Brief Integral Table (continued)

Forms Containing $a + bx$

9. $$\int (a+bx)^n\,dx = \begin{cases} \dfrac{(a+bx)^{n+1}}{(n+1)b} + C, & n \neq -1 \\[2ex] \dfrac{1}{b}\ln|a+bx| + C, & n = -1 \end{cases}$$

10. $$\int x(a+bx)^n\,dx = \begin{cases} \dfrac{1}{b^2(n+2)}(a+bx)^{n+2} - \dfrac{a}{b^2(n+1)}(a+bx)^{n+1} + C, & n \neq -1, -2 \\[2ex] \dfrac{1}{b^2}[a + bx - a\ln|a+bx|] + C, & n = -1 \\[2ex] \dfrac{1}{b^2}\left[\ln|a+bx| + \dfrac{a}{a+bx}\right] + C, & n = -2 \end{cases}$$

11. $$\int x^2(a+bx)^n\,dx = \frac{1}{b^3}\left[\frac{(a+bx)^{n+3}}{n+3} - 2a\frac{(a+bx)^{n+2}}{n+2} + a^2\frac{(a+bx)^{n+1}}{n+1}\right] + C, \quad n \neq -1, n \neq -2, n \neq -3$$

12. $$\int x^m(a+bx)^n\,dx, \quad m > 0, m + n + 1 \neq 0$$

In terms of a lower power of n:

$$\frac{x^{m+1}(a+bx)^n}{m+n+1} + \frac{an}{m+n+1}\int x^m(a+bx)^{n-1}\,dx$$

In terms of a lower power of m:

$$\frac{1}{b(m+n+1)}\left[x^m(a+bx)^{n+1} - ma\int x^{m-1}(a+bx)^n\,dx\right]$$

13. $$\int x^m\sqrt{a+bx}\,dx = \frac{2}{b(2m+3)}\left[x^m\sqrt{(a+bx)^3} - ma\int x^{m-1}\sqrt{a+bx}\,dx\right]$$

If $m = 1$: $$\frac{-2(2a-3bx)\sqrt{(a+bx)^3}}{15b^2} + C$$

If $m = 2$: $$\frac{2(8a^2 - 12abx + 15b^2x^2)\sqrt{(a+bx)^3}}{105b^3}$$

14. $$\int \frac{\sqrt{a+bx}}{x^m}\,dx = \begin{cases} -\dfrac{1}{(m-1)a}\left[\dfrac{\sqrt{(a+bx)^3}}{x^{m-1}} + \dfrac{(2m-5)b}{2}\displaystyle\int \dfrac{\sqrt{a+bx}\,dx}{x^{m-1}}\right], & m \neq 1 \\[2ex] 2\sqrt{a+bx} + \sqrt{a}\ln\left|\dfrac{\sqrt{a+bx}-\sqrt{a}}{\sqrt{a+bx}+\sqrt{a}}\right|, & m = 1, a > 0 \end{cases}$$

(*Table continues inside back cover.*)

Karl J. Smith

College Mathematics and Calculus

with Applications to Management, Life and Social Sciences

2ND EDITION

Brooks/Cole Publishing Company
Pacific Grove, California

Consulting Editor: Robert J. Wisner

Brooks/Cole Publishing Company
A Division of Wadsworth, Inc.

Printed in the United States of America

10 9 8 7 6 5 4 3 2 1

Library of Congress Cataloging-in-Publication Data

Smith, Karl J.
College mathematics and calculus: with applications to management, life, and social sciences/Karl J. Smith.—2nd ed.
p. cm.
Includes index.
ISBN 0-534-16872-8
1. Mathematics. 2. Calculus. I. Title.
QA37.2.S576 1992
510—dc20 91-31581
CIP

Sponsoring Editor: Paula-Christy Heighton
Editorial Assistants: Lainie Giuliano and Carol Ann Benedict
Production Services Coordinator: Joan Marsh
Production: Susan L. Reiland
Manuscript Editor: Adela C. Whitten
Permissions Editor: Mary Kay Hancharick
Interior and Cover Design: Vernon T. Boes
Cover Photo: New York Convention and Visitors Bureau
Interior Illustration: TECHarts; Carl Brown
Typesetting: Polyglot Pte, Ltd, Singapore
Cover Printing: The Lehigh Press
Printing and Binding: R. R. Donnelley & Sons Company

Credits: 1-2-3 is a registered trademark of Lotus Development Corporation. Quattro Pro is a registered trademark of Borland International, Inc. Page 87: Photo courtesy of Sara Hunsaker. Page 88: Dollar Rent-A-Car System trademark reprinted by permission. Avis trademark reprinted by permission. Page 153: Photo courtesy of Sara Hunsaker. Page 235: Photo courtesy of Sara Hunsaker. Pages 237–238: Article reprinted with permission. Copyright © 1991 by Scientific American, Inc. All rights reserved. Photo courtesy of Winnie Klotz. Page 277: Photo courtesy of Sara Hunsaker. Page 323: Photo courtesy of Sara Hunsaker. Page 381: Photo courtesy of Lee Hocker. Page 422: Taken from "Dear Abby" column by Abigail Van Buren. Copyright 1973 Universal Press Syndicate. Reprinted with permission. All rights reserved. Page 425: Photo courtesy of Sara Hunsaker. Page 479: Photo courtesy of Sara Hunsaker. Page 545: Photo courtesy of Sara Hunsaker. Page 575: Photo courtesy of Sara Hunsaker. Page 617: Photo courtesy of New Balance Athletic Shoe, Inc. Page 707: Photo courtesy of Sara Hunsaker. Page 759: Photo courtesy of Sara Hunsaker.

This book is dedicated, with love, to my wife,
Linda Ann Smith

Preface

College Mathematics and Calculus is a textbook for a two-semester or three-quarter course to provide the mathematics background necessary for students in business, management, life sciences, or social sciences. The text is divided into two parts, Finite Mathematics in Chapters 3–8 and Calculus in Chapters 9–16.

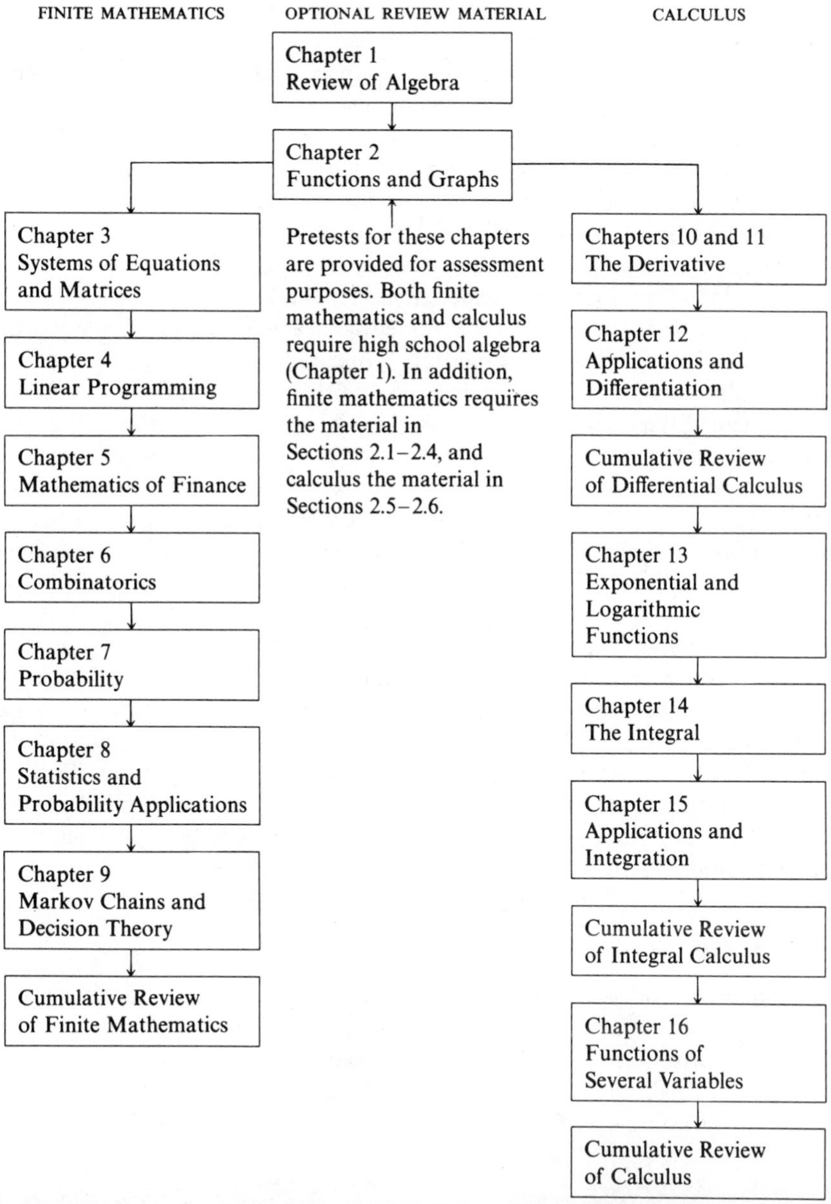

OPTIONAL REVIEW MATERIAL

Chapter 1
Review of Algebra

↓

Chapter 2
Functions and Graphs

Pretests for these chapters are provided for assessment purposes. Both finite mathematics and calculus require high school algebra (Chapter 1). In addition, finite mathematics requires the material in Sections 2.1–2.4, and calculus the material in Sections 2.5–2.6.

FINITE MATHEMATICS

Chapter 3
Systems of Equations and Matrices

↓

Chapter 4
Linear Programming

↓

Chapter 5
Mathematics of Finance

↓

Chapter 6
Combinatorics

↓

Chapter 7
Probability

↓

Chapter 8
Statistics and Probability Applications

↓

Chapter 9
Markov Chains and Decision Theory

↓

Cumulative Review of Finite Mathematics

CALCULUS

Chapters 10 and 11
The Derivative

↓

Chapter 12
Applications and Differentiation

↓

Cumulative Review of Differential Calculus

↓

Chapter 13
Exponential and Logarithmic Functions

↓

Chapter 14
The Integral

↓

Chapter 15
Applications and Integration

↓

Cumulative Review of Integral Calculus

↓

Chapter 16
Functions of Several Variables

↓

Cumulative Review of Calculus

The emphasis in this course is not on algebraic manipulation but rather on an understanding of the modeling process and using mathematics to make statements about real-world applications. The prerequisite for this course is intermediate algebra.

New Edition

It seems as if every new book claims innovation, state-of-the-art production, supplementary materials, readability, abundant problems, and relevant applications. In this new edition I have added over 1,600 problems. In addition, I have included sections on sets, the Fundamental Counting Principle, Riemann sums, and the Fundamental Theorem of Calculus. I have also rearranged much of the material to aid the learning process for the students.

Why should you use my book? I would like you to look at the content, style, and problems in deciding whether this book will fit your needs.

Content

First, every book must cover the appropriate topics, and hopefully in the right order. Finite mathematics has evolved and changed considerably since it was first introduced in the 1960s. Today's course focuses on matrices to solve both linear systems of equations and inequalities (linear programming), sets, combinatorics, and probability, and some supplementary topics such as Markov chains, game theory, and mathematics of finance. New recommendations regarding discrete mathematics are also influencing finite mathematics courses, so I have included appendixes on deductive logic and mathematical induction.

The calculus in Chapters 10–16 is presented by using a variety of real-world applications. In 1986, the Sloan Foundation sponsored a four-day conference to study the calculus curriculum. The conference's work was summarized in the following statement: "We propose a calculus syllabus which emphasizes intuition and conceptual understanding by giving priority to numerical and geometrical methods and approximation techniques." The development of integration that I used in this book follows the recommendations of this conference. The use of tables and numerical integration is emphasized and we do not spend a great deal of time on integration techniques.

Mathematical Models

When a real-life situation is analyzed, it usually does not easily lend itself to mathematical analysis because the real world is far too complicated to be precisely and mathematically defined. It is therefore necessary to develop what is known as a **mathematical model**. This is a body of mathematics based on certain assumptions about the real world, which is modified by experimentation and accumulation of data, and is then used to predict some future occurrence in the real world. A mathematical model is not static or unchanging but is continually being revised and modified as additional relevant information becomes known.

Types of Models

Some mathematical models are quite accurate, particularly in the physical sciences. For example, the path of a projectile, the distance that an object falls in a vacuum, or the time of sunrise tomorrow all have mathematical models that provide very accurate predictions about future occurrences. On the other hand, in the fields of social science, psychology, and management, models provide much less accurate predictions because they must deal with situations that are often random in character. It is therefore necessary to consider two types of models:

A **deterministic model** predicts the exact outcome of a situation because it is based on certain known laws.

A **probabilistic model** deals with situations that are random in character and can predict the outcome within a certain stated or known degree of accuracy.

How do we construct a model? We need to observe a real-world problem and make assumptions about the influencing factors. This is called *abstraction.*

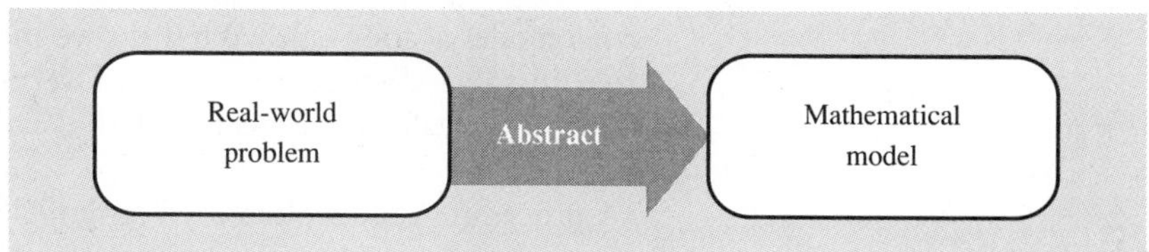

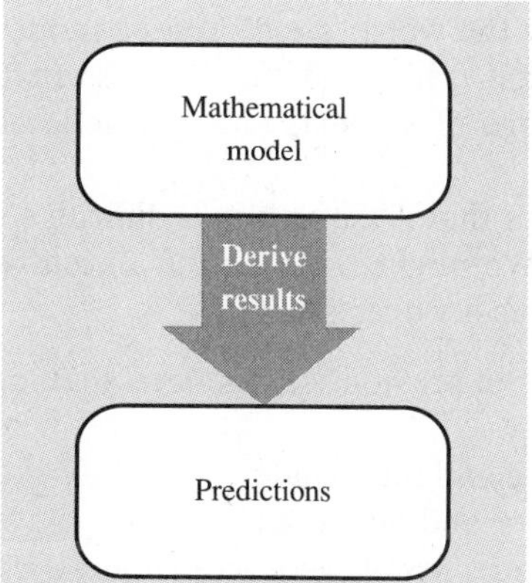

You must know enough about the mechanics of mathematics to *derive results* from the model.

The next step is to gather data. Does the prediction given by the model fit all the known data? If not, we use the data to *modify* the assumptions used to create the model. This is an ongoing process.

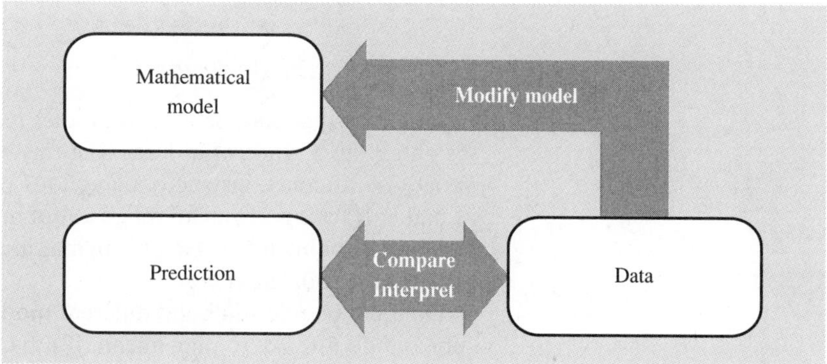

We begin with a rather artificial example of modeling (because at this time we have not developed any content in this course).

A Model for Studying in This Course

How do you plan for success in this course? Suppose we assume that the goal is for you to obtain a grade of C or better in this course. How do you expect to reach this goal? What are your past experiences with college-level courses in general and with mathematics courses in particular? Perhaps your model for studying in this course is quite simple, as shown here:

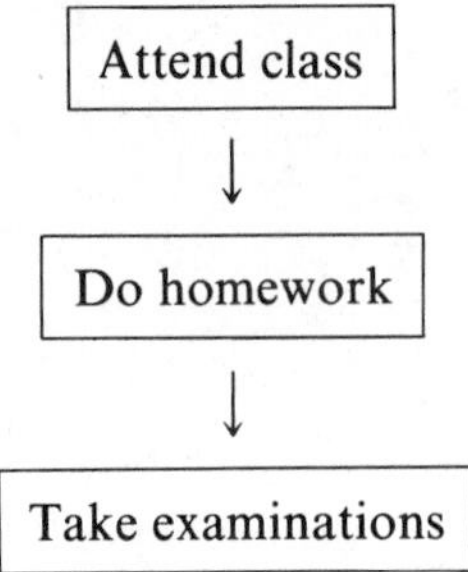

Is this model a good one? What do we mean by the word "good"? Is the model "true" in some absolute sense, or do we mean that it is "valid" in the sense that it has been checked successfully with a wide range of students and college-level courses?

A procedure that is often used in modeling says that if a question is difficult to answer, start by asking some easier questions. We might rephrase our question about this being a "good" model by asking some easier questions:

1. Has this procedure worked with success *for me* in previous college-level mathematics courses?
2. Is there a relationship between class attendance and final grade?
3. Is there a relationship between doing homework and final grade?
4. Will this course be typical of my past experiences in college-level mathematics courses?

Models Should Be Predictive

A good model should be able to predict real-life occurrences. The model we build for obtaining a grade of at least C in this course should be appropriate for a wide variety of students, instructors, methods of instruction, and types of institutions of higher learning. How can we go about making sure that our model is predictive? You are probably interested *only* in making sure that it is predictive for *you* at your school with your instructor.

Different people will build different models. Perhaps you study best under pressure and do not like to plan ahead. If a paper is due on the 15th you will begin on the 14th. Perhaps you are a highly organized person who has already scheduled in advance how much time you have allowed for study and homework in this course.

Perhaps we should try to build a mathematical model. Suppose you count the number of sections you will study in this course. Let this number be n. Also suppose you decide that your grade is determined by the amount of time, t, that you will spend in each section. You might begin by using the following model:

$$G = nt$$

Is *this* a good model for predicting your grade? Not yet. The variables are not well defined, nor are the units for measuring each of the variables. Let t be the time in minutes and let G be the grade as a percent, so $0 \leq G \leq 100$. *Furthermore, suppose that to obtain a C or better requires* $G \geq 70$. Now is this a good model? Not yet. For example, if there are 20 sections and you study each section for 10 minutes, then

$$G = (20)(10) = 200$$

does not make sense. Clearly, we need to refine this mathematical model.

Criticize Your Model

Notice that building a model requires several steps and comparisons between the mathematics and the real world. Do not expect to come up with a working model on the first try. One of the most difficult aspects about teaching and learning how to model mathematics is to deal with the students' impatience. Do not expect to "get it right" on your first attempt. For this model we might consider a scaling factor, k:

$$G = knt$$

Next we could do some research and compare the time spent on each section by a large number of students and the grades obtained by those students. Based on this research, we might determine that the relationship can be described by a line (we consider how to do this in Section 8.6). We could then replace this formula by a more predictive formula, say

$$G = \frac{1}{3}(t - 30) + 50$$

Is this a good model? Not yet. We have not yet built into this model the way we prepare for examinations and other course-related requirements. Perhaps you are beginning to say, "When will I finish this model-building process?" This is a reasonable response, but part of the student impatience is in "getting an answer." The modeling process is never complete. We can simply build better and better models.

In the physical sciences, it is possible to build very accurate models, but in the management, life, and social sciences there are often so many variables to consider that the modeling process is rather complex and will require many simplifying assumptions to come up with a workable model.

You can also see that we need to build some mathematical skills in order to consider some mathematical models. This is the topic of this book. Most books pay lip service to building models but rarely *develop* the skill. True model building cannot be learned in a single lesson, or even in a single course. Learning this skill must be a gradual process, and that is how I approach it in the book—in small steps with realistic applications. I devote an entire section to the nature of linear programming and how to go about the *formulation* of an appropriate model (Section 4.2). In addition, the modeling applications at the beginning of each chapter are included to illustrate the *model-building process* in real-life situations. These applications are presented as suggestions for term papers, which are assignments that are open-ended and require a model-building approach for their development. Even though they are designed to be used as enrichment or for supplementary study, they are also designed to illustrate mathematical modeling. One essay written in response to Modeling Application 1 is given in its entirety in the *Student*

Solutions Manual to show *how* the model building can be developed. Other sample essays are included in the *Instructor's Manual.* These model-building applications, even if not assigned, serve to illustrate, in a very real way, how the material developed in the rest of the chapter can be used to answer some nontrivial questions (see, for example, the Modeling Application on the Cobb–Douglas production function at the end of Chapter 16).

The text is divided into sections of nearly equal size; each takes about one class day to develop. Many sections are marked optional in the Table of Contents. This does not mean that I think these topics are not important; it simply means that the material in these sections is not used in subsequent sections.

Style

The author's writing style is another factor that distinguishes one textbook from another. My writing style is informal, and I always write with the student in mind. I offer study hints along the way and let the students know what is important. Frequent and abundant examples are provided so that the student can understand each step before proceeding to the next. A second color is used to highlight the important steps or particular parts of an equation or formula. The chapter reviews list objectives, practice problems, and also provide a sample test.

Problems

The third factor, which is one of the most important in deciding upon a textbook, is the *number and quality* of the problems presented. This is where I have spent a great deal of effort in writing this book. The problems should help to *develop* the students' understanding of the material, and not inhibit or thwart that understanding by being obscure. The problems start with the simpler ideas, presenting the problems in matched pairs so that for each new problem at least one is provided with an answer in the back and one without the answer. There are about three times more problems than are needed for assignments so that students will have the opportunity to go back and practice additional problems, both for the midterm and for the final. I have provided over 8,000 problems in this text. After the manipulative skills have been developed, each problem set presents a large number of applications to show how the material can be used in business, management, and life and social sciences.

The types of problems include:

Drill. I have included a large number of drill problems to provide adequate practice so that the student can develop a clear understanding of each topic.

Applications. Self-contained applied problems are provided in almost every section to give relevance and practicality to the topic at hand. A list showing the extent and variety of the applications is part of each chapter opening and gives a preview of the material.

Modeling Applications. Each chapter begins with an optional real-life application that allows the material to be applied to a *real* (rather than a textbook) problem, but at a level of difficulty that is manageable for the student. The answers to these questions are open-ended, and a sample essay written in answer to Modeling Application 1 is included in the *Student Solutions Manual.*

CPA, CMA, and Actuary Exam Questions. Actual questions from Actuary, CPA, and CMA exams are scattered throughout the textbook. These test questions provide a link between textbook and profession.

Historical Questions. Historical notes provide insight into the humanness of mathematics, but instead of being superfluous commentaries, these are integrated into the problem sets to give students a taste of some of the ways the topics were originally developed as solutions to mathematical problems.

Supplementary Materials

The final factor often used in selecting a textbook is the type and quality of available supplements. In addition to the answers in the back of the book, I have provided several supplements:

Student Solutions Manual. This provides complete solutions to every odd-numbered problem in the book. It also includes an essay to illustrate the modeling process.

Instructor's Manual. This includes answers to all of the problems in the book, additional questions to accompany the Modeling Applications, as well as sample essays for the Modeling Applications.

Testing Program. This is a computerized test bank with text-editing capabilities that allows you to create an almost unlimited number of tests or retests of the material. This test bank is also available in printed form for those instructors not using a computer.

The Math Lab: Interactive Finite Mathematics. This is a computer supplement prepared by Chris Avery and Charles B. Barker. Appendix D lists the programs available in this supplement.

As the author, I am also available to help you create any other set of supplementary materials that you believe are necessary or worthwhile for your course.

Acknowledgments

Many persons have reviewed either the entire manuscript or parts of this manuscript and have offered many valuable suggestions, and I would like to offer each and every one of them my sincere thanks:

John Alberghini
Manchester Community College
Patricia Bannantine
Marquette University
Craig Benham
University of Kentucky
Thomas Covington
Northwestern State University
Joe S. Evans
Middle Tennessee State University
James L. Forde
Concordia College
Marjorie S. Freeman
University of Houston
Marvin Goodman
Monmouth College
Matthew Gould
Vanderbilt University
Kevin Hastings
University of Delaware
Judith Hector
Walters State Community College
Denise Hennicke
Collin County College
Lou Hoelzle
Bucks County Community College
Susan Kaplan
Western Washington University
Edwin Klein
University of Wisconsin
Ann Barber Megaw
University of Texas, Austin
Norman Mittman
Northeastern Illinois University
Wendell L. Motter
Florida A & M University

Laurence Neises
Spokane Falls Community College
Jacqueline Payton
Virginia State University
William Ramaley
Fort Lewis College
Robert Sharpton
Miami-Dade Community College
Gordon Shilling
University of Texas, Arlington
Rodney Shirey
Jacksonville State University
Roland Sink
Pasadena City College
Harriette J. Stephens
State University of New York, Canton
Ali Tabatabian
San Francisco State University
Mary T. Teegarden
Mesa Junior College
William T. Watkins
Pan American University
Jan Wynn
Brigham Young University
Donald Zalewski
Northern Michigan University

The problem solvers were Nancy Angle (Cerritos College), Gary Gislason (University of Alaska, Fairbanks), Carol Westfall (State University of New York College of Technology and Agriculture), and Harvey Lambert (University of Nevada, Reno), and I would like to thank them for helping to ensure the accuracy of this edition. I continue to thank those who checked the accuracy of the first edition: Pat Bannantine, Gregory Passty, Ernest Ratliff, Terry Shell, John Spellman, Donna Szott, and Richardo Torrejon.

The production of a textbook is a team effort. I would like to thank Paula Heighton for her help and encouragement, Joan Marsh and Susan Reiland for caring so much about the project and for working so hard to meet schedules and deadlines. I would especially like to thank Mary A. McBerty for typing the student's solution manual and the instructor's manual.

And last, but not least, my thanks go to my family, Linda, Missy, and Shannon, for their love, support, and continued understanding of my involvement in writing this book.

Karl J. Smith
Sebastopol, California

Contents

* Optional sections.

* Optional sections.

* Optional sections.

CALCULUS 477

* Optional sections.

* Optional sections.

* Optional sections.

1

Review of Algebra

CHAPTER OVERVIEW

The chapter begins with an algebra pretest. Part I of the pretest reviews general algebraic concepts. If you have difficulty with these questions, take an algebra course before attempting the material in this book. Part II of the pretest reviews more advanced algebraic topics that are discussed in this chapter. If you have difficulty with these questions, review the appropriate section of this chapter.

PERSPECTIVE

The prerequisite for the material in this book is a course in basic algebra. Since it has probably been a while since you studied algebra, this chapter includes a review of the algebraic topics you need to understand the mathematical models presented in this book.

1.1 Algebra Pretests

Part I Basic Algebraic Concepts

Choose the best answer in Problems 1–15.

1. In $5x^2y$, the 5 is
A. a term **B.** a binomial
C. an exponent **D.** a literal factor
E. a numerical coefficient

2. In $6x^2y + 3z$, the 6 and 3 are
A. terms **B.** exponents
C. binomials **D.** coefficients
E. literal factors

3. If $(-2)(-2)(-2) = x$, the value of x is
A. -8 **B.** -6
C. 6 **D.** 8
E. none of these

4. $21 - (-5)$ equals
A. -26 **B.** -16
C. 16 **D.** 26
E. none of these

5. If $a = -3$ and $b = 5$, then $a^2 - 2ab + b^2$ equals
A. -4 **B.** 2
C. 64 **D.** -2
E. none of these

6. The expression $8 + 2 \cdot 3 - 8 \div 2$ equals
A. 11 **B.** 10
C. 3 **D.** -3
E. none of these

7. $3x - 5x$ equals
A. $15x^2$ **B.** $8x$
C. $15x$ **D.** $-2x$
E. none of these

8. -5^2 equals
A. -10 **B.** 25
C. -25 **D.** 10
E. none of these

9. $(-3y^2)^3$ equals
A. $27y^6$ **B.** $-27y^5$
C. $-3y^6$ **D.** $-27y^6$
E. none of these

10. $\dfrac{a+b}{2}$ means
A. $a + b \cdot \frac{1}{2}$ **B.** $a + b \div 2$
C. $a + (b \div 2)$ **D.** $(a + b) \cdot \frac{1}{2}$
E. all of these

11. If $3x + 12 = 6$, x equals
A. 6 **B.** 2
C. -2 **D.** -6
E. none of these

12. If $6 + 3x = x - 4$, x equals
A. 5 **B.** $2\frac{1}{2}$
C. 1 **D.** 2
E. none of these

13. If $2x - 16 = 3x - 9$, x equals
A. $-\frac{7}{5}$ **B.** -7
C. 5 **D.** 25
E. none of these

14. $(32x^8) \div (-2x)$ equals
A. $30x^7$ **B.** $16x^8$
C. $34x^8$ **D.** $-30x^7$
E. none of these

15. Simplify $3(x + 2) - (x - 4y) + (x + y)$.
A. $3x + 11y$ **B.** $3x - 3y + 6$
C. $x + 5y + 6$ **D.** $3x + 5y + 6$
E. none of these

Indicate whether each of the statements in Problems 16–30 is true or false.

16. If $x \neq 0$, then $-x$ always indicates a number that is less than zero.

17. $2(xy) = (2x)(2y)$

18. The domain of a variable is a set from which values of the variable are chosen.

19. x^2 is always positive.

20. $(a + b)^2 = a^2 + b^2$

21. $\dfrac{A + C}{B + C} = \dfrac{A}{B}$

22. $\dfrac{2 + x}{6} = \dfrac{x}{3}$

23. $\frac{0}{5}$ is not defined.

24. $\sqrt{-4}$ is a real number.

25. $\dfrac{A}{B} + \dfrac{C}{B} = \dfrac{1}{B}(A + C)$

26. $\dfrac{-x}{-y} = -\dfrac{x}{y}$

27. $(x - y)^2 = (y - x)^2$

28. $\dfrac{x}{y} + \dfrac{y}{x} = 1$

29. $\dfrac{1}{x}y\dfrac{1}{z} = \dfrac{y}{xz}$

30. $\sqrt{x^2 + y^2} = x + y$

Part II

Algebra Pretest

Choose the best answer for each of the following problems.

[1.2]

1. $|2\pi - 7| =$
A. $2\pi - 7$ **B.** $7 - 2\pi$
C. $-.7168$ **D.** $.7168$
E. The expression may have more than one value.

2. $6.23\overline{4}$ is an example of
A. a natural number
B. a real number
C. an irrational number
D. all the answers in parts **A**, **B**, and **C**
E. none of these

3. The distance between the points (5) and ($\sqrt{10}$) on a real number line is
A. -1.834 **B.** 1.834
C. $5 - \sqrt{10}$ **D.** $\sqrt{10} - 5$
E. none of these

[1.3]

4. If $a = 2$ and $b = -3$, then $2a(a + 2b)$ equals
A. 32 **B.** 16
C. -16 **D.** -48
E. none of these

5. $(x + y)^2$ equals
A. $x^2 + y^2$ **B.** $x^2 + xy + x$
C. $x^2 + 2xy + y^2$ **D.** x^2y^2
E. none of these

6. The result of simplifying $42x - [10 - 2(3x - 4) - 2]$ is
A. $48x + 16$ **B.** $48x - 16$
C. $36x - 20$ **D.** $36x - 16$
E. none of these

[1.4]

7. The greatest common factor of $10x^3$, $5x^2$, and $25x$ is
A. $5x$ **B.** x
C. $5x^2$ **D.** $10x^2$
E. none of these

8. One of the factors of $x^2 - 18x + 80$ is
A. $x + 4$ **B.** $x + 5$
C. $x - 10$ **D.** $x - 16$
E. none of these

9. The complete factorization of $15x^3 - 15xy^2$ is
A. $x(15x^2 - 15y^2)$ **B.** $15x(x + y)(x - y)$
C. $x(15x + y)(x - y)$ **D.** $3xy(5x^2 - 5xy)$
E. none of these

[1.5]

10. Solve $\dfrac{(x + 3)(x - 1)}{x - 1} = 4.$
A. 1 **B.** -1
C. 3 **D.** -3
E. none of these

11. The solution for $-2x \leq 8$ is
A. $x \leq -4$
B. $\{-4, -3, -2, -1, 0, 1, 2, 3, 4, \ldots\}$
C. $x > -4$
D. $x \geq -4$
E. none of these

12. If $3 \leq x + 4 \leq 5$, then
A. $-1 \geq x$ or $x \geq 1$ **B.** $-1 \leq x$ or $x \leq 1$
C. $-1 \leq x \leq 1$ **D.** $7 \leq x \leq 9$
E. none of these

[1.6]

13. Solve $x^2 - 7x + 12 = 0$.
A. $\{-3, -4\}$ **B.** $\{-3, 4\}$
C. $\{3, 4\}$ **D.** $\{2, -6\}$
E. none of these

14. Solve $(x + 1)(2x - 3) = 25$.
A. $\{-1, 3\}$ **B.** $\{4, 8\}$
C. $\{4, -\frac{7}{2}\}$ **D.** $\{6, 2\}$
E. none of these

15. If $(x + 1)(2 - x) < 0$, then
A. $x < -1$ or $x > 2$ **B.** $x < -2$ or $x > 1$
C. $x < -2$ or $x < 1$ **D.** $-1 < x < 2$
E. none of these

16. Solve $x^2 - 2x - 2 = 0$.
A. $-2 \pm \sqrt{3}$ **B.** $4 \pm \sqrt{3}$
C. $2 \pm \sqrt{3}$ **D.** $4 \pm 4\sqrt{3}$
E. none of these

[1.7]

17. Simplify $\dfrac{x^2 - 5x + 4}{x + 3} \cdot \dfrac{x^2 + 2x - 3}{x - 4}$.
A. 1 **B.** $x - 1$
C. $x^2 - 1$ **D.** $(x - 1)^2$
E. none of these

18. Simplify $\dfrac{x}{x^2 - 4} - \dfrac{2}{4 - x^2}$.
A. $x - 2$ **B.** $\dfrac{1}{x - 2}$
C. $\dfrac{1}{2 - x}$ **D.** $\dfrac{x}{x - 2}$
E. none of these

1.2 Real Numbers

In mathematics, we need to be clear about the types of numbers under consideration. For example, if someone asked you to "pick a number," you probably wouldn't choose $\sqrt{5}$ or $\pi/2$, but in this course we want to include such choices—the set of numbers called **real numbers**. The various **sets** of numbers include:

Natural numbers: $N = \{1, 2, 3, 4, \ldots\}$
Whole numbers: $W = \{0, 1, 2, 3, 4, \ldots\}$
Integers: $I = \{\ldots, -3, -2, -1, 0, 1, 2, 3, \ldots\}$
Rationals: $Q = \{$all Quotients p/q where p is an integer and q is a nonzero integer$\}$

Rationals are numbers whose decimal representations either terminate or eventually repeat, such as $\frac{1}{3}$, .5, $\frac{1}{8}$, $.\overline{1}$, $\frac{83}{74}$, or $-\frac{147}{44}$, but the set also includes integers such as 2, 6, $-19, \ldots$ since $2 = 2.0$, $6 = 6.0$, and so on.

Irrationals: $Q' = \{$numbers whose decimal representation does not terminate or repeat$\}$

For example, π, $\sqrt{5}$, $\sqrt[3]{2}$

Real numbers: $R = \{$numbers in Q or $Q'\}$*

A capital letter is usually used to name a set. Thus the set of natural numbers (also called counting numbers) is often referred to by the letter N, the set of integers by I, and the set of rationals by Q (for quotients). One method of designating a set is to enclose the list of its **members**, or **elements**, in braces, $\{\ \}$, and to use three dots, if needed, to indicate that the numbers continue in the shown pattern. The set is said to be **finite** if the number of elements in it is a natural number or less than some natural number. A set that is not finite is said to be **infinite**. A set with no members is called the **empty set**, or **null set**, and is labeled $\{\ \}$ or $\varnothing$.

This course is focused on the set of real numbers, which is easily visualized by using a **coordinate system** called a **number line**, as shown in Figure 1.1. A **one-to-one correspondence** exists between all real numbers and all points on such a number line:

1. Every point on the line corresponds to precisely one real number.
2. For each real number, there corresponds one and only one point.

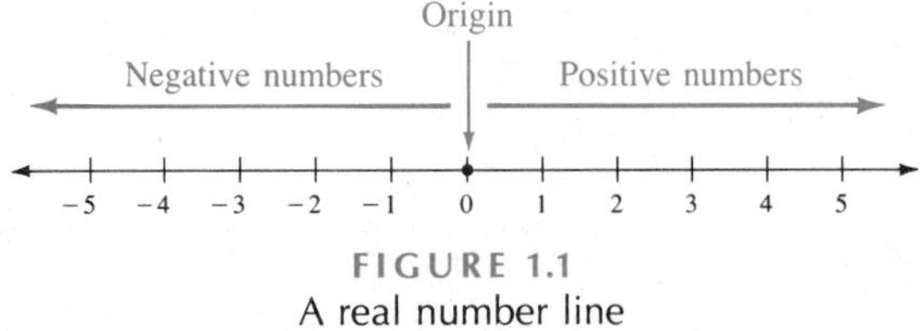

FIGURE 1.1
A real number line

A point associated with a particular number is called the **graph** of that number, and the number is called the **coordinate** of the point. Numbers associated with points to the right of the origin are called **positive real numbers**, and those associated with

* There is another set of numbers used in mathematics, the set of **complex numbers**. In this book, however, we limit ourselves to the set of real numbers.

points to the left are called **negative real numbers**. Thus a number line is also a convenient way of ordering any two real numbers. If a point whose coordinate is a lies to the right of a point whose coordinate is b on a number line, then ***a* is greater than *b***, written $a > b$. For example, $6 > 3$ and $-4 > -10$. If a point's coordinate b is to the left of a point's coordinate a, then we say that ***b* is less than *a***, or $b < a$. For example, $3 < 6$ and $-10 < -4$. The other symbols of comparison are:

$a = b$	a is equal to b
$a \neq b$	a is not equal to b
$a > b$	a is greater than b
$a \geq b$	a is greater than or equal to b
$a < b$	a is less than b
$a \leq b$	a is less than or equal to b

The symbols $>$, $\geq$, $<$, and $\leq$ are referred to as the **inequality symbols**.

A concept called **absolute value** is also associated with points on a number line. The symbol $|a|$ is read "the absolute value of a" and means the distance between the graph of a and the origin. For example:

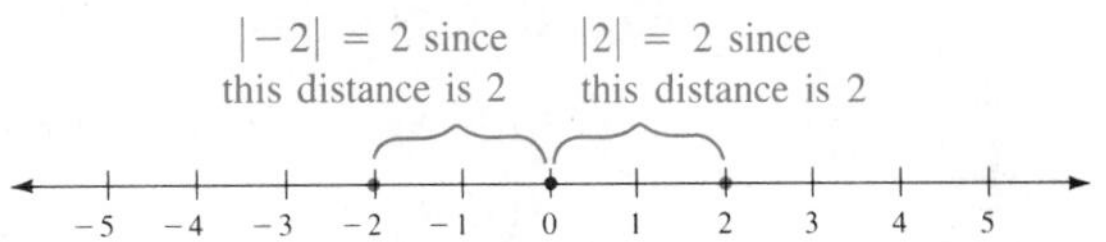

EXAMPLE 1 **a.** $|5| = 5$ **b.** $|-5| = 5$ **c.** $|-\frac{1}{2}| = \frac{1}{2}$

d. $|\sqrt{5}| = \sqrt{5}$ **e.** $-|-5| = -5$ **f.** $-|\pi| = -\pi$

(e) The opposite of the absolute value of -5

(f) The opposite of the absolute value of π ■

Since we will use the notion of absolute value in a variety of contexts, we need a more formal definition:

Absolute Value

The absolute value of a real number a is defined by

$$|a| = \begin{cases} a & \text{if } a \geq 0 \\ -a & \text{if } a < 0 \end{cases}$$

EXAMPLE 2 **a.** $|5| = 5$ since $5 \geq 0$

b. $|-5| = -(-5)$ since $-5 < 0$
$= 5$

c. $-|5| = -5$ since $5 \geq 0$

d. $|-\pi| = -(-\pi)$ since $-\pi < 0$
$= \pi$

e. $|\pi - 3| = \pi - 3$ since $\pi - 3 \geq 0$

f. $|3 - \pi| = -(3 - \pi)$ since $3 - \pi < 0$
$= -3 + \pi$
$= \pi - 3$

g. $|\sqrt{30} - 5| = \sqrt{30} - 5$ since $\sqrt{30} - 5 \geq 0$

h. $|\sqrt{30} - 6| = -(\sqrt{30} - 6)$ since $\sqrt{30} - 6 < 0$
$= 6 - \sqrt{30}$ ■

In algebra, it is necessary to assume certain properties of equality and inequality. The first of these is called the **trichotomy property**, or **property of comparison**:

Property of Comparison

Given any two real numbers a and b, exactly one of the following holds:
1. $a = b$ **2.** $a < b$ **3.** $a > b$

This property tells us that if we are given *any* two real numbers, either they are equal or one of them is greater than the other. It establishes order on the number line. We need this property to derive a formula for the distance between two points on a number line. For example, let P_1 and P_2 be any points on a number line with coordinates x_1 and x_2, usually denoted as $P_1(x_1)$ and $P_2(x_2)$. Then, by the property of comparison, we know that $x_1 = x_2$, $x_1 < x_2$, or $x_1 > x_2$. Consider these possibilities one at a time.

$x_1 = x_2$ That is, the distance between x_1 and x_2 is 0.

$x_1 < x_2$ $x_1 > x_2$

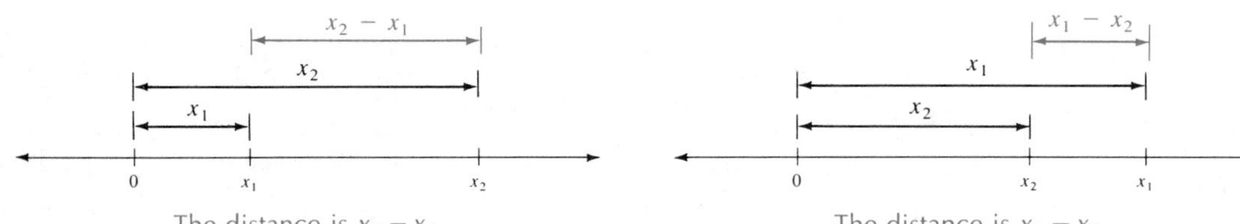

The distance is $x_2 - x_1$. The distance is $x_1 - x_2$.

We can combine these different possibilities into one distance formula by using the idea of absolute value.

Distance on a Number Line

The distance d between points $P_1(x_1)$ and $P_2(x_2)$ is

$$d = |x_2 - x_1|$$

EXAMPLE 3 Let A, B, C, D, and E be points with coordinates as shown on the number line below:

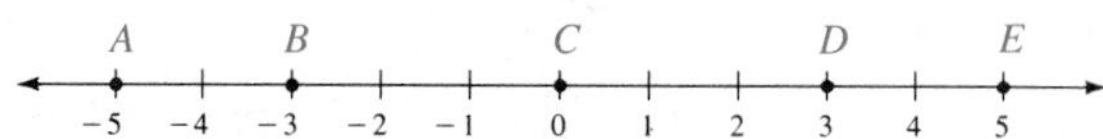

a. The distance from D to E is: $|5 - 3| = |2| = 2$.
b. The distance from E to D is: $|3 - 5| = |-2| = 2$.
c. The distance from A to D is: $|3 - (-5)| = |8| = 8$.
d. The distance from E to B is: $|-3 - 5| = |-8| = 8$. ■

1.2 Problem Set

Classify each example in Problems 1–10 as a natural number (N), whole number (W), integer (I), rational number (Q), irrational number (Q′), or real number (R). Examples may be in more than one set.

1. a. -9 b. $\sqrt{30}$
2. a. $\sqrt{49}$ b. $\frac{17}{2}$
3. a. 5 b. $\sqrt{4}$
4. a. $\frac{0}{5}$ b. $\frac{5}{0}$
5. a. $.\overline{4}$ b. .5252...
6. a. 3.1416 b. π
7. a. .381 b. $\pi/6$
8. a. $\sqrt{121}$ b. $\sqrt{125}$
9. a. 16/0 b. 0/16
10. a. $\pi/3$ b. $\sqrt{4/25}$

Certain relationships among these sets are assumed from your previous courses and are illustrated in Figure 1.2. For Problems 11–40 let

$P = \{\text{primes}\}$
$= \{2, 3, 5, 7, 11, 13, \ldots\}$

Primes are natural numbers with exactly two divisors (itself and 1).

$C = \{\text{composites}\}$
$= \{4, 6, 8, 9, 10, \ldots\}$

Composites are natural numbers greater than 1 that are *not* primes.

$E = \{\text{evens}\}$
$= \{0, 2, 4, 6, 8, 10, \ldots\}$

Even numbers are whole numbers divisible by 2.

$D = \{\text{odds}\}$
$= \{1, 3, 5, 7, 9, 11, \ldots\}$

Odd numbers are natural numbers that can be written as an even number plus 1.

11. Draw Figure 1.2 and show the set *P*.
12. Draw Figure 1.2 and show the set *C*.
13. Draw Figure 1.2 and show the set *E*.
14. Draw Figure 1.2 and show the set *D*.

Indicate whether the statements in Problems 15–40 are true or false. Test your understanding of these sets, both from this section and from previous courses.

15. 0 is a natural number.
16. $\sqrt{3}$ is a real number.
17. .333... is an irrational number.
18. $-\frac{1}{4}$ is an integer.
19. 2 is a real number.
20. 1 is a prime number.
21. $|5|$ is a prime number.
22. The sum of two odd numbers is an odd number.
23. The sum of two even numbers is an even number.
24. The sum of two prime numbers is a prime number.
25. The sum of two composite numbers is a composite number.
26. The product of two even numbers is an even number.
27. The product of two odd numbers is an odd number.
28. The product of two primes is a prime.

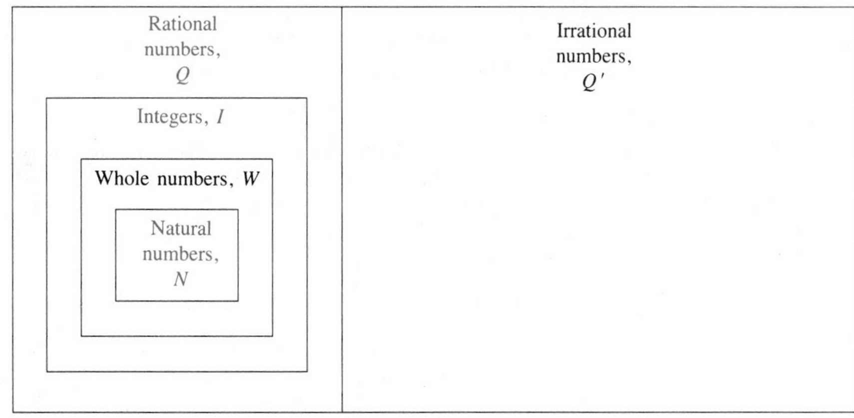

FIGURE 1.2
Relationships within the set of real numbers

29. The product of two rational numbers is a rational number.

30. The product of two irrational numbers is an irrational number.

31. All natural numbers are integers.

32. All irrational numbers are real.

33. All repeating decimals are positive.

34. No even integers are primes.

35. All odd integers are primes.

36. All numbers are either even or odd.

37. Some negative numbers are primes.

38. a/b is a rational number for all numbers a and b.

39. Some real numbers are integers.

40. Some irrational numbers are integers.

Rewrite Problems 41–61 *without the absolute value symbols.*

41. $|9|$ **42.** $-|19|$

43. $|-19|$ **44.** $-|-8|$

45. $-|\pi|$ **46.** $|\pi - 2|$

47. $|2 - \pi|$ **48.** $|\pi - 5|$

49. $|\pi - 10|$ **50.** $|\sqrt{20} - 5|$

51. $|\sqrt{20} - 4|$ **52.** $|\sqrt{50} - 7|$

53. $|\sqrt{50} - 8|$ **54.** $|2\pi - 5|$

55. $|2\pi - 7|$ **56.** $|x + 3|$ if $x \geq 3$

57. $|x + 3|$ if $x \leq -3$ **58.** $|y - 5|$ if $y < 5$

59. $|y - 5|$ if $y \geq 5$ **60.** $|5 - 2s|$ if $s > 10$

61. $|4 + 3t|$ if $t < -10$

Find the distance between the pairs of points whose coordinates are given in Problems 62–70.

62. $A(5)$ and $B(17)$ **63.** $C(-5)$ and $D(-15)$

64. $E(22)$ and $F(-8)$ **65.** $G(-2)$ and $H(9)$

66. $I(\pi)$ and $J(5)$ **67.** $K(\pi)$ and $L(2)$

68. $M(\sqrt{3})$ and $N(1)$ **69.** $P(\sqrt{3})$ and $Q(2)$

70. $R(6)$ and $S(2\pi)$

1.3 Algebraic Expressions

An **algebraic expression** is a grouping of constants and variables obtained by applying a finite number of operations (such as addition, subtraction, multiplication, nonzero division, extraction of roots). **Constants** are symbols with just one possible value (such as 5, 11, 0, .5, or π), and **variables** are symbols (such as x, y, or z) used to represent unspecified numbers selected from some set called the **domain**. In this book, if a domain for a variable is not specified, the domain is the set of real numbers.

Numbers or expressions that are added are called **terms** while those that are multiplied are called **factors**. If a number and a variable are multiplied together, then the factor consisting of the number alone is called the **numerical coefficient**. It is very common to have repeated factors, which are indicated using **exponents**:

Exponent

If b is any real number and n is any natural number, then

$$b^n = \underbrace{b \cdot b \cdot b \cdots b}_{n \text{ factors}}$$

Furthermore, if $b \neq 0$, then

$$b^0 = 1 \qquad \text{and} \qquad b^{-n} = \frac{1}{b^n}$$

The number b is called the **base**, n is called the **exponent**, and b^n is called a **power**.

If an algebraic expression is a finite sum of terms with variables containing whole-number exponents, it is called a **polynomial**. For example,

$$3x, \quad 5, \quad 2x^2 - 3x + 5, \quad \frac{2}{3}x, \quad 0, \quad 5.5, \quad \frac{x-2}{4}$$

are polynomials, but

$$\sqrt{x}, \quad \frac{2}{x}, \quad \frac{x}{0}, \quad x^{1/3}$$

are not. The general form of a polynomial with a single variable is:

General Form of an nth-Degree Polynomial in x

$$a_n x^n + a_{n-1}x^{n-1} + \cdots + a_2 x^2 + a_1 x + a_0 \qquad a_n \neq 0$$

Polynomials are frequently classified according to the number of terms they contain. A **monomial** is a polynomial with one term, a **binomial** has two terms, and a **trinomial** has three. The **degree** of a monomial is the number of variable factors in that monomial. For example, the degree of $3x^2$ is two because $3x^2 = 3xx$ (two variable factors); 3^2x is of degree one because there is one variable factor; and $3x^2y^3$ is of degree five because there are five variable factors ($xxyyy$). The expressions 5, 6^2, and π are all of degree zero since none contains a variable factor. A special case is the monomial 0, to which no degree is assigned. The degree of a simplified polynomial is the degree of the highest-degree term.

From beginning algebra, recall the proper order of operations given in the following box.

Order of Operations Agreement

When simplifying an algebraic expression:

1. Carry out all operations within parentheses (begin with the innermost parentheses).
2. Do exponents next.
3. Complete multiplications and divisions, working from left to right.
4. Finally, do additions and subtractions, working from left to right.

EXAMPLE 1 $2 + 3 \cdot 4 = 2 + 12 = 14$ ■

EXAMPLE 2 $9 + 12 \div 3 + 4 \div 2 + 1 \cdot 2 = 9 + 4 + 2 + 2 = 17$ ■

In algebra we rarely use the symbol $\div$ but, instead, write the expression in Example 2 as

$$9 + \frac{12}{3} + \frac{4}{2} + 1 \cdot 2$$

The fractional bar used for division is also used as a grouping symbol:

$$6 + 4 \div 2 \quad \text{is written as} \quad 6 + \frac{4}{2}$$

whereas

$$(6 + 4) \div 2 \quad \text{is written as} \quad \frac{6+4}{2}$$

EXAMPLE 3 In algebra, $6 + 4 \div 2 + (6 + 4) \div 2 - (6 + 4 \div 2)$ is written as

$$\begin{aligned} 6 + \frac{4}{2} + \frac{6+4}{2} - \left(6 + \frac{4}{2}\right) &= 6 + \frac{4}{2} + \frac{10}{2} - (6 + 2) \\ &= 6 + \frac{4}{2} + \frac{10}{2} - 8 \\ &= 6 + 2 + 5 - 8 \\ &= 5 \end{aligned}$$

■

EXAMPLE 4 Simplify: $2 + 4[1 - 3(3 - 5) - \frac{12}{4}]$

Solution Do the innermost parentheses first:

$$\begin{aligned} 2 + 4\left[1 - 3(3 - 5) - \frac{12}{4}\right] &= 2 + 4\left[1 - 3(-2) - \frac{12}{4}\right] && \text{Multiply and divide inside the brackets.} \\ &= 2 + 4[1 + 6 - 3] && \text{Add and subtract inside the grouping symbols.} \\ &= 2 + 4[4] && \text{Multiply before adding.} \\ &= 2 + 16 \\ &= 18 \end{aligned}$$

■

To **evaluate** an expression means to replace the variables with a given numerical value and then to carry out the order of operations to simplify the resulting numerical expression. This process is illustrated in Example 5.

EXAMPLE 5 Find $-(x + y)[-(s - t)]$ where $x = -2$, $y = -3$, $s = 4$, and $t = 7$.

Solution

$$\begin{aligned} -[(-2) + (-3)][-(4 - 7)] &= -[-5][-(-3)] \\ &= -[-5][3] \\ &= -[-15] = 15 \end{aligned}$$

■

Addition and Subtraction of Polynomials

Terms that are identical, except for their numerical coefficients, are called **similar terms**. That is, similar terms contain the same exponent (or exponents) on the variables. Since the variable parts are identical, they can be simplified as follows:

$$5x + 4x = (5 + 4)x = 9x$$

If you add five apples to four apples, you obtain nine apples; so, too, if you add $5x$'s to $4x$'s, you get $9x$'s. This is a statement of an algebraic property called the **distributive property**:

Distributive Property

For real numbers a, b, and c,

$$a(b + c) = ab + ac$$

You will, of course, use this property and combine similar terms mentally as shown by the following more complicated examples.

EXAMPLE 6

$$(5x^2 + 2x + 1) + (3x^3 - 4x^2 + 3x - 2) = \underset{\text{Third-degree term}}{3x^3} + \underset{\text{Second-degree terms}}{\overset{\text{Similar terms}}{5x^2 + (-4)x^2}} + \underset{\text{First-degree terms}}{\overset{\text{Similar terms}}{2x + 3x}} + \underset{\text{Zero-degree terms}}{\overset{\text{Similar terms}}{1 + (-2)}}$$

$$= 3x^3 + x^2 + 5x - 1$$ ■

EXAMPLE 7 $(4x - 5) + (5x^2 + 2x + 1) = 5x^2 + (4x + 2x) + (-5 + 1) = 5x^2 + 6x - 4$ ■

EXAMPLE 8 $(4x - 5) - (5x^2 + 2x + 1) = 4x - 5 - 5x^2 - 2x - 1 = -5x^2 + 2x - 6$ ■

Note that in Example 7 we added similar terms. In Example 8, notice that the procedure for subtracting a polynomial is to subtract *each* term.

EXAMPLE 9

$$\begin{aligned}(5x^2 + 2x + 1) - (3x^3 - 4x^2 + 3x - 2) &= 5x^2 + 2x + 1 - 3x^3 + 4x^2 - 3x + 2 \\ &= -3x^3 + 9x^2 - x + 3\end{aligned}$$ ■

Multiplication of Polynomials

To understand multiplication of polynomials, it is necessary to first understand multiplication of monomials.

EXAMPLE 10

$x^2(x^3) = xx(xxx) = x^5$ There are five x's used as a factor.

$(x^2)^3 = (x^2)(x^2)(x^2) = (xx)(xx)(xx) = x^6$ There are six x's.

$(x^2y^3)^4 = (x^2y^3)(x^2y^3)(x^2y^3)(x^2y^3) = (x^2x^2x^2x^2)(y^3y^3y^3y^3) = x^8y^{12}$ ■

Example 10 illustrates three laws of exponents that are used to simplify algebraic expressions.

Laws of Exponents

Let a and b be any real numbers, m and n be any integers, and assume that each expression is defined.

FIRST LAW: $b^m \cdot b^n = b^{m+n}$
SECOND LAW: $(b^n)^m = b^{mn}$
THIRD LAW: $(ab)^m = a^m b^m$

These laws of exponents are used along with the distributive property when multiplying polynomials.

EXAMPLE 11 $(4x - 5)(5x^2 + 2x + 1) = (4x - 5)5x^2 + (4x - 5)2x + (4x - 5)1$
$= 20x^3 - 25x^2 + 8x^2 - 10x + 4x - 5$
$= 20x^3 - 17x^2 - 6x - 5$ ■

EXAMPLE 12 $(2x - 3)(x + 4) = (2x - 3)x + (2x - 3)4$
$= 2x^2 - 3x + 8x - 12$
$= 2x^2 + 5x - 12$ ■

Notice that one term, $P = 4x - 5$ in Example 11 and $P = 2x - 3$ in Example 12, is simply distributed to each term in the other factor. We call this **distributive multiplication** to remind us to use the distributive property to do multiplication of polynomials.

Distributive Multiplication

$$P(A_1 + A_2 + \cdots + A_n) = PA_1 + PA_2 + \cdots + PA_n$$

EXAMPLE 13 $(\mathbf{2x + 3y + 1})(5x - 2y - 3)$
$= (\mathbf{2x + 3y + 1})5x + (\mathbf{2x + 3y + 1})(-2y) + (\mathbf{2x + 3y + 1})(-3)$
$= 10x^2 + 15xy + 5x - 4xy - 6y^2 - 2y - 6x - 9y - 3$
$= 10x^2 + 11xy - 6y^2 - x - 11y - 3$ ■

It is frequently necessary to multiply binomials, and even though we use distributive multiplication, we want to be able to carry out the process quickly and efficiently in our heads. Consider Example 14.

EXAMPLE 14 **a.** $(2x + 3)(4x - 5) = (2x + 3)(4x) + (2x + 3)(-5)$
$= \mathbf{8x^2} + \underbrace{\mathbf{12x + (-10x)}} + 3(-5)$

Product of first terms; Sum of inner and outer terms; Product of last terms

$= 8x^2 + 2x - 15$

b. $(5x - 3)(2x + 3) = 10x^2 + (15x - 6x) + (-9)$

Mentally: $10x^2$ — Product of first terms; $(15x - 6x)$ — Sum of outer and inner terms; (-9) — Product of last terms

$= 10x^2 + 9x - 9$

c. $(4x - 3)(3x - 2) = 12x^2 - 17x + 6$

$12x^2$ — Product of first terms; $17x$ — Sum of outer and inner terms; 6 — Product of last terms ■

It is frequently necessary to multiply binomials such as those shown in Example 12; however the process illustrated in that example is too lengthy, so a shortened version, called **FOIL**, is often used instead. Four pairs of terms are multiplied.

Foil

$$(ax + b)(cx + d) = acx^2 + (ad + bc)x + bd$$

① acx^2; ② + ③ $(ad + bc)x$; ④ bd

Do this mentally.

① First terms
② Outer terms
③ Inner terms
④ Last terms

EXAMPLE 15 $(2x - 3)(x + 3) = 2x^2 + 3x - 9$

① $2x^2$; ② + ③: $6x + (-3x)$; ④ -9

$(x + 3)(3x - 4) = 3x^2 + 5x - 12$

$(5x - 2)(3x + 4) = 15x^2 + 14x - 8$ ■

1.3 Problem Set

Simplify the expressions in Problems 1–14.

1. $3 + 2 \cdot 5$
2. $4 + 2 \cdot 3$
3. $5 + 3 \cdot 2$
4. $(-5)^2$
5. $(-6)^2$
6. $(-7)^2$
7. -5^2
8. -6^2
9. -7^2
10. $10 + 6(3 - 5)$
11. $8 - 5(2 - 7)$
12. $3 \cdot 5 - (-2)(-4)$
13. $\dfrac{(-2)6 + (-4)}{-4} + \dfrac{(-3)(-4)}{6}$
14. $(-2) + \dfrac{6}{-3} - 4 + \dfrac{-8}{-4}$

Let $x = -3$, $y = 2$, $z = -1$, and $w = -4$. Evaluate the expressions in Problems 15–27.

15. x^2
16. $-x^2$
17. y^2
18. $-y^2$
19. z^2
20. $-z^2$
21. $x + y - z$
22. $x - (y - z)$
23. $(xy)^2 + xy^2 + x^2y$
24. $5x - (4x + 3w)$
25. $x^2 - w^2(y^2 + z^2)$
26. $\dfrac{x - w^2}{z}$
27. $\dfrac{x - z}{y}$

Mentally multiply the expressions in Problems 28–41.

28. **a.** $(x + 3)(x + 2)$ **b.** $(x + 1)(x + 5)$

29. **a.** $(x - 2)(x + 6)$ **b.** $(x + 5)(x - 4)$

30. **a.** $(x + 1)(x - 2)$ **b.** $(x - 3)(x + 2)$

31. **a.** $(x - 5)(x - 3)$ **b.** $(x + 3)(x - 4)$

32. **a.** $(y + 1)(y - 7)$ **b.** $(y - 3)(y + 5)$

33. **a.** $(2y + 1)(y - 1)$ **b.** $(2y - 3)(y - 1)$

34. **a.** $(y + 1)(3y + 1)$ **b.** $(y + 1)(3y + 2)$

35. **a.** $(2y + 3)(3y - 2)$ **b.** $(2y + 3)(3y + 2)$

36. **a.** $(x + y)(x + y)$ **b.** $(x - y)(x - y)$

37. **a.** $(x + y)(x - y)$ **b.** $(a + b)(a - b)$

38. **a.** $(5x - 4)(5x + 4)$ **b.** $(3y - 2)(3y + 2)$

39. **a.** $(x + 2)^2$ **b.** $(x - 2)^2$

40. **a.** $(x + 4)^2$ **b.** $(x - 3)^2$

41. **a.** $(a + b)^2$ **b.** $(a - b)^2$

Simplify the expressions in Problems 42–68.

42. $(x + y - z) + (2x - 3y + z)$

43. $(x - y - z) + (2x + y - 3z)$

44. $(x + 2y - 3z) + (x - 3y + 5z)$

45. $(x + 3y - 2z) + (3x - 5y + 3z)$

46. $(x + 2y) - (2x - y)$

47. $(2x - y) - (2x + y)$

48. $(x - 3y) - (5x + y)$

49. $(6x - 4y) - (4x - 6y)$

50. $(2x + y + 3) - (x - y + 4)$

51. $(x + y - 5) - (2x - 3y + 4)$

52. $(5x^2 + 3x - 5) - (3x^2 + 2x + 3)$

53. $(6x^2 - 3x + 2) - (2x^2 + 5x + 3)$

54. $(3x^2 + 2x + 6) - (2x^2 + 5x + 3)$

55. $(2x^2 - 3x - 5) - (5x^2 - 6x + 4)$

56. $(x^2 - 1) - (2 - x) + (x^2 - x)$

57. $(x^2 - x) + (x - 3) - (x - x^2)$

58. $(3x - x^2) - (5 - x) - (3x - 2)$

59. $(x^2 - 7) - (3x + 4) - (x - 2x^2)$

60. $(x^2 - 5) - (3x^2 + 2x + 5)$

61. $(x + 1)^3$

62. $(x - 1)^3$

63. $(3x^2 - 5x + 2) + (x^3 - 4x^2 + x - 4)$

64. $(5x + 1) + (x^3 - 4x^2 + x - 4)$

65. $(3x^2 - 5x + 2) - (5x + 1)$

66. $(x^3 - 4x^2 + x - 4) - (3x^2 - 5x + 2)$

67. $(5x + 1)(3x^2 - 5x + 2)$

68. $(3x - 1)(x^2 + 32x - 2)$

1.4 Factoring

A **factor** of a given algebraic expression is an algebraic expression that divides evenly into the given expression. The process of **factoring** involves resolving a given expression into its factors. The procedure we will use is to carry out a series of "tests" for different types of factors. Table 1.1 lists these types in the order in which we should check them when factoring an expression.

TABLE 1.1 Factoring Procedure

Type	Form	Comments
1. Common factors	$ax + ay + az = a(x + y + z)$	Use the distributive property. It can be applied with any number of terms.
2. Difference of squares	$x^2 - y^2 = (x - y)(x + y)$	Remember that the *sum* of two squares cannot be factored in the set of real numbers.
3. FOIL	$x^2 + (c + d)x + cd = (x + c)(x + d)$ $acx^2 + (ad + bc)xy + bdy^2 = (ax + by)(cx + dy)$	Use this trial-and-error procedure with trinomials, after checking for common factors and difference of squares. See the examples in this section.

Common Factoring

The distributive property leads to a very important type of factoring called **common factoring**. When simplifying expressions, we read the distributive property from left to right:

$$a(b + c) = ab + ac$$

When factoring common factors, we read it from right to left, as shown in the following box.

Common Factoring

$$ab + ac = a(b + c)$$
$$ab - ac = a(b - c)$$

EXAMPLE 1 $5x^2 - 25x = (5x)x - (5x)5$
$= 5x(x - 5)$ ■

Common factoring extends to any number of terms.

EXAMPLE 2 $2x^3 - 20x^2 + 6x = 2x(x^2) - 2x(10x) + 2x(3)$
$= 2x(x^2 - 10x + 3)$ ■

If we factor out the *greatest factor* common to each term, or if no other factor can be found, we say the polynomial is **completely factored**.

EXAMPLE 3 $10x^2y + 25xy^2 - 15x^2y^3 = 5xy(2x) + 5xy(5y) + 5xy(-3xy^2)$
$= 5xy(2x + 5y - 3xy^2)$ ■

Common factors refer to the base numerals, not to the exponents, but notice that the smallest exponent (namely, one on the x and y in Example 3) on a common base number leads to the appropriate common factor. For example, $3x^3y^5 + 5x^4y^2$ has a common factor x^3y^2, since the smallest exponent on the common factor x is 3 and on the common factor y is 2.

We usually factor **over the set of integers**, which means that all the numerical coefficients are integers. If the original polynomial has fractional coefficients, however, we factor out the fractional part first.

EXAMPLE 4 $\frac{1}{36}x^2 - 5x + 1 = \frac{1}{36}x^2 - \frac{36}{36}5x + \frac{36}{36}(1)$
$= \frac{1}{36}(x^2 - 180x + 36)$ ■

Common factors do not have to be monomials; they can be any algebraic expression.

EXAMPLE 5 $5x(3a - 2b) + y(3a - 2b) = (5x + y)(3a - 2b)$ The common factor is $(3a - 2b)$. ■

The most important type of factoring used in calculus is finding common factors that are not monomials (as shown in Example 5), except in calculus they are often longer. The following factoring problem is found in Chapter 10.

EXAMPLE 6 Factor $(x^2 + 5x - 8)(3)(5x + 2)^2(5) + (2x + 5)(5x + 2)^3$.

Solution You must first recognize the two terms in this expresssion:

$$(x^2 + 5x - 8)(3)(5x + 2)^2(5) + (2x + 5)(5x + 2)^3$$

Remember that the order of operations groups together the parentheses and the multiplications and the terms are separated by additions or subtractions. The common factor is $(5x + 2)^2$.

$$\begin{aligned}
&(5x + 2)^2\,(x^2 + 5x - 8)(3)(5) + (5x + 2)^2\,(2x + 5)(5x + 2)\\
&= (5x^2 + 2)^2[15(x^2 + 5x - 8) + (2x + 5)(5x + 2)]\\
&= (5x^2 + 2)^2(15x^2 + 75x - 120 + 10x^2 + 29x + 10)\\
&= (5x^2 + 2)^2(25x^2 + 104x - 110)
\end{aligned}$$

■

Difference of Squares

The second type of factoring involves determining whether the expression is a difference of squares.

Difference of Squares

$$x^2 - y^2 = (x - y)(x + y)$$

EXAMPLE 7 **a.** $9x^2 - 25y^2 = (3x)^2 - (5y)^2 = (3x - 5y)(3x + 5y)$

b. $16x^4 - 1 = (4x^2)^2 - (1)^2$ This can be a mental step.

$= (4x^2 - 1)(4x^2 + 1)$ Difference of squares

$= (2x - 1)(2x + 1)(4x^2 + 1)$ Difference of squares again

c. $x^2 - 3$ is irreducible over the set of integers (3 is not a perfect square). Factoring over the set of integers rules out factoring this expression as

$$x^2 - 3 = (x - \sqrt{3})(x + \sqrt{3})$$

since the factors do not have integer coefficients. In this book, an expression is called completely factored if all fractions are eliminated by common factoring and if no further factoring is possible *over the set of integers*.

d. $x^2 + 4$ is irreducible over the set of integers (it is not a difference but a sum of two squares). ■

FOIL—Factoring a Trinomial

Consider the product of two binomials:

$$(2x - 3)(x + 1) = 2x^2 - 3x + 2x - 3 = 2x^2 - x - 3$$

If you understand where the terms of the product came from, you will find it easier to reverse the process. In the following discussion we will concentrate on a particular product, a second-degree polynomial in a single variable, like the one above. You can then see how this case applies to similar products.

First, examine the second-degree (or leading) term and the last term (the constant):

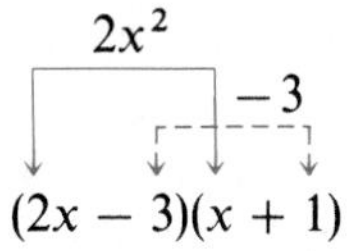

$$2x^2 \qquad -3$$

$$(2x - 3)(x + 1)$$

These terms are the products of the variable terms and the constants of the binomial factors.

Now, recall the origin of the first-degree (or middle) term of the product:

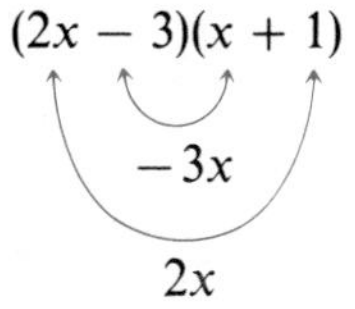

$$(2x - 3)(x + 1)$$

$$-3x$$

$$2x$$

This term is the sum of the products of the variable and constant terms in the binomial factors.

Now consider a product and reverse the multiplication procedure to determine the factors:

$$3x^2 + 13x - 10$$

1. First, find two factors whose product is $3x^2$. These determine the variable terms of the factors and hence the form of the factors:

   $$(x \qquad)(3x \qquad)$$

2. Next, factor the constant term. These factors will yield all possible pairs of factors:

$(x + 2)(3x - 5)$	$(x - 2)(3x + 5)$	$(x + 5)(3x - 2)$
$(x - 5)(3x - 2)$	$(x + 1)(3x - 10)$	$(x - 1)(3x + 10)$
$(x + 10)(3x - 1)$	$(x - 10)(3x + 1)$	

3. Then check each of the possibilities to see which gives the correct middle term:

   $$(x + 5)(3x - 2) = 3x^2 + 13x - 10$$

Thus we factor a polynomial by reversing our knowledge of multiplication. Not all examples are this easy, and, indeed, not all can be factored over the set of integers. If no possibility yields the correct middle term, the polynomial is not factorable over the set of integers. For example, $x^2 + x + 1$ must be of the form

$$(x \quad 1)(x \quad 1)$$

but the only possibilities for the constant terms are

$$(x - 1)(x - 1) \quad \text{and} \quad (x + 1)(x + 1)$$

and no possibility yields the correct middle term. Thus we say that $x^2 + x + 1$ is not factorable over the set of integers.

Procedure for Factoring a Trinomial

1. Find the factors of the leading term and set up the binomials.
2. Find the factors of the constant term, and consider all possible binomials.
3. Determine the factors that yield the correct middle term.
4. If no pair of factors produces the correct full product, then the trinomial is not factorable over the set of integers.

EXAMPLE 8 $x^2 - 8x + 15 = (x - 5)(x - 3)$ ■

EXAMPLE 9 $6x^2 + x - 12 = (2x + 3)(3x - 4)$ ■

EXAMPLE 10 $x^2 - 2xy + y^2 = (x - y)(x - y) = (x - y)^2$

If both the binomial factors are the same, then the expression is called a **perfect square**. ■

Factoring problems are generally not divided into categories as in Examples 1–10. We should always begin by looking for a common factor and then a difference of squares, and finally trying FOIL, as shown in Examples 11–13.

EXAMPLE 11

$$\begin{aligned} 3x^2 - 75 &= 3(x^2 - 25) && \text{Common factor} \\ &= 3(x - 5)(x + 5) && \text{Difference of squares} \end{aligned}$$ ■

EXAMPLE 12

$$\begin{aligned} (x + 3y)^2 - 1 &= [(x + 3y) - 1][(x + 3y) + 1] \\ &= (x + 3y - 1)(x + 3y + 1) \end{aligned}$$ ■

EXAMPLE 13

$$\begin{aligned} 6ax^2 - 21ax - 12a &= 3a(2x^2 - 7x - 4) \\ &= 3a(2x + 1)(x - 4) \end{aligned}$$ ■

In calculus, you must sometimes use common factoring with negative or fractional exponents, as illustrated by the next two examples.

EXAMPLE 14 Factor $-6x^{-4} - x^{-3} + x^{-2}$.

Solution The common factor is found by looking at the smallest power on the common base. It is x^{-4} for this example.

$$-6x^{-4} - x^{-3} + x^{-2} = x^{-4}(-6 - x + x^2)$$ Remember to add exponents when multiplying numbers with the same base.

$$= x^{-4}(x^2 - x - 6)$$ Factor trinomial, if possible.

$$= x^{-4}(x - 3)(x + 2)$$ ■

EXAMPLE 15 Factor $2x^{1/2} + x^{-1/2} + x^{3/2}$.

Solution With a common base, x in this example, the common factor will be that common base with the smallest value of the exponents on that base. The smallest exponent in this example is $-\frac{1}{2}$, so the common factor is $x^{-1/2}$:

$$2x^{1/2} + x^{-1/2} + x^{3/2} = x^{-1/2}(2x + 1 + x^2)$$

$$= x^{-1/2}(x^2 + 2x + 1)$$ Factor trinomial, if possible.

$$= x^{-1/2}(x + 1)^2$$ ■

1.4 Problem Set

Factor the expressions, if possible, in Problems 1–60.

1. $20xy - 12x$
2. $8xy - 6x$
3. $6x - 2$
4. $5y + 5$
5. $xy + xz^2 + 3x$
6. $a^2 - b^2$
7. $a^2 + b^2$
8. $s^2 + 2st + t^2$
9. $m^2 - 2mn + n^2$
10. $u^2 + 2uv + v^2$
11. $x^{-3} + x^{-1} + x^2$
12. $x^{-4} + x^{-1} + x$
13. $(4x - 1)x + (4x - 1)3$
14. $(a + b)x + (a + b)y$
15. $2x^2 + 7x - 15$
16. $x^2 - 2x - 35$
17. $3x^2 - 5x - 2$
18. $6y^2 - 7y + 2$
19. $2x^{-1} - 10x^{-2} - 48x^{-3}$
20. $4x + 1 - 21x^{-1}$
21. $2 - 5x^{-1} + 3x^{-2}$
22. $3x^{-1} + 8x^{-2} - 3x^{-3}$
23. $x^{1/3} + x^{4/3} + x^{7/3}$
24. $x^{1/6} + x^{7/6} + x^{13/6}$
25. $(x - y)^2 - 1$
26. $(2x + 3)^2 - 1$
27. $x^{3/2} + 6x^{1/2} + 9x^{-1/2}$
28. $2x^{3/2} - 7x^{1/2} - 4x^{-1/2}$
29. $(a + b)^2 - (x + y)^2$
30. $(m - 2)^2 - (m + 1)^2$
31. $2x^2 + x - 6$
32. $3x^2 - 11x - 4$
33. $6x^2 + 47x - 8$
34. $6x^2 - 47x - 8$
35. $6x^2 + 49x + 8$
36. $6x^2 - 49x + 8$
37. $4x^2 + 13x - 12$
38. $9x^2 - 43x - 10$
39. $9x^2 - 56x + 12$
40. $12x^2 + 12x - 25$
41. $10x^2 - 9 - x^4$
42. $5x^2 - 4 - x^4$
43. $(x^2 - \frac{1}{4})(x^2 - \frac{1}{9})$
44. $(x^2 - \frac{1}{4})(x^2 - \frac{1}{16})$
45. $(x^2 - 3x - 6)^2 - 4$
46. $(x + y + 2z)^2 - (x - y + 2z)^2$
47. $2(x + y)^2 - 5(x + y)(a + b) - 3(a + b)^2$
48. $2(s + t)^2 + 3(s + t)(s + 2t) - 2(s + 2t)^2$
49. $18x^2(x - 5)^3 + 6x^3(3)(x - 5)^2$
50. $12x^3(x - 8)^4 + 3x^4(4)(x - 8)^3$
51. $10x(3x + 1)^3 + 5x^2(3)(3x + 1)^2(3)$
52. $12x^3(2x - 1)^3 + 3x^4(3)(2x - 1)^2(2)$
53. $2x(4x + 3)^3 + 3x^2(4x + 3)^2(4)$
54. $3x^2(5x - 2)^2 + 2x^3(5x - 2)(5)$
55. $3(x + 1)^2(x - 2)^4 + 4(x + 1)^3(x - 2)^3$
56. $4(x - 5)^3(x + 3)^2 + 2(x - 5)^4(x + 3)$
57. $3(2x - 1)^2(2)(3x + 2)^2 + 2(2x - 1)^3(3x + 2)(3)$
58. $(2x - 3)^3(3)(1 - x)^2(-1) + 3(2x - 3)^2(1 - x)^3(2)$
59. $4(x + 5)^3(x^2 - 2)^3 + (x + 5)^4(3)(x^2 - 2)^2(2x)$
60. $5(x - 2)^4(x^2 + 1)^3 + (x - 2)^5(3)(x^2 + 1)^2(2x)$

1.5 Linear Equations and Inequalities

Linear Equations

A **linear equation in one variable** is an equation that can be written in the form

$$ax + b = 0 \qquad a \neq 0$$

where x is a variable and a and b are any real numbers. An **open** or **conditional equation** is an equation containing a variable that may be either true or false, depending on the replacement for the variable. A **root** or a **solution** is a replacement for the variable that makes the equation true. We also say that the root **satisfies** the equation. The **solution set** of an open equation is the set of all solutions of the equation. To **solve an equation** means to find its solution set. If there are no values for the variable that satisfy the equation, then the solution set is said to be **empty** and is denoted by $\varnothing$. If every replacement of the variable makes the equation true, then the equation is called an **identity**. Two equations with the same solution set are called **equivalent equations**. The process of solving an equation involves finding a sequence of equivalent equations; it ends when the solution or solutions are obvious. There are a few operations that produce equivalent equations, some of which are summarized in the following box.

Properties of Equations

If P and Q are algebraic expressions, and k is a real number, then each of the following is equivalent to $P = Q$:

ADDITION PROPERTY:	$P + k = Q + k$
SUBTRACTION PROPERTY:	$P - k = Q - k$
MULTIPLICATION PROPERTY:	$kP = kQ, \qquad k \neq 0$
DIVISION PROPERTY:	$\dfrac{P}{k} = \dfrac{Q}{k} \qquad k \neq 0$

EXAMPLE 1 Solve $3x + 5 = x - 3$.

Solution

$3x + 5 = x - 3$	
$3x + 5 - x = x - x - 3$	Subtract x from both sides. (Do this step in your head.)
$2x + 5 = -3$	Simplify both sides.
$2x + 5 - 5 = -3 - 5$	Subtract 5 from both sides (mentally).
$2x = -8$	
$x = -4$	Divide both sides by 2 (mental step—not shown).
Check: $3(-4) + 5 = -4 - 3$	
$-7 = -7$	This is true, so the solution checks. ■

EXAMPLE 2 Solve $2x - 5(x - 2) = 3(3 - x)$.

Solution

$2x - 5(x - 2) = 3(3 - x)$

$2x - 5x + 10 = 9 - 3x$ Eliminate the parentheses, then simplify.

$-3x + 10 = 9 - 3x$

$10 = 9$ Add $3x$ to both sides. This results in a false equation.

This is a contradiction, and, since it is equivalent to the original equation, the solution set is empty. ■

EXAMPLE 3 Solve $2x - (7 - x) = x + 1 - 2(4 - x)$.

Solution

$2x - (7 - x) = x + 1 - 2(4 - x)$

$2x - 7 + x = x + 1 - 8 + 2x$ Remove the parentheses first.

$3x - 7 = 3x - 7$ Combine similar terms.

$-7 = -7$ True equation

This is an identity. Because it is equivalent to the original equation, the solution set is the set of all real numbers so that x can be replaced by any real number and the equation will be true. ■

It is frequently necessary to solve variable equations. The following example comes from Section 11.1.

EXAMPLE 4 Solve $2x + xy' + y + 2yy' = 0$ for y'. The variable y' is read "why-prime" and is not the same as y, nor is it y to the first power. Treat y' as a separate variable.

Solution Isolate the terms involving y' on one side.

$xy' + 2yy' = -2x - y$

$(x + 2y)y' = -2x - y$ Isolate the y' by factoring.

$$y' = \frac{-2x - y}{x + 2y}$$ Divide both sides by $x + 2y$. ■

Linear Inequalities

Another type of statement is an **inequality**. Inequalities can be sentences that are always true (for example, $x - 1 < x$ or $5 < 7$), called **absolute inequalities**; always false (for example, $x > x$), called **contradictions**; or sometimes true and sometimes false (for example, $x > 2$), called **conditional inequalities**. The latter can be solved by using a set of properties similar to those for equations.

A notation called **interval notation** is useful in much of the following discussion. Interval notation uses the idea of an ordered pair listing the left and right endpoints of the interval as the first and second components, respectively. The ordered pair is enclosed in brackets or parentheses. Brackets are used when the endpoint is included in the interval, and parentheses are used when the endpoint is excluded from the interval. The smaller number of the pair of numbers a and b must be the first component. The idea is rather simple, as Table 1.2 (page 22) shows.

TABLE 1.2
Interval notation for line segments

	Interval notation	Inequality notation	Line graph
Closed interval	$[a, b]$	$a \le x \le b$	a b
	$[a, b)$	$a \le x < b$	a b
	$(a, b]$	$a < x \le b$	a b
Open interval	(a, b)	$a < x < b$	a b

The symbols ∞ and $-\infty$ are used to denote rays, as shown in Table 1.3.

TABLE 1.3
Interval notation for rays

Interval notation	Inequality notation	Line graph
$(-\infty, b]$	$x \le b$	a b
$(-\infty, b)$	$x < b$	a b
$[a, \infty)$	$x \ge a$	a b
(a, ∞)	$x > a$	a b

Note that the open interval notation ( ) is always used with ∞ and $-\infty$, and that [ ] is never used.

Several properties of inequalities are needed for solving linear inequalities.

Properties of Inequalities

If P and Q are algebraic expressions, and k is a real number, then each of the following is equivalent to $P < Q$:

ADDITION PROPERTY: $P + k < Q + k$

SUBTRACTION PROPERTY: $P - k < Q - k$

MULTIPLICATION PROPERTY:

Positive number k: $kP < kQ$
Negative number k: $kP > kQ$ Note that the inequality is reversed.

DIVISION PROPERTY:

Positive number k: $\frac{P}{k} < \frac{Q}{k}$

Negative number k: $\frac{P}{k} > \frac{Q}{k}$ Note that the inequality is reversed.

These properties also hold for $\le$, $>$, and $\ge$.

Essentially, properties of inequalities allow any operation that is allowed for equations, *except that multiplying or dividing by a negative number reverses the sense of the inequality.*

EXAMPLE 5 Solve $5 - 2(3x - 4) < 4x - 7$.

Solution

$$5 - 2(3x - 4) < 4x - 7$$

$$5 - 6x + 8 < 4x - 7$$

$$-6x + 13 < 4x - 7$$

$$-10x + 13 < -7 \qquad \text{Subtract } 4x \text{ from both sides.}$$

$$-10x < -20 \qquad \text{Subtract 13 from both sides.}$$

$$x > 2 \qquad \text{Divide both sides by } -10\text{; note that the order of the inequality has changed.}$$

We say the solution is the interval $(2, \infty)$. The graph of this solution is

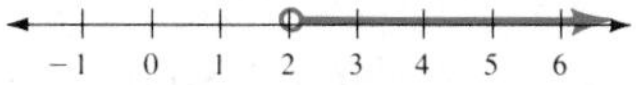

Notice that the open point indicates that $x = 2$ is excluded. ■

EXAMPLE 6

$$5(2x - 3) - 4(x - 2) \leq 3(x - 3)$$

$$10x - 15 - 4x + 8 \leq 3x - 9$$

$$6x - 7 \leq 3x - 9$$

$$3x - 7 \leq -9$$

$$3x \leq -2$$

$$x \leq -\frac{2}{3}$$

The solution is $(-\infty, -\frac{2}{3}]$. The graph of this solution is

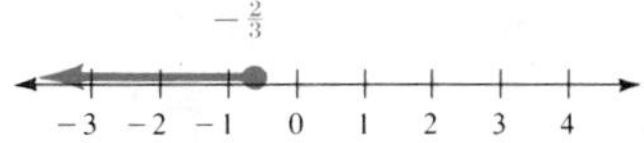

Notice that the closed point indicates that $x = -\frac{2}{3}$ is included. ■

A "string of inequalities" may be used to show the order of three or more quantities. For example, $2 < x < 5$ states that x is a number *between* 2 and 5. The statement is a *compound inequality*, equivalent to $x > 2$ and $x < 5$ at the *same* time, and is graphed as the interval on the number line between 2 and 5:

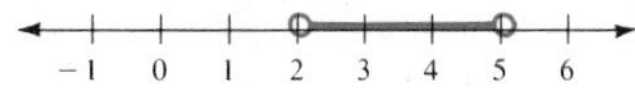

Such inequalities may be solved in a way similar to that used for other inequalities. Note that what is done to one member of the string in each of the following inequalities is done to *all* three members. We try to isolate the x in the middle part of the string.

EXAMPLE 7

$$-3 \le 2x - 5 \le 7$$

$$-3 + 5 \le 2x - 5 + 5 \le 7 + 5$$ Add 5 to all three members.

$$2 \le 2x \le 12$$

$$1 \le x \le 6$$ Divide all three members by 2 (the inequality keeps the same direction).

This says the solution is $[1, 6]$. ■

EXAMPLE 8

$$-2 < 1 - 3x < 7$$

$$-3 < \quad -3x < 6$$ Subtract 1.

$$1 > x > -2$$ Divide by -3; reverse the order of the inequalities.

$$-2 < x < 1$$

The string of inequalities $1 < 2 < 3$ is equivalent to $3 > 2 > 1$. The first states the order of the three integers from smallest to largest, and the second from the largest to smallest. It is standard practice to use the ascending order from smallest to largest, so inequalities are stated with the less than ($<$) relation whenever possible. Notice that $1 > x > -2$ should be rewritten as $-2 < x < 1$ or $(-2, 1)$. ■

1.5 Problem Set

Solve each equation in Problems 1–16.

1. $3x = 5x - 4$ **2.** $2x = 9 - x$

3. $7x + 10 = 5x$ **4.** $9x - 8 = 5x$

5. $3x + 22 = 1 - 4x$ **6.** $5x - 4 = 3x + 8$

7. $2x - 13 = 7x + 2$ **8.** $3x + 32 = 18 - 4x$

9. $7x + 18 = -2x$ **10.** $6x - 1 = 13 - x$

11. $8x - 3 = 15 - x$ **12.** $-5x = 9 - 2x$

13. $2(x - 1) = 1 + 3(x - 2)$ **14.** $11 - x = 2 - 3(x - 1)$

15. $5 - 2x = 1 + 3(x - 2)$ **16.** $1 - 2(x - 3) = 2x - 1$

Solve each inequality in Problems 17–24.

17. $3x - 2 \le 7$ **18.** $2x - 1 \ge 9$

19. $4 - 5x > 29$ **20.** $7 - 4x < 3$

21. $9x + 7 \ge 5x - 9$ **22.** $3x + 7 > 7x - 5$

23. $2(3 - 4x) < 30$ **24.** $3(2x - 5) \le 9$

Solve the compound inequalities in Problems 25–30.

25. $7 < x + 2 < 11$ **26.** $12 < 5 + x < 14$

27. $-2 < x - 1 < 3$ **28.** $-5 \le x - 2 \le 4$

29. $9 < 1 - 2x < 15$ **30.** $-3 \le 1 - 2x \le 7$

Solve each equation in Problems 31–42 for y'.

31. $5x^4 - 5y^4y' = 0$ **32.** $3x^2y^2y' + 2xy^3 = 0$

33. $4x^3y^2 + 2x^4yy' = 0$ **34.** $6x^5y^3 + 3x^6y^2y' = 0$

35. $2x - (xy' + y) + 2yy' = 0$

36. $2y - (x + xy') + 2x = 0$

37. $3x^2 + 4x + (xy' + y) = 0$

38. $2x + (xy + y') + 2yy' = 0$

39. $\frac{1}{25}[2(x + 1)] + \frac{1}{4}[2(y - 1)y'] = 0$

40. $\frac{1}{9}[3(x - 4)] + \frac{1}{12}[3(y + 2)y'] = 0$

41. $3[x^2(3y^2y') + 2xy^3] - 3[x(2yy') + y^2] + 5(xy' + y) = 0$

42. $5[x^2(5y^2y') + 4xy^3] - 5[x(4yy') + y^2] + 6(xy' + y) = 0$

Solve each equation or inequality in Problems 43–60.

43. $2(x - 3) - 5x = 3(1 - 2x)$

44. $3(2x + 7) + 11 = 5(2 - x)$

45. $5(x - 1) + 3(2 - 4x) > 8$

46. $4(1 - x) - 7(2x - 5) < 3$

47. $4(x - 2) + 1 \le 3(x + 1)$

48. $6(2x - 1) \ge 3(x + 4)$

49. $2(4 - 3x) > 4(3 - x)$

50. $3(2 - 5x) \ge 5(x + 2) + 36$

51. $2(3 - 7x) \ge -4 - (5 - x)$

52. $3(7 - x) < 2(x - 2)$

53. $3(1 - x) - 5(x - 2) = 5$

54. $5(5 - 3x) - (x - 8) = 1$

55. $6(2x + 5) = 4(3x + 1)$

56. $9(2 - 3x) = 6(5x + 3)$

57. $3(2x - 5) = 5(4x - 3)$

58. $2(4 - 3x) = 3(x - 2) - (9x - 14)$

59. $5(1 - 2x) = 3(x - 4) - (13x - 17)$

60. $6(1 - 4x) = 3(x - 1) - (5x + 13)$

1.6 Quadratic Equations and Inequalities

Quadratic Equations

A *quadratic equation in one variable* is an equation that can be written in the form

$$ax^2 + bx + c = 0 \qquad a \neq 0$$

where x is a variable, and a, b, and c are real numbers. We will consider three different methods for solving quadratic equations—factoring, square root, and the quadratic formula.

The simplest method can be used if the quadratic expression $ax^2 + bx + c$ is factorable over the integers. The solution then depends on the following property of zero.

Property of Zero

$$AB = 0 \quad \text{if and only if} \quad A = 0 \text{ or } B = 0 \text{ (or both)}$$

Thus, if the product of two factors is zero, then at least one of the factors is zero. If a quadratic equation is factorable, this property provides a method of solution.

EXAMPLE 1 Solve $x^2 = 2x + 15$.

Solution

$x^2 - 2x - 15 = 0$ — Rewrite with a zero on one side (subtract $2x$ and 15 from both sides).

$(x + 3)(x - 5) = 0$ — Factor.

$x + 3 = 0$ or $x - 5 = 0$ — Since the product is zero, one of the factors must be zero.

$x = -3$ $\qquad x = 5$

Solution: $\{-3, 5\}$ ■

You check a quadratic equation just as you check a linear equation, by substitution to see if you obtain a true or false equation. For Example 1,

Check $x = -3$: $(-3)^2 = 2(-3) + 15$, $9 = -6 + 15$, $9 = 9$ ✓

Check $x = 5$: $5^2 = 2(5) + 15$, $25 = 10 + 15$, $25 = 25$ ✓

Solution of Quadratic Equations by Factoring

To solve a quadratic equation that can be expressed as a product of linear factors:

1. Rewrite all nonzero terms on one side of the equation.
2. Factor the expression.
3. Set each of the factors equal to zero.
4. Solve each of the linear equations.
5. Write the solution set that is the union of the solution sets of the linear equations.

EXAMPLE 2 Solve $x^2 - 6x = 0$.

Solution Factor, if possible: $x(x - 6) = 0$

Set each factor equal to zero: $x = 0$ $\quad x - 6 = 0$

$$x = 6$$

Solution is $\{0, 6\}$. However, $x = 0$, $x = 6$ or $x = 0, 6$ is a sufficient answer. ■

When the quadratic equation is not factorable, other methods must be employed. One such method depends on the **square root property**.

Square Root Property

$$P^2 = Q \quad \text{if and only if} \quad P = \pm\sqrt{Q}.$$

For example, the equation $x^2 = 4$ can be rewritten as $x^2 - 4 = 0$, factored, and solved. However, the square root property can also be used:

Square root property	*Factoring*
$x^2 = 4$	$x^2 - 4 = 0$
$x = \pm\sqrt{4}$	$(x - 2)(x + 2) = 0$
$x = \pm 2$	$x = 2, -2$

The square root property can be derived by using the following property of square roots:

Square Root of a Real Number

For all real numbers x,

$$\sqrt{x^2} = |x|$$

This means that, if $x \geq 0$, then $\sqrt{x^2} = x$ and if $x < 0$, then $\sqrt{x^2} = -x$. For example, $\sqrt{2^2} = |2| = 2$ and $\sqrt{(-2)^2} = |-2| = 2$.

The importance of the square root property is that it can be applied to *any* quadratic of the form $ax^2 + bx + c = 0$, $a \neq 0$. The result (which is derived by completing the square and is shown in most algebra textbooks) is called the **quadratic**

formula. This method for solving quadratic equations will solve any quadratic, so in practice we will usually try to solve a quadratic equation by factoring, and if that does not easily work we will go directly to the quadratic formula.

Quadratic Formula

If $ax^2 + bx + c = 0$ with $a \neq 0$, then

$$x = \frac{-b \pm \sqrt{b^2 - 4ac}}{2a}$$

EXAMPLE 3 Solve $5x^2 + 2x - 2 = 0$.

Solution Note that $a = 5$, $b = 2$, and $c = -2$. Thus

$$\begin{aligned} x &= \frac{-2 \pm \sqrt{4 - 4(5)(-2)}}{2(5)} \\ &= \frac{-2 \pm 2\sqrt{1 + 10}}{2(5)} \\ &= \frac{-1 \pm \sqrt{11}}{5} \end{aligned}$$

■

EXAMPLE 4 Solve $5x^2 + 2x + 2 = 0$.

Solution

$$\begin{aligned} x &= \frac{-2 \pm \sqrt{4 - 4(5)(2)}}{2(5)} \\ &= \frac{-2 \pm 2\sqrt{-9}}{10} \end{aligned}$$

Since the square root of a negative number is not a real number, the solution set is *empty over the reals*.* That is, we say the solution set is $\varnothing$. ■

Since the quadratic formula contains a radical, the sign of the radicand (the number under the radical) will determine whether the roots will be real or nonreal. This radicand is called the **discriminant** of the quadratic, and its properties are summarized below:

Discriminant

If $ax^2 + bx + c = 0$ $(a \neq 0)$, then $D = b^2 - 4ac$ is called the **discriminant**.

If $D < 0$, then there are *no real solutions*.
If $D = 0$, then there is *one real solution*.
If $D > 0$, then there are *two real solutions*.

* In algebra, you may have solved this equation by using complex numbers. However, since the domain for variables in this course is the set of real numbers, and since there is no real number whose square is negative, we see that there is no real number that satisfies this equation.

Quadratic Inequalities

A *quadratic inequality in one variable* is an inequality that can be written in the form

$$ax^2 + bx + c < 0 \qquad a \neq 0$$

where x is a variable, and a, b, and c are any real numbers. The symbol $<$ can be replaced by $\leq$, $>$, or $\geq$ and the inequality will still be a quadratic inequality in one variable.

The procedure for solving a quadratic inequality is similar to that for solving a quadratic equality. First, we use the properties of inequality to obtain a zero on one side of the inequality. Then we factor, if possible, the quadratic expression on the left. For example,

$$x^2 - x - 6 \geq 0$$
$$(x + 2)(x - 3) \geq 0$$

A value that causes one of the factors to be zero is called a **critical value** for the inequality; the critical values for the example above are $x = -2$ and $x = 3$. It follows, then, that the inequality must be either positive or negative for every real number that is not a critical value. We therefore examine the intervals defined by the critical values.

EXAMPLE 5 Solve $x^2 - x < 6$.

Solution First obtain a zero on one side, and then factor, if possible:

$$x^2 - x - 6 < 0$$
$$(x + 2)(x - 3) < 0$$

Next, find the critical values (one at a time), plot them on a number line, and label the parts of the number line to the left and right of the critical value "+" or "−" depending on whether the factor is positive or negative in that region.

Factor: $x + 2$ *Critical value:* $x + 2 = 0$

$$x = -2$$

Plot:

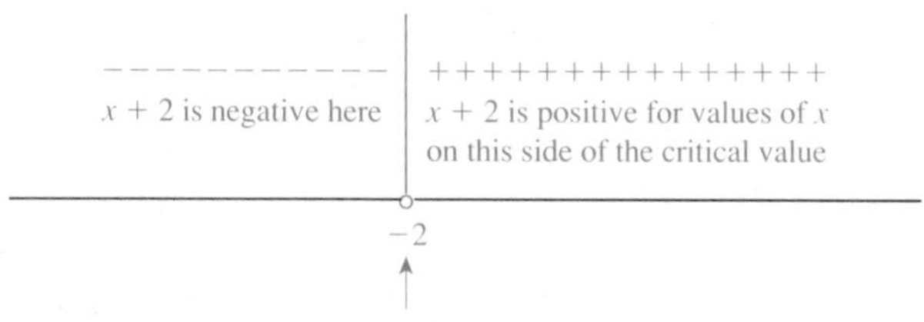

Factor: $x - 3$ *Critical value:* $x - 3 = 0$

$$x = 3$$

Plot (on the same number line):

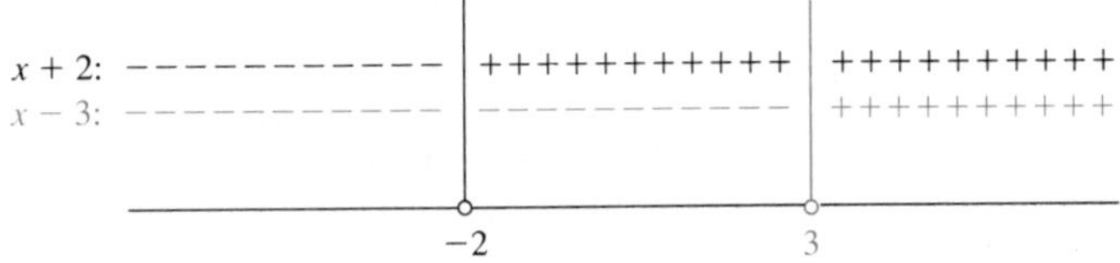

Finally, use the property of products to label the regions of the number line "positive" or "negative" as shown:

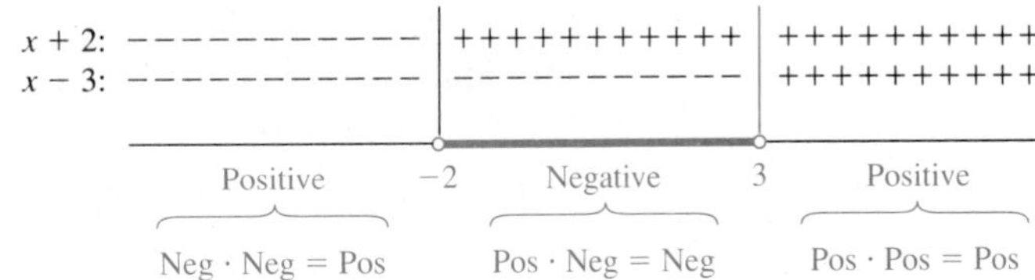

Shade in the appropriate portion of the number line indicated by the original inequality:

$x^2 - x - 6 < 0$ "<0" means you are looking for the portion of the number line labeled "negative."

Answer: $(-2, 3)$ That is, x is between -2 and 3. ■

The previous example seems lengthy because all the steps are shown, but the process is quite easy and will be greatly simplified when you are doing the work. The process is summarized below, while Example 6 shows how your paper should look when doing this type of problem.

Solution of Polynomial Inequalities by Factoring

To solve an inequality that can be expressed as a product less than or greater than zero:

1. Rewrite all nonzero terms on one side of the inequality.
2. Factor the expression.
3. Determine the critical values.
4. Determine the signs of the factors on the intervals between critical values.
5. Select the interval or intervals on which the product has the desired sign.

EXAMPLE 6 Solve $5 + 4x - x^2 \leq 0$.

Solution $(5 - x)(1 + x) \leq 0$ The critical values are 5 and -1. Notice that in this example the critical values are included in the solution since equality is included.

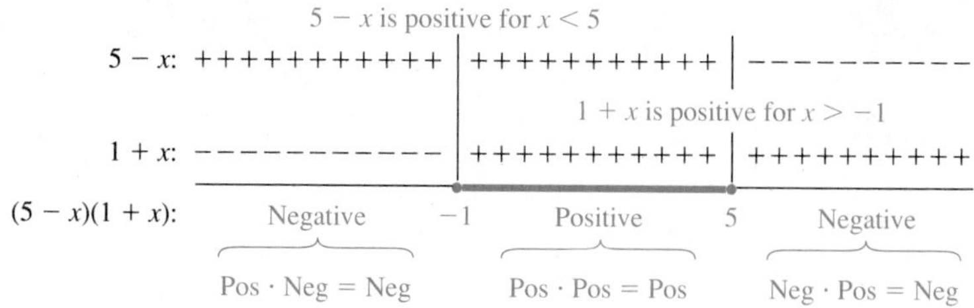

Answer: Since the inequality we are solving is "≤", we look for those portions on the number line where the product is negative ("less than or equal to 0" means the values are negative with the endpoints included). We see that two of the three parts of the number line are labeled negative. This means that any x value in the left one *or* the right one makes the inequality true. The word "or" is written symbolically as "∪". The answer is

$$[-\infty, -1] \cup [5, \infty)$$

■

1.6 Problem Set

Solve each equation in Problems 1–20 over the set of real numbers.

1. $x^2 + 2x - 15 = 0$
2. $x^2 - 8x + 12 = 0$
3. $x^2 + 7x - 18 = 0$
4. $2x^2 + 5x - 12 = 0$
5. $10x^2 - 3x - 4 = 0$
6. $6x^2 + 7x - 10 = 0$
7. $x^2 + 5x - 6 = 0$
8. $x^2 + 5x + 6 = 0$
9. $x^2 - 10x + 25 = 0$
10. $x^2 + 6x + 9 = 0$
11. $12x^2 + 5x - 2 = 0$
12. $2x^2 - 6x + 5 = 0$
13. $5x^2 - 4x + 1 = 0$
14. $2x^2 + x - 15 = 0$
15. $4x^2 - 5 = 0$
16. $3x^2 - 1 = 0$
17. $3x^2 = 7x$
18. $7x^2 = 3$
19. $3x^2 = 5x + 2$
20. $3x^2 - 2 = -5x$

Solve the inequalities in Problems 21–40.

21. $(x - 6)(x - 2) \le 0$
22. $(x + 2)(x - 8) \le 0$
23. $x(x + 3) < 0$
24. $x(x - 3) \ge 0$
25. $(x + 2)(8 - x) \le 0$
26. $(2 - x)(x + 8) \ge 0$
27. $(1 - 3x)(x - 4) < 0$
28. $(2x + 1)(3 - x) > 0$
29. $x^2 \ge 9$
30. $x^2 > 4$
31. $x^2 + 9 \le 0$
32. $x^2 + 2x - 3 < 0$
33. $x^2 - x - 6 > 0$
34. $x^2 - 7x + 12 > 0$
35. $5x - 6 \ge x^2$
36. $4 \ge x^2 + 3x$
37. $5 - 4x \ge x^2$
38. $x^2 + 2x - 1 < 0$
39. $6x^2 - 10 > 59x$
40. $8x^2 < 2 + 15x$

1.7 Rational Expressions

Rational Expressions

If a variable is used in a denominator, then the expression is not called a polynomial. Instead, it is called a **rational expression**.

Rational Expressions

A **rational expression** is an expression that can be written as a polynomial divided by a polynomial. Any values that cause division by zero are excluded from the domain.

The fundamental property used to simplify rational expressions involves factoring both the numerator and denominator and then eliminating common factors according to the property

$$\frac{PK}{QK} = \frac{P}{Q} \quad Q, K \neq 0$$

Some rational expressions are simplified in Examples 1–5. Do not forget that all values for variables that cause division by zero are excluded from the domain.

EXAMPLE 1 $\dfrac{3xyz}{x^2 + 3x} = \dfrac{x(3yz)}{x(x + 3)} = \dfrac{3yz}{x + 3}$ ■

EXAMPLE 2 $\dfrac{x - 2}{x^2 - 4} = \dfrac{1(x - 2)}{(x + 2)(x - 2)} = \dfrac{1}{x + 2}$ ■

Sometimes the factors that are eliminated (as shown in color in Examples 1 and 2) are marked off in pairs as shown in Example 3. The slashes should be viewed as replacing the factor K by the number 1 since

$$\frac{PK}{QK} = \frac{PK}{QK} = \frac{P \cdot 1}{Q \cdot 1} = \frac{P}{Q} \cdot 1 = \frac{P}{Q}$$

EXAMPLE 3 $\dfrac{(x - 5)(x + 2)(x + 1)}{(x + 1)(x - 2)(x - 5)} = \dfrac{(x - 5)(x + 2)(x + 1)}{(x + 1)(x - 2)(x - 5)}$

$= \dfrac{x + 2}{x - 2}$ This is reduced since the x's are terms and *not* factors. You divide factors, not terms. ■

EXAMPLE 4 $\dfrac{6x^2 + 2x - 20}{30x^2 - 68x + 30} = \dfrac{2(3x^2 + x - 10)}{2(15x^2 - 34x + 15)}$ Common factor first.

$= \dfrac{2(3x - 5)(x + 2)}{2(3x - 5)(5x - 3)}$ Complete the factoring; then reduce.

$= \dfrac{x + 2}{5x - 3}$ ■

The laws of exponents can be extended to include rational expressions and negative exponents.

EXAMPLE 5 **a.** $\left(\dfrac{x}{y}\right)^3 = \left(\dfrac{x}{y}\right)\left(\dfrac{x}{y}\right)\left(\dfrac{x}{y}\right) = \dfrac{xxx}{yyy} = \dfrac{x^3}{y^3}$

b. $\dfrac{x^5}{x^3} = \dfrac{xxxxx}{xxx} = xx = x^2$

c. $\dfrac{x^3}{x^5} = \dfrac{xxx}{xxxxx} = \dfrac{1}{x^2} = x^{-2}$ ■

Laws of Exponents

Let a and b be any real numbers, m and n be any integers, and assume that each expression is defined.*

FOURTH LAW: $\left(\frac{a}{b}\right)^m = \frac{a^m}{b^m}$

FIFTH LAW: $\frac{b^m}{b^n} = b^{m-n}$

The procedures for operation on rational expressions are identical to those for operations with fractions. However, with rational expressions, the numerators and denominators are any polynomial (division by zero is excluded) rather than constants. The procedures are summarized below.

Properties of Rational Expressions

Let P, Q, R, S, and K be any polynomials such that all values of the variable that cause division by zero are excluded from the domain.

EQUALITY: $\frac{P}{Q} = \frac{R}{S}$ if and only if $PS = QR$

FUNDAMENTAL PROPERTY: $\frac{PK}{QK} = \frac{P}{Q}$

ADDITION: $\frac{P}{Q} + \frac{R}{S} = \frac{PS + QR}{QS}$

SUBTRACTION: $\frac{P}{Q} - \frac{R}{S} = \frac{PS - QR}{QS}$

MULTIPLICATION: $\frac{P}{Q} \cdot \frac{R}{S} = \frac{PR}{QS}$

DIVISION: $\frac{P}{Q} \div \frac{R}{S} = \frac{PS}{QR}$

Some operations on rational expressions are performed in Examples 6–9. All values of variables that could cause division by zero are excluded.

EXAMPLE 6 $\frac{13}{x - y} + \frac{2}{x - y} = \frac{15}{x - y}$ Common denominator ■

* The first three laws of exponents are in Section 1.3, page 12.

EXAMPLE 7

$$\frac{13}{x-y} - \frac{2}{y-x} = \frac{13}{x-y} + \frac{-2}{y-x}$$

It is helpful to write subtractions as additions.

$$= \frac{13}{x-y} + \frac{-2}{y-x} \cdot \frac{-1}{-1}$$

Multiply by 1 (written as $-1/-1$) in order to obtain a common denominator.

$$= \frac{13}{x-y} + \frac{2}{x-y}$$

$$= \frac{15}{x-y}$$

■

EXAMPLE 8

$$\frac{x+y}{x-y} + \frac{x-2y}{2x+y} = \frac{(x+y)(2x+y)+(x-y)(x-2y)}{(x-y)(2x+y)}$$

$$= \frac{2x^2+3xy+y^2+x^2-3xy+2y^2}{(x-y)(2x+y)}$$

$$= \frac{3x^2+3y^2}{(x-y)(2x+y)}$$

$$= \frac{3(x^2+y^2)}{(x-y)(2x+y)}$$

■

EXAMPLE 9

$$\left(\frac{x^2+5x+6}{2x^2-x-1} \cdot \frac{2x^2-9x-5}{x^2+7x+12}\right) \div \frac{2x^2-13x+15}{x^2+3x-4}$$

$$= \left[\frac{(x+2)(x+3)}{(x-1)(2x+1)} \cdot \frac{(2x+1)(x-5)}{(x+3)(x+4)}\right] \div \frac{(x-5)(2x-3)}{(x-1)(x+4)}$$

$$= \frac{(x+2)\cancel{(x+3)}\cancel{(2x+1)}\cancel{(x-5)}\cancel{(x-1)}\cancel{(x+4)}}{\cancel{(x-1)}\cancel{(2x+1)}\cancel{(x+3)}\cancel{(x+4)}\cancel{(x-5)}(2x-3)}$$

$$= \frac{x+2}{2x-3}$$

■

1.7 Problem Set

Simplify the expressions in Problems 1–30. Values that cause division by zero are excluded. All expressions should be reduced and negative exponents eliminated.

1. $\dfrac{2}{x+y} + 3$
2. $\dfrac{3}{x+y} - 2$
3. $\dfrac{x^2-y^2}{2x+2y}$
4. $\dfrac{x^2-y^2}{3x-3y}$
5. $\dfrac{3x^2-4x-4}{x^2-4}$
6. $\dfrac{x^2+3x-18}{x^2-9}$
7. $\dfrac{3}{x+y} + \dfrac{5}{2x+2y}$
8. $\dfrac{4}{x-y} - \dfrac{3}{2x-2y}$
9. $[7^3 + 2^5(3^3 + 4^4)]^0$
10. $[9^3 + 3^6(5^3 + 8^3)]^0$
11. $(x^2-36)\left(\dfrac{3x+1}{x+6}\right)$
12. $x^2 - 9 \div \dfrac{x+3}{x-3}$
13. $\dfrac{x+3}{x} + \dfrac{3-x}{x^2}$
14. $\dfrac{2}{x-y} + \dfrac{5}{y-x}$
15. $\dfrac{x}{x-1} + \dfrac{x-3}{1-x}$
16. $\dfrac{x+1}{x} + \dfrac{2-x}{x^2}$
17. $\dfrac{2x+3}{x^2} + \dfrac{3-x}{x}$
18. $\dfrac{1}{x^3} + 2xy + \dfrac{x^2}{y^2}$

19. $\frac{x}{y} + 2 + \frac{y}{x}$

20. $\frac{1}{x^3y^2} + \frac{y}{x} + 2xy$

21. $\frac{x}{y} - 2 + \frac{y}{x}$

22. $\frac{1}{2y} + \frac{1}{x} + \frac{y}{2x^2}$

23. $\frac{1}{3xy^2} + xy + \frac{1}{x^3y}$

24. $\frac{2x + y}{(x + y)^2} + \frac{x^2 - 2y^2}{(x + y)^3}$

25. $\frac{4x - 12}{x^2 - 49} \div \frac{18 - 2x^2}{x^2 - 4x - 21}$

26. $\frac{36 - 9x}{3x^2 - 48} \div \frac{15 + 13x + 2x^2}{12 + 11x + 2x^2}$

27. $\frac{1}{x^2 + 1} - \frac{x^2}{x^2 + 1}$

28. $(x^2 + 1) + \frac{x^2}{(x^2 + 1)^2}$

29. $\frac{4x^2}{x^4 - 2x^3} + \frac{8}{4x - x^3} - \frac{-4}{x + 2}$

30. $\frac{6x}{2x + 1} - \frac{2x}{x - 3} + \frac{4x^2}{2x^2 - 5x - 3}$

1.8

*Review

The material of this chapter is reviewed in the following list of objectives. After each objective there are some practice questions. For a sample test select the first question of each set and check your answers. The second question for each objective has no answer given. If you are having trouble with a particular type of problem, look back at the indicated section in the text. When you are finished reviewing these objectives, a sample examination is given at the end of this section.

[1.2]

Objective 1.1: *Classify numbers as natural, whole, integer, rational, irrational, or real.*

1. Classify the following numbers by listing the set(s) into which they fall:

$$-8, \quad \frac{5}{6}, \quad .\overline{23}, \quad \sqrt{169}, \quad \frac{3}{0}$$

2. Draw a diagram showing how the integers, rationals, and reals are related.

Objective 1.2: *Find the absolute value of a number or an expression.*

3. $-|-5|$

4. $|\sqrt{1{,}000} - 35|$ (do not approximate)

5. $|x - 4|$ if $x < -2$

Objective 1.3: *Find the distance between pairs of points on a number line.*

6. $A(-6)$ and $B(11)$

7. $C(2\pi)$ and $D(-4)$

8. $E(4)$ and $F(\sqrt{20})$

[1.3]

Objective 1.4: *Simplify numerical expressions.*

9. **a.** -8^2 **b.** $(-8)^2$ **c.** $-(-8)^2$ **d.** -1^0

10. $6 + 5 \cdot 2 - 8 + 4 + 2$

11. $\frac{6 + 4(-3)}{-2}$

12. $\frac{12 - 2(3)}{-4^2}$

Objective 1.5: *Evaluate algebraic expressions. Let $x = -1$, $y = 2$, and $z = -3$.*

13. $x - (yz - 2z)$

14. $xy - yz + 4xz$

15. **a.** $-x^2$ **b.** $(-x)^2$

16. **a.** $(-y)^2$ **b.** $-y^2$

Objective 1.6: *Mentally multiply binomials using FOIL.*

17. **a.** $(x - 2)(x + 7)$ **b.** $(x + 3)(x - 3)$

18. **a.** $(x + 3)^2$ **b.** $(2x - 1)^2$

19. **a.** $(x + 2y)(x - y)$ **b.** $(x - 3y)(x + 3y)$

20. **a.** $(2x - 3)(x + 5)$ **b.** $(3x - 1)(2x + 3)$

Objective 1.7: *Simplify algebraic expressions.*

21. $(x - y - z) + (2x + y - 3z)$

22. $(5x^2 + 3x - 5) - (2x^2 + 2x + 3)$

23. $2(3 - x^2) - 3(5 - x) - (3x - 2)$

24. $(x + 2)(2x^2 + 3x + 2)$

[1.4]

Objective 1.8: *Factor polynomials.*

25. $x^2 - 5x + 6$

26. $x^2 - 9x + 14$

27. $25 - x^2$

28. $4x^2 - 12x + 8$

29. $9x^4 - 40x^2 + 16$

30. $4x^4 - 13x^3 + 9x^2$

* Optional section.

[1.5]

Objective 1.9: *Solve linear equations.*

31. $90 = 5x - 10$ **32.** $6 - 5x = 5 - 9x$

33. $3 - 4x = 3(5 - 2x)$ **34.** $5(x - 5) + 2 = 3(x - 3)$

Objective 1.10: *Solve linear inequalities.*

35. $3x - 5 > 16$ **36.** $9 - x < -4$

37. $x < 3(2 + x)$ **38.** $2(x - 2) + 3 \le 5(x + 1)$

Objective 1.11: *Solve compound inequalities.*

39. $-8 \le x + 5 \le -3$ **40.** $5 \le x + 1 < 15$

41. $6 < -x < 10$ **42.** $-1 < 5 - 2x \le 1$

[1.6]

Objective 1.12: *Solve quadratic equations.*

43. $(3x - 1)(5x + 2) = 0$ **44.** $(2x + 1)(x - 4) = 11$

45. $6x^2 + 19x = -15$ **46.** $x^2 - 6x + 1 = 0$

47. $x^2 - 6x + 7 = 0$ **48.** $2x^2 + 1 = 2x$

Objective 1.13: *Solve quadratic inequalities.*

49. $(x - 5)(x - 7) < 0$ **50.** $x^2 - 4x \ge 0$

51. $6x^2 > 11x + 10$ **52.** $x^2 - 2x + 3 < 0$

[1.7]

Objective 1.14: *Simplify rational expressions.*

53. $\frac{2x}{15} - \frac{y}{12}$ **54.** $\frac{5x}{x - 1} + 1$

55. $\frac{3x + 2}{9x^2 - 6x + 1} + \frac{1}{3x - 1}$ **56.** $\frac{x^2 - 9}{x + 2} + \frac{x^2 + x - 6}{x^2 - 4}$

SAMPLE TEST

The following sample test (45 minutes) is intended to review the main ideas of this chapter.

1. Draw a diagram showing how the natural numbers, whole number, integers, rationals, and real numbers are related.

2. Write $|2\pi - 10|$ without absolute value symbols (do not approximate).

3. Find the distance between $A(\sqrt{10})$ and $B(5)$.

4. Simplify $\frac{1 + 3 \cdot 5}{-3^2}$.

5. Evaluate $-x^2 + 5x(3 - 2y)$ where $x = -6$ and $y = -3$.

6. Simplify $(3x + 1)(4x - 5)$.

7. Simplify $(x + y)^2$.

8. Simplify $(x + 3)(2x^2 - 5x - 4)$.

9. Factor $6x^2 - 29x - 5$.

10. Factor $1 - 9x^2$.

11. Solve $4 - 3x = 2(5 - 2x)$.

12. Solve $15 < -x$.

13. Solve $2(x + 1) + 5 < 3(1 - x)$.

14. Solve $-3 \le x - 5 < 0$.

15. Solve $x = 5x^2$.

16. Solve $x^2 + 2x - 2 = 0$.

17. Solve $(1 - x)(3x - 2) > 0$.

18. Solve $x^2 - 5 \ge 4x$.

19. Simplify $\frac{7x}{30} - \frac{y}{12}$.

20. Simplify $\frac{x^2 - 25}{x + 1} \div \frac{x^2 - 4x - 5}{x^2 + 2x + 1}$.

2

Functions and Graphs

CHAPTER OVERVIEW

Chapter 2 introduces the building blocks for this course. We begin by reviewing the concept of a function and evaluating functions and functional notation. Then we discuss graphing functions, and in the remainder of the chapter focus on three very important types of functions: linear, quadratic, and rational. The pretest in Section 2.1 previews the material of this chapter.

PERSPECTIVE

This book is divided into two parts, finite mathematics and calculus. Calculus is concerned with infinite processes, and as the name implies, finite mathematics deals with finite, or countable, ideas. This chapter is a prerequisite for both parts of the course.

2.1

Functions and Graphs Pretest

Choose the best answer in Problems 1–20.

1. Which of the following is *false*?
   A. A function is a number.
   B. A function can be defined by a verbal rule.
   C. A function can be defined by a table.
   D. A function can be defined by a graph.
   E. A function can be defined by an algebraic formula.

2. If $f(x) = 3x^2 + 2$, then $f(-2)$ is
   A. y B. -10
   C. $-6x^2 - 4$ D. 14
   E. none of these

3. If $f(x) = 3x^2 + x - 7$, then $f(a - 1) - f(a)$ is
   A. $6x + 2$ B. 2
   C. $2 - 6a$ D. $-4(a + 3)$
   E. none of these

4. If $f(x) = 3x^2 + 2$, then $\dfrac{f(x + h) - f(x)}{h}$ is
   A. $6x + 3h$ B. $\dfrac{6xh + 3h^2 + 4}{h}$
   C. $\dfrac{6x^2 + 6xh + h^2 + 2}{h}$ D. $\dfrac{3x^3 + 3x^2h + 2x + 2h}{h}$
   E. none of these

5. In (h, k), k is the
   A. slope
   B. independent variable
   C. dependent variable
   D. abscissa
   E. domain

6. Which of the following has *no* y-intercept?
   A. $y = 0$ B. $x + y = 0$
   C. $y = x$ D. $x = 2$
   E. $x = y + 2$

7. Consider the function f defined by the figure below. The coordinates of P are
   A. (x, y) B. $(x_1, f(x_1))$
   C. $(x_1, x_1 + h)$ D. about $(1, 2)$
   E. none of these

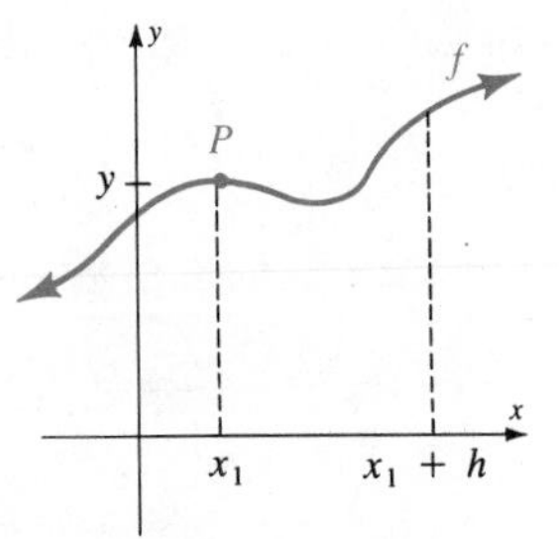

8. What is the slope of the line containing $(5, 2)$ and $(-1, 6)$?
   A. $-\frac{2}{7}$ B. $-\frac{4}{7}$
   C. $-\frac{2}{3}$ D. $-\frac{4}{3}$
   E. none of these

9. If f is a function and $P(x_1, y_1)$ and $Q(x_2, y_2)$ are points on the graph of f, then the slope of the line passing through P and Q is
   A. $\dfrac{x_2 - x_1}{y_2 - y_1}$ B. $f(m)$
   C. $\dfrac{f(x_2) - f(x_1)}{x_2 - x_1}$ D. $\dfrac{f(y_2) - f(y_1)}{x_2 - x_1}$
   E. none of these

10. The equation of the line with y-intercept -2 passing through $(-3, -4)$ is
    A. $2x + y - 2 = 0$ B. $2x + y + 10 = 0$
    C. $2x + y + 11 = 0$ D. $2x - 3y - 6 = 0$
    E. none of these

11. Which of the following is the equation of the line with no slope that passes through $(2, -3)$?
    A. $x = 2$ B. $x = -3$
    C. $y = 2$ D. $y = -3$
    E. none of these

12. The graph of the equation $3x - 2y^2 - 4y = 0$ is
    A. a line
    B. a circle
    C. a parabola
    D. a rational function with at least one asymptote
    E. none of these

13. The graph of the equations $x = 2 + 5t$ and $y = 1 + 2t$ is
    A. a line
    B. a circle
    C. a parabola
    D. a rational function with at least one asymptote
    E. none of these

14. Any number x such that $f(x) = 0$ is called
    A. an x-intercept
    B. a y-intercept
    C. a zero of the function
    D. $x = 0$
    E. none of these

15. A manufacturer selling a product produces x items per day and prices them at $440 - x$ dollars each. The overhead is found to be $9x^2 - 7{,}560x + 1{,}584{,}000$ dollars. Then

$$\begin{aligned}\text{Profit} &= \text{Revenue} - \text{Cost}\\ &= (\text{Number of items})(\text{Price}) - \text{Cost}\end{aligned}$$

$$P = x(440 - x) - (9x^2 - 7{,}560x + 1{,}584{,}000)$$
$$P - 16{,}000 = -10(x - 400)^2$$

What is the maximum profit (in dollars)?

A. 80 **B.** 16,000
C. 400 **D.** 1,658,000
E. none of these

16. The domain for the curve $y^2 + x - 5 = 0$ is
A. $0 < x < 5$ **B.** $0 \le x \le 5$
C. $x < 5$ or $x > 5$ **D.** $-\sqrt{5} < x < \sqrt{5}$
E. none of these

17. If $y = \frac{2}{3}$ is a horizontal asymptote for a curve, then
A. $x = 0$ is a vertical asymptote
B. the degree of the numerator is the same as the degree of the denominator
C. the degree of the numerator is less than the degree of the denominator
D. the degree of the numerator is greater than the degree of the denominator
E. none of these

18. The rational function $y = \dfrac{x^3}{(x-1)(2x+3)}$ has
A. vertical asymptotes at $x = 1$ and $x = -\frac{2}{3}$
B. horizontal asymptotes at $y = 1$ and $y = -\frac{2}{3}$
C. a horizontal asymptote at $y = 0$
D. a vertical asymptote at $x = 0$
E. none of these

19. If $R(x)$ is the revenue function and $D(x)$ is the demand function, then
A. $R(x) > D(x)$ **B.** $R(x) = xD(x)$
C. $D(x) = xR(x)$ **D.** $R(x) = D(x)$
E. none of these

20. If $R(x)$ is the revenue function and $C(x)$ is the cost function, then the break-even point is found by solving
A. $R(x) - C(x) = 0$ **B.** $R(x) + C(x) = 0$
C. $R(x) = xD(x)$ **D.** $xR(x) = D(x)$
E. none of these

2.2 Functions

Definition of a Function

We begin by reviewing an idea from algebra that is fundamental for the study of calculus, namely, the idea of a **function**.

Function

> A **function** of a variable x is a rule f that assigns to each value of x a unique* number $f(x)$, called the *value of the function at* x. A function can be defined by a verbal rule, a table, a graph, or an algebraic formula.

Remember:

1. $f(x)$ is pronounced "f of x."
2. f is a *function* while $f(x)$ is a *number*.
3. The set of replacements for x is called the **domain** of the function.
4. The set of all values $f(x)$ is called the **range** of f.

EXAMPLE 1 *A function defined as a verbal rule.*
Let x be a year from 1990 to 1995, inclusive. We define the function p by the rule that $p(x)$ is the closing price of Xerox stock on January 4 of year x. The domain

* By a unique number, we mean exactly one value.

is the set $\{1990, 1991, 1992, 1993, 1994, 1995\}$, and the range is the set of possible prices for Xerox stock.* For example, $p(1994) = 49\frac{1}{4}$ (or whatever the closing price of Xerox stock was on January 4, 1994). ■

EXAMPLE 2 *A function defined by a table.*

Let g be a function defined so that $g(x)$ is the average price of gasoline in the year x as given by the following table:

Year	Average price per gallon of gasoline on January 4
1944	\$ .21
1954	\$ .29
1964	\$ .30
1974	\$ .53
1984	\$1.24

The domain is $\{1944, 1954, 1964, 1974, 1984\}$.

The range is $\{\$.21, \$.29, \$.30, \$.53, \$1.24\}$.

For example, $g(1984) = \$1.24$. ■

EXAMPLE 3 *A function defined by a graph.*

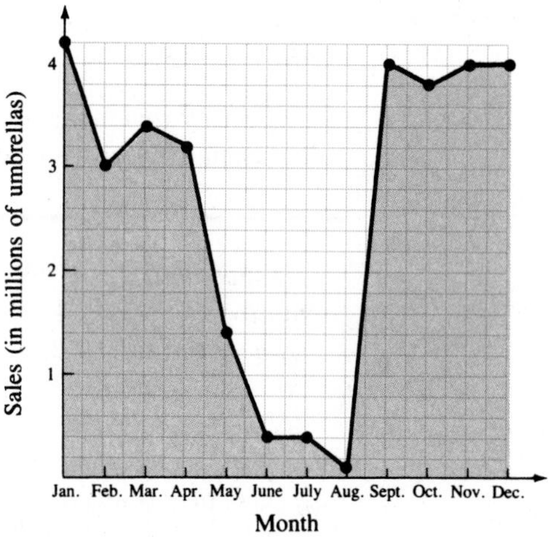

Let s be a function such that $s(x)$ is the sales of umbrellas in millions of units for x, a month in 1992. The domain is {Jan., Feb., ..., Nov., Dec.}, and the range is $.1 \le y \le 4.2$. (Remember that the units on the y-axis are in millions, so this means $100{,}000 \le y \le 4{,}200{,}000$.) For example, $s(\text{Jan.}) \approx 4.2$, $s(\text{Aug.}) \approx .1$. ■

* If you are familiar with the stock market, you know that Xerox stock prices are quoted in eighths of a dollar. The actual prices of Xerox stock over the years 1990–1995 provide the elements of the range.

EXAMPLE 4 *A function defined by an algebraic formula.*
Let f be a function defined by

$$f(x) = 2x - 5$$

This rule says take a number x from the domain, multiply it by 2, and then subtract 5. For example,

$$f(3) = 2(3) - 5 = 1$$

■

Implied Domain of a Function

In this book, unless otherwise specified, the *domain is the set of real numbers for which the given function is meaningful.* For example,

$$f(x) = 2x - 5$$

has a domain consisting of all real numbers, while

$$g(x) = \frac{3}{5x - 5}$$

has a domain that excludes $x = 1$ because division by zero is not defined. For the function g, $5x - 5 = 0$ if $x = 1$. We say $g(1)$ does not exist or we say $g(x)$ is not defined at $x = 1$. Finally, if

$$F(x) = \sqrt{x}$$

then $x \geq 0$ is implied since the square root of a number x is real if and only if $x \geq 0$, so $F(x)$ is defined only if $x \geq 0$.

EXAMPLE 5 Give the domain of each of the following functions.

a. $f(x) = \dfrac{x - 8}{3}$ **b.** $f(x) = \dfrac{3}{x - 8}$ **c.** $f(x) = \sqrt{x + 2}$ **d.** $f(x) = \dfrac{1}{\sqrt{x + 2}}$

Solution **a.** The domain is all real numbers.

b. The domain is all real numbers except where $x - 8 = 0$ or $x = 8$. We state this domain by simply writing $x \neq 8$.

c. The domain is all real numbers for which

$$x + 2 \geq 0$$
$$x \geq -2$$

This can be stated in **set-builder notation** as $\{x \mid x \geq -2\}$, which is read

"The set of all x, such that $x \geq -2$"
"{ x | $x \geq -2$}"

However, set-builder notation is too formal for our work in this course, so we will just write $x \geq -2$ for the domain.

d. This is similar to part **c**, but division by zero must also be excluded, so we see that the domain is $x > -2$.

■

Functional Notation—Evaluating a Function

Example 4 illustrates a very important process called **evaluating a function**. To find $f(3)$ you substitute 3 for every occurrence of x in the formula for $f(x)$. This process is illustrated in Examples 6 and 7.

EXAMPLE 6 Given f and g defined by $f(x) = 2x - 5$ and $g(x) = x^2 + 4x + 3$, find the indicated values:

a. $f(1)$ **b.** $g(2)$ **c.** $g(-3)$ **d.** $f(-3)$

Solution **a.** The number $f(1)$ is found by replacing x by 1 in the expression

$$\begin{aligned} f(x) &= 2x - 5 \\ &\updownarrow \quad \updownarrow \\ f(1) &= 2(1) - 5 \\ &= -3 \end{aligned}$$

WARNING ***This is a very important example.***

b. $g(2)$: $$\begin{aligned} g(x) &= x^2 + 4x + 3 \\ &\updownarrow \quad \updownarrow \quad \updownarrow \\ g(2) &= (2)^2 + 4(2) + 3 \\ &= 4 + 8 + 3 \\ &= 15 \end{aligned}$$

c. $g(-3)$: $$\begin{aligned} g(-3) &= (-3)^2 + 4(-3) + 3 \\ &= 9 - 12 + 3 \\ &= 0 \end{aligned}$$

d. $$\begin{aligned} f(-3) &= 2(-3) - 5 \\ &= -11 \end{aligned}$$

■

A function may be evaluated by using a variable.

EXAMPLE 7 Let F and G be defined by $F(x) = x^2 + 1$ and $G(x) = (x + 1)^2$. Then:

a. $F(w) = w^2 + 1$

b. $$\begin{aligned} G(t) &= (t + 1)^2 \\ &= t^2 + 2t + 1 \end{aligned}$$

c. $$\begin{aligned} F(-a) &= (-a)^2 + 1 \\ &= a^2 + 1 \end{aligned}$$

d. $$\begin{aligned} G(-a) &= (-a + 1)^2 \\ &= a^2 - 2a + 1 \end{aligned}$$

e. $$\begin{aligned} F(w + h) &= (w + h)^2 + 1 \\ &= w^2 + 2wh + h^2 + 1 \end{aligned}$$

f. $$\begin{aligned} F(w^3) &= (w^3)^2 + 1 \\ &= w^6 + 1 \end{aligned}$$

g. $$\begin{aligned} [G(a)]^2 &= [(a + 1)^2]^2 \quad \text{since } G(a) = (a + 1)^2 \\ &= (a + 1)^4 \end{aligned}$$

h. $$\begin{aligned} F[G(x)] &= F[(x + 1)^2] \\ &= [(x + 1)^2]^2 + 1 \\ &= (x + 1)^4 + 1 \end{aligned}$$

i. $F(\sqrt{t}) = (\sqrt{t})^2 + 1 = t + 1$

j. $$\begin{aligned} G(x + h) &= [(x + h) + 1]^2 \\ &= (x + h)^2 + 2(x + h) + 1 \\ &= x^2 + 2xh + h^2 + 2x + 2h + 1 \end{aligned}$$

■

In calculus, functional notation is used to carry out manipulations such as those shown in Example 8.

WARNING ***Because of the extensive later use of this manipulation, be sure you thoroughly understand Example 8.***

EXAMPLE 8 Find $\dfrac{f(x+h)-f(x)}{h}$ for each function given below.

a. $f(x) = x^2$, where $x = 5$. We substitute 5 for every occurrence of x in the formula, but do not substitute for h.

$$\frac{f(5+h)-f(5)}{h} = \frac{(5+h)^2 - 5^2}{h} \qquad \text{Since } f(x) = x^2$$

$$= \frac{25 + 10h + h^2 - 25}{h}$$

$$= \frac{(10+h)h}{h}$$

$$= 10 + h$$

b. $f(x) = 2x^2 + 1$, where $x = 1$:

$$\frac{f(1+h)-f(1)}{h} = \frac{[2(1+h)^2 + 1] - [2(1)^2 + 1]}{h} \qquad \text{Since } f(x) = 2x^2 + 1$$

$$= \frac{[2(1 + 2h + h^2) + 1] - (2 + 1)}{h}$$

$$= \frac{2h^2 + 4h + 3 - 3}{h}$$

$$= 2h + 4$$

c. $f(x) = x^2 + 3x - 2$:

$$\frac{f(x+h)-f(x)}{h} = \frac{[(x+h)^2 + 3(x+h) - 2] - (x^2 + 3x - 2)}{h}$$

$$= \frac{x^2 + 2xh + h^2 + 3x + 3h - 2 - x^2 - 3x + 2}{h}$$

$$= \frac{2xh + h^2 + 3h}{h}$$

$$= 2x + 3 + h$$

■

Functional notation can be used to work a wide variety of applied problems, as shown in Examples 9–12.

EXAMPLE 9 What is the average change per year in the price per gallon of gasoline from 1974 (\$.53) to 1984 (\$1.24)?

Solution The price of gasoline changed by \$.71 per gallon during this 10-year period (\$1.24 − \$.53 = \$.71). Thus the average per-year change in the price is

$$\text{Change in the price of gas} = \frac{\$1.24 - \$.53}{1984 - 1974} = \frac{\$.71}{10} \qquad \longleftarrow \text{10 years}$$

In functional notation, if g is the average price of gasoline, then $g(x)$ is the average price of gas in year x. Thus,

$$\text{Change in the price of gas} = \frac{g(1984) - g(1974)}{1984 - 1974} = \$.071$$

■

EXAMPLE 10 What is the average change per year in the price per gallon of gasoline from 1974 to a year h years later?

Solution Notice that h years after 1974 is $1974 + h$, so the average is

$$\frac{g(1974 + h) - g(1974)}{h}$$

■

EXAMPLE 11 In 1990 the U.S. population was 249.6 million. If we assume a growth rate of 2%, the population (in millions) t years after 1990 can be approximated by the formula

$$P(t) = 249.6(1.02)^t$$

What is the expected population for 1999?

Solution Since $1999 - 1990 = 9$, $t = 9$. Thus:

$$P(9) = 249.6(1.02)^9$$ BY CALCULATOR: [1.02] [y^x] [9] [×] [249.6] [=]

$$\approx 298.3$$ DISPLAY: 298.2951051

The 1999 U.S. population will be about 298.3 million.

■

Composition of Functions

Suppose a farmer sells eggs to Safeway. If the farmer's price for a dozen eggs is x dollars, then there is a function f that can be used to describe $f(x)$, the total cost of those eggs to Safeway. Note that x and $f(x)$ are not the same because Safeway must pay for ordering, shipping, and distributing the eggs. Moreover, in order to determine the price to the consumer, Safeway must add an appropriate markup. Suppose this markup function is called g. Then the price of the eggs to the consumer is

$$g[f(x)] \quad \text{and not} \quad g(x)$$

since the markup must be on Safeway's *total* cost of the eggs and not just on the price the farmer charges for the eggs. This process of evaluating a function of a function illustrates the idea of composition of functions.

Consider two functions f and g such that f is a function from a set X to a set Y and g is a function from set Y to set Z as illustrated by Figure 2.1.

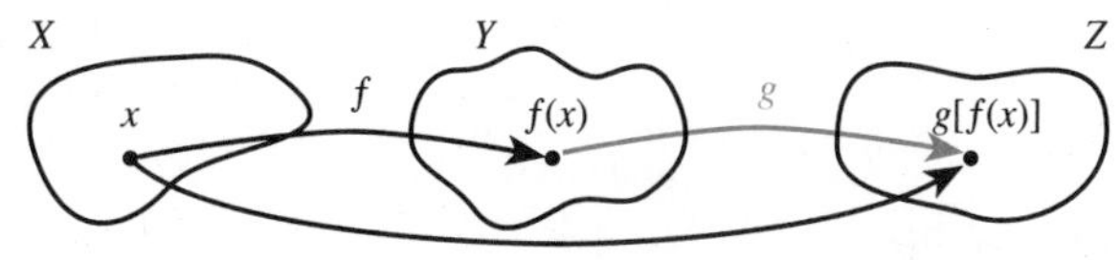

FIGURE 2.1
Composition of f and g

The value in the set Z is the number $g[f(x)]$ and defines a function from X to Z called the **composition of functions f and g**.

Composite Functions

> Let X, Y, and Z be sets of real numbers. Let f be a function from X to Y and g be a function from Y to Z. Then the **composite function** is the function from X to Z defined by
>
> $$g[f(x)]$$

EXAMPLE 12 If $f(x) = x^2$ and $g(x) = x + 4$, find:

a. $g[f(x)]$ **b.** $f[g(x)]$ **c.** $g[f(-1)]$ **d.** $f[g(5)]$

Solution

a. $g[f(x)] = g[x^2]$
$= x^2 + 4$

b. $f[g(x)] = f[x + 4]$
$= (x + 4)^2$
$= x^2 + 8x + 16$

c. $g[f(-1)] = (-1)^2 + 4$
$= 5$

d. $f[g(5)] = f[9]$
$= 9^2$
$= 81$ ■

Notice from Examples 12**a** and 12**b** that $g[f(x)] \neq f[g(x)]$.

2.2 Problem Set

1. What is a mathematical model?
2. Why are mathematical models necessary or useful?
3. What is a function? (Use your own words.)
4. If $y = f(x)$, what is the difference between the symbols f and $f(x)$, if any?
5. One of the following examples expresses y as a function of x and the other does not. Which is a function? Explain your answer.
   a. y is the closing price of IBM stock on March 3 of year x.
   b. x is the closing price of Tandy stock on July 1 of year y.
6. One of the following examples is a function of x and the other is not. Which is a function? Explain your answer.
   a. $y = x^2$ **b.** $x = y^2$
7. Use the table in Example 2 on page 40 to find and interpret each of the following expressions.
   a. $g(1954)$ **b.** $g(1974)$
8. Use the graph in Example 3 on page 40 to find and interpret the following expressions.
   a. $s(\text{March})$ **b.** $s(\text{June})$
9. Use $f(x) = 2x - 5$ (see Example 4) to find
   a. $f(8)$ **b.** $f(-3)$
10. Use $f(x) = 2x - 5$ (see Example 4) to find
   a. $f(-5)$ **b.** $f(7)$

Evaluate the functions in Problems 11–16.

11. Let $f(x) = 5x - 3$. Find
   a. $f(2)$ **b.** $f(10)$
   c. $f(-15)$ **d.** $f(100)$

12. Let $g(x) = 4x - 10$. Find
a. $g(-2)$ **b.** $g(5)$
c. $g(-10)$ **d.** $g(50)$

13. Let $h(x) = 6 - 4x$. Find
a. $h(0)$ **b.** $h(8)$
c. $h(-7)$ **d.** $h(100)$

14. Let $k(x) = 10 - 3x$. Find
a. $k(-4)$ **b.** $k(-5)$
c. $k(6)$ **d.** $k(10)$

15. Let $m(x) = x^2 - 3x + 1$. Find
a. $m(0)$ **b.** $m(1)$
c. $m(2)$ **d.** $m(3)$

16. Let $n(x) = x^2 + x - 3$. Find
a. $n(0)$ **b.** $n(1)$
c. $n(2)$ **d.** $n(3)$

State the domain of each function in Problems 17–20.

17. **a.** $f(x) = 5x^2 - 3x + 2$
b. $f(x) = 6x^2 + 5x - \sqrt{17}$

18. **a.** $f(x) = \dfrac{2x - 5}{3}$ **b.** $f(x) = \dfrac{3}{2x - 5}$

19. **a.** $f(x) = \dfrac{3x + 2}{x + 5}$ **b.** $f(x) = \dfrac{\sqrt{2x - 1}}{2x + 1}$

20. **a.** $f(x) = \sqrt{2x + 6}$ **b.** $f(x) = \dfrac{1}{\sqrt{2x + 6}}$

In Problems 21–36, let $f(x) = 5x - 2$ and $g(x) = 2x^2 - 4x - 5$. Evaluate and simplify:

21. **a.** $f(t)$ **b.** $f(w)$
22. **a.** $g(s)$ **b.** $g(t)$
23. **a.** $f(t + h)$ **b.** $f(s + t)$
24. **a.** $g(t + h)$ **b.** $g(s + t)$
25. $f(t + h + 8)$
26. $g(t - h - 3)$
27. $g(3 + h)$
28. $g(t - 2)$
29. $f(2x^2)$
30. $f(2x^2 - 4x)$
31. $g[f(x)]$
32. $f[g(x)]$
33. $f[g(5x)]$
34. $g[f(2x)]$
35. $g[g(x)]$
36. $f[f(x)]$

Find $[f(x + h) - f(x)]/h$ for each function in Problems 37–44.

37. $f(x) = 2x$
38. $f(x) = 5x$
39. $f(x) = 2x^2$
40. $f(x) = 5x^2$
41. $f(x) = 2x^2 - 3$
42. $f(x) = 5x^2 - 3x$
43. $f(x) = x^2 - 2x + 1$
44. $f(x) = 3x^2 - 2x + 4$

APPLICATIONS

For Problems 45–54, use the following table, which reflects the purchasing power of the dollar from October 1944 to October 1984 (Source: U.S. Bureau of Labor Statistics, Consumer Division). Let x represent the year; let the domain be the set {1944, 1954, 1964, 1974, 1984}.

Year	Round steak (1 lb)	Sugar (5 lb)	Bread (loaf)	Coffee (1 lb)	Eggs (1 doz)	Milk ($\frac{1}{2}$ gal)	Gasoline (1 gal)
1944	$ .45	$ .34	$ .09	$ .30	$ .64	$ .29	$ .21
1954	.92	.52	.17	1.10	.60	.45	.29
1964	1.07	.59	.21	.82	.57	.48	.30
1974	1.78	1.08	.36	1.31	.84	.78	.53
1984	2.15	1.48	1.29	2.69	1.15	1.08	1.52

Let $r(x)$ = price of 1 lb of round steak
$b(x)$ = price of a loaf of bread
$e(x)$ = price of a dozen eggs
$g(x)$ = price of 1 gal of gasoline
$s(x)$ = price of 5 lb of sugar
$c(x)$ = price of 1 lb of coffee
$m(x)$ = price of $\frac{1}{2}$ gal of milk

45. Find: **a.** $r(1954)$ **b.** $m(1954)$
46. Find: **a.** $g(1944)$ **b.** $c(1984)$
47. Find $s(1984) - s(1944)$.
48. Find $b(1984) - b(1944)$.
49. **a.** Find the change in the price of eggs from 1944 to 1984.
b. Use functional notation to write the change in the price of eggs.
50. **a.** Find the change in the price of round steak from 1944 to 1984.
b. Use functional notation to write the change in the price of round steak.

51. a. Find $\dfrac{g(1944 + 40) - g(1944)}{40}$.
 b. What does the expression in part **a** mean?

52. a. Find $\dfrac{m(1944 + 40) - m(1944)}{40}$.
 b. What does the expression in part **a** mean?

53. a. What is the average increase in the price of sugar per year from 1944 to 1954? Use functional notation.
 b. What is the average increase in the price of sugar per year from 1944 to 1964? Use functional notation.
 c. What is the average increase in the price of sugar per year from 1944 to 1974? Use functional notation.
 d. What is the average increase in the price of sugar per year from 1944 to 1984? Use functional notation.
 e. What is the average increase in the price of sugar per year from 1944 to $1944 + h$, where h is an unspecified number of years? Use functional notation.

54. Repeat Problem 53 for coffee instead of sugar.

55. Use $P(t) = 249.6(1.02)^t$ from Example 11 to estimate the U.S. population in the year 2008.

56. Use $P(t) = 249.6(1.02)^t$ from Example 11 to estimate the U.S. population at the turn of the next century.

57. Use $P(t) = 249.6(1.02)^t$ from Example 11 to estimate the population in 1981. How well does this formula model the population if the actual population was 230 million?

58. Use $P(t) = 249.6(1.02)^t$ from Example 11 to estimate the population in 1900.

59. According to the U.S. Public Health Service, the number of marriages in the United States in 1987 was about 2,421,000 and in 1982 it was about 2,495,000. If $M(x)$ represents the number of marriages in year x,
 a. Find $\dfrac{M(1987) - M(1982)}{5}$.
 b. Verbally describe $\dfrac{M(1982 + h) - M(1982)}{h}$.

60. According to the U.S Public Health Service, the number of divorces in the United States in 1987 was about 1,157,000 and in 1982 it was about 1,180,000. If $D(x)$ represents the number of divorces in year x,
 a. Find $\dfrac{D(1987) - D(1982)}{5}$.
 b. Verbally describe $\dfrac{D(1982 + h) - D(1982)}{h}$.

2.3 Graphs

A **two-dimensional coordinate system** consists of two perpendicular coordinate lines in a plane. Usually one of the coordinate lines is horizontal with the positive direction to the right; the other is vertical with the positive direction upward. These coordinate lines are called **coordinate axes**, and the point of intersection is called the **origin**. Note in Figure 2.2 that the axes divide the plane into four parts called the **first**, **second**, **third**, and **fourth quadrants**. This two-dimensional coordinate system is also called a **Cartesian coordinate system** in honor of René Descartes, who was the first to describe it in mathematical detail.

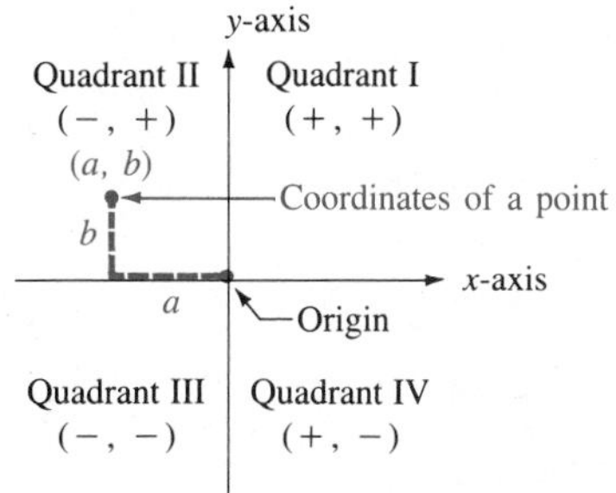

FIGURE 2.2
Cartesian coordinate system

Points in a plane are denoted by ordered pairs. The term *ordered pair* refers to two real numbers represented by (a, b), where a is the **first component** and b is the **second component**. The order in which the components are listed is important since $(a, b) \neq (b, a)$ if $a \neq b$.

The horizontal number line is called the **x-axis** (or the *axis of abscissas*), and x represents the first component of the ordered pair. The vertical number line is called the **y-axis** (or the *axis of ordinates*), and y represents the second component of the ordered pair. The plane determined by the x- and y-axes is called the *coordinate plane*, *Cartesian plane*, or *xy-plane*. When we refer to a point (x, y), we mean a point in the coordinate plane whose abscissa is x and whose ordinate is y. To **plot a point (x, y)** means to locate the point with coordinates (x, y) in the plane and represent its location by a dot. (In this book, if variables other than x and y are used, you will be told which represents the first and which represents the second component.

To **graph a function** means to draw a picture of the ordered pairs that *satisfy* the equation in a one-to-one fashion. The set of x-values permitted for a function is called the **domain** of the function and the set of y-values for the function is called the **range**. We will illustrate graphing functions for three different types of models.

EXAMPLE 1 Graph $f(x) = 2x - 5$.

Solution Let $y = f(x)$ so that we can speak about the ordered pair (x, y) instead of the more notationally cumbersome form $(x, f(x))$. One method for graphing a function is to plot ordered pairs (x, y). That is, *you* choose values for x (the first component) and calculate, or find, the corresponding values for $y = f(x)$ (the second component): Let

$$\begin{aligned}
&x_1 = 0: && f(0) = y_1 = 2 \cdot 0 - 5 = -5 && \text{The ordered pair is } (0, -5).\\
&x_2 = 2: && f(2) = y_2 = 2 \cdot 2 - 5 = -1 && \text{The ordered pair is } (2, -1).\\
&x_3 = 4: && f(4) = y_3 = 2 \cdot 4 - 5 = 3 && \text{The ordered pair is } (4, 3).\\
&x_4 = -2: && f(-2) = y_4 = 2 \cdot (-2) - 5 = -9 && \text{The ordered pair is } (-2, -9).
\end{aligned}$$

Plot the points as shown in Figure 2.3. After you have plotted enough points to see the curve's general shape, connect the points to obtain the curve. ■

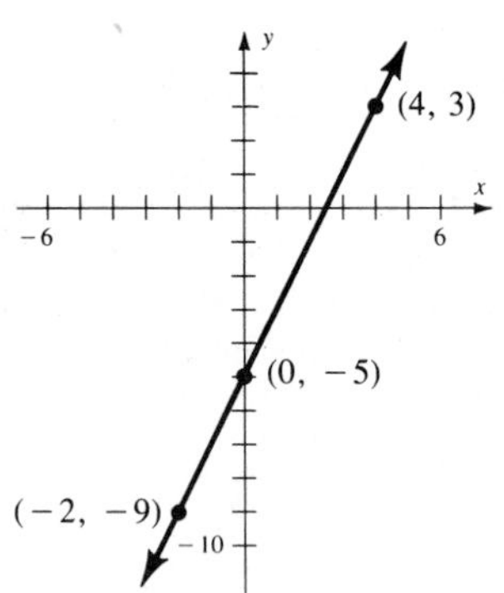

FIGURE 2.3
Graph of $f(x) = 2x - 5$. These points seem to lie on a line. In Section 2.4, we will graph lines much more efficiently than we did here.

EXAMPLE 2 Graph $g(x) = x^2 + 4x + 3$.

Solution Let $y = g(x)$ and

$$\begin{aligned}
&x_1 = 0: && g(0) = y_1 = 0^2 + 4 \cdot 0 + 3 = 3 && \text{Ordered pair } (0, 3)\\
&x_2 = 1: && g(1) = y_2 = 1^2 + 4 \cdot 1 + 3 = 8 && (1, 8)\\
&x_3 = 2: && g(2) = y_3 = 2^2 + 4 \cdot 2 + 3 = 15 && (2, 15)\\
&x_4 = -1: && g(-1) = y_4 = (-1)^2 + 4(-1) + 3 = 0 && (-1, 0)\\
&x_5 = -2: && g(-2) = y_5 = (-2)^2 + 4(-2) + 3 = -1 && (-2, -1)\\
&x_6 = -3: && g(-3) = y_6 = (-3)^2 + 4(-3) + 3 = 0 && (-3, 0)\\
&x_7 = -4: && g(-4) = y_7 = (-4)^2 + 4(-4) + 3 = 3 && (-4, 3)
\end{aligned}$$

Plot the points and draw the curve as shown in Figure 2.4. ■

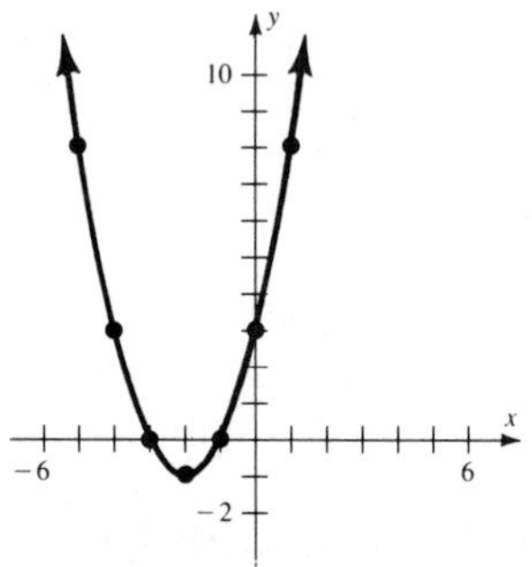

FIGURE 2.4
Graph of $g(x) = x^2 + 4x + 3$. These points do not lie on a line. If the points are connected they form a curve called a **parabola**. We will learn how to efficiently graph parabolas in Section 2.5.

EXAMPLE 3 Sketch the graph of $h(x) = 1/x$.

Solution Let $y = h(x)$ and

$x_1 = 0$:	$y_1 = \frac{1}{0}$	But $y_1 = \frac{1}{0}$ is not defined, so $x = 0$ is not in the domain—*there is no point on the graph corresponding to* $x = 0$.
$x_2 = 1$:	$y_2 = \frac{1}{1} = 1$	The ordered pair is $(1, 1)$.
$x_3 = 2$:	$y_3 = \frac{1}{2}$	$(2, \frac{1}{2})$
$x_4 = 3$:	$y_4 = \frac{1}{3}$	$(3, \frac{1}{3})$
$x_5 = -1$:	$y_5 = 1/(-1) = -1$	$(-1, -1)$
$x_6 = -2$:	$y_6 = 1/(-2) = -\frac{1}{2}$	$(-2, -\frac{1}{2})$
$x_7 = -3$:	$y_7 = 1/(-3) = -\frac{1}{3}$	$(-3, -\frac{1}{3})$

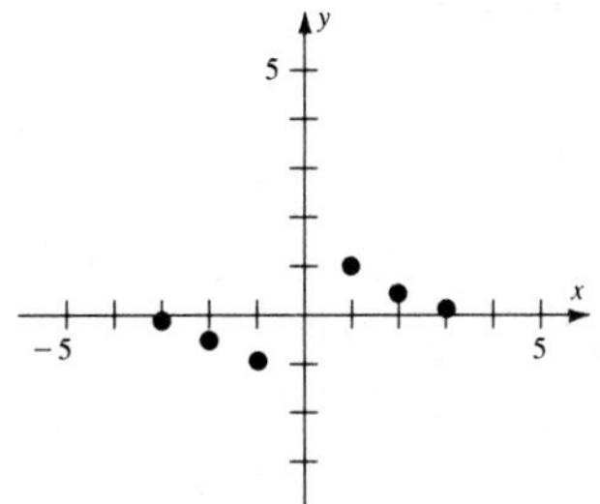

FIGURE 2.5a

If you plot the ordered pairs as shown in Figure 2.5**a**, you see that they cannot be connected with either a line or a parabola. In fact, you cannot connect all of these points with any smooth curve because no point corresponds to $x = 0$. Consider some additional points close to zero:

$$x = \frac{1}{2}: \quad y = \frac{1}{\frac{1}{2}} = 2 \quad \left(\frac{1}{2}, 2\right)$$

$$x = \frac{1}{3}: \quad y = \frac{1}{\frac{1}{3}} = 3 \quad \left(\frac{1}{3}, 3\right)$$

$$\vdots \qquad\qquad\qquad \vdots$$

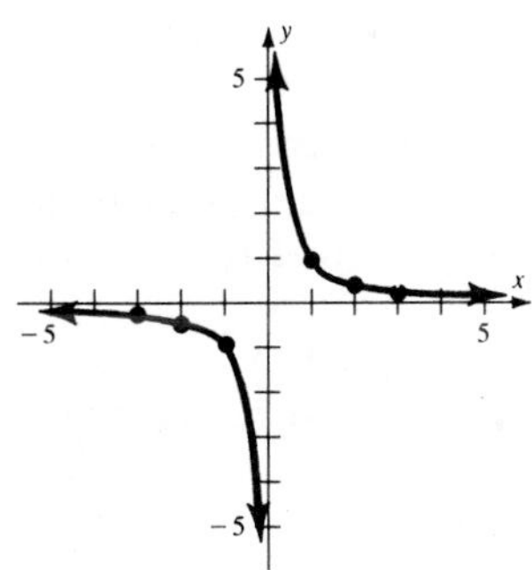

FIGURE 2.5b
Graph of $h(x) = \frac{1}{x}$. This curve is a type of curve that is the graph of what is called a **rational function**. These curves are discussed in Section 2.6.

We can now draw a smooth curve as shown in Figure 2.5**b**. ■

Graphs of Functions

To summarize, the **graph of a function f** is the set of all points $(x, f(x))$ in a coordinate plane, where x is in the domain of f. That is, the graph of f can be described as the set of all points with coordinates (x, y) such that $y = f(x)$. The graph of a typical function f is shown in Figure 2.6.

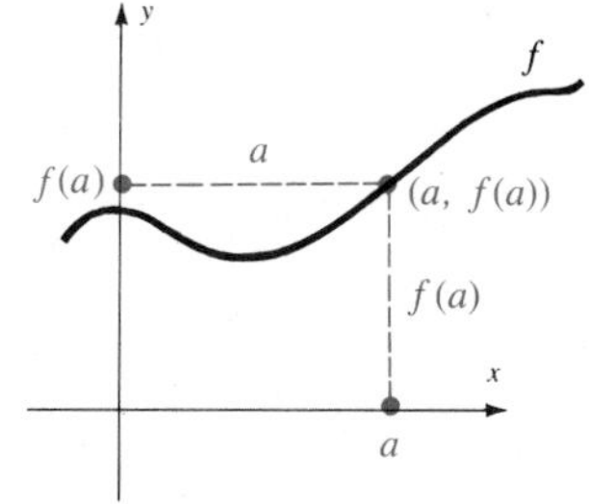

FIGURE 2.6
Graph of a function

Vertical line test Note that the graph of a function is such that for each a in the domain there is only *one point* $(a, f(a))$ on the graph. This means that every vertical line passes through the graph of a function in at most one point. This is the so-called **vertical line test** for the graphs of functions as illustrated in Example 4.

EXAMPLE 4 Which of the following graphs are functions?

Solution

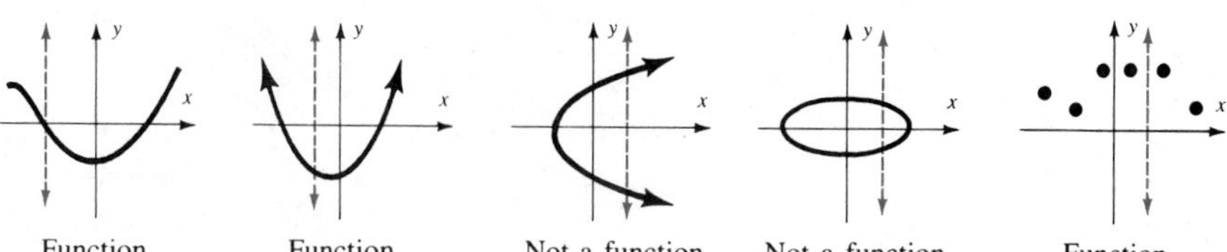

Imagine a vertical line (shown in color) moving from left to right across the plane—if it passes through more than one point of the graph at one time, then the graph does not represent a function. ■

Intercepts There are some points on a graph that are of particular importance to us—the **intercepts**.

Intercepts

If the number zero is in the domain of f, then $f(0)$ is called the **y-intercept** of the graph of f and is the point $(0, f(0))$. This is the point where the graph intersects the y-axis.

If a is a real number in the domain such that $f(a) = 0$, then a is called an ***x*-intercept** and is the point $(a, 0)$. This is the point where the graph intersects the x-axis. Any number x such that $f(x) = 0$ is called a **zero of the function**.

EXAMPLE 5 Find the domain, range, and intercepts for f defined by the graph below.

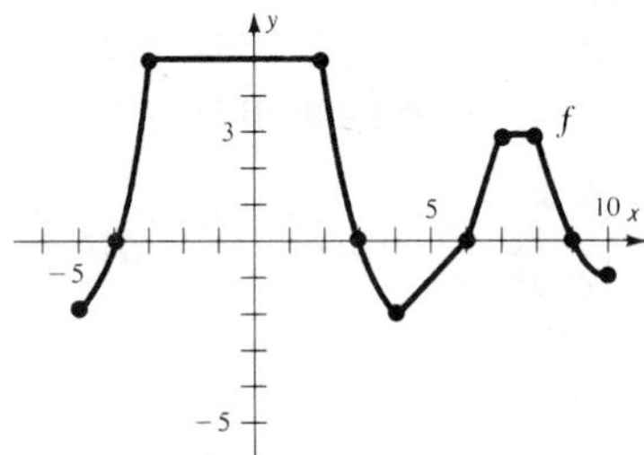

Solution Domain: $-5 \le x \le 10$
Range: $-2 \le y \le 5$
The zeros of the function are -4, 3, 6, and 9.

y-intercept: $(0, 5)$; we usually simply say that the y-intercept is 5. A function will not have more than one y-intercept.
x-intercepts: $(-4, 0)$, $(3, 0)$, $(6, 0)$, and $(9, 0)$. ■

2.3 Problem Set

1. Define the graph of a function f.
2. What does it mean when we say the ordered pair (a, b) satisfies the equation $y = f(x)$?
3. What are the x- and y-intercepts of the graph of a function f?
4. What is a zero of a function f?

Use the figures below to find the coordinates of the points in Problems 5–13. For example, point D has coordinates $(-1, f(-1))$.

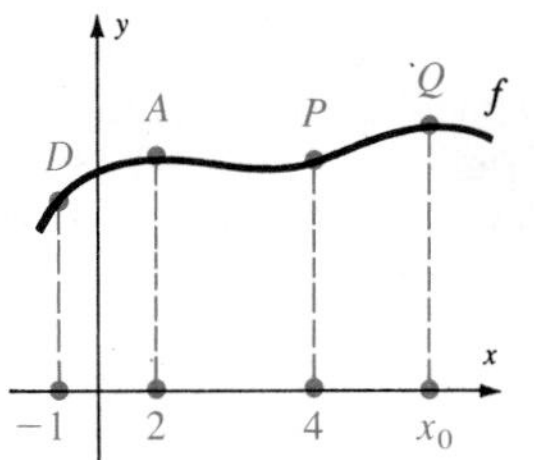

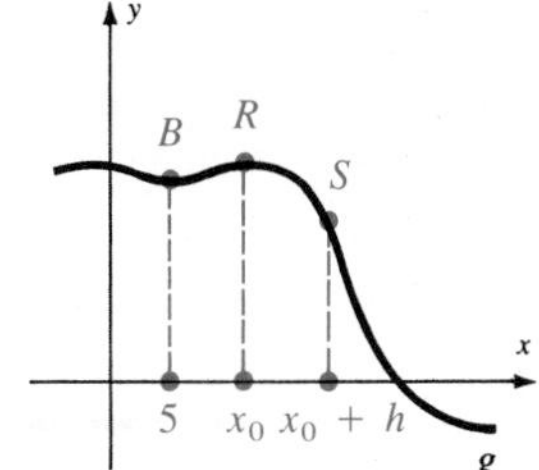

5. Point A
6. Point P
7. Point Q
8. Point B
9. Point R
10. Point S
11. Point C
12. Point T
13. Point U

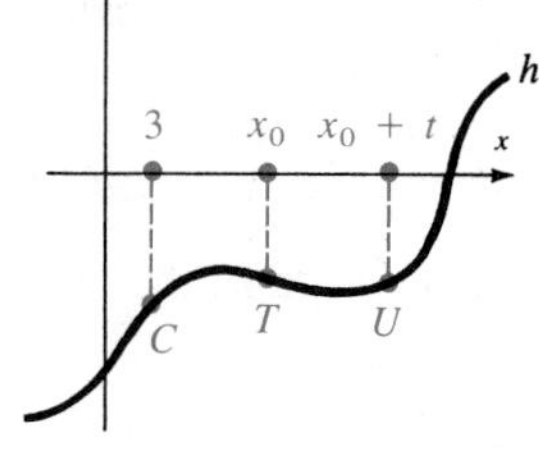

Find the domain, range, and intercepts of the relations defined by the graphs in Problems 14–19. Also state whether the graph defines a function.

14. **15.**

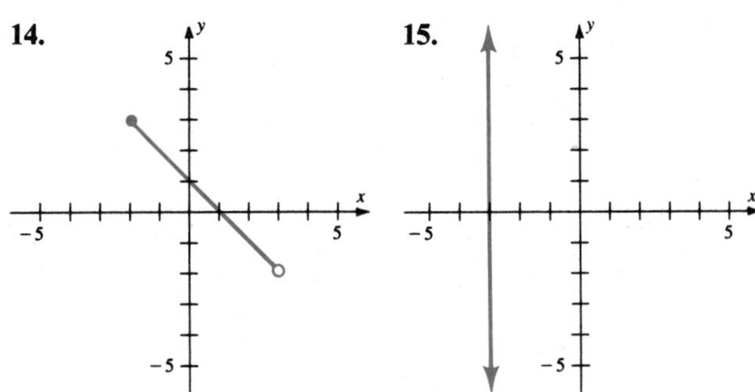

16.

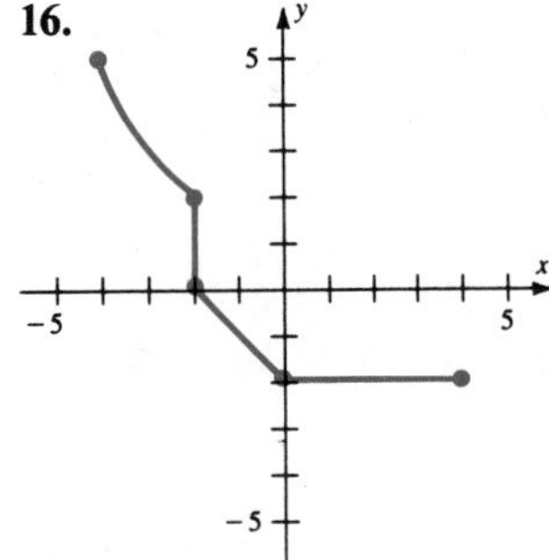

17.

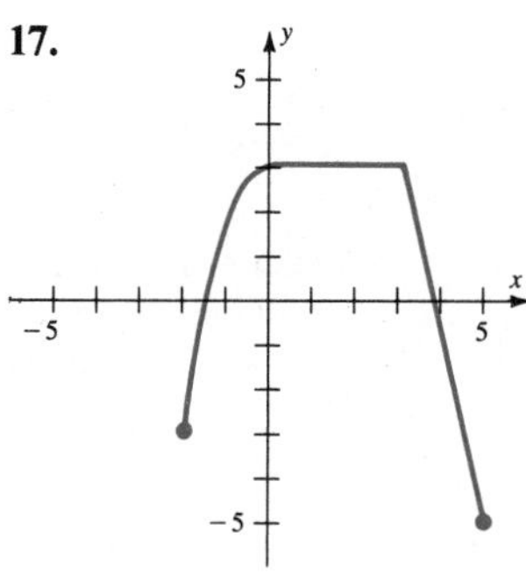

18.

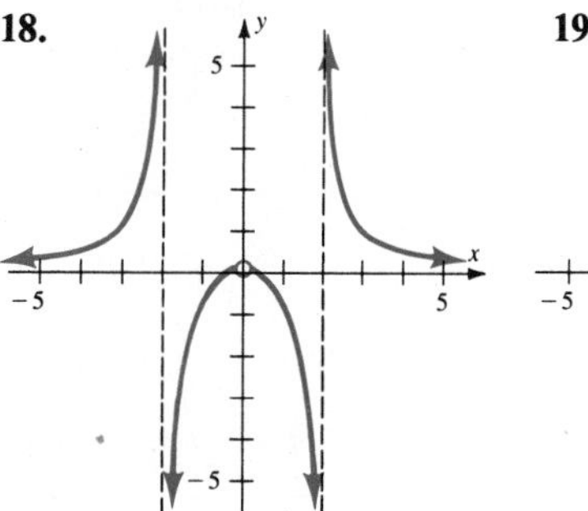

19.

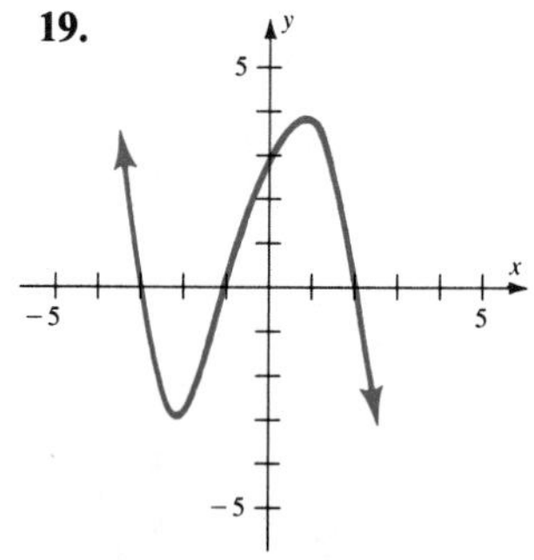

Graph the functions in Problems 20–41 by plotting points.

20. $f(x) = 3x + 1$
21. $f(x) = 2x + 3$
22. $f(x) = x - 4$
23. $f(x) = 6 - 2x$
24. $f(x) = 1 - x$
25. $f(x) = -3x - 1$
26. $g(x) = 2x^2$
27. $g(x) = \frac{1}{2}x^2$
28. $g(x) = \frac{1}{10}x^2$
29. $g(x) = x^2 + 4x + 4$
30. $g(x) = x^2 + 6x + 9$
31. $g(x) = x^2 - 6x + 9$
32. $g(x) = x^2 + 2x - 3$
33. $g(x) = 2x^2 - 4x + 5$
34. $g(x) = 2x^2 - 4x + 4$
35. $h(x) = -1/x$
36. $h(x) = 3/x$
37. $h(x) = -2/x$
38. $h(x) = 1/(x - 1)$
39. $h(x) = 2/(x - 2)$
40. $c(x) = x^3 + 2x - 4$
41. $c(x) = x^3 - 3x + 2$

APPLICATIONS

42. A theater has a capacity of 1,500 seats. The formula relating the number of adult tickets sold, a, and the number of children's tickets sold, c, is $a + c = 1{,}500$. Graph (a, c) satisfying this relationship. Assume that both a and c are positive.

43. A formula from the Tax Rate Schedule X on the 1990 federal income tax for a single person is

$$T = .28x - 3{,}250 \qquad 19{,}451 < x \le 47{,}050$$

where x is the amount on line 37 of Form 1040. Graph (x, T) for $19{,}451 < x \le 47{,}050$.

44. A supply company finds that the number of computer disks sold in year x is given by $s(x) = 5{,}000 + x^2$, where $x = 0$ corresponds to 1980. Graph the sales for the years 1980 to 1990 (inclusive).

45. The pressure, P, in centimeters of mercury, is given as a function of the depth in meters, d, under water, by using the formula $P = .4d + 7.6$. Graph (d, P) for $0 \le d \le 10$.

46. The cost C (in thousands of dollars) of removing p percent of a certain pollutant is given by the formula $C = 20p/(105 - p)$. Graph (p, C) for $0 \le p \le 100$.

47. The number of responses n (per milliseconds) of a nerve is a function of the length of time t (in milliseconds) since the nerve was stimulated. For a certain nerve, this relationship is $n = 150 - (t - 10)^2$. Graph (t, n) for $0 \le t \le 22$.

2.4 Linear Functions

This section reviews material from previous courses and also provides the notation and concepts required in the remainder of this book. Recall that a first-degree equation with two variables is called a *linear equation*.

Linear Function

A function f is a **linear function** if it can be written in the form

$$f(x) = mx + b \quad \text{or} \quad y = mx + b$$

where m and b are real numbers.

If $m = 0$, then $f(x) = b$, and the graph of $f(x) = b$ is a **horizontal line** as shown in Figure 2.7**a**. Functions whose graphs are horizontal lines are called **constant functions**. Now let (x_1, y_1) and (x_2, y_2) be any points on a line so that $x_1 = x_2$. Then this line is parallel to the y-axis and is called a **vertical line**, as shown in Figure 2.7**b**. Note that vertical lines are not the graphs of functions. Vertical lines always have the form $x = c$, where c is some constant ($x = 3$, for example).

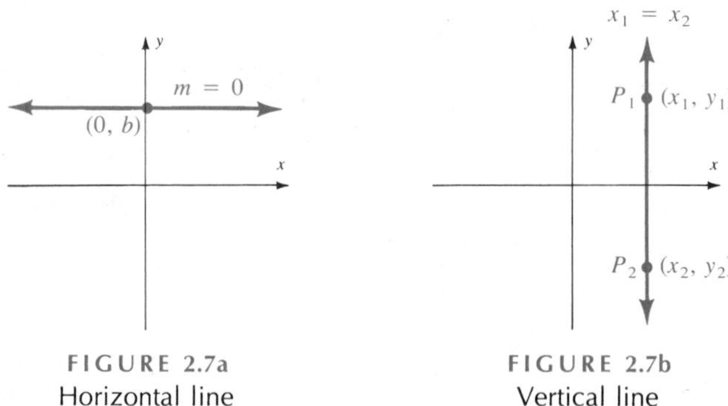

FIGURE 2.7a Horizontal line

FIGURE 2.7b Vertical line

A line is determined by two points, so if we know *any* two points on a line we can draw the line by using a straightedge and the given points. In the last section we evaluated formulas to find those points. However, two points that are usually easy to find are the intercepts:

To find the y-intercept: Let $x = 0$ and solve for y.

To find the x-intercept: Let $y = 0$ and solve for x.

EXAMPLE 1 Graph $f(x) = -\frac{3}{2}x + 3$ by plotting the x- and y-intercepts.

Solution If $f(x) = -\frac{3}{2}x + 3$, write $y = -\frac{3}{2}x + 3$.

Let $x = 0$: $y = -\frac{3}{2}(0) + 3$

$y = 3$ The point (0, 3) is the y-intercept.

Let $y = 0$: $0 = -\frac{3}{2}x + 3$

$\frac{3}{2}x = 3$

$x = 2$ The point (2, 0) is the x-intercept.

Draw the line passing through the plotted intercepts as shown in Figure 2.8. ■

FIGURE 2.8
Graph of $f(x) = -\frac{3}{2}x + 3$

The constants b and m in the form $f(x) = mx + b$ gives us important information about the line we wish to graph. For example, if we let $y = f(x)$, we can write $y = mx + b$. Now we can find the y-intercept. Let $x = 0$:

$$y = m \cdot x + b$$
$$y = b$$

This means that the y-intercept is $(0, b)$. We generally shorten the notation and simply say the y-intercept is b to mean the line passes through the y-axis at the point $(0, b)$.

The second constant m, which is the coefficient of x when the equation is solved for y, tells us the steepness or **slope** of a line. But, first, we need to define what we mean by the slope of a line. This definition requires that we know what is meant by the vertical change (*rise*) relative to a horizontal change (*run*). These ideas are illustrated in Figure 2.9.

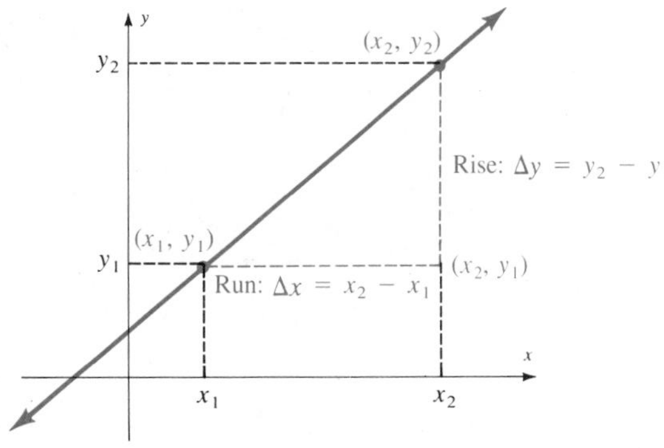

FIGURE 2.9
Slope of a line

Let Δx represent the horizontal change and Δy represent the vertical change.* Note that $\Delta x = x_2 - x_1$ and $\Delta y = y_2 - y_1$.

* Δx is one symbol (not the multiplication of two variables) and is pronounced "delta x" (Δy is pronounced "delta y").

Slope

Let (x_1, y_1) and (x_2, y_2) be points on a line. Then

$$\text{Slope} = \frac{\text{Vertical change}}{\text{Horizontal change}} = \frac{\Delta y}{\Delta x} \quad \text{or} \quad \frac{\text{Rise}}{\text{Run}} = \frac{y_2 - y_1}{x_2 - x_1} = \frac{\Delta y}{\Delta x}$$

If $\Delta x = 0$, then the line is vertical and has *no* slope ($\Delta y/0$ is undefined).
If $\Delta y = 0$, then the line is horizontal and has *zero* slope ($0/\Delta x = 0$).

To show that m in the equation $y = mx + b$ is the slope, consider the line specified by the equation $y = mx + b$, which passes through (x_1, y_1) and (x_2, y_2), with $x_1 \neq x_2$. This means that $y_1 = mx_1 + b$ and $y_2 = mx_2 + b$, so that

$$\begin{aligned}
\text{Slope} = \frac{\Delta y}{\Delta x} &= \frac{y_2 - y_1}{x_2 - y_1} \\
&= \frac{(mx_2 + b) - (mx_1 + b)}{x_2 - x_1} \qquad \text{Substitution} \\
&= \frac{mx_2 - mx_1}{x_2 - x_1} \\
&= \frac{m(x_2 - x_1)}{x_2 - x_1} = m
\end{aligned}$$

This discussion tells us that we can find the y-intercept and slope of linear equations of the form $y = mx + b$ by inspection, as shown by Example 2.

EXAMPLE 2 Find the slope and y-intercept.

	Slope	*y-intercept*	
a. $y = \frac{1}{2}x + 3$	$m = \frac{1}{2}$	$b = 3$; $(0, 3)$	By inspection
b. $y = x - 3$	$m = 1$	$b = -3$; $(0, -3)$	By inspection
c. $y = -\frac{2}{3}x + \frac{5}{2}$	$m = -\frac{2}{3}$	$b = \frac{5}{2}$; $\left(0, \frac{5}{2}\right)$	By inspection

d. $3x + 4y + 8 = 0$

Solve for y: $\quad 4y = -3x - 8$

$$y = -\tfrac{3}{4}x - \tfrac{8}{4}$$

Thus the slope $m = -\frac{3}{4}$, and the y-intercept is $(0, -2)$ since $b = -\frac{8}{4} = -2$. ■

Since the slope and y-intercept are easy to find after the linear equation is solved for y, we give a special name to this form of the equation.

Slope–Intercept Form of the Equation of a Line

The **slope–intercept** form of a linear equation is

$$y = mx + b$$

and the graph of this equation is the line having slope m and y-intercept $(0, b)$.

This form of the equation of a line can be used for graphing certain lines when it is not convenient to plot points. The procedure is summarized in Figure 2.10. Carefully study this procedure.

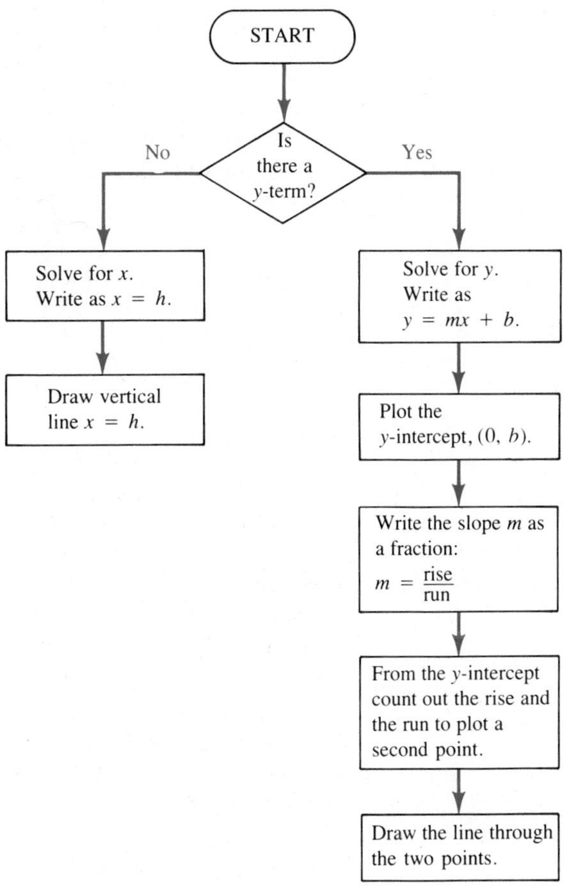

FIGURE 2.10
Procedure for graphing a line by the slope–intercept method

EXAMPLE 3 Graph $y = \frac{1}{2}x + 3$.

Solution By inspection, $b = 3$ and the slope is $\frac{1}{2}$; the line is graphed by first plotting the y-intercept $(0, 3)$ and then finding a second point by counting out the slope: over 2 and up 1.

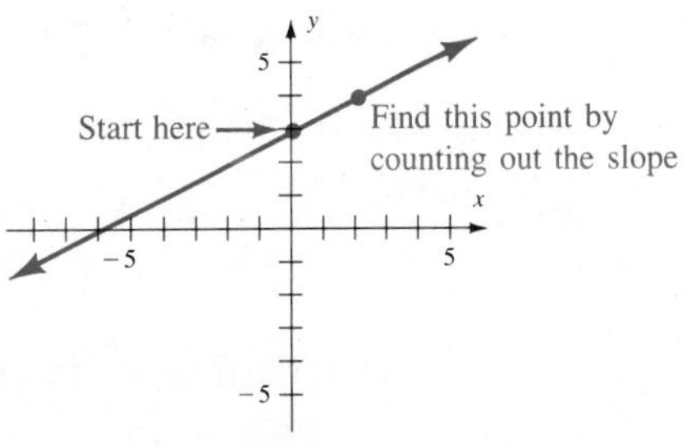

■

EXAMPLE 4 Graph $3x - 2y - 6 = 0$.

Solution Solve for y:

$$2y = 3x - 6$$
$$y = \tfrac{3}{2}x - 3$$

$b = -3$ and the slope is $\frac{3}{2}$; the line is graphed by first plotting the y-intercept $(0, -3)$ and then finding a second point by counting out the slope (over 2, up 3).

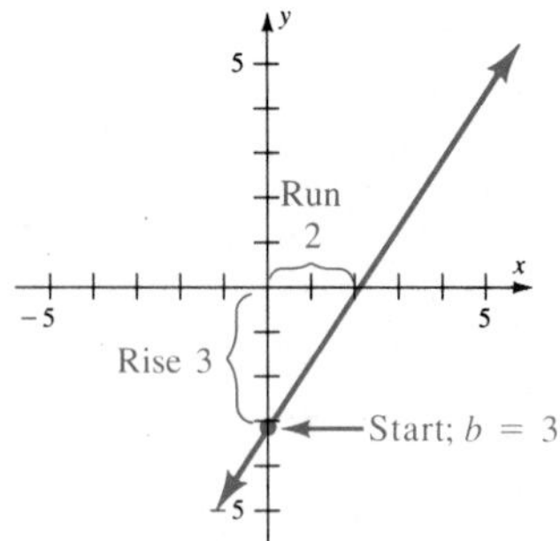

EXAMPLE 5 Graph $4x + 2y - 5 = 0$ for $-1 \le x \le 3$.

Solution Solve for y:

$$2y = -4x + 5$$
$$y = -2x + \tfrac{5}{2}$$

Because of the restriction on the domain, we do not want to graph the entire line, just that part with first components as specified by the restriction, $-1 \le x \le 3$. The usual convention is to show the entire line as a dashed line and the answer as a solid line. For this example $b = \frac{5}{2}$ and the slope is -2; the line is shown as a dashed line. Because of the restriction of the domain, the part of the line with x values between -1 and 3 (inclusive) is shown as a solid line segment.

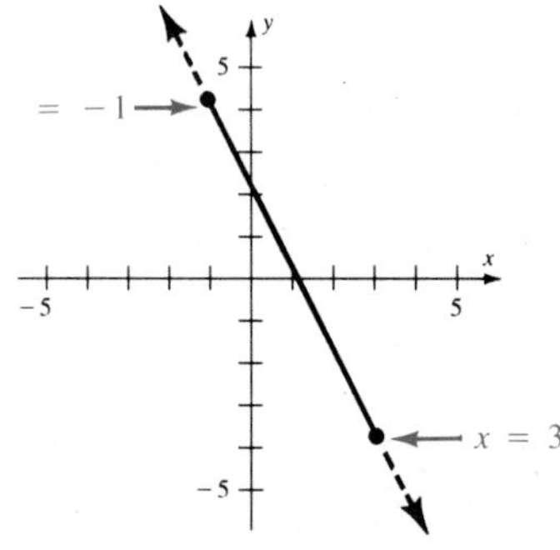

Sometimes a function will have a different equation for different parts of its domain, as shown in Example 6. If each of those equations is linear, then the function is called a **piecewise linear function**.

EXAMPLE 6 Graph $y = x \quad \text{if } x \ge 0$
$y = -x \quad \text{if } x < 0$

Solution Graph each of the line segments as shown in the figure.

Remember from algebra the definition of absolute value:

$$|x| = x \quad \text{if } x \ge 0$$

and

$$|x| = -x \quad \text{if } x < 0$$

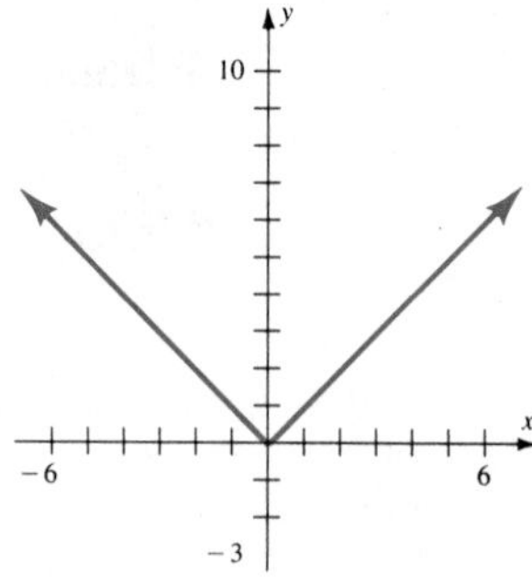

Thus the function in Example 6, the **absolute value function**, can be written $y = |x|$.

EXAMPLE 7 Graph $y = |6 - 2x|$.

Solution We can write this function without absolute value symbols by considering the two parts of the definition of absolute value.

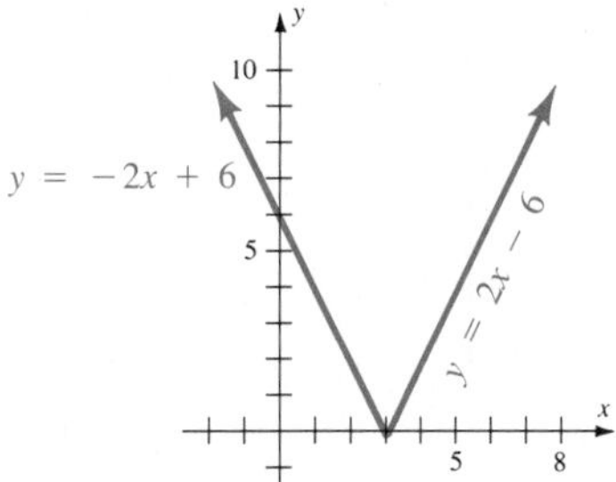

$$y = 6 - 2x \quad \text{if } 6 - 2x \geq 0$$
$$-2x \geq -6$$
$$x \leq 3$$

See Section 1.5 for a review of solving linear inequalities.

And

$$y = -(6 - 2x) \quad \text{if } 6 - 2x < 0$$
$$x > 3$$

Note that $y = -(6 - 2x)$ is the same as $y = 2x - 6$.

This graph is shown at the left. ■

Two relationships make it easy for us to recognize parallel and perpendicular lines.

Parallel Lines
Perpendicular Lines

> Two lines ℓ_1 and ℓ_2 with slopes m_1 and m_2 are
>
> **parallel** if and only if $m_1 = m_2$, and are
> **perpendicular** if and only if $m_1 m_2 = -1$.

EXAMPLE 8 Find the slope and sketch the indicated lines.

a. Passing through $(2, -3)$ and $(-1, 2)$

b. The line passing through the origin which is parallel to the line passing through $(-4, -1)$ and $(1, 3)$

c. The line passing through $(3, 2)$ perpendicular to the line passing through $(-3, 4)$ and $(5, 6)$

d. The line passing through $(-3, 2)$ and $(-3, 4)$

Solution **a.** $m = \dfrac{2 - (-3)}{-1 - 2} = \dfrac{5}{-3} = -\dfrac{5}{3}$ **b.** $m = \dfrac{3 - (-1)}{1 - (-4)} = \dfrac{4}{5}$

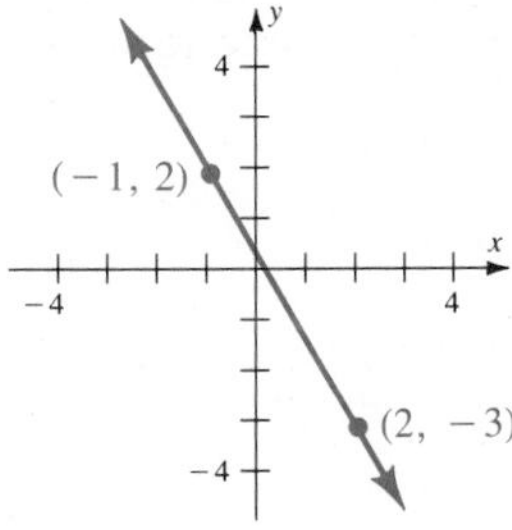

Negative slope

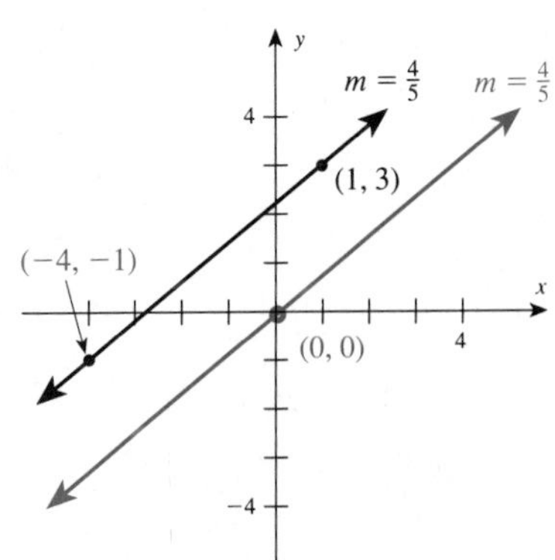

Positive slope

c. $m = \dfrac{6-4}{5-(-3)} = \dfrac{2}{8} = \dfrac{1}{4}$

Slope of perpendicular line is -4

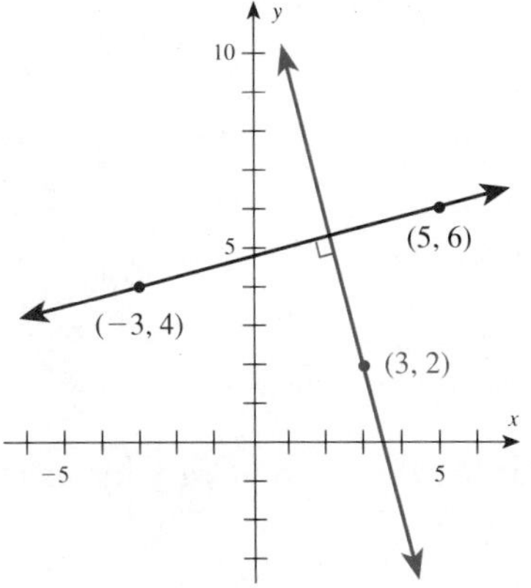

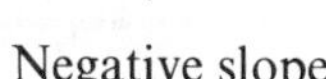
Negative slope

d. 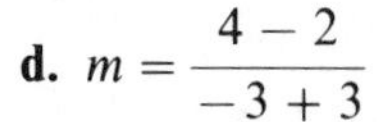 $m = \dfrac{4-2}{-3+3}$ ⟵ This is zero, so the fraction is an undefined number.

Undefined slope; vertical line ■

In constructing mathematical models, it is often necessary to write a linear equation using available or given information about the line. Example 9 shows how to do this if you know the slope and y-intercept.

EXAMPLE 9 Find the equation of the line with y-intercept $(0, 5)$ and slope $-\frac{2}{3}$.

Solution Use the equation $y = mx + b$, where $b = 5$ and $m = -\frac{2}{3}$.

$$y = -\frac{2}{3}x + 5$$

■

More often than not, unfortunately, when you need to find the equation of a line you will not know the y-intercept. You will, however, know a point and the slope or two points. In these cases it is easier to use another form of the equation of a line called the **point–slope form**. It is easy to derive this equation if we remember that

$$m = \frac{y_2 - y_1}{x_2 - x_1}$$

Since (x, y) is any point on the line passing through (x_1, y_1), we have

$$m = \frac{y - y_1}{x - x_1}$$

Thus, by multiplying both sides by $x - x_1$, we obtain

$$m(x - x_1) = y - y_1$$

Point–Slope Form of the Equation of a Line

A nonvertical line having slope m and passing through (x_1, y_1) has the equation

$$y - y_1 = m(x - x_1)$$

EXAMPLE 10 Find an equation of the line with slope 3 passing through $(-2, -5)$.

Solution Use the equation $y - y_1 = m(x - x_1)$, where $m = 3$, $x_1 = -2$, and $y_1 = -5$:

$$y - (-5) = 3[x - (-2)]$$
$$y + 5 = 3(x + 2)$$

■

If you know two points and want the equation, first find the slope and *then* use the point–slope form.

EXAMPLE 11 Find the equation of the line passing through $(-2, 3)$ and $(4, -1)$.

Solution First find the slope:

$$m = \frac{\Delta y}{\Delta x} = \frac{-1 - 3}{4 - (-2)} = \frac{-4}{6} = -\frac{2}{3}$$

Now use the point–slope form (you can use *either* of the given points):

$$(-2, 3){:}\quad y - 3 = -\frac{2}{3}(x + 2) \qquad \text{or} \qquad (4, -1){:}\quad y + 1 = -\frac{2}{3}(x - 4)$$

■

It is not easy to see that the equations in Example 11 are the same. For this reason, we are often asked to algebraically manipulate the answers into the same form. This form is called the **standard form** of the equation of a line.

Standard Form of the Equation of a Line

The **standard form** of the equation of a line is

$$Ax + By + C = 0$$

where (x, y) is any point on the line, and A, B, and C are constants (A and B not both zero).

EXAMPLE 12 Change the point–slope forms given in Example 11 to standard form.

Solution

$y - 3 = -\frac{2}{3}(x + 2)$	Eliminate fractions (multiply by 3).	$y + 1 = -\frac{2}{3}(x + 4)$
$3(y - 3) = -2(x + 2)$	Obtain a 0 on the right.	$3(y + 1) = -2(x - 4)$
$3y - 9 = -2x - 4$	Eliminate parentheses.	$3y + 3 = -2x + 8$
$2x + 3y - 5 = 0$		$2x + 3y - 5 = 0$

■

Note that both equations given in Example 11 are the same in standard form.

EXAMPLE 13 Find the equation of a line passing through $(7, -2)$ with no slope.

Solution Since the line is vertical, the equation has the form $x = h$ when it passes through (h, k). Thus $x = 7$ is the equation. In standard form, $x - 7 = 0$. ■

WARNING ***Do not confuse "no slope" (vertical line) with "zero slope" (horizontal line).***

A mathematical model is often constructed by assuming that the relationship between two variables is linear and then writing an equation using two known data points, as illustrated in Example 14. When relating two values in an applied setting, one value will often depend on the other, so it is customary to designate the dependent one as the **dependent variable**, and the other as the **independent variable**. When x and y are used to represent the variables, x is the independent variable and y the dependent variable.

EXAMPLE 14 A sales executive plotted sales (in millions of dollars) versus the amount spent on advertising (in thousands of dollars) and observed the points as shown in the figure below. It is easy to see that the points lie approximately on a line. By using the points for advertising at \$40,000 and \$80,000, find the equation of this **trend line**. What sales figure can be predicted for an expenditure of \$100,000 on advertising?

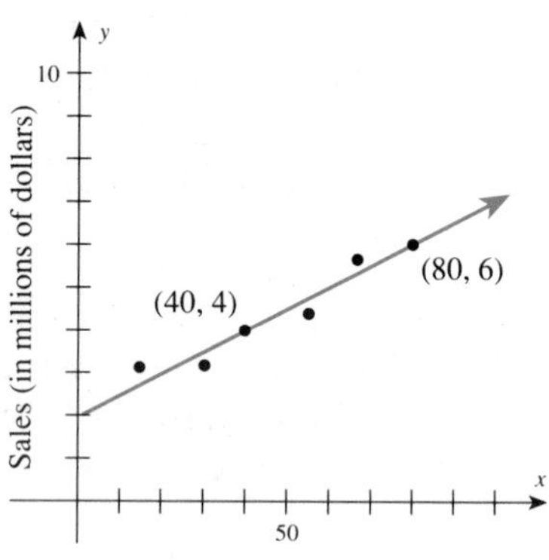

Solution The information gives us points $(40, 4)$ and $(80, 6)$. First find the slope:

$$m = \frac{6 - 4}{80 - 40} = \frac{2}{40} = \frac{1}{20}$$

Next, use the point–slope form using either point:

$$y - y_1 = m(x - x_1)$$
$$y - 4 = \frac{1}{20}(x - 40)$$
$$y = \frac{1}{20}x + 2$$

For advertising expenditures of \$100,000, $x = 100$ so

$$y = \frac{1}{20}(100) + 2$$
$$= 7$$

This means the expected sales are \$7,000,000. ■

Parametric Form of the Equation of a Line

Sometimes it is convenient to define x and y in terms of another variable, say t. We will use this idea in Chapter 3 when we consider systems of equations with more variables than there are equations. For now, let us suppose that the location of a point (x, y) is defined in terms of time, t (in seconds), as follows:

$$x = 1 + t \quad \text{and} \quad y = 2t$$

These equations must be interpreted by saying that although the first component has a 1-second "head start," the rate at which the second component is changing (with respect to time) is twice as fast as the first component. We can tabulate the values of x and y by choosing values for t, where $t \geq 0$ (since time cannot be negative).

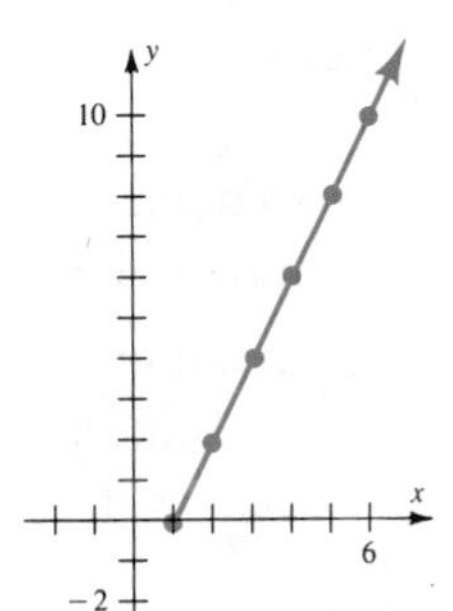

FIGURE 2.11
Graph of $x = 1 + t$, $y = 2t$

t	0	1	2	3	4	5
x	1	2	3	4	5	6
y	0	2	4	6	8	10

The variable t is called a **parameter**, and the equations $x = 1 + t$ and $y = 2t$ are called **parametric equations**. From the graph in Figure 2.11 we see that the data points seem to form a line. We can prove this by *eliminating the parameter*. We solve the first equation for t ($t = x - 1$) and substitute the result into the second equation:

$$y = 2t = 2(x - 1) = 2x - 2$$

which is a linear equation.

Parametric Form of the Equation of a Line

The graph of the parametric equations

$$x = x_1 + at \quad \text{and} \quad y = y_1 + bt$$

is a line passing through (x_1, y_1) with slope $m = b/a \quad (a \neq 0)$.

It is easy to derive this result by solving one of the equations for t and substituting the result into the other equation:

$$x - x_1 = at$$
$$\frac{x - x_1}{a} = t$$

Now, by substitution,

$$y = y_1 + bt = y_1 + b\left(\frac{x - x_1}{a}\right)$$

$$y - y_1 = \frac{b}{a}(x - x_1)$$

This is the equation of a line passing through (x_1, y_1) with slope b/a.

We conclude by summarizing the various forms of linear equations.

Forms of a Linear Equation

STANDARD FORM:	$Ax + By + C = 0$	(x, y) is any point on the line; A, B, and C are constants; A and B are not both zero
SLOPE–INTERCEPT FORM:	$y = mx + b$	m is the slope; $(0, b)$ is the y-intercept
POINT–SLOPE FORM:	$y - y_1 = m(x - x_1)$	m is the slope; (x_1, y_1) is the known point
HORIZONTAL LINE:	$y = k$	(h, k) is a point on the line
VERTICAL LINE:	$x = h$	(h, k) is a point on the line
PARAMETRIC FORM:	$x = x_1 + at$ $y = y_1 + bt$	(x_1, y_1) is a known point; $\frac{b}{a}$ is the slope, $a \neq 0$

2.4 Problem Set

Find the x- and y-intercepts for the lines whose equations are given in Problems 1–8.

1. $y = 2x + 4$
2. $y = 5x - 10$
3. $4x + 3y + 4 = 0$
4. $3x + 2y - 9 = 0$
5. $100x - 250y + 500 = 0$
6. $2x - 5y - 1{,}200 = 0$
7. $y + 2 = 0$
8. $x - 2 = 0$

Find the slope of the line passing through the points in Problems 9–14.

9. $(2, 3)$ and $(5, 4)$
10. $(4, -1)$ and $(-2, 3)$
11. $(5, 2)$ and $(-2, -3)$
12. $(-2, -3)$ and $(4, 5)$
13. $(-2, -3)$ and $(-1, -2)$
14. $(-3, -1)$ and $(-7, -10)$

Find the slope and y-intercept in Problems 15–23.

15. $y = 2x + 4$
16. $y = 5x - 3$
17. $y = 9x + 1$
18. $4x + 3y + 4 = 0$
19. $2x - 3y + 5 = 0$
20. $5x - 2y - 5 = 0$
21. $y - 5 = 0$
22. $y + 9 = 0$
23. $x - 3 = 0$

Graph the lines of the equations in Problems 24–35 by finding the slope and y-intercept.

24. $y = 3x + 4$
25. $y = 2x - 5$
26. $y = -3x + 1$
27. $y = -\frac{1}{4}x + 2$
28. $y = -\frac{2}{3}x - 4$
29. $y = \frac{3}{5}x + \frac{2}{5}$
30. $3x - y + 2 = 0$
31. $x + 3y - 9 = 0$
32. $2x - 3y + 15 = 0$
33. $x = \frac{2}{3}y$
34. $y - 3 = 0$
35. $2x + 5 = 0$

Graph the line segments or piecewise functions given in Problems 36–41.

36. $2x + y + 5 = 0$ if $-3 \leq x \leq 0$
$y + 5 = 0$ if $0 < x < 3$
$x - y - 8 = 0$ if $3 \leq x \leq 10$

37. $y = 3x + 2$ if $0 \leq x \leq 2$
$y - 8 = 0$ if $2 < x < 5$
$x - 2y + 11 = 0$ if $x \geq 5$

38. $y = 2|x|$
39. $y = |3x - 6|$
40. $y = |2x + 4|$
41. $y = -3|x|$

Graph the lines in Problems 42–47 from their parametric equations.

42. $x = -1 + t$
$y = 3 - 2t$

43. $x = -2 + t$
$y = 4 - 5t$

44. $x = -3 + 4t$
$y = -1 - 3t$

45. $x = -2 - 3t$
$y = -2 + 5t$

46. $x = 6 - 2t$
$y = -5 + 3t$

47. $x = 5 - t$
$y = -3 - 2t$

Find the standard form of the equation of the line satisfying the conditions given in Problems 48–59.

48. y-intercept 5; slope 6

49. y-intercept -3; slope -2

50. y-intercept 0; slope 0

51. y-intercept 4; slope 0

52. slope 2; passing through $(4, 3)$

53. slope -1; passing through $(-3, 5)$

54. slope $\frac{1}{2}$; passing through $(5, 3)$

55. slope $\frac{3}{5}$; passing through $(4, -3)$

56. passing through $(-3, -1)$ and $(3, 2)$

57. passing through $(5, 6)$ and $(1, -2)$

58. passing through $(4, -2)$ and $(4, 5)$

59. passing through $(5, 6)$ and $(7, 6)$

APPLICATIONS

Problems 60–65 provide some real-world examples of line graphs. One way of finding the equation of the line is to write two data points from the given information (as shown in Example 14) and then to use those points to write the equation. Use the given information to write an equation in standard form of the line described by the problem.

60. The demand for a certain product is related to the price of the item. Suppose a new line of stationery is tested at two stores. At store A, 25 boxes are sold within a month at \$5 each, and, at store B, 15 boxes priced at \$10 each are sold during the same time. Let x be the price and y be the number of boxes sold. How many boxes would be sold if the price is \$2.00?

61. An important factor related to the demand for a product is its supply. The amount of stationery in Problem 60 that can be supplied is also related to the price. At \$5 each, 10 boxes can be supplied, and, at \$10, 20 boxes can be supplied. Let x be the price and y be the number of boxes supplied. How many boxes could be supplied at \$2.00?

62. The population of Florida in 1980 was roughly 9.7 million, and in 1990 it was 12.6 million. Let x be the year (let 1960 be the base year; that is, $x = 0$ represents 1960, so $x = 10$ represents 1970) and y be the population. Use this equation to predict the population in 2000.

63. The population of Texas in 1980 was roughly 14.2 million, and in 1990 it was 17.2 million. Let x be the year (let 1960 be the base year; that is, $x = 0$ represents 1960, so $x = 10$ represents 1970) and y be the population. Use this equation to predict the population in 2000.

64. It costs \$90 to rent a car if it is driven 100 miles and \$140 if it is driven 200 miles.

65. It costs \$60 to rent a car if it is driven 50 miles and \$60 if it is driven 260 miles.

Many real-world examples have data points that can be approximated by a line. Consider the given data points to find the equation of a trend line, and then answer the question asked in Problems 66–69.

66. If the sales (in thousands of dollars) of a particular item are plotted for the first five years, it can be noticed that these points lie approximately along a straight line.

Year	1	2	3	4	5
Sales	3	4	5	6	6

Using the points $(2, 4)$ and $(5, 6)$, find the equation of a trend line. What sales figure can be predicted for the 8th year?

67. Suppose the cost for maintenance and repairs is plotted as a function of the number of miles (in thousands) the vehicle has been driven.

Miles	10	20	30	40	50
Cost	100	189	309	400	508

Using the points $(10, 100)$ and $(40, 400)$, find the equation of a trend line. What is the expected cost for 53,500 miles?

68. The population of California (in millions) is shown by the following table:

Year	1950	1960	1970	1980	1990
Population	10.6	17.7	19.9	23.7	28.7

Using the points $(0, 10.6)$ and $(30, 23.7)$, find the equation of a trend line that uses 1950 as a base year $(1950 = 0)$. What is the expected population of California in the year 2000?

69. The use of cigarettes among high school seniors is shown by the following table:

Class of	1983	1984	1985	1986	1987
Percentage	70.6	69.7	68.8	67.6	67.2

Using the points $(3, 70.6)$ and $(7, 67.2)$, find the equation of a trend line that uses 1980 as a base year (1980 = 0). What is the expected percentage of cigarette usage for the class of 1992?

Use the following 1990 U.S. Tax Rate Schedule to answer the questions in Problems 70–75.

Schedule Y-1—Use if your filing status is **Married filing jointly or Qualifying widow(er)**

If the amount on Form 1040, line 37, is: Over—	But not over—	Enter on Form 1040, line 38	of the amount over—
\$0	\$32,450	.......15%	\$0
32,450	78,400	\$4,867.50 + 28%	32,450
78,400	185,730	17,733.50 + 33%	78,400
185,730	....	Use **Worksheet** below to figure your tax.	

70. Graph the tax for income from \$0 to \$30,000.

71. Write the equation for the tax, T, for the domain $0 < T \leq 32{,}450$.

72. Write the equation for the tax, T, for the domain $78{,}401 < T \leq 185{,}730$.

73. Graph the tax for income from \$80,000 to \$180,000.

74. Graph the tax for income from \$0 to \$180,000.

75. Write the piecewise linear function for the tax T for amounts up to \$185,730.

2.5 Quadratic and Polynomial Functions

Standard Position Parabola

While many real-world situations can be described by the linear models discussed, many others cannot. It is therefore important to be able to build many *different* types of models. In this section, we discuss a model that requires a second-degree equation for its description.

Quadratic Function

A function f is a **quadratic function** if

$$f(x) = ax^2 + bx + c$$

where a, b, and c are real numbers and $a \neq 0$.

Note that if we write $f(x)$ as y, the quadratic function has one first-degree variable and one second-degree variable. If $b = c = 0$, however, the quadratic function has the form

$$y = ax^2$$

and has a graph called a **standard position parabola**.

EXAMPLE 1 Sketch the graph of $y = 2x^2$.

Solution Begin by finding some ordered pairs that satisfy the equation. Let

$$\begin{array}{lll}
x_1 = 0: & \text{then } y_1 = 2(0)^2 = 0; & \text{the point is } (0, 0) \\
x_2 = 1: & \text{then } y_2 = 2(1)^2 = 2; & \text{the point is } (1, 2) \\
x_3 = -1: & \text{then } y_3 = 2(-1)^2 = 2; & \text{the point is } (-1, 2) \\
x_4 = 2: & \text{then } y_4 = 2(2)^2 = 8; & \text{the point is } (2, 8) \\
x_5 = -2: & \text{then } y_5 = 2(-2)^2 = 8; & \text{the point is } (-2, 8) \\
 & \vdots &
\end{array}$$

These points are plotted and a smooth curve is drawn through them, as shown in Figure 2.12. ■

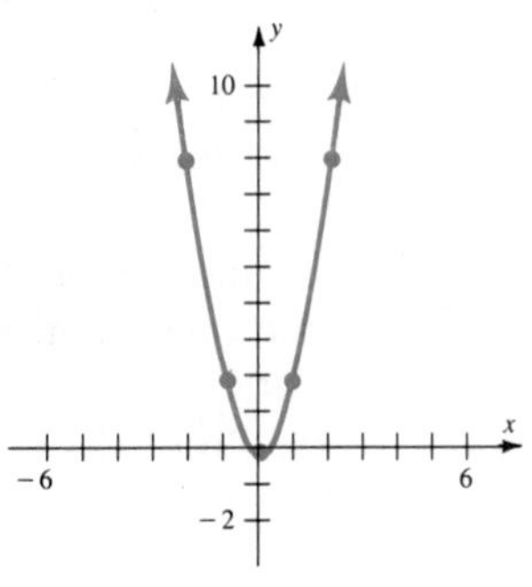

FIGURE 2.12
Graph of $y = 2x^2$

EXAMPLE 2 Sketch the graph of $y = -\frac{1}{2}x^2$.

Solution Plot the points satisfying this equation. If

$$\begin{array}{lll}
x_1 = 0: & \text{then } y_1 = 0; & \text{the point is } (0, 0) \\
x_2 = 1: & \text{then } y_2 = -\dfrac{1}{2}; & \text{the point is } \left(1, -\dfrac{1}{2}\right) \\
x_3 = -1: & \text{then } y_3 = -\dfrac{1}{2}; & \text{the point is } \left(-1, -\dfrac{1}{2}\right) \\
x_4 = 2: & \text{then } y_4 = -2; & \text{the point is } (2, -2) \\
x_5 = -2: & \text{then } y_5 = -2; & \text{the point is } (-2, -2) \\
x_6 = 3: & \text{then } y_6 = -\dfrac{9}{2}; & \text{the point is } \left(3, -\dfrac{9}{2}\right) \\
x_7 = 4: & \text{then } y_7 = -8; & \text{the point is } (4, -8)
\end{array}$$

The plotted points yield the curve shown in Figure 2.13.

CALCULATOR COMMENT

There are several brands of calculators in the \$80–\$100 price range that graph functions. For example, to graph $y = -\frac{1}{2}x^2$ on a TI-81 press the following keys:

[Y=] [(−)] [.5] [X|T] [x^2] [GRAPH]

The coordinates on the curve can be see by pressing [TRACE].

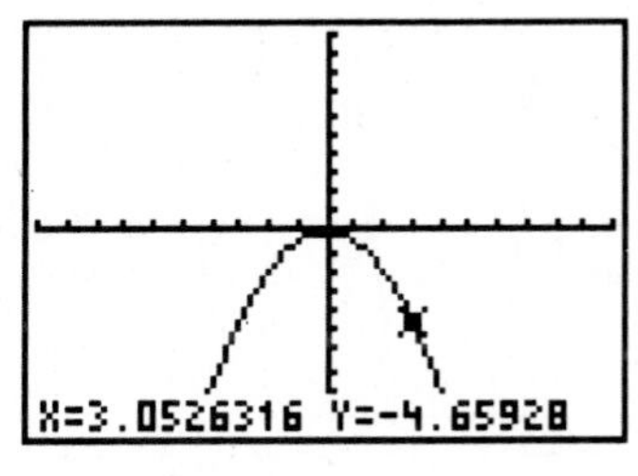

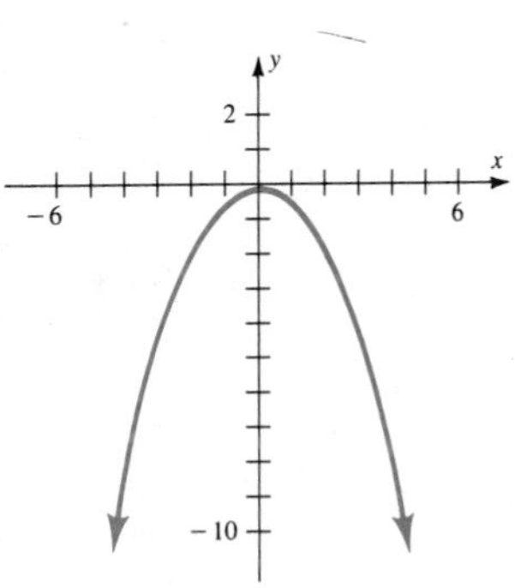

FIGURE 2.13
Graph of $y = -\frac{1}{2}x^2$

■

We can now make some general observations based on the special case $y = ax^2$:

1. The graph has a characteristic shape called a **parabola**.
2. If $a > 0$, the parabola open upward; we say it is **concave upward**. If $a < 0$, the parabola opens downward and is **concave downward**.
3. The point $(0, 0)$ is the lowest point if the parabola opens upward $(a > 0)$; $(0, 0)$ is the highest point if the parabola opens downward $(a < 0)$. This highest or lowest point is called the **vertex**.
4. A parabola is **symmetric** with respect to the vertical line passing through the vertex. This means we can calculate points to the right of the vertex and use symmetry to plot the corresponding points to the left of the vertex.
5. Relative to a fixed scale, the magnitude of a determines the "width" of the parabola; small values of $|a|$ yield "wide" parabolas; large values of $|a|$ yield "narrow" parabolas.

Vertex of a Parabola

For graphs of parabolas of the form

$$y - k = a(x - h)^2$$

the vertex is the point (h, k), and we use this point as the starting point for graphing the parabola. This means that we count out units from the point (h, k) instead of from the origin in order to find new points of the parabola, as illustrated in Example 3.

EXAMPLE 3 Sketch the graph of $y + 3 = 2(x - 2)^2$.

Solution By inspection, $(h, k) = (2, -3)$ is the vertex of the parabola. (Compare this equation with the equation graphed in Example 1.) If the only difference between two equations is the (h, k) point, the graphs will be the same except that they will have different vertices, as shown in Figure 2.14**a**.

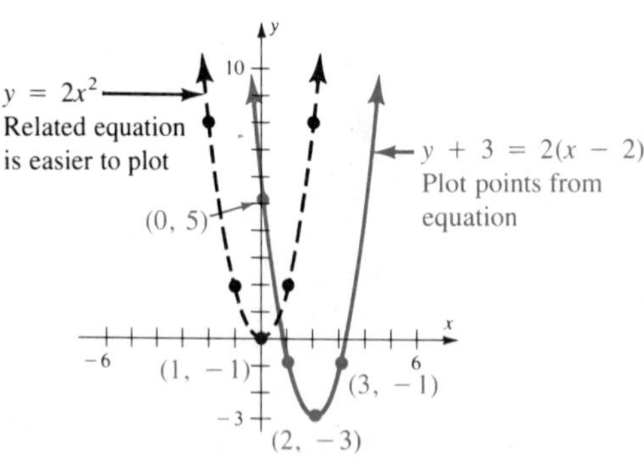

FIGURE 2.14a
Graphs of $y + 3 = 2(x - 2)^2$ and $y = 2x^2$

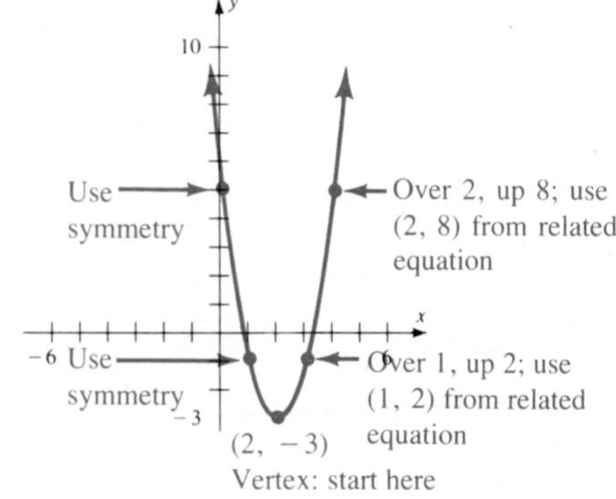

FIGURE 2.14b
Graph of $y = 2x^2$ with points plotted from the vertex $(2, -3)$

Equation: $y + 3 = 2(x - 2)^2$ *Related equation:* $y = 2x^2$

Let $x_1 = 0$:	$(0, 5)$	$(0, 0)$ vertex
$x_2 = 1$:	$(1, -1)$	$(1, 2)$
$x_3 = -1$:	$(-1, 15)$	$(-1, 2)$
$x_4 = 2$:	$(2, -3)$ vertex	$(2, 8)$
$x_5 = -2$:	$(-2, 29)$	$(-2, 8)$
$x_6 = 3$:	$(3, -1)$	$(3, 18)$

Instead of graphing both of these as in Figure 2.14**a**, suppose we count out the points of the related equation *from the point* (h, k) *instead of from the origin*, as shown in Figure 2.14**b**. Note that this graph *is the same as the graph of the more complicated equation.* This leads us to the conclusion that we can graph the simpler related equation if we remember to count out the points from the vertex rather than from the origin. ■

EXAMPLE 4 Sketch the graph of $y - \frac{1}{2} = -2(x + \frac{3}{4})^2$.

Solution We find and plot the vertex $(-\frac{3}{4}, \frac{1}{2})$; we next plot points on the related equation $y = -2x^2$: If

$x_1 = 0$,	then $y_1 = 0$	Vertex
$x_2 = -1$,	then $y_2 = -2$	Count over 1 and down 2 *from the vertex*
$x_3 = -2$,	then $y_3 = -8$	See Figure 2.15

■

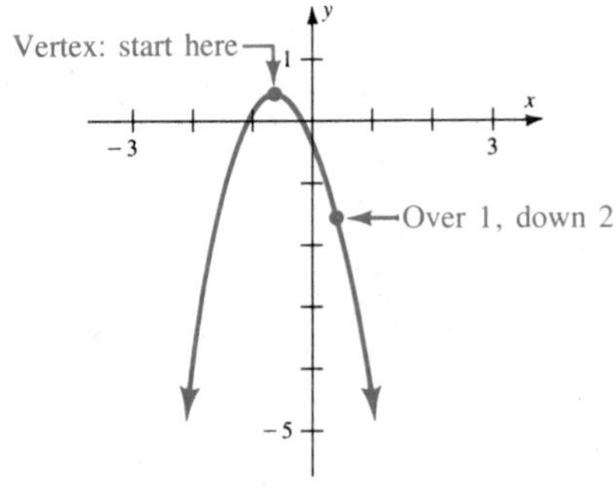

FIGURE 2.15
Graph of $y - \frac{1}{2} = -2(x + \frac{3}{4})^2$

Completing the Square

A general quadratic function

$$y = ax^2 + bx + c$$

can be rewritten in the form

$$y - k = a(x - h)^2$$

by an algebraic procedure called *completing the square.* In order to understand this procedure you need to understand the algebraic form called a *perfect square*:

$$(x + 3)^2 = x^2 + 6x + 9$$
$$(x - 2)^2 = x^2 - 4x + 4$$
$$(x + a)^2 = x^2 + 2ax + a^2$$

A perfect square is a binomial squared. When multiplied out, it is a trinomial in which the first and last terms are the squares of the first and second terms of the binomial. In addition, the middle term is twice the product of the first and second terms of the binomials. This observation gives us a procedure for completing the square for a parabola:

Given: $y = ax^2 + bx + c$ where $a \neq 0$

Step 1: *Subtract the constant term from both sides:*

$$y - c = ax^2 + bx$$

Step 2: *Factor out the coefficient of the squared term:*

$$y - c = a\left(x^2 + \frac{b}{a}x\right)$$

Step 3: *Find $\frac{1}{2}$ the coefficient of the x-term, square it, and add it to both sides. Notice that since you are adding it inside the parentheses on the right, you must* ***add the same number****, which is* $a\left(\frac{b}{2a}\right)^2 = \frac{b^2}{4a}$, *to the left side:*

$$y - c + \frac{b^2}{4a} = a\left[x^2 + \frac{b}{a}x + \left(\frac{b}{2a}\right)^2\right]$$

$$y + \frac{-4ac + b^2}{4a} = a\left(x + \frac{b}{2a}\right)^2$$

This is now in the form $y - y_1 = a(x - x_1)^2$. The algebra here looks terrible, but when you are working with numbers, the process is not so difficult, as shown in Example 5.

EXAMPLE 5 Complete the square for the parabola $y = 2x^2 - 8x + 5$.

Solution

$y - 5 = 2(x^2 - 4x)$	Subtract 5 from both sides and factor out the coefficient of x^2.
$y - 5 + 8 = 2(x^2 - 4x + 4)$	Add 2(4) to both sides.
$y + 3 = 2(x - 2)^2$	Factor.

The graph of this parabola is shown in Example 3. ■

Maximum and Minimum Values

Many applications involve finding the maximum or minimum value of a function f. If f is a quadratic model, then the maximum or minimum value is found by looking at the vertex, (h, k). That is, an equation of the form

$$y - h = a(x - h)^2$$

has vertex (h, k) and can be rewritten in functional form as

$$f(x) = a(x - h)^2 + k$$

If $a > 0$, the parabola opens upward. If $x = h$, then you can see that the minimum value of f is k. You can see that this is true by looking at the graph or noting that $(x - h)^2$ is nonnegative for all x and zero if and only if $x = h$.

If $a < 0$, the parabola opens downward, and the maximum value of f is k.

EXAMPLE 6 A small manufacturer of custom necklaces determines that profit is related to the number of items produced. If x items are produced per day, and the maximum number that can be produced is ten items, the total profit, in dollars, is determined to be

$$P(x) = 360x - 30x^2 - 600$$

How many necklaces should be produced in order to maximize profit?

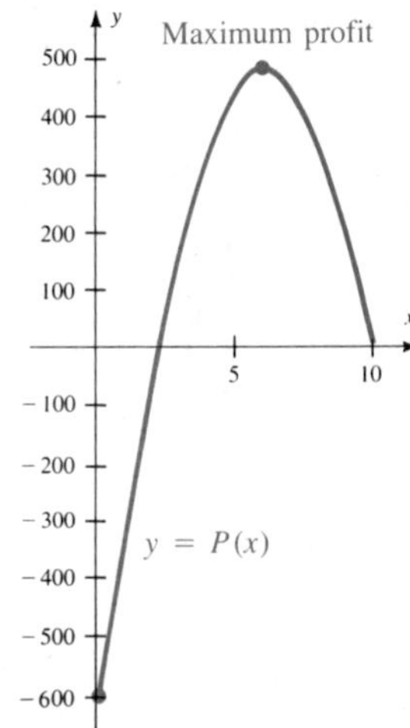

FIGURE 2.16
Graph of $P(x) = -30(x-6)^2 + 480$

Solution The profit function has as its graph a parabola that opens downward. This can be seen by looking at the negative coefficient on the squared term. Next, to find the maximum profit we need to complete the square:

$$P(x) + 600 = -30(x^2 - 12x)$$
$$P(x) + 600 - 1{,}080 = -30(x^2 - 12x + 6^2)$$
$$P(x) - 480 = -30(x-6)^2$$

Note: $6^2(-30) = -1{,}080$

The vertex is (6, 480) so we see that the maximum value of P is 480, which occurs when $x = 6$. Note from the graph shown in Figure 2.16 that if there were a strike and no necklaces were produced, the daily profit would be -600 (this is a \$600-per-day loss). ■

CALCULATOR COMMENT

PRESS: [Y=] [360] [X|T] [−] [30] [X|T] [x^2] [−] [600] [GRAPH]

To answer the question press [TRACE] and move the cursor to the highest point on the curve:

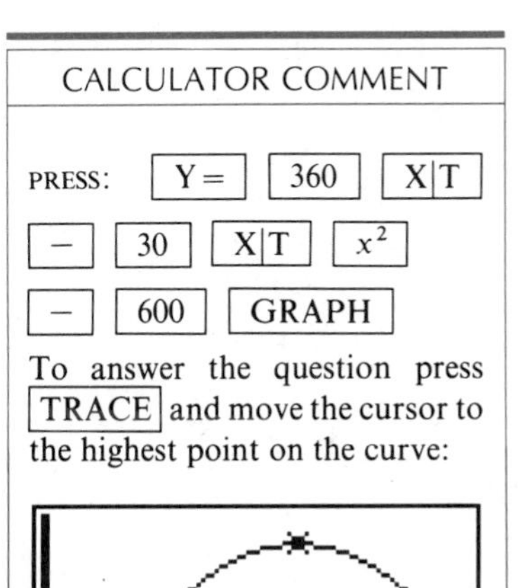

In Example 6 you might ask how it might be possible to come up with the profit function. The **profit function** is found according to the relationship*

PROFIT = REVENUE − COST

Suppose the cost of producing the necklaces in Example 6 is $5x^2 + 40x + 600$. This function,

$$C(x) = 5x^2 + 40x + 600$$

is called a **cost function** (or, to be precise, **total** cost function) because it gives the cost of producing x items. Most cost functions are made up of two parts, a variable that depends on the number of items produced and a fixed part, which does not. The value of $C(0)$ is the **fixed cost**. For this example, $C(0) = 5(0)^2 + 40(0) + 600 = 600$. Suppose also that the price of each necklace is set at $400 - 25x$ dollars. If this function represents the highest price per unit that would sell all x units, it is called the **demand function**. The demand function indicates that the price is determined by the number of necklaces sold. The **revenue function** (i.e., **total** revenue function),

* This is what economists refer to as the **total profit function**; see, for example, Samuelson's *Economics*, 13th ed. (1989, New York: McGraw-Hill, p. 425). In this book, we will simply refer to this as the profit function.

$R(x)$, is defined to be the product of the number of items sold and the price:

REVENUE = (NUMBER OF ITEMS)(PRICE PER ITEM)

For this example, $R(x) = x(400 - 25x)$.

This is also called the demand.

EXAMPLE 7 Draw the graphs of the cost and revenue functions on the same axis and compare with the graph of the profit function in Example 6.

Solution We have:

REVENUE FUNCTION: $R(x) = x(400 - 25x) = -25x^2 + 400x$

$$\text{Complete the square:} \qquad R(x) = -25(x^2 - 16x)$$
$$R(x) - 1{,}600 = -25(x^2 - 16x + 64)$$
$$R(x) - 1{,}600 = -25(x - 8)^2$$

This parabola opens down with vertex at (8, 1600). The domain is [0, 10], and is shown in Figure 2.17.

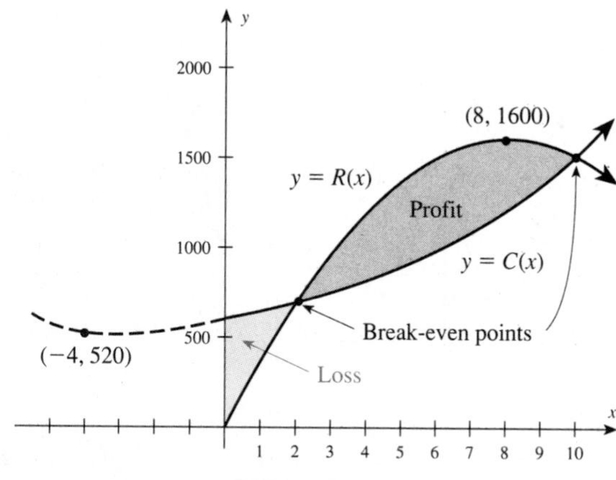

FIGURE 2.17
Graphs of $y = 400x - 25x^2$ and $y = 5x^2 + 40x + 600$

COST FUNCTION: $C(x) = 5x^2 + 40x + 600$

$$\text{Complete the square:} \qquad C(x) - 600 = 5(x^2 + 8x)$$
$$C(x) - 600 + 80 = 5(x^2 + 8x + 16)$$
$$C(x) - 520 = 5(x + 4)^2$$

This parabola opens up with vertex at (−4, 520). The domain is [0, 10], and is shown in Figure 2.17.

PROFIT FUNCTION: $P(x) = R(x) - C(x)$
$$= -25x^2 + 400x - (5x^2 + 40x + 600)$$
$$= -30x^2 + 360x - 600$$

This function was graphed in Example 6. Notice that positive profit is shown as the gray section of Figure 2.17 and negative profit (loss) is shown as the colored section of Figure 2.17. ■

Break-Even Analysis

A company will *break even* (that is, profit will balance loss) if the cost and the revenue are equal. The point (or points) at which this occurs is called a **break-even point**.

EXAMPLE 8 Find the break-even point(s) for Example 7.

Solution Figure 2.17 labels the break-even points. Geometrically, they are the points where the revenue and cost curves intersect. To find these points algebraically, we need to find the x values for which $R(x) = C(x)$.

$$R(x) = C(x)$$
$$400x - 25x^2 = 5x^2 + 40x + 600$$
$$30x^2 - 360x + 600 = 0$$
$$x^2 - 12x + 20 = 0$$
$$(x - 10)(x - 2) = 0$$
$$x = 10, 2$$

The break-even points are at (2, 700) and (10, 1500). ■

Polynomial Functions

Linear and quadratic functions are special types of more general mathematical functions called **polynomial functions**.

Polynomial Function

> A function f is a **polynomial function** in x of degree n if
>
> $$f(x) = a_nx^n + a_{n-1}x^{n-1} + \cdots + a_1x + a_0$$
>
> where $a_0, a_1, a_2, \ldots, a_n$ are real numbers, n is a whole number, and $a_n \neq 0$.

Examples of polynomial functions are

$$f(x) = 5x^3 + 2x^2 - 3x + 5 \qquad g(x) = x^2 - 5x + 1 \qquad h(x) = 6$$

Note that linear and quadratic functions are special types of polynomial functions of degree 1 and 2, respectively. (For a review of the terminology and algebraic simplification of polynomials, see Chapter 1.)

EXAMPLE 9

a. $f(x) = x$ is a polynomial.

b. $f(x) = \frac{1}{x}$ is not a polynomial since $\frac{1}{x} = x^{-1}$ and -1 is not a whole number.

c. $f(x) = \frac{1}{6}x$ is a polynomial ($n = 1$, $a_n = \frac{1}{6}$).

d. $f(x) = \sqrt{x} + 3x^2$ is not a polynomial since $\sqrt{x} = x^{1/2}$ and $\frac{1}{2}$ is not a whole number.

e. $f(x) = x^2 + 3x + 4x^{-2}$ is not a polynomial since the exponent of $4x^{-2}$ is not a whole number. ■

One method of graphing a polynomial function is to plot points as in Section 2.3. However, much better and easier methods are available in calculus, so we will delay the graphing of polynomial functions until we discuss curve sketching in general in Chapter 11. In the meantime we will limit applications to the linear and quadratic functions introduced in this and the previous sections.

2.5 Problem Set

Sketch the graph of each equation in Problems 1–38.

1. $y = x^2$
2. $y = -x^2$
3. $y = -2x^2$
4. $y = 3x^2$
5. $y = -5x^2$
6. $y = 5x^2$
7. $y = \frac{1}{3}x^2$
8. $y = -\frac{1}{3}x^2$
9. $y = \frac{1}{10}x^2$
10. $y = -\frac{1}{10}x^2$
11. $y = \frac{2}{3}x^2$
12. $y = -\frac{2}{3}x^2$
13. $y = (x - 1)^2$
14. $y = -(x + 2)^2$
15. $y = (x + 3)^2$
16. $y = -2(x - 1)^2$
17. $y = \frac{1}{4}(x - 1)^2$
18. $y = -\frac{1}{2}(x + 1)^2$
19. $y - 2 = (x - 1)^2$
20. $y - 2 = 3(x + 2)^2$
21. $y - 2 = -\frac{3}{5}(x - 1)^2$
22. $y + 3 = \frac{2}{3}(x + 2)^2$
23. $y + \frac{2}{3} = (x + \frac{1}{3})^2$
24. $y + \frac{2}{5} = -(x - \frac{1}{3})^2$
25. $y + \frac{2}{5} = (x - \frac{3}{5})^2$
26. $y - .1 = (x + .2)^2$
27. $y = x^2 - 4x + 4$
28. $y = x^2 - 6x + 9$
29. $y = -2x^2 - 2x - 2$
30. $y = 3x^2 + 6x + 3$
31. $y = x^2 + 4x + 5$
32. $y = x^2 - 6x + 11$
33. $y = 2x^2 + 4x$
34. $y = 3x^2 - 6x$
35. $y = -2x^2 - 4x + 1$
36. $y = -2x^2 - 4x + 3$
37. $y = -3x^2 + 12x - 16$
38. $y = 2x^2 - 2x + 5$

Find the maximum or minimum value of y for each of the functions in Problems 39–50.

39. $y = -4(x + 1)^2 + 3$
40. $y = -5(x - 4)^2 + 2$
41. $y = -10(x - 450)^2 + 1{,}250$
42. $y = 12(x + 30)^2 - 140$
43. $y = 25(x - 560)^2 - 1{,}400$
44. $y = -150(x + 2{,}300)^2 + 12{,}000$
45. $y = -3x^2 - 18x - 41$
46. $y = -4x^2 - 40x - 60$
47. $2x^2 + 12x - y + 31 = 0$
48. $3x^2 - 12x - y + 22 = 0$
49. $6x^2 - 12x + 3y + 18 = 0$
50. $5y - 30x^2 - 180x - 370 = 0$

APPLICATIONS

51. After extensive market research, a consulting firm has determined that the demand for a certain item is $1{,}040 - 10x$ dollars, where x is the number of items produced. Since the company has fixed costs of 6,650 dollars, the cost function is found to be $C(x) = 6{,}650 + 500x$.
a. Find the revenue function.
b. Find the break-even point(s).

52. The demand for a certain ratchet flange is $50 - x$ dollars, where x is the number of flanges produced. The cost function is found to be $C(x) = 200 + 20x$.
a. Find the revenue function.
b. Find the break-even point(s).

53. A manufacturer produces quality boats. The profit, in dollars, is determined to be

$$P(x) = -10(x - 375)^2 + 1{,}156{,}250$$

where x is the number of boats.
a. How many boats should be produced in order to produce a maximum profit?
b. What is the profit (or loss) if no boats are produced?
c. What is the maximum profit?

54. Find the maximum profit for Problem 51.

55. Find the maximum profit for Problem 52.

56. Graph the cost and revenue functions of Problem 51 and shade the region representing the company's profit.

57. Graph the cost and revenue functions of Problem 52 and shade the region representing the company's profit.

58. The highest bridge in the world is the bridge over the Royal Gorge of the Arkansas River in Colorado. It is 1,053 ft above the water. If a rock is thrown vertically upward from this bridge with an initial velocity of 64 feet per second, the height h of the rock above the river at time t is described by the function

$$h(x) = -16t^2 + 64t + 1{,}053$$

What is the maximum height possible for a rock thrown vertically upward from the bridge with an initial velocity of 64 feet per second? After how many seconds will it reach that height?

59. In 1974 Evel Knievel attempted a skycycle ride across the Snake River. Suppose the path of the skycycle is given by the equation

$$d(x) = -.0005(x - 2{,}390)^2 + 3{,}456$$

where $d(x)$ is the height above the canyon floor for a horizontal distance of x feet from the launching ramp. What was Knievel's maximum height?

60. **a.** In most states, drivers are required to know approximately how long it takes to stop their cars at various speeds. Suppose you estimate three car lengths for 30 mph and six car lengths for 60 mph. One car length per 10 mph assumes a linear relationship between speed and distance covered. Write a linear equation where x is the speed of the car and y is the distance traveled by the car in feet. (Assume that one car length = 15 ft.)

b. The scheme for the stopping distance of a car given in part **a** is convenient but not accurate. The stopping distance is more accurately approximated by the quadratic equation

$$y = .071x^2$$

This says that a car requires four times as many feet to stop at 60 miles per hour than at 30 mph. That is, doubling the speed quadruples the braking distance. Graph this quadratic equation and the linear equation from part **a** on the same coordinate axes.

c. Comment on the results from part **b**. At about what speed are the two measures the same?

2.6 *Rational Functions*

The third type of function used for mathematical models is called a **rational function.**

Rational Function

A function f is a **rational function** if

$$f(x) = \frac{P(x)}{Q(x)}$$

where P is any polynomial function and Q is a polynomial function whose domain excludes values for which $Q(x) = 0$.

Functions such as

$$f(x) = \frac{1}{x} \qquad y = \frac{1}{x-2} \qquad g(x) = \frac{x^2 + 3x - 1}{x - 1}$$

are examples of rational functions. The values $x = 0$, $x = 2$, and $x = 1$, respectively, are excluded from the domains of these three functions according to the definition of a rational function.

In Section 2.3, we graphed $h(x) = \frac{1}{x}$ (Example 3) by plotting points. It is important to remember the general shape of the graph of this function so we repeat it here for easy reference

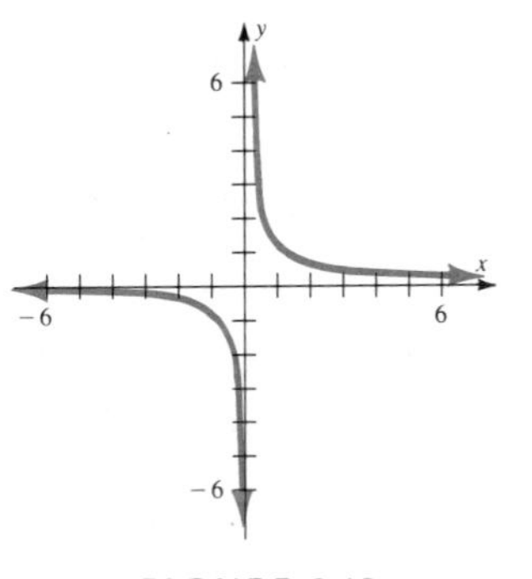

FIGURE 2.18
Graph of $y = \frac{1}{x}$

Note that the domain for the graph in Figure 2.18 consists of all real numbers except $x = 0$. That is, the graph of the function $y = \frac{1}{x}$ does not cross the vertical line $x = 0$. Now consider the graph of the rational function $y = \frac{1}{x-1}$. The definition of rational functions *excludes* from the domain all values for which $x - 1 = 0$, namely, $x = 1$. You can, however, view the equation $x = 1$ as the equation of a vertical line. This line is called a **vertical asymptote** for the curve $y = 1/(x - 1)$ and is illustrated in Example 1**a**. We also say that this function is *unbounded*. In general, a function is **unbounded** if for any number M there is a value of the function whose numerical value is larger than M or smaller than $-M$.

EXAMPLE 1 Graph the following equations by plotting points:

a. $y = \dfrac{1}{x - 1}$ **b.** $y = \dfrac{1}{x + 2}$ **c.** $y = \dfrac{1}{2x - 5}$

Solution The vertical asymptotes are lines for which the denominators are equal to zero. To find the asymptotes, set the denominator equal to zero and solve:

a. $x - 1 = 0$; $x = 1$ **b.** $x + 2 = 0$; $x = -2$ **c.** $2x - 5 = 0$; $x = \frac{5}{2}$

Plot additional points as necessary as shown in Figure 2.19.

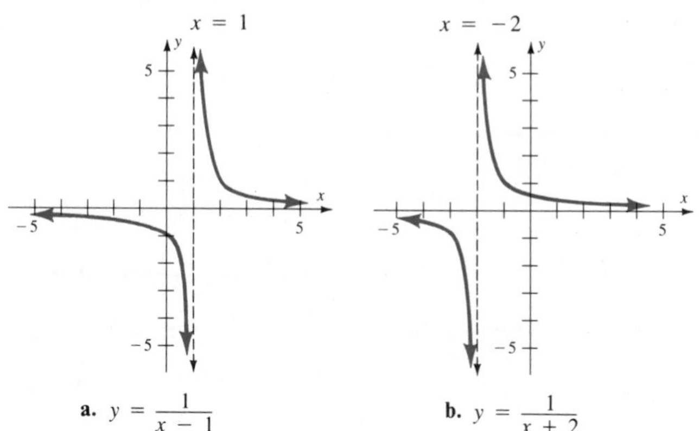

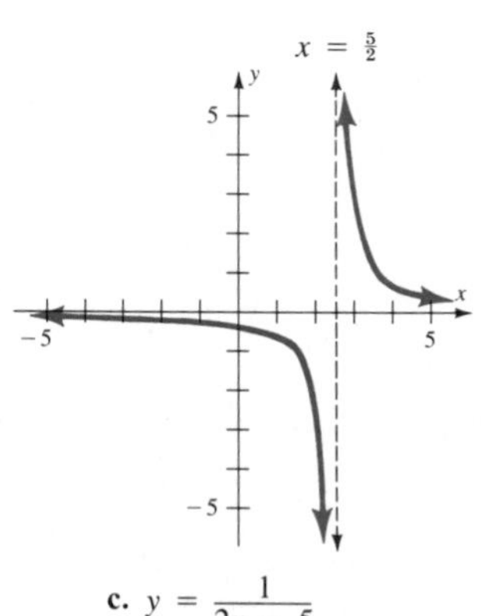

FIGURE 2.19
Vertical asymptotes

A line is an asymptote for a curve if the distance between the line and the curve becomes smaller and smaller within some suitable domain. The graph of a function cannot cross its vertical asymptotes.

EXAMPLE 2 Graph

$$y = \frac{1}{3x^2 - 5x - 2}$$

Solution Find the vertical asymptotes and compare them with those in Figure 2.19:

$$3x^2 - 5x - 2 = 0$$
$$(x - 2)(3x + 1) = 0$$
$$x = 2, -\frac{1}{3}$$

x	y
0	$-\frac{1}{2}$
1	$-\frac{1}{4}$
3	$\frac{1}{10}$
-1	$\frac{1}{6}$
-2	$\frac{1}{20}$

Draw the vertical asymptotes and plot additional points.

See Appendix D for an Introduction to Spreadsheets.

SPREADSHEET PROGRAM

	A	B
1	x	y=1/(3x^2-5x-2)
2	-3	1/(3*A2^2-5*A2-2)
3	+A2+0.5	replicate
4	replicate	

x	y=1/(3x^2-5x-2)
-3	0.025
-2.75	0.029038113
-2.5	0.034188034
-2.25	0.040920716
-2	0.05
-1.75	0.062745098
-1.5	0.081632653
-1.25	0.111888112
-1	0.166666667
-0.75	0.290909091
-0.5	0.8
-0.25	-1.77777778
0	-0.5
0.25	-0.32653061
0.5	-0.26666667
0.75	-0.24615385
1	-0.25
1.25	-0.28070175
1.5	-0.36363636
1.75	-0.64
2	ERR
2.25	0.516129032
2.5	0.235294118
2.75	0.144144144
3	0.1

It is difficult to find all asymptotes with a spreadsheet program.

The error here is that we are requesting division by zero. This indicates a vertical asymptote.

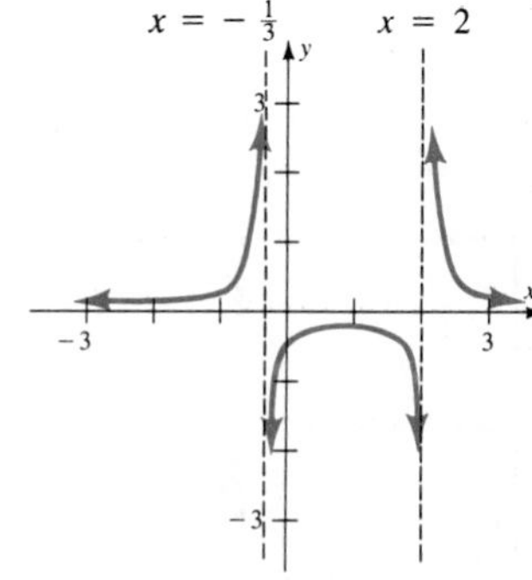

■

It would be convenient if we could say that if $f(x) = P(x)/Q(x)$, then any value for which $Q(x) = 0$ would be a vertical asymptote. Unfortunately, such is not the case, as we see in Example 3.

EXAMPLE 3 Graph

$$y = \frac{x}{x^2 - x}$$

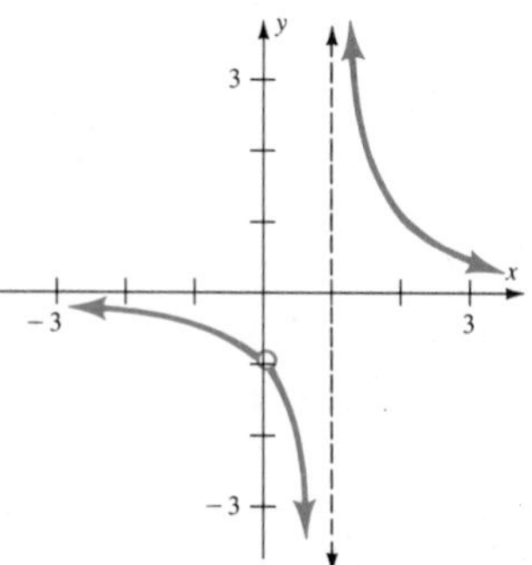

FIGURE 2.20

Graph of $y = \dfrac{x}{x^2 - x}$

Solution Note that if $x^2 - x = 0$ then

$$x(x - 1) = 0$$
$$x = 0, 1$$

There are two values that cause division by zero. However, if we plot points, we see that $x = 1$ is a vertical asymptote and that $x = 0$ is simply a point deleted from the domain. The graph is shown in Figure 2.20.

CALCULATOR COMMENT

If you are using a graphing calculator, you must pay particular attention to Example 3. Note here, if we press

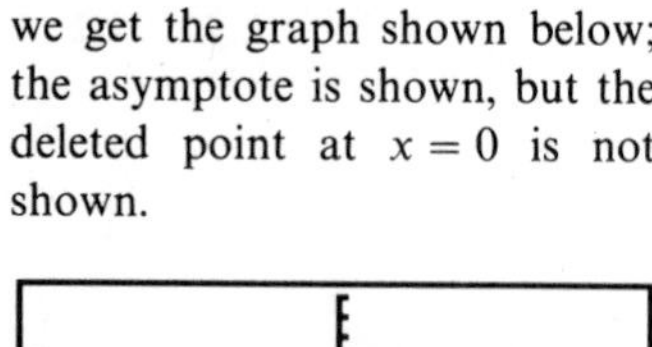

we get the graph shown below; the asymptote is shown, but the deleted point at $x = 0$ is not shown.

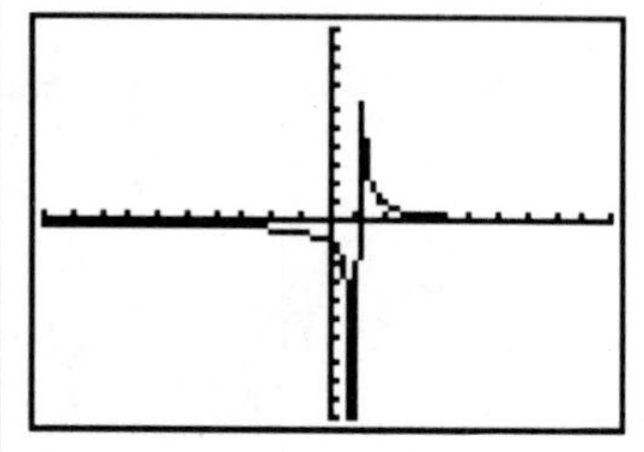

There is a way to decide if a value causing division by zero indicates a deleted point or a vertical asymptote. All we need to do is reduce the fraction and set the resulting denominator equal to zero. Thus

$$\frac{x}{x^2 - x} = \frac{x}{x(x - 1)} = \frac{1}{x - 1}$$
$$x \neq 0, x \neq 1$$

Values that cancel indicate deleted points (that is, $x = 0$).

$x = 1$ is an asymptote — Values that do not cancel indicate asymptotes

EXAMPLE 4 Given the following equations, find the vertical asymptotes.

a. $y = \dfrac{x + 4}{x^2 - 16}$ **b.** $y = \dfrac{x + 4}{x^2 - 15}$ **c.** $y = \dfrac{3x^2 - 5x - 2}{x^2 - 5x + 6}$

Solution **a.** $y = \dfrac{x + 4}{x^2 - 16} = \dfrac{x + 4}{(x - 4)(x + 4)} = \dfrac{1}{x - 4}$

$x \neq 4, x \neq -4$ $x = 4$ is a vertical asymptote.

$x = -4$ is a deleted point on the graph.

b. $y = \dfrac{x + 4}{x^2 - 15}$ is reduced. The vertical asymptotes are found by solving $x^2 - 15 = 0$. The vertical asymptotes have equations $x = \sqrt{15}$, $x = -\sqrt{15}$.

c. $y = \dfrac{3x^2 - 5x - 2}{x^2 - 5x + 6} = \dfrac{(3x + 1)(x - 2)}{(x - 3)(x - 2)} = \dfrac{3x + 1}{x - 3}$

$x = 3$ is a vertical asymptote. ■

Note that the graphs in Figures 2.18 and 2.19 and in Examples 2 and 3 not only have vertical asymptotes but also have a **horizontal asymptote**, $y = 0$. As $|x|$ gets larger, each of the named curves gets closer to the line $y = 0$. There will be a horizontal asymptote $y = 0$ whenever the degree of the numerator is less than the degree of the denominator. Study the variations of asymptotes shown in Figure 2.21**a**–**d**.

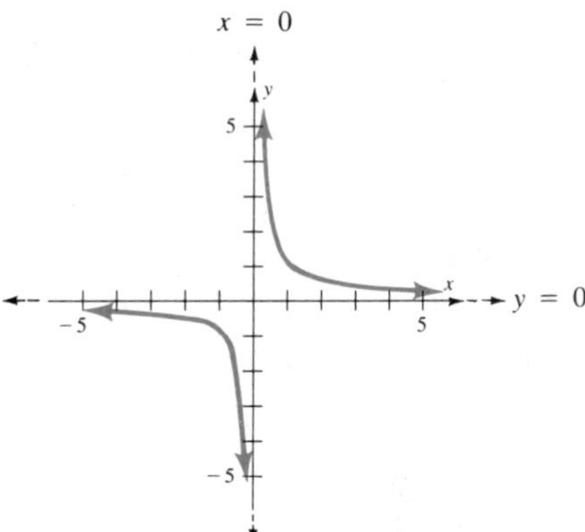

FIGURE 2.21a

Graph of $y = \frac{1}{x}$.

Vertical asymptote, $x = 0$; horizontal asymptote, $y = 0$.

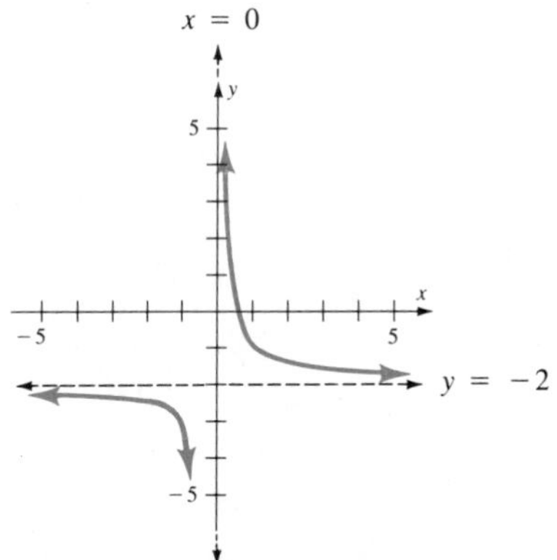

FIGURE 2.21b

Graph of $y = \frac{1}{x} - 2$.

Vertical asymptote, $x = 0$; horizontal asymptote, $y = -2$.

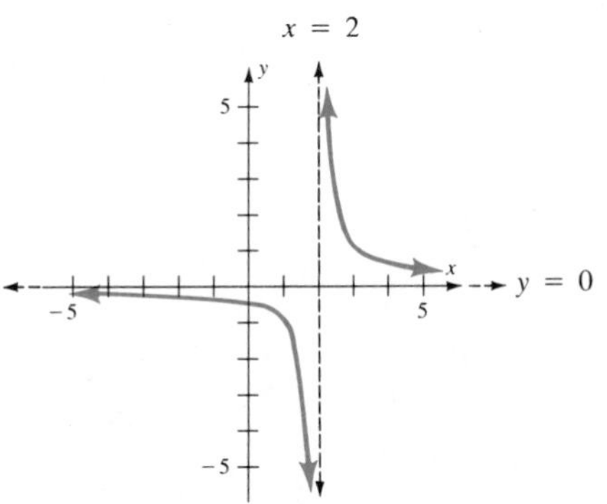

FIGURE 2.21c

Graph of $y = \frac{1}{x - 2}$.

Vertical asymptote, $x = 2$; horizontal asymptote, $y = 0$.

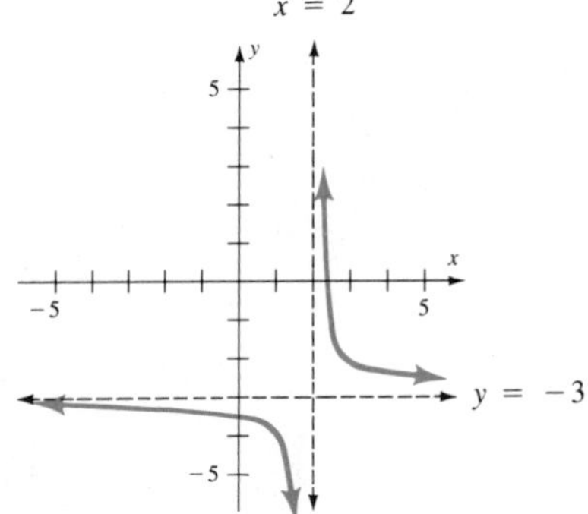

FIGURE 2.21d

Graph of $y = \frac{1}{x - 2} - 3$.

Vertical asymptote, $x = 2$; horizontal asymptote, $y = -3$.

Apparently, horizontal asymptotes depend on the relative degrees of the numerator and denominator polynomials. The following summary is useful when graphing rational functions.

Asymptotes

If $f(x) = P(x)/D(x)$, where $P(x)$ and $D(x)$ are polynomial functions with no common factors, then:

the line $x = r$ is a *vertical asymptote* if $D(r) = 0$;

the line $y = 0$ is a *horizontal asymptote* if the degree of P is less than the degree of D;

the line $y = \dfrac{a_n}{b_n}$ is a *horizontal asymptote* if the degree of P is the same as the degree of D and
$P(x) = a_nx^n + \cdots + a_0$ and
$D(x) = b_nx^n + \cdots + b_0$

EXAMPLE 5 Graph $y = \dfrac{3x - 1}{2x + 5}$.

Solution First find the vertical asymptotes. Set the denominator equal to 0 and solve.

$$2x + 5 = 0$$
$$2x = -5$$
$$x = -\frac{5}{2}$$

Second, find the horizontal asymptotes. The degree of $3x - 1$ and $2x + 5$ is the same, so there is one horizontal asymptote: $y = \frac{3}{2}$. The remaining points are found by calculation.

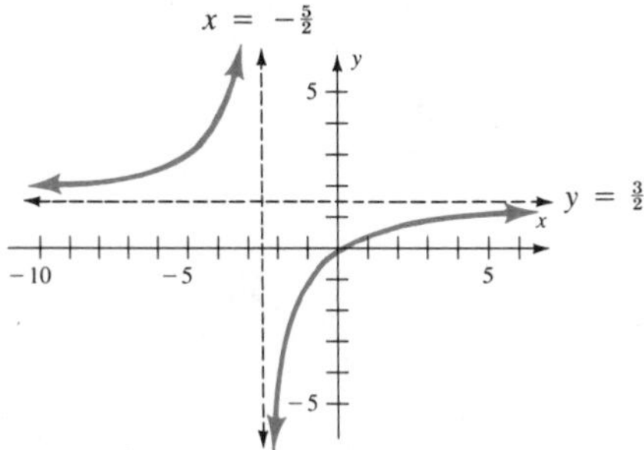

x	y
0	$-\frac{1}{5}$
1	$\frac{2}{7}$
−1	$-\frac{4}{3}$
−2	−7
−3	10
−4	$\frac{13}{3}$
−5	$\frac{16}{5}$

■

Cost-Benefit Models

Many real-life situations involve some desired result, or benefit. However, to achieve this result involves a certain cost and it is often necessary to compare the benefit with the cost. An equation that compares costs and benefits is called a **cost-benefit model**. Suppose that the cost, C, of installing five-field electrostatic precipitators for the removal of p percent of pollutants released into the atmosphere from the emissions of a cement factory is given by the formula

$$C(p) = \frac{200{,}000p}{105 - p}$$

Example 6 demonstrates the use of this cost-benefit model.

EXAMPLE 6 Use the cost-benefit function given above to find:

a. How much would it cost to remove 95% of the pollutants (as required by the Environmental Protection Agency)?

b. How much does it cost to remove 80% of the pollutants (which the company is presently doing)?

c. How much would it cost to remove 100% of the pollutants?

d. Graph the cost-benefit relationship for $0 \le p \le 100$.

Solution **a.** For $p = 95$, $C = \dfrac{200{,}000(95)}{105 - 95} = 1{,}900{,}000$

b. For $p = 80$, $C = \dfrac{200{,}000(80)}{105 - 80} = 640{,}000$

c. For $p = 100$, $C = \dfrac{200{,}000(100)}{105 - 100} = 4{,}000{,}000$

d. Use the points found above (along with additional ones as needed) to graph the function for $0 \le p \le 100$. Note that there is both a horizontal and vertical asymptote.

$$C = \frac{200{,}000p}{105 - p}$$

⟵ Vertical asymptote: $105 - p = 0$
$p = 105$

↑ Horizontal asymptote since the degree of the numerator and denominator is the same; it is

$$C = \frac{200{,}000}{-1} = -200{,}000$$

From the coefficients of p

Note: You will not see the graph approach this asymptote because C approaches $-200{,}000$ as $|p|$ gets larger, but we are asked to graph this curve only for $0 \le p \le 100$.

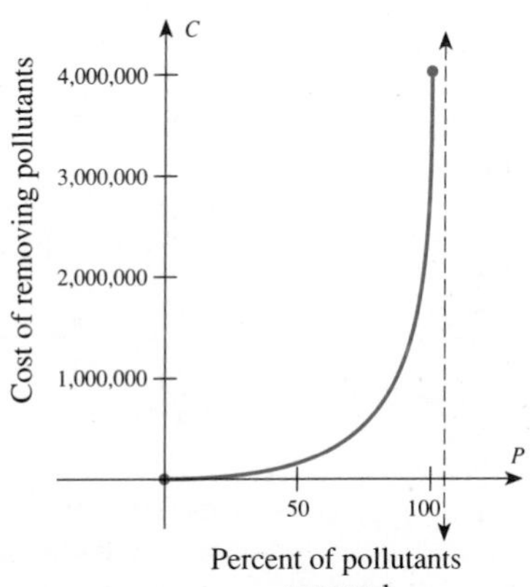

■

2.6
Problem Set

1. What is a rational function?
2. What is a vertical asymptote, and how do you find it?
3. What is a horizontal asymptote, and how do you find it?

Name the horizontal and vertical asymptotes for the curves in Problems 4–15. Also name any deleted points.

4. $f(x) = \dfrac{1}{x+3}$
5. $f(x) = \dfrac{5}{2x-3}$
6. $f(x) = \dfrac{2x-2}{3x^2-x-2}$
7. $g(x) = \dfrac{x+3}{x^2+2x-3}$
8. $g(x) = \dfrac{6}{2x^2+3x-2}$
9. $g(x) = \dfrac{5}{6x^2+5x-6}$
10. $h(x) = \dfrac{2x+5}{3x-4}$
11. $h(x) = \dfrac{5x^2+3x-2}{2x^2+4x+1}$
12. $h(x) = \dfrac{6x^2-3x+1}{2x^2-5}$
13. $y = \dfrac{3x+2}{x^2+x+2}$
14. $y = \dfrac{x-5}{x^2-3x+5}$
15. $y = \dfrac{2x+1}{x^2+x-2}$

Graph the functions in Problems 16–27.

16. $y = \dfrac{2}{x}$
17. $y = -\dfrac{1}{x}$
18. $y = \dfrac{1}{x+3}$
19. $y = \dfrac{1}{x-4}$
20. $y = \dfrac{1}{x} + 4$
21. $y = \dfrac{1}{x} + 3$
22. $y = \dfrac{2x^2+5x+3}{x^2-x-2}$
23. $y = \dfrac{2x^2-3x+1}{x^2+2x-3}$
24. $y = \dfrac{1}{x^2-4}$
25. $y = \dfrac{1}{x^2-9}$
26. $y = \dfrac{1}{3x^2-x-2}$
27. $y = \dfrac{1}{x^2+2x-3}$

APPLICATIONS

In Problems 28–31, use the cost-benefit model

$$C = \frac{40{,}000p}{110-p}$$

where C is the cost (to the nearest dollar) of removing p% of the pollutants.

28. What is the cost of removing 95% of the pollutants?
29. What is the cost of removing 80% of the pollutants?
30. What is the cost of removing 100% of the pollutants?
31. Graph these relationships for $0 \le p \le 100$.

In Problems 32–35, use the cost-benefit model

$$C = \frac{18{,}000p}{100-p}$$

where C is the cost of removing p% of the pollutants.

32. What is the cost of removing 95% of the pollutants?
33. What is the cost of removing 80% of the pollutants?
34. Is it possible to remove 100% of the pollutants?
35. Graph these relationships for $0 \le p < 100$.

Radiologists must deal with three quantities each time an X-ray is taken:

t = time in seconds that the X-ray machine is on
mA = the current, measured in milliamps
FFD = distance from the X-ray machine to the film

Use this information in Problems 36 and 37.

36. If FFD is held constant, the relationship between mA and t is given by

$$mA = \frac{400}{t}$$

Graph this relationship for $0 < t \le 10$.

37. If time is held constant, the relationship between mA and FFD is given by

$$mA = \frac{256}{(FFD)^2}$$

Graph this relationship for $0 < FFD \le 16$.

The pressure-volume relationship is

$$\frac{\text{original pressure}}{\text{new pressure}} = \frac{\text{new volume}}{\text{original volume}}$$

Use this proportion in Problems 38 and 39.

38. If the new pressure is 840 millimeters (mm) of mercury and the new volume is 100 milliliters (ml), write a rational function relating the original pressure and the original volume. Graph this relationship using the pressure as the independent variable.
39. If 400 ml of oxygen is under a pressure of 2,800 mm of mercury, write a rational function relating the new pressure and the new volume. Graph this relationship using the pressure as the independent variable.

40. Suppose that the supply and demand functions for a product are

$$S(p) = p - 20 \quad \text{and} \quad D(p) = \frac{800}{p}$$

The point(s) where $S(p) = D(p)$ are called the **equilibrium point(s)**. This topic will be discussed in Section 3.1.

a. Graph S and D on the same coordinate axes and estimate the equilibrium point(s).
b. Algebraically solve $S(p) = D(p)$.
c. Where does the supply function cross the p-axis? What is the economic significance of this point?

*2.7 Review

The material of this chapter is reviewed in the following list of objectives. After each objective there are some practice questions. For a sample test select the first question of each set and check your answers. The second question for each objective has no answer given. If you are having trouble with a particular type of problem, look back at the indicated section in the text. When you are finished reviewing these objectives, a sample examination is given at the end of this section.

[2.1]
Objective 2.1: *Evaluate a function. Let* $f(x) = 5x - 3$ *and* $g(x) = 2x^2 - 3x + 1$ *and find the requested values.*

1. $f(10)$
2. $f(-3)$
3. $g(-2)$
4. $g(4)$

[2.2]
Objective 2.2: *Evaluate a function and simplify. Let* $f(x) = 3x^2$ *and* $g(x) = x^2 - 5$.

5. $f(s + 3)$
6. $g(t - h)$
7. $f(x + h) - f(x)$
8. $\dfrac{g(x + h) - g(x)}{h}$

Objective 2.3: *Given a graph, find coordinates of indicated points, the domain, range, and intercepts.*

9.

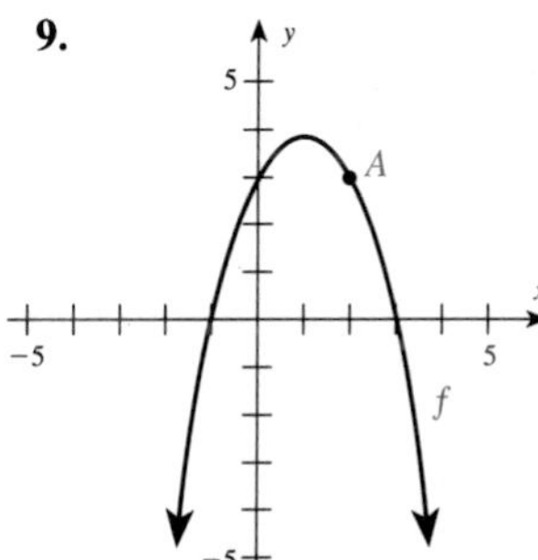

10.

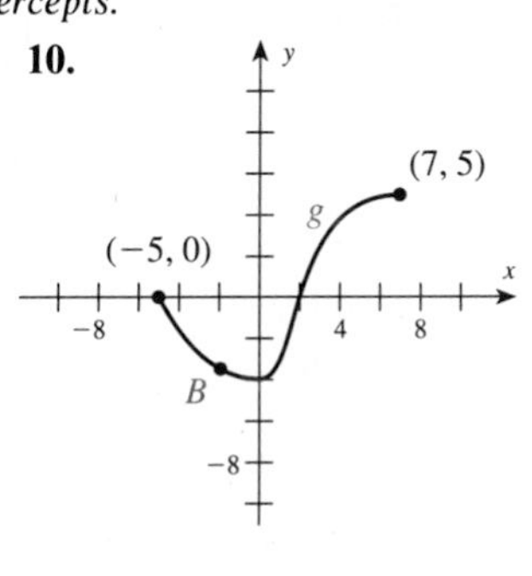

11.

12.

Objective 2.4: *Graph a given function by plotting points.*

13. $f(x) = 9 - 5x$
14. $g(x) = 25 - 5x - x^2$
15. $h(x) = \dfrac{1}{x - 3}$
16. $t(x) = \dfrac{x^2 - 3}{x}$

[2.3]
Objective 2.5: *Find the x- and y-intercepts for a line whose equation is given.*

17. $y = 5x - 4$
18. $3x + 2y - 6 = 0$
19. $3x - 4y + 12 = 0$
20. $50x - 250y = 1{,}000$

Objective 2.6: *Find the slope of the line passing through two given points.*

21. $(4, 1), (-3, -2)$
22. $(-3, -1), (2, 6)$
23. $(-3, 5), (-1, -5)$
24. $(-1, -2), (-6, -8)$

Objective 2.7: *Graph a line by finding the slope and y-intercept.*

25. $y = -\frac{2}{5}x + 2$
26. $y = \frac{2}{3}x - \frac{1}{3}$
27. $x + 2y - 8 = 0$
28. $3x - 5y - 10 = 0$

Objective 2.8: *Graph a line segment or piecewise linear function.*

29. $2x - 3y - 6 = 0; \quad -6 \le x \le 3$
30. $y = -5x + 3; \quad -4 \le x \le 3$
31. $\begin{cases} x - y + 5 = 0 & \text{if } -5 \le x \le 0 \\ x + y + 5 = 0 & \text{if } 0 < x < 5 \end{cases}$
32. $y = 2|x - 1|$

* Optional section.

Objective 2.9: *Find the standard form of the equation of a line satisfying given conditions.*

33. Passing through (2, 10) and (5, 25)

34. Slope $-\frac{2}{3}$; passing through (20, −100)

35. y-intercept −4; slope $\frac{1}{5}$

36. no y-intercept and no slope, passing through (1, 5)

[2.4]

Objective 2.10: *Sketch the graph of a parabola.*

37. $y = 2(x + 3)^2$ **38.** $y + 1 = \frac{1}{2}(x - 2)^2$

39. $y - 4 = -\frac{2}{3}(x + 2)^2$ **40.** $y = 3x^2 - 6x - 1$

Objective 2.11: *Find the maximum or minimum value of y for a quadratic function.*

41. $y + 250 = -\frac{1}{3}(x - 1{,}300)^2$

42. $y = -5(x - 300)^2 + 1{,}100$

43. $y = 2x^2 - 12x + 268$

44. $2y - 3(x + 40)^2 - 100 = 0$

[2.5]

Objective 2.12: *Name the asymptotes for a rational function.*

45. $f(x) = \dfrac{1}{x - 5}$ **46.** $f(x) = \dfrac{3x + 2}{2x - 5}$

47. $g(x) = \dfrac{4}{x^2 - x - 6}$ **48.** $g(x) = \dfrac{3x^2 + 2x + 1}{2x^2 - 7x - 4}$

Objective 2.13: *Graph rational functions.*

49. $f(x) = \dfrac{4}{x}$ **50.** $f(x) = \dfrac{2}{x + 2}$

51. $g(x) = \dfrac{4}{x} + 3$ **52.** $g(x) = \dfrac{2x + 2}{x^2 + 3x + 2}$

Objective 2.14: *Solve applied problems based on the preceding objectives.*

53. *Rate of change.* If $Z(x)$ = the price of Xerox stock on the first trading day in year x, write an expression for the average rate of change in price of Xerox stock from 1985 to 1990.

54. *Revenue.* After market research it was found that the demand equation for a certain product is $4{,}000 - 50x$ dollars, where x is the number of items produced. The cost function for this product is $C(x) = 25{,}000 + 1{,}000x$. Find the revenue function.

55. *Supply.* If a distributor can supply 2,000 items when the cost is \$200 and 8,000 if the cost is \$400, write the supply equation if you know that supply is a linear function. Assume x = cost is the independent variable.

56. *Equilibrium point.* The demand for the item described in Problem 55 is 7,000 items if the price is \$100 and only 1,000 items if the price is \$700. What is the equilibrium point for which the supply and demand are the same provided the demand is linear?

57. *Break-even point.* What is the break-even point for the information described in Problem 54?

58. *Maximum profit.* A manufacturer can produce no more than 200 items and it is determined that the profit (in dollars) is given by the following function:

$$P(x) = -10(x - 170)^2 + 14{,}320$$

What is the maximum profit, and how many items should be manufactured in order to achieve this maximum profit?

59. *Cost-benefit model.* Use the cost-benefit model

$$C(p) = \frac{20{,}000p}{100 - p}$$

to find the cost of removing 95% of the pollutants.

60. *Radiology technology.* Graph the relationship

$$mA = \frac{200}{t}$$

which relates the time that an X-ray machine is on and the current used in the X-ray. You need graph this function only for $0 < t \leq 2$.

SAMPLE TEST

The following sample test (45 minutes) is intended to review the main ideas of this chapter.

In Problems 1–8, let $f(x) = 2x^2 - 3x + 5$ and $g(x) = 1 - 4x$. Find the values requested in Problems 1–6.

1. $f(3)$ **2.** $g(10)$

3. $g(t + 1)$ **4.** $f(t - h)$

5. $\dfrac{g(x + h) - g(x)}{h}$ **6.** $\dfrac{f(x + h) - f(x)}{h}$

7. Graph g.

8. What are the x- and y-intercepts for f?

9. Find the slope of the line passing through (−5, 9) and (−3, −7).

10. Graph $2x + 3y - 12 = 0$ by finding the slope and y-intercept.

11. Graph $y = |x + 3|$.

12. Graph

$$\begin{cases} x + y + 1 = 0 & \text{for } -3 \leq x \leq 1 \\ y + 2 = 0 & \text{for } 1 < x < 5 \end{cases}$$

13. Find the standard form of the equation of a line passing through (30, 20) with slope −5.

14. Find the equation of a line with zero slope and y-intercept -3.

15. Find the equation of a line with no slope passing through $(-3, -2)$.

16. Sketch $y + 3 = \frac{1}{2}(x - 2)^2$.

17. What is the maximum value of the function $y - 550 = -\frac{2}{3}(x - 40)^2$?

18. What are the asymptotes for the function

$$F(x) = \frac{2x + 1}{x^2 + x - 6}$$

19. Use the cost-benefit model

$$C(p) = \frac{500p}{100 - p}$$

to find the cost of removing 90% of the pollutants.

20. Graph C from Problem 19 for $0 \le p < 100$.

Finite Mathematics

Finite mathematics is a recent course in mathematics. It was first proposed in the early 1960s in the now classic book by Kemeny, Snell, and Thompson, *Introduction to Finite Mathematics* (Englewood Cliffs, N.J.: Prentice-Hall, 1956). The topics in that book originally defined the course. Over the years the course has evolved to include the topics that are now included in this part of the book (see list of chapters below), but the topic itself, finite mathematics, is not well defined in the sense that algebra or calculus is well defined. Very loosely defined, finite mathematics is mathematics that does not use calculus. More specifically, finite mathematics means *applying* the elementary mathematics of sets, matrices, linear programming, probability, and statistics to real-life problems.

The prerequisites for this part of the book are at least 2 years of high school algebra (reviewed in Chapter 1) and the concepts of functions, graphs, and linear functions (Sections 2.1–2.4).

3

Systems of Equations and Matrices

CHAPTER OVERVIEW

Here we solve *systems* of equations and explore an extremely powerful concept in mathematical modeling—the *matrix*.

PREVIEW

We first learn what matrices are, then how to manipulate them, and finally, how to use them to solve systems of equations. In 1973 Wassily Leontief (1906–) received the Nobel Prize in Economics for describing the economy in terms of an input–output model. In Section 3.5, we look at two input–output models: a *closed model*, in which we try to find the relative income of each participant within a system, and an *open model*, in which we try to find the amount of production needed to achieve an anticipated demand.

PERSPECTIVE

The most important idea of this chapter—one that must be mastered in order to succeed with the material that follows in the book—is that of *Gauss–Jordan elimination*. It is used over and over again in this course, so work hard to understand the procedure.

MODELING
APPLICATION 1

Gaining a Competitive Edge in Business

Solartex manufactures solar collector panels. During the first year of operation, rent, insurance, utilities, and other fixed costs averaged \$8,500 per month. Each panel sold for \$320 and cost the company \$95 in materials and \$55 in labor. Since company resources are limited, Solartex cannot spend more than \$20,000 in any one month. Sunenergy, another company selling panels, competes directly with Solartex. Last month, Sunenergy manufactured 85 panels at a total cost of \$20,475, but the previous month produced only 60 panels at a total cost of \$17,100.

After you have finished this chapter, write a paper based on this modeling application. This Modeling Application is continued on page 151.

APPLICATIONS

Management *(Business, Economics, Finance, and Investments)*
Supply and demand (3.1, Problems 50–55)
Break-even analysis (3.1, Problems 56–65)
Value of a stock portfolio (3.1, Problems 66–67)
Inventory control (3.2, Problem 5)
Delivery routes (3.2, Problem 53)
Airline routes (3.2, Problem 54)
Shipping of sport wheels (3.2, Problem 55)
Manufacturing construction costs (3.2, Problem 56)
Mixing an alloy in a production process (3.3, Problem 50)
Manufacturing a candy (3.3, Problem 51)
Leontief models (3.5, Problems 7–17; 3.6, Problems 89–92)
Study of the Israeli economy (3.5, Problems 18–20)
Batching process (3.6, Problem 65)
Product allocation (3.6, Problem 66)

Life Sciences *(Biology, Ecology, Health, and Medicine)*
Mixture of a chemical fertilizer (3.1, Problem 69)
Retard soil erosion (3.2, Problem 52)
Preparing a proper cattle-feed mix (3.3, Problem 48; 3.4, Problems 57–59)
Preparing a commercial pesticide (3.3, Problem 49)
Preparing a diet of three basic foods (3.4, Problems 60–62)

Social Sciences *(Demography, Political science, Population, Psychology, Society, and Sociology)*
Voter demographics (3.1, Problem 68; 3.3, Problem 52)
Louvre Tablet from the Babylonian civilization (3.1, Problem 70)
Diplomatic communication channels (3.2, Problem 57)
Population demographics (3.6, Problems 67–68)

General Interest
Best price for a car rental (3.1, Problems 47–49)
Matrix of car rental costs (3.2, Problem 6)
Development of matrix theory (3.2, Problems 58–63)
Ice cream cone with the best value (3.3, Problem 53)

Modeling Application—
Gaining a Competitive Edge in Business

3.1
Systems of Equations

Many situations involve two variables or unknowns related in some specific fashion, as in Example 1.

EXAMPLE 1 **Cost analysis.** Linda is on a business trip and needs to rent a car for a day. The prices at two agencies are different, and she wants to rent the less expensive car.

Agency A: \$10 per day plus 50¢ per mile
Agency B: \$40 per day plus 25¢ per mile

Which agency should she rent from?

Solution The answer to this question depends on the number of miles Linda intends to drive. Let c be the cost (in dollars) of the rental and let m be the number of miles driven. These variables are related as follows:

COST = BASIC CHARGE + MILEAGE CHARGE

where the mileage charge is the cost per mile times the number of miles driven. Thus

Agency A: $c = 10 + .5m$
Agency B: $c = 40 + .25m$

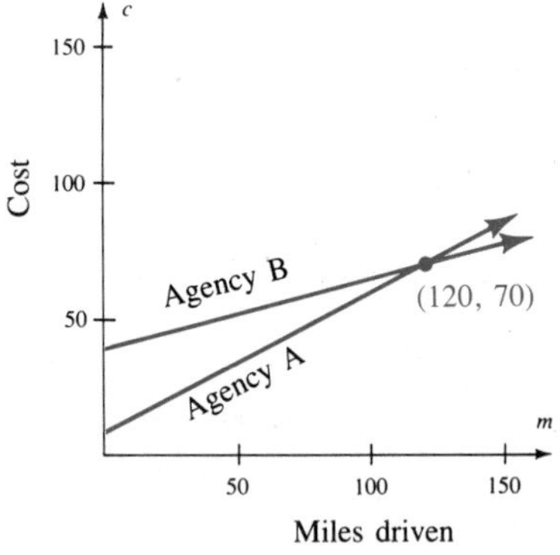

We use a Cartesian graph as a model for this problem and graph the equations for agencies A and B. The point of intersection looks like the point (120, 70). What does this mean? It means that if Linda plans on driving less than 120 miles, she should rent from agency A, and if she plans on driving more than 120 miles, she should rent from agency B. ■

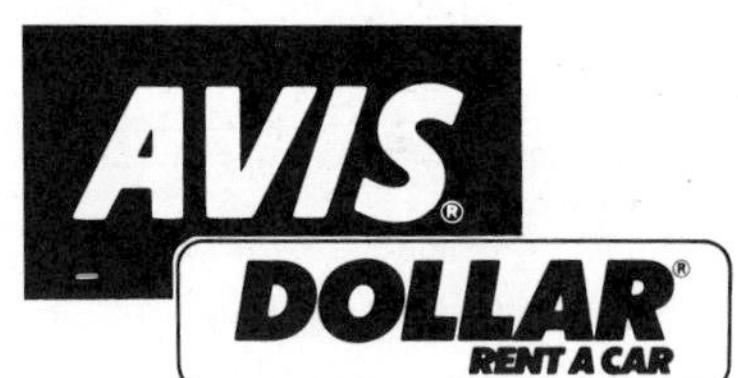

The model in Example 1 is called a **system of equations**. Solving systems of equations is a procedure that is used throughout mathematics. In this text we will solve a variety of systems of equations and inequalities, so we now introduce a notation that will be easy to generalize in later applications. We begin our study of this topic by considering an arbitrary system of two equations with two variables:

$$\begin{cases} a_{11}x_1 + a_{12}x_2 = b_1 \\ a_{21}x_1 + a_{22}x_2 = b_2 \end{cases}$$

The notation may seem strange, but it will prove useful. The variables are x_1 and x_2, and the constants are a_{11}, a_{12}, a_{21}, a_{22}, b_1, and b_2. The subscripts in a_{11}, a_{12}, a_{21}, and a_{22} are called *double subscripts* and indicate *position.* That is, a_{12} should not be read "a sub twelve," but rather as "a sub one-two." This means that it represents the second constant in the first equation:

$$a_{12}$$

Equation number ↗ ↖ Coefficient number in equation

This effort in using notation may seem rather complicated when dealing with only two equations and two variables, but we want to develop a notation that can be used with many variables and many equations.

By a *system of two equations with two variables*, we mean any two equations in those variables. The **simultaneous solution** of a system is the intersection of the solution sets of the individual equations. The brace in front of the equations indicates that this intersection is the desired solution. If all the equations in a system are linear, it is called a **linear system**. We limit our study in this section to linear systems of equations with two unknowns.

Since the graph of each equation in a system of linear equations in two variables is a line, the solution set for the system is the intersection of two lines. In two dimensions, two lines must be related to each other in one of three possible ways:

1. The lines intersect at a single point. They are called **consistent**. In this case there is exactly one solution.
2. The graphs are parallel lines. In this case, the solution set is empty, and the equations are called **inconsistent**.
3. The graphs are the same line. In this case, there are infinitely many points in the solution set, and any solution of one equation is also a solution of the other. The equations of such a system are said to be **dependent**. If the equations of a system are not dependent, then they are called **independent**.

EXAMPLE 2 Relate the system $\begin{cases} 2x - 3y = -8 \\ x + y = 6 \end{cases}$ to the general system $\begin{cases} a_{11}x_1 + a_{12}x_2 = b_1 \\ a_{21}x_1 + a_{22}x_2 = b_2 \end{cases}$ and solve by graphing.

Solution The variables are $x_1 = x$ and $x_2 = y$. The constants are

$$a_{11} = 2 \qquad a_{12} = -3 \qquad b_1 = -8$$
$$a_{21} = 1 \qquad a_{22} = 1 \qquad b_2 = 6$$

The solution is $x = 2$, $y = 4$, or $(2, 4)$, as shown in Figure 3.1.

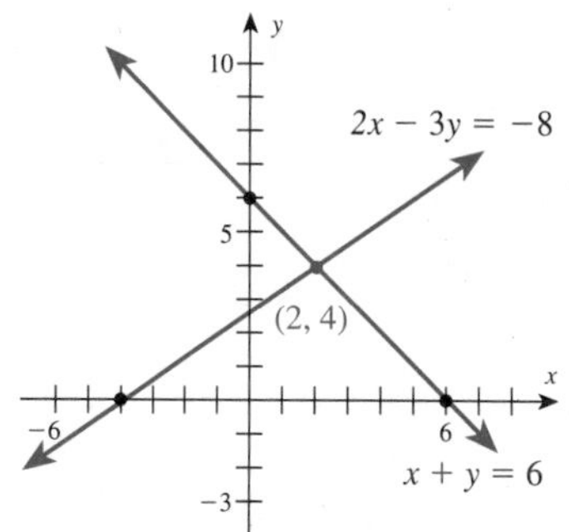

FIGURE 3.1
Graph of the system
$\begin{cases} 2x - 3y = -8 \\ x + y = 6 \end{cases}$

■

EXAMPLE 3 Relate the system $\begin{cases} 2x - 3y = -8 \\ 4x - 6y = -2 \end{cases}$ to the general system and solve by graphing.

Solution The variables are $x_1 = x$, $x_2 = y$, and the constants are

$$a_{11} = 2 \qquad a_{12} = -3 \qquad b_1 = -8$$
$$a_{21} = 4 \qquad a_{22} = -6 \qquad b_2 = -2$$

Since the graphs of the lines are distinct and parallel, as shown in Figure 3.2, there is no point of intersection. These equations are inconsistent.

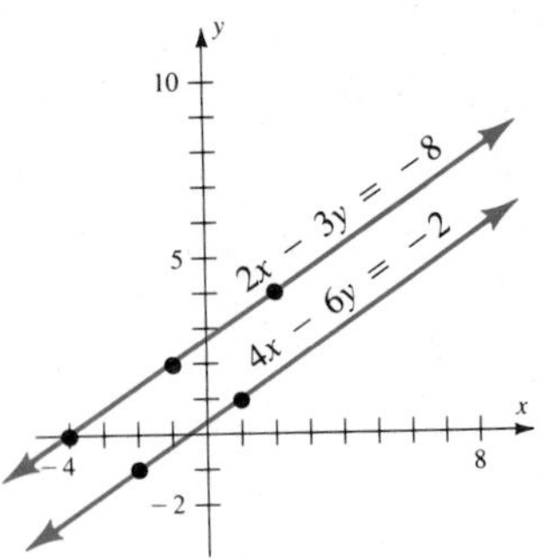

FIGURE 3.2
Graph of the system
$\begin{cases} 2x - 3y = -8 \\ 4x - 6y = -2 \end{cases}$

■

EXAMPLE 4 Relate the system $\begin{cases} 2x - 3y = -8 \\ \quad\;\; y = \frac{2}{3}x + \frac{8}{3} \end{cases}$ to the general system and solve by graphing.

Solution To relate to the general system, both equations must be arranged in the usual form; the second equation needs to be rewritten:

$$y = \frac{2}{3}x + \frac{8}{3}$$
$$3y = 2x + 8 \qquad \text{Multiply by 3}$$
$$-2x + 3y = 8$$

Thus, in standard form, the system is

$$\begin{cases} 2x - 3y = -8 \\ -2x + 3y = 8 \end{cases}$$

This means that the variables are $x_1 = x$, $x_2 = y$, and the constants are

$$a_{11} = 2 \qquad a_{12} = -3 \qquad b_1 = -8$$
$$a_{21} = -2 \qquad a_{22} = 3 \qquad b_2 = 8$$

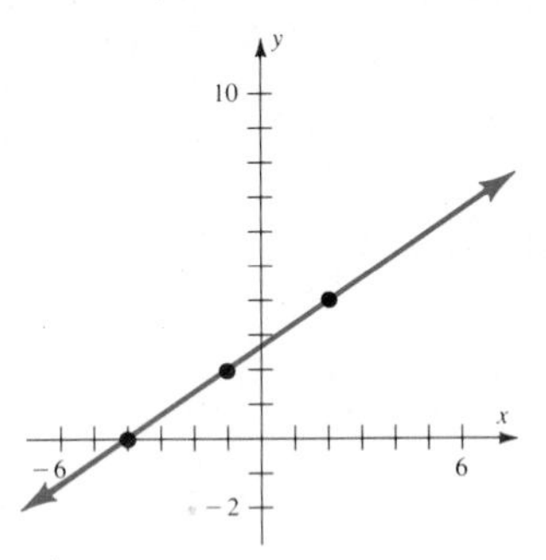

FIGURE 3.3
Graph of the system
$\begin{cases} 2x - 3y = -8 \\ \quad\;\; y = \frac{2}{3}x + \frac{8}{3} \end{cases}$

The equations represent the same line, as shown in Figure 3.3, and are therefore dependent. ■

The **graphing method** can give solutions only as accurate as the graphs we can draw, and, consequently, it is inadequate for most applications. We therefore need more efficient methods.

In general, given a system, the procedure is to write a simpler equivalent system. Two systems are said to be **equivalent** if they have the same solution set. In this chapter we limit ourselves to finding only real roots. There are several ways to go about writing equivalent systems. The first nongraphical method we consider comes from the substitution property of real numbers and leads to a **substitution method** for solving systems.

Substitution Method for Solving Systems of Equations

1. *Solve* one of the equations for one of the variables.
2. *Substitute* the expression obtained into the other equation.
3. *Solve* the resulting equation in a single variable for the value of that variable.
4. *Substitute* that value into either of the original equations to determine the value of the other variable.
5. *State* the solution.

EXAMPLE 5 Solve $\begin{cases} 2p + 3q = 5 \\ \quad\;\; q = -2p + 7 \end{cases}$ by substitution.

Solution Since $q = -2p + 7$, substitute $-2p + 7$ for q in the other equation:

$$\begin{aligned} 2p + 3(-2p + 7) &= 5 \\ 2p - 6p + 21 &= 5 \\ -4p &= -16 \\ p &= 4 \end{aligned}$$

Substitute 4 for p in either of the given equations:

$$\begin{aligned} q &= -2p + 7 \\ &= -2(4) + 7 \\ q &= -1 \end{aligned}$$

The solution is $(p, q) = (4, -1)$. ■

A third method for solving systems is called the **linear combination method**. It involves substitution and the idea that if equal quantities are added to equal quantities, the resulting equation is equivalent to the original system. In general, such addition will not simplify matters unless the numerical coefficients of one or more terms are opposites. However, we can often force them to be opposites by multiplying one or both of the given equations by appropriate nonzero constants.

Linear Combination Method for Solving Systems of Equations

1. *Multiply* one or both of the equations by a (nonzero) constant or constants, so that the coefficients of one of the variables become opposites.
2. *Add* corresponding members of the equations to obtain a new equation in a single variable.
3. *Solve* the derived equation for that variable.
4. *Substitute* the value of the found variable into either of the original equations and solve for the second variable.
5. *State* the solution.

EXAMPLE 6 Solve: $\begin{cases} 3x + 5y = -2 \\ 2x + 3y = 0 \end{cases}$.

Solution Multiply both sides of the first equation by 2 and both sides of the second equation by -3. This procedure forces the coefficients of x to be opposites:

$$\begin{matrix} 2 \\ -3 \end{matrix}\begin{cases} 3x + 5y = -2 \\ 2x + 3y = 0 \end{cases}$$

This means you should add the equations of the system

$$+\begin{cases} 6x + 10y = -4 \\ -6x - 9y = 0 \end{cases}$$

$$[6x + (-6x)] + [10y + (-9y)] = -4 + 0$$

This step should be done mentally

$$y = -4$$

If $y = -4$, then

$$2x + 3y = 0$$
$$2x + 3(-4) = 0$$
$$x = 6$$

The solution is $(6, -4)$. ■

Later we will need to find the corner points for certain regions in a plane and the methods of this section, as illustrated in Example 7, will be helpful.

EXAMPLE 7 Find the corner points labeled A, B, C, and D in the illustrated region.

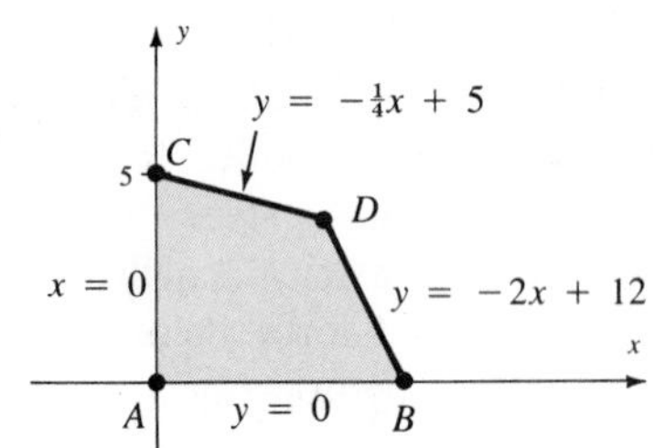

Solution Point A is obvious; it is the origin $(0, 0)$. Point B and C are relatively easy to find because they are the intercepts of the lines.

For point B, the x-intercept is found when $y = 0$ in the equation $y = -2x + 12$. Thus

$$0 = -2x + 12$$
$$-12 = -2x$$
$$6 = x$$

Point B is $(6, 0)$.

For point C, the y-intercept is found by inspection; it is the point $(0, 5)$.

Point D is the intersection of the lines whose equations are given, namely:

$$\begin{cases} y = -\frac{1}{4}x + 5 \\ y = -2x + 12 \end{cases}$$

Solve these by substitution:

$$-2x + 12 = -\frac{1}{4}x + 5$$
$$-8x + 48 = -x + 20$$
$$-7x = -28$$
$$x = 4$$

Finally, $y = -2(4) + 12 = 4$, so D is the point $(4, 4)$. ■

There are two very important business applications of systems of equations: *supply and demand* and *break-even analysis.*

Supply and Demand

In economics, the price, p, and quantity, x (usually called q in economics), of an item are closely related. If these variables are related by ordered pairs (x, p), two relationships between x and p can be considered. The first, called the **supply curve**, expresses the relationship between x and p from the manufacturer's point of view. A *linear* supply curve is shown in Figure 3.4. For every p, the supply curve gives the quantity x that the manufacturer is willing to produce in some time period at the price p. The higher the price, the more the manufacturer is willing to supply. Therefore, the supply curve rises when viewed from left to right. The other curve, the **demand curve**, expresses the relationship between x and p from the consumer's point of view. A *linear* demand curve is shown in Figure 3.4. For every value of p, the demand curve gives the quantity x that the consumer is willing to buy at price p. The higher the price, the less consumers will buy, so demand curves fall when viewed from left to right. It might also be pointed out that, from an economic viewpoint, either the variable x or the variable p can be viewed as the independent variable, but to be consistent with the economic conventions we will plot quantity as the axis of abscissas (horizontal) and price as axis of ordinates (vertical).

If supply and demand curves are drawn on the same coordinate system, the intersection point (if it exists) is called the **equilibrium point** and represents values for which supply equals demand.

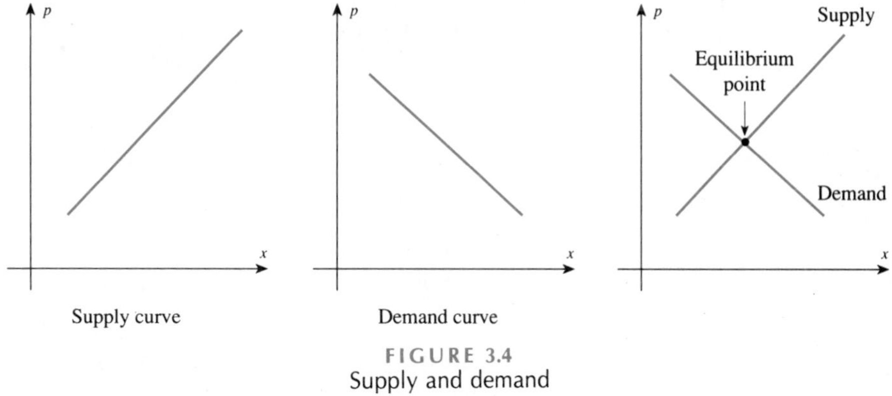

FIGURE 3.4
Supply and demand

In this section, the supply and demand curves are linear (see Figure 3.4), but in most real-life situations they are more general curves.

EXAMPLE 8 A product has a supply curve given by the equation $x = 500p - 8{,}100$ and a demand curve given by the equation $x = -100p + 15{,}000$. Find the equilibrium point.

Solution Find the intersection point by using substitution:

$$\begin{aligned} -100p + 15{,}000 &= 500p - 8{,}100 \\ 600p &= 23{,}100 \\ p &= \frac{23{,}100}{600} = 38.5 \end{aligned}$$

Now, if $p = 38.5$, then

$$x = -100(38.5) + 15{,}000$$
$$= 11{,}150$$

Thus equilibrium is reached when the price is \$38.50 and the quantity is 11,150 items. ■

Break-Even Analysis

In Section 2.5 we introduced four functions which will be important in our study of finite mathematics and calculus in business applications. We will review those ideas here in case you skipped over that section. For this application we will let x be the number of items.

DEMAND FUNCTION This function represents the price per unit that is likely to be paid for an item. It is often used as the price of an item when carrying out a mathematical analysis. We will denote the demand function by

$$p = d(x)$$

REVENUE FUNCTION This is the amount of money received from sales of an item. This number is exclusive of costs and expenses and is denoted by $R(x)$. To find the revenue, use the following formula:

$$\text{REVENUE} = (\text{NUMBER OF ITEMS})(\text{PRICE PER ITEM})$$
$$R(x) = xd(x) \quad \text{or} \quad R(x) = xp$$

COST FUNCTION This is the total cost of producing x items, and is denoted by $C(x)$. It is found according to the formula

$$C(x) = \text{VARIABLE COSTS} + \text{FIXED COSTS}$$

PROFIT FUNCTION This is denoted by $P(x)$ and is defined by

$$P(x) = R(x) - C(x)$$

If the profit function is positive its value is called **profit** and if the profit function is negative its value is called **loss**.

In order to make a profit, revenue must exceed cost. The point at which revenue equals the cost is called the **break-even point**. In Section 2.5 we looked at break-even in general; here we find the break-even point when both the cost and revenue functions are linear.

EXAMPLE 9 A company producing bicycle reflectors has fixed costs of \$2,100 and variable costs of 80¢ per reflector. The selling price of each reflector is \$1.50. What is the break-even point?

Solution Let x be the number of items produced, $C(x)$ be the cost function, and $R(x)$ be the revenue function. Then

$$C(x) = \text{VARIABLE COSTS} + \text{FIXED COSTS}$$
$$= .8x + 2{,}100$$

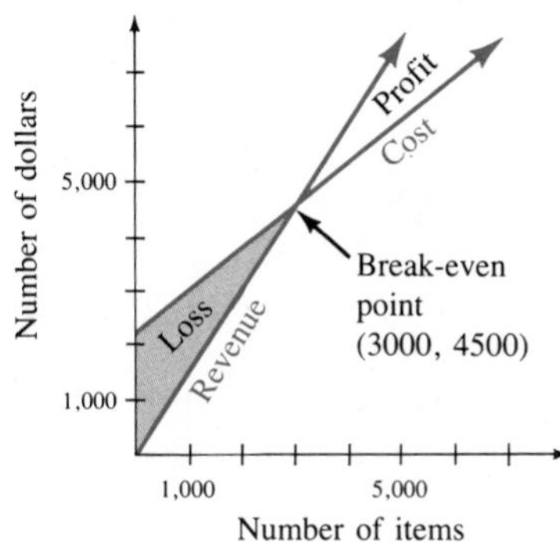

and

$$R(x) = 1.5x$$

Although the graph in the figure illustrates what is meant by break-even analysis and profit and loss, it is often not accurate enough to find the break-even point by inspection. Therefore we will also solve this system of equations by substitution:

$$1.5x = .8x + 2{,}100 \qquad \text{Since } R(x) = C(x) \text{ at the break-even point}$$
$$.7x = 2{,}100$$
$$x = 3{,}000 \qquad \text{If } x = 3{,}000, \text{ then } R(3{,}000) = 1.5(3{,}000) = 4{,}500$$

If 3,000 reflectors are produced, the cost and revenue will both be $4,500. ■

3.1 Problem Set

Solve the systems in Problems 1–6 by graphing.

1. $\begin{cases} y = 3x - 7 \\ y = -2x + 8 \end{cases}$ **2.** $\begin{cases} x + y = 1 \\ 3x + y = -5 \end{cases}$

3. $\begin{cases} y = \frac{2}{3}x - 7 \\ 2x + 3y = 3 \end{cases}$ **4.** $\begin{cases} y = \frac{3}{5}x + 2 \\ 3x - 5y = 10 \end{cases}$

5. $\begin{cases} 2x - 3y = 9 \\ y = \frac{2}{3}x - 3 \end{cases}$ **6.** $\begin{cases} 2x - 3y = 0 \\ y = \frac{2}{3}x - 2 \end{cases}$

Solve the systems in Problems 7–12 by substitution. Relate each system to the general system $\begin{cases} a_{11}x_1 + a_{12}x_2 = b_1 \\ a_{21}x_1 + a_{22}x_2 = b_2 \end{cases}$.

7. $\begin{cases} a = 3b - 7 \\ a = -2b + 8 \end{cases}$ **8.** $\begin{cases} s + t = 1 \\ 3s + t = -5 \end{cases}$

9. $\begin{cases} m = \frac{2}{3}n - 7 \\ 2n + 3m = 3 \end{cases}$ **10.** $\begin{cases} v = \frac{3}{5}u + 2 \\ 3u - 5v = 10 \end{cases}$

11. $\begin{cases} 2p - 3q = 9 \\ q = \frac{2}{3}p - 3 \end{cases}$ **12.** $\begin{cases} 3t_1 + 5t_2 = 1{,}541 \\ t_2 = 2t_1 + 160 \end{cases}$

Solve the systems in Problems 13–18 by linear combinations. Relate each system to the general system $\begin{cases} a_{11}x_1 + a_{12}x_2 = b_1 \\ a_{21}x_1 + a_{22}x_2 = b_2 \end{cases}$.

13. $\begin{cases} c + d = 2 \\ 2c - d = 1 \end{cases}$ **14.** $\begin{cases} 2s_1 + s_2 = 10 \\ 5s_1 - 2s_2 = 16 \end{cases}$

15. $\begin{cases} 3q_1 - 4q_2 = 3 \\ 5q_1 + 3q_2 = 5 \end{cases}$ **16.** $\begin{cases} 9x + 3y = 5 \\ 3x + 2y = 2 \end{cases}$

17. $\begin{cases} 7x + y = 5 \\ 14x - 2y = -2 \end{cases}$ **18.** $\begin{cases} 2x + 3y = 1 \\ 3x - 2y = 0 \end{cases}$

Solve the systems in Problems 19–30 by any method.

19. $\begin{cases} 5x + 4y = 5 \\ 15x - 2y = 8 \end{cases}$ **20.** $\begin{cases} 3x + 2y = 1 \\ 6x + 4y = 2 \end{cases}$

21. $\begin{cases} 4x - 2y = -28 \\ y = \frac{1}{2}x + 5 \end{cases}$ **22.** $\begin{cases} 12x - 5y = -39 \\ y = 2x + 9 \end{cases}$

23. $\begin{cases} y = 2x - 1 \\ y = -3x - 9 \end{cases}$ **24.** $\begin{cases} y = \frac{2}{3}x - 5 \\ y = -\frac{4}{3}x + 7 \end{cases}$

25. $\begin{cases} 2x - 3y - 5 = 0 \\ 3x - 5y + 2 = 0 \end{cases}$ **26.** $\begin{cases} 2x + 3y = a \\ x - 5y = b \end{cases}$

27. $\begin{cases} ax + by = 1 \\ bx + ay = 0 \end{cases}$ **28.** $\begin{cases} .12x + .06y = 210 \\ x + y = 2{,}000 \end{cases}$

29. $\begin{cases} .12x + .06y = 228 \\ x + y = 2{,}000 \end{cases}$ **30.** $\begin{cases} .12x + .06y = 1{,}140 \\ x + y = 10{,}000 \end{cases}$

Find the coordinates of the corner points for each of the regions in Problems 31–46.

31.

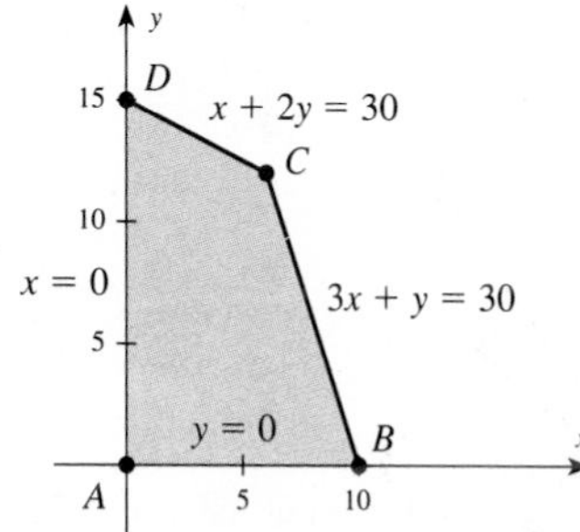

32.

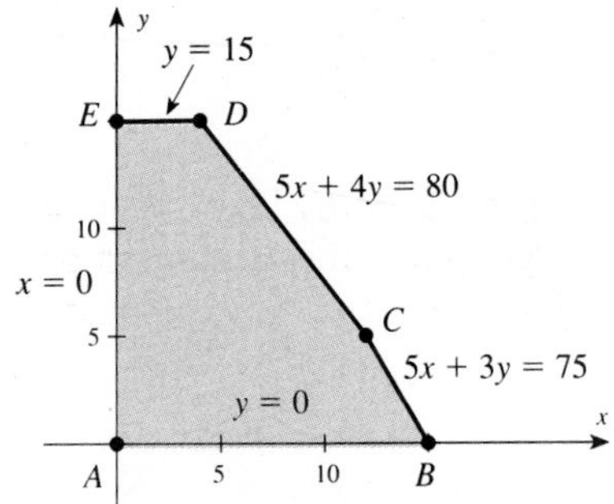

33.

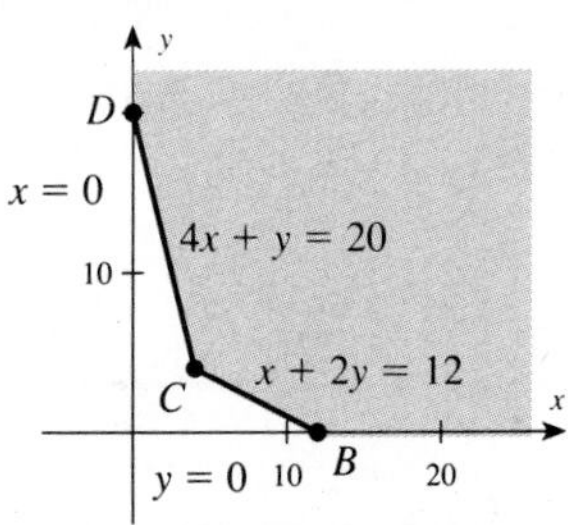

34.

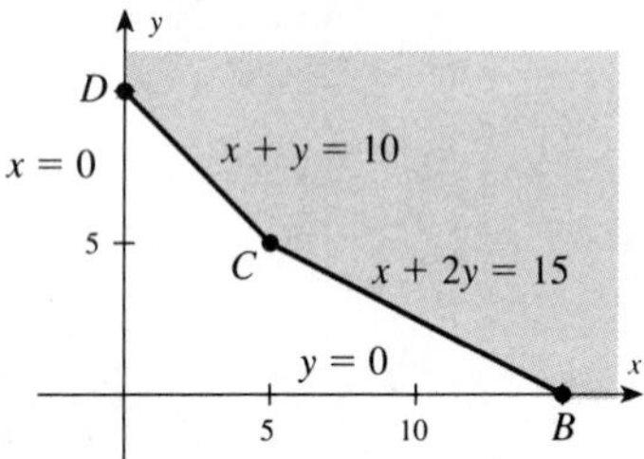

35.

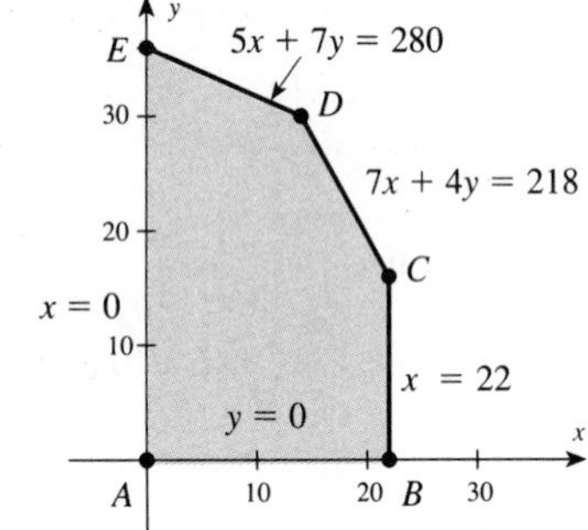

36.

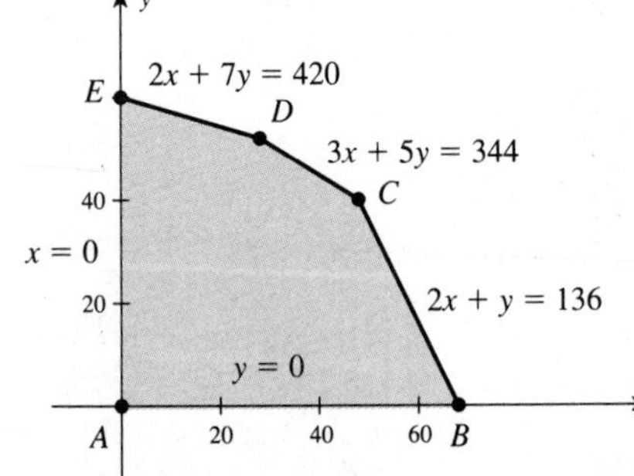

37.

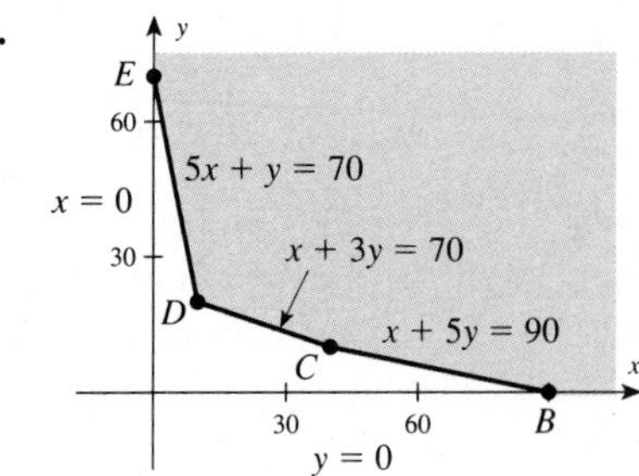

38.

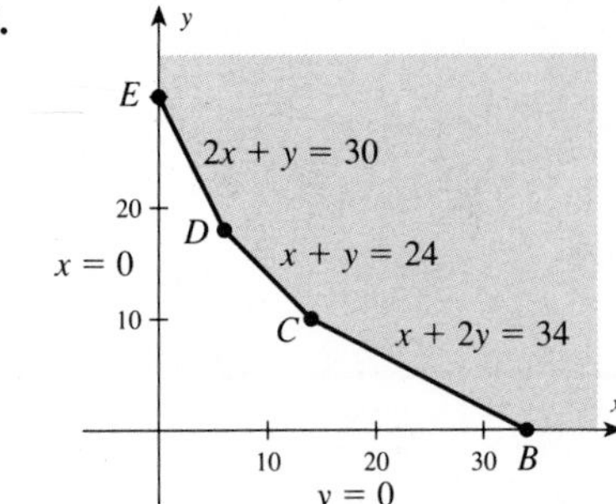

39.

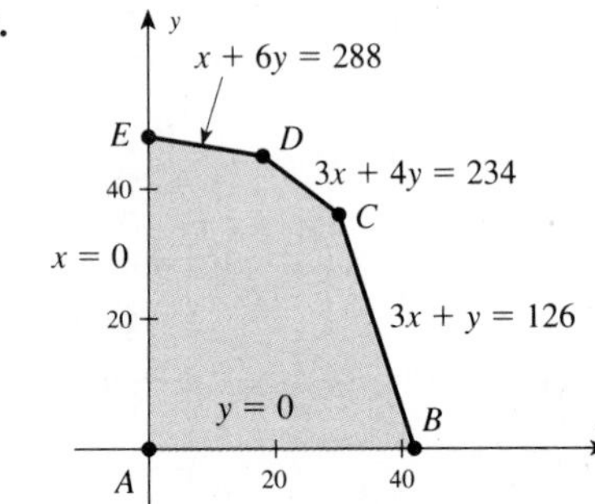

40.

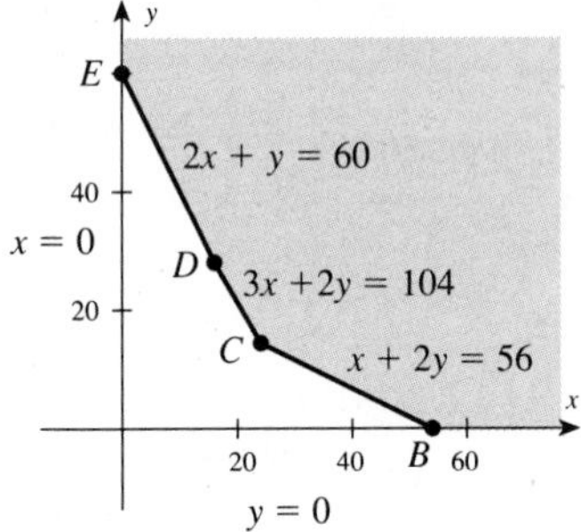

41.

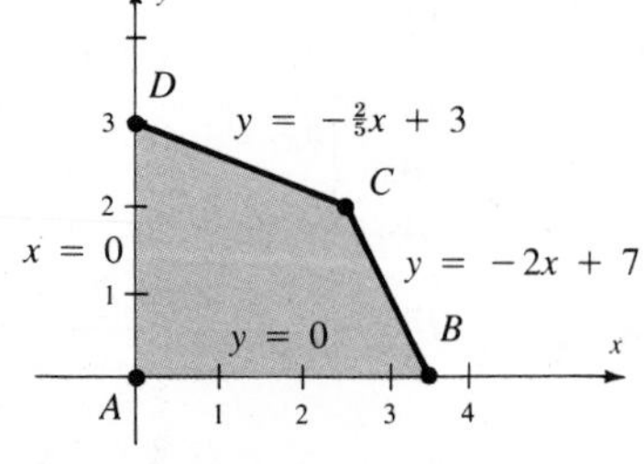

42.

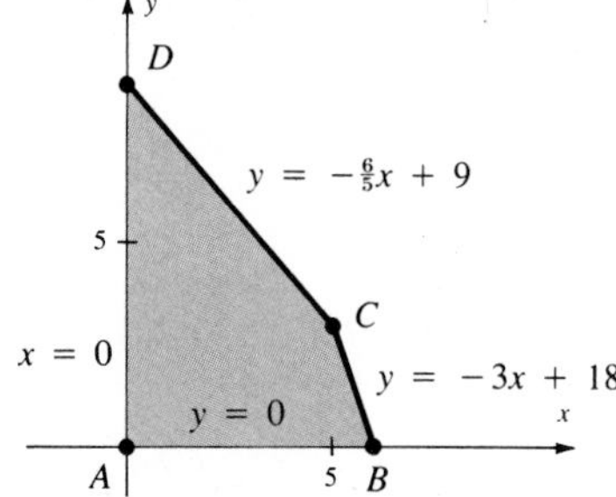

43.

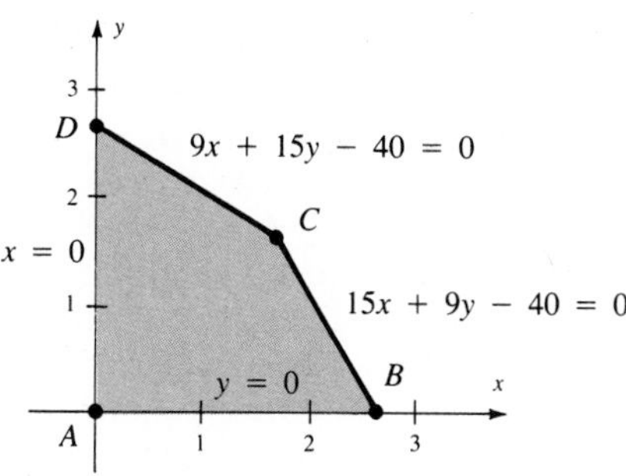

44.

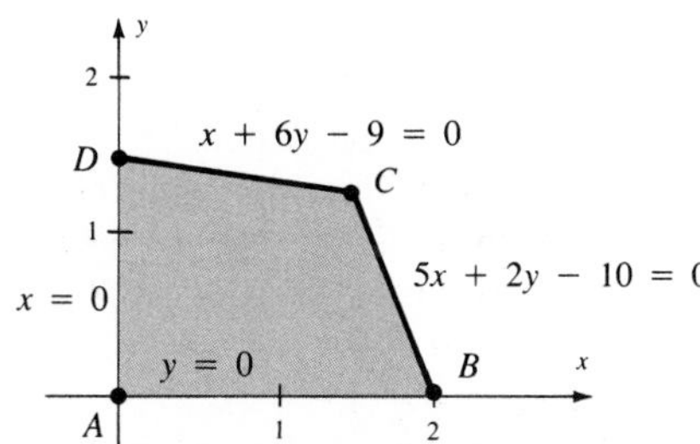

45.

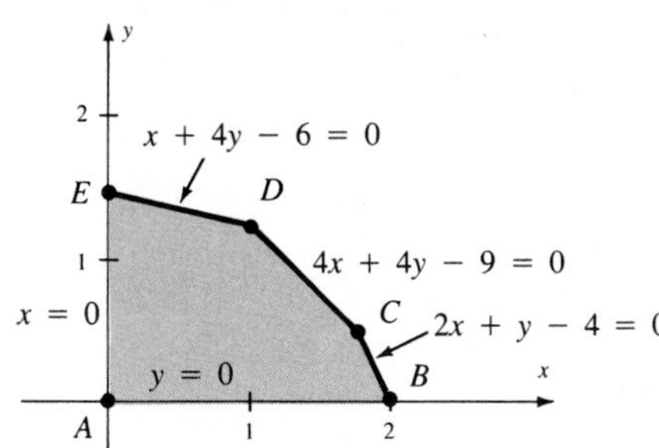

46.

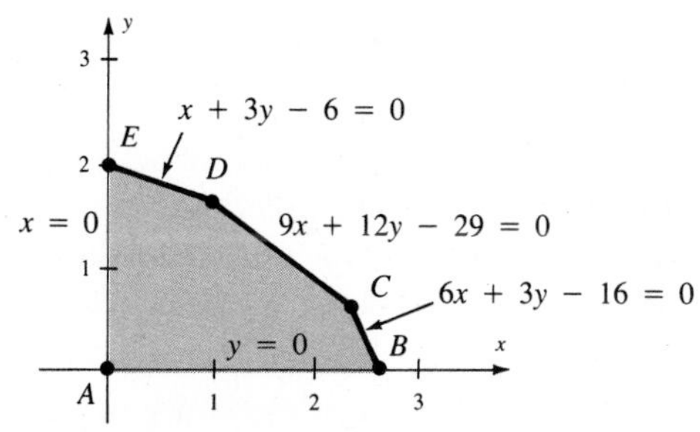

APPLICATIONS

47. Suppose a car rental agency gives the following choices:

Option A:	\$30 per day plus 40¢ per mile
Option B:	Flat \$50 per day with unlimited mileage

At what mileage are both rates the same if you rent the car for three days?

48. Suppose a car rental agency gives the following options:

Option A:	\$40 per day plus 50¢ per mile
Option B:	Flat \$60 per day with unlimited mileage

At what mileage are both rates the same if you rent the car for a week?

49. Two car rental agencies have the following rates:

Agency A:	\$15 per day plus 20¢ per mile
Agency B:	\$25 per day plus 15¢ per mile

At what mileage are both rates the same if you rent the car for four days?

50. The supply for a certain commodity is linear and determined to be $p = .005x + 12$ while the demand is also linear with $p = 150 - .01x$. What is the equilibrium point?

51. The supply for a certain commodity is linear and determined to be $p = .004x + 10$ and a linear demand given by $p = 28 - .005x$. What is the equilibrium point?

52. A certain item has a linear supply curve $p = .0005x - 3$ and a linear demand $p = 8 - .0006x$. What is the equilibrium point?

53. The supply curve for a particular commodity is $x = 250p - 2{,}500$, and the demand curve for the same commodity is $x = 3{,}500 - 150p$. What should the price of the commodity be for the market to be stable?

54. The supply curve for a new product is $x = 2{,}000p + 6{,}000$, and the demand curve is $x = 13{,}000 - 1{,}500p$. What is the equilibrium point for this product?

55. The supply curve for a new software product is $x = 2.5p - 500$, and the demand curve for the same product is $x = 200 - .5p$.
   a. At \$250 for the product, how many items would be supplied? How many would be demanded?
   b. At what price would no items be supplied?
   c. At what price would no items be demanded?
   d. What is the equilibrium price for this product?
   e. How many units will be produced at the equilibrium price?

56. The supply curve for a certain commodity is $x = 2{,}500p - 500$, and the demand curve for the same product is $x = 31{,}500 - 1{,}500p$.
   a. At \$15 per commodity, how many items would be supplied? How many would be demanded?
   b. At what price would no items be supplied?

c. At what price would no items be demanded?
d. What is the equilibrium price for this product?
e. How many units will be produced at the equilibrium price?

57. Suppose you want to sell T-shirts on your campus. You are trying to decide on a price between \$1 and \$7, so you conduct a market-research survey and find that 800 students will buy the shirt for \$1, but none will buy it for \$7. You also find a supplier who can supply 200 shirts if they are sold for \$1 and 600 if they are sold for \$7. At what price will supply equal demand, and how many could you expect to sell if both supply and demand are linear?

58. A new greeting card is introduced in a test market. Within the month, 3,000 cards are sold priced at 50¢ and 1,500 are sold priced at \$2. On the other hand, 2,500 cards could be supplied at a cost of \$2 but only 1,500 could be supplied at a cost of \$1. At what price does the supply equal the demand, assuming that both are linear?

59. A firm producing sunglasses has fixed costs of \$1,200 and variable costs of 80¢ per pair. Find the break-even point if the sunglasses sell for \$2 per pair.

60. Repeat Problem 59 if the sunglasses sell for \$3 per pair.

61. Microdrop Corporation produces floppy computer disks at a cost of \$1.20 per disk with total fixed costs of \$18,000. If the disks sell for \$3 each, what is the break-even point?

62. Repeat Problem 61 if the disks sell for \$2.50 each.

63. Hallmark introduces a new greeting card that has fixed costs of \$2,800 and variable costs of 75¢ per card. If the cards sell for \$1.50 each, what is the break-even point?

64. Repeat Problem 63 if the cards sell for \$1.25 each.

65. Repeat Problem 63 if the cards sell for \$2 each.

66. An investor owns Xerox and Standard Oil stock. The closing prices of each stock for the week are given in the table. If the value of the portfolio was \$19,500 on Monday and \$20,300 on Friday, how many shares of each stock does the investor own?

	Standard Oil	Xerox
Monday	45	60
Tuesday	$45\frac{1}{2}$	$62\frac{1}{8}$
Wednesday	46	63
Thursday	$47\frac{1}{4}$	$61\frac{1}{2}$
Friday	47	62

67. Another investor owns the same stocks as the investor described in Problem 66. This investor's portfolio was valued at \$34,500 on Monday and at \$35,600 on Wednesday. How many shares of each stock are owned by this investor?

68. The total number of registered Democrats and Republicans in a certain community is 100,005. Voter turnout in a recent election showed that 50% of the Democrats voted, 60% of the Republicans voted, and a total of 55,200 votes were cast, which also included Independents and write-in votes. If the registered Democrats outnumber the registered Republicans by 2 to 1, how many registered Democrats and Republicans are there in the community and how many independents or write-in votes were there in the election?

69. Two chemicals are combined to form two grades of fertilizer. One unit of grade A fertilizer requires 10 pounds of potassium and 3 pounds of calcium, while a unit of grade B fertilizer requires 9 pounds of potassium and 4 pounds of calcium. If 780 pounds of potassium and 260 pounds of calcium are available, how many units of each grade of fertilizer can be mixed?

70. ***Historical Question.*** The Louvre Tablet from the Babylonian civilization is dated at about 1500 B.C. It shows a system equivalent to

$$\begin{cases} xy = 1 \\ x + y = a \end{cases}$$

Even though this is not a linear system, you can solve it by substitution if you let $a = 2$. Find x and y.

3.2 *Introduction to Matrices*

One of the biggest problems in applying mathematics to the real world is developing a means of systematizing and handling the great number of variables and large amounts of data that are inherent in real-life situations. One step in handling large amounts of data is the creation of a mathematical model involving what are called **matrices**.

Matrix Representation of Data

Definition of Matrix

A **matrix** is a rectangular array of numbers. (The plural is *matrices.*)

You are already familiar with matrices from everyday experiences. For example, Table 3.1 shows a car rental chart in the form of a matrix.

TABLE 3.1
Example of a Matrix: Costs of Weekly Car Rentals in Europe

Country	Fiat Panda	Ford Fiesta	Opel Kadett	Renault R14	Volkswagen Microbus
Austria	US $149	US $179	US $219	US $269	US $289
Belgium	130	143	222	273	425
Denmark	164	212	269	408	480
France	189	214	257	326	386
Great Britain	160	174	206	215	243
Holland	156	183	213	259	305
Ireland	176	185	198	222	233
Italy	179	213	259	353	408
Luxembourg	130	143	222	273	425
Spain	156	188	206	247	306
Sweden	178	205	246	281	315

In mathematics we enclose the array of numbers in brackets and denote the entire array by using a capital letter. The size of the matrix is called the **order** and is specified by naming the number of **rows** (rows are horizontal) first and then the numbers of **columns** (columns are vertical). A matrix with order $m \times n$ (read "m by n") means that the matrix has m rows and n columns. The order of the matrix in Table 3.1 is 11×5 (eleven by five). The price of the Renault in Denmark is found in row 3, column 4.

EXAMPLE 1 Name the order of each given matrix.

$$A = \begin{bmatrix} 1 & 6 & 3 \\ 4 & 8 & -2 \end{bmatrix} \quad B = \begin{bmatrix} 6 & 1 \\ 4 & -3 \\ 2 & 0 \\ -1 & 4 \end{bmatrix} \quad C = \begin{bmatrix} 5 \\ 8 \\ 1 \end{bmatrix}$$

$$D = [4 \quad 8 \quad 6 \quad 2] \quad E = \begin{bmatrix} 6 & 8 & -2 \\ 5 & 4 & -1 \\ 0 & -5 & 7 \end{bmatrix}$$

Solution The order of A is 2×3; B is 4×2; C is 3×1; D is 1×4; and E is 3×3. ■

If the matrix has only one column (for example, matrix C), then it is called a **column matrix**; if it has only one row (matrix D, for example), then it is called a **row matrix**; and if the number of rows and columns are the same, it is called a **square matrix** (matrix E, for example). The numbers in a matrix are called the **entries**, or **components**, of a matrix. If all the entires of a matrix are zero, it is called a **zero matrix** and is represented by **0**.

Two matrices are said to be **equal** if they have the same order and if the corresponding entries are equal. (Corresponding entries are elements in the same position—that is, same row and column location.)

EXAMPLE 2 Find x and y, if possible, so that the pairs of matrices are equal.

a. $\begin{bmatrix} 2 & x \\ y & -3 \end{bmatrix} = \begin{bmatrix} 2 & 5 \\ -1 & -3 \end{bmatrix}$ By inspection, $x = 5$ and $y = -1$.

b. $\begin{bmatrix} 6 & 4 \\ x & y \end{bmatrix} = \begin{bmatrix} 6 & 0 \\ 4 & -7 \end{bmatrix}$ No values of x and y will make these matrices equal since $4 \neq 0$ in the upper right entries.

c. $\begin{bmatrix} x & 8 \\ y & -4 \end{bmatrix} = \begin{bmatrix} 2 & 8 & -1 \\ 3 & -4 & 0 \end{bmatrix}$ No values of x and y will make these matrices equal since the matrices must have the same order to be equal. ■

EXAMPLE 3 **Inventory control.** Suppose a company produces two models of wireless telephones and manufactures them at four factories. The number of units produced at each factory can be summarized by the following matrix:

	Factory A	B	C	D
Standard model	16	25	15	8
Deluxe model	10	0	14	3

If the entries represent thousands of units, answer the following questions by reading the matrix:

a. How many standard models are produced at factory **D**?
b. How many standard models are produced at factory **A**?
c. How many deluxe models are produced at factory **C**?
d. At which factory are no deluxe models produced?

Solution **a.** 8,000 **b.** 16,000 **c.** 14,000 **d.** Factory **B** ■

EXAMPLE 4 A map (graph) of Mount Monadock is shown. The heavy dots (vertices) represent rest stops, and the lines (edges) represent trails.*

* This problem courtesy of Peter Glidden from an article "From Graphs to Matrices" in the *Mathematics Teacher*, February 1990, p. 128.

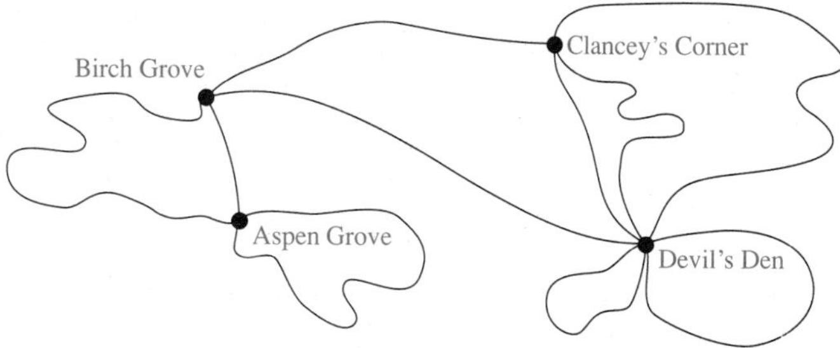

We can represent some characteristics of this graph by using a matrix. We count the number of trails from each rest stop to each other rest stop that do not pass through another stop. Note that there are two routes possible from Aspen Grove to Aspen Grove because you can travel both clockwise and counterclockwise.

	A	**B**	**C**	**D**
Aspen Grove, **A**	2	2	0	0
Birch Grove, **B**	2	0	1	1
Clancey's Corner, **C**	0	1	0	3
Devil's Den, **D**	0	1	3	4

■

If the direction along a route is specified, then the graph is called a **directed graph**. Problem 52 of Problem Set 3.2 reconsiders Example 4 as a directed graph.

Sometimes entries 1 and 0 are used to summarize information in matrix form, as illustrated by Example 5.

EXAMPLE 5 **Communication theory.** Use matrix notation to summarize the following:

The United States has diplomatic relations with the Soviet Union and Mexico, but not with Cuba.

Mexico has diplomatic relations with the United States and the Soviet Union, but not with Cuba.

The Soviet Union has diplomatic relations with the United States, Mexico, and Cuba.

Cuba has diplomatic relations with the Soviet Union, but not with the United States and Mexico.

Solution Draw a graph showing the diplomatic channels.

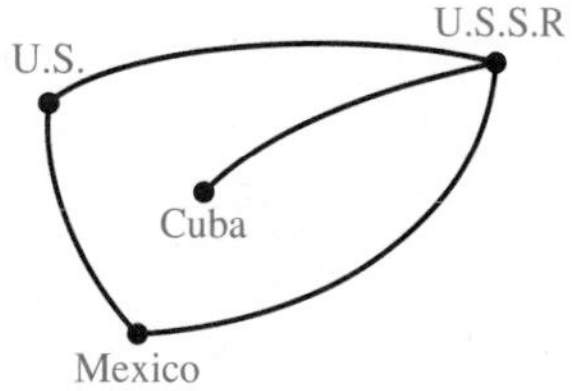

Translate this graph into matrix form by using a 1 to mean that the countries have diplomatic relations and 0 to mean that they do not.

	U.S.	U.S.S.R.	Cuba	Mexico
U.S.	0	1	0	1
U.S.S.R.	1	0	1	1
Cuba	0	1	0	0
Mexico	1	1	0	0

■

The solution to Example 5 is called a **communication matrix** or *incidence matrix.*

Matrix Addition and Subtraction

If two matrices have the same order, we define the **sum of the matrices** as the matrix whose components result from the sum of the corresponding components of the given matrices. We also say the original matrices are **conformable**. If the matrices do not have the same order, then we say the matrices are **not conformable** for addition.

EXAMPLE 6 Let

$$\mathbf{0} = [0 \quad 0 \quad 0] \qquad A = [4 \quad 2 \quad 6]$$

$$B = \begin{bmatrix} 6 \\ 1 \end{bmatrix} \qquad C = [1 \quad 7 \quad 2] \qquad D = \begin{bmatrix} 1 \\ 2 \end{bmatrix}$$

$$E = \begin{bmatrix} 2 & 1 & 0 \\ 4 & 7 & 3 \\ -2 & 0 & 1 \end{bmatrix} \qquad F = \begin{bmatrix} 6 & 1 & 2 \\ 3 & -10 & 4 \\ 1 & 3 & -2 \end{bmatrix}$$

Find (wherever possible):

a. $A + C$ **b.** $B + D$ **c.** $A + \mathbf{0}$ **d.** $A + B$ **e.** $E + F$

Solution

a. $A + C = [4 + 1 \quad 2 + 7 \quad 6 + 2] = [5 \quad 9 \quad 8]$

b. $B + D = \begin{bmatrix} 6 + 1 \\ 1 + 2 \end{bmatrix} = \begin{bmatrix} 7 \\ 3 \end{bmatrix}$

c. $A + \mathbf{0} = [4 + 0 \quad 2 + 0 \quad 6 + 0] = [4 \quad 2 \quad 6]$

d. Not conformable

e. $E + F = \begin{bmatrix} 2 + 6 & 1 + 1 & 0 + 2 \\ 4 + 3 & 7 + (-10) & 3 + 4 \\ -2 + 1 & 0 + 3 & 1 + (-2) \end{bmatrix} = \begin{bmatrix} 8 & 2 & 2 \\ 7 & -3 & 7 \\ -1 & 3 & -1 \end{bmatrix}$ ■

EXAMPLE 7 Let S and T represent the sales (in thousands of dollars) for a company in 1991 and 1992, respectively.

$S =$

	Chicago	Los Angeles	New York
Wholesale	150	200	350
Retail	100	150	50

$T =$

	Chicago	Los Angeles	New York
Wholesale	175	300	400
Retail	110	100	100

What is the combined sales (in thousands of dollars) for the company for the given years?

Solution The answer is found by finding the sum of the matrices:

$$S + T = \begin{bmatrix} 150 + 175 & 200 + 300 & 350 + 400 \\ 100 + 110 & 150 + 100 & 50 + 100 \end{bmatrix}$$

$$= \begin{bmatrix} 325 & 500 & 750 \\ 210 & 250 & 150 \end{bmatrix}$$

■

For the company in Example 7, we might be interested in finding the increase (or decrease) in sales between 1991 and 1992. We see that the difference in wholesale sales in Chicago is found by subtracting:

$$175 - 150 = 25$$

The difference for the other components is found similarly. The **difference of matrices** A and B of the same order, $A - B$, is found by subtracting the entries of B from the corresponding entries of A.

EXAMPLE 8 Find the change in sales between 1991 and 1992 for the company with sales S and T in Example 7.

Solution The answer is found by finding the difference of the matrices:

$$T - S = \begin{bmatrix} 175 - 150 & 300 - 200 & 400 - 350 \\ 110 - 100 & 100 - 150 & 100 - 50 \end{bmatrix}$$

$$= \begin{bmatrix} 25 & 100 & 50 \\ 10 & -50 & 50 \end{bmatrix}$$

The negative entry indicates a decrease in sales. ■

Scalar Multiplication

Continuing with the same example, suppose we want to find the profit generated by the sales in each category and have found that 3% of the gross sales for a given year is actual profit. Then, we find the 1992 profit from wholesale sales in Chicago by multiplying .03 (3%) by 175 (the wholesale sales in thousands of dollars in Chicago in 1992):

$$.03 \cdot 175 = 5.25$$

To find the profit in each category, we multiply the matrix

$$T = \begin{bmatrix} 175 & 300 & 400 \\ 110 & 100 & 100 \end{bmatrix}$$

by .03. This operation of multiplying each entry of a matrix A by a number c is denoted by cA and is called **scalar multiplication**.

EXAMPLE 9 Find the profit (in thousands of dollars) for T.

Solution Calculate $(.03)T$:

$$(.03)T = \begin{bmatrix} (.03)(175) & (.03)(300) & (.03)(400) \\ (.03)(110) & (.03)(100) & (.03)(100) \end{bmatrix} = \begin{bmatrix} 5.25 & 9 & 12 \\ 3.3 & 3 & 3 \end{bmatrix}$$ ■

Matrix Multiplication

You now know how to represent information in matrix form, how to add and subtract matrices, and how to multiply a matrix and a number (scalar multiplication). We now turn our attention to multiplying two matrices.

It might seem natural to multiply two matrices by multiplying corresponding entries. However, such a definition has not proved to be useful, and it turns out that the definition that is useful is rather unusual, so we will consider several examples before formally defining matrix multiplication.

Example 9 illustrates scalar multiplication by considering a 3% profit in three sales locations, as given by

$$(.03)T = \begin{matrix} & \text{Chicago} & \text{Los Angeles} & \text{New York} & \\ & \left[\begin{matrix} 175 \\ 110 \end{matrix}\right. & \begin{matrix} 300 \\ 100 \end{matrix} & \left.\begin{matrix} 400 \\ 100 \end{matrix}\right] & \begin{matrix} \text{Wholesale} \\ \text{Retail} \end{matrix} \end{matrix}$$

Now, however, suppose an inventory tax of 3% is imposed on wholesale sales and a 6% tax is imposed on retail sales. What is the total tax (in thousands of dollars) at each location? First, the necessary calculations to answer this question must be summarized:

	Wholesale tax + Retail tax		Total tax
Chicago tax:	$.03(175) + .06(110)$	$= 5.25 + 6.6$	$= 11.85$
Los Angeles tax:	$.03(300) + .06(100)$	$= 9 + 6$	$= 15$
New York tax:	$.03(400) + .06(100)$	$= 12 + 6$	$= 18$

These calculations summarize the operation of matrix multiplication.

$$\begin{matrix} \text{Wholesale tax} & \text{Retail tax} \\ [.03 & .06] \end{matrix} \begin{matrix} \text{Chicago} & \text{Los Angeles} & \text{New York} \\ \left[\begin{matrix} 175 \\ 110 \end{matrix}\right. & \begin{matrix} 300 \\ 100 \end{matrix} & \left.\begin{matrix} 400 \\ 100 \end{matrix}\right] \end{matrix} \begin{matrix} \text{Wholesale} \\ \text{Retail} \end{matrix}$$

↑ Matrix multiplication

Note size restrictions: $[\ ]_{1\times 2}[\ ]_{2\times 3} = [\ ]_{1\times 3}$

$$= [.03(175) + .06(110) \quad .03(300) + .06(100) \quad .03(400) + .06(100)]$$
$$= [\quad 11.85 \quad 15 \quad 18 \quad]$$

The operation of matrix multiplication has widespread applications, so we will proceed slowly through several examples, each progressing from the previous one.

EXAMPLE 10 $[2 \quad 3 \quad 4 \quad 5]\begin{bmatrix} a \\ b \\ c \\ d \end{bmatrix} = \underbrace{[2a + 3b + 4c + 5d]}$

Order 1×4 — Order 4×1

To multiply matrices, these numbers must be the same.

Multiply corresponding entries and then add the resulting products; the order is 1×1. ■

EXAMPLE 11 $[2 \quad 3 \quad 4 \quad 5]\begin{bmatrix} 1 \\ -3 \\ 0 \\ 2 \end{bmatrix} = [2(1) + 3(-3) + 4(0) + 5(2)]$

$$= [2 - 9 + 0 + 10]$$
$$= [3]$$

In this example a 1×4 matrix and a 4×1 matrix are multiplied and a 1×1 matrix is obtained. ■

EXAMPLE 12 $[5 \quad 3 \quad 2]\begin{bmatrix} 1 \\ 2 \\ 3 \\ 4 \end{bmatrix}$ Not defined

Order 1×3 — Order 4×1

Not the same ■

The process illustrated in Examples 10–12 is repeated over and over for larger-order matrices. If two matrices cannot be multiplied because of their order (as in Example 12), then they are said to be **nonconformable** for multiplication.

You can multiply matrices that are not both row or column matrices. For example,

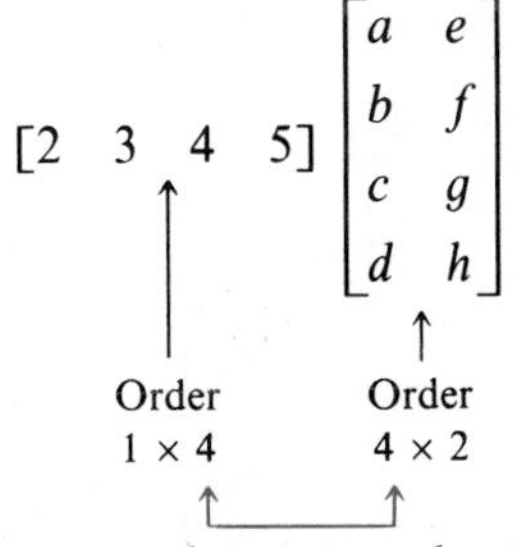

$$[2 \quad 3 \quad 4 \quad 5]\begin{bmatrix} a & e \\ b & f \\ c & g \\ d & h \end{bmatrix}$$

Order 1×4 — Order 4×2

Same, so they are conformable; the answer is a 1×2 matrix.

If there are two columns, then the product is found in two steps:

$$[2 \quad 3 \quad 4 \quad 5]\begin{bmatrix} a & e \\ b & f \\ c & g \\ d & h \end{bmatrix} = [2a + 3b + 4c + 5d \qquad\qquad]$$

Product of row 1 by column 1 is written in position (1, 1).

$$[2 \quad 3 \quad 4 \quad 5]\begin{bmatrix} a & e \\ b & f \\ c & g \\ d & h \end{bmatrix} = [2a + 3b + 4c + 5d \quad 2e + 3f + 4g + 5h]$$

Row 1, column 2 — Answer for position (1, 2)

EXAMPLE 13

$$[2 \quad 3 \quad 4 \quad 5]\begin{bmatrix} 1 & 2 \\ -3 & 1 \\ 0 & -2 \\ 2 & 3 \end{bmatrix} = [2(1) + 3(-3) + 4(0) + 5(2) \qquad]$$

$$[2 \quad 3 \quad 4 \quad 5]\begin{bmatrix} 1 & 2 \\ -3 & 1 \\ 0 & -2 \\ 2 & 3 \end{bmatrix} = [3 \qquad]$$

$$[2 \quad 3 \quad 4 \quad 5]\begin{bmatrix} 1 & 2 \\ -3 & 1 \\ 0 & -2 \\ 2 & 3 \end{bmatrix} = [3 \quad 2(2) + 3(1) + 4(-2) + 5(3)]$$

$$[2 \quad 3 \quad 4 \quad 5]\begin{bmatrix} 1 & 2 \\ -3 & 1 \\ 0 & -2 \\ 2 & 3 \end{bmatrix} = [3 \quad 14]$$

■

EXAMPLE 14

$$[2 \quad -1 \quad 2]\begin{bmatrix} 4 & 2 & -1 \\ 1 & 0 & 2 \\ 3 & -1 & 3 \end{bmatrix} = [2(4) + (-1)1 + 2(3) \qquad]$$

$$[2 \quad -1 \quad 2]\begin{bmatrix} 4 & 2 & -1 \\ 1 & 0 & 2 \\ 3 & -1 & 3 \end{bmatrix} = [13 \quad 2(2) + (-1)(0) + 2(-1) \qquad]$$

$$[2 \quad -1 \quad 2]\begin{bmatrix} 4 & 2 & -1 \\ 1 & 0 & 2 \\ 3 & -1 & 3 \end{bmatrix} = [13 \quad 2 \quad 2(-1) + (-1)(2) + 2(3)]$$

$$= [13 \quad 2 \quad 2]$$

■

Example 14 is written out to clearly demonstrate what is happening. Your work, however, would probably look like this:

Be sure you know where these numbers came from; this is the way work will be shown in subsequent examples.

$$[2 \quad -1 \quad 2]\begin{bmatrix} 4 & 2 & -1 \\ 1 & 0 & 2 \\ 3 & -1 & 3 \end{bmatrix} = [8-1+6 \quad 4+0-2 \quad -2-2+6]$$

$$= [13 \quad 2 \quad 2]$$

EXAMPLE 15

$$\begin{bmatrix} 1 & 3 & 2 \\ -1 & 4 & 3 \end{bmatrix}\begin{bmatrix} 4 & 2 \\ -1 & 3 \\ 2 & -3 \end{bmatrix} = \begin{bmatrix} 4-3+4 & 2+9-6 \\ -4-4+6 & -2+12-9 \end{bmatrix} = \begin{bmatrix} 5 & 5 \\ -2 & 1 \end{bmatrix}$$

■

To summarize this procedure for multiplication of matrices A and B, you multiply the first row of A by the first column of B and place the result in the $(1, 1)$ entry of the product. Then multiply the first row of A by the second column of B and place the result in the $(1, 2)$ entry of the product. Continue multiplying the first row of A by each column of B until you run out of columns of B. Then move to the second row of A and multiply by the first column of B and place the result in the $(2, 1)$ entry of the product. Continue this process until you get to the last row of A and the last column of matrix B. In order to be able to carry out this procedure it is necessary that the number of columns of matrix A be the same as the number of rows of matrix B. This procedure is stated more formally in the following box.

Matrix Multiplication

Let A be an $m \times r$ matrix and let B be an $r \times n$ matrix.
The product matrix AB will be an $m \times n$ matrix. The entry in the ith row and jth column of AB is

the sum of the products formed by multiplying each entry of the ith row of A by the corresponding element in the jth column of B.

EXAMPLE 16 Give the *order* of the following products and give the *position* of the entry found by the highlighted product. If multiplication is not possible, say so.

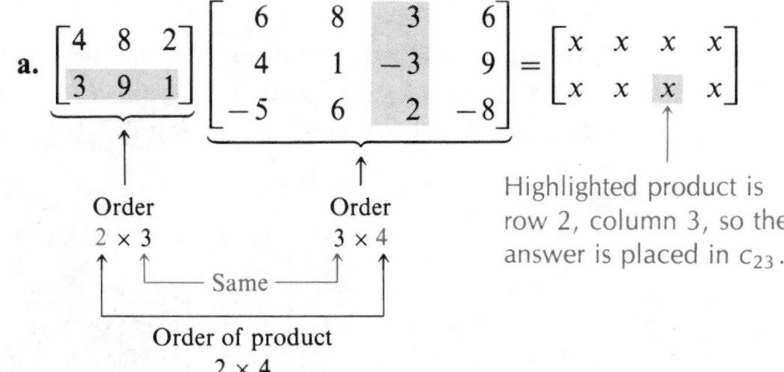

See Appendix D for an Introduction to Spreadsheets.

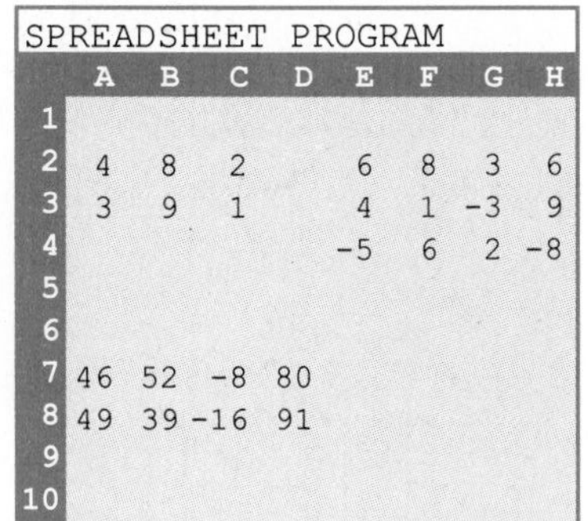

Many spreadsheets have matrix capabilities. In 1-2-3 choose /Data|Matrix|Multiply. In Quattro Pro choose /Tools|Advanced Math|Multiply. All you need to do is insert the entries, tell the program where to find those entries, and then tell the program where to place the answer.

b. $$\begin{bmatrix} 6 & 9 & 2 \\ 3 & 4 & 8 \\ 5 & 1 & 7 \\ -2 & 3 & 4 \end{bmatrix} \begin{bmatrix} 6 & 4 \\ 8 & -4 \\ 1 & 6 \end{bmatrix} = \begin{bmatrix} x & x \\ x & x \\ x & x \\ x & x \end{bmatrix}$$

Order 4×3 — Order 3×2 (the inner numbers are the Same)

Order of product 4×2

c. $$\begin{bmatrix} 2 & 3 & 4 & 1 \\ 6 & 8 & 4 & -3 \end{bmatrix} \begin{bmatrix} 4 & 2 & 6 & -8 \\ 8 & -1 & 0 & 4 \\ 3 & 3 & 4 & 3 \end{bmatrix}$$

Order 2×4 — Order 3×4

Not the same, so not conformable ■

For Examples 17 and 18, let

$$A = \begin{bmatrix} 1 & 2 \\ 3 & 4 \end{bmatrix} \quad \text{and} \quad B = \begin{bmatrix} -1 & 2 \\ 1 & 3 \end{bmatrix}$$

EXAMPLE 17 $$AB = \begin{bmatrix} 1 & 2 \\ 3 & 4 \end{bmatrix} \begin{bmatrix} -1 & 2 \\ 1 & 3 \end{bmatrix} = \begin{bmatrix} -1+2 & 2+6 \\ -3+4 & 6+12 \end{bmatrix} = \begin{bmatrix} 1 & 8 \\ 1 & 18 \end{bmatrix}$$ ■

EXAMPLE 18 $$BA = \begin{bmatrix} -1 & 2 \\ 1 & 3 \end{bmatrix} \begin{bmatrix} 1 & 2 \\ 3 & 4 \end{bmatrix} = \begin{bmatrix} -1+6 & -2+8 \\ 1+9 & 2+12 \end{bmatrix} = \begin{bmatrix} 5 & 6 \\ 10 & 14 \end{bmatrix}$$ ■

WARNING ***AB* ≠ *BA***

Note from Examples 17 and 18 that $AB \neq BA$. That is, matrix multiplication is *not commutative.* This means you must be careful not to switch the order of the matrix factors when working with matrix products.

Matrix Models

This section concludes with some models that use matrix multiplication.

EXAMPLE 19 A developer builds a housing complex featuring two-, three-, and four-bedroom units. Each unit comes in two different floor plans. The matrix P (for production) tells the number of each type of unit for this development.

$$P = \begin{bmatrix} 10 & 5 \\ 25 & 10 \\ 15 & 10 \end{bmatrix} \begin{matrix} \text{2 bedrooms} \\ \text{3 bedrooms} \\ \text{4 bedrooms} \end{matrix}$$

(columns: Plan I, Plan II)

Many materials are used in building these homes, but this model will be simplified to include only lumber, concrete, fixtures, and labor. The matrix M (for materials) gives the amounts of these materials used (in appropriate units) of each).*

$$M = \begin{array}{c} \\ \end{array}\begin{array}{cccc} \text{Lumber} & \text{Concrete} & \text{Fixtures} & \text{Labor} \end{array}$$
$$M = \begin{bmatrix} 7 & 8 & 9 & 20 \\ 8 & 9 & 9 & 22 \end{bmatrix} \begin{array}{l} \text{Plan I} \\ \text{Plan II} \end{array}$$

The matrix product PM gives the amount of material needed for the development.

$$PM = \begin{array}{c} 2 \\ 3 \\ 4 \end{array}\overset{\begin{array}{cc} \text{I} & \text{II} \end{array}}{\begin{bmatrix} 10 & 5 \\ 25 & 10 \\ 15 & 10 \end{bmatrix}} \overset{\begin{array}{cccc} \text{Lumber} & \text{Concrete} & \text{Fixtures} & \text{Labor} \end{array}}{\begin{bmatrix} 7 & 8 & 9 & 20 \\ 8 & 9 & 9 & 22 \end{bmatrix}}$$

$$= \overset{\begin{array}{cccc} \text{Lumber} & \text{Concrete} & \text{Fixtures} & \text{Labor} \end{array}}{\begin{bmatrix} 70 + 40 & 80 + 45 & 90 + 45 & 200 + 110 \\ 175 + 80 & 200 + 90 & 225 + 90 & 500 + 220 \\ 105 + 80 & 120 + 90 & 135 + 90 & 300 + 220 \end{bmatrix}} \begin{array}{l} \text{2 bedrooms} \\ \text{3 bedrooms} \\ \text{4 bedrooms} \end{array}$$

$$= \overset{\begin{array}{cccc} \text{Lumber} & \text{Concrete} & \text{Fixtures} & \text{Labor} \end{array}}{\begin{bmatrix} 110 & 125 & 135 & 310 \\ 255 & 290 & 315 & 720 \\ 185 & 210 & 225 & 520 \end{bmatrix}} \begin{array}{l} \text{2 bedrooms} \\ \text{3 bedrooms} \\ \text{4 bedrooms} \end{array}$$

If the cost of each unit of material is given by the matrix C (for cost), then the total cost of each model for this development is $(PM)C$.

$$C = \overset{\text{Cost per unit}}{\begin{bmatrix} 1{,}800 \\ 190 \\ 2{,}000 \\ 2{,}000 \end{bmatrix}} \begin{array}{l} \text{Lumber} \\ \text{Concrete} \\ \text{Fixtures} \\ \text{Labor} \end{array}$$

$$(PM)C = \begin{bmatrix} 110 & 125 & 135 & 310 \\ 255 & 290 & 315 & 720 \\ 185 & 210 & 225 & 520 \end{bmatrix} \begin{bmatrix} 1{,}800 \\ 190 \\ 2{,}000 \\ 2{,}000 \end{bmatrix} = \begin{bmatrix} 1{,}111{,}750 \\ 2{,}584{,}100 \\ 1{,}862{,}900 \end{bmatrix} \begin{array}{l} \text{2 bedrooms} \\ \text{3 bedrooms} \\ \text{4 bedrooms} \end{array}$$

Since 15 of the units have two bedrooms, the average cost per unit is \$74,117 (\$1,111,750 ÷ 15). For three-bedroom units the average cost per unit is \$73,831, and for four-bedroom units the average cost is \$74,516. ■

* For example, lumber in 1,000 board feet, concrete in cubic yards, etc.

EXAMPLE 20 In Example 5 a matrix was developed to represent the diplomatic relations of the United States, the Soviet Union, Cuba, and Mexico. Let A represent the communication matrix from that example:

$$A = \begin{array}{c} \\ \end{array}\begin{bmatrix} 0 & 1 & 0 & 1 \\ 1 & 0 & 1 & 1 \\ 0 & 1 & 0 & 0 \\ 1 & 1 & 0 & 0 \end{bmatrix}\begin{matrix} \text{U.S.} \\ \text{U.S.S.R.} \\ \text{Cuba} \\ \text{Mexico} \end{matrix}$$

(columns: U.S., U.S.S.R., Cuba, Mexico)

How many channels of communication are open to the various countries if they are willing to speak through an intermediary?

Solution Consider a particular example, say, *Mexico sending a message to Cuba through a single intermediary*. Mexico can send a message to the United States, but the United States cannot pass the message along to Cuba; Mexico can send a message to the Soviet Union and then the Soviet Union can forward it to Cuba; Mexico to Cuba direct does not qualify as sending the message through an intermediary. Therefore, there is only one way that Mexico can send a message to Cuba via an intermediary. *How about the United States sending a message to the United States through an intermediary?* (Perhaps to test for leaks in the communications network!) The United States can send a message to the Soviet Union and then back to the United States, as well as to Mexico and then back. Thus there are two ways that the United States can send a message to itself through an intermediary. Do you see that *there are three ways that the Soviet Union can send a round-trip message through one intermediary*? Summarize this information in matrix form.

U.S. to U.S. 2 ways with an intermediary

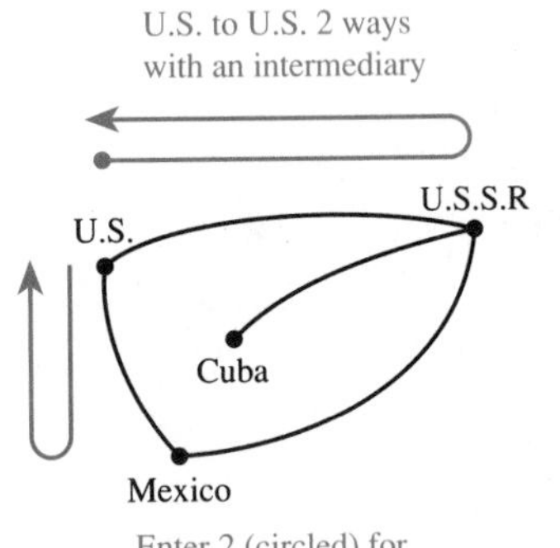

Enter 2 (circled) for U.S. to U.S. in matrix.

	U.S.	U.S.S.R	Cuba	Mexico
U.S.	(2)	1	1	1
U.S.S.R.	1	3	0	1
Cuba	1	0	1	1
Mexico	1	1	1	2

■

Let's compare A^2 for Example 20 with the answer we obtained:

$$A^2 = \begin{bmatrix} 0 & 1 & 0 & 1 \\ 1 & 0 & 1 & 1 \\ 0 & 1 & 0 & 0 \\ 1 & 1 & 0 & 0 \end{bmatrix}\begin{bmatrix} 0 & 1 & 0 & 1 \\ 1 & 0 & 1 & 1 \\ 0 & 1 & 0 & 0 \\ 1 & 1 & 0 & 0 \end{bmatrix}$$

$$= \begin{bmatrix} 0+1+0+1 & 0+0+0+1 & 0+1+0+0 & 0+1+0+0 \\ 0+0+0+1 & 1+0+1+1 & 0+0+0+0 & 1+0+0+0 \\ 0+1+0+0 & 0+0+0+0 & 0+1+0+0 & 0+1+0+0 \\ 0+1+0+0 & 1+0+0+0 & 0+1+0+0 & 1+1+0+0 \end{bmatrix}$$

CALCULATOR COMMENT

Many calculators have built-in matrix operations. For example, on the TI-81 you simply press [MATRX] then you may input three different matrices, called [A], [B], and [C]. After pressing the matrix key, you are asked the size (up to 6 × 6) and then input the entries in each matrix. Matrices are then manipulated using standard notation. For example, after inputting matrix A in Example 4, we press

[A] [x^2] [ENTER]

The display is

```
[ 2  1  2  1 ]
[ 1  3  0  1 ]
[ 1  0  1  1 ]
[ 1  1  1  2 ]
```

$$= \begin{bmatrix} 2 & 1 & 1 & 1 \\ 1 & 3 & 0 & 1 \\ 1 & 0 & 1 & 1 \\ 1 & 1 & 1 & 2 \end{bmatrix} \begin{matrix} \text{U.S.} \\ \text{U.S.S.R.} \\ \text{Cuba} \\ \text{Mexico} \end{matrix}$$

(columns: U.S., U.S.S.R., Cuba, Mexico)

This matrix is the same as $AA = A^2$.

In general, given a communication matrix A:

One intermediary message—one multiplication: $AA = A^2$
Two intermediary messages—two multiplications: $AAA = A^3$
Three intermediary messages—three multiplications: $AAAA = A^4$
And so on.

Another common application of communication matrices is the network of airline routes in the United States. For example, there are several different ways to fly from Kansas City to San Francisco: direct, through one intermediate city, two intermediate cities, and so on. Problem 54 in Problem Set 3.2 involves this type of application.

3.2 Problem Set

State the order of each matrix in Problems 1–2 and whether it is a square matrix, a column matrix, a row matrix, or none of these.

1. a. $A = \begin{bmatrix} 6 & 1 & 4 \\ 7 & 9 & 2 \\ 1 & 5 & 3 \end{bmatrix}$ **b.** $B = \begin{bmatrix} 2 \\ 1 \\ 5 \end{bmatrix}$

c. $C = \begin{bmatrix} 4 & 9 \\ 1 & 6 \\ 7 & 5 \end{bmatrix}$ **d.** $D = \begin{bmatrix} 4 \\ 0 \\ 1 \\ 3 \end{bmatrix}$

e. $E = \begin{bmatrix} 4 & 1 & 7 \\ 9 & 6 & 5 \end{bmatrix}$ **f.** $F = [5 \quad 0 \quad 1 \quad 2]$

2. a. $H = \begin{bmatrix} 6 & 5 \\ 9 & 2 \end{bmatrix}$ **b.** $I = \begin{bmatrix} 1 & 0 & 0 \\ 0 & 1 & 0 \\ 0 & 0 & 1 \end{bmatrix}$

c. $J = \begin{bmatrix} 0 & 1 \\ 1 & 0 \end{bmatrix}$ **d.** $K = \begin{bmatrix} 1 & 0 \\ 0 & 1 \\ 0 & 0 \\ 0 & 0 \\ 0 & 0 \end{bmatrix}$

e. $L = \begin{bmatrix} 1 & 0 & 0 & 0 \\ 0 & 1 & 0 & 0 \end{bmatrix}$ **f.** $M = [5]$

In Problems 3–4 find replacements of the variables that make the matrices equal. If it is not possible to make the matrices equal, state that.

3. a. $\begin{bmatrix} 4 & 8 \\ -1 & x \end{bmatrix} = \begin{bmatrix} y & 8 \\ -1 & 2 \end{bmatrix}$ **b.** $\begin{bmatrix} 6 \\ x \\ y \end{bmatrix} = \begin{bmatrix} z \\ 4 \\ -9 \end{bmatrix}$

4. a. $\begin{bmatrix} 0 & x \\ 0 & 1 \end{bmatrix} = \begin{bmatrix} 1 & 2 \\ 0 & y \end{bmatrix}$ **b.** $\begin{bmatrix} 1 & 0 \\ 0 & 1 \end{bmatrix} = \begin{bmatrix} a & b \\ c & d \end{bmatrix}$

5. A company supplies four parts for General Motors. The parts are manufactured at four factories. Summarize the following production information using a matrix: Factory **A** produces 25 units of part one, 42 units of part two, and 193 units of part three; factory **B** produces 16 units of part one, 39 units of part two, and 150 units of part three; factory **C** produces 50 units of each part; and factory **D** produces 320 units of part four only.

6. Consider Table 3.1 in matrix form.
a. What is the entry in row 2, column 3?
b. What is the entry in row 3, column 2?
c. In which row and column is the price of a Volkswagen Microbus in Sweden?

d. In which row and column is the price of an Opel Kadett in Holland?

e. Would the prices of a Volkswagen Microbus in various countries be represented as a row matrix or a column matrix?

Write the graphs in Problems 7 and 8 in matrix form.

7. **8.**

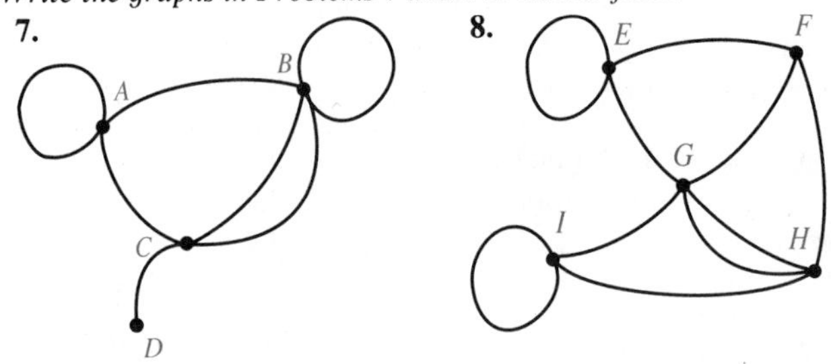

In Problems 9–10 give the order of the product and tell the position of the entry found by the highlighted product. If multiplication is not possible, state that.

9. a. $\begin{bmatrix} 1 & 2 & 3 \\ 8 & 4 & 5 \\ 6 & -1 & 3 \end{bmatrix}\begin{bmatrix} 4 & 6 & 1 \\ 2 & 5 & 1 \\ 4 & 9 & -3 \end{bmatrix}$

b. $\begin{bmatrix} 1 & 2 & 3 \\ 8 & 4 & 5 \\ 6 & -1 & 3 \end{bmatrix}\begin{bmatrix} 4 & 6 & 1 \\ 2 & 5 & 1 \\ 4 & 9 & -3 \end{bmatrix}$

c. $\begin{bmatrix} 1 & 2 & 3 \\ 8 & 4 & 5 \\ 6 & -1 & 3 \end{bmatrix}\begin{bmatrix} 4 & 6 & 1 \\ 2 & 5 & 1 \\ 4 & 9 & -3 \end{bmatrix}$

d. $\begin{bmatrix} 6 & -3 & 2 & -4 & 5 \\ 3 & 5 & 1 & 2 & 3 \end{bmatrix}\begin{bmatrix} 6 & 5 \\ 1 & 1 \\ 2 & 7 \\ 3 & -2 \\ 4 & -1 \end{bmatrix}$

e. $\begin{bmatrix} 6 & -3 & 2 & -4 & 5 \\ 3 & 5 & 1 & 2 & 3 \end{bmatrix}\begin{bmatrix} 6 & 5 \\ 1 & 1 \\ 2 & 7 \\ 3 & -2 \\ 4 & -1 \end{bmatrix}$

10. a. $\begin{bmatrix} 6 & 8 & 1 \\ -2 & 3 & 5 \end{bmatrix}\begin{bmatrix} 5 & 6 & -2 & -9 & -5 \\ 2 & 5 & 1 & 8 & 4 \\ 1 & -3 & 7 & 3 & 12 \end{bmatrix}$

b. $\begin{bmatrix} 6 & 8 & 1 \\ -2 & 3 & 5 \end{bmatrix}\begin{bmatrix} 5 & 6 & -2 & -9 & -5 \\ 2 & 5 & 1 & 8 & 4 \\ 1 & -3 & 7 & 3 & 12 \end{bmatrix}$

c. $\begin{bmatrix} 9 & -8 & 1 & 4 \\ 5 & 2 & 3 & 6 \\ -7 & 6 & -5 & -8 \end{bmatrix}\begin{bmatrix} 1 & 5 & 9 & 13 & 17 \\ 2 & 6 & 10 & 14 & 18 \\ 3 & 7 & 11 & 15 & 19 \\ 4 & 8 & 12 & 16 & 20 \end{bmatrix}$

d. $\begin{bmatrix} 9 & -8 & 1 & 4 \\ 5 & 2 & 3 & 6 \\ -7 & 6 & -5 & -8 \end{bmatrix}\begin{bmatrix} 1 & 5 & 9 & 13 & 17 \\ 2 & 6 & 10 & 14 & 18 \\ 3 & 7 & 11 & 15 & 19 \\ 4 & 8 & 12 & 16 & 20 \end{bmatrix}$

e. $\begin{bmatrix} 6 & 4 & 6 & 3 \\ 9 & -1 & 3 & 4 \end{bmatrix}\begin{bmatrix} 1 & -2 & 4 & 6 \\ 6 & 5 & 8 & -2 \\ 3 & -3 & 7 & 5 \end{bmatrix}$

Find the indicated matrices in Problems 11–12, if possible. Give the order of your answer.

11. a. $[1 \quad -2 \quad 2]\begin{bmatrix} 1 & 2 & 3 \\ 4 & -1 & 3 \\ -3 & 2 & 1 \end{bmatrix}$

b. $[3 \quad 2 \quad -4]\begin{bmatrix} 1 & 2 & -3 \\ -4 & 3 & 2 \\ 2 & 3 & 1 \end{bmatrix}$

12. a. $\begin{bmatrix} 1 & 3 & 2 \\ -1 & 1 & 2 \end{bmatrix}\begin{bmatrix} 6 & 3 & 2 & -3 \\ 1 & 1 & -4 & -1 \\ 2 & 1 & 1 & 2 \end{bmatrix}$

b. $\begin{bmatrix} 6 & 1 & 3 \\ 2 & -3 & 5 \end{bmatrix}\begin{bmatrix} 1 & 0 & 2 & 1 & 1 \\ 2 & 0 & 1 & 1 & 0 \\ 3 & 1 & 4 & 0 & 0 \end{bmatrix}$

Find the indicated matrices in Problems 13–39. Let

$$A = \begin{bmatrix} 1 & 0 & 2 \\ 3 & -1 & 2 \\ 4 & 1 & 0 \end{bmatrix} \quad B = \begin{bmatrix} 1 & 4 & 0 \\ 3 & -1 & 2 \\ -2 & 1 & 5 \end{bmatrix}$$

$$C = \begin{bmatrix} 8 & 1 & 6 \\ 3 & 5 & 7 \\ 4 & 9 & 2 \end{bmatrix} \quad D = \begin{bmatrix} -2 \\ 1 \\ 3 \end{bmatrix} \quad E = \begin{bmatrix} 4 \\ 1 \\ 6 \end{bmatrix}$$

13. $A + B$ **14.** $B + A$

15. $A - B$ **16.** $B - A$

17. $C + D$ **18.** $D + E$

19. $2A + B$ **20.** $3A - 4B$

21. $2C + 3D$ **22.** $C - 2D$

23. $3E - 2D$ **24.** $3D - 2E$

25. $(A + B) + C$ **26.** $3B + 2C$

27. $A + (B + C)$

28. AB

29. BA

30. AC

31. CA

32. BC

33. CB

34. $A(BC)$

35. $(AB)C$

36. $B(AC)$

37. $AB + AC$

38. $A(B + C)$

39. $(B + C)A$

Find the indicated matrices in Problems 40–51. *Let*

$$A = \begin{bmatrix} 1 & 2 \\ 4 & 0 \\ -1 & 3 \\ 2 & 1 \end{bmatrix} \quad B = \begin{bmatrix} 4 & 2 \\ -1 & 3 \end{bmatrix}$$

$$C = \begin{bmatrix} 1 & 0 & 0 & 0 \\ 0 & 1 & 0 & 0 \\ 0 & 0 & 1 & 0 \\ 0 & 0 & 0 & 1 \end{bmatrix} \quad D = \begin{bmatrix} 4 & 1 & 3 & 6 \\ -1 & 0 & -2 & 3 \end{bmatrix}$$

40. AB **41.** BA **42.** B^2 **43.** CA

44. BD **45.** DB **46.** $(B + C)A$ **47.** $BA + CA$

48. CD **49.** A^2 **50.** C^3 **51.** B^3

APPLICATIONS

52. To reduce traffic and slow soil erosion on Mount Monadock (see Example 4), park rangers have decided to designate some ecologically sensitive trails as one-way trails as shown in the figure. This is called a *directed graph.* Write a matrix corresponding to this directed graph.

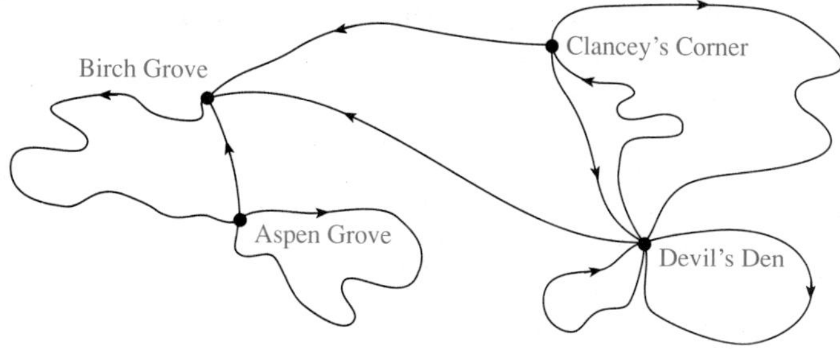

53. A company with five offices in San Francisco operates its own delivery service for office mail. Arrows represent the direction of communication in the figure. Note that A can send mail directly to offices B, D, and E but not to C.

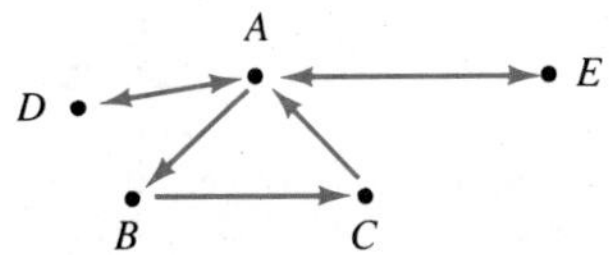

a. Express this delivery network with a communication matrix.

b. Develop a matrix showing the possible two-stop deliveries.

c. Develop a matrix showing the possible three-stop deliveries.

54. Consider the map of airline routes shown below.

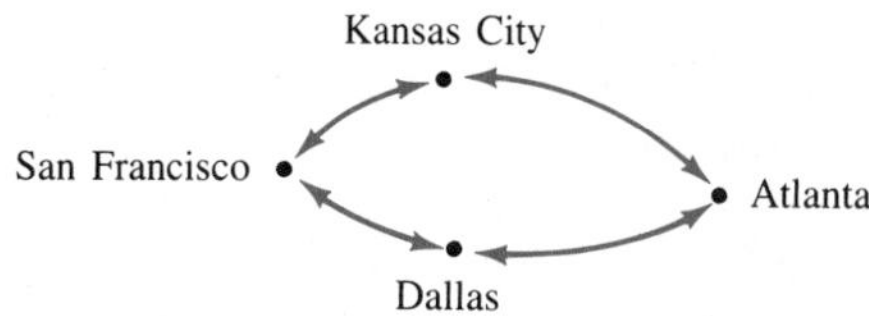

a. Fill in the blanks in the following communication matrix representing the airline routes:

$$T = \begin{matrix} \text{SF} \\ \text{D} \\ \text{A} \\ \text{KC} \end{matrix} \overset{\begin{matrix} \text{SF} & \text{D} & \text{A} & \text{KC} \end{matrix}}{\begin{bmatrix} \quad & \quad & \quad & \quad \\ & & & \\ & & & \\ & & & \end{bmatrix}}$$

b. In how many ways can you travel from Kansas City to San Francisco making exactly one stop?

c. In how many ways can you travel from San Francisco to Kansas City making two stops?

d. Write the communication matrix showing the number of routes among these cities if you make exactly one stop.

e. Write the communication matrix showing the number of routes among these cities if you make exactly two stops.

55. An after-market automotive manufacturer produces several styles of sport wheels. One of the styles is available in two finishes—polished anthracite, A, and black, B—and in three wheel sizes, 13 inch, 14 inch, and 15 inch. The shipments of this particular style to a retailer in the Midwest for the months of January and February can be given by the following matrices:

$$J = \begin{bmatrix} 16 & 24 & 8 \\ 8 & 12 & 4 \end{bmatrix} \quad F = \begin{bmatrix} 12 & 32 & 16 \\ 12 & 20 & 0 \end{bmatrix}$$

Use matrix methods to answer the following questions.

a. How many of each type of wheel were shipped to the retailer during January and February?

b. Suppose the first-quarter sales projections are

$$S = \begin{bmatrix} 40 & 62 & 36 \\ 28 & 32 & 16 \end{bmatrix}$$

What is the necessary March order?

c. The unit price of the wheel depends only on its size. Current prices are \$90, \$110, and \$120 for 13-inch, 14-inch, and 15-inch wheels, respectively. What is the manufacturer's revenue for January?

d. Suppose that unit profit for 13-inch, 14-inch, and 15-inch wheels is \$27, \$33, and \$36, respectively. What matrix could be used to determine both revenue and profit for the month of February?*

56. In Example 19 add 40 units of overhead for the plan I homes and 45 units of overhead for the plan II models. If the cost for overhead is \$1,000 per unit, find the total cost for the housing complex as well as the average cost for two-, three-, and four-bedroom homes assuming all the other information in Example 19 remains unchanged.

57. If the countries in Example 20 are willing to speak through two intermediaries, find the communication matrix and tell how many different ways the United States can communicate with each of the other countries in this fashion.

Historical Question. The English mathematician Arthur Cayley (1821–1895) was responsible for matrix theory. Cayley began his career as a lawyer in 1849, but he was always a mathematician at heart, publishing more than 200 papers in mathematics while practicing law. In 1863 he gave up law to accept a chair in mathematics at Cambridge. He originated the notion of matrices in 1858 and worked on matrix theory over the next several years. His work was entirely theoretical and was not used for any practical purpose until 1925 when matrices were used in quantum mechanics. Since 1950, matrices have played an important role in the social sciences and business. Much of the early work with matrices focused on their properties. In Problems 58–63 we look at some of these properties. Let A, B, and C be square matrices of order 3 and let a and b be real numbers. Give an example to show that each of the following is false or else explain why you think the property is true.

58. $A + B = B + A$

59. $A - B = B - A$

60. $(A - B) - C = A - (B - C)$

61. $(A + B) + C = A + (B + C)$

62. $a(B + C) = aB + aC$

63. $(a + b)C = aC + bC$

3.3 Gauss–Jordan Elimination

In this section we develop a procedure to solve more general linear systems. Consider a system of m equations and n variables:

$$\begin{cases} a_{11}x_1 + a_{12}x_2 + a_{1n}x_n = b_1 \\ a_{21}x_1 + a_{22}x_2 + a_{2n}x_n = b_2 \\ \quad \vdots \qquad\quad \vdots \qquad\quad \vdots \\ a_{m1}x_1 + a_{m2}x_2 + a_{mn}x_n = b_m \end{cases}$$

Now consider what is called the **augmented matrix** for this system:

$$\left[\begin{array}{cccc|c} a_{11} & a_{12} & \cdots & a_{1n} & b_1 \\ a_{21} & a_{22} & \cdots & a_{2n} & b_2 \\ \vdots & \vdots & & \vdots & \vdots \\ a_{m1} & a_{m2} & \cdots & a_{mn} & b_m \end{array}\right]$$

* My thanks to Christian R. Hirsch and Harold L. Schoen for this problem, which was taken from the article "Implementing the Standards," in *Mathematics Teacher*, December 1989, pp. 697–698.

EXAMPLE 1 Write the given systems in augmented matrix form.

a. $\begin{cases} 2x + 3y = 8 \\ 3x + 2y = 7 \end{cases}$ **b.** $\begin{cases} 2x + y = 3 \\ 3x - y = 2 \\ 4x + 3y = 7 \end{cases}$ **c.** $\begin{cases} 3x - 2y + z = -2 \\ 4x - 5y + 3z = -9 \\ 2x - y + 5z = -5 \end{cases}$

d. $\begin{cases} 5x - 3y + z = -3 \\ 2x + 5z = 14 \end{cases}$ **e.** $\begin{cases} x_1 - 3x_3 + x_5 = -3 \\ x_2 + x_4 = -1 \\ x_3 + x_5 = 7 \\ x_1 + x_2 - x_3 + 4x_4 = -8 \\ x_1 + x_2 + x_3 + x_4 + x_5 = 8 \end{cases}$

Solution Note that some coefficients are negative and some are zeros.

a. $\left[\begin{array}{cc|c} 2 & 3 & 8 \\ 3 & 2 & 7 \end{array}\right]$ **b.** $\left[\begin{array}{cc|c} 2 & 1 & 3 \\ 3 & -1 & 2 \\ 4 & 3 & 7 \end{array}\right]$ **c.** $\left[\begin{array}{ccc|c} 3 & -2 & 1 & -2 \\ 4 & -5 & 3 & -9 \\ 2 & -1 & 5 & -5 \end{array}\right]$

d. $\left[\begin{array}{ccc|c} 5 & -3 & 1 & -3 \\ 2 & 0 & 5 & 14 \end{array}\right]$ **e.** $\left[\begin{array}{ccccc|c} 1 & 0 & -3 & 0 & 1 & -3 \\ 0 & 1 & 0 & 1 & 0 & -1 \\ 0 & 0 & 1 & 0 & 1 & 7 \\ 1 & 1 & -1 & 4 & 0 & -8 \\ 1 & 1 & 1 & 1 & 1 & 8 \end{array}\right]$ ■

EXAMPLE 2 Write a system of equations (use $x_1, x_2, \ldots, x_n$ for the variables) that has the given augmented matrix.

a. $\left[\begin{array}{ccc|c} 2 & 1 & -1 & -3 \\ 3 & -2 & 1 & 9 \\ 1 & -4 & 3 & 17 \end{array}\right]$ **b.** $\left[\begin{array}{ccc|c} 1 & 0 & 0 & 3 \\ 0 & 1 & 0 & -2 \\ 0 & 0 & 1 & -5 \end{array}\right]$ **c.** $\left[\begin{array}{cc|c} 1 & 0 & 4 \\ 0 & 1 & 3 \end{array}\right]$

d. $\left[\begin{array}{cc|c} 1 & 0 & -5 \\ 0 & 1 & 2 \\ 2 & 1 & -7 \end{array}\right]$ **e.** $\left[\begin{array}{ccc|c} 1 & 0 & 0 & 3 \\ 0 & 1 & 0 & -3 \\ 0 & 0 & 1 & 6 \\ 0 & 0 & 0 & 1 \end{array}\right]$ **f.** $\left[\begin{array}{ccc|c} 1 & 0 & 0 & 1 \\ 0 & 1 & 0 & 5 \\ 0 & 0 & 1 & -2 \\ 0 & 0 & 0 & 0 \end{array}\right]$

Solution **a.** $\begin{cases} 2x_1 + x_2 - x_3 = -3 \\ 3x_1 - 2x_2 + x_3 = 9 \\ x_1 - 4x_2 + 3x_3 = 17 \end{cases}$ **b.** $\begin{cases} x_1 = 3 \\ x_2 = -2 \\ x_3 = -5 \end{cases}$

c. $\begin{cases} x_1 = 4 \\ x_2 = 3 \end{cases}$ **d.** $\begin{cases} x_1 = -5 \\ x_2 = 2 \\ 2x_1 + x_2 = -7 \end{cases}$

e. $\begin{cases} x_1 = 3 \\ x_2 = -3 \\ x_3 = 6 \\ 0 = 1 \end{cases}$ **f.** $\begin{cases} x_1 = 1 \\ x_2 = 5 \\ x_3 = -2 \\ 0 = 0 \end{cases}$ ■

The goal of this section is to solve a system of m equations with n unknowns. We have already looked at systems of two equations with two unknowns. In high school you may have solved three equations with three unknowns. Now, however, we want to be able to solve problems with two equations and five unknowns, or three equations and two unknowns, or any mixture of linear equations or unknowns. The procedure of this section—**Gauss–Jordan elimination**—is a general method for solving all of these types of systems. We write the system in augmented matrix form (as in Example 1), then carry out a process that transforms the matrix until the solution is obvious. Look back at Example 2—the solutions to parts **b** and **c** are obvious. Part **e** shows $0 = 1$ in the last equation, so this system has no solution (0 cannot equal 1), and part **f** shows $0 = 0$ (which is true for all replacements of the variable), which means that the solution is found by looking at the other equations (namely, $x_1 = 1$, $x_2 = 5$, and $x_3 = -2$). The terms with nonzero coefficients in these examples are arranged on a diagonal, and such a system is said to be in *diagonal form.*

But what process will allow us to transform a matrix into diagonal form? We begin with some steps called **elementary row operations**. Elementary row operations change the *form* of a matrix, but the new form represents an equivalent system. Matrices which represent equivalent systems are called **equivalent matrices** and we introduce these elementary row operations in order to write equivalent matrices. Let us work with a system with three equations and three unknowns (any size will work the same way).

System format *Matrix format*

$$\begin{cases} 3x - 2y + 4z = 11 \\ x - y - 2z = -7 \\ 2x - 3y + z = -1 \end{cases} \qquad \left[\begin{array}{rrr|r} 3 & -2 & 4 & 11 \\ 1 & -1 & -2 & -7 \\ 2 & -3 & 1 & -1 \end{array}\right]$$

Interchanging two equations is equivalent to interchanging two rows in matrix format, and, certainly, if we do this, the solution to the system will be the same:

System format *Matrix format*

$$\begin{cases} x - y - 2z = -7 \\ 3x - 2y + 4z = 11 \\ 2x - 3y + z = -1 \end{cases} \qquad \left[\begin{array}{rrr|r} 1 & -1 & -2 & -7 \\ 3 & -2 & 4 & 11 \\ 2 & -3 & 1 & -1 \end{array}\right]$$

The first and second equations (rows) are interchanged.

Elementary Row Operation 1: *Interchange any two rows.*

Since multiplying or dividing both sides of any equation by any nonzero number does not change the solution, then, in matrix format, the solution will not be changed if any row is multiplied or divided by a nonzero constant. For example, multiply both sides of the first equation by -3:

System format *Matrix format*

$$\begin{cases} -3x + 3y + 6z = 21 \\ 3x - 2y + 4z = 11 \\ 2x - 3y + z = -1 \end{cases} \qquad \left[\begin{array}{rrr|r} -3 & 3 & 6 & 21 \\ 3 & -2 & 4 & 11 \\ 2 & -3 & 1 & -1 \end{array}\right]$$

Both sides of the first equation are multiplied by -3; the first row is multiplied by -3.

Because it is easier to program software by separating multiplication and division into two row operations, we have done the same.

Elementary Row Operation 2: *Multiply all the elements of a row by the same nonzero real number.*

Elementary Row Operation 3: *Divide all the elements of a row by the same nonzero real number.*

> CALCULATOR COMMENT
>
> On the TI-81 calculator, these elementary row operations are found when you press the MATRX key.
>
> *Elementary Row Operation* 1:
> RowSwap
>
> *Elementary Row Operations* 2 *and* 3:
> *Row
>
> *Elementary Row Operation* 4:
> *Row+

The last property we need to carry out Gauss–Jordan elimination rests on the property we used with the linear combination method—namely, adding equations to eliminate a variable. In terms of the system format this means that one equation can be replaced by its sum with another equation in the system. For example, if we add the first equation to the second equation we have:

System format / *Matrix format*

$$\begin{cases} -3x + 3y + 6z = 21 \\ y + 10z = 32 \\ 2x - 3y + z = -1 \end{cases} \qquad \left[\begin{array}{rrr|r} -3 & 3 & 6 & 21 \\ 0 & 1 & 10 & 32 \\ 2 & -3 & 1 & -1 \end{array}\right]$$

In terms of matrix format, any row can be replaced by its sum with some other row.

This process is usually used in conjunction with Elementary Row Operation 2. That is, an equation is changed by adding to it a nonzero multiple of another equation in the system. Go back to the original system:

System format / *Matrix format*

$$\begin{cases} 3x - 2y + 4z = 11 \\ x - y - 2z = -7 \\ 2x - 3y + z = -1 \end{cases} \qquad \left[\begin{array}{rrr|r} 3 & -2 & 4 & 11 \\ 1 & -1 & -2 & -7 \\ 2 & -3 & 1 & -1 \end{array}\right]$$

We can change this system by multiplying the second equation by -3 and adding the result to the first equation. In matrix terminology we would say multiply the second row by -3 and add it to the first row:

System format / *Matrix format*

$$\begin{cases} y + 10z = 32 \\ x - y - 2z = -7 \\ 2x - 3y + z = -1 \end{cases} \qquad \left[\begin{array}{rrr|r} 0 & 1 & 10 & 32 \\ 1 & -1 & -2 & -7 \\ 2 & -3 & 1 & -1 \end{array}\right]$$

Once again, multiply the second row, this time by -2, and add it to the third row. Wait! Why -2? Where did that come from? The idea is the same one we used in the linear combination method—we use a number that will give a zero coefficient to the x in the third equation.

System format / *Matrix format*

$$\begin{aligned} y + 10z &= 32 \\ x - y - 2z &= -7 \\ -y + 5z &= 13 \end{aligned} \qquad \left[\begin{array}{rrr|r} 0 & 1 & 10 & 32 \\ 1 & -1 & -2 & -7 \\ 0 & -1 & 5 & 13 \end{array}\right]$$

Note that the multiplied row is not changed; instead, the changed row is the one to which the multiplied row is added. We call the original row the **pivot row** and the changed row the **target row**.

Elementary Row Operation 4: *Multiply all the entries of a row (the pivot row) by a nonzero real number and add each resulting product to the corresponding entry of another specified row (the target row). Note that this operation changes only the target row.*

There you have it! You should carry out these four elementary row operations until you have a system for which the solution is obvious, as illustrated in Example 3.

EXAMPLE 3 Solve $\begin{cases} 2x - 5y = 5 \\ x - 2y = 1 \end{cases}$.

Solution

System notation

$$\begin{cases} 2x - 5y = 5 \\ x - 2y = 1 \end{cases}$$

Matrix notation

$$\left[\begin{array}{cc|c} 2 & -5 & 5 \\ 1 & -2 & 1 \end{array}\right]$$

Elementary Row Operation 1

Interchange the first and second equations.

$$\begin{cases} x - 2y = 1 \\ 2x - 5y = 5 \end{cases}$$

Interchange the first and second rows.

$$\left[\begin{array}{cc|c} 1 & -2 & 1 \\ 2 & -5 & 5 \end{array}\right] \begin{array}{l} \longleftarrow \text{Pivot row} \\ \longleftarrow \text{Target row} \end{array}$$

Elementary Row Operation 4

Add -2 times the first equation to the second.

$$\begin{cases} x - 2y = 1 \\ \quad - y = 3 \end{cases}$$

Add -2 times the first row to the second row.

$$\left[\begin{array}{cc|c} 1 & -2 & 1 \\ 0 & -1 & 3 \end{array}\right] \begin{array}{l} \longleftarrow \text{Pivot row remains unchanged.} \\ \longleftarrow \text{Target row changes.} \end{array}$$

Elementary Row Operation 2

Multiply both sides of the second equation by -1.

$$\begin{cases} x - 2y = 1 \\ \quad y = -3 \end{cases}$$

Multiply row 2 by -1.

$$\left[\begin{array}{cc|c} 1 & -2 & 1 \\ 0 & 1 & -3 \end{array}\right] \begin{array}{l} \\ \longleftarrow \text{New pivot row} \end{array}$$

Elementary Row Operation 4

Add 2 times the second equation to the first.

$$\begin{cases} x \quad = -5 \\ \quad y = -3 \end{cases}$$

Add 2 times the second row to the first row.

$$\left[\begin{array}{cc|c} 1 & 0 & -5 \\ 0 & 1 & -3 \end{array}\right]$$

This is called the **reduced row-echelon form**.

The solution, $(-5, -3)$, is now obvious. ■

As you study Example 3, first look at how the elementary row operations led to a system equivalent to the first—but one for which the solution is obvious. Next, try to decide *why* a particular row operation was chosen when it was. Most students quickly learn the elementary row operations, but then use a series of (almost random) steps until the obvious solution results. This often works, but is not very efficient. The steps chosen in Example 3 illustrate a very efficient method of using the elementary row operations to determine a system whose solution is obvious. The method was discovered independently by two mathematicians, Karl Friedrich Gauss (1777–1855) and Camille Jordan (1838–1922), so today the process is known as the Gauss–Jordan method. (Since the process was only recently attributed to Jordan, many books refer to the method simply as *Gaussian elimination.*) Before stating the process, however, we will consider a procedure called **pivoting**.

Pivoting

1. Divide all entries in the row in which the pivot appears (called the **pivot row**) by the pivot element so that the pivot entry becomes a 1. This uses Elementary Row Operation 3.
2. Obtain zeros above and below the pivot element by using Elementary Row Operation 4.

EXAMPLE 4 Pivot the given matrix about the circled element.

Solution

$$\left[\begin{array}{cc|c} 15 & \textcircled{5} & 35 \\ 5 & 2 & -3 \end{array}\right] \xrightarrow{R1 \div 5} \left[\begin{array}{cc|c} 3 & 1 & 7 \\ 5 & 2 & -3 \end{array}\right] \xrightarrow{-2R1 + R2} \left[\begin{array}{cc|c} 3 & 1 & 7 \\ -1 & 0 & -17 \end{array}\right]$$

This shorthand notation shows what we did: divide row 1 by 5.

Multiply row 1 (pivot row) by -2 and add it to row 2 (target row).

Consider another matrix:

$$\left[\begin{array}{ccc|c} 2 & 3 & 4 & 1 \\ -1 & \textcircled{2} & 3 & -2 \\ 0 & 1 & -1 & 3 \end{array}\right] \xrightarrow{\frac{1}{2}R2} \left[\begin{array}{ccc|c} 2 & 3 & 4 & 1 \\ -\frac{1}{2} & 1 & \frac{3}{2} & -1 \\ 0 & 1 & -1 & 3 \end{array}\right]$$

$$\xrightarrow{-3R2 + R1} \left[\begin{array}{ccc|c} \frac{7}{2} & 0 & -\frac{1}{2} & 4 \\ -\frac{1}{2} & 1 & \frac{3}{2} & -1 \\ 0 & 1 & -1 & 3 \end{array}\right] \xrightarrow{-1R2 + R3} \left[\begin{array}{ccc|c} \frac{7}{2} & 0 & -\frac{1}{2} & 4 \\ -\frac{1}{2} & 1 & \frac{3}{2} & -1 \\ \frac{1}{2} & 0 & -\frac{5}{2} & 4 \end{array}\right]$$

■

The four elementary row operations are listed below for easy reference.

Elementary Row Operations

There are **four elementary row operations** for producing equivalent matrices:

1. Interchange any two rows.
2. Multiply all the elements of a row by the same nonzero real number.
3. Divide all the elements of a row by the same nonzero real number.
4. Multiply all the entries of a row (*pivot row*) by a nonzero real number and add each resulting product to the corresponding entry of another specified row (*target row*). (Note that this operation changes only the target row.)

COMPUTER COMMENT

The software accompanying this book (see Appendix D) has a program called *Row reduction of a system.* After entering the matrix, you can choose either MANUAL MODE or AUTOMATIC MODE.

You are now ready to see the method worked out by Gauss and Jordan. It efficiently uses the elementary row operations to diagonalize the matrix. That is, the first pivot is the first entry in the first row, first column; the second is the entry in the second row, second column; and so on until the solution is obvious.

Gauss–Jordan Elimination

1. Select the element in the first row, first column, as a pivot.
2. Pivot.
3. Select the element in the second row, second column, as a pivot.
4. Pivot.
5. Repeat the process until you arrive at the last row, or until the pivot element is a zero. If it is a zero and you can interchange that row with a row below it, so that the pivot element is no longer a zero, do so and continue. If it is a zero and you cannot interchange rows so that it is not a zero, continue with the next row.

EXAMPLE 5 Solve $\begin{cases} 3x - 2y + z = -2 \\ 4x - 5y + 3z = -9. \\ 2x - y + 5z = -5 \end{cases}$

Solution We will solve this system by choosing the steps according to the Gauss–Jordan method.

$$\left[\begin{array}{ccc|c} \textcircled{3} & -2 & 1 & -2 \\ 4 & -5 & 3 & -9 \\ 2 & -1 & 5 & -5 \end{array}\right] \xrightarrow{R1 \div 3} \left[\begin{array}{ccc|c} 1 & -\frac{2}{3} & \frac{1}{3} & -\frac{2}{3} \\ 4 & -5 & 3 & -9 \\ 2 & -1 & 5 & -5 \end{array}\right] \xrightarrow{-4R1 + R2} \left[\begin{array}{ccc|c} 1 & -\frac{2}{3} & \frac{1}{3} & -\frac{2}{3} \\ 0 & -\frac{7}{3} & \frac{5}{3} & -\frac{19}{3} \\ 2 & -1 & 5 & -5 \end{array}\right]$$

3 is the pivot.

Obtain a 1 in the pivot position.

Pivot row is 1; target row is 2; −4 is the opposite of the corresponding element in the target row so that a zero is obtained.

$$\xrightarrow{-2R1 + R3} \left[\begin{array}{ccc|c} 1 & -\frac{2}{3} & \frac{1}{3} & -\frac{2}{3} \\ 0 & \textcircled{$-\frac{7}{3}$} & \frac{5}{3} & -\frac{19}{3} \\ 0 & \frac{1}{3} & \frac{13}{3} & -\frac{11}{3} \end{array}\right] \xrightarrow{R2 \div -\frac{7}{3}} \left[\begin{array}{ccc|c} 1 & -\frac{2}{3} & \frac{1}{3} & -\frac{2}{3} \\ 0 & 1 & -\frac{5}{7} & \frac{19}{7} \\ 0 & \frac{1}{3} & \frac{13}{3} & -\frac{11}{3} \end{array}\right] \xrightarrow[\frac{2}{3}R2 + R1]{}$$

Pivot row is 1; target row is 3.

Select a new pivot $(-\frac{7}{3})$.

Pivot

$$\left[\begin{array}{ccc|c} 1 & 0 & -\frac{1}{7} & \frac{8}{7} \\ 0 & 1 & -\frac{5}{7} & \frac{19}{7} \\ 0 & \frac{1}{3} & \frac{13}{3} & -\frac{11}{3} \end{array}\right] \xrightarrow{-\frac{1}{3}R2 + R3} \left[\begin{array}{ccc|c} 1 & 0 & -\frac{1}{7} & \frac{8}{7} \\ 0 & 1 & -\frac{5}{7} & \frac{19}{7} \\ 0 & 0 & \textcircled{$\frac{96}{21}$} & -\frac{96}{21} \end{array}\right] \xrightarrow{R3 \div \frac{96}{21}}$$

Select a new pivot

$$\left[\begin{array}{ccc|c} 1 & 0 & -\frac{1}{7} & \frac{8}{7} \\ 0 & 1 & -\frac{5}{7} & \frac{19}{7} \\ 0 & 0 & 1 & -1 \end{array}\right] \xrightarrow{\frac{5}{7}R3 + R2} \left[\begin{array}{ccc|c} 1 & 0 & -\frac{1}{7} & \frac{8}{7} \\ 0 & 1 & 0 & 2 \\ 0 & 0 & 1 & -1 \end{array}\right] \xrightarrow{\frac{1}{7}R3 + R1} \left[\begin{array}{ccc|c} 1 & 0 & 0 & 1 \\ 0 & 1 & 0 & 2 \\ 0 & 0 & 1 & -1 \end{array}\right]$$

Pivot

The solution, $(1, 2, -1)$, is now obvious.

Note that the Gauss–Jordan method usually introduces (often ugly) fractions. For this reason, many people are turning to readily available computer programs to carry out the drudgery of this method. However, if you do not have access to a computer, you can often reduce the amount of arithmetic by forcing the pivot elements to be 1 by using elementary row operations other than division. Also, you can combine the pivoting steps. In this example we could have forced the first pivot to be 1 by multiplying row 3 by -1 and adding it to the first row instead of dividing by 3. Let us look at the arithmetic in this simplified version:

$$\left[\begin{array}{rrr|r} 3 & -2 & 1 & -2 \\ 4 & -5 & 3 & -9 \\ 2 & -1 & 5 & -5 \end{array}\right] \xrightarrow[-1R3+R1]{} \left[\begin{array}{rrr|r} \textcircled{1} & -1 & -4 & 3 \\ 4 & -5 & 3 & -9 \\ 2 & -1 & 5 & -5 \end{array}\right] \longrightarrow \left[\begin{array}{rrr|r} 1 & -1 & -4 & 3 \\ 0 & -1 & 19 & -21 \\ 0 & 1 & 13 & -11 \end{array}\right] \longrightarrow$$

Pivot; combine two steps in one: $-4R1 + R2$; $-2R1 + R3$

$$\left[\begin{array}{rrr|r} 1 & -1 & -4 & 3 \\ 0 & \textcircled{1} & -19 & 21 \\ 0 & 1 & 13 & -11 \end{array}\right] \longrightarrow \left[\begin{array}{rrr|r} 1 & 0 & -23 & 24 \\ 0 & 1 & -19 & 21 \\ 0 & 0 & \textcircled{32} & -32 \end{array}\right] \longrightarrow \left[\begin{array}{rrr|r} 1 & 0 & -23 & 24 \\ 0 & 1 & -19 & 21 \\ 0 & 0 & 1 & -1 \end{array}\right] \longrightarrow \left[\begin{array}{rrr|r} 1 & 0 & 0 & 1 \\ 0 & 1 & 0 & 2 \\ 0 & 0 & 1 & -1 \end{array}\right]$$

New pivot is $R2 \div (-1)$

Pivot second row (combine two steps): $R2 + R1$; $-1R2 + R3$

New pivot is $R3/32$

Pivot third row: $19R3 + R2$; $23R3 + R1$

The real beauty of Gauss–Jordan elimination is that it works with all sizes of linear systems. Consider the following example consisting of five equations and five unknowns.

EXAMPLE 6 Solve

$$\begin{cases} x_1 - 3x_3 + x_5 = -3 \\ x_2 + x_4 = -1 \\ x_3 + x_5 = 7 \\ x_1 + x_2 - x_3 + 4x_4 = -8 \\ x_1 + x_2 + x_3 + x_4 + x_5 = 8 \end{cases}$$

Solution

$$\left[\begin{array}{rrrrr|r} \textcircled{1} & 0 & -3 & 0 & 1 & -3 \\ 0 & 1 & 0 & 1 & 0 & -1 \\ 0 & 0 & 1 & 0 & 1 & 7 \\ 1 & 1 & -1 & 4 & 0 & -8 \\ 1 & 1 & 1 & 1 & 1 & 8 \end{array}\right] \xrightarrow[\substack{-R1+R4 \\ -R1+R5}]{} \left[\begin{array}{rrrrr|r} 1 & 0 & -3 & 0 & 1 & -3 \\ 0 & \textcircled{1} & 0 & 1 & 0 & -1 \\ 0 & 0 & 1 & 0 & 1 & 7 \\ 0 & 1 & 2 & 4 & -1 & -5 \\ 0 & 1 & 4 & 1 & 0 & 11 \end{array}\right]$$

$$\xrightarrow[\substack{-R2+R4 \\ -R2+R5}]{} \left[\begin{array}{rrrrr|r} 1 & 0 & -3 & 0 & 1 & -3 \\ 0 & 1 & 0 & 1 & 0 & -1 \\ 0 & 0 & \textcircled{1} & 0 & 1 & 7 \\ 0 & 0 & 2 & 3 & -1 & -4 \\ 0 & 0 & 4 & 0 & 0 & 12 \end{array}\right] \xrightarrow[\substack{3R3+R1 \\ -2R3+R4 \\ -4R3+R5}]{} \left[\begin{array}{rrrrr|r} 1 & 0 & 0 & 0 & 4 & 18 \\ 0 & 1 & 0 & 1 & 0 & -1 \\ 0 & 0 & 1 & 0 & 1 & 7 \\ 0 & 0 & 0 & \textcircled{3} & -3 & -18 \\ 0 & 0 & 0 & 0 & -4 & -16 \end{array}\right]$$

$$\xrightarrow[R4 \div 3]{} \left[\begin{array}{ccccc|c} 1 & 0 & 0 & 0 & 4 & 18 \\ 0 & 1 & 0 & 1 & 0 & -1 \\ 0 & 0 & 1 & 0 & 1 & 7 \\ 0 & 0 & 0 & \textcircled{1} & -1 & -6 \\ 0 & 0 & 0 & 0 & -4 & -16 \end{array}\right] \xrightarrow[(-1)R4 + R2]{} \left[\begin{array}{ccccc|c} 1 & 0 & 0 & 0 & 4 & 18 \\ 0 & 1 & 0 & 0 & 1 & 5 \\ 0 & 0 & 1 & 0 & 1 & 7 \\ 0 & 0 & 0 & 1 & -1 & -6 \\ 0 & 0 & 0 & 0 & \textcircled{-4} & -16 \end{array}\right]$$

$$\xrightarrow[R5 \div (-4)]{} \left[\begin{array}{ccccc|c} 1 & 0 & 0 & 0 & 4 & 18 \\ 0 & 1 & 0 & 0 & 1 & 5 \\ 0 & 0 & 1 & 0 & 1 & 7 \\ 0 & 0 & 0 & 1 & -1 & -6 \\ 0 & 0 & 0 & 0 & 1 & 4 \end{array}\right] \xrightarrow[\substack{R5 + R4 \\ -R5 + R3 \\ -R5 + R2 \\ -4R5 + R1}]{} \left[\begin{array}{ccccc|c} 1 & 0 & 0 & 0 & 0 & 2 \\ 0 & 1 & 0 & 0 & 0 & 1 \\ 0 & 0 & 1 & 0 & 0 & 3 \\ 0 & 0 & 0 & 1 & 0 & -2 \\ 0 & 0 & 0 & 0 & 1 & 4 \end{array}\right]$$

The solution is $(x_1, x_2, x_3, x_4, x_5) = (2, 1, 3, -2, 4)$. ■

Example 7 is an example of a system of three equations with two unknowns that has a solution. Example 8 shows what Gauss–Jordan elimination looks like when there are three equations with two unknowns and no solution. (Remember, if there is no solution, the system is inconsistent.)

EXAMPLE 7 Solve $\begin{cases} 2x + \ \ y = 3 \\ 3x - \ \ y = 2. \\ 4x + 3y = 7 \end{cases}$

Solution

$$\left[\begin{array}{cc|c} 2 & 1 & 3 \\ 3 & -1 & 2 \\ 4 & 3 & 7 \end{array}\right] \xrightarrow{-1R1 + R2} \left[\begin{array}{cc|c} 2 & 1 & 3 \\ 1 & -2 & -1 \\ 4 & 3 & 7 \end{array}\right] \xrightarrow[R_1 \leftrightarrow R_2]{\text{Interchange rows 1 and 2}} \left[\begin{array}{cc|c} 1 & -2 & -1 \\ 2 & 1 & 3 \\ 4 & 3 & 7 \end{array}\right]$$

$$\xrightarrow[\substack{-2R1 + R2 \\ -4R1 + R3}]{} \left[\begin{array}{cc|c} \textcircled{1} & -2 & -1 \\ 0 & 5 & 5 \\ 0 & 11 & 11 \end{array}\right] \xrightarrow[R2 \div 5]{} \left[\begin{array}{cc|c} 1 & -2 & -1 \\ 0 & \textcircled{1} & 1 \\ 0 & 11 & 11 \end{array}\right]$$

$$\xrightarrow[\substack{2R2 + R1 \\ -11R2 + R3}]{} \left[\begin{array}{cc|c} 1 & 0 & 1 \\ 0 & 1 & 1 \\ 0 & 0 & 0 \end{array}\right]$$

This final matrix is equivalent to the system: $\begin{cases} x = 1 \\ y = 1. \\ 0 = 0 \end{cases}$

The solution is $(1, 1)$.

Check: $2x + \ \ y = 2(1) + 1 = 3$ √

$3x - \ \ y = 3(1) - 1 = 2$ √

$4x + 3y = 4(1) + 3(1) = 7$ √

■

EXAMPLE 8 Solve $\begin{cases} 2x + y = 3 \\ 3x - y = 2. \\ x - 2y = 4 \end{cases}$

Solution

$$\left[\begin{array}{rr|r} 2 & 1 & 3 \\ 3 & -1 & 2 \\ 1 & -2 & 4 \end{array}\right] \xrightarrow{R1 \leftrightarrow R3} \left[\begin{array}{rr|r} 1 & -2 & 4 \\ 3 & -1 & 2 \\ 2 & 1 & 3 \end{array}\right] \xrightarrow[-2R1 + R3]{-3R1 + R2} \left[\begin{array}{rr|r} 1 & -2 & 4 \\ 0 & 5 & -10 \\ 0 & 5 & -5 \end{array}\right]$$

$$\xrightarrow[R2 \div 5]{} \left[\begin{array}{rr|r} 1 & -2 & 4 \\ 0 & 1 & -2 \\ 0 & 5 & -5 \end{array}\right] \xrightarrow[-5R2 + R3]{2R2 + R1} \left[\begin{array}{rr|r} 1 & 0 & 0 \\ 0 & 1 & -2 \\ 0 & 0 & 5 \end{array}\right]$$

This is equivalent to $\begin{cases} x = 0 \\ y = -2. \\ 0 = 5 \end{cases}$ But, since $0 \neq 5$ regardless of the values of x and y, this is an inconsistent system. ■

A dependent system has infinitely many solutions, but just because a system has infinitely many solutions does not mean that it is satisfied by any set of values. For example, $x + y = 5$ has infinitely many solutions, but not just *any* replacements for x and y will satisfy that equation. In fact, we might say if we choose any value, say, t, for x, then y is fixed to be $5 - t$. In the last chapter we called t a *parameter*, and we could say that any ordered pair of the form $(t, 5 - t)$ will satisfy the equation $x + y = 5$. Example 9 illustrates a dependent system and the use of a parameter in specifying the solution.

EXAMPLE 9 Solve $\begin{cases} 3x - 2y + 4z = 8 \\ x - y - 2z = 5 \\ 4x - 3y + 2z = 13 \end{cases}$.

Solution

$$\left[\begin{array}{rrr|r} 3 & -2 & 4 & 8 \\ 1 & -1 & -2 & 5 \\ 4 & -3 & 2 & 13 \end{array}\right] \xrightarrow[R_1 \leftrightarrow R2]{} \left[\begin{array}{rrr|r} 1 & -1 & -2 & 5 \\ 3 & -2 & 4 & 8 \\ 4 & -3 & 2 & 13 \end{array}\right]$$

$$\xrightarrow{\substack{-3R1 + R2 \\ -4R1 + R3}} \left[\begin{array}{rrr|r} 1 & -1 & -2 & 5 \\ 0 & 1 & 10 & -7 \\ 0 & 1 & 10 & -7 \end{array}\right] \xrightarrow{\substack{R2 + R1 \\ -R2 + R3}} \left[\begin{array}{rrr|r} 1 & 0 & 8 & -2 \\ 0 & 1 & 10 & -7 \\ 0 & 0 & 0 & 0 \end{array}\right]$$

This is equivalent to the system $\begin{cases} x + 8z = -2 \\ y + 10z = -7. \\ 0 = 0 \end{cases}$ If we pick any value for z, say, t, then x and y are, in turn, determined: $x = -2 - 8t$ and $y = -7 - 10t$. This is called a *parametric solution*: $(-2 - 8t, -7 - 10t, t)$. Such a solution means that there are infinitely many ordered triplets satisfying the system (a dependent system), but just *not any* ordered triplet. Let us list a few of the solutions:

Parametric solution: $(-2-8t, -7-10t, t)$

if $t = 0$: $(-2-8\cdot\mathbf{0}, -7-10\cdot\mathbf{0}, \mathbf{0}) = (-2,-7,0)$

Check $(-2,-7,0)$: $3(-2)-2(-7)+4(0)=8$
$(-2)-(-7)-2(0)=5$
$4(-2)-3(-7)+2(0)=13$

if $t = 1$: $(-2-8(\mathbf{1}), -7-10(\mathbf{1}), \mathbf{1}) = (-10,-17,1)$

Check $(-10,-17,1)$: $3(-10)-2(-17)+4(1)=8$
$(-10)-(-17)-2(1)=5$
$4(-10)-3(-17)+2(1)=13$

if $t = \frac{1}{2}$: $(-2-8(\tfrac{1}{2}), -7-10(\tfrac{1}{2}), \tfrac{1}{2}) = (-6,-12,\tfrac{1}{2})$

Check $(-6,-12,\frac{1}{2})$: $3(-6)-2(-12)+4(\tfrac{1}{2})=8$
$(-6)-(-12)-2(\tfrac{1}{2})=5$
$4(-6)-3(-12)+2(\tfrac{1}{2})=13$

Any value for t will give an ordered triplet which satisfies all the equations in the given system. For this reason we call the ordered triplet $(-2-8t, -7-10t, t)$ the **parametric solution**. ■

You may also need a parameter when there are more unknowns than equations, as illustrated by Example 10.

EXAMPLE 10 Solve $\begin{cases} 5x - 3y + z = -3 \\ 2x + 5z = 14 \end{cases}$.

Solution

$$\begin{bmatrix} 5 & -3 & 1 & | & -3 \\ 2 & 0 & 5 & | & 14 \end{bmatrix} \xrightarrow{-2R2+R1} \begin{bmatrix} 1 & -3 & -9 & | & -31 \\ 2 & 0 & 5 & | & 14 \end{bmatrix}$$

$$\xrightarrow{-2R1+R2} \begin{bmatrix} 1 & -3 & -9 & | & -31 \\ 0 & 6 & 23 & | & 76 \end{bmatrix} \xrightarrow[R2 \div 6]{} \begin{bmatrix} 1 & -3 & -9 & | & -31 \\ 0 & 1 & \frac{23}{6} & | & \frac{38}{3} \end{bmatrix}$$

$$\xrightarrow[3R2+R1]{} \begin{bmatrix} 1 & 0 & \frac{5}{2} & | & 7 \\ 0 & 1 & \frac{23}{6} & | & \frac{38}{3} \end{bmatrix}$$

This is equivalent to the system $\begin{cases} x + \frac{5}{2}z = 7 \\ y + \frac{23}{6}z = \frac{38}{3} \end{cases}$.

Here, again, we can choose any z, but then x and y are determined. We could call our choice t (as in the previous examples), but common practice is to let $z = kt$, where k is the least common multiple of the denominators of the fractions. By making this choice, many fractions can be avoided (for those who are not particularly fond of fractions). In this example, we can let $z = 6t$, then

$$x = 7 - \frac{5}{2}(6t) = 7 - 15t$$

$$y = \frac{38}{3} - \frac{23}{6}(6t) = \frac{38}{3} - 23t$$

This gives a parametric solution: $(7-15t, \frac{38}{3}-23t, 6t)$. Remember, *every* different choice of t gives another solution. ■

EXAMPLE 11 Solve $\begin{cases} x + y + 2z + 3w = 5 \\ 4z + 2w = 32 \\ z + 6w = 3 \\ 2z + 12w = 6 \end{cases}$.

Solution Write the augmented matrix:

$$\left[\begin{array}{cccc|c} \textcircled{1} & 1 & 2 & 3 & 5 \\ 0 & 0 & 4 & 2 & -32 \\ 0 & 0 & \boxed{1} & 6 & 3 \\ 0 & 0 & 2 & 12 & 6 \end{array}\right]$$

The first pivot operation (first pivot is circled) is completed. When we move to the second row there is a zero in the pivot position that cannot be eliminated, so we move to the third row. The second pivot is boxed. Carry out the pivot operation.

$$\left[\begin{array}{cccc|c} 1 & 1 & 2 & 3 & 5 \\ 0 & 0 & 4 & 2 & -32 \\ 0 & 0 & \boxed{1} & 6 & 3 \\ 0 & 0 & 2 & 12 & 6 \end{array}\right] \longrightarrow \left[\begin{array}{cccc|c} 1 & 1 & 0 & -9 & -1 \\ 0 & 0 & 0 & -22 & -44 \\ 0 & 0 & 1 & 6 & 3 \\ 0 & 0 & 0 & 0 & 0 \end{array}\right]$$

The final row has all zeros, but the zero in the fourth row, fourth column, can be eliminated by interchanging rows 2 and 4:

$$\longrightarrow \left[\begin{array}{cccc|c} 1 & 1 & 0 & -9 & -1 \\ 0 & 0 & 0 & -22 & -44 \\ 0 & 0 & 1 & 6 & 3 \\ 0 & 0 & 0 & 0 & 0 \end{array}\right] \longrightarrow \left[\begin{array}{cccc|c} 1 & 1 & 0 & -9 & -1 \\ 0 & 0 & 0 & 0 & 0 \\ 0 & 0 & 1 & 6 & 3 \\ 0 & 0 & 0 & -22 & -44 \end{array}\right]$$

$$\longrightarrow \left[\begin{array}{cccc|c} 1 & 1 & 0 & -9 & -1 \\ 0 & 0 & 0 & 0 & 0 \\ 0 & 0 & 1 & 6 & 3 \\ 0 & 0 & 0 & 1 & 2 \end{array}\right] \longrightarrow \left[\begin{array}{cccc|c} 1 & 1 & 0 & 0 & 17 \\ 0 & 0 & 0 & 0 & 0 \\ 0 & 0 & 1 & 0 & -9 \\ 0 & 0 & 0 & 1 & 2 \end{array}\right]$$

This is equivalent to the system

$$\begin{cases} x + y = 17 \\ 0 = 0 \\ z = -9 \\ w = 2 \end{cases}$$

We see that one parameter is needed: $x = 17 - y$ so select $y = t$ to give the following solution: $(x, y, z, w) = (17 - t, t, -9, 2)$. ■

It is possible that you will need more than one parameter for a problem. Suppose you have 2 equations with 4 unknowns. Then you will need at least $4 - 2 = 2$ parameters, as illustrated by Example 12.

EXAMPLE 12 Solve $\begin{cases} -w + 2x - 3y + z = 3 \\ 2w - 3x + y - 2z = 4 \end{cases}$

Solution $\left[\begin{array}{rrrr|r} -1 & 2 & -3 & 1 & 3 \\ 2 & -3 & 1 & -2 & 4 \end{array}\right] \longrightarrow \left[\begin{array}{rrrr|r} 1 & -2 & 3 & -1 & -3 \\ 2 & -3 & 1 & -2 & 4 \end{array}\right] \longrightarrow$

$\left[\begin{array}{rrrr|r} 1 & -2 & 3 & -1 & -3 \\ 0 & 1 & -5 & 0 & 10 \end{array}\right] \longrightarrow \left[\begin{array}{rrrr|r} 1 & 0 & -7 & -1 & 17 \\ 0 & 1 & -5 & 0 & 10 \end{array}\right] \longrightarrow$

This is equivalent to the system

$$\begin{cases} w - 7y - z = 17 & \text{or} \quad w = 17 + 7y + z \\ x - 5y \quad\;\; = 10 & \text{or} \quad x = 10 + 5y \end{cases}$$

You can choose any values for both y and z, and then w and x will be determined; for example, if we use s and t as the parameters so that $y = s$ and $z = t$, then

$$w = 17 + 7s + t$$
$$x = 10 + 5s$$

and the solution is $(17 + 7s + t, 10 + 5s, s, t)$. For example,

if $s = 0$ and $t = 0$, then $(17, 10, 0, 0)$ is a solution;
if $s = 0$ and $t = 1$, then $(18, 10, 0, 1)$ is a solution;
if $s = 1$ and $t = 0$, then $(24, 15, 1, 0)$ is a solution;
if $s = 1$ and $t = 1$, then $(25, 15, 1, 1)$ is a solution; and so on.

Each of these solutions can be checked in the original system to verify that each, in turn, satisfies the system. ■

EXAMPLE 13 A rancher has to mix three types of feed for her cattle. The analysis shown in the table gives the amounts per bag (100 lb) of grain.

Grain	Protein	Carbohydrates	Sodium
A	7 lb	88 lb	1 lb
B	6 lb	90 lb	1 lb
C	10 lb	70 lb	2 lb

How many bags of each type of grain should the rancher mix to provide 71 pounds of protein, 854 pounds of carbohydrates, and 12 pounds of sodium?

Solution Let a, b, and c be the number of bags of grains A, B, and C, respectively, needed for the mixture. Then:

Grain	Protein	Carbohydrates	Sodium
A	$7a$	$88a$	a
B	$6b$	$90b$	b
C	$10c$	$70c$	$2c$
Total	71	854	12

Thus

$$\begin{cases} 7a + 6b + 10c = 71 \\ 88a + 90b + 70c = 854 \\ a + b + 2c = 12 \end{cases}$$

$$\left[\begin{array}{ccc|c} 7 & 6 & 10 & 71 \\ 88 & 90 & 70 & 854 \\ 1 & 1 & 2 & 12 \end{array}\right] \longrightarrow \left[\begin{array}{ccc|c} 1 & 1 & 2 & 12 \\ 88 & 90 & 70 & 854 \\ 7 & 6 & 10 & 71 \end{array}\right]$$

$$\longrightarrow \left[\begin{array}{ccc|c} 1 & 1 & 2 & 12 \\ 0 & 2 & -106 & -202 \\ 0 & -1 & -4 & -13 \end{array}\right] \longrightarrow \left[\begin{array}{ccc|c} 1 & 1 & 2 & 12 \\ 0 & 1 & -53 & -101 \\ 0 & -1 & -4 & -13 \end{array}\right]$$

$$\longrightarrow \left[\begin{array}{ccc|c} 1 & 0 & 55 & 113 \\ 0 & 1 & -53 & -101 \\ 0 & 0 & -57 & -114 \end{array}\right] \longrightarrow \left[\begin{array}{ccc|c} 1 & 0 & 55 & 113 \\ 0 & 1 & -53 & -101 \\ 0 & 0 & 1 & 2 \end{array}\right]$$

$$\longrightarrow \left[\begin{array}{ccc|c} 1 & 0 & 0 & 3 \\ 0 & 1 & 0 & 5 \\ 0 & 0 & 1 & 2 \end{array}\right]$$

The rancher should mix three bags of grain A, five bags of grain B, and two bags of grain C. ■

3.3 Problem Set

Write the systems in Problems 1–2 in augmented matrix form.

1. **a.** $\begin{cases} 4x + 5y = -16 \\ 3x + 2y = 5 \end{cases}$ **b.** $\begin{cases} x + y + z = 4 \\ 3x + 2y + z = 7 \\ x - 3y + 2z = 0 \end{cases}$

c. $\begin{cases} x_1 + 3x_2 + x_3 + x_4 = 3 \\ x_1 - 2x_3 + 2x_4 = 0 \\ x_3 + 5x_4 = -14 \\ x_2 - 3x_3 - x_4 = 2 \end{cases}$

2. **a.** $\begin{cases} 2x + y = 0 \\ 3x - 2y = -7 \\ x - 3y = -1 \end{cases}$ **b.** $\begin{cases} 2x - y + z = 2 \\ 2x - 4z = 32 \end{cases}$

c. $\begin{cases} x_1 - 2x_2 - x_3 - x_4 = 4 \\ 5x_1 + 2x_2 - x_3 + 2x_4 = 23 \\ 3x_1 + 4x_2 - 3x_3 + x_4 = 2 \\ 2x_1 - 2x_2 + x_3 - x_4 = 10 \end{cases}$

Write a system of equations (use $x_1, x_2, x_3, \ldots, x_n$ for the variables) that has the given augmented matrix in Problems 3–4.

3. **a.** $\left[\begin{array}{ccc|c} 2 & 1 & 4 & 3 \\ 6 & 2 & -1 & -4 \\ -3 & -1 & 0 & 1 \end{array}\right]$ **b.** $\left[\begin{array}{ccc|c} 1 & 0 & 0 & 5 \\ 0 & 1 & 0 & -3 \\ 0 & 0 & 1 & 4 \end{array}\right]$

c. $\left[\begin{array}{ccc|c} 1 & 0 & 0 & 3 \\ 0 & 1 & 0 & 2 \\ 0 & 0 & 1 & -8 \\ 0 & 0 & 0 & 1 \end{array}\right]$

4. **a.** $\left[\begin{array}{ccc|c} 1 & 0 & 0 & 5 \\ 0 & 1 & 2 & 4 \end{array}\right]$

b. $\left[\begin{array}{cccc|c} 1 & 0 & 0 & 0 & 3 \\ 0 & 1 & 0 & 0 & -6 \\ 0 & 0 & 1 & 5 & 4 \end{array}\right]$

c. $\left[\begin{array}{cccc|c} 1 & 0 & 0 & 0 & 6 \\ 0 & 1 & 0 & 0 & -3 \\ 0 & 0 & 1 & 0 & 5 \\ 0 & 0 & 0 & 1 & 4 \end{array}\right]$

5. Let $A = \left[\begin{array}{ccc|c} -2 & 2 & 4 & 8 \\ -1 & 3 & -2 & 3 \\ 1 & 5 & 3 & -2 \end{array}\right]$

Work each part separately. Use this matrix for each part and not the results from another part.

a. Interchange two rows to obtain a 1 in the first position of the first row.
b. Multiply each member of row 2 by 2.
c. Multiply each member of row 1 by $\frac{1}{2}$.
d. Multiply row 3 by 2 and add the result to row 1.
e. Multiply row 3 by -1 and add the result to row 2.

6. Let $B = \left[\begin{array}{ccc|c} 2 & -1 & 4 & 8 \\ 3 & -2 & 1 & 4 \\ -5 & 1 & 2 & -7 \end{array}\right]$

Work each part separately. Use this matrix for each part and not the results from another part.

a. Interchange two rows to obtain a 1 in the second position of the second row.
b. Multiply each member of row 1 by $\frac{1}{2}$.
c. Multiply each member of row 3 by -2.
d. Multiply row 1 by 3 and add the result to row 3.
e. Multiply row 2 by 2 and add the result to row 3.

Solve the systems in Problems 7–47 by using Gauss–Jordan elimination.

7. $\begin{cases} x + y = 3 \\ 2x + 3y = 8 \end{cases}$

8. $\begin{cases} x - y = 2 \\ 3x + 2y = 11 \end{cases}$

9. $\begin{cases} 3x - 4y = -2 \\ 2x + 5y = -32 \end{cases}$

10. $\begin{cases} 5x + 2y = 4 \\ 3x + y = -1 \end{cases}$

11. $\begin{cases} 3x - 2y = 10 \\ 6x - 4y = 20 \end{cases}$

12. $\begin{cases} 4x + 5y = 9 \\ 7x + 7y = 14 \end{cases}$

13. $\begin{cases} 2x + y = 0 \\ 3x - 2y = -7 \\ x - 3y = -1 \end{cases}$

14. $\begin{cases} 3x - 5y = 9 \\ 2x - 4y = 8 \\ x + 3y = -11 \end{cases}$

15. $\begin{cases} 2x - 3y = 5 \\ 9x - 7y = 4 \\ x + y = 4 \end{cases}$

16. $\begin{cases} 3x - 2y = 13 \\ 4x + 5y = 2 \\ -2x - 3y = 0 \\ x + y = 1 \end{cases}$

17. $\begin{cases} 4x + 3y = -11 \\ -x - 5y = 24 \\ 2x + 3y = -17 \\ 5x - 2y = 5 \end{cases}$

18. $\begin{cases} 3x - 2y = 6 \\ 5x + 4y = 1 \\ -2x + 3y = 1 \\ x - 4y = -3 \end{cases}$

19. $\begin{cases} x + y + z = 6 \\ 2x - y + z = 3 \\ x - 2y - 3z = -12 \end{cases}$

20. $\begin{cases} 2x - y + z = 3 \\ x - 3y + 2z = 7 \\ x - y - z = -1 \end{cases}$

21. $\begin{cases} x + y + z = 4 \\ x + 3y + 2z = 4 \\ x - 2y + z = 7 \end{cases}$

22. $\begin{cases} x + 2z = 7 \\ x + y = 11 \\ -2y + 9z = -3 \end{cases}$

23. $\begin{cases} x + 2z = 13 \\ 2x + y = 8 \\ -2y + 9z = 41 \end{cases}$

24. $\begin{cases} 6x + y + 20z = 27 \\ x - y = 0 \\ y + z = 2 \end{cases}$

25. $\begin{cases} 4x + y + 2z = 7 \\ x + 2y = 0 \\ 3x - y - z = 7 \end{cases}$

26. $\begin{cases} 2x - y + 4z = 13 \\ 3x + 6y = 0 \\ 2y - 3z = 3 + 3x \end{cases}$

27. $\begin{cases} 3x - 2y + z = 5 \\ 5x - 3y = 24 \\ 2y + z = -5 \end{cases}$

28. $\begin{cases} x + y + z = 4 \\ x - 2y - z = 1 \\ 3x + y - 2z = -1 \end{cases}$

29. $\begin{cases} x + y + z = 4 \\ 3x + 2y + z = 7 \\ x - 3y + 2z = 2 \end{cases}$

30. $\begin{cases} x + y + z = 6 \\ x - 2y - z = 2 \\ 3x - y - 2z = 1 \end{cases}$

31. $\begin{cases} 2x - y + 3z = 7 \\ -x + 3y - 2z = -13 \\ 3x - 4y + 5z = 20 \end{cases}$

32. $\begin{cases} 3x - 2y + z = 13 \\ x - 5y + 2z = 24 \\ 2x + 3y - z = -11 \end{cases}$

33. $\begin{cases} 6x - y - 2z = 7 \\ 5x - 4y - 5z = 5 \\ x + 3y + 3z = 4 \end{cases}$

34. $\begin{cases} 2x - y + z = 2 \\ 2x - 4z = 32 \end{cases}$

35. $\begin{cases} x - 2y + z = -3 \\ 3y - 7z = -6 \end{cases}$

36. $\begin{cases} 2x - 3y = 1 \\ 4y + 6z = -8 \end{cases}$

37. $\begin{cases} w + x - 2y + 3z = 5 \\ 3w - 2x + y - 5z = 8 \end{cases}$

38. $\begin{cases} -2w + 3x - y - z = 6 \\ 3w - 2x + y + 2z = 5 \end{cases}$

39. $\begin{cases} 2w - 3x + z = 5 \\ 3x + 4y - z = 6 \end{cases}$

40. $\begin{cases} x_1 + 2x_2 + x_4 = 3 \\ 3x_1 + x_2 - 2x_4 = -1 \\ x_1 + 3x_3 - x_4 = 2 \end{cases}$

41. $\begin{cases} 6x_1 - 3x_2 + x_4 = -12 \\ 2x_2 + 4x_3 - x_4 = 1 \\ 3x_1 + 2x_2 + 2x_3 = -3 \end{cases}$

42. $\begin{cases} 3x_2 - x_3 - 4x_4 = 1 \\ 2x_1 + x_3 + 3x_4 = -1 \\ 5x_1 - 3x_2 - 5x_3 = -4 \end{cases}$

43. $\begin{cases} x_1 + 3x_2 + x_3 + x_4 = -1 \\ x_1 - 2x_3 + 2x_4 = -4 \\ x_3 + 5x_4 = -14 \\ x_2 - 3x_3 - x_4 = -1 \end{cases}$

44. $\begin{cases} x_1 - 2x_2 - x_3 - x_4 = -1 \\ 5x_1 + 2x_2 - x_3 + 2x_4 = 10 \\ 3x_1 + 4x_2 - 3x_3 + x_4 = -13 \\ 2x_1 - 2x_2 + x_3 - x_4 = 9 \end{cases}$

45. $\begin{cases} x_1 - 3x_2 + x_3 - x_4 = 3 \\ -x_1 + 2x_2 - 2x_3 + x_4 = -3 \\ 3x_1 - 5x_2 - 6x_3 + 4x_4 = 10 \\ 2x_1 + 3x_2 + 4x_3 + 2x_4 = 7 \end{cases}$

46. $\begin{cases} 2x_1 + 3x_2 + 2x_3 - x_4 = 10 \\ x_1 + 2x_2 - 4x_3 + 3x_4 = -3 \\ x_1 + 5x_2 - 2x_3 + 4x_4 = 1 \\ 3x_1 + 3x_2 + 2x_3 - 3x_4 = 14 \\ 5x_1 - 6x_2 - x_3 + 4x_4 = -1 \end{cases}$

47. $\begin{cases} x_1 - x_2 + 2x_3 - x_4 + 2x_5 = 2 \\ 2x_1 + x_2 + x_3 + 2x_4 - 2x_5 = 0 \\ -x_1 - x_2 - x_3 - 3x_4 - x_5 = 3 \\ 3x_1 + 2x_2 - x_3 - x_4 + x_5 = 7 \end{cases}$

APPLICATIONS

48. A farmer must mix three types of cattle feed. The table gives the amounts per bag (100 lb) of grain. How many bags of each type of grain should be mixed to provide 58 pounds of protein, 655 pounds of carbohydrates, and 11 pounds of sodium?

Grain	Protein	Carbohydrates	Sodium
A	9 lb	75 lb	2 lb
B	5 lb	90 lb	1 lb
C	8 lb	80 lb	1 lb

49. In order to control a certain type of disease, it is necessary to use 23 liters of pesticide A and 34 liters of pesticide B. The dealer can order commercial spray I, each container of which holds 5 liters of pesticide A and 2 liters of pesticide B, and commercial spray II, each container of which holds 2 liters of pesticide A and 7 liters of pesticide B. How many containers of each type of commercial spray should be used to attain exactly the right proportion of pesticides needed?

50. In order to manufacture a certain alloy it is necessary to use 33 kg of metal A and 56 kg of metal B. The manufacturer mixes alloy I, each bar of which contains 3 kg of metal A and 5 kg of metal B, with alloy II, each bar of which contains 4 kg of metal A and 7 kg of metal B, to make the desired alloy. How much of the two alloys should be used in order to produce the alloy desired?

51. A candy maker mixes chocolate, milk, and coconut to produce three kinds of candy (I, II, and III) with the following proportions:

Candy I: 7 lb chocolate, 5 gal milk, and 1 oz coconut
Candy II: 3 lb chocolate, 2 gal milk, and 2 oz coconut
Candy III: 4 lb chocolate, 3 gal milk, and 3 oz coconut

If 67 pounds of chocolate, 48 gallons of milk, and 32 ounces of coconut are available, how many units of each kind of candy can be produced?

52. The total number of registered Democrats, Republicans, and Independents in a certain community is 100,000. Voter turnout in a recent election was tabulated as follows:

50% of the Democrats voted
60% of the Republicans voted
70% of the Independents voted
55,200 votes were cast (assume that there were no write-in votes; everyone voted Democratic, Republican, or Independent)

The ratio of registered Democrats to registered Independents is 9 to 1. How many registered Democrats, Republicans, and Independents are there in the community?

53. Baskin-Robbins stores offer three sizes of ice cream cones:

Small scoop: 2.5 oz for $.70
Medium scoop: 4 oz for $.95
Large scoop: 6 oz for $1.40

Are these sizes consistently priced? If not, which size is the best bargain? [*Hint:* You must consider not only the price per ounce for ice cream, but also the price of the cone.]

3.4 Inverse Matrices

In the set of real numbers there are two properties that we now extend to matrices.

Identity Elements

ADDITION:	$a + 0 = 0 + a = a$	0 is the identity for addition in the set of real numbers.
MULTIPLICATION:	$a \cdot 1 = 1 \cdot a = a$	1 is the identity for multiplication in the set of real numbers.

Inverse Elements

ADDITION:	$a + (-a) = (-a) + a = 0$	For each number a there exists an opposite (*additive inverse*) so that the sum of a and its opposite is 0.
MULTIPLICATION:	$a \cdot a^{-1} = a^{-1} \cdot a = 1$	For each nonzero number a there exists a reciprocal (*multiplicative inverse*) so that the product of a and its reciprocal is 1.

For matrices, the identity element for addition is the zero matrix $\mathbf{0}$ so that $A + \mathbf{0} = \mathbf{0} + A = A$ for conformable matrices $\mathbf{0}$ and A. For example,

$$\begin{bmatrix} 1 & 2 & 3 \\ 4 & 5 & 6 \\ 7 & 8 & 9 \end{bmatrix} + \begin{bmatrix} 0 & 0 & 0 \\ 0 & 0 & 0 \\ 0 & 0 & 0 \end{bmatrix} = \begin{bmatrix} 1 & 2 & 3 \\ 4 & 5 & 6 \\ 7 & 8 & 9 \end{bmatrix}$$

Identity matrix for 3 × 3 matrices and addition

Identical

For matrix multiplication, the square matrix I of order $n \times n$ consisting of 1s on the main diagonal and zeros elsewhere is called the **identity matrix** of order n, since $IA = AI = A$ for every conformable matrix A. For example,

$$\begin{bmatrix} 1 & 2 & 3 \\ 4 & 5 & 6 \\ 7 & 8 & 9 \end{bmatrix} \begin{bmatrix} 1 & 0 & 0 \\ 0 & 1 & 0 \\ 0 & 0 & 1 \end{bmatrix} = \begin{bmatrix} 1 & 2 & 3 \\ 4 & 5 & 6 \\ 7 & 8 & 9 \end{bmatrix}$$

Identity matrix for 3 × 3 matrices and multiplication

Identical

The inverse matrix for addition is simply the matrix whose entries are opposites of the corresponding entries of the original matrix. However, it is the inverse for multiplication that is of particular interest to us.

Inverse of a Matrix

If A is a square matrix and if there exists a matrix A^{-1} such that

$$A^{-1}A = AA^{-1} = I$$

then A^{-1} is called the **inverse of** A for multiplication.

Usually, in the context of matrices, when we talk simply of the inverse of A we mean the inverse of A for multiplication.

EXAMPLE 1 Verify that the inverse of $A = \begin{bmatrix} 2 & 1 \\ 3 & 2 \end{bmatrix}$ is $B = \begin{bmatrix} 2 & -1 \\ -3 & 2 \end{bmatrix}$.

Solution

$$AB = \begin{bmatrix} 2 & 1 \\ 3 & 2 \end{bmatrix}\begin{bmatrix} 2 & -1 \\ -3 & 2 \end{bmatrix} = \begin{bmatrix} 4-3 & -2+2 \\ 6-6 & -3+4 \end{bmatrix} = \begin{bmatrix} 1 & 0 \\ 0 & 1 \end{bmatrix} = I$$

$$BA = \begin{bmatrix} 2 & -1 \\ -3 & 2 \end{bmatrix}\begin{bmatrix} 2 & 1 \\ 3 & 2 \end{bmatrix} = \begin{bmatrix} 4-3 & 2-2 \\ -6+6 & -3+4 \end{bmatrix} = \begin{bmatrix} 1 & 0 \\ 0 & 1 \end{bmatrix} = I$$

Thus $B = A^{-1}$. ■

EXAMPLE 2 Show that A and B are inverses of each other, where

$$A = \begin{bmatrix} 0 & 1 & 2 \\ -1 & 1 & 2 \\ 1 & -2 & -5 \end{bmatrix} \quad \text{and} \quad B = \begin{bmatrix} 1 & -1 & 0 \\ 3 & 2 & 2 \\ -1 & -1 & -1 \end{bmatrix}$$

Solution

$$AB = \begin{bmatrix} 0 & 1 & 2 \\ -1 & 1 & 2 \\ 1 & -2 & -5 \end{bmatrix}\begin{bmatrix} 1 & -1 & 0 \\ 3 & 2 & 2 \\ -1 & -1 & -1 \end{bmatrix}$$

$$= \begin{bmatrix} 0+3-2 & 0+2-2 & 0+2-2 \\ -1+3-2 & 1+2-2 & 0+2-2 \\ 1-6+5 & -1-4+5 & 0-4+5 \end{bmatrix} = \begin{bmatrix} 1 & 0 & 0 \\ 0 & 1 & 0 \\ 0 & 0 & 1 \end{bmatrix} = I$$

$$BA = \begin{bmatrix} 1 & -1 & 0 \\ 3 & 2 & 2 \\ -1 & -1 & -1 \end{bmatrix}\begin{bmatrix} 0 & 1 & 2 \\ -1 & 1 & 2 \\ 1 & -2 & -5 \end{bmatrix}$$

$$= \begin{bmatrix} 0+1+0 & 1-1+0 & 2-2+0 \\ 0-2+2 & 3+2-4 & 6+4-10 \\ 0+1-1 & -1-1+2 & -2-2+5 \end{bmatrix} = \begin{bmatrix} 1 & 0 & 0 \\ 0 & 1 & 0 \\ 0 & 0 & 1 \end{bmatrix} = I$$

Since $AB = I = BA$, then $B = A^{-1}$. ■

If a given matrix has an inverse, we say that it is **nonsingular**. The unanswered question, however, is how to *find* an inverse matrix. Consider the following two examples.

EXAMPLE 3 Find the inverse of $A = \begin{bmatrix} 1 & 2 \\ 1 & 4 \end{bmatrix}$.

Solution Find a matrix B (if it exists) so that $AB = I$; since we do not know B, let its entries be variables. That is, let

$$B = \begin{bmatrix} x_1 & x_2 \\ y_1 & y_2 \end{bmatrix}$$

Then

$$\begin{aligned} AB &= \begin{bmatrix} 1 & 2 \\ 1 & 4 \end{bmatrix} \begin{bmatrix} x_1 & x_2 \\ y_1 & y_2 \end{bmatrix} \\ &= \begin{bmatrix} x_1 + 2y_1 & x_2 + 2y_2 \\ x_1 + 4y_1 & x_2 + 4y_2 \end{bmatrix} \\ &= \begin{bmatrix} 1 & 0 \\ 0 & 1 \end{bmatrix} \end{aligned}$$

By the definition of the equality of matrices, we see that

$$\begin{cases} x_1 + 2y_1 = 1 \\ x_1 + 4y_1 = 0 \end{cases} \qquad \begin{cases} x_2 + 2y_2 = 0 \\ x_2 + 4y_2 = 1 \end{cases}$$

Solve these systems by using Gauss–Jordan elimination:

$$\left[\begin{array}{cc|c} 1 & 2 & 1 \\ 1 & 4 & 0 \end{array}\right] \longrightarrow \left[\begin{array}{cc|c} 1 & 2 & 1 \\ 0 & 2 & -1 \end{array}\right] \qquad \left[\begin{array}{cc|c} 1 & 2 & 0 \\ 1 & 4 & 1 \end{array}\right] \longrightarrow \left[\begin{array}{cc|c} 1 & 2 & 0 \\ 0 & 2 & 1 \end{array}\right]$$

$$\longrightarrow \left[\begin{array}{cc|c} 1 & 2 & 1 \\ 0 & 1 & -\frac{1}{2} \end{array}\right] \qquad \longrightarrow \left[\begin{array}{cc|c} 1 & 2 & 0 \\ 0 & 1 & \frac{1}{2} \end{array}\right]$$

$$\longrightarrow \left[\begin{array}{cc|c} 1 & 0 & 2 \\ 0 & 1 & -\frac{1}{2} \end{array}\right] \qquad \longrightarrow \left[\begin{array}{cc|c} 1 & 0 & -1 \\ 0 & 1 & \frac{1}{2} \end{array}\right]$$

$$\begin{cases} x_1 = 2 \\ y_1 = -\frac{1}{2} \end{cases} \qquad \begin{cases} x_2 = -1 \\ y_2 = \frac{1}{2} \end{cases}$$

Therefore the inverse is

$$B = \begin{bmatrix} x_1 & x_2 \\ y_1 & y_2 \end{bmatrix} = \begin{bmatrix} 2 & -1 \\ -\frac{1}{2} & \frac{1}{2} \end{bmatrix}$$

■

> **CALCULATOR COMMENT**
>
> On a calculator such as the TI-81, input the entries of matrix A, then press
>
> [A] x^{-1} ENTER
>
> The output is shown:
>
> [2 -1]
> [-.5 .5]

The solution of the two systems is shown side by side so you can easily see that the steps are identical since the two rows on the left are the same; the third columns are

$$\begin{bmatrix} 1 \\ 0 \end{bmatrix} \begin{bmatrix} 0 \\ 1 \end{bmatrix}.$$

respectively. It would seem that we might combine steps, but, before we do, consider a three-by-three example.

EXAMPLE 4 Find the inverse for the matrix

$$\begin{bmatrix} 1 & -1 & 0 \\ 3 & 2 & 2 \\ -1 & -1 & -1 \end{bmatrix}$$

(This is matrix B from Example 2.)

Solution We need to find a matrix

$$\begin{bmatrix} x_1 & x_2 & x_3 \\ y_1 & y_2 & y_3 \\ z_1 & z_2 & z_3 \end{bmatrix}$$

so that

$$\begin{bmatrix} 1 & -1 & 0 \\ 3 & 2 & 2 \\ -1 & -1 & -1 \end{bmatrix}\begin{bmatrix} x_1 & x_2 & x_3 \\ y_1 & y_2 & y_3 \\ z_1 & z_2 & z_3 \end{bmatrix} = \begin{bmatrix} 1 & 0 & 0 \\ 0 & 1 & 0 \\ 0 & 0 & 1 \end{bmatrix}$$

$$\begin{cases} x_1 - y_1 + 0z_1 = 1 \\ 3x_1 + 2y_1 + 2z_1 = 0 \\ -x_1 - y_1 - z_1 = 0 \end{cases} \quad \begin{cases} x_2 - y_2 + 0z_2 = 0 \\ 3x_2 + 2y_2 + 2z_2 = 1 \\ -x_2 - y_2 - z_2 = 0 \end{cases} \quad \begin{cases} x_3 - y_3 + 0z_3 = 0 \\ 3x_3 + 2y_3 + 2z_3 = 0 \\ -x_3 - y_3 - z_3 = 1 \end{cases}$$

We could solve these as three separate systems using Gauss–Jordan elimination; however, all the steps would be identical since the *variables* on the left of the equal signs are the same in each system. Therefore, suppose we augment the matrix of the coefficients by the *three* rows and do all three at once.

SPREADSHEET PROGRAM

	A	B	C	D	E	F	G	H
1								
2	1	-1	0		0	1	2	
3	3	2	2		-1	1	2	
4	-1	-1	-1		1	-2	-5	
5								
6								

Many spreadsheets find inverse matrices. In 1-2-3 choose /Data|Matrix|Invert. In Quattro Pro choose /Tools|Advanced Math|Invert. All you need to do is insert the entries, tell the program where to find those entries, and then tell the program where to place the inverse.

$$\left[\begin{array}{rrr|rrr} 1 & -1 & 0 & 1 & 0 & 0 \\ 3 & 2 & 2 & 0 & 1 & 0 \\ -1 & -1 & -1 & 0 & 0 & 1 \end{array}\right] \longrightarrow \left[\begin{array}{rrr|rrr} 1 & -1 & 0 & 1 & 0 & 0 \\ 0 & 5 & 2 & -3 & 1 & 0 \\ 0 & -2 & -1 & 1 & 0 & 1 \end{array}\right]$$

$$\xrightarrow{2R3 + R2} \left[\begin{array}{rrr|rrr} 1 & -1 & 0 & 1 & 0 & 0 \\ 0 & 1 & 0 & -1 & 1 & 2 \\ 0 & -2 & -1 & 1 & 0 & 1 \end{array}\right] \longrightarrow \left[\begin{array}{rrr|rrr} 1 & 0 & 0 & 0 & 1 & 2 \\ 0 & 1 & 0 & -1 & 1 & 2 \\ 0 & 0 & -1 & -1 & 2 & 5 \end{array}\right]$$

$$\longrightarrow \left[\begin{array}{rrr|rrr} 1 & 0 & 0 & 0 & 1 & 2 \\ 0 & 1 & 0 & -1 & 1 & 2 \\ 0 & 0 & 1 & 1 & -2 & -5 \end{array}\right]$$

Now, if we relate this back to the original three systems, we see that the inverse is

$$\begin{bmatrix} 0 & 1 & 2 \\ -1 & 1 & 2 \\ 1 & -2 & -5 \end{bmatrix}$$

■

By studying Examples 3 and 4, we are led to a procedure for finding the inverse of a nonsingular matrix:

Procedure for Finding the Inverse of a Matrix

1. Augment the given matrix with I; that is, write $[A \mid I]$, where I is the identity matrix of the same order as the given square matrix A.
2. Perform elementary row operations using Gauss–Jordan elimination in order to change the matrix A into the identity matrix (if possible).
3. If, at any time, you obtain all zeros in a row or column to the left of the dashed line, then there will be no inverse.
4. If steps 1 and 2 can be performed, the result in the augmented part is the inverse of A.

EXAMPLE 5 Find the inverse of the matrix $A = \begin{bmatrix} 1 & 2 \\ 0 & 0 \end{bmatrix}$.

Solution Write the augmented matrix $[A \mid I]$:

$$\left[\begin{array}{cc|cc} 1 & 2 & 1 & 0 \\ 0 & 0 & 0 & 1 \end{array}\right]$$

We want to make the left-hand side look like the corresponding identity matrix. This is impossible since there are no elementary row operations that will put it into the required form. Thus there is no inverse. ■

EXAMPLE 6 Find the inverse of the matrix $A = \begin{bmatrix} 0 & 1 & 2 \\ 2 & -1 & 1 \\ -1 & 1 & 0 \end{bmatrix}$.

Solution Write the augmented matrix $[A \mid I]$ and make the left-hand side look like the corresponding identity matrix (if possible):

$$\left[\begin{array}{ccc|ccc} 0 & 1 & 2 & 1 & 0 & 0 \\ 2 & -1 & 1 & 0 & 1 & 0 \\ -1 & 1 & 0 & 0 & 0 & 1 \end{array}\right] \xrightarrow{R1 \leftrightarrow R3} \left[\begin{array}{ccc|ccc} -1 & 1 & 0 & 0 & 0 & 1 \\ 2 & -1 & 1 & 0 & 1 & 0 \\ 0 & 1 & 2 & 1 & 0 & 0 \end{array}\right]$$

$$\longrightarrow \left[\begin{array}{ccc|ccc} 1 & -1 & 0 & 0 & 0 & -1 \\ 2 & -1 & 1 & 0 & 1 & 0 \\ 0 & 1 & 2 & 1 & 0 & 0 \end{array}\right] \longrightarrow \left[\begin{array}{ccc|ccc} 1 & -1 & 0 & 0 & 0 & -1 \\ 0 & 1 & 1 & 0 & 1 & 2 \\ 0 & 1 & 2 & 1 & 0 & 0 \end{array}\right]$$

$$\longrightarrow \left[\begin{array}{ccc|ccc} 1 & 0 & 1 & 0 & 1 & 1 \\ 0 & 1 & 1 & 0 & 1 & 2 \\ 0 & 0 & 1 & 1 & -1 & -2 \end{array}\right] \longrightarrow \left[\begin{array}{ccc|ccc} 1 & 0 & 0 & -1 & 2 & 3 \\ 0 & 1 & 0 & -1 & 2 & 4 \\ 0 & 0 & 1 & 1 & -1 & -2 \end{array}\right]$$

Thus, $$A^{-1} = \begin{bmatrix} -1 & 2 & 3 \\ -1 & 2 & 4 \\ 1 & -1 & -2 \end{bmatrix}.$$ ■

Inverses, when they are known, can be used to solve matrix equations. For example, suppose

$$A = \begin{bmatrix} 0 & 1 & 2 \\ 2 & -1 & 1 \\ -1 & 1 & 0 \end{bmatrix} \qquad X = \begin{bmatrix} x \\ y \\ z \end{bmatrix} \qquad B = \begin{bmatrix} 0 \\ -1 \\ 1 \end{bmatrix}$$

Then

$$AX = \begin{bmatrix} 0 & 1 & 2 \\ 2 & -1 & 1 \\ -1 & 1 & 0 \end{bmatrix} \begin{bmatrix} x \\ y \\ z \end{bmatrix} = \begin{bmatrix} y + 2z \\ 2x - y + z \\ -x + y \end{bmatrix}$$

and $AX = B$ means

$$\begin{cases} \quad\;\; y + 2z = 0 \\ 2x - y + \;\; z = -1 \\ \;\; -x + \;\; y = 1 \end{cases}$$

since matrices are equal if and only if corresponding entries are equal. If you can write a system in matrix form, then you can solve the system $AX = B$ provided A^{-1} exists:

$AX = B$	Given system
$(A^{-1})AX = A^{-1}B$	Multiply both sides by A^{-1} on the left.
$(A^{-1}A)X = A^{-1}B$	Associative property
$IX = A^{-1}B$	Inverse property
$X = A^{-1}B$	Identity property

This means that to solve a system, all we have to do is multiply A^{-1} and B to find X. Since

$$A = \begin{bmatrix} 0 & 1 & 2 \\ 2 & -1 & 1 \\ -1 & 1 & 0 \end{bmatrix}$$

we see from Example 6,

$$A^{-1} = \begin{bmatrix} -1 & 2 & 3 \\ -1 & 2 & 4 \\ 1 & -1 & -2 \end{bmatrix}$$

so we can solve for X as follows:

$$\begin{aligned} X &= A^{-1}B \\ &= \begin{bmatrix} -1 & 2 & 3 \\ -1 & 2 & 4 \\ 1 & -1 & -2 \end{bmatrix} \begin{bmatrix} 0 \\ -1 \\ 1 \end{bmatrix} = \begin{bmatrix} 0 - 2 + 3 \\ 0 - 2 + 4 \\ 0 + 1 - 2 \end{bmatrix} = \begin{bmatrix} 1 \\ 2 \\ -1 \end{bmatrix} \end{aligned}$$

Thus $x = 1$, $y = 2$, and $z = -1$.

The method of solving a system by using the inverse matrix is very efficient if you know the inverse, as illustrated in Problems 30–56 of Problem Set 3.4. The inverse is also a useful method to use when you have a calculator with matrix operations. Unfortunately, without a calculator or computer, *finding* the inverse for only one system is usually more work than using another method to solve the system. However, there are certain applications that yield the same coefficient matrix over and over. In these cases the inverse method is worthwhile.

3.4

Problem Set

Use multiplication in Problems 1–12 to determine whether the matrices are inverses.

1. $\begin{bmatrix} 1 & 2 \\ 2 & 3 \end{bmatrix}, \begin{bmatrix} -3 & 2 \\ 2 & -1 \end{bmatrix}$

2. $\begin{bmatrix} 2 & -5 \\ -1 & 2 \end{bmatrix}, \begin{bmatrix} -2 & -5 \\ -1 & -2 \end{bmatrix}$

3. $\begin{bmatrix} 3 & 5 \\ 4 & 7 \end{bmatrix}, \begin{bmatrix} 7 & -5 \\ -4 & 3 \end{bmatrix}$

4. $\begin{bmatrix} 4 & 7 \\ 5 & 9 \end{bmatrix}, \begin{bmatrix} 9 & -7 \\ -5 & 4 \end{bmatrix}$

5. $\begin{bmatrix} 4 & 3 \\ 2 & 2 \end{bmatrix}, \begin{bmatrix} 1 & -\frac{3}{2} \\ -1 & 2 \end{bmatrix}$

6. $\begin{bmatrix} 2 & 3 \\ 2 & 1 \end{bmatrix}, \begin{bmatrix} -\frac{1}{4} & \frac{3}{4} \\ \frac{1}{2} & -\frac{1}{2} \end{bmatrix}$

7. $\begin{bmatrix} 0 & 1 & 0 \\ 1 & -1 & 0 \\ -1 & 2 & 1 \end{bmatrix}, \begin{bmatrix} -1 & -1 & 0 \\ -1 & 0 & 0 \\ 1 & -1 & -1 \end{bmatrix}$

8. $\begin{bmatrix} 1 & 0 & 0 \\ 0 & 1 & 1 \\ 2 & 0 & 1 \end{bmatrix}, \begin{bmatrix} 1 & 0 & 0 \\ 2 & 1 & -1 \\ -2 & 0 & 1 \end{bmatrix}$

9. $\begin{bmatrix} 3 & -2 & 4 \\ 2 & 1 & 2 \\ 5 & 3 & 5 \end{bmatrix}, \begin{bmatrix} -1 & 22 & -8 \\ 0 & -5 & 2 \\ 1 & -19 & 7 \end{bmatrix}$

10. $\begin{bmatrix} 6 & -1 & -5 \\ -7 & 1 & 5 \\ -10 & 2 & 11 \end{bmatrix}, \begin{bmatrix} -1 & -1 & 0 \\ -27 & -16 & -5 \\ 4 & 2 & 1 \end{bmatrix}$

11. $\begin{bmatrix} 1 & 1 & 0 & 1 \\ 2 & 0 & -5 & 8 \\ 0 & 1 & 3 & 2 \\ 0 & 0 & 3 & 31 \end{bmatrix}, \begin{bmatrix} 1 & 95 & -1 & -29 \\ 0 & -92 & 1 & 28 \\ 0 & 31 & 0 & -10 \\ 0 & -3 & 0 & 1 \end{bmatrix}$

12. $\begin{bmatrix} 1 & 0 & 0 & 2 \\ 3 & 1 & -1 & 0 \\ 0 & 2 & 1 & 3 \\ 1 & 0 & 0 & 1 \end{bmatrix}, \begin{bmatrix} -2 & 0 & 0 & 2 \\ 0 & \frac{1}{3} & \frac{1}{3} & -1 \\ -3 & -\frac{2}{3} & \frac{1}{3} & 5 \\ 1 & 0 & 0 & -1 \end{bmatrix}$

Find the inverses of the matrices in Problems 13–29, if possible.

13. $\begin{bmatrix} 4 & -7 \\ -1 & 2 \end{bmatrix}$

14. $\begin{bmatrix} 4 & 0 \\ 0 & 5 \end{bmatrix}$

15. $\begin{bmatrix} 3 & 5 \\ 1 & 2 \end{bmatrix}$

16. $\begin{bmatrix} 3 & -1 \\ -4 & 2 \end{bmatrix}$

17. $\begin{bmatrix} 1 & 3 \\ 2 & 0 \end{bmatrix}$

18. $\begin{bmatrix} 2 & 3 \\ 1 & -6 \end{bmatrix}$

19. $\begin{bmatrix} 8 & 6 \\ -2 & 4 \end{bmatrix}$

20. $\begin{bmatrix} 2 & 1 \\ 4 & 3 \end{bmatrix}$

21. $\begin{bmatrix} 1 & -\frac{3}{2} \\ -1 & 2 \end{bmatrix}$

22. $\begin{bmatrix} 1 & 0 & 2 \\ 2 & 1 & 0 \\ 0 & -2 & 9 \end{bmatrix}$

23. $\begin{bmatrix} 6 & 1 & 20 \\ 1 & -1 & 0 \\ 0 & 1 & 3 \end{bmatrix}$

24. $\begin{bmatrix} 4 & 1 & 0 \\ 2 & -1 & 4 \\ -3 & 2 & 1 \end{bmatrix}$

25. $\begin{bmatrix} 1 & -1 & 1 \\ 0 & 2 & -1 \\ 2 & 3 & 0 \end{bmatrix}$

26. $\begin{bmatrix} 15 & 4 & -5 \\ -12 & -3 & 4 \\ -4 & -1 & 1 \end{bmatrix}$

27. $\begin{bmatrix} 1 & 0 & 2 \\ 3 & -1 & 2 \\ 4 & 1 & 0 \end{bmatrix}$

28. $\begin{bmatrix} 1 & 0 & 0 & 1 \\ 0 & 2 & 0 & 0 \\ 0 & 0 & 0 & 1 \\ 2 & 0 & 1 & 0 \end{bmatrix}$

29. $\begin{bmatrix} 0 & 1 & 2 & 0 \\ 0 & 0 & 0 & 1 \\ 1 & 1 & 3 & 0 \\ 2 & 4 & 0 & 0 \end{bmatrix}$

Solve the systems in Problems 30–56 by solving the corresponding matrix equation with an inverse, if possible. That is, if $AX = B$ then $X = A^{-1}B$. Write each system in matrix notation and use the indicated inverse to find a solution.

30. $\begin{cases} 4x - 7y = -1 \\ -x + 2y = 1 \end{cases}$

31. $\begin{cases} 4x - 7y = -2 \\ -x + 2y = 1 \end{cases}$

32. $\begin{cases} 4x - 7y = -65 \\ -x + 2y = 18 \end{cases}$

33. $\begin{cases} 4x - 7y = 48 \\ -x + 2y = -13 \end{cases}$

Problems 34–37 use the inverse found in Problem 19.

34. $\begin{cases} 8x + 6y = 12 \\ -2x + 4y = -14 \end{cases}$

35. $\begin{cases} 8x + 6y = 16 \\ -2x + 4y = 18 \end{cases}$

36. $\begin{cases} 8x + 6y = -6 \\ -2x + 4y = -26 \end{cases}$

37. $\begin{cases} 8x + 6y = 4 \\ -2x + 4y = 32 \end{cases}$

Problems 38–41 use the inverse found in Problem 18.

38. $\begin{cases} 2x + 3y = 2 \\ x - 6y = 16 \end{cases}$

39. $\begin{cases} 2x + 3y = 9 \\ x - 6y = -3 \end{cases}$

40. $\begin{cases} 2x + 3y = 10 \\ x - 6y = 20 \end{cases}$

41. $\begin{cases} 2x + 3y = 2 \\ x - 6y = -14 \end{cases}$

Problems 42–45 use the inverse found in Problem 20.

42. $\begin{cases} 2x + y = 5 \\ 4x + 3y = 9 \end{cases}$

43. $\begin{cases} 2x + y = 16 \\ 4x + 3y = 2 \end{cases}$

44. $\begin{cases} 2x + y = -3 \\ 4x + 3y = 1 \end{cases}$

45. $\begin{cases} 2x + y = -6 \\ 4x + 3y = -8 \end{cases}$

Problems 46–49 use the inverse found in Problem 22.

46. $\begin{cases} x + 2z = -6 \\ 2x + y = 1 \\ -2y + 9z = -24 \end{cases}$

47. $\begin{cases} x + 2z = 7 \\ 2x + y = 16 \\ -2y + 9z = -3 \end{cases}$

48. $\begin{cases} x + 2z = 4 \\ 2x + y = 0 \\ -2y + 9z = 19 \end{cases}$

49. $\begin{cases} x + 2z = 7 \\ 2x + y = 0 \\ -2y + 9z = 31 \end{cases}$

Problems 50–51 use the inverse found in Problem 23.

50. $\begin{cases} 6x + y + 20z = 27 \\ x - y = 0 \\ y + 3z = 4 \end{cases}$

51. $\begin{cases} 6x + y + 20z = 14 \\ x - y = 1 \\ y + 3z = 1 \end{cases}$

Problems 52–53 use the inverse found in Problem 24.

52. $\begin{cases} 4x + y = 7 \\ 2x - y + 4z = -11 \\ -3x + 2y + z = -12 \end{cases}$

53. $\begin{cases} 4x + y = -10 \\ 2x - y + 4z = 20 \\ -3x + 2y + z = 20 \end{cases}$

Problems 54–56 use the inverse found in Problem 29.

54. $\begin{cases} x + 2y = 5 \\ z = 3 \\ w + x + 3y = 9 \\ 2w + 4x = 8 \end{cases}$

55. $\begin{cases} x + 2y = 0 \\ z = -4 \\ w + x + 3y = 4 \\ 2w + 4x = -2 \end{cases}$

56. $\begin{cases} x + 2y = 7 \\ z = -7 \\ w + x + 3y = 16 \\ 2w + 4x = -4 \end{cases}$

APPLICATIONS

57. A rancher has to mix three types of feed. The table gives the amounts per 100-pound bag of grain.

Grain	Protein	Carbohydrates	Sodium
I	10 lb	75 lb	2 lb
II	20 lb	70 lb	3 lb
III	15 lb	80 lb	1 lb

How many bags of each type of grain should be mixed to provide 135 pounds of protein, 740 pounds of carbohydrates, and 22 pounds of sodium?

58. Repeat Problem 57 to provide 350 pounds of protein, 1,885 pounds of carbohydrates, and 48 pounds of sodium.

59. Repeat Problem 57 to provide 545 pounds of protein, 3,090 pounds of carbohydrates, and 79 pounds of sodium.

60. A dietitian is to arrange a diet of three basic foods. The diet is to include 1,800 units of vitamin A, 9,800 units of vitamin C, and 1,420 units of calcium. The number of units per ounce of each of the foods is given in the table.

Food	Vitamin A	Vitamin C	Calcium
I	50 units	300 units	20 units
II	30 units	200 units	40 units
III	40 units	100 units	30 units

How many ounces of each food should be supplied?

61. Repeat Problem 60 to supply 1,200 units of vitamin A, 6,000 units of vitamin C, and 900 units of calcium.

62. Repeat Problem 60 to supply 1,140 units of vitamin A, 7,100 units of vitamin C, and 1,030 units of calcium.

*3.5 Leontief Models

Wassily Leontief received the Nobel Prize in Economics in 1973 for describing the economy in terms of an **input–output model**. We now look at two extended applications of the Leontief model: a **closed model** in which we try to find the relative income of each participant within a system, and an **open model** in which we try to find the amount of production needed to achieve a forecast demand.

A Closed Model

We will illustrate the concept of a closed model with a simple example.

EXAMPLE 1 Suppose that Dave and Bob are contractors who agreed to pool their skills to make repairs on each other's house. Even though they agreed that the amount of work each would do for the other would be about the same, they decided for tax purposes to pay each other about $10,000, but that the pay should be proportional to the amount of work done (including the work done on each one's own house). As it turned out, Dave spent 40% of his time on Bob's house and 60% on his own while Bob spent 65% on Dave's house and 35% on his own. Find the fair wage that each should be paid.

Solution We will place the information about the amount of work done in what is called an **input–output** matrix, T:

PRODUCER
(work done by)

$$T = \begin{bmatrix} .6 & .65 \\ .4 & .35 \end{bmatrix} \begin{matrix} \text{D} \\ \text{B} \end{matrix} \quad \text{USER}$$

(columns: D, B)

In general, an input–output matrix for a closed model consists of n rows and n columns where the sum of each column is 1 and each entry a_{ij} is the fraction (in either common fraction or decimal form) used by i (row heading) and produced by j (column heading). This means

$a_{ij} = a_{11}$ when $i = 1$ (Dave) and $j = 1$ (Dave), so
$a_{11} = .6$ because 60% of Dave's time is used by Dave;

$a_{ij} = a_{12}$ when $i = 1$ (Dave) and $j = 2$ (Bob), so
$a_{12} = .65$ is the amount of time Bob spent on Dave's house;

$a_{ij} = a_{21}$ when $i = 2$ (Bob) and $j = 1$ (Dave), so
$a_{21} = .4$ because Bob used 40% of Dave's work;

$a_{ij} = a_{22}$ when $i = 2$ (Bob) and $j = 2$ (Bob), so
$a_{22} = .35$ since Bob used 35% of his time on his own house.

* Optional section.

Let $X = \begin{bmatrix} x_1 \\ x_2 \end{bmatrix}$ be the column matrix representing the wages of Dave (x_1) and Bob (x_2). The wages include those wages paid to themselves as part of the deal. For the wages to come out fairly (even), the total amount paid out by each must equal the total amount received by each. That is,

$$X = TX$$
$$X - TX = \mathbf{0}$$
$$IX - TX = \mathbf{0} \qquad \text{where } I \text{ is the identity matrix with the same order as } T.$$
$$(I - T)X = \mathbf{0}$$

This matrix equation represents a system of equations in which the right-hand side is always zero. This system is called a **homogeneous system of equations**. There is a trivial solution for which all of the wages (xs) are zero, but that is not the only solution. It can be shown that if the entries in the input–output matrix are positive, and if the sum of each column of T equals 1, then there is a one-parameter solution to the system. This parameter will serve as a scaling factor.

$$I - T = \begin{bmatrix} 1 & 0 \\ 0 & 1 \end{bmatrix} - \begin{bmatrix} .6 & .65 \\ .4 & .35 \end{bmatrix} = \begin{bmatrix} .4 & -.65 \\ -.4 & .65 \end{bmatrix}$$

$$(I - T)X = \begin{bmatrix} .4 & -.65 \\ -.4 & .65 \end{bmatrix} \begin{bmatrix} x_1 \\ x_2 \end{bmatrix} = \begin{bmatrix} .4x_1 - .65x_2 \\ -.4x_1 + .65x_2 \end{bmatrix}$$

This gives the system

$$\begin{cases} .4x_1 - .65x_2 = 0 \\ -.4x_1 + .65x_2 = 0 \end{cases}$$

The solution for this system is shown using Gauss–Jordan elimination.

$$\left[\begin{array}{cc|c} .4 & -.65 & 0 \\ -.4 & .65 & 0 \end{array}\right] \longrightarrow \left[\begin{array}{cc|c} .4 & -.65 & 0 \\ 0 & 0 & 0 \end{array}\right] \longrightarrow \left[\begin{array}{cc|c} 1 & -1.625 & 0 \\ 0 & 0 & 0 \end{array}\right]$$

This system is equivalent to $x_1 = 1.625x_2$. The second variable serves as a parameter, so let $x_2 = \$10{,}000$; then $x_1 = \$16{,}250$. In a closed system, each component pays exactly what it receives, so for income tax purposes Dave received and paid out \$16,250 and Bob received and paid out \$10,000. You can check these amounts by verifying the matrix formula

$$T = TX$$

■

EXAMPLE 2 Suppose a simplified economy consists of three individuals: a farmer, a tailor, and a builder, who provide the essentials of food, clothing, and shelter for each other. Of the food produced by the farmer, 50% is used by the farmer, 20% by the tailor, and 30% by the builder. The builder's production is utilized 40% by the builder, 40% by the farmer, and 20% by the tailor. The tailor's production is divided among them as 45% for the builder, 20% for the farmer, and 35% for the tailor. Suppose that each person's wages are to be about \$1,000 (it could be shells, sheep, or any other convenient unit used as money). How much should each person pay the others?

Solution Write the input–output matrix, which is sometimes called the **technology matrix**.

$$\begin{array}{c} \text{PRODUCER} \\ \begin{array}{cccc} & \text{Farmer} & \text{Builder} & \text{Tailor} \end{array} \end{array}$$

$$T = \begin{bmatrix} .5 & .4 & .2 \\ .3 & .4 & .45 \\ .2 & .2 & .35 \end{bmatrix} \begin{array}{l} \text{Farmer} \\ \text{Builder} \quad \text{USER} \\ \text{Tailor} \end{array}$$

Let X be the column matrix representing the price of each output in the system. That is, let

$$X = \begin{bmatrix} x_1 \\ x_2 \\ x_3 \end{bmatrix} \quad \text{where} \quad \begin{aligned} x_1 &= \text{Farmer's wages} \\ x_2 &= \text{Tailor's wages} \\ x_3 &= \text{Builder's wages} \end{aligned}$$

The wages include those wages paid to themselves as part of the economic system. For the wages to come out even, the total amount paid out by each must equal the total amount received by each. That is,

$$T = TX$$

From Example 1, the solution to this matrix equation is

$$(I - T)X = \mathbf{0}$$

Thus,

$$\left(\begin{bmatrix} 1 & 0 & 0 \\ 0 & 1 & 0 \\ 0 & 0 & 1 \end{bmatrix} - \begin{bmatrix} .5 & .4 & .2 \\ .3 & .4 & .45 \\ .2 & .2 & .35 \end{bmatrix} \right) \begin{bmatrix} x_1 \\ x_2 \\ x_3 \end{bmatrix} = \begin{bmatrix} 0 \\ 0 \\ 0 \end{bmatrix}$$

This gives rise to the system

$$\begin{cases} .5x_1 - .4x_2 - .2x_3 = 0 \\ -.3x_1 + .6x_2 - .45x_3 = 0 \\ -.2x_1 - .2x_2 + .65x_3 = 0 \end{cases}$$

The solution using Gauss–Jordan elimination is shown below:

$$\left[\begin{array}{ccc|c} .5 & -.4 & -.2 & 0 \\ -.3 & .6 & -.45 & 0 \\ -.2 & -.2 & .65 & 0 \end{array}\right] \longrightarrow \left[\begin{array}{ccc|c} 5 & -4 & -2 & 0 \\ -2 & 4 & -3 & 0 \\ -4 & -4 & 13 & 0 \end{array}\right]$$

$$\longrightarrow \left[\begin{array}{ccc|c} 1 & -8 & 11 & 0 \\ -2 & 4 & -3 & 0 \\ -4 & -4 & 13 & 0 \end{array}\right] \longrightarrow \left[\begin{array}{ccc|c} 1 & -8 & 11 & 0 \\ 0 & -12 & 19 & 0 \\ 0 & -36 & 57 & 0 \end{array}\right]$$

$$\longrightarrow \left[\begin{array}{ccc|c} 1 & -8 & 11 & 0 \\ 0 & 1 & -\frac{19}{12} & 0 \\ 0 & 0 & 0 & 0 \end{array}\right] \longrightarrow \left[\begin{array}{ccc|c} 1 & 0 & -\frac{5}{3} & 0 \\ 0 & 1 & -\frac{19}{12} & 0 \\ 0 & 0 & 0 & 0 \end{array}\right]$$

Thus $x_1 = \frac{5}{3}x_3$ and $x_2 = \frac{19}{12}x_3$.

Since the choice of x_3 is arbitrary, choose x_3 so that it is about \$1,000 and also so that it is divisible by 12 (to give an integer solution). Let $x_3 = 1{,}008$:

$$x_1 = \frac{5}{3}(1{,}008) = 1{,}680$$

$$x_2 = \frac{19}{12}(1{,}008) = 1{,}596$$

$$x_3 = 1{,}008$$

The wages to be paid are \$1,680 to the farmer, \$1,596 to the builder, and \$1,008 to the tailor. Note that if you let $x_3 = 636$, then the farmer's wages are \$1,060, the builder's wages \$1,007, and the tailor's wages \$636. No matter what choice is made for x_3, each other participant's share is determined. Sometimes we simply let $x_3 = 12$ so that $x = \frac{5}{3}(12) = 20$, $y = \frac{19}{12}(12) = 19$, which says that an answer is in the ratio of 20:19:12. ■

An Open Model

A more realistic input–output model of an economic system is called an **open model**, in which the economy is composed of several industries that need not only to satisfy their own needs (as in the closed model), but also must satisfy some outside (additional) demands, such as the consumer demand in our society. In order to understand the open model we will begin with an extremely unrealistic and simplified society consisting of only two industries.

EXAMPLE 3 Consider an economy consisting of two industries, gas and electric. Suppose the gas company uses both gas (input) and electricity (input) in the production of gas (output), and the electric company also uses both gas (input) and electricity (input) in the production of electricity (output). Also suppose that the production of a unit of gas uses .1 unit of gas and .2 unit of electricity. Similarly, to produce a unit of electricity requires .15 unit of gas and .25 unit of electricity. The characteristic of an open model is that there is also an outside demand on the industries. Let us suppose that the external demand for both gas and electricity is

$$d_1 = 20 \text{ units for gas}$$
$$d_2 = 60 \text{ units for electricity}$$

How much gas and electricity should be produced in order to meet this final demand?

Solution First, understand the problem. It should be clear that each industry needs to produce more than is necessary for the final demand. For example, if we use the 20 units and 60 units demand, the production processes of the two companies are

GAS TO PRODUCE GAS + GAS TO PRODUCE ELECTRICITY

$$.1(20) \quad + \quad .15(60) \quad = 11 \text{ units of gas}$$

ELECTRICITY TO PRODUCE GAS + ELECTRICITY TO PRODUCE ELECTRICITY

$$.2(20) \quad + \quad .25(60) \quad = 19 \text{ units of electricity}$$

This is called the *internal demand.* The internal demand is the amount necessary for production. After we take out the amounts necessary for internal demand in this example, we see that it would leave only 9 units of gas and 41 units of electricity to satisfy the outside demand. Clearly, the output necessary from the two industries must be larger than the outside demand.

Let x_1 = total output from the gas company
x_2 = total output from the electric company

This gives us the following internal demands:

INTERNAL DEMAND FOR GAS: $.1x_1 + .15x_2$

INTERNAL DEMAND FOR ELECTRICITY: $.2x_1 + .25x_2$

For each industry,

	TOTAL OUTPUT	=	INTERNAL DEMAND	+	OUTSIDE DEMAND
GAS	x_1	=	$(.1x_1 + .15x_2)$	+	d_1
ELECTRICITY	x_2	=	$(.2x_1 + .25x_2)$	+	d_2

We can write this system of equations in matrix form as

$$X = TX + D$$

where

$$X = \begin{bmatrix} x_1 \\ x_2 \end{bmatrix} \text{ is the output matrix;}$$

$$D = \begin{bmatrix} d_1 \\ d_2 \end{bmatrix} = \begin{bmatrix} 20 \\ 60 \end{bmatrix} \text{ is the final (outside) demand matrix}$$

$$T = \text{User} \begin{matrix} \\ G \\ E \end{matrix} \overset{\text{Producer}}{\begin{matrix} G \quad\; E \\ \begin{bmatrix} .1 & .15 \\ .2 & .25 \end{bmatrix} \end{matrix}} \text{ is the input–output matrix.}$$

It is important that you understand how the input–output matrix is put together. Look at the original problem as stated at the top of this example. The production demands for gas are written *down* the column headed GAS and the production demands for electricity are written *down* the column headed ELECTRICITY.

Solve the matrix equation for X:

$$\begin{aligned} X &= TX + D \\ X - TX &= D \\ (I - T)X &= D \\ X &= (I - T)^{-1}D \end{aligned}$$

For this example,

$$\left(\begin{bmatrix} 1 & 0 \\ 0 & 1 \end{bmatrix} - \begin{bmatrix} .1 & .15 \\ .2 & .25 \end{bmatrix}\right)^{-1} = \begin{bmatrix} .9 & -.15 \\ -.2 & .75 \end{bmatrix}^{-1} = \begin{bmatrix} 1.16279 & .23256 \\ .31008 & 1.39535 \end{bmatrix}$$

This means that in matrix form the equation $X = (I - T)^{-1}D$ is

$$\begin{bmatrix} x_1 \\ x_2 \end{bmatrix} = \begin{bmatrix} 1.16279 & .23256 \\ .31008 & 1.39535 \end{bmatrix} \begin{bmatrix} 20 \\ 60 \end{bmatrix}$$

$$\approx \begin{bmatrix} 37.21 \\ 89.22 \end{bmatrix}$$

Therefore, the gas company must have an output of about 37 units and the electric company must have an output of about 89 units in order for both companies to meet both the internal and external demands. ■

Leontief Models

In a Leontief input–output model with input–output (technology) matrix T, to find the unknown output matrix X use one of the following formulas:

A Closed Model
All of the goods are consumed by the existing industries.

Solve

$$(I - T)X = \mathbf{0}$$

An Open Model
There is an external demand on each of the existing industries. Let D be this demand matrix.

Solve

$$X = (I - T)^{-1}D$$

where I is the identity matrix of the same order as T.

EXAMPLE 4 Suppose a simplified economy consists of three industries: farming, clothing, and construction, which provide the essentials of food, clothing, and shelter not only for each other, but also for the consumer.

For each dollar of goods produced by the farming industry, the clothing industry consumes 10¢ and the construction industry consumes 15¢. The farming industry consumes 25¢ worth of its own output for each dollar of goods produced. The consumer demand for the farming industry is $50 billion.

The construction industry consumes 5¢ for each dollar of output, the farming industry consumes 27¢, and the clothing industry consumes 9¢. The demand for the construction industry by the consumer is $65 billion.

The third industry, the clothing industry, uses 4¢ worth of its own production for each dollar of goods produced, while the farming segment uses 8¢ and the construction industry uses 17¢. The consumer demand for the clothing industry is $85 billion.

Construct a model to determine the necessary output for each industry to meet internal consumption and external demand.

Solution Begin by writing the input–output matrix. Remember, the numbers mentioned in the problem form the *column* entries for each industry. Decimal form is usually easier to work with when using a calculator or computer, but when using the decimal form remember that answers are often approximate. The amounts used by each of the industries in the problem are filled in *down* each column headed by that producer. This means the row entries are for the amounts consumed by each of the industries.

		PRODUCER			
	Farming	Construction	Clothing		
$T =$	.25	.27	.08	Farming	
	.15	.05	.17	Construction	USER
	.10	.09	.04	Clothing	

The units in the demand matrix are in billions of dollars.

Let D be the demand matrix: $D = \begin{bmatrix} 50 \\ 65 \\ 85 \end{bmatrix}$ Farming, Construction, Clothing

Let X be the output matrix: $X = \begin{bmatrix} x_1 \\ x_2 \\ x_3 \end{bmatrix}$ Farming, Construction, Clothing

where x_1, x_2, and x_3 are the output from farming, construction, and clothing, respectively.

The solution is given by the matrix equation

$$X = (I - T)^{-1}D$$

$$I - T = \begin{bmatrix} 1 & 0 & 0 \\ 0 & 1 & 0 \\ 0 & 0 & 1 \end{bmatrix} - \begin{bmatrix} .25 & .27 & .08 \\ .15 & .05 & .17 \\ .10 & .09 & .04 \end{bmatrix}$$

$$= \begin{bmatrix} .75 & -.27 & -.08 \\ -.15 & .95 & -.17 \\ -.10 & -.09 & .96 \end{bmatrix}$$

Form the augmented matrix to find the inverse matrix (the details are not shown):

$$(I - T)^{-1} = \begin{bmatrix} 1.4454 & .42942 & .19649 \\ .25952 & 1.14769 & .22486 \\ .17489 & .15233 & 1.08322 \end{bmatrix}$$

Finally,

$$\begin{bmatrix} x_1 \\ x_2 \\ x_3 \end{bmatrix} = \begin{bmatrix} 1.4454 & .42942 & .19649 \\ .25952 & 1.14769 & .22486 \\ .17489 & .15233 & 1.08322 \end{bmatrix} \begin{bmatrix} 50 \\ 65 \\ 85 \end{bmatrix} = \begin{bmatrix} 116.8849 \\ 106.6895 \\ 110.7193 \end{bmatrix}$$

The required outputs for the three industries are about \$117 billion for the farming industry, \$107 billion for the construction industry, and \$111 billion for the clothing industry. ■

CALCULATOR COMMENT

Input I into matrix [A], T into matrix [B], and D into matrix [C] on the TI-81:

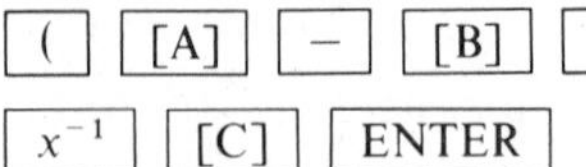

The output is:

```
[ 116.8849486 ]
[ 106.6895023 ]
[ 110.719323  ]
```

3.5

Problem Set

Solve the homogeneous systems in Problems 1–6.

1. $\begin{cases} x + 3y - 5z = 0 \\ 3x + y - 3z = 0 \\ 4x + 6y - 11z = 0 \end{cases}$

2. $\begin{cases} 2x - y + 3z = 0 \\ x + 3y - 4z = 0 \\ 8x + 3y + z = 0 \end{cases}$

3. $\begin{cases} 3x + 6y + 2z = 0 \\ 2x - 3y + 6z = 0 \\ 2x + 3y + 2z = 0 \end{cases}$

4. $\begin{cases} .4x_1 - .5x_2 - .2x_3 = 0 \\ -.1x_1 + .6x_2 - .3x_3 = 0 \\ -.3x_1 - .1x_2 + .5x_3 = 0 \end{cases}$

5. $\begin{cases} .3x_1 - .3x_2 - .65x_3 = 0 \\ -.25x_1 + .5x_2 - .2x_3 = 0 \\ -.05x_1 - .2x_2 + .85x_3 = 0 \end{cases}$

6. $\begin{cases} .7x_1 - .5x_2 - .45x_3 = 0 \\ -.4x_1 + .5x_2 - .35x_3 = 0 \\ -.3x_1 - .25x_2 + .8x_3 = 0 \end{cases}$

APPLICATIONS

7. Suppose each person's wages for the closed model discussed in this section should be around $30,000. Give two possible answers to the question posed if all other information in the problem stays the same.

8. In the closed model discussed in Example 2, suppose the farmer uses 45% of his own production and gives 25% to the tailor. Answer the question if the other information stays the same.

9. Suppose a merchant joins the closed model and allocates her resources as shown by the following input–output matrix:

	Farmer	Builder	Tailor	Merchant	
	% of production				
	.4	.3	.1	.1	Farmer
	.2	.3	.3	.3	Builder
	.2	.2	.4	.3	Tailor
	.2	.2	.2	.3	Merchant

How much should each person pay the others if the wages are to be about $1,000?

Apply the given demand in Problems 10–12 to the open model discussed in Example 4 to find the required output to meet the demand.

10. $D = \begin{bmatrix} 50 \\ 80 \\ 90 \end{bmatrix}$ **11.** $D = \begin{bmatrix} 70 \\ 100 \\ 110 \end{bmatrix}$ **12.** $D = \begin{bmatrix} 100 \\ 180 \\ 170 \end{bmatrix}$

13. Solve the open model discussed in Example 4 for the following input–output matrix, with the other information remaining the same:

$$A = \begin{bmatrix} .3 & .2 & .1 \\ .2 & .4 & .3 \\ .1 & .1 & .2 \end{bmatrix}$$

14. Consider a simple economy with three industries: manufacturing (M), agriculture (A), and transportation (T). Find A and $I - A$ for this system given the information in the table.

Use	Producer M	A	T	External demand	Total output
M	50	20	100	30	200
A	30	80	100	90	300
T	20	10	50	420	500

15. Find $(I - A)^{-1}$ for Problem 14.

16. Find the output in Problem 14 to meet the given demand.

17. Find the output in Problem 14 to meet the external demand if it changes to 200 for M, 90 for A, and 150 for T.

18. ***Historical Question.*** In 1966 Leontief studied a simplified model of the 1958 Israeli economy by dividing it into three sectors—agriculture, manufacturing, and energy—as shown in the table:*

	Agriculture	Manufacturing	Energy
Agriculture	.293	0	0
Manufacturing	.014	.207	.017
Energy	.044	.010	.216

* Wassily Leontief, *Input–Output Economics* (New York: Oxford University Press, 1966), pp. 54–57.

The exports (in thousands of Israeli pounds) are shown in the following table:

Agriculture	138,213
Manufacturing	17,597
Energy	1,786

Find A and $I - A$ for this system.

19. Find $(I - A)^{-1}$ for Problem 18.

20. Find the output in Problem 18 to meet the external demand.

*3.6

Review

The material of this chapter is reviewed in the following list of objectives. After each objective there are some practice questions. For a sample test select the first question of each set and check your answers. The second question for each objective has no answer given. If you are having trouble with a particular type of problem, look back at the indicated section in the text. When you are finished reviewing these objectives, a sample examination is given at the end of this section.

[3.1]

Objective 3.1: *Solve systems of linear equations by graphing.*

1. $\begin{cases} x + y = 4 \\ x - y = 0 \end{cases}$

2. $\begin{cases} y = 2x + 1 \\ y = 10 - x \end{cases}$

3. $\begin{cases} 2x - 3y = -2 \\ 4x - 5y = 0 \end{cases}$

4. $\begin{cases} 5x + 2y = 1 \\ 8x + 3y = 3 \end{cases}$

Objective 3.2: *Solve systems of linear equations by substitution.*

5. $\begin{cases} y = 2x + 1 \\ x + y = 10 \end{cases}$

6. $\begin{cases} y = x - 8 \\ x + 2y + 10 = 0 \end{cases}$

7. $\begin{cases} x + 3y = 7 \\ 2x + 2y = 6 \end{cases}$

8. $\begin{cases} x + 3y = 1 \\ x - 2y = -9 \end{cases}$

Objective 3.3: *Solve systems of linear equations by linear combinations.*

9. $\begin{cases} x + y = 7 \\ x - y = 3 \end{cases}$

10. $\begin{cases} 3x + 2y = 5 \\ 2x + y = 6 \end{cases}$

11. $\begin{cases} x - 3y = -2 \\ 5x - 9y = -5 \end{cases}$

12. $\begin{cases} 5x + y = 4 \\ 9x - 5y = -3 \end{cases}$

[3.2]

Objective 3.4: *State the order of a matrix.*

13. $\begin{bmatrix} 3 & 9 & 0 & 5 & 1 \\ 9 & -1 & 4 & 19 & 0 \end{bmatrix}$

14. $[x + y + z]$

15. $\begin{bmatrix} 1 \\ 4 \\ -8 \\ 0 \\ 0 \end{bmatrix}$

16. $\begin{bmatrix} 16 + 3 + 5 \\ 15 + 9 + 2 \\ 14 + 7 - 11 \end{bmatrix}$

Objective 3.5: *Determine whether the two given matrices are equal. Find the replacements of the variables that make the matrices equal. If it is not possible, so state.*

17. $\begin{bmatrix} 3 & 0 \\ x & -1 \end{bmatrix} = \begin{bmatrix} y & 1 \\ 4 & -1 \end{bmatrix}$

18. $\begin{bmatrix} 1 & 0 & 0 \\ 0 & 1 & 0 \\ 0 & 0 & 1 \end{bmatrix} = \begin{bmatrix} a & b & c \\ d & e & f \\ g & h & i \end{bmatrix}$

19. $[1 + 3 + 4] = [x \quad y \quad z]$

20. $\begin{bmatrix} -1 & 3 \\ 6 & 2 \\ 0 & 4 \end{bmatrix} = \begin{bmatrix} -1 & w \\ x & 2 \\ y & z \end{bmatrix}$

Objective 3.6: *Use a matrix to represent data.*

21. Draw a graph for the matrix

$$\begin{array}{c} \\ A \\ B \\ C \\ D \end{array} \begin{array}{c} \begin{matrix} A & B & C & D \end{matrix} \\ \begin{bmatrix} 0 & 1 & 1 & 0 \\ 0 & 1 & 0 & 0 \\ 1 & 0 & 0 & 0 \\ 0 & 1 & 0 & 0 \end{bmatrix} \end{array}$$

22. What is the entry in row 3, column 4, of the matrix in Problem 21?

* Optional section.

23. Write a matrix for the given directed graph.

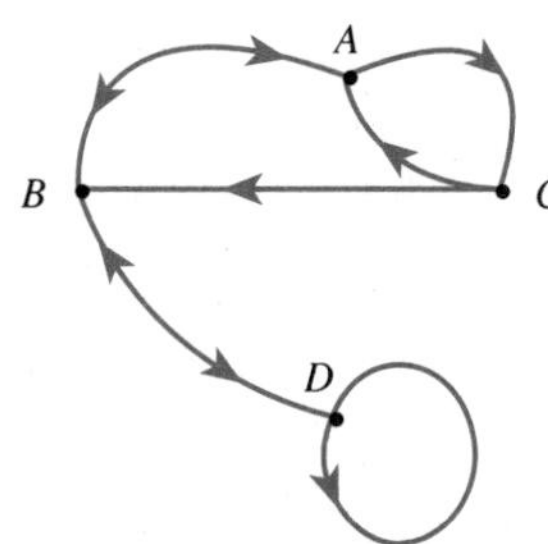

24. Write a matrix for the graph in Problem 23 if you disregard the directed arrows on the routes.

Objective 3.7: *Add conformable matrices.*

25. $\begin{bmatrix} 4 & -2 & 1 \\ -6 & 5 & 10 \end{bmatrix} + \begin{bmatrix} -3 & 5 & 0 \\ -1 & -4 & 41 \end{bmatrix}$

26. $\begin{bmatrix} 6 \\ -1 \\ 3 \end{bmatrix} + \begin{bmatrix} 13 \\ 81 \\ 43 \end{bmatrix}$

27. $\begin{bmatrix} 6 & 0 & 1 \\ -1 & 3 & 4 \\ 2 & 7 & -3 \end{bmatrix} + \begin{bmatrix} 5 & 4 & 1 \\ 6 & -2 & -3 \\ 0 & 7 & 4 \end{bmatrix}$

28. $[3 \quad 2 \quad 11] + [4 \quad 5 \quad 9]$

Objective 3.8: *Subtract conformable matrices.*

29. $[-3 \quad 4 \quad 11] - [-2 \quad -8 \quad 15]$

30. $\begin{bmatrix} 9 & 2 & 1 \\ 6 & 4 & -5 \\ 0 & -2 & 9 \end{bmatrix} - \begin{bmatrix} 10 & 3 & 5 \\ 9 & -9 & 11 \\ 3 & 7 & 0 \end{bmatrix}$

31. $\begin{bmatrix} 4 \\ -3 \\ 1 \\ 7 \end{bmatrix} - \begin{bmatrix} 11 \\ 9 \\ 4 \\ -4 \end{bmatrix}$

32. $\begin{bmatrix} 3 & -3 & 1 \\ 4 & 1 & 0 \end{bmatrix} - \begin{bmatrix} 4 & 5 \\ 2 & 9 \end{bmatrix}$

Objective 3.9: *Perform scalar multiplication.*

33. $3\begin{bmatrix} 4 & 1 & 3 \\ 6 & -2 & 5 \\ 0 & 9 & -6 \end{bmatrix}$

34. $\frac{1}{2}\begin{bmatrix} 6 & 9 & -2 \\ 0 & 3 & 8 \\ 4 & 1 & 2 \end{bmatrix}$

35. $\frac{1}{10}\begin{bmatrix} 5 \\ 9 \\ 10 \end{bmatrix}$

36. $\frac{1}{50}\begin{bmatrix} 150 & -75 \\ 200 & -15 \end{bmatrix}$

Objective 3.10: *Perform mixed operations, if possible. Let*

$A = [2 \quad -3 \quad 4] \qquad B = [-1 \quad 5 \quad 9]$

$C = \begin{bmatrix} 9 & 1 & 8 & -4 \\ 2 & 0 & -3 & -9 \\ 1 & 5 & 2 & -7 \end{bmatrix} \qquad D = \begin{bmatrix} 11 & -2 & 8 & -3 \\ -9 & 5 & -6 & 10 \\ 4 & -8 & 2 & -1 \end{bmatrix}$

37. $2C - 3D$

38. $5A - 9B$

39. $3(C + 2D)$

40. $(A + B) + C$

Objective 3.11: *Multiply matrices, if possible.*

41. $\begin{bmatrix} 2 & -1 & 3 \\ 5 & 2 & -3 \end{bmatrix}\begin{bmatrix} 6 & 1 & 2 \\ -2 & 0 & 4 \\ 5 & 1 & -7 \end{bmatrix}$

42. AC, where A and C are defined in Problems 37–40.

43. $\begin{bmatrix} 3 \\ 4 \\ -2 \\ 1 \end{bmatrix}[6 \quad -3 \quad 6 \quad 8]$

44. CD, where C and D are defined in Problems 37–40.

Objective 3.12: *Set up models requiring matrix multiplication.* On Saturday evenings a favorite pastime of many young residents of a certain small town is to drive around town looking for something to do. A "map" of the town is shown here. Use this map for Problems 45–48. Note that Fourth Street, College Avenue, and Pacific Avenue are two-way streets, while Circle Street and Snark Drive are one-way streets, as shown.

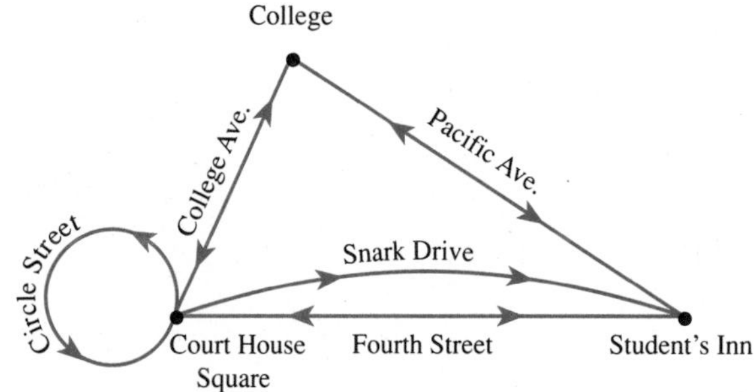

45. Write the information in the map in matrix form. That is, use a matrix to show how many one-block paths there are from one point to another. Define one block to be the distance between any of the given points.

46. Use a matrix to show how many two-block paths there are from one point to another.

47. Use a matrix to show the number of four-block paths from one point to another.

48. Use matrix multiplication to show the number of five-block paths from one point to another.

Objective 3.13: *Write a system of equations in augmented matrix form.*

49. $\begin{cases} 3x_1 - 2x_2 = -3 \\ 4x_3 + x_1 = 7 \end{cases}$

50. $\begin{cases} 4x - 2y + z = 15 \\ 3x + 5y - 2z = 44 \\ 5x - 3y - 4z = -22 \end{cases}$

51. $\begin{cases} 3x + 2y = -1 \\ 5x + 3y = 0 \\ 2x - 4y = 26 \end{cases}$

52. $\begin{cases} 3x_1 - 2x_3 + x_5 = -2 \\ 2x_1 + 5x_2 - x_4 = 5 \\ 3x_2 - 4x_3 + 6x_4 = 15 \\ x_2 + x_4 - x_5 = 4 \\ 2x_3 + x_4 - 2x_5 = 8 \end{cases}$

Objective 3.14: *Write a system of equations given an augmented matrix.*

53. $\left[\begin{array}{ccc|c} 1 & 4 & 9 & -1 \\ 2 & 3 & 1 & 0 \\ 4 & -1 & 2 & 3 \\ 3 & -4 & 0 & 1 \end{array}\right]$

54. $\left[\begin{array}{ccc|c} 1 & 0 & 0 & 3 \\ 0 & 1 & 0 & -2 \\ 0 & 0 & 1 & 4 \end{array}\right]$

55. $\left[\begin{array}{cccc|c} 1 & 0 & 0 & 2 & -3 \\ 0 & 1 & 5 & 0 & -1 \end{array}\right]$

56. $\left[\begin{array}{ccc|c} 1 & 0 & 0 & 3 \\ 0 & 1 & 0 & -2 \\ 0 & 0 & 1 & 5 \\ 0 & 0 & 0 & 2 \end{array}\right]$

[3.3]

Objective 3.15: *Solve systems using Gauss–Jordan elimination.*

57. $\begin{cases} 3x_1 - 2x_2 = -1 \\ 4x_2 + x_1 = 9 \end{cases}$

58. $\begin{cases} 4x - 2y + z = 15 \\ 3x + 5y - 2z = 44 \\ 5x - 3y - 4z = -22 \end{cases}$

59. $\begin{cases} 3x + 2y = -1 \\ 5x + 3y = 0 \\ 2x - 4y = 26 \end{cases}$

60. $\begin{cases} 3x_1 - 2x_3 + x_5 = -2 \\ 2x_1 + 5x_2 - x_4 = 3 \\ 3x_2 - 4x_3 + 6x_4 = 15 \\ x_2 + x_4 - x_5 = 4 \\ 2x_3 + x_4 - 2x_5 = 8 \end{cases}$

Objective 3.16: *Solve systems of equations that require a parameter.*

61. $\begin{cases} x + 3y - z = 3 \\ 2x - 4y + 2z = 16 \\ 3x - y + z = 19 \end{cases}$

62. $\begin{cases} 3x - 2y = -3 \\ 4z + x = 7 \end{cases}$

63. $\begin{cases} 2x + 3y - z = 4 \\ 4x - 2y + 5z = 0 \end{cases}$

64. $\begin{cases} 3x - 4z = 8 \\ 2x - 3y = -6 \end{cases}$

Objective 3.17: *Solve applied problems using Gauss–Jordan elimination.*

65. *Batching process.* A rancher mixes three types of cattle feed. The analysis shown in the table gives the amounts per 100-lb bag of grain.

Grain	Protein	Carbohydrates	Sodium	Other
A	5 lb	85 lb	1 lb	9 lb
B	15 lb	65 lb	10 lb	10 lb
C	8 lb	80 lb	7 lb	5 lb

How many bags of each type of grain should be mixed to provide 316 lb of protein, 1,880 lb of carbohydrates, and 184 lb of sodium?

66. *Product allocation.* A manufacturer is producing two products, I and II. These products each require three ingredients—A, B, and C—to be used as follows:

$$\begin{array}{l} \\ \text{Product I} \\ \text{Product II} \end{array} \begin{array}{c} \begin{array}{ccc} A & B & C \end{array} \\ \begin{bmatrix} 2 & 3 & 10 \\ 3 & 5 & 2 \end{bmatrix} \end{array}$$

If the manufacturer needs to use its supply of 280 units of A, 450 units of B, and 620 units of C, how much of each product can be manufactured?

67. *Population demographics.* Let m_0 and s_0 denote the number of married and single adults (over 18) in a community on January 1, 1990. A study has shown that the number of married and single adults in the following year (m_1 and s_1)

is given by the system of equations

$$\begin{cases} .9m_0 + .3s_0 = m_1 \\ .1m_0 + .7s_0 = s_1 \end{cases}$$

If a town has 38,000 married adults and 14,000 single adults on January 1, 1990, predict the number of married and single adults on January 1, 1991.

68. *Population demographics.* Use the information in Problem 67 to estimate the number of married and single adults on January 1, 1989.

[3.4]

Objective 3.18: *Determine whether the given matrices are inverses.*

69. $\begin{bmatrix} 1 & -2 \\ -3 & 7 \end{bmatrix}, \begin{bmatrix} 7 & 2 \\ 3 & 1 \end{bmatrix}$

70. $\begin{bmatrix} -5 & 3 & 0 \\ -2 & 1 & 0 \\ 0 & 1 & 1 \end{bmatrix}, \begin{bmatrix} 2 & 4 & 3 \\ 0 & 1 & -1 \\ 3 & 5 & 7 \end{bmatrix}$

71. $\begin{bmatrix} 3 & -17 & -20 \\ 3 & -18 & -20 \\ -1 & 6 & 7 \end{bmatrix}, \begin{bmatrix} 6 & 1 & 20 \\ 1 & -1 & 0 \\ 0 & 1 & 3 \end{bmatrix}$

72. $\begin{bmatrix} 4 & -2 & 3 \\ 8 & -3 & 5 \\ 7 & -2 & 4 \end{bmatrix}, \begin{bmatrix} -2 & 2 & -1 \\ 3 & -5 & 4 \\ 5 & -6 & 4 \end{bmatrix}$

Objective 3.19: *Find the inverse of a given matrix.*

73. $\begin{bmatrix} 3 & 3 & -1 \\ -2 & -2 & 1 \\ -4 & -5 & 2 \end{bmatrix}$

74. $\begin{bmatrix} 1 & 1 & 1 \\ 4 & 5 & 0 \\ 0 & 1 & -3 \end{bmatrix}$

75. $\begin{bmatrix} 2 & -3 \\ 4 & 1 \end{bmatrix}$

76. $\begin{bmatrix} 1 & -2 & 0 \\ 2 & 1 & -3 \\ 3 & 5 & -2 \end{bmatrix}$

Objective 3.20: *Solve systems of equations using the inverse matrix.*

77. $\begin{cases} 3x + 3y - z = 5 \\ -2x - 2y + z = 3 \\ -4x - 5y + 2z = -4 \end{cases}$
(See Problem 73 for the inverse.)

78. $\begin{cases} x + y + z = -2 \\ 4x + 5y = 3 \\ y - 3z = -8 \end{cases}$
(See Problem 74 for the inverse.)

79. $\begin{cases} 2x - 3y = -2 \\ 4x + y = 24 \end{cases}$
(See Problem 75 for the inverse.)

80. $\begin{cases} x - 2y = 7 \\ 2x + y - 3z = -11 \\ 3x + 5y - 2z = -11 \end{cases}$
(See Problem 76 for the inverse.)

***Objective 3.21:** *Solve a homogeneous system of equations.*

81. $\begin{cases} .6x_1 - .6x_2 - .5x_3 = 0 \\ -.5x_1 + .8x_2 - .3x_3 = 0 \\ -.1x_1 - .2x_2 + .8x_3 = 0 \end{cases}$

82. $\begin{cases} .3x_1 - .6x_2 - .5x_3 = 0 \\ -.1x_1 + .7x_2 - .2x_3 = 0 \\ -.2x_1 - .1x_2 + .7x_3 = 0 \end{cases}$

83. $\begin{cases} -5x - y + 2z = 0 \\ x + y - z = 0 \\ x - 3y + 2z = 0 \end{cases}$

84. $\begin{cases} 4x - 3y + 2z = 0 \\ 2x + 2y - 3z = 0 \\ 6x - y - z = 0 \end{cases}$

Objective 3.22: *Find the payoffs for a closed economic model. For each of the given closed input–output matrices let x_3 = C's wages = \$30,000.*

85. $\begin{array}{c} \\ A \\ B \\ C \end{array} \begin{array}{c} \begin{matrix} A & B & C \end{matrix} \\ \begin{bmatrix} .5 & .3 & .1 \\ .2 & .5 & .6 \\ .3 & .2 & .3 \end{bmatrix} \end{array}$

86. $\begin{array}{c} \\ A \\ B \\ C \end{array} \begin{array}{c} \begin{matrix} A & B & C \end{matrix} \\ \begin{bmatrix} .4 & .2 & .3 \\ .2 & .4 & .3 \\ .4 & .4 & .4 \end{bmatrix} \end{array}$

87. $\begin{array}{c} \\ A \\ B \\ C \end{array} \begin{array}{c} \begin{matrix} A & B & C \end{matrix} \\ \begin{bmatrix} \frac{1}{2} & \frac{1}{3} & \frac{1}{4} \\ \frac{1}{4} & \frac{1}{3} & \frac{1}{2} \\ \frac{1}{4} & \frac{1}{3} & \frac{1}{4} \end{bmatrix} \end{array}$

88. $\begin{array}{c} \\ A \\ B \\ C \end{array} \begin{array}{c} \begin{matrix} A & B & C \end{matrix} \\ \begin{bmatrix} .25 & .4 & .35 \\ .35 & .2 & .6 \\ .4 & .4 & .05 \end{bmatrix} \end{array}$

***Objective 3.23:** *Find the necessary output to meet demand for an open economic model. Consider the interrelationships among the production of industries M_1, M_2, and M_3 as given in the following table:*

	M_1	M_2	M_3	External demand	Total output
M_1	20	40	80	60	200
M_2	30	50	100	120	300
M_3	60	60	140	240	500

89. Find T and $I - T$.

90. Find $(I - T)^{-1}$.

91. Find the output to meet the external demand.

92. Find the output to meet the external demand if it changes to 190 for M_1, 250 for M_2, and 80 for M_3.

* Optional section.

SAMPLE TEST

The following sample test (45 minutes) is intended to review the main ideas of the chapter.

1. Solve the systems using each of the following methods exactly once: graphing, adding, and substitution.

a. $\begin{cases} 2x + 3y = 2 \\ 5x - 3y = -27 \end{cases}$ **b.** $\begin{cases} y = 2x - 4 \\ y = \frac{1}{3}x + 1 \end{cases}$

c. $\begin{cases} x + y = 13 \\ 3x - 4y = -17 \end{cases}$

Perform the indicated operations in Problems 2–5, if possible.

$$A = \begin{bmatrix} 4 & -1 & 2 \\ 2 & 1 & -1 \\ 0 & 5 & 3 \end{bmatrix} \quad B = \begin{bmatrix} 1 & 3 & -2 \\ 1 & -1 & 1 \\ 2 & -2 & 2 \end{bmatrix}$$

$$C = [1 \quad -2 \quad 3] \quad D = [4 \quad -5 \quad 1]$$

$$E = \begin{bmatrix} 2 & 0 & 1 \\ 1 & -3 & 4 \end{bmatrix} \quad I = \begin{bmatrix} 1 & 0 & 0 \\ 0 & 1 & 0 \\ 0 & 0 & 1 \end{bmatrix}$$

2. a. $2A$ **b.** $3C - 2D$

3. a. EA **b.** AE

4. a. AB **b.** BA

5. a. $A(B + I)$ **b.** $I(A + B)$

6. a. Write the information on this map of airline routes in matrix form.

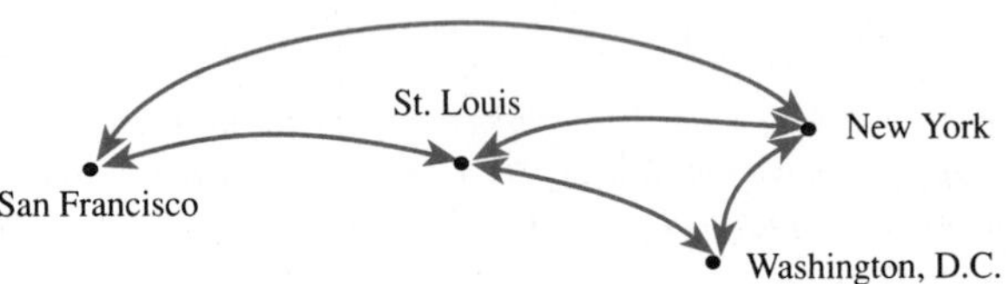

b. Use a matrix to show the number of ways a person could fly making one stop.

c. In how many ways can a person fly from San Francisco to New York making two stops?

Solve the systems in Problems 7–8 by Gauss–Jordan elimination.

7. a. $\begin{cases} 4x - 3y = 18 \\ 5x + 2y = 11 \end{cases}$ **b.** $\begin{cases} 2x - y = 9 \\ x + 5y = -23 \\ 3x - 4y = 26 \end{cases}$

c. $\begin{cases} 3x - y + z = 9 \\ 2x + y - 3z = -2 \end{cases}$

8. $\begin{cases} x - 2y + z = 6 \\ y - 3x + w = -2 \\ w - x - y = 2 \\ 2y - 3w + 2z = -2 \end{cases}$

9. a. Find the inverse of $A = \begin{bmatrix} 3 & 3 & -1 \\ -2 & -2 & 1 \\ -4 & -5 & 2 \end{bmatrix}$.

b. Use the inverse from part **a** to solve

$$\begin{cases} 3x + 3y - z = 8 \\ -2x - 2y + z = -1 \\ -4x - 5y + 2z = 3 \end{cases}$$

10. A manufacturer produces two products, I and II. The products require three ingredients—A, B, and C—to be used as follows:

	A	B	C
Product I	6	3	5
Product II	5	4	7

If the manufacturer has a supply of 3,400 units of items A and C and 2,000 units of item B, how many of each product can be manufactured?

MODELING
APPLICATION 1

Gaining a Competitive Edge in Business

Solartex manufactures solar collector panels. During the first year of operation, rent, insurance, utilities, and other fixed costs averaged \$8,500 per month. Each panel sold for \$320 and cost the company \$95 in materials and \$55 in labor. Since company resources are limited, Solartex cannot spend more than \$20,000 in any one month. Sunenergy, another company selling similar panels, competes directly with Solartex. Last month, Sunenergy manufactured 85 panels at a total cost of \$20,475, but the previous month produced only 60 panels at a total cost of \$17,100.

Mathematical modeling involves creating mathematical equations and procedures to make predictions about the real world. Typical textbook problems focus on limited, specific skills, but when confronted with a real-life example you are often faced with a myriad of "facts" without specific clues on how to fit them together to make predictions. In this book, you will be given a modeling application and asked to write a paper using the given information. You will need to do some research to have adequate data. There are no "right answers" for these papers. In the *Student's Solution Manual*, a paper is presented for this modeling application, but your paper could certainly take a quite different direction.

For this first application, the following questions might help you get started: What is the cost equation for Solartex? How many panels can be manufactured by Solartex given the available capital? What is the minimum number of panels that should be produced? In order to make a profit, revenue must exceed cost; what is Solartex's break-even point? How would you compare Solartex and Sunenergy?

Feel free to supply information that is not given above. For example, a study of market demand could be useful. Suppose you find that at \$75 per panel, Solartex (or a real company you have studied) could sell 200 panels per month, but at \$450, it could sell only 20 panels per month. On the other hand, if Solartex had to sell the panels at \$225 each, it could afford material that would limit it to only 30 panels per month, but at \$450 each, it could afford sufficient material to supply 100 panels per month. What is the equilibrium point for this information?

Write a paper based on this modeling application.

4

Linear Programming

CHAPTER OVERVIEW

This chapter introduces the topic of linear programming. Linear programming is applied when you are interested in maximizing or minimizing a linear function—for example, maximizing profits, minimizing costs, finding the most efficient shipping schedules, minimizing waste in production, securing the proper mixes of ingredients in a product, controlling inventories, or finding the most efficient assignment of personnel.

PREVIEW

Graphing systems of linear inequalities is reviewed in the first section, and then formulating linear programming problems is discussed. These two ideas are then tied together to solve linear programming problems by graphing. Next, an algebraic process for solving linear programming problems called the *simplex method* is introduced. The simplex method is first applied to what is called a "standard" problem. Then we apply it to variations of standard problems. The final section, Duality, ties together, in a very nice and surprising way, maximum- and minimum-type problems.

PERSPECTIVE

This chapter gives us a basic understanding of one of the most important topics in this course—linear programming. However, most real-life applications are not limited to two variables, and since graphing is limited to two dimensions, we also need an algebraic method. The simplex method is this algebraic method and provides a step-by-step procedure for optimization which can easily be adapted to computer solution. Computer solution is absolutely essential for many real-world applications, which may use hundreds of variables and constraints. Many calculators, such as the TI-81, will handle up to 6×6 matrices.

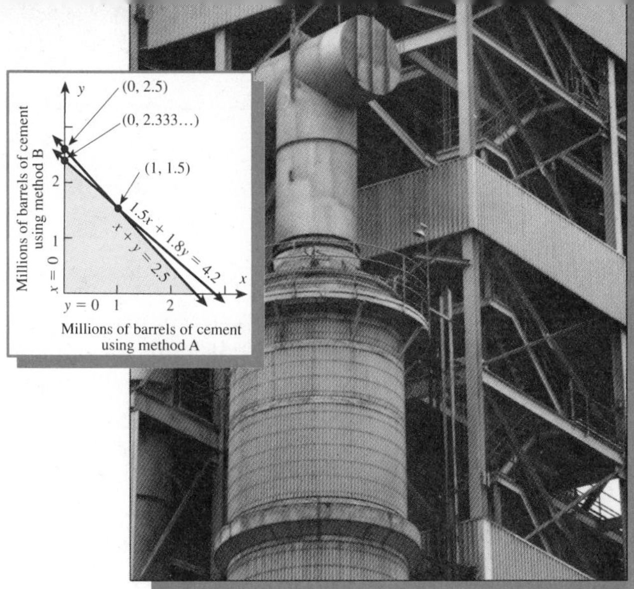

MODELING APPLICATION 2

Air Pollution Control

Alco Cement Company produces cement. The Environmental Protection Agency has ordered Alco to reduce the amount of emissions released into the atmosphere during production. The company wants to comply, but it also wants to do so at the least possible cost. Present production is 2.5 million barrels of cement, and 2 pounds of dust are emitted for every barrel of cement produced. The cement is produced in kilns that are presently equipped with mechanical collectors. However, in order to reduce the emissions to the required level, the mechanical collectors must be replaced by four-field electrostatic precipitators, which would reduce emissions to .5 pound of dust per barrel of cement, or by five-field precipitators, which would reduce emission to .2 pound per barrel. The capital and operating costs for the four-field precipitator are 14¢ per barrel of cement produced, and for the five-field precipitator costs are 18¢ per barrel.*

After you have finished this chapter, write a paper that shows the minimum cost for the control methods that Alco should use in order to reduce particulate emissions by 4.2 million pounds. See page 151 for some general guidelines on preparing a paper.

* This application is adapted from R. E. Kohn, "A Mathematical Programming Model for Air Pollution Control," *School, Science, and Mathematics*, June 1969, pp. 487–499.

APPLICATIONS

Management *(Business, Economics, Finance, and Investments)*
Allocation of resources in production
(4.2, Problems 1–2, 13, 17, and 21; 4.3, Problems 39–41; 4.6, Problem 26; 4.7, Problems 43 and 44)
Allocation of resources in manufacturing
(4.2, Problems 3, 4, and 22; 4.6, Problems 27–29 and 34)
Maximizing yield from an investment
(4.2, Problems 7 and 8; 4.3, Problems 46 and 47; 4.6, Problem 32; 4.7, Problem 42)
Office management
(4.2, Problem 14; 4.6, Problem 33; 4.8, Problems 11, 27)
Maximizing profit
(4.2, Problems 11, 12, 15, 21, and 29; 4.3, Problems 42 and 43; 4.5, Problems 29–32 and 35; 4.8, Problems 29, 57, and 58)
Operating costs (4.2, Problem 16)
Staff utilization (4.2, Problems 24, 25, and 30)
Transportation problem; minimize shipping costs
(4.2, Problems 9, 10, and 26–28; 4.6, Problems 30–31; 4.7, Problems 45–47; 4.8, Problem 12)

Management *(continued)*
Allocation of advertising resources
(4.5, Problems 33 and 34)
Minimizing costs
(4.2, Problem 20; 4.7, Problem 39; 4.8, Problem 30)
Maximize income (4.8, Problems 10, 26, 28)

General Interest
Allocation of time in an exercise program
(4.2, Problem 23; 4.7, Problem 41)

Life Sciences *(Biology, Ecology, Health, and Medicine)*
Maximum number of animals
(4.2, Problems 18–19)
Minimizing exercise time to achieve certain goals
(4.2, Problem 23; 4.7, Problem 41)
Diet problem:
Kellogg's Corn Flakes and Post Honeycombs
(4.2, Problem 5; 4.3, Problem 45)

Life Sciences *(continued)*
Kellogg's Corn Flakes and Kellogg's Raisin Bran
(4.2, Problem 6)
Meeting the minimum daily requirements at the lowest cost
(4.3, Problem 44; 4.8, Problems 9, 25, 28, 79–81)
Determining the proper diet for horses
(4.6, Problem 25)
Food supply (4.8, Problems 55 and 56)
Allocation of oil reserves (low, medium, and high grade)
(4.7, Problem 40)

Society of Actuaries
(4.2, Problem 22; 4.5, Problem 35)

CPA examinations
(4.8, Sample Test Problems 5–10)

Modeling Application—
Air Pollution Control Model

4.1

Systems of Linear Inequalities

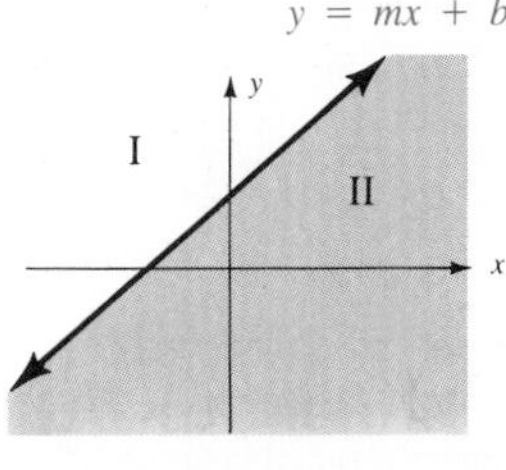

FIGURE 4.1
Half-planes

A **linear inequality** in two variables can be written in one of the following forms:

$$ax + by < c \qquad ax + by > c$$
$$ax + by \le c \qquad ax + by \ge c$$

where a, b, and c are constants (a and b are not both 0). The graphing of a linear inequality is similar to the graphing of a linear equation. Any line divides the plane into three regions, as shown in Figure 4.1. Regions I and II are called **half-planes**, so we can say the three sets determined by a line are the two half-planes and the set of points on the line itself. The line is called the **boundary** of the half-planes. If the boundary is included in the half-plane, the half-plane is said to be **closed**; if the boundary is not included, it is said to be **open**. Half-planes are drawn by showing the boundary as a dashed line, as in Figure 4.2c.

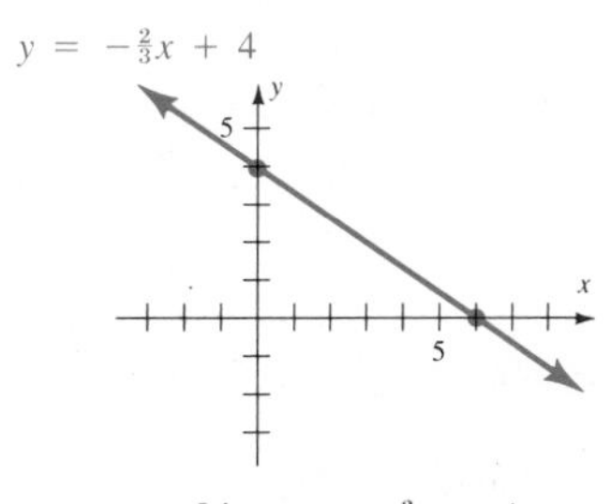

a. Line $y = -\frac{2}{3}x + 4$

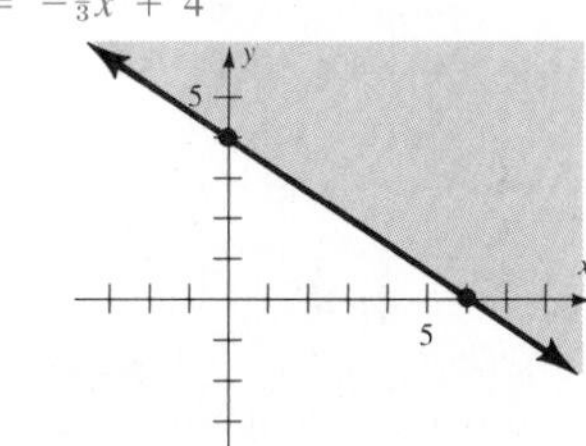

b. Closed half-plane $y \ge -\frac{2}{3}x + 4$
Draw boundary as a solid line

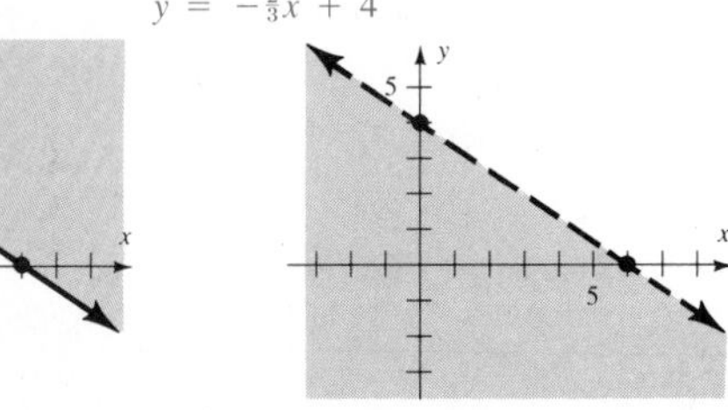

c. Open half-plane $y < -\frac{2}{3}x + 4$
Draw boundary as a dashed line

FIGURE 4.2
Graphing half-planes

It is apparent in Figure 4.2 that every linear inequality has an associated equation that is the boundary of that half-plane. For example, $2x + 3y < 12$ has the boundary line $2x + 3y = 12$. This line is graphed by solving for y: $y = -\frac{2}{3}x + 4$. After you have drawn the boundary (either solid or dashed), the remaining question is: Which half-plane satisfies the inequality? The easiest way to decide is to choose *any* test point not on the boundary line. If it satisfies the inequality, then the solution is the half-plane containing the test point. If the test point does not satisfy the inequality, then the solution is the half-plane that does not contain the test point.

EXAMPLE 1 Graph $5x + 2y - 10 \le 0$.

Solution *Step* 1: *Graph the boundary line.* The boundary is included if the given inequality is of the form $\ge$ or $\le$, so draw $5x + 2y - 10 = 0$ as a solid line. The easiest way to sketch this line is to put it into slope–intercept form: $y = -\frac{5}{2}x + 5$.

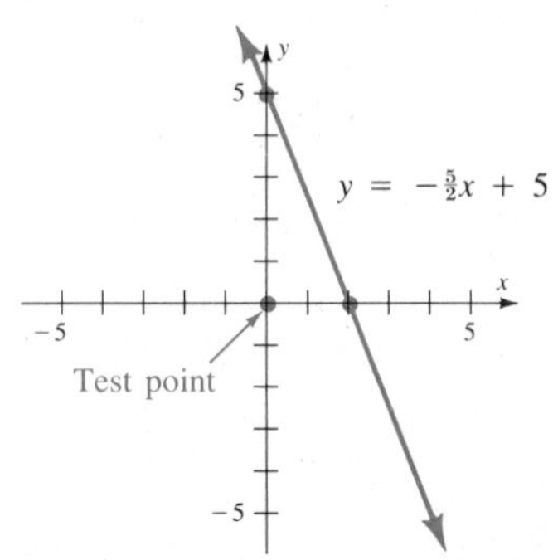

Step 2: *Choose a test point.* The point (0, 0) is usually the best choice because it simplifies the arithmetic:

$$5x + 2y - 10 \leq 0$$

Test (0, 0):

$$5(0) + 2(0) - 10 \leq 0$$

$$-10 \leq 0 \qquad \text{True}$$

Step 3: *Shade the appropriate half-plane.* If the test point satisfies the inequality (it is true, as shown in step 2), shade in the half-plane containing the test point. Otherwise, shade in the other half-plane.

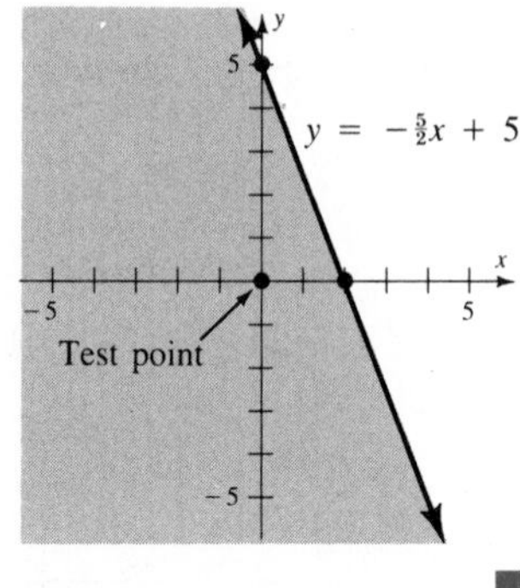

■

Graphing Linear Inequalities

1. Graph the boundary line.
2. Choose a test point not on the boundary.
3. Shade the appropriate half-plane.
   If the coordinates of the test point satisfy the inequality, then shade the half-plane containing the test point; if the coordinates of the test point do not satisfy the inequality, then shade the half-plane not containing the test point.

EXAMPLE 2 Graph $y > -20x + 110$.

Solution *Step 1:* The boundary is $y = -20x + 110$; $m = -\frac{20}{1}$ and $b = 110$. Draw the boundary as a dashed line because it is not included in the half-plane (> symbol).

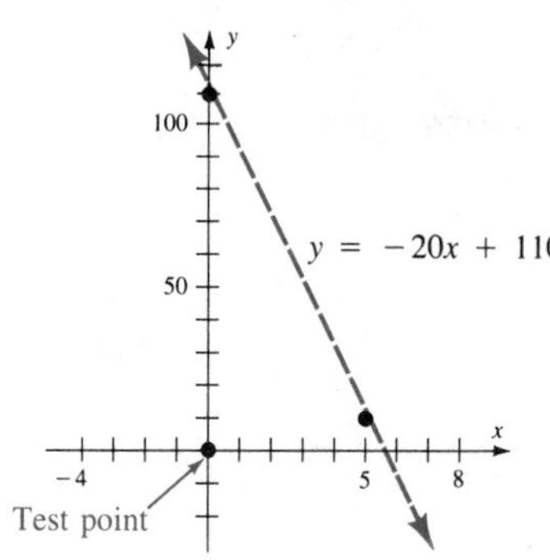

Step 2:

$$y > -20x + 110$$

Test point (0, 0):

$$0 > -20(0) + 110 \qquad \text{False}$$

Step 3: Graph the half-plane that does not contain the test point.

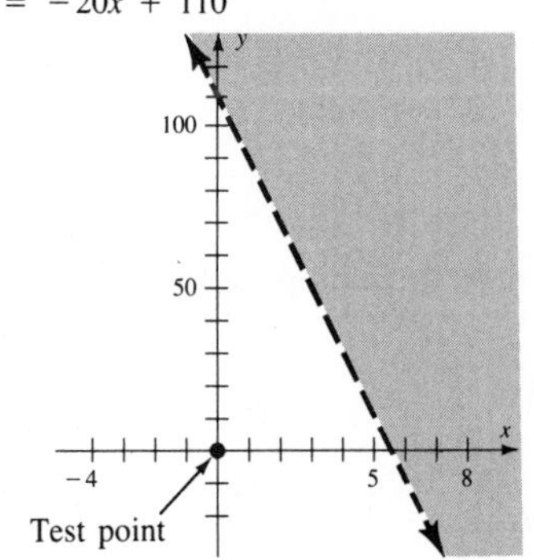

■

> **CALCULATOR COMMENT**
>
> To use the TI-81 for a problem such as Example 2, press
>
> 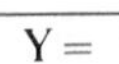
>
> and then input the function:
>
> 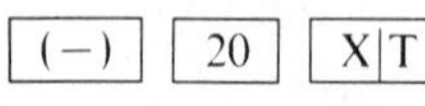
>
> [+] [110]
>
> Finally, draw the graph:
>
> [GRAPH]

There are many applications in which inequalities play an important role. In Chapter 2 we found the *break-even point*—that is, the place where cost and revenue are the same. Sometimes, breaking even for a business takes the form of a linear equation. Suppose a company knows that it will break even if 50 items are sold at \$3 each or if 10 items are sold at \$5 each.* This gives a *break-even linear equation*:

$$y = -20x + 110$$

where x is the price and y is the number of items. Suppose that in order to make a profit it is necessary for

$$y > -20x + 110$$

(This was graphed in Example 2.) However, in this application, it is also necessary that x and y both be positive:

$$x \geq 0$$
$$y \geq 0$$

Furthermore, market research shows that the price must be \$8 or less:

$$x \leq 8$$

This information taken together gives a **system of linear inequalities** similar to a system of equations:

$$\begin{cases} y > -20x + 110 \\ x \geq 0 \\ y \geq 0 \\ x \leq 8 \end{cases}$$

The region determined by the system of inequalities is the solution of the system and is similar to the point of intersection for a system of equations because it is the intersection of the individual inequalities in the system.

EXAMPLE 3 Graph the system $\begin{cases} y > -20x + 110 \\ x \geq 0 \\ y \geq 0 \\ x \leq 8 \end{cases}$

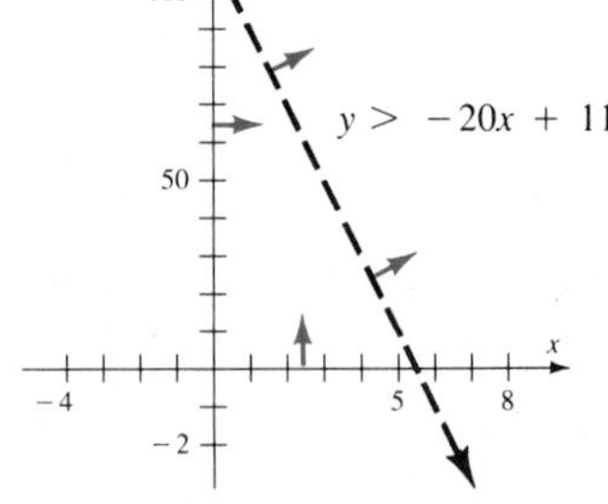

Solution Graph each half-plane, but instead of shading, mark the appropriate regions with arrows. For this example,

$$x \geq 0$$
$$y \geq 0$$

gives the first quadrant, which is marked with arrows as shown.

* In practice, this is more likely to be 50,000 and 10,000 items. We use 50 and 10 here to keep the numbers easy to handle. To find the equation of the line where x is the price and y is the number of items, we use the points (3, 50) and (5, 10). The slope is

$$m = \frac{10 - 50}{5 - 3} = -20$$

and $y = mx + b$, so $50 = -20(3) + b$ or $b = 110$. Thus $y = -20x + 110$.

Next graph $y = -20x + 110$ and use a test point, say, $(0, 0)$, to determine the half-plane. Now draw $x = 8$ and mark it with arrows to show that $x \leq 8$. Finally shade the region of the plane that represents the intersection:

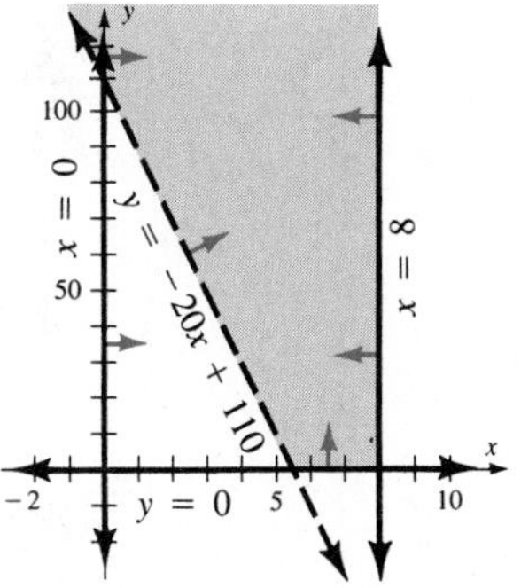

■

EXAMPLE 4 Graph the solution of the system $\begin{cases} 2x + y \leq 3 \\ x - y > 5 \\ x \geq 0 \\ y \geq -10 \end{cases}$

Solution The graphs of the individual inequalities and their intersections are shown in the figure. Note the use of arrows to show the solutions of the individual inequalities. This device replaces the use of a lot of shading, which can be confusing if there are many inequalities in the system.

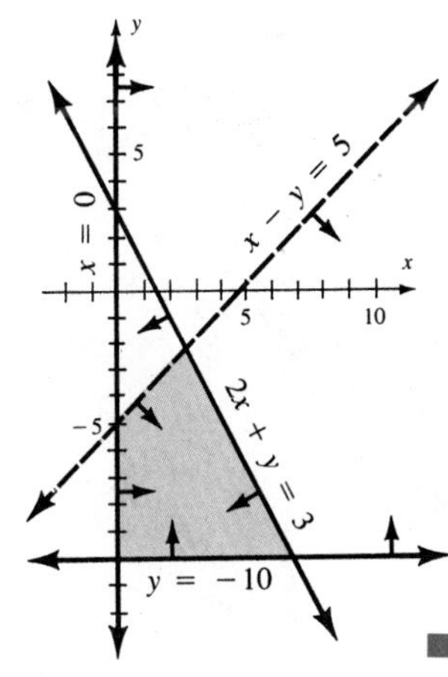

■

4.1 Problem Set

Graph the solution for each of the linear inequalities in Problems 1–24.

1. $y \geq 2x - 3$
2. $y \geq 3x + 2$
3. $y \geq 4x + 5$
4. $y \leq \frac{1}{2}x + 2$
5. $y \leq \frac{2}{3}x - 4$
6. $y \leq \frac{4}{5}x - 3$
7. $x - y > 3$
8. $2x + y < -3$
9. $3x + y < 4$
10. $x + 2y > 4$
11. $x - 3y > 12$
12. $3x - 4y > 8$
13. $y \leq 6$
14. $x \geq 2$
15. $y < -2$
16. $y > 0$
17. $x < 0$
18. $x < 8$
19. $260x - 1{,}040y > 11{,}250$
20. $150x - 450y < 7{,}200$
21. $.06x - .05y < 10{,}000$
22. $.03x - .05y > 10{,}000$
23. $.08x + .10y \geq 500$
24. $.10x - .08y \leq 1{,}500$

Graph the solution of each system in Problems 25–60.

25. $\begin{cases} x \geq 0 \\ y \leq 0 \end{cases}$
26. $\begin{cases} x \geq 0 \\ y \geq 0 \end{cases}$

27. $\begin{cases} x \le 0 \\ y \le 0 \end{cases}$

28. $\begin{cases} x \ge 0 \\ y \ge 0 \\ x \le 5 \\ y \le 6 \end{cases}$

29. $\begin{cases} x \ge 0 \\ y \ge 0 \\ x < 8 \\ y < 5 \end{cases}$

30. $\begin{cases} x \ge 0 \\ y \ge 0 \\ x < 500 \\ y < 1{,}000 \end{cases}$

31. $\begin{cases} y \le 3x - 4 \\ y \ge -2x + 5 \end{cases}$

32. $\begin{cases} 2x + y > 3 \\ 3x - y < 2 \end{cases}$

33. $\begin{cases} 3x - 2y \ge 6 \\ 2x + 3y \le 6 \end{cases}$

34. $\begin{cases} x + 2y \le 18 \\ x + y \ge 4 \end{cases}$

35. $\begin{cases} x + y > 4 \\ x - y > -2 \end{cases}$

36. $\begin{cases} 2x + 3y \ge 3 \\ 2x + 3y \le 9 \end{cases}$

37. $\begin{cases} x - 10 \le 0 \\ x \ge 0 \end{cases}$

38. $\begin{cases} y - 5 \le 0 \\ y \ge 0 \end{cases}$

39. $\begin{cases} y - 25 \le 0 \\ y \ge 0 \end{cases}$

40. $\begin{cases} -5 < x \\ 3 \ge x \\ 5 > y \\ -2 \le y \end{cases}$

41. $\begin{cases} -10 \le x \\ x \le 6 \\ 3 < y \\ y < 8 \end{cases}$

42. $\begin{cases} -5 < x \\ x \le 2 \\ -4 \le y \\ y < 9 \end{cases}$

43. $\begin{cases} x \ge 0 \\ y \ge 0 \\ x + y \le 8 \\ y \le 4 \\ x \le 6 \end{cases}$

44. $\begin{cases} x \ge 0 \\ y \ge 0 \\ x + y \le 9 \\ 2x - 3y \ge -6 \\ x - y \le 3 \end{cases}$

45. $\begin{cases} x \ge 0 \\ y \ge 0 \\ x \le 20 \\ y \le 45 \\ 2x + y \le 65 \end{cases}$

46. $\begin{cases} x \ge 0 \\ y \ge 0 \\ x \le 15 \\ x + 2y \le 120 \\ 4x + y \le 95 \end{cases}$

47. $\begin{cases} x \ge 0 \\ y \ge 0 \\ 4x + y \ge 86 \\ 2x + y \ge 78 \end{cases}$

48. $\begin{cases} x \ge 0 \\ y \ge 0 \\ 7x + 5y \le 140 \\ 2x + y \le 31 \\ 11x + 4y \le 154 \end{cases}$

49. $\begin{cases} x \ge 0 \\ y \ge 0 \\ x + y \ge 35 \\ 2x + y \ge 39 \\ x + 2y \ge 46 \end{cases}$

50. $\begin{cases} x \ge 0 \\ y \ge 0 \\ 3x + 2y \ge 36 \\ x + y \ge 16 \\ x + 2y \ge 25 \end{cases}$

51. $\begin{cases} x \ge 0 \\ y \ge 0 \\ 2x + y \ge 26 \\ 3x + 2y \ge 48 \\ x + y \ge 19 \end{cases}$

52. $\begin{cases} x \ge 0 \\ y \ge 0 \\ y \le 20 \\ x \le 12 \\ 5x + 4y \le 110 \\ 3x + y \le 45 \end{cases}$

53. $\begin{cases} x \ge 0 \\ y \ge 0 \\ 2x + y \ge 8 \\ y \le 5 \\ x - y \le 2 \\ 3x - y \ge 5 \end{cases}$

54. $\begin{cases} 2x + 3y \le 30 \\ 3x + 2y \ge 20 \\ x \ge 0 \\ y \ge 0 \end{cases}$

55. $\begin{cases} 2x - 3y + 30 \ge 0 \\ 3x - 2y + 20 \le 0 \\ x \le 0 \\ y \ge 0 \end{cases}$

56. $\begin{cases} x + y - 10 \le 0 \\ x + y + 4 \ge 0 \\ x - y \le 6 \\ y - x \le 4 \end{cases}$

57. $\begin{cases} 2x - 3y \le 20 \\ 3x + 2y \le 10 \\ x \ge 0 \\ y \le 0 \end{cases}$

58. $\begin{cases} 6x - 3y - 15 \le 0 \\ 4x + 2y - 30 \le 0 \\ x \ge 0 \\ y \ge 0 \end{cases}$

59. $\begin{cases} 4x - 3y + 2 \ge 0 \\ x + y - 3 \le 0 \\ x \ge 0 \\ y \ge 0 \end{cases}$

60. $\begin{cases} 5x - 6y + 3 \ge 0 \\ x + y - 2 \le 0 \\ x \ge 0 \\ y \ge 0 \end{cases}$

4.2 Formulating Linear Programming Models

One of the most difficult tasks in solving a real-world problem is the construction of a mathematical model that simulates the situation. In the 1940s a new math-

Historical Note

George Dantzig developed the simplex method in the 1940s long before the possibility of extensive computer usage was envisioned. However, with computers, the simplex method is extremely useful in many widespread applications.

In 1984, Narendra Karmarkar, a researcher at Bell Laboratories, announced a linear programming technique that supposedly improves the simplex method. It is too early to know if this method is a significant improvement. Computer simulation at the University of California, Berkeley, showed that the Karmarkar technique solved certain selected problems up to four times faster than the simplex method. However, you must be careful in assuming that you do not need to learn the simplex method because you will wait for this new method. In 1979, Leonid Khachian proposed an alternate method, called the ellipsoid method, *which was also purported to be better than the simplex method, but it turned out to be no better than Dantzig's simplex method.*

ematical model called **linear programming** was found to be applicable to a wide range of situations in which the maximum or minimum value of some variables was needed. In this section, you will learn how to formulate or build a model for a variety of applications (see the examples listed in the box), and in the next section you will learn how to solve these models using graphical techniques. (In Section 4.5, an algebraic method, the *simplex method*, for solving linear programming problems is introduced.)

Examples of linear programming models

Allocation of Resources Models
- Maximize profit from a manufacturing process
- Maximize revenue from a manufacturing process
- Minimize costs from a manufacturing process
- Maximize production
- Find the optimal use of land
- Utilization of a sales force
- Utilization of office staff

Nutritional (or Diet) Models
- Minimize cost of a diet with certain nutritional requirements
- Minimize calories given certain nutritional requirements
- Manufacture a processed food to meet government regulations

Investment Strategies
- Maximize the return on an investment
- Allocate funds over several possible investments
- Minimize expenses

Transportation Model
- Minimize shipping costs

EXAMPLE 1 **Allocation of Resources in Production** A farmer has 100 acres on which to plant two crops: corn and wheat. To produce these crops, there are certain expenses, as shown in the table.

Item	Cost per acre
Corn	
Seed	$ 12
Fertilizer	58
Planting/care/harvesting	50
Total	$120
Wheat	
Seed	$ 40
Fertilizer	80
Planting/care/harvesting	90
Total	$210

After the harvest, the farmer must usually store the crops while awaiting favorable market conditions. Each acre yields an average of 110 bushels of corn or

30 bushels of wheat. The limitations of resources are

Available capital:	\$15,000
Available storage facilities:	4,000 bushels

If the net profit (the profit *after* all expenses have been subtracted) per bushel of corn is \$1.30 and for wheat is \$2.00, how should the farmer plant the 100 acres to maximize profits?

Solution First, you might try to solve this problem by using your intuition. If you plant all 100 acres with wheat, the production is $30 \times 100 = 3{,}000$ bushels, for a net profit of $3{,}000 \times 2 = \$6{,}000$. But to plant 100 acres with wheat would cost $\$210 \times 100 = \$21{,}000$, and only \$15,000 is available. On the other hand, if 100 acres of corn are planted, the total cost is $\$120 \times 100 = \$12{,}000$ and the net profit is $110 \times 100 \times \$1.30 = \$14{,}300$. However, the yield of 11,000 bushels (110×100) cannot be stored since there are facilities to store only 4,000 bushels.

To formulate a mathematical model, begin by letting

x = Number of acres to be planted in corn

y = Number of acres to be planted in wheat

There are certain limitations, or **constraints**. The number of acres planted cannot be negative, so

$$\boldsymbol{x \geq 0}$$
$$\boldsymbol{y \geq 0}$$

These constraints apply in almost every model, even though they are not explicitly stated as part of the given problem. Be sure, however, to state them when listing the constraints in the solution.

The amount of available land is 100 acres, so

$$\boldsymbol{x + y \leq 100}$$

Why not $x + y = 100$? It might be more profitable for the farmer to leave some land out of production. That is, it is not *necessary* to plant all the land.

We also know that

$120x$ = Expenses for planting corn

$210y$ = Expenses for planting wheat

The total expenses cannot exceed \$15,000; this is the *available capital*:

$$\boldsymbol{120x + 210y \leq 15{,}000}$$

The yields are

$110x$ = Yield of acreage planted in corn

$30y$ = Yield of acreage planted in wheat

The total yield cannot exceed the storage capacity of 4,000 bushels, so

$$\boldsymbol{110x + 30y \leq 4{,}000}$$

We summarize the constraints (in boldface above) in the following system:

$$\begin{cases} x \geq 0 \\ y \geq 0 \\ x + y \leq 100 \\ 120x + 210y \leq 15{,}000 \\ 110x + 30y \leq 4{,}000 \end{cases}$$

Now let P represent the total profit. The farmer wants to maximize this profit. A function that is to be maximized or minimized is called the **objective function**.

$$\begin{aligned} \text{PROFIT FROM CORN} &= \text{VALUE} \cdot \text{AMOUNT} \\ &= 1.30 \cdot 110x \\ &= 143x \\ \text{PROFIT FROM WHEAT} &= \text{VALUE} \cdot \text{AMOUNT} \\ &= 2.00 \cdot 30y \\ &= 60y \end{aligned}$$

$$\begin{aligned} P &= \text{PROFIT FROM CORN} + \text{PROFIT FROM WHEAT} \\ &= 143x + 60y \end{aligned}$$

The linear programming model is stated as follows:

Maximize: $P = 143x + 60y$

Subject to: $\begin{cases} x \geq 0 \\ y \geq 0 \\ x + y \leq 100 \\ 120x + 210y \leq 15{,}000 \\ 110x + 30y \leq 4{,}000 \end{cases}$ ■

This linear programming model will be solved in the next section.

EXAMPLE 2 **Allocation of resources in manufacturing.** The Wadsworth Widget Company manufactures two types of widgets: regular and deluxe. Each widget is produced at a station consisting of a machine and a person who finishes the widgets by hand. The regular widget requires 3 hours of machine time and 2 hours of finishing time. The deluxe widget requires 2 hours of machine time and 4 hours of finishing time. The profit on the regular widget is \$25, and on the deluxe widget the profit is \$30. If the workday is 8 hours, how many of each type of widget should be produced at each station per day in order to maximize the profit?

Solution We want to maximize the profit, P. There are two types of items, regular and deluxe. Let

x = Number of regular widgets produced

y = Number of deluxe widgets produced

Then

$$\text{Profit from regular widgets} = 25x$$
$$\text{Profit from deluxe widgets} = 30y$$
$$P = \text{PROFIT FROM REGULAR WIDGETS} + \text{PROFIT FROM DELUXE WIDGETS}$$
$$= 25x + 30y$$

Two constraints are

$x \geq 0$ The number of widgets must be nonnegative.
$y \geq 0$

Production time is summarized in the table:

	Machine time (hr)	Manual time (hr)
Regular widget, x	3 each; $3x$ total	2 each; $2x$ total
Deluxe widget, y	2 each; $2y$ total	4 each; $4y$ total
Total workday	$3x + 2y$	$2x + 4y$

Thus

$3x + 2y \leq 8$ Machine workday
$2x + 4y \leq 8$ Manual workday

The linear programming model is:

Maximize: $P = 25x + 30y$

$$\text{Subject to: } \begin{cases} x \geq 0 \\ y \geq 0 \\ 3x + 2y \leq 8 \\ 2x + 4y \leq 8 \end{cases}$$

■

EXAMPLE 3 **Diet problem.** A convalescent hospital wishes to provide, at a minimum cost, a diet that has a minimum of 200 grams of carbohydrates, 100 grams of protein, and 120 grams of fats per day. These requirements can be met with two foods:

	Contents per oz		
Food	Carbohydrates	Protein	Fats
A	10 g	2 g	3 g
B	5 g	5 g	4 g

If food A costs 29¢ per ounce and food B costs 15¢ per ounce, how many ounces of each food should be purchased for each patient per day in order to meet the minimum requirements at the lowest cost?

Solution Let

$x =$ Number of ounces of food A

$y =$ Number of ounces of food B

The *minimum cost*, C, is found by:

Cost of food A $= .29x$

Cost of food B $= .15y$

$C = .29x + .15y$

The constraints are

$x \geq 0$ The amounts of food must be nonnegative.

$y \geq 0$

The table summarizes the nutrients provided:

Food	Amount (in ounces)	Total consumption (in grams)		
		Carbohydrates	Protein	Fats
A	x	$10x$	$2x$	$3x$
B	y	$5y$	$5y$	$4y$
Total		$10x + 5y$	$2x + 5y$	$3x + 4y$

Daily requirements are:

$$10x + 5y \geq 200$$
$$2x + 5y \geq 100$$
$$3x + 4y \geq 120$$

The linear programming model is:

Minimize: $C = .29x + .15y$

$$\text{Subject to: } \begin{cases} x \geq 0 \\ y \geq 0 \\ 10x + 5y \geq 200 \\ 2x + 5y \geq 100 \\ 3x + 4y \geq 120 \end{cases}$$

■

EXAMPLE 4 **Investment.** Brown Bros., Inc., is an investment company analyzing a pension fund for a certain company. A maximum of \$10 million is available to invest in two places. No more than \$8 million can be invested in stocks yielding 12%, and at least \$2 million can be invested in long-term bonds yielding 8%. The stock-to-bond investment ratio cannot be more than 1 to 3. How should Brown Bros. advise its client so that the pension fund will receive the maximum yearly return on investment?

Solution To build this model you need to use the *simple interest formula*:

$$I = Prt$$

where

I = Interest	The amount paid for the use of another's money
P = Principal	The amount invested
r = Interest rate	Write this as a decimal. It is assumed to be an annual interest rate, unless otherwise stated.
t = time	In years, unless otherwise stated

Let

x = Amount invested (in millions) in stocks (12% yield)

y = Amount invested (in millions) in bonds (8% yield)

Then

Stocks: $I = .12x$ $\quad P = x, r = .12$, and $t = 1$

Bonds: $I = .08y$ $\quad P = y, r = .08$, and $t = 1$

The return on investment, R (in millions), is found by

$$R = .12x + .08y$$

The constraints are:

$x \geq 0$	Investments are nonnegative.
$y \geq 0$	
$x + y \leq 10$	Maximum investment is \$10 million; note that the unit chosen for this problem is in millions of dollars.
$x \leq 8$	No more than \$8 million in stocks
$y \geq 2$	No less than \$2 million in bonds
$3x \leq y$	Must invest \$3 million in bonds for every \$1 million invested in stocks for a stock-to-bond ratio of 1 to 3. That is, $\frac{\text{stock}}{\text{bond}} \leq \frac{1}{3}$; 3 stock $\leq$ bond; $3x \leq y$

Thus the linear programming model for this problem is:

Maximize: $R = .12x + .08y$

$$\text{Subject to: } \begin{cases} x \geq 0 \\ y \geq 0 \\ x \leq 8 \\ y \geq 2 \\ x + y \leq 10 \\ 3x \leq y \end{cases}$$

■

EXAMPLE 5 **Transportation problem.** Sears ships a certain air-conditioning unit from factories in Portland, Oregon, and Flint, Michigan, to distribution centers in Los Angeles, California, and Atlanta, Georgia. The shipping costs are summarized in the table:

Source	Destination	Shipping cost
Portland	Los Angeles Atlanta	\$30 \$40
Flint	Los Angeles Atlanta	\$60 \$50

The supply and demand, in number of units, is shown below:

Supply	Demand
Portland, 200 Flint, 600	Los Angeles, 300 Atlanta, 400

How should shipments be made from Portland and Flint to minimize the shipping cost?

Solution The information can be summarized by the following "map."

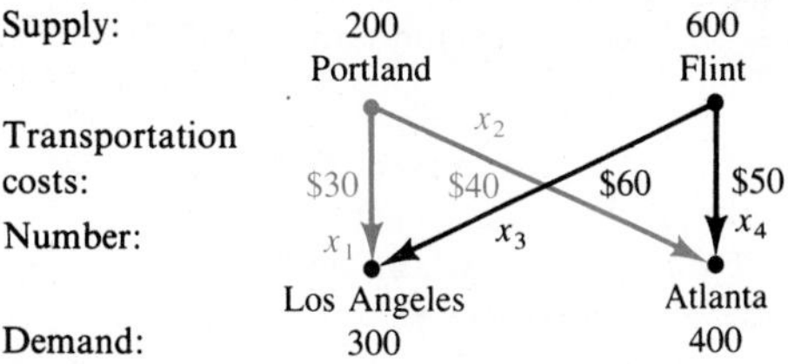

Suppose the following number of units is shipped:

Source	Destination	Number	Shipping cost
Portland	Los Angeles Atlanta	x_1 x_2	$30x_1$ $40x_2$
Flint	Los Angeles Atlanta	x_3 x_4	$60x_3$ $50x_4$

The linear programming problem is then:

$$\text{Minimize: } C = 30x_1 + 40x_2 + 60x_3 + 50x_4$$

$$\text{Subject to: } \begin{cases} x_1 + x_2 \leq 200 \\ x_3 + x_4 \leq 600 \\ x_1 + x_3 \geq 300 \\ x_2 + x_4 \geq 400 \\ x_1 \geq 0, \quad x_2 \geq 0, \quad x_3 \geq 0, \quad x_4 \geq 0 \end{cases}$$

■

Note that as we progressed from Example 1 to Example 5, we were able to formulate the linear programming problem with fewer and fewer words so that the information in the problem was translated almost directly into a linear programming model. Also, notice in Example 5 that we stated the nonnegative constraints together for convenience. In the next section we develop graphical techniques for solving these types of problems.

4.2 Problem Set

Write a linear programming model for Problems 1–30. Do not attempt to solve the problems, but be sure to define all your variables.

APPLICATIONS

1. A farmer has 500 acres on which to plant two crops: corn and wheat. It costs \$120 per acre to produce corn and \$60 per acre to produce wheat, and there is \$24,000 available to pay for this year's production. If the yield per acre is 100 bushels of corn or 40 bushels of wheat and the farmer has contracted to store at least 20,000 bushels, how much should the farmer plant to maximize the profits when the net profit is \$1.20 per bushel for corn and \$2.50 per bushel for wheat?

2. A farmer has 1,500 acres on which to plant three crops: corn, wheat, and soybeans. It costs \$150 per acre to produce corn, \$80 per acre to produce wheat, and \$60 per acre to produce soybeans, and there is \$36,000 available to pay for this year's production. If the yield per acre is 100 bushels of corn, or 40 bushels of wheat, or 60 bushels of soybeans, and the farmer has contracted to store at least 20,000 bushels, how much should the farmer plant to maximize the profits when the net profit is \$1.00 per bushel for corn, \$2 per bushel for wheat, and \$1.50 per bushel for soybeans?

3. The Thompson Company manufactures two industrial products: a standard product (\$45 profit per item) and an economy product (\$30 profit per item). These products are built using machine time and manual labor. The standard product requires 3 hours machine time and 2 hours manual labor. The economy model requires 3 hours machine time and no manual labor. If the week's supply of manual labor is limited to 800 hours and machine time to 1,500 hours, how much of each type of product should be produced each week in order to maximize the profit?

4. A furniture manufacturer makes chairs, sofas, and sofabeds. The production process is divided into carpentry, finishing, and upholstery. Manufacture of a chair requires 3 hours of carpentry, 1 hour of finishing, and 2 hours of upholstery. Manufacture of a sofa requires 5 hours of carpentry, 2 hours of finishing, and 4 hours of upholstery. Manufacture of a sofabed requires 8 hours of carpentry, 3 hours of finishing, and 4 hours of upholstery. The manufacturer has available each day 120 hours of manual labor for carpentry, 24 hours of manual labor for finishing, and 96 hours of manual labor for upholstery. The net profit is \$50 per chair, \$75 per sofa, and \$100 per sofabed. How many of each item should be produced each day in order to maximize profits?

5. The nutritional information in the table is found on the sides of the cereal boxes listed (for 1 ounce of cereal with $\frac{1}{2}$ cup of whole milk). What is the minimum cost in order to receive at least 322 grams of starch (and related carbohydrates) and 119 grams of sucrose (and other sugars) by consuming these two cereals if corn flakes cost 7¢ per ounce and Honeycombs cost 19¢ per ounce?

	Contents per ounce	
Cereal	**Starch and related carbohydrates**	**Sucrose and other sugars**
Kellogg's Corn Flakes	23 g	7 g
Post Honeycombs	14 g	17 g

6. The nutritional information in the table is found on the sides of the cereal boxes listed (for 1 ounce of cereal with $\frac{1}{2}$ cup of whole milk). What is the most cereal a person could consume and not receive more than 322 grams of starch (and related carbohydrates) or more than 126 grams of sucrose (and other sugars)?

Cereal	Contents per ounce: Starch and related carbohydrates	Contents per ounce: Sucrose and other sugars
Kellogg's Corn Flakes	23 g	7 g
Kellogg's Raisin Bran	14 g	18 g

7. Your broker tells you about two investments she thinks worthwhile. She recommends a new issue of Pertec stock, which should yield 20% over the next year, and then to balance your account she recommends Campbell Municipal Bonds with a 10% annual yield. The stock-to-bond ratio should be no less than 1 to 3. If you have no more than $100,000 to invest and do not want to invest more than $70,000 in Pertec or less than $20,000 in bonds, how much should you invest in each to maximize your return?

8. An investment newsletter recommends an investment in zero-coupon bonds paying 8.5% annual interest and in (more risky) junk bonds paying 12% annual interest. Suppose you decide to invest no more than $50,000 in one or both of these investments. You also decide that you want to invest no more than twice as much in junk bonds as you do in zero-coupon bonds. How much should you invest in each of these to receive the maximum yearly return on your investment?

9. Metaltec ships units from New Orleans and Atlanta to Chicago and to Los Angeles. The shipping costs, as well as the supply and demand, are listed in the following table.

Source and supply	Destination and demand	Cost
New Orleans, 1,200	Chicago, 1,500	$120
	Los Angeles, 900	$180
Atlanta, 1,500	Chicago, 1,500	$100
	Los Angeles, 900	$200

How should shipments be made from New Orleans and Atlanta to minimize the shipping cost?

10. The Sebastopol Winery operates two plants for processing grapes, and two warehouses for storing the wine until it is purchased. The costs associated with shipping the wine from the winery (where the processing is done) to the warehouses, as well as the supply and demand, are summarized by the following table.

Winery and supply	Warehouse and demand	Cost
Windsor, 900	Santa Rosa, 600	$.25
	San Rafael, 1,000	$.18
Graton, 700	Santa Rosa, 600	$.25
	San Rafael, 1,000	$.14

How should the shipping from the wineries to the warehouses be made to minimize the cost?

11. Tom's Candy Kitchen makes two types of candy, cream-filled and solid chews. Daily business requires that at least 25 dozen cream-filled and 50 dozen solid chews be made each day. The capacity of the kitchen limits the total amount that can be made each day to no more than 125 dozen. How many dozen of each type should be made in order to maximize profits if the net profit on each cream-filled dozen is $6 and on each dozen solid chews is $4.80?

12. Mama's Home Bakery produces two specialties: German chocolate cake and apple strudel. Daily business requires that at least 50 cakes and 100 strudels be baked every day. The capacity of the ovens limits the total number of items that can be baked each day to no more than 250. How many of each item should be baked in order to maximize profits if the net profit on each cake is $1.50 and on each strudel is $1.20?

13. Karlin Manufacturing has discontinued production of a line of tires that has not been profitable. This has created some excess production time, and management is considering devoting this excess to its Glassbelt or Rainbelt tire lines. The amount of excess production available is 200 hours per month on the composition machine and 400 hours per month on production line #3. The number of hours required for each tire on each of these production lines is given in the table. How many of each tire should Karlin produce to maximize the return if the unit return on each Glassbelt is $52 and on each Rainbelt is $36?

Machine	Glassbelt tire	Rainbelt tire
Composition machine	3 hr	1 hr
Production line #3	2 hr	2 hr

14. An office manager decides to purchase some personal-size computers. Compaq computers cost $3,600 each, will provide 250 MB storage capability, and will require 5 square feet of desk space. IBM/PC computers cost $4,800 each, will provide 612 MB storage, and will require only 3 square feet of desk space. Proper utilization of space requires that a total of not more than 165 square feet be used for this new equipment. If there is $144,000 available to spend on computers, how many of each type of computer should be purchased if the office manager wishes to maximize storage capacity?

15. Foley's Motel has 200 rooms and a restaurant that seats 50 people. Experience shows that 40% of the commercial guests and 20% of the other guests (i.e., noncommercial) eat in the restaurant. There is a net profit of $4.50 per day from each commercial guest and $3.50 per day from each other guest. Find the number of commercial and other guests needed in order to maximize the net profits. Assume one guest per room and that the capacity of the restaurant will not be exceeded.

16. Karlin Enterprises manufactures two games. Standing orders require that at least 24,000 space-battle games and 5,000 football games be produced. The company has two factories: The Gainesville plant can produce 600 space-battle games and 100 football games per day; the Sacramento plant can produce 300 space-battle games and 100 football games per day. If the Gainesville plant costs $20,000 per day to operate and the Sacramento plant costs $15,000 per day, find the number of days per month that each factory should operate to minimize the cost.

17. Alco Company manufactures two products: Alpha and Beta. Each product must pass through two processing operations, and all materials are introduced at the first operation. Alco may produce either one product exclusively or various combinations of both products subject to the constraints given in the table. A shortage of technical labor has limited Alpha production to no more than 700 units per day. There are no constraints on the production of Beta other than the hour constraints shown in the table. How many of each product should be manufactured in order to maximize profits?

Product	Hours required to produce one unit: First process	Second process	Profit per unit
Alpha	1 hr	1 hr	$5
Beta	3 hr	2 hr	$8
Total capacity per day	1,200 hr	1,000 hr	

18. An island is inhabited by two species of animals, A1 and A2, which feed on three types of food, F1, F2, and F3. The first species requires 12 units of food F1, 6 units of F2, and 16 units of F3 to survive. The second species requires 15, 20, and 5 units of foods F1, F2, and F3, respectively. The island can normally supply 1,200 units of F1, 1,200 units of F2, and 800 units of F3. What is the maximum number of animals that this island can support?

19. An isolated geographic region is inhabited by three species of animals, A1, A2, and A3, which feed on three types of food, F1, F2, and F3. The first species requires 12 units of food F1, 6 units of F2, and 16 units of F3 to survive. The second species requires 15, 20, and 5 units of food F1, F2, and F3, respectively, and the third species requires 20, 10, and 10 of food F1, F2, and F3, respectively. The region can supply 8,000 units of F1, 10,000 units of F2, and 6,000 units of F3. What is the maximum number of animals that this region can support?

20. The Cosmopolitan Oil Company requires at least 8,000 barrels of low-grade oil per month; it also needs at least 12,000 barrels of medium-grade oil per month and at least 2,000 barrels of high-grade oil per month. Oil is produced at one of two refineries. The daily production of these refineries is summarized in the table. If it costs $17,000 per day to operate refinery I and $15,000 per day to operate refinery II, how many days per month should each refinery be operated to satisfy the requirements and at the same time minimize the costs?

Refinery	Oil production (barrels): Low grade	Medium grade	High grade
I	200	400	100
II	200	300	100

21. Pell, Inc., manufactures two products: tables and shelves. Each product must be processed in each of three departments: machining, assembling, and finishing. The hours needed to produce one unit of product per department and the maximum possible hours per department are shown in the table. Standing orders require that Pell manufacture

Department	Production time per unit (hr): Tables	Shelves	Maximum capacity (hr)
Machining	2	1	200
Assembling	2	2	240
Finishing	2	3	300

at least 30 tables and 25 shelves. Pell's net profit is $4 per table and $2 per shelf. How many tables should be manufactured to maximize the net profit?

22. A shoe company produces three models of athletic shoes. Shoe parts are first produced in the manufacturing shop and then put together in the assembly shop. The number of hours of labor required per pair in each shop is given in the table. The company can sell as many pairs of shoes as it can produce. During the next month, no more than 1,000 hours can be expended in the manufacturing shop and no more than 800 hours can be expended in the assembly shop. The expected profit from the sale of each pair of shoes is:

Model 1: $12
Model 2: $7
Model 3: $5

The company wants to determine how many pairs of each shoe model to produce to maximize total profits over the next month.*

Shop	Model 1	Model 2	Model 3
Manufacturing	8 hr	5 hr	3 hr
Assembly	5 hr	1 hr	3 hr

23. A woman wants to design a weekly exercise schedule that involves jogging, handball, and aerobic dance. She decides to jog at least 3 hours per week, play handball at least 2 hours per week, and dance at least 5 hours per week. She also wants to devote at least as much time to jogging as to handball because she does not like handball as much as she likes the other exercises. She also knows that jogging consumes 900 calories per hour, handball 600 calories per hour, and dance 800 calories per hour. Her physician told her that she must burn up a total of at least 9,000 calories per week in this exercise program. How many hours should she devote to each exercise if she wishes to minimize her exercise time?

24. To utilize costly equipment efficiently, an assembly line with a capacity of 50 units per shift must operate 24 hours per day. This requires scheduling three shifts with varying labor costs as follows:

Day shift: Labor cost $100 per unit
Swing shift: Labor cost $150 per unit
Graveyard shift: Labor cost $180 per unit

* This problem is taken from a recent examination administered by the Society of Actuaries. Reprinted by permission of the Society of Actuaries.

Each unit produced uses 60 linear feet of sheeting material, and there is a total of 1,800 feet available for the day shift, 1,500 feet for the swing shift, and 3,000 feet for the graveyard shift. The day shift must produce at least 20 units and the other shifts must produce at least 50 units together. How many units should be produced on each shift to maximize the revenue if the units are sold for $850 each?

25. Repeat Problem 24 except minimize labor costs rather than maximize revenue.

26. Suppose a computer dealer has stores in Hillsborough and Palo Alto and warehouses in San Jose and Burlingame. The cost of shipping a computer from one location to another as well as the supply and demand at each location are shown on the following "map":

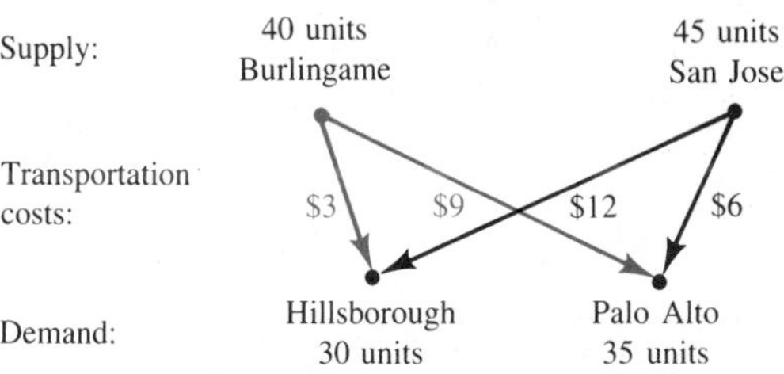

How should the dealer ship the sets to minimize the shipping costs?

27. Camstop, Inc., an importer of computer chips, can ship case lots of components to stores in Dallas and Chicago from ports in either Los Angeles or Seattle. The costs and demand are given in the table. How many cases should be shipped from each port to minimize shipping costs?

Store	Cost to ship (per case): Los Angeles port	Cost to ship (per case): Seattle port	Store demand
Chicago	$9	$7	80 cases
Dallas	$7	$8	110 cases
Available inventory at each port	90 cases	130 cases	

28. A Texas appliance dealer has stores in Fort Worth and in Houston and warehouses in Dallas and San Antonio. The cost of shipping a refrigerator from Dallas to Fort Worth is $10, from Dallas to Houston it is $15, from San Antonio to Fort Worth it is $14, and from San Antonio to Houston it is $18. Suppose that the Fort Worth store has orders for 15 refrigerators and the Houston store has orders for 35. Also suppose that there are 25 refrigerators in stock at Dallas and 45 at San Antonio. What is the most economical way to supply the requested refrigerators to the two stores?

29. One method for separating copper, lead, and zinc from base ores is a flotation separation process. There are three steps to this process: oiling, mixing, and separation. The requirements for this daily process are summarized by the accompanying table. How many units of each metal can be produced in order to maximize the profit?

Metal	Time necessary for production of one unit of metal			Unit profit
	Oiling	**Mixing**	**Separation**	
Copper	2	2	1	\$90
Lead	3	2	1	\$100
Zinc	1	1	3	\$70
Available hours	10	12	8	

30. The manager asked you to increase the output of a division, but also gave you a \$2,000 budget. You have a pool of employees from which to draw. They are of three types: trainees, who are paid \$60 for assembling 36 units; regular employees, who are paid \$90 for assembling 50 units; and supervisors, who are paid \$120 for assembling 60 units. Union regulations require at least one supervisor for every 15 regular employees. Also, there can be no more than 6 trainees working at any one time. How many of each type of employee can you hire in order to maximize the number of units assembled?

4.3 Graphical Solution of Linear Programming Problems

In the previous section we looked at some models called *linear programming problems.* In each case the model had a function called an **objective function**, which was to be maximized or minimized while satisfying several conditions, or **constraints**. If there are only two variables, we can use a graphical method of solution. We begin with the set of constraints and consider them as a system of inequalities. The solution of this system of inequalities is a set of points, S. Each point of the set S is called a **feasible solution**. The objective function can be evaluated for different feasible solutions to obtain maximum or minimum values.

EXAMPLE 1 Maximize: $K = 4x + 5y$

Subject to: $\begin{cases} 2x + 5y \le 25 \\ 6x + 5y \le 45 \\ x \ge 0, \quad y \ge 0 \end{cases}$

Solution The constraints give a set of feasible solutions as graphed in the figure. (You may want to review Section 4.1 on graphing systems of inequalities.)

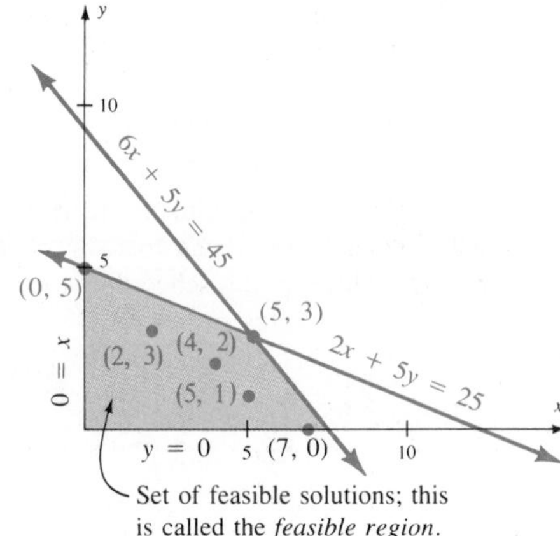

Set of feasible solutions; this is called the *feasible region.*

To solve the linear programming problem, we must now find the feasible solution that makes the objective function as large as possible. Some possible solutions are:

Feasible solution (a point in the solution set of the system)	*Objective function* $K = 4x + 5y$
(2, 3)	$4(2) + 5(3) = 8 + 15 = 23$
(4, 2)	$4(4) + 5(2) = 16 + 10 = 26$
(5, 1)	$4(5) + 5(1) = 20 + 5 = 25$
(7, 0)	$4(7) + 5(0) = 28 + 0 = 28$
(0, 5)	$4(0) + 5(5) = 0 + 25 = 25$

In this list, the point that makes the objective function the largest is (7, 0). But is this the largest for all feasible solutions? How about (6, 1)? Or (5, 3)? It turns out that (5, 3) provides the maximum value:

$$4(5) + 5(3) = 20 + 15 = 35$$

■

In Example 1, how did we know that (5, 3) provides the maximum value for K? Obviously, it cannot be done by trial and error as shown in the example. Let us take a closer look. We want to find the maximum value of K. Suppose we try the value of 40. We can graph the line (see Figure 4.3 on page 172)

$$4x + 5y = 40$$

or, in slope–intercept form,

$$y = -\frac{4}{5}x + 8$$

Notice that the slope of this line is $-\frac{4}{5}$ and the y-intercept is 8; also notice that it lies above the feasible set—that is, it does not intersect the feasible region, so a value of 40 is clearly unrealistic. Try again; try the value of 30 for K. Then

$$4x + 5y = 30 \qquad \text{or} \qquad y = -\frac{4}{5}x + 6$$

(Again, see Figure 4.3.) Now there are many points of intersection between our line and the feasible region. Also, note that both of the lines are parallel since they have the same slope.

Now we begin to formulate a procedure. Think of a "family" of lines, all with slope $-\frac{4}{5}$; some of these lines will intersect the feasible region. If we are maximizing K, we want the point that gives the largest value of K and still has an intersection point (see line N); if we are minimizing K, we want the point that gives the lowest value of K and still has an intersection point (see line n).

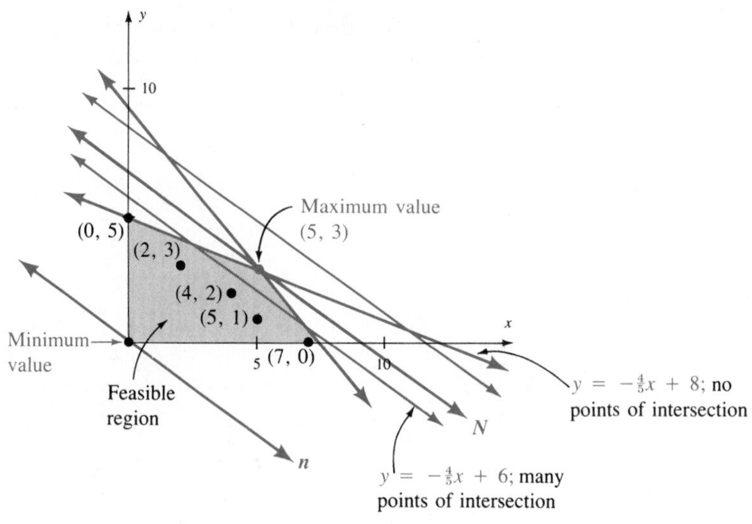

FIGURE 4.3

EXAMPLE 2 Draw (or imagine) lines parallel to the given objective function in the following figures to find the points in the given feasible region that maximize and minimize that function. If there is no such point, state that.

a.

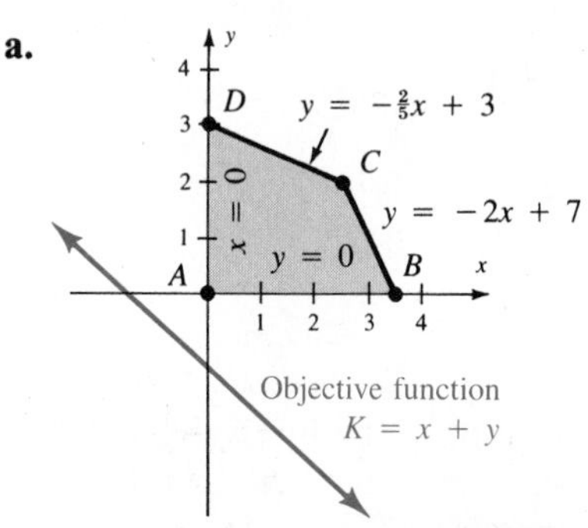

b.

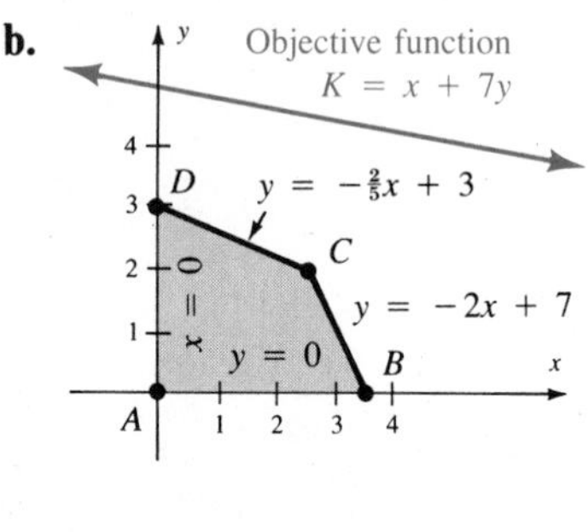

c.

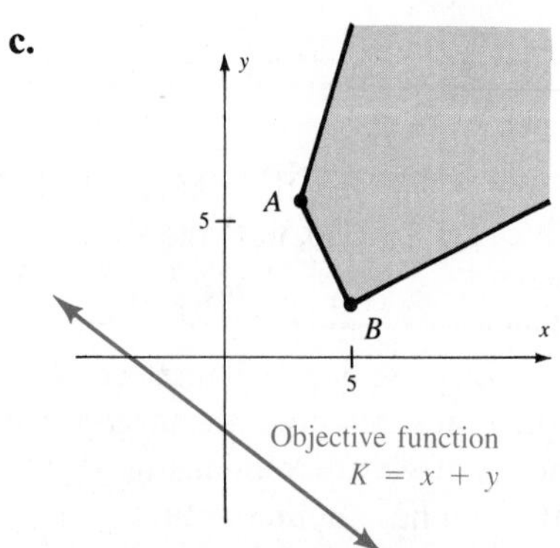

d.

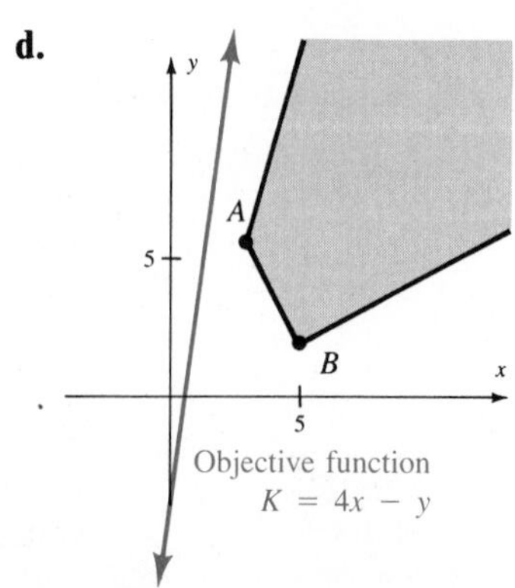

Solution **a.**

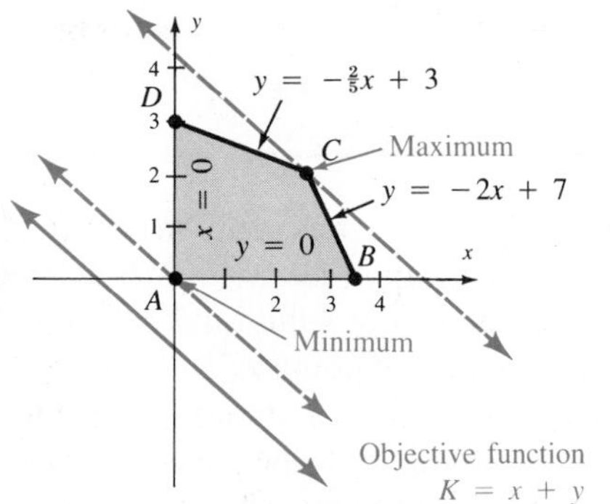

Draw lines parallel to the objective function; the maximum value from the feasible region (shaded) is at point C and the minimum value is at point A.

b.

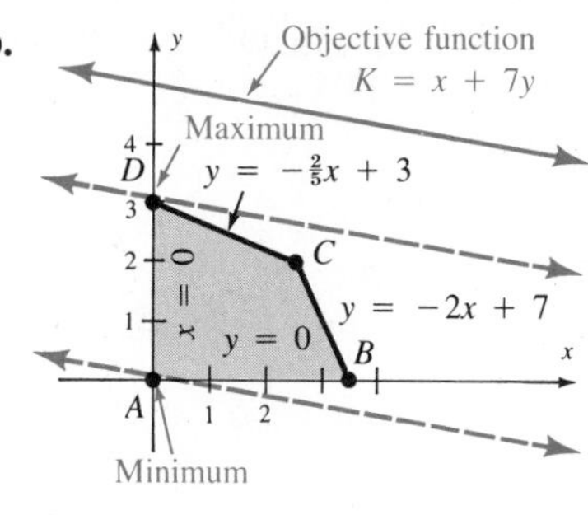

The maximum value from the feasible region is at point D and the minimum value is at point A.

c.

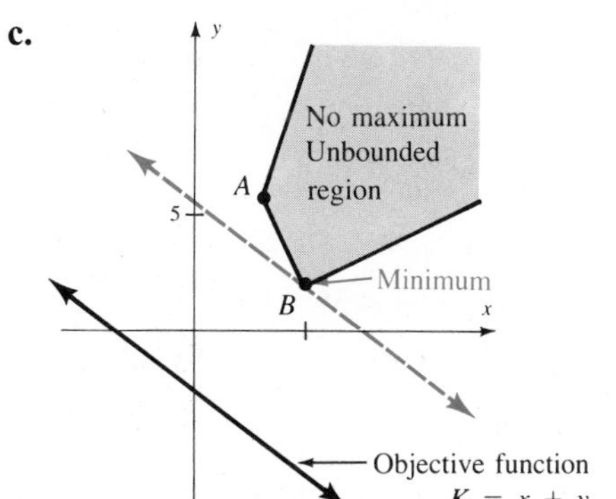

There is no maximum value since the region is unbounded. The minimum value is at point B.

d.

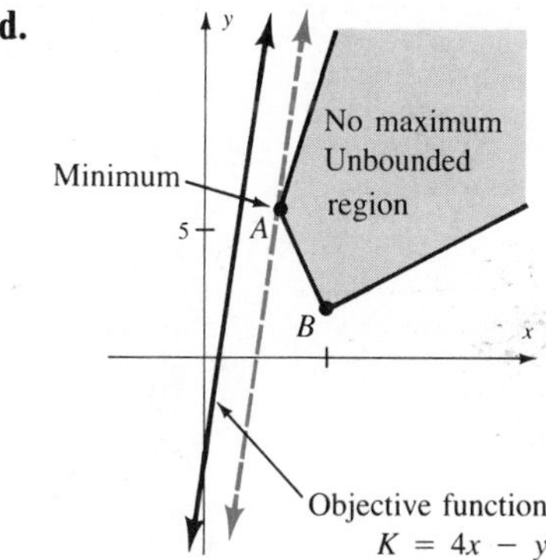

There is no maximum value since the region is unbounded; the minimum value is at point A. ■

Note in Example 2 that all of the optimum values for the objective function occur at **corner points** for the set of feasible solutions. That is, if you consider a family of parallel lines, one member of the family will either coincide with one of the sides of the feasible region or it will intersect it at one point. This follows from the fact that two lines in a plane will either coincide or will intersect at a single point. This fact leads to the **fundamental theorem of linear programming**:

Fundamental Theorem of Linear Programming

A linear expression in two variables,

$$c_1x + c_2y$$

defined over a convex set S* whose sides are line segments, takes on its **maximum value** at a corner point of S and its **minimum value** at a corner point of S. If S is unbounded, there may or may not be an optimum value, but, if there is, then such a value must occur at a corner point.

* A *convex set* is a set that contains the line segment joining any two of its points.

With this theorem, we can now summarize the procedure for solving a linear programming problem by graphing:

Graphing Method for Solving Linear Programming Problems

1. Find the objective expression (the quantity to be maximized or minimized).
2. Find and graph the constraints defined by a system of linear inequalities; the simultaneous solution is called the set S.
3. Find the corner points of S; this may require the solution of a system of two equations with two unknowns, one for each corner point.
4. Find the value of the objective expression for the coordinates of each corner point to find the optimum solutions.

EXAMPLE 3 Solve Example 1 by using the fundamental theorem of linear programming.

Solution Example 1 is repeated below and the graph is shown here with the corner points labeled.

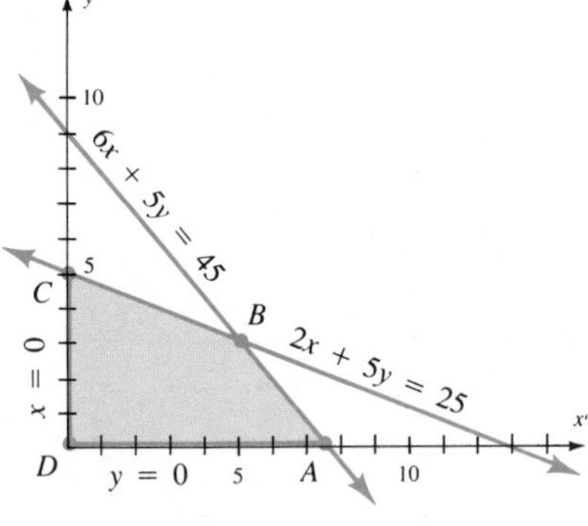

$$\text{Maximize:} \quad K = 4x + 5y$$

$$\text{Subject to:} \quad \begin{cases} 2x + 5y \leq 25 \\ 6x + 5y \leq 45 \\ x \geq 0, \quad y \geq 0 \end{cases}$$

Some corner points can usually be found by inspection. In this case we can see $D = (0, 0)$ and $C = (0, 5)$.

Some corner points may require some work with the boundary *lines* (use equations of boundaries, not the inequalities giving the regions).

Point A System: $\begin{cases} y = 0 \\ 6x + 5y = 45 \end{cases}$

Solve by substitution: $6x + 5(0) = 45$

$$x = \frac{45}{6}$$

$$= \frac{15}{2}$$

Point A is $(\frac{15}{2}, 0)$.

Point B System: $\begin{cases} 2x + 5y = 25 \\ 6x + 5y = 45 \end{cases}$

Solve by adding: $\begin{cases} -2x - 5y = -25 \\ 6x + 5y = 45 \end{cases}$

$$4x = 20$$

$$x = 5$$

If $x = 5$, then

$$2(5) + 5y = 25$$

$$5y = 15$$

$$y = 3$$

Point B is $(5, 3)$.

CALCULATOR COMMENT

You can find the corner points of a linear programming problem by graphing the lines (up to four on the TI-81) and then using the TRACE key to find the point of intersection to any desired degree of accuracy.

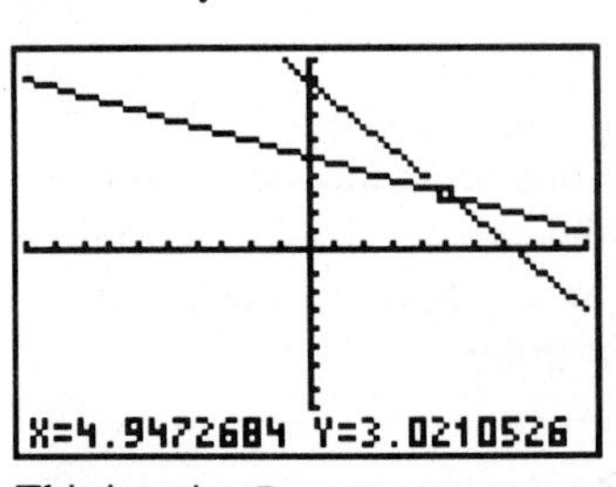

This is point B.

The corner points are thus (0, 0), (0, 5), $(\frac{15}{2}, 0)$, and (5, 3). Check these values.

Corner points	*Objective function* $C = 4x + 5y$
(0, 0)	$4(0) + 5(0) = 0$
(0, 5)	$4(0) + 5(5) = 25$
$(\frac{15}{2}, 0)$	$4(\frac{15}{2}) + 5(0) = 30$
(5, 3)	$4(5) + 5(3) = 35$

↑ Look for largest value of C in this list.

The maximum value for C is 35 at (5, 3). ■

EXAMPLE 4 Maximize $P = x + 5y$ subject to the constraints of Example 1.

Solution The problem has the same corner points as Example 3.

Corner points	*Objective function* $P = x + 5y$
(0, 0)	$0 + 5(0) = 0$
(0, 5)	$0 + 5(5) = 25$
$(\frac{15}{2}, 0)$	$\frac{15}{2} + 5(0) = \frac{15}{2}$
(5, 3)	$5 + 5(3) = 20$

The maximum value for P is 25 at (0, 5). ■

EXAMPLE 5 **Allocation of resources in production.** A farmer has 100 acres on which to plant two crops: corn and wheat. To produce these crops, there are certain expenses, as shown in the table.

Item	Cost per acre
Corn	
Seed	\$ 12
Fertilizer	58
Planting/care/harvesting	50
Total	\$120
Wheat	
Seed	\$ 40
Fertilizer	80
Planting/care/harvesting	90
Total	\$210

After the harvest, the farmer must usually store the crops while awaiting favorable market conditions. Each acre yields an average of 110 bushels of corn or

30 bushels of wheat. The limitations of resources are

Available capital:	\$15,000
Available storage facilities:	4,000 bushels

If the net profit per bushel of corn (after all expenses have been subtracted) is \$1.30 and for wheat is \$2.00, how should the farmer plant the 100 acres to maximize profits?

Solution This linear programming problem was formulated in Section 4.2 (Example 1):

Maximize: $P = 143x + 60y$ — Where x is the number of acres planted in corn and y is the number of acres planted in wheat

$$\text{Subject to: } \begin{cases} x \geq 0 \\ y \geq 0 \\ x + y \leq 100 \\ 120x + 210y \leq 15{,}000 \\ 110x + 30y \leq 4{,}000 \end{cases}$$

First graph the set of feasible solutions by graphing the system of inequalities, as shown in the figure.

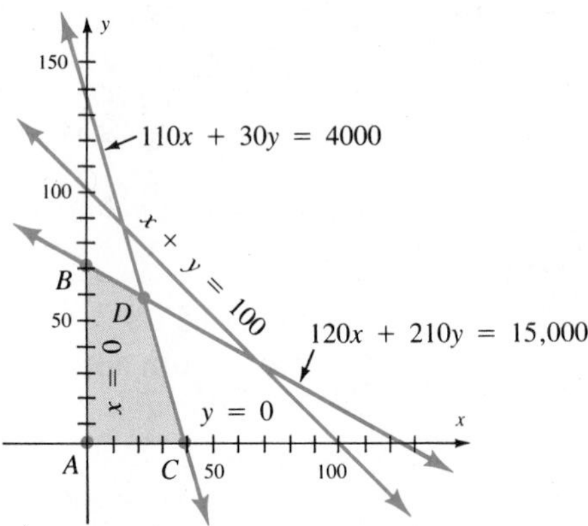

Next find the corner points. By inspection, point $A = (0, 0)$.

$$\text{Point } B: \begin{cases} 120x + 210y = 15{,}000 \\ x = 0 \end{cases}$$

$$120(0) + 210y = 15{,}000$$

$$y = \frac{15{,}000}{210}$$

$$= \frac{500}{7}$$

Point B: $(0, \frac{500}{7})$

Point C: $\begin{cases} 110x + 30y = 4{,}000 \\ y = 0 \end{cases}$

$$110x + 30(0) = 4{,}000$$
$$110x = 4{,}000$$
$$x = \frac{400}{11}$$

Point C: $(\frac{400}{11}, 0)$

Point D: $-7\begin{cases} 110x + 30y = 4{,}000 \\ 120x + 210y = 15{,}000 \end{cases}$ $\quad \begin{cases} -770x - 210y = -28{,}000 \\ 120x + 210y = 15{,}000 \end{cases}$

$$-650x = -13{,}000$$
$$x = 20$$
$$110(20) + 30y = 4{,}000$$
$$30y = 1{,}800$$
$$y = 60$$

Point D: $(20, 60)$

Use the linear programming theorem and check the corner points:

Corner points	*Objective function* $P = 143x + 60y$
A: $(0, 0)$	$143(0) + 60(0) = 0$
B: $(0, \frac{500}{7})$	$143(0) + 60(\frac{500}{7}) \approx 4{,}286$
C: $(\frac{400}{11}, 0)$	$143(\frac{400}{11}) + 60(0) = 5{,}200$
D: $(20, 60)$	$143(20) + 60(60) = 6{,}460$

The maximum value of P is 6,460 at (20, 60). This means that to maximize profits, the farmer should plant 20 acres in corn, plant 60 acres in wheat, and leave 20 acres unplanted. ■

Note from the graph in Example 5 that some of the constraints could be eliminated and everything else would remain unchanged. For example, the boundary $x + y = 100$ was not necessary in finding the maximum value of P. Such a condition is said to be a **superfluous constraint**. It is not uncommon to have superfluous constraints in a linear programming problem. Suppose, however, that the farmer contracted to have the grain stored at a neighboring farm and now the contract calls for *at least* 4,000 bushels to be stored. This change from $110x + 30y \le 4{,}000$ to $110x + 30y \ge 4{,}000$ now makes the condition $x + y \le 100$ important to the solution of the problem. You must be careful about superfluous constraints even though they do not affect the solution at the present time.

It is also possible to have constraints leading to an empty intersection. For example, consider the constraints

$$\begin{cases} x + y \le 1 \\ x + y \ge 2 \\ x \ge 0, \quad y \ge 0 \end{cases}$$

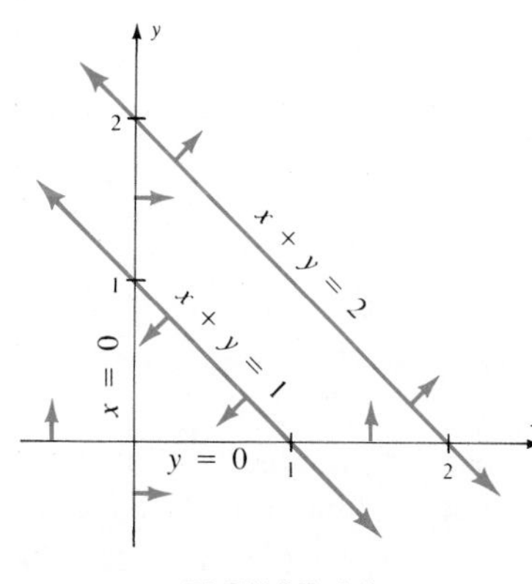

FIGURE 4.4

The graph in Figure 4.4 shows such an empty intersection. We say that this problem has no feasible solution.

EXAMPLE 6 Solve the following linear programming problem.

$$\text{Minimize:} \quad K = 60x + 30y$$

$$\text{Subject to:} \quad \begin{cases} 2x + 3y \geq 120 \\ 2x + y \geq 80 \\ x \geq 0, \quad y \geq 0 \end{cases}$$

Solution Graph:

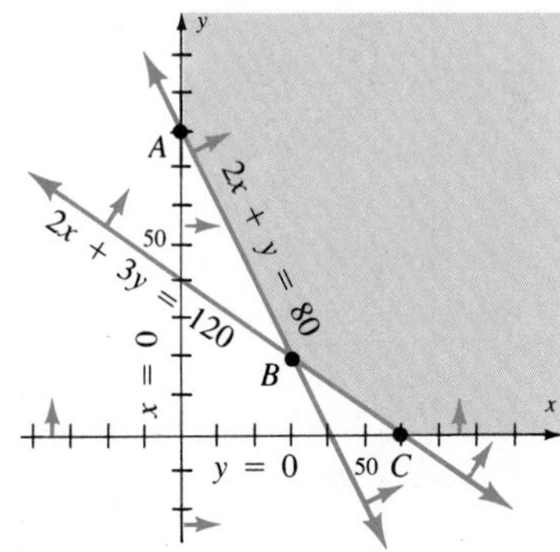

The corner points $A = (0, 80)$ and $C = (60, 0)$ are found by inspection.

$$\text{Point } B: \quad \begin{cases} 2x + 3y = 120 \\ 2x + y = 80 \end{cases}$$

$$2y = 40$$

$$y = 20$$

If $y = 20$, then

$$2x + 20 = 80$$

$$2x = 60$$

$$x = 30$$

Point B: $(30, 20)$

Corner points	*Objective function* $C = 60x + 30y$
$(0, 80)$	$60(0) + 30(80) = 2{,}400$
$(30, 20)$	$60(30) + 30(20) = 2{,}400$
$(60, 0)$	$60(60) + 30(0) = 3{,}600$

From the list above, there are two minimum values for the objective function: $A = (0, 80)$ and $B = (30, 20)$. In this situation, the objective function will have the same minimum value (2,400) at all points along the boundary line segment joining A and B. ■

4.3 Problem Set

Find the points in the given feasible region that (a) maximize the given objective function and (b) minimize the given objective function. If, in either case, there is none, state that.

1. 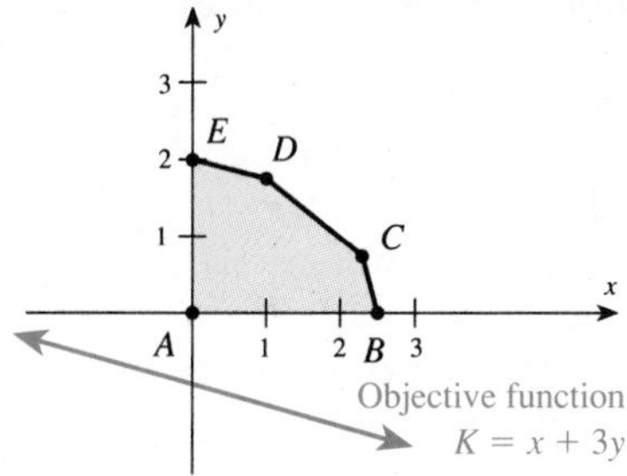

2.  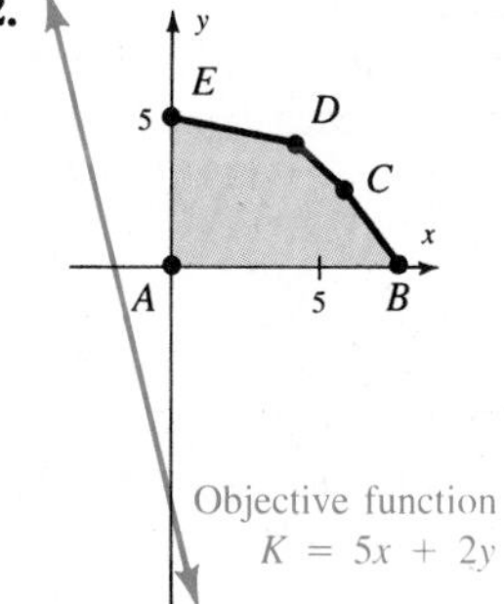

3. 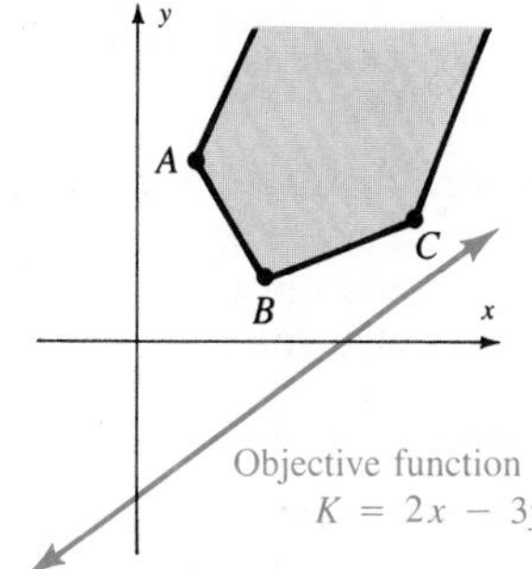

4. 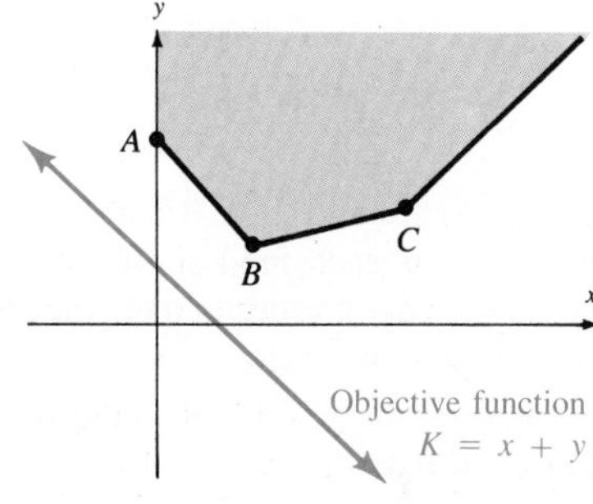

Find the corner points for each set of feasible solutions in Problems 5–18.

5. $\begin{cases} 2x + y \le 12 \\ x + 2y \le 9 \\ x \ge 0, \quad y \ge 0 \end{cases}$

6. $\begin{cases} 2x + 5y \le 20 \\ 2x + y \le 12 \\ x \ge 0, \quad y \ge 0 \end{cases}$

7. $\begin{cases} 3x + 2y \le 12 \\ x + 2y \le 8 \\ x \ge 0, \quad y \ge 0 \end{cases}$

8. $\begin{cases} x \le 10 \\ y \le 8 \\ 3x + 2y \ge 12 \\ x \ge 0, \quad y \ge 0 \end{cases}$

9. $\begin{cases} x + y \le 8 \\ y \le 4 \\ x \le 6 \\ x \ge 0, \quad y \ge 0 \end{cases}$

10. $\begin{cases} x + y \ge 6 \\ -2x + y \ge -16 \\ y \le 9 \\ x \ge 0, \quad y \ge 0 \end{cases}$

11. $\begin{cases} 3x + 2y \le 8 \\ x + 5y \ge 8 \\ x \ge 0, \quad y \ge 0 \end{cases}$

12. $\begin{cases} x \le 8 \\ y \ge 2 \\ x + y \le 10 \\ x \le 3y \\ x \ge 0, \quad y \ge 0 \end{cases}$

13. $\begin{cases} x \le 10 \\ y \le 12 \\ x + y \le 12 \\ y \le x \\ x \ge 0, \quad y \ge 0 \end{cases}$

14. $\begin{cases} x \le 5 \\ y \le 10 \\ 2x + y \le 12 \\ y \le 10x \\ x \ge 0, \quad y \ge 0 \end{cases}$

15. $\begin{cases} 10x + 5y \ge 200 \\ 2x + 5y \ge 100 \\ 3x + 4y \ge 120 \\ x \ge 0, \quad y \ge 0 \end{cases}$

16. $\begin{cases} x + y \le 9 \\ 2x - 3y \ge -6 \\ x - y \le 3 \\ x \ge 0, \quad y \ge 0 \end{cases}$

17. $\begin{cases} 2x + y \ge 8 \\ y \le 5 \\ x - y \le 2 \\ 3x - 2y \ge 5 \\ x \ge 0, \quad y \ge 0 \end{cases}$

18. $\begin{cases} 2x + y \ge 8 \\ x - 2y \le 7 \\ x - y \ge -3 \\ x \le 9 \\ x \ge 0, \quad y \ge 0 \end{cases}$

Find the optimum value for each objective function in Problems 19–38.

19. Maximize $W = 30x + 20y$ subject to the constraints of Problem 5.

20. Maximize $P = 40x + 10y$ subject to the constraints of Problem 5.

21. Maximize $T = 100x + 10y$ subject to the constraints of Problem 6.

22. Maximize $V = 20x + 30y$ subject to the constraints of Problem 6.

23. Maximize $P = 100x + 100y$ subject to the constraints of Problem 7.

24. Maximize $T = 30x + 10y$ subject to the constraints of Problem 7.

25. Minimize $C = 24x + 12y$ subject to the constraints of Problem 8.

26. Minimize $K = 6x + 18y$ subject to the constraints of Problem 8.

27. Maximize $F = 2x - 3y$ subject to the constraints of Problem 9.

28. Maximize $P = 5x + y$ subject to the constraints of Problem 9.

29. Minimize $I = 90x + 20y$ subject to the constraints of Problem 10.

30. Minimize $T = 400x + 100y$ subject to the constraints of Problem 10.

31. Minimize $P = 23x + 46y$ subject to the constraints of Problem 15.

32. Minimize $C = 12x + 15y$ subject to the constraints of Problem 15.

33. Maximize $P = 6x + 3y$ subject to the constraints of Problem 16.

34. Maximize $T = x + 6y$ subject to the constraints of Problem 16.

35. Minimize $X = 5x + 3y$ subject to the constraints of Problem 17.

36. Minimize $A = 2x - 3y$ subject to the constraints of Problem 17.

37. Minimize $K = 140x + 250y$ subject to the constraints of Problem 18.

38. Minimize $C = 640x - 130y$ subject to the constraints of Problem 18.

Solve the linear programming problems in Problems 39–47.

APPLICATIONS

39. Suppose the net profit per bushel of corn in Example 5 increases to \$2.00 and the net profit for wheat drops to \$1.50 per bushel. Maximize the profit if the other conditions of the example remain the same.

40. Suppose the farmer in Example 5 contracted to have the grain stored at a neighboring farm and the contract calls for at least 4,000 bushels to be stored. How many acres should be planted in corn and how many in wheat to maximize profits if the other conditions of the example remain the same?

41. A farmer has 500 acres on which to plant two crops: corn and wheat. It costs \$120 per acre to produce corn and \$60 per acre to produce wheat, and there is \$24,000 available to pay for this year's production. If the yield per acre is 100 bushels of corn or 40 bushels of wheat and the farmer has contracted to store at least 18,000 bushels, how much should the farmer plant to maximize profits when the profit is \$1.20 per bushel for corn and \$2.50 per bushel for wheat?

42. The Wadsworth Widget Company manufactures two types of widgets: regular and deluxe. Each widget is produced at a station consisting of a machine and a person who finishes the widgets by hand. The regular widget requires 2 hours of machine time and 1 hour of finishing time. The deluxe widget requires 3 hours of machine time and 5 hours of finishing time. The profit on the regular widget is \$25, and on the deluxe widget it is \$30. If each workday is 8 hours, how many of each type of widget should be produced at each station per day in order to maximize the profit?

43. The Thompson Company manufactures two industrial products: standard (\$45 profit per item) and economy (\$30 profit per item). These items are built using machine time and manual labor. The standard product requires 3 hours of machine time and 2 hours of manual labor. The economy model requires 3 hours of machine time and no manual labor. If the week's supply of manual labor is limited to 800 hours and machine time to 1,500 hours, how many of each type of product should be produced each week in order to maximize the profit?

44. A convalescent hospital wishes to provide, at a minimum cost, a diet that has a minimum of 200 grams of carbohydrates, 100 grams of protein, and 120 grams of fats per day. These requirements can be met with two foods:

	Contents per oz		
Food	**Carbohydrates**	**Protein**	**Fats**
A	10 g	2 g	3 g
B	5 g	5 g	4 g

If food A costs 29¢ per ounce and food B costs 15¢ per ounce, how many ounces of each food should be purchased for each patient per day in order to meet the minimum requirements at the lowest cost?

45. The nutritional information in the table is found on the sides of the cereal boxes listed (for 1 ounce of cereal with

$\frac{1}{2}$ cup of whole milk). What is the minimum cost in order to receive at least 322 grams of starch (and related carbohydrates) and 119 grams of sucrose (and other sugars) by consuming these two cereals if corn flakes cost 7¢ per ounce and Honeycombs cost 19¢ per ounce?

	Contents per oz	
Cereal	**Starch and related carbohydrates**	**Sucrose and other sugars**
Kellogg's Corn Flakes	23 g	7 g
Post Honeycombs	14 g	17 g

46. Brown Bros., Inc., is an investment company analyzing a pension fund for a certain company. A maximum of \$10 million is available to invest in two places. No more than \$8 million can be invested in stocks yielding 12% and at least \$2 million must be invested in long-term bonds yielding 8%. The stock-to-bond investment ratio cannot be more than 1 to 3. How should Brown Bros. advise its client so that the pension fund will receive the maximum yearly return on investment?

47. Your broker tells you of two investments she thinks are worthwhile. She recommends a new issue of Pertec stock, which should yield 20% over the next year, and then to balance your account she recommends Campbell Municipal Bonds with a 10% annual yield. The stock-to-bond ratio should be no less than 1 to 3. If you have no more than \$100,000 to invest and do not want to invest more than \$70,000 in Pertec or less than \$20,000 in bonds, how much should you invest in each to maximize your return?

4.4 Slack Variables and the Pivot

In Sections 4.1–4.3 we solved linear programming problems using a graphical procedure. However, since graphical methods are inappropriate for more than two variables, we need an algebraic process leading to a solution of the linear programming problem. In 1946 George Dantzig developed an algebraic process called the **simplex method** for solving linear programming problems. Because of the far-reaching applications of this method, some have called it the most important development in applied mathematics of this century. The process itself is easy, but it takes a considerable amount of discussion to set up the proper terminology and notation. In this section we will:

1. Write the linear programming problem in *standard form.*
2. Introduce *slack variables.*
3. Write the linear programming problem in matrix form, called the *initial simplex tableau.*
4. Relate the simplex tableau to the graphing method.
5. Introduce a *pivoting process.*

In the next section we solve linear programming problems using the simplex method.

Standard Form

Throughout this section the objective function is represented by the variable z, and the linear programming variables are represented by $x_1, x_2, x_3, \ldots$. Consider the following linear programming problems using this notation:

Minimize: $z = .08x_1 + .18x_2$

Subject to: $\begin{cases} 23x_1 + 14x_2 \leq 322 \\ 7x_1 + 18x_2 \leq 111 \\ x_1 \geq 0, \quad x_2 \geq 0 \end{cases}$

Maximize: $z = 35x_1 + 18x_2$

Subject to: $\begin{cases} .9x_1 + .06x_2 \leq 1{,}000 \\ 320x_1 + 118x_2 \leq 10{,}000 \end{cases}$

Maximize: $z = 1.2x_1 + 2.5x_2$

Subject to: $\begin{cases} 120x_1 + 200x_2 \leq 24{,}000 \\ 100x_1 + 40x_2 \geq 20{,}000 \\ x_1 \geq 0, \quad x_2 \geq 0 \end{cases}$

Maximize: $z = 45x_1 + 30x_2$

Subject to: $\begin{cases} 2x_1 \leq 800 \\ 2x_1 + 3x_2 \leq 1{,}500 \\ x_1 \geq 0, \quad x_2 \geq 0 \end{cases}$

Standard Linear Programming Problem

A linear programming problem is said to be a *standard linear programming problem* or, more simply, in **standard form**, if the following three conditions are met:

Condition 1. The objective function is to be maximized.
Condition 2. All variables are nonnegative.
Condition 3. All constraints (except those in condition 2) are less than or equal to ($\leq$) a nonnegative constant.

EXAMPLE 1 Determine whether each of the following linear programming problems is in standard form.

a. Minimize: $z = .08x_1 + .18x_2$

Subject to: $\begin{cases} 23x_1 + 14x_2 \leq 322 \\ 7x_1 + 18x_2 \leq 111 \\ x_1 \geq 0, \quad x_2 \geq 0 \end{cases}$

This is not a standard form problem since it is a minimization problem and thus fails to meet condition 1.

b. Maximize: $z = 35x_1 + 18x_2$

Subject to: $\begin{cases} .9x_1 + .06x_2 \leq 1{,}000 \\ 320x_1 + 118x_2 \leq 10{,}000 \end{cases}$

This is not a standard form problem since x_1 and x_2 are not necessarily nonnegative and thus condition 2 has not been met.

c. Maximize: $z = 1.2x_1 + 2.5x_2$

Subject to: $\begin{cases} 120x_1 + 200x_2 \leq 24{,}000 \\ 100x_1 + 40x_2 \geq 20{,}000 \\ x_1 \geq 0, \quad x_2 \geq 0 \end{cases}$

This is not a standard form problem since the second constraint is not less than or equal to a nonzero constant and thus condition 3 has not been met.

d. Maximize: $z = 45x_1 + 30x_2$

Subject to: $\begin{cases} 2x_1 \leq 800 \\ 2x_1 + 3x_2 \leq 1{,}500 \\ x_1 \geq 0, \quad x_2 \geq 0 \end{cases}$

This is a standard linear programming problem.

■

Slack Variables

The goal of the simplex method is to solve a standard linear programming problem by transforming it into a system of linear *equations* that can then be solved using the Gauss–Jordan method. To make this transformation, we introduce new variables to the problems. These variables "take up the slack" created by the inequalities and are consequently called **slack variables**. For example, when we say

$$x \leq 5$$

we are saying that there exists a nonnegative number y, called a *slack variable*, such that

$$x + y = 5$$

That is, slack variables allow us to transform linear programming problems from systems of inequalities to corresponding systems of equations.

Slack Variable

WARNING **Remember that slack variables are nonnegative.**

For every nonnegative constant b and each inequality of the form

$$x \leq b$$

there exists a nonnegative number y, called a **slack variable**, such that

$$x + y = b$$

In this book we denote slack variables by $y_1, y_2, y_3, \ldots$ to distinguish them from the linear programming variables $x_1, x_2, x_3, \ldots$. Also, a statement such as

$$z = 45x_1 + 30x_2$$

is rewritten as

$$-45x_1 - 30x_2 + z = 0$$

In Examples 2 and 3 we rewrite the given standard linear programming problems using slack variables.

EXAMPLE 2 Maximize: $z = 45x_1 + 30x_2$

$$\text{Subject to: } \begin{cases} 2x_1 \leq 800 \\ 2x_1 + 3x_2 \leq 1{,}500 \\ x_1 \geq 0, \quad x_2 \geq 0 \end{cases}$$

Solution Using slack variables,

$$\begin{cases} -45x_1 - 30x_2 + z = 0 \\ 2x_1 + y_1 = 800 \\ 2x_1 + 3x_2 + y_2 = 1{,}500 \\ x_1 \geq 0, \quad x_2 \geq 0 \\ y_1 \geq 0, \quad y_2 \geq 0 \end{cases}$$

These nonnegative requirements are not rewritten using slack variables. ■

EXAMPLE 3 Maximize: $z = 3x_1 + 9x_2 + 12x_3 - 5x_4$

Subject to: $\begin{cases} x_1 + x_2 \le 40 \\ x_3 + x_4 \le 45 \\ x_1 + x_3 \le 30 \\ x_2 + x_4 \le 35 \\ x_1 \ge 0, \quad x_2 \ge 0, \quad x_3 \ge 0, \quad x_4 \ge 0 \end{cases}$

Solution Using slack variables,

$$\begin{cases} -3x_1 - 9x_2 - 12x_3 + 5x_4 + z = 0 \\ x_1 + x_2 + y_1 = 40 \\ x_3 + x_4 + y_2 = 45 \\ x_1 + x_3 + y_3 = 30 \\ x_2 + x_4 + y_4 = 35 \\ x_1 \ge 0, \quad x_2 \ge 0, \quad x_3 \ge 0, \quad x_4 \ge 0 \\ y_1 \ge 0, \quad y_2 \ge 0, \quad y_3 \ge 0, \quad y_4 \ge 0 \end{cases}$$

■

Initial Simplex Tableau and Its Relationship to the Graphing Method

Note that we write the nonnegativity conditions ($x_1 \ge 0, x_2 \ge 0, \ldots$) separately on the bottom lines because the variables are nonnegative for *all* standard linear programming problems. Since we know this, there is no need to include the nonnegative variables in the matrix notation. We now write a matrix called the **initial simplex tableau**. When writing this matrix, it is important that you write the variables in the same order so that the proper coefficients are aligned. The standard form maximization statement at the top is generally written at the bottom in the initial simplex tableau, as shown in Examples 4 and 5.

EXAMPLE 4 Maximize: $z = 45x_1 + 30x_2$

Subject to: $\begin{cases} 2x_1 \le 800 \\ x_2 \le 300 \\ 2x_1 + 3x_2 \le 1{,}500 \\ x_1 \ge 0, \quad x_2 \ge 0 \end{cases}$

Solution Align the slack variables for the tableau form:

$$\begin{array}{rrrrrl} 2x_1 & & + y_1 & & & = 800 \\ & x_2 & & + y_2 & & = 300 \\ 2x_1 + & 3x_2 & & & + y_3 & = 1{,}500 \\ -45x_1 - & 30x_2 & & & & + z = 0 \end{array}$$

The tableau form is:

$$\begin{array}{cccccc|c} x_1 & x_2 & y_1 & y_2 & y_3 & z & \\ \hline 2 & 0 & 1 & 0 & 0 & 0 & 800 \\ 0 & 1 & 0 & 1 & 0 & 0 & 300 \\ 2 & 3 & 0 & 0 & 1 & 0 & 1{,}500 \\ \hdashline -45 & -30 & 0 & 0 & 0 & 1 & 0 \end{array}$$

← These are for reference only and are not part of the tableau.

← This is called the **objective row.**

This is a two-dimensional model since there are two variables (not counting the slack variables). A two-dimensional model can be solved using the graphing procedure of Section 4.3 or the simplex method, which will be introduced in Section 4.5. If there are more than two variables, then only the simplex method can be used. However, for this two-dimensional example it is worthwhile relating the tableau to the graphing method. The graph for this linear programming problem is shown in Figure 4.5.

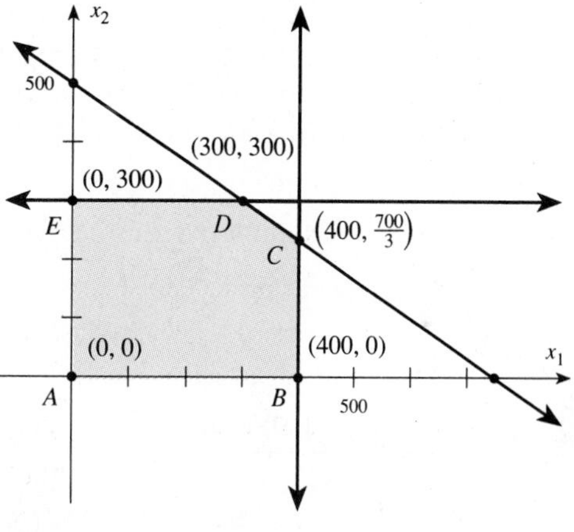

FIGURE 4.5

Now rewrite the tableau back in equation form to see the relationship between the corner points and the tableau values.

System values

	x_1 ↓	x_2 ↓	y_1 ↓	y_2 ↓	y_3 ↓	z ↓	
Coefficients	2	0	1	0	0	0	800
	0	1	0	1	0	0	300
	2	3	0	0	1	0	1,500
	−45	−30	0	0	0	1	0

This means

$$\begin{aligned} \mathbf{2}x_1 + {} & \mathbf{0}x_2 + \mathbf{1}y_1 + \mathbf{0}y_2 + \mathbf{0}y_3 + \mathbf{0}z = \mathbf{800} \\ \mathbf{0}x_1 + {} & \mathbf{1}x_2 + \mathbf{0}y_1 + \mathbf{1}y_2 + \mathbf{0}y_3 + \mathbf{0}z = \mathbf{300} \\ \mathbf{2}x_1 + {} & \mathbf{3}x_2 + \mathbf{0}y_1 + \mathbf{0}y_2 + \mathbf{1}y_3 + \mathbf{0}z = \mathbf{1{,}500} \\ -\mathbf{45}x_1 + {} & -\mathbf{30}x_2 + \mathbf{0}y_1 + \mathbf{0}y_2 + \mathbf{0}y_3 + \mathbf{1}z = \mathbf{0} \end{aligned}$$

The maximum occurs at a corner point, so we will "evaluate" the tableau for each corner point just to see what happens. This will help you understand what the tableau means.

Corner point A on the graph in Figure 4.5 has coordinates (0, 0). We find the values for all the variables as shown. To begin we see $x_1 = 0$ and $x_2 = 0$ (given point).

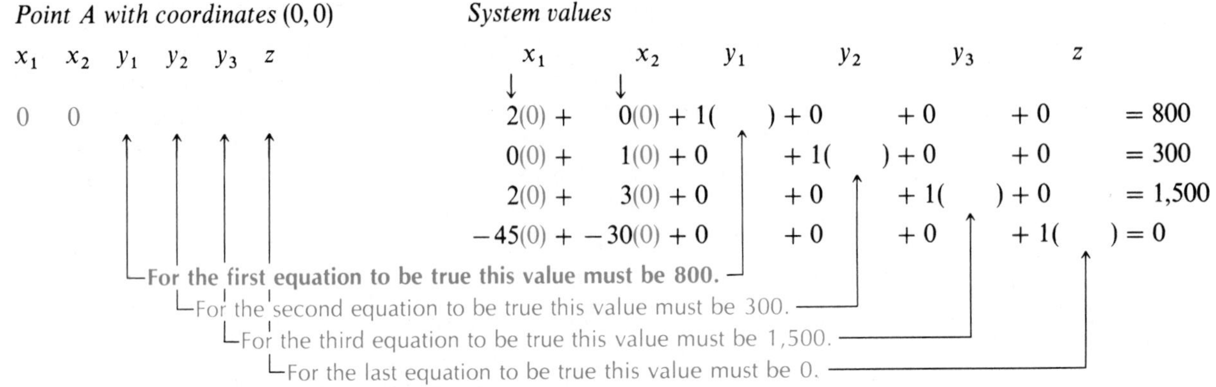

We summarize this below:

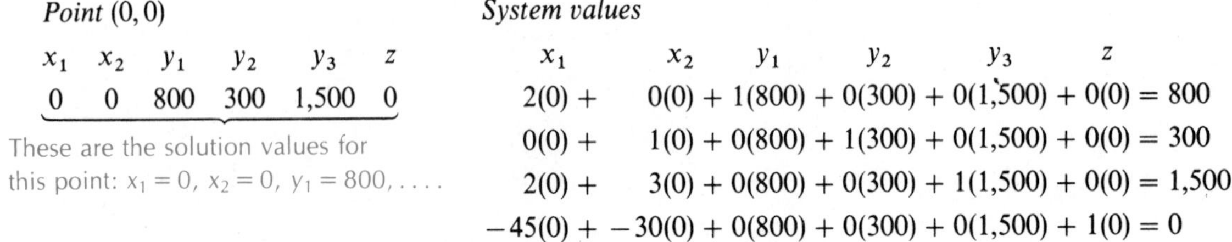

Point (0, 0)

x_1	x_2	y_1	y_2	y_3	z
0	0	800	300	1,500	0

These are the solution values for this point: $x_1 = 0$, $x_2 = 0$, $y_1 = 800, \ldots$.

System values

$$
\begin{aligned}
2(0) + 0(0) + 1(800) + 0(300) + 0(1{,}500) + 0(0) &= 800\\
0(0) + 1(0) + 0(800) + 1(300) + 0(1{,}500) + 0(0) &= 300\\
2(0) + 3(0) + 0(800) + 0(300) + 1(1{,}500) + 0(0) &= 1{,}500\\
-45(0) + -30(0) + 0(800) + 0(300) + 0(1{,}500) + 1(0) &= 0
\end{aligned}
$$

Let's try another corner point, say $B(400, 0)$. Here $x_1 = 400$ and $x_2 = 0$:

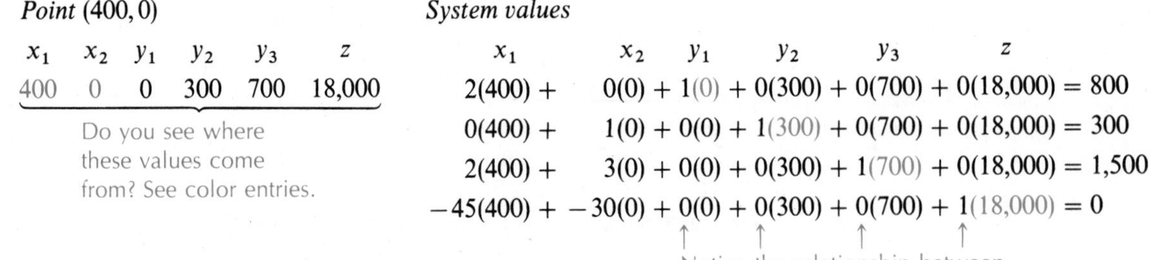

Point (400, 0)

x_1	x_2	y_1	y_2	y_3	z
400	0	0	300	700	18,000

Do you see where these values come from? See color entries.

System values

$$
\begin{aligned}
2(400) + 0(0) + 1(0) + 0(300) + 0(700) + 0(18{,}000) &= 800\\
0(400) + 1(0) + 0(0) + 1(300) + 0(700) + 0(18{,}000) &= 300\\
2(400) + 3(0) + 0(0) + 0(300) + 1(700) + 0(18{,}000) &= 1{,}500\\
-45(400) + -30(0) + 0(0) + 0(300) + 0(700) + 1(18{,}000) &= 0
\end{aligned}
$$

Notice the relationship between the entries in these columns and the color entries.

Consider still another corner point, $E(0, 300)$:

Point (0, 300)

x_1	x_2	y_1	y_2	y_3	z
0	300	800	0	600	9,000

System values

$$\begin{array}{rl} 2(0) + & 0(300) + 1(800) + 0(0) + 0(600) + 0(9{,}000) = 800 \\ 0(0) + & 1(300) + 0(800) + 1(0) + 0(600) + 0(9{,}000) = 300 \\ 2(0) + & 3(300) + 0(800) + 0(0) + 1(600) + 0(9{,}000) = 1{,}500 \\ -45(0) + & -30(300) + 0(800) + 0(0) + 0(600) + 1(9{,}000) = 0 \\ & \uparrow \qquad \uparrow \qquad \uparrow \qquad \uparrow \end{array}$$

The two points checked above are the corner points labeled B and E in Figure 4.5. Now consider the corner point C with coordinates $(400, \frac{700}{3})$.

Point $(400, \frac{700}{3})$

x_1	x_2	y_1	y_2	y_3	z
400	233.33	0	66.67	0	25,000

System values

$$\begin{array}{rl} 2(400) + & 0(233.33) + 1(0) + 0(66.67) + 0(0) + 0(25{,}000) = 800 \\ 0(400) + & 1(233.33) + 0(0) + 1(66.67) + 0(0) + 0(25{,}000) = 300 \\ 2(400) + & 3(233.33) + 0(0) + 0(66.67) + 1(0) + 0(25{,}000) = 1{,}500 \\ -45(400) + & -30(233.33) + 0(0) + 0(66.67) + 0(0) + 1(25{,}000) = 0 \end{array}$$

Finally, consider the corner point D with coordinates (300, 300).

Point (300, 300)

x_1	x_2	y_1	y_2	y_3	z
300	300	200	0	0	22,500

System values

$$\begin{array}{rl} 2(300) + & 0(300) + 1(200) + 0(0) + 0(0) + 0(22{,}500) = 800 \\ 0(300) + & 1(300) + 0(200) + 1(0) + 0(0) + 0(22{,}500) = 300 \\ 2(300) + & 3(300) + 0(200) + 0(0) + 1(0) + 0(22{,}500) = 1{,}500 \\ -45(300) + & -30(300) + 0(200) + 0(0) + 0(0) + 1(22{,}500) = 0 \end{array}$$

Summary

Point	**z-value (no. to be maximized)**
$A(0, 0)$	0
$B(400, 0)$	18,000
$C(400, \frac{700}{3})$	25,000
$D(300, 300)$	22,500
$E(0, 300)$	9,000

The Fundamental Theorem of Linear Programming tells us that the maximum value occurs at one of the corner points. Look at the matrix entries above. The largest value of z at any of the corner points is 25,000. This occurs at the corner point labeled C. Thus, the value of z is a maximum when $x_1 = 400$ and $x_2 = \frac{700}{3} \approx$ 233.33. ■

The simplex method is an **algorithm** that works with the tableau form and has the advantage of moving around the corner points without looking at a graph. This is useful for two-dimensional models, but is necessary with models whose dimension is larger than two. However, before we introduce the simplex method in the next section, we need to practice writing the initial tableau, as well as a process called *pivoting*.

EXAMPLE 5 Write the initial simplex tableau for the following linear programming problem.

$$\text{Maximize:} \quad z = 3x_1 + 9x_2 + 12x_3 - 5x_4$$

$$\text{Subject to:} \quad \begin{cases} x_1 + x_2 \le 40 \\ x_3 + x_4 \le 45 \\ x_1 + x_3 \le 30 \\ x_2 + x_4 \le 35 \\ x_1 \ge 0, \quad x_2 \ge 0, \quad x_3 \ge 0, \quad x_4 \ge 0 \end{cases}$$

Solution Align the slack variables for tableau form:

$$\begin{cases} x_1 + x_2 \qquad\qquad\qquad + y_1 \qquad\qquad\qquad = 40 \\ \qquad\qquad x_3 + x_4 \qquad + y_2 \qquad\qquad = 45 \\ x_1 \qquad + x_3 \qquad\qquad\qquad + y_3 \qquad = 30 \\ \qquad x_2 \qquad + x_4 \qquad\qquad\qquad + y_4 = 35 \\ -3x_1 - 9x_2 - 12x_3 + 5x_4 \qquad\qquad\qquad + z = 0 \end{cases}$$

The tableau form is:

x_1	x_2	x_3	x_4	y_1	y_2	y_3	y_4	z	
1	1	0	0	1	0	0	0	0	40
0	0	1	1	0	1	0	0	0	45
1	0	1	0	0	0	1	0	0	30
0	1	0	1	0	0	0	1	0	35
−3	−9	−12	5	0	0	0	0	1	0

■

Pivoting

Note that the initial simplex tableau looks very much like the augmented matrices we used in Chapter 3 to solve systems of equations. The procedure at that time was to follow a method called *Gauss–Jordan elimination.* For a simplex tableau the process involves a **pivoting operation.**

The *selection* of the pivot elements in the simplex method is not the same as in Gauss–Jordan elimination, as we see in the next section. Example 6, however, reviews the pivoting process. We repeat this pivoting process from page 119 here for easy reference. *The number* 0 *cannot be selected as a pivot element.*

Pivoting

1. Divide all entries in the row in which the pivot appears (called the *pivot row*) by the pivot element so that the pivot entry becomes a 1.
2. Obtain zeros above and below the pivot element.

COMPUTER APPLICATION

You can practice this pivoting process without getting tied up in the arithmetic calculations by using the *Row reduction* program on the software or on a calculator such as the TI-81.

EXAMPLE 6 Pivot about the circled numbers.

a.

$$\begin{array}{ccccc|c} x_1 & x_2 & y_1 & y_2 & z & \\ \left[\begin{array}{ccccc|c} \textcircled{2} & 0 & 1 & 0 & 0 & 800 \\ 2 & 3 & 0 & 1 & 0 & 1{,}500 \\ \hline -45 & -30 & 0 & 0 & 1 & 0 \end{array}\right] \end{array}$$

Step 1:

$$\left[\begin{array}{ccccc|c} \textcircled{1} & 0 & \frac{1}{2} & 0 & 0 & 400 \\ 2 & 3 & 0 & 1 & 0 & 1{,}500 \\ \hline -45 & -30 & 0 & 0 & 1 & 0 \end{array}\right]$$

Divide row 1 by 2: $R_1 \div 2$

Step 2:

$$\left[\begin{array}{ccccc|c} \textcircled{1} & 0 & \frac{1}{2} & 0 & 0 & 400 \\ 0 & 3 & -1 & 1 & 0 & 700 \\ \hline 0 & -30 & \frac{45}{2} & 0 & 1 & 18{,}000 \end{array}\right]$$

← Pivot row

Add $(-2)\cdot$row 1 to row 2: $-2R_1 + R_2$

Add $45\cdot$row 1 to row 3: $45R_1 + R_3$

b.

$$\begin{array}{c} x_1 \quad x_2 \quad x_3 \quad y_1 \quad y_2 \quad z \\ \left[\begin{array}{cccccc|c} 1 & \textcircled{2} & 3 & 1 & 0 & 0 & 100 \\ 2 & 4 & -1 & 0 & 1 & 0 & 200 \\ \hline -14 & -25 & -10 & 0 & 0 & 1 & 0 \end{array}\right] \end{array}$$

Step 1:

$$\left[\begin{array}{cccccc|c} \frac{1}{2} & \textcircled{1} & \frac{3}{2} & \frac{1}{2} & 0 & 0 & 50 \\ 2 & 4 & -1 & 0 & 1 & 0 & 200 \\ \hline -14 & -25 & -10 & 0 & 0 & 1 & 0 \end{array}\right]$$

Divide row 1 by 2: $\frac{1}{2}R_1$

Step 2:

$$\left[\begin{array}{cccccc|c} \frac{1}{2} & \textcircled{1} & \frac{3}{2} & \frac{1}{2} & 0 & 0 & 50 \\ 0 & 0 & -7 & -2 & 1 & 0 & 0 \\ \hline -\frac{3}{2} & 0 & \frac{55}{2} & \frac{25}{2} & 0 & 1 & 1{,}250 \end{array}\right]$$

← Pivot row

Add $(-4)\cdot$row 1 to row 2: $-4R_1 + R_2$

Add $25\cdot$row 1 to row 2: $25R_1 + R_2$

■

4.4 Problem Set

Rewrite the linear programming problems in Problems 1–12 (taken from Problem Set 4.3) using the notation introduced in this section. If one of the three conditions for a standard linear programming problem is violated, tell which one.

1. Maximize: $W = 30x + 20y$

Subject to: $\begin{cases} 2x + y \le 12 \\ 5x + 8y \le 40 \\ x \ge 0, \quad y \ge 0 \end{cases}$

2. Maximize: $T = 100x + 10y$

Subject to: $\begin{cases} 2x + 5y \le 20 \\ 2x + y \le 12 \\ x \ge 0, \quad y \ge 0 \end{cases}$

3. Maximize: $P = 100x + 100y$

Subject to: $\begin{cases} 3x + 2y \le 12 \\ x + 2y \le 8 \\ x \ge 0, \quad y \ge 0 \end{cases}$

4. Minimize: $I = 90x + 20y$

Subject to: $\begin{cases} x + y \ge 6 \\ -2x + y \ge -16 \\ y \le 9 \\ x \ge 0, \quad y \ge 0 \end{cases}$

5. Minimize: $X = 5x + 3y$

Subject to: $\begin{cases} 2x + y \ge 8 \\ y \le 5 \\ x - y \le 2 \\ 3x - 2y \ge 5 \\ x \ge 0, \quad y \ge 0 \end{cases}$

6. Maximize: $P = 23x + 46y$

Subject to: $\begin{cases} x + y \le 6 \\ 2x + y \le -16 \\ y \le 9 \\ x \ge 0, \quad y \ge 0 \end{cases}$

7. Maximize: $I = 90x + 20y$

Subject to: $\begin{cases} x + y \le 6 \\ 2x + y \le -16 \\ y \le 9 \end{cases}$

8. Maximize: $P = 6x + 3y$

Subject to: $\begin{cases} x + y \le 9 \\ 2x - 3y \ge -6 \\ x - y \le 3 \\ x \ge 0, \quad y \ge 0 \end{cases}$

9. Maximize: $T = x + 6y$

Subject to: $\begin{cases} 2x + y \ge 8 \\ y \le 5 \\ x - y \le 2 \\ 3x - 2y \ge 5 \\ x \ge 0, \quad y \ge 0 \end{cases}$

10. Maximize: $P = 160x + 130y$

Subject to: $\begin{cases} 2x + 3y \le 8 \\ 4x + 5y \le 8 \end{cases}$

11. Maximize: $K = 140x + 250y$

Subject to: $\begin{cases} 2x + 5y \le 100 \\ 3x + 4y \le 120 \end{cases}$

12. Minimize: $A = 2x - 3y$

Subject to: $\begin{cases} 3x + 2y \le 12 \\ x + 2y \le 8 \\ x \ge 0, \quad y \ge 0 \end{cases}$

Write the initial simplex tableau for the standard programming problems in Problems 13–22.

13. Maximize: $z = 2x_1 + 3x_2$

Subject to: $\begin{cases} 3x_1 + x_2 \le 300 \\ 2x_1 + 2x_2 \le 400 \\ x_1 \ge 0, \quad x_2 \ge 0 \end{cases}$

14. Maximize: $z = 8x_1 + 16x_2$

Subject to: $\begin{cases} 5x_1 + 3x_2 \le 165 \\ 900x_1 + 1{,}200x_2 \le 36{,}000 \\ x_1 \ge 0, \quad x_2 \ge 0 \end{cases}$

15. Maximize: $z = 45x_1 + 35x_2$

Subject to: $\begin{cases} x_1 + x_2 \le 200 \\ 4x_1 + 2x_2 \le 500 \\ x_1 \ge 0, \quad x_2 \ge 0 \end{cases}$

16. Maximize: $z = 5x_1 + 8x_2$

Subject to: $\begin{cases} x_1 + 3x_2 \le 1{,}200 \\ x_1 + 2x_2 \le 1{,}000 \\ x_1 \le 700 \\ x_1 \ge 0, \quad x_2 \ge 0 \end{cases}$

17. Maximize: $z = x_1 + x_2$

Subject to: $\begin{cases} 12x_1 + 150x_2 \le 1{,}200 \\ 6x_1 + 200x_2 \le 1{,}200 \\ 16x_1 + 50x_2 \le 800 \\ x_1 \ge 0, \quad x_2 \ge 0 \end{cases}$

18. Maximize: $z = 4x_1 + 2x_2$

Subject to: $\begin{cases} 2x_1 + x_2 \le 200 \\ 2x_1 + 2x_2 \le 240 \\ 2x_1 + 3x_2 \le 300 \\ x_1 \ge 0, \quad x_2 \ge 0 \end{cases}$

19. Maximize: $z = 12x_1 + 7x_2 + 5x_3$

Subject to: $\begin{cases} 8x_1 + 5x_2 + 3x_3 \le 1{,}000 \\ 5x_1 + x_2 + 3x_3 \le 800 \\ x_1 \ge 0, \quad x_2 \ge 0, \quad x_3 \ge 0 \end{cases}$

20. Maximize: $z = x_1 + x_2 + x_3$

Subject to: $\begin{cases} 60x_1 \le 1{,}800 \\ 60x_2 \le 1{,}500 \\ 60x_3 \le 3{,}000 \\ x_2 + x_3 \le 50 \\ x_1 \ge 0, \quad x_2 \ge 0, \quad x_3 \ge 0 \end{cases}$

21. Maximize: $z = 9x_1 + 7x_2 + 7x_3 + 8x_4$

Subject to: $\begin{cases} x_1 + x_2 \le 90 \\ x_3 + x_4 \le 130 \\ x_1 + x_3 \le 80 \\ x_2 + x_4 \le 110 \\ x_1 \ge 0, \quad x_2 \ge 0, \quad x_3 \ge 0, \quad x_4 \ge 0 \end{cases}$

22. Maximize: $z = 48x_1 + 61x_2 + 39x_3 + 45x_4$

Subject to: $\begin{cases} x_1 + x_2 \le 5 \\ x_2 + x_3 \le 4 \\ x_3 + x_4 \le 8 \\ x_2 + x_4 \le 7 \\ x_1 \ge 0, \quad x_2 \ge 0, \quad x_3 \ge 0, \quad x_4 \ge 0 \end{cases}$

Perform the pivoting process for the circled pivot in Problems 23–34.

23.
$$\begin{array}{ccccc|c} x_1 & x_2 & y_1 & y_2 & z & \\ \hline 3 & 0 & 1 & 0 & 0 & 90 \\ \textcircled{6} & 2 & 0 & 1 & 0 & 18 \\ \hline -12 & -6 & 0 & 0 & 1 & 0 \end{array}$$

24.
$$\begin{array}{ccccc|c} x_1 & x_2 & y_1 & y_2 & z & \\ \hline \textcircled{5} & 0 & 1 & 0 & 0 & 20 \\ 2 & 3 & 0 & 1 & 0 & 30 \\ \hline -10 & -3 & 0 & 0 & 1 & 0 \end{array}$$

25.
$$\begin{array}{ccccc|c} x_1 & x_2 & y_1 & y_2 & z & \\ \hline 1 & 4 & \frac{1}{3} & 0 & 0 & 430 \\ 0 & \textcircled{2} & 2 & 1 & 0 & 162 \\ \hline 0 & -6 & 4 & 0 & 1 & 360 \end{array}$$

26.
$$\begin{array}{ccccc|c} x_1 & x_2 & y_1 & y_2 & z & \\ \hline 1 & 0 & \frac{1}{5} & 0 & 0 & 4 \\ 0 & \textcircled{3} & -\frac{2}{5} & 1 & 0 & 22 \\ \hline 0 & -3 & 2 & 0 & 1 & 40 \end{array}$$

27.
$$\begin{array}{ccccc|c} x_1 & x_2 & y_1 & y_2 & z & \\ \hline 8 & \textcircled{4} & 1 & 0 & 0 & 40 \\ 12 & 6 & 0 & 1 & 0 & 600 \\ \hline -8 & -10 & 0 & 0 & 1 & 0 \end{array}$$

28.
$$\begin{array}{ccccc|c} x_1 & x_2 & y_1 & y_2 & z & \\ \hline \textcircled{2} & 1 & \frac{1}{4} & 0 & 0 & 10 \\ 0 & 0 & -6 & 1 & 0 & 360 \\ \hline -12 & 0 & \frac{5}{2} & 0 & 1 & 100 \end{array}$$

29.
$$\begin{array}{cccccc|c} x_1 & x_2 & y_1 & y_2 & y_3 & z & \\ \hline 2 & 3 & 1 & 0 & 0 & 0 & 200 \\ \textcircled{2} & 6 & 0 & 1 & 0 & 0 & 100 \\ 3 & 2 & 0 & 0 & 1 & 0 & 300 \\ \hline -10 & -5 & 0 & 0 & 0 & 1 & 0 \end{array}$$

30.
$$\begin{array}{cccccc|c} x_1 & x_2 & y_1 & y_2 & y_3 & z & \\ \hline 20 & \textcircled{10} & 1 & 0 & 0 & 0 & 200 \\ 2 & 4 & 0 & 1 & 0 & 0 & 100 \\ 3 & 2 & 0 & 0 & 1 & 0 & 300 \\ \hline -10 & -30 & 0 & 0 & 0 & 1 & 0 \end{array}$$

31.
$$\begin{array}{cccccc|c} x_1 & x_2 & y_1 & y_2 & y_3 & z & \\ \hline 0 & -5 & 1 & 0 & 0 & 0 & 300 \\ 1 & \textcircled{2} & 0 & 1 & 0 & 0 & 20 \\ 0 & 8 & 0 & -2 & 1 & 0 & 180 \\ \hdashline 0 & -4 & 0 & 6 & 0 & 1 & 300 \end{array}$$

32.
$$\begin{array}{cccccc|c} x_1 & x_2 & y_1 & y_2 & y_3 & z & \\ \hline \textcircled{2} & 1 & \frac{1}{2} & 0 & 0 & 0 & 30 \\ -3 & 0 & 0 & 1 & 0 & 0 & 50 \\ -2 & 0 & 3 & 0 & 1 & 0 & 20 \\ \hdashline -4 & 0 & 2 & 0 & 0 & 1 & 140 \end{array}$$

33.
$$\begin{array}{ccccccc|c} x_1 & x_2 & x_3 & y_1 & y_2 & y_3 & z & \\ \hline 1 & 2 & 4 & 1 & 0 & 0 & 0 & 80 \\ 1 & 4 & 3 & 0 & 1 & 0 & 0 & 60 \\ 8 & 4 & \textcircled{2} & 0 & 0 & 1 & 0 & 10 \\ \hdashline -2 & -3 & -4 & 0 & 0 & 0 & 1 & 0 \end{array}$$

34.
$$\begin{array}{ccccccc|c} x_1 & x_2 & x_3 & y_1 & y_2 & y_3 & z & \\ \hline 1 & 2 & 4 & 1 & 0 & 0 & 0 & 80 \\ \textcircled{1} & 4 & 3 & 0 & 1 & 0 & 0 & 20 \\ 8 & 4 & 2 & 0 & 0 & 1 & 0 & 200 \\ \hdashline -5 & -3 & -4 & 0 & 0 & 0 & 1 & 0 \end{array}$$

In Problems 35–40

a. *Write the simplex tableau.*

b. *Draw the graph showing the set of feasible solutions.*

c. *Find the values of all the variables used in the tableau for each of the corner points, as shown in Example 4.*

35. Maximize: $z = 150x_1 + 80x_2$

Subject to: $\begin{cases} 2x_1 \le 100 \\ 4x_1 + x_2 \le 400 \\ x_1 \ge 0, \quad x_2 \ge 0 \end{cases}$

36. Maximize: $z = 50x_1 + 10x_2$

Subject to: $\begin{cases} 2x_1 \le 100 \\ 4x_1 + x_2 \le 400 \\ x_1 \ge 0, \quad x_2 \ge 0 \end{cases}$

37. Maximize: $z = 15x_1 + 25x_2$

Subject to: $\begin{cases} 5x_2 \le 300 \\ 6x_1 + x_2 \le 240 \\ x_1 \ge 0, \quad x_2 \ge 0 \end{cases}$

38. Maximize: $z = 130x_1 + 20x_2$

Subject to: $\begin{cases} 5x_2 \le 300 \\ 6x_1 + x_2 \le 240 \\ x_1 \ge 0, \quad x_2 \ge 0 \end{cases}$

39. Maximize: $z = 80x_1 + 200x_2$

Subject to: $\begin{cases} x_1 + x_2 \le 50 \\ 2x_1 + 3x_2 \le 120 \\ x_1 \ge 0, \quad x_2 \ge 0 \end{cases}$

40. Maximize: $z = 100x_1 + 140x_2$

Subject to: $\begin{cases} x_1 + x_2 \le 50 \\ 2x_1 + 3x_2 \le 120 \\ x_1 \ge 0, \quad x_2 \ge 0 \end{cases}$

4.5 Maximization by the Simplex Method

In Section 4.4 we set the stage for the **simplex method**. This method requires a standard form linear programming problem written in an initial simplex tableau with slack variables. In this section we discuss four ideas that complete the simplex method:

1. How to read a solution from the simplex tableau
2. How to select a pivot
3. The relationship between simplex and graphing methods
4. How to recognize when a maximum value is found (that is, how to know when you are finished with the simplex method)

Reading Simplex Tableaus and Basic Solutions

Begin by writing Example 5 of Section 4.3 in standard form:

$$\text{Maximize:} \quad z = 143x_1 + 60x_2$$

$$\text{Subject to:} \quad \begin{cases} x_1 + x_2 \le 100 \\ 120x_1 + 210x_2 \le 15{,}000 \\ 110x_1 + 30x_2 \le 4{,}000 \\ x_1 \ge 0, \quad x_2 \ge 0 \end{cases}$$

Write this using slack variables:

$$\begin{cases} x_1 + x_2 + y_1 = 100 \\ 120x_1 + 210x_2 + y_2 = 15{,}000 \\ 110x_1 + 30x_2 + y_3 = 4{,}000 \\ -143x_1 - 60x_2 + z = 0 \end{cases}$$

Now write the initial simplex tableau:

$$\begin{array}{cccccc|c} x_1 & x_2 & y_1 & y_2 & y_3 & z & \\ 1 & 1 & 1 & 0 & 0 & 0 & 100 \\ 120 & 210 & 0 & 1 & 0 & 0 & 15{,}000 \\ 110 & 30 & 0 & 0 & 1 & 0 & 4{,}000 \\ \hdashline -143 & -60 & 0 & 0 & 0 & 1 & 0 \end{array}$$

This vertical line separates the left and right sides of the equations.

This dashed line separates the objective function from the constraints; the last row is called the *objective row*.

Solutions can be read from this matrix tableau. It represents the set of feasible solutions. The problem asks us to maximize an objective function subject to two variables and three constraints (not counting those that restrict the variables to nonnegative values). This means that we have a system of four equations and six variables (three given variables, x_1, x_2, and z, and three slack variables, y_1, y_2, and y_3). From our work with systems of equations we know that any two of the six variables can be chosen arbitrarily and the other four can then be found by solving the remaining system. Also, from the matrix solution of systems of equations, the matrix

$$\begin{array}{cccc|c} y_1 & y_2 & y_3 & z & \\ 1 & 0 & 0 & 0 & 100 \\ 0 & 1 & 0 & 0 & 15{,}000 \\ 0 & 0 & 1 & 0 & 4{,}000 \\ 0 & 0 & 0 & 1 & 0 \end{array}$$

gives the solution

$$\begin{cases} y_1 = 100 \\ y_2 = 15{,}000 \\ y_3 = 4{,}000 \\ z = 0 \end{cases}$$

Thus, if we choose to assign values to x_1 and x_2, we can solve the remaining system. A **basic solution** of a system such as this is one in which we arbitrarily assign values to two variables. A **basic feasible solution** is one in which all the variables are nonnegative. For example, if we let $x_1 = 0$ and $x_2 = 0$, then a basic feasible solution is the one specified above. Since x_1 and x_2 are both zero, this is sometimes called the **initial basic solution**. Notice that $z = 0$ is hardly a maximum value for z, but nevertheless it is a solution to the system. We sometimes write y_1, y_2, y_3, and z to the right of the matrix to help remind us which value in the last column corresponds to which variable.

EXAMPLE 1 Consider the following matrix tableau. Find the initial basic solution.

$$\begin{array}{ccccccc|c} x_1 & x_2 & x_3 & y_1 & y_2 & y_3 & z & \\ \hline 2 & 3 & 4 & 1 & 0 & 0 & 0 & 40 \\ 4 & 5 & 3 & 0 & 1 & 0 & 0 & 30 \\ 6 & 9 & 9 & 0 & 0 & 1 & 0 & 15 \\ \hline -3 & -10 & -5 & 0 & 0 & 0 & 1 & 0 \end{array}$$

Solution The *initial* basic solution chooses the x-values to be 0 and assigns the necessary values to z and the slack variables. Thus, $x_1 = 0$, $x_2 = 0$, and $x_3 = 0$. It follows that $y_1 = 40$, $y_2 = 30$, $y_3 = 15$, and $z = 0$ (you can see this by inspection). From this information *you begin* the simplex method by labeling the right-hand column as shown:

$$\begin{array}{ccccccc|cc} x_1 & x_2 & x_3 & y_1 & y_2 & y_3 & z & & \\ \hline 2 & 3 & 4 & 1 & 0 & 0 & 0 & 40 & y_1 \\ 4 & 5 & 3 & 0 & 1 & 0 & 0 & 30 & y_2 \\ 6 & 9 & 9 & 0 & 0 & 1 & 0 & 15 & y_3 \\ \hline -3 & -10 & -5 & 0 & 0 & 0 & 1 & 0 & z \end{array}$$

The initial basic solution lists the slack variables along with the z-value. ■

The same notation is used to find other basic feasible solutions. For example, to find a basic solution for

$$\begin{array}{ccccccc|c} x_1 & x_2 & x_3 & y_1 & y_2 & y_3 & z & \\ \hline 0 & 8 & 0 & 1 & 4 & 6 & 0 & 30 \\ 1 & 4 & 0 & 0 & 3 & 9 & 0 & 30 \\ 0 & 6 & 1 & 0 & 2 & 3 & 0 & 90 \\ \hline 0 & 9 & 0 & 0 & 4 & 12 & 1 & 4{,}000 \end{array}$$

you can arbitrarily choose some values to be zero. But which ones and how many? Find the columns with a 1 and with all other entries 0, as shown by the shading:

$$\begin{array}{ccccccc|c} x_1 & x_2 & x_3 & y_1 & y_2 & y_3 & z & \\ \hline 0 & 8 & 0 & 1 & 4 & 6 & 0 & 30 \\ 1 & 4 & 0 & 0 & 3 & 9 & 0 & 30 \\ 0 & 6 & 1 & 0 & 2 & 3 & 0 & 90 \\ \hline 0 & 9 & 0 & 0 & 4 & 12 & 1 & 4{,}000 \end{array}$$

If you now look at the remaining columns and choose $x_2 = 0$, $y_2 = 0$, and $y_3 = 0$, then the resulting matrix is

$$\begin{array}{cccc|c} x_1 & x_3 & y_1 & z & \\ \hline 0 & 0 & 1 & 0 & 30 \\ 1 & 0 & 0 & 0 & 30 \\ 0 & 1 & 0 & 0 & 90 \\ \hdashline 0 & 0 & 0 & 1 & 4{,}000 \end{array}$$

From this matrix you can see that $y_1 = 30$, $x_1 = 30$, $x_3 = 90$, and $z = 4{,}000$. Now you can label the column of the original matrix at the right to reflect this information:

$$\begin{array}{ccccccc|cc} x_1 & x_2 & x_3 & y_1 & y_2 & y_3 & z & & \\ \hline 0 & 8 & 0 & 1 & 4 & 6 & 0 & 30 & y_1 \\ 1 & 4 & 0 & 0 & 3 & 9 & 0 & 30 & x_1 \\ 0 & 6 & 1 & 0 & 2 & 3 & 0 & 90 & x_3 \\ \hdashline 0 & 9 & 0 & 0 & 4 & 12 & 1 & 4{,}000 & z \end{array}$$

EXAMPLE 2 Find a basic feasible solution for the given simplex tableau by labeling the column at the right.

$$\begin{array}{ccccccc|c} x_1 & x_2 & x_3 & y_1 & y_2 & y_3 & z & \\ \hline 8 & 0 & 0 & 9 & 1 & 3 & 0 & 90 \\ 5 & 0 & 1 & 19 & 0 & 6 & 0 & 80 \\ 4 & 1 & 0 & 12 & 0 & 9 & 0 & 110 \\ \hdashline -12 & 0 & 0 & 5 & 0 & 6 & 1 & 850 \end{array}$$

Solution Look for the columns with a 1 and 0s:

$$\begin{array}{ccccccc|cc} x_1 & x_2 & x_3 & y_1 & y_2 & y_3 & z & & \\ \hline 8 & 0 & 0 & 9 & 1 & 3 & 0 & 90 & y_2 \\ 5 & 0 & 1 & 19 & 0 & 6 & 0 & 80 & x_3 \\ 4 & 1 & 0 & 12 & 0 & 9 & 0 & 110 & x_2 \\ \hdashline -12 & 0 & 0 & 5 & 0 & 6 & 1 & 850 & z \end{array}$$

A basic feasible solution is $y_2 = 90$, $x_3 = 80$, $x_2 = 110$, and $z = 850$. [This assumes that you let $x_1 = 0$, $y_1 = 0$, and $y_3 = 0$ (the nonshaded columns).] ■

Selecting A Pivot

Is the basic feasible solution in Example 2 an **optimal solution**? To see this let us write the last row (representing the objective function) in equation form:

$$-12x_1 + 5y_1 + 6y_3 + z = 850$$

Solve for z:

$$z = 850 + 12x_1 - 5y_1 - 6y_3$$

Remember we want to make z *as large as possible* and also remember that x_1, y_1, and y_3 are nonnegative. For the terms that are being subtracted ($5y_1$ and $6y_3$ for this example), we see that z will be maximized when $y_1 = y_3 = 0$. On the other hand, if some nonnegative number is added to 850, we do not have a maximum value for z. If you relate this back to the tableau you can see that **as long as there is a negative coefficient in the last row, we have not yet found the maximum value of z.**

Optimal Solution

> The z value in a simplex tableau is maximized when all the entries in the bottom row are nonnegative.

The *process* by which we obtain an optimal solution is called the **pivoting process** and is summarized below.

Pivoting Process

> **Rule 1.** The **pivot column** is the column that has the most negative number as the bottom entry.*
>
> **Rule 2.** The **pivot row** is the row above the line that has the smallest *positive* ratio when the entry in the last column is divided by the corresponding positive entry from the same row of the pivot column.
>
> The **pivot element** is the entry in the intersection of the pivot row and pivot column. Using this pivot element, pivoting is completed in two steps:
>
> 1. Divide the pivot row by the pivot element so that the pivot element becomes a 1.
> 2. Obtain 0s above and below the pivot element.

EXAMPLE 3 Carry out the pivoting process on the given initial simplex tableau.

$$\begin{array}{c} \begin{matrix} x_1 & x_2 & y_1 & y_2 & z & \end{matrix} \\ \left[\begin{array}{rrrrr|r} 2 & 4 & 1 & 0 & 0 & 20 \\ 3 & 2 & 0 & 1 & 0 & 18 \\ \hline -6 & -5 & 0 & 0 & 1 & 0 \end{array}\right] \begin{matrix} y_1 \\ y_2 \\ z \end{matrix} \\ \uparrow \end{array}$$

Solution The pivot column is found by locating the most negative entry.

To see why you select the most negative entry (Rule 1), notice that the last row is the objective function:

$$z = 6x_1 + 5x_2$$

The most negative entry is the opposite of the *largest* coefficient in the objective equation. If we were to set $x_1 = x_2 = 0$, we would obtain $z = 0$ as a first approximation for the largest value of z. It is easy to see that z can be improved by

* If all the entries in this column are negative, the problem is unbounded and there is no solution. Also, if there are duplicate or multiple entries for the most negative, you can choose either.

increasing either x_1 or x_2. Note that for each *unit* x_1 increases, z increases by 6 (because of the $6x_1$ term) and the *per unit* increase in x_2 is 5, so it is easy to see that it is more effective to increase x_1 than x_2. Thus, the pivot column is chosen by looking for the most negative entry.

The variable at the top of the pivot column (x_1 in this example) is sometimes called the **entering variable** because when the pivoting process is finished, the x_1 will appear at the right and will no longer be a zero variable. The reason for this is that this column, when the pivoting process is finished, will consist of a single one (the pivot) and the rest of the entries in the column will be zero.

We now find the pivot row:

$$\begin{array}{c} \\ \\ \\ \\ \end{array}\begin{array}{ccccc|c} x_1 & x_2 & y_1 & y_2 & z & \\ \end{array}$$

$$\left[\begin{array}{ccccc|c} 2 & 4 & 1 & 0 & 0 & 20 \\ 3 & 2 & 0 & 1 & 0 & 18 \\ \hline -6 & -5 & 0 & 0 & 1 & 0 \end{array}\right]\begin{array}{c} y_1 \\ y_2 \\ z \end{array} \qquad \begin{array}{l} 20 \div 2 = 10 \\ 18 \div 3 = 6 \longleftarrow \\ \\ \end{array}$$

Pivot row; this is the smallest positive entry.

↑ Pivot column

In order to see why you select the smallest positive ratio (Rule 2) for the pivot row, look at the tableau in equation form:

$$2x_1 + 4x_2 + y_1 = 20$$
$$3x_1 + 2x_2 + y_2 = 18$$
$$-6x_1 - 5x_2 + z = 0$$

Since the pivot column is the most negative entry (in this example, the coefficient of x_1), the other variable, x_2, is 0.

If $x_2 = 0$, how much can we increase x_1 without causing the other variables to be negative?

$y_1 = 20 - 2x_1$ — Solve the first equation for y_1. The *most* that x_1 can be increased without causing y_1 to be negative is 10; how do you know this? $20 \div 2 = 10$

$y_2 = 18 - 3x_1$ — Solve the second equation for y_2. The *most* that x_1 can be increased without causing y_2 to be negative is 6: $18 \div 3 = 6$

Notice that these operations of division are those that we check when finding the pivot row. What is the *most* that x_1 can be increased without causing *either* y_1 or y_2 to be negative?

If we increase the value of x_1 *beyond the smaller value* it will cause at least one of the variables to be negative. Thus, Rule 2 prevents us from making the variable x_1 too large to be feasible.

The variable labeling the pivot row (y_2 in this example) is sometimes called the **departing variable** because when the pivoting is completed the variable is no longer part of the problem.

To continue with the given tableau, circle the pivot element and carry out the pivot process:

$$\begin{array}{c} \\ \\ \\ \\ \end{array}\begin{array}{ccccc|c} x_1 & x_2 & y_1 & y_2 & z & \\ 2 & 4 & 1 & 0 & 0 & 20 \\ \textcircled{3} & 2 & 0 & 1 & 0 & 18 \\ \hline -6 & -5 & 0 & 0 & 1 & 0 \\ \uparrow & & & & & \end{array}\begin{array}{l} \\ y_1 \\ y_2 \leftarrow \\ z \\ \\ \end{array}$$

Step 1: Divide to make the pivoting element 1:

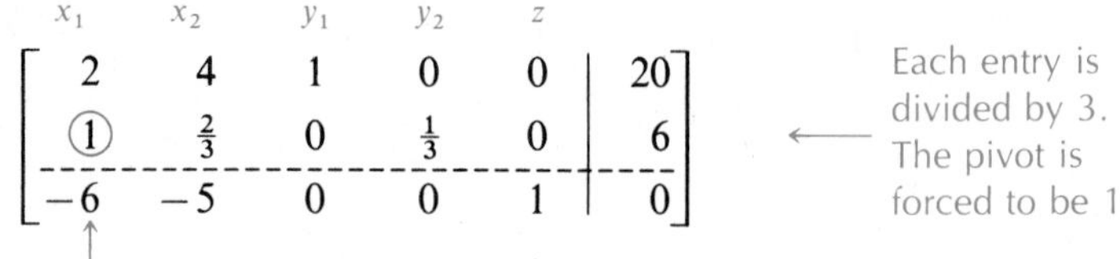

$$\begin{array}{ccccc|c} x_1 & x_2 & y_1 & y_2 & z & \\ 2 & 4 & 1 & 0 & 0 & 20 \\ \textcircled{1} & \frac{2}{3} & 0 & \frac{1}{3} & 0 & 6 \\ \hline -6 & -5 & 0 & 0 & 1 & 0 \\ \uparrow & & & & & \end{array} \leftarrow$$

Each entry is divided by 3. The pivot is forced to be 1.

Step 2: Perform elementary row operations to obtain 0s:

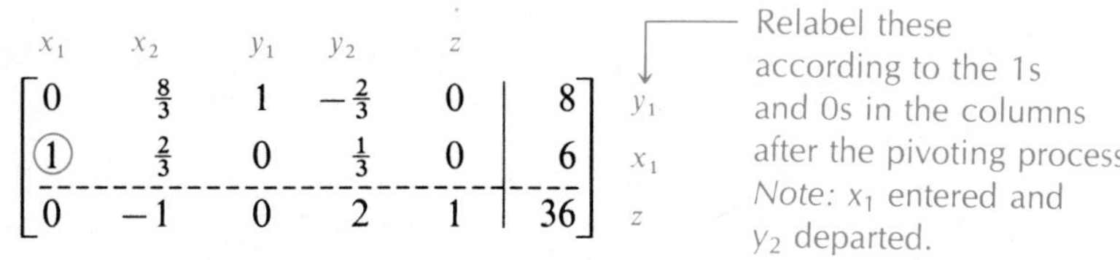

$$\begin{array}{ccccc|c|l} x_1 & x_2 & y_1 & y_2 & z & & \\ 0 & \frac{8}{3} & 1 & -\frac{2}{3} & 0 & 8 & y_1 \\ \textcircled{1} & \frac{2}{3} & 0 & \frac{1}{3} & 0 & 6 & x_1 \\ \hline 0 & -1 & 0 & 2 & 1 & 36 & z \end{array}$$

Relabel these according to the 1s and 0s in the columns after the pivoting process *Note:* x_1 entered and y_2 departed.

Repeat the pivoting process until there are no negative values in the last row.

$$\begin{array}{ccccc|c|l} x_1 & x_2 & y_1 & y_2 & z & & \\ 0 & \frac{8}{3} & 1 & -\frac{2}{3} & 0 & 8 & y_1 \\ 1 & \frac{2}{3} & 0 & \frac{1}{3} & 0 & 6 & x_1 \\ \hline 0 & -1 & 0 & 2 & 1 & 36 & z \\ & \uparrow & & & & & \end{array}$$

$8 \div \frac{8}{3} = 3 \leftarrow$

$6 \div \frac{2}{3} = 9$

Note that we still have a negative value

Step 1:

$$\begin{array}{ccccc|c} x_1 & x_2 & y_1 & y_2 & z & \\ 0 & \textcircled{1} & \frac{3}{8} & -\frac{1}{4} & 0 & 3 \\ 1 & \frac{2}{3} & 0 & \frac{1}{3} & 0 & 6 \\ \hline 0 & -1 & 0 & 2 & 1 & 36 \\ & \uparrow & & & & \end{array}$$

$\leftarrow$ Divide the entries of this row by $\frac{8}{3}$.

Step 2:

$$\begin{array}{ccccc|c|l} x_1 & x_2 & y_1 & y_2 & z & & \\ 0 & 1 & \frac{3}{8} & -\frac{1}{4} & 0 & 3 & x_2 \\ 1 & 0 & -\frac{1}{4} & \frac{1}{2} & 0 & 4 & x_1 \\ \hline 0 & 0 & \frac{3}{8} & \frac{7}{4} & 1 & 39 & z \end{array}$$

$\leftarrow$ All values in the last row are nonnegative, so the process is complete.

Thus the maximum value of z is 39 and it occurs when

$$x_2 = 3$$
$$x_1 = 4$$
$$z = 39$$

There are three equations with five unknowns, so we can arbitrarily choose two variables. We choose $y_1 = 0$ and $y_2 = 0$; the remaining variables are then found by inspection of the last matrix by looking at the columns where a single 1 and 0s elsewhere are found. ■

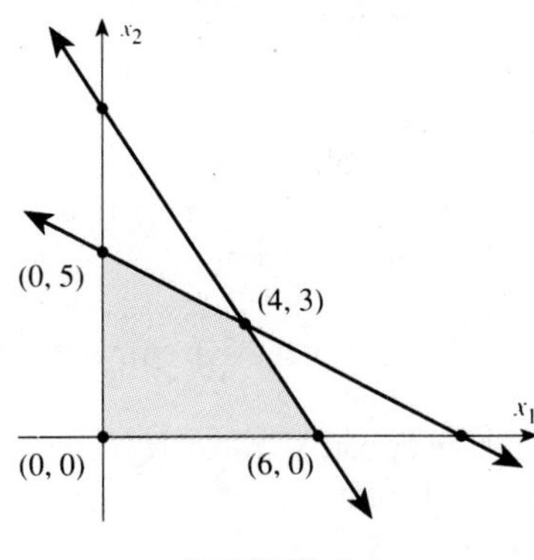

FIGURE 4.6

Relationship Between Simplex and Graphing Methods

In order to see the relationship between the simplex and graphing methods requires that there are no more than two variables. As an example, we will relate the abstract pivoting process we did in Example 3 with the graphical solution.

$$\text{Maximize } z = 6x_1 + 5x_2$$

$$\text{Subject to } \begin{cases} 2x_1 + 4x_2 \le 20 \\ 3x_1 + 2x_2 \le 18 \\ x_1 \ge 0, \quad x_2 \ge 0 \end{cases}$$

The graph is shown in Figure 4.6.
The first step in the pivoting process is repeated here:

Basic feasible solution from the initial tableau:

x_1	x_2	y_1	y_2	z
0	0	20	18	0

This is the corner point (0, 0).

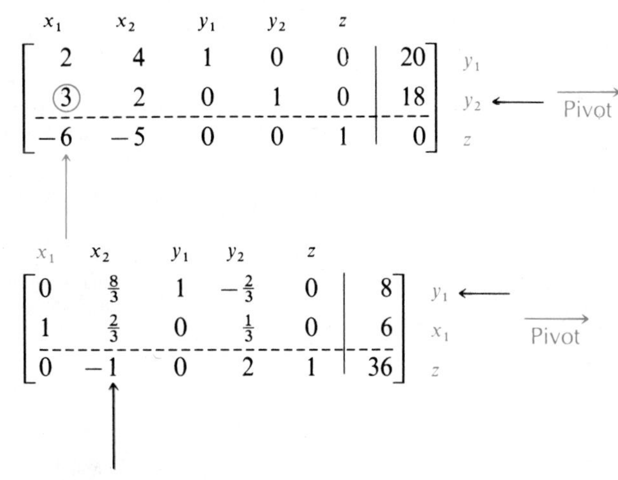

$$\begin{array}{ccccc|c} x_1 & x_2 & y_1 & y_2 & z & \\ 2 & 4 & 1 & 0 & 0 & 20 \\ \textcircled{3} & 2 & 0 & 1 & 0 & 18 \\ \hline -6 & -5 & 0 & 0 & 1 & 0 \end{array} \quad \begin{array}{l} \\ y_1 \\ y_2 \leftarrow \text{Pivot} \\ z \end{array}$$

Basic feasible solution after the first pivot:

x_1	x_2	y_1	y_2	z
6	0	8	0	36

This is the corner point (6, 0).

$$\begin{array}{ccccc|c} x_1 & x_2 & y_1 & y_2 & z & \\ 0 & \frac{8}{3} & 1 & -\frac{2}{3} & 0 & 8 \\ 1 & \frac{2}{3} & 0 & \frac{1}{3} & 0 & 6 \\ \hline 0 & -1 & 0 & 2 & 1 & 36 \end{array} \quad \begin{array}{l} \\ y_1 \leftarrow \\ x_1 \quad \text{Pivot} \\ z \end{array}$$

Note: If you would have chosen the column with the −5 entry you would have obtained the corner point (0, 9). The simplex method forces you to choose the best of the corner points along the x_1 or x_2 axis by forcing you to pick the most negative entry.

Basic feasible solution after the second pivot:

x_1	x_2	y_1	y_2	z
4	3	0	0	39

This is the corner point (4, 3).

$$\begin{array}{ccccc|c} x_1 & x_2 & y_1 & y_2 & z & \\ 0 & 1 & \frac{3}{8} & -\frac{1}{4} & 0 & 3 \\ 1 & 0 & -\frac{1}{4} & \frac{1}{2} & 0 & 4 \\ \hline 0 & 0 & \frac{3}{8} & \frac{7}{4} & 1 & 39 \end{array} \quad \begin{array}{l} \\ x_2 \\ x_1 \\ z \end{array}$$

← All values in the last row are nonnegative, so the process is complete.

As you can see from this comparison, the simplex method moves from one corner point of the feasible region to another, improving z each time until the optimal solution is reached.

Simplex Method

We can now summarize the simplex method.

Simplex Method for Standard Linear Programming Problems

In a standard linear programming problem:

Step 1. Write the initial simplex tableau using slack variables.
Step 2. **Test for maximality:** If all entries in the last row are nonnegative, then the tableau is the final tableau; interpret the solution.
Step 3. Select the pivot element.*
  a. The **pivot column** is the column that has the most negative entry at the bottom.
  b. The **pivot row** is the row that has the smallest positive ratio. If there are no positive ratios, then there is no solution.
Step 4. Carry out the pivoting process and return to step 2.

These steps are written in flowchart form in Figure 4.7.

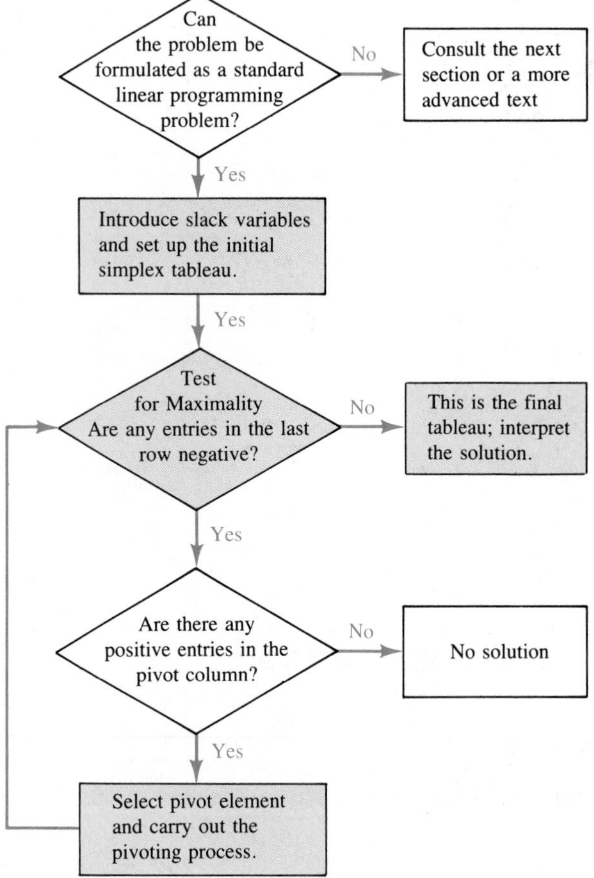

FIGURE 4.7
The simplex method can easily be written as a computer program.

* If two columns have the same entry and this is the most negative entry, then either can be chosen. If two rows give the same positive ratio, then either can be chosen.

EXAMPLE 4 Maximize: $z = 4x_1 + 5x_2$

$$\text{Subject to: } \begin{cases} 2x_1 + 5x_2 \le 25 \\ 6x_1 + 5x_2 \le 45 \\ x_1 \ge 0, \quad x_2 \ge 0 \end{cases}$$

Solution We solved this problem in Examples 1 and 3 of Section 4.3. The maximum value from the graphical method was 35 at (5, 3). We now solve this problem using the simplex method.

$$\begin{array}{c} \begin{array}{cccccc} x_1 & x_2 & y_1 & y_2 & z & \end{array} \\ \left[\begin{array}{ccccc|c} 2 & \textcircled{5} & 1 & 0 & 0 & 25 \\ 6 & 5 & 0 & 1 & 0 & 45 \\ \hline -4 & -5 & 0 & 0 & 1 & 0 \end{array}\right] \begin{array}{c} y_1 \\ y_2 \\ z \end{array} \end{array} \quad \begin{array}{l} 25 \div 5 = 5 \leftarrow \text{Departing variable} \\ 45 \div 5 = 9 \end{array}$$

$\uparrow$ Entering variable

$$\left[\begin{array}{ccccc|c} \frac{2}{5} & \textcircled{1} & \frac{1}{5} & 0 & 0 & 5 \\ 6 & 5 & 0 & 1 & 0 & 45 \\ \hline -4 & -5 & 0 & 0 & 1 & 0 \end{array}\right] \quad \tfrac{1}{5}\text{R1}$$

After the pivoting process you obtain:

$$\left[\begin{array}{ccccc|c} \frac{2}{5} & 1 & \frac{1}{5} & 0 & 0 & 5 \\ 4 & 0 & -1 & 1 & 0 & 20 \\ \hline -2 & 0 & 1 & 0 & 1 & 25 \end{array}\right] \quad \begin{array}{r} -5\text{R1} + \text{R2} \\ 5\text{R1} + \text{R3} \end{array}$$

Select a new pivot:

$$\begin{array}{c} \begin{array}{cccccc} x_1 & x_2 & y_1 & y_2 & z & \end{array} \\ \left[\begin{array}{ccccc|c} \frac{2}{5} & 1 & \frac{1}{5} & 0 & 0 & 5 \\ \textcircled{4} & 0 & -1 & 1 & 0 & 20 \\ \hline -2 & 0 & 1 & 0 & 1 & 25 \end{array}\right] \begin{array}{c} x_2 \\ y_2 \\ z \end{array} \end{array} \quad \begin{array}{l} 5 \div \frac{2}{5} = 12.5 \\ 20 \div 4 = 5 \leftarrow \text{Departing variable} \end{array}$$

$\uparrow$ Entering variable

Pivot again:

$$\left[\begin{array}{ccccc|c} \frac{2}{5} & 1 & \frac{1}{5} & 0 & 0 & 5 \\ \textcircled{1} & 0 & -\frac{1}{4} & \frac{1}{4} & 0 & 5 \\ \hline -2 & 0 & 1 & 0 & 1 & 25 \end{array}\right] \quad \tfrac{1}{4}\text{R2}$$

$$\begin{array}{c} \begin{array}{cccccc} x_1 & x_2 & y_1 & y_2 & z & \end{array} \\ \left[\begin{array}{ccccc|c} 0 & 1 & \frac{3}{10} & -\frac{1}{10} & 0 & 3 \\ 1 & 0 & -\frac{1}{4} & \frac{1}{4} & 0 & 5 \\ 0 & 0 & \frac{1}{2} & \frac{1}{2} & 1 & 35 \end{array}\right] \begin{array}{c} x_2 \\ x_1 \\ z \end{array} \end{array} \quad \begin{array}{l} -\frac{2}{5}\text{R2} + \text{R1} \\ \\ 2\text{R2} + \text{R3} \end{array}$$

This gives the maximum value of 35 at (5, 3). (We have chosen $y_1 = 0$ and $y_2 = 0$.) ■

EXAMPLE 5 Maximize: $z = 3x_1 + 9x_2 + 12x_3 - 5x_4$

$$\text{Subject to: } \begin{cases} x_1 + x_2 \le 40 \\ x_3 + x_4 \le 45 \\ x_1 + x_3 \le 30 \\ x_2 + x_4 \le 35 \\ x_1 \ge 0, \quad x_2 \ge 0, \quad x_3 \ge 0, \quad x_4 \ge 0 \end{cases}$$

Solution Write the initial simplex tableau (this was done as Example 5 of the previous section) and determine the entering and departing variables:

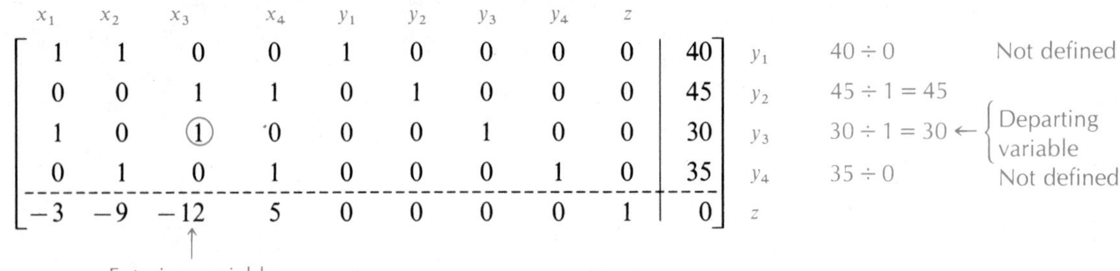

x_1	x_2	x_3	x_4	y_1	y_2	y_3	y_4	z				
1	1	0	0	1	0	0	0	0	40	y_1	$40 \div 0$	Not defined
0	0	1	1	0	1	0	0	0	45	y_2	$45 \div 1 = 45$	
1	0	(1)	0	0	0	1	0	0	30	y_3	$30 \div 1 = 30 \leftarrow$	Departing variable
0	1	0	1	0	0	0	1	0	35	y_4	$35 \div 0$	Not defined
−3	−9	−12	5	0	0	0	0	1	0	z		
		↑ Entering variable										

Next, circle the pivot element and pivot as shown below:

x_1	x_2	x_3	x_4	y_1	y_2	y_3	y_4	z				
1	1	0	0	1	0	0	0	0	40	y_1	$40 \div 1 = 40$	
−1	0	0	1	0	1	−1	0	0	15	y_2	$15 \div 0$	Not defined
1	0	1	0	0	0	1	0	0	30	x_3	$30 \div 0$	Not defined
0	(1)	0	1	0	0	0	1	0	35	y_4	$35 \div 1 = 35 \longleftarrow$	Departing variable
9	−9	0	5	0	0	12	0	1	360	z		
	↑ Entering variable											

The new entering and departing variables and pivot element are found in the previous tableau. Pivot once again:

x_1	x_2	x_3	x_4	y_1	y_2	y_3	y_4	z		
1	0	0	−1	1	0	0	−1	0	5	y_1
−1	0	0	1	0	1	−1	0	0	15	y_2
1	0	1	0	0	0	1	0	0	30	x_3
0	1	0	1	0	0	0	1	0	35	x_2
9	0	0	14	0	0	12	9	1	675	z

This is the final tableau. The solution (by inspection) is

$y_1 = 5$
$y_2 = 15$
$x_3 = 30$
$x_2 = 35$
$z = 675$

This solution assumes that we let $x_1 = 0$, $x_4 = 0$, $y_3 = 0$, and $y_4 = 0$; there are five equations with nine unknowns, so four variables are arbitrarily chosen.

The answer, however, is given in terms of the original variables in the problem, so we say that the objective function has a maximum value of 675 when $x_1 = 0$, $x_2 = 35$, $x_3 = 30$, and $x_4 = 0$. ■

4.5 Problem Set

Find the initial basic solution for each matrix tableau in Problems 1–3.

1. a.

x_1	x_2	y_1	y_2	z	
4	8	1	0	0	30
6	5	0	1	0	50
−10	−20	0	0	1	0

b.

x_1	x_2	y_1	y_2	z	
9	12	1	0	0	120
5	18	0	1	0	180
−40	−30	0	0	1	0

2. a.

x_1	x_2	y_1	y_2	y_3	z	
5	9	1	0	0	0	18
3	12	0	1	0	0	35
7	21	0	0	1	0	49
−5	−9	0	0	0	1	0

b.

x_1	x_2	y_1	y_2	y_3	z	
9	5	1	0	0	0	19
6	12	0	1	0	0	35
12	1	0	0	1	0	48
−10	−25	0	0	0	1	0

3. a.

x_1	x_2	x_3	y_1	y_2	y_3	z	
8	12	40	1	0	0	0	60
5	9	12	0	1	0	0	30
6	15	8	0	0	1	0	40
−5	−12	−3	0	0	0	1	0

b.

x_1	x_2	x_3	y_1	y_2	y_3	y_4	z	
5	9	11	1	0	0	0	0	80
6	8	4	0	1	0	0	0	50
9	18	1	0	0	1	0	0	60
1	8	1	0	0	0	1	0	90
−8	−12	−5	0	0	0	0	1	0

Find a basic feasible solution for each simplex tableau in Problems 4–6 and determine whether it is the final tableau. You do not need to pivot these problems.

4. a.

x_1	x_2	y_1	y_2	z	
3	1	4	0	0	60
4	0	8	1	0	24
−3	0	12	0	1	80

b.

x_1	x_2	y_1	y_2	z	
0	1	6	2	0	12
1	0	3	6	0	80
0	0	−3	9	1	120

5. a.

x_1	x_2	y_1	y_2	y_3	z	
1	8	0	3	0	0	10
0	4	1	1	0	0	12
0	1	0	0	1	0	20
0	3	0	2	0	1	32

b.

x_1	x_2	x_3	y_1	y_2	y_3	z	
0	2	0	1	2	4	0	20
0	4	1	0	8	8	0	80
1	3	0	0	9	3	0	120
0	−5	0	0	4	−2	1	360

6. a.

x_1	x_2	x_3	y_1	y_2	y_3	z	
0	1	0	4	0	0	0	90
1	0	0	6	1	0	0	70
0	0	1	3	3	0	0	65
0	0	0	5	2	1	0	20
0	0	0	19	10	0	1	250

b.

x_1	x_2	x_3	y_1	y_2	y_3	z	
1	2	1	0	0	1	0	5
1	−1	3	0	1	0	0	10
6	5	−2	1	0	0	0	8
−3	−5	−6	0	0	0	1	0

Carry out the pivoting process on each initial simplex tableau in Problems 7–12.

7.
$$\begin{array}{ccccc|c} x_1 & x_2 & y_1 & y_2 & z & \\ \hline 6 & 3 & 1 & 0 & 0 & 20 \\ 2 & 4 & 0 & 1 & 0 & 4 \\ \hdashline -4 & -20 & 0 & 0 & 1 & 0 \end{array}$$

8.
$$\begin{array}{ccccc|c} x_1 & x_2 & y_1 & y_2 & z & \\ \hline 3 & 6 & 1 & 0 & 0 & 60 \\ 12 & 6 & 0 & 1 & 0 & 18 \\ \hdashline -1 & -30 & 0 & 0 & 1 & 0 \end{array}$$

9.
$$\begin{array}{cccccc|c} x_1 & x_2 & y_1 & y_2 & y_3 & z & \\ \hline 2 & 3 & 1 & 0 & 0 & 0 & 50 \\ 45 & 15 & 0 & 1 & 0 & 0 & 30 \\ 4 & 3 & 0 & 0 & 1 & 0 & 70 \\ \hdashline -5 & -20 & 0 & 0 & 0 & 1 & 0 \end{array}$$

10.
$$\begin{array}{cccccc|c} x_1 & x_2 & y_1 & y_2 & y_3 & z & \\ \hline 30 & 15 & 1 & 0 & 0 & 0 & 60 \\ 5 & 3 & 0 & 1 & 0 & 0 & 50 \\ 2 & 5 & 0 & 0 & 1 & 0 & 100 \\ \hdashline -8 & -20 & 0 & 0 & 0 & 1 & 0 \end{array}$$

11.
$$\begin{array}{ccccccc|c} x_1 & x_2 & x_3 & y_1 & y_2 & y_3 & z & \\ \hline 5 & -1 & -3 & 1 & 0 & 0 & 0 & 20 \\ 2 & 1 & 9 & 0 & 1 & 0 & 0 & 10 \\ 1 & 3 & 9 & 0 & 0 & 1 & 0 & 2 \\ \hdashline -4 & -5 & -9 & 0 & 0 & 0 & 1 & 0 \end{array}$$

12.
$$\begin{array}{ccccccc|c} x_1 & x_2 & x_3 & y_1 & y_2 & y_3 & z & \\ \hline 3 & 9 & 6 & 1 & 0 & 0 & 0 & 6 \\ 5 & 9 & 8 & 0 & 1 & 0 & 0 & 15 \\ 3 & 6 & 5 & 0 & 0 & 1 & 0 & 30 \\ \hdashline -5 & -30 & -25 & 0 & 0 & 0 & 1 & 0 \end{array}$$

In Problems 13–28 solve the linear programming problems by using the simplex method.

13. Maximize: $z = 2x_1 + 3x_2$

Subject to: $\begin{cases} 3x_1 + x_2 \le 300 \\ 2x_1 + 2x_2 \le 400 \\ x_1 \ge 0, \quad x_2 \ge 0 \end{cases}$

14. Maximize: $z = 5x_1 - 3x_2$

Subject to: $\begin{cases} 2x_1 + x_2 \le 200 \\ 5x_1 + 2x_2 \le 100 \\ x_1 \ge 0, \quad x_2 \ge 0 \end{cases}$

15. Maximize: $z = 8x_1 + 16x_2$

Subject to: $\begin{cases} 5x_1 + 3x_2 \le 165 \\ 900x_1 + 1{,}200x_2 \le 36{,}000 \\ x_1 \ge 0, \quad x_2 \ge 0 \end{cases}$

16. Maximize: $z = 240x_1 + 100x_2$

Subject to: $\begin{cases} 300x_1 + 900x_2 \le 39{,}000 \\ 3x_1 + x_2 \le 150 \\ x_1 \ge 0, \quad x_2 \ge 0 \end{cases}$

17. Maximize: $z = 45x_1 + 35x_2$

Subject to: $\begin{cases} x_1 + x_2 \le 200 \\ 4x_1 + 2x_2 \le 500 \\ x_1 \ge 0, \quad x_2 \ge 0 \end{cases}$

18. Maximize: $z = 90x_1 + 55x_2$

Subject to: $\begin{cases} x_1 + x_2 \le 100 \\ 3x_1 + 4x_2 \le 400 \\ x_1 \ge 0, \quad x_2 \ge 0 \end{cases}$

19. Maximize: $z = 5x_1 + 8x_2$

Subject to: $\begin{cases} x_1 + 3x_2 \le 1{,}200 \\ x_1 + 2x_2 \le 1{,}000 \\ x_1 \le 700 \\ x_1 \ge 0, \quad x_2 \ge 0 \end{cases}$

20. Maximize: $z = 8x_1 + 10x_2$

Subject to: $\begin{cases} 2x_1 + x_2 \le 1{,}000 \\ x_1 + 3x_2 \le 1{,}500 \\ x_2 \le 500 \\ x_1 \ge 0, \quad x_2 \ge 0 \end{cases}$

21. Maximize: $z = 2x_1 + x_2$

Subject to: $\begin{cases} 12x_1 + 150x_2 \le 1{,}200 \\ 6x_1 + 200x_2 \le 1{,}200 \\ 16x_1 + 48x_2 \le 800 \\ x_1 \ge 0, \quad x_2 \ge 0 \end{cases}$

22. Maximize: $z = 140x_1 + 80x_2$

Subject to: $\begin{cases} 100x_1 + 5x_2 \le 1{,}500 \\ 200x_1 + 800x_2 \le 1{,}200 \\ 150x_1 + 3x_2 \le 1{,}050 \\ x_1 \ge 0, \quad x_2 \ge 0 \end{cases}$

23. Maximize: $z = 4x_1 + 2x_2$

Subject to: $\begin{cases} 2x_1 + x_2 \le 200 \\ 2x_1 + 2x_2 \le 240 \\ 2x_1 + 3x_2 \le 300 \\ x_1 \ge 0, \quad x_2 \ge 0 \end{cases}$

24. Maximize: $z = 3x_1 + 2x_2$

Subject to: $\begin{cases} 2x_1 + 4x_2 \le 80 \\ 2x_1 + x_2 \le 100 \\ 2x_1 + 2x_2 \le 120 \\ x_1 \ge 0, \quad x_2 \ge 0 \end{cases}$

25. Maximize: $z = 12x_1 + 7x_2 + 5x_3$

Subject to: $\begin{cases} 2x_1 + 4x_2 + 6x_3 \le 1{,}000 \\ x_1 + x_2 + 3x_3 \le 800 \\ x_1 \ge 0, \quad x_2 \ge 0, \quad x_3 \ge 0 \end{cases}$

26. Maximize: $z = 2x_1 + 3x_2 + x_3$

Subject to: $\begin{cases} 60x_1 \le 1{,}800 \\ 60x_2 \le 1{,}500 \\ 60x_3 \le 960 \\ x_2 + x_3 \le 50 \\ x_1 \ge 0, \quad x_2 \ge 0, \quad x_3 \ge 0 \end{cases}$

27. Maximize: $z = 9x_1 + 7x_2 + 7x_3 + 8x_4$

Subject to: $\begin{cases} x_1 + x_2 \le 90 \\ x_3 + x_4 \le 130 \\ x_1 + x_3 \le 80 \\ x_2 + x_4 \le 110 \\ x_1 \ge 0, \quad x_2 \ge 0, \quad x_3 \ge 0, \quad x_4 \ge 0 \end{cases}$

28. Maximize: $z = 48x_1 + 61x_2 + 39x_3 + 45x_4$

Subject to: $\begin{cases} x_1 + x_2 + x_3 \le 5 \\ x_2 + x_3 \le 4 \\ x_3 + x_4 \le 2 \\ x_2 + x_4 \le 7 \\ x_1 \ge 0, \quad x_2 \ge 0, \quad x_3 \ge 0, \quad x_4 \ge 0 \end{cases}$

Answer Problems 29–35 by using the simplex method.

APPLICATIONS

29. Alco Company manufactures two products: Alpha and Beta. Each product must pass through two processing operations, and all materials are introduced at the first operation. Alco may produce either one product exclusively or various combinations of both products subject to the constraints given in the table. A shortage of technical labor has limited Alpha production to no more than 700 units per day. There are no constraints on the production of Beta other than the hour constraints in the table. How many of each product should be manufactured in order to maximize the profit?

	Hours required to produce 1 unit		
Product	**First process**	**Second process**	**Profit per unit**
Alpha	1 hr	1 hr	\$5
Beta	3 hr	2 hr	\$8
Total capacity per day	1,200 hr	1,000 hr	

30. Repeat Problem 29 using the following table. Also assume that, instead of Alpha production being limited, Beta production has been limited to 500 units per day.

	Hours required to produce 1 unit		
Product	**First process**	**Second process**	**Profit per unit**
Alpha	2 hr	1 hr	\$8
Beta	1 hr	3 hr	\$10
Total capacity per day	1,000 hr	1,500 hr	

31. If each Alpha costs \$2 to manufacture and each Beta costs \$3, rework Problem 29 with the additional restriction that only \$2,000 per day is available to pay for these manufacturing costs.

32. If each Alpha costs \$2 to manufacture and each Beta costs \$3, rework Problem 30 with the additional restriction that only \$2,000 per day is available to pay for these manufacturing costs.

33. Rosenberg's Department store has \$150,000 to spend on newspaper advertising, and will select up to three newspapers. An advertisement in the *Press Democrat* costs \$12,000 and reaches 80,000 readers, an advertisement in the *News Herald* costs \$2,000 and reaches 10,000 readers, and an advertisement in *U.S.A. Today* costs \$20,000 and reaches 4 million readers, but only 100,000 of these are in the purchasing area for store shoppers. The *Press Democrat* will not accept more than three advertisements from one business, and Rosenberg's management will not allow more than five advertisements with any one paper. How should Rosenberg's advertise to maximize the number of readers?

34. Rework Problem 33 but suppose *U.S.A. Today* raises the price to $24,000.

35. A shoe company produces three models of athletic shoes. Shoe parts are first produced in the manufacturing shop and then put together in the assembly shop. The number of hours of labor required per pair in each shop is given in the table. The company can sell as many pairs of shoes as it can produce. However, during the next month, no more than 1,000 hours can be expended in the manufacturing shop and no more than 800 hours can be expended in the assembly shop. The expected profit from the sale of each pair of shoes is $12 for model 1, $7 for model 2, and $5 for model 3. The company wants to determine how many pairs of each shoe model to produce to maximize total profits over the next month.*

Shop	Model 1	Model 2	Model 3
Manufacturing	8 hr	5 hr	3 hr
Assembly	5 hr	1 hr	3 hr

* This problem is taken from a recent examination administered by the Society of Actuaries. Reprinted by permission of the Society of Actuaries.

*4.6 Nonstandard Linear Programming Problems

The standard linear programming problem is subject to three conditions:

Condition 1. The objective function is to be maximized.
Condition 2. All variables are nonnegative.
Condition 3. All constraints (except those in condition 2) are less than or equal to a nonnegative constant.

In this section we learn how to handle certain types of nonstandard problems.

Mixed Constraints

If there are mixed constraints of the type $\leq$ and $\geq$, then you can reverse the inequality by multiplying both sides by -1. This forces all of the constraints to be less than or equal to constraints, but then the requirement that all variables be nonnegative may be violated. For example, if

$$2x_1 + 3x_2 \geq 4$$

then multiplying both sides by -1 yields

$$-2x_1 - 3x_2 \leq -4$$

This type of constraint violates condition 3:

Linear polynomial $\leq$ nonnegative constant

We now introduce a second phase to the linear programming process. This phase is required to transform the problem to one that meets the criteria of a standard linear programming problem.

* This section is not required for subsequent textual development.

Nonstandard Linear Programming Problems, Phase I/Phase II

PHASE I Are there negative values in the right-hand column (ignore the objective row)? If not, proceed to Phase II. If there are negative entries in the right-hand column, you must use the following procedure for selecting the pivot:

Select pivot row first; it is the row (except for the value of z in the last row) with the most negative entry.*

Select pivot column second; form ratios using the right-hand column entry as the denominator. *The largest ratio gives the pivot column.*

Pivot; after pivoting, return to ask the Phase I question.

PHASE II This is the procedure for the standard linear programming problem. This phase requires that there is a basic feasible solution.

Select pivot column first; it is the column (except for the value of z in the last column) with the most negative entry.

Select pivot row second; form ratios using the right-hand column entry as the denominator. *The smallest positive ratio gives the pivot row.*

Pivot; after pivoting, return to ask the Phase I question.

EXAMPLE 1 Maximize: $z = 8x_1 + 12x_2$

$$\text{Subject to: } \begin{cases} x_1 + x_2 \le 10 \\ x_1 - x_2 \ge 5 \\ x_1 \ge 0, \quad x_2 \ge 0 \end{cases}$$

Solution Multiply the second constraint by -1:

$$\begin{cases} x_1 + x_2 \le 10 \\ -x_1 + x_2 \le -5 \\ x_1 \ge 0, \quad x_2 \ge 0 \end{cases}$$

$$\begin{array}{ccccc|c} x_1 & x_2 & y_1 & y_2 & z & \\ 1 & 1 & 1 & 0 & 0 & 10 \\ \textcircled{-1} & 1 & 0 & 1 & 0 & -5 \\ \hdashline -8 & -12 & 0 & 0 & 1 & 0 \\ \uparrow & \uparrow & \uparrow & \uparrow & \uparrow & \end{array}$$

$\longleftarrow$ **Phase I** Since there is a negative entry in the right-hand column, the pivot row is selected first.

Ratios: $-1 \div (-5) = \frac{1}{5}$, $1 \div (-5) = -\frac{1}{5}$, $0 \div (-5) = 0$, $1 \div (-5) = -\frac{1}{5}$, $0 \div (-5) = 0$

The largest ratio is $\frac{1}{5}$, so the first pivot is circled.

* If there is a tie for the most negative element, you can choose either. If all entries in the pivot row are nonnegative, there are no feasible solutions and the linear programming problem has no solution.

Pivot:

$$\begin{array}{c} \\ \\ \\ \end{array}\begin{array}{ccccc|c} x_1 & x_2 & y_1 & y_2 & z & \\ \hline 1 & 1 & 1 & 0 & 0 & 10 \\ \textcircled{1} & -1 & 0 & -1 & 0 & 5 \\ \hline -8 & -12 & 0 & 0 & 1 & 0 \end{array}$$

$$\left[\begin{array}{ccccc|c} 0 & 2 & 1 & 1 & 0 & 5 \\ 1 & -1 & 0 & -1 & 0 & 5 \\ \hline 0 & -20 & 0 & -8 & 1 & 40 \end{array}\right] \quad \begin{array}{l} -\text{R2} + \text{R1} \\ \\ 8\text{R2} + \text{R3} \end{array}$$

Phase I There are no negative entries in the right-hand column, so proceed with Phase II.

Phase II Select the pivot column first, as in the usual simplex procedure.

$$\begin{array}{ccccc|c} x_1 & x_2 & y_1 & y_2 & z & \\ \hline 0 & 2 & 1 & 1 & 0 & 5 \\ 1 & -1 & 0 & -1 & 0 & 5 \\ \hline 0 & -20 & 0 & -8 & 1 & 40 \\ & \uparrow & & & & \end{array} \quad \begin{array}{l} 5 \div 2 = \frac{5}{2} \\ 5 \div (-1) \text{ not positive} \\ \\ \\ \end{array}$$

Pivot column

$$\left[\begin{array}{ccccc|c} 0 & \textcircled{1} & \frac{1}{2} & \frac{1}{2} & 0 & \frac{5}{2} \\ 1 & -1 & 0 & -1 & 0 & 5 \\ \hline 0 & -20 & 0 & -8 & 1 & 40 \end{array}\right] \quad \begin{array}{l} \frac{5}{2} \div 1 = \frac{5}{2} \\ 5 \div (-1) \text{ not positive} \\ \\ \end{array}$$

$$\begin{array}{ccccc|c|c} x_1 & x_2 & y_1 & y_2 & z & & \\ \hline 0 & 1 & \frac{1}{2} & \frac{1}{2} & 0 & \frac{5}{2} & x_2 \\ 1 & 0 & \frac{1}{2} & -\frac{1}{2} & 0 & \frac{15}{2} & x_1 \\ \hline 0 & 0 & 10 & 2 & 1 & 90 & z \end{array}$$

The maximum value is $z = 90$, and the optimal solution is $(x_1, x_2) = (\frac{15}{2}, \frac{5}{2})$. ■

Equality Constraints

If one of the constraints is an equality, you can eliminate one of the variables from the problem as illustrated by Example 2.

EXAMPLE 2 Maximize: $z = 3x_1 + 5x_2 + x_3$

Subject to:
$$\begin{cases} 2x_1 + 2x_2 + x_3 \le 150 \\ 4x_1 + 5x_2 + 3x_3 \le 350 \\ x_1 + x_2 + x_3 = 90 \\ x_1 \ge 0, \quad x_2 \ge 0, \quad x_3 \ge 0 \end{cases}$$

Solution Since $x_1 + x_2 + x_3 = 90$, we can solve for any of these three variables. For example,

$$x_3 = 90 - x_2 - x_1$$

SPREADSHEET PROGRAM

In 1-2-3 choose /Date|Optimization.
In Quattro Pro choose
/Tools|Advanced Math|Optimization.

These programs will solve linear programming problems with <, <=, >, >=, and = constraints, as well as solving both maximum and minimum values. They will also solve the dual problem discussed in the next section.

Substitute this into all of the other statements:

$$\text{Maximize:} \quad z = 3x_1 + 5x_2 + (\mathbf{90} - \boldsymbol{x_2} - \boldsymbol{x_1})$$

$$\text{Subject to:} \quad \begin{cases} 2x_1 + 2x_2 + \ \ (\mathbf{90} - \boldsymbol{x_2} - \boldsymbol{x_1}) \le 150 \\ 4x_1 + 5x_2 + 3(\mathbf{90} - \boldsymbol{x_2} - \boldsymbol{x_1}) \le 350 \\ x_1 \ge 0, \quad x_2 \ge 0 \end{cases}$$

Algebraically simplify:

$$\text{Maximize:} \quad z = 2x_1 + 4x_2 + 90$$

$$\text{Subject to:} \quad \begin{cases} x_1 + \ \ x_2 \le 60 \\ x_1 + 2x_2 \le 80 \\ x_1 \ge 0, \quad x_2 \ge 0 \end{cases}$$

We can now begin the simplex procedure:

$$\begin{array}{c} \begin{matrix} x_1 & x_2 & y_1 & y_2 & z & \end{matrix} \\ \left[\begin{array}{ccccc|c} 1 & 1 & 1 & 0 & 0 & 60 \\ 1 & 2 & 0 & 1 & 0 & 80 \\ \hline -2 & -4 & 0 & 0 & 1 & 90 \end{array}\right] \begin{matrix} y_1 \\ y_2 \\ \\ \end{matrix} \end{array}$$

Phase I is complete (there are no negative entries in the right column), so move to Phase II. The pivot column is -4 (entering variable is x_2) and the pivot row is 80 (departing variable is y_2).

After the first pivot, we obtain the final tableau:

$$\begin{array}{c} \begin{matrix} x_1 & x_2 & y_1 & y_2 & z & \end{matrix} \\ \left[\begin{array}{ccccc|c} \frac{1}{2} & 0 & 1 & -\frac{1}{2} & 0 & 20 \\ \frac{1}{2} & 1 & 0 & \frac{1}{2} & 0 & 40 \\ \hline 0 & 0 & 0 & 2 & 1 & 250 \end{array}\right] \begin{matrix} y_1 \\ x_2 \\ \\ \end{matrix} \end{array}$$

If $x_2 = 40$ and $x_1 = 0$, then $x_3 = 90 - x_1 - x_2 = 50$.

The variable z obtains its maximum value of 250 when $x_1 = 0$, $x_2 = 40$, $x_3 = 50$. ■

Minimization Problems

For the rest of this section (as well as the next section), we turn our attention to *minimization* problems. We will first consider minimization problems that have only less than constraints, then we will consider those with mixed constraints, and in the next section will consider those with only greater than constraints.

Less Than Constraints

The first, and easiest, type of minimization problem is one in which all of the constraints are of the $\le$ type. (The constraints that require the variables to be nonnegative are excepted, of course.)

EXAMPLE 3 Minimize: $z = -4x_1 - 5x_2$

Solution Subject to: $\begin{cases} 2x_1 + 5x_2 \le 25 \\ 6x_1 + 5x_2 \le 45 \\ x_1 \ge 0, \quad x_2 \ge 0 \end{cases}$

This is a minimization problem with $\le$ constraints. It is *not* a standard linear programming problem. Let $z' = -z$. This means that the smallest value for z is the largest value for z'. To see this more clearly, write down any set of positive numbers, say,

$$S = \{3, 9, 18, 20\}$$

Opposites: $S' = \{-3, -9, -18, -20\}$
Minimize S: 3 is the smallest element in S
Maximize S': -3 is the largest element in S'

This means that the objective function for this example can be rewritten as:

Maximize: $z' = -z = 4x_1 + 5x_2$

Subject to: $\begin{cases} 2x_1 + 5x_2 \le 25 \\ 6x_1 + 5x_2 \le 45 \\ x_1 \ge 0, \quad x_2 \ge 0 \end{cases}$

Solving the problem using the simplex method is the same as with the standard linear programming problem except that now the coefficient of z is negative. For this example, the simplex tableau is

$$\begin{array}{c} \begin{matrix} x_1 & x_2 & y_1 & y_2 & z' & \end{matrix} \\ \left[\begin{array}{rrrrr|r} 2 & 5 & 1 & 0 & 0 & 25 \\ 6 & 5 & 0 & 1 & 0 & 45 \\ \hline -4 & -5 & 0 & 0 & 1 & 0 \end{array}\right] \begin{matrix} y_1 \\ y_2 \\ \\ \end{matrix} \end{array}$$

The simplex method offers a straightforward (2 pivots) way to obtain the final simplex tableau:

$$\begin{array}{c} \begin{matrix} x_1 & x_2 & y_1 & y_2 & z' & \end{matrix} \\ \left[\begin{array}{rrrrr|r} 0 & 1 & \frac{3}{10} & -\frac{1}{10} & 0 & 3 \\ 1 & 0 & -\frac{1}{4} & \frac{1}{4} & 0 & 5 \\ \hline 0 & 0 & \frac{1}{2} & \frac{1}{2} & 1 & 35 \end{array}\right] \begin{matrix} x_2 \\ x_1 \\ \\ \end{matrix} \end{array}$$

WARNING ***We are focusing on the simplex method, but, in general, when solving linear programming problems, do not forget that graphing is a very good procedure when dealing with two variables.***

However, since this example has only two variables the most expedient method of solution is by graphing. From Example 1 in Section 4.3, the solution is $z' = 35$ when $x_1 = 5$ and $x_2 = 3$. This means that the minimum value of z is -35 when $x_1 = 5$ and $x_2 = 3$. ■

Mixed Constraints

The second type of minimization problem in this section is one that has *both* $\le$ and $\ge$ type constraints. For $\ge$ constraints, we multiply both sides by -1 and treat the tableau like the one in Example 1. We illustrate this procedure with an applied problem from Section 4.2.

EXAMPLE 4 **Transportation problem.** Sears ships a certain air-conditioning unit from factories in Portland, Oregon, and Flint, Michigan, to distribution centers in Los Angeles, California, and Atlanta, Georgia. Shipping costs are summarized in the table:

Source	Destination	Shipping cost
Portland	Los Angeles Atlanta	\$30 \$40
Flint	Los Angeles Atlanta	\$60 \$50

Supply and demand, in number of units, are:

Supply	Demand
Portland, 200 Flint, 600	Los Angeles, 300 Atlanta, 400

How should shipments be made from Portland and Flint to minimize the shipping cost?

Solution This model was built on that of Example 5 of Section 4.2, which we copy here for convenience. Let

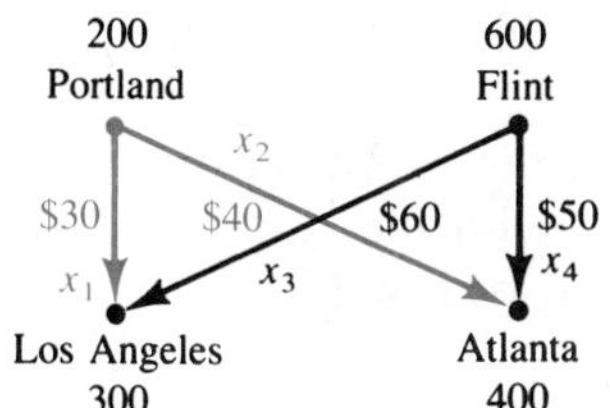

x_1 = Number shipped from Portland to Los Angeles
x_2 = Number shipped from Portland to Atlanta
x_3 = Number shipped from Flint to Los Angeles
x_4 = Number shipped from Flint to Atlanta

Minimize: $C = 30x_1 + 40x_2 + 60x_3 + 50x_4$

Subject to: $\begin{cases} x_1 + x_2 \leq 200 \\ x_3 + x_4 \leq 600 \\ x_1 + x_3 \geq 300 \\ x_2 + x_4 \geq 400 \end{cases}$

Thus the problem in matrix form is:

Maximize $z' = -z = 30x_1 + 40x_2 + 60x_3 + 50x_4$

$$\begin{array}{ccccccccc|c} x_1 & x_2 & x_3 & x_4 & y_1 & y_2 & y_3 & y_4 & z & \\ 1 & 1 & 0 & 0 & 1 & 0 & 0 & 0 & 0 & 200 \\ 0 & 0 & 1 & 1 & 0 & 1 & 0 & 0 & 0 & 600 \\ -1 & 0 & -1 & 0 & 0 & 0 & 1 & 0 & 0 & -300 \\ 0 & -1 & 0 & \textcircled{-1} & 0 & 0 & 0 & 1 & 0 & -400 \\ \hline 30 & 40 & 60 & 50 & 0 & 0 & 0 & 0 & 1 & 0 \end{array}$$

Since there are negative elements in the right-hand column, this is Phase I. Select the most negative entry (−400) for the pivot row. There are two choices for the pivot column (second and fourth column). We select the circled one shown in row 4, column 4. Although not necessary, we choose the one with the larger entry in the objective row. Multiply row 4 by -1 and pivot to obtain:

$$
\begin{array}{ccccccccc|c}
x_1 & x_2 & x_3 & x_4 & y_1 & y_2 & y_3 & y_4 & z' & \\
1 & 1 & 0 & 0 & 1 & 0 & 0 & 0 & 0 & 200 \\
0 & -1 & 1 & 0 & 0 & 1 & 0 & 1 & 0 & 300 \\
-1 & 0 & -1 & 0 & 0 & 0 & 1 & 0 & 0 & -300 \\
0 & 1 & 0 & 1 & 0 & 0 & 0 & -1 & 0 & 400 \\
\hdashline
30 & -10 & 60 & 0 & 0 & 0 & 0 & 50 & 1 & -20{,}000
\end{array}
$$

We are still in a Phase I problem. Select row 3 as the pivot row (it is the only negative element). Again, the largest ratio is a tie between the first and third columns. We choose the one with the larger number in the objective row; this pivot element is circled. Divide row 3 by -1 and pivot:

$$
\begin{array}{ccccccccc|c}
x_1 & x_2 & x_3 & x_4 & y_1 & y_2 & y_3 & y_4 & z' & \\
1 & 1 & 0 & 0 & 1 & 0 & 0 & 0 & 0 & 200 \\
-1 & -1 & 0 & 0 & 0 & 1 & 1 & 1 & 0 & -100 \\
1 & 0 & 1 & 0 & 0 & 0 & -1 & 0 & 0 & 300 \\
0 & 1 & 0 & 1 & 0 & 0 & 0 & -1 & 0 & 400 \\
\hdashline
-30 & -10 & 0 & 0 & 0 & 0 & 60 & 50 & 1 & -38{,}000
\end{array}
$$

This is still a Phase I problem and the pivot row is row 2. To break the tie we choose the element in the column with the number with the largest magnitude (-30) in the objective row.* This element is circled. Divide row 2 by -1 and pivot. The result is:

$$
\begin{array}{ccccccccc|cl}
x_1 & x_2 & x_3 & x_4 & y_1 & y_2 & y_3 & y_4 & z' & & \\
0 & 0 & 0 & 0 & 1 & 1 & 1 & 1 & 0 & 100 & \leftarrow \\
1 & 1 & 0 & 0 & 0 & -1 & -1 & -1 & 0 & 100 & \\
0 & -1 & 1 & 0 & 0 & 1 & 0 & 1 & 0 & 200 & \\
0 & 1 & 0 & 1 & 0 & 0 & 0 & -1 & 0 & 400 & \\
\hdashline
0 & 20 & 0 & 0 & 0 & -30 & 30 & 20 & 1 & -35{,}000 & \\
 & & & & & \uparrow & & & & &
\end{array}
$$

Now that the negative elements in the right-hand column (except for the $-35{,}000$ in the objective row) are gone, we can now apply the Phase II procedure, which means we pick the column first. The most negative element is -30, so the pivot column is column 6; the smallest positive ratio is 100, so the pivot row is row 1.

* If you choose the element in the column with -10 in the objective row, the problem will cycle back to this tableau. If your problem cycles back to the same tableau when you are performing the simplex procedure, select the other choice to break the tie the second time around.

After pivoting, we obtain:

$$\begin{array}{ccccccccc|c} x_1 & x_2 & x_3 & x_4 & y_1 & y_2 & y_3 & y_4 & z' & \\ 0 & 0 & 0 & 0 & 1 & 1 & 1 & 1 & 0 & 100 \\ 1 & 1 & 0 & 0 & 1 & 0 & 0 & 0 & 0 & 200 \\ 0 & -1 & 1 & 0 & -1 & 0 & -1 & 0 & 0 & 100 \\ 0 & 1 & 0 & 1 & 0 & 0 & 0 & -1 & 0 & 400 \\ \hline 0 & 20 & 0 & 0 & 30 & 0 & 60 & 50 & 1 & -32{,}000 \end{array}$$

The pivoting is complete. The solution is $x_1 = 200$, $x_3 = 100$, and $x_4 = 400$. This means that the maximum value for $z' = -32{,}000$, so the corresponding minimum value for $z = -z' = 32{,}000$. The minimum value of \$32,000 is achieved by shipping 200 air-conditioning units from Portland to Los Angeles and none from Portland to Atlanta. Flint ships 100 units to Los Angeles, and 400 to Atlanta. ■

4.6

Problem Set

1. Use the simplex method to maximize $z = 20x_1 + 50x_2$

Subject to $\begin{cases} 8x_1 + 3x_2 \le 24 \\ x_1 + x_2 \ge 5 \\ x_1 \ge 0, \quad x_2 \ge 0 \end{cases}$

2. Use the simplex method to maximize $z = 30x_1 + 40x_2$

Subject to $\begin{cases} x_1 + 2x_2 \le 8 \\ 3x_1 + x_2 \ge 9 \\ x_1 \ge 0, \quad x_2 \ge 0 \end{cases}$

3. Use the graphing method to solve the linear programming problem in Problem 1. Reconcile the answer from the graphing method and the simplex method.

4. Use the graphing method to solve the linear programming problem in Problem 2. Reconcile the answer from the graphing method and the simplex method.

5. Use the simplex method to minimize $z = 30x_1 + 90x_2$

Subject to $\begin{cases} 8x_1 + 3x_2 \le 24 \\ 2x_1 + 3x_2 \ge 12 \\ x_1 \ge 0, \quad x_2 \ge 0 \end{cases}$

6. Use the simplex method to maximize $z = 10x_1 + 200x_2$

Subject to $\begin{cases} x_1 + x_2 \ge 6 \\ x_1 + 4x_2 \le 12 \\ x_1 \ge 0, \quad x_2 \ge 0 \end{cases}$

7. Use the graphing method to solve the linear programming problem in Problem 5. Reconcile the answer from the graphing method and the simplex method.

8. Use the graphing method to solve the linear programming problem in Problem 6. Reconcile the answer from the graphing method and the simplex method.

Use the simplex method to solve Problems 9–24.

9. Maximize: $z = 10x_1 + 40x_2$

Subject to: $\begin{cases} 3x_1 + 5x_2 \le 50 \\ 2x_1 + 3x_2 \ge -10 \\ x_1 \ge 0, \quad x_2 \ge 0 \end{cases}$

10. Maximize: $z = 500x_1 + 300x_2$

Subject to: $\begin{cases} 9x_1 + 5x_2 \ge -5 \\ 15x_1 + 3x_2 \le 75 \\ x_1 \ge 0, \quad x_2 \ge 0 \end{cases}$

11. Maximize: $z = 30x_1 + 40x_2$

Subject to: $\begin{cases} 5x_1 + 3x_2 \ge -5 \\ 2x_1 + 3x_2 \le 40 \\ x_1 \ge 0, \quad x_2 \ge 0 \end{cases}$

12. Maximize: $z = 5x_1 + 4x_2$

Subject to: $\begin{cases} 3x_1 + 6x_2 \le 90 \\ 2x_1 + 5x_2 \le 100 \\ x_1 \ge 0, \quad x_2 \ge 0 \end{cases}$

13. Maximize: $z = 30x_1 - 20x_2$

Subject to: $\begin{cases} 2x_1 - x_2 \le 12 \\ -2x_2 \le 9 \\ x_1 \ge 0, \quad x_2 \le 0 \end{cases}$

14. Maximize: $z = 100x_1 - 10x_2$

Subject to: $\begin{cases} 2x_1 - 5x_2 \le 20 \\ 2x_1 - 2x_2 \le 12 \\ x_1 \ge 0, \quad x_2 \le 0 \end{cases}$

15. Maximize: $z = 50x_1 + 40x_2$

Subject to: $\begin{cases} 3x_1 + 2x_2 \le 25 \\ x_1 + 5x_2 \le 8 \\ x_1 \ge 0, \quad x_2 \ge 0 \end{cases}$

16. Maximize: $z = 3x_1 + 2x_2$

Subject to: $\begin{cases} 2x_1 + 3x_2 \le 105 \\ 5x_1 + 10x_2 \ge 15 \\ x_1 \ge 0, \quad x_2 \ge 0 \end{cases}$

17. Maximize: $z = 5x_1 + 3x_2$

Subject to: $\begin{cases} 12x_1 + 4x_2 \le 16 \\ 2x_1 + 5x_2 \ge 10 \\ x_1 \ge 0, \quad x_2 \ge 0 \end{cases}$

18. Minimize: $z = -x_1 - 3x_2$

Subject to: $\begin{cases} 6x_1 + 2x_2 \le 5 \\ 2x_1 + 3x_2 \le 6 \\ x_1 \ge 0, \quad x_2 \ge 0 \end{cases}$

19. Minimize: $z = -20x_1 - 5x_2$

Subject to: $\begin{cases} 3x_1 + 5x_2 \le 10 \\ 3x_1 + 2x_2 \le 6 \\ x_1 \ge 0, \quad x_2 \ge 0 \end{cases}$

20. Minimize: $z = x_1 + 3x_2$

Subject to: $\begin{cases} x_1 + x_2 \le 10 \\ 5x_1 + 2x_2 \ge 20 \\ -x_1 + 2x_2 \ge 0 \\ x_1 \ge 0, \quad x_2 \ge 0 \end{cases}$

21. Minimize: $z = 35x_1 + 10x_2$

Subject to: $\begin{cases} -2x_1 + 3x_2 \ge 0 \\ 8x_1 + x_2 \le 52 \\ -2x_1 + x_2 \le 2 \\ x_1 \ge 3 \\ x_1 \ge 0, \quad x_2 \ge 0 \end{cases}$

22. Minimize: $z = 140x_1 - 60x_2$

Subject to: $\begin{cases} -2x_1 + 3x_2 \ge 0 \\ 8x_1 + x_2 \le 52 \\ -2x_1 + x_2 \le 2 \\ x_1 \ge 3 \\ x_1 \ge 0, \quad x_2 \ge 0 \end{cases}$

23. Maximize: $z = 2x_1 + 3x_2 + 5x_3$

Subject to: $\begin{cases} x_1 + 3x_2 + x_3 \le 46 \\ 2x_1 + x_2 + x_3 \le 40 \\ x_2 + x_3 = 10 \\ x_1 \ge 0, \quad x_2 \ge 0, \quad x_3 \ge 0 \end{cases}$

24. Maximize: $z = x_1 + x_2 + 5x_3$

Subject to: $\begin{cases} x_1 - 3x_3 \le 30 \\ x_2 - 2x_3 \le 20 \\ x_1 + 5x_2 \le 40 \\ x_1 + x_2 - x_3 = 8 \\ x_1 \ge 0, \quad x_2 \ge 0, \quad x_3 \ge 0 \end{cases}$

APPLICATIONS

25. Tony's veterinarian prescribes three food supplements for his horses. Tony must feed them at least 114 kilograms of oats and 50 kilograms of alfalfa, but no more than 200 kilograms of grain. The amounts of each of these in the supplements are summarized in the table.

	Supplement A	Supplement B	Supplement C
Oats	1 kg	2 kg	3 kg
Alfalfa	2 kg	1 kg	1 kg
Grain	1 kg	0 kg	1 kg
Cost per kg:	$1	$1	$4

How many units of A, B, and C should be mixed to minimize the cost?

26. Helmer's apple farm grows apples. Helmer's cannot grow more than 3,000 bins (but can grow less). The crop is divided into two types of apples: eating and applesauce. Helmer's must supply at least 80 bins of eating apples and 800 bins of apples for sauce. The cost of producing these apples is $4 per bin of eating apples and $3.25 per bin of applesauce. How many bins of each should be produced to minimize the cost?

27. Pell, Inc., manufactures two products: tables and shelves. Each product must be processed in each of three departments: machining, assembling, and finishing. The hours needed to produce one unit of product per department and the maximum possible hours per department are shown in the table. Standing orders require that Pell manufacture at least 30 tables and 25 shelves. Pell's net profit is \$4 per table and \$2 per shelf. How many tables should be manufactured to maximize the profit? Solve this problem graphically.

	Production time per unit (hr)		Maximum capacity (hr)
Department	**Tables**	**Shelves**	
Machining	2	1	200
Assembling	2	2	240
Finishing	2	3	300

28. Solve Problem 27 by using the simplex method.

29. The answer to Problem 27 using the graphical method gives some integral coordinates, but the answer to Problem 28 using the simplex method does not. Look at the graphical solution and reconcile the answers to Problems 27 and 28.

30. Metaltec ships units from New Orleans and Atlanta to Chicago and to Los Angeles. The shipping costs, as well as the supply and demand, are listed in the following table.

Source and supply	Destination and demand	Cost
New Orleans, 1,200	Chicago, 1,500	\$120
	Los Angeles, 900	\$180
Atlanta, 1,500	Chicago, 1,500	\$100
	Los Angeles, 900	\$200

How should shipments be made from New Orleans and Atlanta to minimize the shipping cost?

31. The Sebastopol Winery operates two plants for processing grapes, and two warehouses for storing the wine until it is purchased. The costs associated with shipping the wine from the winery (where the processing is done) to the warehouses, as well as the supply and demand, are summarized by the following table.

Winery and supply	Warehouse and demand	Cost
Windsor, 900	Santa Rosa, 600	\$.25
	San Rafael, 1,000	\$.18
Graton, 700	Santa Rosa, 600	\$.25
	San Rafael, 1,000	\$.14

How should the shipping from the wineries to the warehouses be made to minimize the cost?

32. An investment newsletter recommends an investment in zero-coupon bonds paying 8.5% annual interest and in (more risky) junk bonds paying 12% annual interest. Suppose you decide to invest no more than \$48,000 in one or both of these investments. You also decide that you want to invest not more than twice as much in junk bonds as you do in zero-coupon bonds. How much should you invest in each of these to receive the maximum yearly return on your investment?

33. The manager asked you to increase the output of your division, but also gave you a \$1,830 budget. You have a pool of employees from which to draw. The three types of employees are trainees who are paid \$60 for assembling 36 units, regular employees who are paid \$90 for assembling 50 units, and supervisors who are paid \$120 for assembling 60 units. Union regulations require at least one supervisor for every 15 regular employees. Also, there can be no more than 6 trainees working at any one time. How many of each type of employee can you hire in order to maximize the number of units assembled?

34. Tom's Candy Kitchen makes three types of candy: cream-filled, nuts, and solid chews. Daily business requires that at least 25 dozen cream-filled, 40 dozen candy with nuts, and 50 dozen solid chews be made each day. The capacity of the kitchen limits the total amount that can be made each day to no more than 200 dozen. How many dozen of each type should be made in order to maximize profits if the net profit on each cream-filled dozen is \$6, on each dozen of nuts is \$4, and on each dozen solid chews is \$4.80?

*4.7 Duality

The most common type of minimization problem is one in which all of the constraints are of the $\geq$ type. In this section we use a procedure for solving these problems. The process was first introduced by John von Neumann (1903–1957), who has been described as one of the greatest geniuses of this century. It involves solving a related maximization problem called the **dual** of the minimization problem. We begin with a two-dimensional example so that we can work both the original problem and the dual problem graphically to illustrate the plausibility of the method. The dual can be used with any number of variables.

EXAMPLE 1 Minimize: $z = 50x_1 + 60x_2$

Subject to: $\begin{cases} 7x_1 + 2x_2 \geq 14 \\ 3x_1 + 5x_2 \geq 20 \\ x_1 \geq 0, \quad x_2 \geq 0 \end{cases}$

Solution We begin by using the graphical method:

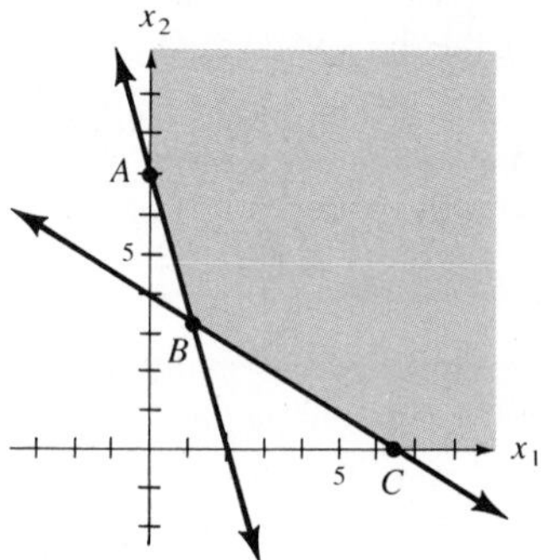

Corner points: A: $(0, 7)$
B: $(\frac{30}{29}, \frac{98}{29})$
C: $(\frac{20}{3}, 0)$

Check the corner points and find:

$(0, 7)$: $z = 50(0) + 60(7) = 420$
$(\frac{30}{29}, \frac{98}{29})$: $z = 50(\frac{30}{29}) + 60(\frac{98}{29}) \approx 254.48$
$(\frac{20}{3}, 0)$: $z = 50(\frac{20}{3}) + 60(0) \approx 333.33$

The minimum value is 254.48, which occurs at $(\frac{30}{29}, \frac{98}{29})$.

Next, write the problem using an augmented matrix (the objective function is written in the last row):

$$\left[\begin{array}{cc|c} 7 & 2 & 14 \\ 3 & 5 & 20 \\ \hline 50 & 60 & 0 \end{array}\right]$$

* This section is not required for subsequent textual development.

If we interchange the rows and columns of this matrix (it is called the **transpose** of the given matrix and we will discuss it below), we obtain

$$\left[\begin{array}{cc|c} 7 & 3 & 50 \\ 2 & 5 & 60 \\ \hline 14 & 20 & 0 \end{array}\right]$$

John von Neumann discovered that if the problem specified by this matrix is maximized using $\leq$ constraints, the same answer as the corresponding minimum problem will always be obtained. Check it out; we will use y's instead of x's so we can keep the two problems separate.

Maximize: $z = 14y_1 + 20y_2$

Subject to: $\begin{cases} 7y_1 + 3y_2 \leq 50 \\ 2y_1 + 5y_2 \leq 60 \\ y_1 \geq 0, \quad y_2 \geq 0 \end{cases}$

Here again, solve the linear programming problem by graphing.

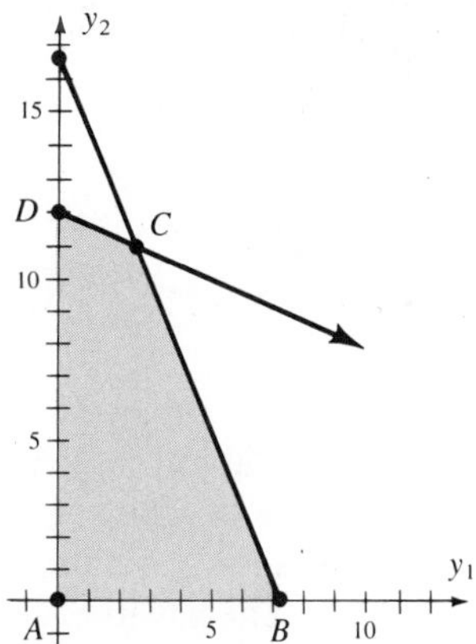

Corner points: A: $(0, 0)$
B: $(\frac{50}{7}, 0)$
C: $(\frac{70}{29}, \frac{320}{29})$
D: $(0, 12)$

Check values:
$(0, 0)$: $z' = 14(0) + 20(0) = 0$
$(\frac{50}{7}, 0)$: $z' = 14(\frac{50}{7}) + 20(0) = 100$
$(\frac{70}{29}, \frac{320}{29})$: $z' = 14(\frac{70}{29}) + 20(\frac{320}{29}) \approx 254.48$
$(0, 12)$: $z' = 14(0) + 20(12) = 240$

The maximum value is 254.48 at the point $(\frac{70}{29}, \frac{320}{29})$.

WARNING ***Take notice of this sentence.***

Note that the minimum value of the original problem is the same as the maximum value of the dual: *This is always true.* ■

The feasible regions of the two problems are different, and the corner points are different, but the *extreme values* of the objective functions are the same. You should note an even closer connection between a problem and its dual as we use the simplex method to solve the dual (the maximum problem).

$$\begin{array}{ccccc|c} y_1 & y_2 & u & v & z' & \\ 7 & 3 & 1 & 0 & 0 & 50 \\ 2 & \textcircled{5} & 0 & 1 & 0 & 60 \\ \hline -14 & -20 & 0 & 0 & 1 & 0 \end{array}$$

↑ Pivot column

For now, let us use slack variables u and v (you will soon see these turn out to be x_1 and x_2).
$50 \div 3 = 16.666\ldots$
$60 \div 5 = 12$

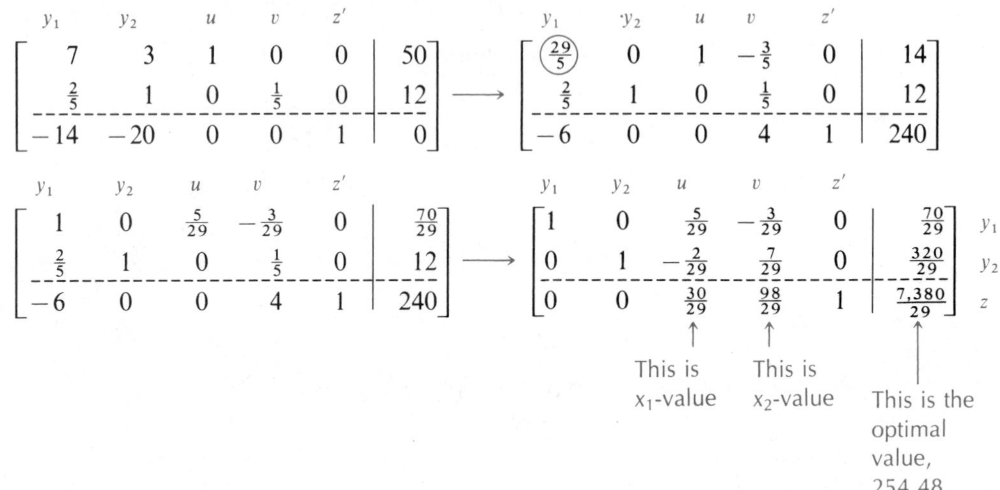

Note that the solution to the *minimum problem* is found in the bottom row of the tableau under the columns u and v (the slack variables); that is, $x_1 = \frac{30}{29}$ and $x_2 = \frac{98}{29}$. This suggests that when solving the dual, you can use the original x-values as slack variables.

The key to changing a minimization problem to its dual maximization problem is the interchanging of the rows and columns in the augmented matrix of the minimization problem. We called this the *transpose*:

Transpose of a Matrix

The **transpose**, M^T, of a matrix M is the matrix formed by interchanging the corresponding rows and columns of M.

EXAMPLE 2 Find the transpose of the given matrices.

$$A = \begin{bmatrix} 1 & 2 \\ 3 & 4 \end{bmatrix} \quad B = \begin{bmatrix} 6 & 8 & 9 \\ 4 & 1 & 7 \\ 3 & 2 & 1 \end{bmatrix} \quad C = [4 \quad 8]$$

$$D = \begin{bmatrix} 6 \\ 1 \\ 2 \end{bmatrix} \quad E = \begin{bmatrix} 1 & 3 & 6 \\ 4 & 9 & 2 \end{bmatrix}$$

Solution $A^{\mathrm{T}} = \begin{bmatrix} 1 & 3 \\ 2 & 4 \end{bmatrix}$ $B^{\mathrm{T}} = \begin{bmatrix} 6 & 4 & 3 \\ 8 & 1 & 2 \\ 9 & 7 & 1 \end{bmatrix}$ $C^{\mathrm{T}} = \begin{bmatrix} 4 \\ 8 \end{bmatrix}$

$D^{\mathrm{T}} = [6 \quad 1 \quad 2]$ $E^{\mathrm{T}} = \begin{bmatrix} 1 & 4 \\ 3 & 9 \\ 6 & 2 \end{bmatrix}$ ■

We can now summarize this process.

Von Neumann's Duality Principle

The optimum value of a minimum linear programming problem, if a solution exists, is the same as the optimum value of its dual. That is, the maximum value of z' is the same as the minimum value of z.

Linear Programming Minimization Problems with $\geq$ Constraints—Dual Procedure

To find the dual:

1. The given problem should be a minimization problem consisting only of $\geq$ constraints.* (You may have to use some of the techniques of the last section if the order of some of the constraints is not correct.)
2. Write the simplex problem in matrix form, with each constraint on a different row and with the objective function as the bottom row.
3. Find the transpose of the problem; this gives the dual problem (the dual of a minimum problem is a maximum problem). The constraints of the dual are formed from the rows of the transpose; the objective function is at the bottom; and the inequality signs are the reverse of those in the original problem.
4. The optimal solution is given by the entries in the bottom row of the columns corresponding to the slack variables, and the minimum value of the objective function of the minimization problem is the same as the maximum value of the objective function of the dual.

EXAMPLE 3 Minimize: $z = 50x_1 + 60x_2$

Subject to: $\begin{cases} 7x_1 + 2x_2 \geq 14 \\ 3x_1 + 5x_2 \geq 20 \\ 6x_1 + 10x_2 \geq 30 \\ x_1 \geq 0, \quad x_2 \geq 0 \end{cases}$

* The dual can also be used to transform a maximization problem into a minimization problem. In this book, however, we will use the dual only to transform minimization problems into maximization ones.

Solution Solve this by solving the dual problem. First, write in augmented matrix form:

$$\left[\begin{array}{cc|c} 7 & 2 & 14 \\ 3 & 5 & 20 \\ 6 & 10 & 30 \\ \hline 50 & 60 & 0 \end{array}\right]$$

Next, find the transpose:

$$\left[\begin{array}{ccc|c} 7 & 3 & 6 & 50 \\ 2 & 5 & 10 & 60 \\ \hline 14 & 20 & 30 & 0 \end{array}\right]$$

The transpose leads to the dual problem:

Maximize: $z' = 14y_1 + 20y_2 + 30y_3$

Subject to: $\begin{cases} 7y_1 + 3y_2 + 6y_3 \le 50 \\ 2y_1 + 5y_2 + 10y_3 \le 60 \end{cases}$

Solve using the simplex method:

$$\begin{array}{c} \begin{array}{cccccc} y_1 & y_2 & y_3 & x_1 & x_2 & z' \end{array} \\ \left[\begin{array}{cccccc|c} 7 & 3 & 6 & 1 & 0 & 0 & 50 \\ 2 & 5 & \textcircled{10} & 0 & 1 & 0 & 60 \\ \hline -14 & -20 & -30 & 0 & 0 & 1 & 0 \end{array}\right] \end{array} \quad \begin{array}{l} 50 \div 6 = 8.333 \\ 60 \div 10 = 6 \end{array}$$

$\uparrow$ Pivot column

$$\begin{array}{c} \begin{array}{cccccc} y_1 & y_2 & y_3 & x_1 & x_2 & z' \end{array} \\ \left[\begin{array}{cccccc|c} 7 & 3 & 6 & 1 & 0 & 0 & 50 \\ \frac{1}{5} & \frac{1}{2} & 1 & 0 & \frac{1}{10} & 0 & 6 \\ \hline -14 & -20 & -30 & 0 & 0 & 1 & 0 \end{array}\right] \end{array}$$

$$\begin{array}{c} \begin{array}{cccccc} y_1 & y_2 & y_3 & x_1 & x_2 & z' \end{array} \\ \left[\begin{array}{cccccc|c} \textcircled{\frac{29}{5}} & 0 & 0 & 1 & -\frac{3}{5} & 0 & 14 \\ \frac{1}{5} & \frac{1}{2} & 1 & 0 & \frac{1}{10} & 0 & 6 \\ \hline -8 & -5 & 0 & 0 & 3 & 1 & 180 \end{array}\right] \end{array} \quad \begin{array}{l} 14 \div \frac{29}{5} = 2.413\ldots \\ 6 \div \frac{1}{5} = 30 \end{array}$$

$\uparrow$ Pivot column

$$\begin{array}{c} \begin{array}{cccccc} y_1 & y_2 & y_3 & x_1 & x_2 & z' \end{array} \\ \left[\begin{array}{cccccc|c} 1 & 0 & 0 & \frac{5}{29} & -\frac{3}{29} & 0 & \frac{70}{29} \\ \frac{1}{5} & \frac{1}{2} & 1 & 0 & \frac{1}{10} & 0 & 6 \\ \hline -8 & -5 & 0 & 0 & 3 & 1 & 180 \end{array}\right] \end{array}$$

$$\begin{array}{c} \begin{array}{cccccc} y_1 & y_2 & y_3 & x_1 & x_2 & z' \end{array} \\ \left[\begin{array}{cccccc|c} 1 & 0 & 0 & \frac{5}{29} & -\frac{3}{29} & 0 & \frac{70}{29} \\ 0 & \textcircled{\frac{1}{2}} & 1 & -\frac{1}{29} & \frac{35}{290} & 0 & \frac{160}{29} \\ \hline 0 & -5 & 0 & \frac{40}{29} & \frac{63}{29} & 1 & \frac{5{,}780}{29} \end{array}\right] \end{array}$$

$$\begin{array}{c}
\begin{array}{cccccccc} y_1 & y_2 & y_3 & x_1 & x_2 & z' & & \end{array} \\
\left[\begin{array}{cccccc|c}
1 & 0 & 0 & \frac{5}{29} & -\frac{3}{29} & 0 & \frac{70}{29} \\
0 & 1 & 2 & -\frac{2}{29} & \frac{70}{290} & 0 & \frac{320}{29} \\
\hline
0 & -5 & 0 & \frac{40}{29} & \frac{63}{29} & 1 & \frac{5{,}780}{29}
\end{array}\right]
\end{array}$$

$$\begin{array}{c}
\begin{array}{cccccccc} y_1 & y_2 & y_3 & x_1 & x_2 & z' & & \end{array} \\
\left[\begin{array}{cccccc|c}
1 & 0 & 0 & \frac{5}{29} & -\frac{3}{29} & 0 & \frac{70}{29} \\
0 & 1 & 2 & -\frac{2}{29} & \frac{70}{290} & 0 & \frac{320}{29} \\
\hline
0 & 0 & 10 & \frac{30}{29} & \frac{98}{29} & 1 & \frac{7{,}380}{29}
\end{array}\right]
\end{array}$$

The maximum is $z' = \frac{7{,}380}{29} \approx 254.5$, so von Neumann's duality principle tells us that the minimum value of z is also 254.5. In addition to giving the minimum value of z, the duality process also gives the values of x_1 and x_2. These values appear at the bottom of the columns labeled x_1 and x_2, respectively. Thus the objective function is minimized when $x_1 = \frac{30}{29}$ and $x_2 = \frac{98}{29}$. ■

COMPUTER COMMENT

Most linear programming problems require an extensive amount of tedious arithmetic calculations and for that reason most mathematicians and financial consultants use computer programs of the simplex method to carry out the actual "work" of this method. One of the things you will notice when using a computer program is that instead of exact results, the work is rounded to a specified number of decimal places. Example 3 may look like this when a computer program is used:

$$\begin{array}{c}
\begin{array}{cccccccc} y_1 & y_2 & y_3 & x_1 & x_2 & z' & & \end{array} \\
\left[\begin{array}{cccccc|c}
\boxed{5.80} & 0 & 0 & 1 & -.60 & 0 & 14 \\
.20 & .50 & 1 & 0 & .10 & 0 & 6 \\
\hdashline
-8 & -5 & 0 & 0 & 3 & 1 & 180
\end{array}\right]
\end{array}$$

This is the step in Example 3 where $\frac{29}{5}$ is circled.

$$\begin{array}{c}
\begin{array}{cccccccc} y_1 & y_2 & y_3 & x_1 & x_2 & z' & & \end{array} \\
\left[\begin{array}{cccccc|c}
1 & 0 & 0 & .17 & -.10 & 0 & 2.41 \\
0 & \boxed{.50} & 1 & -.03 & .12 & 0 & 5.52 \\
\hdashline
0 & -5 & 0 & 1.38 & 2.17 & 1 & 199.31
\end{array}\right]
\end{array}$$

This is the step in Example 3 where $\frac{1}{2}$ is circled.

$$\begin{array}{c}
\begin{array}{cccccccc} y_1 & y_2 & y_3 & x_1 & x_2 & z' & & \end{array} \\
\left[\begin{array}{cccccc|c}
1 & 0 & 0 & .17 & -.10 & 0 & 2.41 \\
0 & 1 & 2 & -.07 & .24 & 0 & 11.03 \\
\hdashline
0 & 0 & 10 & 1.03 & 3.38 & 1 & 254.48
\end{array}\right]
\end{array}$$

This is the final step.

A final note on integer solutions: In this chapter we have been requiring nonnegative variables and have not worked problems requiring integer solutions because linear programming problems that impose integer constraints involve techniques that are beyond the scope of this text. Unfortunately, you cannot simply solve such a problem by the techniques discussed here and

(continued)

(continued)

then round your results to the nearest integer. The rounded solution *may not* be the best integer solution. However, you can use the following result: Suppose you are maximizing z and

z_0 = Maximum value for the continuous problem (the answer found using the simplex method in this chapter)

z_1 = Maximum found by rounding z_0 to the nearest integer

z_2 = Maximum integer solution (found by techniques not discussed in this text)

The best we can say is that

$$z_1 \le z_2 \le z_0$$

Linear programming provides extremely useful mathematical models, but there are many applied applications for which the model developed here is not sufficient. For more advanced study of linear programming, you will need to consult a linear algebra or a linear programming textbook.

4.7 Problem Set

Find the transpose of the matrices in Problems 1–4.

1. a. $A = \begin{bmatrix} 6 & 9 \\ 4 & 8 \end{bmatrix}$ **b.** $B = \begin{bmatrix} 5 & 6 \\ 3 & 8 \end{bmatrix}$

c. $C = \begin{bmatrix} 4 & 9 & 1 \\ 6 & 1 & 4 \end{bmatrix}$

2. a. $D = \begin{bmatrix} 8 & 1 \\ 6 & 2 \\ 4 & 5 \end{bmatrix}$ **b.** $E = \begin{bmatrix} 6 & 8 & 1 \\ 2 & 3 & 4 \\ 5 & 9 & 7 \end{bmatrix}$

c. $F = \begin{bmatrix} 1 & 0 & 3 \\ 4 & 9 & 7 \\ 8 & 6 & 5 \end{bmatrix}$

3. a. $G = [1 \quad 3 \quad 5]$ **b.** $H = \begin{bmatrix} 4 \\ 9 \\ 6 \end{bmatrix}$

c. $J = \begin{bmatrix} 1 \\ 0 \\ 3 \\ 2 \end{bmatrix}$

4. a. $K = [1 \quad 8 \quad 7 \quad 4]$

b. $L = \begin{bmatrix} 4 & 8 & 0 & 3 & 2 \\ 6 & 1 & 4 & 7 & 9 \end{bmatrix}$ **c.** $M = \begin{bmatrix} 1 & 4 & 3 \\ 6 & 9 & 2 \\ 4 & 7 & 1 \\ 5 & 7 & 11 \end{bmatrix}$

Write the dual of Problems 5–8.

5. Minimize: $z = 3x_1 + 4x_2$

Subject to: $\begin{cases} 2x_1 + 8x_2 \ge 10 \\ 3x_1 + 5x_2 \ge 30 \\ x_1 \ge 0, \quad x_2 \ge 0 \end{cases}$

6. Minimize: $z = 50x_1 + 30x_2$

Subject to: $\begin{cases} 5x_1 + 5x_2 \ge 25 \\ 6x_1 - 2x_2 \ge 10 \\ x_1 \ge 0, \quad x_2 \ge 0 \end{cases}$

7. Minimize: $z = 3x_1 + 2x_2 + 5x_3$

Subject to: $\begin{cases} x_1 + x_2 + x_3 \ge 10 \\ 2x_1 + 3x_3 \ge 2 \\ x_2 + 2x_3 \ge 1 \\ 3x_1 + 5x_3 \ge 15 \\ x_1 \ge 0, \quad x_2 \ge 0, \quad x_3 \ge 0 \end{cases}$

8. Minimize: $z = x_1 + 2x_2 + 3x_3$

Subject to: $\begin{cases} x_1 + x_2 + x_3 \geq 50 \\ 6x_1 + 5x_3 \geq 10 \\ 7x_2 + 5x_3 \geq 35 \\ 9x_1 + 4x_2 \geq 18 \\ x_1 \geq 0, \quad x_2 \geq 0, \quad x_3 \geq 0 \end{cases}$

Solve Problems 9–38 by using the dual.

9. Minimize: $z = 3x_1 + 4x_2$

Subject to: $\begin{cases} 2x_1 + 8x_2 \geq 10 \\ 3x_1 + 5x_2 \geq 30 \\ x_1 \geq 0, \quad x_2 \geq 0 \end{cases}$

10. Minimize: $z = 50x_1 + 30x_2$

Subject to: $\begin{cases} 5x_1 + 5x_2 \geq 25 \\ 6x_1 - 2x_2 \geq 10 \\ x_1 \geq 0, \quad x_2 \geq 0 \end{cases}$

11. Minimize: $z = 18x_1 + 24x_2$

Subject to: $\begin{cases} 3x_1 + x_2 \geq 24 \\ x_1 + 2x_2 \geq 21 \\ x_1 + x_2 \geq 18 \\ x_1 \geq 0, \quad x_2 \geq 0 \end{cases}$

12. Minimize: $z = 120x_1 + 150x_2$

Subject to: $\begin{cases} 2x_1 + x_2 \geq 18 \\ x_1 + 2x_2 \geq 21 \\ x_1 + x_2 \geq 15 \\ x_1 \geq 0, \quad x_2 \geq 0 \end{cases}$

13. Minimize: $z = 27x_1 + 24x_2$

Subject to: $\begin{cases} 3x_1 + x_2 \geq 24 \\ x_1 + x_2 \geq 18 \\ x_1 + 3x_2 \geq 21 \\ x_1 \geq 0, \quad x_2 \geq 0 \end{cases}$

14. Minimize: $z = 1{,}000x_1 + 1{,}200x_2$

Subject to: $\begin{cases} x_1 + x_2 \geq 18 \\ 2x_1 + x_2 \geq 20 \\ x_1 + 2x_2 \geq 24 \\ x_1 \geq 0, \quad x_2 \geq 0 \end{cases}$

15. Minimize: $z = 32x_1 + 36x_2$

Subject to: $\begin{cases} 4x_1 + x_2 \geq 24 \\ x_1 + x_2 \geq 16 \\ x_1 + 2x_2 \geq 20 \\ x_1 \geq 0, \quad x_2 \geq 0 \end{cases}$

16. Minimize: $z = 6x_1 + 8x_2$

Subject to: $\begin{cases} 2x_1 + x_2 \geq 18 \\ x_1 + 2x_2 \geq 20 \\ x_1 + x_2 \geq 15 \\ x_1 \geq 0, \quad x_2 \geq 0 \end{cases}$

17. Minimize: $z = 24x_1 + 32x_2$

Subject to: $\begin{cases} 2x_1 + x_2 \geq 20 \\ x_1 + 4x_2 \geq 24 \\ x_1 \geq 0, \quad x_2 \geq 0 \end{cases}$

18. Minimize: $z = 10x_1 + 12x_2$

Subject to: $\begin{cases} x_1 + x_2 \geq 20 \\ 2x_1 + x_2 \geq 24 \\ x_1 + 2x_2 \geq 28 \\ x_1 \geq 0, \quad x_2 \geq 0 \end{cases}$

19. Minimize: $z = 10x_1 + 12x_2$

Subject to: $\begin{cases} x_1 + x_2 \geq 20 \\ 2x_1 + x_2 \geq 30 \\ x_1 + 2x_2 \geq 25 \\ x_1 \geq 0, \quad x_2 \geq 0 \end{cases}$

20. Minimize: $z = 10x_1 + 12x_2$

Subject to: $\begin{cases} x_1 + x_2 \geq 24 \\ 2x_1 + x_2 \geq 28 \\ x_1 + 2x_2 \geq 32 \\ x_1 \geq 0, \quad x_2 \geq 0 \end{cases}$

21. Minimize: $z = 12x_1 + 16x_2 + 18x_3$

Subject to: $\begin{cases} x_1 + 2x_2 + x_3 \geq 20 \\ x_1 + x_2 + 2x_3 \geq 16 \\ 2x_1 + x_2 + x_3 \geq 24 \\ x_1 \geq 0, \quad x_2 \geq 0, \quad x_3 \geq 0 \end{cases}$

22. Minimize: $z = 36x_1 + 48x_2 + 48x_3$

Subject to: $\begin{cases} 2x_1 + x_2 + x_3 \geq 48 \\ x_1 + x_2 + 2x_3 \geq 40 \\ x_1 + 2x_2 + x_3 \geq 44 \\ x_1 \geq 0, \quad x_2 \geq 0, \quad x_3 \geq 0 \end{cases}$

23. Minimize: $z = 36x_1 + 36x_2 + 24x_3$

Subject to: $\begin{cases} 2x_1 + x_2 + x_3 \geq 36 \\ x_1 + x_2 + x_3 \geq 24 \\ x_1 + 2x_2 + x_3 \geq 30 \\ x_1 \geq 0, \quad x_2 \geq 0, \quad x_3 \geq 0 \end{cases}$

24. Minimize: $z = 24x_1 + 30x_2 + 36x_3$

Subject to: $\begin{cases} x_1 + 2x_2 + x_3 \geq 36 \\ x_1 + x_2 + 2x_3 \geq 30 \\ 2x_1 + x_2 + x_3 \geq 42 \\ x_1 \geq 0, \quad x_2 \geq 0, \quad x_3 \geq 0 \end{cases}$

25. Minimize: $z = 40x_1 + 35x_2 + 30x_3$

Subject to: $\begin{cases} 2x_1 + x_2 + x_3 \geq 30 \\ x_1 + x_2 + x_3 \geq 25 \\ x_1 + 2x_2 + x_3 \geq 35 \\ x_1 \geq 0, \quad x_2 \geq 0, \quad x_3 \geq 0 \end{cases}$

26. Minimize: $z = 40x_1 + 36x_2 + 24x_3$

Subject to: $\begin{cases} 2x_1 + x_2 + x_3 \geq 32 \\ x_1 + x_2 + x_3 \geq 28 \\ x_1 + 3x_2 + x_3 \geq 36 \\ x_1 \geq 0, \quad x_2 \geq 0, \quad x_3 \geq 0 \end{cases}$

27. Minimize: $z = 3x_1 + 2x_2 + 5x_3$

Subject to: $\begin{cases} x_1 + x_2 + x_3 \geq 10 \\ 2x_1 + 3x_3 \geq 2 \\ x_2 + 2x_3 \geq 1 \\ 3x_1 + 5x_3 \geq 15 \\ x_1 \geq 0, \quad x_2 \geq 0, \quad x_3 \geq 0 \end{cases}$

28. Minimize: $z = x_1 + 2x_2 + 3x_3$

Subject to: $\begin{cases} x_1 + x_2 + x_3 \geq 50 \\ 6x_1 + 5x_3 \geq 10 \\ 7x_2 + 5x_3 \geq 35 \\ 9x_1 + 4x_2 \geq 18 \\ x_1 \geq 0, \quad x_2 \geq 0, \quad x_3 \geq 0 \end{cases}$

29. Minimize: $z = 50x_1 + 10x_2 + 20x_3$

Subject to: $\begin{cases} x_1 + x_2 + x_3 \geq 25 \\ 2x_1 + x_2 + 3x_3 \geq 99 \\ x_1 \geq 0, \quad x_2 \geq 0, \quad x_3 \geq 0 \end{cases}$

30. Minimize: $z = 10x_1 + 20x_2 + 30x_3$

Subject to: $\begin{cases} x_1 + 3x_2 + x_3 \geq 75 \\ 2x_1 + x_2 + 5x_3 \geq 80 \\ x_1 \geq 0, \quad x_2 \geq 0, \quad x_3 \geq 0 \end{cases}$

31. Minimize: $z = 24x_1 + 12x_2$

Subject to: $\begin{cases} x_1 \leq 10 \\ x_2 \leq 8 \\ 3x_1 + 2x_2 \geq 12 \\ x_1 \geq 0, \quad x_2 \geq 0 \end{cases}$

32. Minimize: $z = 6x_1 + 18x_2$

Subject to: $\begin{cases} x_1 \leq 10 \\ x_2 \leq 8 \\ 3x_1 + 2x_2 \geq 12 \\ x_1 \geq 0, \quad x_2 \geq 0 \end{cases}$

33. Minimize: $z = 90x_1 + 20x_2$

Subject to: $\begin{cases} x_1 + x_2 \geq 6 \\ -2x_1 + x_2 \geq -16 \\ x_2 \leq 9 \\ x_1 \geq 0, \quad x_2 \geq 0 \end{cases}$

34. Minimize: $z = 400x_1 + 100x_2$

Subject to: $\begin{cases} x_1 + x_2 \geq 6 \\ -2x_1 + x_2 \geq -16 \\ x_2 \leq 9 \\ x_1 \geq 0, \quad x_2 \geq 0 \end{cases}$

35. Minimize: $z = 5x_1 + 3x_2$

Subject to: $\begin{cases} 2x_1 + x_2 \geq 9 \\ x_2 \leq 5 \\ x_1 - x_2 \leq 3 \\ 3x_1 - 2x_2 \geq 5 \\ x_1 \geq 0, \quad x_2 \geq 0 \end{cases}$

36. Minimize: $z = 2x_1 - 3x_2$

Subject to: $\begin{cases} 2x_1 + x_2 \geq 8 \\ x_2 \leq 5 \\ x_1 - x_2 \leq 2 \\ 3x_1 - 2x_2 \geq 5 \\ x_1 \geq 0, \quad x_2 \geq 0 \end{cases}$

37. Minimize: $z = 140x_1 + 250x_2$

Subject to: $\begin{cases} 2x_1 + x_2 \geq 8 \\ x_1 - x_2 \leq 7 \\ x_1 - x_2 \geq -3 \\ x_1 \leq 9 \\ x_1 \geq 0, \quad x_2 \geq 0 \end{cases}$

38. Minimize: $z = 640x_1 - 130x_2$

Subject to: $\begin{cases} 2x_1 + x_2 \geq 9 \\ x_1 - x_2 \leq 7 \\ x_1 - x_2 \geq -3 \\ x_1 \leq 9 \\ x_1 \geq 0, \quad x_2 \geq 0 \end{cases}$

APPLICATIONS

39. Karlin Enterprises manufactures two electronic games. Standing orders require that at least 24,000 space-battle games and 5,000 football games be produced. The company has two factories: the Gainesville plant can produce 600 space-battle games and 100 football games per day; the Sacramento plant can produce 300 space-battle games and 100 football games per day. If the Gainesville plant costs $20,000 per day to operate and the Sacramento factory costs $15,000 per day, find the number of days each factory should operate to minimize the cost.

40. The Cosmopolitan Oil Company requires at least 4,000 barrels of low-grade oil. It also needs at least 12,000 barrels of medium-grade oil and at least 8,000 barrels of high-grade oil. Oil is produced at one of two refineries. The daily production of these refineries is summarized in the table. If it costs $17,000 per day to operate refinery I and $15,000 per day to operate refinery II, how many days per month should each refinery be operated to satisfy the requirements and at the same time minimize the costs?

	Oil production (barrels)		
Refinery	**Low grade**	**Medium grade**	**High grade**
I	100	400	400
II	100	200	100

41. A woman wants to design a weekly exercise schedule that involves jogging, handball, and aerobic dance. She decides to jog at least 3 hours per week, play handball at least 2 hours per week, and dance at least 5 hours per week. She also wants to devote at least as much time to jogging as to handball because she does not like handball as much as she likes the other exercises. She also knows that jogging consumes 900 calories per hour, handball 600 calories per hour, and dance 800 calories per hour. Her physician told her that she must burn up a total of at least 9,700 calories per week in this exercise program. How many hours should she devote to each exercise if she wishes to minimize her exercise time?

42. Brown Brothers is an investment company analyzing the pension fund of a certain company. A maximum of $10 million is available to invest in two places. No more than $8 million can be invested in stocks yielding 12%, and at least $2 million can be invested in long-term bonds yielding 8%. The stock-to-bond investment ratio cannot be more than 1 to 3. How should Brown Brothers advise its client so that the pension fund will receive the maximum yearly return on its investment?

43. To utilize costly equipment efficiently, an assembly line with a capacity of 50 units per shift must operate 24 hours per day. This requires scheduling three shifts with varying labor costs as follows:

Day shift:	Labor costs are $100 per unit
Swing shift:	Labor costs are $150 per unit
Graveyard shift:	Labor costs are $180 per unit

Each unit produced uses 60 linear feet of sheeting material, and there is a total of 1,800 feet available for the day shift, 1,500 feet for the swing shift, and 3,000 feet for the graveyard shift. The day shift must produce at least 20 units and the other shifts must produce at least 50 units together. How many units should be produced on each shift to maximize the revenue if the units are sold for $850 each?

44. Repeat Problem 43 except minimize labor costs rather than maximize revenue.

45. Suppose a computer dealer has stores in Hillsborough and Palo Alto and warehouses in San Jose and Burlingame. The cost of shipping a computer from one location to another as well as the supply and demand at each location are shown on the following "map":

Supply: 40 units Burlingame; 45 units San Jose

Transportation Costs: $3, $9, $12, $6

Demand: Hillsborough 30 units; Palo Alto 35 units

How should the dealer ship the computers to minimize the shipping costs?

46. Camstop, an importer of computer chips, can ship case lots of components to stores in Dallas and Chicago from ports in either Los Angeles or Seattle. The costs and demand are given in the table. How many cases should be shipped from each port to minimize shipping costs?

Store	Cost to ship (each case)		
	Los Angeles port	**Seattle port**	**Store demand**
Chicago	$9	$7	80 cases
Dallas	$7	$8	110 cases
Available inventory at each port	90 cases	130 cases	

47. A Texas appliance dealer has stores in Fort Worth and in Houston and warehouses in Dallas and San Antonio. The cost of shipping a refrigerator from Dallas to Fort Worth is $14, and from Dallas to Houston shipping is $15; from San Antonio to Fort Worth the cost is $10, and from San Antonio to Houston it is $18. Suppose that the Fort Worth store has orders for 15 refrigerators and the Houston store has orders for 35. Also suppose that there are 25 refrigerators in stock at Dallas and 45 at San Antonio. What is the most economical way to supply the requested refrigerators to the two stores?

*4.8 Review

The material of this chapter is reviewed in the following list of objectives. After each objective there are some practice questions. For a sample test select the first question of each set and check your answers. The second question for each objective has no answer given. If you are having trouble with a particular type of problem, look back at the indicated section in the text. When you are finished reviewing these objectives, a sample examination is given at the end of this section.

[4.1]
Objective 4.1: *Graph linear inequalities.*

1. $50x + 30y < 150$
2. $y > \frac{2}{3}x - 7$
3. $2x - 3y < 48$
4. $.05x + .08y \geq 160$

Objective 4.2: *Graph a system of inequalities.*

5. $\begin{cases} 7y \leq 2x + 420 \\ 2x + 2y \leq 500 \\ x \geq 50 \\ y \leq 80 \end{cases}$

6. $\begin{cases} x \leq 0 \\ y \geq 0 \\ 3x + 2y \geq -3 \\ x - y \leq 0 \end{cases}$

7. $\begin{cases} x - y - 8 \leq 0 \\ x - y + 4 \geq 0 \\ x + y \geq -5 \\ y + x \leq 4 \end{cases}$

8. $\begin{cases} 2x + 3y \leq 600 \\ x + 2y \leq 360 \\ 3x - 2y \leq 240 \\ x \geq 0, \quad y \geq 0 \end{cases}$

[4.2]
Objective 4.3: *Formulate linear programming problems.*

9. *Diet problem.* A convalescent hospital wishes to provide for its patients a diet that has a minimum of 100 g of carbohydrates, 60 g of proteins, and 40 g of fats per day. These requirements can be met with two foods, as shown in the table. It is also important to minimize costs, and food A costs 14¢ per ounce and food B costs 6¢ per ounce. How many ounces of each food should be bought for each patient per day in order to meet the minimum requirements at the lower cost?

Food	Carbohydrates	Proteins	Fats
A	6 g	3 g	1 g
B	2 g	2 g	2 g

* Optional section.

10. *Maximize income.* Regal Products, Inc. has on hand 120, 150, and 200 units of raw materials, A, B, and C, respectively. The company produces two products requiring the number of units of each ingredient listed in the table. Product I sells for $25 and product II for $55. If the company wishes to maximize the gross income, how many of each product should be produced?

Product	Raw materials A	B	C
I	2	3	10
II	3	5	2

11. *Allocation of human resources.* Bradbury Bros. Realty plans to open several new branch offices and has $1,275,000 in capital for this expansion. It will open either a four-person office or a six-person branch office. It will take an initial cash outlay of $175,000 for a four-person branch and $200,000 for a six-person branch. Bradbury Bros. has also decided not to hire more than 32 new employees and will not open more than 10 branch offices. How many of each type of branch office should be opened to maximize cash inflow if it is expected to be $50,000 for four-person branches and $65,000 for six-person branch offices?

12. *Transportation problem.* St. Jean Winery has warehouses in Kenwood and in Graton. It also has two tasting outlets, one in Calistoga and the other in Napa. The costs, as well as the supply and demand, are given in the following "map":

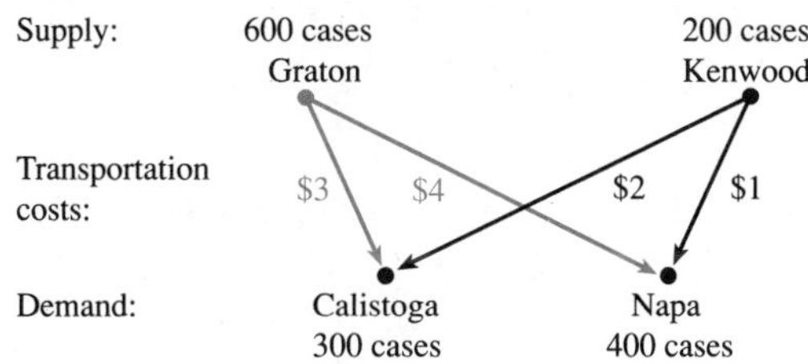

How should the winery schedule the shipping to minimize shipping costs?

[4.3]

Objective 4.4: *Graph the set of feasible solutions.*

13. Maximize: $P = 5x + 7y$

Subject to: $\begin{cases} 4x + y \geq 7 \\ x + 2y \leq 12 \\ x \leq 7 \\ y \geq 0 \end{cases}$

14. Minimize: $C = 8x + 5y$

Subject to: $\begin{cases} 2x + y \geq 8 \\ x - y \leq 2 \\ x \geq 0, \quad y \geq 0 \end{cases}$

15. Maximize: $M = 5x + 4y$

Subject to: $\begin{cases} 2x + y \leq 420 \\ 2x + 2y \leq 500 \\ 2x + 3y \leq 600 \\ x \geq 0, \quad y \geq 0 \end{cases}$

16. Maximize: $S = x + y$

Subject to: $\begin{cases} 12x + 150y \leq 1{,}200 \\ 6x + 200y \leq 1{,}200 \\ 26x + 50y \leq 800 \\ x \geq 0, \quad y \geq 0 \end{cases}$

Objective 4.5: *Find the corner points for a feasible set of solutions.*

17. Maximize: $P = 5x + 7y$

Subject to: $\begin{cases} 4x + y \geq 7 \\ x + 2y \leq 12 \\ x \leq 7 \\ y \geq 0 \end{cases}$

18. Minimize: $C = 8x + 5y$

Subject to: $\begin{cases} 2x + y \geq 8 \\ x - y \leq 2 \\ x \geq 0, \quad y \geq 0 \end{cases}$

19. Maximize: $M = 5x + 4y$

Subject to: $\begin{cases} 2x + y \leq 420 \\ 2x + 2y \leq 500 \\ 2x + 3y \leq 600 \\ x \geq 0, \quad y \geq 0 \end{cases}$

20. Maximize: $S = x + y$

Subject to: $\begin{cases} 12x + 150y \leq 1{,}200 \\ 6x + 200y \leq 1{,}200 \\ 16x + 50y \leq 800 \\ x \geq 0, \quad y \geq 0 \end{cases}$

Objective 4.6: *Find the optimum value for a given objective function and set of constraints.* (*See Problems* 13–20.)

21. Maximize: $P = 5x + 7y$

Subject to: $\begin{cases} 4x + y \geq 7 \\ x + 2y \leq 12 \\ x \leq 7 \\ y \geq 0 \end{cases}$

22. Minimize: $C = 8x + 5y$

Subject to: $\begin{cases} 2x + y \geq 8 \\ x - y \leq 2 \\ x \geq 0, \quad y \geq 0 \end{cases}$

23. Maximize: $M = 5x + 4y$

Subject to: $\begin{cases} 2x + y \leq 420 \\ 2x + 2y \leq 500 \\ 2x + 3y \leq 600 \\ x \geq 0, \quad y \geq 0 \end{cases}$

24. Maximize: $S = x + y$

Subject to: $\begin{cases} 12x + 150y \leq 1{,}200 \\ 6x + 200y \leq 1{,}200 \\ 16x + 50y \leq 800 \\ x \geq 0, \quad y \geq 0 \end{cases}$

Objective 4.7: *Solve applied linear programming problems by graphing.*

25. *Diet problem.* Solve Problem 9.

26. *Maximize income.* Solve Problem 10.

27. *Allocation of human resources.* Solve Problem 11.

28. *Maximize income.* How many of each product of Problem 10 should be produced if product II sells for $30 and everything else remains the same?

29. *Maximize profit.* Foley's Motel has 200 rooms and a restaurant that seats 50 people. Experience shows that 40% of the commercial guests and 20% of the other guests eat in the restaurant. Suppose there is a net profit of $4.50 per day from each commercial guest and $3.50 per day from each other guest. Find the number of commercial and other guests needed in order to maximize the net profits. Assume one guest per room and that the capacity of the restaurant will not be exceeded.

30. *Minimize cost.* Karlin Enterprises manufactures two electronic games. Standing orders require that at least 20,000 space-battle games and 5,000 football games be produced. The company has two factories: the Gainesville plant can produce 600 space-battle games and 100 football games per day; the Sacramento plant can produce 300 space-battle games and 100 football games per day. If the Gainesville plant costs $20,000 per day to operate and the Sacramento factory costs $15,000 per day, find the number of days per month each factory should operate to minimize the cost.

[4.4]

Objective 4.8: *Write linear programming problems in standard form using the proper notation. If the problem is not written in standard form, determine which condition is violated.*

31. Maximize: $P = 60x + 20y$

Subject to: $\begin{cases} 3x + 2y \leq 52 \\ 6x - 4y \leq 100 \\ x \geq 0, \quad y \geq 0 \end{cases}$

32. Minimize: $C = 2x + 3y$

Subject to: $\begin{cases} 19x - 3y \leq 60 \\ 6x + 2y \leq 100 \\ x \geq 0, \quad y \geq 0 \end{cases}$

33. Maximize: $T = .05x + .08y$

Subject to: $\begin{cases} 3x - 2y \geq 10 \\ 2x + 5y \leq 20 \\ 2x + 9y \leq 40 \\ x \geq 0, \quad y \geq 0 \end{cases}$

34. Maximize: $I = 190x + 250y$

Subject to: $\begin{cases} 6x + 2y \leq 10 \\ 3x - 4y \leq 20 \\ 6x + 5y \leq 30 \\ x \geq 0, \quad y \geq 0 \end{cases}$

Objective 4.9: *Write the initial simplex tableau for a standard linear programming problem.*

35. Maximize: $z = 6x_1 + 25x_2 + 3x_3$

Subject to: $\begin{cases} 2x_1 + x_3 \leq 50 \\ 4x_2 + x_3 \leq 90 \\ 3x_1 + 4x_2 \leq 100 \\ x_1 \geq 0, \quad x_2 \geq 0, \quad x_3 \geq 0 \end{cases}$

36. Maximize: $z = 2x_1 + 3x_2 + 5x_3$

Subject to: $\begin{cases} x_1 + x_2 + 5x_3 \leq 10 \\ 2x_1 + 2x_2 + 3x_3 \leq 20 \\ x_1 + 5x_2 + x_3 \leq 30 \\ x_1 \geq 0, \quad x_2 \geq 0, \quad x_3 \geq 0 \end{cases}$

37. Maximize: $z = 15x_1 + 5x_2 + x_3$

Subject to: $\begin{cases} 3x_1 + 2x_2 \leq 120 \\ 3x_2 + x_3 \leq 90 \\ x_3 \leq 100 \\ x_1 \geq 0, \quad x_2 \geq 0, \quad x_3 \geq 0 \end{cases}$

38. Maximize: $z = x_1 + 3x_2$

Subject to: $\begin{cases} x_1 + 5x_2 \leq 100 \\ 2x_1 + 3x_2 \leq 180 \\ 3x_1 + x_2 \leq 50 \\ x_1 \leq 40 \\ x_2 \leq 90 \\ x_1 \geq 0, \quad x_2 \geq 0 \end{cases}$

Objective 4.10: *Perform a pivot operation for a designated pivot.*

39.
$$\begin{array}{cccccc|c} x_1 & x_2 & y_1 & y_2 & y_3 & z & \\ \hline 4 & 5 & 1 & 0 & 0 & 0 & 36 \\ \textcircled{3} & 6 & 0 & 1 & 0 & 0 & 12 \\ 1 & 3 & 0 & 0 & 1 & 0 & 20 \\ \hdashline 10 & -5 & 0 & 0 & 0 & 1 & 0 \end{array}$$

40.
$$\begin{array}{cccccc|c} x_1 & x_2 & y_1 & y_2 & y_3 & z & \\ \hline 3 & -5 & 1 & 0 & 0 & 0 & 25 \\ 2 & 4 & 0 & 1 & 0 & 0 & 60 \\ 10 & \textcircled{5} & 0 & 0 & 1 & 0 & 70 \\ \hdashline -5 & -10 & 0 & 0 & 0 & 1 & 0 \end{array}$$

41.
$$\begin{array}{ccccccc|c} x_1 & x_2 & x_3 & y_1 & y_2 & y_3 & z & \\ \hline 2 & 3 & 4 & 1 & 0 & 0 & 0 & 60 \\ 3 & 6 & \textcircled{3} & 0 & 1 & 0 & 0 & 40 \\ 5 & 1 & 1 & 0 & 0 & 1 & 0 & 30 \\ \hdashline -5 & -6 & -8 & 0 & 0 & 0 & 1 & 0 \end{array}$$

42.
$$\begin{array}{ccccccc|c} x_1 & x_2 & x_3 & y_1 & y_2 & y_3 & z & \\ \hline 1 & -2 & 2 & 1 & 0 & 0 & 0 & 20 \\ 2 & \textcircled{3} & 4 & 0 & 1 & 0 & 0 & 9 \\ 3 & -3 & -1 & 0 & 0 & 1 & 0 & 15 \\ \hdashline -10 & -12 & -5 & 0 & 0 & 0 & 1 & 0 \end{array}$$

[4.5]
Objective 4.11: *Find the initial basic solution for a given matrix tableau.*

43.
$$\begin{array}{ccccccc|c} x_1 & x_2 & x_3 & y_1 & y_2 & y_3 & z & \\ \hline 3 & -2 & 1 & 1 & 0 & 0 & 0 & 30 \\ 4 & 8 & 0 & 0 & 1 & 0 & 0 & 20 \\ 2 & 4 & 9 & 0 & 0 & 1 & 0 & 10 \\ \hdashline -10 & -20 & -30 & 0 & 0 & 0 & 1 & 0 \end{array}$$

44.
$$\begin{array}{cccccc|c} x_1 & x_2 & y_1 & y_2 & y_3 & z & \\ \hline 0 & 0 & 3 & 1 & -3 & 0 & 14 \\ 0 & 1 & 2 & 0 & 9 & 0 & 36 \\ 1 & 0 & 3 & 0 & 6 & 0 & 10 \\ \hdashline 0 & 0 & 9 & 0 & 41 & 1 & 180 \end{array}$$

45.
$$\begin{array}{ccccccc|c} x_1 & x_2 & x_3 & y_1 & y_2 & y_3 & z & \\ \hline 6 & 0 & 0 & 1 & 3 & 6 & 0 & 60 \\ 4 & 0 & 1 & 0 & 9 & -3 & 0 & 19 \\ 8 & 1 & 0 & 0 & 4 & -4 & 0 & 38 \\ \hdashline -10 & 0 & 0 & 0 & -5 & 6 & 1 & 120 \end{array}$$

46.
$$\begin{array}{ccccccc|c} x_1 & x_2 & y_1 & y_2 & y_3 & y_4 & z & \\ \hline 0 & -5 & 0 & 1 & 6 & 0 & 0 & 10 \\ 0 & 9 & 1 & 0 & -5 & 0 & 0 & 12 \\ 1 & 2 & 0 & 0 & 9 & 0 & 0 & 140 \\ 0 & 1 & 0 & 0 & 1 & 1 & 0 & 280 \\ \hdashline 0 & 10 & 0 & 0 & 11 & 0 & 1 & 360 \end{array}$$

Objective 4.12: *Carry out the pivoting process on the given initial tableau.*

47.
$$\begin{array}{ccccc|c} x_1 & x_2 & y_1 & y_2 & z & \\ \hline 4 & 6 & 1 & 0 & 0 & 24 \\ 2 & 4 & 0 & 1 & 0 & 10 \\ \hdashline -3 & -1 & 0 & 0 & 1 & 0 \end{array}$$

48.
$$\begin{array}{cccccc|c} x_1 & x_2 & y_1 & y_2 & y_3 & z & \\ \hline 3 & 8 & 1 & 0 & 0 & 0 & 40 \\ 4 & 9 & 0 & 1 & 0 & 0 & 45 \\ 8 & 4 & 0 & 0 & 1 & 0 & 12 \\ \hdashline -5 & -10 & 0 & 0 & 0 & 1 & 0 \end{array}$$

49.
$$\begin{array}{ccccccc|c} x_1 & x_2 & x_3 & y_1 & y_2 & y_3 & z & \\ \hline 1 & 1 & 3 & 1 & 0 & 0 & 0 & 2 \\ 3 & 1 & 4 & 0 & 1 & 0 & 0 & 30 \\ 2 & 0 & 3 & 0 & 0 & 1 & 0 & 20 \\ \hdashline -5 & -3 & -2 & 0 & 0 & 0 & 1 & 85 \end{array}$$

50.
$$\begin{array}{ccccccc|c} x_1 & x_2 & x_3 & y_1 & y_2 & y_3 & z & \\ \hline 0 & 9 & 6 & 1 & 0 & 0 & 0 & 36 \\ 0 & 3 & 5 & 0 & 1 & 0 & 0 & 25 \\ 45 & 27 & 9 & 0 & 0 & 1 & 0 & 18 \\ \hdashline -1 & -5 & -10 & 0 & 0 & 0 & 1 & 0 \end{array}$$

Objective 4.13: *Solve a standard linear programming problem, if possible.*

51. Maximize: $z = 6x_1 + 25x_2 + 3x_3$

Subject to: $\begin{cases} 2x_1 + x_3 \le 50 \\ 4x_2 + x_3 \le 88 \\ 3x_1 + 4x_2 \le 118 \\ x_1 \ge 0, \quad x_2 \ge 0, \quad x_3 \ge 0 \end{cases}$

52. Maximize: $z = 2x_1 + 3x_2 + 5x_3$

Subject to: $\begin{cases} x_1 + x_2 + 5x_3 \le 10 \\ 2x_1 + 2x_2 + 3x_3 \le 20 \\ x_1 + 5x_2 + x_3 \le 30 \\ x_1 \ge 0, \quad x_2 \ge 0, \quad x_3 \ge 0 \end{cases}$

53. Maximize: $z = 15x_1 + 5x_2 + x_3$

Subject to: $\begin{cases} 3x_1 + 2x_2 \le 120 \\ 3x_2 + x_3 \le 90 \\ x_3 \le 100 \\ x_1 \ge 0, \quad x_2 \ge 0, \quad x_3 \ge 0 \end{cases}$

54. Maximize: $z = x_1 + 3x_2$

Subject to: $\begin{cases} x_1 + 5x_2 \le 100 \\ 2x_1 + 3x_2 \le 74 \\ 3x_1 + x_2 \le 50 \\ x_1 \le 40 \\ x_2 \le 90 \\ x_1 \ge 0, \quad x_2 \ge 0 \end{cases}$

Objective 4.14: *Solve applied linear programming problems using the simplex method.*

55. *Food supply.* An island is inhabited by two species of animals, A1 and A2, which feed on three types of food, F1, F2, and F3. The first species of animals requires 12 units of food F1, 14 units of F2, and 16 units of F3 to survive. The second species requires 15, 20, and 5, units of foods F1, F2, and F3, respectively. The island can normally supply 1,200 units of F1, 1,200 units of F2, and 800 units of F3. What is the maximum number of animals that this island can support?

56. *Food supply.* Suppose a new animal, A3, is introduced to the island described in Problem 55. If A3 requires 5 units of F1, 8 units of F2, and 6 units of F3, how many animals will the island now support?

57. *Maximize profit.* Pell, Inc., manufactures two products: tables and shelves. Each product must be processed in each of three departments: machining, assembling, and finishing. The hours needed to produce one unit of product per department and the maximum possible hours per department are shown in the table. Pell's net profit is \$50 per table and \$40 per shelf. How many tables should be manufactured to maximize the profit?

	Production time per unit (hr)		
Department	**Tables**	**Shelves**	**Maximum capacity (hr)**
Machining	2	1	200
Assembling	2	2	240
Finishing	1	2	300

58. *Maximize profit.* Suppose that the net profits in Problem 57 change so that each shelf brings a net profit of \$70 and each table \$80. Now, how many of each should be manufactured to maximize the profit?

[4.6]

Objective 4.15: *Determine whether each programming problem is in standard form, and write the initial simplex tableau, if possible.*

59. Maximize: $z = 5x_1 + 2x_2$

Subject to: $\begin{cases} x_1 + 3x_2 \le 30 \\ 2x_1 - x_2 \le 24 \\ x_1 \ge 0, \quad x_2 \ge 0 \end{cases}$

60. Maximize: $z = 35x_1 + 90x_2$

Subject to: $\begin{cases} x_1 + x_2 \ge 10 \\ 3x_1 + x_2 \le 30 \\ x_1 + 2x_2 \le 20 \\ x_1 \ge 0, \quad x_2 \ge 0 \end{cases}$

61. Maximize: $z = x_1 + 2x_2 + 5x_3$

Subject to: $\begin{cases} x_1 + x_3 \le 10 \\ x_2 + x_3 \le 30 \\ x_1 + 2x_2 \ge -10 \\ x_1 \ge 0, \quad x_2 \ge 0, \quad x_3 \ge 0 \end{cases}$

62. Maximize: $z = 3x_1 + 2x_2 + x_3$

Subject to: $\begin{cases} x_2 + x_3 \le 100 \\ 3x_1 + x_3 \le 60 \\ x_1 + 2x_2 \le 50 \\ x_1 + x_2 + x_3 = 200 \\ x_1 \ge 0, \quad x_2 \ge 0, \quad x_3 \ge 0 \end{cases}$

Objective 4.16: *Solve linear programming problems that are not in standard form.*

63. Maximize: $z = 5x_1 + 3x_2$

Subject to: $\begin{cases} -3x_1 + 7x_2 \leq 49 \\ 2x_1 - x_2 \geq 4 \\ x_1 \geq 0, \quad x_2 \geq 0 \end{cases}$

64. Maximize: $z = 10x_1 - 5x_2$

Subject to: $\begin{cases} 4x_1 + x_2 \leq 19 \\ 3x_1 - x_2 \leq 12 \\ x_1 \geq 0, \quad x_2 \leq 0 \end{cases}$

65. Maximize: $z = 20x_1 + 60x_2$

Subject to: $\begin{cases} 2x_1 + 3x_2 \leq 18 \\ 2x_1 + x_2 \geq 10 \\ x_1 \geq 0, \quad x_2 \geq 0 \end{cases}$

66. Minimize: $z = 25x_1 + 10x_2$

Subject to: $\begin{cases} 3x_1 - x_2 \leq 25 \\ x_1 + 2x_2 \geq 10 \\ x_1 \geq 0, \quad x_2 \geq 0 \end{cases}$

[4.7]

Objective 4.17: *Write the transpose of a given matrix.*

67. $\begin{bmatrix} 6 & 8 & 1 & 9 \\ -4 & 3 & -2 & 0 \\ 3 & 4 & 5 & 7 \end{bmatrix}$

68. $\begin{bmatrix} 6 \\ -1 \\ 2 \\ 7 \end{bmatrix}$

69. $[8 \quad 1 \quad 4]$

70. $\begin{bmatrix} 6 & 0 & 9 \\ 4 & -3 & 8 \\ 4 & 1 & -6 \end{bmatrix}$

Objective 4.18: *Write the dual of a given linear programming problem.*

71. Minimize: $z = 30x_1 + 25x_2$

Subject to: $\begin{cases} x_1 + 5x_2 \geq 100 \\ 2x_1 + 3x_2 \geq 600 \\ x_1 \geq 0, \quad x_2 \geq 0 \end{cases}$

72. Minimize: $z = 10x_1 + 15x_2$

Subject to: $\begin{cases} 3x_1 + 8x_2 \geq 24 \\ 5x_1 + 2x_2 \geq 30 \\ x_1 \geq 0, \quad x_2 \geq 0 \end{cases}$

73. Minimize: $z = 100x_1 + 200x_2$

Subject to: $\begin{cases} 3x_1 + 5x_2 \geq 70 \\ 2x_1 + 3x_2 \geq 60 \\ 5x_1 + 2x_2 \geq 50 \\ x_1 \geq 0, \quad x_2 \geq 0 \end{cases}$

74. Minimize: $z = 50x_1 + 80x_2$

Subject to: $\begin{cases} 9x_1 + x_2 \geq 18 \\ 3x_1 + 12x_2 \geq 36 \\ 2x_1 + 3x_2 \geq 30 \\ x_1 \geq 0, \quad x_2 \geq 0 \end{cases}$

Objective 4.19: *Solve a minimization problem by using the simplex method. (The duals of these were found in Problems* 71–74.)

75. Minimize: $z = 30x_1 + 25x_2$

Subject to: $\begin{cases} x_1 + 5x_2 \geq 100 \\ 2x_1 + 3x_2 \geq 600 \\ x_1 \geq 0, \quad x_2 \geq 0 \end{cases}$

76. Minimize: $z = 10x_1 + 15x_2$

Subject to: $\begin{cases} 3x_1 + 8x_2 \geq 26 \\ 5x_1 + 2x_2 \geq 32 \\ x_1 \geq 0, \quad x_2 \geq 0 \end{cases}$

77. Minimize: $z = 100x_1 + 200x_2$

Subject to: $\begin{cases} 3x_1 + 5x_2 \geq 70 \\ 2x_1 + 3x_2 \geq 60 \\ 5x_1 + 2x_2 \geq 50 \\ x_1 \geq 0, \quad x_2 \geq 0 \end{cases}$

78. Minimize: $z = 50x_1 + 80x_2$

Subject to: $\begin{cases} 9x_1 + x_2 \geq 18 \\ 3x_1 + 12x_2 \geq 36 \\ 2x_1 + 3x_2 \geq 30 \\ x_1 \geq 0, \quad x_2 \geq 0 \end{cases}$

Objective 4.20: *Solve an applied minimization problem.*

79. *Diet problem.* Of the twenty or so amino acids making up protein, eight cannot be synthesized by humans and are known as the *essential amino acids.* A diet that provides adequate amounts of the three amino acids tryptophan (TRP), lysine (LYS), and methionine (MET) will also generally provide enough of the other essential amino acids. The table at the top of page 232 shows three sources for these acids and the approximate number of grams of TRP, LYS, and MET per pound.

Food	TRP	LYS	MET
Beef	1.2	8.0	2.4
Peanuts	1.5	5.0	1.2
Cashews	2.0	3.0	1.5

If peanuts cost \$1 per pound and cashews cost \$3 per pound, find the least expensive combination of peanuts and cashews that contains an amount of each of the three amino acids equal to or greater than the amount in a pound of beef.

80. *Diet problem.* One pound of hamburger contains 90 g of fat and 100 IU of vitamin A. One pound of chicken contains 15 g of fat and 400 IU of vitamin A. If it is necessary to provide a diet containing at least 180 g of fat and 800 IU vitamin A, how many pounds of each type of meat are necessary to minimize the cost? Assume that the current price of hamburger is \$4.00/lb and chicken is \$3.00/lb.

81. *Crop management.* Niels is studying the utility of his garden crops, consisting of cabbage and peanuts. He needs to provide 60 g of protein and 40 g of carbohydrates for himself for dinner and knows that each pound of cabbage provides 15 g of protein and 25 g of carbohydrates, whereas a pound of peanuts provides 30 g of protein and 10 g of carbohydrates. It takes Niels 30 minutes to collect a pound of cabbage and 20 minutes to collect a pound of peanuts. What is the least time Niels can spend collecting cabbage and peanuts to provide at least the basic nutrients of protein and carbohydrates?

SAMPLE TEST

The following sample test (45 minutes) is intended to review the main ideas of the chapter.

*In Problems 1–4 set up the initial tableau for the linear programming problems and indicate the location of the first pivot. You do not need to solve.**

1. Maximize: $z = 2x_1 + 20x_2 + 3x_3$

Subject to: $\begin{cases} 2x_1 + x_3 \le 50 \\ 4x_2 + x_3 \le 100 \\ 3x_1 + 4x_2 \le 109 \\ x_1 \ge 0, \quad x_2 \ge 0, \quad x_3 \ge 0 \end{cases}$

* For extra practice you can solve Problems 1–4, but to do so will require more time.

2. Maximize: $z = 5x_1 + 3x_2$

Subject to: $\begin{cases} 7x_1 - 3x_2 \ge 3 \\ 2x_1 + x_2 \le 12 \\ x_1 \ge 0, \quad x_2 \ge 0 \end{cases}$

3. Minimize: $z = 50x_1 + 80x_2$

Subject to: $\begin{cases} 9x_1 + x_2 \ge 18 \\ 3x_1 + 12x_2 \ge 36 \\ 2x_1 + 3x_2 \ge 30 \\ x_1 \ge 0, \quad x_2 \ge 0 \end{cases}$

4. Maximize: $z = x_1 + 3x_2 + 2x_3$

Subject to: $\begin{cases} x_1 + x_2 + x_3 \le 16 \\ 4x_1 + 2x_2 + 3x_3 \le 48 \\ x_1 + 2x_2 + x_3 \le 22 \\ x_1 \ge 0, \quad x_2 \ge 0, \quad x_3 \ge 0 \end{cases}$

Problems 5–10 are reprinted with the permission of the American Institute of Certified Public Accountants, Inc. The questions are multiple-choice questions taken from recent CPA examinations. Choose the best, or most appropriate, response for each question.

CPA Exam May 1973

The Ball Company manufactures three types of lamps: A, B, and C. Each lamp is processed in two departments—I and II. Total available hours of labor per day for departments I and II are 400 and 600, respectively. No additional labor is available. Time requirements and profit per unit for each lamp type are:

	A	B	C
Hours required in department I	2	3	1
Hours required in department II	4	2	3
Profit per unit (sales price less all variable costs)	\$5	\$4	\$3

The company has assigned you, as the accounting member of its profit planning committee, to determine the number of types of A, B, and C lamps that it should produce in order to maximize its total profit from the sale of lamps. The following questions concern the linear programming model your group has developed.

5. The coefficients of the objective function would be

A. 4, 2, 3 **B.** 2, 3, 1
C. 5, 4, 3 **D.** 400, 600

6. The constraints in the model would be
   A. 2, 3, 1 **B.** 5, 4, 3
   C. 4, 2, 3 **D.** 400, 600
7. The constraint imposed by the available hours in department I could be expressed as
   A. $4x_1 + 2x_2 + 3x_3 \leq 400$
   B. $4x_1 + 2x_2 + 3x_3 \geq 400$
   C. $2x_1 + 3x_2 + 1x_3 \leq 400$
   D. $2x_1 + 3x_2 + 1x_3 \geq 400$
8. The most types of lamps that would be included in the optimal solution would be
   A. 2 **B.** 1 **C.** 3 **D.** 0

CPA Exam May 1979

9. The Pauley Company plans to expand its sale force by opening several new branch offices. Pauley has $10,400,000 in capital available for new branch offices. Pauley will consider opening only two types of branches: 20-person branches (type A) and 10-person branches (type B). Expected initial cash outlays are $1,300,000 for a type A branch and $670,000 for a type B branch. The expected annual profit is $92,000 for a type A branch and $36,000 for a type B branch. Pauley will hire no more than 200 employees for the new branch offices and will not open more than 20 branch offices. Use linear programming to help decide how many branch offices should be opened.

   In a system of equations for a linear programming model, which of the following equations would **not** represent a constraint (restriction)?
   A. $A + B \leq 20$
   B. $20A + 10B \leq 200$
   C. $\$92{,}000A + \$36{,}000B \leq \$128{,}000$
   D. $\$1{,}300{,}000A + \$670{,}000B \leq \$10{,}400{,}000$

CPA Exam November 1975

10. Patsy, Inc., manufactures two products, X and Y. Each product must be processed in each of three departments: machining, assembling, and finishing. The hours needed to produce one unit of product per department and the maximum possible hours per department are:

	Production hours per unit		Maximum
Department	X	Y	**capacity in hours**
Machining	2	1	420
Assembling	2	2	500
Finishing	2	3	600

Other restrictions are:

$$X \geq 50$$
$$Y \geq 50$$

The objective function is to maximize profits where profit $= \$4X + \$2Y$. Given the objective and constraints, what is the most profitable number of units of X and Y, respectively, to manufacture?

5

Mathematics of Finance

CHAPTER OVERVIEW

The notion of simple interest is introduced, which leads directly to the concept of compound interest. This chapter uses a unified notation, so you should learn the meanings of the variables used early in the chapter, especially the difference between present value, P, and future value, A. Pay special attention to the relationship between annual rate, r, and the rate per period, i; and between the total time, t (in years), and the number of periods, N. One of the important skills to be learned in this chapter is the ability to look at a particular situation and decide what type of financial formula applies (that is, what model to use).

PREVIEW

One of the cornerstones upon which business is built is the payment of interest for the temporary use of another's money. In this chapter we discuss simple and compound interest, present value, annuities, and sinking funds. We will be concerned with comparing the present value, P, with its value, A, at some future time. We can make payments or deposits in two ways: deposit a lump sum or make periodic payments. Consider the following simplified chapter overview:

	Present value, *P*	**Future value, *A***	**Method for finding unknown value**
Lump-sum payment or deposit	Known	Unknown	Compound interest
	Unknown	Known	Present value
Periodic payment	Known	Unknown	Annuity
	Unknown	Known	Sinking fund

PERSPECTIVE

Intelligent handling of money goes hand in hand with financial success. While understanding compound interest, annuities, amortization, and sinking funds does not guarantee success, such understanding is necessary when handling money and investments.

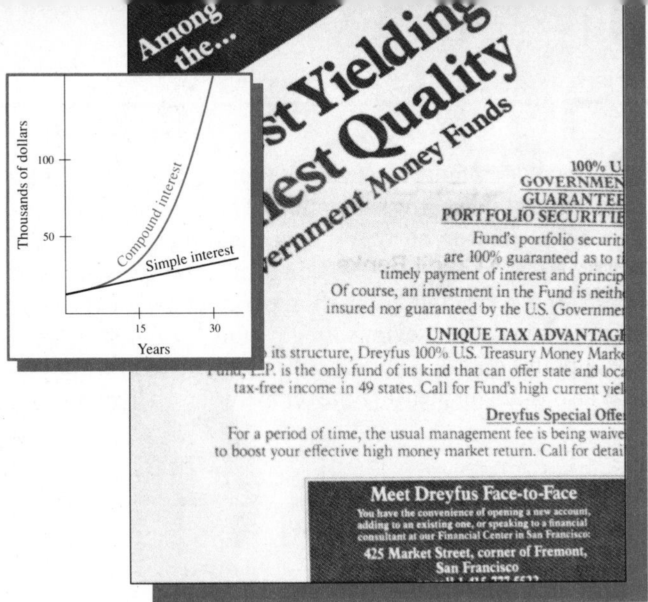

MODELING
APPLICATION 3

Investment Opportunities

You have just inherited $30,000 and need to decide what to do with the money. You check around and find that the current interest rate for deposits is 13.58%, while for loans it ranges from 14% to 21%. Inflation has been around 4%, but long-range estimates are that it will average about 10%. For this problem, assume that you have no unusual debts, are 30 years old, and have a home worth $60,000 with a $30,000 mortgage at an 8% interest rate with payments of $225 per month for the next 25 years.

Write a paper discussing your investment alternatives. Omit any discussion of tax considerations. For general guidelines about writing this essay, see the commentary for Modeling Application 1 on page 151.

APPLICATIONS

Management *(Business, Economics, Finance, and Investments)*
Salary increases (5.1, Problems 50 and 51)
Comparing simple and compound interest (5.2, Problems 1–26)
Finding the future value of bank accounts (5.2, Problems 27–31; 5.6, Problems 17–28; 5.6, Test Problems 1–2)
Finding the effective yield of an investment (5.2, Problems 32–35)
Present value problem (5.2, Problems 36–40, 53–58; 5.5, Problems 25–26, 70; 5.6, Problems 33–36; Test Problem 11)
Effect of inflation (5.2, Problems 41–52; 5.5, Problem 23; 5.6, Problem 67; Test Problem 7)
Annuity values (5.3, Problems 1–24; 5.5, Problems 22, 29–30; 5.6, Problems 37–44, 53–56, 62, 65, 68–69, 71)
Setting up a retirement fund (5.3, Problems 25–29 and 31)
Value of Social Security deposits (5.3, Problems 30 and 32)
Monthly loan payment (5.4, Problems 13–24; 5.5, Problem 27; 5.6, Problems 49–52; Test Problem 16)
Present value of an annuity (5.4, Problems 1–12, 27; 5.6,

Management *(continued)*
Problems 45–48, 66, 72)
Sinking fund (5.5, Problems 1–12, 24)
Sinking funds to pay off notes (5.5, Problems 13–17, 21–22; 5.6, Problems 57–60, 73)
Repayment of a bond (5.5, Problem 28; 5.6, Problem 60, Test Problem 17)
Effective yield (5.6, Problems 29–32)
Value of a savings certificate (5.6, Problem 61)
Value of a lump-sum insurance payment (5.6, Problems 63–64)

Life Sciences *(Biology, Ecology, Health, and Medicine)*
Cell division (5.1, Problem 57)
Growth patterns for rabbits (5.1, Problem 58)
Price of wheat as determined by last year's production (5.1, Problem 60)

Social Sciences *(Demography, Political Science, Population, Psychology, Society, and Sociology)*
Breaking news story (5.1, Problem 55)
Spread of rumors (5.1, Problem 56)

General Interest
Triangular numbers (5.1, Problem 59)
National debt (5.2, Problem 60)
Calculating total interest paid for a consumer's loan (5.4, Problems 25–26)
Present value of a lottery prize (5.2, Problem 59; 5.4, Problem 28; 5.6, Test Problem 15)
Deposit to a Keogh account (5.3, Problem 25)
Value of an insurance annuity (5.4, Problems 29–30; 5.6, Test Problem 3)
Saving for a purchase (5.4, Problems 31–38; 5.5, Problems 18, 21, 24–26)
Investing an inheritance (5.5, Problem 19)
Savings account for a child (5.5, Problem 20)

Management Accounting Exam
(5.6, Test Problem 18)

CPA Exam
(5.6, Test Problems 19 and 20)

Modeling Application—
Investment Opportunities

5.1 Difference Equations

A frequent real-life situation involves a finite sequence of steps. For example, next year's population is dependent on this year's population. The balance in a bank account earning interest over some period of time is dependent on the amount of money in the account in the previous period of time. The age of an object is based on the amount of carbon present, and the amount of carbon present now is dependent on the amount of carbon present in some previous time period.

For example, suppose you pour yourself a cup of coffee and the temperature of the coffee is 200°F. Suppose also that you measure the temperature each minute for the first 5 minutes.

SPREADSHEET PROGRAM

	A	B
1	Elapsed time	Temperature
2	0	200
3	+A2+1	+B2*(.95)+5
4	replicate	replicate

Elapsed time	Temperature
0	200
1	195
2	190.25
3	185.7375
4	181.450625
5	177.3780938
6	173.5091891
7	169.8337296
8	166.3420431
9	163.024941
10	159.8736939
11	156.8800092
12	154.0360088
13	151.3342083
14	148.7674979
15	146.329123
16	144.0126669
17	141.8120335
18	139.7214318
19	137.7353603
20	135.8485922

Elapsed Time (in minutes)	Temperature (in degrees Fahrenheit)
0	200
1	195
2	190
3	185
4	181
5	177

Notice that the temperatures—call them T_0, T_1, T_2, T_3, T_4, and T_5—form what is called a **sequence** or **progression**. There is a formula which gives these temperatures. Suppose we assume that each temperature T_n is related to the previous temperature T_{n-1}. In general, an equation that gives each number of a sequence in terms of preceding numbers is called a **difference equation**. The goal of this section is to find such a formula. After you know such a formula, you can use it to forecast future events. For example, suppose you know that the formula for the temperature T_n is

$$T_n = .95T_{n-1} + 5$$

Then you can predict the temperature for the sixth minute (if you assume that the temperature $T_5 = 177$),

$$T_6 = .95T_5 + 5 = .95(177) + 5 = 173.15$$

A spreadsheet output for the first 20 minutes is shown in the margin.

First-Order Difference Equation

Consider the set of even numbers

$$2, 4, 6, 8, 10, \ldots$$

They form a sequence. The individual numbers are called *terms* of the sequence. The number x_0 is the starting value of the sequence and is called the *initial value*. In terms of investments and interest, it is usually the amount of money you have now (initially). The number x_1 is the value after the first change. Think of the subscript as a measurement of time—the second value of the sequence, x_1, is the value at the end of the *first* year. Thus, x_5 is the value at the end of the fifth year (which is the sixth term of the sequence).

Sequences are usually defined by giving what is called the *general term*, which is a formula such as

$$x_n = 3n + 2$$

where $n = 0, 1, 2, 3, 4, \ldots$. You can then find any term by evaluation:

$$x_0 = 3(0) + 2 = 2 \quad \text{initial value}$$
$$x_1 = 3(1) + 2 = 5 \quad \text{value after first change}$$
$$x_2 = 3(2) + 2 = 8 \quad \text{value after second change}$$
$$x_3 = 3(3) + 2 = 11 \quad \ldots$$

Thus, the sequence is $2, 5, 8, 11, \ldots$.

Although sequences are used throughout mathematics in a variety of contexts and applications, we will use them in connection with financial formulas by considering general terms of the type

$$x_{n+1} = ax_n + b$$

where a and b are constants. An equation of this type is called a **first-order linear difference equation**.

Yale Set Designers Turn to Finite Element Analysis

This spring a group of students at Yale School of Drama will learn how to use a mathematical technique that helped to build the Stealth bomber. Instead of making low-profile planes, however, these graduates will move on to Broadway or scenery studios to erect 35-foot-high robotic reptiles or construct super-structures that weigh 50,000 pounds when loaded with lights and equipment.

The formal introduction of a course in finite element analysis (FEA)—it was taught once before on a trial basis—is an attempt to move beyond the methodology for technical set design that has prevailed since stagehands cobbled together balconies at Shakespeare's Globe Theater. "You put something in, it broke, then you put in something bigger or stronger," says Bronislaw J. Sammler, production supervisor and chairman of the technical design and production department of the Yale drama school.

A more incisive look at stage construction has been called for because blockbuster Broadway shows, Las Vegas-style extravaganzas, opera and even regional theater must be loaded with special effects and labyrinthine sets to entice audiences jaded by movies and television.

Finite element analysis is used today to analyze point stresses in automobiles, jet fighters and even knee braces for professional football players. In set design the technique uses matrices of simultaneous equations to derive deflections and stresses when a load—a 250-pound opera singer, for example—is applied to a series of finite elements: the tubing, I-beams and other components that make up the steel frame for an overhanging ledge on which the diva stands.

Prospective technical designers will not be required to wrestle with equation solving to learn the technique, however. A personal computer software program,

(*continued*)

(continued)

FINITE ELEMENT ANALYSIS helped design the Metropolitan Opera's set for the Flying Dutchman. Photo: Winnie Klotz

called Algor, will do the calculations. The students will learn the program, in part, from a tutorial written as a master's thesis by David L. Sword, the assistant professor who will teach FEA as an elective course.

The software simplifies the process by requiring a user to trace out a three-dimensional model on the screen and input a few variables—type of material, thickness and tensile strength, among others. It then solves the equations and supplies a contour map that indicates in red areas of highest stress.

Yale alumni have already become finite element proselytizers, and New York City's Metropolitan Opera is one beneficiary. Geoff Webb was snapped up by the opera house as a technical designer after completing a master's thesis in 1988 at Yale on finite element analysis. In working on a set for Wagner's *Flying Dutchman*, Webb had to make adjustments to the specifications for a 50-foot staircase of steel tubing that descended from the Dutchman's floating ship. The software showed that the steps might have flexed up and down slightly underfoot when the singer made his entrance on it.

Webb is looking at still other software from the design engineer's toolbox, so computerized methods for designing concealed aircraft may get quite visible use by those working hidden behind the curtain.

—*Gary Stix*

EXAMPLE 1 Find the first four terms of the sequence that satisfies the difference equation

$$x_{n+1} = 6x_n + 50$$

and has an initial value of 100.

Solution Given $x_0 = 100$; then

$$x_1 = 6x_0 + 50 = 6(100) + 50 = 650$$
$$x_2 = 6x_1 + 50 = 6(650) + 50 = 3{,}950$$
$$x_3 = 6x_2 + 50 = 6(3{,}950) + 50 = 23{,}750$$

The sequence is 100, 650, 3,950, 23,750, ... ■

EXAMPLE 2 Suppose you deposit \$10,000 in a bank account. Let $A_0 = 10{,}000$. This is called the **initial condition**. Let $A_1, A_2, A_3, \ldots$ be the amount in the account after one year, two years, three years, . . . , respectively. That is, A_n is the amount in the bank account after n years. Also suppose that the difference formula is

$$A_{n+1} = 1.08A_n$$

Find the sequence of bank balances for the first five years.

Solution Since $A_0 = 10{,}000$,

$$A_1 = 1.08A_0 = 1.08(10{,}000) = 10{,}800$$
$$A_2 = 1.08A_1 = 1.08(10{,}800) = 11{,}664$$
$$A_3 = 1.08A_2 = 1.08(11{,}664) = 12{,}597.12$$
$$A_4 = 1.08A_3 = 1.08(12{,}597.12) \approx 13{,}604.89$$
$$A_5 = 1.08A_4 = 1.08(13{,}604.89) \approx 14{,}693.28$$

■

A **solution** of a linear difference equation is a sequence in which the successive terms will satisfy the equation. The difference between finding the solution and finding the general term is that the solution allows us to calculate any term without first calculating all the preceding terms. Our goal, therefore, is to find this solution for the first-order linear difference equation. First, however, we will solve two special cases, called *arithmetic* and *geometric* sequences.

Arithmetic Sequence

If $a = 1$, then the general difference equation becomes

$$x_{n+1} = x_n + b$$

EXAMPLE 3 Let $b = 2$ in the difference equation

$$x_{n+1} = x_n + 2$$

with initial value 1. Find the first four terms of this sequence.

Solution This difference equation is $x_{n+1} = x_n + 2$ with $x_0 = 1$:

$$x_0 = 1$$
$$x_1 = x_0 + 2 = 1 + 2 = 3$$
$$x_2 = x_1 + 2 = 3 + 2 = 5$$
$$x_3 = x_2 + 2 = 5 + 2 = 7$$

The sequence is 1, 3, 5, 7, ... ■

A difference equation with $a = 1$, namely

$$x_{n+1} = x_n + b$$

is called an **arithmetic sequence** or **progression**. It is characterized by the fact that there is a *common difference*, b. Notice that the other terms of the sequence are

$$\begin{aligned}
&x_0 \\
&x_1 = x_0 + b \\
&x_2 = x_1 + b = x_0 + b + b = x_0 + 2b \\
&x_3 = x_2 + b = x_0 + 2b + b = x_0 + 3b \\
&x_4 = x_3 + b = x_0 + 3b + b = x_0 + 4b \\
&\quad\vdots \\
&x_n = x_0 + nb
\end{aligned}$$

Arithmetic Sequence

The solution of the arithmetic sequence with first term x_0 and common difference b is

$$x_n = x_0 + nb$$

EXAMPLE 4 Find the x_{10} term of the arithmetic sequence whose first term is -3 and whose common difference is 4.

Solution $x_0 = -3$; $b = 4$, so

$$\begin{aligned}
x_n &= x_0 + nb \\
x_{10} &= -3 + 10(4) = 37
\end{aligned}$$

■

Geometric Sequence

If $b = 0$, then the general difference equation becomes

$$x_{n+1} = ax_n$$

EXAMPLE 5 Find the first four terms of the sequence $x_{n+1} = ax_n$ for the initial value of 1 and $a = 2$.

Solution The difference equation is $x_{n+1} = 2x_n$:

$$\begin{aligned}
&x_0 = 1 \\
&x_1 = 2x_0 = 2(1) = 2 \\
&x_2 = 2x_1 = 2(2) = 4 \\
&x_3 = 2x_2 = 2(4) = 8 \\
&\quad\vdots
\end{aligned}$$

The sequence is 1, 2, 4, 8, ...

■

A difference equation with $b = 0$, namely

$$x_{n+1} = ax_n$$

is called a **geometric sequence** or **progression.** It is characterized by the fact that there is a *common ratio, a.* Notice that the terms of the sequence are

$$\begin{aligned}
&x_0\\
&x_1 = ax_0\\
&x_2 = ax_1 = a(ax_0) = a^2x_0\\
&x_3 = ax_2 = a(a^2x_0) = a^3x_0\\
&x_4 = ax_3 = a(a^3x_0) = a^4x_0\\
&\quad\vdots\\
&x_n = a^nx_0
\end{aligned}$$

Geometric Sequence

The solution of the geometric sequence with first term x_0 and common ratio a is

$$x_n = a^nx_0$$

EXAMPLE 6 Find the x_5 term of the geometric sequence whose first term is 100 and whose common ratio is $\frac{1}{2}$.

Solution $x_0 = 100$, $a = \frac{1}{2}$, so

$$x_n = a^nx_0$$

$$x_5 = \left(\frac{1}{2}\right)^5(100) = 3.125$$

■

General Solution of a Difference Equation

We can now find the general solution of the difference equation

$$x_{n+1} = ax_n + b$$

Begin by looking for a pattern. Let x_0 be the initial value.

$$\begin{aligned}
x_1 &= ax_0 + b\\
x_2 &= ax_1 + b = a(ax_0 + b) + b = a^2x_0 + ab + b\\
x_3 &= ax_2 + b = a(a^2x_0 + ab + b) = a^3x_0 + a^2b + ab + b\\
x_4 &= ax_3 + b = a(a^3x_0 + a^2b + ab + b) + b = a^4x_0 + a^3b + a^2b + ab + b\\
&\quad\vdots\\
x_n &= a^nx_0 + a^{n-1}b + a^{n-2}b + \cdots + a^2b + ab + b
\end{aligned}$$

Even though this is the general solution, it is not particularly useful because of its algebraic form. We now change the algebraic form so that it is easier to use.

$$\begin{aligned}
x_n &= a^nx_0 + a^{n-1}b + a^{n-2}b + \cdots + a^2b + ab + b\\
&= a^nx_0 + b(a^{n-1} + a^{n-2} + \cdots + a^2 + a + 1)
\end{aligned}$$

Now, consider the sum

$$S_n = 1 + a + a^2 + \cdots + a^n$$

and

$$aS_n = a + a^2 + a^3 + \cdots + a^{n+1}$$ Multiply both sides by a.

$$S_n - aS_n = 1 - a^{n+1}$$ Subtract (notice the arrows in the lines above; all of those terms are zero when you subtract).

$$(1-a)S_n = 1 - a^{n+1}$$ Factor.

$$S_n = \frac{1-a^{n+1}}{1-a}$$ $a \neq 1$

Thus,

$$a^{n-1} + a^{n-2} + \cdots + a^2 + a + 1 = \frac{1-a^{(n-1)+1}}{1-a} = \frac{1-a^n}{1-a}$$

The general solution is now stated (by substitution in the second step)

$$x_n = a^n x_0 + b(a^{n-1} + a^{n-2} + \cdots + a^2 + a + 1)$$

$$x_n = a^n x_0 + b\left(\frac{1-a^n}{1-a}\right)$$

$$= a^n x_0 + \frac{b}{1-a}(1-a^n)$$

$$= a^n x_0 + \frac{b}{1-a} - \frac{b}{1-a}a^n$$

$$= \frac{b}{1-a} + a^n x_0 - \frac{b}{1-a}a^n$$

$$= \frac{b}{1-a} + \left(x_0 - \frac{b}{1-a}\right)a^n$$

General Solution of a Difference Equation

The difference equation $x_{n+1} = ax_n + b$ with $a \neq 1$ has solution

$$x_n = \frac{b}{1-a} + \left(x_0 - \frac{b}{1-a}\right)a^n$$

EXAMPLE 7 Solve each difference equation and find x_4 for each.

a. $x_{n+1} = 5x_n + 8, \quad x_0 = 1$ **b.** $x_{n+1} = 3x_n, \quad x_0 = 4$

c. $x_{n+1} = x_n + 4, \quad x_0 = 3$

Solution **a.** $a = 5, b = 8, x_0 = 1$, so using the general solution,

$$\frac{b}{1-a} = \frac{8}{1-5} = \frac{8}{-4} = -2$$

$$x_n = -2 + (1 + 2)5^n$$
$$= -2 + 3 \cdot 5^n \qquad \text{This is the solution.}$$

Also, $x_4 = -2 + 3 \cdot 5^4 = 1{,}873$.

b. $a = 3, b = 0, x_0 = 4$, so using the general solution,

$$\frac{b}{1-a} = \frac{0}{1-3} = 0$$
$$x_n = 0 + (4 + 0)3 = 4 \cdot 3^n \qquad \text{This is the solution.}$$

Notice, however, since $b = 0$, this is a geometric sequence, so using the solution for geometric sequences we find (directly)

$$x_n = 4 \cdot 3^n$$

Also, $x_4 = 4 \cdot 3^4 = 324$.

c. $a = 1,\ b = 4,\ x_0 = 3$. This is an arithmetic sequence. From the solution for arithmetic sequences, we find (directly)

$$x_n = 3 + 4n$$

Also, $x_4 = 3 + 4 \cdot 4 = 19$. ■

EXAMPLE 8 Suppose the average beginning salary for a recent college graduate is \$25,000 and that this graduate can expect annual salary increases of \$1,000 a year plus a 5% cost-of-living increase. What is the graduate's salary for the fifth year?

Solution Let x_n be the salary in the nth year. Now,

$$x_n = \text{last year's salary} + .05\,(\text{last year's salary}) + 1{,}000$$
$$= x_{n-1} + .05x_{n-1} + 1{,}000$$
$$= 1.05x_{n-1} + 1{,}000$$

Note that the general solution is given for a difference equation of the form $x_{n+1} = ax_n + b$ and we have modeled this example in terms of x_n (instead of x_{n+1}):

$$x_n = 1.05x_{n-1} + 1{,}000$$

These equations are the same since the relationship between x_n and x_{n+1} is the same as that between x_{n+1} and x_n (that is, one year change). Therefore, we can use the general solution formula for a difference equation where $a = 1.05$, $b = 1{,}000$, and $x_0 = 25{,}000$:

$$x_n = \frac{1{,}000}{1 - 1.05} + \left(25{,}000 - \frac{1{,}000}{1 - 1.05}\right)(1.05)^n$$

In particular, we want $n = 5$ (fifth year):

$$x_5 = \frac{1{,}000}{-.05} + \left(25{,}000 - \frac{1{,}000}{-.05}\right)(1.05)^5 \approx 37{,}432.67$$

The graduate's salary in five years should be \$37,433. ■

Sometimes models using difference equations are developed using the idea of **proportionality**. We say that two quantities are proportional if one is equal to a constant times the other. For example, if x and y are proportional then $x = ky$ for some constant k, which is called the *constant of proportionality.*

EXAMPLE 9 Suppose that a major news story breaks on the four national television networks at the same time. The number of people learning the news each hour after it broke is proportional to the number who have not heard it by the end of the preceding hour. If we assume that the population is 220 million, how many people would hear of the news 24 hours after it broke if the constant of proportionality is .2?

Solution Let x_n = number of people who have heard the news after n hours. If P is the population in millions (220 in this example), then the number of people who have not heard the news after n hours is $P - x_n$. Thus,

$$\begin{pmatrix}\text{number who have}\\ \text{heard the news}\\ \text{after } n+1 \text{ hours}\end{pmatrix} = \begin{pmatrix}\text{number who know}\\ \text{after } n \text{ hours}\end{pmatrix} + \begin{pmatrix}\text{number who learn}\\ \text{news during the}\\ (n+1)\text{st hour}\end{pmatrix}$$

$$x_{n+1} \qquad = \qquad x_n \qquad + \qquad k(P - x_n)$$

For this example, $k = .2$ and $P = 220$:

$$\begin{aligned} x_{n+1} &= x_n + .2(220 - x_n) \\ &= x_n + 44 - .2x_n \\ &= .8x_n + 44 \end{aligned}$$

If we assume that no one heard the news initially ($x_0 = 0$) we can find the solution where $a = .8$, $b = 44$, and $x_0 = 0$:

$$\begin{aligned} x_n &= \frac{b}{1-a} + \left(x_0 - \frac{b}{1-a}\right)a^n \\ &= \frac{44}{1-.8} + \left(0 - \frac{44}{1-.8}\right)(.8)^n \\ &= 220 - 220(.8)^n \end{aligned}$$

For $n = 24$,

$$x_{24} = 220 - 220(.8)^{24} \approx 219$$

Thus, after one day (24 hours) approximately 1 million people would not have heard the news. ■

Sigma Notation

Sometimes we want to write a sum of terms of a sequence using a shorthand notation. This notation uses the uppercase Greek letter sigma, so is often referred to as **sigma notation**. For example, we wrote

$$S_n = 1 + a + a^2 + a^3 + \cdots + a^n$$

Using sigma notation this sum could be written as

$$S_n = \sum_{k=0}^{n} a^k \quad \text{since} \quad \sum_{k=0}^{n} a^k = 1 + a + a^2 + a^3 + \cdots + a^n$$

In order words, the sigma notation evaluates the expression immediately following the sigma (a^k in this example) first for $k = 0$, then for $k = 1$, next for $k = 2$, and so on, where k counts up (one unit at a time) until it reaches the final value of n. The value 0 in this example is called the *lower limit of summation* and n the *upper limit of summation*, while the variable k is called the *index of summation*.

EXAMPLE 10 Find $\sum_{i=1}^{4} i(i + 3)$.

Solution We have
$$\sum_{i=1}^{4} i(i+3) = \overbrace{1(1+3)}^{i=1} + \overbrace{2(2+3)}^{i=2} + \overbrace{3(3+3)}^{i=3} + \overbrace{4(4+3)}^{i=4}$$
$$= 1 \cdot 4 + 2 \cdot 5 + 3 \cdot 6 + 4 \cdot 7$$
$$= 4 + 10 + 18 + 28$$
$$= 60$$
■

5.1 Problem Set

Compute x_1, x_2, and x_3 for the linear difference equations given in Problems 1–12.

1. $x_{n+1} = x_n + 10;\ x_0 = 6$
2. $x_{n+1} = x_n - 5;\ x_0 = 3$
3. $x_{n+1} = 4x_n;\ x_0 = 1$
4. $x_{n+1} = 10x_n;\ x_0 = 0$
5. $x_{n+1} = -5x_n;\ x_0 = 10$
6. $x_{n+1} = 3x_n;\ x_0 = 4$
7. $x_{n+1} = 2x_n + 3;\ x_0 = 4$
8. $x_{n+1} = -3x_n - 4;\ x_0 = 5$
9. $x_{n+1} = 2x_n + 3;\ x_0 = 1$
10. $x_{n+1} = 3x_n - 2;\ x_0 = 4$
11. $x_{n+1} = 5x_n - 2;\ x_0 = 3$
12. $x_{n+1} = 4x_n + 3;\ x_0 = 2$

Find x_4 for each difference equation given in Problems 13–24.

13. $x_{n+1} = x_n + 8;\ x_0 = 0$
14. $x_{n+1} = x_n + 100;\ x_0 = 100$
15. $x_{n+1} = 3x_n;\ x_0 = 1$
16. $x_{n+1} = 2x_n;\ x_0 = 1$
17. $x_n = (\frac{1}{2})x_{n-1} + 2;\ x_0 = 100$
18. $x_n = (\frac{1}{10})x_{n-1} + 2;\ x_0 = 1{,}000$
19. $x_{n+1} = 5x_n + 2;\ x_0 = 0$
20. $x_{n+1} = 2x_n - 3;\ x_0 = 10$
21. $x_{n+1} = 2x_n + 1;\ x_0 = 8$
22. $x_{n+1} = 1 - 2x_n;\ x_0 = 0$
23. $x_{n+1} = 1 - (\frac{1}{2})x_n;\ x_n = 0$
24. $x_{n+1} = 10 - (\frac{1}{10})x_n;\ x_0 = 0$

Solve the difference equations in Problems 25–40.

25. $x_{n+1} = x_n + 25;\ x_0 = 5$
26. $x_{n+1} = x_n - 20;\ x_0 = 20$
27. $x_{n+1} = x_n - 2;\ x_0 = 4$
28. $x_{n+1} = x_n + 1;\ x_0 = 0$
29. $x_n = x_{n-1} + 4;\ x_0 = 0$
30. $x_n = x_{n-1} - 5;\ x_0 = 5$
31. $x_n = .3x_{n-1};\ x_0 = 1$
32. $x_n = (\frac{1}{5})x_{n-1};\ x_0 = 5$
33. $x_n = 2x_{n-1};\ x_0 = 1$
34. $x_n = 4x_{n-1};\ x_0 = 1$
35. $x_n = 2x_{n-1} - 3;\ x_0 = 0$
36. $x_n = -2x_{n-1} + 15;\ x_0 = 0$
37. $x_n = 5x_{n-1} + 9;\ x_0 = 2$
38. $x_n = -3x_{n-1} + 5;\ x_0 = 1$
39. $x_n = 4x_{n-1} + 2;\ x_0 = 1$
40. $x_n = 3x_{n-1} - 4;\ x_0 = 10$

In Problems 41–49 evaluate the given expression.

41. $\sum_{k=2}^{6} k$
42. $\sum_{m=1}^{4} m^2$
43. $\sum_{n=0}^{6} (2n + 1)$
44. $\sum_{k=2}^{5} (10 - 2k)$

45. $\sum_{k=1}^{5} (-2)^{k-1}$

46. $\sum_{k=0}^{4} 3(-2)^k$

47. $\sum_{k=0}^{3} 2(3^k)$

48. $\sum_{k=2}^{5} (100 - 5k)$

49. $\sum_{k=0}^{5} (20 - 3k)$

APPLICATIONS

50. Suppose a job pays a starting salary of \$20,000 with a \$500 per year increase and a 3% cost-of-living increase. Write the difference equation and find the salary (to the nearest dollar) in the fifth year.

51. Suppose a job pays a starting salary of \$38,000 with a \$2,500 per year increase and a 5% cost-of-living increase. Write the difference equation and find the salary (to the nearest dollar) in the fourth year.

52. Solve the difference equation for Problem 50.

53. Solve the difference equation for Problem 51.

54. Suppose a glass of cola at 35°F warms to room temperature according to the difference equation

$$T_n = 1.05T_{n-1} + 2$$

where T_n is the temperature of the cola after n minutes (up to 10 minutes). Find the temperature (to the nearest degree) for each of the first five minutes.

55. Suppose a news story breaks in a town of 200,000 and the number of people learning of the news after the story broke is proportional to the number who have not heard it by the end of the preceding hour. How many will have heard the story in 8 hours if the constant of proportionality is .3?

56. Suppose the number of persons in a company of 3,000 employees who have heard a particular piece of gossip is proportional to the number of persons who have not heard it by the end of the previous day. How many will have heard the gossip one week after the rumor started if the constant of proportionality is .4?

57. Suppose a cell divides every 20 minutes. If there is initially one cell, how many cells will there be in 8 hours?

58. ***Historical Question.*** In 1202, Leonardo Fibonacci wrote a book entitled *Liber Abaci,* in which he discussed the advantages of the Hindu–Arabic numbers over Roman numerals. In this book, he posed a problem to find the number of rabbits alive after a given number of generations. Assume:
   i. You start with one pair of rabbits of opposite sex.
   ii. It takes a newborn rabbit two months to mature into an adult rabbit.
   iii. A pair of adult rabbits of opposite sex produces one pair of newborn rabbits also of opposite sex each month.
   iv. Rabbits never die.

   Find a difference equation for the number of rabbits alive after n months.

59. ***Historical Question.*** The Greeks studied a group of numbers called *triangular numbers.* These are numbers that can be arranged in the form of a triangle. For example, if you have a pile of pennies, you could form triangles with 1, 3, 6, or 10 pennies.

Let T_n be the nth triangular number. Notice that $T_0 = 1$, $T_1 = 3$, $T_2 = 6, \ldots$. Find a difference equation for the sequence of triangular numbers.

60. According to the Department of Agriculture (Statistical Reporting Service), the U.S. wheat production in 1989 was approximately 2.0 billion bushels. It is known that the current wheat crop affects next year's level of production and next year's price. Let p_n denote the price of wheat (in dollars per bushel) and q_n denote the quantity of wheat produced (in billions of bushels), and suppose that p_n and q_n are related by the equations

$$p_n = 10 - .01q_n$$
$$q_{n+1} = .75p_n - 5$$

Solve the difference equation for wheat production.

5.2 Interest

When Albert Einstein was asked what was the most amazing formula he knew, you might think he would have given his famous $E = mc^2$ formula, but he did not. He thought the most amazing formula was the compound interest formula.

Indeed, one of the most fundamental mathematical concepts for business people and consumers is the idea of interest. Simply stated, **interest** is rent paid for the use of another's money. That is, we receive interest when we let others use our money (when we deposit money in a savings account, for example), and we pay interest when we use the money of others (for example, when we borrow from a bank).

Simple Interest

The amount of the deposit or loan is called the **principal**, or **present value**, and the **interest rate** is stated as a percentage of the principal over a given period of time.

Simple Interest Formula

The **simple interest formula** is

$I = Prt$ where I = Amount of interest
P = Principal or Present value
r = Annual interest rate
t = Time (in years)

EXAMPLE 1 How much interest does a \$73 deposit earn in 3 years if the interest rate is 8%?

Solution

$$
\begin{aligned}
I &= Prt \\
&= 73(.08)(3) \\
&= 17.52
\end{aligned}
$$

Notice that interest rates are written as decimals when substituted into formulas.

The amount of interest is \$17.52. ■

Future Value

The **future value**, A, of a deposit is the amount of money on deposit after a given amount of time. For Example 1, the future value of the account in 3 years is $A = \$73 + \$17.52 = \$90.52$.

Future Value (simple interest)

The **future value**, A, is

$$
A = P + I \quad \text{or} \quad A = P + Prt = P(1 + rt)
$$

where P = Principal
I = Amount of interest

Example 1 is an example of simple interest, but banks and businesses pay **compound interest**. For compound interest, after some designated period of time, the earned interest is added to the account so that future calculations include this earned interest as part of the principal. For Example 1, suppose that interest is *compounded annually*. This means that at the end of the first year the value of the account is found as follows:

$$\begin{aligned} I &= Prt \\ &= 73(.08)(1) \qquad t = 1 \text{ at the end of the first year.} \\ &= 5.84 \end{aligned}$$

Therefore,

$$\begin{aligned} A &= P + I \\ &= 73 + 5.84 \\ &= 78.84 \end{aligned}$$

This amount then becomes the principal for the second year:

$$\begin{aligned} I &= Prt \\ &= 78.84(.08)(1) \\ &= 6.3072 \qquad \text{This is \$6.31 interest for the second year.} \end{aligned}$$

At the end of the second year,

$$\begin{aligned} A &= P + I \\ &= 78.84 + 6.31 \\ &= 85.15 \end{aligned}$$

For the third year,

$$\begin{aligned} I &= 85.15(.08)(1) \qquad \text{and} \qquad & A &= 85.15 + 6.81 \\ &= 6.812 & &= 91.96 \end{aligned}$$

Note that with simple interest (Example 1) the future value in 3 years is \$90.52 and with compound interest it is \$91.96.

The process just described with numbers is easy to follow, but to find the general formula we need to repeat the steps algebraically.

First Year:

$$\begin{aligned} A &= P + I \\ &= P + Pr \\ &= P(1 + r) \end{aligned}$$

$I = Prt = Pr$ when $t = 1$.
Factor out the common factor P; this number becomes the principal for the second year.

Second Year:

$$\begin{aligned} A &= P(1 + r) + I \\ &= P(1 + r) + P(1 + r)r \\ &= P(1 + r)[1 + r] \\ &= P(1 + r)^2 \end{aligned}$$

The second year principal is $P(1 + r)$. Also,

$$\begin{aligned} I &= \text{principal} \times \text{rate} \times \text{time} \\ &= \underbrace{P(1 + r)}_{\text{Second-year principal}}r(1) \end{aligned}$$

This becomes the principal for the third year.

Third Year:

$$\begin{aligned} A &= P(1 + r)^2 + I \\ &= P(1 + r)^2 + P(1 + r)^2 r \\ &= P(1 + r)^2[1 + r] \\ &= P(1 + r)^3 \end{aligned}$$

If you continue in the same fashion for t years at a rate of r compounded annually, you can see that the formula is

$$A = P(1 + r)^t$$

We can also derive this formula using a difference equation.

$$\text{BALANCE AT BEGINNING OF NEXT YEAR} = \text{BALANCE AT BEGINNING OF THIS YEAR} + \text{INTEREST}$$

$$A_{n+1} = A_n + I$$

$$A_{n+1} = A_n + A_n r \qquad \text{Since } I = Prt = A_n r(1)$$

$$A_{n+1} = A_n(1 + r)$$

This is a geometric sequence whose first term is P and whose common ratio is $(1+r)$, so after t years the formula is

$$A_t = (1 + r)^t P \quad \text{or} \quad A = P(1 + r)^t$$

EXAMPLE 2 Find the compound interest for a \$73 deposit at 8% for 3 years compounded annually.

Solution

$$A = 73(1 + .08)^3$$
$$= 73(1.08)^3$$

On a calculator,

[73] [×] [1.08] [×] [1.08] [×] [1.08] [=]

or, if your calculator has an exponent key,

[73] [×] [1.08] [y^x] [3] [=]

The rest of this chapter assumes that calculators have an exponent key; if yours does not, you will need to use repeated multiplication as shown on the first line. The result is 91.958976 or \$91.96; the interest is \$91.96 − \$73 = \$18.96. ■

If you wish, you may use a table for compound interest problems. Table 5 in Appendix E shows the compound interest for \$1 for N periods. For Example 2, find the column headed 8% and look in the row $N = 3$. The entry is 1.259712. Multiply this number by the principal (73) to obtain 91.958976 or \$91.96.

EXAMPLE 3 You are considering a 6-year \$1,000 certificate of deposit paying 12% compounded annually. How much money will you have at the end of 5 years?

Solution

By calculator

$A = 1{,}000(1.12)^5$

[1.12] [y^x] [5] [×] [1000] [=]

The result 1762.341683 rounded to the nearest cent is \$1,762.34.

By Table 5, Appendix E

Find the column headed 12% and the row labeled $N = 5$ to find the entry 1.762342. Multiply:

$$1{,}000(1.762342) \approx \$1{,}762.34$$

■

The disadvantages of working with tables are (1) a different table is needed for each different rate; (2) tables are not as readily available as calculators; and (3) even when using a table, a final multiplication by the principal is necessary.

If you have deposited money or seen savings and loan advertisements lately, you know that most financial institutions compound interest more frequently than annually. This is to your advantage because the shorter the compounding period, the sooner you earn interest on your interest.

Since n is the number of times compounded per year we say

$n = 2$ if compounded semiannually,
$n = 4$ if quarterly,
$n = 12$ if monthly, and
$n = 360$ if compounded daily.

Note that when paying interest, banks use 360 for the number of days in a year; this is called **ordinary interest**. Unless instructed otherwise, use 360 for the number of days in a year. If 365 days are used, it is called **exact interest**.

Future Value (compound interest)

$$A = P(1 + i)^N$$

where A = Future value
P = Present value
r = Annual interest rate
t = Number of years
n = Number of times compounded per year
i = Rate per period $= \frac{r}{n}$
N = Number of periods $= nt$

EXAMPLE 4 Reconsider Example 3 for 12% compounded monthly.

Solution $P = 1{,}000$, $r = 12\%$, $t = 5$, $n = 12$; $i = \frac{r}{n} = \frac{.12}{12} = .01$, $N = nt = 12(5) = 60$

$$A = P(1 + i)^N = 1{,}000(1 + .01)^{60} = 1{,}000(1.01)^{60}$$

$= 1816.696699$ PRESS: [1.01] [y^x] [60] [×] [1000] [=]

Answer rounded to the nearest cent: $1,816.70 ■

We can also use Table 5 in Appendix E to work problems like Example 4. Notice that the table is set up to use $i = r/n$ and $N = nt$.

EXAMPLE 5 Use Table 5 in Appendix E to find the amount you would have if you invested $1,000 for 10 years at 8% interest compounded:

a. Annually **b.** Semiannually **c.** Quarterly

Solution **a.** $N = nt = 1(10) = 10$, and $i = \frac{r}{1} = 8\%$: $A = 1{,}000 \times 2.158925$

$= \$2{,}158.93$

b. $N = nt = 2(10) = 20$, and $i = \frac{r}{2} = \frac{8\%}{2} = 4\%$: $A = 1{,}000 \times 2.191123$

$= \$2{,}191.12$

c. $N = nt = 4(10) = 40$, and $i = \frac{r}{4} = \frac{8\%}{4} = 2\%$: $A = 1{,}000 \times 2.208040$

$= \$2{,}208.04$ ■

Table 5 is not extensive enough to find monthly or daily compounding of \$1,000 at an 8% annual interest; these problems require a calculator. Also, keep in mind the difference between interest and future value for both the simple and compound interest formulas. With simplex interest, you find the interest first and then the future value. For compound interest, you first find the future value and then find the interest. Table 5.1 summarizes this comparison.

TABLE 5.1
Comparison of simple and compound interest

	Interest, I	**Future value, A**
Simple interest formula	$I = Prt$ This is found first for simple interest.	$A = P + I$ This is found after using the simple interest formula.
Compound interest	$I = A - P$ This is found after using Table 5.	$A = P(1 + i)^N$ or $A = P$ (Table 5 number) This is found first for compound interest.

Effective Yield

You have no doubt seen advertisements that say

8.00% = 8.33%*

* Effective annual yield when principal and interest are left in the account.

In order to understand the phrase **effective annual yield**, look at the earnings of \$1 at 8% compounded annually and quarterly:

Annually — $1(1 + .08) = 1.08$

Quarterly — $1\left(1 + \frac{.08}{4}\right)^4 = (1.02)^4 = 1.08243216$

The \$1 compounded quarterly at 8% is the same as \$1 compounded annually at a rate of 8.243216%. Therefore 8.24% is called the *effective annual yield* of 8% compounded quarterly. That is, the effective yield Y is the rate compounded annually that equals the same amount compounded n times per year at rate r. That is,

$$\underbrace{\left(1 + \frac{r}{n}\right)^{nt}}_{\text{compounded } n \text{ times per year}} = \underbrace{\left(1 + \frac{Y}{1}\right)^{t}}_{n = 1}$$

For $t = 1$ we have

$$\left(1 + \frac{r}{t}\right)^n = 1 + Y$$

$$Y = \left(1 + \frac{r}{t}\right)^n - 1$$

Effective Annual Yield

The **effective annual yield**, Y, or **effective rate**, for an account paying $r\%$ compounded n times per year is the simple annual interest rate that would pay an equivalent amount. It is found by the following formula:

$$Y = \left(1 + \frac{r}{n}\right)^n - 1$$

EXAMPLE 6 Verify the claims of the bank advertisement shown. (You need a calculator with an exponent key for this example.)

Solution The compounding is on a daily basis ($n = 360$), so:

$$\left(1 + \frac{.08}{360}\right)^{360} \approx 1.08327744$$

$$\text{Effective rate} \approx 1.08327744 - 1$$

$$\approx .0833 \quad \text{or} \quad 8.33\%$$

■

Present Value

Businesses sometimes need to have a given amount of money on hand at some time in the future. That is, suppose a business has a \$1,000 note payable and due in 5 years. The business would like to know the amount of money that must be deposited *today* so that in 5 years the total amount of principal *and* interest will be \$1,000. This sum to be deposited is called the **present value** and is the same variable P that appeared in the future value formulas. That is, for present value, the interest formulas are solved for P since the variable A is known.

Present Value

SIMPLE INTEREST: $P = A - I$

COMPOUND INTEREST: By formula, $P = A(1 + i)^{-N}$

By table, $P = A \div$ (Table 5, Appendix E, entry)

(continued)

(continued)

The variables in this formula have the same meaning they had in the compound interest formula:

P = Present value (principal)
A = Future value
r = Annual interest rate
t = Number of years
n = Number of times compounded per year

$i = \frac{r}{n}$ (rate per period)
$N = nt$ (number of periods)

It is easy to see where the simple interest formula comes from (subtract I from both sides of the future value formula). The compound interest formula is also easy to derive when you remember the meaning of a negative exponent:

$$A = P(1 + i)^N \quad \text{Future value formula}$$

$$P = \frac{A}{(1 + i)^N} \quad \text{Divide both sides by } (1 + i)^N.$$

$$= A(1 + i)^{-N} \quad (1 + i)^{-N} = \frac{1}{(1 + i)^N}$$

EXAMPLE 7 Don's Shoe Repair needs \$1,000 for a note payable in 5 years, and Don wants to make a deposit now to pay for the note. What is the present value if the money is deposited in a savings account paying 8% compounded quarterly?

Solution $i = \frac{r}{n} = \frac{.08}{4} = .02$

$N = nt = 4(5) = 20$

By calculator

$A = 1{,}000(1.02)^{-20}$

[1.02] [y^x] [20] [+/−] [×] [1000] [=]

The display shows: 672.9713331
Don should deposit \$672.97.

By Table 5, Appendix E

Find the column headed $i = 2\%$ and the row labeled $N = 20$ to find the entry 1.485947.

$A = 1{,}000 \div 1.485947$
$\approx \$672.97$ ■

EXAMPLE 8 Tom Mikalson has built a good reputation for his sporting goods store, and the business is growing at an annual rate of 21%. Tom plans on selling out and retiring in 5 years. Yesterday a developer offered Tom \$240,000 for the business, and Tom is trying to decide whether he should sell now.

The business was recently appraised for \$198,000, so Tom uses this as the present value of the business.

If he sells, he will invest the money with his stockbroker in an account paying 12% compounded quarterly.

He will sell if the price today is better than the one he can expect in 5 years. What should Tom do?

Solution First, find the expected value of the business in 5 years, then find the present value of that amount.

Future value

$$A = P(1 + i)^N \qquad n = 1 \text{ (compounded annually)}$$
$$= 198{,}000(1 + .21)^5 \qquad r = \text{Growth rate } (21\%)$$
$$= 513{,}561.0071 \qquad N = nt = 1(5) = 5$$
$$i = r/n = \tfrac{.21}{1} = .21$$

Present value

$$P = A(1 + i)^{-N} \qquad n = 4 \text{ (compounded quarterly)}$$
$$= 513{,}561.0071(1 + .03)^{-20} \qquad r = \text{Interest rate } (12\%)$$
$$= 284{,}346.2779 \qquad N = nt = 4(5) = 20$$
$$i = \tfrac{.12}{4} = .03$$

This means that Tom should not accept less than $284,346.28 for his business (even though it is presently appraised at $198,000). ■

Inflation

Example 8 involves a very important application of present and future value—inflation. **Inflation** indicates an increase in the amount of currency in circulation, which leads to a fall in the currency's value and a consequent rise in prices. Inflation is usually specified as a percent, and for our purposes we will use the future and present value formulas with $n = 1$ so that $N =$ the number of years and $i =$ the annual inflation rate.

EXAMPLE 9 The inflation rate in 1990 was about 6%. A person earning a $30,000 salary wants to know what salary to expect in 2000 (to the nearest thousand dollars) if this rate of inflation continues and if her salary keeps pace with inflation.

Solution This is a future value problem; use Table 5 in Appendix E or a calculator.

$$A = 30{,}000(1 + .06)^{10} \qquad \text{Future value formula}$$
$$\approx 53{,}725.43 \qquad P = 30{,}000$$
$$i = .02 \text{ (inflation rate)}$$
$$N = 6 \text{ (number of years)}$$

By table, $A = 30{,}000(1.79084)$ In Table 5, $N = 10$, $i = 6\% = 1.79084$

$\approx 53{,}725.44$

The expected wage will be about $54,000.* ■

EXAMPLE 10 An insurance agent wants to sell you a policy that will pay you $20,000 in 30 years. If you assume an average rate of inflation of 9% over the next 30 years, what is the value of that insurance payment in terms of today's dollars?

* When working with inflation you are guessing the future inflation rate over several years. Since this rate is an estimate, do not try to force exact answers when working with inflation problems. For this reason, we will usually specify the accuracy for your answers.

Solution Use a calculator (or Table 5) to find the present value ($N = 30$, $i = 9\%$):

$$P = 20{,}000(1 + .09)^{-30} \approx 1{,}507.42$$

The present value of the \$20,000 is about \$1,500. ■

The effects of compound interest can sometimes be rather dramatic. It is not uncommon to find some investments that pay 16% (although such investments are much riskier than insured savings accounts).

EXAMPLE 11 Your first child has just been born. You want to give her a million dollars when she retires at age 65. If you invest your money at 16% compounded quarterly, how much do you need to invest today so your child will have a million dollars in 65 years?

Solution Find the present value of 1,000,000 when $n = 4$, $r = 16\%$, $t = 65$, $i = \frac{.16}{4} = .04$, and $N = 4(65) = 260$. Use the present value formula:

$$P = 1{,}000{,}000(1 + .04)^{-260} \approx 37.26763062$$

This means that if you invest \$37.27 at 16% compounded quarterly in your child's name, she will have \$1,000,000 at age 65. ■

5.2 Problem Set

APPLICATIONS

In Problems 1–6 compare the amount of simple interest and the interest if the investment is compounded annually.

1. \$1,000 at 8% for 5 years
2. \$5,000 at 10% for 3 years
3. \$2,000 at 12% for 3 years
4. \$2,000 at 12% for 5 years
5. \$5,000 at 12% for 20 years
6. \$1,000 at 14% for 30 years

In Problems 7–12 compare the future amount you would have if the money were invested at simple interest or invested with interest compounded annually.

7. \$1,000 at 8% for 5 years
8. \$5,000 at 10% for 3 years
9. \$2,000 at 12% for 3 years
10. \$2,000 at 12% for 5 years
11. \$5,000 at 12% for 20 years
12. \$1,000 at 14% for 30 years

Find i, N, A, and I for the information given in Problems 13–26.

	Compounding period, n	Principal, P	Yearly rate, r	Time, t
13.	Annually, 1	\$1,000	9%	5 years
14.	Semiannually, 2	\$1,000	9%	5 years
15.	Annually, 1	\$500	8%	3 years
16.	Semiannually, 2	\$500	8%	3 years
17.	Quarterly, 4	\$500	8%	3 years
18.	Semiannually, 2	\$3,000	18%	3 years
19.	Quarterly, 4	\$5,000	18%	10 years
20.	Quarterly, 4	\$624	16%	5 years
21.	Quarterly, 4	\$5,000	20%	10 years
22.	Monthly, 12	\$350	12%	5 years
23.	Monthly, 12	\$4,000	24%	5 years
24.	Quarterly, 4	\$800	12%	90 days
25.	Quarterly, 4	\$1,250	16%	450 days
26.	Quarterly, 4	\$1,000	12%	900 days

Answer the questions in Problems 27–35. A calculator with an exponent key is required for these exercises.

27. What is the future amount of $12,000 invested for 5 years at 14% compounded monthly?

28. What is the future amount of $800 invested for 1 year at 10% compounded daily?

29. What is the future amount of $9,000 invested for 4 years at 20% compounded monthly?

30. If $5,000 is compounded annually at $5\frac{1}{2}$% for 12 years, what is the total interest received at the end of that time?

31. If $10,000 is compounded annually at 8% for 18 years, what is the total interest received at the end of that time?

32. What is the effective yield of 6% compounded quarterly?

33. What is the effective yield of 6% compounded monthly?

34. What is the effective yield of 8% compounded monthly?

35. What is the effective yield of 12% compounded daily?

Find the present value in Problems 36–40.

	Amount needed	Time	Interest	Compounded
36.	$7,000	5 years	8%	Annually
37.	$25,000	5 years	8%	Semiannually
38.	$165,000	10 years	12%	Semiannually
39.	$500,000	10 years	12%	Quarterly
40.	$3,000,000	20 years	12%	Semiannually

In Problems 41–52 calculate the expected price in the year 2012 if you assume a 10% inflation rate and use the given 1992 price. Answers should be rounded to the nearest dollar.

41. Cup of coffee, $.75

42. Small car, $6,000

43. Big Mac, $1.85

44. Movie admission, $5.00

45. Monthly rent, $400

46. Sunday paper, $1.00

47. Textbook, $38.00

48. Tuition, $14,000

49. Yearly salary, $25,000

50. Condominium, $85,000

51. House, $175,000

52. Business, $675,000

53. You owe $5,000 due in 3 years, but you would like to pay the debt today. If the present interest rate is compounded annually at 11%, how much should you pay today so that the present value is equivalent to the $5,000 payment in 3 years?

54. Ricon Bowling Alley will need $20,000 in 5 years to resurface the lanes. What deposit should be made today in an account that pays 9% compounded semiannually?

55. An accounting firm agrees to purchase a computer for $150,000. It will be delivered in 270 days. How much do the owners need to deposit in an account paying 18% compounded quarterly?

56. The Fair View Market must be remodeled in 3 years. Remodeling will cost $200,000. How much should be deposited now (to the nearest dollar) in order to pay for this remodeling if the account pays 10% compounded monthly?

57. A laundromat will need seven new washing machines in $2\frac{1}{2}$ years for a total cost of $2,900. How much money (to the nearest dollar) should be deposited at the present time to pay for these machines? The interest rate is 11% compounded semiannually.

58. A computerized checkout system is planned for Able's Grocery Store. The system will be delivered in 18 months at a cost of $560,000. How much should be deposited today into an account paying 7.5% compounded daily?

59. A contest offers the winner $50,000 now or $10,000 now and $45,000 in 1 year. Which is the best choice if the current interest rate is 10% compounded monthly and the winner does not intend to use any of the money for 1 year?

60. In 1990 the national debt was $2 trillion. If the annual interest rate is 12%, what is the interest on the national debt *every second*? (Use a 365-day year.)

5.3 Annuities

In Section 5.2 we discussed the present and future value of a lump-sum deposit bearing compound interest. However, it is far more typical to make monthly or other periodic payments into an account. For example, if you put $2,200 in the bank each year for 7 years at age 18, and then *make no further deposits*, you will have over $1 million at age 65 if the annual interest rate is 10%. On the other hand, if you wait until you are age 30 and *then* make *annual* deposits *each year* until you are 65 you will have less than $656,000. It is easy to see the power of the compounding of interest over long periods of time. In this section, we will develop formulas that verify these amounts.

Suppose you decide to give up smoking and save the \$2 per day you spend on cigarettes. How much will you save in 5 years? If you save the money without earning any interest, you will have

$$\$2 \times 365 \times 5 = \$3{,}650$$

However, let us assume that you save \$2 per day and at the end of each month you deposit the \$60 (assume all months are 30 days) into a savings account earning 12% interest compounded monthly. Now, how much will you have in 5 years?

TABLE 5.2

Time	Amount saved	Interest earned	Total amount
Start	\$0	\$0	\$0
1 month	\$60	\$0	\$60
2 months	\$120	\$.60	\$120.60
3 months	\$180	\$1.21	\$181.81
4 months	\$240	\$1.82	\$243.63
	⋮		

If we continue to calculate the entries in Table 5.2 month by month, we will tire long before we reach 5 years (that would be 60 entries to calculate). Instead, let us shorten the problem to 6 months and try to notice a pattern so we can use a formula. At the same time, we will use the variables of the last section to mean the same thing: P = present value; A = future value; r = annual rate; t = number of years; n = number of times compounded per year; $i = \frac{r}{n}$; and $N = nt$. We will also introduce a new variable, m, as the periodic payment (monthly payment in this example).

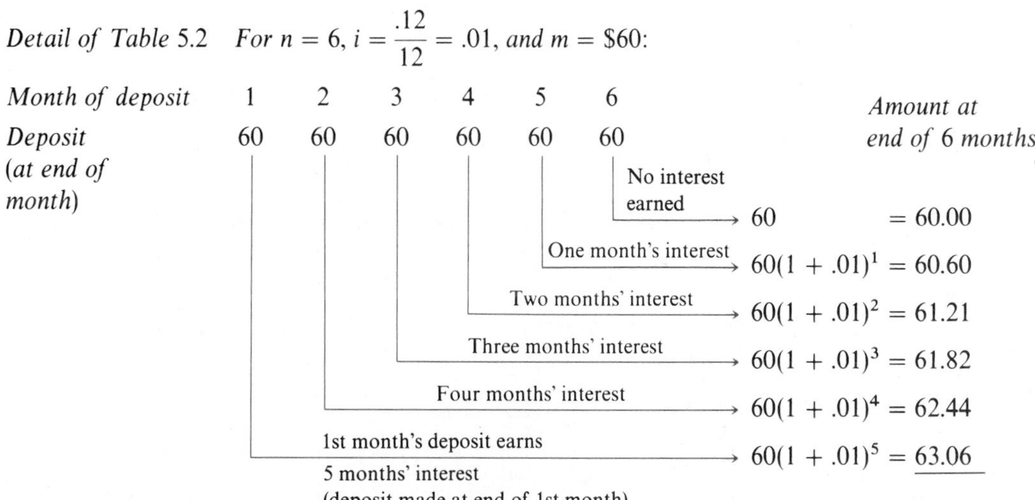

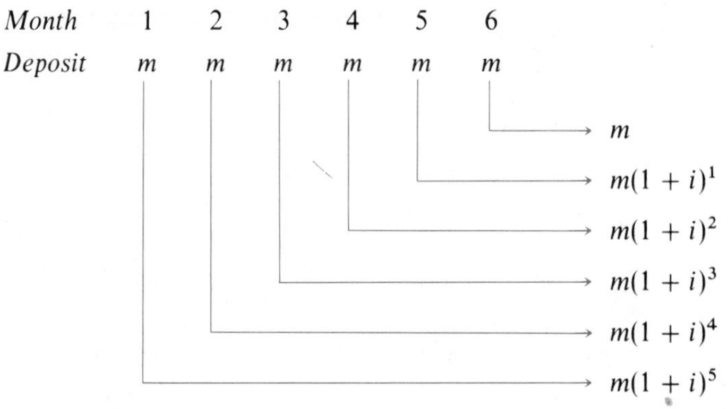

Month	1	2	3	4	5	6
Deposit	m	m	m	m	m	m

Month 6 deposit → m
Month 5 deposit → $m(1+i)^1$
Month 4 deposit → $m(1+i)^2$
Month 3 deposit → $m(1+i)^3$
Month 2 deposit → $m(1+i)^4$
Month 1 deposit → $m(1+i)^5$

Total after 6 months: $A = m + m(1+i)^1 + m(1+i)^2 + m(1+i)^3 + m(1+i)^4 + m(1+i)^5$

Ordinary Annuity

A sequence of payments into an interest-bearing account is called an **annuity**. If the payments are made at the end of the time period, and if the frequency of payments is the same as the frequency of compounding, the annuity is called an **ordinary annuity**. The amount of an annuity is the sum of all payments made plus all accumulated interest.

To find a general equation for the amount of an annuity, let

A_N = The amount of money in the annuity after N periods
i = Interest rate per period of time (i.e., $i = r/n$)
m = Monthly deposit

Also assume that $A_0 = 0$ (that is, there is nothing in the account when the account is opened). Now,

CURRENT AMOUNT = PREVIOUS AMOUNT + INTEREST + DEPOSIT

$$A_N \quad = \quad A_{N-1} \quad + \quad iA_{N-1} \quad + \quad m$$

$$A_N = (1+i)A_{N-1} + m$$

The solution of this difference equation is found by noticing that $a = 1 + i$, $b = m$, and $A_0 = 0$, so

$$\begin{aligned} A_N &= \frac{m}{1-(1+i)} + \left[0 - \frac{m}{1-(1+i)}\right](1+i)^N \\ &= \frac{m}{-i} + \frac{m}{i}(1+i)^N \\ &= m\left[\frac{-1}{i} + \frac{(1+i)^N}{i}\right] \\ &= m\left[\frac{(1+i)^N - 1}{i}\right] \end{aligned}$$

Ordinary Annuity

By formula	*By table*
Let $i = \frac{r}{n}$ and $N = nt$ with monthly payment m; then	Use Table 6, Appendix E:
	Find this table entry as usual using i and N
$$A = m\left[\frac{(1+i)^N - 1}{i}\right]$$	$A = m(\text{Table 6 entry})$

EXAMPLE 1 How much do you save in 5 years if you deposit \$60 at the end of each month into an account paying 12% compounded monthly?

Solution The rate i is $\frac{.12}{12} = .01$ and $N = 5(12) = 60$. From Table 6 in Appendix E,

$$A = 60(81.669670) = \$4{,}900.18$$

■

EXAMPLE 2 Repeat Example 1 for 2 years.

Solution This time $N = 2(12) = 24$. Table 6 does not have an entry for $N = 24$, so you must use the formula:

$$i = \frac{.12}{12} = .01 \qquad A = 60\left[\frac{(1+.01)^{24} - 1}{.01}\right]$$

PRESS: [1.01] [y^x] [24] [−] [1] [=] [÷] [.01] [=] [×] [60] [=]

DISPLAY: 1618.4079

You would have \$1,618.41. ■

Annuity Due

The annuities described thus far, in which the deposit is made at the end of each period, are ordinary annuities. Another type of annuity is one in which the payments are made at the *beginning* of each period. These are called **annuities due**. To derive a formula for an annuity due, look at the detail for Table 5.2 and note that if the periodic deposit is put into the account at the beginning of the period instead of at the end, the only difference will be that each exponent is increased by 1 (for one more period). This leads to the formula shown below:

Annuity Due

By formula	*By table*
Let $i = \frac{r}{n}$ and $N = nt$ with monthly payment m; then	Use Table 7, Appendix E:
$$A = m\left[\frac{(1+i)^{N+1} - 1}{i} - 1\right]$$	$A = m(\text{Table 7 entry})$

EXAMPLE 3 What is the value of an annuity due for which \$100 per month is deposited into an account paying 18% compounded monthly for $7\frac{1}{2}$ years?

Solution Here, $i = \frac{18}{12} = .015$ and $N = 7\frac{1}{2}(12) = 90$. Use Table 7 in Appendix E to find: 190.748849. Thus

$$A = \$100(190.748849)$$
$$= \$19{,}074.88$$

■

EXAMPLE 4 Compare the amounts you will have at age 65 if you are presently 18 years old and deposit \$2,200 per year for 7 years at 10% annual interest with what you would have if you wait until you are 30 years old and then make annual \$2,200 deposits until age 65.

Solution First, \$2,200 per year for 7 years is an annuity due:

$$A = \$2{,}200\left[\frac{(1+.1)^{7+1}-1}{.1} - 1\right] = \$22{,}958.95$$

Now, find the future value of this amount from age 25 to 65 (40 years):

$$A = \$22{,}958.95(1 + .1)^{40} \approx \$1{,}039{,}105$$

On the other hand, if you wait until you are 30 years old, then it is an annuity due for 35 years (age 30 to 65):

$$A = \$2{,}200\left[\frac{(1+.1)^{35+1}-1}{.1} - 1\right] = \$655{,}878.97$$

In the first case, you are depositing 7(\$2,200) = \$15,400 to have \$1,039,105 at age 65; in the second, you are depositing 35(\$2,200) = \$77,000 to have \$655,879 at age 65. You can see the power of compounding over long periods of time. ■

5.3 Problem Set

Find the value of each annuity in Problems 1–24 at the end of the indicated number of years. Assume that the interest is compounded with the same frequency as the deposits. Give your anwers to the nearest dollar.

	Amount of deposit	Frequency compounded	Rate	Number of years	Type of annuity
1.	\$500	Annually	8%	30	Ordinary
2.	\$500	Annually	6%	30	Ordinary
3.	\$250	Semiannually	8%	30	Ordinary
4.	\$600	Semiannually	12%	10	Ordinary
5.	\$300	Quarterly	12%	10	Ordinary
6.	\$100	Monthly	12%	5	Ordinary
7.	\$500	Annually	8%	30	Due
8.	\$500	Annually	6%	30	Due

	Amount of deposit	Frequency compounded	Rate	Number of years	Type of annuity
9.	\$250	Semiannually	8%	30	Due
10.	\$600	Semiannually	12%	10	Due
11.	\$300	Quarterly	12%	10	Due
12.	\$100	Monthly	12%	5	Due
13.	\$50	Monthly	18%	5	Due
14.	\$200	Quarterly	18%	20	Ordinary
15.	\$400	Quarterly	16%	20	Ordinary
16.	\$2,500	Semiannually	18%	30	Due
17.	\$30	Monthly	18%	5	Due
18.	\$30	Monthly	18%	5	Ordinary
19.	\$5,000	Annually	8%	10	Ordinary

	Amount of deposit	Frequency compounded	Rate	Number of years	Type of annuity
20.	\$5,000	Annually	8%	10	Due
21.	\$100	Monthly	15%	7	Due
22.	\$2,500	Semiannually	8.5%	20	Ordinary
23.	\$1,250	Quarterly	10%	20	Ordinary
24.	\$125	Monthly	10%	30	Due

APPLICATIONS

25. Self-employed persons can make contributions for their retirement into a special tax-deferred account called a **Keogh account**. Suppose you are able to contribute \$20,000 into this account each year. How much will you have at the end of 20 years if you make your first deposit today and are paid 8% annual interest?

26. The owner of Sebastopol Tree Farm deposits \$650 at the beginning of each quarter for 5 years into an account paying 8% compounded quarterly. What is the value of the account at the end of 5 years?

27. The owner of Oak Hill Squirrel Farm deposits \$1,000 at the end of each quarter for 5 years into an account paying 8% compounded quarterly. What is the value at the end of 5 years?

28. John and Rosamond want to retire in 5 years and can save \$150 every three months. They plan to deposit the money at the beginning of each quarter into an account paying 8% compounded quarterly. How much will they have at the end of 5 years?

29. You want to retire at age 65. You decide to make a deposit to yourself at the beginning of each year into an account paying 13%, compounded annually. Assuming you are now 25 and can spare \$1,200 per year, how much will you have (to the nearest dollar) when you retire?

30. In 1988 the maximum Social Security deposit was \$3,818.10. Suppose you are 25 and make a deposit of this amount into an account at the end of each year. How much would you have (to the nearest dollar) when you retire if the account pays 8% compounded annually and you retire at age 65?

31. Repeat Problem 29 using your own age.

32. Repeat Problem 30 using your own age.

5.4 Amortization

Present Value of an Annuity

Suppose that instead of making monthly payments at periodic intervals, we want to deposit a lump sum today that will have the same value as an annuity at the end of some time period.

EXAMPLE 1 Chen and Mai are partners who have decided to set aside a fund for their secretary who will retire in 10 years. Chen wants to make a \$50 monthly deposit into this account while Mai wants to make a lump-sum deposit today, but both want to deposit equal amounts. How much will be in the account in 10 years if the interest rate is 9% compounded monthly, and how much should Mai deposit in order to equal Chen's contribution?

Solution Chen's deposits are an ordinary annuity with $m = \$50$, $r = 9\%$, $t = 10$, $i = \frac{.09}{12}$, and $N = 10(12) = 120$:

$$A = m\left[\frac{(1+i)^N - 1}{i}\right] = \$50\left[\frac{(1+\frac{.09}{12})^{120} - 1}{\frac{.09}{12}}\right] = \$9{,}675.71$$

Mai's contribution is a present value problem. When we speak of the **present value of an annuity**, we mean the lump-sum deposit today that will equal the future value of a given annuity. Using the present value formula from Section 5.2, we find

$$P = A(1+i)^{-N} = 9{,}675.71\left(1 + \frac{.09}{12}\right)^{-120} = \$3{,}947.08$$

For Chen and Mai to contribute equal amounts, Chen will deposit \$50 per month for 10 years and Mai will make a lump-sum contribution of \$3,947.08. The total value in the account when the secretary retires will be

From Chen's annuity (includes principal and interest) ↓ — From Mai's contribution: Amount + Interest ↓ ↓

$$\$9{,}675.71 + (\$3{,}947.08 + \$5{,}728.63) = \underbrace{\$19{,}351.42}_{\text{Total}}$$

■

The goal of this section is to develop a formula that will allow us to calculate a lump-sum deposit (see Example 1) *without first finding the value of an annuity*. Recall that $\$P$ deposited today will amount to $A = P(1 + i)^N$ (as before, $i = \frac{r}{n}$ and $N = nt$). Substitute this into the ordinary annuity formula with monthly payment m to obtain

$$P(1 + i)^N = m\left[\frac{(1 + i)^N - 1}{i}\right]$$

Solve this formula for P and simplify to obtain a formula for the present value of an annuity:

$$P = \frac{m}{i}\left[\frac{(1 + i)^N - 1}{(1 + i)^N}\right] = \frac{m}{i}\left[1 - \frac{1}{(1 + i)^N}\right] = \frac{m}{i}[1 - (1 + i)^{-N}]$$

Present Value of an Annuity

By formula	*By table*
$P = m\left[\frac{1 - (1 + i)^{-N}}{i}\right]$	Use Table 8, Appendix E: $P = m(\text{Table 8 entry})$

EXAMPLE 2 Use the formula for the present value of an annuity to calculate Mai's deposit as described in Example 1. That is, find the present value of an annuity where $m = \$50$, $r = 9\%$, and $t = 10$.

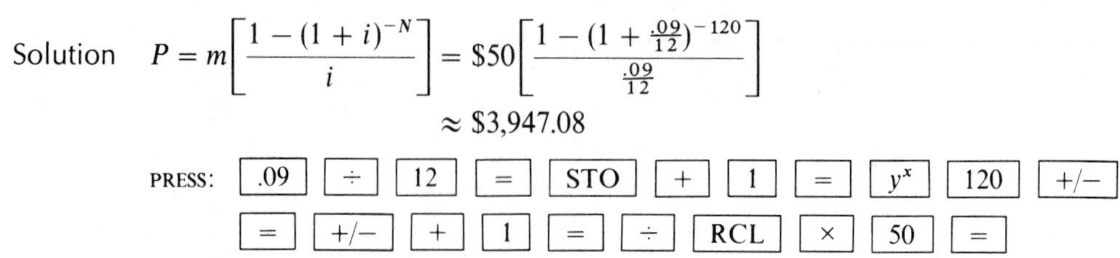

Solution
$$P = m\left[\frac{1 - (1 + i)^{-N}}{i}\right] = \$50\left[\frac{1 - (1 + \frac{.09}{12})^{-120}}{\frac{.09}{12}}\right]$$
$$\approx \$3{,}947.08$$

PRESS: [.09] [÷] [12] [=] [STO] [+] [1] [=] [y^x] [120] [+/−] [=] [+/−] [+] [1] [=] [÷] [RCL] [×] [50] [=] ■

The calculator steps are rather complicated, so you might want to use Table 8, Appendix E. This procedure is illustrated in Example 3.

EXAMPLE 3 What is the present value of an annuity of \$30 per month for 5 years deposited into an account paying 12% compounded monthly?

Solution We have $i = \frac{.12}{12} = .01$ and $N = 5(12) = 60$. From Table 8 in Appendix E,

$$\begin{aligned} P &= 30(44.955038) \\ &= \$1{,}348.65114 \end{aligned}$$

It would require a deposit of \$1,348.65. ■

EXAMPLE 4 Theresa's Social Security benefit is \$450 per month if she retires at age 62 instead of age 65. What is the present value of an annuity that would pay \$450 per month for 3 years if the interest rate is 10% compounded monthly?

Solution Here, $i = \frac{.10}{12} = .008\overline{3}$ and $N = 12(3) = 36$. These values are not included in Table 8, so you must use the formula:

$$P = 450\left[\frac{1 - (1 + i)^{-36}}{i}\right]$$

To use this formula you need a calculator. First, store i in memory:

[.10] [÷] [12] [=] [STO]

[+] [1] [=] — This is $(1 + i)$.

[y^x] [36] [+/−] [=] — This is $(1 + i)^{-36}$.

[+/−] [+] [1] [=] — This gives the numerator.

(↑ [+/−]: This key changes sign; it allows you to subtract without reentering the values.)

[÷] [RCL] [×] [450] [=] — This divides by i and multiplies by 450 for the final result.

DISPLAY: 13946.056

The present value is \$13,946.06. ■

The most common use for the present value of an annuity is to tell you the amount you can borrow with a given monthly payment on an installment loan.

EXAMPLE 5 You look at your budget and decide that you can afford \$250 per month for a car. What is the maximum loan you can afford if the interest rate is 13% and you want to repay the loan in four years?

Solution This is the present value of an annuity: $i = \frac{.13}{12}$, $N = 12(4) = 48$, $m = 250$ and we want to find P, the amount you can borrow. We see that these numbers are not to be found in Table 8, so we must use the formula:

$$P = m\left[\frac{1 - (1 + i)^{-N}}{i}\right] = 250\left[\frac{1 - (1 + \frac{.13}{12})^{-48}}{\frac{.13}{12}}\right]$$

This is a substantial calculator problem, similar in type to that described in Example 4:

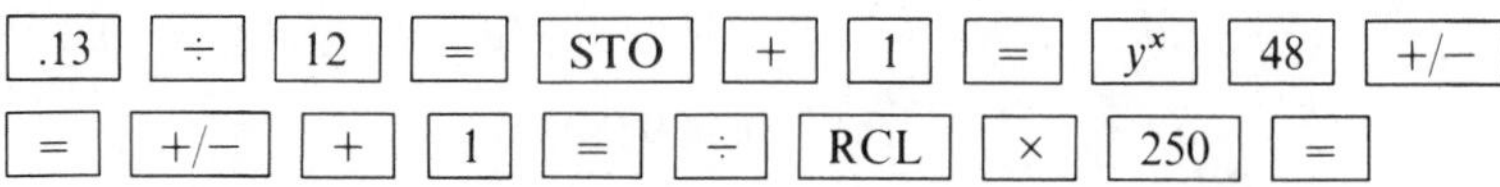

DISPLAY: 9318.7974

This says that you can afford to finance about \$9,300. If you add to this amount the down payment that you can afford (perhaps your trade-in) you will have the amount you can spend for your car. ■

Amortization

Installment loans are one of the most common examples of the present value of an annuity. The process of paying off a debt by systematically making partial payments until the debt (principal) and the interest are repaid is called **amortization**. If the loan is paid off in regular equal installments, then we can use the formula for the present value of an annuity to find the monthly payments by algebraically solving for m:

$$P = m\left[\frac{1-(1+i)^{-N}}{i}\right]$$

Multiply both sides by i:

$$Pi = m[1-(1+i)^{-N}]$$

Divide by $[1-(1-i)^{-N}]$ to solve for m:

$$\frac{Pi}{1-(1+i)^{-N}} = m$$

Installment Payments

The monthly payment, m, for a fully amortized installment loan of P dollars with annual percentage rate r is found as follows:

By formula	*By table*
	Use the present value of an annuity table, Table 8 in Appendix E:
$m = \dfrac{Pi}{1-(1+i)^{-N}}$	$m = P \div (\text{Table 8 entry})$
where $i = \dfrac{r}{12}$.	

EXAMPLE 6 In 1990 the average price of a new home was \$162,000 and the interest rate was 11%. If this amount is financed for 30 years at 11% interest, what is the monthly payment?

Solution We have $i = \frac{.11}{12}$, $N = 12(30) = 360$, and $P = 162{,}000$. Thus

$$m = \frac{162{,}000i}{1 - (1+i)^{-360}}$$

PRESS: [.11] [÷] [12] [=] [STO] — Store i for future use.

[+] [1] [=] — Continue to find $(1 + i)$.

[y^x] [360] [+/−] [=] — This gives $(1 + i)^{-360}$.

[+/−] [+] [1] [=] — This gives the denominator.

(+/−: Subtracts without reentering the number)

[1/x] [×] [162000] [×] [RCL] [=]

(1/x: By inverting the denominator and then multiplying by the numerator, you do not need to reenter the number.)

DISPLAY: 1542.7639

The monthly payment is $1,542.76. This is the payment for interest and principal to pay off a 30-year 11% loan of $162,000. ■

The schedule of payments on a loan showing how much of each payment goes for interest and how much goes to repay the principal is called an **amortization schedule**. For a home loan this schedule is rather long, so we show the first and last year of the loan in Example 6 in Table 5.3.

TABLE 5.3

Amortization Schedule

End of period	Payment	Interest	Principal	Outstanding balance	End of period	Payment	Interest	Principal	Outstanding balance
0				$162,000.00	348	$1,542.76	$172.66	$1,370.10	$17,465.46
1	$1,542.76	$1,485.00	$57.76	$161,942.24	349	$1,542.76	$160.10	$1,382.66	$16,082.80
2	$1,542.76	$1,484.47	$58.29	$161,883.95	350	$1,542.76	$147.43	$1,395.33	$14,687.47
3	$1,542.76	$1,483.94	$58.82	$161,825.13	351	$1,542.76	$134.64	$1,408.12	$13,279.34
4	$1,542.76	$1,483.40	$59.36	$161,765.76	352	$1,542.76	$121.73	$1,421.03	$11,858.31
5	$1,542.76	$1,482.85	$59.91	$161,705.86	353	$1,542.76	$108.70	$1,434.06	$10,424.25
6	$1,542.76	$1,482.30	$60.46	$161,645.40	354	$1,542.76	$95.56	$1,447.20	$8,977.05
7	$1,542.76	$1,481.75	$61.01	$161,584.39	355	$1,542.76	$82.29	$1,460.47	$7,516.58
8	$1,542.76	$1,481.19	$61.57	$161,522.82	356	$1,542.76	$68.90	$1,473.86	$6,042.72
9	$1,542.76	$1,480.63	$62.13	$161,460.69	357	$1,542.76	$55.39	$1,487.37	$4,555.35
10	$1,542.76	$1,480.06	$62.70	$161,397.98	358	$1,542.76	$41.76	$1,501.00	$3,054.35
11	$1,542.76	$1,479.48	$63.28	$161,334.70	359	$1,542.76	$28.00	$1,514.76	$1,539.59
12	$1,542.76	$1,478.90	$63.86	$161,270.85	360	$1,553.70	$14.11	$1,539.59	$0.00

It is interesting to see how much interest is paid for the home loan of Example 6. There are 360 payments of $1,542.76, so the total repaid is

$$360(1{,}542.76) = 555{,}393.60$$

Since the loan was for $162,000, the interest is

$$\$555{,}393.60 - \$162{,}000 = \$393{,}393.60$$

The amount of interest paid can be reduced by making a large down payment. For Example 6, if a 20% down payment (which is fairly standard) was made, then $162,000(.80) = $129,600 is the amount to be financed. If this amount is amortized over 20 years (instead of 30), the monthly payments are $1,337.72 with total interest of $191,452.80. The total amount paid on this loan is

$$\underset{\text{Payments}}{\underset{\uparrow}{240(1{,}337.72)}} + \underset{+\text{ Down payment}}{\underset{\uparrow}{32{,}400}} = 353{,}452.80$$

The savings yielded over the term of the loan by making a 20% down payment is $201,940.80 ($555,393.60 – $353,452.80).

5.4 Problem Set

Find the present value of the ordinary annuities in Problems 1–12.

	Amount of deposit	Frequency compounded	Rate	Number of years
1.	$500	Annually	8%	30
2.	$500	Annually	6%	30
3.	$250	Semiannually	8%	30
4.	$600	Semiannually	12%	10
5.	$300	Quarterly	12%	10
6.	$100	Monthly	12%	5
7.	$200	Quarterly	18%	20
8.	$400	Quarterly	16%	20
9.	$30	Monthly	18%	5
10.	$75	Monthly	10%	10
11.	$50	Monthly	11%	30
12.	$100	Monthly	13%	40

APPLICATIONS

Find the monthly payment for the loans in Problems 13–24.

13. $500 loan for 12 months at 12%

14. $100 loan for 18 months at 18%

15. $4,560 loan for 20 months at 21%

16. $3,520 loan for 30 months at 19%

17. Used-car financing of $2,300 for 24 months at 15%

18. New car financing of 2.9% on a 30-month $12,450 loan

19. Furniture financed at $3,456 for 36 months at 23%

20. A refrigerator financed for $985 at 17% for 15 months

21. A $112,000 home bought with a 20% down payment and the balance financed for 30 years at 11.5%

22. A $108,000 home bought with a 30% down payment and the balance financed for 30 years at 12.05%

23. Finance $450,000 for a warehouse with a 12.5% 30-year loan

24. Finance $859,000 for an apartment complex with a 13.2% 20-year loan

25. How much interest would be saved in Problem 21 if the home were financed for 15 rather than for 30 years?

26. How much interest would be saved in Problem 22 if the home were financed for 15 rather than for 30 years?

27. Pat agrees to contribute $500 to the alumni fund at the end of each year for the next 5 years. Karl wants to match Pat's gift, but he wants to make a lump-sum contribution. If the current interest rate is 12.5% compounded annually, how much should Karl contribute to equal Pat's gift?

28. A $1,000,000 lottery prize pays $50,000 per year for the next 20 years. If the current rate of return is 12.25%, what is the present value of this prize?

29. Suppose you have an annuity from an insurance policy and you have the option of being paid \$250 per month for 20 years or having a lump-sum payment of \$25,000. Which has more value if the current rate of return is 10% compounded monthly?

30. An insurance policy offers you the option of being paid \$750 per month for 20 years or a lump sum of \$50,000. Which has more value if the current rate of return is 9% compounded monthly and you expect to live for 20 years?

31. You look at your budget and decide that you can afford \$250 per month for a car. What is the maximum loan you can afford if the interest rate is 13% and you want to repay the loan in four years?

32. Suppose your gross monthly income is \$5,500 and your current monthly payments are \$625. If the bank will allow you to pay up to 36% of your gross monthly income (less current monthly payments) for house payments, what is the maximum loan you can obtain if the rate for a 30-year mortgage is 9.65%?

33. Suppose your gross monthly income is \$4,550 and your spouse's gross monthly salary is \$3,980. Your monthly bills are \$1,235. The home you wish to purchase costs \$355,000 and the loan is an 11.85% 30-year loan. How much down payment (rounded to the nearest hundred dollars) is necessary to be able to afford this home? This down payment is what percent of the cost of the home? Assume the bank will allow you to pay up to 36% of your gross monthly income (less current monthly payments) for house payments.

34. Suppose you want to purchase a home for \$225,000 with a 30-year mortgage at 10.24% interest. Suppose also that you can put down 25%. What are the monthly payments? What is the total amount of interest? What is the amount saved if this home is financed for 15 years instead of for 30 years?

35. The McBertys have \$30,000 in savings to use as a down payment on a new home. They also have determined that they can afford between \$1,500 and \$1,800 per month for mortgage payments. If the mortgage rates are 11% per year compounded monthly, what is the price range for houses they should consider for a 30-year loan?

36. Rework Problem 35 for a 20-year instead of a 30-year loan.

37. Rework Problem 35 if interest rates go up to 12.5%.

38. Rework Problem 35 if interest rates go down to 10.2%.

*5.5 Sinking Funds

The last financial application we consider is the situation in which we need to have a lump sum of money in a certain period of time. The present value formula will tell us how much we need to have today, but we frequently do not have that amount available. Suppose your goal is \$10,000 in 5 years. You can obtain 8% compounded quarterly, so the present value is \$6,729.71. However, this is more than you can afford to put into the bank now. The next choice is to make a series of small equal investments to accumulate at 8% compounded quarterly, so that the end result is the same, namely, \$10,000 in 5 years. The account you set up to receive those investments is called a **sinking fund**.

To find a formula for a sinking fund, we begin with the formula for an ordinary annuity and solve for m:

$$A = m\left[\frac{(1+i)^N - 1}{i}\right]$$

$$m[(1+i)^N - 1] = Ai$$

$$m = \frac{Ai}{(1+i)^N - 1}$$

* This section is not required for subsequent textual development.

Sinking Fund	By formula	By table
	$m = \dfrac{Ai}{(1+i)^N - 1}$	Use the ordinary annuity table, Table 6, Appendix E: $m = A \div (\text{Table 6 entry})$

EXAMPLE 1 A business needs to raise \$1,000,000 in 20 years by making equal quarterly payments into an account paying 12% interest compounded quarterly. What is the required amount of each deposit?

Solution Here, $i = \frac{.12}{4} = .03$ and $N = 4(20) = 80$. From Table 6 in Appendix E, we obtain 321.363019. Thus

$$\begin{aligned} m &= 1{,}000{,}000 \div 321.363019 \\ &= \$3{,}111.745723 \end{aligned}$$

The business needs to make quarterly deposits of \$3,112 (rounded to the nearest dollar). ■

The primary reason that businesses set up a sinking fund is to pay off bonds. A **bond** is a certificate (a written promise) of a business to repay a certain amount at some future time. It is often how a business, corporation, or government agency borrows to buy new equipment or raise money for construction. Usually, each bond has a face value of \$1,000 and a specified interest rate, called the *contract rate*. This interest is paid to be bondholders at specified intervals (usually twice a year). The bonds are usually sold to an underwriter, who sells them to investors. Since the contract rate is fixed but prevailing interest rates are not, bonds are bought and sold for either more or less than the face value. This, in effect, changes the interest rate actually received (because of the changes in the principal). This interest rate is called the *market rate* or *effective rate*. Bond prices are quoted as a percent of their face amounts. A quote of 103 is 103% of the face amount, and a quote of $85\frac{1}{2}$ is $85\frac{1}{2}\%$ of the face amount.

EXAMPLE 2 The Packard-Hue Corporation issues \$50,000,000 worth of bonds for a new plant. These bonds are 20-year bonds with interest payable semiannually at a contract rate of 6%. In addition to paying interest on the bonds, the company sets up a sinking fund into which they will make semiannual payments and receive 8% interest compounded semiannually. What is the amount of each semiannual payment necessary to pay the interest on the bonds and for the sinking fund?

Solution *Bond interest:* This is simple interest because interest is paid to bondholders and does not accumulate.

$$\begin{aligned} I &= Prt \\ &= \$50{,}000{,}000(.06)\left(\frac{1}{2}\right) \quad \leftarrow \text{Semiannual payment} \\ &= \$1{,}500{,}000 \end{aligned}$$

Sinking fund payment: $i = \frac{.08}{2} = .04$ and $N = 2(20) = 40$; thus

$$m = \$50{,}000{,}000 \div 95.025516 \longleftarrow \text{Table 6 entry}$$
$$\approx \$526{,}174 \quad \text{Rounded to the nearest dollar}$$

The total semiannual cost is $1,500,000 + $526,174 = $2,026,174 ■

Summary

For many students, the most difficult part of working with business formulas is determining *which* formula or *which* table to use. To make the processes easier for you, consistent notation and consistently constructed tables are used here. Your main task, therefore, is one of classification, summarized in Table 5.4 on pages 270–271.

5.5 Problem Set

Find the amount of payment necessary for each deposit to a sinking fund in Problems 1–12. Assume that the deposit is made at the end of each compounding period.

	Amount needed	Time	Interest rate	Compounded
1.	$7,000	5 years	8%	Annually
2.	$25,000	5 years	11%	Annually
3.	$25,000	5 years	12%	Semiannually
4.	$50,000	10 years	14%	Semiannually
5.	$165,000	10 years	12%	Semiannually
6.	$3,000,000	20 years	12%	Semiannually
7.	$500,000	10 years	12%	Quarterly
8.	$55,000	5 years	16%	Quarterly
9.	$100,000	8 years	10%	Quarterly
10.	$35,000	12 years	18%	Quarterly
11.	$45,000	6 years	18%	Monthly
12.	$120,000	30 years	14%	Monthly

13. Clearlake Optical has a $50,000 note that comes due in 4 years. The owners wish to create a sinking fund to pay this note. If the fund earns 8% compounded semiannually, how much should each semiannual deposit be?

14. A business must raise $70,000 in 5 years. What should be the size of the owners' quarterly payment to a sinking fund paying 8% compounded quarterly?

15. Clearlake Optical has developed a new lens. The owners plan on issuing a $4,000,000 30-year bond with a contract rate of $5\frac{1}{2}$% paid annually to raise capital to market this new lens. To pay off the debt, they will also set up a sinking fund paying 8% interest compounded annually. What size annual payment is necessary for interest and sinking fund combined? (Answers to the nearest dollar.)

16. The owners of Bardoza Greeting Cards wish to introduce a new line of cards but need to raise $200,000 to do it. They decide to issue 10-year bonds with a contract rate of 6% paid semiannually. They also set up a sinking fund paying 8% interest compounded semiannually. How much money will they need to make the semiannual interest payments as well as payments to the sinking fund? (Answer to the nearest dollar.)

Problems 17–30 provide a mixture of financial problems. For each problem: ***a.*** *Classify the type.* ***b.*** *Answer the questions. (Round to the nearest dollar.)*

17. Rincon Bowling Alley will need $80,000 in 4 years to resurface the lanes. What lump sum would be necessary today if the owner of the business can deposit it in an account that pays 9% compounded semiannually?

18. Rita wants to save for a trip to Tahiti, so she puts $2.00 per day into a jar. After 1 year she has saved $730 and puts the money into a bank account paying 10% compounded annually. She continues to save in this manner and makes her annual $730 deposit for 15 years. How much does she have at the end of that time period?

19. Karen receives a $12,500 inheritance that she wants to save until she retires in 22 years. If she deposits the money in a fixed 11% account, compounded daily, how much will she have when she retires?

20. You want to give your child $1,000,000 when he retires at age 65. How much money do you need to deposit to an account paying 9% compounded monthly if your child is now 10 years old?

TABLE 5.4 Classification and Formulas for Financial Problems

DEFINITION OF VARIABLES:

P = Present value (sometimes called principal)
A = Future value
I = Amount of interest
r = Annual interest rate
t = Number of years
n = Periods; that is, the number of times per year that the interest is compounded
m = Periodic payment
i = Rate per period $= \frac{r}{n}$
N = Number of periods $= nt$

FINANCIAL TABLES (Appendix E): *For all tables use i and N to find entry.*

Future value (Table 5): Lump sum known, to find future lump sum
Present value (Table 5): Future lump sum known, to find present value
Ordinary annuity (Table 6): Periodic payments at the end of each period, to find the future value
Annuity due (Table 7): Periodic payments at the beginning of each period, to find the future value
Present value of an annuity (Table 8): Find the present lump sum necessary to deposit to equal the future value of periodic payments for a given period of time.
Installment payments (Table 8): Find the periodic payments to pay off both principal and interest in a given period of time.
Sinking fund (Table 6): Find the amount of periodic payments in order to have a given amount at some future date.

21. An accounting firm agrees to purchase a computer for \$150,000 (cash on delivery) and the delivery date is in 270 days. How much do the owners need to deposit in an account paying 18% compounded quarterly?

22. For 5 years, Thompson Cleaners deposit \$900 at the beginning of each quarter into an account paying 8% compounded quarterly. What is the value of the account to the nearest dollar at the end of 5 years?

23. In 1980 the inflation rate hit 16%. Suppose that the average cost of a textbook in 1980 was \$15. What is the expected cost in the year 2000 if we project this rate of inflation on the cost?

24. What is the necessary amount of monthly payments to an account paying 18% compounded monthly in order to have \$100,000 in $8\frac{1}{3}$ years if the deposits are made at the end of the month?

25. Thomas' Grocery Store is going to be remodeled in 5 years, and the remodeling will cost \$300,000. How much should be deposited now (to the nearest dollar) in order to pay for this remodeling if the account pays 12% compounded monthly?

26. If an apartment complex will need painting in $3\frac{1}{2}$ years and the job will cost \$45,000, what amount needs to be deposited into an account now in order to have the neces-

Classification of Types:

LUMP SUM

Present value, P	Future value, A	Classification	Formula	Table
Known	Unknown	Future value	$A = P(1+i)^N$	$A = P$(Table 5 entry)
Unknown	Known	Present value	$P = \dfrac{A}{(1+i)^N}$	$P = A \div$ (Table 5 entry)

PERIODIC PAYMENTS

Periodic payment, m	Present value, P	Future value, A	Classification	Formula	Table
Known		Unknown	Ordinary annuity (end of each period)	$A = m\left[\dfrac{(1+i)^N - 1}{i}\right]$	$A = m$(Table 6 entry)
Known		Unknown	Annuity due (start of each period)	$A = m\left[\dfrac{(1+i)^{N+1} - 1}{i} - 1\right]$	$A = m$(Table 7 entry)
Known	Unknown*		Present value of an annuity	$P = m\left[\dfrac{1-(1+i)^{-N}}{i}\right]$	$P = m$(Table 8 entry)
Unknown	Known		Installment payment	$m = \dfrac{Pi}{1-(1+i)^{-N}}$	$m = P \div$ (Table 8 entry)
Unknown		Known	Sinking fund	$m = \dfrac{Ai}{(1+i)^N - 1}$	$m = A \div$ (Table 6 entry)

* Amount needed to deposit today to equal the future value of an annuity; this is the amount you can borrow with a given monthly payment on an installment loan.

sary funds? The account pays 12% interest compounded semiannually.

27. Teal and Associates needs to borrow \$45,000. The best loan they can find is one at 12% that must be repaid in monthly installments over the next $3\frac{1}{2}$ years. How much are the monthly payments?

28. A city issues \$20 million in tax-exempt 25-year bonds with 8% interest payable quarterly. In addition to paying interest on these bonds, the city sets up an account into which quarterly payments are made and 12% interest compounded quarterly is received. How much needs to be paid into this account to pay off the \$20 million in 25 years?

29. Certain Concrete Company deposits \$4,000 at the end of each quarter into an account paying 10% interest compounded quarterly. What is the value of the account at the end of $7\frac{1}{2}$ years?

30. Major Magic Corporation deposits \$1,000 at the beginning of each month into an account paying 18% interest compounded monthly. What is the value of the account at the end of $8\frac{1}{3}$ years?

*5.6

Review

The material of this chapter is reviewed in the following list of objectives. After each objective there are some practice questions. For a sample test select the first question of each set and check your answers. The second question for each objective has no answer given. If you are having trouble with a particular type of problem, look back at the indicated section in the text. When you are finished reviewing these objectives, a sample examination is given at the end of this section.

[5.1]

Objective 5.1: *Find terms of a linear difference equation.*

1. Find the first four terms of the difference equation $x_{n+1} = 3x_n$ where $x_0 = 10$.
2. Find the first four terms of the difference equation $x_{n+1} = 5x_n - 500$; $x_0 = 1{,}000$.
3. Find x_8 for the difference equation $x_{n+1} = 450x_n + 200$; $x_7 = 10$.
4. Find x_{10} for the difference equation $x_{n+1} = .5x_n + 400$; $x_9 = 4{,}700$.

Objective 5.2: *Solve a difference equation.*

5. Solve $x_{n+1} = x_n + 250$; $x_0 = 50$
6. Solve $x_{n+1} = 4x_n$; $x_0 = 1$
7. Solve $x_n = x_{n-1} + 10$; $x_0 = 5$
8. Solve $x_{n+1} = .5x_n + 100$; $x_0 = 0$

[5.2]

Objective 5.3: *Find the amount of simple interest.*

9. \$400 invested at 13% for 3 years
10. \$725 invested at 9% for 5 years
11. \$1,000,000 invested at 12% for 3 years
12. \$6,000 invested at 25% for 2 years

Objective 5.4: *Find the amount of interest compounded annually.*

13. \$400 invested at 13% for 3 years
14. \$725 invested at 9% for 5 years
15. \$1,000,000 invested at 12% for 3 years
16. \$6,000 invested at 25% for 2 years

Objective 5.5: *Find the future value for both simple interest and interest compounded annually.*

17. \$400 invested at 13% for 3 years
18. \$725 invested at 9% for 5 years
19. \$1,000,000 invested at 12% for 3 years
20. \$6,000 invested at 25% for 2 years

* Optional section.

Objective 5.6: *Find the future value of an account compounded n times per year. You will need a calculator with an exponent key for these exercises.*

21. \$400 invested at 13% for 3 years compounded monthly
22. \$725 invested at 9% for 5 years compounded quarterly
23. \$1,000,000 invested at 12% for 3 years compounded daily (1 year = 365 days)
24. \$6,000 invested at 25% for 2 years compounded semiannually

Objective 5.7: *Find the future value of an account compounded n times per year using Table 5 in Appendix E.*

25. \$120 invested at 7% for 25 years compounded annually
26. \$3,200 invested at 14% for 6 years compounded semiannually
27. \$900 invested at 16% for 3 years compounded quarterly
28. \$1,250 invested at 12% for $7\frac{1}{2}$ years compounded monthly

Objective 5.8: *Find the effective yield.*

29. 10% compounded monthly
30. 12% compounded monthly
31. 10% compounded daily
32. 11% compounded daily

Objective 5.9: *Find the present value for a given amount.*

33. Need \$500 in 2 years; 8% compounded annually
34. Need \$12,000 in 5 years; 10% compounded semiannually
35. Need \$5,000 in 3 years; 12% compounded quarterly
36. Need \$6,000,000 in 20 years; 12% compounded monthly

[5.3]

Objective 5.10: *Find the value of an ordinary annuity.*

37. \$25 monthly payment for 10 years into an account paying 14% compounded monthly
38. \$50,000 annual payment for 25 years into an account paying 10% compounded annually
39. \$300 quarterly payment for 12 years into an account paying 9% compounded quarterly
40. \$4,000 semiannual payment for 12 years into an account paying 17% compounded semiannually

Objective 5.11: *Find the value on an annuity due.*

41. \$25 monthly payment for 10 years into an account paying 14% compounded monthly
42. \$50,000 annual payment for 25 years into an account paying 10% compounded annually

43. $300 quarterly payment for 12 years into an account paying 9% compounded quarterly

44. $4,000 semiannual payment for 12 years into an account paying 17% compounded semiannually

Objective 5.12: *Find the present value of an annuity.*

45. $25 monthly payment for 10 years into an account paying 14% compounded monthly

46. $50,000 annual payment for 25 years into an account paying 10% compounded annually

47. An insurance policy offers you a choice of $800 per month for 6 years or a lump-sum payment. What lump-sum payment would equal the monthly payments if the current interest rate is 12% compounded monthly?

48. A lottery offers you a choice of $1,000,000 per year for 5 years or a lump-sum payment. What lump-sum payment would equal the annual payment if the current interest rate is 14% compounded annually?

[5.4]

Objective 5.13: *Find the monthly payment for an installment loan.*

49. The amount to be financed on a home loan is $85,000 for a 14% 30-year loan.

50. The amount to be financed on a new car is $9,500. The terms are 17% for 4 years.

51. Purchase a car for $12,500 with 25% down and finance the balance for 4 years at 11.9%.

52. Purchase a home for $125,000 with 20% down and finance the balance for 30 years at 13.4%.

Objective 5.14: *Solve applied annuity problems and be able to distinguish ordinary annuities from annuities due.*

53. Phyllis Niklas decides to save $50 per month starting now and deposits this amount to an account paying 18%. If she makes these deposits for 5 years, how much will she have?

54. Joan Marsh drops 25¢ into a jar each day. At the end of each month (assume 30-day months) she deposits this amount into an account paying 12%. If she makes these deposits for 5 years, how much will she have?

55. Carol will retire in 15 years and is depositing $2,000 into an IRA account at the end of each year. If the IRA is paying 13% compounded annually, how much will Carol have in this account in 15 years?

56. Hal agrees to give the building fund $2,000 at the end of each year for the next 3 years. Amber wants to match Hal's gift but wants to make a lump-sum contribution today. If the current rate is 9% compounded annually, how much should Amber contribute?

[5.5]

Objective 5.15: *Find the amount of payment necessary for a deposit to a sinking fund. Assume that the deposit is made at the end of each compounding period.*

57. Alcar Repair will need $12,000 in 5 years. What is the monthly deposit necessary to raise this amount if it is deposited into an account paying 12% compounded monthly?

58. A business must raise $30,000 in 7 years, so the owners open a sinking fund paying 11% compounded semiannually. What should be the size of their semiannual payment to the fund?

59. What is the quarterly payment to a sinking fund paying 14% compounded quarterly in order to raise $10,000 in 5 years?

60. Thompson International decides to raise $100,000,000 in order to develop a new geothermal energy source. It issues a 50-year bond with a contract rate of 6% paid annually. It also sets up a sinking fund to repay the debt. This sinking fund pays 8% compounded annually. What is the size of the annual payment for both the interest and the bond repayment?

Objective 5.16: *Be able to classify the various types of financial problems of this chapter. These include compounded interest, present value, ordinary annuity, annuity due, present value of an annuity, installment payment, and sinking fund. With the given information for the following problems, classify each type. [Note: The directions tell you only to classify these problems. However, for additional practice you can answer the questions by assuming a 12% interest rate compounded annually.]*

61. Find the value of a $1,000 certificate in 3 years.

62. Deposit $300 at the end of each year. What is the total in the account in 10 years?

63. An insurance policy pays $10,000 in 5 years. What lump-sum deposit today will yield $10,000 in 5 years?

64. A 5-year term policy has an annual payment of $300, and at the end of 5 years, all payments and interest are refunded. What lump-sum deposit today will yield $10,000 in 5 years?

65. What annual deposit is necessary to give $10,000 in 5 years?

66. A $5,000,000 apartment complex is to be paid off in 10 years by making 10 equal annual payments. How much is each payment?

67. The prices of automobiles have increased at 6.25% per year. How much would you expect a $10,000 automobile to cost in 5 years?

68. Deposit $450 at the beginning of each year. What is the total amount in the account in 25 years?

69. Deposit $825 into an account at the end of each year. What is the total in the account in 23 years?

70. What lump-sum deposit today will yield $1,000,000 in 37 years?

71. What annual deposit is necessary to give $4,000 in 17 years?

72. What lump-sum deposit today is equal to 33 annual deposits of $500?

73. You want to have $10,000 in 6 years by making a deposit at the beginning of each year.

SAMPLE TEST

The following Sample Test (45 minutes) is intended to review the main ideas of this chapter.

In Problems 1–17 classify the type of financial problem, state the appropriate formula, and identify the variables and constants. For extra practice (outside the time limit), answer the question.

1. Find the value of a $1,000 certificate in $2\frac{1}{2}$ years if the interest rate is 12% compounded monthly.
2. You deposit $300 at the end of each year into an account paying 12% compounded annually. How much is in the account in 10 years?
3. An insurance policy pays $10,000 in 5 years. What lump-sum deposit today will yield $10,000 in 5 years if it is deposited at 12% compounded quarterly?
4. A 5-year term policy has an annual premium of $300, and at the end of 5 years, all payments and interest are refunded. What lump-sum deposit is necessary to equal this amount if you assume an interest rate of 10% compounded annually?
5. What annual deposit is necessary to give $10,000 in 5 years if the money is deposited at 9% interest compounded annually?
6. A $5,000,000 apartment complex loan is to be paid off in 10 years by making 10 equal annual payments. How much is each payment if the interest rate is 14% compounded annually?
7. The prices of automobiles have increased at 6.25% per year. How much would you expect a $10,000 automobile to cost in 5 years?
8. The amount to be financed on a new car is $9,500. The terms are 17% for 4 years. What is the monthly payment?
9. At the beginning of each year, you deposit $450 into an account paying 11% compounded annually. How much is in the account in 25 years?
10. At the end of each year, you deposit $825 into an account paying 7.5% compounded annually. How much is in the account in 23 years?
11. What lump-sum deposit today will yield $1,000,000 in 37 years if it is deposited at 12% compounded annually?
12. What deposit today is equal to 33 annual deposits of $500 into an account paying 8% compounded annually?
13. What annual deposit into an account paying 11.5% is necessary to give $5,000 in 15 years?
14. You want to have $10,000 in 6 years by making a deposit at the end of each year into an account paying 12% interest compounded annually. How much should your annual deposit be?
15. A lottery offers you a choice of $1,000,000 per year for 5 years or a lump-sum payment. What lump-sum payment would equal the annual payments if the current interest rate is 14% compounded annually?
16. What is the monthly payment for a home costing $125,000 with a 20% down payment and the balance financed for 30 years at 12%?
17. Western Electric decides to raise $150,000,000 in order to develop a new geothermal energy source. It issues a 50-year bond with a contract rate of 9% paid annually. It also sets up a fund to repay the debt. This fund pays 10% compounded annually. What is the size of the annual payment for both the interest and the bond repayment?

The following multiple-choice questions have appeared on recent Management Accounting and CPA examinations. The questions are reprinted with permission of the National Association of Accountants and the American Institute of Certified Public Accountants, Inc.

Management Accounting Exam June 1983

18. A firm has daily cash receipts of $200,000. A commercial bank has offered to reduce the collection time by 3 days. The bank requires a monthly fee of $4,000 for providing this service. If money market rates will average 12% during the year, the additional annual income (loss) of having the service is
    A. $(24,000) **B.** $24,000
    C. $66,240 **D.** $68,000
    E. Some amount other than those given above

CPA Exam November 1976

19. A businesswoman wants to withdraw $3,000 (including principal) from an investment fund at the end of each year for 5 years. How should she compute her required initial investment at the beginning of the first year if the fund earns 6% compounded annually?
    A. $3,000 times the amount of an annuity of $1 at 6% at the end of each year for 5 years
    B. $3,000 divided by the amount of an annuity of $1 at 6% at the end of each year for 5 years

C. \$3,000 times the present value of an annuity of \$1 at 6% at the end of each year for 5 years
D. \$3,000 divided by the present value of an annuity of \$1 at 6% at the end of each year for 5 years

CPA Exam November 1976

20. A businessman wants to invest a certain sum of money at the end of each year for 5 years. The investment will earn 6% compounded annually. At the end of 5 years, he will need a total of \$30,000 accumulated. How should he compute his required annual investment?
A. \$30,000 times the amount of an annuity of \$1 at 6% at the end of each year for 5 years
B. \$30,000 divided by the amount of an annuity of \$1 at 6% at the end of each year for 5 years
C. \$30,000 times the present value of an annuity of \$1 at 6% at the end of each year for 5 years
D. \$30,000 divided by the present value of an annuity of \$1 at 6% at the end of each year for 5 years

6

Combinatorics

CHAPTER OVERVIEW

This chapter focuses on sets and counting techniques. In addition, you will learn about the fundamental counting principle and two very useful counting models, permutations and combinations. This chapter concludes with the binomial theorem.

PREVIEW

The idea of a set is one of those very simple, and profound, concepts that is used to simplify and unify mathematics. After looking at sets, set operations, and set relationships in the first section, we move to the notion of counting and develop the ideas of permutations, combinations, and the fundamental counting principle. Make a special note of Table 6.1 on page 309 because being able to tell *how* to proceed is just as important as carrying out the actual process.

PERSPECTIVE

We all learned to count in a straightforward "one, two, three, . . ." procedure. But in mathematics, especially when working with probabilities, it is necessary to count the number of elements in involved sets, and straightforward counting is often difficult or even impossible. This chapter involves preliminary, but very necessary, work for the next chapter on probability.

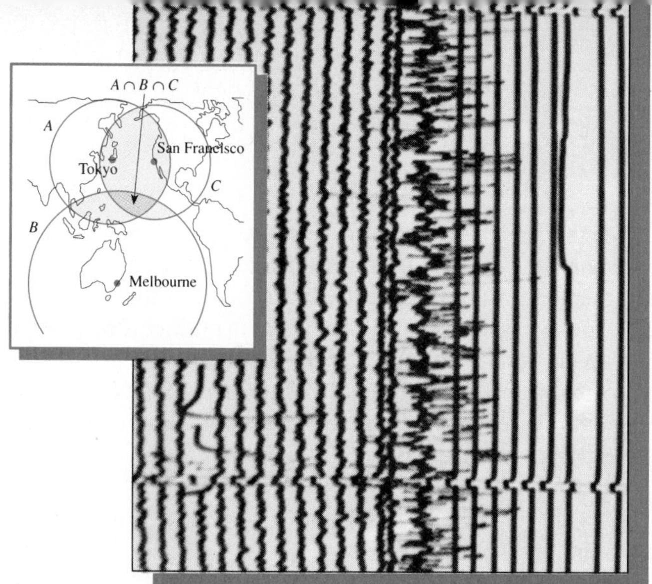

MODELING APPLICATION 4

Earthquake Epicenter

During an earthquake, vibrations emanate in all directions from an origin called the *epicenter*. Two kinds of vibrations of importance in the study of earthquakes are called *primary* (P waves) and *secondary* (S waves). These names are due to the waves' order of arrival at a seismographic laboratory. The P wave is the fastest type of wave generated by an earthquake, with an average speed of about 8 kilometers per second. Concurrently, an S wave is produced that travels at approximately two-thirds the velocity of a P wave, so that it arrives at the recording station after the initial P wave. The greater the distance between the epicenter of the earthquake and the seismograph, the greater the time interval between the arrival of the two waves. Suppose an earthquake is recorded at three recording stations, one in Tokyo, another in Melbourne, and a third in San Francisco. The arrival times of the P and S waves are given in the table in the margin.

After you have finished this chapter, write a paper that determines the probable location of the epicenter by using Venn diagrams.. For general guidelines about writing this essay, see the commentary for Modeling Application* 1 *on page* 151.

*The idea and material for this extended application are derived from Joseph Di Carlucci, "Earthquakes and Venn Diagrams," *The Mathematics Teacher*, September 1979, pp. 428–433.

Recording stations

I. Tokyo
II. Melbourne
III. San Francisco

Arrival times

	P wave	S wave
I.	6:25:15 P.M.	6:32:19 P.M.
II.	7:25:15 P.M.	7:34:07 P.M.
III.	7:25:15 A.M.	7:29:15 A.M.

APPLICATIONS

Social Sciences (*Demography, Political science, Population, Psychology, Society, and Sociology*)
Surveys (6.2, Problems 32–41)
Social Security numbers (6.3, Problem 31)
License plate problem (6.3, Problems 32–35, 44)
Election problem (6.4, Problem 19)
Committee problem (6.4, Problem 32; 6.5, Problems 41–43; 6.6, Problems 34–35; 6.7, Test Problems 8–9)
Grouping individuals in a psychology experiment (6.5, Problem 35)
Psychology experiment (6.7, Problems 51, 57)

General Interest
Historical questions dealing with John Venn (6.1, Problem 57)
Deciding on proper number of buses to charter (6.2, Problem 31)

General Interest (continued)
Likelihood of purchasing a defective tire (6.2, Problem 35)
Number of selections possible (6.3, Problems 19–27, 37–38; 6.4, Problems 19–40; 6.5, Problems 12–17)
ATM access codes (6.3, Problems 39–40)
Combination locks (6.3, Problems 36, 43)
Ways of obtaining different poker hands (6.5, Problems 14–17, 22)
Ways of answering a multiple choice test (6.5, Problem 38)
Pascal's triangle (6.6, Problems 39–42)
Survey about traffic tickets (6.7, Problems 33–36)
Dealing cards to bridge players (6.7, Test Problem 6)

Modeling Application—
Earthquake Epicenter

Management (*Business, Economics, Finance, and Investments*)
Analysis of sales representatives (6.1, Problem 55)
Different categories of employees within the Ampex Corp. (6.1, Problem 56)
Survey of a national board (6.2, Problem 34)
Ways of rating businesses (6.5, Problem 34)
Grouping stocks (6.5, Problem 37)
Likelihood of an IRS investigation (6.5, Problem 39)

Life Sciences (*Biology, Ecology, Health, and Medicine*)
Menu selections (6.3, Problems 28, 29, 41, and 42)
Public opinion poll (6.3, Problem 27)
Human blood types (6.7, Problem 50, Test Problem 15)

6.1

Sets and Set Operations

Terminology

The concept of sets and set theory is attributed to Georg Cantor (1845–1918), a great German mathematician. Cantor's ideas were not adopted as fundamental underlying concepts in mathematics until the twentieth century, but today the idea of sets is used to unify and explain nearly every other concept in mathematics. Recall that a counting number is an element of the set $\{1, 2, 3, \ldots\}$.

Sets are specified by roster or by description. In the **roster method**, sets are specified by listing the elements and enclosing those elements between braces: $\{\ \ \}$. In the **description method**, sets are described in words, such as

The set of subscribers to the *Wall Street Journal*

or by using **set-builder notation**, such as

$\{s \mid 0 < s < 100, s \text{ a counting number}\}$

The proper way to read the set-builder notation here is "the set of all counting numbers, s, such that s is between 0 and 100."

EXAMPLE 1 Let U be the set of counting numbers. Write the given sets by roster.

$$A = \{x \mid (x-2)(x-3) = 0\}$$
$$B = \{y \mid 3 \le y < 7\}$$
$$C = \{y \mid y^2 = 4\}$$

Solution By roster,

$$A = \{2, 3\}$$
$$B = \{3, 4, 5, 6\}$$
$$C = \{2\}$$

Note: The number -2 also makes the statement $y^2 = 4$ true, but -2 is not in the universe for this problem.

If S is a set, we write $a \in S$ if a is a member of the set S and $b \notin S$ if b is not a member of S. For example, let C be the set of cities in California. If a represents the city of Anaheim and b stands for the city of Berlin, Germany, we would write

$a \in C$ and $b \notin C$

Two special sets require attention. The first is the set that contains no elements and the second is a set that contains every element under consideration:

Empty and Universal Sets

The **empty set** contains no elements and is denoted by $\{\ \ \}$ or $\varnothing$.

The **universal set** contains all the elements under consideration in a given discussion and is denoted by U.

For example, if we agree that $U = \{1, 2, 3, 4, 5, 6, 7, 8, 9\}$, then all sets that are considered would have elements only among the elements of U. No set could contain the number 10 since 10 is not included in what we have decided is the universe. For every problem, a universal set must be specified or implied, and it must remain fixed for that problem. However, when a new problem is begun, a new universal set can be specified.

The number of elements in a set is called the **cardinality** of the set. The cardinality of the empty set is 0. If a set is denoted by A, then the notation $|A|$ is used to symbolize the cardinality of A.

Two sets can have the same cardinality and be different sets, such as $\{1, 2, 3\}$ and $\{7, 8, 9\}$. However, if two sets are the same, such as $\{1, 2, 3\}$ and $\{2, 3, 1\}$, we say they do not only have the same cardinality, but are also equal.

Equal Sets

Set A and B are equal, denoted by $A = B$, if they have exactly the same elements.

Notice that equal sets must have the same cardinality, but sets with the same cardinality are not necessarily equal.

Remember that when set notation is used, the *order* of the elements of a subset is not important. If the *order* of the elements is important, then a different notation is used. Let $A = \{2, 4, 6, 8\}$ and pick a pair of elements from A.

Order not important: $\{2, 6\} = \{6, 2\}$
Order important: $(2, 6) \neq (6, 2)$

Parentheses are used to indicate order. If there are two elements, it is called an **ordered pair**; three are called an **ordered triplet**; if there are n ordered elements, it is called an **ordered *n*-tuple**.

Subsets

To deal effectually with sets it is necessary to understand certain relationships among sets.

Subsets

A set A is a **subset** of a set B, denoted by $A \subseteq B$, if every element of A is an element of B.

Consider the following sets:

$$U = \{1, 2, 3, 4, 5, 6, 7, 8, 9\}$$
$$A = \{2, 4, 6, 8\}$$
$$B = \{1, 3, 5, 7\}$$
$$C = \{5, 7\}$$

A, B, and C are subsets of the universal set U (all sets considered are subsets of the universe, by definition of the universal set). Also, $C \subseteq B$ since every element in C is also an element of B.

However, C is not a subset of A. To substantiate this claim, we must show that there is some member of C that is not a member of A. In this case we merely note that $5 \in C$ and $5 \notin A$.

Do not confuse the notions of "element" and "subset." That is, 5 is an *element* of C since it is listed in C; $\{5\}$ is a *subset* of C but is not an element since we do not find $\{5\}$ contained in C—if we did, C might look like $\{5, \{5\}, 7\}$.

To find all the possible subsets of the set $C = \{5, 7\}$ we use what is called a **tree diagram**, which is a device that helps us enumerate a list of possibilities:

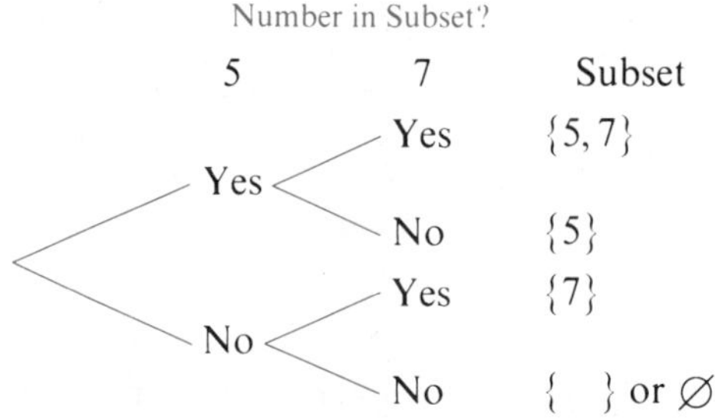

The subsets of C are: $\{5, 7\}$, $\{5\}$, $\{7\}$, $\varnothing$. Note that the set C has only two elements, but it has four subsets. Certainly, then, you must be careful to distinguish between a subset and an element.

The subsets of C can be classified into two categories: **proper** and **improper**.

Proper subsets of C	Improper subset of C
$\{5\}$ $\{7\}$ $\{\ \}$	$\{5, 7\}$

U

B

A

a. $A \subseteq B$

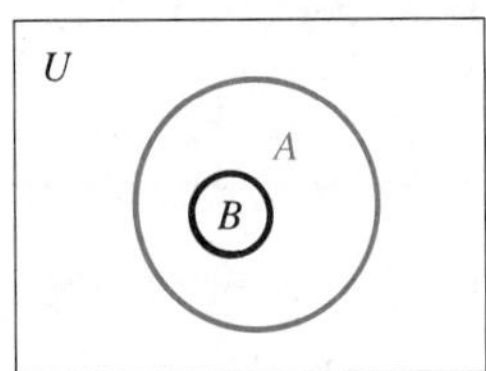

b. $B \subseteq A$

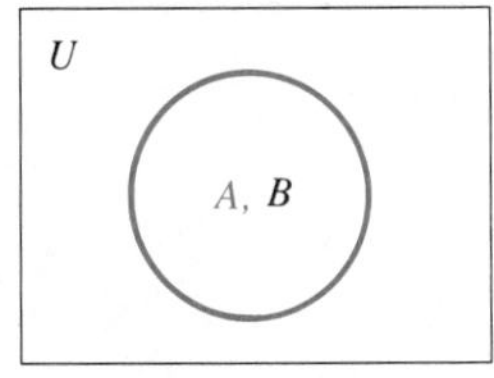

c. $A = B$

FIGURE 6.1
Venn diagrams showing subset and equal relationships

Every set has just one improper subset, and that is the set itself. All other subsets are proper subsets. The notation for a proper subset (as distinguished from an improper subset) does not include the small equality symbol; that is,

$A \subset B$ ← Used to denote the idea that A is a *proper subset* of B

Venn Diagrams

A useful way to depict relationships among sets is to represent the universal set by a rectangle, with the proper subsets in the universe represented by circular or oval-shaped regions, as in Figure 6.1.

These figures are called **Venn diagrams**, after John Venn (1834–1923). The Swiss mathematician Leonhard Euler (1707–1783) also used circles to illustrate principles of logic, so sometimes these diagrams are called *Euler circles*. However, Venn was the first person to use them in a general way.

We can also illustrate other relationships between two sets: A and B may have no elements in common, in which case they are **disjoint**, or they may overlap and have some elements in common (Figure 6.2).

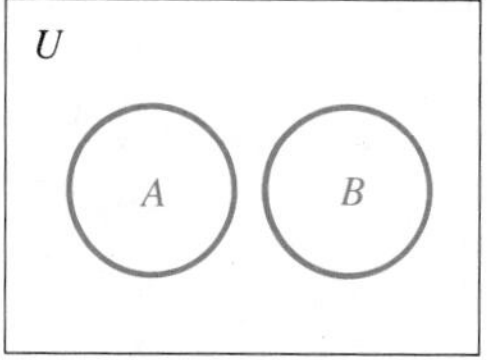

a. Disjoint sets

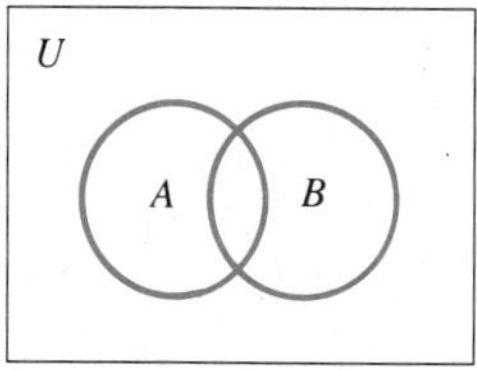

b. Overlapping sets

FIGURE 6.2
Venn diagrams for disjoint and overlapping sets

FIGURE 6.3
Venn diagram showing two general sets

Sometimes we are given two sets X and Y and we know nothing about how they are related. In this situation we draw a general figure, such as the one shown in Figure 6.3.

Note that the two circles divide the universe into four disjoint regions. When sets are drawn in this manner it does not mean that the only possibility is overlapping sets. For example:

If $X \subseteq Y$, then region 2 is empty.
If $Y \subseteq X$, then region 4 is empty.
If $X = Y$, then regions 2 and 4 are both empty.
If X and Y are disjoint, then region 3 is empty.

Set Operations

Four common operations performed on sets are Cartesian product, union, intersection, and complementation.

Cartesian Product

Cartesian Product

The **Cartesian product** of two sets A and B, denoted by $A \times B$, is defined by

$$A \times B = \{\text{all ordered pairs } (a, b) \text{ where } a \in A \text{ and } b \in B\}$$

EXAMPLE 2 Let $A = \{a, b, c\}$ and $B = \{x, y\}$. Then

$$A \times B = \{a, x), (a, y), (b, x), (b, y), (c, x), (c, y)\}$$ ■

EXAMPLE 3 A club is to select a president from the following set:

$$A = \{\text{Alfie, Bogie, Calvin, Doug, Ernie}\}$$

and a vice-president from the following set:

$$B = \{\text{Fred, Gail, Hazel}\}$$

List all possible results of the election.

Solution We can systematically list these by finding $A \times B$, and to do this we use an array as shown:

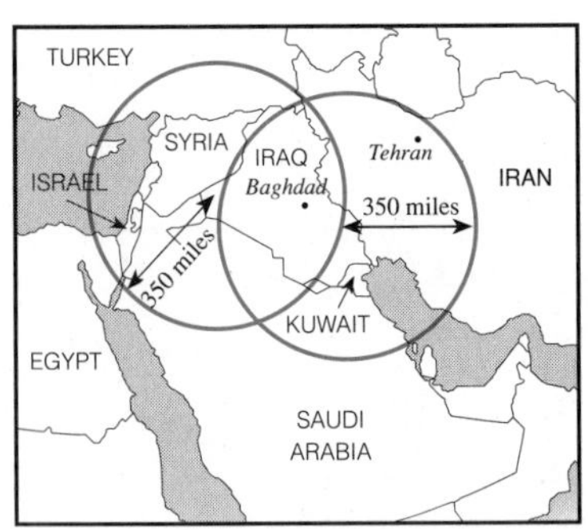

A 1990 news article described the potential range of an immense artillery gun by using Venn diagrams.

$A \times B$	f Fred	g Gail	h Hazel
a Alfie	(a, f)	(a, g)	(a, h)
b Bogie	(b, f)	(b, g)	(b, h)
c Calvin	(c, f)	(c, g)	(c, h)
d Doug	(d, f)	(d, g)	(d, h)
e Ernie	(e, f)	(e, g)	(e, h)

■

Notice that $A \times B$ has 15 elements. In general, $|A \times B| = |A| \cdot |B|$. Also notice that since the Cartesian product gives **ordered** pairs, the order is important, so

$$A \times B \neq B \times A$$

Union

Union is an operation for sets A and B in which a set is formed that consists of all the elements in A or B or both. The symbol for the operation of union is $\cup$, and we write $A \cup B$.

Union of Sets

> The **union** of sets A and B, denoted by $A \cup B$, is the set consisting of all elements of A or B or both.
>
> $$A \cup B = \{x \mid x \in A \text{ or } x \in B \text{ or } x \in (\text{both } A \text{ and } B)\}$$

EXAMPLE 4 Let $U = \{1, 2, 3, 4, 5, 6, 7, 8, 9\}$, $A = \{2, 4, 6, 8\}$, $B = \{1, 3, 5, 7\}$, $C = \{5, 7\}$. Then:

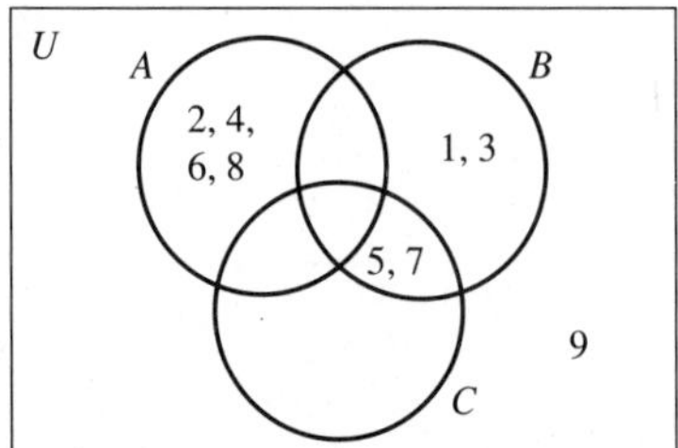

a. $A \cup C = \{2, 4, 5, 6, 7, 8\}$; that is, the union of A and C is the set consisting of all elements in A or in C or in both.

b. $B \cup C = \{1, 3, 5, 7\}$; note that, even though the elements 5 and 7 appear in both sets, they are listed only once. That is, the sets $\{1, 3, 5, 7\}$ and $\{1, 3, 5, 5, 7, 7\}$ are equal (exactly the same).

c. $A \cup B = \{1, 2, 3, 4, 5, 6, 7, 8\}$.

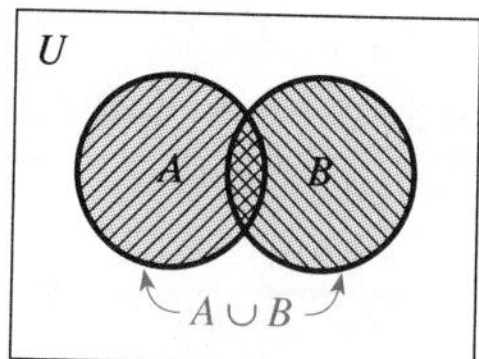

FIGURE 6.4
Venn diagram showing the union of two sets (color shading)

d. $(A \cup B) \cup \{9\} = \{1, 2, 3, 4, 5, 6, 7, 8, 9\} = U$; here we are considering the union of three sets; however, the parentheses indicate the operation that should be performed first. Note also that, because the solution $\{1, 2, 3, 4, 5, 6, 7, 8, 9\}$ has a name, we write down the name rather than the set. That is, $(A \cup B) \cup \{9\} = U$ is the simpler representation. ■

We can use Venn diagrams to illustrate union. In Figure 6.4 we first shade A and then shade B. *The union is all parts that have been shaded at least once.*

Intersection

A third operation for sets is called **intersection.**

Intersection of Sets

The **intersection** of sets A and B, denoted by $A \cap B$, is the set consisting of all elements common to A and B.

$$A \cap B = \{x \mid x \in A \textit{ and } x \in B\}$$

EXAMPLE 5 Let $U = \{a, b, c, d, e\}$, $A = \{a, c, e\}$, $B = \{c, d, e\}$, $C = \{a\}$, and $D = \{e\}$.

Note: Sets are U, A, B, C, D and E. Elements are a, b, c, d, and e. (Capital and lowercase letters represent different things.)

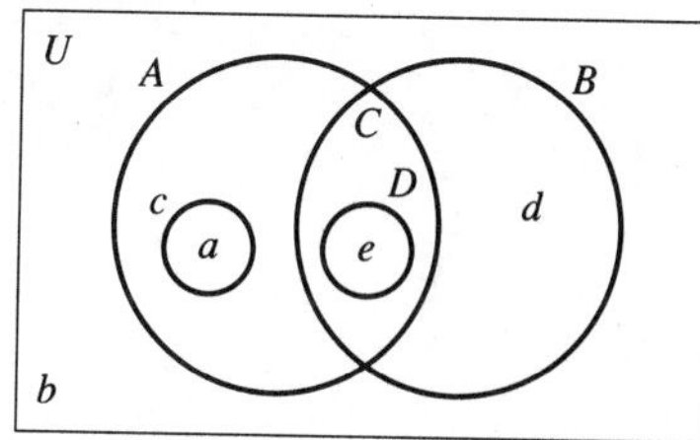

a. $A \cap B = \{c, e\}$; that is, the intersection of A and B is the set consisting of elements in both A and B.

b. $A \cap C = C$ since $A \cap C = \{a\}$ and $\{a\} = C$; we write down the name for the set that is the intersection.

c. $B \cap C = \varnothing$ since B and C have no elements in common (they are disjoint).

d. Parentheses tell us which operation to do first:

$$(A \cap B) \cap D = \{c, e\} \cap \{e\}$$

$$= \{e\}$$ {e} is the set of elements common to the sets {c, e} and {e}.

$$= D$$ ■

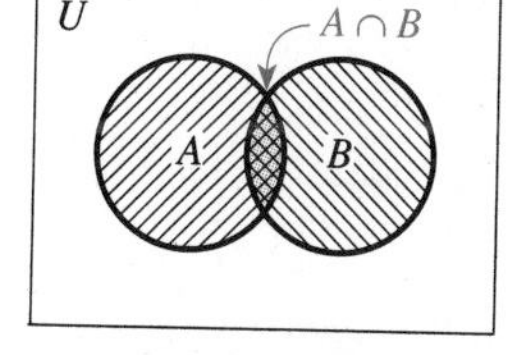

FIGURE 6.5
Venn diagram showing the intersection of two sets (color shading)

The intersection of sets can also be easily shown in a Venn diagram. To find the intersection of two sets A and B, first shade A and then shade B; *the intersection is all parts shaded twice* (Figure 6.5).

Complementation

Complementation is an operation on a set that must be performed in relation to the universal set.

Complementation

> The **complement** of a set A, denoted by $\bar{A}$, is the set of all elements of U that are not in the set A.
>
> $$\bar{A} = \{x \mid x \notin A\}$$

EXAMPLE 6 Let $U = \{\text{People in California}\}$, $A = \{\text{People who are over 30}\}$, $B = \{\text{People who are 30 or under}\}$, and $C = \{\text{People who own a car}\}$. Then:

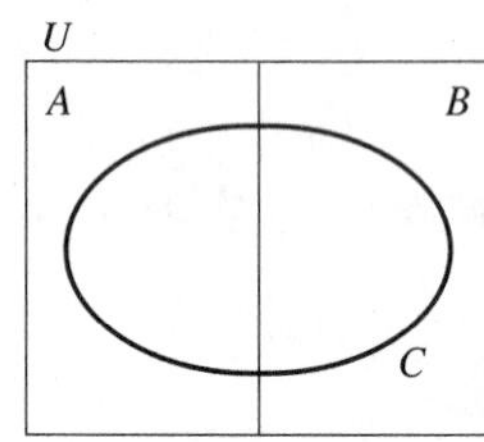

a. $\bar{C} = \{\text{Californians who do not own a car}\}$

b. $\bar{A} = B$

c. $\bar{B} = A$

d. $\bar{U} = \varnothing$

e. $\bar{\varnothing} = U$ ■

Complementation can be shown in a Venn diagram. The color shading in Figure 6.6 shows the complement of A. In a Venn diagram, the complement is everything in U that is not in the given set (in this case, everything not in A).

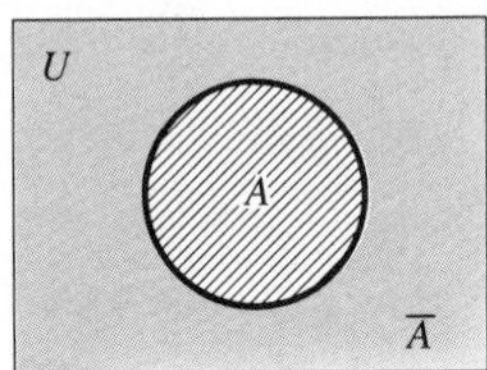

FIGURE 6.6
Venn diagram showing the complement of a set A (color shading)

6.1
Problem Set

1. Explain the difference between *element of a set* and *subset of a set*. Give examples.
2. Explain the difference between *subset* and *proper subset*.
3. Explain the difference between *universal set* and *empty set*.
4. In your own words, define *union*, *intersection*, and *complement*. Illustrate with examples.

List all subsets of each set given in Problems 5–10; also give the cardinality of each set.

5. $\{\ \}$
6. $\{2\}$
7. $\{2,4\}$
8. $\{2,4,6\}$
9. $\{2,4,6,8\}$
10. $\{m,a,t,h\}$
11. How many subsets are there for $\{1,2,3,4,5,6,7,8,9,10\}$?

12. How many subsets are there for $\{a, b, c, d, e, f\}$?

13. What is the cardinality of the set of current U.S. senators? How many subsets can be formed which consist only of U.S. senators?

Let $A = \{1, 3, 5\}$ *and* $B = \{a, b\}$ *for Problems* 14–17.

14. $A \times B$ **15.** $B \times A$

16. $A \times A$ **17.** $B \times B$

18. Give an example of a set with cardinality 0.

19. Give an example of a set with cardinality larger than a million.

Perform the set operations in Problems 20–29. *Let* $U = \{1, 2, 3, 4, 5, 6, 7, 8, 9, 10\}$.

20. $\{2, 6, 8\} \cup \{6, 8, 10\}$

21. $\{2, 6, 8\} \cap \{6, 8, 10\}$

22. $\{1, 2, 3, 4, 5\} \cap \{3, 4, 5, 6, 7\}$

23. $\{1, 2, 3, 4, 5\} \cup \{3, 4, 5, 6, 7\}$

24. $\{2, 5, 8\} \cup \{3, 6, 9\}$ **25.** $\{2, 5, 8\} \cap \{3, 6, 9\}$

26. $\overline{\{2, 8, 9\}}$ **27.** $\overline{\{1, 2, 5, 7, 9\}}$

28. $\overline{\{8\}}$ **29.** $\overline{\{6, 7, 8, 9, 10\}}$

Let $U = \{1, 2, 3, 4, 5, 6, 7\}$, $A = \{1, 2, 3, 4\}$, $B = \{1, 2, 5, 6\}$, *and* $C = \{3, 5, 7\}$. *List all the members of each of the sets in Problems* 30–38.

30. $A \cup B$ **31.** $A \cup C$

32. $B \cup C$ **33.** $A \cap B$

34. $A \cap C$ **35.** $B \cap C$

36. $\bar{A}$ **37.** $\bar{B}$

38. $\bar{C}$

39. *Determine whether each of the following is true or false*:

a. $\varnothing \in \varnothing$ **b.** $\varnothing \subseteq \varnothing$
c. $\varnothing \in \{\varnothing\}$ **d.** $\varnothing = \varnothing$
e. $\varnothing = \{\varnothing\}$

40. *Determine whether each of the following is true or false*:

a. $\varnothing \subseteq \{\varnothing\}$
b. $\varnothing \in A$ for all sets A
c. $\varnothing = 0$
d. $\varnothing = \{0\}$
e. $\varnothing \subseteq \{0\}$

Draw a Venn diagram for each relationship in Problems 41–46.

41. $X \cup Y$ **42.** $Y \cup Z$

43. $X \cap Z$ **44.** $X \cap Y$

45. $\bar{X}$ **46.** $\bar{Z}$

Graph the intersection of the sets in Problems 47–50 *on a number line. Assume the universal set is the set of real numbers.*

47. $A = \{x \geq 5\}$, $B = \{x \leq 8\}$

48. $C = \{x < 4\}$, $D = \{x > -3\}$

49. $E = \{x \geq -8\}$, $F = \{x < -1\}$

50. $G = \{-2, 0, 2, 5, 8\}$, $H = \{-4, 2, 6, 8, 10\}$

Find the intersection of the two-dimensional sets in Problems 51–54. *Assume the universal set is the set of all ordered pairs of real numbers.*

51. $A = \{x + 2y = 5\}$, $B = \{3x - y = 8\}$

52. $C = \{x + y = 5\}$, $D = \{x - y = 1\}$

53. $E = \{-x + y = -3\}$, $F = \{3x + 2y = 4\}$

54. $G = \{2x - 3y = 6\}$, $H = \{x + 2y = 10\}$

APPLICATIONS

55. Let A represent the top sales representatives of the year 1991:

{Bob Wisner, Joan Marsh, Craig Barth, Phyllis Niklas}

and let B represent the top representatives of the year 1992:

{Phyllis Niklas, Craig Barth, Shannon Smith, Christy Anton}

a. What is the set of top sales representatives for 1991 or 1992? What set operation should be used to find this answer?

b. What is the set of top sales representatives for the 2 years in a row? What set operation should be used to find this answer?

56. Let

U = {All employees of Ampex Corporation}
A = Set of executives = {Employees with salaries of \$60,000 or more}
B = Set of junior executives = {Employees with salaries between \$40,000 and \$60,000}
C = Set of nonexecutives = {Employees with salaries of \$20,000 and less}
D = Set of white-collar workers = {Employees with salaries of \$12,000 or more}

a. Draw a Venn diagram showing how these sets are related.

Determine whether each of the following is true or false:

b. $A \subseteq U$ **c.** $A \cup B = U$
d. $D \subseteq A$ **e.** $\varnothing \subseteq A$
f. $C \cap D = \varnothing$ **g.** $\varnothing \in A$
h. $B \cap C = \varnothing$ **i.** $\overline{A \cup B \cup C} = D$

57. ***Historical Question.*** John Venn (1834–1923) used diagrams to illustrate logic in his *Symbolic Logic*, published in 1881. Venn was an ordained priest, but he left the clergy in 1883 to spend all his time teaching and studying logic. In the text we noted that three intersecting circles divide the universe into eight regions. Venn also considered the case of four intersecting circles. Show what this case might look like, and tell how many disjoint regions are produced by four general intersecting circles.

6.2 Combined Operations with Sets

Venn Diagrams

In the last section we defined Cartesian product, union, intersection, and complementation of sets. However, the real payoff for studying these relationships comes when dealing with combined operations or with several sets at the same time.

EXAMPLE 1 Illustrate $\overline{A \cup B}$ in a Venn diagram.

Solution This is a combined operation. Since it is the complement of the union, you must *first* find $A \cup B$; *then* find the complement.

Step 1:

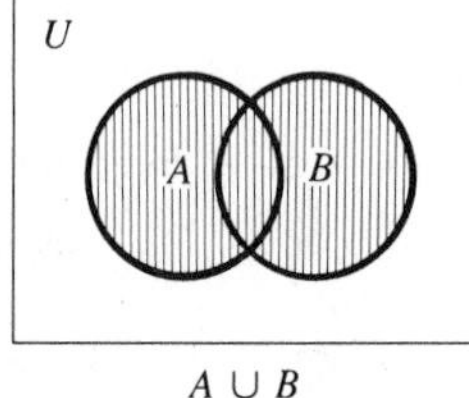

$A \cup B$

Step 2:

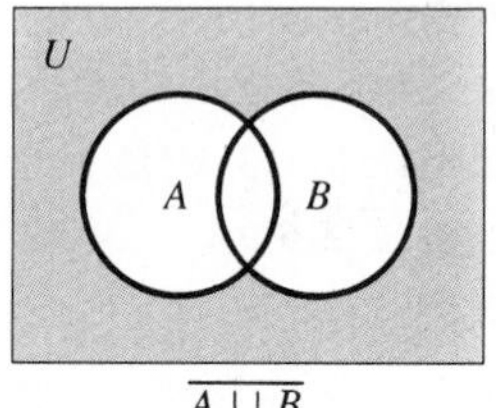

$\overline{A \cup B}$

These steps are generally combined into one diagram:

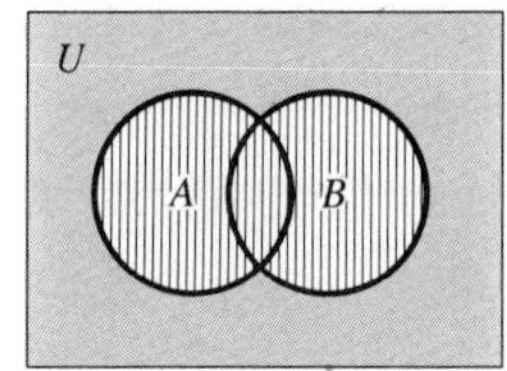

The answer is shown by color shading. The vertical lines show the preliminary step. It is generally a good idea to show the final answer (the part in color) by using a highlighter pen. ■

EXAMPLE 2 Illustrate $\bar{A} \cup \bar{B}$ in a Venn diagram.

Solution Compare this statement with Example 1. This is the union of the complements, so here we *first* find $\bar{A}$ and $\bar{B}$; *then* find the union.

Step 1:

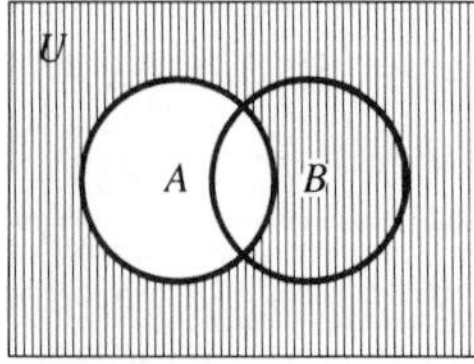

$\bar{A}$ (vertical lines)

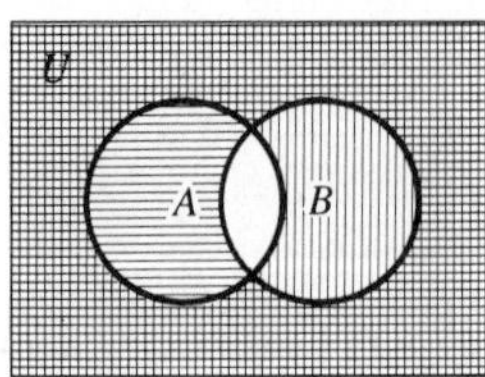

$\bar{A}$ with $\bar{B}$ (horizontal lines)

Step 2:

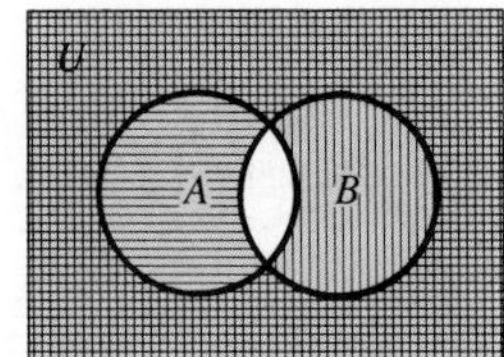

$\bar{A} \cup \bar{B}$ (color shading; the union is all parts that have horizontal or vertical lines or both)

Your work will normally show the above steps with one diagram. ■

Notice that $\overline{A \cup B} \neq \bar{A} \cup \bar{B}$. If they were equal, the final *shaded* color portions of the Venn diagrams from Examples 1 and 2 would be the same.

EXAMPLE 3 Prove that $\overline{A \cup B} = \bar{A} \cap \bar{B}$.

Step 1: Draw a Venn diagram for the left side of the equal sign.

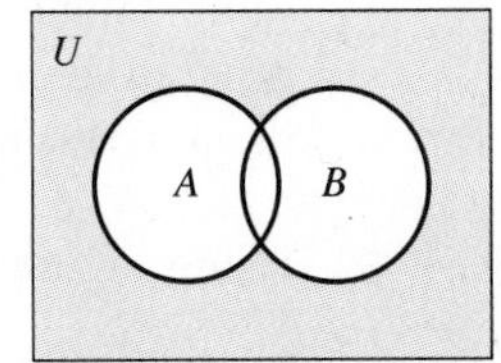

$\overline{A \cup B}$ (see Example 1 for details)

Step 2: Draw a separate Venn diagram for the right side of the equal sign.

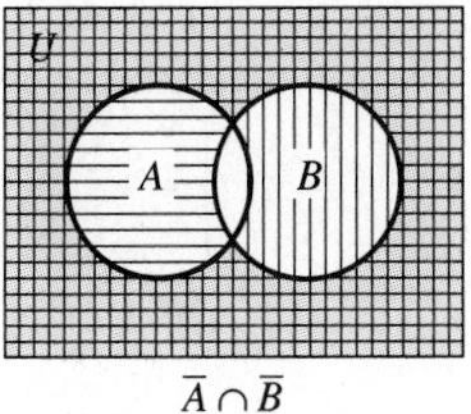

$\bar{A} \cap \bar{B}$

Detail: $\bar{A}$: vertical lines
$\bar{B}$: horizontal lines
$\bar{A} \cap \bar{B}$
↑
The intersection of horizontal and vertical lines is the shaded portion—it consists of all parts having both horizontal and vertical lines.

Step 3: Compare shaded portions of the two Venn diagrams. They are the same, so we have proved $\overline{A \cup B} = \bar{A} \cap \bar{B}$. ■

The result proved in Example 3 is called **De Morgan's Law**. In the problem set you are asked to prove the second part of De Morgan's Law.

De Morgan's Laws

$$\overline{X \cup Y} = \bar{X} \cap \bar{Y} \qquad \overline{X \cap Y} = \bar{X} \cup \bar{Y}$$

Order of Operations

The *order* of operations for sets is left to right, except if there are parentheses. Operations within parentheses are performed first, as shown by Example 4.

EXAMPLE 4 Illustrate $(A \cup \bar{C}) \cap \bar{C}$ in a Venn diagram.

Solution

$(A \cup C) \cap \bar{C}$

Detail:

$A \cup C$: Vertical lines

$\bar{C}$: Horizontal lines

$(A \cup C) \cap \bar{C}$: The intersection of vertical and horizontal lines is the part shaded in color. ■

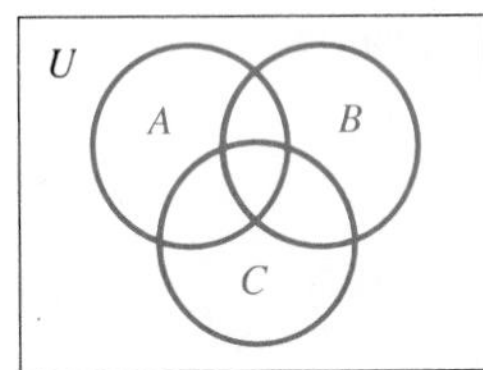

FIGURE 6.7
Three general sets

Sometimes you will be asked to consider relationships among three sets. The general Venn diagram for this is shown in Figure 6.7. Note that three sets divide the universe into eight regions. Can you number each?

EXAMPLE 5 Using Figure 6.7 as a guide, shade in each of the following sets:

a. $A \cup B$ **b.** $A \cap C$ **c.** $B \cap C$ **d.** $\bar{A}$

e. $\overline{A \cup B}$ **f.** $A \cap B \cap C$ **g.** $A \cup B \cup C$ **h.** $\overline{A \cup B \cup C}$

Solution

a.

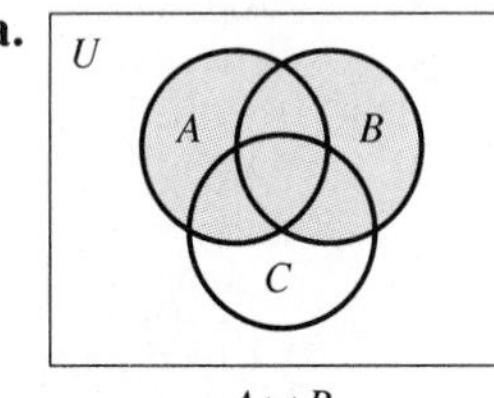

$A \cup B$

b.

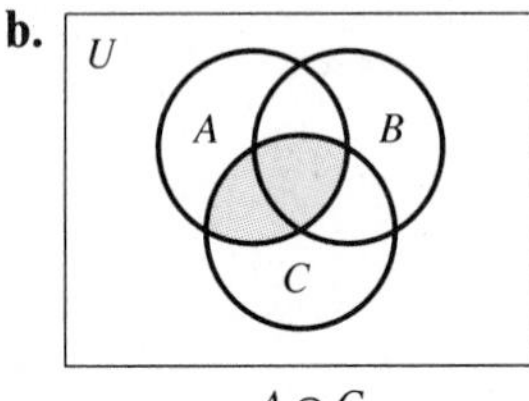

$A \cap C$

c.

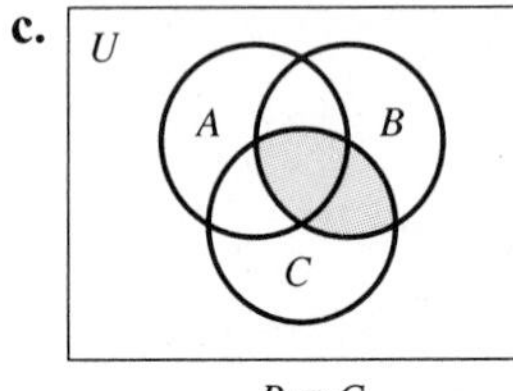

$B \cap C$

d.

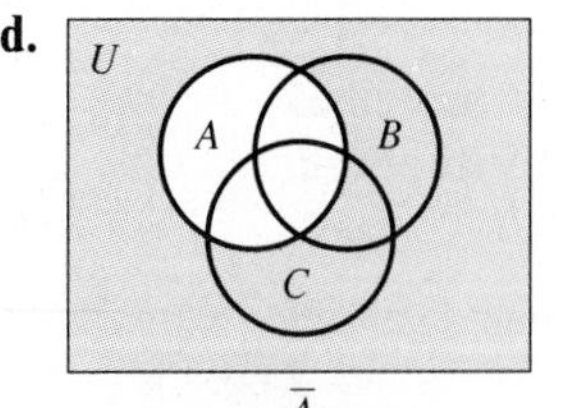

$\bar{A}$

e.

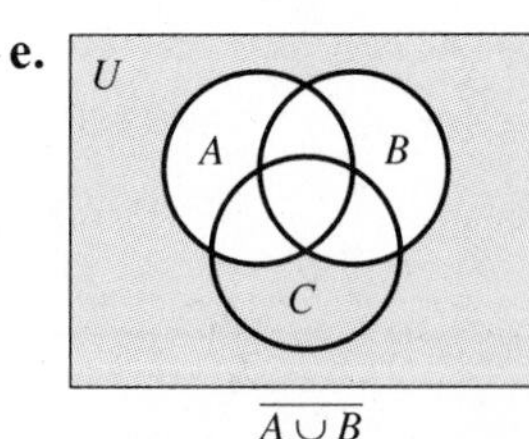

$\overline{A \cup B}$

f.

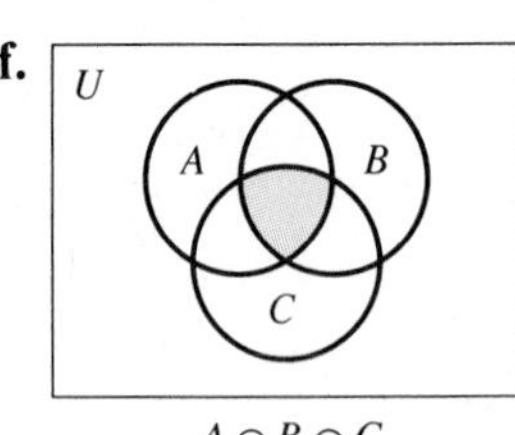

$A \cap B \cap C$

g.

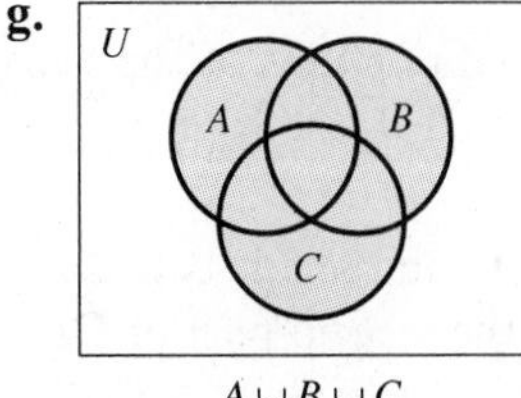

$A \cup B \cup C$

h.

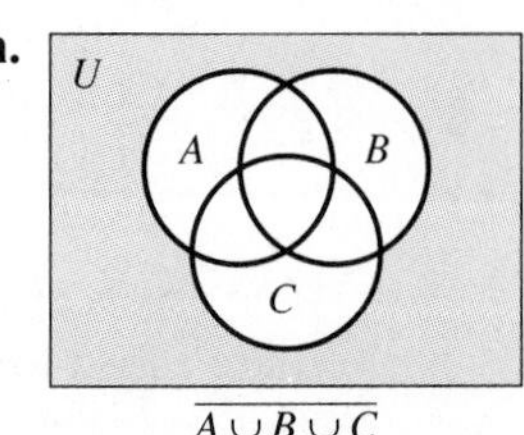

$\overline{A \cup B \cup C}$

■

EXAMPLE 6 Draw a Venn diagram for $A \cup (B \cap C)$.

Solution The Venn diagram is shown in the margin.

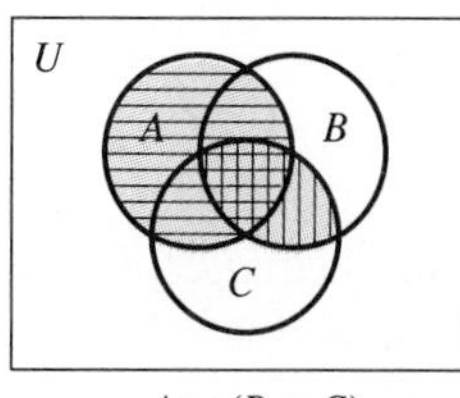

$A \cup (B \cap C)$

Detail:

Step 1: Parentheses first (vertical lines for $B \cap C$)

Step 2: Set A (horizontal lines)

Step 3: The union is all parts that show either vertical lines or horizontal lines, or both—this part is shaded in color. ■

EXAMPLE 7 Draw a Venn diagram for $\overline{A \cup B} \cap C$.

Solution

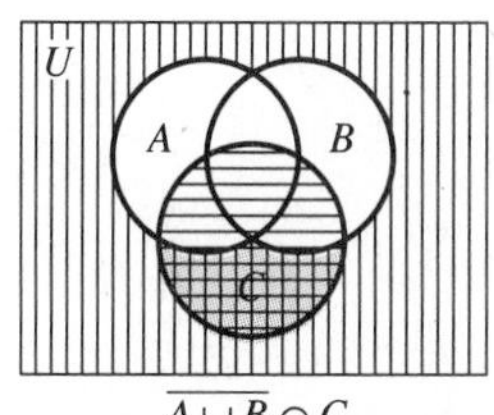

$\overline{A \cup B} \cap C$

Detail:

Step 1: $\overline{A \cup B}$ first (vertical)

Step 2: C (horizontal)

Step 3: The intersection is all parts that show both vertical and horizontal lines—this part is shaded in color. ■

Cardinality of a Union

U

$|X| + |Y|$ adds this region twice

X $X \cap Y$ Y

FIGURE 6.8
Venn diagram for the number of elements in the union of two sets

Consider the problem of finding the cardinality of the union of two sets. If the sets are small enough, you can simply find the union and then count the number of elements. For example, if $U = \{1, 2, 3, 4, 5, 6, 7, 8, 9, 10\}$, $A = \{2, 4, 6, 8, 10\}$, and $B = \{1, 3, 4, 5, 6, 9\}$, then $|A| = 5$, $|B| = 6$, and $|A \cup B|$ can be found by first noting that $A \cup B = \{1, 2, 3, 4, 5, 6, 8, 9, 10\}$, so $|A \cup B| = 9$. However, if the sets are larger, then it would be convenient to find $|A \cup B|$ without first finding the set $A \cup B$. Some students might want to find $|A \cup B|$ by adding $|A|$ and $|B|$, but you can see from this example that $|A| + |B| = 5 + 6 \neq 9$. However, if you look at the Venn diagram for the number of elements in the union of two sets, the situation becomes quite clear, as shown in Figure 6.8.

Cardinality of a Union of Sets

$$|X \cup Y| = \underbrace{|X| + |Y|} - \underbrace{|X \cap Y|}$$

↑ The elements in the intersection are counted twice here.

↑ This corrects for the error introduced by counting those elements in the intersection twice.

This result generalizes for more sets. For example, for three sets,

$$|A \cup B \cup C| = |A| + |B| + |C| - |A \cap B| - |B \cap C| - |C \cap A| + |A \cap B \cap C|$$

The proof of this is left as a problem (see Problem 42 of Problem Set 6.2).

$$\begin{aligned} \text{Thus, } |A \cup B| &= |A| + |B| - |A \cap B| \\ &= 5 + 6 - 2 \quad \text{since} \quad A \cap B = \{4, 6\} \\ &= 9 \end{aligned}$$

Survey Problems

EXAMPLE 8 Suppose a survey indicates that 45 students are taking mathematics and 41 are taking English. How many students are taking math or English?

Solution At first, it might seem that all you need to do is add 41 and 45, but such is not the case.

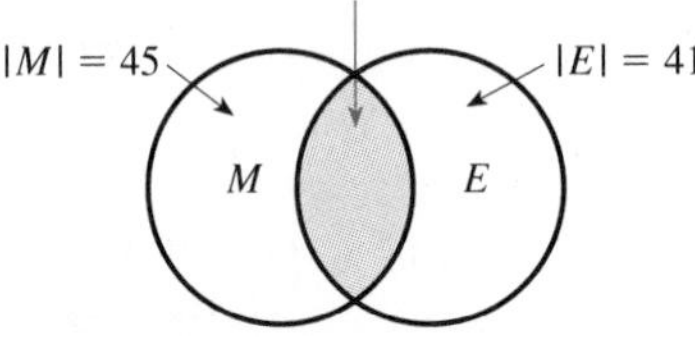

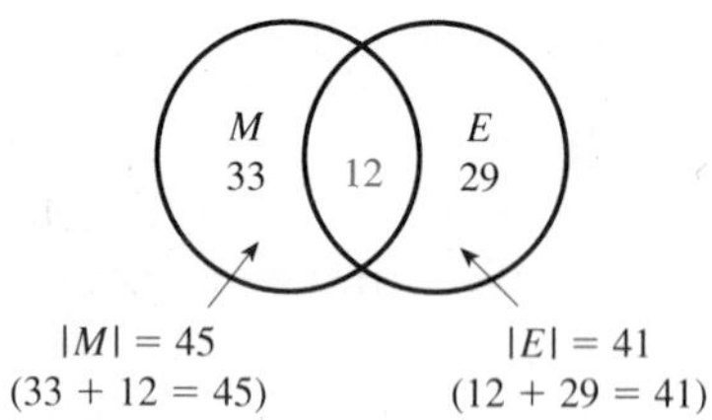

As you can see, you need further information. Suppose 12 students are taking both math and English. In this case, we see that

$$\begin{aligned} |M \cup E| &= |M| + |E| - |M \cap E| \\ &= 45 + 41 - 12 \\ &= 74 \end{aligned}$$

■

For three sets, the situation is a little more involved. There is a formula for the number of elements, but it is easier to use Venn diagrams, as illustrated by Example 8. Remember, the overall procedure is to fill in the number in the innermost region first and work your way out through the Venn diagram using subtraction.

EXAMPLE 9 **Survey.** A survey of 470 students gives the following information:

45 students are taking finite math
41 students are taking statistics
40 students are taking computer programming
15 students are taking finite math and statistics
18 students are taking finite math and computer programming
17 students are taking statistics and computer programming
7 students are taking all three

a. How many students are taking only finite math?

b. How many students are taking only statistics?

c. How many students are taking only computer programming?

d. How many students are not taking any of these courses?

Solution The method of solution is to draw a Venn diagram and fill in the various regions. Let

$$F = \{x \mid x \text{ is taking finite math}\}$$
$$S = \{x \mid x \text{ is taking statistics}\}$$
$$C = \{x \mid x \text{ is taking computer programming}\}$$

Step 1: Fill in $F \cap S \cap C$, which is given: $|F \cap S \cap C| = 7$.

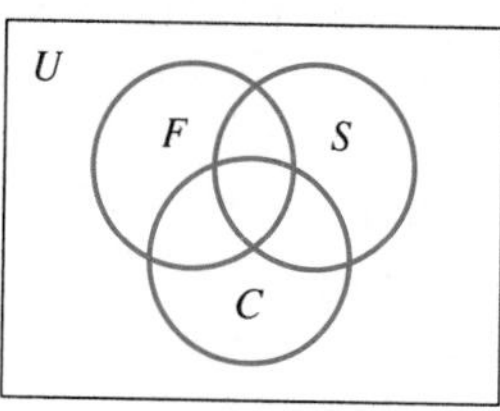

Sets F, S, and C

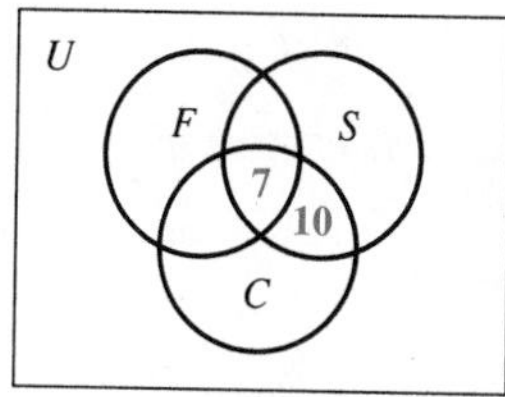

Steps 1 and 2

Step 2: Fill in $S \cap C$, which is given: $|S \cap C| = 17$. But 7 have previously been accounted for in this region, so we need only account for 10 additional members.

Step 3: Fill in $|F \cap C| = 18$. Need to fill in only 11 additional members.

Step 4: Fill in $|F \cap S| = 15$. Need to fill in only 8 additional members.

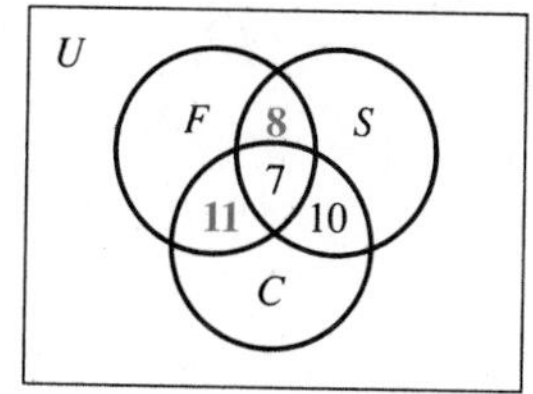

Steps 3 and 4

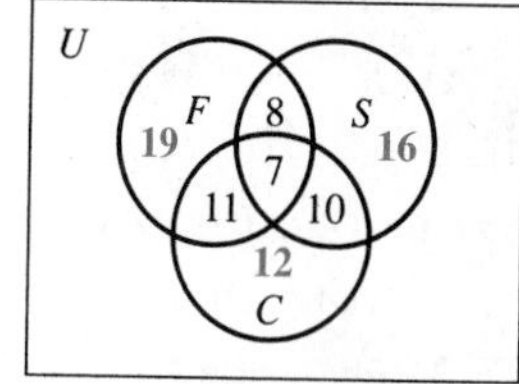

Steps 5, 6, and 7

Step 5: Fill in $|C| = 40$. Note that 28 have previously been filled in, so we need an additional 12 members.

Step 6: Fill in $|S| = 41$. Need to fill in 16 members.

Step 7: Fill in $|F| = 45$. Need to fill in 19 members.

Step 8: Adding all the numbers in the diagram, we have accounted for 83 members. Since 470 students were surveyed, we see that 387 are not taking any of the three courses.

We now have all the answers to the questions directly from the Venn diagram:

a. 19 **b.** 16 **c.** 12 **d.** 387 ■

6.2 Problem Set 6.2

Draw a Venn diagram to illustrate each relationship given in Problems 1–15.

1. $A \cap \bar{B}$

2. $\bar{A} \cup B$

3. $\bar{A} \cap C$

4. $\overline{B \cap C}$

5. $\overline{A \cap C}$

6. $\overline{A \cup B}$

7. $\overline{B \cup C}$

8. $A \cap (B \cup C)$

9. $A \cup (B \cup C)$

10. $\overline{(A \cup B) \cup C}$

11. $(A \cap B) \cap (A \cap C)$ **12.** $(A \cap B) \cup (A \cap C)$

13. $\overline{(A \cap B) \cup C}$ **14.** $A \cap \overline{B \cup C}$

15. $\overline{A \cup B} \cup C$

In Problems 16–24, let $U = \{1, 2, 3, 4, 5, 6, 7, 8, 9, 10\}$

$$A = \{2, 4, 6, 8\}$$
$$B = \{5, 9\}$$
$$C = \{2, 5, 8, 9, 10\}$$

List all the members of each set.

16. $(A \cup B) \cup C$ **17.** $(A \cap B) \cap C$

18. $A \cup (B \cap C)$ **19.** $A \cap (\bar{B} \cup C)$

20. $\bar{A} \cap (B \cup C)$ **21.** $A \cap \overline{B \cup C}$

22. $\overline{A \cap (B \cup C)}$ **23.** $\bar{A} \cup (B \cap C)$

24. $\overline{A \cup (B \cap C)}$

In Problems 25–30, use Venn diagrams to prove or disprove each expression. Remember to draw a diagram for the left side of the equation and another for the right side. If the final shaded portions are the same, then you have proved the result. If the final shaded portions are not identical, then you have disproved the result.

25. $\overline{A \cup B} = \bar{A} \cup \bar{B}$

26. $\bar{A} \cap \bar{B} = \overline{A \cup B}$

27. $(A \cup B) \cup C = A \cup (B \cup C)$

28. $A \cup (B \cup C) = (A \cup B) \cup (A \cup C)$

29. $A \cap (B \cup C) = (A \cap B) \cap (A \cap C)$

30. Prove De Morgan's Law: $\overline{X \cap Y} = \bar{X} \cup \bar{Y}$

APPLICATIONS

31. Montgomery College has a 50-piece band and a 36-piece orchestra. If 14 people are members of both the band and the orchestra, can the band and orchestra travel in two 40-passenger buses?

32. In a survey of a TriDelt chapter with 50 members, 18 were taking mathematics and 35 were taking English while 6 were taking both. How many were not taking either of these subjects?

33. Of the senior class at Rancho Cotati High School there were 25 football players and 16 basketball players. If 7 persons played both sports, how many different people played in these sports?

34. In a survey of the executive board of a national charity, it is found that 15 members earn more than \$100,000 per year and 9 own more than \$1,000,000 in negotiable securities. If 7 earn more than \$100,000 and own more than \$1,000,000 in negotiable securities, how many members are on the executive board, if all the members fall into one of the two categories?

35. In a sample of defective tires, 72 have defects in materials, 89 have defects in workmanship, and 17 have defects of both types. How many tires are in the sample of defective tires (the sample consists of only tires with defects in materials or workmanship)?

36. A survey of 100 persons at Plimbo Corporation shows that 40 jog, 25 swim, and 15 both swim and jog. How many do neither?

37. A survey of executives of *Fortune* 500 companies finds that 520 have MBA degrees, 650 have business degrees, and 450 have both degrees. How many executives with MBAs have nonbusiness undergraduate degrees? (*Note:* The MBA is a graduate degree and requires an undergraduate degree.)

38. A survey of 100 persons at Better Widgets finds that 40 jog, 25 swim, 16 cycle, 15 swim and jog, 10 swim and cycle, 8 jog and cycle, and 3 jog, swim, and cycle. Let

$$J = \{\text{People who jog}\}$$
$$S = \{\text{People who swim}\}$$
$$C = \{\text{People who cycle}\}$$

Use a Venn diagram to show how many are in each of the eight possible categories.

39. A survey of 100 women finds that 59 use shampoo A, 41 use shampoo B, 35 use shampoo C, 24 use shampoos A and B, 19 use shampoos A and C, 13 use shampoos B and C, and 11 use all three. Let

$$A = \{\text{Women who use shampoo A}\}$$
$$B = \{\text{Women who use shampoo B}\}$$
$$C = \{\text{Women who use shampoo C}\}$$

Use a Venn diagram to show how many are in each of the eight possible categories.

40. To see if Shannon Smith can handle the job he has applied for, the personnel manager sends him out to poll 100 people about their favorite types of TV shows. Shannon obtains the following data: 59 prefer comedies, 38 prefer variety shows, 42 prefer serious drama, 18 prefer comedies and variety programs, 12 prefer variety and serious drama, 16 prefer comedies and serious drama, 7 prefer all types, and 2 do not like any TV shows. If you were the personnel manager, would you hire Shannon on the basis of this survey?

41. In an interview of 50 students, 12 liked Proposition 8 and Proposition 13, 18 liked Proposition 8 but not Proposition 2, 4 liked Proposition 8, Proposition 13, and Proposition 2, 25 liked Proposition 8, 15 liked Proposition 13, 10 liked Proposition 2 but not Proposition 8 or Proposition 13, and 1 liked Proposition 13 and Proposition 2 but not Proposition 8.

a. Of those surveyed, how many did not like any of the three propositions?
b. How many liked Proposition 8 and Proposition 2?
c. Show the completed Venn diagram.

42. Show that

$$|A \cup B \cup C| = |A| + |B| + |C| - |A \cap B| - |A \cap C| - |B \cap C| + |A \cap B \cap C|$$

6.3 Fundamental Counting Principle

In the previous section we found a formula for counting the number of elements in a union of sets. However, sometimes we want to count in more general settings. In this section we will introduce the **Multiplication Principle**, which is so important to our study of combinatorics that it is also called the **Fundamental Counting Principle**. In order to understand this principle, we begin with an example.

EXAMPLE 1 Consider a club with five members:

$$\{\text{Alfie, Bogie, Calvin, Doug, Ernie}\}$$

In how many ways can they elect a president and secretary? (One person cannot hold both positions.)

Solution There are at least two ways to solve this problem. The first, and perhaps the easiest, is by drawing a tree diagram:

President	Secretary	Result (*Note*: Order is important—use parentheses.)
① a	① b	(a, b)
	② c	(a, c)
	③ d	(a, d)
	④ e	(a, e)
② b	① a	(b, a)
	② c	(b, c)
	③ d	(b, d)
	④ e	(b, e)
③ c	① a	(c, a)
	② b	(c, b)
	③ d	(c, d)
	④ e	(c, e)
④ d	① a	(d, a)
	② b	(d, b)
	③ c	(d, c)
	④ e	(d, e)
⑤ e	① a	(e, a)
	② b	(e, b)
	③ c	(e, c)
	④ d	(e, d)

a = Alfie
b = Bogie
c = Calvin
d = Doug
e = Ernie

By counting the number of branches in the tree, we see that there are 20 possibilities. This method is effective for small numbers, but the technique quickly gets out of hand. For example, if we wished to see how many ways the club could elect a president, secretary, and treasurer, this technique would be very lengthy.

A second method of solution is by using boxes or "pigeonholes" to represent each choice separately. For example:

Ways of choosing a president	*Ways of choosing a secretary*
5	4

↑ Since we have chosen a president, only four members remain.

If we multiply the numbers in the pigeonholes, we get

$$5 \cdot 4 = 20$$

and we see that the result is the same as that from the tree diagram. ■

This pigeonhole procedure is called the **multiplication principle** or the **fundamental counting principle**.

Multiplication Principle

> If task A can be performed in m ways, and, after task A is performed, a second task B can be performed in n ways, then task A followed by task B can be performed in $m \cdot n$ ways.

The multiplication principle can be used repeatedly in a particular problem so that it can be applied not only to two tasks, but also to three tasks, four tasks, or any number of tasks.

EXAMPLE 2 How many ways can the club in Example 1 select a president, secretary, and treasurer?

Solution Continue from where we left off in Example 1 and use the fundamental counting principle a second time:

$$5 \cdot 4 \cdot 3 = 60$$

President (5), Secretary (4) — From Example 1; 3 — Number of ways of selecting a treasurer

Can you draw a tree diagram for this example? ■

EXAMPLE 3 **a.** A coin is tossed three times and the sequence of heads and tails is recorded. What are the possible outcomes?

b. Three coins are tossed simultaneously. What are the possible outcomes?

Solution **a.** Since a single coin can land in two ways, the number of different ways (from the multiplication principle) is

$$2 \cdot 2 \cdot 2 = 8$$

We can show this sequence using a tree diagram:

First toss	*Second toss*	*Third toss*	*Outcomes*
H	H	H	(H, H, H)
		T	(H, H, T)
	T	H	(H, T, H)
		T	(H, T, T)
T	H	H	(T, H, H)
		T	(T, H, T)
	T	H	(T, T, H)
		T	(T, T, T)

b. If you toss three coins simultaneously, you might be inclined to say the possibilities are 3 heads, 2 heads and a tail, 2 tails and a head, and three tails, so there are 4 possibilities, but such an analysis is not correct. Even though you are tossing the coins simultaneously, there is no reason for the results to be any different from tossing a single coin three times. Look at the tree diagram for part **a** and replace the words "first toss," "second toss," and "third toss," with the words "first coin," "second coin," and "third coin," and you should see that the answer to this part is identical to the answer in part **a**. This says, for example, that if you toss five coins simultaneously, the number of possibilities is $2^5 = 32$. ■

The State of California "ran out" of codes that could be issued on license plates. The scheme in California is practiced in many states; that is, license plate codes consist of three letters followed by three numerals. For example, CWB 072 is a license plate code. When the state ran out of available new numbers, a change had to take place. The decision was made to leave the old numbers in circulation and to issue new plates in the order of three numerals followed by three letters, such as

The problem is to determine how many plates were available before the switch.* The solution to this problem is found by using the Fundamental Counting Principle.

* Certain plates are never issued because the combinations of letters produce obscene or confusing words, such as CHP, which might be misinterpreted as a car of the California Highway Patrol. But in our discussion here, we will assume that all possible license plates were issued.

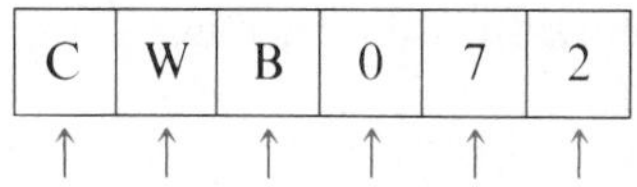

There are 26 possibilities for each of these letters. There are 10 possibilities for each of these numerals.

We have extended the Fundamental Counting Principle for performing 6 tasks.

Thus the total number of possible license plates that could be issued is:

$$26 \cdot 26 \cdot 26 \cdot 10 \cdot 10 \cdot 10 = 17{,}576{,}000$$

EXAMPLE 4 How many license plates can be formed if repetitions of letters or digits are not allowed?

Solution The result is given by

$$26 \cdot 25 \cdot 24 \cdot 10 \cdot 9 \cdot 8 = 11{,}232{,}000$$

■

As you can see, the multiplication principle is a very efficient general method of counting that breaks a larger number down into smaller tasks that can easily be enumerated. It works if the number of choices is independent of the choices made. Repetitions are allowed with this principle. This means that the same choice may be made twice.

EXAMPLE 5 It was recently reported that Wendy's spent nearly $1 million on thousands of taste tests to develop a better hamburger and came up with "The Big Classic." They tested 9 different buns, 40 sauces, 3 types of lettuce, 2 sizes of tomatoes, 4 different boxes, and 500 names. How many different hamburgers are possible if one choice is made from each category?

Solution The multiplication principle applies:

$$9 \times 40 \times 3 \times 2 \times 4 \times 500 = 4{,}320{,}000^*$$

■

In our work with combinatorics, we will frequently encounter products such as

$$6 \cdot 5 \cdot 4 \cdot 3 \cdot 2 \cdot 1$$

$$10 \cdot 9 \cdot 8 \cdot 7 \cdot 6 \cdot 5 \cdot 4 \cdot 3 \cdot 2 \cdot 1$$

$$52 \cdot 51 \cdot 50 \cdot 49 \cdot \cdots \cdot 4 \cdot 3 \cdot 2 \cdot 1$$

Since these are tedious to write out, **factorial notation** is used:

Definition of Factorial

$n!$ is called n *factorial* and is defined by

$$n! = n(n-1)(n-2) \cdot \cdots \cdot 3 \cdot 2 \cdot 1$$

for n a natural number. Also, $0! = 1$.

* The news article did not report the number of different hamburgers, but if the taste test was done at one location and each hamburger took 10 minutes to eat, this number of hamburgers would take over 80 years of nonstop eating to test each possibility! Notice that Wendy's also thought a different name changed the burger. If you do not count a different name as a different burger, then there are

$$9 \times 40 \times 3 \times 2 \times 4 = 8{,}640$$

different hamburgers.

EXAMPLE 6 Find the factorial of each whole number up to 10.

$$
\begin{aligned}
0! &= \mathbf{1}\\
1! &= \mathbf{1}\\
2! &= 1 \cdot 2 = \mathbf{2}\\
3! &= 1 \cdot 2 \cdot 3 = \mathbf{6}\\
4! &= 1 \cdot 2 \cdot 3 \cdot 4 = \mathbf{24}\\
5! &= 1 \cdot 2 \cdot 3 \cdot 4 \cdot 5 = \mathbf{120}\\
6! &= 1 \cdot 2 \cdot 3 \cdot 4 \cdot 5 \cdot 6 = \mathbf{720}\\
7! &= 1 \cdot 2 \cdot 3 \cdot 4 \cdot 5 \cdot 6 \cdot 7 = \mathbf{5{,}040}\\
8! &= 1 \cdot 2 \cdot 3 \cdot 4 \cdot 5 \cdot 6 \cdot 7 \cdot 8 = \mathbf{40{,}320}\\
9! &= 1 \cdot 2 \cdot 3 \cdot \cdots \cdot 8 \cdot 9 = \mathbf{362{,}880}\\
10! &= 1 \cdot 2 \cdot 3 \cdot \cdots \cdot 9 \cdot 10 = \mathbf{3{,}628{,}800}
\end{aligned}
$$

■

Calculator note: Look for a [!] or [$x!$] key on your calculator. This is a factorial key. Verify one of the entries above by using your calculator.

EXAMPLE 7 $5! - 4! = 120 - 24$ PRESS: [5] [$x!$] [−] [4] [$x!$] [=]

$= 96$

■

EXAMPLE 8 $(5 - 4)! = 1!$ PRESS: [5] [−] [4] [=] [$x!$]

$= 1$

■

EXAMPLE 9 $(2 \cdot 3)! = 6!$

$= 720$

■

EXAMPLE 10 $2!3! = 2 \cdot 1 \cdot 3 \cdot 2 \cdot 1$

$= 12$

■

EXAMPLE 11 $\dfrac{8!}{4!} = \dfrac{8 \cdot 7 \cdot 6 \cdot 5 \cdot \not{4} \cdot \not{3} \cdot \not{2} \cdot \not{1}}{\not{4} \cdot \not{3} \cdot \not{2} \cdot \not{1}}$

$= 8 \cdot 7 \cdot 6 \cdot 5$

$= 1{,}680$

■

EXAMPLE 12 $\dfrac{100!}{98!} = \dfrac{100 \cdot 99 \cdot \not{98!}}{\not{98!}}$ Note that $100! = 100 \cdot 99!$
$= 100 \cdot 99 \cdot 98!$

$= 9{,}900$

■

As the problems use larger numbers, you will find it necessary to simplify first and then use your calculator, if necessary. Example 12, for example, cannot be worked on a calculator without first simplifying.

EXAMPLE 13 $\dfrac{8!}{3!(8-3)!} = \dfrac{8!}{3!5!}$

$$= \frac{8 \cdot 7 \cdot \not{6} \cdot \not{5}!}{\not{3} \cdot \not{2} \cdot 1 \cdot \not{5}!}$$

$$= 56$$

■

Examples 12 and 13 illustrate a useful property of factorial.

Multiplication Property of Factorial

$$n! = n(n-1)!$$

This property was used in Example 12:

$$100! = 100 \cdot 99! \qquad \text{and} \qquad 99! = 99 \cdot 98!$$

Thus $100! = 100 \cdot 99 \cdot 98!$. This means that when using factorial notation, we can "count down" to any convenient number and then affix a factorial symbol. See Example 13, where we used $8! = 8 \cdot 7 \cdot 6 \cdot 5!$.

6.3

Problem Set 6.3

Evaluate each expression in Problems 1–18.

1. $6! - 4!$ **2.** $7! - 3!$

3. $8! - 5!$ **4.** $(6 - 4)!$

5. $(7 - 3)!$ **6.** $(8 - 5)!$

7. $4!3!$ **8.** $(4 \cdot 3)!$

9. $(5 \cdot 2)!$ **10.** $5!2!$

11. $\dfrac{9!}{7!}$ **12.** $\dfrac{10!}{6!}$

13. $\dfrac{12!}{8!}$ **14.** $\dfrac{10!}{4!6!}$

15. $\dfrac{9!}{5!4!}$ **16.** $\dfrac{12!}{3!(12-3)!}$

17. $\dfrac{11!}{4!(11-4)!}$ **18.** $\dfrac{52!}{3!(52-3)!}$

APPLICATIONS

19. How many outfits consisting of a skirt and a blouse can a woman select if she has three skirts and five blouses?

20. How many outfits consisting of a suit and a tie can a man select if he has two suits and eight ties?

21. In how many ways can a group of 15 people elect a president and a vice president?

22. In how many ways can a group of 10 people elect a president, a vice president, and a secretary?

23. In how many ways can a group of four people elect a president, vice president, and secretary? Find your answer using the multiplication principle and then substantiate your answer with a tree diagram.

24. A coin is tossed four times and the sequence of heads and tails is recorded. Use the multiplication principle to determine the number of different possibilities and then substantiate your answer with a tree diagram.

25. There are two routes connecting Ferndale and Centerville, and four routes connecting Centerville and Weymouth. How many different routes are there between Ferndale and Weymouth?

26. An apartment complex has studio, one-, and two-bedroom units. Each of these units comes with and without a view. How many different types of apartments can be rented?

27. A public opinion poll asks the following questions:

(1) Sex (M or F)
(2) Age (under 20; 20–50; over 50)
(3) Political affiliation (Democrat, Republican, Independent)

How many different classifications are possible?

28. A restaurant allows you to pick one item from Column A (which has 5 items from which to choose), one item from Column B (which has 4 items from which to choose), and one item from Column C (which has 6 items from which to choose). How many different meals can be chosen?

29. Foley's Village Inn offers the following menu in its restaurant:

Main course	Dessert	Beverage
Prime rib	Ice cream	Coffee
Steak	Sherbet	Tea
Chicken	Cheesecake	Milk
Ham		Sanka
Shrimp		

In how many different ways can someone order a meal consisting of one choice from each category?

30. How many seven-digit telephone numbers are possible if the first two digits cannot be ones or zeros?

31. A typical Social Security number is 555-38-4459; the first digit cannot be zero. How many Social Security numbers are possible?

32. Some California license plates consist of one digit, followed by three letters, followed by three digits. How many such license plates are possible?

33. If a state issues license plates that consist of one letter followed by five digits, how many different plates are possible?

34. Repeat Problem 33 if the first letter cannot be O, Q, or I.

35. Repeat Problem 33 without repetition of digits.

36. A certain lock has five tumblers, each of which can assume six positions. How many different positions are possible?

37. How many different answers are possible for a true–false test with 10 questions?

38. How many different answers are possible for a multiple-choice test with 10 questions and five possible answers for each question?

39. Suppose an automated teller machine (ATM) card is accessed by pressing in a secret four-digit code. How many codes are possible if
    a. the code consists of digits only?
    b. the code consists of letters only?
    c. the code consists of digits or letters?

40. Answer the questions in Problem 39 if repetitions are not allowed.

41. A restaurant allows you to pick one item from Column A (which has 5 items from which to choose), one item from Column B (which has 4 items from which to choose), and one item from Column C (which has 6 items from which to choose). How many different meals can be chosen?

42. A restaurant allows you to pick one item from Column A (which has 9 items from which to choose), one item from Column B (which has 6 items from which to choose), and one item from Column C (which has 10 items from which to choose). How many different meals can be chosen?

43. A combination lock is unlocked by turning to the left, then to the right, and then to the left again. If there are 30 digits on the dial, how many possible combinations are there?

44. New York license plates consist of three letters followed by three digits. It is also known that 245 specific arrangements of three letters are not allowed because they are considered obscene. How many license plates are possible?

45. Suppose you flip a coin and keep a record of the results. In how many ways could you obtain at least one head if you flip the coin four times?

46. Repeat Problem 45 if you flip the coin ten times.

6.4 *Permutations*

Consider the set

$$A = \{a, b, c, d, e\}$$

Remember, when set symbols are used, the order in which the elements are listed is not important. Suppose now that we wish to select elements from A by taking them *in a certain order*. For example, suppose set A is a club and the members are holding an election for president and secretary. If the first person selected is the president, and the second person selected is the secretary, then selecting members a and b is not the same as selecting members b and a for these positions. Since set notation signifies that the order of the elements is not important, another notation must be used when the order is important. This notation uses parentheses rather than braces. For this example, (a, b) or (b, a) represents (president, secretary) and $(a, b) \neq (b, a)$. If two elements are selected, then the notation (a, b) is called an **ordered pair**; if three are selected, it is called an **ordered triplet**; if four are selected, it is called an **ordered four-tuple**; and so on. These selections are called **arrangements** of the given set, A in this example. Arrangements are said to be selections from the given set *without repetitions*, since a symbol cannot be selected from the set twice.

EXAMPLE 1 Find the number of permutations (arrangements) of the elements a, c, and d from the set A. List these arrangements by using a tree diagram.

Solution From the multiplication principle we see that the number of permutations is

$$3 \cdot 2 \cdot 1 = 6$$

In listing a permutation, be sure to use ordered n-tuple notation. In this case, an ordered triplet, such as (a, d, c), means the *order is important*.

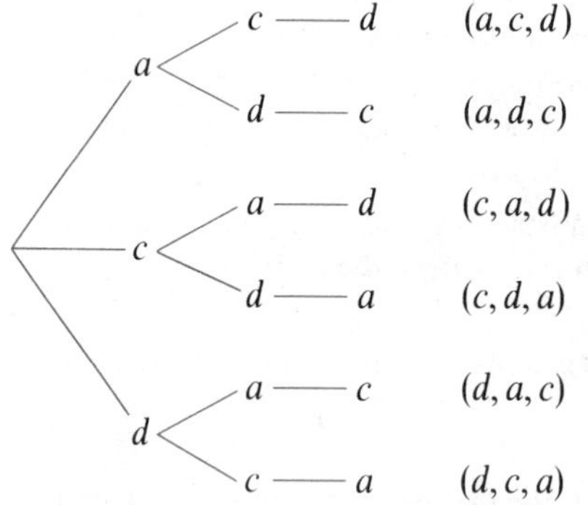

■

Permutation

> A **permutation** of r elements of a set S with n distinct elements is an ordered arrangement of those r elements selected without repetitions.

EXAMPLE 2 How many permutations of two elements can be selected from a set of six elements?

Solution Let $B = \{a, b, c, d, e, f\}$ and select two elements:

(a, b)	(a, c)	(a, d)	(a, e)	(a, f)
(b, a)	(b, c)	(b, d)	(b, e)	(b, f)
(c, a)	(c, b)	(c, d)	(c, e)	(c, f)
(d, a)	(d, b)	(d, c)	(d, e)	(d, f)
(e, a)	(e, b)	(e, c)	(e, d)	(e, f)
(f, a)	(f, b)	(f, c)	(f, d)	(f, e)

There are 30 permutations of two elements selected from a set of six elements.

■

Example 2 brings up two difficulties. The first is the lack of notation for the phrase

the number of permutations of two elements selected from a set of six elements and the second is the inadequacy of relying on direct counting, especially if the sets are very large.

Notation for Permutations

$_nP_r$ is a symbol used to denote the *number of permutations* of r elements selected from a set of n elements. Another way of saying this is to say we are selecting n distinct objects taken r at a time.

Example 2 can now be shortened by writing

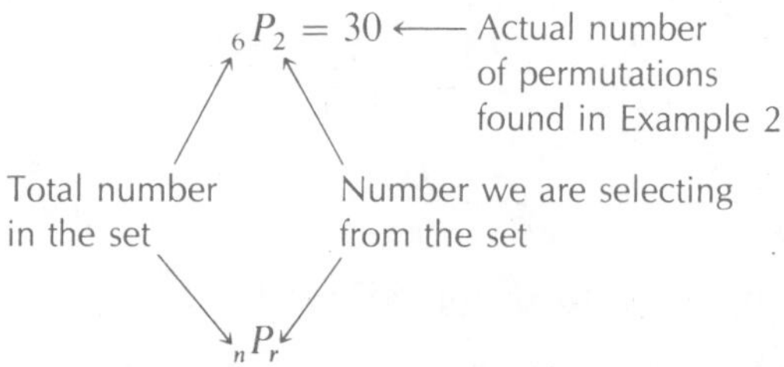

Next, to find a formula for $_nP_r$, we turn to the *multiplication principle*. The number of ways of permuting n distinct objects taken r at a time, written $_nP_r$, is given by

$$_nP_r = \underbrace{n(n-1)(n-2)\cdot\cdots\cdot(n-r+1)}_{r \text{ factors}}$$

Using factorials, the formula for $_nP_r$ can be written more simply as follows:

$$\begin{aligned} _nP_r &= n(n-1)(n-2)\cdot\cdots\cdot(n-r+1) \\ &= n(n-1)(n-2)\cdot\cdots\cdot(n-r+1)\cdot\frac{(n-r)!}{(n-r)!} \\ &= \frac{n(n-1)(n-2)\cdot\cdots\cdot(n-r+1)(n-r)(n-r-1)\cdot\cdots\cdot 3\cdot 2\cdot 1}{(n-r)!} \\ &= \frac{n!}{(n-r)!} \end{aligned}$$

This is the general formula for $_nP_r$.

Permutation Formula

$$_nP_r = \frac{n!}{(n-r)!}$$

EXAMPLE 3

$$\begin{aligned} _{10}P_2 &= \frac{10!}{(10-2)!} \\ &= \frac{10\cdot 9\cdot 8!}{8!} \\ &= 90 \end{aligned}$$

■

EXAMPLE 4 $${}_nP_0 = \frac{n!}{(n-0)!} = 1$$ ■

EXAMPLE 5 Find the number of license plates possible in a state using only three letters, if none of the letters can be repeated.

Solution This is a permutation of 26 objects taken 3 at a time. Thus the solution is given by

$$\begin{aligned} {}_{26}P_3 &= \frac{26!}{23!} \\ &= \frac{26 \cdot 25 \cdot 24 \cdot \cancel{23!}}{\cancel{23!}} \\ &= 15{,}600 \end{aligned}$$ ■

EXAMPLE 6 Repeat Example 5 if repetitions are allowed.

Solution Remember, permutations do not allow repetitions, so this is *not* a permutation. However, the multiplication principle still applies:

$$26 \cdot 26 \cdot 26 = 17{,}576$$ ■

EXAMPLE 7 Find the number of arrangements of letters in the word MATH.

Solution This is a permutation of four objects taken 4 at a time:

$$\begin{aligned} {}_4P_4 &= \frac{4!}{0!} \\ &= \frac{24}{1} \\ &= 24 \end{aligned}$$ ■

EXAMPLE 8 How many permutations are there of the letters in the word HATH?

Solution If you try to solve this problem as you did Example 7, you would have

$${}_4P_4 = 4! = 24$$

However, if you list the possibilities, you would find only 12 *different permutations.* The difficulty here is that two of the letters in the word HATH are indistinguishable. If you label them as H_1ATH_2, you would find such possibilities as

H_1ATH_2 H_2ATH_1 H_1AH_2T

If you complete this list, you would indeed find $4! = 24$ possibilities. But since there are *two indistinguishable* letters, you divide the total, 4!, by 2 (different orderings of two letters) to find the result:

$$\frac{4!}{2} = \frac{24}{2} = 12$$ ■

EXAMPLE 9 How many permutations are there of the letters in the word ASSIST?

Solution There are six letters, and if you consider the letters as distinguishable, as in

$$AS_1S_2IS_3T$$

there are ${}_6P_6 = 6! = 720$ possibilities. However,

$$\begin{array}{lll} AS_1S_2IS_3T & AS_1S_3IS_2T & AS_2S_1IS_3T \\ AS_2S_3IS_1T & AS_3S_1IS_2T & AS_3S_2IS_1T \end{array}$$

are all indistinguishable, so you must divide the total by $3! = 6$.

$$\frac{6!}{3!} = \frac{6 \cdot 5 \cdot 4 \cdot 3!}{3!} = 120$$

There are 120 permutations of the letters in the word ASSIST. ■

Examples 8 and 9 suggest a general result:

Permutations of n Objects, Not All Distinct

Given a set of n objects in which n_1 are of one kind, n_2 are of another kind, . . . , and n_k are of a further kind so that

$$n = n_1 + n_2 + \cdots + n_k$$

Then the number of *distinguishable* permutations is given by the formula $\dfrac{n!}{n_1!n_2! \cdot \cdots \cdot n_k!}$. This is called an **ordered partition** and is denoted by

$$\binom{n}{n_1, n_2, \ldots, n_k}$$

EXAMPLE 10 The number of permutations of the letters in the word ATTRACT is 7! (since there are seven letters) divided by factorials of the number of subcategories of repeated letters:

$$\frac{7!}{3!2!1!1!}$$

← Total number of objects. (7!)

3! — Letter T occurs three times.

2! — Letter A occurs twice.

1!1! — Letters R and C occur once.

This number can now be simplified:

$$\frac{7 \cdot 6 \cdot 5 \cdot \overset{2}{\cancel{4}} \cdot \cancel{3!}}{\cancel{3!} \cdot \cancel{2}} = 420$$

■

EXAMPLE 11 Wohlert's Realty received ten calls about a listing. In how many ways can these inquiries be routed to the company brokers if Walt Wohlert is to receive four of them, Joel Simpson is to receive three of them, Al Rommer is to receive two of them, and Tim Selbo will receive the remaining inquiry?

Solution The number of ways of assigning the inquiries is

$$\frac{10!}{4!3!2!1!} = \frac{10 \cdot 9 \cdot 8 \cdot 7 \cdot 6 \cdot 5 \cdot 4!}{4! \cdot 3 \cdot 2 \cdot 2} = 12{,}600 \text{ ways}$$ ■

EXAMPLE 12 Suppose you have 8 history books, 5 English books, and 4 math books. You have room on your bookshelf for 10 books and decide to select 5 history books, 3 English books, and 2 math books. How many ways can these books be arranged on the shelf if they are arranged in this order: English books first, math books in the middle, and history books on the right?

Solution Since the order is important, this is a permutation problem. We think of this arrangement as three separate tasks, and begin by using the multiplication principle:

$$\begin{pmatrix}\text{WAYS OF ARRANGING}\\ \text{ENGLISH BOOKS}\end{pmatrix} \cdot \begin{pmatrix}\text{WAYS OF ARRANGING}\\ \text{MATH BOOKS}\end{pmatrix} \cdot \begin{pmatrix}\text{WAYS OF ARRANGING}\\ \text{HISTORY BOOKS}\end{pmatrix}$$

Now the number of ways each of these tasks can be accomplished is a permutation, so we have

$$_5P_3 \cdot {_4P_2} \cdot {_8P_5} = \frac{5!}{(5-3)!} \cdot \frac{4!}{(4-2)!} \cdot \frac{8!}{(8-5)!} = \frac{5! \cdot 4! \cdot 8!}{2! \cdot 2! \cdot 3!} = 4{,}838{,}400 \text{ ways}$$ ■

6.4 Problem Set

Evaluate each of the numbers in Problems 1–6.

1. a. $_9P_1$ b. $_9P_2$ c. $_9P_3$ d. $_9P_4$ e. $_9P_0$
2. a. $_5P_4$ b. $_{52}P_3$ c. $_7P_2$ d. $_4P_4$ e. $_{100}P_1$
3. a. $_{12}P_5$ b. $_5P_3$ c. $_8P_4$ d. $_8P_0$ e. $_gP_h$
4. a. $_{92}P_0$ b. $_{52}P_1$ c. $_7P_5$ d. $_{16}P_3$ e. $_nP_4$
5. a. $_7P_3$ b. $_5P_5$ c. $_{50}P_{48}$ d. $_{25}P_1$ e. $_mP_3$
6. a. $_8P_3$ b. $_{12}P_0$ c. $_{10}P_2$ d. $_{11}P_4$ e. $_nP_5$

How many permutations are there for the letters in the words in Problems 7–16?

7. HOLIDAY
8. FORMULA
9. ESCHEW
10. ANNEX
11. MISSISSIPPI
12. OBFUSCATION
13. BOOKKEEPING
14. CONCENTRATION
15. APOSIOPESIS
16. GRAMMATICAL

17. Find the number of permutations of the elements $\{a, b, c\}$ from the set $\{a, b, c, d, e\}$. List these arrangements by using a tree diagram.
18. Find the number of permutations of the elements $\{b, c, d, e\}$ from the set $\{a, b, c, d, e\}$. List these arrangements by using a tree diagram.

APPLICATIONS

Answer the questions in Problems 19–40 by using the multiplication principle, the permutation formulas (with and without distinct objects), or both.

19. In how many ways can a group of five people elect a president, vice president, secretary, and treasurer?
20. In how many different ways can eight books be arranged on a shelf?
21. In how many ways can you select and read three books from a shelf of eight books?
22. In how many ways can ten toys be distributed to three children so that each child receives one toy?

23. In how many ways can seven tennis players be seeded (ranked) in a tournament?

24. In how many ways can a row of 3 contestants for a TV game show be selected from an audience of 362 people?

25. What is the number of ways a baseball team consisting of nine players can be arranged for batting order?

26. A couple knows they will be giving birth to a girl and have a list of seven first names and five middle names for the new bundle of joy. How many different first and middle names are possible if they choose from their list?

27. In how many ways can McDonald's select two sites for construction if there are eight possible sites from which to choose?

28. If five people arrive at a bank at the same time, in how many different ways can they arrange themselves in the same line?

29. In how many ways can six half-hour prime time slots be filled with 15 different programs?

30. A night security person must choose a route and visit eight sites. If the security person works five nights a week for 50 weeks a year, how long will it take the security person to be required to repeat a route if a different route is chosen each evening?

31. A truck driver has seven different deliveries to make each day. If a different route is chosen each day, how long will it take the truck driver to complete all different routes for the deliveries? Assume that deliveries are made Monday–Saturday for 50 weeks a year.

32. A senate investigative team is to be formed. One member is to come from among 12 Republicans, one from among 15 Democrats, and one from the special prosecutor's staff of 5. In how many ways can this investigative team be formed?

33. A group of three drivers and two persons who do not drive form a carpool. If the car seats six persons, in how many ways can the five persons arrange themselves in the car?

34. Common sense would dictate that the empty seat in Problem 33 be the middle front seat. If this is the case, answer the question posed in Problem 33.

35. Seven calls are routed to three salespersons. One of these salespersons receives three of the calls, and each of the other two salespersons receives two calls each. In how many different ways can these calls be routed to the three salespersons?

36. If nine calls are routed to three salespersons by giving three to each, in how many ways can this be done?

37. Ten seniors are to line up for a group photograph. If the order of students in the line is important, how many different arrangements are possible?

38. **a.** In how many ways can five race car drivers be assigned to two cars?
    b. In how many ways can five race car drivers be assigned to seven cars?

39. Suppose you have six math books and nine computer science books to place on a shelf with five slots. In how many ways can you put the books on the shelf if the first two slots are filled with math books and the next three slots with computer science books?

40. Suppose you have six math books, eight business books, and three accounting books to place on a shelf with nine slots. In how many ways can you put the books on the shelf if the first three slots are filled with math books, the next four slots with business books, and the last two slots with accounting books?

6.5 Combinations

Consider the set

$$A = \{a, b, c, d, e\}$$

If two elements are selected from A in a certain order, they are represented by an *ordered pair* and the ordered pair is called a *permutation*. On the other hand, if two elements are selected from A *without regard to the order in which they are selected*, they are represented as a subset of A.

EXAMPLE 1 Select two elements from A:

Permutations—Order important			
(a, b)	(a, c)	(a, d)	(a, e)
(b, a)	(b, c)	(b, d)	(b, e)
(c, a)	(c, b)	(c, d)	(c, e)
(d, a)	(d, b)	(d, c)	(d, e)
(e, a)	(e, b)	(e, c)	(e, d)

Notation: ${}_5P_2 = 20$

There are 20 permutations—order is important.

Subsets—Order not important			
$\{a, b\}$	$\{a, c\}$	$\{a, d\}$	$\{a, e\}$
	$\{b, c\}$	$\{b, d\}$	$\{b, e\}$
		$\{c, d\}$	$\{c, e\}$
			$\{d, e\}$

Do not list $\{b, a\}$ since $\{b, a\} = \{a, b\}$

There are 10 subsets; note that set notation is used. ■

We use the following name and notation when listing different subsets of a given set.

Definition of Combination

A **combination** of r elements of a finite set S is a subset of S that contains r distinct elements.

Remember when listing the elements of a subset that the order in which those elements are listed is not important. A notation similar to that used for permutations is used to denote the number of combinations.

Notation for Combinations

$\binom{n}{r}$ and ${}_nC_r$ are symbols used to denote the number of combinations of r elements selected from a set of n elements ($r \leq n$).

The notation ${}_nC_r$ is similar to the notation used for permutations, but since $\binom{n}{r}$ is used in later work, we will use that notation for combinations. The notation $\binom{n}{r}$ is read as "n choose r."

The formula for the number of permutations leads directly to a formula for the number of combinations since each subset of r elements has $r!$ permutations of its members. Thus, by the multiplication principle,

$$\binom{n}{r} \cdot r! = {}_nP_r$$

so

$$\binom{n}{r} = \frac{{}_nP_r}{r!} = \frac{n!}{r!(n-r)!}$$

Combination Formula

$$\binom{n}{r} = \frac{n!}{r!(n-r)!}$$

EXAMPLE 2 $\binom{10}{3} = \frac{10!}{3!7!} = \frac{\overset{5}{\cancel{10}} \cdot \overset{3}{\cancel{9}} \cdot 8 \cdot \cancel{7!}}{\cancel{3} \cdot \cancel{2} \cdot \cancel{7!}} = 120$ ■

EXAMPLE 3 $\binom{n}{0} = \frac{n!}{0!(n-0)!} = 1$ ■

EXAMPLE 4
$$\binom{m-1}{2} = \frac{(m-1)!}{2!(m-1-2)!}$$
$$= \frac{(m-1)(m-2)(m-3)!}{2 \cdot 1 \cdot (m-3)!}$$
$$= \frac{(m-1)(m-2)}{2}$$
■

EXAMPLE 5 In how many ways can a club of five members select a three-person committee?

Solution $\binom{5}{3} = \frac{5!}{3!2!} = \frac{5 \cdot 4 \cdot 3}{3!} = 10$ ■

EXAMPLE 6 Find the number of five-card hands that can be drawn from an ordinary deck of cards.

Solution
$$\binom{52}{5} = \frac{52!}{5!47!}$$
$$= \frac{52 \cdot 51 \cdot 50 \cdot 49 \cdot 48}{5 \cdot 4 \cdot 3 \cdot 2 \cdot 1}$$
$$= 2{,}598{,}960$$

CALCULATOR COMMENT

Some calculators are very efficient at working with permutations, combinations, and factorials. The TI-81, for example, will evaluate the expression in Example 6 if you press

[52] [MATH] [PRB] [nCr]

[ENTER] [5] [ENTER]

■

EXAMPLE 7 In how many ways can a diamond flush be drawn in poker? (A diamond flush is a hand of five diamonds.)

Solution This is a combination of 13 objects (diamonds) taken 5 at a time. Thus, the solution is given by

$$\binom{13}{5} = \frac{13!}{5!8!} = \frac{13 \cdot 12 \cdot 11 \cdot 10 \cdot 9}{5!} = 1{,}287$$
■

EXAMPLE 8 In how many ways can a flush be drawn in poker?*

Solution Begin with the multiplication principle:

$$\underbrace{4}_{\substack{\text{Number} \\ \text{of suits}}} \cdot \underbrace{1{,}287}_{\substack{\text{Number of ways of} \\ \text{drawing a flush in} \\ \text{a particular suit} \\ \text{(from Example 7)}}} = 5{,}148$$

■

EXAMPLE 9 In how many ways can a full house of three aces and two kings be dealt?

Solution Begin with the multiplication principle:

$$\begin{pmatrix} \text{WAYS OF OBTAINING} \\ \text{ACES} \end{pmatrix} \cdot \begin{pmatrix} \text{WAYS OF OBTAINING} \\ \text{KINGS} \end{pmatrix}$$

Each of these numbers is a combination (since the order in which you are dealt the cards is not important):

$$\begin{aligned} \begin{pmatrix} \text{WAYS OF OBTAINING} \\ \text{ACES} \end{pmatrix} \cdot \begin{pmatrix} \text{WAYS OF OBTAINING} \\ \text{KINGS} \end{pmatrix} &= \binom{4}{3}\binom{4}{2} \\ &= \frac{4!}{3!1!} \cdot \frac{4!}{2!2!} \\ &= \frac{4 \cdot 3 \cdot 4 \cdot 3}{3 \cdot 2} = 24 \end{aligned}$$

■

Which Method?

We have now looked at several counting schemes: tree diagrams, the multiplication principle, permutations, combinations, and ordered partitions. In practice, you will generally not be told what type of counting problem you are dealing with—you will need to decide. Table 6.1 should help with that decision.

EXAMPLE 10 What is the number of license plates possible in Florida if each license plate consists of three letters followed by three digits, and we add the condition that repetition of letters or digits is not permitted?

Solution This is a permutation problem since the *order* in which the elements are arranged is important. That is, CWB072 and BCW072 are different plates. The number is found by using permutations along with the multiplication principle:

$$\begin{aligned} {}_{26}P_3 \cdot {}_{10}P_3 &= 26 \cdot 25 \cdot 24 \cdot 10 \cdot 9 \cdot 8 \\ &= 11{,}232{,}000 \end{aligned}$$

■

* A flush in poker is five cards from one suit (see Figure 7.1 on page 331).

TABLE 6.1

Multiplication principle	Permutations	Combinations	Ordered partitions
Counting total of separate tasks	Number of ways of selecting r items out of n items		Partition n elements into k categories
Repetitions allowed	Repetitions are not allowed		Repetitions are not allowed
If tasks $1, 2, 3, \ldots, k$ can be performed in $n_1, n_2, n_3, \ldots, n_k$ ways, respectively, then the total number of ways the k tasks can be performed is $n_1 \cdot n_2 \cdot n_3 \cdot \cdots \cdot n_k$ ways	Order is important (select and arrange)	Order is not important (just select)	Order is not important
	Arrangements of r items from a set of n items	Subsets of r items from a set of n items	Place n items into k categories where $n = n_1 + n_2 + n_3 + \cdots + n_k$
	${}_nP_r = \dfrac{n!}{(n-r)!}$	$\dbinom{n}{r} = \dfrac{n!}{r!(n-r)!}$	$\dbinom{n}{n_1, n_2, \ldots, n_k} = \dfrac{n!}{n_1!n_2!\cdots n_k!}$

EXAMPLE 11 A quartet is to be selected from a choir. There is to be one soprano selected from a group of six sopranos, two tenors selected from a group of five tenors, and a bass selected from three basses.

a. In how many ways can the quartet be formed?

b. In how many ways can the quartet be formed if one of the tenors is designated lead tenor?

Solution **a.** Begin with the multiplication principle:

$$\begin{aligned}
&\left(\begin{matrix}\text{WAYS OF SELECTING}\\ \text{SOPRANO}\end{matrix}\right)\cdot\left(\begin{matrix}\text{WAYS OF SELECTING}\\ \text{TENORS}\end{matrix}\right)\cdot\left(\begin{matrix}\text{WAYS OF SELECTING}\\ \text{BASS}\end{matrix}\right)\\
&= \binom{6}{1} \cdot \binom{5}{2} \cdot \binom{3}{1}\\
&= 6 \cdot 10 \cdot 3\\
&= 180
\end{aligned}$$

b. Since the order of selecting the tenors is important, the middle factor from part **a** is replaced by

$$_5P_2 = \frac{5!}{3!} = 20$$

The number of ways of selecting the quartet is $6 \cdot 20 \cdot 3 = 360$. ■

EXAMPLE 12 Suppose that the Sharp Investment Company has 15 sales representatives who are to be reassigned into three geographical areas as follows: 4 in the North, 5 in the South, and 6 in the West. In how many ways could the sales representatives be assigned to the geographical areas?

Solution Use the multiplication principle to fill three pigeonholes:

North: 15 choose 4

South: 11 choose 5; there are 11 persons left after 4 are removed for the northern region.

$$\binom{15}{4} \cdot \binom{11}{5} \cdot \binom{6}{6} = \frac{15!}{4!\,11!} \cdot \frac{11!}{5!\,6!} \cdot \frac{6!}{6!\,0!} = \frac{15!}{4!\,5!\,6!}$$

West: 6 choose 6; there are 6 persons left after 5 more are removed from the 11 for the southern region.

Note the connection between combinations and ordered partitions. We could model this problem by

$$\binom{15}{4,\,5,\,6} = \frac{15!}{4!\,5!\,6!}$$

The fractional form $\frac{15!}{4!5!6!}$ is acceptable; but if you have a calculator you can state the answer as 630,630. ■

EXAMPLE 13 Find the number of bridge hands (13 cards) consisting of six hearts, four spades, and three diamonds.

Solution The order in which the cards are received is unimportant, so finding the number of bridge hands is a combination problem. Also use the multiplication principle:

Number of ways of obtaining hearts · Number of ways of obtaining spades · Number of ways of obtaining diamonds

$$\binom{13}{6} \cdot \binom{13}{4} \cdot \binom{13}{3}$$

$$= \frac{13!}{6!(13-6)!} \cdot \frac{13!}{4!(13-4)!} \cdot \frac{13!}{3!(13-3)!}$$

$$= \frac{13!13!13!}{6!7!4!9!3!10!}$$ ⟵ This answer is acceptable.

$$= 350{,}904{,}840$$ ■

EXAMPLE 14 A club with 42 members wants to elect a president, a vice president, and a treasurer. From the other members, an advisory committee of 5 people is to be selected. In how many ways can this be done?

Solution This is both a permutation and a combination problem, with the final result calculated by using the multiplication principle:

$$\underbrace{\begin{matrix}\text{Number of ways}\\ \text{of selecting}\\ \text{officers}\end{matrix}}\quad\underbrace{\begin{matrix}\text{Number of ways}\\ \text{of selecting}\\ \text{committee}\end{matrix}}$$

$$
\begin{aligned}
&{}_{42}P_3 \cdot \binom{39}{5}\\
&= 42 \cdot 41 \cdot 40 \cdot \frac{39!}{5!(39-5)!}\\
&= 39{,}658{,}142{,}160 \quad \longleftarrow \text{Found with a calculator}
\end{aligned}
$$

■

EXAMPLE 15 At Baker Naval Air Station there are five officers, 15 petty officers, and 30 enlisted personnel. In how many ways could the following five-person committees be formed?

a. five officers

b. five petty officers

c. one officer, 2 petty officers, and 2 enlisted members

d. at least three officers

Solution **a.** $\binom{5}{5} = \frac{5!}{5!0!} = 1$

b. $\binom{15}{5} = \frac{15!}{5!10!} = \frac{15 \cdot 14 \cdot 13 \cdot 12 \cdot 11 \cdot 10!}{5 \cdot 4 \cdot 3 \cdot 2 \cdot 10!} = 3{,}003$

c. This requires both the multiplication principle and combinations:

$$
\begin{aligned}
\binom{5}{1}\binom{15}{2}\binom{30}{2} &= \frac{5!}{1!4!} \cdot \frac{15!}{2!13!} \cdot \frac{30!}{2!28!}\\
&= \frac{5 \cdot 4! \cdot 15 \cdot 14 \cdot 13! \cdot 30 \cdot 29 \cdot 28!}{4! \cdot 2 \cdot 13! \cdot 2 \cdot 28!}\\
&= 228{,}375
\end{aligned}
$$

d. To have at least three officers we need to find the sum of the numbers in each of the following committees:

$$\begin{pmatrix}\text{NO. OF COMMITTEES}\\ \text{WITH 3 OFFICERS}\end{pmatrix} + \begin{pmatrix}\text{NO. OF COMMITTEES}\\ \text{WITH 4 OFFICERS}\end{pmatrix} + \begin{pmatrix}\text{NO. OF COMMITTEES}\\ \text{WITH 5 OFFICERS}\end{pmatrix}$$

We now need to find the number in each of these committees by using the multiplication principle:

$$\begin{pmatrix}\text{NO. OF COMMITTEES}\\ \text{WITH 3 OFFICERS}\end{pmatrix} = \begin{pmatrix}\text{NO. OF WAYS OF}\\ \text{PICKING 3 OFFICERS}\end{pmatrix} \cdot \begin{pmatrix}\text{NO. OF WAYS OF PICK-}\\ \text{ING 2 NON-OFFICERS}\end{pmatrix}$$

$$= \binom{5}{3}\binom{45}{2} = \frac{5!}{3!2!} \cdot \frac{45!}{2!43!} = \frac{5 \cdot 4 \cdot 3! \cdot 45 \cdot 44 \cdot 43!}{3! \cdot 2 \cdot 2 \cdot 43!} = 9{,}900$$

$$\begin{pmatrix}\text{NO. OF COMMITTEES}\\ \text{WITH 4 OFFICERS}\end{pmatrix} = \begin{pmatrix}\text{NO. OF WAYS OF}\\ \text{PICKING 4 OFFICERS}\end{pmatrix} \cdot \begin{pmatrix}\text{NO. OF WAYS OF PICK-}\\ \text{ING 1 NON-OFFICER}\end{pmatrix}$$

$$= \binom{5}{4}\binom{45}{1} = \frac{5!}{4!1!} \cdot \frac{45!}{1!44!} = \frac{5 \cdot 4! \cdot 45 \cdot 44!}{4! \cdot 44!} = 225$$

$$\begin{pmatrix}\text{NO. OF COMMITTEES}\\ \text{WITH 5 OFFICERS}\end{pmatrix} = \begin{pmatrix}\text{NO. OF WAYS OF}\\ \text{PICKING 5 OFFICERS}\end{pmatrix}$$

$$= \binom{5}{5} = \frac{5!}{5!0!} = 1$$

Thus,

$$\begin{pmatrix}\text{NO. OF COMMITTEES}\\ \text{WITH 3 OFFICERS}\end{pmatrix} + \begin{pmatrix}\text{NO. OF COMMITTEES}\\ \text{WITH 4 OFFICERS}\end{pmatrix} + \begin{pmatrix}\text{NO. OF COMMITTEES}\\ \text{WITH 5 OFFICERS}\end{pmatrix}$$

$$= 9{,}900 + 225 + 1 = 10{,}126$$ ■

6.5

Problem Set

Evaluate each of the numbers in Problems 1–11.

1. a. $\binom{9}{1}$ b. $\binom{9}{2}$ c. $\binom{9}{3}$ d. $\binom{9}{4}$ e. $\binom{9}{0}$

2. a. $\binom{5}{4}$ b. $\binom{52}{3}$ c. $\binom{7}{2}$ d. $\binom{4}{4}$ e. $\binom{100}{1}$

3. a. $\binom{7}{3}$ b. $\binom{5}{5}$ c. $\binom{50}{48}$ d. $\binom{25}{1}$ e. $\binom{g}{h}$

4. a. ${}_7P_3$ b. ${}_5P_5$ c. ${}_{52}P_2$ d. ${}_{25}P_1$ e. ${}_gP_h$

5. a. ${}_7P_5$ b. $\binom{8}{0}$ c. $\binom{10}{2}$ d. ${}_nP_4$ e. $\binom{n}{4}$

6. a. $\binom{92}{0}$ b. ${}_{52}P_3$ c. ${}_mP_n$ d. $\binom{16}{3}$ e. $\binom{m}{n}$

7. a. ${}_{92}P_0$ b. $\binom{53}{3}$ c. $\binom{7}{7}$ d. ${}_7P_7$ e. ${}_nP_5$

8. a. $\binom{4}{1}\binom{3}{1}\binom{2}{2}$ b. $\binom{6}{3}\binom{3}{2}\binom{1}{1}$

9. a. $\binom{5}{1}\binom{4}{3}\binom{1}{1}$ b. $\binom{8}{2}\binom{6}{4}\binom{2}{2}$

10. a. $\binom{9}{1}\binom{8}{4}\binom{4}{4}$ b. $\binom{7}{1}\binom{6}{4}\binom{2}{2}$

11. a. $\binom{10}{6}\binom{4}{2}\binom{2}{1}\binom{1}{1}$ b. $\binom{8}{2}\binom{6}{3}\binom{3}{1}\binom{2}{2}$

APPLICATIONS

12. A bag contains 12 pieces of candy. In how many ways can 5 pieces be selected?
13. If the Senate is to form a new committee of 5 members, in how many different ways can the committee be chosen if all 100 senators are available to serve on this committee?
14. In how many ways can three aces be drawn from a deck of cards?
15. In how many ways can two kings be drawn from a deck of cards?
16. In how many ways can a heart flush be obtained? (A heart flush is a hand of five hearts.)
17. In how many ways can a spade flush be obtained? (A spade flush is a hand of five spades.)

In Problems 18–31 decide whether you would use the multiplication principle, a permutation, a combination, an ordered partition, or none of these. Next, write the solution using permutation, combination, or ordered partition notation, if possible; and finally, answer the question asked.

18. How many different arrangements are there of letters in the word CORRECT?
19. At Artist's Dance Studio, every man must dance the last dance. If there are five men and eight women, in how many ways can dance couples be formed for the last dance?
20. Martin's Ice Cream Store sells sundaes with chocolate, strawberry, butterscotch, or marshmallow toppings, nuts, and whipped cream. If you can choose exactly three of these extras, how many possible sundaes are there?
21. Five people are to dine together at a rectangular table, but the host cannot decide on a seating arrangement. In how many ways can the guests be seated, assuming fixed chair positions?
22. In how many ways can three hearts be drawn from a deck of cards?
23. A shipment of a hundred TV sets is received. Six sets are to be chosen at random and tested for defects. In how many ways can the six sets be chosen?
24. A night watchman visits 15 offices every night. To prevent others from knowing when he will be at a particular office,

he varies the order of his visits. In how many ways can this be done?

25. A certain manufacturing process calls for six chemicals to be mixed. One liquid is to be poured into the vat, and then the others are to be added in turn. All possible combinations must be tested to see which gives the best results. How many tests must be performed?

26. There are three boys and three girls at a party. In how many ways can they be seated in a row if they can sit down four at a time?

27. There are three boys and three girls at a party. In how many ways can they be seated in a row if they want to sit alternating boy, girl, boy, girl?

28. If there are ten people in a club, in how many ways can they choose a dishwasher and a bouncer?

29. How many subsets of size three can be formed from a set of five elements?

30. In how many ways can you be dealt two cards from an ordinary deck of cards?

31. In how many ways can five taxi drivers be assigned to six cars?

32. A fraternity contains 8 sophomores, 15 juniors, and 12 seniors. In how many ways can a committee consisting of 1 sophomore, 3 juniors, and 3 seniors be chosen?

33. A sorority has 2 freshmen, 5 sophomores, 25 juniors, and 18 seniors as members. In how many ways can a committee consisting of 2 sophomores, 5 juniors, and 5 seniors be chosen?

34. A rating service rates 25 businesses as A, AA, or AAA. Suppose that 5 companies will be rated AAA, 4 companies will be rated A, and the rest rated AA. In how many ways can this be done?

35. A psychology experiment will group 50 persons into one of three categories: dominant, passive, neutral. In how many ways can 15 be dominant, 10 passive, with the rest neutral? (You can leave your answer in factorial notation.)

36. In a state lottery, 20 people will be chosen as winners. Of these 20, 5 will be awarded $3 million, 5 will win $100,000, and the rest will win $10,000. In how many ways can this be done?

37. Suppose that in a study of 50 stocks, 7 are expected to go up, 5 to go down, and the rest will stay about the same. In how many different ways (to the nearest billion) can the 50 stocks be divided up so that the conditions of this problem are satisfied?

38. A group of students found that on a certain multiple-choice test of 15 questions, the answer sheet shows response "**a**" occurs 3 times, "**b**" occurs 4 times, "**c**" occurs 3 times, "**d**" occurs 2 times, and "**e**" occurs 3 times. In how many ways can the questions be answered? Leave your answer in factorial notation.

39. The investigative division of the Internal Revenue Service has a list of 10 companies cited for violations of proper tax procedures. A spy for one of the companies found out that the IRS intends to bring 4 of these companies to trial, will refer 5 cases for further investigation, and the other case will be dropped. In how many ways can the disposition of the 10 cases be made according to the known information?

40. A sample of 5 watches is selected at random from a lot of 20 watches. The entire lot will be rejected if 2 or more watches are found to be defective.
    a. How many different samples can be selected?
    b. If 2 of the watches are defective, how many of the samples will lead to rejection?

41. A group of 10 Americans, 6 Russians, and 5 Finns must choose a committee of 3 persons. In how many ways can a committee be chosen that consists of:
    a. All Americans?
    b. 2 Americans?
    c. 1 American?
    d. 1 of each nationality?
    e. 2 Russians and 1 Finn?

42. How many committees of 2 men and 3 women can be chosen from 10 men and 8 women?

43. How many committees of 3 Americans and 3 Russians can be chosen from 10 Americans and 12 Russians?

44. Show that $\binom{n}{n-r} = \binom{n}{r}$.

45. Show that $\binom{n}{k-1} + \binom{n}{k} = \binom{n+1}{k}$.

46. Prove the formula for the number of ordered partitions for the case where $k = 3$.

6.6 Binomial Theorem

Combinations can be used to help us form a pattern for the expansion of a binomial $(a + b)^n$. This problem occurs frequently in mathematics, and direct calculation is too tedious for a very large exponent n. Also, we sometimes need to find only one term in the expansion, and a pattern will help us find only that term without having to find all the others.

Consider the powers of $(a + b)$ listed below, which are found by direct multiplication:

$$\begin{aligned}
(a+b)^0 &= 1 \\
(a+b)^1 &= 1\cdot a + 1\cdot b \\
(a+b)^2 &= 1\cdot a^2 + 2\cdot ab + 1\cdot b^2 \\
(a+b)^3 &= 1\cdot a^3 + 3\cdot a^2b + 3\cdot ab^2 + 1\cdot b^3 \\
(a+b)^4 &= 1\cdot a^4 + 4\cdot a^3b + 6\cdot a^2b^2 + 4\cdot ab^3 + 1\cdot b^4 \\
(a+b)^5 &= 1\cdot a^5 + 5\cdot a^4b + 10\cdot a^3b^2 + 10\cdot a^2b^3 + 5\cdot ab^4 + 1\cdot b^5 \\
&\vdots
\end{aligned}$$

First, ignore the coefficients (shown in color) and focus on the variables:

$$\begin{array}{llllllll}
(a+b)^1: & a & b \\
(a+b)^2: & a^2 & ab & b^2 \\
(a+b)^3: & a^3 & a^2b & ab^2 & b^3 \\
(a+b)^4: & a^4 & a^3b & a^2b^2 & ab^3 & b^4 \\
(a+b)^5: & a^5 & a^4b & a^3b^2 & a^2b^3 & ab^4 & b^5 \\
 & \vdots
\end{array}$$

Do you see a pattern? As you read from left to right, the powers of a decrease and the powers of b increase. Note that the sum of the exponents for each term is the same as the original exponent:

$$(a+b)^n: \quad a^nb^0 \quad a^{n-1}b^1 \quad a^{n-2}b^2 \quad \cdots \quad a^{n-r}b^r \quad \cdots \quad a^2b^{n-2} \quad a^1b^{n-1} \quad a^0b^n$$

Next, consider the numerical coefficients:

$$\begin{array}{lccccccccccc}
(a+b)^0: & & & & & & 1 \\
(a+b)^1: & & & & & 1 & & 1 \\
(a+b)^2: & & & & 1 & & 2 & & 1 \\
(a+b)^3: & & & 1 & & 3 & & 3 & & 1 \\
(a+b)^4: & & 1 & & 4 & & 6 & & 4 & & 1 \\
(a+b)^5: & 1 & & 5 & & 10 & & 10 & & 5 & & 1 \\
 & & & & & & \vdots
\end{array}$$

Do you see the pattern? This arrangement of numbers is called **Pascal's triangle**. The rows and columns of Pascal's triangle are usually numbered, as shown in Figure 6.9.

Note that the rows of Pascal's triangle are numbered to correspond to the exponent n on $(a + b)^n$. There are many relationships associated with this pattern, but, for now, we are concerned with an expression representing the entries in the pattern. Do you see how to generate additional rows of the triangle? Look at Figure 6.9.

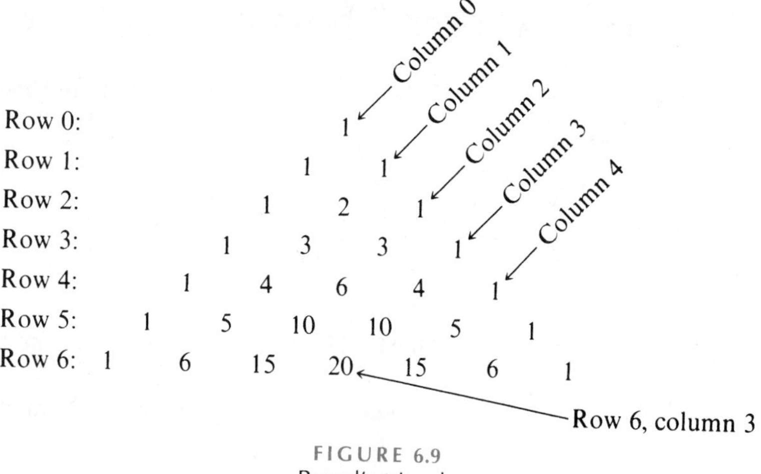

FIGURE 6.9
Pascal's triangle

Notice the following:

1. Each row begins and ends with a 1.
2. We began counting the rows with row 0. This is because after row 0, the second entry in the row is the same as the row number. Thus, row 7 begins 1 7....
3. The triangle is symmetric about the middle. This means that the entries of each row are the same at the beginning and the end. Thus row 7 ends with...7 1. (This property is proved in Problem 41.)
4. To find new entries, we simply add the two entries just above in the preceding row. Thus, row 7 is found by looking at row 6:

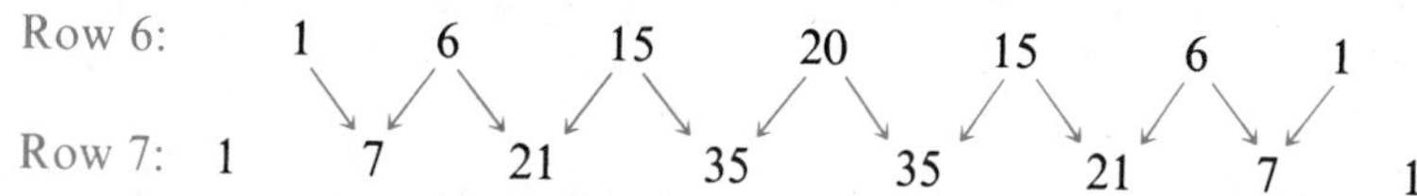

(This property is proved in Problem 40.)

Also note that the entries in Pascal's triangle are the combinations we computed in Section 6.5. That is, the entry in the third column of the fourth row is $\binom{4}{3}$, as shown:

$$\binom{0}{0} = 1$$

$$\binom{1}{0} = 1 \quad \binom{1}{1} = 1$$

$$\binom{2}{0} = 1 \quad \binom{2}{1} = 2 \quad \binom{2}{2} = 1$$

$$\binom{3}{0} = 1 \quad \binom{3}{1} = 3 \quad \binom{3}{2} = 3 \quad \binom{3}{3} = 1$$

$$\binom{4}{0} = 1 \quad \binom{4}{1} = 4 \quad \binom{4}{2} = 6 \quad \binom{4}{3} = 4 \quad \binom{4}{4} = 1$$

EXAMPLE 1 Find $\binom{3}{2}$ both by Pascal's triangle and by formula.

Solution Third row, second column; entry in Pascal's triangle is 3. By formula,

$$\binom{3}{2} = \frac{3!}{2!(3-2)!} = \frac{3!}{2!1!} = 3$$

■

EXAMPLE 2 Find $\binom{6}{4}$ both by Pascal's triangle and by formula.

Solution Row 6, column 4; entry is 15. By formula,

$$\binom{6}{4} = \frac{6!}{4!(6-4)!} = \frac{6!}{4!2!} = \frac{6 \cdot 5 \cdot 4!}{2 \cdot 1 \cdot 4!} = 15$$

■

Using this notation we can now state a very important theorem in mathematics—the **binomial theorem.**

Binomial Theorem

For any positive integer n,

$$(a+b)^n = \binom{n}{0}a^n + \binom{n}{1}a^{n-1}b + \binom{n}{2}a^{n-2}b^2 + \cdots + \binom{n}{k}a^{n-k}b^k + \cdots + \binom{n}{n-2}a^2b^{n-2} + \binom{n}{n-1}ab^{n-1} + \binom{n}{n}b^n = \sum_{k=0}^{n}\binom{n}{k}a^{n-k}b^k$$

Pascal's triangle is efficient for finding the numerical coefficients for exponents that are relatively small, as shown in Figure 6.10. However, for larger exponents we need to use the binomial theorem to find the coefficients.

EXAMPLE 3 Find $(x+y)^8$.

Solution Use Pascal's triangle (Figure 6.10) to obtain the coefficients in the expansion. Thus

$$(x+y)^8 = x^8 + 8x^7y + 28x^6y^2 + 56x^5y^3 + 70x^4y^4 + 56x^3y^5 + 28x^2y^6 + 8xy^7 + y^8$$

■

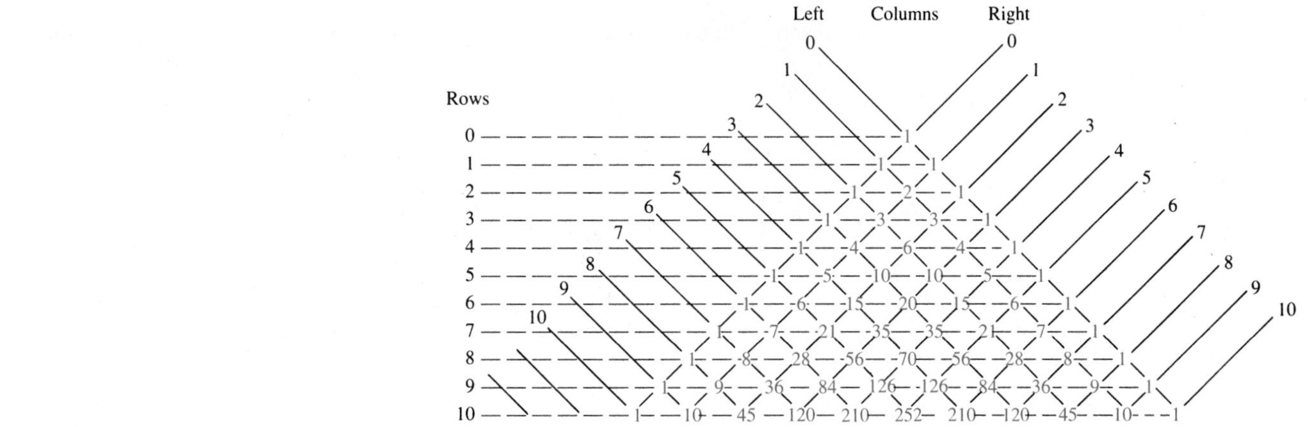

Left Columns Right

Rows	
0	1
1	1 1
2	1 2 1
3	1 3 3 1
4	1 4 6 4 1
5	1 5 10 10 5 1
6	1 6 15 20 15 6 1
7	1 7 21 35 35 21 7 1
8	1 8 28 56 70 56 28 8 1
9	1 9 36 84 126 126 84 36 9 1
10	1 10 45 120 210 252 210 120 45 10 1

FIGURE 6.10
Pascal's triangle—an extended version can be found in Table 1, Appendix E

EXAMPLE 4 Find $(x - 2y)^4$.

Solution In this example, $a = x$ and $b = -2y$ and the coefficients are shown in Figure 6.10.

$$\begin{aligned}(x - 2y)^4 &= [x + (-2y)]^4\\ &= 1 \cdot x^4 + 4 \cdot x^3(-2y) + 6 \cdot x^2(-2y)^2 + 4 \cdot x(-2y)^3 + 1 \cdot (-2y)^4\\ &= x^4 - 8x^3y + 24x^2y^2 - 32xy^3 + 16y^4\end{aligned}$$

■

EXAMPLE 5 Find $(a + b)^{15}$.

Solution The power is rather large, so use the binomial theorem or Table 1, Appendix E, to find the coefficients:

$$\begin{aligned}(a + b)^{15} &= \binom{15}{0}a^{15} + \binom{15}{1}a^{14}b + \binom{15}{2}a^{13}b^2 + \cdots + \binom{15}{14}ab^{14} + \binom{15}{15}b^{15}\\ &= \frac{15!}{0!15!}a^{15} + \frac{15!}{1!14!}a^{14}b + \frac{15!}{2!13!}a^{13}b^2 + \cdots + \frac{15!}{14!1!}ab^{14} + \frac{15!}{15!0!}b^{15}\\ &= a^{15} + 15a^{14}b + 105a^{13}b^2 + \cdots + 15ab^{14} + b^{15}\end{aligned}$$

■

EXAMPLE 6 Find the coefficient of the term x^2y^{10} in the expansion of $(x + 2y)^{12}$.

Solution Here, $n = 12$, $r = 10$, $a = x$, and $b = 2y$; thus

$$\binom{12}{10}x^2(2y)^{10} = \frac{12!}{10!2!}(2)^{10}x^2y^{10}$$

The coefficient is 66(1,024) = 67,584. ■

EXAMPLE 7 How many subsets of $\{1, 2, 3, 4, 5\}$ are there?

Solution In Section 6.1 we found that a set containing two elements has four subsets, and we did this by direct enumeration with a tree diagram. To follow the same procedure for this example would be too tedious. Instead, consider the following argument: There are $\binom{5}{5}$ subsets of five elements; $\binom{5}{4}$ subsets of four elements; $\binom{5}{3}$ subsets of three elements; $\binom{5}{2}$ subsets of two elements; $\binom{5}{1}$ subsets of a single element; and $\binom{5}{0}$ subsets of zero elements (the empty set). Thus the *total* number of subsets is

$$\binom{5}{0}+\binom{5}{1}+\binom{5}{2}+\binom{5}{3}+\binom{5}{4}+\binom{5}{5}$$

Now note that

$$(a+b)^5=\binom{5}{0}a^5+\binom{5}{1}a^4b+\binom{5}{2}a^3b^2+\binom{5}{3}a^2b^3+\binom{5}{4}ab^4+\binom{5}{5}b^5$$

so that if we let $a = 1$ and $b = 1$,

$$\underbrace{(1+1)^5}_{2^5}=\underbrace{\binom{5}{0}+\binom{5}{1}+\binom{5}{2}+\binom{5}{3}+\binom{5}{4}+\binom{5}{5}}_{\text{Total number of subsets}}$$

We see that there are 2^5 subsets of a set containing five elements. ■

EXAMPLE 8 How many subsets are there of a set containing 10 elements?

Solution Using the ideas of Example 7, there are 2^{10} subsets. ■

Examples 7 and 8 suggest a general result:

Number of Subsets of a Finite Set

A set of n distinct elements has 2^n subsets.

6.6 Problem Set

Evaluate the expressions in Problems 1–12 using both Pascal's triangle and the binomial theorem.

1. $\binom{8}{1}$ 2. $\binom{5}{4}$ 3. $\binom{8}{2}$ 4. $\binom{7}{5}$

5. $\binom{8}{3}$ 6. $\binom{9}{5}$ 7. $\binom{12}{1}$ 8. $\binom{15}{0}$

9. $\binom{20}{20}$ 10. $\binom{32}{31}$ 11. $\binom{18}{2}$ 12. $\binom{46}{2}$

Expand the binomial expressions in Problems 13–27 using the binomial theorem or Pascal's triangle.

13. $(x + y)^5$
14. $(x + y)^6$
15. $(x + y)^4$
16. $(x - y)^4$
17. $(x - y)^5$
18. $(x - y)^6$
19. $(x + 2)^5$
20. $(x - 3)^5$
21. $(2x + 3y)^4$
22. $(1 - x)^8$
23. $(1 - x)^{10}$
24. $(x - 2y)^8$
25. $(x - y)^{15}$
26. $(x + 1)^{18}$
27. $(x + y)^{20}$

Find the requested coefficient in Problems 28–33.

28. a^5b^6 term in $(a + b)^{11}$
29. a^4b^7 term in $(a - b)^{11}$
30. $a^{14}b$ term in $(a - 2b)^{15}$
31. $a^{10}b^4$ term in $(a + 3b)^{14}$
32. $x^{14}y^2$ term in $(2x^2 + y)^9$
33. $x^{14}y$ term in $(x^2 - 2y)^8$

APPLICATIONS

34. How many different subsets can be chosen from a set of seven elements?
35. How many different subsets can be chosen from the U.S. Senate? (There are 100 U.S. senators.)

36. Suppose a coin is tossed eight times. How many possible outcomes will there be of five heads and three tails? [*Hint:* Denote the possibilities of a single toss by H + T. Since there are eight tosses, we can denote the set of all possibilities by $(\mathrm{H} + \mathrm{T})^8$. In this problem, you are looking for the coefficients of $\mathrm{H}^5\mathrm{T}^3$; that is, heads five times (H^5) and tails three times (T^3).]

37. Suppose a coin is tossed ten times. How many possible outcomes will there be of four heads and six tails? [See the hint in Problem 36.]

38. Suppose a coin is tossed nine times. How many possible outcomes will there be of two heads and seven tails? [See the hint in Problem 36.]

39. Generalize the argument of Example 7 to show that $\binom{n}{0} + \binom{n}{1} + \binom{n}{2} + \cdots + \binom{n}{n-1} + \binom{n}{n} = 2^n$. (This says that the sum of the entries of the nth row of Pascal's triangle is 2^n.)

40. Show that $\binom{n-1}{r-1} + \binom{n-1}{r} = \binom{n}{r}$.

[This says that to find any entry in Pascal's triangle (except the first and last), simply add the two entries directly above.]

41. Show that Pascal's triangle is symmetric.

42. ***Historical Question.*** Blaise Pascal (1623–1662) has been described as "the greatest 'might-have-been' in the history of mathematics." Pascal was frail, and, because he needed to conserve his energy, he was forbidden to study mathematics. This aroused his curiosity and forced him to acquire most of his knowledge of the subject by himself. At 16, he wrote an essay on conic sections that astounded the mathematicians of his time. At 18, he had already invented one of the first calculating machines. However, at 27, because of his health, he promised God that he would abandon mathematics and spend his time in religious study. Three years later he broke this promise and wrote *Traité du Triangle Arithmétique*, in which he investigated what we today call Pascal's triangle. The very next year he was almost killed when his runaway horse jumped an embankment. He took this to be a sign of God's displeasure with him and again gave up mathematics—this time permanently. Answer the following questions about Pascal's triangle:

a. What is the sum of the entries in the nth row?

b. How are the powers of 11 related to Pascal's triangle?

c. Find a formula for the largest number in the nth row.

*6.7 Review

The material of this chapter is reviewed in the following list of objectives. After each objective there are some practice questions. For a sample test select the first question of each set and check your answers. The second question for each objective has no answer given. If you are having trouble with a particular type of problem, look back at the indicated section in the text. When you are finished reviewing these objectives, a sample examination is given at the end of this section.

[6.1]

Objective 6.1: *Write sets using the roster method when the sets are given using the description method or set-builder notation. Let U be the set of counting numbers.*

1. $\{x \mid x^2 = 9\}$
2. $\{y \mid -3 \le y \le 0\}$
3. $\{t \mid 4 \le t \le 6\}$
4. $\{m \mid (m-3)(2m+1) = 0\}$

Objective 6.2: *List the subsets of a given set.*

5. $\{1, 2, 3\}$
6. $\{c, a, t\}$
7. $\{h, m\}$
8. $\{6, 8, 10, 12\}$

Objective 6.3: *Find the union of given sets. Let*

$$U = \{1, 2, 3, 4, 5, 6, 7, 8, 9, 10\}$$
$$A = \{5, 10\}$$
$$B = \{6, 7, 8, 9, 10\}$$
$$C = \{4, 6\}$$

9. $A \cup B$
10. $A \cup C$
11. $B \cup C$
12. $\varnothing \cup B$

Objective 6.4: *Find the intersection of given sets. Let U, A, B, and C be defined as in Objective 6.3.*

13. $A \cap B$
14. $A \cap C$
15. $B \cap C$
16. $\varnothing \cap B$

Objective 6.5: *Find the complement of a given set. Let U, A, and C be defined as in Objective 6.3.*

17. $\bar{A}$
18. $\bar{B}$
19. $\bar{C}$
20. $\bar{\varnothing}$

Objective 6.6: *Draw a Venn diagram for union, intersection, and complementation.*

21. $R \cup S$
22. $R \cap S$
23. $\bar{R}$
24. $\bar{S}$

* Optional section.

[6.2]

Objective 6.7: *Draw a Venn diagram for mixed operations.*

25. $\overline{S \cup T}$

26. $\overline{S} \cap \overline{T}$

27. $\overline{R \cup (S \cup T)}$

28. $(R \cup S) \cap (R \cup T)$

Objective 6.8: *Find the union, intersection, and complement for mixed operations.*

$$U = \{2, 4, 6, 8, 10, 12, 14, 16\}$$
$$A = \{6, 8, 12, 16\}$$
$$B = \{4, 6, 8, 10\}$$
$$C = \{8, 10, 12, 14, 16\}$$

29. $(A \cup B) \cap (A \cup C)$

30. $\overline{A \cap (B \cup C)}$

31. $\bar{A} \cap \bar{C}$

32. $\overline{A \cup C}$

Objective 6.9: *Use Venn diagrams to solve survey questions. Use the results of the following survey: In a survey of* 500 *motorists it was found that*

140 had received a ticket for speeding
120 had received a ticket for failure to yield the right of way
98 had received a ticket for failure to stop at a stop sign
41 had received tickets for speeding and failure to yield
35 had received tickets for speeding and failure to stop at a stop sign
47 had received tickets for failure to yield and failure to stop at a stop sign
24 had received tickets for all three violations

33. How many motorists did not receive any tickets?

34. How many motorists received only one ticket?

35. How many motorists received exactly two tickets?

36. How many motorists received only a ticket for speeding?

[6.3]

Objective 6.10: *Evaluate expressions with factorial notation.*

37. $\dfrac{10!}{4!(10-4)!}$

38. $\dfrac{12!}{9!(12-9)!}$

39. $\dfrac{135!}{133!}$

40. $6! - 5!$

Objective 6.11: *Answer applied problems involving the multiplication principle.*

41. How many four-letter code words are possible using the letters *A, E, I, O, U,* and *Y* if no letter can be repeated?

42. How many four-letter code words are possible using the letters *X, Y, Z, R, S, T,* and *U* if adjacent letters cannot be alike?

43. A new car comes with your choice of 8 colors, 6 interiors, and 3 models. How many variations for this car are possible?

44. A personalized license plate allows any numeral, space, or letter in any one of six positions. How many plates, regular or personalized, are possible if the numbering scheme is three letters followed by three numerals?

[6.4]

Objective 6.12: *Evaluate permutations.*

45. ${}_8P_5$

46. ${}_{18}P_0$

47. ${}_{40}P_2$

48. ${}_sP_3$

Objective 6.13: *Answer applied problems involving permutations.*

49. Find the number of arrangements of letters in the word ERROR.

50. If blood can be typed as A, B, AB, or O, and also as Rh+ or Rh−, how many possible blood types are there according to these classifications?

51. A psychologist classifies her subject according to one of six personality types, one of five educational groups, and one of two sexes. How many categories are possible?

52. In how many ways can a teacher select a roll keeper, a blackboard cleaner, and a hall monitor from a class of 20 pupils?

[6.5]

Objective 6.14: *Evaluate combinations.*

53. $\dbinom{8}{5}$

54. $\dbinom{18}{0}$

55. $\dbinom{40}{2}$

56. $\dbinom{s}{3}$

Objective 6.15: *Answer counting problems involving the multiplication principle, permutations, combinations, or distinguishable permutations.*

57. *Psychology experiment.* A psychologist has an experiment consisting of five colored boxes. If the possible colors are white, red, blue, or green, how many ways can the five boxes be painted?

58. In how many ways can a pair of hearts be drawn from a deck of cards?

59. *Manufacturing.* A sample of five automobiles is selected from an assembly line producing 100 cars. In how many ways can this sample be selected?

60. *Sports.* In how many ways can uniforms be distributed to five players if there are 20 uniforms from which to choose?

[6.6]

Objective 6.16: *Expand binomial expressions using Pascal's triangle or the binomial theorem for the numerical coefficients.*

61. $(x + y)^7$

62. $(x - y)^4$

63. $(2x + y)^5$

64. $(x^2 - 2y)^6$

Objective 6.17: *Find a particular coefficient in a binomial expansion.*

65. x^5y^9 term in $(x - y)^{14}$

66. x^2y^3 term in $(x + 2y)^5$

67. x^4y^3 term in $(2x - y)^7$

68. x^6y^3 term in $(x^2 + y^3)^4$

Objective 6.18: *Answer counting problems involving the binomial theorem or the number of subsets.*

69. How many subsets can be chosen from a set of six elements?

70. Suppose a coin is tossed five times. How many possible outcomes will there be for three heads and two tails?

71. How many outcomes are there of two girls and two boys for a family having four children?

72. What are the total number of boy-girl outcomes for a family having four children?

SAMPLE TEST

The following sample test (45 minutes) is intended to review the main ideas of this chapter.

1. Let $U = \{2, 3, 4, 5, 6, 7, 8, 9, 10, 11, 12\}$, $A = \{5, 6, 7\}$, $B = \{7, 11\}$, $C = \{2, 7, 12\}$.

a. Find $\overline{A \cup B}$. **b.** Find $\bar{A} \cup \bar{B}$.
c. Find $A \cap \overline{(B \cup C)}$. **d.** Find $\overline{A \cap (B \cup C)}$.

2. Use Venn diagrams to prove or disprove $\overline{A \cup B} = \bar{A} \cup \bar{B}$.

Evaluate the expressions in Problems 3–5.

3. **a.** $7!$ **b.** $7! - 4!$
c. $(7 - 4)!$ **d.** $\dfrac{100!}{98!}$

4. **a.** ${}_6P_3$ **b.** ${}_6C_3$

5. **a.** $\binom{8}{2}$ **b.** $\binom{8}{5, 2, 1}$

6. In how many ways can a deck of 52 cards be dealt to four bridge players? Leave your answer in factorial notation.

7. In how many ways can you choose two books to read from a bookshelf containing seven books?

8. A club consists of 17 men and 19 women. In how many ways can a president, vice president, treasurer, and secretary be chosen, along with an advisory committee of 6 people, if no 2 persons can serve in more than one capacity?

9. Repeat Problem 8, but now add that exactly two of the officers must be women.

10. In how many ways can the letters in the word WORRY be arranged?

11. A mathematics test contains ten questions. In how many ways can the test be answered if the possible answers are true and false?

12. Answer Problem 11 if the possible answers are **a**, **b**, **c**, **d**, and **e**—that is, if the test is a multiple-choice test.

13. Expand $(T + H)^6$.

14. Suppose a coin is tossed six times. How many outcomes will there be for four heads and two tails?

15. Human blood can be classified into types A, B, AB, or O. Furthermore, each of these types is classified as positive or negative depending on whether the antigen called Rh is present. These possibilities are summarized in the Venn diagram. Note that O blood type refers to the absence of either the A or B blood types.

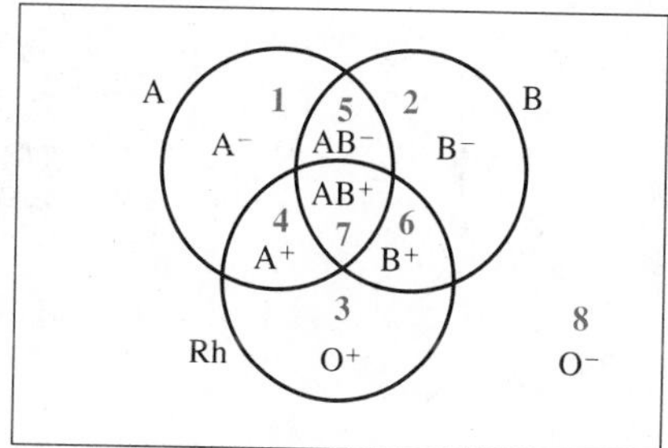

Region	*Blood type*
1	A^-
2	B^-
3	O^+
4	A^+
5	AB^-
6	B^+
7	AB^+
8	O^-

All people with the A antigen are put in circle A, those with the B antigen are put in circle B, and those with the Rh antigen are put in circle Rh. Now suppose that a hospital reports the following data:

60 patients have the A antigen
50 patients have the B antigen
65 patients have the Rh antigen
15 patients have both the A and B antigens
20 patients have both the B and Rh antigens
30 patients have both the A and Rh antigens
5 patients have all three antigens
10 patients have none of the antigens

How many patients
a. Were surveyed?
b. Have AB^+ blood?
c. Have A^+ blood?
d. Have O^- blood?
e. Have exactly one antigen?

7

Probability

CHAPTER OVERVIEW

We begin by discussing the notion of a probability model; then consider the probability of simple, equally likely events; and finally construct more complicated models by relating them back to the simpler cases.

PREVIEW

This chapter introduces probabilistic models. Real-world events can either be predicted with certainty or not. Of course, very little in the real world is *certain*, but events can be predicted with more or less certainty. Mathematical models are used to measure the amount of certainty, or probability. As you can imagine, it is an extremely important idea, not only in mathematics, but in almost all fields of study.

PERSPECTIVE

The first major topic of finite mathematics was solving systems (of both equations and inequalities) using the mathematics of matrices. The second major topic of finite mathematics is probability and it is introduced in this chapter.

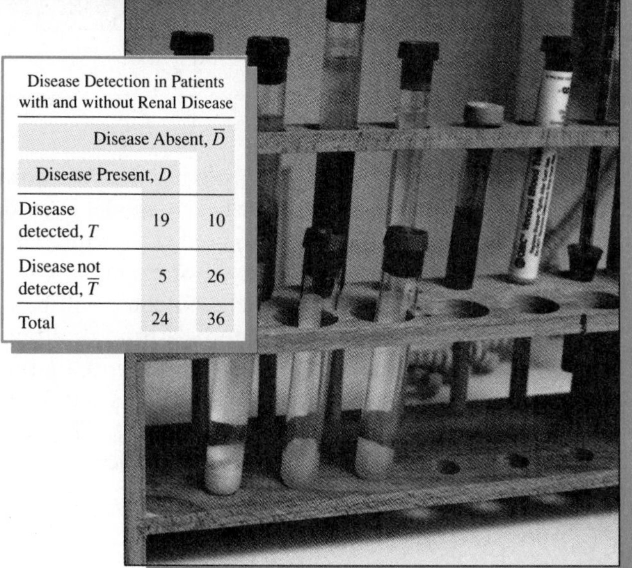

Disease Detection in Patients with and without Renal Disease	Disease Present, D	Disease Absent, $\overline{D}$
Disease detected, T	19	10
Disease not detected, $\overline{T}$	5	26
Total	24	36

MODELING APPLICATION 5

Medical Diagnosis

After a patient with observable symptoms sees a doctor, the doctor must decide which of several possible diseases is the most probable cause of the symptoms. The doctor may order further tests before making a diagnosis. Suppose a new test for detecting renal disease is being developed. The test will be given in a double-blind study to 24 patients with confirmed renal disease and 36 patients who are free of the disease. The results are as follows: of those with renal disease, 19 test positively and 5 test negatively; for those free of the disease, 10 test positively and 26 show negative results.*

After you have finished this chapter, write a paper based on Bayes' theorem to develop a model to answer the question: "Are the results of the test described above conclusive enough for a doctor to make a diagnosis?" The development of this model should show that two probabilities are small. One of these is a false positive result, which is a positive test result when in fact the patient does not have the disease. The other is a false negative result, which is a negative test result when in fact the patient does have the disease. For general guidelines about writing this essay, see the commentary for Modeling Application 1 on page 151.

* This modeling application is adapted from J. S. Milton and J. J. Corbet, "Conditional Probability and Medical Tests: An Exercise in Conditional Probability," *The UMAP Journal*, Vol. III, No. 2, 1982. © 1982 Education Development Center, Inc.

APPLICATIONS

Management *(Business, Economics, Finance, and Investments)*
Industrial testing
(7.1, Problems 6, 10; 7.2, Problems 14, 20; 7.3, Problems 1–6, 34–39, 46–51; 7.4, Problems 41, 45–46, 48–49; 7.6, Problems 31–36; 7.7, Problems 1, 4, 67–68)
Analysis of accident risk (7.6, Problem 70)
Probability that a shipment has a defect
(7.6, Problem 71)

Life Sciences *(Biology, Ecology, Health, and Medicine)*
Probabilities of family makeup—male/female (7.3, Problems 14, 24–28)
Fruit fly experiment
(7.4, Problems 25–34; 7.7, Test Problems 5–7)
Medical diagnosis
(7.6, Problem 72; 7.7, Problem 70)

Modeling Application—
Medical Diagnosis

Social Sciences *(Demography, Political science, Population, Psychology, Society, and Sociology)*
Public opinion polling
(7.1, Problems 8, 17–19)
Psychological study (7.1, Problem 9)
Banking survey (7.2, Problem 19)
Survey of persons dropping a college course
(7.4, Problem 42)
Cancer Society smoking survey
(7.5, Problem 70)
Qualifying for graduate school
(7.6, Problem 66)
Probability that a criminal is lying
(7.6, Problem 67)
Demographics of college students
(7.6, Problems 68, 69)

General Interest
Sample space for the results of the World Series (7.1, Problem 5)
Probability of winning a raffle
(7.2, Problem 17)

General Interest *(continued)*
Probability of an A on an examination
(7.2, Problem 15)
Probability of a red light at an intersection
(7.2, Problem 18)
Dice probabilities
(7.2, Problems 23–40; 7.3, Problem 33; 7.4, Problems 15–18)
Dungeons and Dragons (7.2, Problem 41)
Historical dice games (7.2, Problems 42–43)
Racetrack odds
(7.3, Problem 32; 7.7, Problem 11)
Poker probabilities
(7.3, Problems 59–60; 7.4, Problems 43–44; 7.7, Problem 33)
Birthday problem (7.3, Problems 61–62)
Historical probability problem
(7.3, Problem 63)
Roulette probabilities
(7.5, Problem 46; 7.7, Problem 40)
Accuracy when working math problems
(7.5, Problems 71–72)
Graduate school examination
(7.6, Problem 66)

7.1

Probability Models

The models we have considered so far have been deterministic models. We now turn to a different type of model called a **probabilistic model**. This model is used with situations that are random in character and attempts to predict the outcomes of events with a certain stated or known degree of accuracy. For example, if we toss a coin, it is impossible to predict in advance whether the outcome will be a head or a tail. Our intuition tells us that it is equally likely to be a head or a tail, and somehow we sense that if we repeat the experiment of tossing a coin a large number of times, heads will occur "about half the time." To check this out, I recently flipped a coin 1,000 times and obtained 460 heads and 540 tails. The percentage of heads is $\frac{460}{1,000} = .46 = 46\%$, which is called the **relative frequency**:

Relative Frequency of a Repeated Experiment

If an experiment is repeated n times and an event occurs m times, then

$$\frac{m}{n}$$

is called the **relative frequency** of the event.

Our task in this chapter is to create a model that will assign a number p, called the *probability of an event*, which will predict the relative frequency. This means that for a *sufficiently large number of repetitions* of an experiment

$$p \approx \frac{m}{n}$$

Probabilities can be obtained in one of three ways:

1. **Theoretical probabilities** (also called *a priori models*) are obtained by logical reasoning. For example, the probability of rolling a die and obtaining a 3 is $\frac{1}{6}$ because there are six possible outcomes, each with an equal chance of occurring, so a 3 should appear $\frac{1}{6}$ of the time.
2. **Empirical probabilities** (also called *a posteriori models*) are obtained from experimental data. For example, an assembly line producing brake assemblies for General Motors produces 1,500 items per day. The probability of a defective brake can be obtained by experimentation. Suppose the 1,500 brakes are tested and 3 are found to be defective. Then the relative frequency, or probability, is

$$\frac{3}{1,500} = .002 \quad \text{or} \quad .2\%$$

3. **Subjective probabilities** are obtained by experience and indicate a measure of "certainty." For example, a TV reporter studies the satellite maps and issues a prediction about tomorrow's weather based on past experience under similar circumstances: 80% chance of rain tomorrow.

We must define a probability measure so that it conforms to these different ways of using the word *probability*.

We begin by formalizing our terminology. We have been speaking about experiments and events. An **experiment** is the observation of any physical occurrence. A

sample space of an experiment is the set of all possible outcomes. An **event** is a subset of the sample space. If an event is the empty set, it is called the **impossible event**; if it has only one element, it is called a **simple event**.

EXAMPLE 1 **a.** The sample space S for the experiment of tossing a coin is

$$S = \{\text{head, tail}\}$$

The events $H = \{\text{head}\}$ and $T = \{\text{tail}\}$ are simple events.

b. The sample space S for the experiment of rolling a die is

$$S = \{1, 2, 3, 4, 5, 6\}$$

The events $F = \{4\}$ and $W = \{2\}$ are simple events, whereas the events $E = \{\text{an even is rolled}\} = \{2, 4, 6\}$ and $F = \{1, 6\}$ are not simple events.

c. Finally, the sample space S for the experiment of simultaneously tossing a coin and rolling a die (which is a little more complicated to find than the one in parts **a** and **b**) can be found by using a tree diagram:

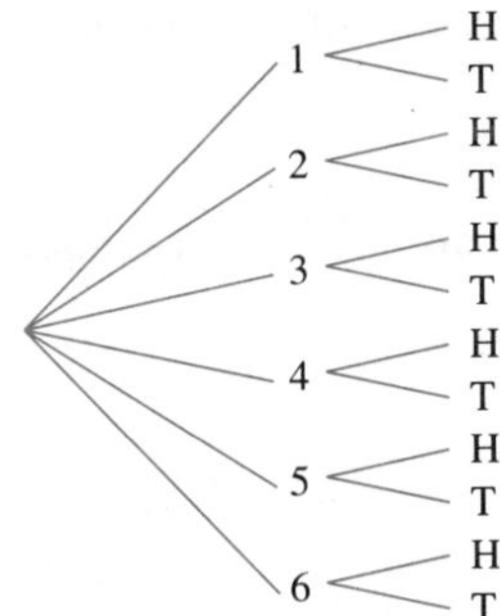

$S = \{1H, 1T, 2H, 2T, 3H, 3T, 4H, 4T, 5H, 5T, 6H, 6T\}$

An example of a simple event for this sample space is $\{5H\}$, which is to simultaneously roll a 5 and a head. The event $H = \{\text{head}\}$ is not a simple event since $H = \{1H, 2H, 3H, 4H, 5H, 6H\}$. Remember, a simple event is a subset of the sample space that has only one element. ■

EXAMPLE 2 A computer chip is operated until it fails, and its lifetime, t (in hours), is recorded. The sample space S is

$$S = \{t \mid t \geq 0\}$$

An example of an event E is that the chip lasts more than 1,000 hours; that is,

$$E = \{t \mid t > 1{,}000\}.$$ ■

EXAMPLE 3 A UCLA alumnus records the results of a UCLA football game. The sample space S is

$$S = \{w, l, t\}$$

where w, l, and t denote win, lose, and tie, respectively. An example of an event E is that UCLA wins; that is, $E = \{w\}$. This is also an example of a simple event, whereas the events in Examples 1 and 2 are not simple events. ■

Two events E and F are said to be **mutually exclusive** if $E \cap F = \varnothing$.

EXAMPLE 4 Suppose you perform an experiment of rolling a die. Then the sample space S is

$$S = \{1, 2, 3, 4, 5, 6\}$$

Let $E = \{1, 3, 5\}$, $F = \{2, 4, 6\}$, $G = \{1, 3, 6\}$, and $H = \{2, 4\}$. Then:

E and F are mutually exclusive
G and H are mutually exclusive

But

F and H are *not* mutually exclusive
E and G are *not* mutually exclusive ■

We can now define a probability measure:

Probability Measure

Let S be a sample space associated with some experiment. With each event E we associate a real number called the *probability of* E, denoted by $P(E)$, satisfying the following properties:

1. $0 \leq P(E) \leq 1$
2. $P(S) = 1$
3. If E and F are mutually exclusive events, then

$$P(E \cup F) = P(E) + P(F)$$

Note that this definition does not tell us how to compute $P(E)$; it simply gives some general properties of $P(E)$ from which we can build a variety of probability models. Also note that set notation is used in property 3. We often use set notation and terminology to describe probability problems. Some of the commonly used terms are given in Table 7.1.

TABLE 7.1

S Sample space	U Universal set $\varnothing$ Empty set $A \cup B$ Union of sets A and B $A \cap B$ Intersection of sets A and B $\bar{A}$ Complement of a set A

Probabilistic statement	**Set notation**
Events A and B	Sets A, B
A and B are mutually exclusive	$A \cap B = \varnothing$
A and B occur	$A \cap B$
A or B occurs	$A \cup B$
A does not occur	$\bar{A}$
Neither A nor B occurs	$\overline{A \cup B}$, or (equivalently) $\bar{A} \cap \bar{B}$
A and B are equally likely	$P(A) = P(B)$
A is more likely than B	$P(A) > P(B)$
A is less likely than B	$P(A) < P(B)$

In this chapter our discussion will be limited to experiments with a finite number of outcomes. Example 2 illustrates an infinite sample space. When the sample space is finite, it can be written as

$$S = \{s_1, s_2, s_3, \ldots, s_k\}$$

for some counting number k. Here $s_1, s_2, \ldots, s_k$ are the simple events or outcomes of our experiment. It follows from property 2 that

$$P(s_1) + P(s_2) + P(s_3) + \cdots + P(s_k) = 1$$

which means that all the probabilities of the simple events in a probabilistic model must add up to 1.

EXAMPLE 5 Suppose a die is rolled and the number of the top face is recorded. Then $S = \{1, 2, 3, 4, 5, 6\}$. There are many possible models:

Model 1

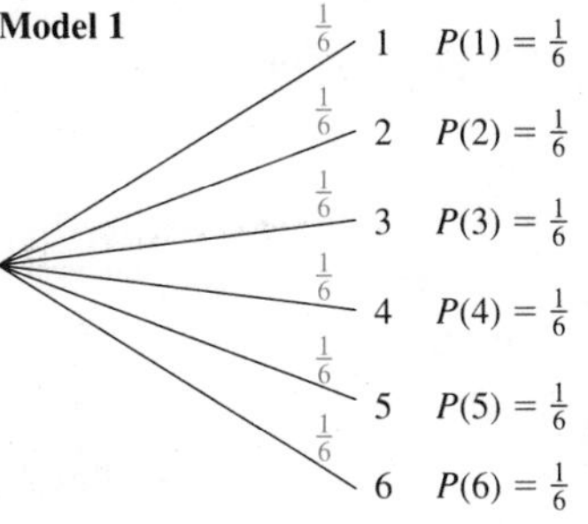

This is a model consisting of equally likely outcomes. Notice that the probabilities of each simple event are often shown on the branches of a tree diagram. A die is called **fair** if these outcomes are equally likely. We will assume fair dice unless otherwise noted.

Model 2

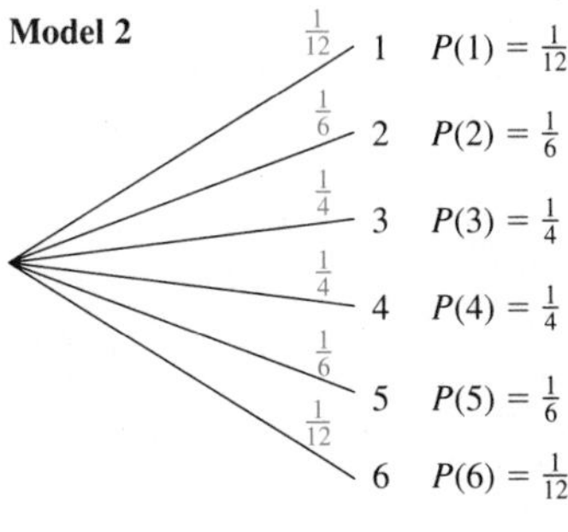

This is a model that favors the outcomes of 3 and 4. We say these simple events are not equally likely. This is an example of a "loaded" die—one that is not fair.

Model 3

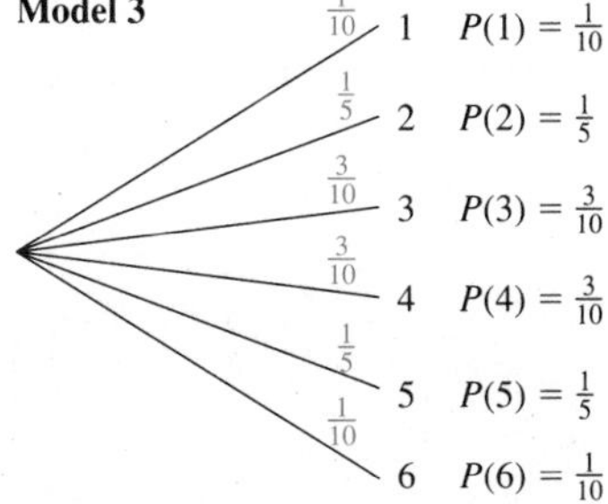

This is *not a probability model* because the sum of the probabilities of the simple events is not 1. This means that a die could not even be loaded to fit this model since the sum is not 1.

Other models for this experiment are also possible. ■

If an event is not a simple event, but has finitely many elements, then its probability can be found using the **addition principle**:

Addition Principle

> If $E = \{s_1, s_2, \ldots, s_n\}$, for simple events $s_1, s_2, \ldots$, then
>
> $$P(E) = P(s_1) + P(s_2) + \cdots + P(s_n) \qquad \text{or} \qquad P(E) = \sum_{i=1}^{n} P(s_i)$$
>
> where $P(s_1), P(s_2), \ldots, P(s_n)$ are the probabilities of the simple events.

EXAMPLE 6 Suppose a die is loaded so that the probabilities of the simple events are $P(1) = \frac{1}{12}$, $P(2) = \frac{2}{12}$, $P(3) = \frac{3}{12}$, $P(4) = \frac{3}{12}$, $P(5) = \frac{2}{12}$, and $P(6) = \frac{1}{12}$. Find the probabilities for rolling the die once and obtaining:

a. An even number

b. A prime number

c. A number greater than 3

d. A 2 or a 6

e. A 2 and a 6

Solution **a.**

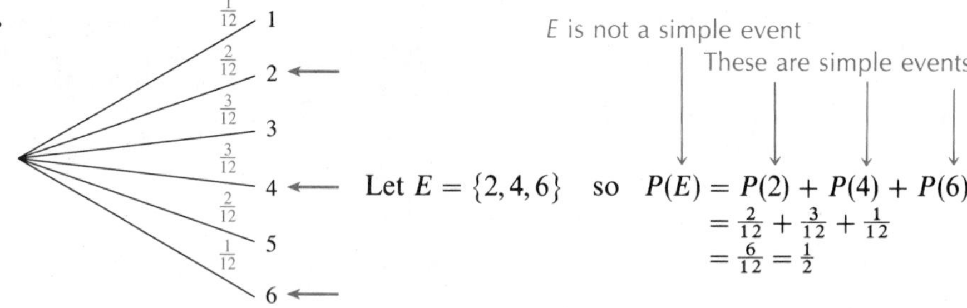

b.

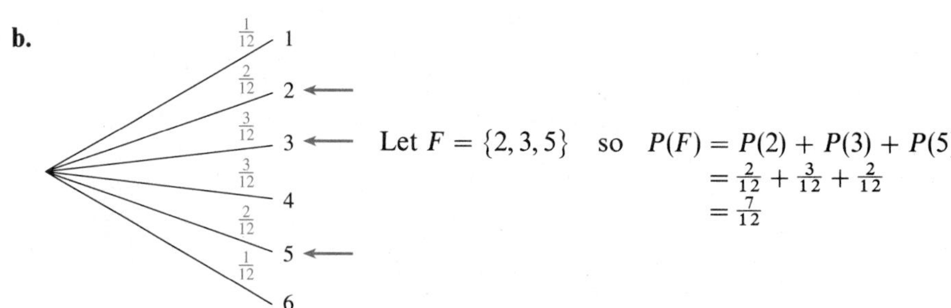

c.

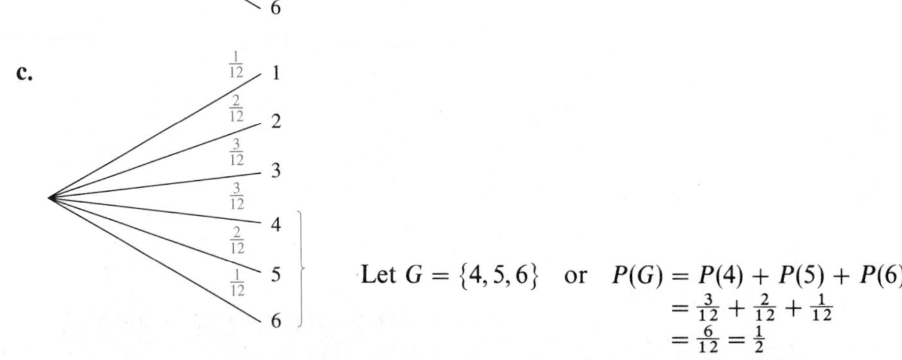

d. Let $H = \{2, 6\}$ so $P(H) = P(2) + P(6)$

$$= \tfrac{2}{12} + \tfrac{1}{12}$$
$$= \tfrac{3}{12} = \tfrac{1}{4}$$

e. $I = \varnothing$ so $P(I) = 0$ ■

In Example 6e, our intuition tells us that the probability of the empty event should be 0, but the following argument proves this fact by using only the definition of a probability measure:

$P(E) = P(E \cup \varnothing)$ Since $E = E \cup \varnothing$ for any set E

$P(E) = P(E) + P(\varnothing)$ Property 3 of a probability measure since E and $\varnothing$ are mutually exclusive

$0 = P(\varnothing)$ Subtract $P(E)$ from both sides.

Probability of the Empty Set

$$P(\varnothing) = 0$$

In the next section we discuss finding a model for simple events, and in Section 7.4 we find a model for the probabilities of events that are not simple.

7.1 Problem Set

Describe the sample space for the experiments in Problems 1–10.

1. A pair of dice is rolled and the sum of the top faces is recorded.
2. A coin is tossed three times and the sequence of heads and tails is recorded.
3. Ten people are asked if they graduated from college and the number of people responding "yes" is recorded.
4. A jar contains five red balls, four white balls, and three green balls. One ball is drawn and the result is recorded.

APPLICATIONS

5. The number of games necessary to complete the World Series for the years 1960–1990 is recorded. (The World Series consists of playing until one team wins four games.)
6. An Eveready Long-Life® battery is tested and the life of the battery is recorded.
7. The number of words in which an error is made on a typing test consisting of 500 words is recorded.
8. A person is randomly selected for a public opinion poll. The person is asked two questions:

   Sex: male (m), female (f)
   Political party: Democrat (d), Republican (r), Independent (i), not registered (n)

   The responses are recorded.
9. A psychologist is studying sibling relationships in families with three children. A family is asked about the sex of their children and the result is recorded. If a family has a girl, then a boy, and finally another boy, this would be recorded as {gbb}.
10. A sample of five radios is selected from an assembly line and tested. The number of defective radios is recorded.

In Problems 11–19 an experiment is described and some events are listed. Write each event using set notation and determine whether the given events are mutually exclusive.

11. A pair of dice is rolled and the sum of the numbers on the top faces is recorded.
    a. Event E is rolling an even number.
    b. Event F is rolling a prime number.
12. A pair of dice is rolled and the sum of the numbers on the top faces is recorded.
    a. Event M is rolling a 7 or an 11.
    b. Event N is rolling a 2, 3, or 12.
13. A pair of dice is rolled and the sum of the numbers on the top faces is recorded.
    a. Event G is rolling a 2 (snake eyes).
    b. Event H is rolling a sum greater than 2.

14. A coin is tossed three times and the sequence of heads and tails is recorded.
   a. Event E is tossing one head.
   b. Event F is tossing two heads.

15. A coin is tossed three times and the sequence of heads and tails is recorded.
   a. Event G is tossing a head on the first roll.
   b. Event H is tossing a head on the second roll.

16. A coin is tossed three times and the sequence of heads and tails is recorded.
   a. Event I is tossing at least one head.
   b. Event J is tossing no heads.

17. Consider the experiment described in Problem 8.
   a. Event F is that the selected person is female.
   b. Event D is that the selected person is a Democrat.

18. Consider the experiment described in Problem 8.
   a. Event I is that the selected person is a registered Independent.
   b. Event N is that the selected person is not registered.

19. Consider the experiment described in Problem 8.
   a. Event E is that the selected person is either female or Republican.
   b. Event R is that the selected person is a registered voter.

In Problems 20–28 consider the experiment of rolling a single die and recording the number on the top face. Let N be the event of rolling an odd number; E be the event of rolling an even number; and L be the event of rolling a number less than 3. Describe each event in words and list the events in each set.

20. $N \cup E$ **21.** $L \cup E$

22. $N \cap E$ **23.** $E \cap L$

24. $\bar{N}$ **25.** $\bar{L}$

26. $\overline{N \cup L}$ **27.** $\bar{E} \cap \bar{L}$

28. $\bar{N} \cap \bar{E}$

In Problems 29–38 suppose a pair of dice are loaded so that the probabilities of the simple events associated with rolling these dice and recording the sum on the two top faces are

$P(2) = \frac{1}{11}$ $P(3) = \frac{1}{11}$ $P(4) = \frac{1}{11}$ $P(5) = \frac{1}{11}$
$P(6) = \frac{1}{11}$ $P(7) = \frac{1}{11}$ $P(8) = \frac{1}{11}$ $P(9) = \frac{1}{11}$
$P(10) = \frac{1}{11}$ $P(11) = \frac{1}{11}$ $P(12) = \frac{1}{11}$

Let $C = \{2, 3, 12\}$, $E = \{7, 11\}$, *and* $F = \{8, 9, 10\}$.

29. Is this a probability model? If it is, explain why.

30. Find $P(E)$. **31.** Find $P(C)$.

32. Find $P(F)$. **33.** Find $P(E \cup F)$.

34. Find $P(E \cup C)$. **35.** Find $P(E \cap C)$.

36. Find $P(\bar{F})$. **37.** Find $P(\bar{C})$.

38. Find $P(\overline{E \cup F})$.

In Problems 39–48 suppose a pair of dice are loaded so that the probabilities of the simple events associated with rolling these dice and recording the sum on the two top faces are

$P(2) = \frac{1}{36}$ $P(3) = \frac{2}{36}$ $P(4) = \frac{3}{36}$ $P(5) = \frac{4}{36}$
$P(6) = \frac{5}{36}$ $P(7) = \frac{6}{36}$ $P(8) = \frac{5}{36}$ $P(9) = \frac{4}{36}$
$P(10) = \frac{3}{36}$ $P(11) = \frac{2}{36}$ $P(12) = \frac{1}{36}$

Let $C = \{2, 3, 12\}$, $E = \{7, 11\}$, *and* $F = \{8, 9, 10\}$.

39. Is this a probability model? If it is, explain why.

40. Find $P(E)$. **41.** Find $P(C)$.

42. Find $P(F)$. **43.** Find $P(E \cup F)$.

44. Find $P(E \cup C)$. **45.** Find $P(E \cap C)$.

46. Find $P(\bar{F})$. **47.** Find $P(\bar{C})$.

48. Find $P(\overline{E \cup F})$.

7.2 Probability of Equally Likely Events

In this section we begin to *calculate* theoretical probabilities. Suppose a sample space is divided into simple events that are **equally likely**. For example, the experiment of flipping a coin has a sample space

$$S = \{\text{H}, \text{T}\}$$

and the simple events H = {heads}, T = {tails} are equally likely. In this text, coins are considered *fair* (equally likely heads and tails) unless otherwise noted. This means that the coin is perfectly balanced and symmetrical and the events H and T are equally likely to occur.

EXAMPLE 1 Consider two dice, each with faces labeled 1, 2, 3, 4, 5, 6. An experiment consists of rolling the dice and noting the numbers on the top face.

For die A, $P(1) = \frac{1}{6}$, $P(2) = \frac{1}{6}$, $P(3) = \frac{1}{6}$, $P(4) = \frac{1}{6}$, $P(5) = \frac{1}{6}$, and $P(6) = \frac{1}{6}$.
For die B, $P(1) = \frac{1}{12}$, $P(2) = \frac{2}{12}$, $P(3) = \frac{3}{12}$, $P(4) = \frac{3}{12}$, $P(5) = \frac{2}{12}$, and $P(6) = \frac{1}{12}$.

For die A, each of the six possible outcomes are equally likely. Die A is called a *fair die*. The outcomes for die B are not equally likely, so die B is called a *loaded die*. All dice used in this book are considered to be fair dice unless otherwise noted. ■

Die A in Example 1 has a sample space consisting of six equally likely simple events, so we will assign each simple event probability $\frac{1}{6}$. In so doing we are creating a probability model for rolling a single die called a **uniform probability model**.

Uniform Probability Model

> If an experiment has a sample space consisting of n mutually exclusive and equally likely simple events, then a *uniform probability model* assigns the probability of $1/n$ to each simple event.

Spades (black cards)

Hearts (red cards)

Clubs (black cards)

Diamonds (red cards)

FIGURE 7.1
A deck of 52 cards

EXAMPLE 2 Suppose a single card is chosen from a deck of cards. What is the probability it is an ace of spades?

Solution In this section, when we refer to a deck of cards we are assuming the standard bridge deck shown in Figure 7.1. Since there are 52 equally likely outcomes for this experiment, we assign a probability of $\frac{1}{52}$ to the event of drawing the ace of spades. We write this as

$$P(\text{ace of spades}) = \frac{1}{52}$$ ■

EXAMPLE 3 Find $P(\text{ace})$ for the experiment in Example 2.

Solution Let $A = \{\text{ace of spades, ace of clubs, ace of hearts, ace of diamonds}\}$. Then

$$\begin{aligned} P(A) &= P(\text{ace of spades}) + P(\text{ace of clubs}) + P(\text{ace of hearts}) \\ &\quad + P(\text{ace of diamonds}) \\ &= \frac{1}{52} + \frac{1}{52} + \frac{1}{52} + \frac{1}{52} \\ &= \frac{4}{52} \\ &= \frac{1}{13} \quad \text{or} \quad .08 \end{aligned}$$

State probability as a reduced fraction or as a decimal rounded to the nearest hundredth. ■

The procedure illustrated by Example 3 is unnecessarily lengthy. Let us consider a general result. Suppose

$$E = \{k_1, k_2, k_3, \ldots, k_s\}$$

Then

$$\begin{aligned} P(E) &= P(k_1) + P(k_2) + P(k_3) + \cdots + P(k_s) \\ &= \underbrace{\frac{1}{n} + \frac{1}{n} + \frac{1}{n} + \cdots + \frac{1}{n}}_{s \text{ simple events}} \\ &= \frac{s}{n} \end{aligned}$$

Each simple event has probability of $1/n$.

This leads us to the following important probability model:

Probability of an Event That Can Occur in Any One of n Mutually Exclusive and Equally Likely Ways

If an experiment can occur in any of n ($n \geq 1$) mutually exclusive and equally likely ways and if s of these ways are considered favorable to E, then the probability of the event E, denoted by $P(E)$, is

$$P(E) = \frac{s}{n} = \frac{\text{Number of outcomes favorable to } E}{\text{Number of all possible outcomes}}$$

For Examples 4–8, suppose a single card is selected from a deck of cards (see Figure 7.1).

EXAMPLE 4 Find $P(\text{heart})$.

Solution There are 52 elements in the sample space and 13 of these are hearts, or successes. Therefore

$$P(\text{heart}) = \frac{13}{52} = \frac{1}{4}$$ ■

EXAMPLE 5 $P(\text{heart or an ace}) = \frac{16}{52}$ 13 hearts + 3 *additional* aces (be careful not to count the ace of hearts twice)

$= \frac{4}{13}$ ■

EXAMPLE 6 $P(\text{heart and an ace}) = \frac{1}{52}$ The ace of hearts is the only such card. ■

EXAMPLE 7 $P(\text{ace or a 2}) = \frac{8}{52} = \frac{2}{13}$ ■

EXAMPLE 8 $P(\text{ace and a 2}) = \frac{0}{52}$ There is no way of drawing a single card and obtaining an ace *and* a 2.

$= 0$ ■

Finding the Probability of an Event

In summary, to find the probability of some event:

1. Describe and identify the sample space, and then count the number of simple events (these should be equally likely). Call this number n.
2. Count the number of occurrences that are favorable to the event; call this the *number of successes* and denote it by s.
3. Compute the probability of the event: $P(E) = \frac{s}{n}$.

Do not forget that the simple events must be equally likely. Consider the following situation for Examples 9–11: A jar contains three red marbles, two black marbles, and five white marbles. Conduct an experiment of drawing out a single marble from the jar.

EXAMPLE 9 Find $P(\text{red})$.

Solution A possible sample space is {red, black, white}, but the simple events {red}, {black}, {white} are *not* equally likely. The problem, then, is to create a sample space that *has* equally likely simple events. For this example it seems that the individual marbles each have an equal chance of being chosen. Consider the sample space consisting of the individual marbles labeled as follows:

$$\{R_1, R_2, R_3, B_1, B_2, W_1, W_2, W_3, W_4, W_5\}$$

where R_1, R_2, R_3 represent the red marbles; B_1, B_2 represent the black marbles; and W_1, W_2, W_3, W_4, W_5 represent the white marbles. Then the sample space, as now described, consists of equally likely outcomes; so we use this model. Thus

$$P(\text{red}) = \frac{\text{Number of favorable outcomes}}{\text{Number of simple events}} = \frac{3}{10}$$ ■

EXAMPLE 10 $P(\text{black or white}) = \frac{7}{10}$ ■

EXAMPLE 11 $P(\text{black and white}) = \dfrac{0}{10} = 0$ The number of favorable outcomes is 0. You cannot draw a black *and* a white marble with only one draw. ■

Sometimes probabilities are found empirically by using the model developed in this section, as shown in Example 12.

EXAMPLE 12 Suppose that in a certain study, 46 out of 155 people showed a certain kind of behavior. Assign a probability to this behavior.

Solution $P(\text{behavior of interest}) = \dfrac{46}{155} \approx .30$ By calculator ■

EXAMPLE 13 Suppose we conduct an experiment by rolling a pair of dice 50 times and recording the sum. This experiment is repeated 3 times for a total of 150 rolls of the dice. The results of these experiments are recorded below:

Outcome	First trial	Second trial	Third trial	Total
2	|		|	2
3	|||	||	||	7
4	𝍸	||||	𝍸	14
5	||	𝍸 ||	𝍸 |	15
6	𝍸 ||||	𝍸 |||	𝍸 |	23
7	𝍸 |||	𝍸 𝍸 |	𝍸 ||	26
8	𝍸 |	𝍸 |||	𝍸 ||||	23
9	𝍸 |	|||	𝍸 ||	16
10	||||	||||	||	10
11	𝍸	||	𝍸	12
12	|	|		2
Total	50	50	50	150

The empirical probabilities can now be calculated:

$P(2) = \frac{2}{150} \approx .01$ $P(8) = \frac{23}{150} \approx .15$

$P(3) = \frac{7}{150} \approx .05$ $P(9) = \frac{16}{150} \approx .11$

$P(4) = \frac{14}{150} \approx .09$ $P(10) = \frac{10}{150} \approx .07$

$P(5) = \frac{15}{150} = .10$ $P(11) = \frac{12}{150} = .08$

$P(6) = \frac{23}{150} \approx .15$ $P(12) = \frac{2}{150} \approx .01$

$P(7) = \frac{26}{150} \approx .17$ ■

EXAMPLE 14 Find the theoretical probabilities for the sum of the two top faces when rolling a pair of dice.

Solution The sample space is $\{2, 3, 4, 5, 6, 7, 8, 9, 10, 11, 12\}$. If we assume that each simple event is equally likely, then

$$P(2) = \frac{1}{11} \approx .09$$

This same result should then hold for any of the numbers 2–12. But these theoretical probabilities do not correspond to the probabilities found in Example 13 by actually conducting an experiment of rolling a pair of dice. We know these two numbers should be approximately the same. The difficulty here is our assumption that the outcomes in this sample space are equally likely. You *must* find a sample space of

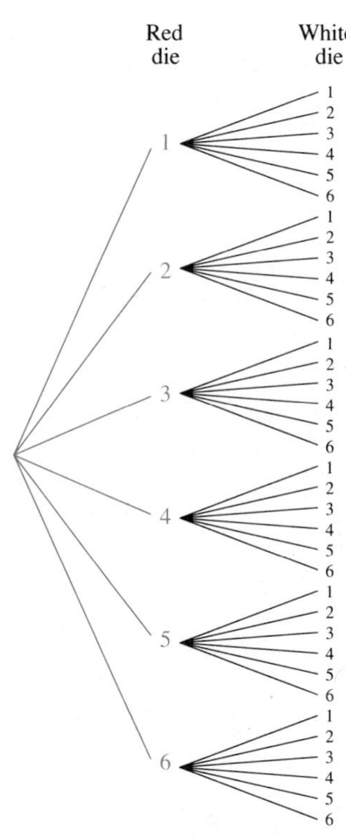

equally likely possibilities to use the model given in this section. To find the correct sample space, let us use a tree diagram. If you picture the dice as two different colors, you can see the arrangements in Figure 7.2. Notice that this is the Cartesian product of the sample spaces for the individual dice.

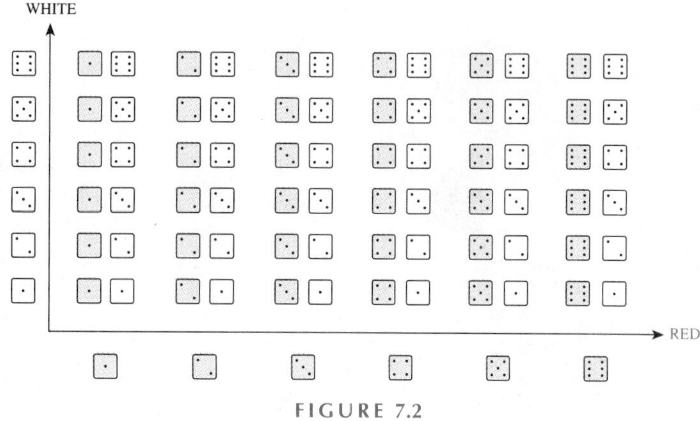

FIGURE 7.2
Sample space for tossing a pair of dice

Using this model, we find

$$P(2) = \frac{1}{36}$$ ⟵ Number of successful possibilities in sample space
⟵ Number of equally likely possibilities; look at Figure 7.2.

$$P(3) = \frac{2}{36}$$

$$P(4) = \frac{3}{36}$$ ■

After calculating the other probabilities in Example 14 (you will be asked to do some of these in Problem Set 7.2), we can compare *these* theoretical probabilities with the empirical probabilities and find consistent results, as shown in Table 7.2.

TABLE 7.2
Comparison of the empirical and theoretical probabilities for rolling a pair of dice

Outcome	Theoretical probability	Empirical probability
2	.0278	.0133
3	.0556	.0467
4	.0833	.0933
5	.1111	.1000
6	.1389	.1533
7	.1667	.1733
8	.1389	.1533
9	.1111	.1067
10	.0833	.0667
11	.0556	.0800
12	.0278	.0133

7.2

Problem Set

Use the spinners shown for Problems 1–4 and find the requested probabilities. Assume that the pointer can never lie on a border line.

1. a. $P(\text{white})$
   b. $P(\text{black})$
   c. $P(\text{red})$

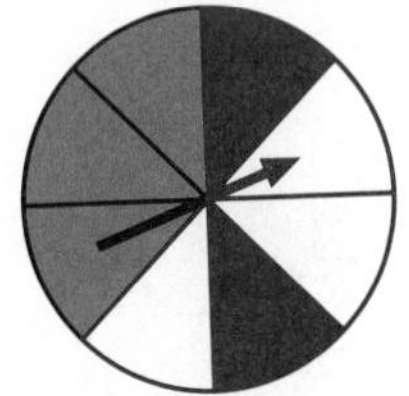

2. a. $P(A)$
   b. $P(B)$
   c. $P(C)$

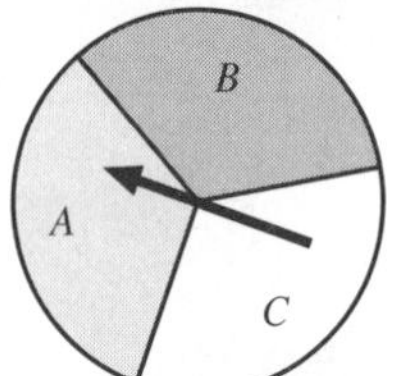

3. a. $P(D)$
   b. $P(E)$
   c. $P(F)$

4. a. $P(G)$
   b. $P(H)$
   c. $P(I)$
   d. $P(J)$

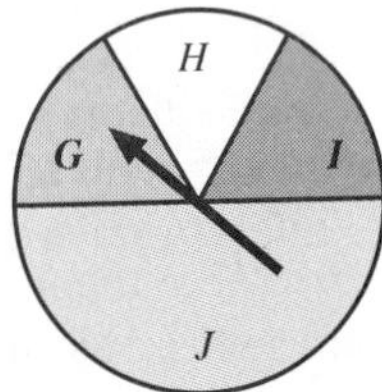

Consider the jar containing marbles shown here. Suppose each marble has an equal chance of being picked from the jar. Find the probabilities in Problems 5–7.

5. a. $P(\text{white})$
   b. $P(\text{black})$
   c. $P(\text{red})$

6. a. $P(\text{white or black})$
   b. $P(\text{red or black})$
   c. $P(\text{white and black})$

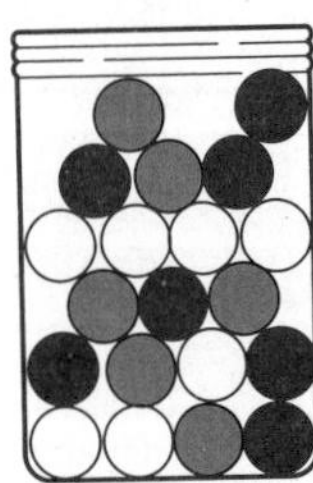

7. If you were to pick a marble from the jar, note its color, return the marble to the jar, and then repeat the experiment 100 times, how many times would you expect to pick a black marble?

A single card is selected from a deck of 52 cards. Find the probabilities in Problems 8–10.

8. a. $P(\text{five})$
   b. $P(\text{jack})$
   c. $P(\text{five and a jack})$
   d. $P(\text{five or a jack})$

9. a. $P(\text{three})$
   b. $P(\text{diamond})$
   c. $P(\text{three and a diamond})$
   d. $P(\text{three or a diamond})$

10. a. $P(\text{seven})$
    b. $P(\text{heart})$
    c. $P(\text{seven and a heart})$
    d. $P(\text{seven or a heart})$

For the dart boards shown in Problems 11 and 12, assume that a dart is randomly thrown at the board and it strikes the board each time. Find the requested probabilities.

11. a. $P(A)$
    b. $P(B)$
    c. $P(C)$
    d. $P(A \text{ or } B)$
    e. $P(A \text{ and } B)$

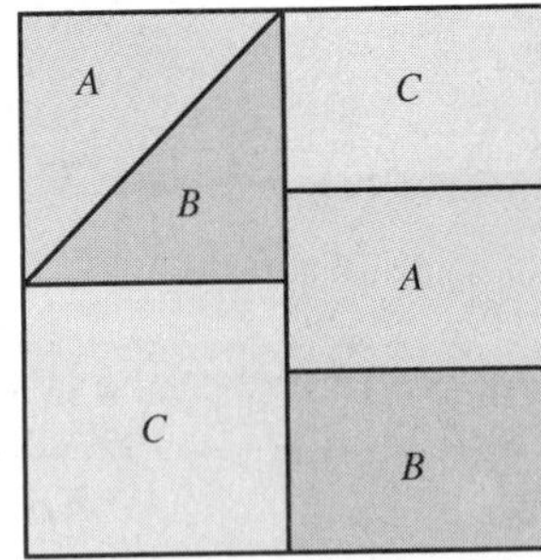

12. a. $P(R)$
    b. $P(B)$
    c. $P(G)$
    d. $P(Y)$
    e. $P(R \text{ or } B)$
    f. $P(G \text{ or } Y)$

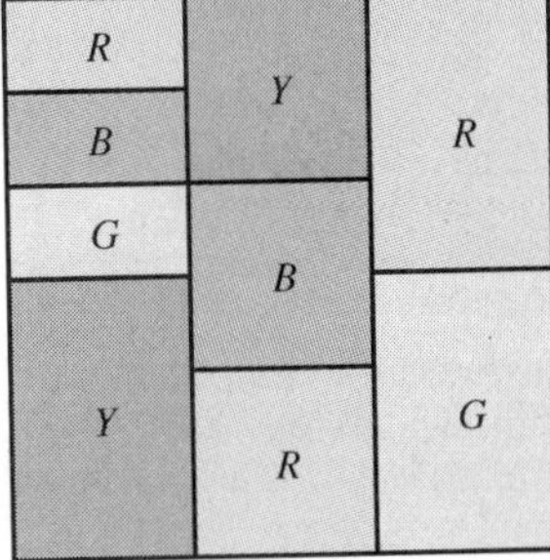

13. Create a square dart board with areas marked red (R), green (G), and black (B) so that $P(R) = \frac{1}{3}$ and $P(G) = \frac{1}{6}$. What is $P(B)$?

APPLICATIONS

Give the probabilities in Problems 14–20 in decimal form (correct to two decimal places).

14. Last year, 1,485 calculators were returned to the manufacturer. If 85,000 were produced, assign a number to specify the probability that a particular calculator would be returned.

15. Last semester, Professor Math gave 13 A's out of 285 grades. If the grades were assigned randomly, what is the probability of an A?

16. Last year, it rained on 85 days in a certain city. What is the probability of rain on a day selected at random?

17. A campus club is having a raffle and needs to sell 1,500 tickets. If the people on your floor of the dorm buy 285 of the tickets, what is the probability that someone on your floor will hold the winning ticket?

18. If the traffic signal at a certain intersection is green for 45 seconds, yellow for 5 seconds, and red for 30 seconds, what is the probability that the signal will be red when you arrive at the intersection?

19. A bank did a survey of the number of people waiting in line at different times during the business day and found the following probabilities:

Number in line:	0	1	2	3	4+
Probability:	.29	.15	.22	.18	.16

a. What is the probability that there will be at most two people in line?
b. What is the probability that there will be at least two people in line?

20. In an experiment to study the effectiveness of an automatic brake on a power lawn mower in reducing injuries, 500 of 1,000 lawn mowers were equipped with the automatic brake. At the end of a one-year trial period, the records showed that of those equipped with an automatic brake, there were 20 injuries, and of those not equipped with the brake there were 45 injuries. Based on these data, what is the probability that a person using the lawn mower with an automatic brake will be injured within a one-year period? What is the probability that a person using the lawn mower without an automatic brake will be injured within a one-year period?

Perform the experiments described in Problems 21–22, tally your results, and calculate the empirical probabilities for each outcome (to the nearest hundredth).

21. Flip three coins simultaneously 50 times and note the results. The possible outcomes are (1) three heads, (2) two heads and one tail, (3) two tails and one head, and (4) three tails. Do these appear to be equally likely outcomes?

22. Prepare three cards that are identical except for the color. One card is black on both sides, one is white on both sides, and one is black on one side and white on the other. One card is selected at random and placed flat on the table. You will see either a black or a white card; record the color of the face. This is not the event with which we are concerned; rather, we are interested in finding the probability of the *other* side being black or white. Record the color of the underside, as shown in the table:

Color of face	Frequency	Outcome (color of the underside)	Frequency	Probability
White		White		
		Black		
Black		White		
		Black		

Repeat the experiment 50 times and find the probability of occurrence with respect to the known color. Do these appear to be equally likely outcomes?

Use the sample space shown in Figure 7.2 to find the probabilities that the sums of the top faces on the dice are the numbers requested in Problems 23–34.

23. $P(5)$
24. $P(6)$
25. $P(7)$
26. $P(8)$
27. $P(9)$
28. $P(10)$
29. $P(11)$
30. $P(12)$
31. $P(4 \text{ or } 5)$
32. $P(4 \text{ and } 5)$
33. $P(\text{even number})$
34. $P(\text{odd number})$

35. Dice is a popular game in gambling casinos. Two dice are tossed, and various amounts are paid according to the outcome. If a 7 or 11 occurs on the first roll, the player wins. What is the probability of winning on the first roll?

36. In the game of dice, the player loses if the outcome of the first roll is a 2, 3, or 12. What is the probability of losing on the first roll?

37. In the game of dice, a pair of 1s is called *snake eyes*. What is the probability of losing a dice game by rolling snake eyes?

38. Consider a die with only four sides, marked 1, 2, 3, and 4: Write out a sample space similar to the one shown in Figure 7.2 for rolling a pair of these dice.

39. Using the sample space you found in Problem 38, find the probability that the sum of the dice is the given number. Assume equally likely outcomes.
a. $P(2)$
b. $P(3)$
c. $P(4)$

40. Using the sample space you found in Problem 38, find the probability that the sum of the dice is the given number. Assume equally likely outcomes.
a. $P(5)$ **b.** $P(6)$ **c.** $P(7)$

41. The game of Dungeons and Dragons uses nonstandard dice. Consider a die with eight sides marked 1, 2, 3, 4, 5, 6, 7, and 8. Write out a sample space similar to the one shown in Figure 7.2 for rolling a pair of these dice.

42. ***Historical Question.*** The Romans played many dice games using a stone with 14 faces marked with the Roman numerals I to XIV. Assuming that each face of this die has an equally likely chance of occurring, find the requested probabilities when one such stone is tossed.
a. $P(\text{V})$ **b.** $P(\text{VII})$
c. $P(\text{X})$ **d.** $P(\text{XV})$

43. ***Historical Question.*** The Romans played a game of chance that required the participants to roll a pair of dice like the one described in Problem 42. Find the probability that the sum of the faces is the given number.
a. $P(20)$ **b.** $P(2)$
c. $P(15)$ **d.** $P(25)$

7.3
Calculated Probabilities

In Section 7.2 our focus was on deciding whether the events were mutually exclusive and equally likely. Then we applied the formula

$$P(E) = \frac{\text{Number of favorable outcomes}}{\text{Number of simple events}}$$

In this section we focus on different ways of counting the number of simple events (all possible outcomes) and the number of favorable outcomes (successes). You will need to use the multiplication principle, permutations, and combinations. It will also be helpful to have a calculator to put your answers in decimal form.

EXAMPLE 1 Assume that in an assortment of 20 electronic calculators there are 5 with defective switches. If one machine is selected at random, what is the probability of picking one with a defective switch?

Solution The solution is given by

$$\frac{\text{Number of ways of selecting 1 defective from the 5}}{\text{Number of ways of selecting 1 machine from the 20}} = \frac{5}{20} = \frac{1}{4}$$

Now, to use the terminology of combinations in anticipation of more complicated problems, we can rework this problem:

$$\frac{\binom{5}{1}}{\binom{20}{1}} = \frac{5}{20} = .25$$

■

EXAMPLE 2 Assume that in an assortment of 20 electronic calculators there are 5 with defective switches. If two machines are selected at random, what is the probability they both have defective switches?

Solution The solution is given by

$$\frac{\begin{pmatrix}\text{Number of ways of selecting}\\ \text{2 defectives from the 5}\end{pmatrix}}{\begin{pmatrix}\text{Number of ways of selecting}\\ \text{2 machines from the 20}\end{pmatrix}} = \frac{\binom{5}{2}}{\binom{20}{2}} = \frac{\frac{5 \cdot 4}{2}}{\frac{20 \cdot 19}{2}} = \frac{1}{19} \approx .05$$

■

EXAMPLE 3 Assume that in an assortment of 20 electronic calculators there are 5 with defective switches. If three machines are selected at random, what is the probability that exactly one of those selected has a defective switch?

Solution In this problem we use the multiplication principle to determine the number of successes. Picking one machine with a defective switch:

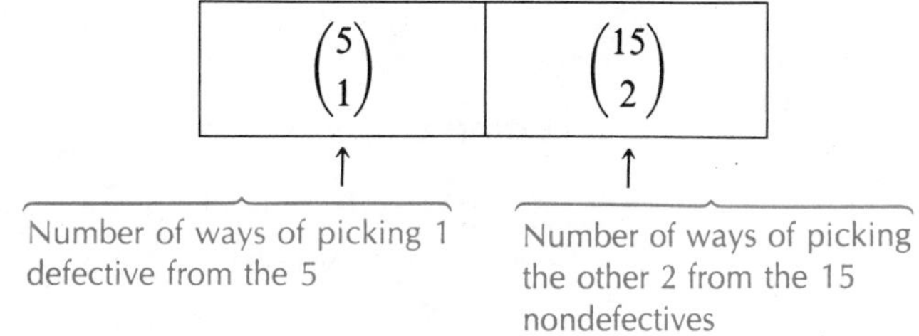

The probability we are seeking is

$$\frac{\begin{pmatrix}\text{Number of ways of picking}\\ \text{1 defective from the 5}\end{pmatrix} \cdot \begin{pmatrix}\text{Number of ways of picking the}\\ \text{other 2 from the 15 nondefectives}\end{pmatrix}}{\text{Number of ways of picking 3 from 20}}$$

$$= \frac{\binom{5}{1} \cdot \binom{15}{2}}{\binom{20}{3}} = \frac{\frac{5 \cdot 15 \cdot 14}{2}}{\frac{20 \cdot 19 \cdot 18}{3 \cdot 2 \cdot 1}}$$

$$= \frac{5 \cdot 15 \cdot 7}{20 \cdot 19 \cdot 3} = \frac{35}{76} \approx .46$$

■

EXAMPLE 4 Extrasensory perception (ESP) can be tested by using five colored cards. The subject is asked to arrange the red, blue, green, black, and white cards in the same order in which a person in another room has arranged them. What is the probability that the subject would arrange the cards in the same order by chance?

Solution Let $E = \{\text{proper arrangement}\}$.

$$P(E) = \frac{\text{Number of proper arrangements}}{\text{Number of simple events}} = \frac{1}{{}_5P_5}$$

Permutation because order is specified

$$= \frac{1}{5!} = \frac{1}{120} \approx .008$$

■

EXAMPLE 5 What is the probability that a family that has five children has exactly two girls?

Solution Assume that the probability of a boy or a girl is the same.

$$P(\text{exactly 2 girls}) = \frac{\text{Number of families with 5 children having exactly 2 girls}}{\text{Number of all possible 5-children families}}$$

$$= \frac{\binom{5}{2}}{2^5}$$

← Ways of selecting 2 out of 5

← Multiplication principle, 2 possibilities for each of the 5 children

$$= \frac{10}{32}$$

$$= .3125$$

■

Complementary Probabilities

We have been using the formula

$$P(E) = \frac{\text{Number of favorable outcomes}}{\text{Number of simple events}}$$

Now we wish to expand our scope. Let

s = Number of favorable outcomes (success)

f = Number of unfavorable outcomes (failure)

and

n = Number of possibilities (simple events); note that $n = s + f$

Then the probability **for** an event E is $P(E) = \frac{s}{n}$.

The probability **against** an event E, called the **complement** of E, written $\bar{E}$ is $P(\bar{E}) = \frac{f}{n}$.

An important property of probability is found by adding these probabilities:

$$P(E) + P(\bar{E}) = \frac{s}{n} + \frac{f}{n} = \frac{s+f}{n} = \frac{n}{n} = 1$$

The property that $P(E) + P(\bar{E}) = 1$ is usually written in a more useful form as shown in the following box.

Probability of Complements

$$P(E) = 1 - P(\bar{E}) \qquad \text{or} \qquad P(\bar{E}) = 1 - P(E)$$

This result extends to any number of events in a sample space. For example, Table 7.3 shows all of the possible poker hands and their probabilities. Notice that

TABLE 7.3 Probabilities of Poker Hands

	Hand	Number of favorable events	Probability
Royal flush		4	.00000153908
Other straight flush		36	.00001385169
Four of a kind		624	.00024009604
Full house		3,744	.00144057623
Flush		5,108	.00196540155
Straight		10,200	.00392464682
Three of a kind		54,912	.02112845138
Two pair		123,552	.04753901561
One pair		1,098,240	.42256902761
Other hands		1,302,540	.50117739403
		2,598,960	1.00000000000

the sum of all the probabilities is 1. We will calculate some of the probabilities on this table in the examples and problems.

EXAMPLE 6 From the table of probabilities of poker in Table 7.3, the probability of obtaining one pair is about .42. What is the probability of not obtaining one pair? (*Note:* In poker, a pair in a better hand—for example, "two pairs" or "three of a kind"—does not count as a pair.)

Solution Since $P(\text{pair}) \approx .42$,

$$\begin{aligned} P(\text{not a pair}) &= 1 - P(\text{pair}) \\ &\approx 1 - .42 = .58 \end{aligned}$$

■

EXAMPLE 7 What is the probability of being dealt a flush in poker?

Solution First we will work out this problem using Table 7.3, then we verify the entries in Table 7.3 for this example.

$$P(\text{flush}) = \frac{\text{Number of ways of obtaining a flush}}{\text{Number of possible poker hands}} = \frac{5{,}108}{2{,}598{,}960} \approx .0019654$$

These numbers were found in the table. Now, we need to verify these table numbers:

$$\begin{aligned}
&\text{Number of ways of obtaining a flush}\\
&\quad = (\text{Number of suits}) \cdot (\text{Number of flushes in a particular suit})\\
&\quad = 4 \cdot \binom{13}{5}\\
&\quad = 4 \cdot \frac{13!}{5!8!}\\
&\quad = 4 \cdot 1{,}287 = 5{,}148
\end{aligned}$$

This is *not* the number shown in Table 7.3. What we have just calculated here is the number of *all* possible flushes, but in poker there are special flushes that are better than regular flushes. These are royal flushes and straight flushes. From Table 7.3 you can see that there are 4 possible royal flushes (one for each suit) and 36 straight flushes (see Problem 59). Thus, the number of regular flushes is

$$5{,}148 - 4 - 36 = 5{,}108$$

$$\begin{aligned}
&\text{Number of possible poker hands}\\
&\quad = \text{Number of ways of drawing 5 cards from 52}\\
&\quad = \binom{52}{5}\\
&\quad = \frac{52!}{5!47!} = 2{,}598{,}960
\end{aligned}$$

■

EXAMPLE 8 What is the probability that a random 7-digit telephone number has one or more repeated digits?*

Solution There are $10^7 = 10{,}000{,}000$ distinct telephone numbers. This is the number of simple events in the sample space. If

$$E = \{\text{there is one or more repeated digits}\}$$

then

$$\bar{E} = \{\text{there are no repeated digits}\}$$

By the multiplication principle, the number of 7-digit numbers that have no repeated digits is

$$10 \cdot 9 \cdot 8 \cdot 7 \cdot 6 \cdot 5 \cdot 4 = 604{,}800$$

* Ignore the fact that certain arrangements of digits do not represent "real" phone numbers.

Thus,

$$P(\bar{E}) = \frac{604{,}800}{10{,}000{,}000} = .06048$$

Finally,

$$P(E) = 1 - P(\bar{E}) = 1 - .06048 = .93952$$

■

Odds

s = Number of successes
f = Number of failures
n = Number of possibilities

Related to probability is the notion of **odds**. Instead of forming the ratios s/n and f/n for the probabilities for and against an event, we form the following ratios to find the odds:

Odds for an event E: $\frac{s}{f}$ (ratio of success to failure)

Odds against an event E: $\frac{f}{s}$ (ratio of failure to success)

EXAMPLE 9 If a jar has 7 quarters, 3 dimes, and 5 nickels, what are the odds against choosing a coin at random and picking a nickel?

Solution A success is drawing a nickel, so $s = 5$; a failure is drawing a quarter or a dime, so $f = 10$. The odds against drawing a nickel are found by writing the ratio of failures to successes:

$$\frac{f}{s} = \frac{10}{5} = \frac{2}{1}$$

Odds are stated as ratios and not fractions, whereas probabilities are usually stated as fractions or decimals. Thus, we say that the odds against drawing a nickel are 2 to 1. ■

EXAMPLE 10 Compare the probability of drawing a heart from an ordinary deck of cards and the odds in favor of drawing a heart.

Solution

$$\text{Probability of drawing a heart} = \frac{13}{52} = \frac{1}{4}$$

← 13 hearts in the deck / 52 is the total no. of cards

$$\text{Odds in favor of drawing a heart} = \frac{13}{39} = \frac{1}{3}$$

← 13 cards are successes / 39 cards are failures

We would say the probability of drawing a heart is $\frac{1}{4}$ or .25 and the odds in favor of drawing a heart are 1 to 3. ■

Sometimes you will know the probability and want to find the odds, or you may know the odds and want to find the probability. The relationship is easy to see if you study Figure 7.3.

Remember,
s = Number of successes
f = Number of failures
n = Number of possibilities

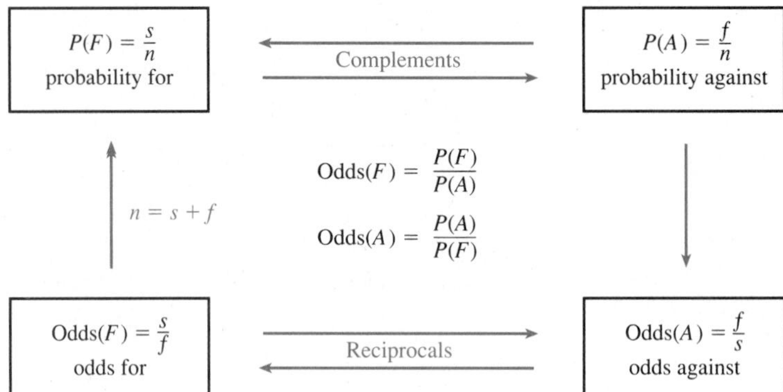

FIGURE 7.3
Relationship between probability and odds

EXAMPLE 11 If the probability of an event is .45, what are the odds in favor of the event?

Solution $P(E) = .45 = \frac{45}{100} = \frac{9}{20}$

$$P(\bar{E}) = 1 - P(E) = 1 - \frac{9}{20} = \frac{11}{20}$$

$$\text{Odds}(F) = \frac{P(E)}{P(\bar{E})} = \frac{\frac{9}{20}}{\frac{11}{20}} = \frac{9}{20} \div \frac{11}{20} = \frac{9}{20} \cdot \frac{20}{11} = \frac{9}{11}$$

The odds in favor are 9 to 11. ■

EXAMPLE 12 If the odds against you are 100 to 1, what is the probability of the event?

Solution $\text{Odds}(A) = \frac{100}{1}$ Since this is odds against, $s = 1$ and $f = 100$.

$$n = s + f = 100 + 1 = 101$$

$$P(E) = \frac{s}{n} = \frac{1}{101}$$ ■

EXAMPLE 13 If the odds in favor of some event are 2 to 5, what is the probability of the event?

Solution $\text{Odds}(F) = \frac{2}{5}$ This is odds for, so $s = 2$ and $f = 5$.

$$n = s + f = 7$$

$$P(E) = \frac{s}{n} = \frac{2}{7}$$ ■

Birthday Problem

There is an interesting, and counterintuitive, result known as the *birthday problem.* Consider the set of people in your classroom or the set of U.S. presidents. What would you guess is the probability that two persons of the group are born on the same day of the year? As it turns out, Polk and Harding were both born on November 2. If any 23 or more people are selected at random, the probability that two or more of them will have the same birthday is greater than 50%. This is a seemingly paradoxical situation that will fool most people. In order to substantiate this claim, we calculate the probability of a birthday match for four people in Example 14.

EXAMPLE 14 A group of 4 people is selected at random. What is the probability that at least 2 of them have the same birthday?

Solution Let us first determine the number of simple events in the sample space.

There are 365 possibilities for each person's birthday (exclude leap years). Use the multiplication principle to find 365^4—the total number of simple events in the sample space.

Let $M = \{2$ or more of the 4 people have birthdays that match$\}$. We need to compute $P(M)$. It is easier to find $P(\bar{M})$ because

$$\bar{M} = \{\text{no 2 of the people have a birthday that matches}\}$$

$$\begin{aligned} P(M) &= 1 - P(\bar{M}) \\ &= 1 - \frac{\text{Number of ways that there are no matches}}{\text{Number of simple events in the sample space}} \\ &= 1 - \frac{{}_{365}P_4}{365^4} \\ &= 1 - \frac{365 \cdot 364 \cdot 363 \cdot 362}{365^4} \\ &\approx 1 - .9836441 = .0163559 \end{aligned}$$

■

If the result of Example 14 is extended to n people, the result is

$$P(M) = 1 - P(\bar{M}) = 1 - \frac{{}_{365}P_n}{365^n}$$

For $n = 23$, $P(M) \approx .507$. Figure 7.4 shows the birthday probabilities for $n = 0$ where $P(M) = 0$ to $n = 60$ where $P(M) \approx 1$. [Note that $P(M)$ does not *equal* 1 until $n = 366$.]

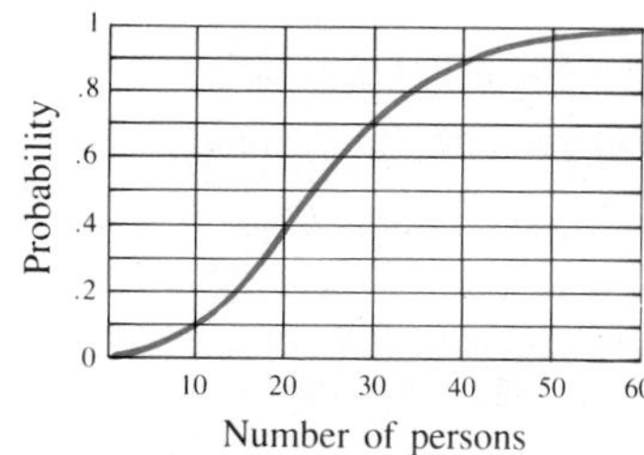

FIGURE 7.4
Probabilities of a birthday match

7.3 Problem Set

APPLICATIONS

It is known that a company has 10 pieces of machinery with no defects, 4 with minor defects, and 2 with major defects. Thus each of the 16 machines falls into one of these three categories. The inspector is about to come, and it is her policy to choose one machine and check it. Find the requested probabilities in Problems 1–6.

1. $P(\text{no defects})$

2. $P(\text{major defect})$

3. P(minor defect)
4. P(no major defects)
5. P(no minor defects)
6. P(defect)

Suppose a jar contains six red balls, four white balls, and three black balls. One ball is drawn at random. Find the requested probabilities in Problems 7–12.

7. P(red)
8. P(white)
9. P(black)
10. P(black or white)
11. P(red or white)
12. P(black and white)
13. What are the odds in favor of drawing a heart from an ordinary deck of 52 cards?
14. What are the odds against a family with four children having four boys?
15. Suppose the odds that a man will be bald by the time he is 60 are 9 to 1. State this as a probability.
16. Suppose the odds are 33 to 1 that someone will lie to you at least once in the next 7 days. State this as a probability.
17. What are the odds in favor of drawing an ace from an ordinary deck of 52 cards?

A certain magic trick with cards requires that five cards be randomly placed side by side in a row. Find the requested probabilities in Problems 18–23.

18. P(first card at the left is the ace of spades)
19. P(middle card is a heart)
20. P(first and last cards are diamonds)
21. P(left 3 cards are clubs and the next two are not)
22. P(second and fourth cards are kings)
23. P(middle 3 cards are red)

Suppose that a family has four children. Find the requested probabilities in Problems 24–28.

24. P(exactly 2 boys)
25. P(exactly 1 girl)
26. P(all boys)
27. P(2 girls and 2 boys)
28. P(3 girls and 1 boy)
29. What is the probability of flipping a coin five times and obtaining exactly two heads?
30. What is the probability of flipping a coin five times and obtaining three heads and two tails?
31. What is the probability of flipping a coin six times and obtaining exactly three heads?
32. Racetracks quote the approximate odds for a particular race on a large display board called a tote board. Some examples from a particular race are:

Horse number	Odds
1	2 to 1
2	15 to 1
3	3 to 2
4	7 to 5
5	1 to 1

The odds stated are for the horse losing. Thus

$$P(\text{horse 1 losing}) = \frac{2}{2+1} = \frac{2}{3}$$

$$P(\text{horse 1 winning}) = 1 - \frac{2}{3} = \frac{1}{3}$$

What would be the probability of winning for each of these horses?*

33. What are the odds in favor of rolling a 7 or 11 on a single roll of a pair of dice?

For Problems 34–39 assume that the inspector described in Problems 1–6 selects 2 machines at random. Find the requested probabilities as decimals rounded to the nearest hundredth.

34. P(both defective)
35. P(both nondefective)
36. P(both have major defects)
37. P(both have minor defects)
38. P(one with no defects and one with major defects)
39. P(one with a major defect and one with a minor defect)

For Problems 40–45 suppose two balls are drawn at random from the jar described in Problems 7–12. Find the requested probabilities as decimals rounded to the nearest hundredth.

40. P(both red)
41. P(both white)
42. P(both black)
43. P(1 red and 1 white)
44. P(1 black and 1 white)
45. P(1 red and 1 black)

For Problems 46–51 assume that the inspector described in Problems 1–6 selects 3 machines at random. Find the requested probabilities as decimals rounded to the nearest hundredth.

46. P(all have major defects)
47. P(all have minor defects)

* Racetrack odds are based on betting rather than probability so the sum of the probabilities for each horse winning is not necessarily 1.

48. P(all are nondefective)

49. P(exactly one major defect)

50. P(exactly one minor defect)

51. P(one of each type)

For Problems 52–58 suppose three balls are drawn at random from the jar described in Problems 7–12. Find the requested probabilities as decimals rounded to the nearest hundredth.

52. P(3 red)

53. P(3 white)

54. P(3 black)

55. P(2 red and 1 white)

56. P(2 white and 1 black)

57. P(2 black and 1 red)

58. P(one of each color)

Find the requested probabilities in Problems 59–61 as decimals correct to eight decimal places.

59. P(straight flush).

60. P(full house of three aces and a pair of twos)

61. ***Birthday Problem (continued).*** To calculate the probabilities graphed in Figure 7.4, suppose you have a group of n people. Let M be the event that at least two of the people have the same birthday. Complete the following table of probabilities using the formula given in the text.

n	5	10	20	22	23	24	30	40	50
$P(M)$	**a.**	**b.**	**c.**	**d.**	**e.**	**f.**	**g.**	**h.**	**i.**

62. Rework Example 14 but do not exclude leap years. Is 23 still the number of people for which the probability of a match first exceeds 50%?

63. ***Historical Question.*** The mathematical theory of probability arose in France in the seventeenth century when a gambler, the Chevalier de Méré, was interested in adjusting the stakes so that he could be certain of winning if he played long enough. He was betting that he could get at least one 6 in four rolls of a die. He also bet that in 24 tosses of a pair of dice, he would get at least one 12. He found that he won more often than he lost with the first bet, but not with the second. He did not know why, so he wrote to the mathematician Blaise Pascal (1623–1662) to find out why. Pascal sent these questions to another mathematician, Pierre de Fermat (1601–1665), and together they developed the first theory of probability. What are the probabilities for winning in the two games described by de Méré?

7.4 Conditional Probability

The probability of an event depends on what information is known about that event. For example, suppose you know that a family has two children and you are interested in the probability that both the children are boys. This probability depends on additional information, as illustrated by Example 1.

EXAMPLE 1 What is the probability that a family with two children has two boys if:

a. You have no additional information.

b. You know that there is at least one boy.

c. You know that the youngest child is a boy.

Solution Let

$$D = \{2 \text{ boys}\}$$
$$E = \{\text{at least 1 boy}\}$$
$$F = \{\text{youngest is a boy}\}$$

a. Consider the sample space:

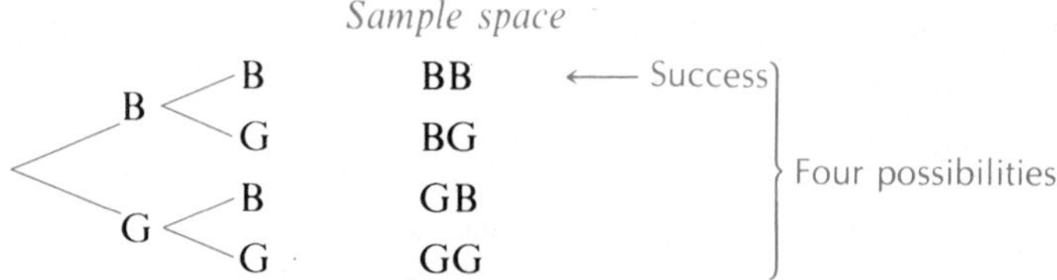

Thus $P(D) = \frac{1}{4}$.

b. The sample space from part **a** is *altered* because of the additional information:

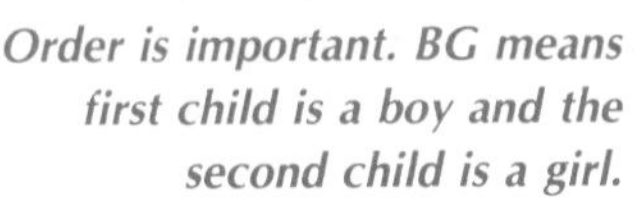

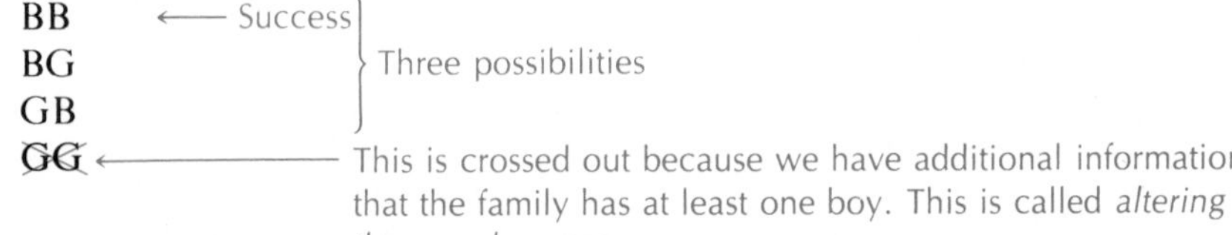

We write $P(D|E)$ to mean the probability of D *given the additional information* that E has occurred. Thus $P(D|E) = \frac{1}{3}$.

c. Again consider the altered sample space:

Sample space

BB
GB
~~BG~~
~~GG~~ } These are crossed out because you know that the youngest child is a boy.

Thus $P(D|F) = \frac{1}{2}$. ■

In Example 1 we needed a notation to represent the probability of an event E *given that an event F has occurred*. This is the idea of **conditional probability**. We write $P(E|F)$ to denote the probability of an event E *given* that an event F has occurred. One way of finding a conditional probability is to consider an altered sample space.

EXAMPLE 2 Consider rolling a pair of dice.

a. What is the probability that the sum is 6?

b. What is the probability that the sum is 6 if you know that one of the dice shows a 5?

c. What is the probability that at least one of the numbers showing is a 6?

d. What is the probability of rolling a sum of 2, 3, or 12 (this is known as *craps*) if you know that at least one of the dice shows a 6?

Solution The sample space is shown in Figure 7.2 (page 335).

a. Out of a sample space consisting of 36 equally likely pairs (x, y), there are five ways of obtaining a 6 (shown in color):

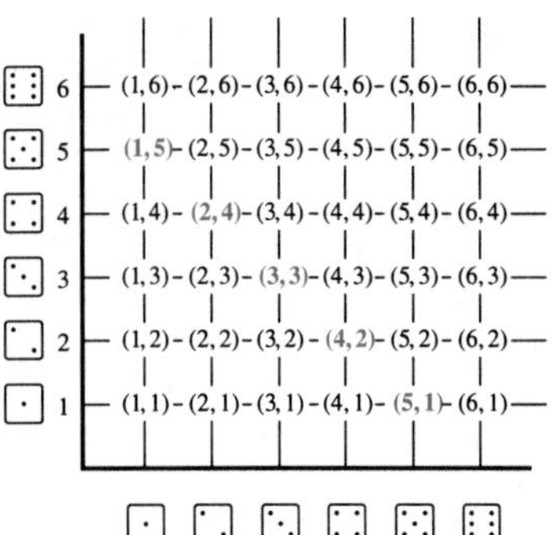

Thus $P(6) = \frac{5}{36}$.

b. The sample space is now reduced to 11 possibilities:

Altered sample space is circled.

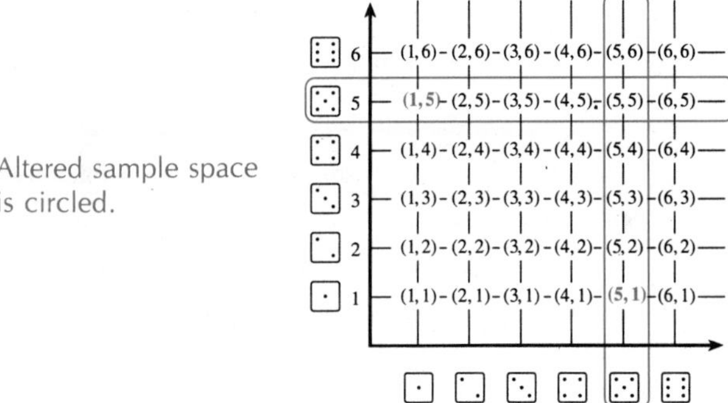

In the reduced sample space, two (shown in color) are considered successes. Thus,

$$P(\text{sum is } 6 \,|\, \text{one die is a } 5) = \frac{2}{11}$$

c. There are 11 ways to have at least one 6 (shown in color).

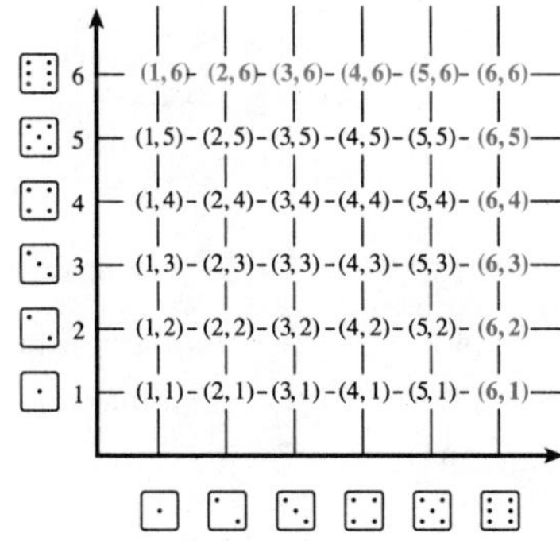

$$P(\text{at least one number shown is a } 6) = \frac{11}{36}$$

d. The altered sample space is circled. Within this altered sample space, there is only one possibility (shown in color) that gives a sum of 2, 3, or 12. Thus,

$$P(\text{roll a craps} \mid \text{one die is a six}) = \frac{1}{11}$$

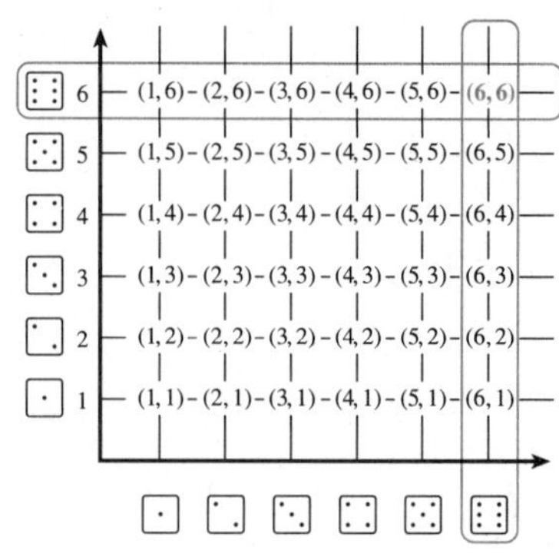

■

EXAMPLE 3 Before an advertising campaign for a new product is launched, a survey is conducted to determine the present usage for a certain brand of hair dryer. The results are shown in the table. Assume that one respondent is chosen at random from this sample of 1,400. Let

$M = \{\text{male}\}$ $D = \{\text{daily use}\}$

$F = \{\text{female}\}$ $C = \{\text{occasional use}\}$

	Daily use	**Occasional use**	**Total**
Male	142	258	400
Female	619	381	1,000
Total	761	639	1,400

Find each of the following (to the nearest hundredth) and interpret:

a. $P(M)$ **b.** $P(D)$ **c.** $P(M \mid D)$ **d.** $P(D \mid M)$

Solution **a.** $P(M)$ is the probability that the respondent is male:

	Daily use	**Occasional use**	**Total**	
Male	142	258	400	← $s = 400$
Female	619	381	1,000	
Total	761	639	1,400	
			↑ $n = 1{,}400$	

$$P(M) = \frac{400}{1{,}400} \begin{array}{l} \leftarrow \text{Number of males} \\ \leftarrow \text{Total number of respondents} \end{array}$$

$$\approx .29$$

b. $P(D)$ is the probability that the respondent uses a hair dryer daily:

	Daily use	Occasional use	Total
Male	142	258	400
Female	619	381	1,000
Total	761	639	1,400

← $n = 1{,}400$ (Total row); ↑ $s = 761$ (Daily use column)

$$P(D) = \frac{761}{1{,}400} \approx .54$$

← Number of daily users
← Total number of respondents

c. $P(M \mid D)$ is the probability that the respondent is a male if it is known that the respondent uses a hair dryer daily:

	Daily use	Occasional use	Total
Male	142	258	400
Female	619	381	1,000
Total	761	639	1,400

← $s = 142$ (Male row); ↑ $n = 761$ (Daily use column, circled)

Note that the sample space (the table) is altered as shown by part circled.

$$P(M \mid D) = \frac{142}{761} \approx .19$$

← Number of males
← Total number
} Altered sample space; daily users

d. $P(D \mid M)$ is the probability that the respondent uses a hair dryer daily if it is known that the respondent is a male:

	Daily use	Occasional use	Total
Male	142	258	400
Female	619	381	1,000
Total	761	639	1,400

← $n = 400$ (Male row, circled); ↑ $s = 142$ (Daily use column)

Note the altered sample space.

$$P(D \mid M) = \frac{142}{400} \approx .36$$

← Number of daily users
← Total number
} Reduced sample space; males

■

Examples 2 and 3 were designed to help you understand the idea of conditional probability, but these are simple examples with simple altered sample spaces. It is not always practical to list the sample space and then the altered sample space, so we seek to define $P(E|F)$ in such a way as to tell us how to calculate $P(E|F)$ without listing the sample space.

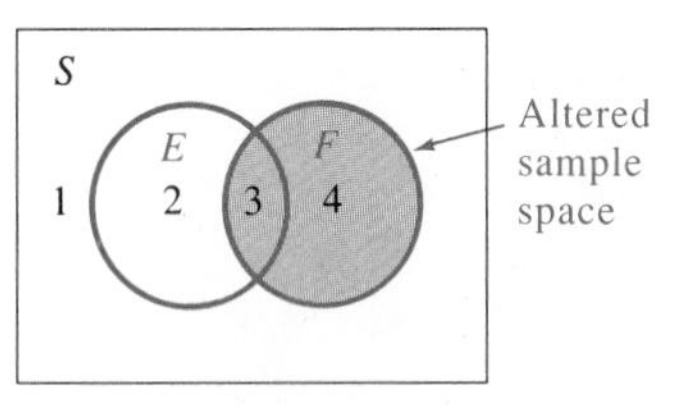

FIGURE 7.5

Consider the Venn diagram for two events E and F in Figure 7.5. We seek $P(E|F)$. Let $|E|$, $|F|$, and $|E \cap F|$ be the number of times that events E, F, and $E \cap F$ have occurred among the n repetitions of the experiment. Now $P(E|F)$ can be found by focusing our attention on region 3. We calculate the number

$$\frac{|E \cap F|}{|F|}$$

which represents the number of times E has occurred among those outcomes in which F has occurred. Now

$$\frac{|E \cap F|}{|F|} = \frac{\frac{|E \cap F|}{n}}{\frac{|F|}{n}} = \frac{P(E \cap F)}{P(F)}$$

This leads to the following definition:

Conditional Probability

The **conditional probability** that an event E has occurred given that event F has occurred is denoted by $P(E|F)$ and defined by

$$P(E|F) = \frac{P(E \cap F)}{P(F)} \qquad \text{provided} \qquad P(F) \neq 0$$

We will now rework parts of Examples 2 and 3 by using this formula.

EXAMPLE 4 Rework Example 2**b** by using the formula. That is, find

$$P(\text{sum is } 6 \mid \text{one of the dice is a } 5)$$

Solution Let E = sum is 6 and F = one of the dice is a 5. Look at the sample space. Event E is shown in color, namely

$$E = \{(1,5), (2,4), (3,3), (4,2), (5,1)\}$$

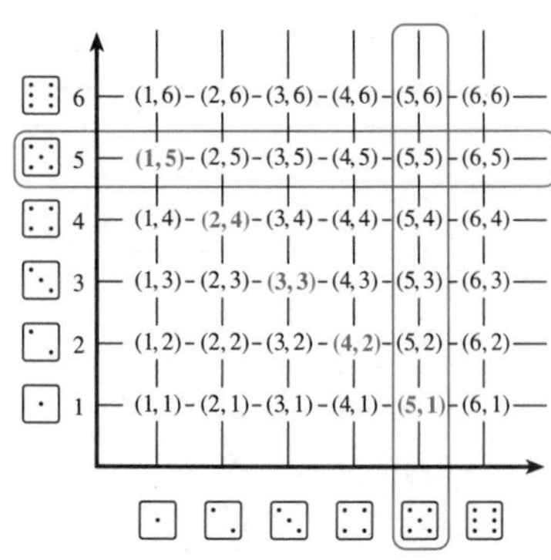

Event F is circled, namely

$$F = \{(1,5), (2,5), (3,5), (4,5), (5,5), (6,5), (5,6), (5,4), (5,3), (5,2), (5,1)\}$$

$E \cap F$ are those simple events that are in both E and F, namely

$$E \cap F = \{(1,5), (5,1)\}$$

Thus, from the sample space we see that

$$P(E) = \frac{5}{36} \qquad P(F) = \frac{11}{36} \qquad P(E \cap F) = \frac{2}{36}$$

so

$$P(E \mid F) = \frac{P(E \cap F)}{P(F)} = \frac{\frac{2}{36}}{\frac{11}{36}} = \frac{2}{11}$$

The formula gives the same answer we found in Example **2b**. Notice that we can also find

$$P(F \mid E) = \frac{P(F \cap E)}{P(E)} = \frac{\frac{2}{36}}{\frac{5}{36}} = \frac{2}{5}$$

■

EXAMPLE 5 Rework Example **3c** using the formula.

Solution

$$P(M \mid D) = \frac{P(M \cap D)}{P(D)} = \frac{\frac{142}{1{,}400}}{\frac{761}{1{,}400}} = \frac{142}{761}$$

This result agrees with the answer from Example **3c**. ■

EXAMPLE 6 A tire manufacturer has found that 10% of the tires produced have cosmetic defects and 2% have both cosmetic and structural defects. What is the probability that one tire selected at random is structurally defective if it is known that it has a cosmetic defect?

Solution Let $C = \{\text{cosmetic defect}\}$ and $S = \{\text{structural defect}\}$; find $P(S \mid C)$:

$$P(S \mid C) = \frac{P(S \cap C)}{P(C)}$$

$$= \frac{\frac{2}{100}}{\frac{10}{100}} = \frac{1}{5}$$

■

Product Rule

Conditional probability is often used to find the probability of an intersection of two events. If both sides of the equation for conditional probability are multiplied by the number $P(F)$, we obtain an alternate form of the conditional probability formula that is known as the product rule.

Product Rule

$$P(E \cap F) = P(E \mid F) \cdot P(F)$$

EXAMPLE 7 Suppose you take a sample of two calculators from a batch of 20 calculators, of which it is known that 5 have defective switches. Find the probability that the first one selected is defective and the second one is also defective.

Solution Let $A = \{$the first calculator picked has a defective switch$\}$
$B = \{$the second calculator picked has a defective switch$\}$
Begin by finding $P(A) = \frac{5}{20} = \frac{1}{4}$.

If A occurred, then there are 19 calculators left from which to choose and 4 of the remaining calculators are defective, so

$$P(B \mid A) = \frac{4}{19}$$

Thus, by the product rule,

$$P(A \cap B) = P(B \cap A) = P(B \mid A) \cdot P(A) = \frac{4}{19} \cdot \frac{1}{4} = \frac{1}{19}$$ ■

Stochastic Processes and Tree Diagrams

Many experiments, such as the one in Example 7, consist of a sequence of two or more events in which the probability at any state depends on the outcomes and associated probabilities of the preceding states. This chain of experiments is called a **finite stochastic process**. The easiest way to analyze stochastic processes is to look at a tree diagram.

Look at the tree diagram for Example 7.

Step 1: Label the paths with appropriate probabilities. The sum of these should be 1.

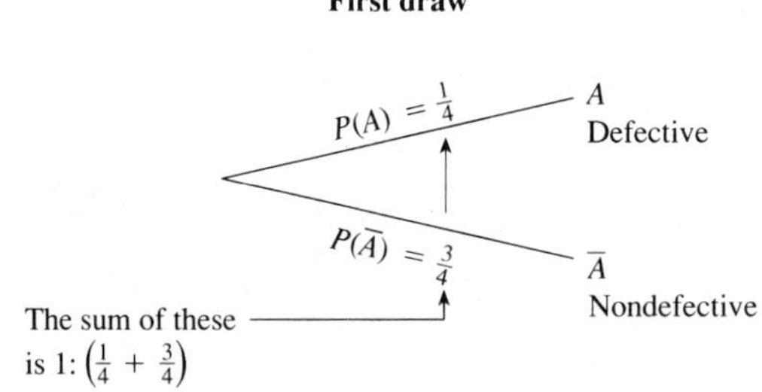

Step 2:

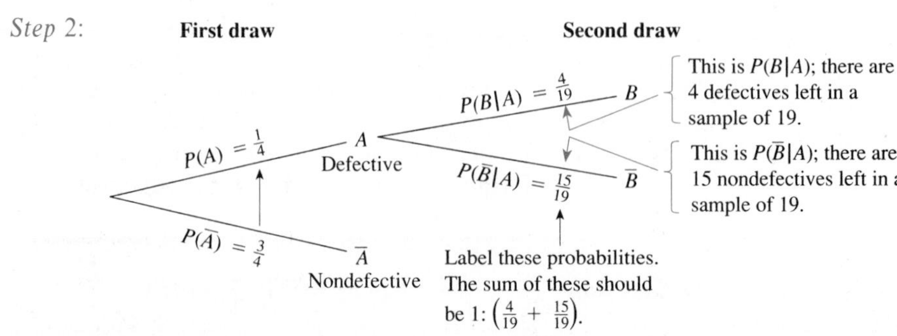

Step 3:

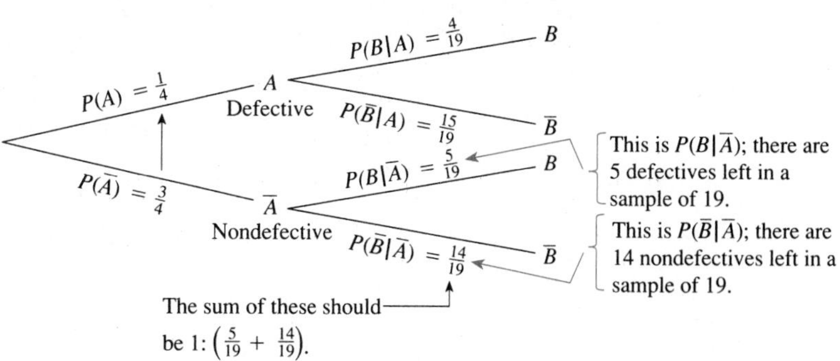

Step 4:

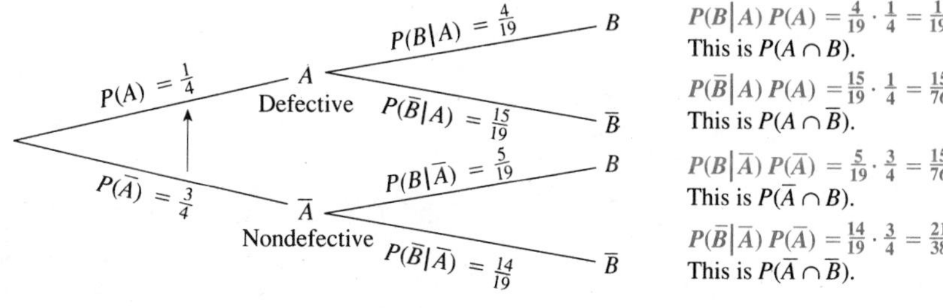

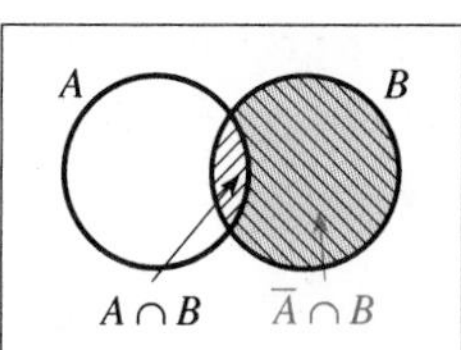

If we want to find $P(B)$ we just need to notice that for any set B,

$$B = (A \cap B) \cup (\bar{A} \cap B)$$

Thus, since $(A \cap B)$ and $(\bar{A} \cap B)$ are mutually exclusive,

$$P(B) = P(A \cap B) + P(\bar{A} \cap B)$$

What does this mean? It means that if you want to find $P(B)$, just look at the above tree diagram and add the numbers at the ends of the respective branches for the event B.

From step 4 of the tree diagram we have

$$P(B) = \frac{1}{19} + \frac{15}{76} = \frac{19}{76} = \frac{1}{4}$$

Notice for this example that the probability of drawing a defective switch on the second draw is the same as it is for drawing it on the first draw, if we assume that we do not know the results of the first draw.

EXAMPLE 8 Consider an experiment consisting of drawing two cards in a row from a deck consisting of tens, jacks, and kings. Use a tree diagram to find the requested probabilities. Let

$$T_1 = \{\text{drawing a ten on the first draw}\}$$
$$J_2 = \{\text{drawing a jack on the second draw}\}$$
$$K_2 = \{\text{drawing a king on the second draw}\}$$

Find each of the following probabilities.

a. $P(T_1)$ **b.** $P(J_2 \mid J_1)$

c. $P(K_1 \cap J_2)$ **d.** $P(\text{one jack and one king})$

Solution

First draw	Second draw	Path number	Probabilities
T_1 ($\frac{4}{12}=\frac{1}{3}$)	T_2 ($\frac{3}{11}$)	1	$T_1T_2: \frac{1}{3}\cdot\frac{3}{11}=\frac{1}{11}$
	J_2 ($\frac{4}{11}$)	2	$T_1J_2: \frac{1}{3}\cdot\frac{4}{11}=\frac{4}{33}$
	K_2 ($\frac{4}{11}$)	3	$T_1K_2: \frac{1}{3}\cdot\frac{4}{11}=\frac{4}{33}$
J_1 ($\frac{4}{12}=\frac{1}{3}$)	T_2 ($\frac{4}{11}$)	4	$J_1T_2: \frac{1}{3}\cdot\frac{4}{11}=\frac{4}{33}$
	J_2 ($\frac{3}{11}$)	5	$J_1J_2: \frac{1}{3}\cdot\frac{3}{11}=\frac{1}{11}$
	K_2 ($\frac{4}{11}$)	6	$J_1K_2: \frac{1}{3}\cdot\frac{4}{11}=\frac{4}{33}$
K_1 ($\frac{4}{12}=\frac{1}{3}$)	T_2 ($\frac{4}{11}$)	7	$K_1T_2: \frac{1}{3}\cdot\frac{4}{11}=\frac{4}{33}$
	J_2 ($\frac{4}{11}$)	8	$K_1J_2: \frac{1}{3}\cdot\frac{4}{11}=\frac{4}{33}$
	K_2 ($\frac{3}{11}$)	9	$K_1K_2: \frac{1}{3}\cdot\frac{3}{11}=\frac{1}{11}$

a. Find T_1 under the first draw: $P(T_1) = \frac{4}{12} = \frac{1}{3}$

b. This is the last branch of path 5 labeled J_2: $P(J_2 \mid J_1) = \frac{3}{11}$

c. This is path 7: $P(K_1 \cap J_2) = P(J_2 \mid K_1)P(K_1) = \frac{4}{11}\cdot\frac{1}{3} = \frac{4}{33}$

d. Paths 6 and 8 each have one jack and one king; add these probabilities since they are mutually exclusive:

$$\begin{aligned} P(\text{one jack and one king}) &= P(K_2 \mid J_1)P(J_1) + P(J_2 \mid K_1)P(K_1) \\ &= \frac{4}{11}\cdot\frac{1}{3} + \frac{4}{11}\cdot\frac{1}{3} \\ &= \frac{8}{33} \\ &\approx .24 \end{aligned}$$

■

7.4 Problem Set

In Problems 1–8 consider a family with three children and find the requested probabilities.

1. $P(\text{exactly 2 boys})$
2. $P(\text{exactly 1 girl})$
3. $P(\text{exactly 3 girls})$
4. $P(\text{exactly 2 boys given that at least 1 child is a boy})$
5. $P(\text{exactly 1 girl given that at least 1 child is a girl})$
6. $P(\text{exactly 3 girls given that at least 1 child is a girl})$
7. $P(\text{at least 1 girl})$
8. $P(\text{all girls given that there is at least 1 girl})$

Suppose a coin is flipped four times. Find the requested probabilities in Problems 9–14.

9. $P(\text{exactly 3H})$
10. $P(\text{exactly 2H})$
11. $P(\text{exactly 3T})$
12. $P(\text{exactly 3H given that there are at least 2H})$
13. $P(\text{exactly 2H given that there is at least 1H})$
14. $P(\text{exactly 3T given that there is at least 1T})$

Consider rolling a pair of dice. Find the probabilities requested in Problems 15–18.

15. P(sum is 7)

16. P(at least one 5 is rolled)

17. P(sum is 7 | 2 is on one of the dice)

18. P(sum is 2 or 3 | at least one 5 is rolled)

Use the following tree diagram to answer the questions in Problems 19–24.

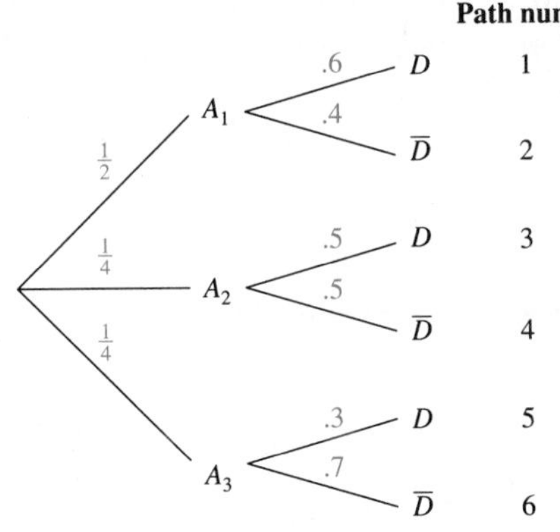

19. Which path number represents $P(D \mid A_2)$?

20. Which path number represents $P(\bar{D} \cap A_3)$?

21. Describe path 1 as a probability.

22. Find $P(\bar{D} \mid A_2)$.

23. Find $P(D \mid A_1)P(A_1)$.

24. Find $P(A_2 \cap D)$.

APPLICATIONS

In an experiment it is necessary to examine fruit flies and determine their sex and whether they have mutated after exposure to radiation. For 1,000 *fruit flies examined, there are* 643 *females and* 357 *males. Also,* 403 *of the females are normal and* 240 *are mutated, while* 190 *of the males are normal and* 167 *are mutated. In Problems* 25–34 *assume a single fruit fly is chosen at random from the sample of* 1,000 *and calculate the requested probabilities to the nearest hundredth.*

25. P(male)

26. P(female)

27. P(normal male)

28. P(mutated male)

29. P(normal | male)

30. P(mutated | male)

31. P(male | normal)

32. P(male | mutated)

33. P(female | mutated)

34. P(mutated | female)

In Problems 35–40 *suppose that E and F are events with the given probabilities. Use the definition of conditional probability to calculate both* $P(E \mid F)$ *and* $P(F \mid E)$.

35. $P(E) = .5$, $P(F) = .2$, $P(E \cap F) = .1$

36. $P(E) = .85$, $P(F) = .45$, $P(E \cap F) = .3$

37. $P(E) = \frac{1}{3}$, $P(F) = \frac{1}{2}$, $P(E \cap F) = \frac{1}{6}$

38. $P(E) = \frac{3}{5}$, $P(F) = \frac{7}{10}$, $P(E \cap F) = \frac{1}{2}$

39. $P(E) = .5$, $P(F) = .8$, $P(E \cap F) = .4$

40. $P(E) = \frac{6}{7}$, $P(F) = \frac{2}{3}$, $P(E \cap F) = \frac{4}{7}$

41. The Ross Light Bulb Company has found that 5% of the bulbs it manufactures have defective filaments and 3% have both defective filaments and defective workmanship. What is the probability of defective workmanship in a particular bulb if you know it has a defective filament?

42. At Southeastern University a survey of students taking both college algebra and statistics found that 25% dropped college algebra, 30% dropped statistics, and 10% dropped both courses. If a person dropped college algebra, what is the probability that the person also dropped statistics?

Use a tree diagram to help you answer the questions in Problems 43–50.

43. Two cards are drawn without replacement from a deck of 52 cards. Let

$$J_1 = \{\text{drawing a jack on the first draw}\}$$
$$J_2 = \{\text{drawing a jack on the second draw}\}$$
$$K_1 = \{\text{drawing a king on the first draw}\}$$

Find:

a. $P(J_1)$
b. $P(J_2 \mid J_1)$
c. $P(K_1 \cap J_2)$
d. $P(\text{one jack and one king})$

44. Two cards are drawn without replacement from a deck of 52 cards. Let

H_1 = a heart is drawn on the first draw
H_2 = a heart is drawn on the second draw

Find

a. $P(H_1)$
b. $P(H_2 \mid H_1)$
c. $P(H_1 \cap \bar{H}_2)$
d. $P(\text{two hearts})$

45. An experiment consists of taking samples from a collection of 10 items, three of which are known to be defective. The experiment continues until a nondefective item is found. What is the probability that the experiment will conclude after two items are selected?

46. An experiment consists of taking samples from a collection of 10 items, three of which are known to be defective. The experiment continues until a nondefective item is found. What is the probability that the experiment will conclude after three items are selected?

47. A game consists of drawing a card from a deck of cards. If a card other than an ace is drawn, the game stops; but if an ace is drawn, the game continues. What is the probability (rounded to the nearest thousandth) that the experiment will conclude after three cards are selected? Draw a tree diagram for the entire game.

48. Karlin Enterprises produces radar detectors at three plants. Nashville supplies 50% of the detectors, New Orleans supplies 30% of the detectors, and Los Angeles supplies 20% of the detectors. It is also known that 1%, 2%, and 3% of the detectors are defective when produced at Nashville, New Orleans, and Los Angeles, respectively. What is the probability that a randomly selected detector will be defective?

49. Karlin Enterprises produces radar detectors at three plants. Nashville supplies 40% of the detectors, New Orleans supplies 25% of the detectors, and Los Angeles supplies 35% of the detectors. It is also known that 3%, 2%, and 1% of the detectors are defective when produced at Nashville, New Orleans, and Los Angeles, respectively. What is the probability that a randomly selected detector will be defective?

50. Two equally matched students are deciding among offers from three schools. The probability of either student selecting UCLA is 40%, Vanderbilt is 50%, and Stanford is 10%. What is the probability that one will select UCLA and one will select Vanderbilt?

7.5 Independent Events, Intersections, and Unions

Independent Events

The conditional probability $P(E|F)$ is the probability that E occurs when it is known that F occurs. Sometimes, though, the occurrence of F does not affect $P(E)$. In such circumstances we say that E and F are **independent**. We will first illustrate this concept with an example.

EXAMPLE 1 Suppose a coin is tossed twice. Let

$$H_1 = \{\text{head on first toss}\}$$
$$H_2 = \{\text{head on second toss}\}$$

These events seem to be independent since the occurrence of one event does not affect the occurrence of the other. We can verify this by using the definition.

Sample space

HH
HT
TH
TT

$$\left.\begin{aligned} P(H_1) &= \tfrac{2}{4} = \tfrac{1}{2} \\ P(H_2) &= \tfrac{2}{4} = \tfrac{1}{2} \\ P(H_1 \cap H_2) &= \tfrac{1}{4} \end{aligned}\right\}$$

These probabilities are found by looking at the sample space.

Thus

$$P(H_2|H_1) = \frac{P(H_2 \cap H_1)}{P(H_1)}$$
$$= \frac{\frac{1}{4}}{\frac{1}{2}} \longleftarrow \textit{Note: } P(H_2 \cap H_1) = P(H_1 \cap H_2)$$
$$= \frac{1}{2} = P(H_1)$$

■

Notice from Example 1 that the probability of H_1 did not change by the knowledge that H_2 occurred. That means that the occurrence of H_2 does not affect the probability of the event H_1. In this case, we call H_1 and H_2 independent.

Independent Events

The events E and F are said to be **independent** if

$$P(E \mid F) = P(E)$$

That is, the occurrence of E is not affected by the occurrence or nonoccurrence of F.

EXAMPLE 2 An experiment consists of drawing two consecutive cards from an ordinary deck of cards. Let $E = \{\text{first card is red}\}$ and $F = \{\text{second card is black}\}$. Are these events independent if:

a. The cards are drawn with replacement?

b. The cards are drawn without replacement?

Solution To verify independence we need to calculate $P(F \mid E)$ and compare this with $P(F)$; the events are independent if

$$P(F \mid E) = P(F)$$

a. $P(F \mid E) = \dfrac{26}{52}$ ← Number of black cards in deck / ← Number of cards in deck

$P(F) = \dfrac{1}{2}$ There are the same number of outcomes with the second card black as there are with the second card red.

Thus $P(F \mid E) = P(F)$, so these are independent events. This should be consistent with your intuition about these events.

b. $P(F \mid E) = \dfrac{26}{51}$ ← Number of black cards in deck / ← Number of cards in deck (remember, this time the card is not replaced)

$P(F) = \dfrac{1}{2}$ This event, by itself, has the same probability whether or not the experiment is done with replacement.

Thus $P(F \mid E) \neq P(F)$, so these are not independent events. ■

There is an alternate way to check for independence. Suppose that E and F are independent. Then

$$P(E \mid F) = P(E)$$

Suppose we rewrite the expression:

$$P(E) = P(E \mid F)$$
$$= \frac{P(E \cap F)}{P(F)}$$

Therefore, by multiplication of both sides of the equation,

$$P(E) \cdot P(F) = P(E \cap F)$$

provided E and F are independent. We can also carry out the same calculation in reverse to arrive at the following test for independence:

Test for Independent Events

> Events E and F are independent if and only if
>
> $$P(E \cap F) = P(E) \cdot P(F)$$

EXAMPLE 3 Toss a single die twice. Let

$$E = \{\text{first toss is a prime}\}$$
$$F = \{\text{first toss is a 3}\}$$
$$G = \{\text{second toss is a 2}\}$$
$$H = \{\text{second toss is a 3}\}$$

Decide which of these events are independent.

Solution The sample space for both tosses has 36 possibilities. Using this sample space, we find:

$$P(E) = \frac{18}{36} = \frac{1}{2} \qquad P(E \cap F) = \frac{6}{36} = \frac{1}{6}$$

$$P(F) = \frac{6}{36} = \frac{1}{6} \qquad P(E \cap G) = \frac{3}{36} = \frac{1}{12}$$

$$P(G) = \frac{6}{36} = \frac{1}{6} \qquad P(E \cap H) = \frac{3}{36} = \frac{1}{12}$$

$$P(H) = \frac{6}{36} = \frac{1}{6} \qquad P(F \cap G) = \frac{1}{36}$$

$$P(F \cap H) = \frac{1}{36}$$

$$P(G \cap H) = 0$$

6	(1, 6)	(2, 6)	(3, 6)	(4, 6)	(5, 6)	(6, 6)
5	(1, 5)	(2, 5)	(3, 5)	(4, 5)	(5, 5)	(6, 5)
4	(1, 4)	(2, 4)	(3, 4)	(4, 4)	(5, 4)	(6, 4)
3	(1, 3)	(2, 3)	(3, 3)	(4, 3)	(5, 3)	(6, 3)
2	(1, 2)	(2, 2)	(3, 2)	(4, 2)	(5, 2)	(6, 2)
1	(1, 1)	(2, 1)	(3, 1)	(4, 1)	(5, 1)	(6, 1)

First toss

Therefore:

***WARNING* This is an important example—but you must think through these conclusions and not just read this.**

E and F are not independent since $P(E \cap F) \neq P(E) \cdot P(F)$.
E and G are independent since $P(E \cap G) = P(E) \cdot P(G)$.
F and G are independent since $P(F \cap G) = P(F) \cdot P(G)$.
E and H are independent since $P(E \cap H) = P(E) \cdot P(H)$.
F and H are independent since $P(F \cap H) = P(F) \cdot P(H)$.
G and H are not independent since $P(G \cap H) \neq P(G) \cdot P(H)$. ■

Probability of Intersections and Unions

Some probability models can be built by breaking down the events being considered into simpler ones by using the words *and* (intersection) or *or* (union).

For a formula for the **probability of the intersection of two events**, recall that events E and F are independent if the occurrence of one of these events in no way affects the occurrence of the other; that is, events E and F are independent if $P(E|F) = P(E)$. Therefore, from the conditional probability formula,

$$P(E|F) = \frac{P(E \cap F)}{P(F)}$$

$$P(E|F) \cdot P(F) = P(E \cap F)$$

If E and F are independent, $P(E \cap F) = P(E)P(F)$. This formula is now used to derive a formula for the **probability of a union of two events**. From the definition of a probability measure (Section 7.1), we obtain a formula for

$$P(E \cup F) = P(E) + P(F) \qquad \text{if } E \text{ and } F \text{ are mutually exclusive}$$

If E and F are not mutually exclusive, then Figure 7.5 (page 352) leads us to a generalization of this formula:

$$P(E \cup F) = P(E) + P(F) - P(E \cap F)$$

We can now summarize these formulas for the probability of the intersection or union of two events.

Probability of Intersections and Unions

INTERSECTION:	$P(E \cap F) = P(E) \cdot P(F)$	Provided E and F are independent
UNION:	$P(E \cup F) = P(E) + P(F) - P(E \cap F)$	

Suppose a coin is tossed and a die is simultaneously rolled. Let $T = \{\text{tail is tossed}\}$, $F = \{4 \text{ is rolled}\}$, and $N = \{\text{odd number is rolled}\}$. Find the probabilities in Examples 4 and 5.

EXAMPLE 4 $P(T \cap F) = P(T) \cdot P(F)$ — This is the probability of a tail *and* a 4. Note that T and F are independent.

$$= \frac{1}{2} \cdot \frac{1}{6}$$

$$= \frac{1}{12}$$

■

EXAMPLE 5 $P(T \cup F) = P(T) + P(F) - P(T \cap F)$ — This is the probability of a tail *or* a 4.

$$= \frac{1}{2} + \frac{1}{6} - \frac{1}{12}$$

$$= \frac{7}{12}$$

■

In Examples 6–9 suppose a die is rolled twice and let

$A = \{\text{first toss is a prime}\}$ $\quad B = \{\text{first toss is a 3}\}$

$C = \{\text{second toss is a 2}\}$ $\quad D = \{\text{second toss is a 3}\}$

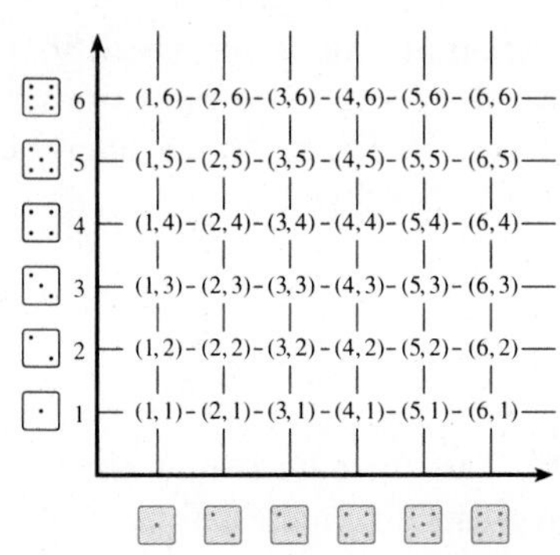

The sample space for this experiment has 36 possible outcomes. By considering this sample space, we see that

$$P(A) = \frac{18}{36} = \frac{1}{2}$$

$$P(B) = P(C) = P(D) = \frac{6}{36} = \frac{1}{6}$$

$$P(A \cap B) = \frac{6}{36} = \frac{1}{6}$$

Events A and B are not independent because $P(A \cap B) \neq P(A) \cdot P(B)$.

$$P(A \cap C) = P(A \cap D) = \frac{3}{36} = \frac{1}{12}$$

Events A and C *are* independent because $P(A \cap C) = P(A) \cdot P(C)$.

$$P(B \cap C) = P(B \cap D) = \frac{1}{36}$$

$$P(C \cap D) = 0$$

This means that C and D are mutually exclusive.

Can you name other pairs that are or are not independent? We will use these calculations in Examples 6–9.

EXAMPLE 6 $P(A \cup B) = P(A) + P(B) - P(A \cap B)$

$$= \frac{1}{2} + \frac{1}{6} - \frac{1}{6}$$

$$= \frac{1}{2}$$

This is the probability that a prime *or* a 3 is obtained on the first toss. ■

EXAMPLE 7 $P(A \cup C) = P(A) + P(C) - P(A \cap C)$

$$= \frac{1}{2} + \frac{1}{6} - \frac{1}{12}$$

$$= \frac{7}{12}$$

This is the probability that the first toss is a prime *or* the second toss is a 2. ■

EXAMPLE 8 $P(B \cup D) = P(B) + P(D) - P(B \cap D)$

$$= \frac{1}{6} + \frac{1}{6} - \frac{1}{36}$$

$$= \frac{11}{36}$$

This is the probability that the first toss is a 3 *or* the second toss is a 3. ■

EXAMPLE 9 $P(C \cup D) = P(C) + P(D) - P(C \cap D)$

$$= \frac{1}{6} + \frac{1}{6} - 0$$

$$= \frac{1}{3}$$

This is the probability that the second toss is a 2 *or* a 3. Note that C and D are mutually exclusive. ■

Experiments with and without Replacement

Some experiments are said to be performed *with replacement* and others *without replacement*. Consider the following examples:

Suppose cards are drawn from a deck of 52 cards. Let

$$S_1 = \{\text{draw a spade on the first draw}\}$$
$$H_1 = \{\text{draw a heart on the first draw}\}$$
$$H_2 = \{\text{draw a heart on the second draw}\}$$

To draw **with replacement** means that the first card is drawn, the result is noted, and then the card is replaced in the deck before the second card is drawn. To draw **without replacement** means that one card is drawn, the result is noted, and then a second card is drawn without replacing the first card. Find the probabilities in Examples 10–13. Compare the following tree diagrams and note that only the probabilities are different (not the tree diagrams).

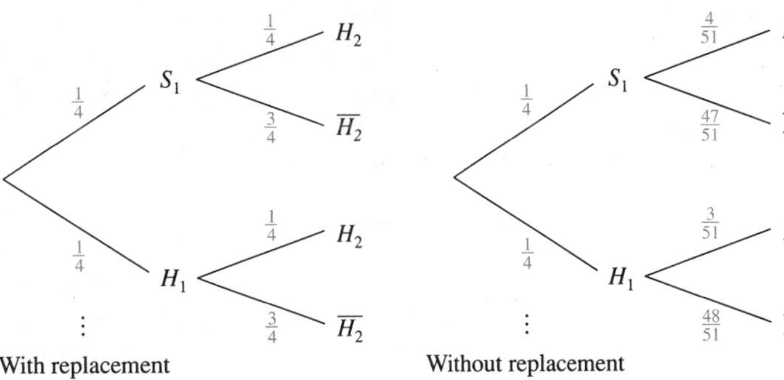

EXAMPLE 10 Find $P(S_1 \cap H_2)$, with replacement.

Solution S_1 and H_2 are independent since the cards are drawn with replacement.

$$P(S_1 \cap H_2) = P(S_1) \cdot P(H_2) = \frac{1}{4} \cdot \frac{1}{4} = \frac{1}{16}$$ ■

EXAMPLE 11 Find $P(S_1 \cap H_2)$, without replacement.

Solution In this experiment the events are not independent since the probability of drawing the second card depends on what was drawn on the first card. This problem is equivalent to drawing two cards from a deck of cards. The order in which the cards are drawn is important since the question specifies that the *first* card is a spade and the *second* card is a heart; thus this is a permutation problem.

$$P(S_1 \cap H_2) = \frac{{}_{13}P_1 \cdot {}_{13}P_1}{{}_{52}P_2} = \frac{13 \cdot 13}{52 \cdot 51} = \frac{13}{204}$$ ■

EXAMPLE 12 What is the probability of drawing two hearts with replacement?

Solution H_1 and H_2 are independent.

$$P(H_1 \cap H_2) = P(H_1) \cdot P(H_2) = \frac{1}{4} \cdot \frac{1}{4} = \frac{1}{16}$$ ■

EXAMPLE 13 What is the probability of drawing two hearts without replacement?

Solution H_1 and H_2 are not independent, but the order in which the cards are drawn is not important since the question simply asks for two hearts; thus this is a combination problem.

$$P(\text{two hearts}) = \frac{{}_{13}C_2}{{}_{52}C_2} = \frac{\frac{13 \cdot 12}{2}}{\frac{52 \cdot 51}{2}} = \frac{1}{17} \quad \blacksquare$$

Summary of Probability Formulas

We conclude this section by summarizing the probability formulas:

Summary of Probability Formulas

1. $P(E) = \frac{s}{n}$ where event E can occur in n mutually exclusive and equally likely ways, and where s of them are considered favorable.
2. $P(S) = 1$ where S is the sample space.
3. $P(\varnothing) = 0$
4. $P(E) = 1 - P(\bar{E})$
5. $P(E|F) = \frac{P(E \cap F)}{P(F)}$ provided $P(F) \neq 0$.
6. $P(E \text{ and } F) = P(E \cap F) = P(E) \cdot P(F)$ provided E and F are independent.
7. $P(E \text{ or } F) = P(E \cup F) = P(E) + P(F) - P(E \cap F)$
8. $P(E \text{ or } F) = P(E \cup F) = P(E) + P(F)$ provided E and F are mutually exclusive.

Note: When doing probability problems, assume that experiments are performed *without replacement* unless it is otherwise stated.

7.5
Problem Set

Use the definition in Problems 1–4 to determine whether events E and F are independent.

1. $P(E) = .5,\ P(F) = .2,\quad P(E \cap F) = .1$
2. $P(E) = \frac{3}{5},\ P(F) = \frac{7}{10},\quad P(E \cap F) = \frac{1}{2}$
3. $P(E) = \frac{6}{7},\ P(F) = \frac{2}{3},\quad P(E \cap F) = \frac{4}{7}$
4. $P(E) = .5,\ P(F) = .8,\quad P(E \cap F) = .4$

Suppose a die is rolled twice. Let

$A = \{$*first toss is an even*$\}$
$B = \{$*first toss is a 6*$\}$
$C = \{$*second toss is a 2*$\}$
$D = \{$*second toss is a 3*$\}$

Determine whether the events in Problems 5–10 are independent.

5. A and B
6. A and C
7. A and D
8. B and C
9. B and D
10. C and D

In Problems 11–16, consider a family with four children. Let

$E = \{$*2 boys and 2 girls*$\}$
$F = \{$*exactly 1 boy*$\}$
$G = \{$*at most 1 boy*$\}$
$H = \{$*at least 1 child of each sex*$\}$

Determine whether the events are independent.

11. E and F
12. E and G
13. E and H
14. F and G
15. F and H
16. G and H

Suppose that events A, B, and C are all independent so that

$P(A) = \frac{1}{2}$ $\quad P(B) = \frac{1}{3}$ $\quad P(C) = \frac{1}{6}$

Find the probabilities in Problems 17–34:

17. $P(\bar{A})$
18. $P(\bar{B})$
19. $P(\bar{C})$
20. $P(A \cap B)$
21. $P(A \cap C)$
22. $P(B \cap C)$
23. $P(A \cup B)$
24. $P(A \cup C)$
25. $P(B \cup C)$
26. $P(\overline{A \cap B})$
27. $P(\overline{A \cap C})$
28. $P(\overline{B \cap C})$
29. $P(\overline{A \cup B})$
30. $P(\overline{A \cup C})$
31. $P(\overline{B \cup C})$
32. $P(A \cap B \cap C)$
33. $P(\overline{A \cap B \cap C})$
34. $P[(A \cap B) \cup C]$

In Problems 35–40 suppose a coin is tossed twice. Find the requested probabilities.

35. P(2H)
36. P(2T)
37. P(3T)
38. P(1H and 1T)
39. P(match)*
40. P(No match)

Suppose A, B, and C are independent events so that

$P(A) = \frac{1}{2}$ $\quad P(B) = \frac{2}{3}$ $\quad P(C) = \frac{5}{6}$

Find the probabilities in Problems 41–44.

41. **a.** $P(\bar{A})$ **b.** $P(\bar{B})$ **c.** $P(\bar{C})$
42. **a.** $P(A \cap B)$ **b.** $P(A \cap C)$ **c.** $P(B \cap C)$
43. **a.** $P(A \cup B)$ **b.** $P(A \cup C)$ **c.** $P(B \cup C)$
44. **a.** $P(\overline{A \cap C})$ **b.** $P(\overline{A \cup B})$ **c.** $P(\overline{B \cup C})$

APPLICATIONS

45. The probability of tossing a coin four times and obtaining four heads in a row is $P(4\text{H}) = \frac{1}{2} \cdot \frac{1}{2} \cdot \frac{1}{2} \cdot \frac{1}{2} = \frac{1}{16}$. What is the probability of tossing a coin and obtaining a head if we know that heads have occurred on the previous four flips of the coin?

* This means P(both the same).

46. One "system" used by roulette players is to watch a game of roulette until a large number of reds occur in a row (say, 10). After 10 successive reds, they reason that black is "due" to occur and begin to bet large sums on black. Where is the fallacy in the reasoning of this "system" betting?

In Problems 47–59 suppose a coin is tossed and simultaneously a die is rolled. Let $H = \{$head is tossed$\}$, $S = \{$6 is rolled$\}$, and $E = \{$even number is rolled$\}$. Find the requested probabilities.

47. **a.** $P(H)$ **b.** $P(S)$ **c.** $P(E)$
48. $P(H \cap S)$
49. $P(H \cap E)$
50. $P(S \cap E)$
51. $P(S \cup E)$
52. $P(H \cup E)$
53. $P(H \cup S)$
54. $P(H \mid S)$
55. $P(S \mid H)$
56. $P(S \mid E)$
57. $P(E \mid S)$
58. $P(H \mid E)$
59. $P(E \mid H)$

In Problems 60–64 assume a box has five red cards and three black cards, and two cards are drawn with replacement. Find the requested probabilities.

60. P(2 red cards)
61. P(2 black cards)
62. P(1 red and 1 black card)
63. P(red on first draw and black on second draw)
64. P(red on first draw or black on second draw)

Problems 65–69 repeat the experiment of Problems 60–64 except that the cards are drawn without replacement.

65. P(2 red cards)
66. P(2 black cards)
67. P(1 red and 1 black card)
68. P(red on first draw and black on second draw)
69. P(red on first draw or black on second draw)
70. A survey conducted by the American Cancer Society found that of the 5,000 persons surveyed, 35% were heavy smokers and 3% had emphysema. Of those who had emphysema, 95% were also heavy smokers. Using this information, decide whether being a heavy smoker and having emphysema are independent events.
71. Suppose that a certain mathematics student can do a factoring problem correctly 85% of the time, can do a fraction problem correctly 80% of the time, and can combine similar terms correctly 95% of the time. What is the probability that this student will correctly work a problem that requires factoring, combining fractions, and combining similar terms twice if you assume these skills are independent?
72. Assume the information provided in Problem 71 and give the number of questions you would expect to get correct on a test consisting of 20 questions like those described in Problem 71.

7.6 Bayes' Theorem

Probabilities with a Partition of the Sample Space

In Section 7.4 we noted that for any event B

$$P(B) = P(A \cap B) + P(\bar{A} \cap B)$$

We would now like to generalize this property. The product rule allows us to rewrite this as

$$P(B) = P(B \mid A)P(A) + P(B \mid \bar{A})P(\bar{A})$$

We say that A and $\bar{A}$ form a **partition** of the sample space.

Partition of a Set

A set S is said to be **partitioned** if S is divided into r subsets satisfying the following two conditions:

1. If A and B are any two subsets, then $A \cap B = \varnothing$; that is, the members of the subsets are pairwise disjoint.
2. The union of all the subsets is S; that is, there are no elements of S that are not included in one of the subsets.

EXAMPLE 1 Suppose we partition a subspace as follows. We divide a room of people into two groups as follows:

Members: 15 men; 20 women; 0 children
Nonmembers: 10 men; 8 women; 12 children

Suppose a prize is given to one person who is selected at random by choosing from one group or the other with the members being twice as likely to be chosen as nonmembers. After the group is selected, then one person in that group is selected with every person in the group having an equal chance of being chosen. If $N =$ {a man is chosen}, $W =$ {a woman is chosen}, and $C =$ {a child is chosen}, find $P(N)$.

Solution The groups of members, M, and nonmembers, $\bar{M}$, form a partition of the sample space. Using the product rule for this partition, we have

$$P(N) = P(N \mid M)P(M) + P(N \mid \bar{M})P(\bar{M})$$

Since $P(M)$ is twice as likely as $P(\bar{M})$ we see*

$$P(M) = \frac{2}{3} \quad \text{and} \quad P(\bar{M}) = \frac{1}{3}$$

* $P(M) + P(\bar{M}) = 1$. Let $x = P(\bar{M})$. Then $P(M) = 2x$, so $2x + x = 1 \Rightarrow x = \frac{1}{3}$.

Also,

$$P(N \mid M) = \frac{15}{35} = \frac{3}{7} \quad \text{and} \quad P(N \mid \bar{M}) = \frac{10}{30} = \frac{1}{3}$$

(15 men members; 35 members) (10 men nonmembers; 30 nonmembers)

Thus,

$$\begin{aligned} P(N) &= P(N \mid M)P(M) + P(N \mid \bar{M})P(\bar{M}) \\ &= \frac{3}{7} \cdot \frac{2}{3} + \frac{1}{3} \cdot \frac{1}{3} \\ &= \frac{25}{63} \\ &\approx .40 \end{aligned}$$

By calculator, .3968254 ■

We now want to generalize the result illustrated by Example 1 to a partition of the sample space into more than two parts. The difficulty with the solution of Example 1 is that when it is generalized it is much more difficult to visualize. Instead of an algebraic generalization, we can reconsider it using a tree diagram:

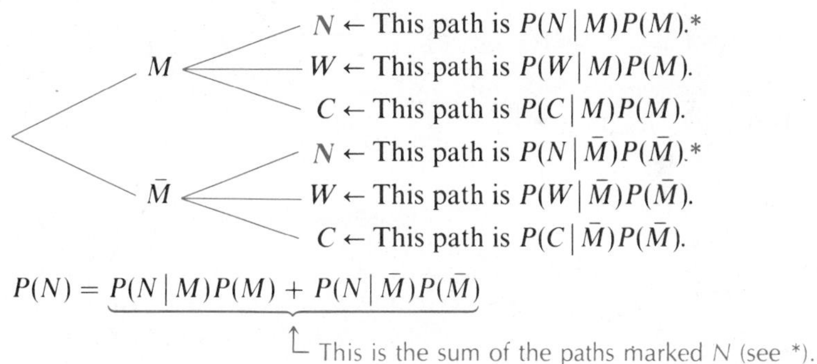

$$P(N) = \underbrace{P(N \mid M)P(M) + P(N \mid \bar{M})P(\bar{M})}$$

This is the sum of the paths marked N (see *).

The probability of some event N is the sum of the entries at the end of each path labeled N on the associated tree diagram.

If a partition has more than two possibilities, the probability of some event E is the sum of the numbers at the end of each path labeled E. This is stated algebraically as follows:

Probability of a Partitioned Event

If $A_1, A_2, \ldots, A_n$ form a partition of a sample space and E is any event, then

$$P(E) = P(E \mid A_1)P(A_1) + P(E \mid A_2)P(A_2) + \cdots + P(E \mid A_n)P(A_n)$$

This algebraic statement is usually interpreted in terms of a tree diagram, as shown in Figure 7.6 on page 368.

Partition with two sets

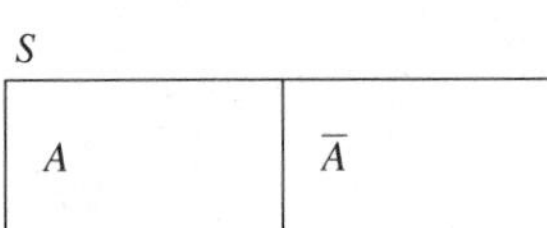

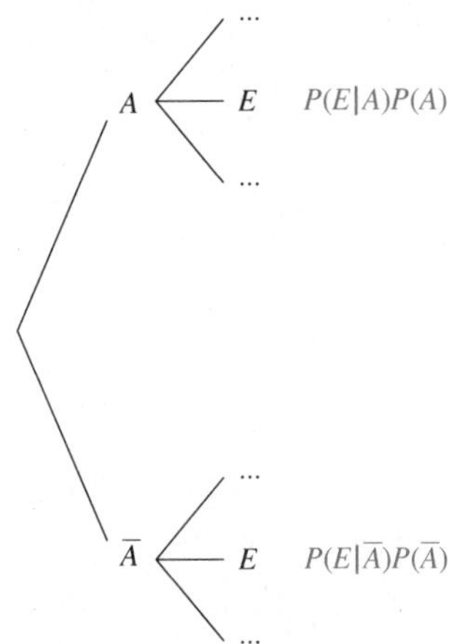

$P(E)$ is the sum of the branches labeled E

$$P(E) = P(E|A)P(A) + P(E|\overline{A})P(\overline{A})$$

Partition with three sets

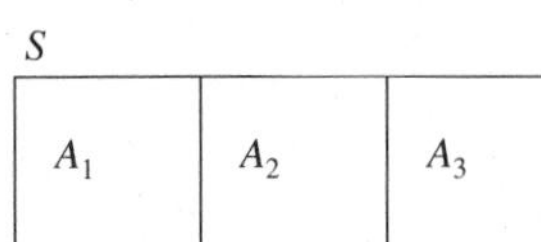

A_1 — E — $P(E|A_1)P(A_1)$

A_2 — E — $P(E|A_2)P(A_2)$

A_3 — E — $P(E|A_3)P(A_3)$

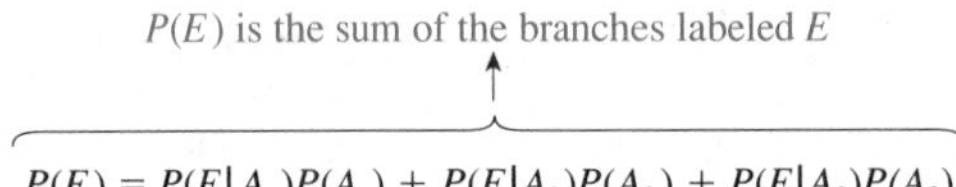

$P(E)$ is the sum of the branches labeled E

$$P(E) = P(E|A_1)P(A_1) + P(E|A_2)P(A_2) + P(E|A_3)P(A_3)$$

Partition with n sets

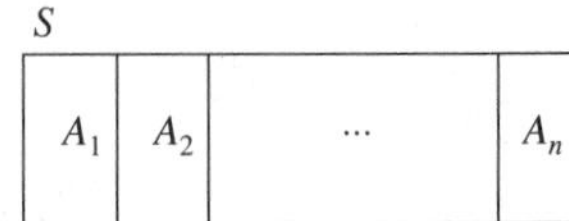

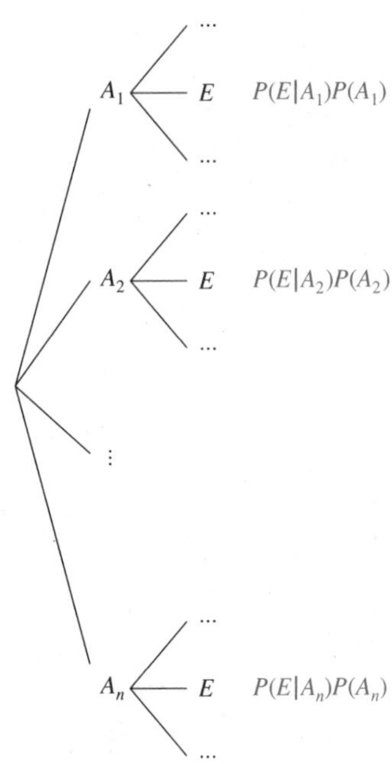

Look for the branches labeled E.

$$P(E) = P(E|A_1)P(A_1) + P(E|A_2)P(A_2) + \cdots + P(E|A_n)P(A_n)$$

FIGURE 7.6
Partitions, tree diagrams, and probabilities

EXAMPLE 2 Three cards are drawn (without replacement) from a deck of cards. What is the probability that the third one drawn is a heart?

Solution Construct the following tree diagram. Be sure you understand how the probabilities at each step are found.

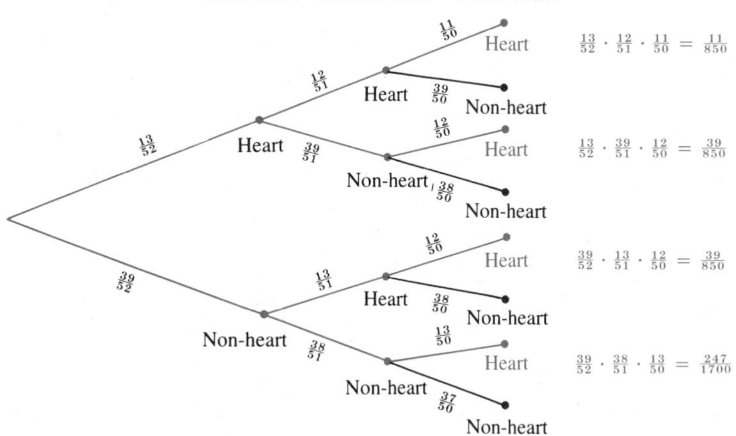

Thus the probability of a heart on the third draw is the sum of the branches labeled "heart" on the third draw:

$$P(\text{heart on the third draw}) = \frac{11}{850} + \frac{39}{850} + \frac{39}{850} + \frac{247}{1{,}700}$$

$$= \frac{425}{1{,}700} = \frac{1}{4}$$

■

EXAMPLE 3 Three cards are drawn (without replacement) from a deck of cards. What is the probability that exactly two cards are hearts?

Solution Use the tree diagram from Example 2, and trace out different paths to indicate success.

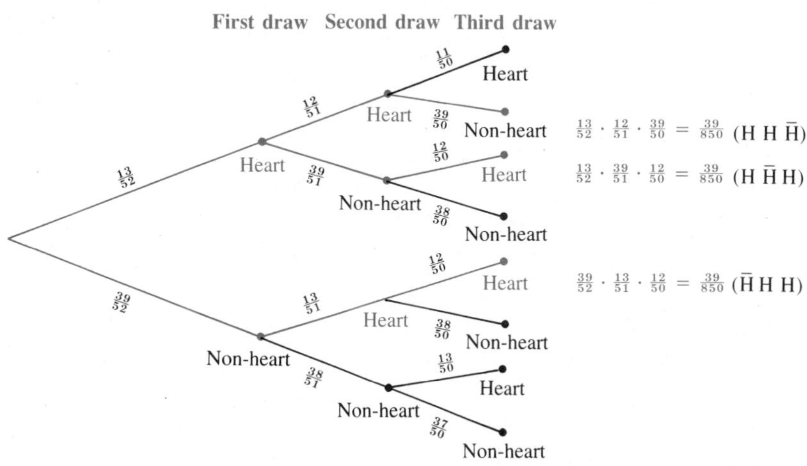

Thus

$$P(\text{exactly two hearts}) = \frac{39}{850} + \frac{39}{850} + \frac{39}{850} = \frac{117}{850} \approx .138$$

■

A Posteriori Probabilities

The tree diagrams in Examples 2 and 3 are concerned with calculating probabilities that give the likelihood that an event will occur. These are sometimes called ***a priori* probabilities** because they are calculated *prior* to observing the result. On the other hand, sometimes it is necessary to consider probabilities that are calculated *after* the outcomes of experiments have been observed, such as in medical diagnosis. Such probabilities are called ***a posteriori* probabilities.**

In order to see what we mean, let us go back to Example 1, in which we calculated the probability of a man being selected from one of two groups, members and nonmembers. Suppose we perform the experiment and do indeed select a man. Now, we are interested in knowing the probability that that man is a member.

EXAMPLE 4 Suppose some people are divided into two groups as follows:

Members: 15 men; 20 women; 0 children
Nonmembers: 10 men; 8 women; 12 children

First one group is selected with the member group being twice as likely to be chosen as the nonmember group. Then one person is selected from that group, with every person in the group having an equal chance of being chosen. If $N = \{\text{a man is chosen}\}$, $W = \{\text{a woman is chosen}\}$, and $C = \{\text{a child is chosen}\}$, what is the probability that a selected man is a member?

Solution Notice that this is an *a posteriori* probability. We wish to find the probability that the selected person is a member, given that the person is a man—namely,

$$P(M \mid N)$$

Compare with Example 1. We know $P(N \mid M)$, but we need to find $P(M \mid N)$ so we begin with the definition of conditional probability:

$$\begin{aligned}
P(M \mid N) &= \frac{P(M \cap N)}{P(N)} \\
&= \frac{P(N \cap M)}{P(N)} \\
&= \frac{P(N \mid M)P(M)}{P(N)} \quad \begin{array}{l} \longleftarrow \text{We know these numbers.} \\ \longleftarrow \text{We found this number in Example 1.} \end{array} \\
&= \frac{P(N \mid M)P(M)}{P(N \mid M)P(M) + P(N \mid \bar{M})P(\bar{M})} \\
&= \frac{\frac{3}{7} \cdot \frac{2}{3}}{\frac{25}{63}} \\
&= \frac{18}{25} = .72
\end{aligned}$$

■

Notice that we derived a formula in Example 4 for *a posteriori probability*. The general formula was first published in 1763 by Thomas Bayes (1702–1763) and is known as Bayes' theorem.

Bayes' Theorem

If $A_1, A_2, \ldots, A_n$ form a partition of the sample space S and E is an event associated with S, then the *a posteriori* probability $P(A_i \mid E)$ is given by

$$P(A_i \mid E) = \frac{P(E \mid A_i)P(A_i)}{P(E \mid A_1)P(A_1) + P(E \mid A_2)P(A_2) + \cdots + P(E \mid A_n)P(A_n)} \qquad i = 1, 2, \ldots, n$$

For a partition with two sets A and $\bar{A}$:

$$P(A \mid E) = \frac{P(E \mid A)P(A)}{P(E \mid A)P(A) + P(E \mid \bar{A})P(\bar{A})}$$

and

$$P(\bar{A} \mid E) = \frac{P(E \mid \bar{A})P(\bar{A})}{P(E \mid A)P(A) + P(E \mid \bar{A})P(\bar{A})}$$

For a partition with three sets A_1, A_2, and A_3:

$$P(A_1 \mid E) = \frac{P(E \mid A_1)P(A_1)}{P(E \mid A_1)P(A_1) + P(E \mid A_2)P(A_2) + P(E \mid A_3)P(A_3)}$$

$$P(A_2 \mid E) = \frac{P(E \mid A_2)P(A_2)}{P(E \mid A_1)P(A_1) + P(E \mid A_2)P(A_2) + P(E \mid A_3)P(A_3)}$$

$$P(A_3 \mid E) = \frac{P(E \mid A_3)P(A_3)}{P(E \mid A_1)P(A_1) + P(E \mid A_2)P(A_2) + P(E \mid A_3)P(A_3)}$$

In practice, however, we use tree diagrams to calculate these probabilities as illustrated by the following example. A tree diagram used to calculate an *a posteriori* probability is called a *Bayes' tree.*

EXAMPLE 5 Thompson Associates, a management consulting firm, has agreed to run a quality control check on three assembly lines at Karlin Manufacturing. In checking the inventory, Thompson found a defective item. What is the probability that it came from a particular assembly line?

Analysis In checking the records, Thompson Associates gathers the following information:

Number of items in inventory	*Previous testing results*
50% from assembly line A	2% of items from assembly line A are defective
30% from assembly line B	3% of items from assembly line B are defective
20% from assembly line C	4% of items from assembly line C are defective

Let

$A = \{\text{item tested is from assembly line A}\}$
$B = \{\text{item tested is from assembly line B}\}$
$C = \{\text{item tested is from assembly line C}\}$
$D = \{\text{item tested is defective}\}$

Thompson is now able to calculate the following probabilities from the given information:

$$P(A) = .5 \qquad P(B) = .3 \qquad P(C) = .2$$
$$P(D \mid A) = .02 \qquad P(D \mid B) = .03 \qquad P(D \mid C) = .04$$

Find $P(A \mid D)$, $P(B \mid D)$, and $P(C \mid D)$.

Solution Find $P(A \mid D)$ first by formula:

$$\begin{aligned} P(A \mid D) &= \frac{P(D \mid A)P(A)}{P(D \mid A)P(A) + P(D \mid B)P(B) + P(D \mid C)P(C)} \\ &= \frac{(.02)(.5)}{(.02)(.5) + (.03)(.3) + (.04)(.2)} \\ &= \frac{.01}{.027} \approx .37 \end{aligned}$$

Since it can often be confusing to use the formula, we draw a *Bayes' tree*:

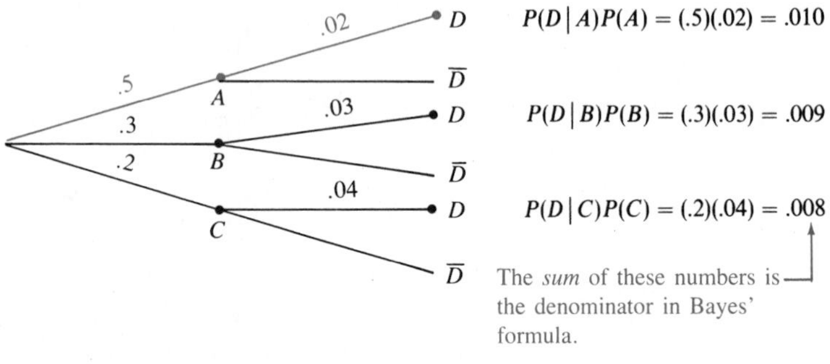

Using Bayes' tree, find the quotient of the desired path (shown in color) divided by the sum of all path possibilities:

$$P(A \mid D) = \frac{.010}{.027} \approx .37$$

Use Bayes' tree to find the other probabilities:

$$P(B \mid D) = \frac{.009}{.027} \approx .33 \qquad P(C \mid D) = \frac{.008}{.027} \approx .30$$

■

EXAMPLE 6 Suppose a test for AIDS is given to a patient. Let

$$A = \{\text{person has AIDS}\}$$
$$R = \{\text{test shows a positive result}\}$$

If the person has AIDS, the test shows a positive result 93% of the time; if the person does not have AIDS, the test shows a positive result 4% of the time. It is also known that .1% of the population has undetected AIDS. Find the probability that if the test shows a positive result, the person actually has AIDS.

Solution Given: $P(R\,|\,A) = .93$, $P(R\,|\,\bar{A}) = .04$, and $P(A) = .001$. Find $P(A\,|\,R)$.

$$\begin{aligned} P(A\,|\,R) &= \frac{P(R\,|\,A)P(A)}{P(R\,|\,A)P(A) + P(R\,|\,\bar{A})P(\bar{A})} \\ &= \frac{(.93)(.001)}{(.93)(.001) + (.04)(.999)} \\ &= \frac{.00093}{.00093 + .03996} \\ &\approx .0227439 \end{aligned}$$

This means that even though the test is highly reliable and will detect AIDS in 93% of the cases, a positive result indicates that the person actually has AIDS in only 2.3% of the cases. ■

7.6 Problem Set

Problems 1–8 refer to the following tree diagram for a two-stage experiment. Find the requested probabilities.

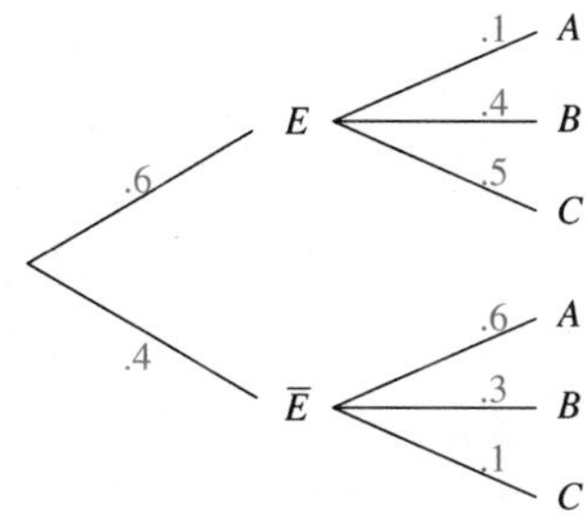

1. $P(A\,|\,E)P(E)$
2. $P(B\,|\,E)P(E)$
3. $P(B\,|\,\bar{E})P(\bar{E})$
4. $P(C\,|\,\bar{E})P(\bar{E})$
5. $P(A)$
6. $P(B)$
7. $P(E\,|\,A)$
8. $P(\bar{E}\,|\,B)$

Problems 9–16 refer to the following tree diagram for a two-stage experiment. Find the requested probabilities.

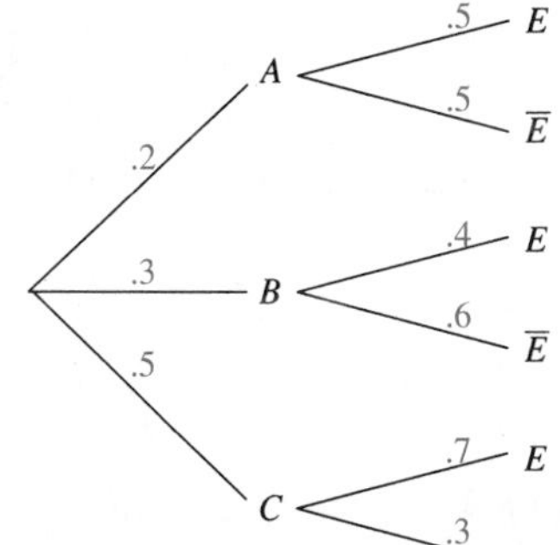

9. $P(E\,|\,A)P(A)$
10. $P(E\,|\,B)P(B)$
11. $P(\bar{E}\,|\,B)P(B)$
12. $P(\bar{E}\,|\,C)P(C)$

13. $P(E)$ **14.** $P(\bar{E})$

15. $P(A \mid E)$ **16.** $P(B \mid E)$

This experiment has two mutually exclusive events, A and $\bar{A}$, that form a partition of the sample space S. The number of elements in each set is shown in each region. Find the requested probabilities in Problems 17–22 by referring to the Venn diagram.

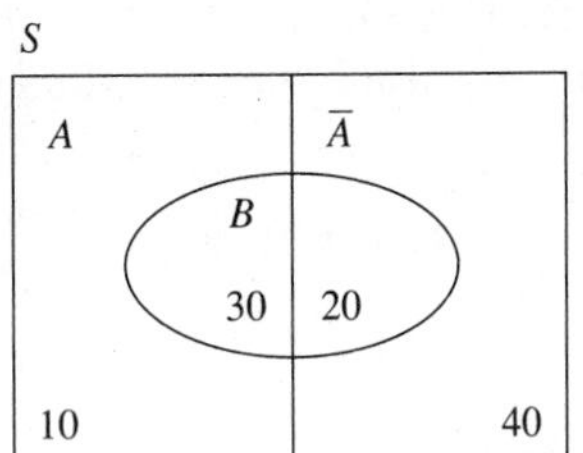

17. $P(B)$ **18.** $P(A \mid B)$

19. $P(\bar{A} \mid B)$ **20.** $P(\bar{B})$

21. $P(B \mid A)$ **22.** $P(B \mid \bar{A})$

23. Draw a tree diagram using the information in the Venn diagram.

This experiment has two mutually exclusive events, A and $\bar{A}$, that form a partition of the sample space S. The number of elements in each set is shown in each region. Find the requested probabilities in Problems 24–29 by referring to the Venn diagram.

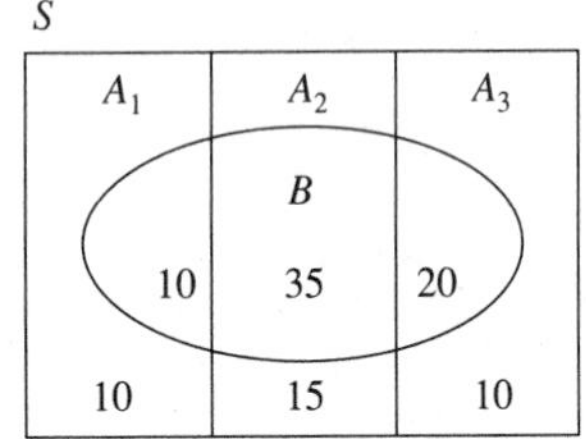

24. $P(B)$ **25.** $P(A_1 \mid B)$

26. $P(A_2 \mid B)$ **27.** $P(\bar{B})$

28. $P(B \mid A_3)$ **29.** $P(B \mid A_2)$

30. Draw a tree diagram using the information in the Venn diagram.

APPLICATIONS

In Problems 31–36 consider the experiment of selecting 2 items (without replacement) from a sample space of 100, of which 5 items are defective. Let $A = \{$first item selected is defective$\}$ and $B = \{$second item selected is defective$\}$. Find the probabilities correct to two decimal places.

31. $P(A)$ **32.** $P(\bar{A})$

33. $P(B \mid A)$ **34.** $P(B \mid \bar{A})$

35. $P(B)$ **36.** $P(\bar{B})$

Two cards are drawn in succession from a deck of 52 cards (without replacement). Let $D_1 = \{$diamond is drawn on first draw$\}$ and $D_2 = \{$diamond is drawn on second draw$\}$. Find the probabilities in Problems 37–42.

37. $P(D_1)$ **38.** $P(\overline{D_1})$

39. $P(D_2 \mid D_1)$ **40.** $P(D_2 \mid \overline{D_1})$

41. $P(D_2)$ **42.** $P(\overline{D_2})$

For the experiment in Problems 31–36, draw 3 items and let $C = \{$third item selected is defective$\}$. Find the probabilities in Problems 43–48 correct to the nearest hundredth.

43. $P(C \mid B)$ **44.** $P(C \mid \bar{B})$

45. $P(C \mid A)$ **46.** $P(C \mid \bar{A})$

47. $P(C)$ **48.** $P(\bar{C})$

For the experiment in Problems 37–42, draw three cards and let $D_3 = \{$diamond is drawn on third draw$\}$. Find the probabilities in Problems 49–53 correct to the nearest hundredth.

49. $P(D_3)$

50. $P(D_3 \mid D_2)$

51. P(exactly 1 diamond is drawn}

52. P(3 diamonds)

53. P(at least 1 diamond)

Suppose the people in a room are divided into two groups as follows:

Members: 15 *men;* 20 *women;* 0 *children*
Nonmembers: 10 *men;* 8 *women;* 12 *children*

Suppose a prize is given to one person who is selected at random by first choosing one group or the other, with the members M being twice as likely to be chosen as nonmembers $\bar{M}$. After the group is selected, then one person in that group is selected with each person in the group having an equal chance of being chosen. Let $N = \{$a man is chosen$\}$, $W = \{$a woman is chosen$\}$, and $C = \{$a child is chosen$\}$.

Use this information for Problems 54–59.

54. $P(W)$ **55.** $P(C)$

56. $P(\bar{M} \mid N)$ **57.** $P(M \mid W)$

58. $P(M \mid C)$ **59.** $P(\bar{M} \mid W)$

Given two urns, numbered Urn I and Urn II.
Contents of Urn I: 5 *red marbles,* 6 *blue marbles, and* 10 *green marbles*
Contents of Urn II: 30 *red marbles,* 20 *blue marbles, and* 10 *green marbles.*

The probability of selecting Urn I is .6 and of selecting Urn II is .4. After one of the urns is selected, a marble is selected from that urn. Let $I = \{$Urn I is selected$\}$, $\bar{I} = \{$Urn II is selected$\}$,

R = {a red marble is selected}, B = {a blue marble is selected}, and G = {a green marble is selected}. Find the probabilities in Problems 60–65 correct to the nearest hundredth.

60. $P(R)$ **61.** $P(B)$

62. $P(G)$ **63.** $P(I \mid R)$

64. $P(\bar{I} \mid B)$ **65.** $P(I \mid G)$

Write your answers to Problems 66–72 correct to three decimal places.

66. A test is given to candidates for graduate school. If the person is qualified, the probability of passing the test is 95%; if the person is not qualified, the probability of passing the test is 5%. Assume that the probability that a person taking the test is actually qualified is 70%. What is the probability that a person who passes the test is actually qualified?

67. Two persons are questioned by the police. One man tells the truth half of the time and the other always tells the truth. One man is chosen at random and asked one test question (a question to which the answer is known). The man answers truthfully. What is the probability that this man is the one who always tells the truth?

68. The student body of Oak Ridge University consists of 55% men and 45% women. A survey finds that 60% of the men and 30% of the women have outside jobs. What is the probability that a student who has a job is:
a. A man? **b.** A woman?

69. At Oak Ridge University 60% of the students come from the west, 15% from the midwest, and 25% from the east. A surveys finds that 80% of the students from the west, 40% of the students from the midwest, and 60% of the students from the east had attended at least one rock concert. What is the probability that a student chosen at random who has attended a rock concert comes from:
a. The east? **b.** The midwest?
c. The west?

70. Evergreen Insurance Company classifies its policyholders according to the following table:

Policyholder	Percent of policyholders	Probability of an accident-free year
AA(good risk)	60%	.999
A (average)	30%	.995
B (poor risk)	10%	.99
X(bad risk)	0%	.75

If Cliff Wilson buys a policy and subsequently has an accident, what is the probability he is a class B policyholder?

71. An electronic calculator has eight essential components that a manufacturer purchases from a supplier. Three components are selected at random and are tested. One is found to be defective. A search of the records gives the following information about the supplier:

Supplier's name	Number of shipments	Number of defects per shipment			
		0	1	2	More than 2
Abdullah	30	21	6	3	0

Given the sample result, what is the probability that the shipment has two defects?

72. Suppose a test for cancer is given. If a person has cancer, the test will detect it in 96% of the cases; and if the person does not have cancer, the test will show a positive result 1% of the time. If we assume that 12% of the population taking the test actually has cancer, what is the probability that a person taking the test and obtaining a positive result actually has cancer?

*7.7 Review

The material of this chapter is reviewed in the following list of objectives. After each objective there are some practice questions. For a sample test select the first question of each set and check your answers. The second question for each objective has no answer given. If you are having trouble with a particular type of problem, look back at the indicated section in the text. When you are finished reviewing these objectives, a sample examination is given at the end of this section.

[7.1]
Objective 7.1: *Write a sample space for a given experiment.*

1. *Industrial testing.* The reliability of a floppy disk is tested by the number of bits incorrectly recorded on a disk consisting of 1.44 MB.

* Optional section.

2. *Survey.* A person is randomly selected for a public opinion poll. The person is asked three questions:

Sex: male (m), female (f)
Income: low (l), medium (d), high (h)
Support president: yes (y), no (n)

A sample response is recorded as "mdy."

3. A drawer contains blue, black, and brown socks. The experiment consists of reaching into the drawer and selecting two socks at random.

4. *Industrial testing.* The life of a flashlight battery is tested by selecting a battery at random, using it continuously, and noting the time to failure. If it is still operating at 1,000 hours, the testing is terminated and the battery is assumed to have a life of 1,000 hours.

Objective 7.2: *Determine whether events are mutually exclusive.*

5. A coin and a die are simultaneously tossed onto a table. Let

$$H = \{\text{Head on the coin}\}$$
$$F = \{5 \text{ on die}\}$$

6. Consider the experiment described in Problem 2. Let

$$M = \{\text{Male}\}$$
$$S = \{\text{Support president}\}$$

7. Two dice are rolled and the sum of the numbers on the top faces recorded. Let

$$T = \{2 \text{ or } 3\}$$
$$F = \{4 \text{ or } 5\}$$

8. Two dice are rolled and the sum of the numbers on the top faces recorded. Let

$$E = \{\text{Even number}\}$$
$$P = \{\text{Prime number}\}$$

Objective 7.3: *Know the definition of a probability measure.*

9. Define a probability measure.

10. Consider the experiment of rolling a loaded die with the probabilities

$$P(1) = \frac{1}{20}, \quad P(2) = \frac{1}{5}, \quad P(3) = \frac{1}{4},$$
$$P(4) = \frac{1}{4}, \quad P(5) = \frac{1}{5}, \quad P(6) = \frac{1}{20}$$

Tell why this can (or cannot) be used as a probability model.

11. *Gambling.* At a horse race the following probabilities were calculated from the tote board:

$P(\text{Horse 1 winning}) = .13$ $\quad P(\text{Horse 2 winning}) = .24$
$P(\text{Horse 3 winning}) = .35$ $\quad P(\text{Horse 4 winning}) = .28$
$P(\text{Horse 5 winning}) = .19$ $\quad P(\text{Horse 6 winning}) = .10$

Tell why this can (or cannot) be used as a probability model.

12. Suppose a weighted coin and a loaded die are simultaneously tossed onto a table. Let

$$E = \{\text{Head}\}$$
$$S = \{6\}$$
$$F = \{\text{Head or } 6\}$$
$$G = \{\text{Head and } 6\}$$

It is known that

$$P(E) = .4 \qquad P(S) = .2 \qquad P(F) = .5 \qquad P(G) = .6$$

Tell why this can (or cannot) be used as a probability model.

Objective 7.4: *Use the addition principle with simple events. Suppose a box contains seven flavors of jelly beans, designated by color. Let*

$$A = \{\textit{Red, Green, Black}\}$$
$$B = \{\textit{Blue, Purple}\}$$
$$C = \{\textit{White, Pink}\}$$

If each color has a probability of $\frac{1}{7}$ of being chosen, find the probabilities in Problems 13–20.

13. $P(A)$ **14.** $P(B)$

15. $P(\bar{C})$ **16.** $P(\bar{B})$

17. $P(A \cup B)$ **18.** $P(A \cap B)$

19. $P(\bar{A} \cup \bar{B})$ **20.** $P(\overline{A \cap B})$

[7.2]

Objective 7.5: *Find the probability of mutually exclusive and equally likely events. Suppose a card is drawn from a deck of 52 cards. Find the probabilities.*

21. $P(7)$

22. $P(\text{Club})$

23. $P(7 \text{ or club})$

24. $P(7 \text{ and club})$

Objective 7.6: *Find empirical probabilities. Flip two coins simultaneously 50 times and note the results. The outcomes are*

$A = \{\text{Two heads}\}$

$B = \{\text{A head and a tail}\}$

$C = \{\text{Two tails}\}$

25. $P(A)$

26. $P(B)$

27. $P(C)$

28. Do the results of this experiment seem to indicate that these events are equally likely?

[7.3]

Objective 7.7: *Find probabilities that require combinations and permutations. Suppose three cards are drawn from a deck of cards (without replacement). Find the indicated probabilities as a decimal correct to three places.*

29. P(All hearts)

30. P(All one suit)

31. P(Spade, heart, and club)

32. P(Spade first, heart next, then a club last)

Objective 7.8: *Find probabilities by using the complement.*

33. Find the probability (to the nearest hundredth) of being dealt at least one heart in seven consecutive cards dealt from a deck of 52 cards.

34. What is the probability (to the nearest hundredth) of tossing at least one head by flipping a single coin five times?

35. If $P(A) = .6$ and $P(B) = .3$ for independent events A and B, what is $P(\overline{A \cup B})$?

36. Suppose that two cards are drawn without replacement from a deck of 52 cards. Find the probability that neither card is a spade.

[7.4]

Objective 7.9: *Find conditional probabilities.*

37. At a certain fast-food outlet, it was shown that 10% of the hamburgers were overdone, 3% had a missing ingredient, and 1% were overdone with a missing ingredient. What is the probability that the hamburger is overdone if the tomato is missing?

38. What is the probability that the hamburger in Problem 37 is missing an ingredient if it is overdone?

39. What is the probability of tossing at least two heads by flipping a single coin five times if it is known that at least one head occurred?

40. *Gambling.* In the American version of roulette, there are 38 individual slots into which a steel ball has an equal chance of falling; 18 are red, 18 are black, and 2 are green. What is the probability of red on the fifth consecutive play if it is known that red occurred on the first four plays?

[7.5]

Objective 7.10: *Find probabilities of unions and intersections. For Problems 41–48 consider the following table:*

Party	Opinion of President's Policies: Satisfied, S	Not satisfied, $\bar{S}$	Total
Republican, R	225	175	400
Democrat, D	300	200	500
Independent, I	25	75	100
Total	550	450	1,000

41. $P(R \cap S)$

42. $P(R \cup D)$

43. $P(S \cup \bar{S})$

44. $P(I \cap \bar{S})$

Problems 45–58 also use Objective 7.9.

45. $P(R \mid S)$

46. $P(S \mid R)$

47. $P(D \mid \bar{S})$

48. $P(\bar{S} \mid D)$

Problems 49–62 use Objective 7.7 in addition to Objective 7.10. A lot of 11 TV sets contains 5 black-and-white sets and 6 color sets. Two sets are drawn at random. Let

$B_1 = \{\text{Black-and-white set is drawn on the first draw}\}$

$B_2 = \{\text{Black-and-white set is drawn on the second draw}\}$

$C_1 = \{\text{Color set is drawn on the first draw}\}$

$C_2 = \{\text{Color set is drawn on the second draw}\}$

Compute the probabilities in Problems 49–55 by drawing with replacement.

49. P(Both color)

50. P(Both black-and-white)

51. $P(C_1 \cap B_2)$

52. $P(B_1 \cap C_2)$

53. $P(C_1 \cup B_2)$

54. $P(B_1 \cup C_2)$

55. P(One black-and-white set and one color set)

Repeat the experiment described above without replacement in Problems 56–62.

56. P(Both color)

57. P(Both black-and-white)

58. $P(C_1 \cap B_2)$

59. $P(B_1 \cap C_2)$

60. $P(C_1 \cup B_2)$

61. $P(B_1 \cup C_2)$

62. P(One black-and-white set and one color set)

Objective 7.11: *Find probabilities of a stochastic process. An armored car transports money between two locations and travels along one of three routes: A_1, B_1, C_1. On one day, the route* *is randomly selected according to the probabilities* $P(A_1) = .4$, $P(B_1) = .2$, $P(C_1) = .4$. *However, on the following day, the probabilities change so that the probability of the previously chosen route is .2, with the other two routes equally probable. Denote these second-day routes by* A_2, B_2, *and* C_2. *Find the probabilities.*

63. $P(A_2 \mid A_1)$

64. $P(A_2 \mid B_1)$

65. $P(A_2)$

66. $P(B_2)$

[7.7]

Objective 7.12: *Apply Bayes' theorem.*

67. *Industrial testing.* It is discovered that electric blenders of a certain company have defective switches. After an investigation it is found that the defective blenders came from three plants, as listed in the table. If a blender is selected at random and it has a defective switch, what is the probability that it came from Detroit if all three plants produce the same number of blenders?

Location of plant	Blenders with defective switches
I. Los Angeles	10%
II. Houston	40%
III. Detroit	20%

68. *Industrial testing.* Repeat Problem 67 except find the probability that the blender came from Los Angeles.

69. *Geology.* Geologists predict the presence of underground water by noting the presence of certain types of permeable surface rock. Records show that when there is subsurface water the probability of permeable surface rock is .8, and when there is no water the probability of permeable surface rock is .3. It is also known that the probability of hitting water at a feasible cost by sinking a test well at random is .4. What is the probability of water beneath permeable surface rock?

70. *Medical testing.* Suppose a test for hypoglycemia is given. If a person has hypoglycemia, the test will detect it in 90% of the cases; and if the person does not have hypoglycemia, the test will show a positive result 3% of the time. If we assume that 20% of the population taking the test actually have hypoglycemia, what is the probability that a person taking the test and obtaining a positive result actually has hypoglycemia?

SAMPLE TEST

The following sample test (45 *minutes*) *is intended to review the main ideas of the chapter.*

1. **a.** Define a probability measure.
b. What is meant by mutually exclusive events?
c. What is meant by independent events?

2. Suppose a box contains five flavors of jelly beans: 10 raspberry, 20 coconut, 25 blueberry, 30 watermelon, and 15 mint. The outcomes of selecting one jelly bean at random from the box are: $A = \{\text{coconut}\}$; $B = \{\text{raspberry, blueberry}\}$; and $C = \{\text{watermelon, mint}\}$. Find:
a. $P(A)$
b. $P(B)$
c. $P(\bar{C})$
d. $P(A \cup B)$
e. $P(\overline{A \cap B})$

3. Suppose you flip two coins simultaneously 100 times and note the results according to the following list of possibilities: $A = \{2H\}$, $B = \{2T\}$, and $C = \{1H \text{ and } 1T\}$. Will the experiment show these to be equally likely events? Why or why not?

4. Suppose two cards are drawn from a deck of cards (without replacement). Find the probabilities (to the nearest hundredth) of the following events.
a. P(both spades)
b. P(spade first, heart second)
c. P(1 spade and 1 heart)
d. P(not both spades)

5. The following table shows the results of an examination of 1,000 fruit flies exposed to massive amounts of radiation.

Sex of fruit fly	Normal, N	Mutated, $\bar{N}$	Total
Male, M	125	475	600
Female, $\bar{M}$	250	150	400
Total	375	625	1,000

Find:
a. $P(M)$
b. $P(\bar{N})$
c. $P(M \cup N)$
d. $P(M \cap \bar{N})$

6. **a.** Draw a tree diagram for the table in Problem 5.
b. From the table, find $P(N)$.
c. Find $P(N)$ using the tree diagram and show that the result is the same as the answer you found in part **b**.
d. From the table, find $P(N \mid \bar{M})$.
e. Use the tree diagram and Bayes' theorem to find $P(N \mid \bar{M})$, and verify that the answer is the same as the answer you found in part **d**.

7. Using the information in Problem 6, find:
a. $P(M \mid N)$
b. $P(N \mid M)$
c. $P(\bar{M} \mid \bar{N})$
d. $P(N \mid \bar{M})$

8. A shipment of six Broncos and four Thunderbirds arrives at a dealership, and two are selected at random and tested for defects. Let

$$B_1 = \{\text{first choice is a Bronco}\}$$
$$B_2 = \{\text{second choice is a Bronco}\}$$
$$T_1 = \{\text{first choice is a Thunderbird}\}$$
$$T_2 = \{\text{second choice is a Thunderbird}\}$$

Compute the requested probabilities by drawing with replacement.

a. $P(\text{both Broncos})$ **b.** $P(T_1 \cap B_2)$
c. $P(T_1 \cup B_2)$

9. Repeat Problem 8 by drawing without replacement.

10. A test for glaucoma will detect the disease in 95% of the cases and give a false positive result 2% of the time. If we assume that 10% of the population taking the test actually has glaucoma, what is the probability that a person taking the test and obtaining a positive result actually has glaucoma?

8

Statistics and Probability Applications

CHAPTER OVERVIEW

We are now introduced to the concept of a random variable. We use random variables to discuss two important probability distributions: the binomial distribution and the normal distribution. We also discuss some statistical measures needed to analyze data. These are the measures of central tendency (often called *averages*)—the mean, median, and mode. We also discuss the measures of dispersion—the range, variance, and standard deviation. Knowledge of these statistical measures enables us to make general statements and comparisons when dealing with sets of data or large populations.

PREVIEW

When an experiment with two possible outcomes is repeated, the binomial distribution is used to analyze the results. For example, it is used to determine the failure rate on an assembly line, to determine the likelihood of certain occurrences in genetics, and in determining the reliability of predictions based on psychology experiments. The normal distribution is a continuous distribution that is used to characterize many naturally occurring phenomena or as an approximation to certain binomial distributions. Together these distributions comprise a powerful tool in making probabilistic statements in a great variety of circumstances.

PERSPECTIVE

The central thread tying the material of this course together is that of a mathematical model. One of the difficult aspects of building a model is being able to handle the large amounts of data available in real-world settings. Matrices are one type of model used for analyzing data. In this chapter other types of mathematical models are considered: those used to measure central tendency and dispersion, and those used to describe binomial and normal distributions.

MODELING APPLICATION 6

USSR: The Impact of Climate on Grain Production

Suppose we wish to investigate crop yield over a period of years. Yield is expected to be related to a variety of factors: rainfall, temperature, fertilizer, humidity, and new technology, to name just a few. For this modeling application restrict your concern to rainfall as a predictor of crop yield. The data in the accompanying table show the precipitation, X (in centimeters) and crop yield Y (in centners per hectare*) for 15 growing seasons in the Soviet Union.

After you have finished this chapter, write a paper based on finding the best fitting line using this data.

* A centner is a measure used in many European countries; it is roughly equal to 50 kg. A hectare is 10,000 square meters, about 2.471 acres.

Crop Production Data[a]

Precipitation, X	Yield, Y
29	12
32	16
30	17
34	13
33	14
30	16
44	20
35	18
34	18.5
30	19
46	21
32	23
30	18
37	27
35	35

[a] This table adapted from *USSR: The Impact of Recent Climate Change on Grain Production.* Washington, D.C.: Central Intelligence Agency, 1976.

APPLICATIONS

Management *(Business, Economics, Finance, and Investments)*

Waiting/transaction times at a bank (8.1, Problems 7–8)
Quality control (8.1, Problems 14–15; 8.3, Problems 1–6, 38–39; 8.7, Problems 25, 29, 42, 45–46)
Salary comparisons (8.2, Problems 15, 23, 34)
Analysis of telephone usage (8.2, Problem 29)
Mean life of light bulbs (8.4, Problems 31–36)
Reliability of a computer system (8.5, Problem 50)
Overbooking of airline flights (8.5, Problem 51)
Testing a computer circuit (8.6, Problem 25)

Life Sciences *(Biology, Ecology, Health, and Medicine)*

Probability of white-faced calves (8.3, Problems 7–12)
Physical characteristics (8.3, Problems 30–31; 8.4, Problems 17–21)
Germination rate of a seed (8.3, Problem 41)
Rainfall in Ferndale (8.4, Problems 29–30)
Death rate from cancer (8.5, Problems 48–49)
Side effects of a drug (8.7, Problem 44)
Probability of a 310-day pregnancy (8.7, Test Problem 9)
Radioactive contamination and cancer mortality (8.7, Test Problems 10–11)

Social Sciences *(Demography, Political science, Population, Psychology, Society, and Sociology)*

Analysis of test scores (8.2, Problems 17–22; 32–33; 8.3, Problem 40)
Probability of a missile reaching target (8.3, Problem 37)
Testing for ESP (8.3, Problems 42–43)
Grading on a curve (8.4, Problems 22–26)
Parapsychology experiment (8.5, Problem 46)
Correlation of IQ scores and high school grades (8.6, Problem 26)
Correlation (8.6, Problems 25–32)

General Interest

Probability that a team will win the World Series (8.3, Problems 34–35)
Probability of a hit in baseball (8.3, Problem 36; 8.5, Problem 47)
Game of odd person out (8.3, Problems 44–47)
Breaking strength of materials (8.4, Problems 27, 28, 38–40)
Correlation between education and salary (8.6, Problem 29)

Actuary Exam Questions

(8.7, Test Problems 12–15)

Modeling Application—

World Running Records

8.1

Random Variables

The management at Chrysler wants to introduce a "zero defects" advertising campaign on a new line of automobiles, so they decide to do extensive testing of parts coming off the assembly line. Suppose 150 parts come off the assembly line every day, and inspectors keep track of the number of items with any type of defect. If X is the number of defective items, then X can obviously assume any of the values 0, 1, 2, 3,..., 149, 150. This variable X is an example of a **discrete random variable**.

Random Variable

> A **random variable** X associated with a probability space S is a function that assigns a real number to each simple event in S. If S has a finite number of outcomes, then X is called a **finite discrete random variable**; if X can assume infinitely many values arranged in a sequence, it is called an **infinite discrete random variable**; and if X can assume any real value on an interval, then it is called a **continuous random variable**.

In finite mathematics we focus on discrete random variables, but we can often convert continuous values to discrete values by rounding off to a given number of decimal places. For example, if X represents the heights of individuals, then X is a continuous random variable. However, if we round to the nearest eighth of an inch, then it is a discrete random variable. We will look at relationships such as this and other continuous random variables in Sections 8.4 and 8.5. In the meantime, we limit our discussion to discrete random variables.

In statistics, random variables are represented by using capital letters, such as X or Y, while lowercase letters, such as x or y, are used to denote a particular value of a random variable.

EXAMPLE 1 Suppose the length of time (in days) that a person waited to respond to a parking ticket is summarized by the list of tallies in the margin. Write out a table showing the relative frequencies.

Solution

0	卌 卌 卌 ||
1	|||
2	||
3	卌 卌 卌 卌 |
4	|
5	卌
6	
7	
8	|

Time (days)	Frequency	Relative frequency
0	17	$\frac{17}{50} = .34$
1	3	$\frac{3}{50} = .06$
2	2	$\frac{2}{50} = .04$
3	21	$\frac{21}{50} = .42$
4	1	$\frac{1}{50} = .02$
5	5	$\frac{5}{50} = .10$
6	0	$\frac{0}{50} = 0$
7	0	$\frac{0}{50} = 0$
8	1	$\frac{1}{50} = .02$
Total	50	1.00

If X is the number of days a person waited to respond to a parking ticket, then X is a discrete random variable where $X = 0, 1, 2, \ldots, 8$. ■

Notice in Example 1 that since 0, 1, 2, ..., 8 is the entire sample space,

$$P(X = 0) + P(X = 1) + P(X = 2) + \cdots + P(X = 8) = 1$$

Also notice that each $P(X = x_i)$ for $i = 1, 2, \ldots, 8$ is a number between 0 and 1, since each is a probability. We summarize these observations in the following box.

Probability Distribution

A **probability distribution** is the collection of all values that a random variable assumes along with the probabilities that correspond to these values. Furthermore,

1. $\sum_{i=1}^{n} P(X = x_i) = 1$

2. $0 \leq P(X = x_i) \leq 1 \quad$ for every i, $\quad 1 \leq i \leq n$

It is often useful to display the information in a probability distribution graphically in a special kind of bar graph called a **histogram**.

EXAMPLE 2 Represent the information from Example 1 in a histogram.

Solution Use the horizontal axis to delineate the values of the random variable and the vertical axis to represent the relative frequencies.

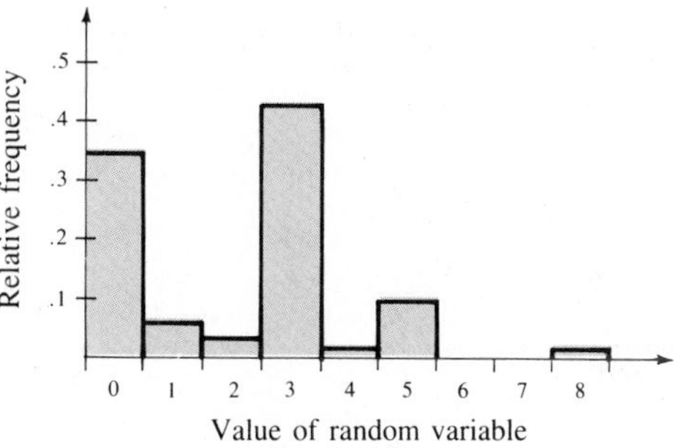

■

There is a very important relationship between the area of the rectangles in a histogram and the probabilities. Note that since the width of each bar is 1 unit, the **area of each bar is the probability of occurrence for that value of the random variable.** Therefore the **total area of the rectangles is 1**. This property is very important in our study of probability distributions.

WARNING ***Note the sentence in color.***

EXAMPLE 3 Draw a histogram for the probability distribution of rolling a single die and noting the number on top. Shade the portion that represents $P(X \geq 5)$.

Solution Let $X = 1, 2, 3, 4, 5, 6$; then $P(X = x_i) = \frac{1}{6}$, for $1 \le i \le 6$, and

$$P(X \ge 5) = P(X = 5) + P(X = 6)$$

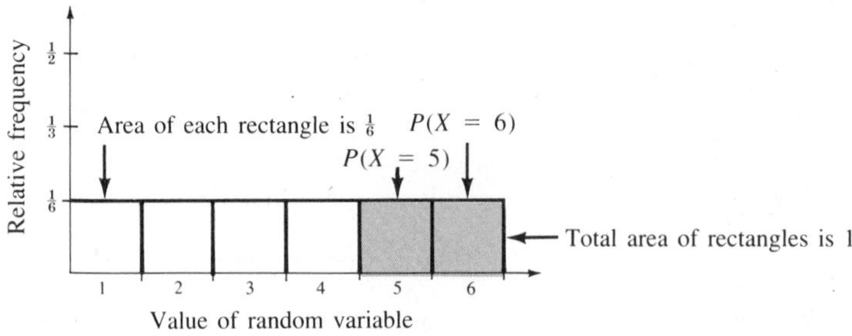

■

EXAMPLE 4 Let X denote the random variable for the probability distribution of rolling a pair of dice and noting the sum of the numbers on top. Draw a histogram for this probability distribution and shade the portion that represents $P(6 \le X \le 8)$.

Solution $P(6 \le X \le 8) = P(X = 6) + P(X = 7) + P(X = 8)$, so shade in the portions representing these random variables as shown in the figure.

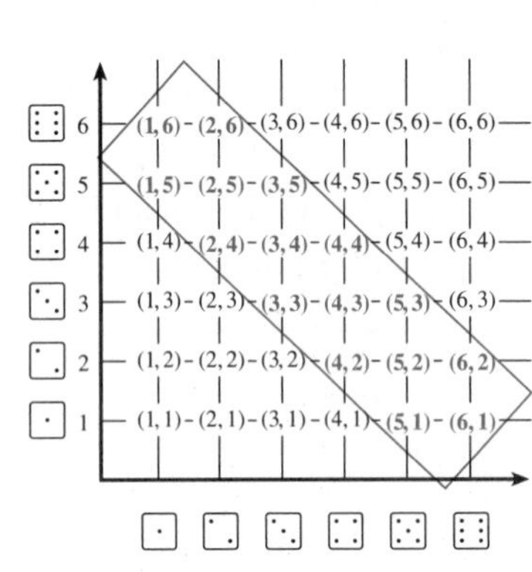

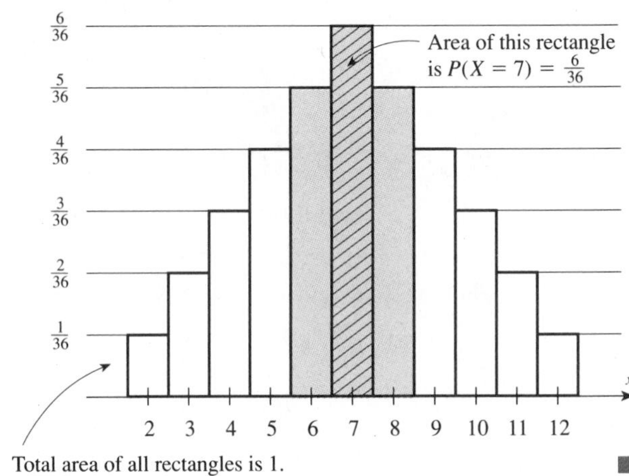

■

8.1 Problem Set

Give the frequency distribution and define the random variables in Problems 1–3.

1. The heights of 30 students are as follows (figures rounded to the nearest inch):

68	67	66	68	64	70	67	67	68	64	65	66	66	69	69
67	65	64	70	72	71	69	64	63	70	71	63	69	65	67

2. The numbers of years several leading batters played in the major leagues are as follows:

Ty Cobb	24	Paul Waner	20
Rogers Hornsby	23	Stan Musial	22
Ed Delehanty	16	Henie Manush	17
Dan Brouthers	19	Honus Wagner	21
Sam Thompson	15	Willie Keeler	19
Al Simmons	20	Ted Williams	19
Joe DiMaggio	13	Tris Speaker	22
Jimmy Foxx	20	Billy Hamilton	14
Lou Gehrig	17	Harry Heilmann	17
George Sisler	16	Babe Ruth	22
Nap Lajoie	21	Jesse Burkett	16
Cap Anson	22	Bill Terry	14
Eddie Collins	25		

3. A pair of dice are rolled and the sum of the spots on the tops of the dice are recorded as follows:

3	2	6	5	3	8	8	7	10	9
7	5	12	9	6	11	8	11	11	8
7	7	7	10	11	6	4	8	8	7
6	4	10	7	9	7	9	6	6	9
4	4	6	3	4	10	6	9	6	11

Find the probability distribution (that is, the relative frequencies) and draw a histogram for Problems 4–9.

4. Use the data in Problem 1.

5. Use the data in Problem 2.

6. Use the data in Problem 3.

APPLICATIONS

7. Blane, Inc., a consulting firm, is employed to perform an efficiency study at National City Bank. As part of the study, the number of times per day that people are waiting in line is summarized.

Number waiting	Frequency
2	20
3	15
4	7
5	5
6	2
7	1
8	0

8. Blane, Inc., also notes the transaction times (rounded to the nearest minute) for customers at National City Bank, as listed in the table:

Time	Frequency
1	10
2	12
3	18
4	25
5	16
6	10
7	6
8	1
9	0
10	2

9. Three coins are tossed onto a table and the following frequencies are noted:

Number of heads	Frequency
3	18
2	56
1	59
0	17

Use your knowledge of probability from Chapter 7 *to find the probability distributions in Problems* 10–19. *Let X be the random variable.*

10. Three coins are tossed onto a table and the number of tails is noted.

11. Four coins are tossed onto a table and the number of heads is noted.

12. Two dice are rolled and the total number of spots on the top faces is recorded.

13. Two cards are simultaneously drawn from a deck of 52 cards and the number of aces is noted.

14. From a lot containing 25 items, 5 of which are defective, 3 items are chosen at random and the number of defectives is noted.

15. Repeat Problem 14 except choose the 3 items one at a time, note the result, and then return the chosen item before choosing again.

16. Three cards are drawn from a deck of 52 cards and the number of hearts is noted.

17. Four cards are drawn from a deck of 52 cards and the number of hearts is noted.

18. A bag of Halloween candy contains 25 Hershey bars, 10 Rocky Roads, and 15 Almond Joys. Two pieces of candy are randomly selected and the number of Rocky Roads selected is noted.

19. Repeat Problem 18 except note the number of Almond Joys.

For Problems 20–29 *draw a histogram and shade in the region indicated for each problem.*

20. $P(X \geq 2)$ for the experiment in Problem 10.

21. $P(X$ is at least 3) for the experiment in Problem 11.

22. $P(X = 7 \text{ or } X = 11)$ for the experiment in Problem 12.

23. $P(X = 1)$ for the experiment in Problem 13.

24. $P(X \geq 1)$ for the experiment in Problem 14.

25. $P(X = 2)$ for the experiment in Problem 15.

26. $P(X = 3)$ for the experiment in Problem 16.

27. $P(2 \le X \le 3)$ for the experiment in Problem 17.

28. $P(X = 2)$ for the experiment in Problem 18.

29. $P(X$ is at least 1) for the experiment in Problem 19.

30. Draw a histogram for $P(X = x_i)$, where $P(X = x_i) = \dfrac{x_i}{10}$ for $0 \le i \le 4$, i a counting number. For example, if $x_i = 4$, then $P(X = 4) = \frac{4}{10} = \frac{2}{5}$.

31. Draw a histogram for $P(X = x_i)$, where $P(X = x_i) = \dfrac{x_i}{20}$ for $2 \le i \le 6$, i a counting number. For example, if $x_i = 2$, then $P(X = 2) = \frac{2}{20} = \frac{1}{10}$.

8.2 Analysis of Data

Measures of Central Tendency

The word *average* is frequently used in everyday language, but in mathematics it can take on a variety of meanings. For example, suppose we compare the "averages" of two bowlers in a tournament consisting of seven games:

Andrew: 185 average
Bob: 170 average

It would appear that Andrew did better in the tournament, but suppose the winner is found by looking at the individual games, as listed below:

	Andrew	**Bob**	**Winner (Better score)**
Game 1	175	180	Bob
Game 2	150	130	Andrew
Game 3	160	161	Bob
Game 4	180	185	Bob
Game 5	160	163	Bob
Game 6	183	185	Bob
Game 7	287	186	Andrew
Total	1,295	1,190	Andrew

To find each average, we divide the total by the number of games:

$$\text{Andrew: } \frac{1{,}295}{7} = 185 \qquad \text{Bob: } \frac{1{,}190}{7} = 170$$

However, if we consider the games separately, Bob beat Andrew in *five out of seven games.* Clearly, the "average" we calculated above does not completely describe the situation.

The first statistical measures we consider are called **measures of central tendency**, or **averages**. We already used the *summation notation* shown below:

$$\sum_{i=1}^{n} x_i = x_1 + x_2 + \cdots + x_n$$

In statistics this is sometimes abbreviated as Σx to mean *the sum of all values that x can assume*. Similarly, Σx^2 means *square each value that x can assume and then add the results.*

Averages, or Measures of Central Tendency

The *mean* is the most sensitive average. It reflects the entire distribution and is probably the most important of the averages.

The *median* gives the middle value. It is especially useful when there are a few extraordinary values that distort the mean.

The *mode* is the average that measures "popularity." It is possible to have no mode or more than one mode.

Given a set of data, there are three common *measures of central tendency*:

1. The **mean** (or **arithmetic mean**) is found by adding the data and then dividing by the number of data items, n:

$$\text{Mean} = \frac{\Sigma x}{n}$$

The mean of a set of sample scores is denoted by $\bar{x}$.

2. The **median** is the middle number when the data are arranged in order of size. If there is an even number of data items, then the median is the mean of the two middle numbers.
3. The **mode** is the value that occurs most frequently. If there is no number that occurs more than once, there is no mode. It is possible to have more than one mode.

EXAMPLE 1 Find the mean, median, and mode for Andrew's and Bob's bowling scores.

Solution Mean:

Andrew	*Bob*
$\bar{x} = \dfrac{1{,}295}{7} = 185$	$\bar{x} = \dfrac{1{,}190}{7} = 170$

There is a rather nice physical example that illustrates the idea of the mean. Consider a seesaw that consists of a plank and a movable support (called a fulcrum).

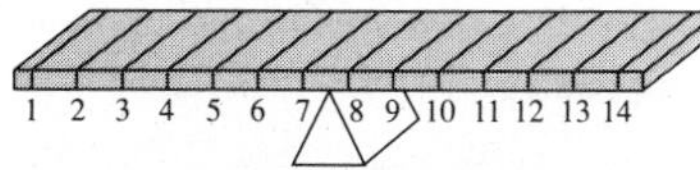

We assume that the plank has no weight and is marked off into units as shown. Now let's place some 1-pound weights in the positions of the numbers in our distribution. The balance point is the mean. In Example 1a, we place weights at 3, 5 (two weights), 8, and 9.

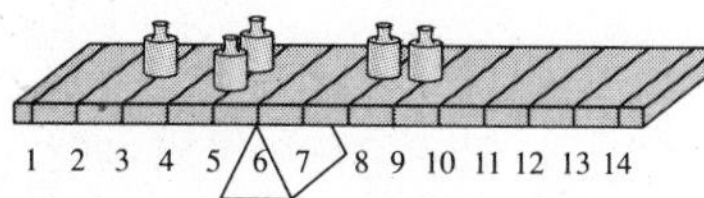

If we place the fulcrum at the mean, 6, the seesaw will balance.

Median:

Andrew		*Bob*
150		130
160		161
160		163
175	← The middle number is the median. →	180
180		185
183		185
287		186

Mode:

Andrew		*Bob*
150		130
160 }		161
160 }	The mode is the number that occurs most frequently.	163
175		180
180		{ 185
183		{ 185
287		186

Summary:

	Andrew	*Bob*	*Winner (Better Average)*
Mean	185	170	Andrew
Median	175	180	Bob
Mode	160	185	Bob

■

Find the mean, median, and mode for the sets of data in Examples 2–4.

EXAMPLE 2 3, 5, 5, 8, 9

Solution Mean: $\dfrac{\text{Sum of terms}}{\text{Number of terms}} = \dfrac{3 + 5 + 5 + 8 + 9}{5} = \dfrac{30}{5} = 6$

Median: Arrange in order: 3, 5, 5, 8, 9
The middle term, 5, is the median.

Mode: The term that occurs most frequently is the mode, which is 5. ■

EXAMPLE 3 4, 10, 9, 8, 9, 4, 5

Solution Mean: $\dfrac{4 + 10 + 9 + 8 + 9 + 4 + 5}{7} = \dfrac{49}{7} = 7$

Median: 4, 4, 5, 8, 9, 9, 10
The median is 8.

Mode: This set of data is **bimodal**; that is, it has two modes.
The modes are 4 and 9. ■

EXAMPLE 4 6, 5, 4, 7, 1, 9

Solution Mean: $\dfrac{6 + 5 + 4 + 7 + 1 + 9}{6} = \dfrac{32}{6} \approx 5.33$

Median: $1, 4, \underbrace{5, 6}, 7, 9$

$$\frac{5 + 6}{2} = \frac{11}{2} = 5.5$$

Mode: There is no mode. ■

Suppose the data are presented in a frequency distribution and there are more data than can be handled conveniently by the method used in Examples 2–4. Consider Example 5, which shows how you might proceed.

EXAMPLE 5 The following table gives the number of days one must wait for a marriage license in various states. Find the mean, median, and mode.

Wait for marriage license (days)	Frequency (number of states)
0	18
1	3
2	2
3	23
4	1
5	3
Total	50

Solution Mean: To find the mean, we could, of course, add all 50 individual numbers. But, instead, notice that

0 occurs 18 times, so we write $0 \cdot 18$
1 occurs 3 times, so we write $1 \cdot 3$
2 occurs 2 times, so we write $2 \cdot 2$
3 occurs 23 times, so we write $3 \cdot 23$
4 occurs 1 time, so we write $4 \cdot 1$
5 occurs 3 times, so we write $5 \cdot 3$

Thus the mean is

$$\begin{aligned}\bar{x} &= \frac{0 \cdot 18 + 1 \cdot 3 + 2 \cdot 2 + 3 \cdot 23 + 4 \cdot 1 + 5 \cdot 3}{50} \\ &= \frac{0 + 3 + 4 + 69 + 4 + 15}{50} \\ &= \frac{95}{50} = 1.90\end{aligned}$$

Median: Since there are 50 values, the mean of the twenty-fifth and twenty-sixth largest values is the median. From the table, we see that the twenty-fifth term is 3 and the twenty-sixth term is 3, so the median is

$$\begin{aligned}\frac{3 + 3}{2} &= \frac{6}{2} \\ &= 3\end{aligned}$$

Mode: The mode is the value that occurs most frequently, which is 3. ■

The mean from a frequency distribution is called the **weighted mean.**

Weighted Mean

If the scores $x_1, x_2, x_3, \ldots, x_n$ occur $w_1, w_2, \ldots, w_n$ times, respectively, then

$$\bar{x} = \frac{\sum_{i=1}^{n} w_i \cdot x_i}{\sum_{i=1}^{n} w_i} \qquad \text{or simply} \qquad \bar{x} = \frac{\Sigma w \cdot x}{\Sigma w}.$$

We have been using $\bar{x}$ to denote the mean of a set of *sample scores.* Now we let μ (Greek letter mu) denote the mean of *all scores* in some population. A **population** is the complete and entire collection of elements to be studied, so a sample must be a subset of a population. A sample obtained by selecting from the population is a **random sample** if each item in the population has an equal chance of being drawn. In the formula for a weighted mean, if we consider the entire population, then $\Sigma wx/\Sigma w$ can be rewritten as

$$\mu = \Sigma x \cdot P(x)$$

if we consider $P(x)$ to be the relative frequency with which x occurs.

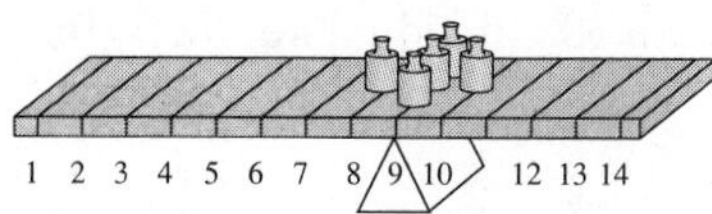

(a) First example (don't forget that the plank has no weight)

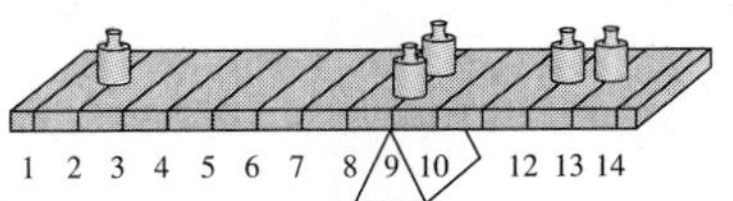

(b) Second example

Measures of Dispersion

The measures we have been discussing can help us interpret information, but they do not give the whole story. For example, consider the following two sets of data:

Data set 1: 8, 10, 9, 9, 9
Data set 2: 9, 9, 2, 12, 13

For both these data sets the mean, median, and mode are 9. It would seem that some additional analysis is in order. Note that the data in the second set are more spread out than the data in the first set. The amount that the data are spread out is called the **dispersion**. We now consider three measures of dispersion: the **range**, **variance**, and **standard deviation**. The simplest measure of dispersion is the range:

Range

The **range** in a set of data is the difference between the largest and the smallest numbers in the set.

The ranges for the above data sets are:

Range of data set 1: $10 - 8 = 2$
Range of data set 2: $13 - 2 = 11$

Note that the range is determined only by the largest and the smallest numbers in the set; it does not give us any information about the other numbers. It thus seems reasonable to invent some other measures of dispersion that take into account all the numbers in the data.

The variance and standard deviation are measures that give information about the dispersion. The **variance** uses all the numbers in the data set to measure the dispersion. When finding the variance, we must make a distinction between the variance of the entire population and the variance of a sample chosen randomly from that population. When the variance is based on a set of sample scores, it is denoted by s^2; and when it is based on all scores in a population, it is denoted by σ^2 (σ is the lowercase Greek letter sigma). The variance is found by

$$s^2 = \frac{\Sigma(x - \bar{x})^2}{n - 1} \qquad \sigma^2 = \frac{\Sigma(x - \mu)^2}{n}$$

If n is large (say, greater than 30), s^2 and σ^2 will be almost the same. The formula for s^2 looks a little intimidating, but it is based on arithmetic procedures, which, if taken one at a time, are quite simple. Understanding the variance formula is easy if you systematically deal with the data as indicated by the following procedure:

Procedure for Finding the Variance for a Set of Sample Scores

Step 1: Determine the mean of the numbers. (Find $\bar{x}$.)

Step 2: Subtract the mean from each number. (Find $x - \bar{x}$.)

Step 3: Square each of these differences. [This is $(x - \bar{x})^2$.]

Step 4: Find the sum of the squares of these differences. [This is $\Sigma(x - \bar{x})^2$.]

Step 5: Divide this sum by 1 less than the number of pieces of data.

EXAMPLE 6 Find the variance for each set of data.

a. Data set 1: 8, 9, 9, 9, 10 **b.** Data set 2: 2, 9, 9, 12, 13

Solution From the preceding discussion, we know that the mean, median, and mode for both these examples are the same (9). The range for data set 1 is 2, and for data set 2 it is 11. Now we want to find the second measure of dispersion, the variance. The procedure is lengthy, so make sure to follow each step carefully.

Step 1: Find the mean.

a. $\bar{x} = 9$ **b.** $\bar{x} = 9$

Step 2: Subtract each number from the mean.

a.

Data, x	Difference from the mean, $x - \bar{x}$
8	−1
9	0
9	0
9	0
10	1

b.

Data, x	Difference from the mean, $x - \bar{x}$
2	−7
9	0
9	0
12	3
13	4

Step 3: Square each of these differences. Notice that some of the differences in step 2 are positive and others are negative. Remember, we wish to find a measure of total dispersion. But if we add all these differences, we will not obtain the total variability. Indeed, if we simply add the differences for either example, the sum is zero. But we do not wish to say there is no dispersion. To resolve this difficulty with positive and negative differences, we square each difference so the result will always be nonnegative.

a.

Data, x	Difference from the mean, $x - \bar{x}$	Square of the difference, $(x - \bar{x})^2$
8	−1	1
9	0	0
9	0	0
9	0	0
10	1	1

b.

Data, x	Difference from the mean, $x - \bar{x}$	Square of the difference, $(x - \bar{x})^2$
2	−7	49
9	0	0
9	0	0
12	3	9
13	4	16

Step 4: Find the sum of these squares.

a. $1 + 0 + 0 + 0 + 1 = 2$ **b.** $49 + 0 + 0 + 9 + 16 = 74$

Step 5: Divide this sum by 1 less than the number of terms. In each of these examples there are five pieces of data, so divide by 4 to find the variance:

a. Variance $= \frac{2}{4} = .5$ **b.** Variance $= \frac{74}{4} = 18.5$ ■

The larger the variance, the more dispersion there is in the original data, and a smaller variance means that the data are more closely clustered around the mean.

The **standard deviation** for a sample is s and for the entire population is σ. That is,

$$\text{Standard deviation} = \sqrt{\text{variance}}$$

EXAMPLE 7 Find the standard deviations for the data of Example 6.

a. Data set 1: 8, 9, 9, 9, 10 **b.** Data set 2: 2, 9, 9, 12, 13

Solution We did most of the work in finding the standard deviation in Example 6. Now, all we need to do is to find the square root of the answers to Example 6.

a. $\sqrt{.5} \approx .71$ **b.** $\sqrt{18.5} \approx 4.30$ ■

If the variance is based on a random variable of a probability distribution, it can be written as

$$\sigma^2 = \Sigma(X - \mu)^2 \cdot P(X)$$

This formula can be manipulated into an equivalent form to help us carry out the calculations more easily:

$$\sigma^2 = \Sigma[X^2 \cdot P(X)] - \mu^2$$

EXAMPLE 8 Roll a pair of dice and let the random variable represent the total on the tops of the dice. Find the mean, variance, and standard deviation of the population.

Solution Organize the computations in table form, as shown.

X	$P(X)$	$X \cdot P(X)$	X^2	$X^2 \cdot P(X)$
2	$\frac{1}{36}$	$\frac{2}{36}$	4	$\frac{4}{36}$
3	$\frac{2}{36}$	$\frac{6}{36}$	9	$\frac{18}{36}$
4	$\frac{3}{36}$	$\frac{12}{36}$	16	$\frac{48}{36}$
5	$\frac{4}{36}$	$\frac{20}{36}$	25	$\frac{100}{36}$
6	$\frac{5}{36}$	$\frac{30}{36}$	36	$\frac{180}{36}$
7	$\frac{6}{36}$	$\frac{42}{36}$	49	$\frac{294}{36}$
8	$\frac{5}{36}$	$\frac{40}{36}$	64	$\frac{320}{36}$
9	$\frac{4}{36}$	$\frac{36}{36}$	81	$\frac{324}{36}$
10	$\frac{3}{36}$	$\frac{30}{36}$	100	$\frac{300}{36}$
11	$\frac{2}{36}$	$\frac{22}{36}$	121	$\frac{242}{36}$
12	$\frac{1}{36}$	$\frac{12}{36}$	144	$\frac{144}{36}$
Total		$\frac{252}{36} = 7$		$\frac{1,974}{36}$

Mean: $\mu = \Sigma X \cdot P(X) = 7$

Variance: $\sigma^2 = \Sigma[X^2 \cdot P(X)] - \mu^2$

$$= \frac{1,974}{36} - 7^2$$

$$= 5.8$$

Standard deviation: $\sigma = \sqrt{5.8} \approx 2.4$

■

CALCULATOR COMMENT

After the data points are input, most calculators will find the mean, variance, and standard deviation with a single key. Check your owner's manual.

If your calculator does not have statistical functions, you can arrange the calculations as shown in Example 8 and use the square root table given in Appendix E.

8.2
Problem Set

Find the mean, median, mode, range, variance, and standard deviation for each set of values in Problems 1–12.

1. 1, 2, 3, 4, 5
2. 17, 18, 19, 20, 21
3. 103, 104, 105, 106, 107
4. 765, 766, 767, 768, 769
5. 4, 7, 10, 7, 5, 2, 7
6. 15, 13, 10, 7, 6, 9, 10
7. 3, 5, 8, 13, 21
8. 1, 4, 9, 16, 25
9. 79, 90, 95, 95, 96
10. 70, 81, 95, 79, 85
11. 1, 2, 3, 3, 3, 4, 5
12. 0, 1, 1, 2, 3, 4, 16
13. Compare Problems 1–4. What do you notice about the mean and standard deviation?
14. By looking at Problems 1–4 and discovering a pattern, find the mean and standard deviation of the following set of numbers:

217,850 217,851 217,852 217,853 217,854

APPLICATIONS

15. Find the mean, median, and mode of the following sample of salaries of the employees of Green Lawn Landscaping Company:

Salary	Frequency
$10,000	4
16,000	3
20,000	2
30,000	1

16. Linda Foley, the leading salesperson for the Green Lawn Landscaping Company, turned in the following summary of sales contacts for the week of October 23–28. Find the mean, median, and mode.

Date	Number of clients contacted
Oct. 23	12
Oct. 24	9
Oct. 25	10
Oct. 26	16
Oct. 27	10
Oct. 28	21

17. Find the mean, median, and mode of the following sample test scores:

Test score	Frequency
90	1
80	3
70	10
60	5
50	2

18. The following sample scores were obtained on a test:

Score	Frequency
90	1
80	6
70	10
60	4
50	3
40	1

Find the mean, median, mode, and range for the class.

19. A sample of class scores on a test is shown below.

Score	Frequency
90	2
80	4
70	9
60	5
50	3
40	1
30	2
0	4

Find the mean, median, mode, and range for the class.

Find the variance (s^2) and standard deviation (s) for Problems 20–24.

20. Problem 17
21. Problem 18
22. Problem 19
23. Problem 15
24. Problem 16

Find the mean, variance, and standard deviation (σ) *of the probability distributions in Problems* 25–28.

25.

X	$P(X)$
0	$\frac{1}{2}$
1	$\frac{1}{4}$
2	$\frac{1}{4}$

26.

X	$P(X)$
1	.1
2	.8
3	.1

27.

X	$P(X)$
2	.2
3	.3
4	.4
5	.1

28.

X	$P(X)$
5	$\frac{1}{6}$
10	$\frac{1}{3}$
20	$\frac{1}{3}$
30	$\frac{1}{6}$

29. A survey shows that the number of times a nonbusiness telephone rings before it is answered is:

X	$P(X)$
0	.00
1	.20
2	.31
3	.16
4	.18
5	.10
6	.02
7	.03

Find the mean, variance, and standard deviation for the number of rings.

30. The probability for the number of customers arriving at a bank in a given period of time is summarized below:

X	$P(X)$
0	.25
1	.43
2	.18
3	.09
4	.03
5	.01
6	.01

Find the mean, variance, and standard deviation for the number of arrivals.

31. Suppose a variance is zero. What can you say about the data?

32. A professor gives five exams. The scores for two students have the same mean, although one student seemed to do better on all the tests except one. Give an example of such scores.

33. A professor gives six exams. The scores of two students have the same mean, although one student's scores have a small standard deviation and the other student's scores have a large standard deviation. Give an example of such scores.

34. The salaries for the executives of a small company are shown below. Find the mean, median, and mode. Which measure seems to best describe the average executive salary for the company?

Position	Salary
President	\$90,000
1st VP	40,000
2nd VP	40,000
Supervising manager	34,000
Accounting manager	30,000
Personnel manager	30,000

35. Roll a pair of dice 20 times and record the sum of the numbers on the two top faces. Find the mean, variance, and standard deviation for your data. Compare your results with Example 8.

36. Repeat Problem 35 for 100 rolls of the dice.

37. Roll a pair of dice until all 11 outcomes (the sum of the numbers on the two top faces) occur at least once. Repeat the experiment 20 times. Find the mean, variance, and standard deviation for the number of tosses.

38. The *harmonic mean* (or H.M.) is found by dividing the number of scores, n, by the sum of the reciprocals of all scores:

$$\text{H.M.} = \frac{n}{\Sigma \frac{1}{x}}$$

Find the harmonic mean for the data in Problem 5.

39. Repeat Problem 38 for the data in Problem 6.

40. The *geometric mean* (or G.M.) is found by taking the nth root of the product of the n scores. Find the geometric mean for the data in Problem 5.

41. Repeat Problem 40 for the data in Problem 6.

42. The *quadratic mean*, or *root mean square* (R.M.S.), is found by squaring each score; adding the results; dividing by the number of scores, n; and then taking the square root of the result:

$$\text{R.M.S.} = \sqrt{\frac{\Sigma x^2}{n}}$$

Find the quadratic mean for the data in Problem 5.

43. Repeat Problem 42 for the data in Problem 6.

8.3
The Binomial Distribution

Consider now a common type of experiment—one with only two outcomes, A and $\bar{A}$. Suppose that $P(A) = p$ and $P(\bar{A}) = q = 1 - p$. We are interested in n repetitions of the experiment. If $P(A)$ remains the same for each repetition and we let X represent the number of times that event A has occurred, then we call X a **binomial random variable.**

EXAMPLE 1 Toss a coin four times. The sample space is shown below:

Number of heads	Outcomes					
4	HHHH					
3	HHHT	HHTH	HTHH	THHH		
2	HHTT	HTHT	HTTH	THTH	THHT	TTHH
1	TTTH	TTHT	THTT	HTTT		
0	TTTT					

Let the random variable X represent the number of heads that have occurred. That is, $X = 0$ if we obtain no heads; $X = 1$ means one head is obtained; $X = 2$ means two heads are obtained; $X = 3$, $X = 4$ mean three and four heads are obtained, respectively. Since there are 16 possibilities,

$$P(X = 4) = \frac{1}{16}$$

$$P(X = 3) = \frac{4}{16} = \frac{1}{4}$$

$$P(X = 2) = \frac{6}{16} = \frac{3}{8}$$

$$P(X = 1) = \frac{4}{16} = \frac{1}{4}$$

$$P(X = 0) = \frac{1}{16}$$

Note that the sum is one:

$$\frac{1}{16} + \frac{4}{16} + \frac{6}{16} + \frac{4}{16} + \frac{1}{16} = 1$$

■

This example illustrates a common discrete probability distribution called the **binomial distribution**, which is a list of outcomes and probabilities for a **binomial experiment**.

Binomial Experiment

A **binomial experiment** is an experiment that meets four conditions:

1. There must be a fixed number of trials. Denote this number by n.
2. There must be two possible mutually exclusive outcomes for each trial. Call them *success* and *failure*.
3. Each trial must be independent. That is, the outcome of a particular trial is not affected by the outcome of any other trial.
4. The probability of success and failure must remain constant for each trial.

Consider the manufacture of computer chips. Suppose three items are chosen at random from a day's production and are classified as defective (F) or nondefective (S). We are interested in the number of successes obtained. (A "success" is the occurrence of the event we are considering—in this case, nondefectives.) Suppose that an item has a probability of .1 of being defective and therefore a probability of .9 of being nondefective. We will assume that these probabilities remain the same throughout the experiment and that the classification of any particular item is independent of the classification of any other item. The sample space, along with the probabilities, is listed below:

Sample space	*Associated probabilities*
1. FFF	$(.1)(.1)(.1) = (.1)^3$
2. FFS	$(.1)(.1)(.9) = (.9)(.1)^2$
3. FSF	$(.1)(.9)(.1) = (.9)(.1)^2$
4. SFF	$(.9)(.1)(.1) = (.9)(.1)^2$
5. FSS	$(.1)(.9)(.9) = (.9)^2(.1)$
6. SFS	$(.9)(.1)(.9) = (.9)^2(.1)$
7. SSF	$(.9)(.9)(.1) = (.9)^2(.1)$
8. SSS	$(.9)(.9)(.9) = (.9)^3$

If we let $X =$ the number of successes obtained, then $P(X = 0)$ is found on line 1 above:

$P(X = 0) = (.1)^3$

$P(X = 1)$ is found from lines 2, 3, and 4:

2. FFS $\quad (.9)(.1)^2$
3. FSF $\quad (.9)(.1)^2$
4. SFF $\quad (.9)(.1)^2$

Total: $\quad P(X = 1) = 3(.9)(.1)^2$

$P(X = 2)$ is found by

5. FSS $(.9)^2(.1)$
6. SFS $(.9)^2(.1)$
7. SSF $(.9)^2(.1)$

Total: $P(X = 2) = 3(.9)^2(.1)$

$P(X = 3)$ is found in line 8:

$P(X = 3) = (.9)^3$

Note that the same results can be achieved by simply considering the following binomial expansion:

$$(.1 + .9)^3 = (.1)^3 + 3(.1)^2(.9) + 3(.1)(.9)^2 + (.9)^3$$

This leads us to the following theorem:

Binomial Distribution Theorem

Let X be a random variable for the number of successes in n independent and identical repetitions of an experiment with two possible outcomes, success and failure. If p is the probability of success, then

$$P(X = k) = \binom{n}{k} p^k (1 - p)^{n-k} \qquad k = 0, 1, \ldots, n$$

A sequence of independent trials for which there are only two possible outcomes is also sometimes called a sequence of **Bernoulli trials**, after Jacob Bernoulli (1654–1705).

EXAMPLE 2 Suppose a sociology teacher always gives true–false tests with 10 questions.

a. What is the probability of getting exactly 70% by guessing?

b. What is the probability of getting 70% or better by guessing?

c. If a student can be sure of getting five questions correct, but must guess at the others, what is the probability of getting 70% or better?

Solution **a.**

$$P(X = 7) = \binom{10}{7}\left(\frac{1}{2}\right)^7\left(\frac{1}{2}\right)^3 = \frac{10!}{7!3!} \cdot \frac{1}{2^{10}} = \frac{120}{1{,}024} = \frac{15}{128}$$

b.

$$P(X = 8) = \binom{10}{8}\left(\frac{1}{2}\right)^8\left(\frac{1}{2}\right)^2 = \frac{45}{1{,}024}$$

$$P(X = 9) = \binom{10}{9}\left(\frac{1}{2}\right)^9\left(\frac{1}{2}\right)^1 = \frac{10}{1{,}024} = \frac{5}{512}$$

$$P(X = 10) = \binom{10}{10}\left(\frac{1}{2}\right)^{10} = \frac{1}{1{,}024}$$

$$P(X \geq 7) = \frac{120}{1{,}024} + \frac{45}{1{,}024} + \frac{10}{1{,}024} + \frac{1}{1{,}024} = \frac{176}{1{,}024} = \frac{11}{64}$$

c. $$P(2 \le X \le 5) = \sum_{k=2}^{5} \binom{5}{k}\left(\frac{1}{2}\right)^k\left(\frac{1}{2}\right)^{5-k}$$
$$= \binom{5}{2}\left(\frac{1}{2}\right)^5 + \binom{5}{3}\left(\frac{1}{2}\right)^5 + \binom{5}{4}\left(\frac{1}{2}\right)^5 + \binom{5}{5}\left(\frac{1}{2}\right)^5$$
$$= 10 \cdot \frac{1}{32} + 10 \cdot \frac{1}{32} + 5 \cdot \frac{1}{32} + 1 \cdot \frac{1}{32}$$
$$= \frac{5}{16} + \frac{5}{16} + \frac{5}{32} + \frac{1}{32} = \frac{13}{16} = .8125$$ ■

As you can see from the above examples, the calculations for binomial probabilities can become rather tedious. For this reason, tables of binomial distributions have been compiled. If you wish to calculate the probability of X successes in n independent Bernoulli trials with the probability of success on a single trial equal to p, you can use Table 3 in Appendix E. For example, the probability

$$P(X = 1) = \binom{5}{1}\left(\frac{1}{10}\right)^1\left(\frac{9}{10}\right)^4$$

$n = 5$

$k = 1$ $p = \frac{1}{10} = .1$

can be found in Table 3, where $n = 5$, $k = 1$, and $p = .1$:

$$P(X = 1) = .328$$

Also, we find $P(X = 0) = .590$.

Table 3 is particularly useful when calculating sums of binomial probabilities. For Example 2**c**, we find from Table 3 (where $n = 5$, $p = .5$, and $k = 2, 3, 4$, and 5):

$$P(2 \le X \le 5) = .313 + .312 + .156 + .031 = .812$$

For simple calculations, a calculator will probably be more efficient for finding binomial probabilities, as shown in Example 3.

EXAMPLE 3 A missile has a probability of $\frac{1}{10}$ of penetrating enemy defenses and reaching its target. If five missiles are aimed at the same target, what is the probability that exactly one will hit its target? What is the probability that at least one will hit its target?

Solution
$$P(X = 1) = \binom{5}{1}\left(\frac{1}{10}\right)^1\left(\frac{9}{10}\right)^4$$
$$= \frac{5 \cdot 9^4}{10^5} = \frac{6{,}561}{20{,}000} = .32805$$

$$P(X \ge 1) = 1 - P(X = 0)$$
$$= 1 - \binom{5}{0}\left(\frac{9}{10}\right)^5 = 1 - \frac{9^5}{10^5}$$
$$= 1 - .59049 = .40951$$ ■

EXAMPLE 4 A typist can type a page accurately (with no errors) with a probability of .25. What is the probability that there will be exactly one page with at least one error in a report of five pages?

Solution The probability of at least one error on one page is .75, so the solution is given by

$$P(X = 1) = \binom{5}{1}(.75)^1(.25)^4 = .0146484375$$ ■

We can also find the mean, variance, and standard deviation for a biomial distribution.

EXAMPLE 5 Find the mean, variance, and standard deviation for the number of heads obtained from tossing a coin four times.

Solution Use the information in Example 1 along with the formulas for mean and variance from Section 8.2:

$$\mu = \Sigma X \cdot P(X) \qquad \text{and} \qquad \sigma^2 = \Sigma[X^2 \cdot P(X)] - \mu^2$$

X	$P(X)$	μ $X \cdot P(X)$	X^2	$X^2 \cdot P(X)$
4	$\frac{1}{16} = .0625$	.25	16	1
3	$\frac{1}{4} = .25$	.75	9	2.25
2	$\frac{3}{8} = .375$	.75	4	1.5
1	$\frac{1}{4} = .25$	.25	1	.25
0	$\frac{1}{16} = .0625$	0	0	0
Total		2		5

From the table, $\mu = 2$, and

$$\sigma^2 = 5 - 4 = 1 \qquad \sigma = \sqrt{1} = 1$$ ■

The computations in Example 5 use the formulas for mean, variance, and standard deviation given in Section 8.2. However, if we apply some complicated algebraic manipulations to these formulas (the details are too involved to show here), we arrive at some very simple formulas for the mean, variance, and standard deviation of a binomial distribution.

Mean, Variance, and Standard Deviation of a Binomial Distribution

A binomial distribution of n independent trials, each with a probability of success p and failure $q = 1 - p$, has mean μ, variance σ^2, and standard deviation σ, given by

$$\mu = np \qquad \sigma^2 = npq \qquad \sigma = \sqrt{npq}$$

We can verify these formulas by comparison with the results found in Example 5. We had four repetitions of tossing a coin, so $n = 4$; $p = \frac{1}{2}$ since the probability of success (a head) is $\frac{1}{2}$; and $q = 1 - p = \frac{1}{2}$. Thus

$$\mu = np \qquad \sigma^2 = npq \qquad \sigma = \sqrt{npq}$$
$$= 4 \cdot \tfrac{1}{2} \qquad = 4 \cdot \tfrac{1}{2} \cdot \tfrac{1}{2} \qquad = \sqrt{1}$$
$$= 2 \qquad = 1 \qquad = 1$$

EXAMPLE 6 The probability of a transmission failure in the first year on a new automobile is .005. Find the mean number of failures for 10,000 cars if we assume that the failures are independent. Also find the standard deviation.

Solution $n = 10{,}000$; $q = .005$; $p = 1 - .005 = .995$

Mean number of failures

Standard deviation

$$\mu = nq \qquad \sigma = \sqrt{npq}$$
$$= 10{,}000(.005) \qquad = \sqrt{(10{,}000)(.995)(.005)}$$
$$= 50 \qquad = \sqrt{49.75}$$
$$\approx 7.05$$

■

8.3 Problem Set

APPLICATIONS

A manufacturer of disk drives chooses three items from the day's production and classifies each as defective or nondefective. Suppose that an item has a probability of .2 of being defective. In Problems 1–5 find the requested probabilities.

1. No defective disk drives are found.
2. All three are defective.
3. Two are found to be defective.
4. At least two are defective.
5. One or more are defective.
6. Numerically, how are your answers for Problems 1 and 5 related?

All the cows in a certain herd are white-faced. The probability that a white-faced calf will be born from the mating with a certain bull is .9. Suppose four cows are bred to the same bull. Find the probabilities in Problems 7–12.

7. Four white-faced calves
8. Exactly three white-faced calves
9. No white-faced calves
10. One white-faced calf
11. At least three white-faced calves
12. Not more than two white-faced calves

Use Table 3 (Appendix E) or a calculator to find the binomial probabilities in Problems 13–20.

13. $n = 5, \quad X = 3, \quad p = .30$
14. $n = 4, \quad X = 3, \quad p = .25$
15. $n = 12, \quad X = 6, \quad p = .65$
16. $n = 10, \quad X = 4, \quad p = .80$
17. $n = 6, \quad X = 6, \quad p = \frac{1}{2}$
18. $n = 8, \quad X = 8, \quad p = \frac{3}{4}$
19. $n = 7, \quad X = 5, \quad p = \frac{1}{10}$
20. $n = 15, \quad X = 13, \quad p = \frac{2}{5}$

In Problems 21–29 find the mean, variance, and standard deviation for the given values of n and p. Assume that the binomial conditions are satisfied in each case.

21. $n = 5, \quad p = .3$
22. $n = 4, \quad p = .25$
23. $n = 12, \quad p = .65$
24. $n = 10, \quad p = .8$
25. $n = 6, \quad p = \frac{1}{2}$
26. $n = 8, \quad p = \frac{3}{4}$
27. $n = 7, \quad p = \frac{1}{10}$
28. $n = 15, \quad p = \frac{2}{5}$
29. $n = 10, \quad p = \frac{3}{5}$
30. The probability that children in a specific family will have blond hair is $\frac{3}{4}$. If there are four children in the family, what is the probability that two of them are blond?

31. The probability that the children in a specific family will have brown eyes is $\frac{3}{8}$. If there are four children in the family, what is the probability that at least two of them have brown eyes?

32. It is known that 85% of the graduates of Foley's School of Motel Management are placed in a job within 6 months of graduation. If a class has 20 graduates, what is the probability that 15 will be placed within 6 months?

33. If a graduating class of Foley's School in Problem 32 has 10 graduates, what is the probability that at least 9 will be placed within 6 months?

34. Suppose the National League team has a probability of $\frac{3}{5}$ of winning the World Series and the American League team has a probability of $\frac{2}{5}$. The series is over as soon as one team wins four games. Find the probability that the series is over in four games.

35. In Problem 34 what is the probability that the series will go seven games?

36. The batting average of the star baseball player Brian Thompson is .300 (that is, $P(\text{hit}) = .3$). What is the probability that Thompson will get at least three hits for four times at bat?

37. Suppose that research has shown that the probability that a missile penetrates enemy defenses and reaches its target is $\frac{1}{10}$. Find the smallest number of identical missiles that are necessary in order to be 80% certain of hitting the target at least once.

38. Eighty percent of the widgets of the Ampex Widget Company meet the specifications of its customers. If a sample of six widgets is tested, what is the probability that three or more of them would fail to meet the specifications?

39. Find the mean number of failures for the widgets described in Problem 38. Also find the standard deviation.

40. Ten percent of the students fail the final exam in Math 9. If the class size is 30, find the mean and standard deviation for the number of failures in each class.

41. A certain type of seed has a 60% germination rate. If the seeds are planted in rows of 50 seeds each, find the mean and standard deviation for the number of seeds that germinate in each row.

42. Suppose a person claims to have ESP and says he can read mental images 75% of the time. We set up an experiment where he must determine whether we are looking at a picture of a circle or a square. We agree that if he can correctly identify the mental image at least five out of six times we will grant his claim.
    a. What is the probability of granting his claim if he is in fact only guessing?
    b. What is the probability of denying his claim if he really does have the ability he claims?

43. Repeat Problem 42 where the person claims to read mental images 90% of the time.

44. In a certain office, three men determine who will pay for coffee by each flipping a coin. If one of them has an outcome that is different from the other two, he must pay for the coffee. What is the probability that in any play of the game there will be an odd man out?

45. In the game of Problem 44, what is the probability that there will be an odd man out in a particular play if there are five players?

46. In Problem 44, what is the probability that it will require more than two tosses to produce an odd man out?

47. Generalize Problem 44 for n men playing the game of odd man out.

8.4
The Normal Distribution

Suppose we survey the results of 20 children's scores on an IQ test. The scores (rounded to the nearest 5 points) are: 115, 90, 100, 95, 105, 105, 95, 105, 105, 95, 125, 120, 110, 100, 100, 90, 110, 100, 115, and 80. The mean is 103, the standard deviation is 10.93, and the frequency is shown in Figure 8.1a on page 402.

If we consider 100,000 IQ scores instead of only 20, we might obtain the frequency distribution shown in Figure 8.1b. As you can see, the frequency distribution approximates a curve. If we connect the endpoints of the bars in Figure 8.1b, we obtain a curve that is very close to a curve called the **normal distribution curve**, or simply the **normal curve**, as shown in Figure 8.2.

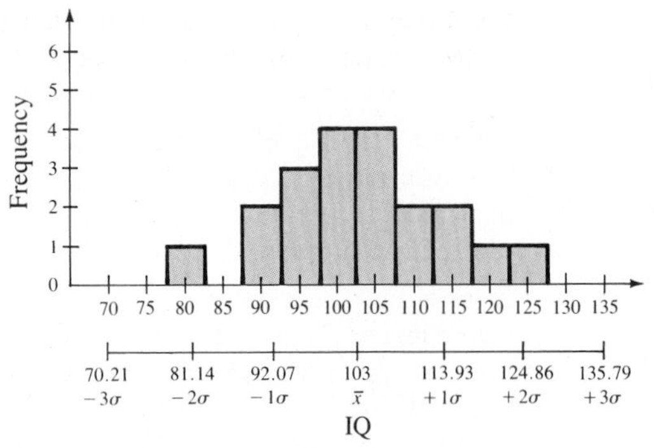

a. Frequencies of IQs for a sample of 20 children

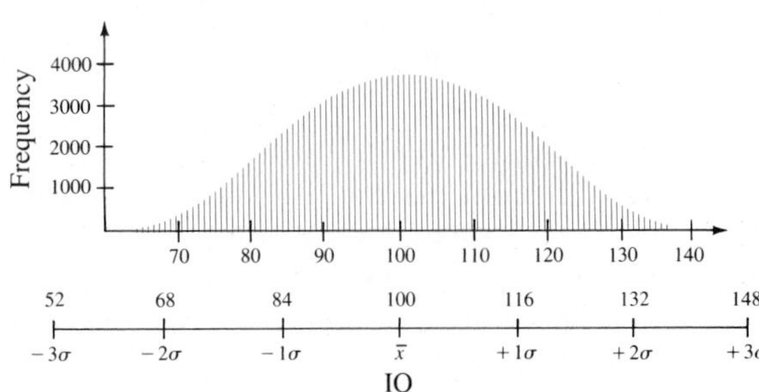

b. Frequencies of IQs for a sample of 100,000 children

FIGURE 8.1
Frequencies of IQ scores

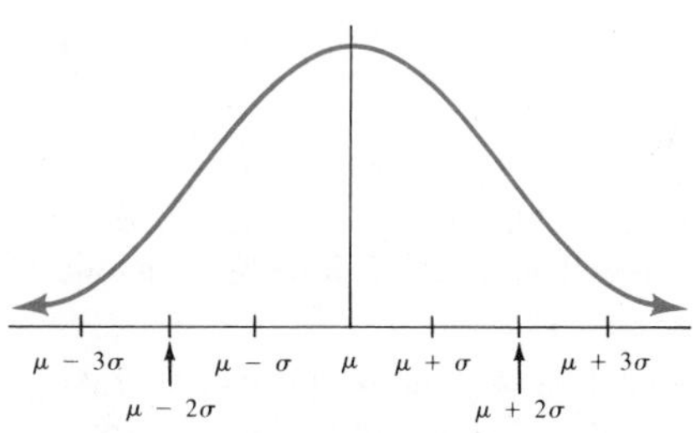

FIGURE 8.2
Normal distribution curve

The normal distribution is a **continuous** (rather than discrete) **distribution**, and it extends indefinitely in both directions, never touching the x-axis. It is symmetric about a vertical line drawn through the mean, μ. The equation of this curve is

$$y = \frac{e^{-(x-\mu)^2/(2\sigma^2)}}{\sigma\sqrt{2\pi}} \qquad \text{where } \mu = \text{Mean}$$

$$\sigma = \text{Standard deviation}$$

$$\pi \approx 3.1416$$

$$e \approx 2.7183$$

Graphs of this curve for several choices of σ are shown in Figure 8.3. Using calculus, it can be shown that the "curvature" of the normal curve changes at $\mu + \sigma$ and $\mu - \sigma$. In calculus this point is called a *point of inflection.*

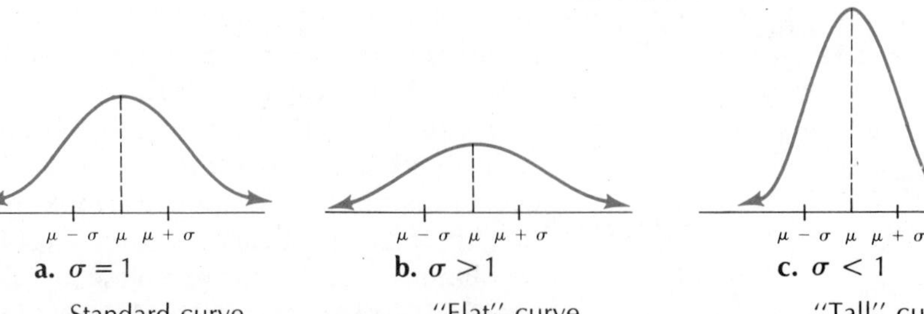

a. $\sigma = 1$ Standard curve

b. $\sigma > 1$ "Flat" curve

c. $\sigma < 1$ "Tall" curve

FIGURE 8.3
Variations of normal curves

Since the normal distribution is a probability distribution, we know the area under this curve is 1. Therefore we can relate the area to probabilities:

Probabilities of a Normal Probability Distribution

Let X be a random variable with a normal probability distribution. Then:

$P(a \leq X \leq b)$ is the area under the associated normal curve between a and b.

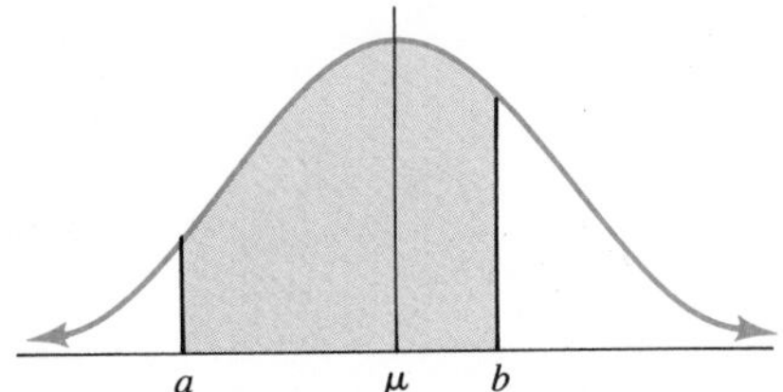

$P(X > \mu) = P(X < \mu) = \frac{1}{2}$; that is, the curve is symmetric about the mean.
$P(X = x) = 0$ for any real number x. (Since there are *infinitely* many possibilities, the probability of a particular value is 0.)
$P(X < x) = P(X \leq x)$ for any real number x.
$P(X > x) = 1 - P(X \leq x)$ for any real number x.

If $\mu = 0$ and $\sigma = 1$, the curve is called the **standard normal curve**.

Since it requires calculus to find particular areas under the standard normal curve, extensive tables have been compiled to determine the area under this curve without the necessity of going through actual computations. We know, for example, that approximately 68% (.6826) of the area lies between $\mu + \sigma$ and $\mu - \sigma$. Also, approximately 14% (.1359) is between $\mu + \sigma$ and $\mu + 2\sigma$, and 2% (.0215) between $\mu + 2\sigma$ and $\mu + 3\sigma$. These areas are depicted in Figure 8.4.

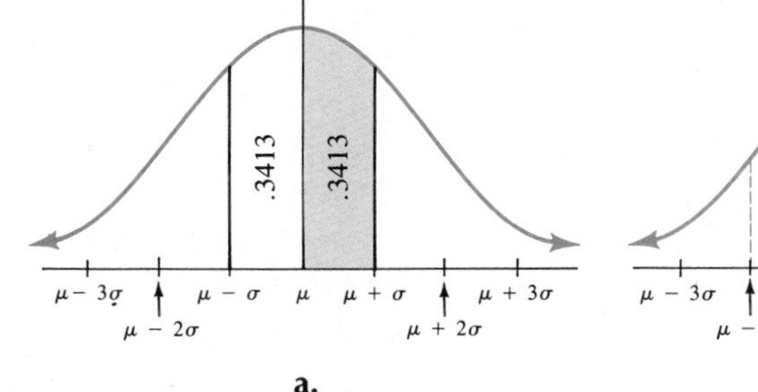

a.

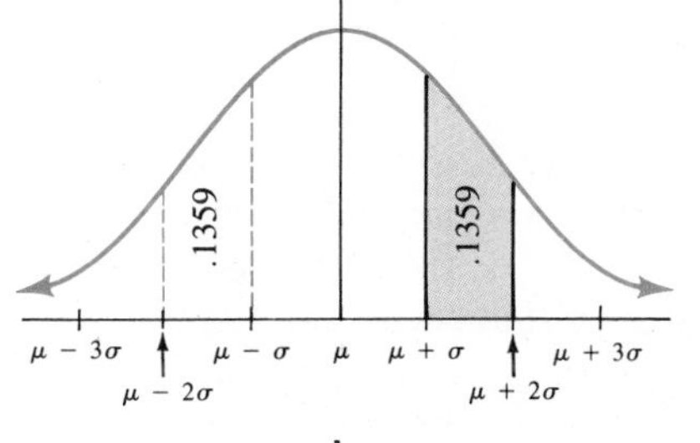

b.

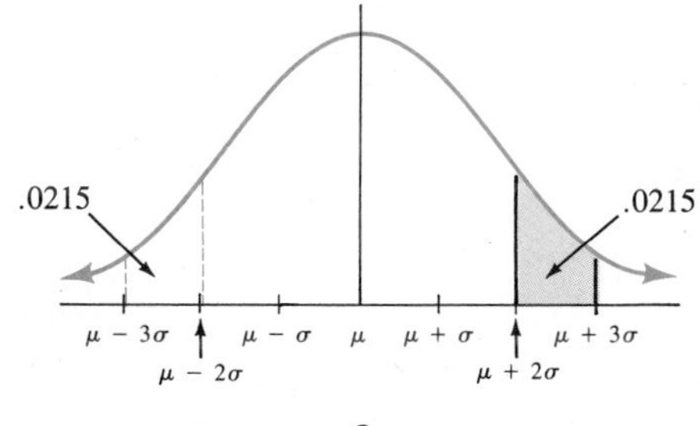

c.

FIGURE 8.4
Areas under a normal curve

For additional areas, use Table 4 in Appendix E. The table is arranged to give the area under the **standard normal curve** to the left of a vertical line through some number z, where $\mu = 0$ and $\sigma = 1$. Thus, to verify the area between $\mu + \sigma$ and $\mu + 2\sigma$ (shaded in color in Figures 8.4b and 8.5), you need to find the area to the left of $z = 2$ and then subtract the area to the left of $z = 1$. From the table, the area to the left of $z = 2$ is .9772 and the area to the left of $z = 1$ is .8413.

Therefore the area between $z = 2$ and $z = 1$ is

$$.9772 - .8413 = .1359$$

or about 14% of the area under the curve (as stated earlier).

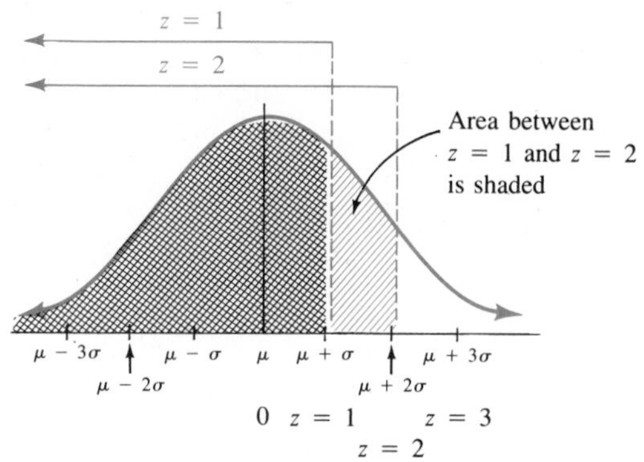

FIGURE 8.5
Standard normal curve

EXAMPLE 1 Find the area under the standard normal curve to the left of $z = .57$.

Solution From Table 4 in Appendix E, it is .7157. ■

EXAMPLE 2 Find the area under the standard normal curve to the right of $z = -.13$.

Solution From Table 4, the area to the *left* of $-.13$ is .4483, so the area to the right is

$$1 - .4483 = .5517$$

■

EXAMPLE 3 Find the area under the standard normal curve between $z = -.05$ and $z = .93$.

Solution The area to the left of $-.05$ is .4801 and to the left of .93 it is .8238, so the area between those values is

$$.8238 - .4801 = .3437$$

■

What if we are not working with a standard normal curve? That is, suppose the curve is a normal curve, but it does not have a mean of 0. We can use the information in Figure 8.4, which, for convenience, is summarized in Figure 8.6.

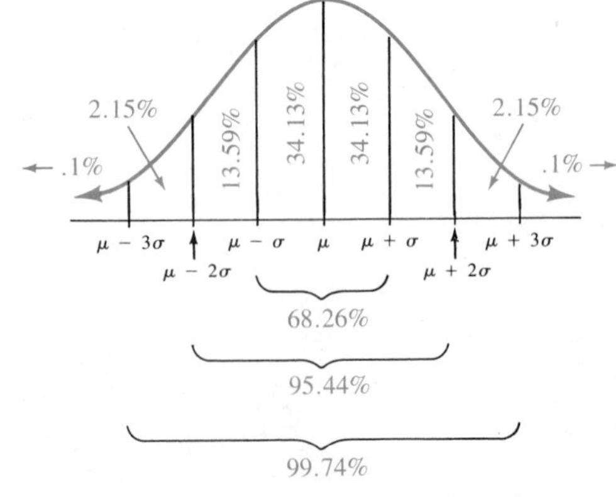

FIGURE 8.6

EXAMPLE 4 A teacher claims to grade "on a curve." This means the teacher believes that the scores on a given test are normally distributed. If 200 students take the exam, with mean 73 and standard deviation 9, how would the teacher grade the students?

Solution First, draw a normal curve with a mean 73 and standard deviation 9, as shown below:

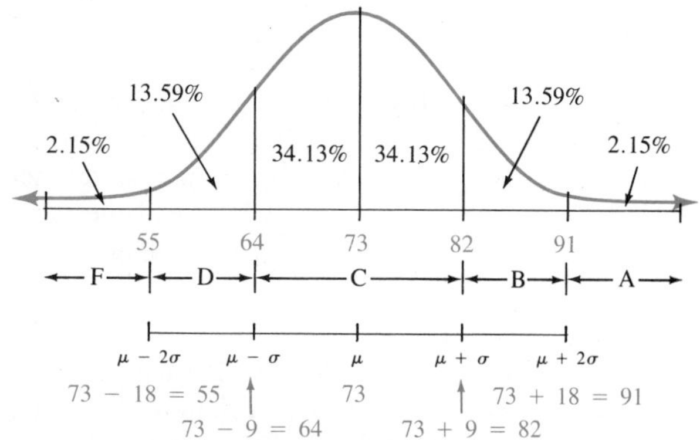

The range of 73 to 82 will contain about 34% of the class, and the range of 82 to 91 will contain about 14% of the class. Finally, about 2% of the class will score higher than 91. The teacher would therefore give grades according to the following table:

Grade on final	Letter grade	Number receiving grade	Percentage of class
Above 91	A	4	2%
83–91	B	28	14%
64–82	C	136	68%
55–63	D	28	14%
54 or below	F	4	2%

■

EXAMPLE 5 The Ridgemont Light Bulb Company tested a new line of light bulbs and found their life to be normally distributed, with a mean life of 98 hours and a standard deviation of 13.

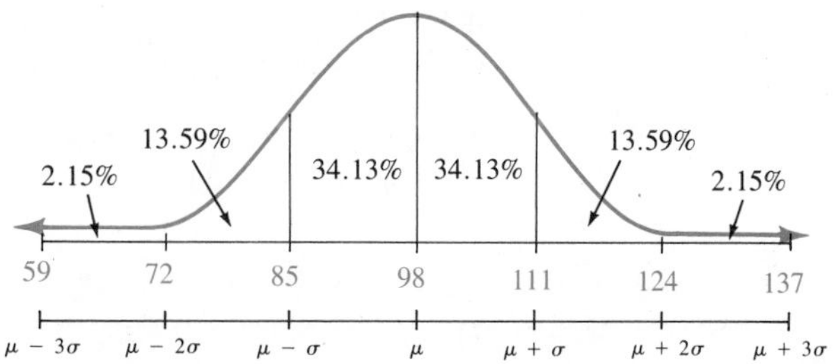

a. What percentage of bulbs with last less than 72 hours?

b. What is the probability that a bulb selected at random will last more than 111 hours?

Solution Draw a normal curve with mean 98 and standard deviation 13.

a. About 2% will last less than 72 hours.

b. We see that about 16% of the bulbs will last longer than 111 hours, so

$$P(X > 111) \approx .16$$

■

z-Scores

If you need to use a finer division of standard deviations than 1, 2, or 3, you will have to use calculus, or have a table available for that particular μ and σ, or use a number called a **z-score**. The z-score essentially translates any normal curve into a standard normal curve by use of the simple calculation given below. (This is why Table 4 in Appendix E uses z values.)

z-Score

The area to the left of a value x under a normal curve with mean μ and standard deviation σ is the same as the area under a standard normal curve to the left of the following value of z:

$$z = \frac{x - \mu}{\sigma}$$

EXAMPLE 6 Find the probability that one of the light bulbs described in Example 5 will last between 110 and 120 hours.

Solution From Example 5, $\mu = 98$ and $\sigma = 13$. Let $x = 110$; then

$$z = \frac{110 - 98}{13} \approx .92$$

Now, look up $z \approx .92$ in Table 4 (Appendix E). The area to the left of $x = 110$ is .8212.

For $x = 120$,

$$z = \frac{120 - 98}{13} \approx 1.69$$

Again, from Table 4, the area to the left of $x = 120$ is .9545 (find $z \approx 1.69$ in the table).

The area between $x = 110$ and $x = 120$ is

$$.9545 - .8212 = .1333$$

Thus $P(110 < X < 120) \approx .13$. ■

8.4 Problem Set

Find the area under the standard normal curve satisfying the conditions in Problems 1–16.

1. Left of -2
2. Left of 0
3. Left of 1
4. Left of 1.23
5. Left of $-.61$
6. Left of .81

7. Right of -2
8. Right of 1
9. Right of -1
10. Right of -1.73
11. Right of 1.69
12. Right of $-.11$
13. Between .5 and 1.61
14. Between $-.4$ and .4
15. Between -1.03 and 1.59
16. Between -2.8 and $-.46$

APPLICATIONS

In Problems 17–21 suppose that people's heights (in centimeters) are normally distributed, with a mean of 170 and a standard deviation of 5. We take a random sample of 50 persons.

17. How many would you expect to be between 165 and 175 centimeters tall?
18. How many would you expect to be taller than 160 centimeters?
19. How many would you expect to be taller than 175 centimeters?
20. If a person is selected at random, what is the probability that he or she is taller than 165 centimeters?
21. What is the variance (s^2) for this sample?

In Problems 22–26 suppose that, for a certain exam, a teacher grades on a curve. It is known that the mean is 50 and the standard deviation is 5. There are 45 students in the class.

22. How many students would receive a C?
23. How many students would receive an A?
24. What score would be necessary to obtain an A?
25. If an exam paper is selected at random, what is the probability that it will be a failing paper?
26. What is the variance for this exam?
27. The breaking strength of a rope (in pounds) is normally distributed, with a mean of 100 pounds and a standard deviation of 16. What is the probability that a certain rope will break with a force of 132 pounds?
28. The diameter of an electric cable is normally distributed, with a mean of .9 inch and a standard deviation of .01. What is the probability that the diameter will exceed .91 inch?
29. The annual rainfall in Ferndale, California, is known to be normally distributed, with a mean of 35.5 inches and a standard deviation of 2.5. About 2.15% of the time, the rainfall will exceed how many inches?
30. In Problem 29, what is the probability that the rainfall will exceed 30.5 inches?

If the life of a light bulb is normally distributed with a mean life of 250 hours and a standard deviation of 25 hours, find the probabilities requested in Problems 31–36.

31. $P(X > 250)$
32. $P(X > 300)$
33. $P(X < 220)$
34. $P(200 < X < 300)$
35. $P(220 \le X \le 320)$
36. $P(230 \le X \le 240)$
37. The diameter of a pipe is normally distributed, with a mean of .4 inch and a variance of .0004. What is the probability that the diameter will exceed .44 inch?
38. The breaking strength of a certain new synthetic material is normally distributed, with a mean of 165 pounds and a variance of 9. The material is considered defective if the breaking strength is less than 159 pounds. What is the probability that a sample of one chosen at random will be defective?
39. Repeat Problem 37 to find the probability that the diameter will exceed .43 inch.
40. Repeat Problem 38 to find the probability that the breaking strength of a sample is between 170 and 180 pounds.

8.5 Normal Approximation to the Binomial

The normal distribution in Section 8.4 was introduced by looking at a histogram showing IQ frequencies. In this section we will look at the relationship between the binomial and normal distributions.

EXAMPLE 1 Consider an experiment of tossing a coin 16 times. What is the probability of obtaining exactly x heads for $x = 0, 1 2, \ldots, 15, 16$?

Solution This is a binomial distribution for which $n = 16$, $p = \frac{1}{2}$, and $q = \frac{1}{2}$. The model is

$$P(X = x) = \binom{n}{x} p^x q^{n-x} = \binom{16}{x}\left(\frac{1}{2}\right)^x\left(\frac{1}{2}\right)^{16-x}$$

X	P(X)
0	.00002
1	.0002
2	.0018
3	.0085
4	.0278
5	.0667
6	.1222
7	.1746
8	.1964
9	**.1746**
10	.1222
11	.0667
12	.0278
13	.0085
14	.0018
15	.0002
16	.00002

The table in the margin shows the results of all the calculations for $x = 0, 1, 2, \ldots, 15$, and 16. Let us consider the procedure for one of these, say, $x = 9$ (see boldface entry):

$$
\begin{aligned}
P(X = 9) &= \binom{16}{9}\left(\frac{1}{2}\right)^9\left(\frac{1}{2}\right)^7 \\
&= \frac{16!}{(16-9)!9!}\left(\frac{1}{2}\right)^{16} \\
&= \frac{16!}{7!9!2^{16}} \\
&\approx .1746
\end{aligned}
$$

■

We can construct a histogram showing the results for Example 1. In Figure 8.7, which shows this, we use $2^{16} = 65{,}536$ for the total number of possibilities (2^{16} by the multiplication principle).

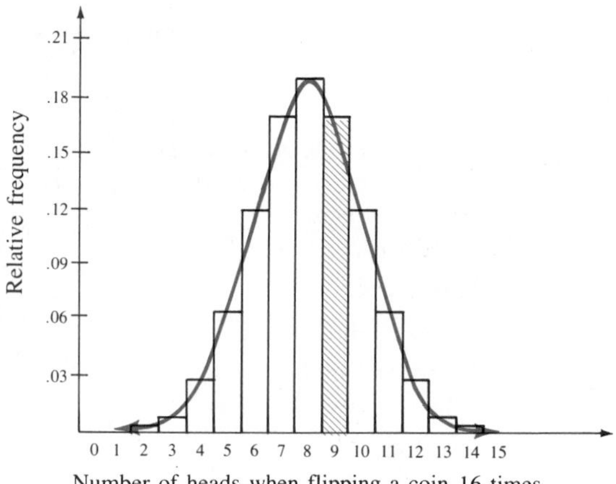

FIGURE 8.7
Histogram for the binomial distribution in Example 1

Note that we have superimposed a normal curve over the histogram in Figure 8.7. Under certain conditions a normal curve can be used to approximate a binomial distribution. This relationship was first noted by the mathematician Abraham De Moivre in 1718. Suppose we use the normal curve to approximate the binomial distribution in Example 1. We first find the mean and the standard deviation:

$$
\begin{aligned}
\mu &= np & \sigma &= \sqrt{npq} \\
&= 16\left(\frac{1}{2}\right) & &= \sqrt{16\left(\frac{1}{2}\right)\left(\frac{1}{2}\right)} \\
&= 8 & &= \sqrt{4} \\
& & &= 2
\end{aligned}
$$

The normal curve shown in Figure 8.7 has mean 8 and standard deviation 2. The important difference between the binomial and normal distributions is that the binomial is *discrete* (or finite) and the normal is *continuous* (or infinite). If we were to find the probability of $x = 9$ for a normal distribution, we would find it to be 0. However, take a closer look at Figure 8.7, as shown in Figure 8.8.

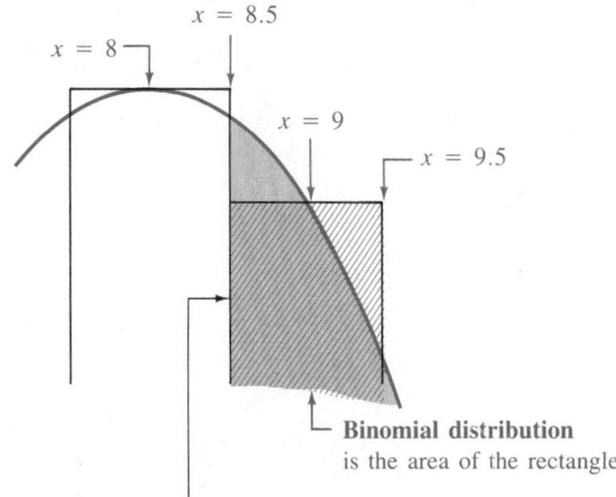

Normal probability is the area from $x = 8.5$ to $x = 9.5$ under the curve. If this is used to approximate the area of the rectangle, it is called the **split unit cutoff method.**

FIGURE 8.8
Detail of Figure 8.7

The z-score for $x = 9.5$ is

$$z = \frac{x - \mu}{\sigma} = \frac{9.5 - 8}{2} = .75$$

The Table 4 entry for $z = .75$ is .7734.

The z-score for $x = 8.5$ is

$$z = \frac{8.5 - 8}{2} = .25$$

The Table 4 entry for $z = .25$ is .5987.

The area between $x = 9.5$ and $x = 8.5$ is therefore

$$.7734 - .5987 = .1747$$

Note that this is a *very* good approximation for $P(X = 9)$ in the binomial distribution found in Example 1. This method of approximating the binomial probabilities is called the **split unit cutoff method**.

The Normal Distribution as an Approximation to the Binomial Distribution

If X is a binomial random variable of n independent trials, each with probability of success p and failure $q = 1 - p$, and if

$$np \geq 5 \quad \text{and} \quad nq \geq 5$$

then the binomial random variable is approximated by a normal distribution with mean and standard deviation given by

$$\mu = np \qquad \sigma = \sqrt{npq}$$

To show the tremendous value of this approximation, consider another, more realistic example.

EXAMPLE 2 Experimental data indicate that the cure rate of a new drug is 85%. If the drug is given to 100 patients, find the probability that at least 90 of them will be cured.

Solution This is a binomial distribution with $p = .85$, $q = .15$, and $n = 100$. We want to find $P(X \geq 90)$.

$$P(X \geq 90) = P(X = 90) + P(X = 91) + \cdots + P(X = 99) + P(X = 100)$$

Now,

$$P(X = 90) = \binom{100}{90}(.85)^{90}(.15)^{10}$$

$$P(X = 91) = \binom{100}{91}(.85)^{91}(.15)^{9}$$

$$P(X = 100) = \binom{100}{100}(.85)^{100}(.15)^{0}$$

These calculations are certainly out of hand (they are even out of the range of what most calculators can handle). However,

$$np = 100(.85) = 85 \geq 5 \qquad \text{and} \qquad nq = 100(.15) = 15 \geq 5$$

so we can use the normal distribution as an approximation.

$$\begin{aligned} \mu &= np & \sigma &= \sqrt{npq} \\ &= 85 & &= \sqrt{100(.85)(.15)} \\ & & &\approx 3.571 \end{aligned}$$

Now, $P(X \geq 90) = 1 - P(X < 90)$, so for $x = 90$, we use the z-score for 89.5 (split unit cutoff method) and Table 4 (Appendix E):

$$\begin{aligned} z &= \frac{x - \mu}{\sigma} \\ &= \frac{89.5 - 85}{3.571} \\ &\approx 1.26 \end{aligned}$$

From Table 4, $P(z < 1.26) = .8962$.

Then

$$\begin{aligned} P(X \geq 90) &\approx 1 - .8962 \\ &= .1038 \end{aligned}$$

Note that for $P(X < 90)$ we used the split unit 89.5. For $P(X \leq 90)$ we would have used the split unit 90.5. ■

TABLE 8.1 Minimum Sample Required to Approximate a Binomial Distribution by a Normal Distribution

p	n must be at least
.001	5,000
.01	500
.1	50
.2	25
.3	17
.4	13
.5	10
.6	13
.7	17
.8	25
.9	50
.99	500
.999	5,000

In Example 2 we calculated np and nq in order to see if we could apply a normal approximation. Table 8.1 provides a quick reference as to when we may use this approximation. For Example 2, we could have used $p = .9$ and the table to find that n must be at least 50 to conclude that the normal is an acceptable approximation.

EXAMPLE 3 In certain grades in elementary school, students are given a vision test. Experience has shown that 18% of the students do not pass the test. Find the probability that if the test is given to 100 students, *exactly* 18 will fail the test.

Solution This is a binomial distribution with $n = 100$ and $p = .18$.

$$P(X = 18) = \binom{100}{18}(.18)^{18}(.82)^{82}$$

This is a formidable calculation (even with a calculator; on my TI-81 I obtain .1033420394). However, Table 8.1 indicates that we can use a normal approximation with

$$\begin{aligned} \mu &= np & \sigma &= \sqrt{npq} \\ &= (100)(.18) & &= \sqrt{100(.18)(.82)} \\ &= 18 & &\approx 3.84 \end{aligned}$$

We cannot find $x = 18$ exactly, so we find the difference of $x = 18.5$ and $x = 17.5$:

For $x = 18.5$: *For* $x = 17.5$:

$$z = \frac{18.5 - 18}{3.84} \approx .13 \qquad z = \frac{17.5 - 18}{3.84} \approx -.13$$

From Table 4,

$$P(X \le 18.5) \approx .5517 \qquad P(X \le 17.5) \approx .4483$$

Thus

$$\begin{aligned} P(17.5 \le X \le 18.5) &\approx .5517 - .4483 \\ &= .1034 \end{aligned}$$

This means that there is about a 10% probability that exactly 18 out of the 100 students will fail the test. ■

8.5 Problem Set

For the given values associated with a binomial experiment in Problems 1–9, determine whether the normal distribution is a suitable approximation.

1. $n = 1{,}000$, $p = .01$
2. $n = 20$, $p = .5$
3. $n = 10$, $p = .6$
4. $n = 15$, $p = .6$
5. $n = 100$, $p = .4$
6. $n = 50$, $p = .1$
7. $n = 1{,}000$, $p = .04$
8. $n = 12$, $p = .6$
9. $n = 40$, $p = .09$

A coin is tossed 15 times. Let X be a random variable representing the number of tails. Find the probabilities in Problems 10–21.

10. $P(X < 8)$
11. $P(X < 10)$
12. $P(X < 11)$
13. $P(X > 13)$
14. $P(X > 6)$
15. $P(X > 9)$
16. $P(7 < X < 9)$
17. $P(4 \le X \le 8)$
18. $P(5 \le X \le 7)$
19. $P(X = 5)$
20. $P(X = 11)$
21. $P(X = 14)$

A die is tossed 1,000 times. Let X be the number of 6s tossed. Find the probabilities in Problems 22–33.

22. $P(X < 200)$
23. $P(X \le 170)$
24. $P(X \le 190)$
25. $P(X \ge 150)$
26. $P(X \ge 140)$
27. $P(X > 160)$

28. $P(140 \le X \le 150)$

29. $P(150 < X < 160)$

30. $P(140 < X < 150)$

31. $P(X = 140)$

32. $P(X = 150)$

33. $P(X = 160)$

APPLICATIONS

A new drug has a cure rate of 92%. The drug is administered to 1,000 patients. Let X be the number who are cured by the drug. Find the probabilities in Problems 34–45.

34. $P(X < 900)$

35. $P(X < 950)$

36. $P(X \le 850)$

37. $P(X \ge 920)$

38. $P(X > 900)$

39. $P(X > 910)$

40. $P(900 \le X \le 1{,}000)$

41. $P(850 < X < 950)$

42. $P(800 < X < 950)$

43. $P(X = 900)$

44. $P(X = 920)$

45. $P(X = 930)$

46. An experiment in parapsychology consists of correctly identifying one of five shapes. If the test is repeated 10 times, what is the probability that the subject will correctly identify the object more than 8 times?

47. A baseball player's batting average is .250. What is the probability of more than six hits in the next ten times at bat?

48. In 1990 the yearly death rate from cancer was 186.3 per 100,000. If a person associates with roughly 500 people, estimate the probability that more than 1 out of these 500 will die of cancer in a year.

49. Repeat Problem 48 to find the probability that more than 2 of the 500 will die in a year.

50. A computer system is made up of 100 components, each with a reliability of .95 (that is, the probability that the component operates properly is .95). If these components function independently of one another, and the computer requires at least 80 components to function properly, what is the reliability of the whole system?

51. An airline is penalized for overbooking, but loses money with empty seats. Suppose the records show that for a certain flight, 8% of the advance reservations do not show up. If the plane seats 300 people, find the probability that if the airline takes 315 advance reservations, more than 300 will show up.

*8.6 Correlation and Regression

Correlation

Mathematical modeling relates numerical data to assumptions about the relationship between two variables. For example:

IQ and salary
Study time and grades
Age and heart disease
Runner's speed and runner's brand of shoes
Teachers' salaries and beer consumption

All are attempts to relate two variables in some way or another. If there is a relationship between those variables we say that there is a **correlation**. Once a correlation is established, then the next step in the modeling process is to identify the nature of the relationship. This is called **regression analysis**. We now turn our attention to finding a *linear relationship* that is called the *best-fitting line* by a technique called the **least squares method**. The derivations of the results and the formulas given in this section are, for the most part, based on the calculus developed later in Section 16.3 and will therefore not be done here. We will, however,

* The material of this section is not required for understanding of subsequent material.

focus on how to use the formulas as well as discuss what they mean and how they are interpreted.

We begin with correlation. We want to know if two variables are related—that is, depend on one another. Let us call one variable x and the other y. These variables can be represented as ordered pairs (x, y) in a graph called a **scatter diagram**.

EXAMPLE 1 A survey (random sample) of 20 students compared the grade received on an examination with the length of time the student studied. Draw a scatter diagram to represent the data in the table.

Student Number	1	2	3	4	5	6	7	8	9	10
Length of Study Time (to nearest 5 minutes)	30	40	30	35	45	15	15	50	30	0
Grade (100 possible)	72	85	75	78	89	58	71	94	78	10
Student Number	11	12	13	14	15	16	17	18	19	20
Length of Study Time (to nearest 5 minutes)	20	10	25	25	25	30	40	35	20	15
Grade (100 possible)	75	43	68	60	70	68	82	75	65	62

Solution Let x be the study time (in minutes) and let y be the grade (in points). The graph is shown below:

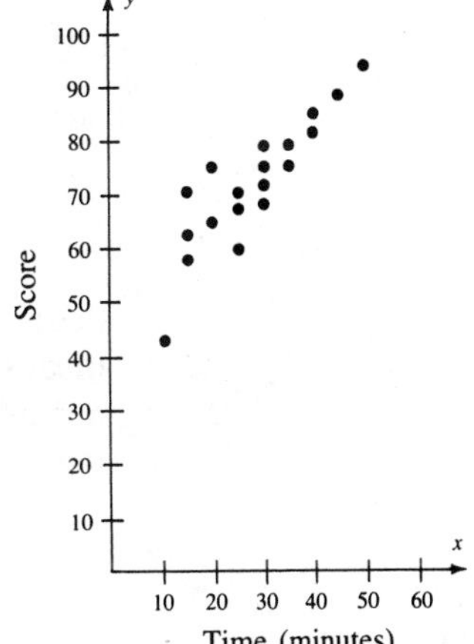

■

Correlation is a measure to determine whether there is a statistically significant relationship between two variables. Intuitively, it should assign a measure consistent with the scatter diagrams shown in Figure 8.9 on page 414.

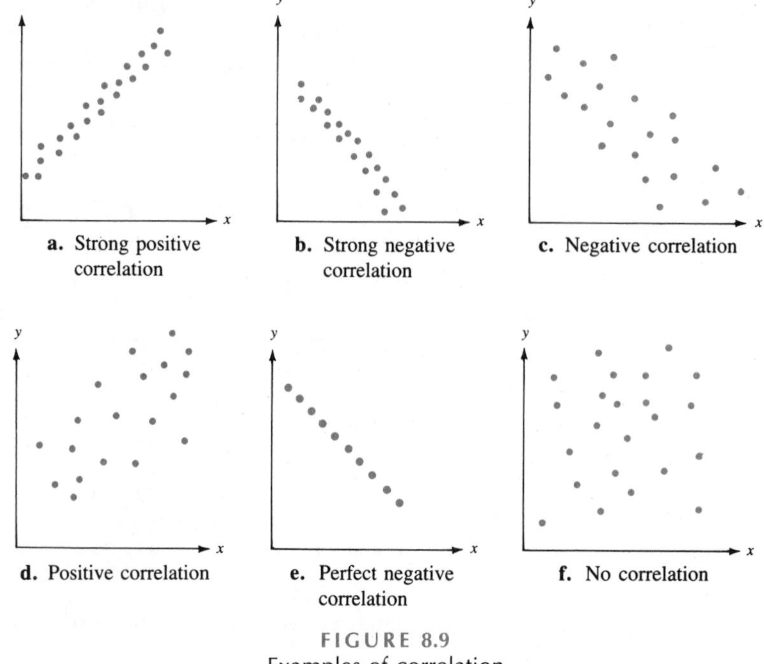

FIGURE 8.9
Examples of correlation

Such a measure, called the **linear correlation coefficient,** r, is defined so that it has the following properties:

1. r measures the correlation between x and y.
2. r is between -1 and 1.
3. If r is close to 0, it means there is little correlation.
4. If r is close to 1, it means there is a strong positive correlation.
5. If r is close to -1, it means there is a strong negative correlation.

To write a formula for r, we let n denote the number of pairs of data present; and, as before:

Σx	denotes the sum of the x values
Σx^2	means square the x values and then sum
$(\Sigma x)^2$	means sum the x values and then square
Σxy	means multiply each x value by the corresponding y value and then sum
$n\Sigma xy$	means multiply n by Σxy
$(\Sigma x)(\Sigma y)$	means multiply Σx by Σy

Correlation Coefficient

The **linear correlation coefficient** r is

$$r = \frac{n\Sigma xy - (\Sigma x)(\Sigma y)}{\sqrt{n(\Sigma x^2) - (\Sigma x)^2}\sqrt{n(\Sigma y^2) - (\Sigma y)^2}}$$

EXAMPLE 2 Find r for the data in Example 1.

Solution

	Study time, x	Score, y	xy	x^2	y^2
	30	72	2,160	900	5,184
	40	85	3,400	1,600	7,225
	30	75	2,250	900	5,625
	35	78	2,730	1,225	6,084
	45	89	4,005	2,025	7,921
	15	58	870	225	3,364
	15	71	1,065	225	5,041
	50	94	4,700	2,500	8,836
	30	78	2,340	900	6,084
	0	10	0	0	100
	20	75	1,500	400	5,625
	10	43	430	100	1,849
	25	68	1,700	625	4,624
	25	60	1,500	625	3,600
	25	70	1,750	625	4,900
	30	68	2,040	900	4,624
	40	82	3,280	1,600	6,724
	35	75	2,625	1,225	5,625
	20	65	1,300	400	4,225
	15	62	930	225	3,844
Total	535 ↑ Σx	1,378 ↑ Σy	40,575 ↑ Σxy	17,225 ↑ Σx^2	101,104 ↑ Σy^2

$$
\begin{aligned}
r &= \frac{n\Sigma xy - (\Sigma x)(\Sigma y)}{\sqrt{n(\Sigma x^2) - (\Sigma x)^2}\sqrt{n(\Sigma y^2) - (\Sigma y)^2}} \\
&= \frac{20(40{,}575) - (535)(1{,}378)}{\sqrt{20(17{,}225) - (535)^2}\sqrt{20(101{,}104) - (1{,}378)^2}} \\
&= \frac{74{,}270}{\sqrt{58{,}275}\sqrt{123{,}196}} \\
&\approx .8765
\end{aligned}
$$

■

TABLE 8.2 Correlation Coefficient

n	$\alpha = .05$	$\alpha = .01$
4	.950	.999
5	.878	.959
6	.811	.917
7	.754	.875
8	.707	.834
9	.666	.798
10	.632	.765
11	.602	.735
12	.576	.708
13	.553	.684
14	.532	.661
15	.514	.641
16	.497	.623
17	.482	.606
18	.468	.590
19	.456	.575
20	.444	.561
25	.396	.505
30	.361	.463
35	.335	.430
40	.312	.402
45	.294	.378
50	.279	.361
60	.254	.330
70	.236	.305
80	.220	.286
90	.207	.269
100	.196	.256

NOTE: The derivation of this table is beyond the scope of this course. It shows the critical values of the *Pearson correlation coefficient*.

Example 2 shows a very strong positive correlation. But if r for Example 2 had been .46, would we still be able to assume that there is a strong correlation? This question is a topic of major concern in statistics. The term **significance level** is used to denote the cutoff between results attributed to chance and the results attributed to significant differences. Table 8.2 gives **critical values** for determining whether two variables are correlated. If $|r|$ is greater than the given table value, then we may assume that a correlation exists between the variables. If we use the column labeled $\alpha = .05$, then the significance level is 5%. This means that the probability is .05 that we will say the variables are correlated when, in fact, the results are attributed to chance. This is also true for a significance level of 1% ($\alpha = .01$). For Example 2, since $n = 20$, we see in Table 8.2 that $r = .46$ would show a linear correlation at a 5% significance level, but not at a 1% level.

EXAMPLE 3 Find the critical value of the linear correlation coefficient for 10 pairs of data and a significance level of .05.

Solution From Table 8.2, the critical value is $r = .632$. For $n = 10$, any value greater than $r = .632$ is considered linearly correlated. ■

EXAMPLE 4 If $r = -.85$ and $n = 10$, are the variables correlated at a significance level of 1%?

Solution For $n = 10$ and $\alpha = .01$, the Table 8.2 entry is .765. Since r is negative and since $|r| > .765$, we see that there is a negative linear correlation. ■

EXAMPLE 5 The following table is a sample of some past annual mean salaries for teachers in elementary and secondary schools, along with the annual per capita beer consumption (in gallons) for Americans. Find the correlation coefficient.*

Year	1960	1965	1970	1972	1973	1983
Mean Teacher Salary	\$5,000	\$6,200	\$8,600	\$9,700	\$10,200	\$16,400
Per Capita Beer Consumption (gal)	24.02	25.46	28.55	29.43	29.68	35.2

Solution

$$n = 6$$
$$\Sigma x = 56{,}100$$
$$\Sigma y = 172.34$$
$$\Sigma x^2 = 604{,}490{,}000$$
$$\Sigma y^2 = 5{,}026.3418$$
$$(\Sigma x)^2 = 3{,}147{,}210{,}000$$
$$(\Sigma y)^2 = 29{,}701.0756$$
$$\Sigma xy = 1{,}688{,}969$$
$$r = \frac{6(1{,}688{,}969) - (56{,}100)(172.34)}{\sqrt{6(604{,}490{,}000) - (56{,}100)^2}\sqrt{6(5{,}026.3418) - (172.34)^2}}$$
$$\approx .994$$

The number $r \approx .994$ is so close to 1 that we hardly need to consult Table 8.2 to know that there is a strong positive linear correlation between teachers' salaries and per capita beer drinking. ■

The significance of the correlation implies that teachers use their raises to buy more beer, right? Wrong. Perhaps increases in teachers' salaries precipitate higher

* Example 5 and the paragraph following are from Mario F. Triola, *Elementary Statistics*, 2nd ed. (Menlo Park, Calif.: Benjamin/Cummings, 1983). © 1983 by The Benjamin/Cummings Publishing Company, Inc. Reprinted by permission.

taxes, which in turn cause taxpayers to drown their sorrows and forget their financial difficulties by drinking more beer. Or perhaps higher teachers' salaries and greater beer consumption are both manifestations of some other factor, such as a general improvement in the standard of living. In any event, the techniques in this chapter can be used only to establish a *statistical* linear relationship. *We cannot establish the existence or absence of any inherent cause-and-effect relationship.*

Best-Fitting Line

The final step in our discussion is to find the best-fitting line. That is, we want to find a line $y' = mx + b$ so that the sum of the distances of the data points from this line will be as small as possible. (We use y' instead of y to distinguish between the actual second component, y, and the predicted y value, y'.) Since some of these distances may be positive and some negative and since we do not want large opposites to "cancel each other out," we minimize the sum of the *squares* of these distances. Therefore, the **regression line** is sometimes called the **least squares line**.

Least Squares, or Regression, Line

The **least squares**, or **regression, line** $y' = mx + b$ is the line of best fit when

$$m = \frac{n(\Sigma xy) - (\Sigma x)(\Sigma y)}{n(\Sigma x^2) - (\Sigma x)^2} \qquad b = \frac{\Sigma y - m(\Sigma x)}{n}$$

EXAMPLE 6 Find the best-fitting line for the data in Example 1.

Solution From Example 2, $n = 20$, and

$$\Sigma xy = 40{,}575$$
$$\Sigma x = 535$$
$$\Sigma y = 1{,}378$$
$$\Sigma x^2 = 17{,}225$$
$$(\Sigma x)^2 = (535)^2$$

Since m is used as part of the formula for b, you must first find m:

$$m = \frac{20(40{,}575) - (535)(1{,}378)}{20(17{,}225) - (535)^2} = \frac{74{,}270}{58{,}275} \approx 1.27447$$

Now you can use this m when finding b:

$$b = \frac{1{,}378 - 1.27447(535)}{20} \approx 34.8078$$

Thus we may approximate the least squares line as $y' = 1.3x + 35$. This line is shown in the figure in the margin. ■

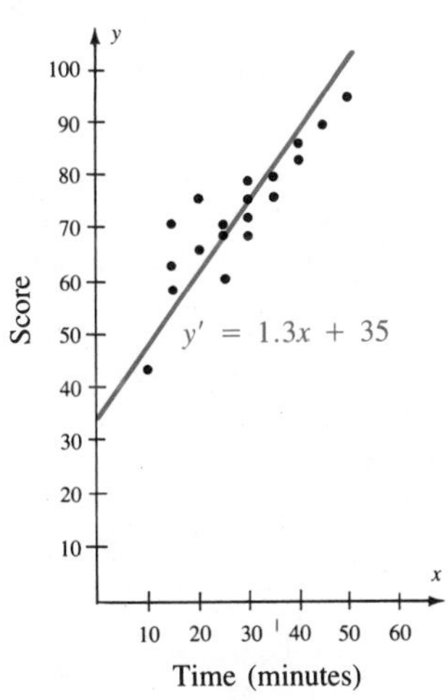

Regression line for the data in Example 1

EXAMPLE 7 Use the regression line of Example 6 to predict the score of a person who studied $\frac{1}{2}$ hour.

Solution $x = 30$ minutes, so $y' = 1.3(30) + 35 = 74$ ■

A final word of caution: Use the regression line only if r indicates that there is a significant linear correlation, as given in Table 8.2.

8.6 Problem Set

In Problems 1–12 a sample of paired data gives a linear correlation coefficient r. In each case use Table 8.2 to determine whether there is a significant linear correlation.

1. $n = 10,\quad r = .7,$ significance level 5%
2. $n = 10,\quad r = .7,$ significance level 1%
3. $n = 30,\quad r = .4,$ significance level 1%
4. $n = 30,\quad r = .4,$ significance level 5%
5. $n = 15,\quad r = -.732,$ significance level 5%
6. $n = 35,\quad r = -.4127,$ significance level 1%
7. $n = 50,\quad r = -.3416,$ significance level 1%
8. $n = 100,\quad r = -.41096,$ significance level 5%
9. $n = 23,\quad r = .501,$ significance level 1%
10. $n = 38,\quad r = .416,$ significance level 5%
11. $n = 28,\quad r = -.214,$ significance level 5%
12. $n = 55,\quad r = -.14613,$ significance level 1%

Draw a scatter diagram and find r for the data in each table in Problems 13–18 and determine whether there is a linear correlation at either the 5% or 1% level.

13.

x	1	2	3	4
y	1	5	8	13

14.

x	4	5	10	10
y	0	−10	−10	−20

15.

x	0	1	2	3	4
y	25	19	16	12	10

16.

x	1	3	3	5	8
y	30	22	19	15	10

17.

x	10	20	30	30	50	60
y	50	48	60	58	70	75

18.

x	85	90	100	102	105	110
y	80	40	30	28	25	15

In Problems 19–24 find the regression line for the indicated problem.

19. Problem 13
20. Problem 15
21. Problem 17
22. Problem 14
23. Problem 16
24. Problem 18

APPLICATIONS

25. A new computer circuit was tested and the times (in nanoseconds) required to carry out different subroutines were recorded as follows:

Difficulty Level	1	2	2	3	4	5	5	5
Time	10	11	13	8	15	18	21	19

Find r and determine whether it is statistically significant at the 1% level.

26. Ten people are given a standard IQ test. Their scores were then compared with their high school grades:

IQ	Grade (GPA)
117	3.1
105	2.8
111	2.5
96	2.8
135	3.4
81	1.9
103	2.1
99	3.2
107	2.9
109	2.3

Find r and determine whether it is statistically significant at the 1% level.

27. Find the regression line for the data in Problem 25.

28. Find the regression line for the data in Problem 26.

29. The following data represent the number of years of full-time education (x) and the annual salary in thousands of dollars (y) for 15 persons. Is there a correlation between these variables, and if so, is it at the 5% or the 1% significance level?

Education:	20	27	28	18	13	18	9	16
Salary:	35.2	24.6	23.7	33.3	24.4	33.4	11.2	32.3

Education:	16	12	12	19	16	14	13
Salary:	25.1	22.1	18.9	37.8	25.9	28.4	29.6

30. The following data show measurements of temperature (°F) and chirping frequency (in chirps per second) for the striped ground cricket. Is there a correlation between these variables, and if so, is it at the 5% or the 1% significance level?

Temperature:	31.4	22.0	34.1	29.1	27.0	24.0	20.9	27.8
Frequency:	20.0	16.0	19.8	18.4	17.1	15.5	14.7	17.1

Temperature:	20.8	28.5	26.4	28.1	27.0	28.6	24.6
Frequency:	15.4	16.2	15.0	17.2	16.0	17.0	14.4

31. A bank records the number of mortgage applications and its own prevailing interest rate (at the first of the month) for each of 16 consecutive months. Is there a correlation between these variables, and if so, it is at the 5% or the 1% significance level?

Interest:	9.5	9.9	10.0	10.5	11.0	11.5	11.0	12.0
Number:	27	29	25	25	19	20	17	13

Interest:	12.0	12.5	13.0	13.5	13.0	12.5	11.5	11.5
Number:	15	10	10	6	5	5	11	14

32. A researcher chooses and interviews a group of 15 male workers in an automobile plant. The researcher then gives a score ranging from 1 to 20 based on a scale of patriotism—the higher the score, the more patriotic the person appeared to be. Each person is then given a written test and is scored on their patriotism. Is there a correlation between the researcher score and the test score, and if so, is it at the 5% or the 1% significance level?

Researcher:	10	14	15	17	17	18	18	19
Test:	15	12	19	8	9	16	17	6

Researcher:	16	18	20	12	14	9	17
Test:	11	14	12	11	10	12	6

33. Find the best-fitting line for the data in Problem 29.

34. Find the best-fitting line for the data in Problem 30.

35. Find the best-fitting line for the data in Problem 31.

36. Find the best-fitting line for the data in Problem 32.

*8.7 Review

The material of this chapter is reviewed in the following list of objectives. After each objective there are some practice questions. For a sample test select the first question of each set and check your answers. The second question for each objective has no answer given. If you are having trouble with a particular type of problem, look back at the indicated section in the text. When you are finished reviewing these objectives, a sample examination is given at the end of this section.

* Optional section.

[8.1]
Objective 8.1: *Prepare a frequency distribution.*

1. Twenty-five students obtained the following scores on a recent mathematics examination:

75 70 70 65 75 80 70 75 60 65 65 90 40
70 60 65 65 75 85 80 55 75 85 55 60

2. The following outcomes were obtained when rolling a single die 50 times.

3 2 5 6 6 1 4 3 2 1 4 1 2 1 3 6 5
6 3 1 4 1 2 6 5 6 5 1 2 1 2 3 6 4
4 3 1 5 2 2 1 6 4 3 6 6 3 6 6 3

3. The wages of the employees of a small accounting firm are as follows:

\$40,000	\$45,000	\$45,000	\$45,000
\$20,000	\$20,000	\$25,000	\$40,000
\$30,000	\$18,000	\$18,000	\$18,000
\$18,000	\$18,000	\$16,000	\$14,000

4. The divorce rates by state in a given year were as follows:

Alabama	7.0	Kentucky	4.5	North Dakota	3.2
Alaska	8.6	Louisiana	3.8	Ohio	5.5
Arizona	8.2	Maine	5.6	Oklahoma	7.9
Arkansas	9.3	Maryland	4.1	Oregon	7.0
California	6.1	Massachusetts	3.0	Pennsylvania	3.4
Colorado	6.0	Michigan	4.8	Rhode Island	3.9
Connecticut	4.5	Minnesota	3.7	South Carolina	4.7
Delaware	5.3	Mississippi	5.6	South Dakota	3.9
D. C.	6.8	Missouri	5.7	Tennessee	6.8
Florida	7.9	Montana	6.5	Texas	6.9
Georgia	6.5	Nebraska	4.0	Utah	5.6
Hawaii	5.5	Nevada	16.8	Vermont	4.6
Idaho	7.1	New Hampshire	5.9	Virginia	4.5
Illinois	4.6	New Jersey	3.2	Washington	6.9
Indiana	7.7	New Mexico	8.0	West Virginia	5.3
Iowa	3.9	New York	3.7	Wisconsin	3.6
Kansas	5.4	North Carolina	4.9	Wyoming	7.8

Prepare a frequency distribution by rounding to the nearest percent (7.0 means 7.0%; 8.6% means 8.6% or 9% rounded to the nearest percent).

Objective 8.2: *Find the probability distribution for a given set of data.*

5. Use the data from Problem 1.

6. Use the data from Problem 2.

7. Use the data from Problem 3.

8. Use the data from Problem 4.

Objective 8.3: *Represent data using a histogram.*

9. Use the data from Problem 1.

10. Use the data from Problem 2.

11. Use the data from Problem 3.

12. Use the data from Problem 4.

[8.2]

Objective 8.4: *Find the mean, median, and mode for a set of data.*

13. 8, 12, 19, 24, 10, 24, 15

14. 816, 817, 818, 819, 820, 821, 822

15. 15,000; 21,000; 45,000; 40,000; 21,000

16. 18, 22, 24, 18, 25, 24, 28, 30

Objective 8.5: *Find the mean, median, and mode for a frequency distribution.*

17. Use the data from Problem 1.

18. Use the data from Problem 2.

19. Use the data from Problem 3.

20. Use the data from Problem 4.

Objective 8.6: *Find the range, variance, and standard deviation for a set of data.*

21. Use the data from Problem 13.

22. Use the data from Problem 2.

23. Roll a die and let a random variable represent the number of dots on the top of the die.

24. Use the following probability distribution:

X	$P(X)$
0	.1
2	.3
4	.4
6	.2

[8.3]

Objective 8.7: *Find probabilities using a binomial probability distribution.*

25. *Quality control.* An aircraft has built-in redundancy. Suppose that at least two of five similar components must be functioning in order to operate the aircraft. If the probability of failure of any of these independently operating components is .01, what is the probability that the aircraft will fail due to failure of these components?

26. Use Table 3 in Appendix E to compute a binomial probability where $n = 15$, $k = 4$, and $p = .7$.

27. What is the probability that a baseball player with a .250 batting average gets exactly two hits for four times at bat? (Assume that the events are independent.)

28. *Psychology experiment.* Twenty subjects participated in a study to determine whether a graphologist could distinguish a normal person's handwriting from that of a psychotic. Ten normals were matched by age, sex, and education to ten psychotics, and then handwriting samples of each matched pair were placed in ten folders. The graphologist was presented with the ten folders and was to pick out the sample in each folder that had been written by a psychotic. If

$$p = P(\text{Any particular psychotic/normal pair is correctly identified})$$

then $p = \frac{1}{2}$ assumes no special ability on the part of the graphologist. If X is the number of correctly identified psychotic/normal pairs and if $p = \frac{1}{2}$, find $P(X > 7)$.

Objective 8.8: *Find the mean, variance, and standard deviation for a binomial distribution.*

29. *Quality control.* If 10,000 aircraft of the type described in Problem 25 are in service, what is the mean number of aircraft failures? Also find the variance and standard deviation.

30. Find the mean, variance, and standard deviation for the distribution described in Problem 26.

31. Find the mean number of hits for 50 times at bat for the baseball player described in Problem 27. Also find the variance and standard deviation.

32. Find the mean, variance, and standard deviation for the number of correctly identified pairs in the experiment described in Problem 28.

[8.4]

Objective 8.9: *Find the area under different parts of the standard normal curve.*

33. Right of 1.5

34. Left of -2.5

35. Between $-.5$ and $.5$

36. Between -1.2 and 1.7

Objective 8.10: *Find the area under different parts of a normal curve given the mean and standard deviation.*

37. $\mu = 55, \sigma = 10$; to the right of 60

38. $\mu = 75, \sigma = 8$; to the left of 70

39. $\mu = 130, \sigma = 15$; between 120 and 140

40. $\mu = 67.4, \sigma = 6.37$; between 60 and 70

Objective 8.11: *Solve applied problems using the normal distribution.*

41. Suppose a teacher grades on a curve. If the mean is 75 and the standard deviation is 8, make up a grading distribution. If there were 50 people in the class, how many could expect to get As, Bs, etc.?

42. *Quality control.* The mean life expectancy for a certain pair of shoes is 15 months, and the standard deviation is 4 months. If the manufacturer is contemplating offering a 1-year money-back guarantee, how many pairs could be expected to wear out if 10,000 pairs are sold?

[8.5]

Objective 8.12: *Use the normal approximation as an approximation for a binomial distribution.*

43. Suppose two dice are rolled 500 times. What is the probability that a sum of 7 is tossed more than 100 times?

44. *Side effects of a drug.* A drug has side effects for 1 person out of 1,000 taking the drug. If the drug is administered to 155,000 people, what is the probability that more than 180 people will show side effects?

45. *Quality control.* A manufacturer of microwave ovens produces 5,000 units per day. Quality control studies have shown that 6% are defective and cannot be sold. A random sample of 100 ovens are selected from each day's production and are tested. What is the probability that there are more than 10 defective ovens in the sample?

46. *Quality control.* Repeat Problem 45 to find the probability that there is at least 1 but not more than 6 defective ovens.

[8.6]

Objective 8.13: *Draw a scatter diagram for a set of data.*

47. A study was conducted to test the relationship between speed (mph) and fuel consumption (mpg). The following information was obtained:

Speed (mph)	20	30	40	50	60	70
Fuel Consumption (mpg)	24	28	29	27	23	19

48. Compare IQ with productivity on an assembly line.

IQ	105	87	110	101	85
Productivity	83	110	81	90	103
IQ	90	92	109	95	97
Productivity	109	111	80	105	103

49. Compare age with blood pressure.

Age	20	25	30	40	50	35	68	55
Blood Pressure	85	91	84	93	100	86	94	92

Objective 8.14: *Find the linear correlation coefficient and determine whether the variables are significantly correlated at either the 1% or the 5% level.*

50. Use the data in Problem 47.

51. Use the data in Problem 48.

52. Use the data in Problem 49.

Objective 8.15: *Find the regression line for a set of data.*

53. Use the data in Problem 47.

54. Use the data in Problem 48.

55. Use the data in Problem 49.

SAMPLE TEST

The following sample test (45 minutes) is intended to review the main ideas of the chapter.

Use the following outcomes, obtained from rolling a pair of dice 50 *times and recording the sum of the two top faces, in Problems* 1–5.

4, 3, 6, 10, 8, 9, 2, 4, 7, 4, 6, 7, 11, 7, 8, 6, 4,
8, 3, 9, 7, 8, 7, 9, 5, 9, 6, 6, 10, 7, 3, 7, 10, 6,
11, 5, 9, 10, 6, 11, 8, 11, 7, 5, 6, 11, 12, 7, 8, 9

1. Prepare a frequency distribution.
2. Find the probability distribution.
3. Represent the data in a histogram.
4. Find the mean, median, and mode.
5. Find the range, variance, and standard deviation.
6. Use Table 3 in Appendix E to compute a binomial probability where $n = 10$, $k = 4$, and $p = .8$.
7. Find the mean, variance, and standard deviation for the distribution described in Problem 6.
8. Find the area under a normal curve with $\mu = 86$ and $\sigma = 5$ between 80 and 90.
9. The following item once appeared in Dear Abby's column:

> **Dear Abby:** You wrote in your column that a woman is pregnant for 266 days. Who said so? I carried my baby for ten months and five days, and there is no doubt about it because I know the exact date my baby was conceived. My husband is in the Navy and it couldn't have possibly been conceived any other time because I saw him only once for an hour, and I didn't see him again until the day before the baby was born.
>
> I don't drink or run around, and there is no way this baby isn't his, so please print a retraction about that 266-day carrying time because otherwise I am in a lot of trouble.
>
> *San Diego Reader*
>
> Abby's answer was consoling and gracious but not very statistical:
>
> **Dear Reader:** The average gestation period is 266 days. Some babies come early. Others come late. Yours was late.

Assume that pregnancy durations have a normal distribution. If the mean pregnancy duration is 266 days and the standard deviation is 16 days, what is the probability of having a pregnancy longer than 310 days?

10. Since World War II, plutonium for use in atomic weapons has been produced at an Atomic Energy Commission facility in Hanford, Washington. One of the major safety problems encountered there has been the storage of radioactive wastes. Over the years, significant quantities of these substances have leaked from their open-pit storage areas into the nearby Columbia River, which flows through parts of Oregon and eventually empties into the Pacific Ocean. To measure the health consequences of this contamination, an index of exposure was calculated for each of the nine Oregon counties having frontage on either the Columbia River or the Pacific Ocean. The cancer mortality rate for each of these counties was also determined. The data are listed in the table, where higher index values represent higher levels of contamination.*

Radioactive Contamination and Cancer Mortality in Oregon Counties

County	Index of exposure	Cancer mortality per 100,000
Clatsop	8.34	210.3
Columbia	6.41	177.9
Gilliam	3.41	129.9
Hood River	3.83	162.3
Morrow	2.57	130.1
Portland	11.64	207.5
Sherman	1.25	113.5
Umatilla	2.49	147.1
Wasco	1.62	137.5

Find the linear correlation coefficient and determine whether the variables are significantly correlated at either the 1% or 5% level.

11. Draw a scatter diagram and the regression line for the set of data in Problem 10.

The following multiple-choice questions dealing with probability and random variables are from actual actuarial exams and are reprinted with permission of the Society of Actuaries, 208 *South LaSalle Street, Chicago, Illinois* 60604.

Actuary Exam May 1982

12. Suppose Q and S are independent events such that the probability that at least one of them occurs is $\frac{1}{3}$ and the probability that Q occurs but S does not occur is $\frac{1}{9}$. What is $P(S)$?
A. $\frac{4}{9}$ **B.** $\frac{1}{3}$ **C.** $\frac{2}{9}$ **D.** $\frac{1}{7}$
E. $\frac{1}{9}$

* From Richard J. Larsen and Donna Fox Stroup, *Statistics in the Real World* (New York: Macmillan, 1976).

13. A fair coin is tossed until a head appears. Given that the first head appeared on an even-numbered toss, what is the conditional probability that the head appeared on the fourth toss?
A. $\frac{1}{16}$ **B.** $\frac{1}{8}$ **C.** $\frac{3}{16}$ **D.** $\frac{1}{4}$
E. $\frac{15}{16}$

14. A card is drawn at random from an ordinary deck of 52 cards and replaced. This is done a total of 5 independent times. What is the conditional probability of drawing the ace of spades exactly 4 times, given that this ace is drawn at least 4 times?
A. $\frac{1}{2}$ **B.** $\frac{12}{13}$ **C.** $\frac{13}{14}$ **D.** $\frac{60}{61}$
E. $\frac{255}{256}$

15. Suppose an experiment consists of tossing a fair coin until three heads occur. What is the probability that the experiment ends after exactly six flips of the coin with a head on the fifth toss as well as on the sixth?
A. $\frac{1}{16}$ **B.** $\frac{1}{8}$ **C.** $\frac{5}{32}$ **D.** $\frac{1}{4}$
E. $\frac{10}{32}$

9

Markov Chains and Decision Theory

CHAPTER OVERVIEW

This chapter presents three mathematical models. This first, introduced in Chapter 7, involves a stochastic process. It is a repeated stochastic process that is called a Markov chain. Two types of Markov chains are considered in this chapter—absorbing and nonabsorbing. The second model considered is mathematical expectation, which places numerical values on various courses of action and allows us to make decisions based on expected return. The third mathematical model in this chapter is game theory, which enables us to analyze not only our own choices in decision making, but also the possible alternatives open to our opponent with the ultimate goal of maximizing our return.

PREVIEW

As we move from the mechanics of mathematics to more realistic problem solving, the models we must use become more complicated because real-life situations have many variables and component parts. Models must accommodate these complexities, so the methods rely to a great extent on the aid of computers. In this chapter we combine the ideas of probability and matrices to study experiments whose outcomes depend only on the outcome of the previous experiment. A stochastic process refers to a random event that depends on previous random events. After looking at Markov chains, we turn to the mathematics of decision making and relate it to mathematical expectation and game theory.

PERSPECTIVE

In business, as well as in almost all endeavors, it is frequently necessary to choose between opposing alternatives. Much of this decision making is done intuitively because the choices are not usually well defined or clear-cut. The goal of this chapter is to aid in the decision making process. Markov chains are used to make stock market predictions, to study living patterns in metropolitan areas, to follow the purchases of consumers in certain markets, to study the ability of parents to pass certain physical traits to their offspring, and to monitor the outcome of some gambling situations that fall into a category called *financial ruin problems.* Decision theory provides a systematic way to attack problems of decision making. This problem solving skill is an essential one.

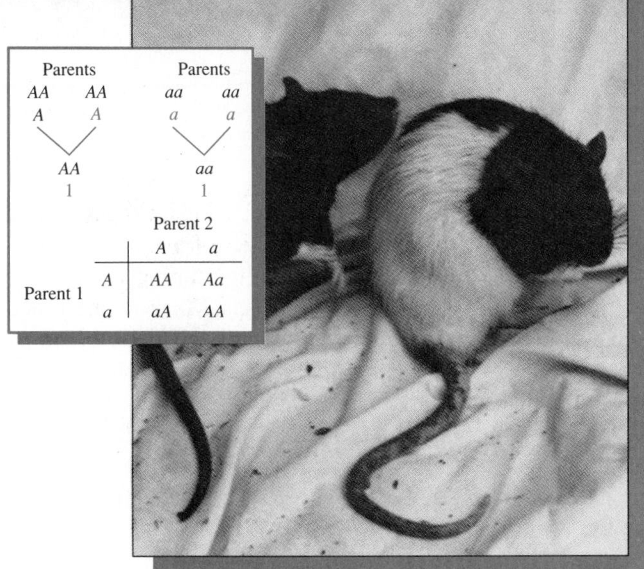

MODELING APPLICATION 7

Genetics

A researcher interested in studying the ability of parents to pass certain physical traits to their offspring sets up a classic *brother–sister mating model* on a family of rats. In this experiment, two rats are mated, and from among their direct descendants two individuals of the opposite sex are selected at random and are mated. Then the process is repeated indefinitely and the target traits are studied. What are the expected results from such experimentation?

After you have finished this chapter, write a paper based on this modeling application.

This Modeling Application is continued on page 469.

APPLICATIONS

Management *(Business, Economics, Finance, and Investments)*
Stock analysis (9.1, Problem 15)
Distribution of rental cars in San Francisco (9.1, Problems 17, 19–20)
Homeowners insurance purchasing patterns in Dallas–Fort Worth area (9.1, Problems 18, 21–22)
Computer "learning program" probabilities (9.1, Problems 31–38)
Purchasing patterns for buying dog food (9.2, Problem 17)
Advertising strategies (9.2, Problem 18)
Purchasing patterns for coffee (9.2, Problems 19–22)
Probability that a defective monitor will be approved (9.3, Problem 27)
Life insurance costs (9.4, Problems 11–16; 9.7, Problem 40)
Rental car distribution (9.4, Problems 34–35; 9.7, Problem 44)
Military strategies (9.5, Problems 21, 24; 9.6, Problem 13; 9.7, Problem 58)
Location for a McDonald's (9.5, Problem 26)
Market share (9.7, Problems 2, 9)

Management *(continued)*
Contract bidding (9.7, Problem 41)
Competition between greeting card companies (9.7, Problems 9–10, 21)

Life Sciences *(Biology, Ecology, Health, and Medicine)*
Distribution of a cattle herd in future generations (9.2, Problems 25–26)
Probability that a substance will be in a liquid or gaseous state in the future (9.7, Problem 3)
Distribution of a chemical (9.7, Problem 3)
Cross-pollination of plants (9.7, Problems 35–36)

Social Sciences *(Demography, Political science, Population, Psychology, Society, and Sociology)*
Living patterns in a metropolitan area (9.1, Problem 16)
Long-range voting patterns (9.2, Problems 23–24)
Campaign strategies in a presidential election (9.2, Problems 28–30; 9.5, Problem 23)
Rat maze problem (9.3, Problems 29–30)

General Interest
Likelihood of a quiz (9.1, Problems 23–30)
Financial ruin in a game (9.3, Problems 25–26; 9.7, Problems 33–34)
Probability of a revoked credit card (9.3, Problem 28)
Contest expectations (9.4, Problems 7–10; 9.7, Problems 37, 42)
U.S. roulette (9.4, Problems 17–25)
Real estate listings (9.4, Problem 26; 9.7, Problem 43)
Oil drilling sites (9.4, Problems 27–28, 32)
Sherlock Holmes' "Final Problem" (9.5, Problem 25)
CPA Examination (Cumulative Review, Problems 24–27, 43)
Actuary Examination (Cumulative Review, Problems 28–34)
CMA Examination (Cumulative Review, Problems 35–37)

Modeling Application—
Genetics

9.1

Introduction to Markov Chains

Markov Chain

In the first part of this chapter, we build a probability model using a process called a **Markov chain**. A Markov chain process is used when a series of events or experiments consists of a finite number of trials, each with a finite number of possible outcomes having a fixed probability of occurrence. That is, a Markov chain is a probabilistic process in which the future development is completely determined by the present state, regardless of how that present state arose. It is a natural extension of the tree diagrams of stochastic processes first introduced in Chapter 7.

For example, suppose you own 100 shares of CBS stock and are keeping a record of its progress at the close of each trading day. Each day there are three possibilities: up, down, or unchanged. After a lengthy analysis you determine that the following probabilities apply:

Description	*Tree diagram*	*Conditional probability*
If the stock increases one day, the probabilities for the following day are given by:	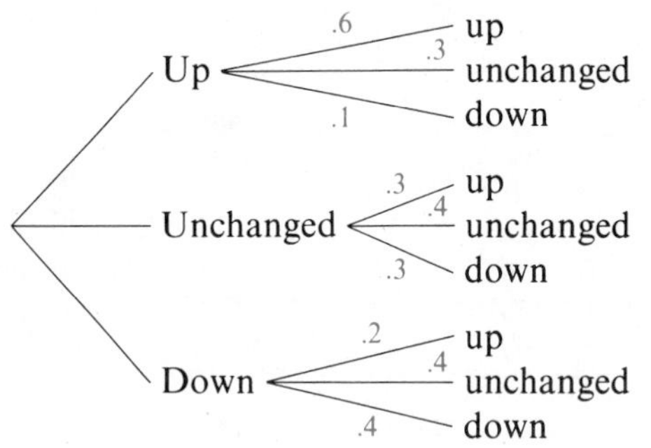	$P(\text{up} \mid \text{up}) = .6$ $P(\text{unchanged} \mid \text{up}) = .3$ $P(\text{down} \mid \text{up}) = .1$
If the stock remains unchanged one day, the probabilities for the following day are given by:		$P(\text{up} \mid \text{unchanged}) = .3$ $P(\text{unchanged} \mid \text{unchanged}) = .4$ $P(\text{down} \mid \text{unchanged}) = .3$
If the stock decreases one day, the probabilities for the following day are given by:		$P(\text{up} \mid \text{down}) = .2$ $P(\text{unchanged} \mid \text{down}) = .4$ $P(\text{down} \mid \text{down}) = .4$

Transition Matrix

All of these probabilities can be readily summarized in matrix form:

		Next state		
		Up	Unchanged	Down
Present state	Up	.6	.3	.1
	Unchanged	.3	.4	.3
	Down	.2	.4	.4

The matrix representation for these probabilities is called a **transition matrix** because it gives the probability of moving from a present state to the next state. A transition matrix must have the following properties:

1. *It is square* because all possible states are used both as rows and columns.
2. *All entries are between* 0 *and* 1 *inclusive* because all entries represent probabilities.
3. *The sum of the entries in any row is* 1 because the numbers in the row give the probability of changing from the state listed at the left to one of the states listed across the top.

Probability Vector

A **probability vector** is a $1 \times n$ matrix for which the sum of the n entries is 1. If $s_1, s_2, \ldots, s_n$ are the states of a Markov chain and p_{ij} is the probability that an experiment will next be in state s_j if it is in state s_i now, then the row matrices

$$[p_{11} \quad p_{12} \quad \cdots \quad p_{1n}]$$
$$[p_{21} \quad p_{22} \quad \cdots \quad p_{2n}]$$
$$\vdots$$
$$[p_{n1} \quad p_{n2} \quad \cdots \quad p_{nn}]$$

Transition Matrix

are the probability vectors, and the matrix T formed by these probability vectors is called the **transition matrix**:

$$T = \begin{bmatrix} p_{11} & p_{12} & \cdots & p_{1n} \\ p_{21} & p_{22} & \cdots & p_{2n} \\ & & \vdots & \\ p_{n1} & p_{n2} & \cdots & p_{nn} \end{bmatrix}$$

EXAMPLE 1 A mathematics instructor gives surprise quizzes. She never gives a quiz 2 days in a row, but if she does not give a quiz one day, she is just as likely to give a quiz the following day as she is not to give a quiz. This is an example of a Markov chain with two *states*: s_1 = quiz and s_2 = no quiz. The probability of a quiz the next day depends on the present state. We can summarize the probabilities in matrix form:

	Quiz next day	No quiz next day
Quiz one day	0	1
No quiz one day	$\frac{1}{2}$	$\frac{1}{2}$

■

Suppose the mathematics instructor in Example 1 uses the same rules for giving quizzes day after day. If today is Monday, what are the possibilities for Wednesday if the class meets daily? Let Q = quiz given and let N = no quiz given. Then we can draw a tree diagram, as shown. The probabilities for each outcome are found by multiplying along each branch of the tree; $P(NNN) = \frac{1}{4}$ means that there is a probability of $\frac{1}{4}$ that no quiz will be given on any of the 3 days.

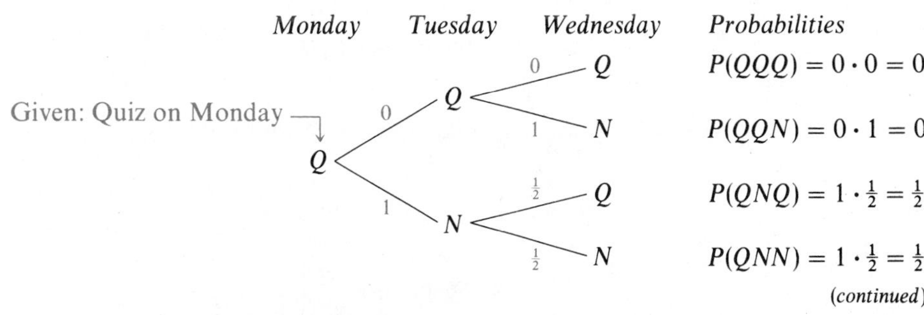

(*continued*)

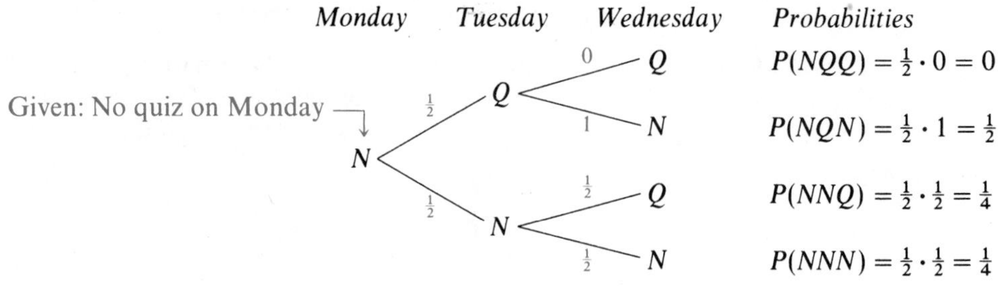

EXAMPLE 2 What is the probability that the instructor described in Example 1 will give a quiz on Wednesday, given the following information?

a. There is a quiz on Monday.

b. There is no quiz on Monday.

What is the probability that there is no quiz on Wednesday, given the following information?

c. There is a quiz on Monday.

d. There is no quiz on Monday.

Solution We use the previous tree diagram and the addition principle for probability.

a. $P(Q \text{ on Wednesday} \mid Q \text{ on Monday}) = P(QQQ) + P(QNQ) = 0 + \frac{1}{2} = \frac{1}{2}$

b. $P(Q \text{ on Wednesday} \mid N \text{ on Monday}) = P(NQQ) + P(NNQ) = 0 + \frac{1}{4} = \frac{1}{4}$

c. $P(N \text{ on Wednesday} \mid Q \text{ on Monday}) = P(QQN) + P(QNN) = 0 + \frac{1}{2} = \frac{1}{2}$

d. $P(N \text{ on Wednesday} \mid N \text{ on Monday}) = P(NQN) + P(NNN) = \frac{1}{2} + \frac{1}{4} = \frac{3}{4}$ ■

EXAMPLE 3 If T is the transition matrix given in Example 1, find T^2.

Solution

$$T^2 = \begin{bmatrix} 0 & 1 \\ \frac{1}{2} & \frac{1}{2} \end{bmatrix} \begin{bmatrix} 0 & 1 \\ \frac{1}{2} & \frac{1}{2} \end{bmatrix} = \begin{bmatrix} 0 + \frac{1}{2} & 0 + \frac{1}{2} \\ 0 + \frac{1}{4} & \frac{1}{2} + \frac{1}{4} \end{bmatrix} = \begin{bmatrix} \frac{1}{2} & \frac{1}{2} \\ \frac{1}{4} & \frac{3}{4} \end{bmatrix}$$ ■

Now compare the results of Examples 2 and 3:

$$P(Q \text{ on Wed.} \mid Q \text{ on Mon.}) = \frac{1}{2} \qquad P(N \text{ on Wed.} \mid Q \text{ on Mon.}) = \frac{1}{2}$$

$$P(Q \text{ on Wed.} \mid N \text{ on Mon.}) = \frac{1}{4} \qquad P(N \text{ on Wed.} \mid N \text{ on Mon.}) = \frac{3}{4}$$

$$\begin{array}{ccc} & & \text{Wednesday} \\ & & \begin{array}{cc} Q & N \end{array} \\ \text{Monday} & \begin{array}{c} Q \\ N \end{array} & \begin{bmatrix} \frac{1}{2} & \frac{1}{2} \\ \frac{1}{4} & \frac{3}{4} \end{bmatrix} \end{array}$$

WARNING ***Note the important property illustrated by Example 3 and summarized in this paragraph.***

Thus, in this example we see that the *square* of the transition matrix gives the probabilities for moving from one state to another in *two* stages. This is true, in general, and higher powers of the transition matrix carry similar interpretations. If T is the transition matrix of a Markov chain, then the (i, j) entry of T^n gives the probability of passing from state i to state j in n stages.

EXAMPLE 4 What are the probabilities that the instructor described in Example 1 will or will not give a quiz on Friday (4 days after Monday)?

Solution If the above result can be applied twice, we need to find T^4:

$$T^4 = \begin{bmatrix} 0 & 1 \\ \frac{1}{2} & \frac{1}{2} \end{bmatrix}^4 = \begin{bmatrix} 0 & 1 \\ \frac{1}{2} & \frac{1}{2} \end{bmatrix}^2 \begin{bmatrix} 0 & 1 \\ \frac{1}{2} & \frac{1}{2} \end{bmatrix}^2 = \begin{bmatrix} \frac{1}{2} & \frac{1}{2} \\ \frac{1}{4} & \frac{3}{4} \end{bmatrix} \begin{bmatrix} \frac{1}{2} & \frac{1}{2} \\ \frac{1}{4} & \frac{3}{4} \end{bmatrix}$$

$$= \begin{bmatrix} \frac{1}{4} + \frac{1}{8} & \frac{1}{4} + \frac{3}{8} \\ \frac{1}{8} + \frac{3}{16} & \frac{1}{8} + \frac{9}{16} \end{bmatrix} = \begin{bmatrix} \frac{3}{8} & \frac{5}{8} \\ \frac{5}{16} & \frac{11}{16} \end{bmatrix}$$

The row 1, column 1 entry in T^4 signifies that the probability of a quiz on Friday, given a quiz on Monday, is $\frac{3}{8}$. ■

Eight (school) days after the initial Monday the probabilities are found by considering T^8:

$$T^8 = \begin{bmatrix} 0 & 1 \\ \frac{1}{2} & \frac{1}{2} \end{bmatrix}^4 \begin{bmatrix} 0 & 1 \\ \frac{1}{2} & \frac{1}{2} \end{bmatrix}^4 = \begin{bmatrix} \frac{3}{8} & \frac{5}{8} \\ \frac{5}{16} & \frac{11}{16} \end{bmatrix} \begin{bmatrix} \frac{3}{8} & \frac{5}{8} \\ \frac{5}{16} & \frac{11}{16} \end{bmatrix}$$

$$= \begin{bmatrix} \frac{9}{64} + \frac{25}{128} & \frac{15}{64} + \frac{55}{128} \\ \frac{15}{128} + \frac{55}{256} & \frac{25}{128} + \frac{121}{256} \end{bmatrix}$$

$$= \begin{bmatrix} \frac{43}{128} & \frac{85}{128} \\ \frac{85}{256} & \frac{171}{256} \end{bmatrix}$$

$$\approx \begin{bmatrix} .3359 & .6641 \\ .3320 & .6680 \end{bmatrix}$$

This suggests that as we take higher and higher powers of T, the result gets closer and closer to

$$\begin{bmatrix} \frac{1}{3} & \frac{2}{3} \\ \frac{1}{3} & \frac{2}{3} \end{bmatrix}$$

Note that the same probability vector appears in both rows. Suppose we multiply this vector $[\frac{1}{3} \quad \frac{2}{3}]$ times the original transition matrix:

$$[\tfrac{1}{3} \quad \tfrac{2}{3}] \begin{bmatrix} 0 & 1 \\ \frac{1}{2} & \frac{1}{2} \end{bmatrix} = [0 + \tfrac{1}{3} \quad \tfrac{1}{3} + \tfrac{1}{3}] = [\tfrac{1}{3} \quad \tfrac{2}{3}]$$

The answer is still $[\frac{1}{3} \quad \frac{2}{3}]$! This is no coincidence, and if $T, T^2, T^3, \ldots, T^n$, approach a matrix all of whose rows are V, then $VT = V$. For this reason we call V a **fixed probability vector**. We discuss the process for finding a fixed probability vector in the next section.

9.1 Problem Set

Are the matrices in Problems 1–3 probability vectors?

1. a. $\begin{bmatrix} \frac{1}{2} & \frac{1}{2} & \frac{1}{2} \end{bmatrix}$ b. $\begin{bmatrix} \frac{1}{2} & \frac{1}{2} \end{bmatrix}$
   c. $\begin{bmatrix} \frac{2}{3} & \frac{1}{5} \end{bmatrix}$
2. a. $[1 \quad 0 \quad 0]$ b. $[.6 \quad .3 \quad .1]$
   c. $[.1 \quad .1 \quad .1]$
3. a. $[.5 \quad .6 \quad -.1]$
   b. $[.05 \quad .15 \quad .63 \quad .17]$
   c. $[0 \quad 0 \quad 0]$

Are the matrices in Problems 4–6 transition matrices?

4. a. $\begin{bmatrix} .3 & .2 & .5 \\ 0 & .1 & .9 \\ .8 & .1 & .1 \end{bmatrix}$ b. $\begin{bmatrix} .5 & 0 & .5 \\ .1 & .1 & .1 \\ 1 & 0 & 0 \end{bmatrix}$
5. a. $\begin{bmatrix} \frac{1}{3} & \frac{2}{3} \\ \frac{1}{5} & \frac{4}{5} \end{bmatrix}$ b. $\begin{bmatrix} \frac{2}{3} & \frac{4}{5} \\ \frac{5}{6} & \frac{1}{6} \end{bmatrix}$
6. a. $\begin{bmatrix} \frac{1}{4} & \frac{1}{2} & \frac{1}{4} \\ \frac{1}{5} & \frac{3}{5} & \frac{2}{5} \\ 0 & 0 & 0 \end{bmatrix}$ b. $\begin{bmatrix} 1 & 0 & 0 \\ 0 & 1 & 0 \\ 0 & 0 & 1 \end{bmatrix}$

If T is the matrix in Problems 7–12, find T^2 and T^3.

7. $\begin{bmatrix} \frac{1}{2} & \frac{1}{2} \\ \frac{1}{4} & \frac{3}{4} \end{bmatrix}$
8. $\begin{bmatrix} .1 & .2 & .7 \\ .1 & .8 & .1 \\ .2 & .3 & .5 \end{bmatrix}$
9. $\begin{bmatrix} .6 & .4 \\ .23 & .77 \end{bmatrix}$
10. $\begin{bmatrix} 1 & 0 & 0 \\ 0 & 1 & 0 \\ .4 & .3 & .3 \end{bmatrix}$
11. $\begin{bmatrix} .15 & .25 & .6 \\ .5 & .35 & .15 \\ .1 & .8 & .1 \end{bmatrix}$
12. $\begin{bmatrix} .29 & .53 & .18 \\ .01 & .99 & 0 \\ .6 & .3 & .1 \end{bmatrix}$
13. If a certain experiment has two states 0 and 1 with the probabilities $P(0|0) = .3$, $P(1|0) = .7$, $P(0|1) = .6$, and $P(1|1) = .4$, write the transition matrix.
14. If a certain experiment has three states 0, 1, and 2 with the probabilities $P(0|0) = .4$, $P(1|0) = .3$, $P(2|0) = .3$, $P(0|1) = .1$, $P(1|1) = .5$, $P(2|1) = .4$, $P(0|2) = .7$, $P(1|2) = .2$, $P(2|2) = .1$, write the transition matrix.

APPLICATIONS

15. A stock broker has been watching the price changes of a particular stock. Over the past year, the broker has found that on a day the stock is traded,

$$P(\text{increase} \mid \text{increase on previous day}) = .3$$
$$P(\text{decrease} \mid \text{increase on previous day}) = .7$$
$$P(\text{increase} \mid \text{decrease on previous day}) = .8$$
$$P(\text{decrease} \mid \text{decrease on previous day}) = .2$$

Write a transition matrix for this information.

16. A city planner has studied the living patterns within a certain greater metropolitan area. Over the last 10 years the study shows that

$$P(\text{move to city} \mid \text{lived in suburbs previous year}) = .1$$
$$P(\text{stay in suburbs} \mid \text{lived in suburbs previous year}) = .6$$
$$P(\text{leave area} \mid \text{lived in suburbs previous year}) = .3$$
$$P(\text{stay in city} \mid \text{lived in city previous year}) = .8$$
$$P(\text{move to suburbs} \mid \text{lived in city previous year}) = .1$$
$$P(\text{leave area} \mid \text{lived in city previous year}) = .1$$
$$P(\text{move to city} \mid \text{lived out of the area previous year}) = .4$$
$$P(\text{move to suburbs} \mid \text{lived out of the area previous year}) = .6$$

Write a transition matrix for this information.

17. San Francisco is served by three airports, San Francisco International (SFO), Oakland (OAK), and San Jose (SJC). A car rental company plans to begin operations in San Francisco by setting up rental and car storage facilities at these three airports. The probabilities that a person will return the car to a particular location are given by the following matrix, T:

		Returned to		
		SFO	OAK	SJC
Rented from	SFO	.8	.1	.1
	OAK	.2	.7	.1
	SJC	.3	.2	.6

   a. Is this a transition matrix? Why or why not?
   b. What does the entry .8 signify?
   c. Which airport has the lowest percentage of returns?

18. A survey of the residents in the greater Dallas–Fort Worth area shows that they purchased their homeowners insurance from one of three companies. The probabilities that a resident will purchase homeowners insurance from one of these companies in the subsequent year are sum-

marized by the following matrix:

		Company the subsequent year		
		A	B	C
Present company	A	.4	.4	.2
	B	.1	.8	.1
	C	.1	.2	.7

a. Is this a transition matrix? Why or why not?
b. What does the entry .8 signify?
c. Which company looks like it has the lowest percentage of satisfied customers?

19. Find T^2 for the matrix in Problem 17.

20. Find T^4 for the matrix in Problem 17.

21. Find T^2 for the matrix in Problem 18.

22. Find T^4 for the matrix in Problem 18.

In Problems 23–30 suppose that the transition matrix in Example 1 is changed to

		Next day	
		Q	N
One day	Q	$\frac{1}{2}$	$\frac{1}{2}$
	N	$\frac{3}{4}$	$\frac{1}{4}$

Assume that the instructor does not give a quiz on the first day of class.

23. What is the probability of a quiz on the second day of class?

24. What is the probability of a quiz on the third day of class?

25. What is the probability of a quiz on the fourth day of class?

26. What is $P(Q \text{ on Wednesday} \mid Q \text{ on Monday})$?

27. What is $P(Q \text{ on Wednesday} \mid N \text{ on Monday})$?

28. What is $P(N \text{ on Wednesday} \mid Q \text{ on Monday})$?

29. What is $P(N \text{ on Wednesday} \mid N \text{ on Monday})$?

30. Write a matrix that will answer Problems 26–29.

A computer program is called a "learning program" if it can learn by making mistakes and then correcting those mistakes. Suppose the probability that the computer will be functioning properly on a given day depends on whether it was functioning properly on the previous day. These probabilities are summarized by the following transition matrix, where C means functioning properly and E means malfunctioning:

		State next day (Tuesday)	
		C	E
Present state (Monday)	C	.99	.01
	E	.9	.1

In Problems 31–38 assume that the program is operating properly on Monday and give answers correct to four decimal places.

31. What is the probability that the computer will be in a correct state on Wednesday (i.e., after two transitions)?

32. What is the probability that the computer will be in a correct state on Thursday (i.e., after three transitions)?

33. What is the probability that the computer will be in a correct state on Friday (i.e., after four transitions)?

34. What is $P(C \text{ on Wednesday} \mid E \text{ on Monday})$?

35. What is $P(C \text{ on Wednesday} \mid C \text{ on Monday})$?

36. What is $P(E \text{ on Wednesday} \mid E \text{ on Monday})$?

37. What is $P(E \text{ on Wednesday} \mid C \text{ on Monday})$?

38. Write a matrix that will answer Problems 34–37.

9.2 Regular Markov Chains

In Section 9.1 we discussed transition matrices that summarized the probabilities for changing from one state to another. We also were introduced to the idea of a *fixed probability vector*, which is now defined:

Fixed Probability Vector

A probability vector V such that $VT = V$ is called a **fixed probability vector** for the matrix T.

To understand this concept, consider a marketing analysis for Tide®. Suppose the following transition matrix is given:

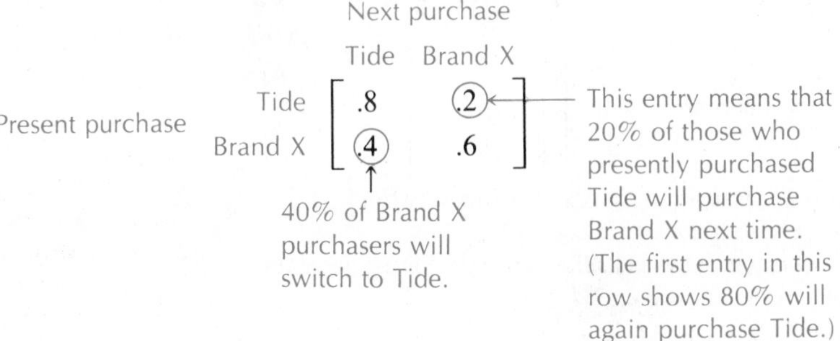

Now, suppose that there are 10 million people who purchase a detergent each month and that 25% of the consumers purchase Tide and 75% purchase Brand X (all other brands lumped together). Also, suppose that every purchaser makes one purchase per month. What share of the market will use Tide in the next month, given these assumptions?

$$\begin{matrix} \text{Tide} & \text{Brand X} \\ [.25 & .75\ ] \end{matrix} \begin{bmatrix} .8 & .2 \\ .4 & .6 \end{bmatrix} = [.5 \quad .5]$$

Second month:

$$[.5 \quad .5] \begin{bmatrix} .8 & .2 \\ .4 & .6 \end{bmatrix} = [.6 \quad .4]$$

Third month:

$$[.6 \quad .4] \begin{bmatrix} .8 & .2 \\ .4 & .6 \end{bmatrix} = [.64 \quad .36]$$

Fourth month:

$$[.64 \quad .36] \begin{bmatrix} .8 & .2 \\ .4 & .6 \end{bmatrix} = [.656 \quad .344]$$

This means that at the end of the fourth month, Tide could expect 65.6% of the market, or 6.56 million customers. If we continue this process, will the market share for Tide and Brand X stabilize at some specific share of the market for each? If so, this share is represented by the fixed probability vector.

Let $V = [v_1 \quad v_2]$. Now find V so that

$$[v_1 \quad v_2] \begin{bmatrix} .8 & .2 \\ .4 & .6 \end{bmatrix} = [v_1 \quad v_2]$$

$$\begin{cases} .8v_1 + .4v_2 = v_1 \\ .2v_1 + .6v_2 = v_2 \end{cases} \quad \text{or} \quad \begin{cases} .2v_1 - .4v_2 = 0 \\ .2v_1 - .4v_2 = 0 \end{cases}$$

In addition, $[v_1 \quad v_2]$ is a probability vector, so $v_1 + v_2 = 1$, giving the system

$$\begin{cases} .2v_1 - .4v_2 = 0 \\ .2v_1 - .4v_2 = 0 \\ v_1 + \quad v_2 = 1 \end{cases}$$

$$\left[\begin{array}{cc|c} .2 & -.4 & 0 \\ .2 & -.4 & 0 \\ 1 & 1 & 1 \end{array}\right] \longrightarrow \left[\begin{array}{cc|c} 1 & 1 & 1 \\ 0 & -.6 & -.2 \\ 0 & 0 & 0 \end{array}\right] \longrightarrow \left[\begin{array}{cc|c} 1 & 1 & 1 \\ 0 & 1 & \frac{1}{3} \\ 0 & 0 & 0 \end{array}\right] \longrightarrow \left[\begin{array}{cc|c} 1 & 0 & \frac{2}{3} \\ 0 & 1 & \frac{1}{3} \\ 0 & 0 & 0 \end{array}\right]$$

You do not need to use Gauss–Jordan. Substituting $v_2 = 1 - v_1$ into $.2v_1 - .4v_2 = 0$ would work just as well.

Thus $v_1 = \frac{2}{3}$, $v_2 = \frac{1}{3}$, and $[\frac{2}{3} \quad \frac{1}{3}]$ is the fixed probability vector. This means that, in the long run, Tide can expect to capture $\frac{2}{3}$ of the market.

The next example ties together the concepts of a fixed probability vector and the successive powers of a transition matrix.

EXAMPLE 1 Suppose a professor gives quizzes (Q) or no quizzes (N) according to the transition matrix

$$\begin{array}{c} \\ Q \\ N \end{array} \begin{array}{c} \begin{array}{cc} Q & N \end{array} \\ \begin{bmatrix} 0 & 1 \\ \frac{1}{2} & \frac{1}{2} \end{bmatrix} \end{array}$$

Find the fixed probability vector $V = [v_1 \quad v_2]$.

Solution

$$[v_1 \quad v_2] \begin{bmatrix} 0 & 1 \\ \frac{1}{2} & \frac{1}{2} \end{bmatrix} = [v_1 \quad v_2]$$

We solve this equation for v_1 and v_2:

$$[v_1 \quad v_2] \begin{bmatrix} 0 & 1 \\ \frac{1}{2} & \frac{1}{2} \end{bmatrix} = [0 + \tfrac{1}{2}v_2 \quad v_1 + \tfrac{1}{2}v_2] = [\tfrac{1}{2}v_2 \quad v_1 + \tfrac{1}{2}v_2]$$

If $[\frac{1}{2}v_2 \quad v_1 + \frac{1}{2}v_2] = [v_1 \quad v_2]$, the corresponding components are equal:

$$\begin{cases} \frac{1}{2}v_2 = v_1 \\ v_1 + \frac{1}{2}v_2 = v_2 \end{cases}$$

Using this relationship, along with the fact that $v_1 + v_2 = 1$, gives the matrix system

$$\left[\begin{array}{cc|c} 1 & 1 & 1 \\ -1 & \frac{1}{2} & 0 \\ 1 & -\frac{1}{2} & 0 \end{array}\right] \longrightarrow \left[\begin{array}{cc|c} 1 & 1 & 1 \\ 0 & \frac{3}{2} & 1 \\ 0 & -\frac{3}{2} & -1 \end{array}\right] \longrightarrow \left[\begin{array}{cc|c} 1 & 1 & 1 \\ 0 & 1 & \frac{2}{3} \\ 0 & 0 & 0 \end{array}\right] \longrightarrow \left[\begin{array}{cc|c} 1 & 0 & \frac{1}{3} \\ 0 & 1 & \frac{2}{3} \\ 0 & 0 & 0 \end{array}\right]$$

Thus $v_1 = \frac{1}{3}$ and $v_2 = \frac{2}{3}$, or $V = [\frac{1}{3} \quad \frac{2}{3}]$. ■

Now let us compare the solution of Example 1 with the powers of T found in Section 9.1:

$$T^2 = \begin{bmatrix} \frac{1}{2} & \frac{1}{2} \\ \frac{1}{4} & \frac{3}{4} \end{bmatrix} \qquad T^4 = \begin{bmatrix} \frac{3}{8} & \frac{5}{8} \\ \frac{5}{16} & \frac{11}{16} \end{bmatrix} \qquad T^8 \approx \begin{bmatrix} .3359 & .6641 \\ .3320 & .6680 \end{bmatrix}$$

It is no coincidence that the higher powers of T approach the matrix consisting of the fixed probability vector $[\frac{1}{3} \;\; \frac{2}{3}]$:

$$\begin{bmatrix} \frac{1}{3} & \frac{2}{3} \\ \frac{1}{3} & \frac{2}{3} \end{bmatrix}$$

This relationship between the higher powers of T and the fixed probability vector are summarized in the **transition matrix theorem**. However, before we can state this theorem, we need to consider some terminology that will help us describe when the transition matrix theorem applies.

Consider the matrix

$$\begin{bmatrix} 1 & 0 \\ \frac{1}{5} & \frac{4}{5} \end{bmatrix}$$

The powers of this matrix will always have $p_{12} = 0$. We can see this by considering

$$\begin{bmatrix} a & 0 \\ b & c \end{bmatrix}\begin{bmatrix} a & 0 \\ b & c \end{bmatrix}$$

The entry in the first row, second column of the product is

$$a \cdot 0 + 0 \cdot c = 0$$

Thus, regardless of the other entries, the powers of this matrix will always have $p_{12} = 0$. This means that the powers of this matrix will never approach a matrix whose rows are all a probability vector V. However, if for some power of T all the entries are positive, then the powers of the matrix will approach a matrix whose rows are all a probability vector. If this is the case, we call the transition matrix **regular**.

Regular Transition Matrix

A transition matrix is **regular** if there exists some power of it that contains only positive elements.

EXAMPLE 2 Is $\begin{bmatrix} \frac{1}{5} & \frac{4}{5} \\ 1 & 0 \end{bmatrix}$ regular?

Solution

$$\begin{bmatrix} \frac{1}{5} & \frac{4}{5} \\ 1 & 0 \end{bmatrix}\begin{bmatrix} \frac{1}{5} & \frac{4}{5} \\ 1 & 0 \end{bmatrix} = \begin{bmatrix} \frac{1}{25} + \frac{4}{5} & \frac{4}{25} + 0 \\ \frac{1}{5} + 0 & \frac{4}{5} + 0 \end{bmatrix}$$

$$= \begin{bmatrix} \frac{21}{25} & \frac{4}{25} \\ \frac{1}{5} & \frac{4}{5} \end{bmatrix}$$

Since the second power of the matrix has all positive elements, we see that it is regular. ■

How many powers do you need to test in order to determine whether a Markov chain is regular? It depends on the size of T, but it is known that if T is the transition matrix for a regular Markov chain with n rows, then one of the first $(n-1)^2 + 1$ powers contains only positive entries. Thus, if each of the first $(n-1)^2 + 1$ powers of T contains at least one zero entry, k, then the Markov chain is not regular. Also, when finding successive powers of T it is more efficient to calculate T^2, T^4, $T^8, \ldots$ than it is to calculate T^2, T^3, T^4, $T^5, \ldots$. You do not need to worry about missing a power which had all positive entries, say T^7, because any time one power has only positive entries, all higher powers will have only positive entries. These results are summarized by the following transition matrix theorems:

Transition Matrix Theorems

Theorem for Determining whether a Transition Matrix with n Rows Is Regular
If the first $(n-1)^2 + 1$ powers of a transition matrix with n rows all contain at least one zero, then the Markov chain is not regular.

Theorem of Nonreturning Zeros
A zero can never return to a transition matrix after all zeros are gone when raising T to higher powers.

Consequences of a Regular Transition Matrix
If T is a regular transition matrix, then:

1. T has a unique fixed probability vector, V, whose elements are all positive.
2. $T, T^2, T^3, \ldots, T^n$ approach the matrix all of whose rows are V. In practice, it is more efficient to consider $T, T^2, T^4, T^8, \ldots$.
3. If W is *any* probability vector, then $WT, WT^2, WT^3, \ldots, WT^n$ approach the fixed probability vector V.

These theorems give us a procedure; first determine whether T is regular. To do this, find $T^2, T^4, \ldots$ until you find a power with nonzero entries or until you get to the $(n-1)^2 + 1$ power. Thus, for a transition matrix

with two rows, you need only to check up to T^2;
with three rows, you need to check up to T^5;
with four rows, you need to check up to T^{17}.

After you know that it is regular, let V be the fixed probability vector. Let

$$VT = V$$

This will give a dependent system whose variables are the entries of the fixed probability vector. Add the additional constraint that the sum of the entries of this fixed probability vector is 1 and the result is a system of equations that can be solved for the fixed probability vector V. This gives the long-run trend for the transition matrix.

These steps are illustrated by Example 3.

EXAMPLE 3 Suppose that General Motors (GM), Ford (F), and Chrysler (C) each introduce a front-wheel-drive compact with a mileage rating better than 50 mpg. Assume that each company initially captures about one-third of the market. During the year:

1. General Motors keeps 85% of its customers but loses 10% to Ford and 5% to Chrysler.

2. Ford keeps 80% of its customers but loses 10% to General Motors and 10% to Chrysler.
3. Chrysler keeps 60% of its customers but loses 25% to General Motors and 15% to Ford.

If this trend continues, what share of the market is each likely to have at the end of next year, and what is the long-run prediction if this trend continues?

Solution The transition matrix is

$$\begin{array}{c} \\ GM \\ F \\ C \end{array}\begin{array}{c} \begin{array}{ccc} GM & F & C \end{array} \\ \begin{bmatrix} \frac{17}{20} & \frac{1}{10} & \frac{1}{20} \\ \frac{1}{10} & \frac{4}{5} & \frac{1}{10} \\ \frac{1}{4} & \frac{3}{20} & \frac{3}{5} \end{bmatrix} \end{array}$$

Note: $85\% = \frac{85}{100} = \frac{17}{20}$; do the same for all the percents.

Since the entries are all positive, the matrix is a regular transition matrix and the transition matrix theorem applies. To determine what share of the market each company will have at the end of next year we multiply the original share vector $[\frac{1}{3} \quad \frac{1}{3} \quad \frac{1}{3}]$ by T:

$$[\tfrac{1}{3} \quad \tfrac{1}{3} \quad \tfrac{1}{3}]\begin{bmatrix} \frac{17}{20} & \frac{1}{10} & \frac{1}{20} \\ \frac{1}{10} & \frac{4}{5} & \frac{1}{10} \\ \frac{1}{4} & \frac{3}{20} & \frac{3}{5} \end{bmatrix}$$

$$= [\tfrac{17}{60} + \tfrac{1}{30} + \tfrac{1}{12} \quad \tfrac{1}{30} + \tfrac{4}{15} + \tfrac{1}{20} \quad \tfrac{1}{60} + \tfrac{1}{30} + \tfrac{1}{5}]$$

$$= [\tfrac{2}{5} \quad \tfrac{7}{20} \quad \tfrac{1}{4}]$$

We see that at the end of next year GM will have $\frac{2}{5}$ or 40% of the market, Ford will have 35% of the market, and Chrysler will have 25% of the market.

For the long-range prediction we find the fixed probability vector:

$$[v_1 \quad v_2 \quad v_3]\begin{bmatrix} \frac{17}{20} & \frac{1}{10} & \frac{1}{20} \\ \frac{1}{10} & \frac{4}{5} & \frac{1}{10} \\ \frac{1}{4} & \frac{3}{20} & \frac{3}{5} \end{bmatrix} = [v_1 \quad v_2 \quad v_3]$$

By the third part of the transition matrix theorem, any probability vector $[\frac{1}{3} \quad \frac{1}{3} \quad \frac{1}{3}]T$ approaches this fixed probability vector. Multiplying, we get

$$[\tfrac{17}{20}v_1 + \tfrac{1}{10}v_2 + \tfrac{1}{4}v_3 \quad \tfrac{1}{10}v_1 + \tfrac{4}{5}v_2 + \tfrac{3}{20}v_3 \quad \tfrac{1}{20}v_1 + \tfrac{1}{10}v_2 + \tfrac{3}{5}v_3] = [v_1 \quad v_2 \quad v_3]$$

By the equality of vectors, we obtain the following system:

$$\begin{cases} \frac{17}{20}v_1 + \frac{1}{10}v_2 + \frac{1}{4}v_3 = v_1 \\ \frac{1}{10}v_1 + \frac{4}{5}v_2 + \frac{3}{20}v_3 = v_2 \\ \frac{1}{20}v_1 + \frac{1}{10}v_2 + \frac{3}{5}v_3 = v_3 \end{cases} \quad \text{or} \quad \begin{cases} 3v_1 - 2v_2 - 5v_3 = 0 \\ 2v_1 - 4v_2 + 3v_3 = 0 \\ v_1 + 2v_2 - 8v_3 = 0 \end{cases}$$

> **CALCULATOR COMMENT**
>
> If your calculator has matrix operations, it is easy to find the fixed probability vector by using the transition matrix theorem. For Example 3 we entered the transition matrix in our TI-81 (called it [A]) and carried out the following steps:
>
> 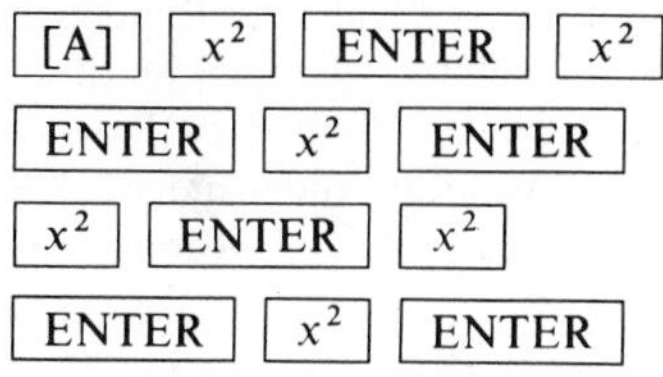
>
>
> This gives two-place accuracy. Continue with
>
> x^2 ENTER x^2
>
> ENTER
>
> to obtain:
>
> [.4905660377 .358490566 .1509433962]

If you solve this system of equations, you will find that it is a dependent system. Add to this system the fact that V is a probability vector, namely, $v_1 + v_2 + v_3 = 1$, to obtain the matrix:

$$\left[\begin{array}{rrr|r} 3 & -2 & -5 & 0 \\ 2 & -4 & 3 & 0 \\ 1 & 2 & -8 & 0 \\ 1 & 1 & 1 & 1 \end{array}\right] \longrightarrow \left[\begin{array}{rrr|r} 1 & 1 & 1 & 1 \\ 0 & -5 & -8 & -3 \\ 0 & -6 & 1 & -2 \\ 0 & 1 & -9 & -1 \end{array}\right] \longrightarrow \left[\begin{array}{rrr|r} 1 & 1 & 1 & 1 \\ 0 & 1 & -9 & -1 \\ 0 & 0 & -53 & -8 \\ 0 & 0 & -53 & -8 \end{array}\right]$$

$$\longrightarrow \left[\begin{array}{rrr|r} 1 & 1 & 1 & 1 \\ 0 & 1 & -9 & -1 \\ 0 & 0 & 1 & \frac{8}{53} \\ 0 & 0 & 0 & 0 \end{array}\right] \longrightarrow \left[\begin{array}{rrr|r} 1 & 0 & 0 & \frac{26}{53} \\ 0 & 1 & 0 & \frac{19}{53} \\ 0 & 0 & 1 & \frac{8}{53} \\ 0 & 0 & 0 & 0 \end{array}\right]$$

Therefore, $v_1 = \frac{26}{53}$, $v_2 = \frac{19}{53}$, and $v_3 = \frac{8}{53}$. Thus the fixed probability vector is

$$\begin{matrix} & \text{GM} & \text{F} & \text{C} \\ [\frac{26}{53} & \frac{19}{53} & \frac{8}{53}] \approx [.49 & .36 & .15] \end{matrix}$$

This means that the long-range prediction is that General Motors will have about 49% of the market, Ford 36%, and Chrysler 15%. Note that the transition matrix theorem tells us that this is the long-term prediction *regardless of the original market share.* Even if GM started with 10% of the market, Ford 20%, and Chrysler 70%, the long-term prediction, if the same trend continues, will still be 49% for GM, 36% for Ford, and 15% for Chrysler. ■

9.2 Problem Set

In Problems 1–12 determine whether the matrix is regular. For each matrix that is regular, find the matrix that it approaches as it is raised to higher and higher powers.

1. $\begin{bmatrix} 0 & 1 \\ \frac{1}{3} & \frac{2}{3} \end{bmatrix}$

2. $\begin{bmatrix} \frac{1}{2} & \frac{1}{2} \\ 1 & 0 \end{bmatrix}$

3. $\begin{bmatrix} \frac{1}{3} & \frac{2}{3} \\ 0 & 1 \end{bmatrix}$

4. $\begin{bmatrix} 1 & 0 \\ \frac{1}{2} & \frac{1}{2} \end{bmatrix}$

5. $\begin{bmatrix} \frac{1}{4} & \frac{3}{4} \\ \frac{1}{3} & \frac{2}{3} \end{bmatrix}$

6. $\begin{bmatrix} 1 & 0 \\ \frac{2}{5} & \frac{3}{5} \end{bmatrix}$

7. $\begin{bmatrix} \frac{2}{5} & \frac{3}{5} \\ 1 & 0 \end{bmatrix}$

8. $\begin{bmatrix} .6 & .4 \\ .2 & .8 \end{bmatrix}$

9. $\begin{bmatrix} 0 & 1 & 0 \\ \frac{1}{2} & \frac{1}{4} & \frac{1}{4} \\ 0 & \frac{1}{3} & \frac{2}{3} \end{bmatrix}$

10. $\begin{bmatrix} \frac{1}{3} & \frac{1}{3} & \frac{1}{3} \\ 0 & 0 & 1 \\ \frac{3}{10} & \frac{1}{2} & \frac{1}{5} \end{bmatrix}$

11. $\begin{bmatrix} .6 & .2 & .2 \\ .1 & .8 & .1 \\ .1 & .2 & .7 \end{bmatrix}$

12. $\begin{bmatrix} 0 & .3 & .7 \\ .4 & 0 & .6 \\ .5 & .5 & 0 \end{bmatrix}$

13. **a.** Is the following matrix regular?

$$\begin{bmatrix} 0 & 1 \\ 1 & 0 \end{bmatrix}$$

b. Does this matrix have a unique fixed probability vector?
c. Does the transition matrix theorem apply? Why or why not?

14. **a.** Is the following matrix regular?

$$\begin{bmatrix} 1 & 0 \\ \frac{1}{2} & \frac{1}{2} \end{bmatrix}$$

b. Does this matrix have a unique fixed probability vector?
c. Does the transition matrix theorem apply? Why or why not?

15. Consider the following matrix:

$$\begin{bmatrix} a & 1-a \\ 1-a & a \end{bmatrix}$$

a. What conditions on a are necessary if this is to be a transition matrix?

b. Find the fixed probability vector (assuming that the conditions you found in part **a** apply).

16. Let

$$T = \begin{bmatrix} 1 & 0 & 0 \\ \frac{1}{2} & 0 & \frac{1}{2} \\ 0 & 0 & 1 \end{bmatrix}$$

a. Find the matrix that T^n approaches.

b. Does T have more than one fixed probability vector?

APPLICATIONS

17. A housewife buys three kinds of dog food: canned, packaged, and dry. Her dog likes variety so she never buys the same kind on successive days. If she buys canned food one day, then the next day she buys packaged food. But if she buys packaged or dry food, then the next day she is twice as likely to buy canned food as the other types. Find the transition matrix. In the long run, how often does she buy each of the three varieties?

18. A publishing company has three books to promote. It is company policy to take out an advertisement for one of the books each month in a national journal. The author of book A is the brother-in-law of an editor and has a 60% chance of being picked for advertising each month. The other books will not be picked two months in a row. On the other hand, if book A is picked one month and then in the following month book A is not chosen, then books B and C have an equal chance of being picked. Find the transition matrix. In the long run, how often will each book be advertised?

19. A retailer stocks three brands of coffee. Each brand has about one-third of the market. A survey is taken and finds that each month the following occurs:

a. Alpha Coffee Concern keeps 70% of its customers but loses 10% to Better Bean Coffee and 20% to Carolyn's Certified Coffee.

b. Better Bean Coffee keeps 80% of its customers but loses 10% to Alpha Coffee Concern and 10% to Carolyn's Certified Coffee.

c. Carolyn's Certified Coffee keeps 50% of its customers but loses 10% to Alpha Coffee and 40% to Better Bean Coffee.

Write the transition matrix for this information.

20. What is the market share of each brand of coffee in Problem 19 after 1 month? After 2 months?

21. What is the market share of each brand of coffee in Problem 19 in the long run?

22. What is the market share of each brand of coffee in Problem 19 in the long run if Alpha Coffee starts with 10%, Better Bean with 30%, and Carolyn's Certified Coffee with 60% of the market? What is the market share after 1 month?

23. In an eastern state a survey finds that if a person's father is a Democrat, the person will vote Democratic 70% of the time and Independent the rest of the time. If a person's father is a Republican, the person will vote Republican 50% of the time, Democratic 40% of the time, and Independent 10% of the time. Of the children of Independents, 40% of the time they will vote Independent, 40% Democratic, and 20% Republican. What is the long-range voting pattern for each party?

24. What is the probability that the grandchild of a Democrat will vote Independent in the eastern state described in Problem 23?

25. On Niels' cattle ranch it has been shown that the probability that a Jersey cow will have a Jersey calf is .6, that it will have a Guernsey calf is .2, or that it will have a white-faced calf is .2. The probabilities that a Guernsey cow will have a Jersey, Guernsey, or white-faced calf are .1, .7, and .2, respectively; and the probabilities that a white-faced cow will have a Jersey, Guernsey, or white-faced calf are .1, .1, and .8, respectively. Assume that the herd is now 30% Jersey, 20% Guernsey, and 50% white-faced. What will the distribution of the herd be in two generations?

26. What will the distribution of the herd of cattle described in Problem 25 be after a large number of generations?

27. Two companies compete against each other, and the transition matrix for people changing from one company to another each month is given by the following transition matrix:

		To company I	To company II
From company	I	.7	.3
	II	.2	.8

What is the long-range expectation for the companies if this trend continues?

28. In a recent close presidential race a poll finds that the incumbent would have to carry California, New York, and Texas to be reelected, so the campaign manager decides to devote the last week of the campaign to these states. She also decides that the incumbent will not campaign in the same state 2 days in a row and that if he campaigns in California, he will be twice as likely to campaign in New York the next day. Also, if he campaigns in New York, he will be twice as likely to campaign in California the next day. If he campaigns in Texas, he will be equally likely to campaign in California or New York the next day. What is the transition matrix? Is the transition matrix regular?

29. If the incumbent of Problem 28 spoke in Texas on Wednesday, what is the probability that he will be in California on Friday?

30. If the incumbent of Problem 28 spoke in California on Wednesday, what is the probability that he will be in California again on Friday?

*9.3
Absorbing Markov Chains

Under certain conditions a stochastic process will give rise to a transition matrix for a Markov chain with the property that once a given state is reached it is impossible to move out of that state. For example, suppose two people each have two quarters and decide to flip one coin each and call "match" or "no match." The winner takes both coins. If they play long enough, common sense tells us that eventually one player will go broke and the other player will end up with all four coins to put an end to the game. The states of having no or four coins are examples of what are called **absorbing states**.

Absorbing States

A state i of a Markov chain is an **absorbing state** if once the system reaches state i on some trial, the system remains in that state on all future trials. This will occur whenever $p_{ii} = 1$.

EXAMPLE 1 Let

$$T = \begin{matrix} \\ 1 \\ 2 \\ 3 \end{matrix} \begin{matrix} \begin{matrix} 1 & 2 & 3 \end{matrix} \\ \begin{bmatrix} .1 & .2 & .7 \\ 0 & 1 & 0 \\ .5 & .3 & .2 \end{bmatrix} \end{matrix}$$

Then $p_{12} = .2$ is the probability of going from state 1 to state 2. Since $p_{22} = 1$, the probability of going from state 2 to state 2 (that is, remaining in state 2) is 1. This says that state 2 is an absorbing state. Note also that if one entry in any row is 1, then the other entries must be 0s, since each row of a transition matrix is a probability vector. ■

EXAMPLE 2 Suppose two people each have two coins and decide to play match–no match. Write the transition matrix for the first player. The position p_{ij} is the probability of changing from state i to state j. In this example, state i means having i coins.

* Optional section.

Number of ending coins of player 1

Number of beginning coins of player 1

$$\begin{array}{c} \\ 0 \\ 1 \\ 2 \\ 3 \\ 4 \end{array} \begin{array}{c} \begin{array}{ccccc} 0 & 1 & 2 & 3 & 4 \end{array} \\ \begin{bmatrix} 1 & 0 & 0 & 0 & 0 \\ \frac{1}{2} & 0 & \frac{1}{2} & 0 & 0 \\ 0 & \frac{1}{2} & 0 & \frac{1}{2} & 0 \\ 0 & 0 & \frac{1}{2} & 0 & \frac{1}{2} \\ 0 & 0 & 0 & 0 & 1 \end{bmatrix} \end{array}$$

Note that states 0 and 4 are absorbing. Also note that at each state the player's bankroll will increase or decrease by one coin, each with a probability of $\frac{1}{2}$. ■

Suppose we want to examine the long-term trend for a matrix with one or more absorbing states. Let

$$T = \begin{array}{c} \\ 1 \\ 2 \\ 3 \end{array} \begin{array}{c} \begin{array}{ccc} 1 & 2 & 3 \end{array} \\ \begin{bmatrix} 1 & 0 & 0 \\ .1 & .5 & .4 \\ 0 & 0 & 1 \end{bmatrix} \end{array}$$

Both states 1 and 3 are absorbing states—once these states are entered, the system will remain in that state. If the Markov chain begins in state 2, it too must eventually end up in either state 1 or state 3. Now consider the powers of T (correct to two decimal places):

$$T^2 = \begin{bmatrix} 1 & 0 & 0 \\ .15 & .25 & .6 \\ 0 & 0 & 1 \end{bmatrix} \qquad T^4 = \begin{bmatrix} 1 & 0 & 0 \\ .19 & .06 & .75 \\ 0 & 0 & 1 \end{bmatrix}$$

$$T^8 = \begin{bmatrix} 1 & 0 & 0 \\ .20 & .00 & .80 \\ 0 & 0 & 1 \end{bmatrix}$$

It appears that if the system begins in state 2, the probability that it will end in absorbing state 1 is $\frac{1}{5}$ (entry p_{21}) and in absorbing state 3, $\frac{4}{5}$ (entry p_{23}). Now remember that with regular Markov chains the final result is independent of the initial state. This is not true for absorbing Markov chains, as we can see from this example.

Absorbing Markov Chain Theorem

If T is a transition matrix for an absorbing Markov chain, then:

1. $T, T^2, T^3, \ldots$ approach some particular matrix.

2. In a finite number of steps the chain will enter an absorbing state and stay there.

3. The long-term trend depends on the initial state.

The next step is to learn a procedure for finding the probabilities that a particular initial nonabsorbing state will lead to a particular absorbing state. The following theorem outlines this procedure:

Transition Theorem for Absorbing Markov Chains

Let T be a transition matrix for an absorbing Markov chain.

Step 1: Rearrange the rows and columns of T so that the absorbing states come first:

$$T = \left[\begin{array}{c|c} I_n & \mathbf{0} \\ \hline P & Q \end{array}\right]$$

where I_n is an identity matrix and $\mathbf{0}$ is a matrix consisting of all 0s. Matrices P and Q are the matrices consisting of the remaining entries shown in the indicated positions.

Step 2: Let $F = [I_m - Q]^{-1}$, where I_m is an identity matrix of the same order as Q. Remember, $[I_m - Q]^{-1}$ is the inverse of the matrix $[I_m - Q]$. Call F the **fundamental matrix**.

Step 3: FP is a matrix whose entries p_{ij} represent the probability that nonabsorbing state i will lead to absorbing state j.

EXAMPLE 3 Let

$$T = \begin{array}{c} 1 \\ 2 \\ 3 \end{array} \overset{\begin{array}{ccc} 1 & 2 & 3 \end{array}}{\begin{bmatrix} 1 & 0 & 0 \\ .1 & .5 & .4 \\ 0 & 0 & 1 \end{bmatrix}}$$

Find and interpret FP.

Solution *Step* 1:

$$\text{interchange rows 2 and 3: } \begin{array}{c} 1 \\ 3 \\ 2 \end{array} \overset{\begin{array}{ccc} 1 & 2 & 3 \end{array}}{\left[\begin{array}{ccc} 1 & 0 & 0 \\ 0 & 0 & 1 \\ \hline .1 & .5 & .4 \end{array}\right]} \qquad \text{interchange columns: } \begin{array}{c} 1 \\ 3 \\ 2 \end{array} \overset{\begin{array}{ccc} 1 & 3 & 2 \end{array}}{\left[\begin{array}{cc|c} 1 & 0 & 0 \\ 0 & 1 & 0 \\ \hline .1 & .4 & .5 \end{array}\right]}$$

$$I_2 = \begin{bmatrix} 1 & 0 \\ 0 & 1 \end{bmatrix} \qquad \mathbf{0} = \begin{bmatrix} 0 \\ 0 \end{bmatrix} \qquad P = [.1 \quad .4] \qquad Q = [.5]$$

Step 2: $F = [I_1 - Q]^{-1} = [1 - .5]^{-1} = [.5]^{-1} = [2]$

Step 3: $FP = [2][.1 \quad .4] = \overset{\begin{array}{cc} 1 & 3 \end{array}}{[.2 \quad .8]}\ 2$

This means that if the system begins with nonabsorbing state 2, the probability that it will end up in absorbing state 1 is .2 and in absorbing state 3 is .8. This is consistent with the arithmetic we did by finding T^2, T^4, and T^8 earlier. ■

EXAMPLE 4 Find and interpret FP for the game of match–no match and the matrix

$$T = \begin{array}{c} \\ 0 \\ 1 \\ 2 \\ 3 \\ 4 \end{array} \begin{array}{c} \begin{array}{ccccc} 0 & 1 & 2 & 3 & 4 \end{array} \\ \begin{bmatrix} 1 & 0 & 0 & 0 & 0 \\ \frac{1}{2} & 0 & \frac{1}{2} & 0 & 0 \\ 0 & \frac{1}{2} & 0 & \frac{1}{2} & 0 \\ 0 & 0 & \frac{1}{2} & 0 & \frac{1}{2} \\ 0 & 0 & 0 & 0 & 1 \end{bmatrix} \end{array}$$

Solution

$$\begin{array}{c} \\ 0 \\ 4 \\ 1 \\ 2 \\ 3 \end{array} \begin{array}{c} \begin{array}{ccccc} 0 & 1 & 2 & 3 & 4 \end{array} \\ \left[\begin{array}{ccccc} 1 & 0 & 0 & 0 & 0 \\ 0 & 0 & 0 & 0 & 1 \\ \hline \frac{1}{2} & 0 & \frac{1}{2} & 0 & 0 \\ 0 & \frac{1}{2} & 0 & \frac{1}{2} & 0 \\ 0 & 0 & \frac{1}{2} & 0 & \frac{1}{2} \end{array}\right] \end{array} \qquad \begin{array}{c} \\ 0 \\ 4 \\ 1 \\ 2 \\ 3 \end{array} \begin{array}{c} \begin{array}{ccccc} 0 & 4 & 1 & 2 & 3 \end{array} \\ \left[\begin{array}{cc|ccc} 1 & 0 & 0 & 0 & 0 \\ 0 & 1 & 0 & 0 & 0 \\ \hline \frac{1}{2} & 0 & 0 & \frac{1}{2} & 0 \\ 0 & 0 & \frac{1}{2} & 0 & \frac{1}{2} \\ 0 & \frac{1}{2} & 0 & \frac{1}{2} & 0 \end{array}\right] \end{array}$$

$$P = \begin{bmatrix} \frac{1}{2} & 0 \\ 0 & 0 \\ 0 & \frac{1}{2} \end{bmatrix} \qquad Q = \begin{bmatrix} 0 & \frac{1}{2} & 0 \\ \frac{1}{2} & 0 & \frac{1}{2} \\ 0 & \frac{1}{2} & 0 \end{bmatrix}$$

$$I_3 - Q = \begin{bmatrix} 1 & 0 & 0 \\ 0 & 1 & 0 \\ 0 & 0 & 1 \end{bmatrix} - \begin{bmatrix} 0 & \frac{1}{2} & 0 \\ \frac{1}{2} & 0 & \frac{1}{2} \\ 0 & \frac{1}{2} & 0 \end{bmatrix} = \begin{bmatrix} 1 & -\frac{1}{2} & 0 \\ -\frac{1}{2} & 1 & -\frac{1}{2} \\ 0 & -\frac{1}{2} & 1 \end{bmatrix}$$

$$F = (I_3 - Q)^{-1} = \begin{bmatrix} \frac{3}{2} & 1 & \frac{1}{2} \\ 1 & 2 & 1 \\ \frac{1}{2} & 1 & \frac{3}{2} \end{bmatrix}$$

Note: See Section 3.4 for the method for finding the inverse of a matrix.

$$FP = \begin{bmatrix} \frac{3}{2} & 1 & \frac{1}{2} \\ 1 & 2 & 1 \\ \frac{1}{2} & 1 & \frac{3}{2} \end{bmatrix} \begin{bmatrix} \frac{1}{2} & 0 \\ 0 & 0 \\ 0 & \frac{1}{2} \end{bmatrix}$$

$$= \begin{array}{c} \\ 1 \\ 2 \\ 3 \end{array} \begin{array}{c} \begin{array}{cc} 0 & 4 \end{array} \\ \begin{bmatrix} \frac{3}{4} & \frac{1}{4} \\ \frac{1}{2} & \frac{1}{2} \\ \frac{1}{4} & \frac{3}{4} \end{bmatrix} \end{array}$$

This means that if the system was originally in state 1 (player 1 having one coin), then the probability of financial ruin (0 coins) is $\frac{3}{4}$ and the probability that all four coins would be obtained is $\frac{1}{4}$. If the system began in state 2 (player 1 having two coins), then there is a probability of $\frac{1}{2}$ of either winning or losing both coins. If the system began in state 3 (player 1 having three coins), then there is a probability of $\frac{1}{4}$ of losing all three coins and $\frac{3}{4}$ of winning all four coins. ■

CALCULATOR COMMENT

Enter matrix T and repeat the following steps:

[A] x^2 ENTER x^2

ENTER

Repeat

x^2 ENTER

eight more times to find:

```
[  1   0  0  0  0   ]
[ .75  0  0  0 .25  ]
[ .5   0  0  0 .5   ]
[ .25  0  0  0 .75  ]
[  0   0  0  0  1   ]
```

Compare this with the answer obtained using the transition theorem.

9.3 Problem Set

Determine whether each matrix in Problems 1–12 is a transition matrix for an absorbing Markov chain. If it is, find the absorbing states.

1. $\begin{bmatrix} .3 & .5 & .2 \\ .1 & .7 & .2 \\ .2 & .6 & .2 \end{bmatrix}$

2. $\begin{bmatrix} .1 & .6 & .3 \\ 1 & 0 & 0 \\ .4 & .3 & .3 \end{bmatrix}$

3. $\begin{bmatrix} \frac{1}{2} & \frac{1}{3} & \frac{1}{6} \\ \frac{2}{3} & \frac{1}{6} & \frac{1}{12} \\ \frac{1}{6} & \frac{1}{3} & \frac{1}{2} \end{bmatrix}$

4. $\begin{bmatrix} \frac{1}{3} & \frac{1}{3} & \frac{1}{3} \\ 0 & 1 & 0 \\ \frac{1}{6} & \frac{1}{3} & \frac{1}{2} \end{bmatrix}$

5. $\begin{bmatrix} 0 & 1 & 0 \\ .3 & .2 & .5 \\ .5 & .5 & 0 \end{bmatrix}$

6. $\begin{bmatrix} 1 & 0 & 0 \\ .3 & .2 & .5 \\ .5 & .5 & 0 \end{bmatrix}$

7. $\begin{bmatrix} 1 & 0 & 0 \\ 0 & 1 & 0 \\ 0 & 0 & 1 \end{bmatrix}$

8. $\begin{bmatrix} 0 & 1 & 0 \\ 1 & 0 & 0 \\ 0 & 0 & 0 \end{bmatrix}$

9. $\begin{bmatrix} 1 & 0 & 0 \\ 0 & 1 & 0 \\ .41 & .34 & .25 \end{bmatrix}$

10. $\begin{bmatrix} \frac{1}{2} & \frac{1}{3} & \frac{1}{4} & 0 \\ 0 & 1 & 0 & 0 \\ \frac{1}{4} & \frac{1}{4} & \frac{1}{4} & \frac{1}{4} \\ 0 & 0 & 0 & 1 \end{bmatrix}$

11. $\begin{bmatrix} \frac{1}{2} & \frac{1}{4} & \frac{1}{5} & \frac{1}{20} \\ \frac{1}{3} & 0 & \frac{1}{3} & \frac{1}{3} \\ 0 & 0 & 1 & 0 \\ \frac{1}{5} & \frac{1}{5} & \frac{1}{5} & \frac{2}{5} \end{bmatrix}$

12. $\begin{bmatrix} 0 & 1 & 0 & 0 \\ 0 & 0 & 1 & 0 \\ 0 & 0 & 0 & 1 \\ 1 & 0 & 0 & 0 \end{bmatrix}$

Find and interpret FP for the matrices in Problems 13–24. If approximate answers are necessary, give them correct to two decimal places.

13. $\begin{bmatrix} .2 & 3 & .5 \\ 0 & 1 & 0 \\ 0 & 0 & 1 \end{bmatrix}$

14. $\begin{bmatrix} 1 & 0 & 0 \\ 1 & 8 & 1 \\ 0 & 0 & 1 \end{bmatrix}$

15. $\begin{bmatrix} 1 & 0 & 0 \\ 0 & 1 & 0 \\ .15 & .35 & .5 \end{bmatrix}$

16. $\begin{bmatrix} \frac{1}{3} & \frac{1}{3} & \frac{1}{3} \\ \frac{1}{2} & \frac{1}{4} & \frac{1}{4} \\ 0 & 0 & 1 \end{bmatrix}$

17. $\begin{bmatrix} 1 & 0 & 0 \\ .1 & .9 & 0 \\ .3 & .6 & .1 \end{bmatrix}$

18. $\begin{bmatrix} \frac{1}{4} & \frac{1}{3} & \frac{5}{12} \\ 0 & 1 & 0 \\ \frac{1}{2} & 0 & \frac{1}{2} \end{bmatrix}$

19. $\begin{bmatrix} 1 & 0 & 0 & 0 \\ \frac{1}{3} & \frac{1}{3} & 0 & \frac{1}{3} \\ \frac{1}{4} & \frac{1}{4} & \frac{1}{4} & \frac{1}{4} \\ 0 & 0 & 0 & 1 \end{bmatrix}$

20. $\begin{bmatrix} \frac{1}{2} & \frac{1}{2} & 0 & 0 \\ 0 & 1 & 0 & 0 \\ 0 & 0 & 1 & 0 \\ 0 & \frac{1}{3} & \frac{1}{3} & \frac{1}{3} \end{bmatrix}$

21. $\begin{bmatrix} 1 & 0 & 0 & 0 \\ .4 & .2 & .1 & .3 \\ 0 & 0 & 1 & 0 \\ .1 & .5 & .4 & 0 \end{bmatrix}$

22. $\begin{bmatrix} \frac{1}{5} & \frac{1}{5} & \frac{1}{5} & \frac{1}{5} & \frac{1}{5} \\ 0 & 1 & 0 & 0 & 0 \\ \frac{1}{3} & \frac{1}{3} & \frac{1}{3} & 0 & 0 \\ 0 & 0 & 0 & 1 & 0 \\ 0 & 0 & 0 & 0 & 1 \end{bmatrix}$

23. $\begin{bmatrix} 1 & 0 & 0 & 0 & 0 \\ .1 & .2 & .3 & .3 & .1 \\ .2 & .2 & .2 & .2 & .2 \\ 0 & 0 & 1 & 0 & 0 \\ 0 & 0 & 0 & 0 & 1 \end{bmatrix}$

24. $\begin{bmatrix} 1 & 0 & 0 & 0 & 0 \\ \frac{1}{3} & 0 & \frac{2}{3} & 0 & 0 \\ 0 & \frac{1}{3} & 0 & \frac{2}{3} & 0 \\ 0 & 0 & \frac{1}{3} & 0 & \frac{2}{3} \\ 0 & 0 & 0 & 0 & 1 \end{bmatrix}$

APPLICATIONS

25. **a.** Write a transition matrix for a match–no match game for two players, each with a single coin.
b. Find *FP*.
c. What is the probability of financial ruin for the first player?

26. **a.** Write a transition matrix for a match–no match game for two players with a total of six coins between them.
b. Find *FP*.
c. What is the probability of financial ruin if the first player begins with three coins?
d. What is the probability of financial ruin if the first player begins with two coins?
e. What is the probability of financial ruin if the first player begins with four coins?

27. Noxin, Inc., produces color computer monitors. Each monitor is tested and categorized before it is shipped. If it passes the test, it is shipped; if it needs minor repairs, it is returned to the assembly room for repair and retesting; if it then fails the test, it is destroyed. The probabilities for these events are summarized in the following transition matrix:

		Subsequent test		
		Failed	Returned	Passed
First test	Failed	1	0	0
	Returned	.1	.2	.7
	Passed	0	0	1

What is the probability that a returned item will eventually be passed?

28. Rennolds Department Store issues credit cards and has a policy of revoking any cards with accounts delinquent for 3 months. The auditor calculates the probabilities for the following transition matrix by looking at the records of 5,000 credit card customers:

First month	Subsequent month: Paid up	Delinquent 1 month	Delinquent 2 months	Revoke card
Paid up	.65	.35	0	0
Delinquent 1 month	.70	0	.30	0
Delinquent 2 months	.3	.4	0	.3
Revoke card	0	0	0	1

a. What is the probability that a paid-up customer will eventually have a revoked card?

b. What is the probability that a customer who is 1 month delinquent will eventually have a revoked card?

29. A rat is placed at random in one of the compartments of the maze shown here. When the rat is in some room, there are equal probabilities that it will choose any door in that room. When the rat reaches room 1, it is rewarded with food, and it no longer leaves that room. Room 3 has one-way doors that do not permit the rat to leave the room once it enters. What is the probability that the rat will end up in room 1 if it was originally placed in room 5?

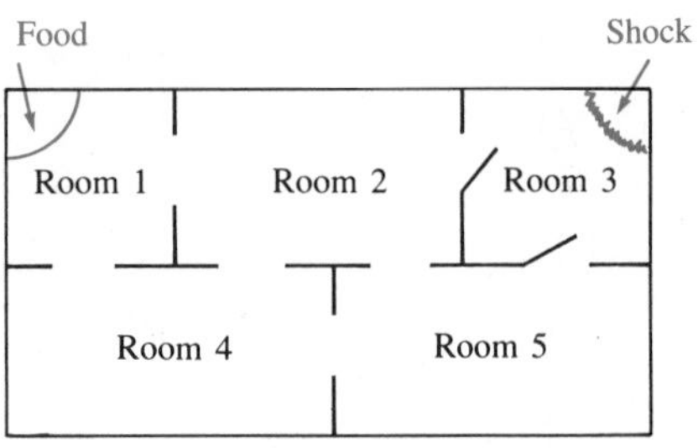

30. The experiment in Problem 29 is repeated. What is the probability that the rat will end up in room 1 if it is placed in room 2?

9.4 Expectation

Markov chains can be used in a management setting to make more intelligent decisions than might be made without using the theory. However, Markov chains are only one tool that can be used in making decisions. Another tool to use is **operations research**, which is the science of making optimal decisions. Operations research uses linear programming as well as a concept called **expected value**.

If we apply the notion of a weighted mean to the average value of a random variable, we obtain a measure called the *expected value* or **mathematical expectation**. This concept began as a means of measuring probable winnings in gambling situations, but today it has widespread application to the decision-making process in business, economics, and operations research. We will begin, however, by applying it to some simple gambling situations.

EXAMPLE 1 Suppose $1 is bet on a single number in a game of roulette. The roulette wheel has 38 numbered slots, and a ball is rotated around the wheel until it falls into one of the slots. There are many ways to place a bet, but if a single number is chosen and the ball lands on the chosen number, the bet is returned along with 35 times the original bet. If the ball lands on any other number, the bet is lost. Let the random variable X represent the amount won by the player. Then X has either of two values, 35 or -1, and

$$P(X = +35) = \frac{1}{38} \quad \text{and} \quad P(X = -1) = \frac{37}{38}$$

Suppose now that a player is a compulsive gambler who always bets on a single number, and makes n bets. The player should win approximately $\frac{n}{38}$ times and lose approximately $\frac{37n}{38}$ times. Since each win is \$35, and each loss is \$1, we have

$$\text{Total winnings:} \quad 35 \cdot \left(\frac{n}{38}\right) = \frac{35n}{38}$$

$$\text{Total losses:} \quad 1 \cdot \left(\frac{37n}{38}\right) = \frac{37n}{38}$$

$$\text{Net winnings:} \quad \frac{35n}{38} - \frac{37n}{38} = -\frac{2n}{38}$$

Negative winnings are interpreted as a net loss. If we are concerned with the *average* winnings per game, we divide the total by the number of trials, n, to obtain the expected value:

$$\text{Expected value} = -\frac{2}{38} = -\frac{1}{19}$$

This is a loss of about 5.263¢ per game.

■

In general:

Expected Value of a Random Variable

> If a random variable X has possible values $x_1, x_2, \ldots, x_n$ occurring with probabilities $p_1, p_2, \ldots, p_n$ $(\sum p_i = 1)$, we define the **expected value of X**, denoted by $E(X)$, by
>
> $$E(X) = x_1p_1 + x_2p_2 + \cdots + x_np_n$$

EXAMPLE 2 Suppose you are going to play a game by rolling a pair of dice. You will be paid \$5.00 every time you roll a pair of 6s. You will not receive anything for any other outcome. It costs \$.50 to play. What is the expected value for this game?

Solution Compare the wording of this example and Example 1. In Example 1 you would not lose the dollar you bet until after the game is played. In this example, there is an admission price of \$.50, which means that this amount is paid, win or lose. In situations for which there is an "admission price" this amount must be subtracted from the proposed winnings.

Let X be the amount won. Then

$$X = 5.00 - .50 = \$4.50 \quad \text{if you win the game}$$

$$X = -\$.50 \quad \text{if you lose the game}$$

and

$$P(X = \$4.50) = \frac{1}{36} \quad \text{and} \quad P(X = -\$.50) = \frac{35}{36}$$

Using the notation of the above definition,

$$E(X) = x_1p_1 + x_2p_2 \quad \text{where} \begin{cases} x_1 = 4.50 \\ x_2 = -.50 \end{cases} \quad \text{and} \quad \begin{matrix} p_1 = \frac{1}{36} \\ p_2 = \frac{35}{36} \end{matrix}$$

Thus

$$\begin{aligned} E(X) &= 4.50\left(\frac{1}{36}\right) + (-.50)\left(\frac{35}{36}\right) \\ &\approx .125 - .486 \\ &\approx -.36 \end{aligned}$$

When we say that the expectation is a loss of \$.36 per game it does *not* mean that you will lose \$.36 every time you play the game. (Indeed, you will *never* lose \$.36; you will either win \$5.00 or nothing—less the \$.50 admission charge.) But, if you were to play this game a large number of times, you could expect to lose an *average* of \$.36 per game. ■

We say that a game is **fair** if $E(X)$ is zero; if $E(X)$ is positive, the game is favorable to the player; and if $E(X)$ is negative, the game is unfavorable to the player.

EXAMPLE 3 A contest offers the following prizes:

Prize	Value	Probability of winning
Grand prize trip	$\$1{,}500 = x_1$	$.000026 = p_1$
Weber Kettle	$110 = x_2$	$.000032 = p_2$
Magic Chef Range	$279 = x_3$	$.000016 = p_3$
Murray Bicycle	$191 = x_4$	$.000021 = p_4$
Lawn Boy Mower	$140 = x_5$	$.000026 = p_5$
Samsonite Luggage	$183 = x_6$	$.000016 = p_6$

What is the expected value for this contest?

Solution

$$\begin{aligned} E(X) &= x_1p_1 + x_2p_2 + \cdots + x_6p_6 \\ &= 1{,}500(.000026) + 110(.000032) + 279(.000016) \\ &\quad + 191(.000021) + 140(.000026) + 183(.000016) \\ &= .057563 \end{aligned}$$

The expected value is a little less than 6¢. ■

Expectation is obviously useful in decision-making processes. For example, should you enter the contest in Example 3? What is the cost of entering? Do you need to mail in your entry? From a mathematical standpoint, you should not be willing to pay more than 5¢ to enter the contest. Let us now turn to a more practical problem in Example 4.

EXAMPLE 4 Karlin, Inc., manufactures automated vending machines and must choose between bidding on two contracts. Contract I will cost \$250 in preparation costs to make the bid, but if Karlin is selected, there will be a net profit of \$5,000. Contract II will cost \$100 to prepare, with a net profit of \$3,000 if the bid is won. The actual probability of winning contract I is 20% and of winning contract II is 25%. Should Karlin bid on contract I or contract II?

Solution The solution can be found by calculating the expectation:

$$\begin{aligned} E(\text{contract I}) &= 5{,}000(.20) + (-250)(.80) \\ &= 1{,}000 + (-200) \\ &= 800 \\ E(\text{contract II}) &= 3{,}000(.25) + (-100)(.75) \\ &= 750 + (-75) \\ &= 675 \end{aligned}$$

Thus, on the basis of the given information, Karlin should bid on contract I. ■

EXAMPLE 5 A \$150,000 house has an unstable foundation and is sliding off a hillside. The only hope for the house is to drill a horizontal well to tap a reservoir of stored water that is causing the slippage. If the cost of drilling each well is \$1,000, and if the probability that a particular well will hit the water reservoir is .8, how many wells should be drilled?

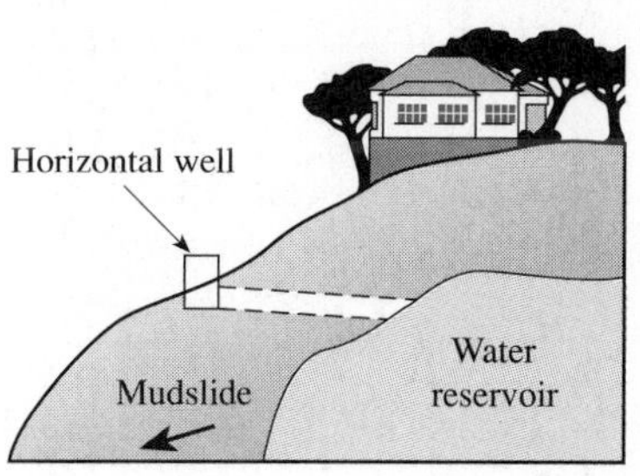

Solution It is obvious that drilling no well is too few and drilling 150 wells is too many. Our task is to decide on the number of wells, x, to be drilled. The probability that the x wells will be unsuccessful is $.2^x$ (assume that the probability of hitting water for each well is independent). Thus the probability that at least one well will hit the water reservoir is

$$1 - .2^x$$

The expected gain is $\$150{,}000(1 - .2^x)$, while the cost of drilling the wells is $\$1{,}000x$, so

$$E(x) = \$150{,}000(1 - .2^x) - \$1{,}000x$$

We want to find the maximum value of $E(x)$ for the various values of x.

$E(0) = 0$ If no wells are tried, the house will slide off the hill and be lost.

$$\begin{aligned} E(1) &= \$150{,}000(1 - .2^1) - \$1{,}000 \cdot 1 \\ &= \$119{,}000 \\ E(2) &= \$150{,}000(1 - .2^2) - \$1{,}000 \cdot 2 \\ &= \$142{,}000 \\ E(3) &= \$150{,}000(1 - .2^3) - \$1{,}000 \cdot 3 \\ &= \$145{,}800 \end{aligned}$$

CALCULATOR COMMENT

You might find it interesting to graph the curve

$$y = 150{,}000(1 - .2^x) - 1{,}000x$$

The graph on a TI-81 looks like:

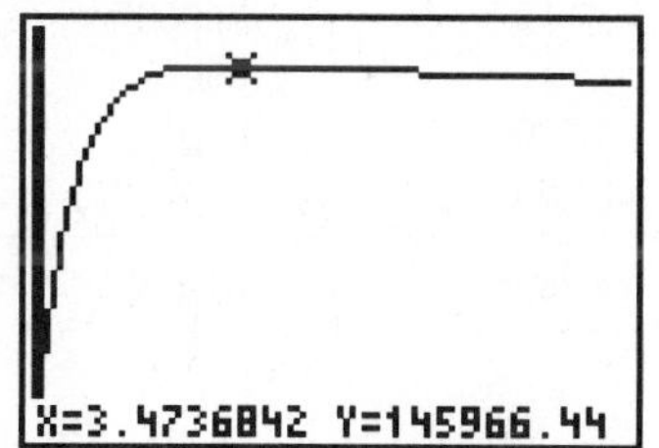

The trace tells us the maximum occurs near $x = 3.5$, but since x must be a positive integer, the maximum (using the trace) shows: X=3 Y=145800

$$E(4) = \$150{,}000(1 - .2^4) - \$1{,}000 \cdot 4$$
$$= \$145{,}760$$
$$E(5) = \$150{,}000(1 - .2^5) - \$1{,}000 \cdot 5$$
$$= \$144{,}952$$
$$E(6) = \$150{,}000(1 - .2^6) - \$1{,}000 \cdot 6$$
$$= \$143{,}990.40$$

Note that sinking additional wells does not increase the expected gain. The expected value is greatest when three wells are drilled, so operations research tells us to drill three wells. ■

9.4 Problem Set

Given the probability functions in Problems 1–6, find the expected value of X.

1.

X	1	2	3
$P(X)$	$\frac{1}{4}$	$\frac{1}{2}$	$\frac{1}{4}$

2.

X	5	10	21
$P(X)$	.2	.5	.3

3.

X	3	6	9	12
$P(X)$	.4	.3	.2	.1

4.

X	4	9	12	18
$P(X)$	$\frac{1}{4}$	$\frac{1}{3}$	$\frac{1}{4}$	$\frac{1}{6}$

5.

X	0	1	2	3	4
$P(X)$	.10	.25	.30	.20	.15

6.

X	1	4	9	16	25
$P(X)$	.5	.25	.10	.10	.05

APPLICATIONS

7. A magazine subscription service is having a contest in which the prize is \$80,000. If the company receives 1 million entries, what is the mathematical expectation of entering the contest?

8. Krinkles potato chips is having a "Lucky Seven Sweepstakes." The grand prize is \$70,000; 7 second prizes each pay \$7,000; 77 third prizes each pay \$700; and 777 fourth prizes each pay \$70. How much is the expectation of entering this contest if we assume that there are 10 million entries?

9. A company presents the following table for its contest:

Prize	Prizes available	Approximate probability of winning
\$1,000.00	13	.000005
100.00	52	.00002
10.00	520	.0002
1.00	28,900	.010989
Total	29,485	.011111*

* This is $\frac{1}{90}$.

What is the expectation for playing this game one time?

10. A company holds a bingo card contest where the following chances of winning are given:

	One card 1 time	One card 7 times	One card 13 times
$25 prize	1 in 21,252	1 in 3,036	1 in 1,630
$ 3 prize	1 in 2,125	1 in 304	1 in 163
$ 1 prize	1 in 886	1 in 127	1 in 68
Any prize	1 in 609	1 in 87	1 in 47

What is the expectation in dollars for playing one card 13 times?

Life insurance policies use mathematical expectation. The "winning" value is the value of the policy and the "winning" probability is the probability of dying during the life of the policy. The probability of dying is found from an actuarial table similar to Table 9 in Appendix E. Note that the table gives p_x (probability of living) and q_x (probability of dying). Find the expected value for the 1-year policies described in Problems 11–16.

11. $10,000 issued at age 10

12. $10,000 issued at age 38

13. $10,000 issued at age 65

14. $25,000 issued at age 19

15. $50,000 issued at age 19

16. $125,000 issued at age 35

A U.S. roulette wheel has 38 numbered slots (1–36, 0, and 00, as shown in the figure). Some of the more common bets and payoffs are shown in the figure. If the payoff is listed as 6 to 1, the player would receive $6 plus the $1 bet. Use the figure to find the expectation in Problems 17–25 for playing $1 one time. One play consists of the dealer spinning the wheel and a small white ball in opposite directions. As the ball slows to a stop it lands in one of the 38 numbered slots, which are also colored black, red, or green.

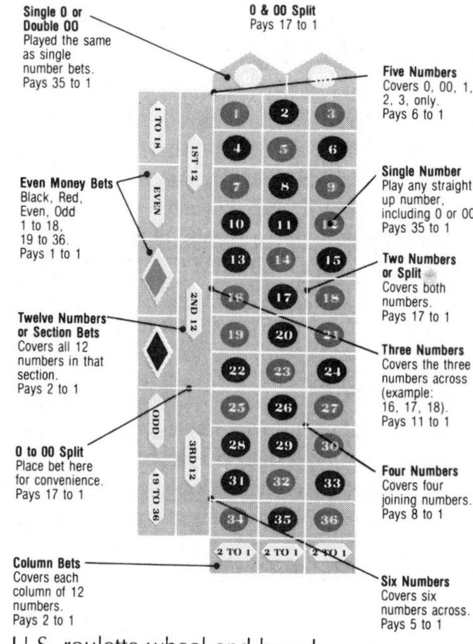

U.S. roulette wheel and board

17. Black

18. Odd

19. Single-number bet

20. Double-number bet

21. Three-number bet

22. Four-number bet

23. Five-number bet

24. Six-number bet

25. Column bet

26. A realtor who takes the listing on a house to be sold knows that she will spend $800 trying to sell the house. If she sells it herself, she will earn 6% of the selling price. If another realtor sells a house from her list, our realtor will earn only 3% of the price. If the house is unsold in 6 months, she will lose the listing. Suppose the probabilities are as follows:

	Probability
Sell by herself	.50
Sell by another realtor	.30
Not sell in 6 months	.20

What is the expected profit for listing an $85,000 house? (Be sure to subtract the $800 from the commission because the profit is 6% of the selling price *less* selling costs.)

27. An oil drilling company knows that it will cost $25,000 to sink a test well. If oil is hit, the income for the drilling company will be $425,000. If only natural gas is hit, the income will be $125,000. If nothing is hit, the company will have no income. If the probability of hitting oil is $\frac{1}{40}$ and if the probability of hitting gas is $\frac{1}{20}$, what is the expectation for the drilling company? Should the test well be sunk? (Do not forget to subtract the cost from the income in determining the payoffs.)

28. In Problem 27, suppose the income for hitting oil is changed to $825,000 and the income for gas to $225,000. Now, what is the expectation for the drilling company? Should the test well be sunk?

29. Suppose you roll one die. You are paid $5 if you roll a 1, and you pay $1 otherwise. What is the expectation?

30. A game involves drawing a single card from an ordinary deck. If an ace is drawn, the player receives 50¢; if a face card is drawn, the player receives 25¢; if the two of spades is drawn, the player receives $1. If the cost of playing is 10¢, should you play?

31. Consider the following game where a player rolls a single die: If a prime (2, 3, or 5) is rolled, the player wins $2. If a square (1 or 4) is rolled, the player wins $1. However, if the player rolls a perfect number (6), then it costs the player $11. Is this a good deal for the player or not?

32. Suppose the probability of hitting the reservoir in Example 5 is .6 instead of .8. Now, how many wells should be tried?

33. Suppose each well in Example 5 costs $5,000 to drill instead of $1,000. Now, how many wells should be tried?

34. Kingston and Associates, a consulting firm, is hired by a national car rental agency to determine the number of cars to purchase for a new outlet. After appropriate research, the consulting firm finds that the daily demand will be from 12 to 17 cars, distributed as follows:

Number of customers	12	13	14	15	16	17
Probability	.05	.1	.35	.2	.2	.1

Records also show that it costs the agency $20 per day per car whether the car is rented or not. If the rental price per car is $32, then the profit is $12 for each day the car is rented. What is the optimal number of cars that the consulting firm should recommend the agency to obtain?

35. Repeat Problem 34 for the following values:

Number of customers	10	11	12	13	14	15
Probability	.1	.2	.3	.2	.1	.1

9.5 Game Theory

Another application of decision theory is concerned with the analysis of certain types of conflict that may be either real or artificial—for example, a game of cards or chess played between friends, union and management at a bargaining table, a presidential election, opposing armies on a battlefield, an oil company drilling for oil (oil company versus nature).

The outcomes in the games of chess and poker, for example, depend on more than mere chance. A player must play according to some set of strategies and in anticipation of what the opponent will do. The analysis of these conflict situations and their corresponding strategies for decision making are developed in a relatively recent branch of mathematics called **game theory**. Game theory is primarily concerned with the logic of strategy and was first envisioned by the great German mathematician Gottfried Leibniz (1646–1716). However, the theory of games as we know it today was developed in the 1920s by John von Neumann and Emile Borel. It gained wide acceptance in 1944 in a book titled *Theory of Games and Economic Behavior* by von Neumann and Oskar Morgenstern.

Let us begin our study of game theory by considering a rather simple game. Suppose a friend of ours, Linda, likes to play games and suppose we will act as her consultant in matters relating to game theory. Her opponent, therefore, is our opponent.

The first game Linda decides to play is a variation of the game of two-finger mora. The game consists of two persons simultaneously holding out either one or two fingers from a closed fist. Linda's opponent makes her the following offer: "Let's play two-finger mora. If the sum of the fingers is two, I'll pay you 5¢. If it is four, you pay me 5¢. If the sum is odd, we break even."

This is an example of a two-person **zero-sum game**. If the payoffs for a game are such that a win for one player results in a corresponding loss for the other player, then we describe the situation as a zero-sum game. The game of two-finger mora can be summarized by a 2×2 **game matrix**:

$$\begin{array}{cc} & \text{Opponent} \\ & \begin{array}{cc} 1 & 2 \end{array} \\ \text{Linda} \begin{array}{c} 1 \\ 2 \end{array} & \begin{bmatrix} 5 & 0 \\ 0 & -5 \end{bmatrix} \end{array}$$

Here is how to read this game matrix:

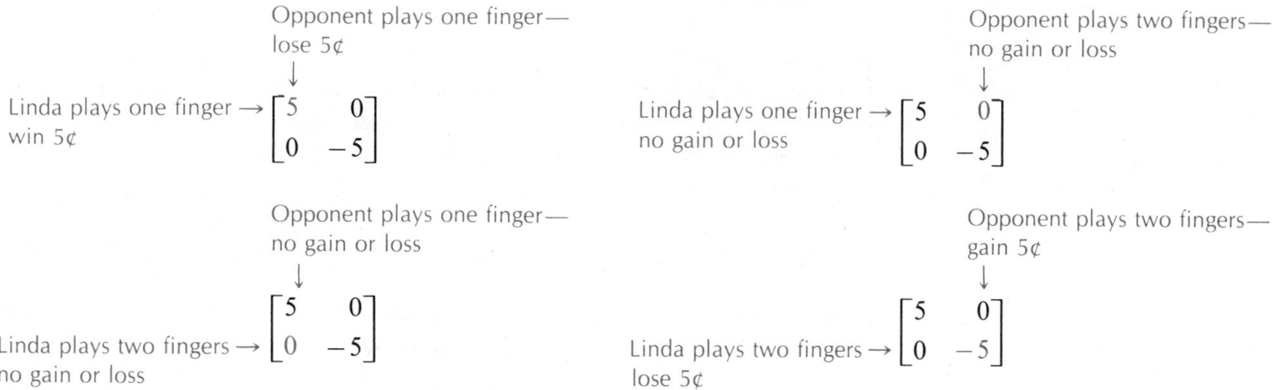

Notice that a gain for Linda is a loss for the opponent and a loss for Linda is a gain for the opponent. This is what we mean by a zero-sum game.

We would like to develop a **strategy** for this game. Each row of the game matrix represents a strategy for Linda, while each column represents a strategy for her opponent. If we examine the game matrix, we see that Linda's best strategy is to show one finger since the *worst* she can do is break even. On the other hand, the best strategy for her opponent is to show two fingers since the worst he can do is break even. The payoff when they both play their best strategies is called the **value of the game**. We see that the value of this game is 0, and whenever the value of a game is 0 we call the game **fair**.

EXAMPLE 1 Determine the best strategies and value of the game of three-finger mora where player II agrees to pay player I the difference between the fingers shown (number of fingers shown by player I minus the number shown by player II).

Solution First we write the game matrix:

		Player II 1	2	3
	1	0	−1	−2
Player I	2	1	0	−1
	3	2	1	0

We see that player I can reason as follows: "If I play one finger, then I stand to lose 2; if I play two fingers, I stand to lose 1; if I play three fingers, the *worst* I can do is break even." We see that player I looks for the *minimum entry* in each row. On the other hand, player II looks for the *maximum entry* in each column (since the negative values indicate a win for player II). He reasons, "If I play column 1, I stand to lose 2; column 2, I can lose 1; but if I play column 3, the *worst* I can do is break even." In this game we see that player I would play three fingers and player II would also play three fingers. The value of the game is 0. ■

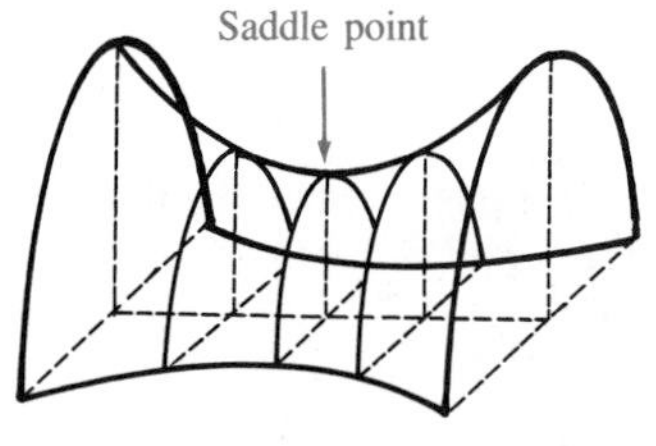

FIGURE 9.1
Saddle point

Games such as this are said to be **strictly determined** since a knowledge of our opponent's strategy would not alter our own strategy. In a strictly determined game the smallest entry in some row is also the largest entry in its column. This point is called a **saddle point**. The reason it is called a saddle point is because it can be visualized by thinking of the surface of a saddle. The same point is a maximum for one person and at the same time a minimum for another. There is such a point on a saddle, as we can see in Figure 9.1. The **value of a strictly determined game** is the value of this saddle point.

If a matrix game has a saddle point, then the row containing the saddle point is the best strategy for player I, and the column containing the saddle point is the best strategy for player II.

Let us now look at the matrices in Examples 2–4 and decide if the game determined by each matrix is strictly determined. We will use the following procedure:

Step 1: Place an asterisk (*) next to the minimum of each row.

Step 2: Check to see if each element marked by an * is the maximum in its column. If so, circle it. This is a saddle point and the game is strictly determined.

EXAMPLE 2

				Row minimum
	1	−2	−3*	−3
	2	4	(1*)	1
	−3	−4*	−2	−4
Column maximum	2	4	1	

The saddle point is circled, and the value of the game is 1. The game is strictly determined. ■

EXAMPLE 3

$$
\begin{array}{r}
\\
\\
\\
\text{Column maximum}
\end{array}
\begin{array}{c}
\begin{bmatrix} 2 & 1 & -3^* \\ -2^* & 0 & 2 \\ 3 & -1^* & 1 \end{bmatrix} \\
\begin{array}{ccc} 3 & \;\;1 & \;\;\;2 \end{array}
\end{array}
\begin{array}{c}
\text{Row minimum} \\ -3 \\ -2 \\ -1 \\ \\
\end{array}
$$

There is no saddle point. In this case, we say that the game is **nonstrictly determined**. ■

EXAMPLE 4

$$
\begin{array}{r}
\\
\\
\\
\\
\text{Column maximum}
\end{array}
\begin{array}{c}
\begin{bmatrix} 0 & 5 & -2^* & 3 \\ \enclose{circle}{3^*} & 4 & \enclose{circle}{3^*} & 5 \\ 2 & 0 & 1 & -1^* \\ \enclose{circle}{3^*} & 5 & \enclose{circle}{3^*} & 6 \end{bmatrix} \\
\begin{array}{cccc} 3 & \;\;5 & \;\;3 & \;\;\;6 \end{array}
\end{array}
\begin{array}{c}
\text{Row minimum} \\ -2 \\ 3 \\ -1 \\ 3 \\ \\
\end{array}
$$

The saddle points are circled, and the value of the game is 3. We see that a game can have more than one saddle point, but all saddle points must be numerically the same. The game is strictly determined. ■

Now consider a game that is nonstrictly determined. Linda's friend says he is tired of playing two-finger mora and has thought of another game. He says it is played with a dime and a nickel. Each of them will select one of two coins and put it in a hand. On a given signal they will open their hands together. If the coins match, Linda will win both. If the coins do not match, her friend will win both coins. We advise Linda to play this silly game, provided we can first write out the game matrix.

$$
\begin{array}{cc}
 & \text{Opponent} \\
 & \begin{array}{cc} \text{Nickel} & \text{Dime} \end{array} \\
\text{Linda} \begin{array}{r} \text{Nickel} \\ \text{Dime} \end{array} & \begin{bmatrix} 5 & -5 \\ -10 & 10 \end{bmatrix}
\end{array}
$$

We quickly check for a saddle point and see that there is none, so this is not a strictly determined game. Game theory will not help Linda play this game if she plays only once, but if she plays over and over, we can try to develop a **mixed strategy** that will tell her how often to play a nickel and how often to play a dime.

Let p be the probability that Linda will choose row 1 and q the probability that her opponent will choose column 1 (that is, both choose nickels). Then p and q are numbers between 0 and 1 (inclusive). Also, since Linda must choose either row 1 or row 2, the probability she will choose row 2 is $1 - p$. Likewise, column 2 for her opponent will be chosen with the probability $1 - q$. Now, if we represent Linda's strategy by the row matrix

$$[p \quad 1 - p]$$

and her opponent's strategy by the column matrix

$$\begin{bmatrix} q \\ 1-q \end{bmatrix}$$

we mean that Linda will choose row 1 of the game matrix with probability p and her opponent will choose column 1 with probability q.

Since Linda and her opponent are making their choices independently, the probability that the payoff to Linda is 5¢ is pq:

$$\underset{\text{Linda picks row 1}\ \downarrow\ \downarrow\ \text{Opponent picks column 1}}{pq}$$

If Linda's opponent plays column 1, then Linda's expectation is

$$5p - 10(1-p)$$

If her opponent plays column 2, then her expectation is

$$-5p + 10(1-p)$$

These expectations can be summarized by using matrix multiplication:

$$\begin{aligned}[p \quad 1-p]\begin{bmatrix} 5 & -5 \\ -10 & 10 \end{bmatrix} &= [5p - 10(1-p) \quad -5p + 10(1-p)] \\ &= [5p - 10 + 10p \quad -5p + 10 - 10p] \\ &= [15p - 10 \quad -15p + 10]\end{aligned}$$

In 1928 John von Neumann proved that if a game is not strictly determined, then the best strategy for the first player (Linda) is to choose p so that the entries of this product matrix are equal. That is,

$$\begin{aligned} 15p - 10 &= -15p + 10 \\ 30p &= 20 \\ p &= \frac{2}{3} \end{aligned}$$

Thus Linda's best strategy for this game is to pick row 1 two-thirds of the time and row 2 one-third of the time. For her opponent, find

$$\begin{aligned}\begin{bmatrix} 5 & -5 \\ -10 & 10 \end{bmatrix}\begin{bmatrix} q \\ 1-q \end{bmatrix} &= [5q - 5(1-q) \quad -10q + 10(1-q)] \\ &= [5q - 5 + 5q \quad -10q + 10 - 10q] \\ &= [10q - 5 \quad -20q + 10]\end{aligned}$$

Now we solve:

$$\begin{aligned} 10q - 5 &= -20q + 10 \\ 30q &= 15 \\ q &= \frac{1}{2} \end{aligned}$$

The best strategy for Linda's opponent is to choose each column one-half of the time.

Von Neumann also proved that if both players select the optimum strategy, then the expectation of winning for the row player is the same as the expectation of losing for the column player. This is called Von Neumann's **minimax theorem**. This common expectation is called the **value of the nonstrictly determined game**. Thus, to calculate the value of the game, we can use either player's optimum strategy:

$$\begin{bmatrix} \frac{2}{3} & \frac{1}{3} \end{bmatrix} \begin{bmatrix} 5 & -5 \\ -10 & 10 \end{bmatrix} \quad \text{or} \quad \begin{bmatrix} 5 & -5 \\ -10 & 10 \end{bmatrix} \begin{bmatrix} \frac{1}{2} \\ \frac{1}{2} \end{bmatrix}$$

We can use the product in *any* row or column to find the value (we choose column 1 of the first product):

$$\frac{2}{3}(5) + \frac{1}{3}(-10) = \frac{10}{3} - \frac{10}{3} = 0$$

The value of this game is 0, which means that it is a fair game.

Optimal Strategy Theorem

In the zero-sum two-person game

$$\begin{bmatrix} a & b \\ c & d \end{bmatrix}$$

with strategies

$$\begin{bmatrix} p & 1-p \end{bmatrix} \quad \text{and} \quad \begin{bmatrix} q \\ 1-q \end{bmatrix}$$

the optimal strategy for the row player occurs when the entries of the product

$$\begin{bmatrix} p & 1-p \end{bmatrix} \begin{bmatrix} a & b \\ c & d \end{bmatrix}$$

are equal, provided that the game matrix has no saddle point. The optimal strategy for the column player occurs when the entries of the product

$$\begin{bmatrix} a & b \\ c & d \end{bmatrix} \begin{bmatrix} q \\ 1-q \end{bmatrix}$$

are equal, provided that the game matrix has no saddle point.

EXAMPLE 5 Find the optimal strategies and value for the game

$$\begin{bmatrix} -10 & 5 \\ 8 & -10 \end{bmatrix}$$

Solution There is no saddle point. Next, find the products. For the row player:

$$\begin{bmatrix} p & 1-p \end{bmatrix} \begin{bmatrix} -10 & 5 \\ 8 & -10 \end{bmatrix} = \begin{bmatrix} -10p + 8(1-p) & 5p - 10(1-p) \end{bmatrix}$$

Solve:

$$\begin{aligned}
-10p + 8(1-p) &= 5p - 10(1-p) \\
-10p + 8 - 8p &= 5p - 10 + 10p \\
-18p + 8 &= 15p - 10 \\
-33p &= -18 \\
p &= \frac{18}{33}
\end{aligned}$$

For the column player:

$$\begin{bmatrix} -10 & 5 \\ 8 & -10 \end{bmatrix} \begin{bmatrix} q \\ 1-q \end{bmatrix} = \begin{bmatrix} -10q + 5(1-q) \\ 8q - 10(1-q) \end{bmatrix}$$

Solve:

$$\begin{aligned}
-10q + 5(1-q) &= 8q - 10(1-q) \\
-10q + 5 - 5q &= 8q - 10 + 10q \\
-15q + 5 &= 18q - 10 \\
-33q &= -15 \\
q &= \frac{15}{33}
\end{aligned}$$

Thus the optimal strategies are $[\frac{18}{33} \;\; \frac{15}{33}]$ and $\begin{bmatrix} \frac{15}{33} \\ \frac{18}{33} \end{bmatrix}$. The value of the game is

$$-10\left(\frac{18}{33}\right) + 8\left(1 - \frac{18}{33}\right) = -\frac{60}{33}$$

The game is biased in favor of the second player since the value of the game is negative. ■

Sometimes the original game is not 2×2, but can be reduced to a 2×2 game matrix by eliminating certain rows or columns. Consider the following game matrix:

Opponent

$$\text{Linda} \quad \begin{bmatrix} 2 & 4 & -3 \\ -1 & 3 & 2 \\ 0 & -1 & -4 \end{bmatrix}$$

In finding the optimum strategies we first look for a saddle point and see that there is none. Suppose Linda is playing this game with an opponent. Certainly she would never choose row 3 since every entry in row 3 is smaller than the corresponding entry in row 1. She can therefore eliminate row 3 from any further consideration. We say that row 1 **dominates** row 3. *We can delete* **dominated rows**. The result is a 2×3 *subgame* of the original game:

Opponent

$$\text{Linda} \quad \begin{bmatrix} 2 & 4 & -3 \\ -1 & 3 & 2 \end{bmatrix}$$

On the other hand, Linda's opponent wants to make his choices as small as possible (negative values are good for him). He would not pick column 2 as a strategy since each entry in column 1 is smaller than the corresponding entry in column 2. We say that column 2 dominates column 1 and the opponent would *delete the* **dominating column** from further consideration. The result is a 2×2 subgame of the original game:

$$\text{Linda} \overset{\text{Opponent}}{\begin{bmatrix} 2 & -3 \\ -1 & 2 \end{bmatrix}}$$

The following principle can be applied to any matrix game:

Dominating Principle

> In a given game matrix, *dominated rows* can be eliminated (*eliminate the smaller row*). *Dominating columns* can be eliminated (*eliminate the larger column*).

EXAMPLE 6 Find the strategies for the game matrix:

$$\text{Linda} \overset{\text{Opponent}}{\begin{bmatrix} 2 & 4 & -3 \\ -1 & 3 & 2 \\ 0 & -1 & -4 \end{bmatrix}}$$

Solution Eliminate dominated rows (row 3) and dominating columns (column 2), leaving the following subgame:

$$\begin{bmatrix} 2 & -3 \\ -1 & 2 \end{bmatrix}$$

Solve this subgame (the details are left for you) to obtain:

$$\text{Linda's strategy} = \begin{bmatrix} \frac{3}{8} & \frac{5}{8} \end{bmatrix} \qquad \text{Opponent's strategy} = \begin{bmatrix} \frac{5}{8} \\ \frac{3}{8} \end{bmatrix}$$

Next, it is necessary to state the strategy for the *original game*. Use a 0 to denote the eliminated row and column. Thus, since row 3 was deleted from Linda's strategy, we write

$$\begin{bmatrix} \frac{3}{8} & \frac{5}{8} & 0 \end{bmatrix}$$

and since column 2 was deleted, the opponent's strategy is

$$\begin{bmatrix} \frac{5}{8} \\ 0 \\ \frac{3}{8} \end{bmatrix}$$

■

9.5

Problem Set

Determine the optimal strategies and the value of the game in Problems 1–18.

1. $\begin{bmatrix} 4 & 0 \\ 3 & -2 \end{bmatrix}$

2. $\begin{bmatrix} 0 & -3 \\ -2 & 0 \end{bmatrix}$

3. $\begin{bmatrix} 1 & -1 \\ -1 & 1 \end{bmatrix}$

4. $\begin{bmatrix} 1 & -2 \\ 0 & 2 \end{bmatrix}$

5. $\begin{bmatrix} 3 & -2 \\ 1 & -3 \end{bmatrix}$

6. $\begin{bmatrix} 10 & 0 \\ -10 & 5 \end{bmatrix}$

7. $\begin{bmatrix} 5 & -1 \\ 3 & -2 \end{bmatrix}$

8. $\begin{bmatrix} 1.5 & -.5 \\ .5 & 2.5 \end{bmatrix}$

9. $\begin{bmatrix} -1 & 0 \\ \frac{1}{4} & -\frac{1}{4} \end{bmatrix}$

10. $\begin{bmatrix} 2 & 1 & 3 \\ -2 & 0 & 3 \\ 4 & -2 & -3 \end{bmatrix}$

11. $\begin{bmatrix} 2 & 3 & 3 \\ 1 & 0 & -1 \\ 0 & 0 & 4 \end{bmatrix}$

12. $\begin{bmatrix} 2 & 1 & 3 \\ 0 & 1 & -2 \\ -1 & 0 & -3 \end{bmatrix}$

13. $\begin{bmatrix} 4 & -2 & 3 \\ 3 & -3 & 1 \\ 2 & 0 & -1 \end{bmatrix}$

14. $\begin{bmatrix} -1 & -8 & -5 \\ 1 & 5 & -4 \\ -5 & 7 & 2 \end{bmatrix}$

15. $\begin{bmatrix} 5 & -3 & -4 \\ 4 & -1 & 0 \\ 3 & -3 & 6 \end{bmatrix}$

16. $\begin{bmatrix} 0 & 1 & 2 & 0 \\ 1 & 2 & 2 & 1 \\ 3 & 0 & -1 & 0 \\ 2 & -1 & 0 & -2 \end{bmatrix}$

17. $\begin{bmatrix} 2 & -1 & 4 & 3 \\ -4 & -5 & -2 & 1 \\ -1 & 0 & 3 & 0 \end{bmatrix}$

18. $\begin{bmatrix} 9 & 10 & 11 & 9 \\ 10 & 11 & 11 & 10 \\ 12 & 9 & 8 & 9 \\ 11 & 8 & 9 & 8 \end{bmatrix}$

APPLICATIONS

19. A friend suggests the following game: He will hide a quarter, dime, or half-dollar in one hand behind his back. You are to guess which coin he has. If you guess correctly, you get the coin. If you guess incorrectly, he gets the difference between your guess and the coin held. Write the game matrix and say if the game is strictly determined.

20. A friend suggests the following four-finger mora: If the sum is even, you win an amount equal to the sum of the fingers shown; if the sum is odd, your friend wins an amount equal to the sum of the fingers shown. Write the game matrix and say if the game is strictly determined.

21. A general with two regiments is trying to capture a city. He can attack from the north with both regiments, from the south with both regiments, or from the north with one regiment and from the south with one regiment. The defending forces also have two regiments that can be deployed for protection in the north, in the south, or one in the north and one in the south. The general of the attacking forces will gain 1 point if the attack succeeds, −1 if the attack fails, and 0 if the forces are held at a standoff. Whichever force has more regiments in an area wins the battle, and if one regiment is deployed against one regiment, the armies are held at a standoff. Write the game matrix and state whether this is a strictly determined game.

22. A friend proposes the following game: "We each flip an imaginary coin and call out heads or tails without knowing each other's choice. If there is a match, I win $1, and if there is no match, I lose $1." What are the strategies for the friend and her opponent, and what is the value of the game?

23. In a presidential campaign the Democratic and Republican candidates can campaign in either urban or rural areas. The units assigned to each choice are in gains or losses of thousands of votes:

		Republican	
		Urban	Rural
Democratic	Urban	−5	3
	Rural	4	2

What are the strategies, and what is the value of the game?

24. In submarine warfare, a submarine can attack enemy ships from shallow depths or can launch rockets toward enemy ships from deep water. The shallow-depth attack gives more accurate results but is also more dangerous for the submarine. The surface ships can also drop depth charges set for shallow detonation or drop charges set for deep detonation. If a submarine avoids the depth charges, we credit it with 50 points if it was deep and 80 points if it was shallow (since it is more effective if it is shallow). If it is hit with a depth charge, it is credited with −100 points. What are the strategies, and what is the value of the game?

25. In Sir Arthur Conan Doyle's story "The Final Problem," Sherlock Holmes is pursuing his archenemy Professor Moriarty. The professor is out to kill Holmes, whose only chance for escape is to flee to the Continent by taking a train from London to Dover. Just as the train is leaving London, the two men see each other and the professor is left at the station. Holmes knows that if he meets Moriarty again it means certain death. Let us assign this occurrence the value of -100 (for Holmes). Holmes can stay on the train until he reaches Dover or he can get off at Canterbury. What should he do? If he eludes Moriarty and makes it to Dover, we will assign this occurrence the value of 50 (for Holmes). If he eludes Moriarty but only reaches Canterbury, we will call this a draw and assign this occurrence the value of 0. On the other hand, Moriarty can catch Holmes by chartering a special train, but must decide whether to go to Canterbury or to Dover. Find the strategies for Holmes and Moriarty. Assume that the results are for repeated plays of the "game."

26. McDonald's is planning to build a restaurant in a large city. It can build uptown, downtown, or in the suburbs. Burger King also plans to build a restaurant in the same city. The payoff matrix in thousands of dollars is:

		Burger King		
		Uptown	Downtown	Suburbs
	Uptown	4	1	−3
McDonald's	Downtown	−2	0	−5
	Suburbs	4	2	3

What are the strategies, and what is the value of the game?

*9.6 $m \times n$ Matrix Games

We now discuss the solution to certain types of $m \times n$ matrix games. If the $m \times n$ matrix is strictly determined or if it can be reduced to a 2×2 game, we know how to find optimum strategies. Remember, the first steps in solving *any* matrix game are:

1. Check for a saddle point; if the matrix has one, then the game is strictly determined and the saddle point is the value of the game.
2. Eliminate dominated rows (eliminate the smaller row).
3. Eliminate dominating columns (eliminate the larger column).

The discussion that follows assumes that these steps are carried out first.

$2 \times n$ Matrix Games

Suppose we check for a saddle point and dominated rows or dominating columns and the resulting game matrix is a $2 \times n$ game, where n is some positive integer greater than 2. For a $2 \times n$ matrix game, player I has two strategies and player II has n strategies. For example, suppose Linda is playing the following game with an opponent:

$$\text{Linda}\ \overset{\text{Opponent}}{\begin{bmatrix} 5 & -1 & 1 \\ -1 & 3 & 0 \end{bmatrix}}$$

Recall that V is the value of the game if both Linda and her opponent use optimum strategies on repeated playing of the same game. Therefore, if Linda uses her

* The material in this section is not required for the remainder of the text.

optimal strategy and her opponent does not, Linda's winnings will be greater than or equal to V since positive entries indicate payoffs to her. On the other hand, if her opponent plays an optimum strategy and Linda does not, his winnings will be less than or equal to V since negative entries indicate payoffs to her opponent.

First consider the situation where we find Linda's optimum strategies first. We do this in the case where we have a $2 \times n$ matrix game. We denote the value of the game by V and denote her strategy by

$$[p \quad 1-p]$$

Next we calculate the payoffs if her opponent plays each of the three columns. Given

$$[p \quad 1-p]\begin{bmatrix} 5 & -1 & 1 \\ -1 & 3 & 0 \end{bmatrix}$$

for the first column,

$$[p \quad 1-p]\begin{bmatrix} 5 \\ -1 \end{bmatrix} = 5p + (-1)(1-p) = 5p - 1 + p = 6p - 1$$

For the second column,

$$[p \quad 1-p]\begin{bmatrix} -1 \\ 3 \end{bmatrix} = -p + 3(1-p) = -p + 3 - 3p = -4p + 3$$

For the third column,

$$[p \quad 1-p]\begin{bmatrix} 1 \\ 0 \end{bmatrix} = p$$

These must all be greater than or equal to V since we wish to make Linda's payoffs as great as possible:

$$\begin{aligned} 6p - 1 &\geq V \\ -4p + 3 &\geq V \\ p &\geq V \end{aligned}$$

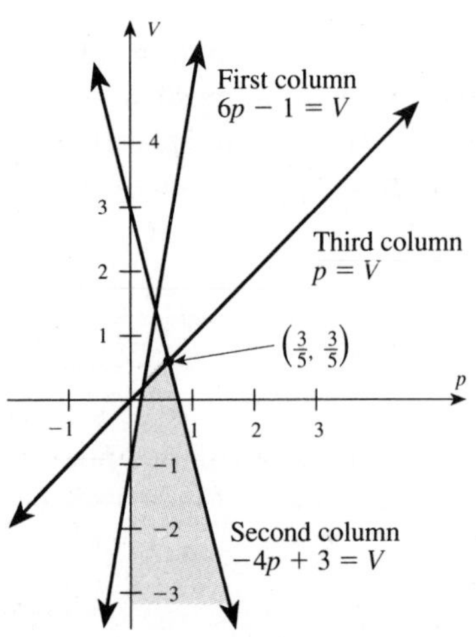

FIGURE 9.2

We wish to maximize V subject to the inequalities above (also remember that $0 \leq p \leq 1$). We do this graphically as shown in Figure 9.2. The color portion shows the simultaneous solution of the system of inequalities and represents Linda's possible winnings, depending on the column her opponent plays. We see that we can *maximize* V at the intersection of the lines

$$p = V \qquad \text{and} \qquad -4p + 3 = V$$

This point of intersection occurs when $p = \frac{3}{5}$ and $V = \frac{3}{5}$—that is, at the point $(\frac{3}{5}, \frac{3}{5})$. Therefore Linda's optimal strategy is when $p_1 = \frac{3}{5}$ and $p_2 = 1 - p = \frac{2}{5}$ or

$$[\tfrac{3}{5} \quad \tfrac{2}{5}]$$

and the value of the game is $\frac{3}{5}$.

The optimal strategy for Linda's opponent is found by again considering the game matrix

$$\begin{bmatrix} 5 & -1 & 1 \\ -1 & 3 & 0 \end{bmatrix}$$

Her opponent can see that Linda's optimum strategy is determined by the intersection of the lines from the second and third columns, so he will concentrate his attention on these two columns:

$$\begin{bmatrix} -1 & 1 \\ 3 & 0 \end{bmatrix}$$

$$\begin{bmatrix} -1 & 1 \\ 3 & 0 \end{bmatrix} \begin{bmatrix} q \\ 1-q \end{bmatrix} = \begin{bmatrix} -q+1-q \\ 3q+0 \end{bmatrix} = \begin{bmatrix} 1-2q \\ 3q \end{bmatrix}$$

Thus

$$1 - 2q = 3q$$

$$q = \frac{1}{5}$$

So Linda's opponent's strategy for the subgame is

$$\begin{bmatrix} \frac{1}{5} \\ \frac{4}{5} \end{bmatrix}$$

Therefore, for the original game, his strategy is

$$\begin{bmatrix} 0 \\ \frac{1}{5} \\ \frac{4}{5} \end{bmatrix}$$

m × 2 *Matrix Games*

When Linda is playing an $m \times 2$ matrix game we use a similar method of solution except that for this game we *first* find the strategies for her opponent. For an $m \times 2$ matrix game, player I has m strategies and player II has two strategies. Consider the following game:

$$\begin{array}{cc} & \text{Opponent} \\ \text{Linda} & \begin{bmatrix} 1 & 2 \\ 3 & -2 \\ -3 & 3 \end{bmatrix} \end{array}$$

Linda's opponent wants to play a strategy that results in a payoff that is as small as possible. If the value of the game is denoted by V and her opponent's strategy is

$$\begin{bmatrix} q \\ 1-q \end{bmatrix}$$

where $q_1 = q$ and $q_2 = 1 - q$, we calculate her opponent's payoffs for each of the three rows that Linda might play.

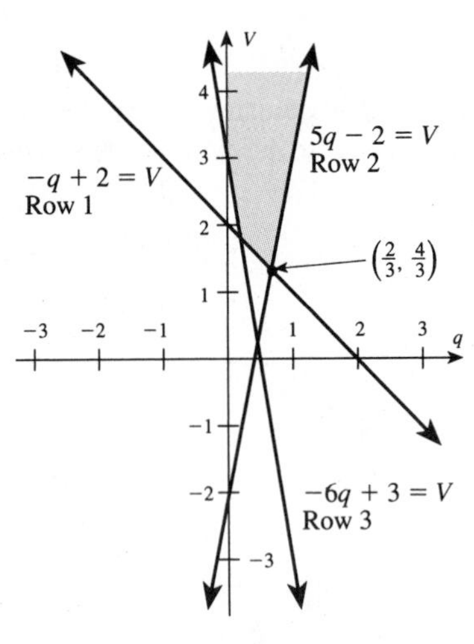

FIGURE 9.3

$$\text{Row 1:} \quad [1 \quad 2]\begin{bmatrix} q \\ 1-q \end{bmatrix} = q + 2(1-q) = -q + 2$$

$$\text{Row 2:} \quad [3 \quad -2]\begin{bmatrix} q \\ 1-q \end{bmatrix} = 3q - 2(1-q) = 5q - 2$$

$$\text{Row 3:} \quad [-3 \quad 3]\begin{bmatrix} q \\ 1-q \end{bmatrix} = -3q + 3(1-q) = -6q + 3$$

These must be less than or equal to V since her opponent wants to make the payoffs as small as possible:

$$\begin{aligned} -q + 2 &\le V \\ 5q - 2 &\le V \\ -6q + 3 &\le V \end{aligned}$$

The value V is minimized subject to the inequalities above and $q \geq 0$, as shown in Figure 9.3.

Linda's opponent wishes to *minimize* V so we see that the color area is minimized at the point of intersection of the lines

$$-q + 2 = V \qquad \text{and} \qquad 5q - 2 = V$$

The point of intersection is $(\frac{2}{3}, \frac{4}{3})$ and the optimal strategy for her opponent is

$$\begin{bmatrix} \frac{2}{3} \\ \frac{1}{3} \end{bmatrix}$$

For Linda's strategy, we concentrate on the first and second rows (why?), and solve the matrix game

$$\begin{bmatrix} 1 & 2 \\ 3 & -2 \end{bmatrix}$$

The solution is found by multiplying

$$\begin{aligned} [p \quad 1-p]\begin{bmatrix} 1 & 2 \\ 3 & -2 \end{bmatrix} &= [p + 3(1-p) \quad 2p - 2(1-p)] \\ &= [-2p + 3 \quad 4p - 2] \end{aligned}$$

We equate the entries and solve to obtain:

$$\begin{aligned} -2p + 3 &= 4p - 2 \\ -6p &= -5 \\ p &= \frac{5}{6} \end{aligned}$$

Linda's strategy for the subgame is $[\frac{5}{6} \quad \frac{1}{6}]$ and for the original game is $[\frac{5}{6} \quad \frac{1}{6} \quad 0]$. The value of the game is $\frac{8}{6} = \frac{4}{3}$.

$m \times n$ Matrix Games

The general solution to a 3×3, or even higher-order game, is a problem in linear programming and is beyond the scope of this course.

A word of caution is in order: The theory of games is still being developed, and at the present time we can solve only the simplest types of games. Most practical problems in game theory are too complicated and remain yet unsolved. The difficulty is in constructing a mathematical model that adequately accounts for all the interrelationships that may be present in a real game between buyer and seller, competing companies, people and nature, and so on. Nevertheless, game theory can, and does, help us begin to intelligently analyze conflict situations.

9.6
Problem Set

Find the optimal strategies and game values for the games in Problems 1–12.

1. $\begin{bmatrix} 2 & 1 \\ -3 & 2 \\ 1 & -4 \end{bmatrix}$

2. $\begin{bmatrix} 0 & -2 \\ 2 & 1 \\ -3 & -1 \end{bmatrix}$

3. $\begin{bmatrix} 1 & -1 \\ 0 & 1 \\ -1 & 0 \\ 1 & -2 \end{bmatrix}$

4. $\begin{bmatrix} 4 & 3 \\ -1 & 2 \\ -5 & 6 \\ 2 & -3 \end{bmatrix}$

5. $\begin{bmatrix} 1 & 4 & -3 \\ -2 & 1 & 2 \end{bmatrix}$

6. $\begin{bmatrix} -2 & 1 & 3 \\ 1 & -3 & 4 \end{bmatrix}$

7. $\begin{bmatrix} 1 & 0 & 3 \\ -3 & 2 & -4 \\ 2 & 3 & 1 \end{bmatrix}$

8. $\begin{bmatrix} 1 & -2 & 3 \\ 2 & -3 & 2 \\ 4 & 1 & -3 \end{bmatrix}$

9. $\begin{bmatrix} 2 & 5 & -4 \\ 1 & 0 & -1 \\ 7 & -2 & -3 \end{bmatrix}$

10. $\begin{bmatrix} 4 & -3 & 1 \\ 2 & -1 & 0 \\ 3 & -2 & 1 \end{bmatrix}$

11. $\begin{bmatrix} 5 & 4 & 2 & 1 \\ 1 & -2 & 4 & 3 \\ 4 & 3 & 2 & 2 \\ 1 & 0 & -1 & -2 \end{bmatrix}$

12. $\begin{bmatrix} -3 & 1 & 0 & -2 \\ 2 & 3 & -4 & 1 \\ 1 & -1 & -5 & 0 \\ -2 & 2 & 1 & -1 \end{bmatrix}$

APPLICATION

13. A general with two regiments is trying to capture a city. He can attack from the north with both regiments, from the south with both regiments, or from the north with one regiment and from the south with one regiment. The defending forces also have two regiments that can be deployed for protection in the north, in the south, or one in the north and one in the south. The general of the attacking forces will gain 1 point if the attack succeeds, -1 if the attack fails, and 0 if the forces are held at a standoff. Whichever force has more regiments in an area wins the battle, and if one regiment is deployed against one regiment, the armies are held at a standoff. Find the generals' strategies.

*9.7
Review

The material of this chapter is reviewed in the following list of objectives. After each objective there are some practice questions. For a sample test select the first question of each set and check your answers. The second question for each objective has no answer given. If you are having trouble with a particular type of problem, look back at the indicated section in the text. When you are finished reviewing these objectives, a sample examination is given at the end of this section.

* Optional section.

[9.1]

Objective 9.1: *Write a transition matrix.*

1. If a certain experiment has three states 0, 1, and 2 with the probabilities

$P(0\|0) = .12$	$P(0\|1) = .47$	$P(0\|2) = .17$
$P(1\|0) = .63$	$P(1\|1) = .15$	$P(1\|2) = .42$
$P(2\|0) = .25$	$P(2\|1) = .38$	$P(2\|2) = .41$

write the transition matrix.

2. *Market share.* Hallmark and Buzza-Cardoza are competing greeting card companies in a certain geographical area. A survey of 240 persons showed that the last time they purchased greeting cards, 180 purchased a Hallmark card and 60 purchased a Buzza-Cardoza card. The respondents were then shown two sample cards, and 144 of the Hallmark users said they would stay with Hallmark while the rest said they would switch. Of the Buzza-Cardoza buyers, 15 would switch to Hallmark and the remainder would stay with Buzza-Cardoza. Write a transition matrix.

3. Suppose a certain chemical can exist in a solid, liquid, or gaseous state. An experiment is performed and the following information is obtained:

   If the beginning state is solid, there is a 60% probability it will go to a liquid state after 1 hour and a 40% probability it will go to a gaseous state.
   If the beginning state is a liquid, there is a 20% chance it will be in a solid state after 1 hour, a 30% probability it will be in a gaseous state, and a 50% probability it will remain in a liquid state.
   If the beginning state is a gas, there is a 50% probability it will be in a liquid state in 1 hour and a 50% probability it will stay in a gaseous state.

   Write the hourly transition matrix for this problem.

4. A TV anchor person has a wardrobe that can be divided into three color groups: maroon, blue, and green. If that person wears blue or green one day, that color cannot be worn the next day. In this case the anchor person will select maroon half the time and the other color the other half the time. However, since maroon is the favorite color, if maroon is worn one day, there is a 20% probability that it will be worn again the next day, a 40% probability that blue will be worn the next day, and a 40% probability that green will be worn the next day. Write the transition matrix for this problem.

Objective 9.2: *Use matrix multiplication to find probabilities after n trials. A transition matrix for a chemical substance being in a liquid or a gaseous state after 1 hour is*

$$T = \begin{array}{c} \\ L \\ G \end{array} \begin{array}{c} \begin{array}{cc} L & G \end{array} \\ \begin{bmatrix} .8 & .2 \\ .4 & .6 \end{bmatrix} \end{array}$$

5. What is the probability that a substance presently in a liquid state will be in a gaseous state in 3 hours?
6. What is the probability that a substance presently in a gaseous state will be in a liquid state in 2 hours?
7. What is the probability that a substance presently in a liquid state will be in a liquid state in 2 hours?
8. What is the probability that a substance presently in a gaseous state will be in a gaseous state in 3 hours?

[9.2]

Objective 9.3: *Given a transition matrix and the present share of a market, find the short-term expectations for the share of the market for a given set of products.*

9. *Market share.* Suppose the yearly transition matrix for customers of Hallmark and Buzza-Cardoza cards is

$$T = \begin{array}{c} \\ H \\ B \end{array} \begin{array}{c} \begin{array}{cc} H & B \end{array} \\ \begin{bmatrix} .6 & .4 \\ .2 & .8 \end{bmatrix} \end{array}$$

   If the present market is 5,000,000 customers and Hallmark has 75% of the market while Buzza-Cardoza has the other 25%, what is the expected number of Buzza-Cardoza customers at the end of the year?

10. Repeat Problem 9 for the end of 2 years.

11. The probability that a manufacturer will buy raw materials from three companies each month is summarized with the following transition matrix:

$$\begin{array}{c} \\ A \\ B \\ C \end{array} \begin{array}{c} \begin{array}{ccc} A & B & C \end{array} \\ \begin{bmatrix} .60 & .30 & .10 \\ .20 & .60 & .20 \\ .20 & .20 & .60 \end{bmatrix} \end{array}$$

   Notice that if the manufacturer buys from a company one time, it is more probable that the manufacturer will buy from the same company the next time. If the manufacturer is currently giving one-third of the business to each of the three companies, how much business will the manufacturer do with each of the three companies after 2 months?

12. If the company in Problem 11 is giving 80% of its business to company *A* and 10% to each of companies *B* and *C*, how much of its business will the manufacturer do with each of companies *A*, *B*, and *C* after 2 months?

Objective 9.4: Determine whether a given matrix is a regular transition matrix.

13. $\begin{bmatrix} 0 & 1 & 0 \\ .4 & .2 & .4 \\ .7 & 0 & .3 \end{bmatrix}$

14. $\begin{bmatrix} .6 & .2 & .2 \\ .7 & .1 & .1 \\ .5 & .2 & .3 \end{bmatrix}$

15. $\begin{bmatrix} \frac{3}{5} & \frac{2}{5} \\ 0 & 1 \end{bmatrix}$

16. $\begin{bmatrix} \frac{4}{5} & \frac{1}{5} & 0 \\ \frac{2}{3} & \frac{1}{3} & 0 \\ \frac{7}{10} & \frac{3}{10} & 0 \end{bmatrix}$

Objective 9.5: Find the fixed probability vector for a given transition matrix.

17. Find the fixed probability vector for Problem 9.
18. Find the fixed probability vector for Problems 5–8.
19. Find the fixed probability vector for Problem 3.
20. Find the fixed probability vector for Problem 13.

Objective 9.6: Given a transition matrix and the present share of a market, find the long-term expectations for the share of the market for a given set of products.

21. What is the long-term expectation for Problem 9?
22. Use the results of Problems 9, 10, and 21 to devise an investment strategy for someone wishing to invest in either Hallmark or Buzza-Cardoza cards, assuming that the most reward goes to the company with the most customers.
23. Find the long-range percentage of business the manufacturer in Problem 11 gives to each company after a long period of time.
24. What is the long-range probable composition for the chemical in Problem 3?

[9.3]

Objective 9.7: Determine whether a given matrix is a transition matrix for an absorbing Markov chain. If it is, list the absorbing states.

25. $\begin{bmatrix} \frac{2}{3} & \frac{1}{4} & \frac{1}{12} \\ \frac{1}{2} & \frac{1}{3} & \frac{1}{6} \\ 0 & 0 & 1 \end{bmatrix}$

26. $\begin{bmatrix} .4 & .4 & .2 \\ 1 & 0 & 0 \\ .8 & .1 & .1 \end{bmatrix}$

27. $\begin{bmatrix} 1 & 0 & 1 \\ \frac{1}{3} & \frac{1}{3} & \frac{1}{3} \\ 0 & 0 & 1 \end{bmatrix}$

28. $\begin{bmatrix} \frac{1}{2} & 0 & \frac{1}{2} & 0 \\ 0 & \frac{1}{2} & 0 & \frac{1}{2} \\ 0 & 0 & 1 & 0 \\ 1 & 0 & 0 & 0 \end{bmatrix}$

Objective 9.8: Find and interpret FP for a given transition matrix.

29. $\begin{bmatrix} .2 & .3 & .5 \\ 0 & 1 & 0 \\ .5 & .3 & .2 \end{bmatrix}$

30. $\begin{bmatrix} 1 & 0 & 0 \\ .3 & .2 & .5 \\ 0 & 0 & 1 \end{bmatrix}$

31. $\begin{bmatrix} 1 & 0 & 0 & 0 \\ .5 & 0 & .5 & 0 \\ 0 & .5 & 0 & .5 \\ 0 & 0 & 0 & 1 \end{bmatrix}$

32. $\begin{bmatrix} 1 & 0 & 0 & 0 \\ .2 & .3 & .4 & .1 \\ 0 & 1 & 0 & 0 \\ 0 & 0 & 0 & 1 \end{bmatrix}$

Objective 9.9: Solve applied absorbing Markov chain problems.

33. Suppose two players start a game with $4 between them. They play a game with the following transition matrix for the first player:

$$\begin{array}{c} \\ 0 \\ 1 \\ 2 \\ 3 \\ 4 \end{array} \begin{array}{c} \begin{array}{ccccc} 0 & 1 & 2 & 3 & 4 \end{array} \\ \begin{bmatrix} 1 & 0 & 0 & 0 & 0 \\ \frac{3}{5} & 0 & \frac{2}{5} & 0 & 0 \\ 0 & \frac{3}{5} & 0 & \frac{2}{5} & 0 \\ 0 & 0 & \frac{3}{5} & 0 & \frac{2}{5} \\ 0 & 0 & 0 & 0 & 1 \end{bmatrix} \end{array}$$

a. What are the absorbing states?
b. Does this game favor the first player or the other player?

34. If the first player in Problem 33 begins with $2, what is the probability of financial ruin (to the nearest hundredth)?
35. A cross-pollination of plants gives the following transition matrix for 6 states:

$$\begin{array}{c} \\ 1 \\ 2 \\ 3 \\ 4 \\ 5 \\ 6 \end{array} \begin{array}{c} \begin{array}{cccccc} 1 & 2 & 3 & 4 & 5 & 6 \end{array} \\ \begin{bmatrix} 1 & 0 & 0 & 0 & 0 & 0 \\ \frac{1}{16} & \frac{1}{4} & \frac{1}{4} & \frac{1}{4} & \frac{1}{8} & \frac{1}{16} \\ 0 & \frac{1}{4} & \frac{1}{2} & 0 & 0 & \frac{1}{4} \\ \frac{1}{4} & \frac{1}{4} & 0 & \frac{1}{2} & 0 & 0 \\ 0 & 1 & 0 & 0 & 0 & 0 \\ 0 & 0 & 0 & 0 & 0 & 1 \end{bmatrix} \end{array}$$

a. What are the absorbing states?
b. Find Q. **c.** Find $I_4 - Q$.

36. Find and interpret FP for the transition matrix in Problem 35.

[9.4]

Objective 9.10: Find the expected value for a random variable.

37. A contest offered a $10,000 first prize, 10 second prizes worth $1,000 each, and 1,000 third prizes worth $100 each. What is the expectation for this contest if we assume that there are one million entries?
38. Find the expected value of X:

X	0	1	5	10	25
$P(X)$	$\frac{1}{2}$	$\frac{3}{20}$	$\frac{1}{10}$	0	$\frac{1}{4}$

39. If a pair of dice is rolled and you obtain a 7 or an 11, you will receive \$2. Otherwise you will receive nothing. What is the expectation?

40. If the probability of dying in the next year is .002, what is the expectation from a \$25,000 life insurance policy? (See Table 9 in Appendix E.)

Objective 9.11: *Make decisions based on mathematical expectation. This is an application of operations research.*

41. *Contract bidding.* An electrical contractor must choose which of three contracts to bid upon. The costs and profits for the bids are summarized in the accompanying table. For which contract should the contractor submit a bid? (Subtract preparatory costs from profits when making your decisions.)

Contract	Preparatory costs	Profit	Probability of winning contract
I	\$350	\$3,000	$\frac{2}{3}$
II	\$2,500	\$50,000	$\frac{1}{10}$
III	\$2,400	\$25,000	$\frac{1}{5}$

42. A lottery offers a prize of a color TV (value \$500), and 800 tickets are sold. What is the expectation if you buy three tickets? If the tickets cost \$1 each, should you buy them?

43. Walt, who is a realtor, knows that if he takes a listing to sell a house, it will cost him \$1,000. However, if he sells the house, he will receive 6% of the selling price. If another realtor sells the house, Walt will receive 3% of the selling price. If the house is unsold in 3 months, he will lose the listing and receive nothing. Suppose Walt must decide upon one of the following listings:

		Probability of selling the house		
House	Value	Listing agent	Another agent	Not selling
I	\$85,000	.6	.3	.1
II	\$100,000	.5	.4	.1
III	\$190,000	.2	.3	.5

Which listing should Walt take if he must choose only one?

44. It is determined that the daily demand at a car rental agency is between 5 and 9 cars as follows:

Number of Customers	5	6	7	8	9
Probability	.2	.3	.3	.1	.1

If it costs \$5 per day per car whether a car is rented or not, and if the profit is \$10 per day on the days the car is rented, what is the optimal number of cars that the agency should have on hand?

[9.5]

Objective 9.12: *Decide whether a game is strictly determined, and if it is, state the value of the game.*

45. $\begin{bmatrix} 1 & 0 & -1 & 2 \\ 0 & -2 & 3 & -1 \\ 1 & 2 & -1 & 0 \\ -1 & 0 & 1 & -2 \end{bmatrix}$

46. $\begin{bmatrix} 2 & 1 & 3 \\ 0 & 1 & -2 \\ -1 & 2 & -3 \end{bmatrix}$

47. $\begin{bmatrix} 2 & 1 & 3 \\ -2 & 0 & 3 \\ 4 & -2 & -3 \end{bmatrix}$

48. $\begin{bmatrix} 0 & 1 & 2 & 0 \\ 1 & 2 & 2 & 1 \\ 3 & 0 & -1 & 0 \\ 2 & -1 & 0 & -2 \end{bmatrix}$

Objective 9.13: *Find the optimal strategies and value for a* 2×2 *game.*

49. $\begin{bmatrix} 4 & -2 \\ 2 & -3 \end{bmatrix}$

50. $\begin{bmatrix} 4 & -1 \\ 2 & -2 \end{bmatrix}$

51. $\begin{bmatrix} -4 & 0 \\ 1 & -1 \end{bmatrix}$

52. $\begin{bmatrix} -3 & 1 \\ -1 & 5 \end{bmatrix}$

Objective 9.14: *Find dominated rows and dominating columns and find the optimal strategies and value for a game.*

53. $\begin{bmatrix} 3 & -3 & 2 \\ 2 & -4 & 0 \\ 1 & -1 & -2 \end{bmatrix}$

54. $\begin{bmatrix} 40 & 10 & -30 \\ -20 & 0 & -50 \\ 40 & 20 & 30 \end{bmatrix}$

55. $\begin{bmatrix} -10 & 5 & -2 \\ 8 & -10 & 10 \\ 4 & -12 & 6 \end{bmatrix}$

56. $\begin{bmatrix} -1 & 1 & 0 \\ 0 & -2 & 3 \\ 1 & -1 & 4 \end{bmatrix}$

Objective 9.15: *Solve problems using game theory.*

57. Suppose we play the following game: A friend has a dime and a quarter, and we have a quarter and a half-dollar. We each select a coin and put it on the table without knowing the choice of our opponent. If the total is 50¢ or 60¢, we win the amount on the table; otherwise we lose the amount on the table. What are the strategies and value for this game?

58. A secret courier for the SSO must carry a coded message from New York to London. She can either fly or take a ship. Her opponent, a special agent of a leading government, must prevent her from delivering the message and must guess which route she will take so that he can intercept the courier. If the special agent intercepts her on the ship, he will dispose of her and her message, resulting in a score of -100 for the SSO courier. On the other hand, if he intercepts her on the plane, he will only be able to intercept the message, resulting in a score of -25 for the SSO courier. If the courier delivers the message, her score is 10. Find the strategies for the courier and the value of this game.

59. Suppose we are planning to market a new pretzel and we can manufacture either a stick or a figure-8 type of pretzel. Our competitor can manufacture five varieties of pretzels, and the resulting gain or loss in thousands of dollars of sales is shown by the following payoff matrix:

$$\begin{array}{l} \\ \text{Stick} \\ \text{Figure-8} \end{array} \begin{array}{c} \begin{array}{ccccc} A & B & C & D & E \end{array} \\ \begin{bmatrix} 10 & 5 & 1 & 0 & -3 \\ 2 & -1 & 2 & 5 & 1 \end{bmatrix} \end{array}$$

If we must choose to manufacture only one type each month and our competitor can also choose only one type each month, what are our strategies and what is the value of the game?

60. Consider the following game: I hold a 2 of spades and an 8 of hearts. My opponent holds a 2 of diamonds and an 8 of clubs. We each select a card and place it face up on the table. If they are both the same color, I win an amount equal to the sum of the values of the cards on the table. If they are different colors, my opponent wins an amount equal to the sum of the values of the cards. What are my strategies, and what is the value of the game?

[9.6]

Objective 9.16: *Determine the strategies and the value of an $m \times n$ matrix game.*

61. $\begin{bmatrix} 1 & -3 & 2 & 1 \\ 4 & 2 & -5 & 0 \\ 0 & -5 & 1 & 0 \end{bmatrix}$

62. $\begin{bmatrix} 1 & -1 & -3 \\ 5 & -6 & 3 \\ 2 & 1 & -2 \end{bmatrix}$

63. $\begin{bmatrix} 2 & -1 & 3 \\ -1 & 2 & 0 \\ 0 & -5 & 2 \end{bmatrix}$

64. $\begin{bmatrix} 3 & 4 & -1 & 1 \\ 2 & 3 & 4 & 5 \\ 4 & 5 & 0 & 1 \\ -1 & 0 & 5 & 6 \\ -3 & 2 & -1 & 0 \end{bmatrix}$

SAMPLE TEST

The following sample test (45 minutes) is intended to review the main ideas of the chapter.

1. A certain chemical can exist in a solid, liquid, or gaseous state. An experiment is performed and the following information is obtained:

If the beginning state is solid, there is a 75% probability that the chemical will go to a liquid state after 1 hour, a 5% probability that it will go to a gaseous state after 1 hour, and a 20% probability that it will remain in a solid state.

If the beginning state is liquid, there is a 10% probability that the chemical will be in a solid state after 1 hour, a 30% probability that it will be in a gaseous state, and a 60% probability that it will remain in a liquid state.

If the beginning state is gaseous, there is a 40% probability that the chemical will be in a liquid state in 1 hour and a 60% probability that it will stay in a gaseous state.

Write the hourly transition matrix.

2. The yearly transition matrix for customers of Hallmark and Buzza-Cardoza cards is

$$T = \begin{array}{c} \\ H \\ B \end{array} \begin{array}{c} \begin{array}{cc} H & B \end{array} \\ \begin{bmatrix} .7 & .3 \\ .5 & .5 \end{bmatrix} \end{array}$$

If the present market is 2 million customers and Hallmark has 80% of the market while Buzza-Cardoza has the other 20%, what is the expected number of Buzza-Cardoza customers at the end of 1 year?

3. Find the fixed probability vector for the transition matrix in Problem 2.

4. What is the probable long-range composition for the chemical in Problem 1?

5. Two players start a game with $5 between them. The game has the following transition matrix for the first player:

$$\begin{array}{c} \\ 0 \\ 1 \\ 2 \\ 3 \\ 4 \\ 5 \end{array} \begin{array}{c} \begin{array}{cccccc} 0 & 1 & 2 & 3 & 4 & 5 \end{array} \\ \begin{bmatrix} 1 & 0 & 0 & 0 & 0 & 0 \\ .5 & 0 & .5 & 0 & 0 & 0 \\ 0 & .5 & 0 & .5 & 0 & 0 \\ 0 & 0 & .5 & 0 & .5 & 0 \\ 0 & 0 & 0 & .5 & 0 & .5 \\ 0 & 0 & 0 & 0 & 0 & 1 \end{bmatrix} \end{array}$$

a. What are the absorbing states?

b. State P and Q for this Markov chain. If

$$F = (I - Q)^{-1} = \begin{bmatrix} 1 & -.5 & 0 & 0 \\ -.5 & 1 & -.5 & 0 \\ 0 & -.5 & 1 & -.5 \\ 0 & 0 & -.5 & 1 \end{bmatrix}^{-1} = \begin{bmatrix} 1.6 & 1.2 & .8 & .4 \\ 1.2 & 2.4 & 1.6 & .8 \\ .8 & 1.6 & 2.4 & 1.2 \\ .4 & .8 & 1.2 & 1.6 \end{bmatrix}$$

does this game favor the first player? Why or why not?

c. If the first player begins with \$4, what is the probability of financial ruin (to the nearest hundredth)?

6. Find the optimal strategies and value of each game, and determine whether the game is strictly determined.

a. $\begin{bmatrix} 3 & 5 \\ 2 & -5 \end{bmatrix}$ **b.** $\begin{bmatrix} 2 & 1 \\ 0 & 1 \end{bmatrix}$

c. $\begin{bmatrix} 4 & -5 \\ 2 & 3 \end{bmatrix}$

d. $\begin{bmatrix} 50 & 20 & 20 \\ -10 & 0 & -50 \\ 40 & 20 & 30 \end{bmatrix}$

7. Find the optimal strategies and value of the following game.

$$\begin{bmatrix} 2 & 1 & 3 \\ 0 & 1 & -2 \\ -1 & 2 & -3 \end{bmatrix}$$

8. *Contract bidding.* A contractor must choose one of three contracts on which to bid. The costs and profits for the bids are:

Contract	Preparation costs	Profit	Probability of winning contract
I	\$100	\$10,000	.9
II	\$5,000	\$100,000	.1
III	\$4,500	\$135,000	.1

If the contractor has time to submit only one bid, which should it be? Base your answer on mathematical expectation and remember to subtract preparation costs from the profits when making your decision.

9. The daily demand at a car rental agency is between 20 and 25 cars as follows:

Demand	20	21	22	23	24	25
Probability	.1	.2	.3	.2	.1	.1

If it costs the agency \$10 per day whether the car is rented or not, and if the net profit is \$30 per day on the days the car is rented, what is the optimal number of cars that the agency should have on hand?

10. Suppose you are producing two types of cereal. One type appeals to children and the other with granola is for adults. Your competitor markets five varieties of cereals, and the resulting gain or loss in thousands of dollars of sales is shown below:

	Competitor's cereals				
	A	*B*	*C*	*D*	*E*
Children	20	10	2	0	−6
Adults	4	−2	4	10	1

If you must choose to manufacture only one type each month and your competitor can also choose only one type each month, what are your best strategies, and what is the value of this game?

MODELING
APPLICATION 7

Genetics

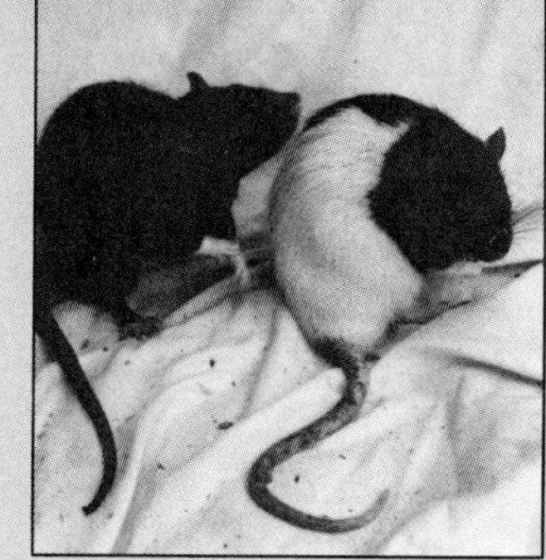

A researcher interested in studying the ability of parents to pass certain physical traits to their offspring sets up a classic *brother–sister mating model* on a family of rats. In this experiment, two rats are mated, and from among their direct descendants two individuals of the opposite sex are selected at random and are mated. Then the process is repeated indefinitely and the target traits are studied. What are the expected results from such experimentation?* Assume that traits are determined by *genes* that are passed from parents to their offspring. Each parent has a pair of genes, and the basic assumption is that each offspring inherits one gene from each parent to form the offspring's own pair. The genes are selected in a random, independent way. In your model, assume that the researcher is studying a trait that is both easily identifiable (such as color of a rat's fur) and determined by a pair of genes consisting of a *dominant* gene, denoted by A, and a *recessive* gene, denoted by a. The possible pairings of genes are called *genotypes*:

AA is *dominant*, or homozygous.
Aa is *hybrid*, or heterozygous; genetically, the genotype aA is the same as Aa.
aa is called *recessive*.

Write a paper considering the possibilities for the offspring. Also, suppose one parent has genotype Aa (heterozygous) and is mated with a parent whose genotype is unknown. The offspring is mated with another heterozygous genotype (Aa) and this process is continued. Build a model to predict the genotype of the offspring over time. For general guidelines about writing this essay, see the commentary accompanying Modeling Application 1 on page 151.

*This modeling application is based on the work of Gregor Mendel (1822–1884), an Austrian monk, who formulated the laws of heredity and genetics. Mendel's work was later amplified and explained by a mathematician, G. H. Hardy (1877–1947), and a physician, Wilhelm Weinberg (1862–1937). For years Mendel taught science without any teaching credentials because he had failed the biology portion of the licensing examination! His work, however, laid the foundation for the very important branch of biology known today as genetic science.

1. Let

$$A = \begin{bmatrix} 4 & 0 & -1 \\ 3 & -2 & 1 \\ 0 & 3 & -2 \end{bmatrix} \qquad B = \begin{bmatrix} 4 & -1 & 3 \\ -3 & 1 & 2 \\ 2 & -3 & 0 \end{bmatrix}$$

$$I = \begin{bmatrix} 1 & 0 & 0 \\ 0 & 1 & 0 \\ 0 & 0 & 1 \end{bmatrix}$$

Perform the indicated operations.

a. AB **b.** $A(B + I)$
c. $I(2A + B)$ **d.** A^{-1}

Solve the systems of equations or inequalities in Problems 2–7.

2. $\begin{cases} x + y = 3 \\ 2x - y = 9 \end{cases}$

3. $\begin{cases} y = 2x + 1 \\ y = 3x - 5 \\ y = x + 7 \end{cases}$

4. $\begin{cases} x + 2y + 3z = 2 \\ 2x - y + z = -1 \end{cases}$

5. $\begin{cases} 3x + y + z = -6 \\ x + 2y - z = 9 \\ 5x + y - 3z = -4 \end{cases}$

6. $\begin{cases} 3x + 2y < -3 \\ x - y > 0 \\ x \le 0 \end{cases}$

7. $\begin{cases} 2x + 3y \le 60 \\ x + 2y \le 36 \\ 3x - 2y \le 24 \\ x \ge 0, \quad y \ge 0 \end{cases}$

8. Maximize: $23x_1 + 24x_2$

Subject to: $\begin{cases} 4x_1 + 2x_2 \le 1{,}800 \\ 2x_1 + 3x_2 \le 1{,}200 \\ 5x_1 + 4x_2 \le 2{,}400 \\ x_1 \ge 0, \quad x_2 \ge 0 \end{cases}$

9. Let $U = \{1, 2, 3, \ldots, 8, 9, 10\}$, $A = \{2, 3, 5, 7\}$, $B = \{1, 3, 5, 7, 9\}$, and $C = \{2, 4, 6, 8, 10\}$. Find

a. $A \cap C$ **b.** $B \cup \bar{C}$
c. $U \cap C$ **d.** $\bar{U}$
e. $\overline{A \cap (B \cup C)}$

APPLICATIONS

10. In a survey of 600 college students:

250 had tried marijuana
350 had tried alcohol
175 had tried cocaine
110 had tried both cocaine and alcohol
140 had tried both marijuana and alcohol
100 had tried both marijuana and cocaine
70 had tried all three

a. How many had not tried any of these drugs?
b. How many had tried only one of these drugs?
c. How many had tried exactly two of these drugs?

11. **a.** How many subsets can be chosen from five elements?
b. In how many ways can six hats be distributed to six persons?

12. Suppose two cards are drawn from a deck of cards without replacement. Find the requested probabilities.
a. P(both hearts)
b. P(1 heart and 1 diamond)
c. P(heart on first draw and diamond on second draw)
d. P(diamond on second draw given a heart on first draw)
e. P(ace of hearts drawn both times)

13. Repeat Problem 12 assuming that the cards are drawn with replacement.

14. A sample of 100 ball bearings is drawn from a day's production and 15 are found to be defective.
a. What is the probability of drawing one ball bearing and finding that it is defective?
b. Is this problem an example of empirical or theoretical probability?

15. At a fast-food outlet, 15% of the hamburgers sold were cold, 5% had a missing ingredient, and .5% were both cold and had a missing ingredient. What is the probability that your hamburger is cold if the pickle is missing?

16. A messenger transports money between two locations and travels along one of three routes, A Street, B Street, or C Street. One day, the route is randomly selected according to the probabilities $P(A) = .3$, $P(B) = .5$, and $P(C) = .2$. On the following day, the probabilities change so that the probability of the previously chosen route is .2, with the other two routes being equally probable. Find the indicated probabilities (subscripts are used to indicate the day on which the particular route is taken; for example, A_2 means that A was chosen on the second day).
a. $P(A_2 | A_1)$ **b.** $P(A_2 | B_1)$
c. $P(A_2)$ **d.** $P(C_3)$

17. Consider the following test scores:

85, 70, 75, 90, 65, 40, 70, 95, 80, 70, 55, 65, 70, 80, 95

a. Prepare a frequency distribution.
b. Find the probability distribution and draw a histogram.
c. Find the mean, median, and mode.
d. Find the range, variance, and standard deviation using the frequency distribution.

18. A psychology teacher grades on a curve. If the mean is 65 and the standard deviation is 10, how would the grades be distributed? If there are 100 students in the class, how many could expect to get As, Bs, etc.?

19. A study is conducted to test the relationship between speed (mph) and fuel consumption (mpg). The following information is obtained:

Speed	20	30	40	50	60
Fuel Consumption	35	38	40	34	29

a. Find the linear correlation coefficient and determine whether the variables are significantly correlated at either the 1% or the 5% level.
b. Find the regression line for this set of data.

Solve the linear programming problems in Problems 20–23.

20. A hospital wishes to provide a diet for its patients that has a minimum of 50 grams of carbohydrates, 30 grams of proteins, and 40 grams of fats per day. These requirements can be met with two foods, A and B. Food A costs $.14 per ounce and supplies 6 grams of carbohydrates, 3 grams of proteins, and 1 gram of fats per ounce. Food B costs $.06 per ounce and supplies 2 grams of carbohydrates, 2 grams of proteins, and 2 grams of fats per ounce. How many ounces of each food should be bought for each patient per day in order to meet the minimum requirements at the lowest cost?

21. An island is inhabited by three species of animals, A, B, and C, which feed on three types of food, F1, F2, and F3. The amount of each type of food (in ounces) to sustain one animal of each type is summarized here:

	Food		
Animal	F1	F2	F3
A	12	14	16
B	15	20	5
C	5	8	6

If the island contains 2,000 ounces of food F1, 12,000 ounces of food F2, and 8,000 ounces of food F3, what is the maximum number of animals that it can support?

22. Karlin Manufacturing produces two types of children's toys, wagons and carts. Each product must be processed in each of three departments: machining, assembling, and finishing. The hours needed to produce one toy per department and the maximum possible hours available per department are shown here:

	Production Time per Unit (hours)		Maximum Capacity
Department	Wagons	Carts	(hours)
Machining	2	1	2,000
Assembling	2	2	2,400
Finishing	1	2	3,000

Karlin's net profit is $5 per wagon and $4 per cart. How many wagons and carts should be manufactured to maximize the profit?

23. Of the 20 or so amino acids comprising protein, 8 cannot be synthesized by humans and are known as the *essential amino acids.* A diet that provides an adequate amount of the three amino acids tryptophan (TRP), lysine (LYS), and methionine (MET) will also generally provide enough of the other essential amino acids. The table shows three sources for these acids and the approximate number of grams of TRP, LYS, and MET per pound.

	Amino Acid		
Food	TRP	LYS	MET
Beef	1.2	8.0	2.4
Peanuts	1.5	5.0	1.2
Cashews	2.0	3.0	1.5

If peanuts cost $2.50 per pound and cashews cost $5 per pound, find the cheapest combination of peanuts and cashews that contains an amount of each of the three amino acids equal to or greater than the amount in a pound of beef.

The following multiple-choice questions are from an actual CPA examination given by the American Institute of Certified Public Accountants, Inc. They are reproduced with permission.

CPA Exam November 1974

The Golden Hawk Manufacturing Company wants to maximize the profits on products A, B, and C. The contribution margin for each product follows:

Product	Contribution margin
A	$2
B	$5
C	$4

The production requirements and departmental capacities, by departments, are:

Department	Production requirements by product (hours) A	B	C	Departmental capacity (total hours)
Assembling	2	3	2	30,000
Painting	1	2	2	38,000
Finishing	2	3	1	28,000

24. What is the profit maximization formula for the Golden Hawk Company?
A. $\$2A + \$5B + \$4C = X$, where X = profit
B. $5A + 8B + 5C \leq 96{,}000$
C. $\$2A + \$5B + \$4C \leq X$, where X = profit
D. $\$2A + \$5B + \$4C = 96{,}000$

25. What is the constraint for the painting department of the Golden Hawk Company?
A. $1A + 2B + 2C \geq 38{,}000$
B. $\$2A + \$5B + \$4C \geq 38{,}000$
C. $1A + 2B + 2C \leq 38{,}000$
D. $2A + 3B + 2C \leq 30{,}000$

CPA Exam November 1980

26. Duguid Company is considering a proposal to introduce a new product, XPL. An outside marketing consultant prepares the following table describing the relative likelihood of monthly sales volume levels and related income (loss) for XPL:

Monthly sales volume	Probability	Income (loss)
\$3,000	.10	\$(35,000)
\$6,000	.20	\$5,000
\$9,000	.40	\$30,000
\$12,000	.20	\$50,000
\$15,000	.10	\$70,000

If Duguid decides to market XPL, the expected value of the added monthly income will be
A. \$24,000 **B.** \$26,500
C. \$30,000 **D.** \$120,000

CPA Exam May 1976

27. Your client wants your advice on which of two alternatives he should choose. One alternative is to sell an investment now for \$10,000. Another alternative is to hold the investment 3 days, after which he can sell it for a certain selling price based on the following probabilities:

Selling price	Probability
\$5,000	.4
\$8,000	.2
\$12,000	.3
\$30,000	.1

Using probability theory, which of the following is the most reasonable statement?
A. Hold the investment 3 days because the expected value of holding exceeds the current selling price.
B. Hold the investment 3 days because of the chance of getting \$30,000 for it.
C. Sell the investment now because the current selling price exceeds the expected value of holding.
D. Sell the investment now because there is a 60% chance that the selling price will fall in 3 days.

Actuary Exam November 1981

28. A family has five children. Assuming that the probability of a girl on each birth was $\frac{1}{2}$ and that the five births were independent, what is the probability that the family has at least one girl, given that they have at least one boy?
A. $\frac{31}{32}$ **B.** $\frac{30}{31}$
C. $\frac{15}{16}$ **D.** $\frac{5}{31}$
E. $\frac{5}{32}$

29. A calculator has a random number generator key that, when pushed, displays a random digit (0, 1, . . . , 9). The key is pushed four times. Assuming the numbers generated are independent, what is the probability of obtaining one 0, one 5, and two 9s in any order?
A. $\frac{10!}{2!}\left(\frac{1}{10}\right)^{10}$ **B.** $\frac{10!}{2!}\left(\frac{1}{10}\right)^{4}$
C. $\frac{4!}{2!}\left(\frac{1}{10}\right)^{4}$ **D.** $\frac{9!}{4!}\left(\frac{1}{9}\right)^{4}$
E. $\left(\frac{1}{10}\right)^{4}$

30. Mr. Flowers plants ten rose bushes in a row. Eight of the bushes are white and two are red, and he plants them in random order. What is the probability that he will consecutively plant seven or more white bushes?
A. $\frac{1}{10}$ **B.** $\frac{1}{9}$
C. $\frac{2}{15}$ **D.** $\frac{7}{15}$
E. $\frac{1}{5}$

31. Events S and T have probabilities $P(S) = P(T) = \frac{1}{3}$ and $P(S \mid T) = \frac{1}{6}$. What is $P(\bar{S} \cap \bar{T})$?

A. $\frac{1}{6}$ **B.** $\frac{1}{3}$
C. $\frac{7}{18}$ **D.** $\frac{4}{9}$
E. $\frac{1}{2}$

32. A jar has three red marbles and one white marble. A shoebox has one red marble and one white marble. Three marbles are chosen at random without replacement from the jar and placed in the shoebox. Then two marbles are chosen at random and without replacement from the shoebox. What is the probability that both marbles chosen from the shoebox are red?

A. $\frac{9}{10}(\frac{3}{4})^3$ **B.** $\frac{43}{100}$
C. $\frac{3}{8}$ **D.** $(\frac{3}{4})^3$
E. $\frac{9}{40}$

33. A jar contains n black balls and n white balls. Three balls are chosen at random and without replacement. What is the value of n if the probability is $\frac{1}{12}$ that all three balls are white?

A. 4 **B.** 5
C. 8 **D.** 10
E. 12

34. An automobile manufacturing company produces three different car models. The table presents data on sales and average gasoline consumption for these three models. What is the mean miles per gallon (mpg) for the cars sold by the company, assuming that each car uses the same number of gallons of gasoline?

Model	Number of cars sold	Gas consumption (mpg)
I	2,000	15
II	4,000	20
III	4,000	25

A. 25 **B.** 21
C. 20 **D.** 15
E. Cannot be determined from the given information

The next three questions are taken from an exam given by the Institute of Management Accounting of the National Association of Accountants. The questions are reproduced by permission.

CMA Exam December 1979

The Elon Co. manufactures two industrial products: X-10, which sells for $90 a unit, and Y-12, which sells for $85 a unit. Each product is processed through both of the company's manufacturing departments. The limited availability of labor, material, and equipment capacity has restricted the ability of the firm to meet the demand for its products. The production department believes that linear programming can be used to routinize the production schedule for the two products.

The following data are available to the production department:

	Amount required per unit	
	X-10	**Y-12**
Direct material: Weekly supply is limited to 1,800 pounds at $12 per pound	4 lb	2 lb
Direct labor:		
Department 1—weekly supply is limited to 10 people at 40 hours each at an hourly cost of $6	$\frac{2}{3}$ hr	1 hr
Department 2—weekly supply is limited to 15 people at 40 hours each at an hourly rate of $8	$1\frac{1}{4}$ hr	1 hr
Machine time:		
Department 1—weekly capacity is limited to 250 hours	$\frac{1}{2}$ hr	$\frac{1}{2}$ hr
Department 2—weekly capacity is limited to 300 hours	0 hr	1 hr

The overhead costs for Elon are accumulated on a plantwide basis. The overhead is assigned to products on the basis of the number of direct labor hours required to manufacture the product. This base is appropriate for overhead assignment because most of the variable overhead costs vary as a function of labor time. The estimated overhead cost per direct labor hour is:

Variable overhead cost	$ 6
Fixed overhead cost	6
Total overhead cost per direct labor hour	$12

The production department formulated the following equations for the linear programming statement of the problem:

A = number of units of X-10 to be produced
B = number of units of Y-12 to be produced

Objective function to minimize costs:

Minimize: $Z = 85A + 62B$

Constraints:

Material	$4A + 2B \le 1{,}800$ lb
Department 1 labor	$\frac{2}{3}A + 1B \le 400$ hr
Department 2 labor	$1\frac{1}{4}A + 1B \le 600$ hr
Nonnegativity	$A \ge 0, \quad B \ge 0$

35. The formulation of the linear programming equations as prepared by Elon Co.'s production department is incorrect. Explain what errors have been made in the formulation prepared by the production department.

36. Formulate and label the proper equations (i.e., the simplex tableau) for the linear programming statement of Elon Co.'s production problem.

37. Explain how linear programming could help Elon Co. determine how large a change in the price of direct materials would have to be to change the optimum production mix of X-10 and Y-12.

38. Suppose the probability that a manufacturer will buy raw materials from three companies each month is summarized by the following transition matrix:

$$\begin{array}{c} \\ A \\ B \\ C \end{array} \begin{array}{c} \begin{array}{ccc} A & B & C \end{array} \\ \begin{bmatrix} .50 & .30 & .20 \\ .30 & .50 & .20 \\ .30 & .30 & .40 \end{bmatrix} \end{array}$$

If the manufacturer buys from a company one time, it is more probable that the manufacturer will buy from that company the next time. If the manufacturer is currently giving one-third of the business to each of the three companies, how much business will the manufacturer do with each of the companies after 1 month?

39. Repeat Problem 38 for 2 months.

40. What are the long-term prospects for doing business with companies A, B, and C in Problem 38?

41. Find the fixed probability vector for

$$\begin{bmatrix} \frac{2}{3} & \frac{1}{3} \\ \frac{2}{5} & \frac{3}{5} \end{bmatrix}$$

42. **a.** What is an absorbing Markov chain?
b. Find and interpret FP for

$$\begin{array}{c} \\ 1 \\ 2 \\ 3 \end{array} \begin{array}{c} \begin{array}{ccc} 1 & 2 & 3 \end{array} \\ \begin{bmatrix} .1 & .5 & .4 \\ .5 & .3 & .2 \\ 0 & 0 & 1 \end{bmatrix} \end{array}$$

The following multiple-choice question is reprinted with permission of the American Institute of Certified Public Accountants, Inc.

CPA Exam May 1975

43. The Stat Company wants more information on the demand for its products. The following data are relevant:

Units demanded	Probability of unit demand	Total cost of units demanded
0	.10	\$0
1	.15	1.00
2	.20	2.00
3	.40	3.00
4	.10	4.00
5	.05	5.00

What is the total expected value or payoff with perfect information?
A. \$2.40 **B.** \$7.40
C. \$9.00 **D.** \$9.15

44. Martha, a realtor, knows that the expenses for selling a house are \$1,500 per listing. If she sells the house, she will receive 6% of the selling price. If another realtor sells the house, Martha will receive 3% of the selling price. If the house is unsold in 3 months, she will lose the listing and receive nothing. Now she must decide on only one of the following listings. Which should she take if she wants to maximize her mathematical expectation?

		Probability of selling the house		
House	Value	Listing agent	Another agent	Not selling
A	\$125,000	.5	.4	.1
B	\$168,000	.4	.5	.1
C	\$225,000	.2	.3	.5

45. Determine the value of the following game, as well as both strategies.

$$\begin{bmatrix} -3 & 1 & 2 \\ -5 & 0 & 2 \\ -1 & 5 & 6 \end{bmatrix}$$

46. A secret courier for the SSO must carry a coded message from London to Paris. This courier can take a train or can fly. Her opponent from Scotland Yard must prevent her (the courier) from delivering the message and must guess which route she will take so that he (the Scotland Yard inspector) can intercept the courier. If he intercepts her on the train, he will dispose of her and her message, resulting in a "score" of 100 for Scotland Yard. On the other hand, if he intercepts her on the plane, he will only be able to intercept the message, resulting in a "score" of 25 for Scotland Yard. If the courier delivers the message, her "score" is -10. Find the strategy for the Scotland Yard inspector, and also find the value of the game.

47. What is the future value of \$5,680 deposited at 5.85% interest compounded daily for $2\frac{1}{2}$ years?

48. Carol will retire in 20 years and is depositing \$2,000 into an IRA account at the end of each year. If the IRA is paying 9.5% interest compounded annually, how much will she have in 20 years?

49. Dale wants to retire in 20 years with the same amount of money in his retirement fund that Carol (Problem 48) has in her account, but he wants to make a one-time deposit now (because he just received an inheritance and wants to invest the money). He will place the money into a bond paying a guaranteed 9.5% interest compounded annually for the next 20 years. How much must he deposit?

50. Carlos buys a \$10,000 car and agrees to finance it for 3 years at 12% interest (APR). What is his monthly payment?

Calculus

The ideas of calculus were first considered toward the end of the sixteenth century, but the theory was not developed until the second half of the seventeenth century when Gottfried Leibniz (1646–1716) and Isaac Newton (1642–1727) simultaneously, and independently, invented the calculus we use today. Calculus extends the range of problems we can work with. For example:

Elementary mathematics	**Calculus**
1. Slope of a line 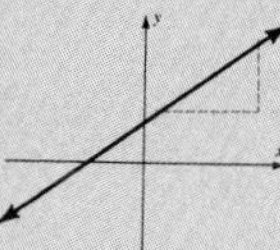	1. Slope of a curve 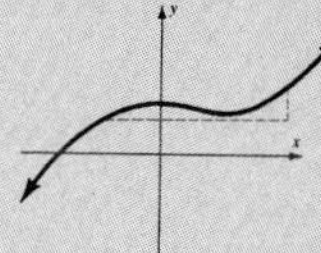
2. Tangent line to a circle 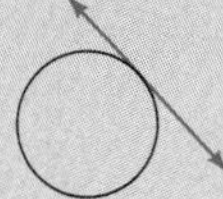	2. Tangent line to a general curve 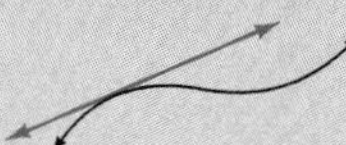
3. Area of a region bounded by line segments 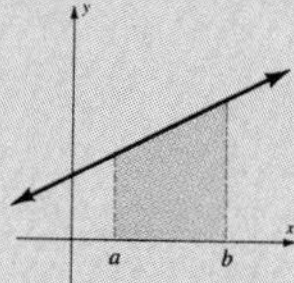	3. Area of a region bounded by curves 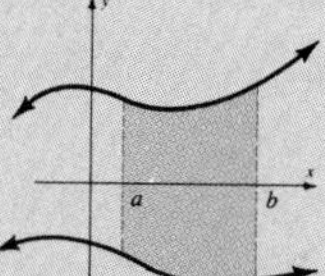
4. Average changes, velocity, or acceleration	4. Instantaneous change, velocity, or acceleration
5. Average of a finite collection of numbers	5. Average value of a function over an interval

Up to now, your study of mathematics has focused on the *mechanics* of mathematics (such as solving equations, drawing graphs, and manipulating symbols), but now you will not only reinforce such skills and learn new ones but you will also apply mathematics to particular real-life situations.

The prerequisites for this part of the text are at least 2 years of high school mathematics (reviewed in Chapter 1) and the concepts of functions and graphs (Chapter 2).

10

The Derivative

CHAPTER OVERVIEW

We focus here on the concept of a derivative and introduce efficient ways of finding the derivative. The derivative is one of the fundamental ideas in all of calculus and is the cornerstone of more advanced mathematics.

PREVIEW

We first introduce the derivative as a rate of change, but soon show that it has many additional useful applications. After just a glimpse of these applications (to be continued in the next chapter), we concentrate on finding derivatives.

PERSPECTIVE

The skills learned in this chapter are used in the next two chapters to develop applications of the derivative. Then the integral is introduced as an "antiderivative"—which will again use the knowledge of the derivatives introduced in this chapter. Two of the single most revolutionary concepts in all of mathematics are the ideas of limits and derivatives. Remember, it took some of the greatest minds in the history of mathematics many years to formulate these ideas, so do not despair if you have trouble understanding them in one evening, or even in one course. Hard work and perseverance will pay off.

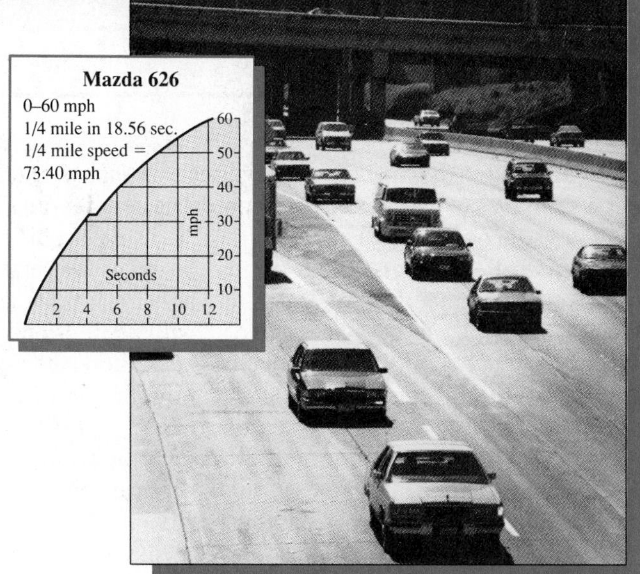

MODELING
APPLICATION 8

Instantaneous Acceleration: A Case Study of the Mazda 626

Consider the performance of the Mazda 626 Sport Coupe. The graph shown here, taken from an advertisement, shows the car's acceleration.*

*After you have finished this chapter, write a paper discussing the average rate of travel, the distance traveled, the velocity, and the acceleration for this car.***

For general guidelines abou writing this essay, See Modeling Application 1 on page 151.

* From *Time*, January 4, 1982.

** This modeling application is adapted from Peter A. Lindstrom, "A Beginning Calculus Project," UMAP, Vol. 5, No. 3, pp. 271–276.

APPLICATIONS

Management (*Business, Economics, Finance, and Investments*)
Cost of manufacturing (10.1, Problem 69; 10.5, Problem 55)
Rental charges for a fleet of trucks (10.2, Problem 46)
Output as a function of the number of workers (10.3, Problems 5–8)
Rate of change of the Gross National Product (10.3, Problems 13–16)
Cost of a per unit increase in production (10.3, Problems 47–49; 10.8, Sample Test, Problem 19)
Marginal profit (10.3, Problem 50; 10.5, Problem 56; 10.8, Sample Test, Problem 20)
Average rate of change of cost (10.3, Problems 51–55)
Marginal cost (10.3, Problems 58–59; 10.4, Problems 28–37; 10.5, Problem 55)
Change in the rate of earnings in a corporation (10.5, Problem 59)
Consumer Price Index (10.5, Problem 65)

Management (*continued*)
Demand for a commodity in a free market (10.6, Problems 41–44)
Advertising to influence purchasing (10.6, Problem 45)
Rate of change of prices in a free market (10.6, Problems 46–48)
Enrollment projections (10.7, Problem 59)

Life Sciences (*Biology, Ecology, Health, and Medicine*)
Temperatures at Death Valley (10.2, Problem 48)
Rate at which the number of bacteria in a culture change (10.4, Problems 38–40; 10.7, Problem 53)
Relationship between current and time on an X-ray machine (10.5, Problems 57–58)
The relationship between a population of foxes and rabbits (10.5, Problems 62–63)
Rate of a liquid flowing into a reservoir (10.5, Problem 64)
Effect of a drug in the bloodstream (10.6, Problems 49–50)

Social Sciences (*Demography, Political science, Population, Psychology, Society, and Sociology*)
The number of animals available for a psychology experiment (10.1, Problem 70)
Learning theory (10.1, Problem 71; 10.5, Problems 60–61; 10.6, Problems 51–52; 10.7, Problems 54, 60)
SAT scores of first-year college students (10.3, Problems 9–12)

General Interest
Postal charges (10.2, Problem 47)
Height of a projectile after t seconds (10.3, Problems 1–4)
Average speeds in a daily commute (10.3, Problems 17–20)
Velocity of an object moving in a straight line (10.4, Problems 41–44)
Interest rate changes (10.7, Problems 55–57)

Modeling Application—
Instantaneous Acceleration: A Case Study of the Mazda 626

10.1 Limits

Mathematical analysis can be divided into two broad categories: *continuous* and *discrete*. Let us consider a few examples to clarify the distinction. The counter on a turnstile may look very similar to the odometer on a car, but the turnstile is a discrete counting device while the odometer is a continuous device. The set of integers is a discrete set, whereas the set of real numbers is not. Calculus was being invented as mathematicians all over the world realized they needed to deal with new notions concerning the transition from discrete to continuous. Calculus is based on a continuous model, and the central key to understanding calculus is the notion of a limit.

Calculus was originally developed intuitively. Over time, every concept was subjected to the most meticulous scrutiny. It was felt that all mathematical thinking should eventually lead to the ideas of calculus and be continuous in nature. Then, in 1956, a landmark text by Kemeny, Snell, and Thompson called *Introduction to Finite Mathematics* was published. It dealt with discrete ideas not contained in calculus. For over 20 years this course has been offered at various colleges and universities but it was never meant to be a replacement for calculus. However, as computers became more and more a part of mathematics and mathematical development, the necessity of treating the continuous ideas of calculus from a discrete standpoint became increasingly important. Many mathematicians are now suggesting that continuous and discrete be accepted as fundamental classifications in mathematics and the mathematics curriculum and that courses in calculus be offered alongside courses in discrete mathematics.

Intuitive Notion of a Limit

We will now turn to the notion of a limit. Imagine a child throwing a ball at a brick wall:

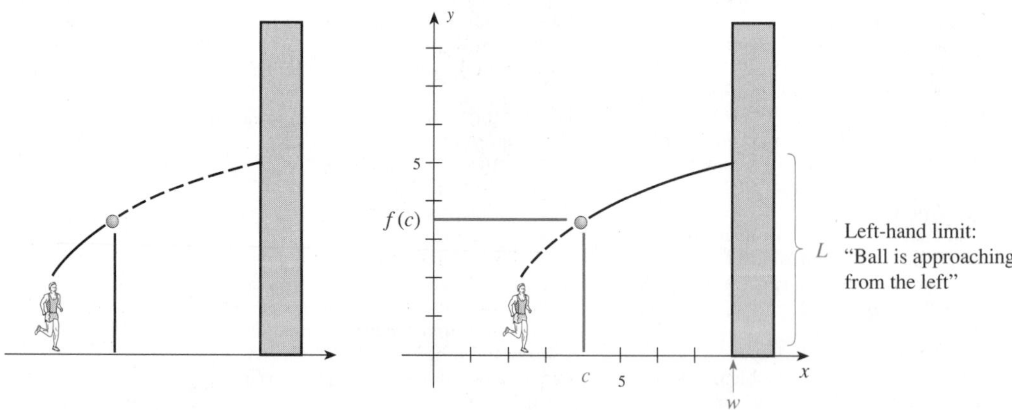

We might say that the ball is getting closer and closer to the wall, and that the wall is a limiting position for the ball.

Can we develop a mathematical model for this situation? In mathematical notation, we can describe the path of the ball by some function, say f, which gives the position of the ball. We see that the point $(c, f(c))$ is a point on the ball's path when x has the value of c. Suppose also that the wall is located at some point $(w, 0)$ on the x-axis. If we write $x \to w^-$ we mean that x approaches the wall from the left (the negative means it is approaching from the left side). If we write $x \to w^+$ we means that x is approaching the wall from the right (the positive means it is approaching from the right side).

The next question that we ask is how high is the ball when it hits the wall? If we mark the location on the wall using the coordinate system, we see the first component must be w. Thus, if L represents the height, then the ball strikes the wall at (w, L).

The mathematical model for this situation uses a notation called *limit notation* as follows:

$$\lim_{x \to w^-} f(x) = L$$

This is read as "the limit of f as x approaches w from the left equals L." In this case, L is called the "left-hand limit." This means that we predict that as x approaches w from the left along the x-axis, the path described by the function f will hit the wall at a height L.

If the projectile approaches from the right, we write

$$\lim_{x \to w^+} f(x) = R$$

The number R is called the "right-hand limit."

If we throw a projectile at a wall from both the left and the right, there is no reason to expect that it will hit at exactly the same height as shown by the following figure. Notice that our wall is now drawn as a line with no thickness:

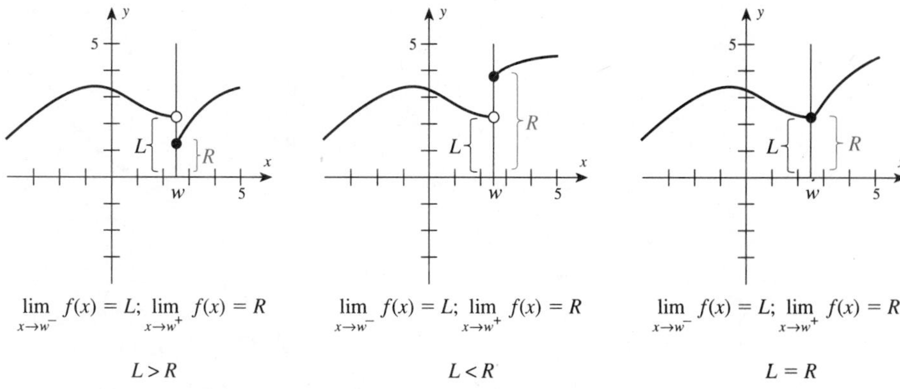

If the left- and right-hand limits are the same, then we write

$$\lim_{x \to w} f(x) = L$$

Notice that this notation does not specify a right- or left-hand limit. That is, there is no plus or minus associated with the number w. If the left- and right-hand limits are not the same, we see that *the limit does not exist at* w.

Limit Notation

> The notation $\lim_{x \to c} f(x) = L$ is read
>
> "the limit of f as x approaches c is L"
>
> and means that for all values of x in the domain of f, the values of $f(x)$ get closer and closer to the number L as x gets closer and closer to (but remains different from) c.

This intuitive definition of limit will suffice for this course. In a more formal course we would define what we mean by "closer and closer."* Notice that the limit as $x \to c$ does not require that the number $f(c)$ exists. Functions with the property that

$$\lim_{x \to c} f(x) = f(c)$$

are said to be **continuous at $x = c$**. We will consider this in the next section.

Limits by Graphing

Figure 10.1 shows the graph of a function f and the number $c = 3$. The arrowheads show possible sequences of values of x approaching $c = 3$ from both the left and the right. As x approaches $c = 3$, the $f(x)$ values get closer and closer to 5. We write this as $\lim_{x \to 3} f(x) = 5$.

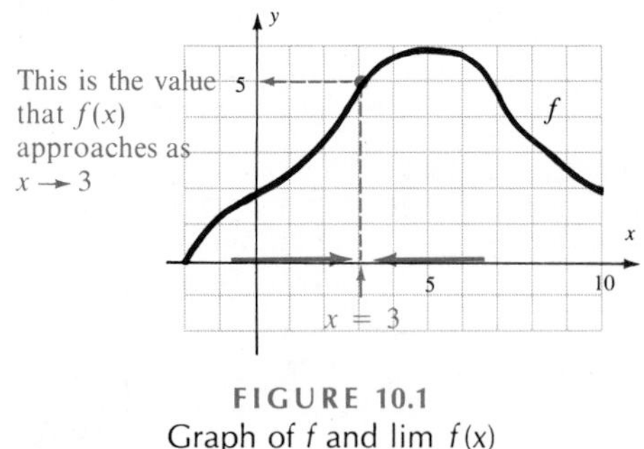

FIGURE 10.1
Graph of f and $\lim_{x \to c} f(x)$

* To make this intuitive notion precise, we must define what is meant by "closer and closer." For each $c > 0$ there exists a $d > 0$ such that $|f(x) - L| < c$ whenever $0 < |x - c| < d$. Notice also that we say the limit **is equal to L**. The statement that $0 < |x - c| < d$ implies that $x \neq c$. This exclusion ($x \neq c$) will be particularly important when we evaluate certain limits of rational functions.

EXAMPLE 1 Given the function defined by the graph below, find the following limits:

a. $\lim_{x \to 3^-} f(x)$ **b.** $\lim_{x \to -2^+} f(x)$ **c.** $\lim_{x \to 0} f(x)$ **d.** $\lim_{x \to 5} f(x)$

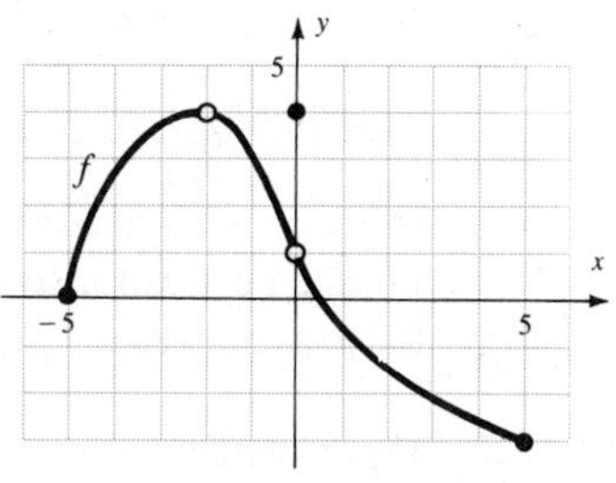

Solution Remember that $f(c)$ does not need to be defined in order to consider a limit. Also remember that an open circle on a graph indicates an excluded point. The following limits are found by inspection:

a. $\lim_{x \to 3^-} f(x) = -2$ **b.** $\lim_{x \to -2^+} f(x) = 4$

c. $\lim_{x \to 0^-} f(x) = 1$ and $\lim_{x \to 0^+} f(x) = 1$, so $\lim_{x \to 0} f(x) = 1$

d. $\lim_{x \to 5^-} f(x) = -3$, so $\lim_{x \to 5} f(x) = -3$. Because the domain of f is the closed interval $[-5, 5]$, we do not need to consider the right-hand limit, $\lim_{x \to 5^+} f(x)$, which implies $x > 5$ (which is not in the domain). ■

EXAMPLE 2 Find the requested limits on $[-6, 5)$.

a. $\lim_{x \to 3^-} f(x)$ **b.** $\lim_{x \to -2^-} f(x)$

c. $\lim_{x \to -2^+} f(x)$ **d.** $\lim_{x \to 5} f(x)$

Solution **a.** $\lim_{x \to 3^-} f(x) = 6$ **b.** $\lim_{x \to -2^-} f(x) = 2$

c. $\lim_{x \to -2^+} f(x) = 4$

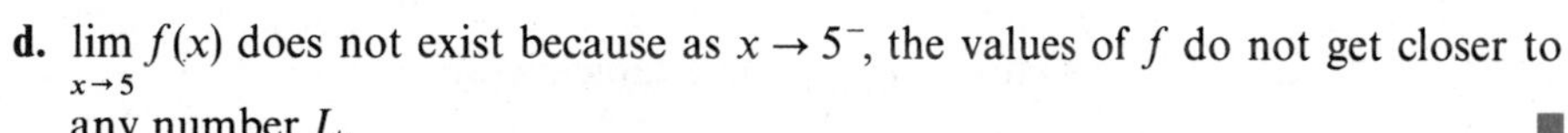

d. $\lim_{x \to 5} f(x)$ does not exist because as $x \to 5^-$, the values of f do not get closer to any number L. ■

EXAMPLE 3 Find $\lim_{x \to 2} 5$.

Solution Look at the graph of $f(x) = 5$. It is easy to see that regardless of what x is approaching, the value of $f(x)$ is 5. This means that $\lim_{x \to 2} 5 = 5$.

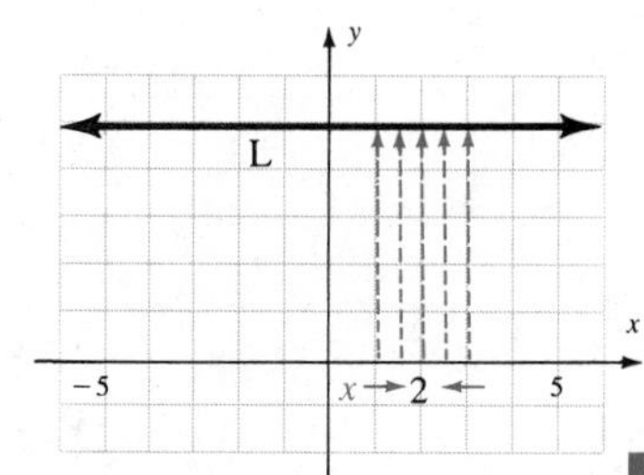

■

Limit of a Constant

> If $f(x) = k$ is a constant, then
> $$\lim_{x \to c} f(x) = k$$

Limits by Table

It is not always convenient (or even possible) to first draw a graph in order to find limits. You can also use a calculator or a computer to construct a table of values for f as $x \to c$.

EXAMPLE 4 Find $\lim_{x \to 4} x^2$.

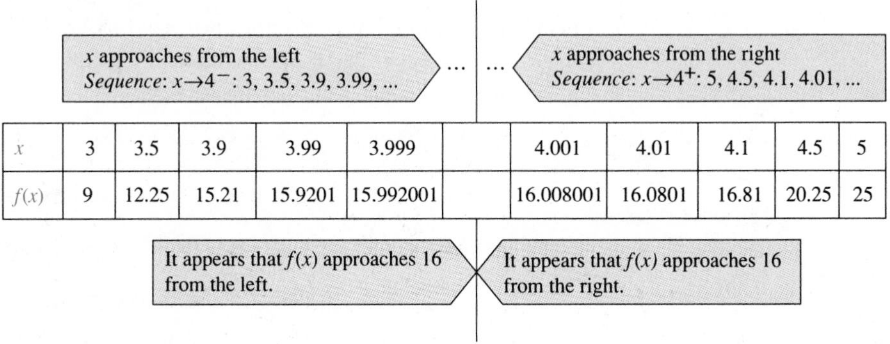

x	3	3.5	3.9	3.99	3.999		4.001	4.01	4.1	4.5	5
$f(x)$	9	12.25	15.21	15.9201	15.992001		16.008001	16.0801	16.81	20.25	25

Thus: $\lim_{x \to 4} x^2 = 16$. ■

After Example 4 you might be saying to yourself, why not just substitute $x = 4$ in x^2 to obtain the limit 16? Well, it is not that easy, as Example 5 illustrates.

EXAMPLE 5 Find $\lim_{x \to 2} \dfrac{x^2 - x - 6}{x - 2}$.

Notice that $f(2)$ is not defined: $f(2) = \dfrac{2^2 + (2) - 6}{2 - 2}$ ⟵ Division by 0

Solution Construct a table of values.

				$x \to 2^-$ Left-hand limit ⟶			⟵ Right-hand limit $x \to 2^+$					
x	1	1.5	1.9	1.99	1.999	1.9999	2.0001	2.001	2.01	2.1	2.5	3
$f(x)$	4	4.5	4.9	4.99	4.999	4.9999	5.0001	5.001	5.01	5.1	5.5	6

Since the left- and right-hand limits appear to be 5, we say the limit of $f(x)$ as x approaches 2 is 5.

A spreadsheet can be a very valuable and easy-to-use tool when evaluating limits by table (see Appendix D). The following illustration shows this example using a spreadsheet.

SPREADSHEET PROGRAM

	A	B	C	D	E
1	What is c?	[input c here]			
2					
3	n	left approach	f(x)	right approach	f(x)
4					
5	1	+B1-1	(B5^2+B5-6)/(B5-2)	+B1+1	(B5^2+B5-6)/(B5-2)
6	+A5+1	+$B5+(1-(0.1)^A6)	replicate	+$B5+(1-(0.1)^A6)	replicate
7	replicate	replicate		replicate	

OUTPUT:

What is c?	2			
n	left approach	f(x)	right approach	f(x)
1	1	4	3	6
2	1.9	4.9	2.1	5.1
3	1.99	4.99	2.01	5.01
4	1.999	4.999	2.001	5.001
5	1.9999	4.9999	2.0001	5.0001
6	1.99999	4.99999	2.00001	5.00001
7	1.999999	4.999999	2.000001	5.000001
8	1.9999999	4.9999998999975	2.0000001	5.0000000999981
9	1.99999999	4.999999989982	2.00000001	5.0000000099747

■

Limits by Using Algebra

In Example 5 we saw that substitution of x for 2 leads to undefined values for the function. This occurs because the domain of f in Example 5 excludes the limiting value for x. Instead of substitution, suppose we try algebraic simplification.

EXAMPLE 6 Find $\lim_{x \to 2} \dfrac{x^2 + x - 6}{x - 2}$.

Solution Instead of finding a table of values (as we did in Example 5 and in the spreadsheet following Example 5), let us simplify the given expression:

$$\lim_{x \to 2} \frac{x^2 + x - 6}{x - 2} = \lim_{x \to 2} \frac{(x + 3)(x - 2)}{x - 2} = \lim_{x \to 2}(x + 3)$$

The above simplification is valid only if $x \neq 2$; but as $x \to 2$ we know that $x \neq 2$, so the simplification is valid. What about the limit of $x + 3$ as x approaches 2? You can complete this by graphing or by a table of values to see that when x is close to 2, $x + 3$ is close to 5. Thus,

$$\lim_{x \to 2} \frac{x^2 + x - 6}{x - 2} = 5$$

■

Clearly, we need a theorem to help us evaluate limits such as the one in Example 6. As long as f is a polynomial function, the limit can be found by direct evaluation. On the other hand, if f is not a polynomial function, then substitution *may* cause undefined values for f. The procedure is to algebraically simplify the given expression, and then *if the simplified expression is a polynomial*, evaluate it by direct substitution. This is summarized in the following box.

Limit of a Polynomial

> If f is any polynomial function, then
> $$\lim_{x \to c} f(x) = f(c) \qquad \text{for any real number } c$$

EXAMPLE 7 Find:

a. $\lim_{x \to 1}(4x^3 - 2x^2 + x - 1)$ **b.** $\lim_{x \to 3}(x^2 - x)$

c. $\lim_{x \to 3} \frac{x - 3}{x - 3}$ **d.** $\lim_{x \to 1} \frac{x^2 - 1}{x - 1}$

Solution **a.** $\lim_{x \to 1}(4x^3 - 2x^2 + x - 1) = 4(1)^3 - 2(1)^2 + 1 - 1 = 2$
since $4x^3 - 2x^2 + x - 1$ is a polynomial

b. $\lim_{x \to 3}(x^2 - x) = 3^2 - 3 = 6$ since $x^2 - x$ is a polynomial

c. $\lim_{x \to 3} \frac{x - 3}{x - 3} = \lim_{x \to 3} 1 = 1 \qquad x \neq 3$ Limit of a constant

d. $\lim_{x \to 1} \frac{x^2 - 1}{x - 1} = \lim_{x \to 1} \frac{(x - 1)(x + 1)}{x - 1} = \lim_{x \to 1}(x + 1) = 2 \qquad x \neq 1$ ■

EXAMPLE 8 Find $\lim_{x \to 0} f(x)$ where $f(x) = \begin{cases} x + 5 & \text{if } x > 0 \\ x & \text{if } x < 0 \end{cases}$.

Solution Notice that $f(0)$ is not defined, but since the limit is as x approaches 0, it is not necessary that $f(0)$ be defined. It is necessary, however, to consider left- and right-hand limits for this problem.

$$\lim_{x \to 0^-} f(x) = \lim_{x \to 0^-} x$$ As x approaches from the left of 0, it must be the case that $x < 0$, so $f(x) = x$ when $x < 0$.

$$= 0$$

$$\lim_{x \to 0^+} f(x) = \lim_{x \to 0^+} (x + 5)$$ As x approaches from the right of 0, $x > 0$.

$$= 5$$

Since the left- and right-hand limits are not the same, we say $\lim_{x \to 0} f(x)$ does not exist. ■

It is instructive to compare the results of Example 8 with the graph of the function, as shown in Figure 10.2. After it is graphed, it is easy to see that the left- and right-hand limits are not the same. Compare with the graph of g in the following example.

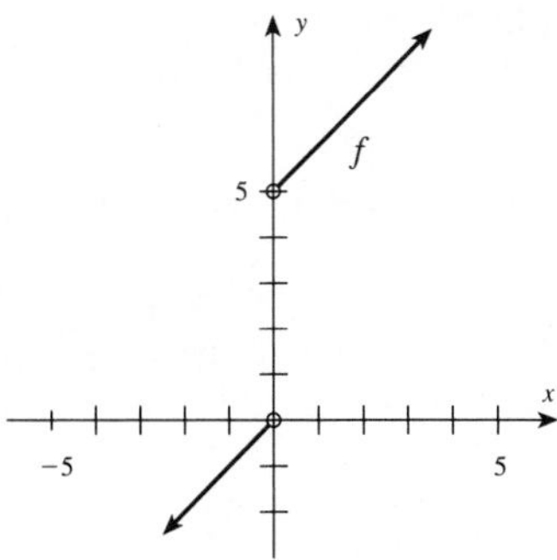

Graph of $f(x) = \begin{cases} x + 5 & \text{if } x > 0 \\ x & \text{if } x < 0 \end{cases}$

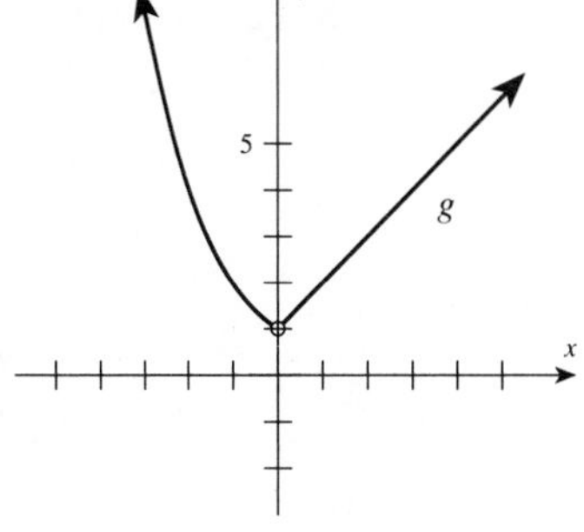

Graph of $g(x) = \begin{cases} x + 1 & \text{if } x > 0 \\ x^2 + 1 & \text{if } x < 0 \end{cases}$

FIGURE 10.2
Graphs of the functions in Examples 8 and 9

EXAMPLE 9 Find $\lim_{x \to 0} g(x)$ where $g(x) = \begin{cases} x + 1 & \text{if } x > 0 \\ x^2 + 1 & \text{if } x < 0 \end{cases}$.

Solution

$$\lim_{x \to 0^-} g(x) = \lim_{x \to 0^-} (x^2 + 1) = 1$$

$$\lim_{x \to 0^+} g(x) = \lim_{x \to 0^+} (x + 1) = 1$$

Since the left- and right-hand limits are equal, $\lim_{x \to 0} g(x) = 1$. ■

Sometimes the function is a rational function that cannot be simplified to a polynomial. Sometimes such an expression will have a limit (Example 10) and sometimes it will not (Example 11). In order to evaluate such limits we need a result called the Limit of a Quotient Theorem.

Limit of a Quotient Theorem

Let f and g be two functions whose limits as $x \to c$ exist and $\lim_{x \to c} g(x) \neq 0$. Then

$$\lim_{x \to c} \frac{f(x)}{g(x)} = \frac{\lim_{x \to c} f(x)}{\lim_{x \to c} g(x)}$$

The theorem says that if the limit of the denominator function is not zero, the limit of the ratio of the two functions is the ratio of their limits. In particular, if f and g are polynomial functions, then you can evaluate a rational function by substitution.

EXAMPLE 10 Find $\lim_{x \to 2} \frac{x - 2}{x + 2}$.

Solution

$$\lim_{x \to 2} \frac{x - 2}{x + 2} = \frac{\lim_{x \to 2}(x - 2)}{\lim_{x \to 2}(x + 2)} = \frac{0}{4} = 0$$

■

EXAMPLE 11 Find $\displaystyle\lim_{x\to -2}\frac{x-2}{x+2}$.

Solution $\displaystyle\lim_{x\to -2}\frac{x-2}{x+2}$ does not exist. You cannot apply the Limit of a Quotient Theorem because the limit of $x + 2$ (the denominator) is 0 as $x \to -2$. In order to see that this limit does not exist, you can use a table of values to see that the quotient increases without limit as $x \to -2^-$ and decreases as $x \to -2^+$, or you can draw the graph as shown in the margin. ■

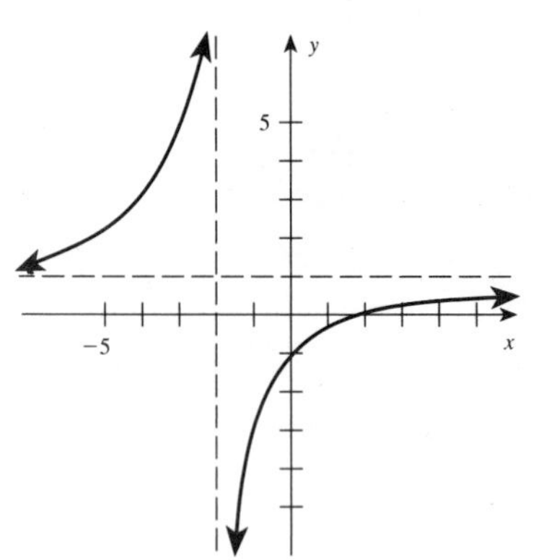

COMPUTER APPLICATION

See the program *Limit of f(x)* on the computer disk accompanying this book. It sets up a table of values (either left- or right-hand limits). You can also draw the graph. From the graph and the table of values, you will be able to see if the limit exists, and in most cases will also be able to find the limit.

Several other limit theorems that we will need to use occasionally are stated here for completeness. We have also included the other limit theorems in this list.

Limit Theorems

Let f and g be two functions whose limits as $x \to c$ exist.

Limit of a constant: $\displaystyle\lim_{x\to c} k = k$ for any constant k

Limit of a polynomial: $\displaystyle\lim_{x\to c} f(x) = f(c)$ for any polynomial function f

Limit of a sum: $\displaystyle\lim_{x\to c}[f(x) + g(x)] = \lim_{x\to c} f(x) + \lim_{x\to c} g(x)$

Limit of a difference: $\displaystyle\lim_{x\to c}[f(x) - g(x)] = \lim_{x\to c} f(x) - \lim_{x\to c} g(x)$

Limit of a product: $\displaystyle\lim_{x\to c}[f(x)\cdot g(x)] = \left[\lim_{x\to c} f(x)\right]\left[\lim_{x\to c} f(x)\right]$

Limit of a quotient: $\displaystyle\lim_{x\to c}\frac{f(x)}{g(x)} = \frac{\lim_{x\to c} f(x)}{\lim_{x\to c} g(x)}$ if $\displaystyle\lim_{x\to c} g(x) \neq 0$

Limit of a power: $\displaystyle\lim_{x\to c}[f(x)]^n = \left[\lim_{x\to c} f(x)\right]^n$ where n is a positive integer

Limit of a root: $\displaystyle\lim_{x\to c}\sqrt[n]{f(x)} = \sqrt[n]{\lim_{x\to c} f(x)}$ where $n \geq 2$ is a positive integer, and both roots are defined.

The expression $0/0$ is not a real number; in calculus we call this an **indeterminate form**. As long as an indeterminate form is not obtained, you may evaluate the limit as $x \to c$ by substituting c into the given expression. These limit theorems apply for one-sided limits as well as for two-sided limits.

EXAMPLE 12 Find $\lim_{x \to 1} \frac{\sqrt{x} - 1}{x - 1}$.

Solution Try substitution: $\lim_{x \to 1} \frac{\sqrt{x} - 1}{x - 1} = \frac{\sqrt{1} - 1}{1 - 1} = \frac{1 - 1}{1 - 1} = \frac{0}{0}$ Indeterminate form

WARNING ***Do not assume when you obtain an indeterminate form that the limit does not exist.***

You could proceed with a table of values for this example but, instead, we will **rationalize the numerator**. Remember, from algebra, the process of *rationalizing the denominator*. The process here is the same—namely, multiply both numerator and denominator by $\sqrt{x} + 1$:

$$\begin{aligned}\frac{\sqrt{x} - 1}{x - 1} \cdot \frac{\sqrt{x} + 1}{\sqrt{x} + 1} &= \frac{\sqrt{x} \cdot \sqrt{x} - \sqrt{x} + \sqrt{x} - 1}{(x - 1)(\sqrt{x} + 1)} \\ &= \frac{x - 1}{(x - 1)(\sqrt{x} + 1)} \\ &= \frac{1}{\sqrt{x} + 1}\end{aligned}$$

Thus

$$\lim_{x \to 1} \frac{\sqrt{x} - 1}{x - 1} = \lim_{x \to 1} \frac{1}{\sqrt{x} + 1} = \frac{1}{1 + 1} = \frac{1}{2}$$

■

Limits at Infinity

In Section 2.6 we considered asymptotes as an aid for graphing rational functions. The idea of a horizontal asymptote seeks to find the limiting value of a function as x becomes very large or very small. Consider Examples 13 and 14.

EXAMPLE 13 Find the value of $\frac{1}{x}$, if it exists, as x increases without bound. This is called a *limit at infinity*.

Solution The limit theorems are not stated for limits at infinity so we cannot assume they apply. If the limit theorems do not apply, we must rely on a graph or a table. For clarity we will do both.

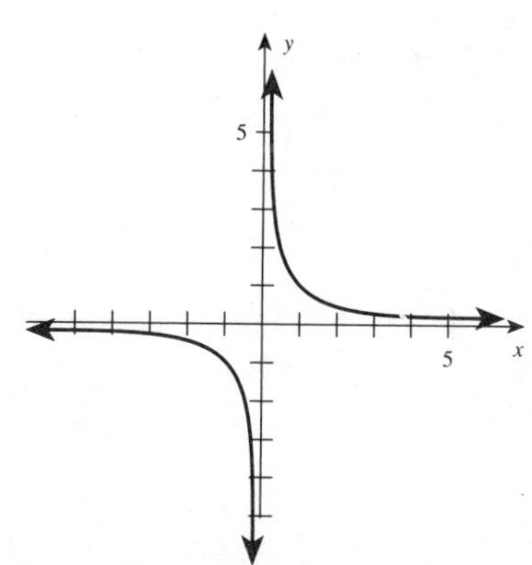

x	1	2	10	100	1,000	10,000
$f(x)$	1	.5	.1	.01	.001	.0001

We say that $\frac{1}{x}$ approaches 0 as x increases without bound, and we symbolize this by $\lim_{x \to \infty} \frac{1}{x} = 0$.

■

EXAMPLE 14 Find the value of $\frac{1}{x}$ as x decreases without bound.

x	-1	-2	-10	-100	$-1{,}000$	$-10{,}000$
$f(x)$	-1	$-.5$	$-.1$	$-.01$	$-.001$	$-.0001$

We say that $\frac{1}{x}$ approaches 0 as x decreases without bound, and we symbolize this by $\lim_{x\to-\infty} \frac{1}{x} = 0$. ■

Limits at Infinity

The statement

$$\lim_{x\to\infty} f(x) = L$$

means that $f(x)$ is close to L for very large positive values of x. Also,

$$\lim_{x\to-\infty} f(x) = L$$

means that $f(x)$ is close to L for negative values of x with very large absolute values. Finally,

$$\lim_{|x|\to\infty} f(x) = L$$

means that $f(x)$ is close to L for both positive and negative values of x with very large absolute values.

In Examples 13 and 14 we see that $\lim_{|x|\to\infty} \frac{1}{x} = 0$. It is easy to show that the limit is also 0 for any constant k divided by x, and for $\frac{1}{x^n}$ where n is a positive real number. This is summarized by the following limit theorem.

Limit to Infinity

$$\lim_{|x|\to\infty} \frac{1}{x^n} = 0 \quad \text{and} \quad \lim_{|x|\to\infty} \frac{k}{x^n} = 0 \quad \text{for any constant } k \text{ and } n > 0.$$

EXAMPLE 15 Find $\lim_{x\to\infty} \frac{x}{2x+1}$.

Solution We could construct a table of values. Instead, suppose we multiply the rational expression by 1, written as $\frac{\frac{1}{x}}{\frac{1}{x}}$:

$$\frac{x}{2x+1} \cdot \frac{\frac{1}{x}}{\frac{1}{x}} = \frac{1}{2+\frac{1}{x}}$$

Now, since $\lim_{x\to\infty} \frac{1}{x} = 0$, we see that

$$\lim_{x\to\infty} \frac{1}{2+\frac{1}{x}} = \frac{1}{2+0} = \frac{1}{2}$$

■

EXAMPLE 16 Find $\lim_{x\to\infty} \frac{3x^2 - 7x + 2}{7x^2 + 2x + 5}$.

Solution Note that the largest power of x in the expression is x^2, so we multiply the numerator and denominator by $\frac{1}{x^2}$:

$$\lim_{x\to\infty} \frac{3x^2 - 7x + 2}{7x^2 + 2x + 5} \cdot \frac{\frac{1}{x^2}}{\frac{1}{x^2}}$$

Now, since $\frac{k}{x}$ and $\frac{k}{x^2}$ both approach 0 as x increases without bound, we have

$$= \lim_{x\to\infty} \frac{3 - \frac{7}{x} + \frac{2}{x^2}}{7 + \frac{2}{x} + \frac{5}{x^2}} = \frac{3 - 0 + 0}{7 + 0 + 0}$$

Thus $\lim_{x\to\infty} f(x) = \frac{3}{7}$. ■

Asymptotes

We can now substantiate the result about horizontal asymptotes stated in Section 2.6. If

$$f(x) = \frac{P(x)}{D(x)}$$

where $P(x)$ and $D(x)$ are polynomial functions with no common factors, then the line $y = 0$ is a *horizontal asymptote* if the degree of $P(x) = a_n x^n + \cdots + a_0$ is less than the degree of $D(x) = b_m x^m + \cdots + b_0$. If $y = f(x)$, then

$$y = \lim_{x\to\infty} \frac{P(x)}{D(x)} = \lim_{x\to\infty} \frac{a_n x^n + \cdots + a_0}{b_m x^m + \cdots + b_0} \qquad \text{where } n < m$$

$$= \lim_{x\to\infty} \frac{a_n x^n + \cdots + a_0}{b_m x^m + \cdots + b_0} \cdot \frac{\frac{1}{x^m}}{\frac{1}{x^m}} \qquad \text{where } n < m$$

$$= \lim_{x\to\infty} \frac{\frac{a_n}{x^{m-n}} + \cdots + \frac{a_0}{x^m}}{b_m + \frac{b_{m-1}}{x} + \cdots + \frac{b_0}{x^m}}$$

$$= \frac{0 + 0 + \cdots + 0}{b_m + 0 + 0 + \cdots + 0} = 0$$

You can similarly show that if the degree of P is equal to the degree of D then the horizontal asymptote is

$$y = \frac{a_n}{b_n}$$

This is left as a problem.

EXAMPLE 17 Graph $y = \dfrac{x^2}{x^2 + 1}$.

Solution There are no vertical asymptotes (no values of x that cause division by zero). The horizontal asymptote(s) are found by taking the limit at infinity:

$$y = \lim_{|x| \to \infty} \frac{x^2}{x^2 + 1} = \lim_{|x| \to \infty} \frac{1}{1 + \dfrac{1}{x^2}} = 1$$

The graph is shown in Figure 10.3.

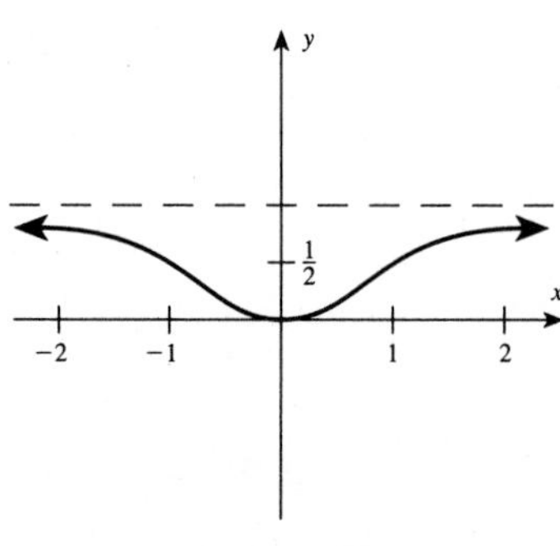

FIGURE 10.3

10.1 Problem Set

Given the functions defined by the graphs in Figure 10.4, *find the limits in Problems* 1–12.

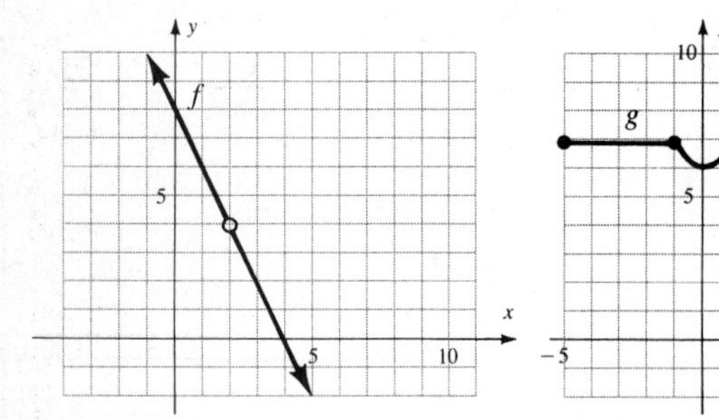

FIGURE 10.4
Functions f, g, and t

1. $\lim_{x \to 3} f(x)$
2. $\lim_{x \to 2} f(x)$
3. $\lim_{x \to 0} f(x)$
4. $\lim_{x \to -3} g(x)$
5. $\lim_{x \to -1^-} g(x)$
6. $\lim_{x \to 2^+} g(x)$
7. $\lim_{x \to 3^+} g(x)$
8. $\lim_{x \to 2^-} t(x)$
9. $\lim_{x \to 2^+} t(x)$
10. $\lim_{x \to 2} t(x)$
11. $\lim_{x \to 4} t(x)$
12. $\lim_{x \to -4} t(x)$

Find the limits by filling in the appropriate values in the tables in Problems 13–16.

13. $\lim_{x\to 5^-} (4x - 5)$

x	2	3	4	4.5	4.9	4.99
$f(x)$	3					

14. $\lim_{x\to 2} \frac{x^3 - 8}{x^2 + 2x + 4}$

x	1	1.5	1.9	1.99	1.999	2.5	2.01	2.001
$f(x)$								

15. $\lim_{x\to 2} \frac{x^2 + 2x + 4}{x^3 - 8}$

x	1	1.5	1.9	1.99	1.999	2.5	2.01	2.001
$f(x)$								

16. $\lim_{x\to\infty} \frac{x^2 + 6x + 9}{x + 3}$

x	1	10	100	1,000	10,000	100,000	1,000,000
$f(x)$							

Find the limits in Problems 17–62.

17. $\lim_{x\to 0} x^8$

18. $\lim_{x\to 2}(x^2 - 4)$

19. $\lim_{x\to 3}(x^2 - 4)$

20. $\lim_{x\to 1} \frac{1}{x - 3}$

21. $\lim_{x\to -3} \frac{1}{x - 3}$

22. $\lim_{x\to 3} \frac{1}{x - 3}$

23. $\lim_{x\to 0} \frac{1}{x^2 + 1}$

24. $\lim_{x\to 1} \frac{3x + 2}{x - 2}$

25. $\lim_{x\to\infty} \frac{1}{x^2 + 1}$

26. $\lim_{x\to\infty} 2x$

27. $\lim_{x\to -1} \frac{1}{x^2 + 1}$

28. $\lim_{x\to 2} \frac{1}{x^2 - 4}$

29. $\lim_{x\to 2} \frac{x^2 - 4}{x - 2}$

30. $\lim_{x\to\infty} (3x - 4)$

31. $\lim_{x\to 3} \frac{x^2 + 3x - 10}{x - 2}$

32. $\lim_{x\to 3} \frac{x^2 - 8x + 15}{x - 3}$

33. $\lim_{x\to 4} \frac{\sqrt{x} - 4}{x - 16}$

34. $\lim_{x\to -5} \frac{x^2 + 3x - 10}{x + 5}$

35. $\lim_{x\to 9} \frac{\sqrt{x} - 3}{x - 3}$

36. $\lim_{x\to 9} \frac{\sqrt{x} - 3}{x - 9}$

37. $\lim_{x\to 2} \frac{x^2 - 1}{x - 2}$

38. $\lim_{x\to 4} \frac{\sqrt{x} - 2}{x - 2}$

39. $\lim_{x\to 2} \frac{x^3 - 8}{x^2 + 2x + 4}$

40. $\lim_{x\to 4} \frac{2x^2 - 5x - 12}{x - 4}$

41. $\lim_{x\to 2} \frac{x + 2}{x^3 + 8}$

42. $\lim_{x\to 2} \frac{x^2 + 2x + 4}{x^3 - 8}$

43. $\lim_{|x|\to\infty} \frac{2x^2 - 5x - 3}{x^2 - 9}$

44. $\lim_{x\to 2} \frac{6 - x}{2x - 15}$

45. $\lim_{|x|\to\infty} \frac{3x - 1}{2x + 3}$

46. $\lim_{|x|\to\infty} \frac{x^2 + 6x + 9}{x + 3}$

47. $\lim_{x\to -\infty} \frac{5x + 10{,}000}{x - 1}$

48. $\lim_{|x|\to\infty} \frac{6x^2 - 5x + 2}{2x^2 + 5x + 1}$

49. $\lim_{x\to\infty} \left(x + 2 + \frac{3}{x - 1}\right)$

50. $\lim_{x\to -\infty} \frac{4x + 10^6}{x + 1}$

51. $\lim_{x\to 0} \frac{1 - \frac{1}{x + 1}}{x}$

52. $\lim_{x\to 1} \frac{1 - \frac{1}{x}}{x - 1}$

53. $\lim_{x\to -\infty} \frac{3x^2 - 5x + 15}{x + 3}$

54. $\lim_{x\to -\infty} \left(2x - 3 + \frac{4}{x + 2}\right)$

55. $\lim_{x\to -\infty} \frac{4x^4 - 3x^3 + 2x + 1}{3x^4 - 9}$

56. $\lim_{x\to 1} \frac{x^2 + x + 1}{x^3 - 1}$

57. $\lim_{x\to 0} f(x) = \begin{cases} 2x + 3 & \text{if } x > 0 \\ x^2 + 3 & \text{if } x < 0 \end{cases}$

58. $\lim_{x\to 0} f(x) = \begin{cases} x^2 - 1 & \text{if } x > 0 \\ 2x^2 + 1 & \text{if } x < 0 \end{cases}$

59. $\lim_{x\to 1} f(x) = \begin{cases} 2x + 3 & \text{if } x > 0 \\ x^2 + 3 & \text{if } x < 0 \end{cases}$

60. $\lim_{x\to -3} f(x) = \begin{cases} x^2 - 1 & \text{if } x > 0 \\ 2x^2 + 1 & \text{if } x < 0 \end{cases}$

61. $\lim_{x\to 3} f(x) = \begin{cases} x + 7 & \text{if } x > 3 \\ x^2 + 1 & \text{if } x \le 3 \end{cases}$

62. $\lim_{x\to -2} f(x) = \begin{cases} 5x & \text{if } x \ge -2 \\ x^2 + 1 & \text{if } x < -2 \end{cases}$

APPLICATIONS

Use limits to find the horizontal asymptotes for the curves whose equations are given in Problems 63–68.

63. $y = \frac{3x^2}{x^2 + 2}$

64. $y = \frac{-2x^2}{x^2 + 1}$

65. $y = \frac{2x^2 + 5}{3x^2 + 2}$

66. $y = \frac{2x^2 + 1}{5x^2}$

67. $y = \dfrac{2x^2}{x^2 - 4}$

68. $y = \dfrac{x^2}{x^2 - 1}$

69. The cost of manufacturing a specialized machine tool is a function of the number of items manufactured. This cost is graphed below. Note that there is a jump in the cost after 10,000 items because at that point it is necessary to add a second shift.

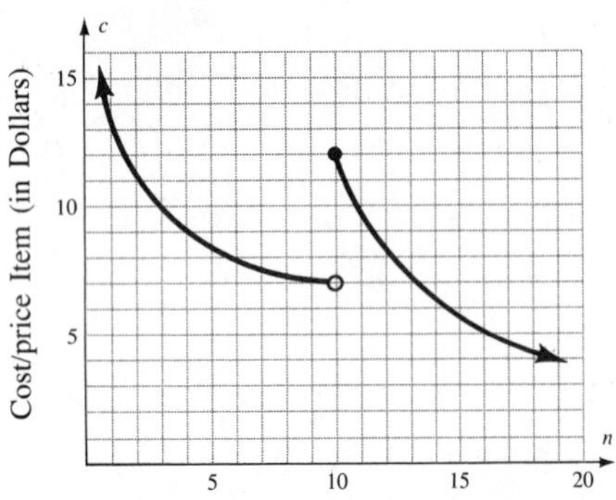

Find the following limits.

a. $\lim_{n \to 10^-} c(n)$ **b.** $\lim_{n \to 10^+} c(n)$

c. $\lim_{n \to 13} c(n)$ **d.** $\lim_{n \to 10} c(n)$

70. The number of live animals during a psychology experiment is shown in the graph. The experiment begins with 2 animals, and after time t_4 there are 14 animals. Find the following limits.

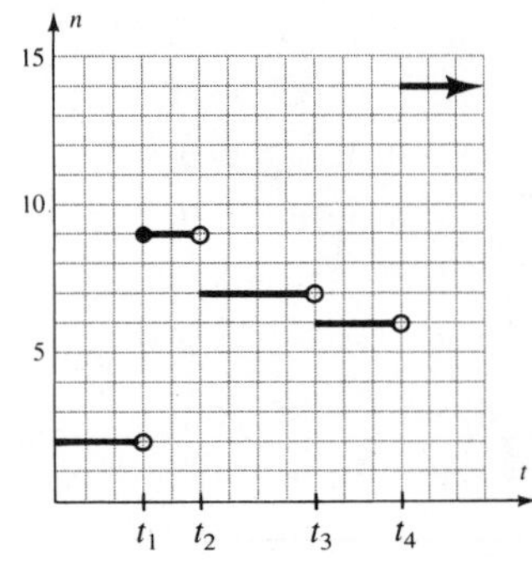

a. $\lim_{t \to t_1^-} n(t)$ **b.** $\lim_{t \to t_1^+} n(t)$

c. $\lim_{t \to t_1} n(t)$ **d.** $\lim_{t \to t_4} n(t)$

71. Learning theory measures the percentage of mastery of a subject as a function of time. The learning curve for a particular learning task is shown below. Note that at time t_1 there is a jump in mastery. Find the following limits.

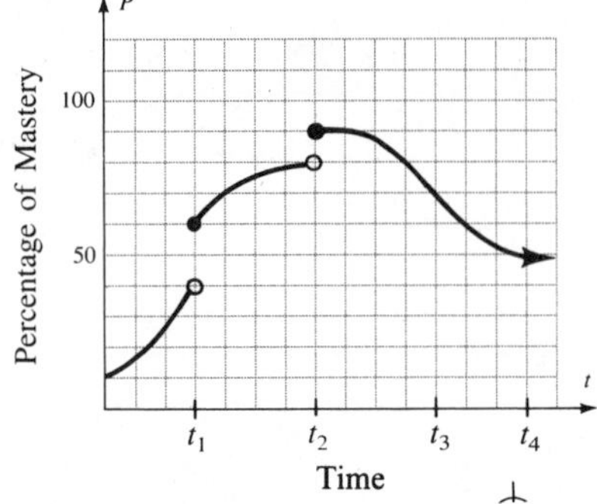

a. $\lim_{t \to t_1} P(t)$ **b.** $\lim_{t \to t_2} P(t)$

c. $\lim_{t \to t_3} P(t)$ **d.** $\lim_{t \to \infty} P(t)$

72. If $f(x) = \dfrac{P(x)}{D(x)}$ where $P(x)$ and $D(x)$ are polynomial functions with no common factors, and degree n, show that the horizontal asymptote is

$$y = \frac{a_n}{b_n}$$

73. If $f(x) = \sqrt{x}$, find $\lim_{h \to 0} \dfrac{f(1 + h) - f(1)}{h}$.

74. If $f(x) = x^2 + 1$, find $\lim_{h \to 0} \dfrac{f(2 + h) - f(2)}{h}$.

75. If $f(x) = \dfrac{1}{x}$, find $\lim_{h \to 0} \dfrac{f(3 + h) - f(3)}{h}$.

76. If $f(x) = x^3$, find $\lim_{h \to 0} \dfrac{f(-1 + h) - f(-1)}{h}$.

10.2 Continuity

Introduction

You may remember the puzzle in the margin at the top of page 495 from elementary school: the challenge is to draw the figure without lifting your pencil from the paper or retracing any of the lines. In calculus we are concerned with figures that

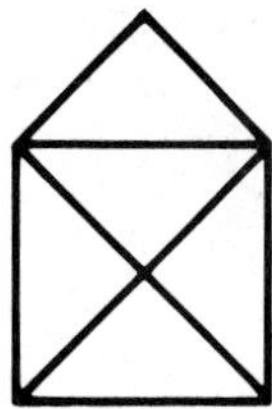

can be drawn without lifting a pencil from the paper, but we focus our attention on functions. The idea of *continuity* evolved from the notion of a curve "without breaks or jumps" to a rigorous definition given by Karl Weierstrass (1815–1897). Galileo and Leibniz had thought of continuity in terms of the density of points on a curve, but they were in error since the rational numbers have this property of denseness, yet do not form a continuous curve. Another mathematician, J. W. R. Dedekind (1831–1916), took an entirely different approach and concluded that continuity is due to the division of a curve into two parts so that there is one and only one point that makes this division. As Dedekind wrote, "By this commonplace remark, the secret of continuity is to be revealed."* We begin with a discussion of *continuity at a point*. It may seem strange to talk about continuity *at a point*, but it should seem natural to talk about a curve being "discontinuous at a point," as illustrated by Example 1.

EXAMPLE 1 Which of the following curves appear to have a discontinuity at the point $x = 1$?

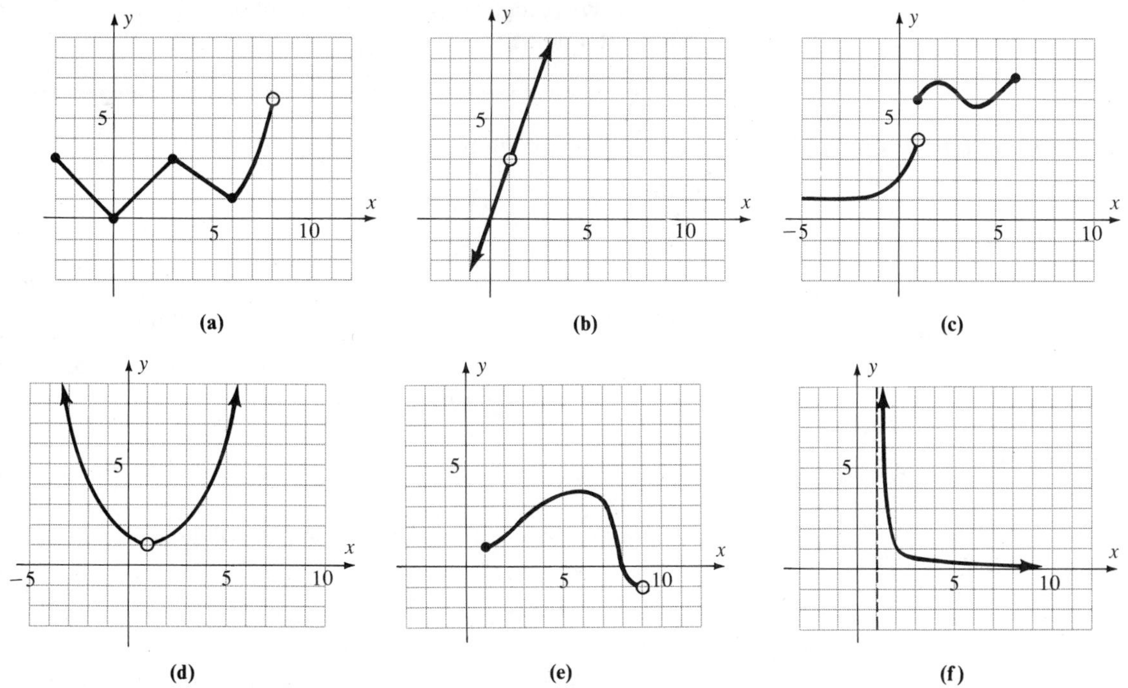

Solution The functions whose graphs are shown in parts **b**, **c**, **d**, and **f** are obviously discontinuous at $x = 1$. If, however, $x = 1$ happens to be the endpoint of an interval (part **e**) we do not say the curve is discontinuous at $x = 1$. ■

The solution to Example 1 illustrates the idea of continuity from an intuitive standpoint, but we need to define and formalize this idea more completely. There are two essential conditions for a function f to be continuous at a point c. First, $f(c)$ must be defined. For example, the curve in part **e** of Example 1 is not continuous at

* From Carl Boyer, *A History of Mathematics* (New York: Wiley, 1968), p. 607.

its right endpoint because it is not defined at $x = 9$ (the open dot indicates an excluded point).

EXAMPLE 2 Which, if any, of the given functions is continuous at $x = 0$?

a. $f(x) = \dfrac{3x^2 - x + 2}{x}$ **b.** $g(x) = \dfrac{x^2 - 5x}{x}$

c. $s(x) = \begin{cases} x + 3 & \text{if } x > 0 \\ x^2 & \text{if } x < 0 \end{cases}$

Solution None of these functions is defined at $x = 0$, so none is continuous at $x = 0$. ■

Definition of Continuity at a Point

A second condition for continuity at a point $x = c$ is that the function makes no jumps at this point. This means that if "x is close to c," then "$f(x)$ must be close to $f(c)$." Looking at Example 1, we see that the graphs in parts **b**, **c**, and **d** jump at the point $x = 1$. We recognize this as the concept of limit and now define the concept of continuity at a point:

Definition of Continuity at a Point

A function f is continuous at a point $x = c$ if

1. $f(c)$ exists and
2. $\lim_{x \to c} f(x) = f(c)$

The conditions of this definition are summarized in Table 10.1.

Test the continuity of each function in Examples 3–6 at the point $x = 1$. If it is not continuous at $x = 1$, tell why.

EXAMPLE 3 $f(x) = \dfrac{x^2 + 2x - 3}{x - 1}$

Solution Not continuous at $x = 1$ (hole), because f is not defined at this point. ■

EXAMPLE 4 $g(x) = \dfrac{x^2 + 2x - 3}{x - 1}$ if $x \neq 1$ and $g(x) = 6$ if $x = 1$

Solution Note that g is very similar to f in Example 3 except that g is defined at $x = 1$. Now, to test the second condition of continuity, $g(1) = 6$ and

$$\begin{aligned} \lim_{x \to 1} g(x) &= \lim_{x \to 1} \frac{x^2 + 2x - 3}{x - 1} \\ &= \lim_{x \to 1} \frac{(x - 1)(x + 3)}{x - 1} \\ &= \lim_{x \to 1} (x + 3) \\ &= 4 \end{aligned}$$

Since $\lim_{x \to 1} g(x) \neq g(1)$, we see that g is not continuous at $x = 1$ (jump). ■

TABLE 10.1
Holes, Poles, Jumps, and Continuity

$f(c)$ is defined	$f(c)$ is not defined
Continuous at $x = c$ $\lim_{x\to c} f(x)$ exists and is equal to $f(c)$ 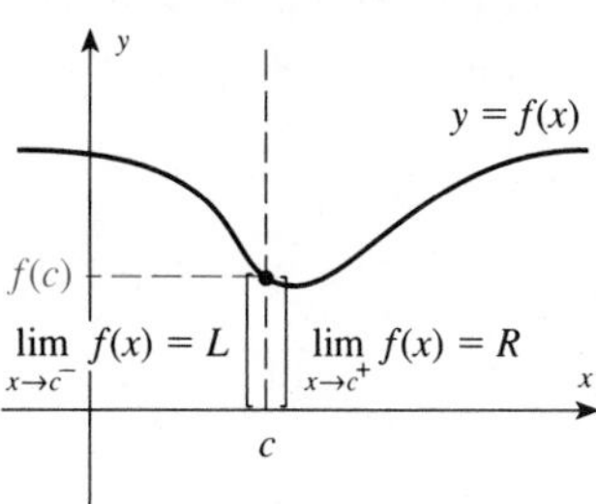	
Not continuous at $x = c$: Hole $\lim_{x\to c} f(x)$ exists and is not equal to $f(c)$ 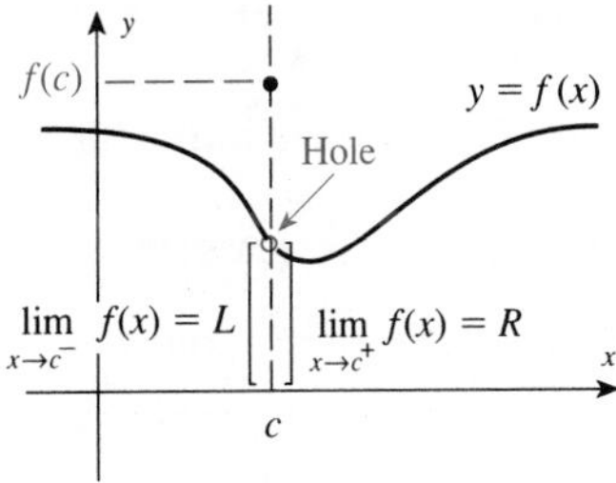	*Not continuous at $x = c$: Hole* $\lim_{x\to c} f(x)$ exists and $f(c)$ is not defined 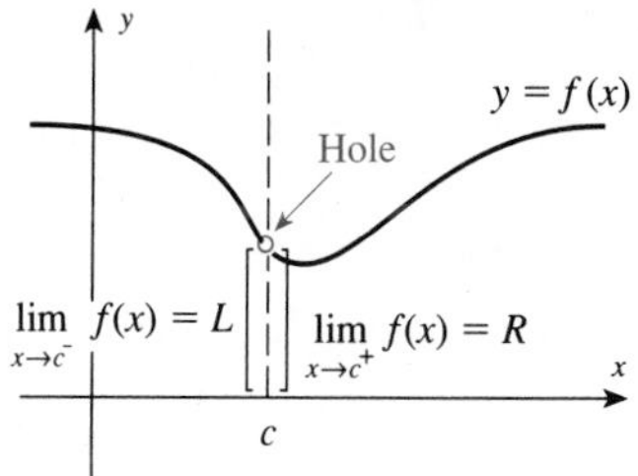
Not continuous at $x = c$: Pole $\lim_{x\to c} f(x)$ does not exist; $f(c)$ is defined 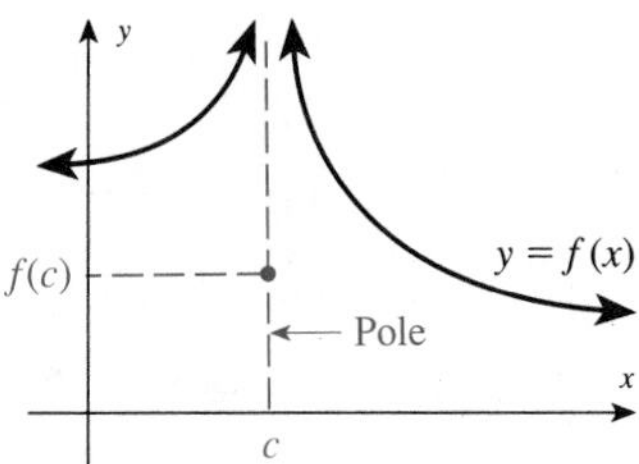	*Not continuous at $x = c$: Pole* $\lim_{x\to c} f(x)$ does not exist; $f(c)$ is not defined 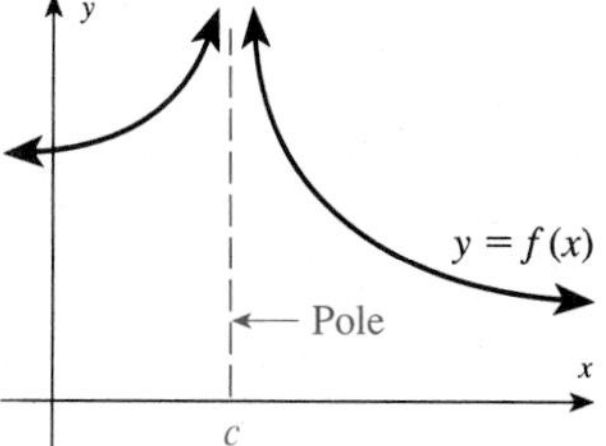
Not continuous at $x = c$: Jump $\lim_{x\to c} f(x)$ does not exist; $f(c)$ is defined 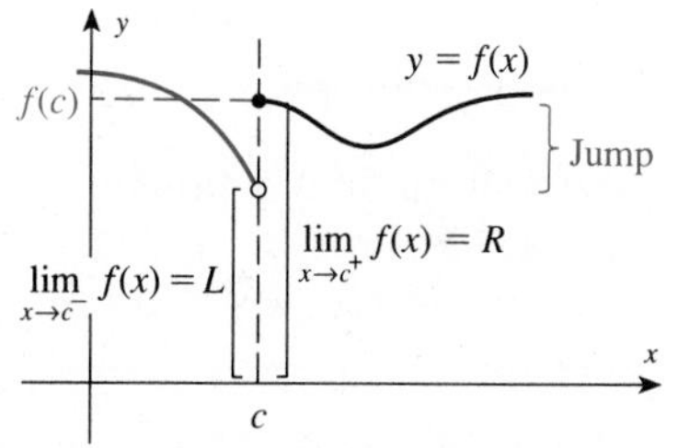	*Not continuous at $x = c$: Jump* $\lim_{x\to c} f(x)$ does not exist; $f(c)$ is not defined 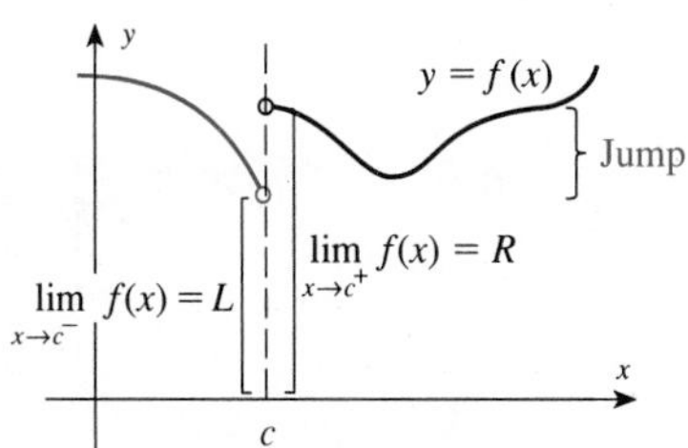

EXAMPLE 5 $h(x) = \dfrac{x^2 + 2x - 3}{x - 1}$ if $x \neq 1$ and $h(x) = 4$ if $x = 1$

Solution Compare h with g of Example 4. We see that both conditions of continuity are satisfied, which means that h is continuous at $x = 1$. ■

EXAMPLE 6 $m(x) = 3x^3 + 5x^2 - 4x + 1$

Solution $m(1) = 3(1)^3 + 5(1)^2 - 4(1) + 1 = 5$, so m is defined at $x = 1$. Also, $\lim_{x\to 1} m(x) = m(1)$, so the function is continuous at $x = 1$. ■

Continuity Theorems

Note that Example 6 is a polynomial function, and from the previous section we know that if f is any polynomial function, then

$$\lim_{x\to c} f(x) = f(c)$$

for any real number c. Thus we immediately have the following result:

Continuity of a Polynomial

> Every polynomial function is continuous at every point in its domain.

If a function is continuous for every point in a given open interval, then we say that the function is continuous over that interval. A function that is not continuous is said to be a **discontinuous function**. For example, polynomials are continuous for all real numbers. On the other hand,

$$f(x) = \frac{x + 1}{x - 1}$$

is continuous for $-1 \le x \le 0$ but not for $0 \le x \le 2$ since f is undefined at $x = 1$. The function h from Example 5,

$$h(x) = \begin{cases} \dfrac{x^2 + 2x - 3}{x - 1} & \text{if } x \neq 1 \\ 4 & \text{if } x = 1 \end{cases}$$

is continuous for all real numbers. As you can see from the above examples, we are usually concerned with finding points of discontinuity. The following result summarizes the properties of continuity:

Continuity Theorem

> Let f and g be continuous functions at $x = c$. Then the following functions are also continuous at $x = c$:
> **1.** $f + g$ **2.** $f - g$ **3.** fg **4.** f/g $(g(c) \neq 0)$

Since most of the functions we will discuss in this book are continuous over certain intervals, our task will be to look for *points of discontinuity.* These points may be values for which the definition of the function changes or values that cause division by zero. We call such points **suspicious points**. We then check for the continuity at the suspicious points and use the continuity theorem for all other points in the interval.

EXAMPLE 7 Let $f(x) = \dfrac{x^2 + 3x - 10}{x - 2}$. Check continuity.

Solution The suspicious point is $x = 2$ since this is a value for which the function $(x - 2)$ is equal to zero. The function f is not defined at this point, so it is *discontinuous at* $x = 2$. By the continuity theorem (part 4), it is continuous at all other points. To graph f, notice that

$$f(x) = \frac{x^2 + 3x - 10}{x - 2} = \frac{(x + 5)(x - 2)}{x - 2} = x + 5 \qquad x \neq 2$$

The graph is shown in Figure 10.5.

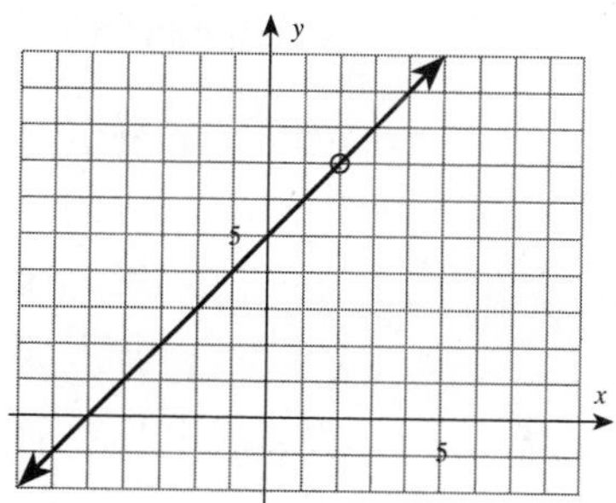

FIGURE 10.5
Graph of $f(x) = \dfrac{x^2 + 3x - 10}{x - 2}$ ■

EXAMPLE 8 Let $g(x) = \begin{cases} \dfrac{x^2 + 3x - 10}{x - 2} & \text{if } x \neq 2 \\ 6 & \text{if } x = 2 \end{cases}$. Check continuity.

Solution The suspicious point is $x = 2$ since it is a value for which the definition of the function changes. Check this value:

Step 1: $g(2) = 6$, so the function is defined at $x = 2$.

Step 2:
$$\begin{aligned} \lim_{x \to 2} g(x) &= \lim_{x \to 2} \frac{x^2 + 3x - 10}{x - 2} \\ &= \lim_{x \to 2} \frac{(x + 5)(x - 2)}{x - 2} \\ &= \lim_{x \to 2} (x + 5) \\ &= 7 \end{aligned}$$

But $g(2) = 6$ and thus $\lim_{x \to 2} g(x) \neq g(2)$, so the function is *discontinuous at* $x = 2$. The graph of g is the same as f in Figure 10.5 except the point $(2, 6)$ is included. ■

EXAMPLE 9 Let $G(x) = \begin{cases} \dfrac{x^2 + 3x - 10}{x - 2} & \text{if } x \neq 2 \\ 7 & \text{if } x = 2 \end{cases}$. Check continuity.

Solution Note that we have simply redefined the functional value of g of Example 8 at a single point. Thus

Step 1: $G(2) = 7$ and

Step 2: $\lim_{x \to 2} G(x) = 7$ (from Example 8) and $\lim_{x \to 2} G(x) = G(2)$. Thus the function is continuous for all real numbers.

The graph of G is the same as f in Figure 10.5 except the point $(2, 7)$ is included. Compare this with Example 8. In this example, the point "plugs the hole" to force continuity. ■

Note in Example 9 that you do not need to check the continuity at other points in the domain. The other points are not suspicious points and you only need to apply the continuity theorem for all of these points.

EXAMPLE 10 Let $f(x) = \begin{cases} 3 - x & \text{if } -5 \leq x < 2 \\ x - 2 & \text{if } 2 \leq x \leq 5 \end{cases}$. Check continuity.

Solution The suspicious point is $x = 2$ since it is a value for which the definition of the function changes. Check continuity by applying the definition: that is, f is defined at $x = 2$ since $f(2) = 0$. In order to find the limit as $x \to 2$ you need to check both the left- and right-hand limits (since the function is defined according to a different rule depending on from which direction x approaches 2).

$$\lim_{x \to 2^-} f(x) = \lim_{x \to 2^-} (3 - x) = 1$$
$$\lim_{x \to 2^+} f(x) = \lim_{x \to 2^+} (x - 2) = 0$$

Since the left- and right-hand limits are different, $\lim_{x \to 2} f(x)$ does not exist, so the function is discontinuous at $x = 2$.

If you draw the graph of f you can see the discontinuity at $x = 2$. Note that f is continuous over $-5 \leq x < 2$ and over $2 \leq x \leq 5$ but not over $-5 \leq x \leq 5$.

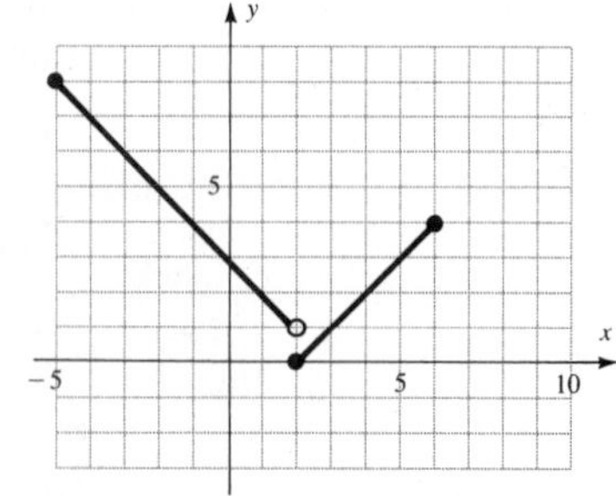

■

EXAMPLE 11 Let $g(x) = \begin{cases} 2 - x & \text{if } -5 \leq x < 2 \\ x - 2 & \text{if } 2 \leq x < 5 \end{cases}$. Check continuity.

Solution The suspicious point is $x = 2$. g is defined at $x = 2$ since $g(2) = 0$.

$$\lim_{x \to 2^-} g(x) = \lim_{x \to 2^-} (2 - x) = 0$$

$$\lim_{x \to 2^+} g(x) = \lim_{x \to 2^+} (x - 2) = 0$$

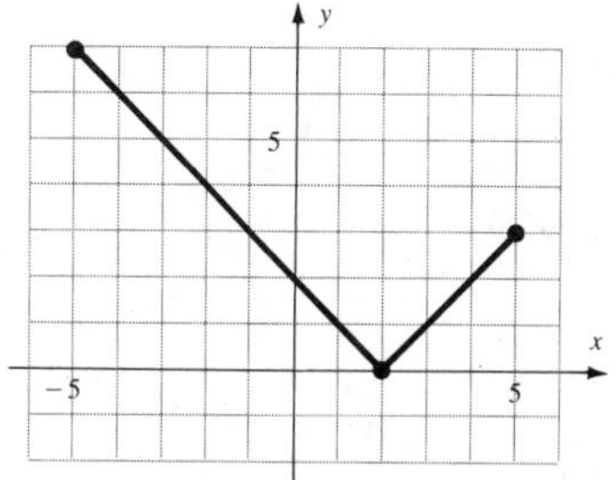

By looking at the graph of g you can see that it is continuous over $-5 \leq x \leq 5$. Note that $g(x) = |x - 2|$. Since $\lim_{x \to 2^+} g(x) = \lim_{x \to 2^-} g(x) = 0$, we can write

$$\lim_{x \to 2} g(x) = 0$$

The function is continuous at the suspicious point, so we conclude that g is continuous for all real numbers over $-5 \leq x \leq 5$. ■

You will note that the graphs of the continuous functions in the last few examples can be sketched without lifting pencil from paper. The graphs of the discontinuous functions cannot be so sketched.

10.2 Problem Set

In Problems 1–12 find all suspicious points and tell which of those are points of discontinuity.

1.

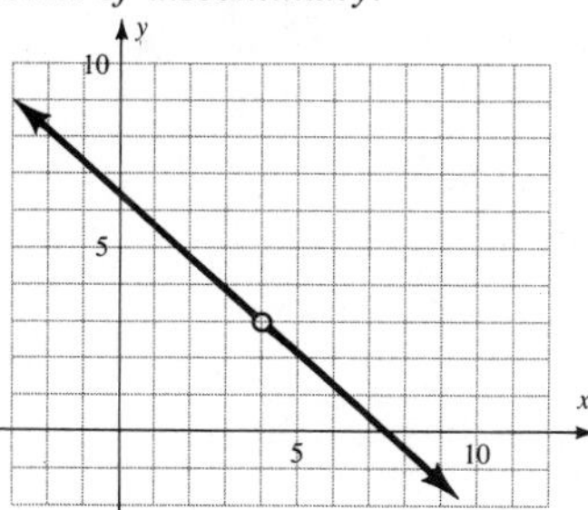

2.

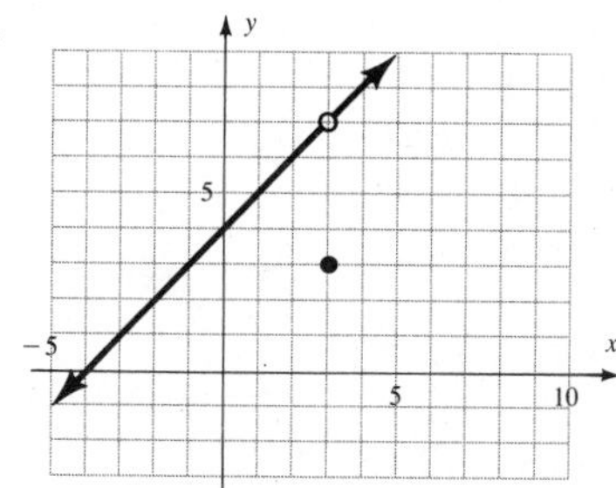

3.

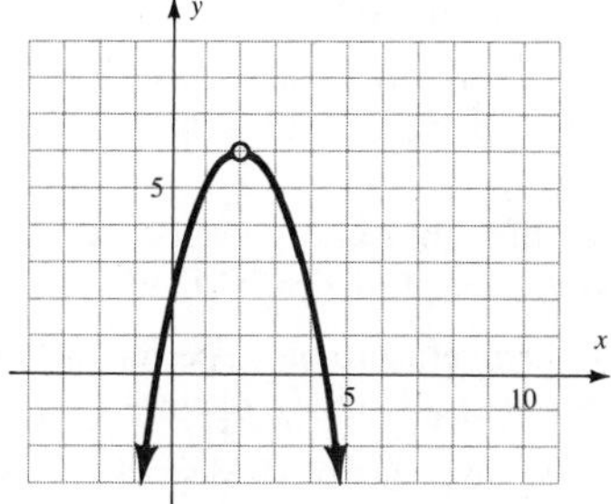

4.

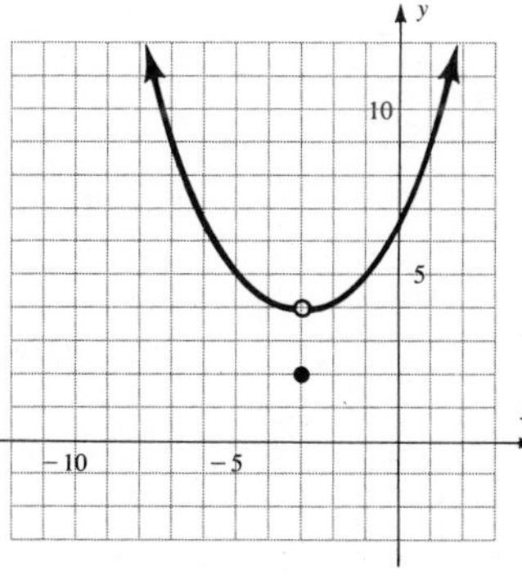

5.

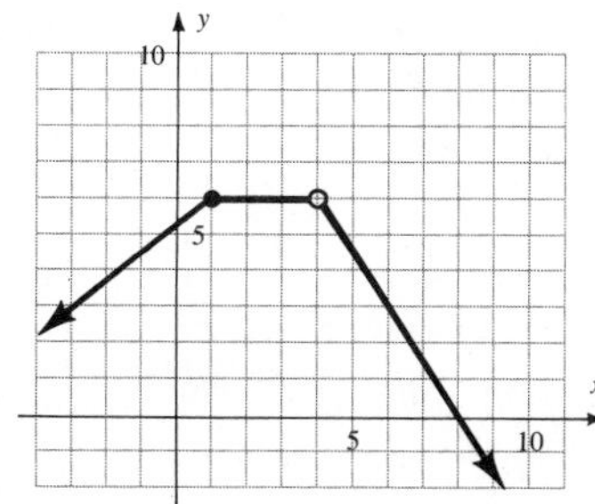

6.

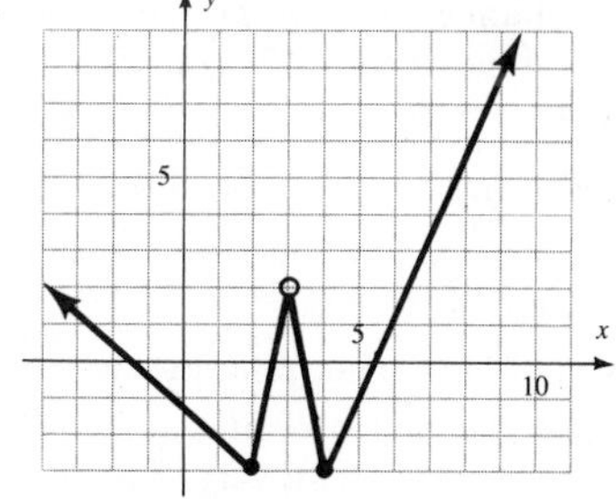

7. 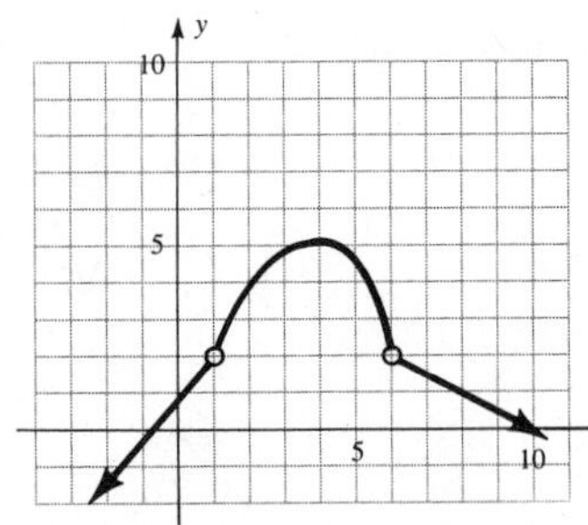

8. 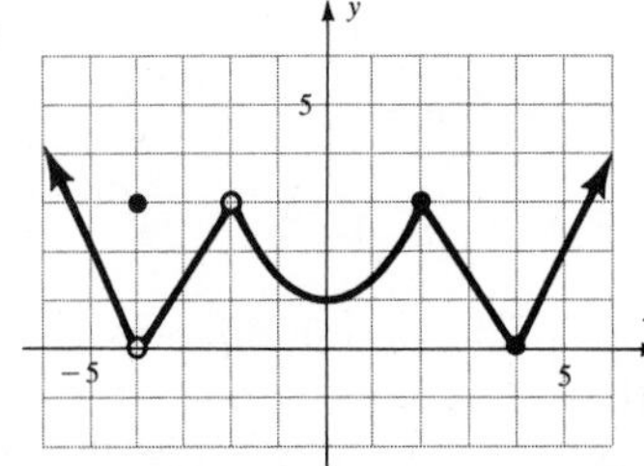

9. 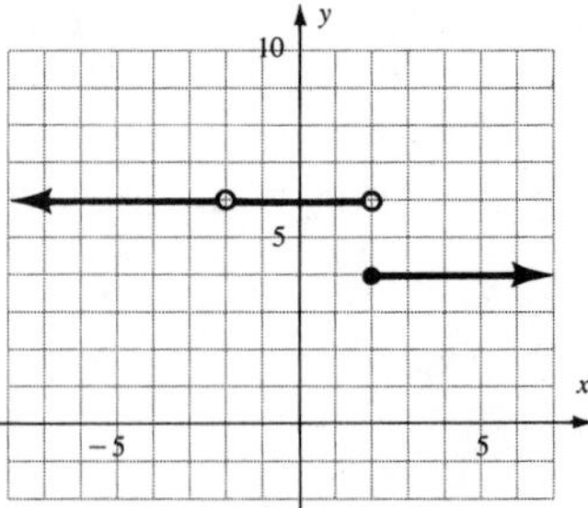

10. 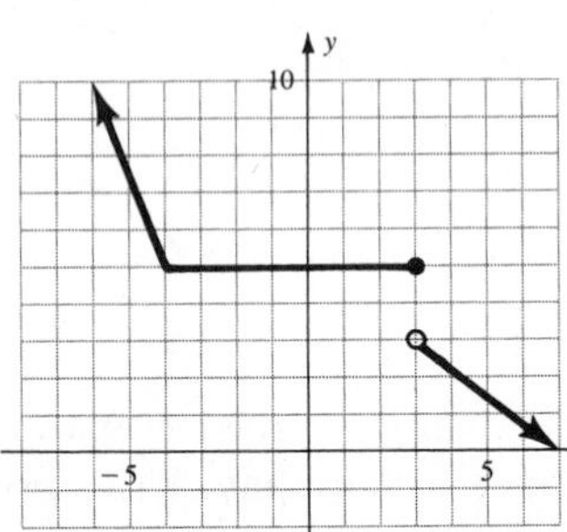

11. 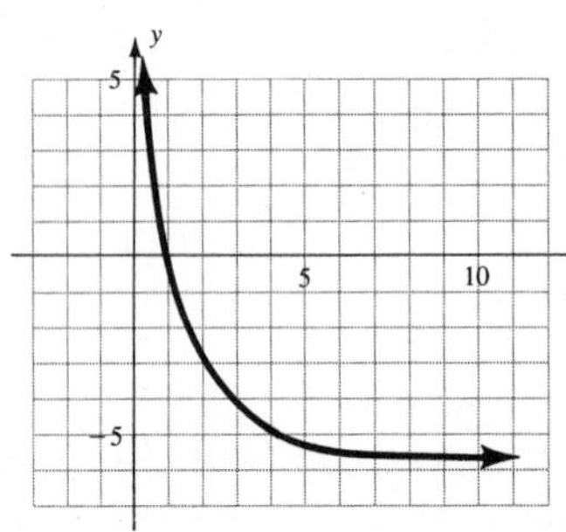

12. 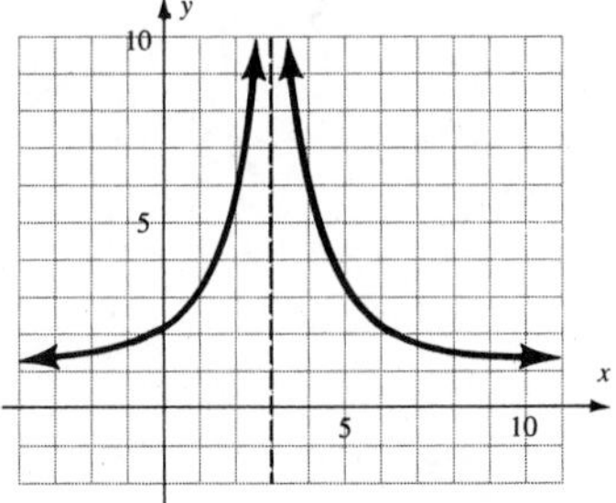

13. Is the graph in Problem 11 continuous at $x = 1$? Explain your answer.

14. Is the graph in Problem 12 continuous at $x = 2.5$? Explain your answer.

15. Is the graph in Problem 11 continuous at $x = .001$? Explain your answer.

16. Is the graph in Problem 12 continuous at $x = 2.9999$? Explain your answer.

Which of the functions described in Problems 17–22 represent continuous functions? State the domain, if possible, for each example.

17. The humidity on a specific day at a given location considered as a function of time

18. The temperature on a specific day at a given location considered as a function of time

19. The selling price of IBM stock on a specific day considered as a function of time

20. The number of unemployed people in the United States during December 1989 considered as a function of time

21. The charges for a telephone call from Los Angeles to New York considered as a function of time

22. The charges for a taxi ride across town considered as a function of mileage

In Problems 23–45 a function along with an interval is given. State whether the function is continuous at all points in this interval. Give the points of discontinuity, if any.

23. $f(x) = \dfrac{1}{x^2 + 5}, \quad -5 \le x \le 5$

24. $f(x) = \dfrac{1}{x^2 - 5}, \quad -5 \le x \le 5$

25. $f(x) = \dfrac{1}{x^2 - 9}, \quad 5 \le x \le 10$

26. $f(x) = \dfrac{1}{x^2 - 9}, \quad -5 \le x \le 5$

27. $f(x) = \dfrac{x - 1}{x^2 - 1}, \quad -5 \le x \le 5$

28. $f(x) = \dfrac{3x}{x^3 - x}, \quad -5 \le x \le 5$

29. $f(x) = \dfrac{x + 2}{x^2 - 6x - 16}, \quad -5 \le x \le 5$

30. $f(x) = \dfrac{x + 2}{x^2 - 6x - 16}, \quad -10 \le x \le 10$

31. $f(x) = \dfrac{1}{x - 3}, \quad 0 \le x \le 2$

32. $f(x) = \dfrac{1}{x - 3}, \quad 0 \le x \le 5$

33. $f(x) = \dfrac{x}{(x - 8)^2}, \quad 0 \le x \le 5$

34. $f(x) = \dfrac{x}{(x - 3)^2}, \quad 0 \le x \le 5$

35. $f(x) = \begin{cases} \dfrac{1}{x - 3} & -5 \le x \le 5, x \ne 3 \\ 4 & x = 3 \end{cases}$

36. $f(x) = \dfrac{x^2 - x - 6}{x + 2}, \quad -5 \le x \le 5$

37. $f(x) = \begin{cases} \dfrac{x^2 - x - 6}{x + 2} & -5 \le x \le 5, x \ne -2 \\ -4 & x = -2 \end{cases}$

38. $f(x) = \begin{cases} \dfrac{x^2 - x - 6}{x + 2} & -5 \le x \le 5, x \ne -2 \\ -5 & x = -2 \end{cases}$

39. $f(x) = \dfrac{x^2 - 3x - 10}{x + 2}, \quad 0 \le x \le 5$

40. $f(x) = \dfrac{x^2 - 3x - 10}{x + 2}, \quad -5 \le x \le 5$

41. $f(x) = \begin{cases} \dfrac{x^2 - 3x - 10}{x + 2} & -5 \le x \le 5, x \ne -2 \\ -3 & x = -2 \end{cases}$

42. $f(x) = \begin{cases} \dfrac{x^2 + x + 1}{x^3 - 1} & 0 \le x \le 5, x \ne 1 \\ 1 & x = 1 \end{cases}$

43. $f(x) = |x|, \quad -5 \le x \le 5$

44. $f(x) = |x - 2|, \quad -5 \le x \le 5$

45. $f(x) = |x + 3|, \quad -5 \le x \le 5$

APPLICATIONS

46. A rental agency will lease trucks at a cost C of \$.55 per mile if the annual mileage is under 10,000 miles, but the rate is lowered to \$.40 per mile for mileage over 10,000 or more miles.
   a. Is C a continuous function? What is the domain?
   b. If m is the number of miles, we can write $C(m)$ to represent the mileage charge. What is $C(9{,}000)$? What is $C(11{,}000)$?
   c. Graph the function C.

47. Postal charges are \$.29 for the first ounce and \$.23 for each additional ounce or fraction thereof. Let C be the cost function for mailing a letter weighing w ounces.
   a. Is C a continuous function? What is the domain?
   b. What is $C(1.9)$? $C(2.01)$? $C(2.89)$?
   c. Graph the function C for $[0, 10]$.

48. On July 10, 1913, the following temperatures were recorded at Death Valley, California:

Time	8	10	12	1	3	5	7
°F	90	115	123	134	130	128	105

   Let T represent the temperature in degrees Fahrenheit and assume that T is a function of time (measured on a 24-hour clock) so that $T(8) = 90, \ldots, T(12) = 123, \ldots, T(19) = 105$.
   a. Is T a continuous function? What is the domain?
   b. On the day in question, what is the minimum number of times during the day that the temperature was 100°F? 110°F? 80°F?

49. Give an example of a function defined on $1 \le x \le 3$ that is discontinuous at $x = 2$ and continuous elsewhere on the interval.

50. Give an example of a function defined on $3 \le x \le 7$ that is discontinuous at $x = 4$ and $x = 5$ and continuous elsewhere on the interval.

10.3 Rates of Change

Average Rate of Change

Elementary mathematics focuses on formulas and relationships among variables but does not include the analysis of quantities that are in a state of constant change. For example, the following problem might be found in elementary mathematics. If you drive at 55 miles per hour (mph) for a total of 3 hours, how far did you travel? This example uses a formula, $d = rt$ (distance = rate · time), and the answer is $d = 55(3) = 165$. However, this model does not adequately describe the situation in the real world. You could drive for 3 hours at an *average rate* of 55 mph, but you probably could not drive for 3 hours at a *constant rate* of 55 mph—in reality, the rate would be in a state of constant change. Other applications in which rates may not remain constant quickly come to mind:

Profits changing with sales
Population changing with the growth rate

TABLE 10.2 Distance and Time for a Commuter Car

Time	Distance from home
6:09 A.M.	0 miles
6:25	16
6:30	21
6:34	25
6:36	26.7
6:37	27.7
6:39	28.5
7:01	34
7:03	35
7:09	38
7:15	43
7:28	50

Property taxes changing with the tax rate
Tumor sizes changing with chemotherapy
The speed of falling objects changing over time

The rate at which one quantity changes relative to another is mathematically described by using a concept called a **derivative**. We will see that the rate of change of one quantity relative to another is mathematically determined by finding the slope of a line drawn tangent to a curve.

We begin with a simple example. Consider the speed of a moving object, say, a car. By *speed* we mean the rate at which the distance traveled varies with time. It has magnitude, but no direction. We can measure speed in two ways: *average speed* and *instantaneous speed.* We begin with the average speed of a commuter driving from home to the office.

To find the average speed we use the formula $d = rt$ or $r = \frac{d}{t}$; that is, we divide the distance traveled by the elapsed time:

$$\text{Average speed} = \frac{\text{distance traveled}}{\text{elapsed time}}$$

EXAMPLE 1 A commuter left home one morning and set the odometer to zero. The times and distances were noted as shown in Table 10.2. Find the average speed for the requested time intervals.

a. 6:09 A.M. to 6:39 A.M. (0–30 min)

b. 6:39 A.M. to 7:09 A.M. (30–60 min)

c. 7:01 A.M. to 7:28 A.M. (52–79 min)

Solution **a.** 0–30 min: $\text{Average speed} = \dfrac{28.5 - 0}{\frac{1}{2} - 0} = 57 \text{ mph}$

b. 30–60 min: $\text{Average speed} = \dfrac{38 - 28.5}{1 - \frac{1}{2}} = 19 \text{ mph}$

c. 52–79 min: $\text{Average speed} = \dfrac{50 - 34}{\frac{79}{60} - \frac{52}{60}} \approx 35.6 \text{ mph}$ ■

We can generalize the work done in Example 1 by writing the following formula:

$$\text{Average speed} = \frac{d_2 - d_1}{t_2 - t_1} \text{ mph}$$

where the car travels d_1 miles in t_1 hours and d_2 miles in t_2 hours. The relationship is shown graphically in Figure 10.6, where the average speed is the slope of the line joining the points (d_1, t_1) and (d_2, t_2). This line is sometimes called the **secant line.** We see that the average rate of change is $\dfrac{f(t_2) - f(t_1)}{t_2 - t_1}$.

It is worthwhile to state this formula in terms of the usual variable x. Let $x = t_1$ and let $h = t_2 - t_1$, which is the length of the time interval over which we are finding an average. From these substitutions it follows that

$$h = t_2 - x$$

$$h + x = t_2$$

This leads to a general statement of the average rate of change.

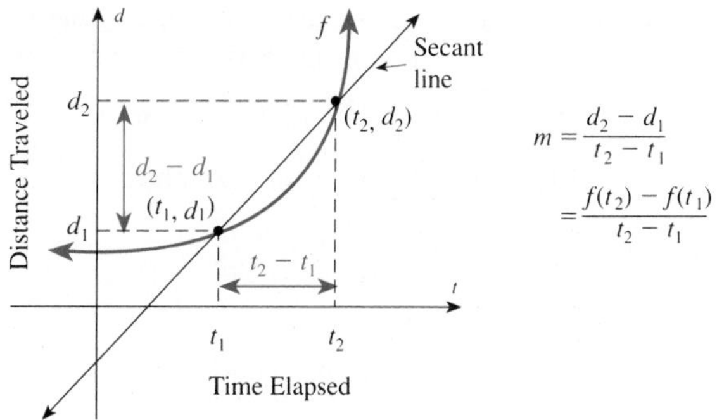

FIGURE 10.6
Geometrical interpretation of average speed

Average Rate of Change

The **average rate of change** of a function f with respect to x over an interval $[x, x + h]$ is given by the formula

$$\frac{f(x + h) - f(x)}{h}$$

EXAMPLE 2 Find the average rate of change of f with respect to x between $x = 3$ and $x = 5$ if $f(x) = x^2 - 4x + 7$.

Solution Given $x = 3$ and $h = 5 - 3 = 2$:

$$\begin{aligned} f(x + h) &= f(3 + 2) \\ &= f(5) \\ &= 5^2 - 4(5) + 7 \\ &= 12 \\ f(x) &= f(3) \\ &= 3^2 - 4(3) + 7 \\ &= 4 \end{aligned}$$

$$\begin{aligned} \text{Average rate of change} &= \frac{f(x + h) - f(x)}{h} \\ &= \frac{12 - 4}{2} \\ &= 4 \end{aligned}$$

The slope of the secant line is 4, as shown in Figure 10.7.

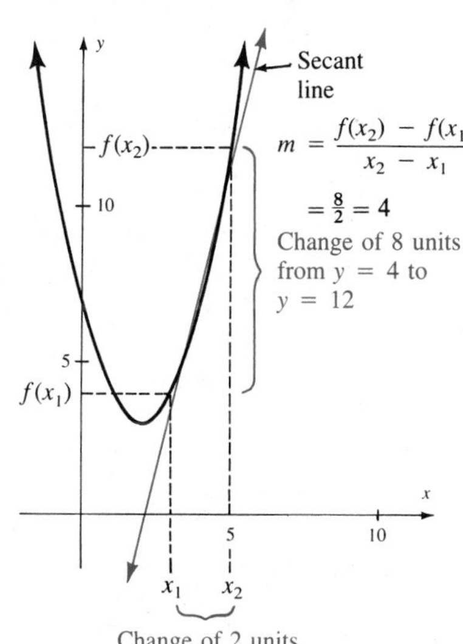

FIGURE 10.7
Average rate of change of f, where $f(x) = x^2 - 4x + 7$ between $x = 3$ and $x = 5$ ■

EXAMPLE 3 Table 10.3 shows the number of divorces for selected years from 1970 to 1987. Find the average divorce rate for the following periods of time:

a. 1970–1980 **b.** 1970–1987 **c.** 1980–1987 **d.** 1985–1987

TABLE 10.3 Number of U.S. Divorces

Year	Number (in millions)
1970	.708
1975	1.036
1980	1.182
1985	1.187
1987	1.157

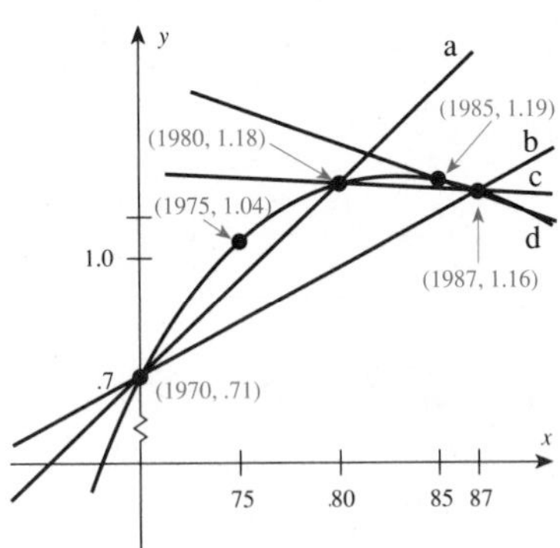

Solution Average rate of change for f defined by Table 10.3:

$$\frac{f(x+h)-f(x)}{h}$$

a. $x = 1970$ and $h = 10$: $\dfrac{f(1980)-f(1970)}{h} = \dfrac{1.182-.708}{10}$

$= .0474$

The average divorce rate (rate of change of divorces) between the years 1970 and 1980.

b. $x = 1970$ and $h = 17$: $\dfrac{f(1987)-f(1970)}{17} = \dfrac{1.157-.708}{17}$

$\approx .0264$

c. $x = 1980$ and $h = 7$: $\dfrac{f(1987)-f(1980)}{7} = \dfrac{1.157-1.182}{7}$

$\approx -.0036$

A negative rate of change indicates a decrease in the rate at which divorces are occurring between 1980 and 1987.

d. $x = 1985$ and $h = 2$: $\dfrac{f(1987)-f(1985)}{2} = \dfrac{1.157-1.187}{2}$

$= -.015$ ■

Instantaneous Rate of Change

Note in Examples 2 and 3 that the slopes of the secant lines we found when calculating the average rate of change can vary greatly and really do not tell us much about the rate *at a particular time.* Imagine that a curve called f is defined by a roller coaster track. Consider a fixed location on this track, call it $(x, f(x))$. Now imagine the front car of the roller coaster at some point on the track a horizontal distance of h from x. The slope of the secant line between these points is the average rate of change of the function f between the two points.

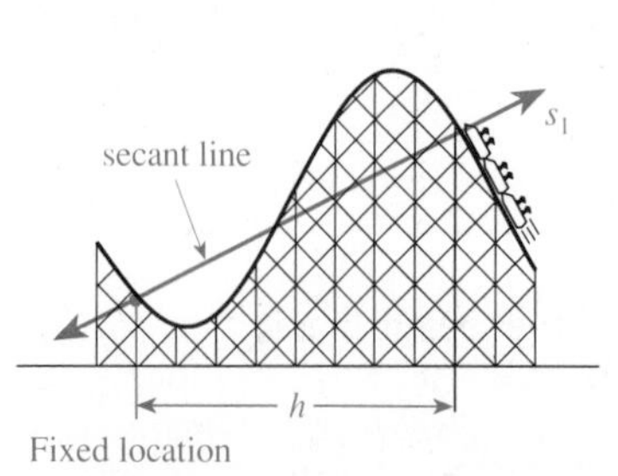

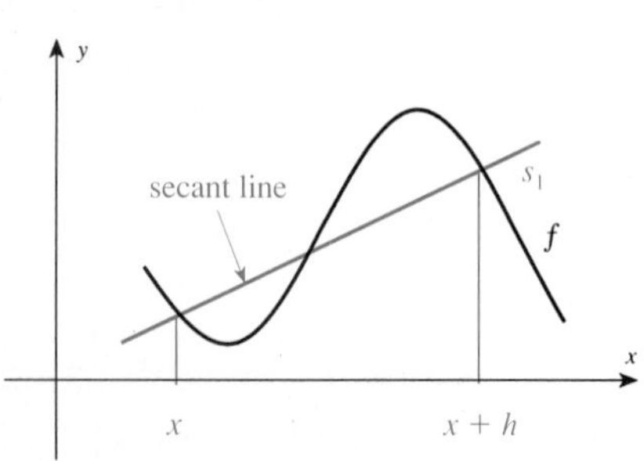

Also imagine the roller coaster car moving along the track toward the fixed location. That is, let $h \to 0$. As the car moves along the track we obtain a sequence of secant lines. The roller coaster is shown at the left and the curve with secant lines is shown at the right.

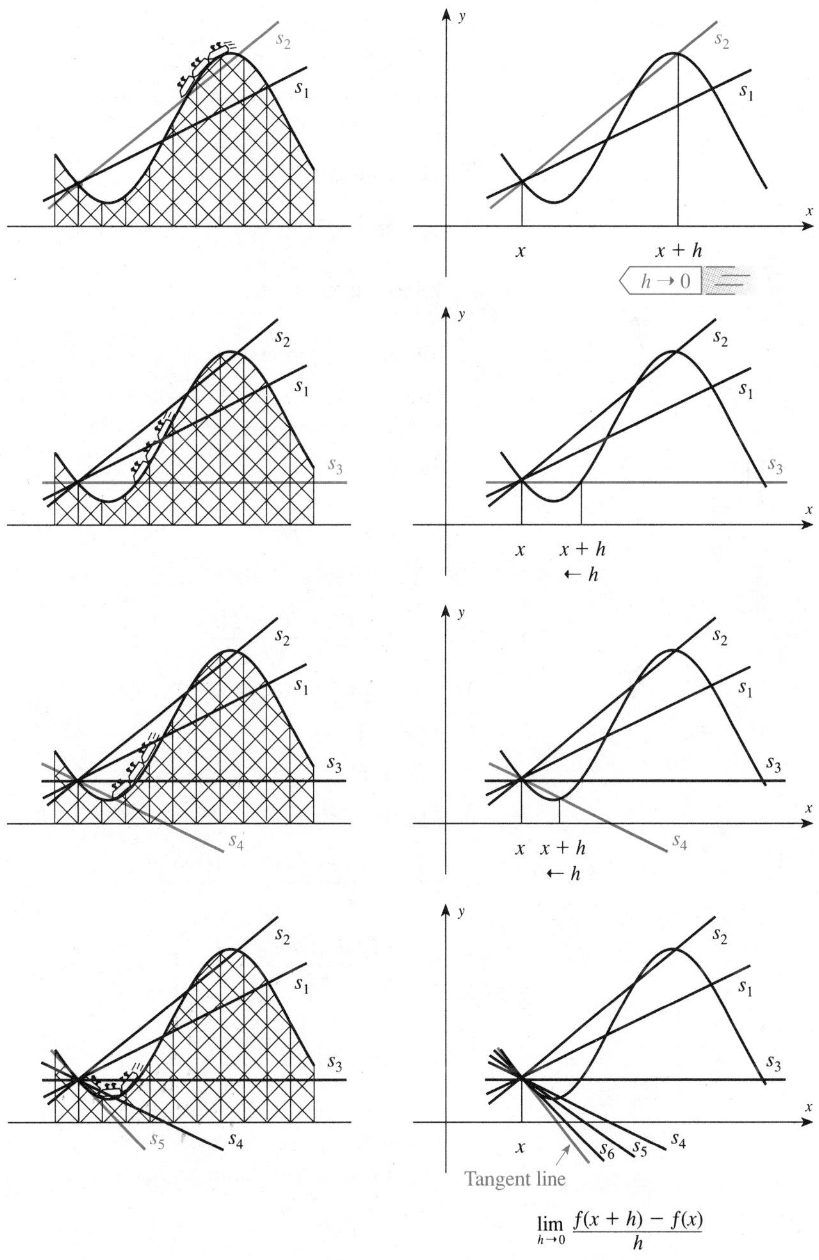

As the car gets close to the fixed location, the secant lines approach a limiting line (shown in color). This limiting line for which $h \to 0$ is called the **tangent line**.

Instantaneous Rate of Change

Given a function f and the graph of $y = f(x)$, the *tangent line* at the point $(x, f(x))$ is the line that passes through this point with slope

$$\lim_{h \to 0} \frac{f(x+h) - f(x)}{h}$$

if this limit exists. The slope of the tangent line is also referred to as the *instantaneous rate of change* of the function f with respect to x.

To summarize:

Velocity interpretation	**Average rate of change**	**Instantaneous rate of change**
Algebraic interpretation	$\dfrac{f(x+h) - f(x)}{h}$	$\lim_{h \to 0} \dfrac{f(x+h) - f(x)}{h}$
Geometric interpretation	Slope of the secant line through the points $[x, f(x)]$ and $[x + h, f(x + h)]$	Slope of the tangent line at the point $[x, f(x)]$

EXAMPLE 4 Find the instantaneous rate of change of the function $f(x) = 5x^2$ with respect to x.

Solution Find $\lim_{h \to 0} \dfrac{f(x+h) - f(x)}{h}$.

We evaluate this expression in small steps:

1. $f(x)$ is given.
2. $f(x+h) = 5(x+h)^2 = 5x^2 + 10xh + 5h^2$
3. $f(x+h) - f(x) = (5x^2 + 10xh + 5h^2) - 5x^2 = 10xh + 5h^2$
4. $\dfrac{f(x+h) - f(x)}{h} = \dfrac{10xh + 5h^2}{h} = 10x + 5h \quad (h \neq 0)$
5. $\lim_{h \to 0} \dfrac{f(x+h) - f(x)}{h} = \lim_{h \to 0}(10x + 5h) = 10x$ ■

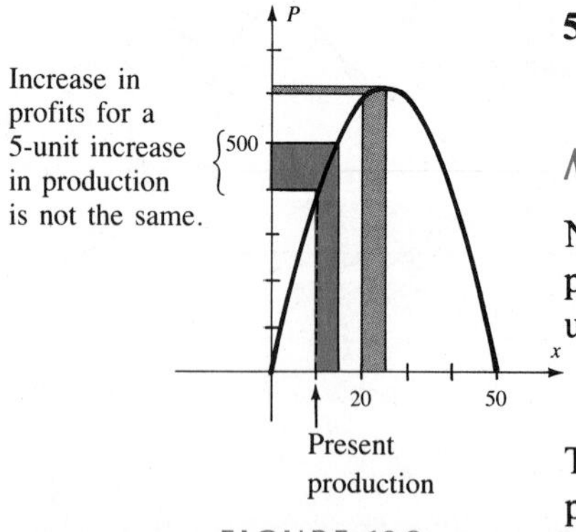

Marginal Profit

Now we will use the concept of instantaneous rate of change in a business application. Suppose the profit, P, for a manufacturer is a function of the number of units produced and behaves according to the model

$$P(x) = 50x - x^2$$

Increase in profits for a 5-unit increase in production is not the same.

P 500 x 20 50 Present production

FIGURE 10.8
Profit function

The graph of this profit function is shown in Figure 10.8. Now suppose that present production is 10 units. What is the profit?

$$P(10) = 50(10) - 10^2 = 400$$

What is the per unit increase in profit if production is increased from 10 to 20 units?

$$P(20) = 50(20) - 20^2 = 600$$

Increased profit: $P(20) - P(10) = 600 - 400 = 200$

Per unit increase in profit: $\dfrac{P(20) - P(10)}{20 - 10} = \dfrac{200}{10} = 20$

EXAMPLE 5 What is the per unit increase in profit if production is increased from 10 units to:

a. 15 units? **b.** 11 units?

Solution **a.** $\dfrac{P(15) - P(10)}{15 - 10} = \dfrac{525 - 400}{5} = \dfrac{125}{5} = 25$ This is the average rate of change of profit as x increases from 10 to 15.

b. $\dfrac{P(11) - P(10)}{11 - 10} = \dfrac{429 - 400}{1} = \dfrac{29}{1} = 29$ ■

From Example 5 we see that the per unit increase in profit is different if we increase from 10 to 15 units than if we increase from 10 to 11 units. We would therefore talk about the average per unit increase in profit from 10 to 15 or from 10 to 11 units.

Let us consider this situation in general. An increase in production from x units to $x + h$ units would produce an average per unit increase in profit for a function P as follows:

$$\frac{P(x + h) - P(x)}{x + h - x} = \frac{P(x + h) - P(x)}{h}$$

For the instantaneous rate of change we can let $h \to 0$. *This represents the per unit increase in profit at a production level of x units.* In business this per unit increase is called the **marginal profit** and is defined by

$$\lim_{h \to 0} \frac{P(x + h) - P(x)}{h}$$

if this limit exists.

EXAMPLE 6 Find the marginal profit for the above model; that is, find the marginal profit for

$$P(x) = 50x - x^2$$

Solution

1. $\mathbf{P(x)} = 50x - x^2$ is given.
2. $\mathbf{P(x + h)} = 50(x + h) - (x + h)^2 = 50x + 50h - x^2 - 2xh - h^2$
3. $\mathbf{P(x + h) - P(x)} = (50x + 50h - x^2 - 2xh - h^2) - (50x - x^2)$
   $= 50h - 2xh - h^2$
4. $\dfrac{\mathbf{P(x + h) - P(x)}}{\mathbf{h}} = \dfrac{(50 - 2x - h)h}{h} = 50 - 2x - h \quad (h \neq 0)$
5. $\displaystyle\lim_{h \to 0} \frac{\mathbf{P(x + h) - P(x)}}{\mathbf{h}} = \lim_{h \to 0}(50 - 2x - h) = 50 - 2x \quad (h \neq 0)$ ■

We check the results of Example 6 with our previous results:

1. Production level of 10 units where $x = 10$

$$50 - 2(10) - h = 30 - h$$

2. Increases: From 10 to 20 ($h = 10$): $30 - h = 30 - 10 = 20$
   From 10 to 15 ($h = 5$): $30 - h = 30 - 5 = 25$
   From 10 to 11 ($h = 1$): $30 - h = 30 - 1 = 29$
3. Marginal profit ($h \neq 0$): $\lim_{h \to 0}(30 - h) = 30$

Velocity

The final application of the concept of an instantaneous rate of change is **velocity**.

EXAMPLE 7 At an amusement park there is a "free fall" ride called "The Edge." The ride involves falling 100 feet in 2.5 seconds. As you are falling, you pass the 16-foot mark at one second and the 64-foot mark at two seconds. What is the velocity at the *instant* the ride passes the 100-foot mark (measured from the top)?

Solution Let $s(t)$ be the distance in feet from the top t seconds after release. Also assume that $s(t) = 16t^2$.*

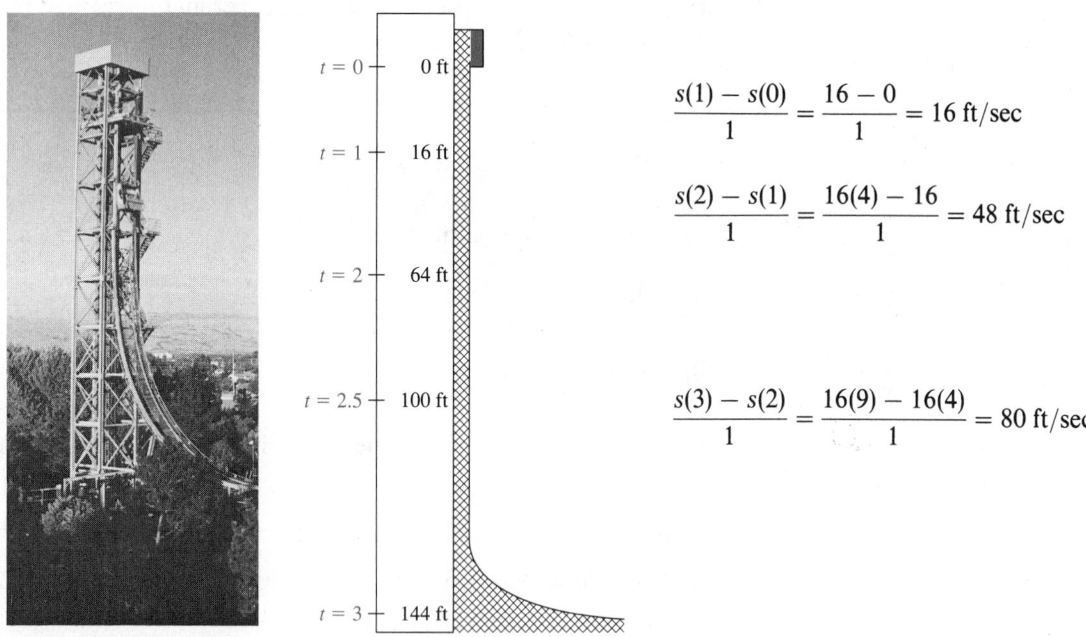

Courtesy of Great America Theme Park, Santa Clara, CA

* This is the formula for free fall in a vacuum; it is sufficiently accurate for our purposes in this problem.

In general, the average velocity from time $t = t_1$ to $t = t_1 + h$ is given by the following formula (where $h \neq 0$).

$$\begin{aligned}\text{Average velocity} &= \frac{\text{Change in position}}{\text{Change in time}}\\ &= \frac{s(t_1 + h) - s(t_1)}{h}\\ &= \frac{16(t_1 + h)^2 - 16t_1^2}{h}\\ &= \frac{16t_1^2 + 32t_1h + 16h^2 - 16t_1^2}{h}\\ &= \frac{32t_1h + 16h^2}{h}\\ &= 32t_1 + 16h\end{aligned}$$

Now, to find the velocity at a particular instant in time, we simply need to consider the limit as $h \to 0$:*

$$\begin{aligned}\text{Instantaneous velocity} &= \lim_{h\to 0} \frac{s(t_1 + h) - s(t_1)}{h}\\ &= \lim_{h\to 0}(32t_1 + 16h)\\ &= 32t_1\end{aligned}$$

We can now use these formulas to find the average and instantaneous velocities:

Time interval	**Average velocity** $\mathbf{32t + 16h}$	**Instantaneous velocity** $\mathbf{32t}$
At $t = 0$		$32(0) = 0$
From $t = 0$ to $t = 1$ ($h = 1$)	$32(0) + 16(1) = 16$	
At $t = 1$		$32(1) = 32$
From $t = 1$ to $t = 2$ ($h = 1$)	$32(1) + 16(1) = 48$	
From $t = 1$ to $t = 3$ ($h = 2$)	$32(1) + 16(2) = 64$	
At $t = 2$		$32(2) = 64$
⋮	⋮	⋮

* Note that as $h \to 0$ it must be true that $h \neq 0$.

We are given that the ride passes 100 feet at 2.5 seconds, so the instantaneous velocity at that instant is

$$32\left(2\frac{1}{2}\right) = 80 \text{ ft/sec}$$

(By the way, 80 ft/sec is approximately 54 mph.) ■

10.3 Problem Set

APPLICATIONS

The graph in Figure 10.9 shows the height h of a projectile after t seconds. Find the average rate of change of height (in ft) with respect to the changes in time t (in sec) in Problems 1–4.

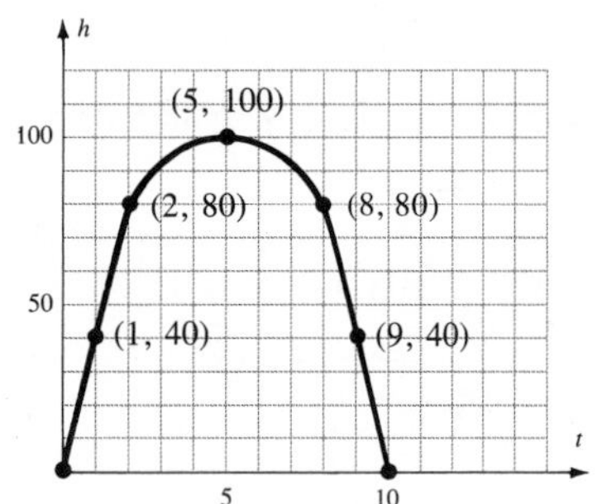

FIGURE 10.9
Height of a projectile in feet t seconds after fired

1. 1 to 7
2. 1 to 5
3. 1 to 2
4. 2 to 9

The graph in Figure 10.10 shows company output as a function of the number of workers. Find the average rate of change of output for the changes in the number of workers in Problems 5–8.

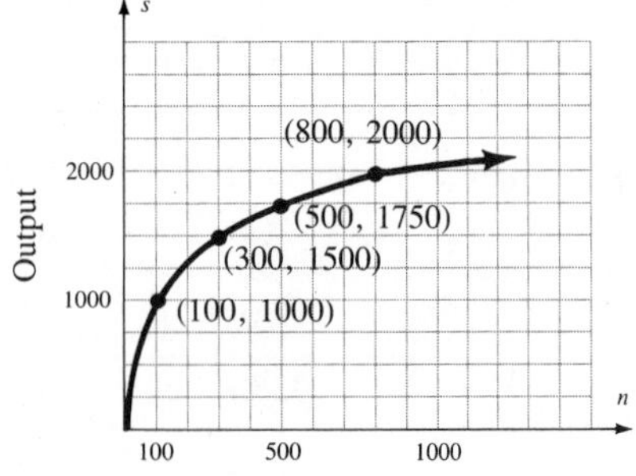

FIGURE 10.10
Output in production relative to the number of employees at Kampbell Construction

5. 100 to 800
6. 300 to 800
7. 500 to 800
8. 300 to 500

The SAT scores of entering first-year college students are shown in Figure 10.11. Find the average yearly rate of change of the scores for the periods of time in Problems 9–12.

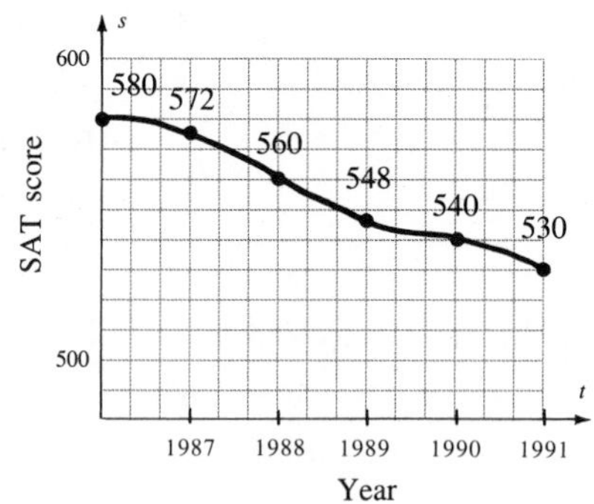

FIGURE 10.11
SAT scores at Riveria College

9. 1986 to 1991
10. 1987 to 1991
11. 1989 to 1991
12. 1990 to 1991

Table 10.4 shows the Gross National Product (GNP) in trillions of dollars for the years 1960–1987. Find the average yearly rate of change of the GNP for the years given in Problems 13–16.

TABLE 10.4 Gross National Product

Year	Dollars (in trillions)
1960	.5153
1970	1.0155
1975	1.5984
1980	2.7320
1986	4.2403
1987	4.5267

13. 1960 to 1987

14. 1970 to 1987

15. 1980 to 1987

16. 1986 to 1987

Table 10.2 gives some distances and commute times for a typical daily commute. Find the average speed for the requested time intervals.

17. 6:09 A.M. to 6:36 A.M.

18. 6:36 A.M. to 7:03 A.M.

19. 7:03 A.M. to 7:28 A.M.

20. 6:09 A.M. to 7:28 A.M.

Trace the curves in Problems 21–26 onto your own paper and draw the secant line passing through P and Q. Next, imagine $h \to 0$ and draw the tangent line at P assuming that Q moves along the curve to the point P. Finally, estimate the slope of the curve at P using the slope of the tangent line you have drawn.

21.

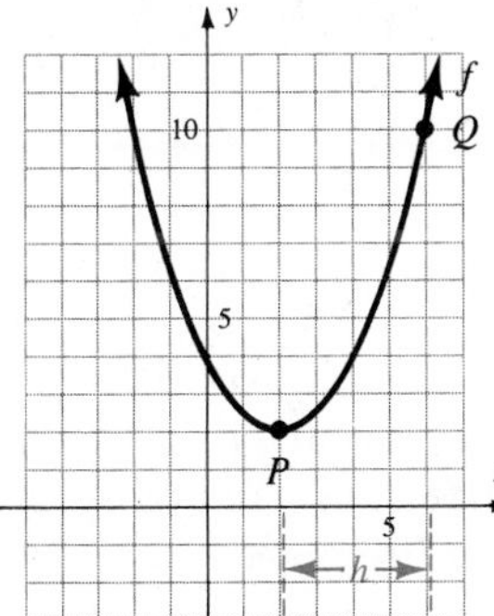

22.

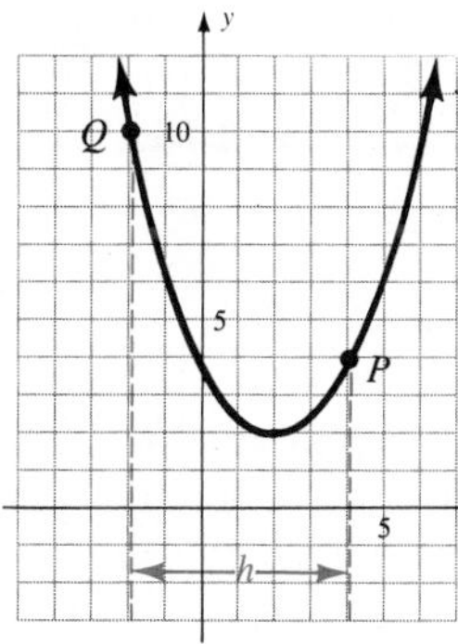

23.

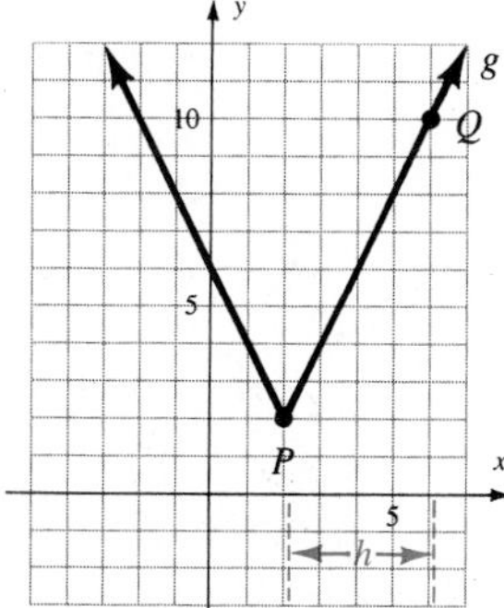

24.

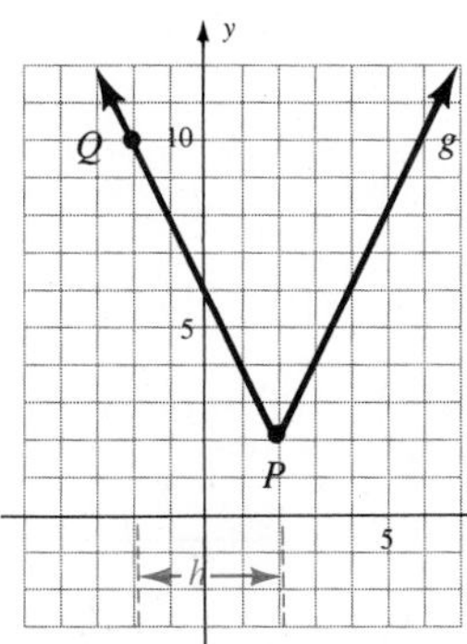

25.

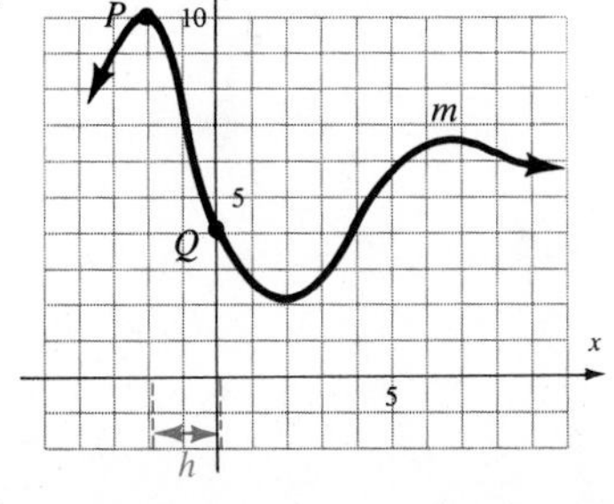

26.

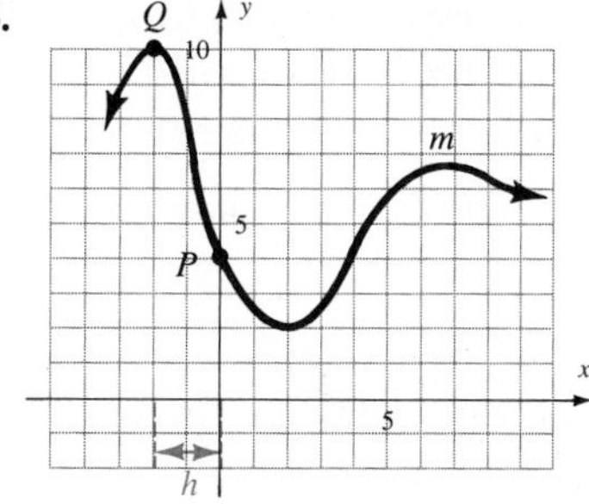

Find the average rate of change for the functions in Problems 27–36.

27. $f(x) = 4 - 3x$ for $x = -3$ to $x = 2$

28. $f(x) = 5x - 1$ for $x = -5$ to $x = -1$

29. $f(x) = 5$ for $x = -3$ to $x = 3$

30. $f(x) = -2$ for $x = 0$ to $x = 5$

31. $f(x) = 3x^2$ for $x = 1$ to $x = 3$

32. $f(x) = x^2 - 3x$ for $x = 0$ to $x = 3$

33. $y = -2x^2 + x + 4$ for $x = 1$ to $x = 4$

34. $y = \sqrt{x}$ for $x = 4$ to $x = 9$

35. $y = \dfrac{-2}{x+1}$ for $x = 1$ to $x = 5$

36. $y = 2x^3 + 7x$ for $x = 1$ to $x = 3$

Find the instantaneous rate of change for the functions in Problems 37–46. These are the same functions as those given in Problems 27–36.

37. $f(x) = 4 - 3x$ for $x = -3$

38. $f(x) = 5x - 1$ for $x = -5$

39. $f(x) = 5$ for $x = -3$

40. $f(x) = -2$ for $x = 0$

41. $f(x) = 3x^2$ for $x = 1$

42. $f(x) = x^2 - 3x$ for $x = 0$

43. $y = -2x^2 + x + 4$ for $x = 1$

44. $y = \sqrt{x}$ for $x = 4$

45. $y = \dfrac{-2}{x+1}$ for $x = 0$

46. $y = 2x^3 + 7x$ for $x = 1$

APPLICATIONS

Suppose the profit, P, for a manufacturer is a function of the number of units produced and behaves according to the model

$$P(x) = 50x - x^2$$

Also suppose that the present production is 20 units. Use this model for Problems 47–50.

47. What is the per unit increase in profit if production is increased from 20 to 30 units?

48. Repeat Problem 47 for an increase from 20 to 25 units.

49. Repeat Problem 47 for an increase from 20 to 21 units.

50. What is the marginal profit for P at $x = 20$?

The cost, C, in dollars for producing x items is given by

$$C(x) = 30x^2 - 100x$$

Use this model for Problems 51–57.

51. Find the average rate of change of cost as x increases from 100 to 200 items.

52. Repeat Problem 51 for an increase from 100 to 110 items.

53. Repeat Problem 51 for an increase from 100 to 101 items.

54. Repeat Problem 51 for an increase from 100 to $(100 + h)$ items.

55. Repeat Problem 51 for an increase from x to $(x + h)$ items.

56. Find $\lim\limits_{h \to 0} \dfrac{C(100 + h) - C(100)}{h}$.

57. Find $\lim\limits_{h \to 0} \dfrac{C(x + h) - C(x)}{h}$.

58. Attach a possible meaning for the result of Problem 56. In business the result of Problem 56 is called the *marginal cost* for the production level of 100. We will discuss this concept later in the text.

59. Attach a possible meaning for the result of Problem 57. In business the result of Problem 57 is called the *marginal cost* for the production level of x. We will discuss this concept later in the text.

10.4 Definition of Derivative

Derivative

The concept of the derivative is a very powerful mathematical idea, and the variety of applications is almost unlimited. In the last section we investigated the instantaneous rate of change of a function f per unit change in x (marginal profit, instantaneous rate of change, velocity, and the slope of a line tangent to a curve at a particular point). All of these ideas can be summarized by a single concept, called the **derivative**:

Definition of Derivative

For a given function f, we define the *derivative of f at x*, denoted by $f'(x)$, to be

$$f'(x) = \lim_{h \to 0} \frac{f(x + h) - f(x)}{h}$$

provided this limit exists. If the limit exists we say f is a *differentiable function at x.*

If the limit does not exist, then we say that f is *not differentiable* at x. We have now developed all of the techniques necessary to apply the definition of derivative to a variety of functions. Also note that as $h \to 0$, $h \neq 0$ so we will assume $h \neq 0$ without stating it when using the five-step process for finding the derivative.

EXAMPLE 1 Use the definition to find the derivative of $y = x^2$.

Solution We carry out the five-step process of the last section:

1. $f(x) = x^2$ is given
2. $f(x + h) = (x + h)^2$ Evaluate f at $x + h$.
   $= x^2 + 2xh + h^2$

3. $f(x+h) - f(x) = (x^2 + 2xh + h^2) - x^2$
$= 2xh + h^2$

4. $\dfrac{f(x+h) - f(x)}{h} = \dfrac{2xh + h^2}{h}$
$= \dfrac{h(2x+h)}{h}$
$= 2x + h$

5. $\lim\limits_{h\to 0} \dfrac{f(x+h) - f(x)}{h} = \lim\limits_{h\to 0}(2x + h)$
$= 2x$

The derivative of $y = x^2$ is $2x$. Sometimes we write $y' = 2x$ or $f'(x) = 2x$. ■

Tangent Line

EXAMPLE 2 Find the equation of the line tangent to the curve $y = x^2$ at $x = 3$.

Solution From Example 1, $y' = 2x$, so the slope of the tangent line at any point is $2x$. When $x = 3$, then $y = 3^2$, so we are looking for the tangent line passing through $(3, 9)$ with slope equal to the derivative. Remember, the slope of a curve at a point is the value of the derivative at that point. Thus, at $x = 3$, $f'(3) = 2(3) = 6$. Now we can use the point–slope form:

WARNING ***y' or f' means derivative and does not represent an exponent.***

$$y - 9 = 6(x - 3)$$
$$y = 6x - 9 \quad \text{or, in standard form,} \quad 6x - y - 9 = 0$$ ■

EXAMPLE 3 Find the instantaneous rate of change of profit $P(x) = 80x - 25x^2$, if the present level of production is 50 units.

Solution Evaluate the derivative at $x = 50$ (because that is the present production level). First, find $P'(x)$:

1. $P(x) = 80x - 25x^2$

2. $P(x+h) = 80(x+h) - 25(x+h)^2$
$= 80x + 80h - 25x^2 - 50xh - 25h^2$

3. $P(x+h) - P(x) = (80x + 80h - 25x^2 - 50xh - 25h^2) - (80x - 25x^2)$
$= 80h - 50xh - 25h^2$

4. $\dfrac{P(x+h) - P(x)}{h} = \dfrac{80h - 50xh - 25h^2}{h}$
$= 80 - 50x - 25h$

5. $\lim\limits_{h\to 0} \dfrac{P(x+h) - P(x)}{h} = \lim\limits_{h\to 0}(80 - 50x - 25h)$
$= 80 - 50x$

Thus $P'(x) = 80 - 50x$ and $P'(50) = -2{,}420$. ■

We use tangent lines to sketch functions in the next chapter. It is worth noting here, though, that if the derivative of a function is positive, then the curve is rising at that point (since the slope of the tangent line is also positive); and if the derivative is negative at a point, then the curve is falling at that point. If the derivative is 0, then the tangent is horizontal. In Example 2, we found the equation of the tangent line of the function $y = x^2$ at $(2, 4)$. This tangent line and others are shown in Figure 10.12.

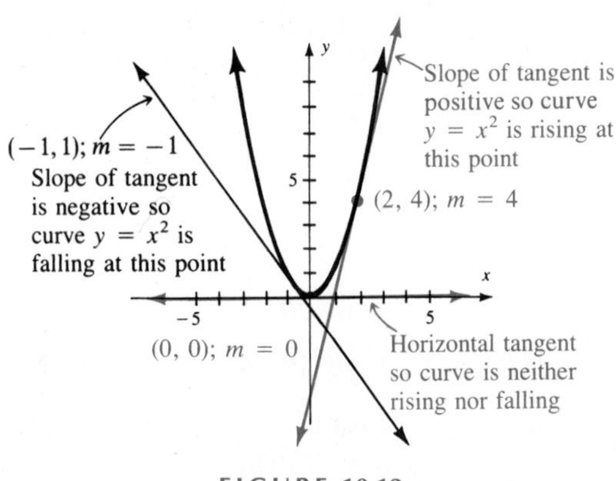

FIGURE 10.12
Graph of $y = x^2$ with tangent lines

Functions That Are Not Differentiable

A derivative may not exist at a particular point. If the limit does not exist at $x = a$, then we say that the function is not differentiable at $x = a$. The concept of differentiability is more easily understood if you relate it to tangent lines. Geometrically, a tangent line will not exist when the graph of the function "has a sharp point," as shown in Figure 10.13a.

In Figure 10.13a, the tangent lines as $h \to 0^-$ and $h \to 0^+$ are not the same, so limit does not exist.

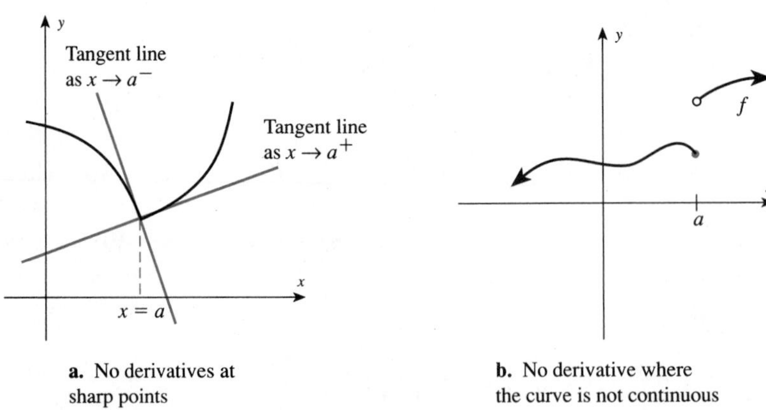

a. No derivatives at sharp points

b. No derivative where the curve is not continuous

FIGURE 10.13
Points on a curve for which the derivative is not defined

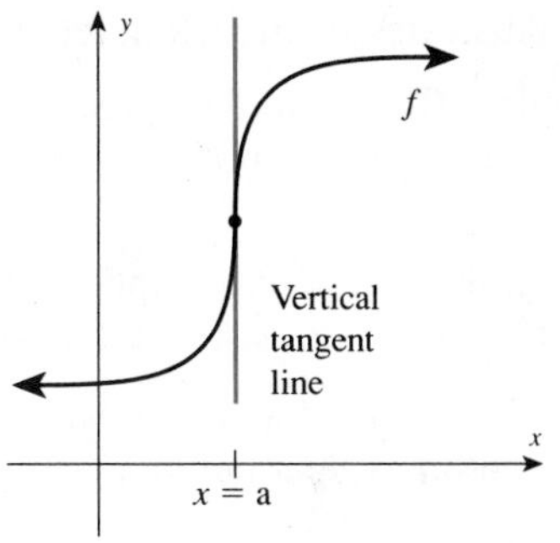

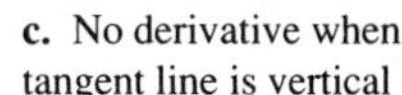

c. No derivative when tangent line is vertical

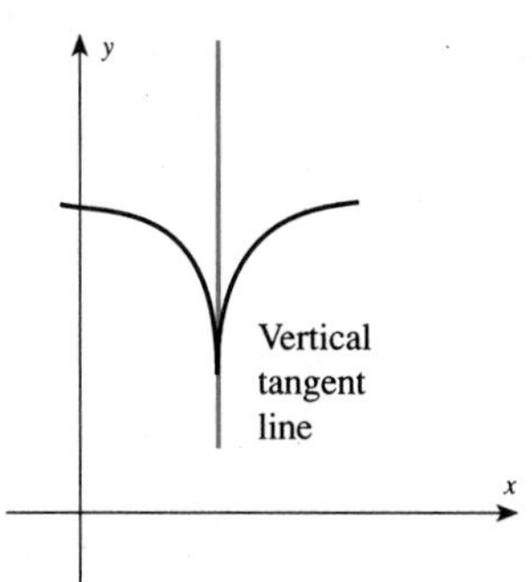

d. No derivative when tangent line is vertical

FIGURE 10.13 *(continued)*

The slope of a curve does not exist at a point for which the derivative is not defined. This means that there might be a vertical tangent line, and when this occurs we also say there is no slope (see Figures 10.13**c** and **d**). A point for which a function is not continuous cannot have a derivative at that point (see Figure 10.13**b**). However, be careful with this statement because a **function may be continuous at $x = a$ but still not be differentiable there** (see Figures 10.13**a**, **c**, and **d**).

EXAMPLE 4 Let $y = |x|$. Show that y is continuous at $x = 0$ but not differentiable at $x = 0$.

Solution First, $f(x) = |x|$ is continuous at $x = 0$ since $f(0) = |0| = 0$ is defined and $\lim_{x\to 0}|x| = 0$ (see Figure 10.14). Next, find $f'(x)$:

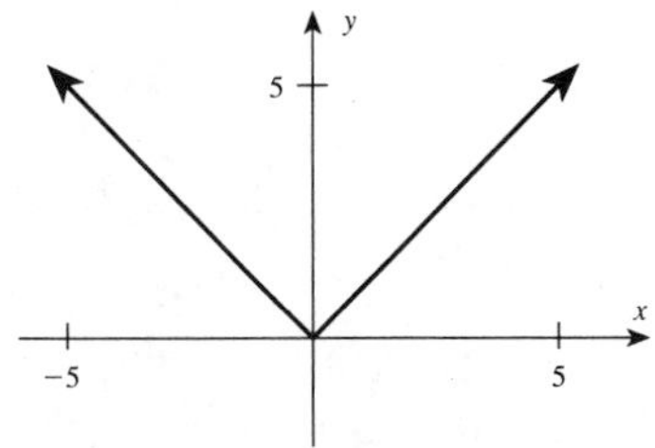

FIGURE 10.14
Graph of $y = |x|$

1. $\boldsymbol{f(x)} = |x|$
2. $\boldsymbol{f(x+h)} = |x+h|$
3. $\boldsymbol{f(x+h) - f(x)} = |x+h| - |x|$
4. $\dfrac{\boldsymbol{f(x+h) - f(x)}}{\boldsymbol{h}} = \dfrac{|x+h| - |x|}{h}$
5. $\displaystyle \lim_{h\to 0} \frac{f(x+h) - f(x)}{h} = \lim_{h\to 0} \frac{|x+h| - |x|}{h} = \lim_{h\to 0} \frac{|0+h| - |0|}{h}$ at $x = 0$

$$= \lim_{h\to 0} \frac{|h|}{h}$$

Now consider the left- and right-hand limits:

$$\lim_{h\to 0^+} \frac{|h|}{h} = \lim_{h\to 0^+} \frac{h}{h} = 1 \qquad \text{If } h\to 0^+, \text{ then } h \text{ is positive, so } |h| = h.$$

$$\lim_{h\to 0^-} \frac{|h|}{h} = \lim_{h\to 0^-} \frac{-h}{h} = -1 \qquad \text{If } h\to 0^-, \text{ then } h \text{ is negative, so } |h| = -h.$$

Since the left- and right-hand limits are not the same, we see that this limit does not exist. Therefore the derivative does not exist. Thus $f(x) = |x|$ is continuous at $x = 0$ but is not differentiable at $x = 0$. ■

Example 4 illustrates an important relationship between derivatives and continuity:

If a function f has a derivative at $x = c$, then it is continuous at $x = c$.
If a function f is continuous at $x = c$, then it does *not necessarily* have a derivative at $x = c$.

To show that **differentiability implies continuity** it is necessary to establish (1) $f(c)$ is defined, (2) $\lim_{x \to c} f(x)$ exists, and (3) $\lim_{x \to c} f(x) = f(c)$. These conditions all follow from the definition of derivative (we will not show the details here). Example 4 shows that **continuity does not imply differentiability**.

Marginal Cost

In the previous section *marginal profit* was defined. In business and economics the adjective *marginal* means rate of change. Mathematically this means that it is a derivative. Another example of this usage is **marginal cost**, which means the rate of change in cost per unit change in production at an output level of x units. This idea is illustrated with the following example.

EXAMPLE 5 Suppose the total cost in thousands of dollars for manufacturing x thousand items is described by the following equation:

$$C(x) = 2x^2 + 4x + 2{,}500$$

Also suppose that current production is 50,000 items ($x = 50$). Derive the formula for marginal cost and then find the marginal cost for $x = 50$. Also find the actual cost for producing one additional item at this production level.

Solution The formula for the marginal cost is

$$C'(x) = \lim_{h \to 0} \frac{C(x + h) - C(x)}{h}$$

Carry out the five-step procedure.

1. $\mathbf{C(x)} = 2x^2 + 4x + 2{,}500$
2. $\mathbf{C(x + h)} = 2(x + h)^2 + 4(x + h) + 2{,}500$
3. $\begin{aligned}\mathbf{C(x + h) - C(x)} &= [2(x + h)^2 + 4(x + h) + 2{,}500] - [2x^2 + 4x + 2{,}500] \\ &= 2x^2 + 4xh + 2h^2 + 4x + 4h + 2{,}500 - 2x^2 - 4x - 2{,}500 \\ &= 4xh + 2h^2 + 4h\end{aligned}$
4. $\dfrac{\mathbf{C(x + h) - C(x)}}{\mathbf{h}} = \dfrac{4xh + 2h^2 + 4h}{h} = 4x + 2h + 4$
5. $\lim_{\mathbf{h \to 0}} \dfrac{\mathbf{C(x + h) - C(x)}}{\mathbf{h}} = \lim_{h \to 0}(4x + 2h + 4) = 4x + 4$

The marginal cost for $x = 50$ (current production is 50,000) is

$$C'(50) = 4(50) + 4 = 204$$

We also want to find the *actual cost* of producing the 51st unit. We need to find

$$\frac{C(51) - C(50)}{1} = 4(50) + 2(1) + 4 = 206$$

This is the number we calculated in step 4 where $x = 50$ and $h = 1$. ■

The actual cost of producing one unit is very close to the marginal cost. It is usually easier to find the derivative (especially after studying the next two sections) than it is to calculate the actual cost of producing one additional unit. For this reason economists often use marginal cost to approximate the actual cost of producing one additional unit.

10.4 Problem Set

Find the x-value for all points where the functions in Problems 1–6 do not have derivatives.

1. 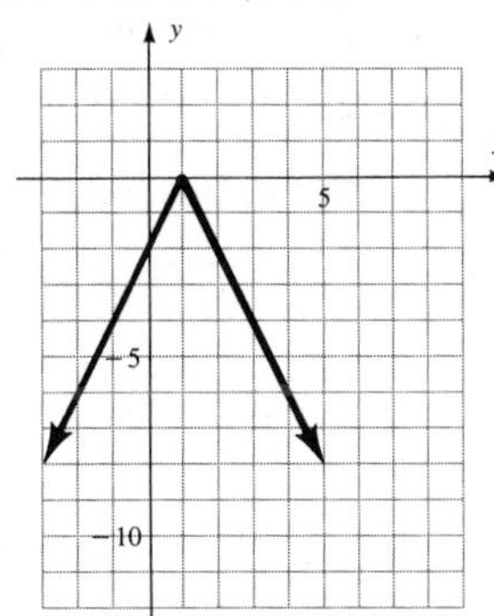

2. 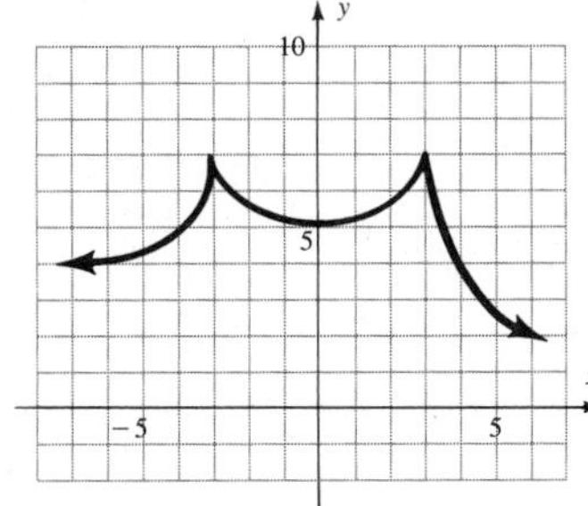

3. 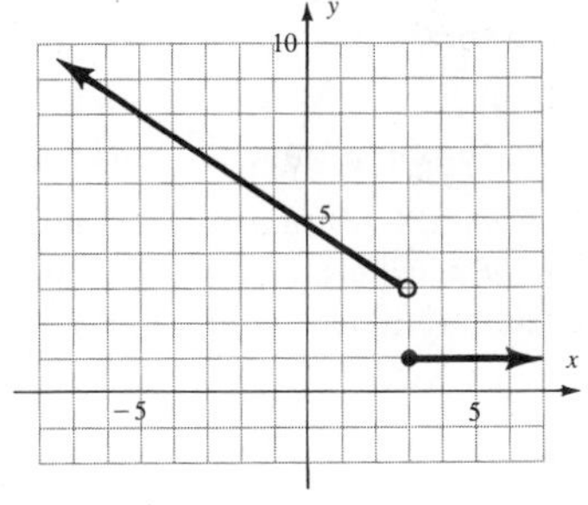

4. 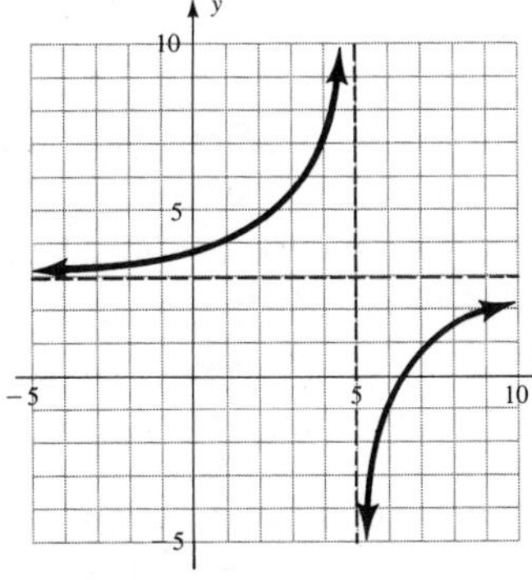

5. 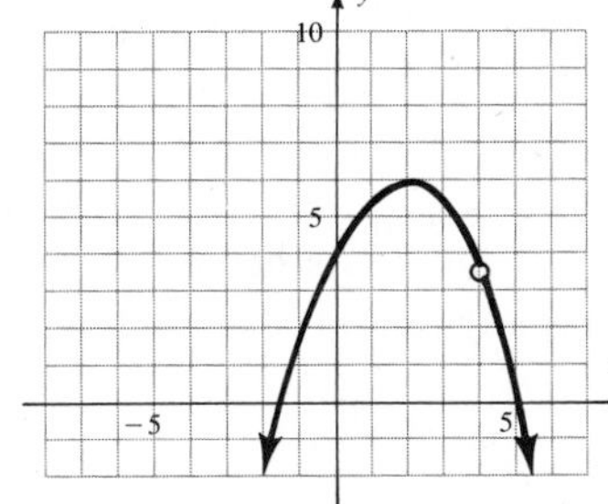

6. 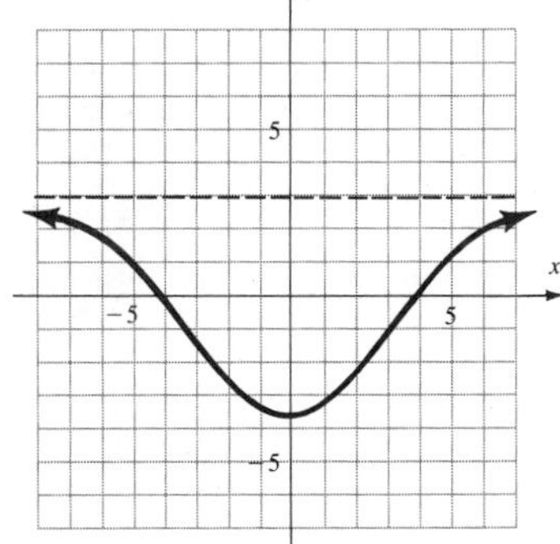

Find the derivative, $f'(x)$, of each of the functions in Problems 7–22 by using the derivative definition.

7. $f(x) = 2x^2$
8. $f(x) = 5x^2$
9. $f(x) = -3x^2$
10. $y = 2x + 1$
11. $y = 4 - 5x$
12. $y = 25 - 250x$
13. $y = 3x^2 + 4x$
14. $y = 3 + 2x - 3x^2$
15. $y = -3x^2 - 50x + 125$
16. $y = x^3$
17. $f(x) = 2x^3$
18. $g(x) = x^3 + x^2$
19. $f(x) = -\dfrac{3}{x}$
20. $f(x) = -\dfrac{2}{x+1}$
21. $y = 2\sqrt{x}$
22. $y = \sqrt{3x}$

Find the equation of the line tangent to the curves in Problems 23–27 at the given point.

23. $y = 5x^2$ at $x = -3$
24. $y = 2x^2$ at $x = 4$
25. $y = 4 - 5x$ at $x = -2$
26. $y = 3x^2 + 4x$ at $x = 0$
27. $y = 3 + 2x - 3x^2$ at $x = -1$

APPLICATIONS

In Problems 28–32 let $C(x) = x^2 - 40x + 2{,}000$ be the total cost function for producing x items where $20 \leq x \leq 40$.

28. Derive the formula for the marginal cost of this function.

29. Find the marginal cost for a production level of 20 items.
30. Find the marginal cost for a production level of 30 items.
31. Find the actual cost of producing one additional item if the current production level is 20 items.
32. Find the actual cost of producing one additional item if the current production level is 30 items.

In Problems 33–37 let $C(x) = x^2 - 100x + 3{,}000$ *be the total cost function for producing x items where* $50 \le x \le 100$.

33. Derive the formula for the marginal cost of this function.
34. Find the marginal cost for a production level of 50 items.
35. Find the marginal cost for a production level of 60 items.
36. Find the actual cost of producing one additional item if the current production level is 50 items.
37. Find the actual cost of producing one additional item if the current production level is 60 items.

In Problems 38–40 suppose the number (in millions) of bacteria present in a culture at time t is given by the formula $N(t) = 2t^2 - 200t + 1{,}000$.

38. Derive the formula for the instantaneous rate of change of the number of bacteria with respect to time.
39. Find the instantaneous rate of change of the number of bacteria with respect to time at time $t = 3$.
40. Find the instantaneous rate of change of the number of bacteria with respect to time at the beginning of this experiment.
41. An object moving in a straight line travels d miles in t hours according to the formula $d(t) = \frac{1}{5}t^2 + 5t$. What is the object's velocity when $t = 3$?
42. Derive the formula for the instantaneous velocity in Problem 41.
43. An object moving in a straight line travels m miles in t minutes according to the formula $m(t) = t^2 + 10t$. What is the object's velocity when $t = 5$?
44. Derive the formula for the instantaneous velocity in Problem 43.

10.5 Differentiation Techniques, Part I

Differentiation is one of the most powerful and useful concepts in mathematics but it would indeed be cumbersome if we had to apply the definition of derivative as we did in the last section every time we wanted to use this concept. Luckily there are many shortcuts to the process, and these are presented as differentiation techniques in this and the following two sections. In stating and working with these derivative formulas it is helpful to use some alternative notations for derivative. We have already used y', f' and $f'(x)$. Some other notations include

$$\frac{dy}{dx} \qquad \frac{d}{dx}y \qquad \frac{d}{dx}f(x) \qquad D_x y \qquad D_x[f(x)]$$

These are usually pronounced "dee y, dee x," "derivative of y with respect to x," "derivative of f with respect to x," "derivative of y with respect to x," and "derivative of f of x with respect to x," respectively. We will use these notations interchangeably.

In this section we develop five very important derivative formulas, in the next section two more, and in Section 10.7 a derivative formula called the *chain rule.* We begin by finding a formula for the derivative of a power; that is, if $y = x^n$, then what is y'? In the last section (Example 1) we found that if $y = x^2$, then $f'(x) = 2x$. How about $y = x^3$?

EXAMPLE 1 If $f(x) = x^3$, find $f'(x)$.

Solution
1. $f(x) = x^3$
2. $f(x + h) = (x + h)^3 = x^3 + 3x^2h + 3xh^2 + h^3$

3. $f(x + h) - f(x) = x^3 + 3x^2h + 3xh^2 + h^3 - x^3$
$= 3x^2h + 3xh^2 + h^3$

4. $\dfrac{f(x + h) - f(x)}{h} = \dfrac{h(3x^2 + 3xh + h^2)}{h}$
$= 3x^2 + 3xh + h^2$

5. $\lim\limits_{h\to 0} \dfrac{f(x + h) - f(x)}{h} = \lim\limits_{h\to 0}(3x^2 + 3xh + h^2) = 3x^2$ ■

If $y = x^4$, then, if we carry out the steps shown in Example 1, we will find $f'(x) = 4x^3$. The details are left as an exercise. Finally, look for a pattern:

$$y = x^2 \quad \rightarrow \quad y' = 2x$$
$$y = x^3 \quad \rightarrow \quad y' = 3x^2$$
$$y = x^4 \quad \rightarrow \quad y' = 4x^3$$
$$y = x^5 \quad \rightarrow \quad ?$$

Can you replace the question mark by the pattern you found?

$$y = x^5 \quad \rightarrow \quad y' = 5x^4$$
$$y = x^6 \quad \rightarrow \quad y' = 6x^5$$
$$\vdots$$
$$y = x^n \quad \rightarrow \quad y' = nx^{n-1}$$

We have simply demonstrated the plausibility of the following **power rule**. The general proof for any real number n requires algebra beyond the scope of this course, but we can look at several specific cases to get a feel for the proof. The proof for n a natural number is requested in Problem 69, Problem Set 10.5.

Power Rule

If f is a differentiable function, and if $f(x) = x^n$, then

$$f'(x) = nx^{n-1}$$

for any real number n.

EXAMPLE 2 If $y = x$, then $n = 1$, and the power rule asserts that $y' = 1 \cdot x^{1-1} = x^0 = 1$. Verify this result by using the definition of a derivative.

Solution

1. $f(x) = x$

2. $f(x + h) = x + h$

3. $f(x + h) - f(x) = h$

4. $\dfrac{f(x + h) - f(x)}{h} = 1$

5. $\lim\limits_{h\to 0} \dfrac{f(x + h) - f(x)}{h} = \lim\limits_{h\to 0} 1 = 1$ ■

EXAMPLE 3 Find derivatives of the following functions by using the power rule.

a. x^4 **b.** x^5 **c.** x^6 **d.** x^0 **e.** x^{-1}
f. x^{-2} **g.** $x^{1/3}$ **h.** $\sqrt{x}$ **i.** $x^{-2.5}$

Solution

a. If $y = x^4$, then $y' = 4x^3$.
b. If $y = x^5$, then $y' = 5x^4$.
c. If $y = x^6$, then $y' = 6x^5$.
d. If $y = x^0$, then $y' = 0x^{-1} = 0$.
e. If $y = x^{-1}$, then $y' = (-1)x^{-1-1} = -x^{-2}$ or $\frac{-1}{x^2}$.
f. If $y = x^{-2}$, then $y' = (-2)x^{-2-1} = -2x^{-3}$ or $\frac{-2}{x^3}$.
g. If $y = x^{1/3}$, then $y' = (\frac{1}{3})x^{1/3-1} = \frac{1}{3}x^{-2/3}$.
h. If $y = \sqrt{x}$, write (or think of it) as $y = x^{1/2}$; now apply the power rule: $y' = (\frac{1}{2})x^{1/2-1} = \frac{1}{2}x^{-1/2}$. Since the problem was given in radical notation, we state the answer in radical notation: $y' = \dfrac{1}{2\sqrt{x}}$.
i. If $y = x^{-2.5}$, then $y' = -2.5x^{-3.5}$. ■

A constant rule can be easily derived using the derivative definition:

1. $f(x) = k$ is given.
2. $f(x + h) = k$
3. $f(x + h) - f(x) = 0$
4. $\dfrac{f(x + h) - f(x)}{h} = 0$
5. $\lim\limits_{h \to 0} \dfrac{f(x + h) - f(x)}{h} = \lim\limits_{h \to 0} \dfrac{k - k}{h} = \lim\limits_{h \to 0} 0 = 0$

Constant Rule

If $f(x) = k$ for some constant k, then

$$f'(x) = 0$$

EXAMPLE 4 Use the constant rule to find the derivatives of the given functions.

a. $y = 5$ **b.** $y = -18$ **c.** $y = \sqrt{3}$ **d.** $y = \pi$

Solution The derivative of any constant is 0; thus

a. $y' = 0$ **b.** $y' = 0$ **c.** $y' = 0$ **d.** $y' = 0$ ■

The power rule and the constant rule for derivatives can be combined to prove the following result:

Constant Times a Function

If $y = kf(x)$ for a differentiable function f, then

$$y' = kf'(x)$$

This derivative formula for a constant times a function is easy to derive using the definition of a derivative. (The procedure is left as an exercise). Example 5 illustrates how to use the formula.

EXAMPLE 5 Find the derivatives of the given functions.

a. $y = 3x^2$ **b.** $y = 6x^{-5}$ **c.** $y = -\frac{3}{5}x^{10}$ **d.** $y = -2\sqrt{x}$

e. $y = 5$

Solution **a.** $y' = 3[(2)x^1]$
$= 6x$

b. $y' = 6[(-5)x^{-6}]$
$= -30x^{-6}$

c. $y' = -\frac{3}{5}[(10)x^9]$
$= -6x^9$

d. $y' = -2[(\frac{1}{2})x^{1/2-1}]$
$= -x^{-1/2}$

e. $y' = 0$ by the constant rule. However, note that we can write $y = 5 = 5x^0$ so $y' = 5 \cdot 0x^{-1} = 0$. This says that the power rule and the constant rule give the same results. ■

The next two derivative formulas are for sums or differences of functions:

Sum Rule
Difference Rule

Suppose f and g are differentiable functions of x.

If $y = f + g$, then $y' = f' + g'$.
If $y = f - g$, then $y' = f' - g'$.

We assume that if f and g are differentiable at a point $x = a$, then the sum rule implies that the sum $f + g$ is also differentiable at $x = a$.

The following proof of the sum formula is optional, and you may skip over to Example 6 if you are not interested in deriving these formulas. We begin by writing y as $u(x)$ so that the sum formula can be written as $u(x) = f(x) + g(x)$. We want to show that $u'(x) = f'(x) + g'(x)$. We use the definition of derivative:

1. $\boldsymbol{u(x)} = f(x) + g(x)$
2. $\boldsymbol{u(x+h)} = f(x+h) + g(x+h)$
3. $\boldsymbol{u(x+h) - u(x)} = [f(x+h) + g(x+h)] - [f(x) + g(x)]$
$= [f(x+h) - f(x)] + [g(x+h) - g(x)]$
4. $\dfrac{\boldsymbol{u(x+h) - u(x)}}{\boldsymbol{h}} = \dfrac{[f(x+h) - f(x)] + [g(x+h) - g(x)]}{h}$
$= \dfrac{f(x+h) - f(x)}{h} + \dfrac{g(x+h) - g(x)}{h}$
5. $\displaystyle\lim_{h\to 0} \frac{\boldsymbol{u(x+h) - u(x)}}{\boldsymbol{h}} = \lim_{h\to 0}\left[\frac{f(x+h) - f(x)}{h} + \frac{g(x+h) - g(x)}{h}\right]$
$\displaystyle = \lim_{h\to 0} \frac{f(x+h) - f(x)}{h} + \lim_{h\to 0} \frac{g(x+h) - g(x)}{h}$

Thus $u'(x) = f'(x) + g'(x)$. The proof for the difference formula is identical to the one for the sum formula.

These last two derivative formulas, along with the ones already discussed, allow us to easily find the derivatives of polynomials, as shown in Example 6.

EXAMPLE 6 Find the derivatives of the given functions.

a. $y = 5x^2 - 3x + 15$ **b.** $y = 9x^3 - 4x^2 + 5x - 12$

Solution **a.** $y' = 5(2)x - 3(1) + 0$
$= 10x - 3$

b. $y' = 9(3)x^2 - 4(2)x + 5(1) - 0$
$= 27x^2 - 8x + 5$ ■

EXAMPLE 7 If $f(x) = 6x^8 - 5\sqrt{x} + \dfrac{4}{x}$, find $f'(x)$.

Solution Write (or think of) this as $f(x) = 6x^8 - 5x^{1/2} + 4x^{-1}$. Then

$$f'(x) = 6(8)x^7 - 5\left(\frac{1}{2}\right)x^{-1/2} + 4(-1)x^{-2}$$

$$= 48x^7 - \frac{5}{2\sqrt{x}} - \frac{4}{x^2}$$ ■

EXAMPLE 8 If $f(x) = \dfrac{x^4 + 5x^3 + 1}{\sqrt{x}}$ find $f'(x)$.

Solution Write or think of this as

$$x^{-1/2}(x^4 + 5x^3 + 1) = x^{7/2} + 5x^{5/2} + x^{-1/2}$$

Thus, if $f(x) = x^{7/2} + 5x^{5/2} + x^{-1/2}$, then

$$f'(x) = \frac{7}{2}x^{5/2} + \frac{25}{2}x^{3/2} - \frac{1}{2}x^{-3/2}$$ ■

EXAMPLE 9 Suppose the amount of calcium remaining in a person's bloodstream is given by the formula $A = t^{-3/2}$ where t is the number of days after the calcium was injected ($t \geq \frac{1}{2}$). How fast is the body removing calcium from the blood exactly 48 hours (or $t = 2$) after the calcium was injected?

Solution The rate of change (per day) of calcium in the blood is given by the derivative

$$A'(t) = \frac{dA}{dt} = -\frac{3}{2}t^{-5/2}$$

When $t = 2$, this rate is $-\frac{3}{2}(2)^{-5/2}$. This rate can be approximated on a calculator by pressing

[2] [x^y] [2.5] [+/−] [×] [1.5] [+/−] [=] −0.2651650429

The amount of calcium in the blood is changing at the rate of about −.27 unit per day when $t = 2$. The negative sign tells us that the amount of calcium is decreasing. ■

Note in Example 9 that the variable is not x and that the symbol used is dA/dt. This symbol is read as "derivative of A with respect to t," or as "dee A, dee t." Even though the variables x and y are used most of the time, keep in mind that any variables may be substituted for x and y.

The derivative rules of this section are summarized here for easy reference:

Differentiation Formulas

POWER RULE: If $y = x^n$, then $y' = nx^{(n-1)}$ for any real number n.
CONSTANT RULE: If $y = k$, then $y' = 0$.
CONSTANT TIMES A FUNCTION RULE: If $y = kf$, then $y' = kf'$.
SUM RULE: If $y = f + g$, then $y' = f' + g'$.
DIFFERENCE RULE: If $y = f - g$, then $y' = f' - g'$.

10.5 Problem Set

Find the derivatives of the functions in Problems 1–42 and simplify them algebraically.

1. $y = x^7$
2. $y = x^{16}$
3. $y = x^{12}$
4. $y = x^{-3}$
5. $y = x^{-5}$
6. $y = 25$
7. $y = -130$
8. $y = 5x^{-4}$
9. $y = -4x^{-8}$
10. $y = 5\sqrt{x}$
11. $y = -\frac{1}{2}\sqrt{x}$
12. $y = \frac{3}{4}\sqrt{x}$
13. $y = 5x^{-8}$
14. $y = -3x^{-2}$
15. $y = 12x^{5/4}$
16. $y = 5x^2 - 9$
17. $x = 3x^2 + x$
18. $y = 5x^3 - 9x^2$
19. $y = 2x^2 - 5x - 6$
20. $y = 5x^2 - 5x + 12$
21. $y = 5x^3 - 5x^2 + 4x - 5$
22. $y = 6x^3 - 25x^2 - 6x + 45$
23. $y = x^{-3} + x^2 + x^{-1}$
24. $y = x^{-5} - x^{-3} - x^{-1}$
25. $y = (x^4 + 2)^2$
26. $y = (x^5 - 1)^2$
27. $y = \dfrac{2}{x} + \dfrac{5}{x^2}$
28. $y = \dfrac{3}{x} - \dfrac{2}{x^3}$
29. $f(x) = -5x^7 + 2\sqrt{x} - 3x^{-1}$
30. $f(x) = 4x^{-3/4} - 5\sqrt{x} + 2x^{-3}$
31. $f(x) = \dfrac{x^3 + 4x^2 + 1}{x^2}$
32. $f(x) = \dfrac{3x^3 - 5x + 6}{2x^2}$
33. $f(x) = \dfrac{x^5 + 3x^3}{\sqrt{x}}$
34. $f(x) = \dfrac{2x^4 - 5x^2}{2\sqrt{x}}$
35. $\dfrac{d}{dx}\left(\sqrt[3]{x} + \dfrac{5}{8}\right)$
36. $\dfrac{d}{dy}(4y^{1/2} - 5)$
37. $\dfrac{d}{dD}(\pi D)$
38. $\dfrac{d}{dr}(\pi r^2)$
39. $\dfrac{dV}{dr}$ where $V = \dfrac{4}{3}\pi r^3$
40. $\dfrac{dV}{dR}$ where $V = \dfrac{10}{3}\pi R^2$
41. $\dfrac{dK}{dC}$ where $K = \dfrac{C^2}{4\pi}$
42. $\dfrac{dS}{dR}$ where $S = 4\pi R^2$

Find the slope of the tangent to the graph of f at the given point in Problems 43–48.

43. $f(x) = x^5$ at $(1, 1)$
44. $f(x) = x^5$ at $(2, 32)$
45. $f(x) = \dfrac{1}{\sqrt{x}}$ at $(1, 1)$
46. $f(x) = \dfrac{1}{\sqrt{x}}$ at $\left(4, \dfrac{1}{2}\right)$
47. $f(x) = 3x^2 - 4x + 10$ at $(0, 10)$
48. $f(x) = 3x^2 - 4x + 10$ at $(3, 25)$

Find the equation of the tangent line of f at the given point in Problems 49–54.

49. $f(x) = x^2 + 2x + 1$ at $(0, 1)$
50. $f(x) = x^2 + 2x + 1$ at $(-2, 1)$
51. $f(x) = 2 - \dfrac{2}{x^2}$ at $(1, 0)$
52. $f(x) = 2 - \dfrac{2}{x^2}$ at $(0, 2)$
53. $f(x) = \sqrt[3]{x}$ at $(1, 1)$
54. $f(x) = \sqrt[3]{x}$ at $(8, 2)$

APPLICATIONS

55. The cost of producing a certain type of boat is $C(x) = 20x^2 + 500x + 250{,}000$. What is the marginal cost?

56. The profit function for a certain item is $P(x) = 45x - 3x^3$. What is the marginal profit?

57. The relationship between the current and the time on an X-ray machine is given by the formula $m = 400/t$. Find dm/dt.

58. The relationship between the current and the distance on an X-ray machine is given by the formula $m = 256/s^2$. Find dm/ds.

59. The earnings (in thousands of dollars) of Amdex Corporation are a function of the number of years since its founding in 1986 according to the formula $A(t) = .05t^2 + 25t + 5$. At what rate are the earnings changing in 1992?

60. In a certain experiment, subjects are found to learn according to the model $N = 25\sqrt{x}$, where N is the number of tasks learned in x hours. How fast are the subjects learning at the end of the second hour?

61. Repeat Problem 60 except now determine how fast the subjects are learning at the end of the fifth hour.

62. In a wildlife reserve the population P of foxes depends on the population x of rabbits according to the formula $P = .0005x + .00001x^2$. What is the rate at which the population of foxes is changing when the number of rabbits is 100,000?

63. Repeat Problem 62 for a population of 1 million rabbits.

64. The number of gallons, g, of water pumped into a reservoir after t minutes is given by the formula $g(t) = 2t + \sqrt{t}$. At what rate is water flowing into the reservoir when $t = 10$?

65. Assume that the consumer price index (CPI) of an economy is described by the function

$$P(t) = -.05t^2 + 5t + 250$$

where t is the number of years after 1990 (i.e., $t = 0$ corresponds to 1990).

a. What was the average rate of change in the CPI from the years 1990 to 1994?

b. At what rate was the consumer price index changing in the year 1991?

66. Show that if $y = x^4$, then $y' = 4x^3$ by using the derivative definition.

67. Derive the formula for a constant times a function by using the derivative definition.

68. Prove the formula for the difference of functions by using the derivative definition.

69. Derive the power rule for a natural number n by using the binomial theorem.

10.6 Differentiation Techniques, Part II

There are two additional shortcut differentiation formulas that simplify work with calculus. These are the *product* and *quotient* formulas. Since the sum and difference formulas are easy to remember and are intuitively obvious, students often incorrectly try to generalize to product and quotient formulas. The derivative of a sum or difference is the sum or difference of the derivatives. But this is not true of the product and quotient formulas.

WARNING *The derivative of a product is not the product of the derivatives.*

The following simple example will illustrate that the derivative of a product is not the product of the derivatives. Let $f(x) = x^2$ and $g(x) = x^3$. Consider $y = f(x)g(x) = x^5$:

$$f'(x) = 2x \qquad g'(x) = 3x^2 \qquad y' = 5x^4$$

and

$$f'(x)g'(x) = 2x(3x^2) = 6x^3 \neq 5x^4$$

Note that $y' \neq f'g'$. The product formula tells us how to find the derivative of a product without the necessity of first multiplying them together.

Product Rule

If f and g are differentiable functions such that $y = f(x)g(x)$, then

$$y' = f(x)g'(x) + g(x)f'(x)$$

This formula can more easily be remembered in the following form: If $y = fg$, then

$$y' = fg' + gf'$$

First function times the derivative of the second plus second times the derivative of the first

This differentiation formula is proved in Example 4, but we will first work through a couple of examples to make sure you understand how it works.

EXAMPLE 1 If $f(x) = x^2$, $g(x) = x^3$, and $y = fg$, verify that $y' = fg' + gf' = 5x^4$.

Solution $f'(x) = 2x$ and $g'(x) = 3x^2$, so

$$\begin{aligned} fg' + f'g &= (x^2)(3x^2) + (x^3)(2x) \\ &= 3x^4 + 2x^4 \\ &= 5x^4 \end{aligned}$$

Thus $y' = fg' + f'g = 5x^4$. ■

EXAMPLE 2 If $y = 2x^3(x^2 - 3x)$, find $\dfrac{dy}{dx}$ and simplify.

Solution You could multiply the factors together to obtain a polynomial, then use the results of the last section to find the derivative, or you can use the product rule. This example illustrates the product rule.

$$\begin{aligned} \frac{dy}{dx} = \frac{d}{dx}[\overbrace{2x^3}^{\text{first}}\,\overbrace{(x^2 - 3)}^{\text{second}}] &= \overbrace{2x^3}^{\text{first}} \cdot \overbrace{\frac{d}{dx}(x^2 - 3)}^{\text{derivative of second}} \;+\; \overbrace{(x^2 - 3)}^{\text{second}} \cdot \overbrace{\frac{d}{dx}(2x^3)}^{\text{derivative of first}} \\ &= 2x^3(2x) + (x^2 - 3)(6x^2) \\ &= 4x^4 + 6x^4 - 18x^2 \\ &= 10x^4 - 18x^2 \\ &= 2x^2(5x^2 - 9) \end{aligned}$$

■

EXAMPLE 3 If $y = (\sqrt{x} + 1)(2\sqrt{x} - 3)$, find dy/dx and simplify.

Solution

$$\begin{aligned} \frac{dy}{dx} &= \frac{d}{dx}[\overbrace{(x^{1/2} + 1)}^{\text{first}}\overbrace{(2x^{1/2} - 3)}^{\text{second}}] \\ &= \overbrace{(x^{1/2} + 1)}^{\text{first}}\overbrace{\frac{d}{dx}(2x^{1/2} - 3)}^{\text{derivative of second}} + \overbrace{(2x^{1/2} - 3)}^{\text{second}}\overbrace{\frac{d}{dx}(x^{1/2} + 1)}^{\text{derivative of first}} \\ &= (x^{1/2} + 1)[2(\tfrac{1}{2})x^{-1/2}] + (2x^{1/2} - 3)[\tfrac{1}{2}x^{-1/2}] \\ &= (x^{1/2} + 1)x^{-1/2} + (2x^{1/2} - 3)[\tfrac{1}{2}x^{-1/2}] \\ &= x^0 + x^{-1/2} + x^0 - \tfrac{3}{2}x^{-1/2} \\ &= 2 - \tfrac{1}{2}x^{-1/2} \\ &= 2 - \frac{1}{2\sqrt{x}} \cdot \frac{\sqrt{x}}{\sqrt{x}} \\ &= 2 - \frac{\sqrt{x}}{2x} \\ &= \frac{4x - \sqrt{x}}{2x} \end{aligned}$$

■

WARNING Example 3 shows that the product rule is not always the most economical method. Notice that if you first multiply the factors to obtain

$$y = 2x - \sqrt{x} - 3$$

You can find the derivative by using the power rule:

$$y' = 2 - \frac{1}{2}x^{-1/2} = 2 - \frac{\sqrt{x}}{2x}$$

$$= \frac{4x - \sqrt{x}}{2x}$$

EXAMPLE 4 Prove the product rule. That is, if f and g are differentiable functions such that $y = f(x)g(x)$, then prove that

$$y' = f(x)g'(x) + g(x)f'(x)$$

Solution Use the definition of derivative:

$$y' = \lim_{h \to 0} \frac{f(x+h)g(x+h) - f(x)g(x)}{h}$$

Add and subtract $f(x)g(x+h)$ in the numerator:

$$y' = \lim_{h \to 0} \frac{f(x+h)g(x+h) - f(x)g(x) + f(x)g(x+h) - f(x)g(x+h)}{h}$$

$$= \lim_{h \to 0} \frac{[f(x+h)g(x+h) - f(x)g(x+h)] + [f(x)g(x+h) - f(x)g(x)]}{h}$$

$$= \lim_{h \to 0} \left[g(x+h) \cdot \frac{f(x+h) - f(x)}{h} + f(x) \cdot \frac{g(x+h) - g(x)}{h} \right]$$

$$= \lim_{h \to 0} \left[g(x+h) \cdot \frac{f(x+h) - f(x)}{h} \right] + \lim_{h \to 0} \left[f(x) \cdot \frac{g(x+h) - g(x)}{h} \right]$$

Since $g(x)$ is differentiable, it is continuous, and $\lim_{h \to 0} g(x+h) = g(x)$. Using this and the definition of derivative, we have

$$y' = g(x)f'(x) + f(x)g'(x)$$ ■

WARNING The derivative formula for quotients also does *not* follow the pattern of simply taking the quotient of the derivatives. The quotient formula is stated in the following box, and you are asked to prove it in the problems. It is very easy to prove because you can write f/g as fg^{-1} and use the product rule together with the chain rule, which is discussed in the next section.

Quotient Rule

If f and g are differentiable functions such that $y = f(x)/g(x)$ then $y' = [g(x)f'(x) - f(x)g'(x)]/[g(x)]^2$. This formula can be more easily stated and remembered by using the following notation:

If $y = \dfrac{f}{g}$, then

$$y' = \frac{gf' - fg'}{g^2}$$

The *d*enominator function times the derivative of the *n*umerator minus the *n*umerator times the derivative of the *d*enominator all divided by the *d*enominator squared (Remember that *d* comes before *n* alphabetically.) You can easily remember this by: "low dee high minus high dee low, over the square of the denominator must go."

EXAMPLE 5 If $y = \dfrac{2x^3}{3x+1}$, find $\dfrac{dy}{dx}$ and simplify.

Solution

$$y' = \frac{d}{dx}\left[\frac{2x^3}{3x+1}\right] \begin{array}{l} \longleftarrow \text{top} \\ \longleftarrow \text{bottom} \end{array}$$

$$y' = \frac{\overbrace{(3x+1)}^{\text{bottom}}\overbrace{(2x^3)'}^{\text{derivative of top}} - \overbrace{(2x^3)}^{\text{top}}\overbrace{(3x+1)'}^{\text{derivative of bottom}}}{\underbrace{(3x+1)^2}_{\text{bottom squared}}}$$

"Low dee high minus high dee low over the square of the denominator must go. . ."

$$= \frac{(3x+1)(6x^2) - (2x^3)(3)}{(3x+1)^2}$$

$$= \frac{18x^3 + 6x^2 - 6x^3}{(3x+1)^2}$$

$$= \frac{12x^3 + 6x^2}{(3x+1)^2}$$

Leave this in factored form—do *not* expand. ■

EXAMPLE 6 If $y = \dfrac{3x^2+2}{5x^2-1}$, find $\dfrac{dy}{dx}$ and simplify.

Solution

$$y' = \frac{(5x^2-1)(6x) - (3x^2+2)(10x)}{(5x^2-1)^2}$$

$$= \frac{30x^3 - 6x - 30x^3 - 20x}{(5x^2-1)^2}$$

$$= \frac{-26x}{(5x^2-1)^2}$$

■

EXAMPLE 7 An apple factory can produce x thousand gallons of apple juice per week at an average weekly cost of

$$C(x) = \frac{x + 15}{x} \text{ per gallon on the domain } [1, 20].$$

Show that this average is always decreasing.

Solution A function is decreasing when its derivative is negative. Thus we need to show the inequality $C'(x) < 0$ on $[1, 20]$. First find $C'(x)$.

Method I: Quotient rule

$$C'(x) = \frac{x(1) - (x + 15)(1)}{x^2} = -\frac{15}{x^2}$$

Method II: Power rule. Write $C(x) = 1 + 15x^{-1}$. Then

$$C'(x) = -15x^{-2}$$

WARNING Do not forget that quotients can often be written as products and that the problem may be easier to work as a product than as a quotient. However, in either case, you will obtain the same results if you do not make any mistakes. Since $C'(x) = -15x^{-2}$ we see that $C'(x) < 0$ for all nonzero x. Thus, the total cost is always decreasing on $[1, 20]$. ■

10.6 Problem Set

Find the derivative for each of the functions in Problems 1–28 and simplify.

1. $f(x) = 5x^2(x^2 - 6)$
2. $f(x) = 3x^3(x^2 + 7)$
3. $f(x) = (x + 1)(x - 2)$
4. $f(x) = (2x - 5)(x + 7)$
5. $g(x) = (3x^2 + 5)(2x^2 - 5)$
6. $g(x) = (9x^2 - 8)(2x^2 - 5)$
7. $g(x) = 5x^4(2x^2 - 5x + 1)$
8. $g(x) = 9x^5(3x^2 - 2x + 15)$
9. $y = \dfrac{x}{x - 3}$
10. $y = \dfrac{x}{3x - 1}$
11. $y = \dfrac{x + 5}{x - 3}$
12. $y = \dfrac{3x + 1}{15 - x}$
13. $y = \dfrac{x^2 + 3}{x^2 - 5}$
14. $y = \dfrac{5 - 7x^3}{1 + x^3}$
15. $y = \dfrac{x^2 + x - 6}{x^2 - 4}$
16. $y = \dfrac{2x^2 - x}{x^2 + 6x}$
17. $f(x) = (2x^2 - 1)(x^3 + 2x^2 - 3)$
18. $f(x) = (3x^2 + 2)(x^3 - 5x^2 + 4)$
19. $f(x) = (4x^2 + x)(x^3 - 3x^2 + 13)$
20. $f(x) = (x^2 + 3x)(x^3 + 4x^2 + 25)$
21. $f(x) = (3x^3 - 2x + 5)(2x^4 + 5x - 9)$
22. $f(x) = (7x^4 - 8x^2 + 144)(5x^3 + 25)$
23. $f(x) = 5x^{2/3}(5x^{-1} + 3x)$
24. $f(x) = 3x^2(x^{-2} + x)$
25. $f(x) = 6\sqrt{x}(2x^2 - 5)$
26. $f(x) = 2(\sqrt{x} + 3x)(\sqrt{x} - x)$
27. $g(x) = \dfrac{2x}{5x^2 - 11x + 3}$
28. $g(x) = \dfrac{5x}{125 - 25x - 3x^2}$

Find the slope of the tangent to the graph of f at the value of x given in Problems 29–34.

29. $f(x) = \dfrac{5x^2 + 5x}{x - 5}$ at $x = 1$
30. $f(x) = \dfrac{x + 2}{4x - 7x^2}$ at $x = -1$
31. $f(x) = x^{1/2}(x^2 + 3)$ at $x = 4$

32. $f(x) = x^{2/3}(x^2 - 5)$ at $x = 8$

33. $f(x) = \dfrac{x^2 - 5x + 1}{x^2 + 4x + 4}$ at $x = 0$

34. $f(x) = \dfrac{x^3 + 1}{x^3 - 1}$ at $x = 0$

Find an equation of the tangent line to the graph of f at the given value of x in Problems 35–40.

35. $f(x) = \dfrac{5x^2 + 5x}{x - 5}$ at $x = 0$

36. $f(x) = \dfrac{x + 2}{4x - 7x^2}$ at $x = 1$

37. $f(x) = x^{1/2}(x^2 + 3)$ at $x = 4$

38. $f(x) = x^{2/3}(x^2 - 5)$ at $x = 8$

39. $f(x) = \dfrac{x^2 - 5x + 1}{x^2 + 4x + 4}$ at $x = 1$

40. $f(x) = \dfrac{x^3 + 1}{x^3 - 1}$ at $x = 0$

APPLICATIONS

41. Classical economic theory predicts that in a free market the demand, D, for a commodity must decrease as the price, x, increases. Such a demand curve is drawn in Figure 10.15.

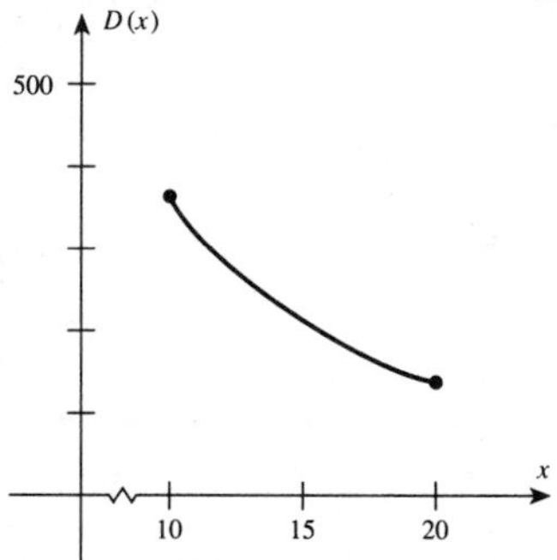

FIGURE 10.15
Demand curve
$$D(x) = \frac{100{,}000}{x^2 + 15x + 25}$$
on the domain [10, 20]

Find the rate of change of demand with respect to the price change for the demand curve shown in Figure 10.15.

42. If the demand curve is
$$D(x) = \frac{25{,}000}{x^2 + 3x + 20} \text{ on the domain } [10, 20]$$
find the rate of change of demand with respect to a price change.

43. Find the rate of change of demand for $x = 15$ using the demand curve in Figure 10.15, and interpret this answer in terms of the graph. Is the demand increasing or decreasing at that point?

44. Find the rate of change of demand for $x = 15$ using the demand curve in Problem 42 and interpret this answer. Is the demand increasing or decreasing at that point?

45. Dollars spent on advertising will influence the number of items purchased, up to some saturation point as shown by the graph in Figure 10.16.

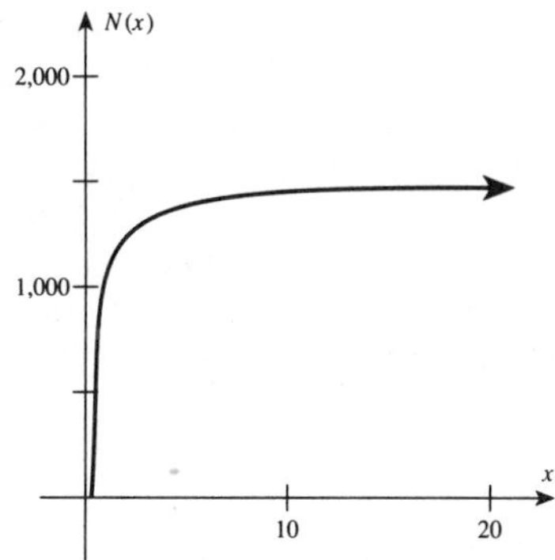

FIGURE 10.16
Number of items sold (in thousands) for dollars spent (in millions) according to the formula
$$N(x) = 1{,}500 - \frac{575}{x}$$

Find the rate of change of the number of items with respect to the number of dollars spent.

46. Find the rate of change of the number of items with respect to the number of dollars spent for the function
$$N(x) = 2{,}500 - \frac{400}{x}$$
where N is measured in thousands of items and x in thousands of dollars.

47. Find the rate of change of the number of items if \$10,000,000 is spent for the function defined in Figure 10.16. Interpret your answer in terms of this graph. Is the number of items purchased increasing or decreasing at this point?

48. Find the rate of change of the number of items if \$10,000 is spent for the function defined in Problem 46, and interpret this answer. Is the number of items purchased increasing or decreasing at this point?

49. Many drugs injected into the bloodstream have a dramatic effect over a very short period of time and then stabilize as shown in Figure 10.17.

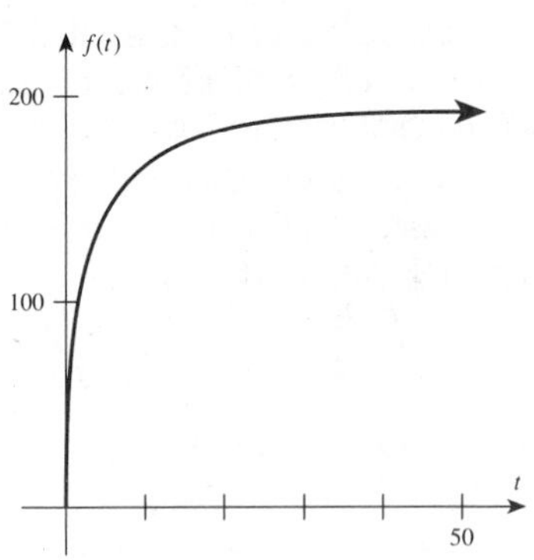

FIGURE 10.17
Amount of a drug (in number of milligrams) injected into the bloodstream after t minutes according to the formula

$$f(t) = \frac{t}{.01 + .005t}$$

Find the rate at which the amount of the drug is changing relative to the time t.

50. At what rate is the drug being absorbed into the system (in milligrams per minute) after 5 minutes for the drug shown in Figure 10.17?

51. Psychologists tell us that there is a certain interest value in learning a certain task, but after the task has been mastered, interest (and consequently, learning) decreases. Suppose an experiment is set up in which learning is tested by counting the number of items, N, learned by a subject in a given amount of time, t (in minutes), according the formula given in Figure 10.18.

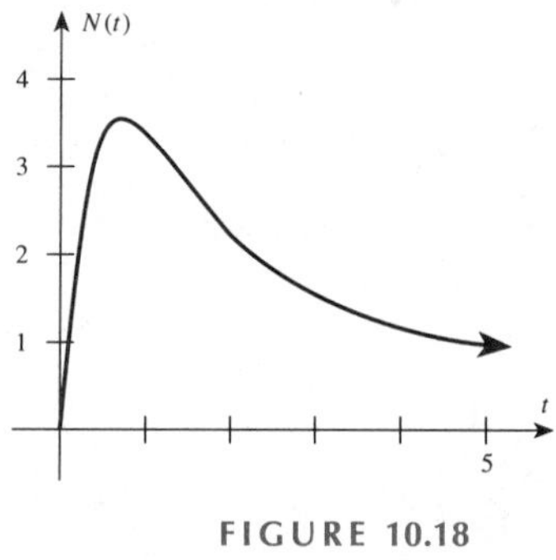

FIGURE 10.18
Learning curve

$$N(t) = \frac{10t}{2t^2 + 1}$$

What is the rate of change of number of items learned with respect to time, t?

52. Find the rate of learning for the learning curve shown in Figure 10.18 after 2 minutes. Interpret this answer in terms of the graph.

10.7 The Chain Rule

We now come to the last of the differentiation techniques we will consider, the **chain rule**. The chain rule, along with the other derivative theorems we have discussed, will enable us to efficiently find the derivative of almost every function we will consider. Remember, however, as you progress throughout this or other courses, that if you encounter a function that does not "fit" any of the differentiation formulas we have developed, you can resort to using the derivative definition.

Introduction to the Chain Rule

Suppose it is known that the carbon monoxide pollution in the air is .02 ppm (parts per million) for each person in a town whose population is growing at the rate of 1,000 people per year. To find the rate at which the level of pollution is increasing with respect to time, we form the product

$$(.02 \text{ ppm/person})(1{,}000 \text{ people/year}) = 20 \text{ ppm/year}$$

In this example, the level of pollution, L, is a function of the population, P, which is itself a function of time, t. Thus, L is a function of t, and

$$\begin{bmatrix}\text{RATE OF CHANGE OF } L \\ \text{WITH RESPECT TO } t\end{bmatrix} = \begin{bmatrix}\text{RATE OF CHANGE OF } L \\ \text{WITH RESPECT TO } P\end{bmatrix}\begin{bmatrix}\text{RATE OF CHANGE OF } P \\ \text{WITH RESPECT TO } t\end{bmatrix}$$

Expressing each of these rates in terms of an appropriate derivative, we obtain the following equation:

$$\frac{dL}{dt} = \frac{dL}{dP}\frac{dP}{dt}$$

These observations suggest the following differentiation rule.

Chain Rule

Suppose y is a differentiable function of u, and u, in turn, is a differentiable function of x. Then

$$\frac{dy}{dx} = \frac{dy}{du} \cdot \frac{du}{dx}$$

EXAMPLE 1 If $y = u^6$ and $u = 2x^3 - 5x$, find

a. $\dfrac{dy}{du}$ **b.** $\dfrac{du}{dx}$ **c.** $\dfrac{dy}{dx}$

Write your answer as a function of x.

Solution

a. $\dfrac{dy}{du} = \dfrac{d}{du}(u^6) = 6u^5 = 6(2x^3 - 5x)^5$

b. $\dfrac{du}{dx} = \dfrac{d}{dx}(2x^3 - 5x) = 6x^2 - 5$

c. $\dfrac{dy}{dx} = \dfrac{dy}{du} \cdot \dfrac{du}{dx}$ This is the chain rule.

$= 6u^5(6x^2 - 5)$ Answers from parts **a** and **b**

$= 6(2x^3 - 5x)^5(6x^2 - 5)$ Write u in terms of x. ■

EXAMPLE 2 If $y = 2u^5$ and $u = 2x^3 - 5x$, find $\dfrac{dy}{dx}$.

Solution When asking for the derivative of y with respect to x, it will be assumed that the answer should be given as a function of x, unless stated otherwise.

$\dfrac{dy}{dx} = \dfrac{dy}{du} \cdot \dfrac{du}{dx}$ This is the chain rule.

$= 10u^4(6x^2 - 5)$ Substitute $u = 2x^3 - 5x$ to write this as a function of x.

$= 10(2x^3 - 5x)^4(6x^2 - 5)$ ■

More often than not, you will need to identify the function u, rather than being given the function u, as you were in Examples 1 and 2.

EXAMPLE 3 If $y = (4x + 1)^5$, find y'.

Solution **Think:** $u = 4x + 1$ so $y = \square^5$

$$y' = \frac{dy}{du} \cdot \frac{du}{dx}$$ Chain rule; remember $y' = \frac{dy}{dx}$.

$$= 5(4x + 1)^4 \, (4)$$ **Think:** $y' = 5\,\square^4$ times derivative of $\square$.

$$= 20(4x + 1)^4$$ ■

Because you will frequently use the chain rule, it is important that the process become natural and that you do not let the notation become too cumbersome. For this reason, the next example will show you in black what you would write down and in color what you should be thinking as you use the chain rule.

EXAMPLE 4 If $f(x) = (2x^2 + 3x - 5)^3$, find $f'(x)$.

Solution

$$f(x) = \overbrace{(2x^2 + 3x - 5)}^{u}{}^{\overbrace{3}^{n}}$$ **Think:** $f(x) = \square^3$

$$f'(x) = \overbrace{3}^{n}\ \overbrace{(2x^2 + 3x - 5)^2}^{u^{n-1}}\ \overbrace{(4x + 3)}^{u'}$$ $f'(x) = 3\,\square^2 \cdot \frac{d\,\square}{dx}$

$$= 3(4x + 3)(2x^2 + 3x - 5)^2$$ ■

Differentiation of Composite Functions

The chain rule is actually a rule for differentiating composite functions. In particular, if $y = f(u)$ and $u = u(x)$, then y is the **composite function** $y = (f \circ u)(x) = f[u(x)]$ and the chain rule can be rewritten as follows.

Chain Rule for Composite Functions

> If u is differentiable at x and f is differentiable at $u(x)$, then the composite function $f \circ u$ is differentiable at x and
>
> $$(f \circ u)'(x) = f'(u)u'(x) \quad \text{or} \quad \frac{d}{dx} f[u(x)] = f'(u)\frac{du}{dx}$$

EXAMPLE 5 Differentiate $y = (3x^4 - 7x + 5)^3$.

Solution $y' = \boxed{3}(3x^4 - 7x + 5)^{\boxed{3-1}}(3x^4 - 7x + 5)'$ **Think:** $y = \square^3$ so $y' = 3\,\square^2 \cdot \frac{d\,\square}{dx}$

This step is usually done mentally. Note that we are thinking of $u(x) = 3x^4 - 7x + 5$. What you write down is:

$$y' = 3(3x^4 - 7x + 5)^2(12x^3 - 7)$$ ■

You could, with a lot of work, have found the derivative in Example 5 without using the chain rule either by expanding the polynomial or by using the product rule. The answer would be the same, but would involve much more algebra. In order to compare these methods, however, consider the following problem with a simpler function.

EXAMPLE 6 Differentiate $y = (3x + 2)^2$.

a. By expansion **b.** By the product rule

c. By the chain rule

Solution **a.** $y = (3x + 2)^2 = 9x^2 + 12x + 4$

$y' = 18x + 12 = 6(3x + 2)$

b. $y = (3x + 2)^2 = (3x + 2)(3x + 2)$

$$\begin{aligned} y' &= (3x + 2)(3x + 2)' + (3x + 2)(3x + 2)' \\ &= (3x + 2)(3) + (3x + 2)(3) \\ &= 6(3x + 2) \end{aligned}$$

c. $y = (3x + 2)^2$

$y' = 2(3x + 2)(3) = 6(3x + 2)$ ■

The chain rule allows us to find the derivative of functions that would otherwise be very difficult to handle. To efficiently use the chain rule, you must recognize when it is required and when it is not.

Chain rule not required	*Chain rule required*	*Function of x*	*Function of u*	*Derivative*
$y = 5x + 1$	$y = (5x + 1)^3$	$u = 5x + 1$	$y = u^3$	$y' = 3(5x + 1)^2(5)$
$y = 3x + 1$	$y = (3x + 1)^5$	$u = 3x + 1$	$y = u^5$	$y' = 5(3x + 1)^4(3)$
$y = 4x + 1$	$y = \sqrt{4x + 1}$	$u = 4x + 1$	$y = \sqrt{u}$	$y' = \frac{1}{2}(4x + 1)^{-1/2}(4)$
$y = 6x + 1$	$y = \frac{1}{6x + 1}$	$u = 6x + 1$	$y = \frac{1}{u}$	$y' = (-1)(6x + 1)^{-2}(6)$
$y = 2 - 7x$	$y = \sqrt[3]{(2 - 7x)^2}$	$u = 2 - 7x$	$y = \sqrt[3]{u^2}$	$y' = \frac{2}{3}(2 - 7x)^{-1/3}(-7)$
$y = \frac{3x + 1}{4x + 1}$	$y = \left(\frac{3x + 1}{4x + 1}\right)^2$	$u = \frac{3x + 1}{4x + 1}$	$y = u^2$	$y' = 2\left(\frac{3x + 1}{4x + 1}\right)\left[\frac{3x + 1}{4x + 1}\right]'$

$$\begin{aligned} &= 2\left(\frac{3x + 1}{4x + 1}\right)\left[\frac{(4x + 1)(3) - (3x + 1)(4)}{(4x + 1)^2}\right] \\ &= 2\left(\frac{3x + 1}{4x + 1}\right)\left[\frac{-1}{(4x + 1)^2}\right] \\ &= \frac{-2(3x + 1)}{(4x + 1)^3} \end{aligned}$$

Do you see the difference between the functions that do not require the chain rule and those that do? The chain rule is required when operating on a composite function.

The previous examples illustrate the most common type of composite function—namely, the one for which

$$y = [u(x)]^n$$

The rule for differentiating such functions is called the **generalized power rule**, and is a special case of the chain rule. In this course, this will be the most common application of the chain rule.

Generalized Power Rule

Let u be some differentiable function of x and $y = [u(x)]^n$, where n is a rational number. Then

$$y' = nu^{n-1}u' \qquad \text{or} \qquad \frac{dy}{dx} = n[u(x)]^{(n-1)}\frac{du}{dx}$$

EXAMPLE 7 Let $y = 15\sqrt{3x^2 + x}$. Find $\frac{dy}{dx}$ and simplify.

Solution

$$y' = 15(\overbrace{3x^2 + x}^{u})^{\overbrace{1/2}^{n}}$$

$$= 15\overbrace{\left(\frac{1}{2}\right)}^{n}\overbrace{(3x^2 + x)^{-1/2}}^{u^{n-1}}\overbrace{(6x + 1)}^{u'}$$

$$= \frac{15(6x + 1)}{2\sqrt{3x^2 + x}}$$

■

You should notice that we simplified in Example 7. Sometimes it is as much of a problem to simplify as it is to use the chain rule and find the derivative. You might wish to review Section 1.4, in particular simplifying by finding the common factor. The next two examples show that the chain rule can be used with the product and quotient rules as well as with the power rule.

EXAMPLE 8 Let $y = x^2(5x + 2)^3$. Find y' and simplify.

Solution This is a product, so begin with the product rule:

$$y' = \overbrace{x^2}^{\text{first}}\overbrace{[(5x + 2)^3]'}^{\text{derivative of second}} + \overbrace{(5x + 2)^3}^{\text{second}}\ \overbrace{[x^2]'}^{\text{derivative of first}}$$

Prime marks indicate that derivatives are yet to be found.

$$= \underbrace{x^2(3)(5x + 2)^2\overbrace{(5)}^{u'}}_{\text{chain rule (generalized power rule)}} + (5x + 2)^3(2x)$$

$$= x(5x + 2)^2[x(3)(5) + 2(5x + 2)]$$ Remove common factors.

$$= x(5x + 2)^2(25x + 4)$$ Simplify expression in brackets.

EXAMPLE 9 Let $y = \dfrac{(5x-3)^4}{-2x}$. Find $\dfrac{dy}{dx}$ and simplify.

Solution This is a quotient, so begin with the quotient rule:

$$\frac{dy}{dx} = \frac{-2x[(5x-3)^4]' - (5x-3)^4[-2x]'}{(-2x)^2}$$

Prime marks indicate that derivatives are yet to be found.

$$= \frac{-2x(4)(5x-3)^3(5) - (5x-3)^4(-2)}{(-2x)^2}$$

Note the factor (5) from the generalized power rule.

$$= \frac{-2(5x-3)^3[20x - (5x-3)]}{(-2x)^2}$$

Common factor in numerator

$$= \frac{-2(5x-3)^3(15x+3)}{4x^2}$$

Simplify algebraically.

$$= \frac{-(5x-3)^3 3(5x+1)}{2x^2}$$

$$= \frac{-3(5x-3)^3(5x+1)}{2x^2}$$

■

EXAMPLE 10 Let $y = [(2x-1)(5x^3+2)]^9$. Find dy/dx.

Solution This is a power, so begin with the power rule:

$$\begin{aligned} y' &= 9[(2x-1)(5x^3+2)]^8[(2x-1)(5x^3+2)]' && \text{Chain rule} \\ &= 9[(2x-1)(5x^3+2)]^8[(2x-1)(15x^2) + (5x^3+2)(2)] && \text{Product rule} \\ &= 9[(2x-1)(5x^3+2)]^8[30x^3 - 15x^2 + 10x^3 + 4] && \text{Simplify bracket.} \\ &= 9[(2x-1)(5x^3+2)]^8[40x^3 - 15x^2 + 4] \\ &= 9(40x^3 - 15x^2 + 4)[(2x-1)(5x^3+2)]^8 \end{aligned}$$

■

EXAMPLE 11 Find the standard form of the equation of the line tangent to the graph of f where $f(x) = \dfrac{24}{3x+2}$ at $x = 2$.

Solution We will use the point–slope form of the equation of a line—namely,

$$\boldsymbol{y - y_1 = m(x - x_1)}$$

First, find (x_1, y_1), which is the point on the curve at $x = 2$:

$$f(2) = \frac{24}{3(2)+2} = 3; \quad \text{so the point is } (2, 3)$$

The slope of the tangent line at a point is the same as the value of the derivative at that point. Write $f(x) = 24(3x+2)^{-1}$.

$$\begin{aligned} f'(x) &= -24(3x+2)^{-2}(3) \\ &= -72(3x+2)^{-2} \end{aligned}$$

Do not forget the factor 3 (because of the chain rule).

Thus,

$$f'(2) = -72[3(2) + 2]^{-2} = \frac{-72}{64} = \frac{-9}{8}$$

This value is the slope, m, in the equation:

$$y - 3 = \frac{-9}{8}(x - 2)$$

$$8y - 24 = -9x + 18$$

$$9x + 8y - 42 = 0$$

■

EXAMPLE 12 Find the point where the tangent line is horizontal for the function defined by $f(x) = \dfrac{x}{(3x + 2)^2}$.

Solution The tangent line will be horizontal when its slope is 0, and the slope of the tangent line is the same as the slope of the curve at a particular point.

$$f'(x) = \frac{(3x + 2)^2(1) - x(2)(3x + 2)(3)}{(3x + 2)^4}$$

$$= \frac{(3x + 2)[(3x + 2) - 6x]}{(3x + 2)^4}$$

$$= \frac{(3x + 2)(2 - 3x)}{(3x + 2)^4}$$

$$= \frac{2 - 3x}{(3x + 2)^3}$$

We need to find the value(s) of x for which

$$\frac{2 - 3x}{(3x + 2)^3} = 0$$

$$2 - 3x = 0 \qquad x \neq -\frac{2}{3}$$

$$x = \frac{2}{3}$$

We need to find the point on the curve for which $x = \frac{2}{3}$:

$$f\left(\frac{2}{3}\right) = \frac{\frac{2}{3}}{\left[3\left(\frac{2}{3}\right) + 2\right]^2}$$

$$= \frac{\frac{2}{3}}{16} = \frac{1}{24}$$

The point is $(\frac{2}{3}, \frac{1}{24})$.

■

10.7 Problem Set

Find the derivatives of the functions in Problems 1–30 and simplify.

1. $f(x) = (3x + 2)^3$
2. $f(x) = (6x + 5)^3$
3. $f(x) = (5x - 1)^4$
4. $f(x) = (4 - 2x)^5$
5. $y = (2x^2 + x)^3$
6. $y = (3x^2 - 2x)^3$
7. $y = (2x^2 - 3x + 2)^2$
8. $y = (3x^2 + 5x - 1)^2$
9. $g(x) = (x^3 + 5x)^4$
10. $g(x) = (2x^3 - 7x^2)^4$
11. $m(x) = (2x^2 - 5x)^{-2}$
12. $m(x) = (6x^2 - 3x)^{-3}$
13. $t(x) = (x^4 + 3x^3)^{-1}$
14. $t(x) = (5x^3 - 2x^2)^{-2}$
15. $y = 5(4x^3 + 3x^2)^3$
16. $y = 6(x^3 + 4x^2)^4$
17. $y = (x^2 - 3x)^{1/4}$
18. $y = (2x^2 + 5x)^{2/3}$
19. $y = \sqrt{x^2 + 16}$
20. $y = \sqrt{x^2 - 25}$
21. $y = 5\sqrt{x^3 + 8}$
22. $y = 9\sqrt{x^4 + 3x}$
23. $y = x\sqrt{3x + 1}$
24. $y = 2x\sqrt{5x - 3}$
25. $y = \sqrt[3]{2x + 5}$
26. $y = \sqrt[3]{7 - 5x}$
27. $y = \dfrac{1}{5x + 3}$
28. $y = \dfrac{1}{3x - 2}$
29. $y = \dfrac{1}{(x^2 + 3)^2}$
30. $y = \dfrac{1}{(x^3 - 2)^3}$

In Problems 31–36 find the standard form of the equation of the line tangent to the graph of f at the indicated value of x.

31. $f(x) = (3x^2 + 2)^2$ at $x = 1$
32. $f(x) = \dfrac{1}{2x - 3}$ at $x = 2$
33. $f(x) = \dfrac{8}{5 - 3x}$ at $x = -1$
34. $f(x) = x^2(2 - x)^3$ at $x = 1$
35. $f(x) = x\sqrt{x - 4}$ at $x = 5$
36. $f(x) = x\sqrt{x + 3}$ at $x = 1$

In Problems 37–42 find the value(s) of x for which the tangent line of the given function is horizontal.

37. $f(x) = x^2(x - 3)^2$
38. $f(x) = x^3(x - 2)^3$
39. $f(x) = \dfrac{x}{(2x + 3)^2}$
40. $f(x) = \dfrac{x - 1}{(x - 2)^2}$
41. $f(x) = \sqrt{x^2 - 5x + 20}$
42. $f(x) = \sqrt{x^2 + 2x + 3}$

In Problems 43–52

a. *Find the "unsimplified" derivative.*

b. *Find the simplified version of the derivative.*

43. $f(x) = (2x + 1)^2(3x + 2)^3$
44. $f(x) = (1 - 3x)^3(4 - x)^2$
45. $y = (5x + 1)^2(4x + 3)^{-1}$
46. $y = (9x + 2)^2(2x - 5)^{-1}$
47. $g(x) = \dfrac{(2x - 5)^2}{5x + 3}$
48. $g(x) = \dfrac{(7x + 11)^3}{x^2 + 3}$
49. $t(x) = \dfrac{(x + 5)^4}{(2x - 5)^2}$
50. $t(x) = \dfrac{(x^2 - 3x + 1)^2}{(5x + 2)^2}$
51. $f(x) = \dfrac{3x}{\sqrt{x^2 + 1}}$
52. $f(x) = \dfrac{-2x}{\sqrt{4 - x^2}}$

APPLICATIONS

53. A colony of bacteria that has been sprayed by a bacterial agent in a controlled situation can be measured in terms of N, the number of viable bacteria remaining after t minutes, according to the formula

$$N(t) = 10^9(50 - 2t)^3$$

What is the rate of decrease at the end of 10 minutes?

54. A psychologist finds that after t minutes $(0 \le t \le 10)$ a person is able to score

$$s(t) = \frac{100t^2}{(2t + 10)^2}$$

on a performance test. Is the person's score increasing or decreasing at time $t = 10$ minutes?

55. If \$10,000 is invested at an annual rate r compounded annually, the amount in the account at the end of 10 years is

$$A = 10{,}000(1 + r)^{10}$$

Find the rate of change of A with respect to r $\left(\text{i.e., find } \dfrac{dA}{dr}\right)$.

56. If the interest in Problem 55 is compounded quarterly (instead of annually), the formula is

$$A = 10{,}000\left(1 + \frac{r}{4}\right)^{40}$$

Find $\dfrac{dA}{dr}$.

57. If you are to receive \$10,000 in 10 years, the present value of this amount, if it is invested at an annual rate r, is given by the formula

$$P = \frac{10{,}000}{(1 + r)^{10}}$$

Find the rate of change of P with respect to r $\left(\text{i.e., find } \dfrac{dP}{dr}\right)$.

58. If the amount in Problem 57 is compounded monthly instead of annually, the formula is

$$P = \frac{10{,}000}{(1 + \frac{r}{12})^{120}}$$

Find $\frac{dP}{dr}$.

59. The registrar of an eastern agricultural college estimates that the total student enrollment t years from now is given by the formula

$$N = 8{,}000 - \frac{4{,}000}{\sqrt{1 + .2t}} \quad (t \geq 0)$$

a. What is the enrollment now?
b. What is the enrollment in 10 years?
c. What is the rate of change in enrollment with respect to time?
d. How fast is the enrollment changing today?
e. How fast is the enrollment changing in 10 years?

60. The psychologist L. L. Thurstone studied learning by asking subjects to memorize a list of n words. One such formula he used to predict learning time, T, was

$$T = 4n\sqrt{n - 3}$$

a. Find the rate of change in time with respect to the length of the list n.
b. Find $\frac{dT}{dn}$ for $n = 20$ and interpret your result.

61. Prove the quotient rule.

*10.8

Review

The material of this chapter is reviewed in the following list of objectives. After each objective there are some practice questions. For a sample test select the first question of each set and check your answers. The second question for each objective has no answer given. If you are having trouble with a particular type of problem, look back at the indicated section in the text. When you are finished reviewing these objectives, a sample examination is given at the end of this section.

[10.1]
Objective 10.1: *Given a function defined by a graph, find a limit as $x \to c$.*

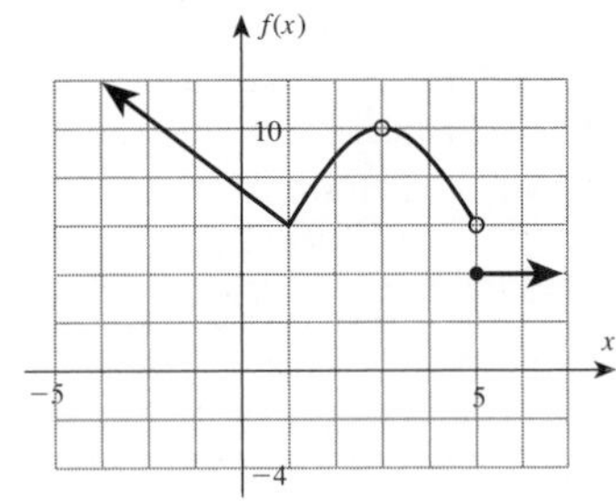

1. $\lim_{x\to 3} f(x)$

2. $\lim_{x\to 1} f(x)$

3. $\lim_{x\to 5} f(x)$

4. $\lim_{x\to \infty} f(x)$

Objective 10.2: *Find limits by using a calculator and filling in values on a table.*

5. $\lim_{x\to 3} \frac{5x + 1}{3 - x}$

6. $\lim_{x\to 0} \frac{5x + 1}{3 - x}$

7. $\lim_{|x|\to \infty} \frac{5x + 1}{3 - x}$

8. $\lim_{|x|\to \infty} \frac{145 - 2{,}000x + 15x^3}{2x^2 - 3x^3}$

Objective 10.3: *Evaluate limits.*

9. $\lim_{x\to 1} \frac{x^2 + 5x - 6}{x - 1}$

10. $\lim_{x\to 1} \frac{x^2 + 7x + 6}{x - 1}$

11. $\lim_{x\to -1} \frac{x^2 + 7x + 6}{x - 1}$

12. $\lim_{x\to \infty} \frac{3x^2 - 5x + 10}{8x^2 + 2x - 5}$

* Optional section.

[10.2]

Objective 10.4: *From a graph, find all suspicious points and tell which of them are points of discontinuity.*

13.

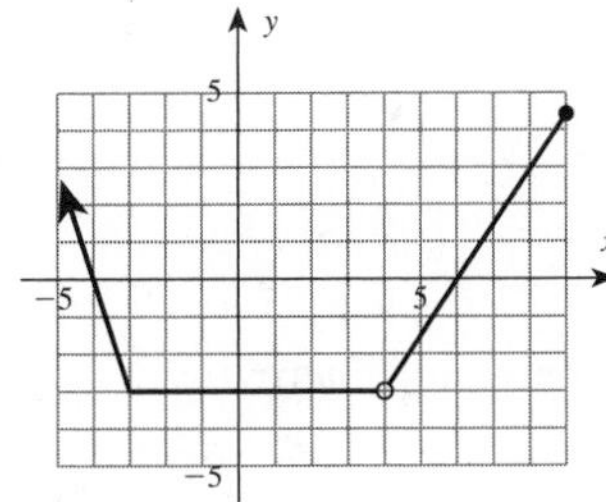

14.

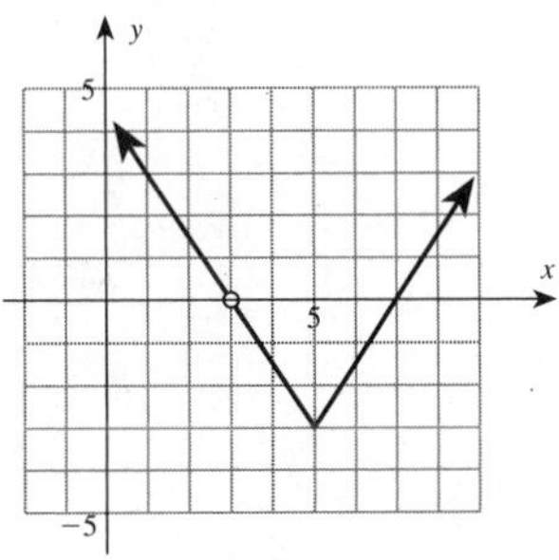

15.

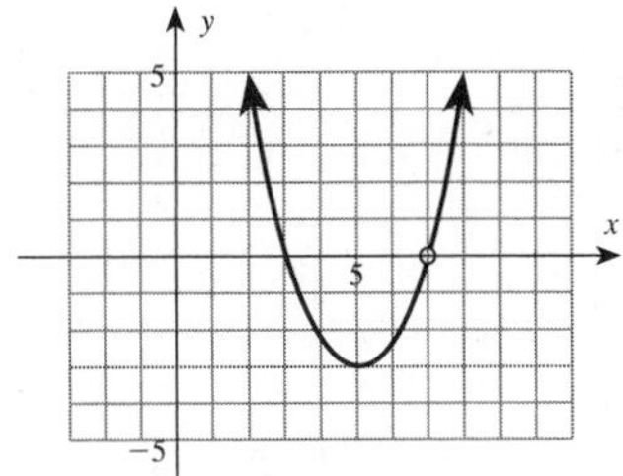

16.

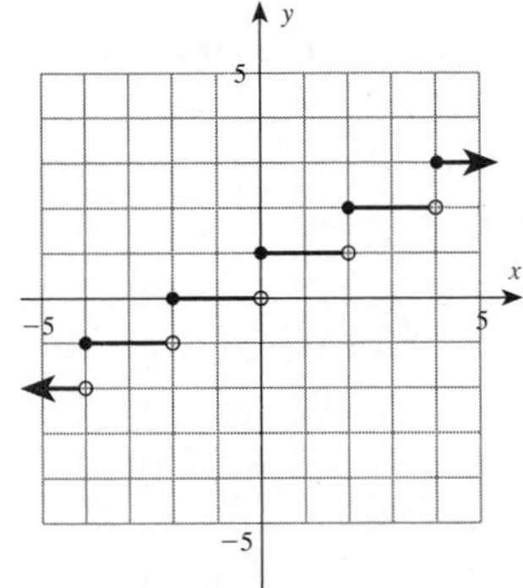

Objective 10.5: *Decide whether a given applied situation describes a continuous or a discontinuous function.*

17. The number of bacteria in a culture as a function of time

18. The distance that a skydiver falls as a function of the time since leaving the aircraft

19. The odometer reading on an automobile as a function of the distance traveled

20. Distance a skydiver has fallen as shown on frames of a motion picture film

Objective 10.6: *Given a function defined over a certain interval, determine whether the function is continuous at all points in the interval. Give the points of discontinuity.*

21. $f(x) = \dfrac{x^2 - 15x + 56}{x - 8}, \quad -5 \le x \le 5$

22. $f(x) = \dfrac{x^2 - 15x + 56}{x - 8}, \quad 0 \le x \le 10$

23. $f(x) = \begin{cases} \dfrac{x^2 - 15x + 56}{x - 8} & 0 \le x \le 10, x \ne 8 \\ 4 & x = 8 \end{cases}$

24. $f(x) = \begin{cases} \dfrac{x^2 - 15x + 56}{x - 8} & 0 \le x \le 10, x \ne 8 \\ 1 & x = 8 \end{cases}$

[10.3]

Objective 10.7: *Find the average rate of change for a given function on some interval.*

25. $y = 3x^2 + 4x$ for $x = 1$ to $x = 3$

26. $y = 3 + 2x - x^2$ for $x = -1$ to $x = 1$

27. $y = \sqrt{3x}$ for $x = 0$ to $x = 6$

28. $y = \dfrac{1}{x} - 5$ for $x = 1$ to $x = 4$

Objective 10.8: *Find the instantaneous rate of change for a given function at some point.*

29. $y = 3x^2 + 4x$ at $x = 1$

30. $y = 3 + 2x - x^2$ at $x = -1$

31. $y = \sqrt{3x}$ at $x = 3$

32. $y = \dfrac{1}{x} - 5$ at $x = 1$

[10.4]

Objective 10.9: *Know the definition of derivative and also use the definition to find a derivative.*

33. In your own words, state the definition of derivative.

34. $y = 3 - 8x^2$

35. $y = \dfrac{1}{x - 5}$

36. $y = \sqrt{x^2 + 1}$

[10.5]
Objective 10.10: *Use the power rule, constant rule, constant times a function rule, sum rule, and difference rule to find the derivative of a given function.*

37. $y = x^{14}$ **38.** $y = x^{-8}$

39. $y = x^{-7/9}$ **40.** $y = 2x$

41. $y = 150$ **42.** $y = -\dfrac{23}{25}$

43. $y = 2x^3 - 5x^2 + 12$

44. $f(x) = 45 - 13x^2 - 5x^3$

[10.6]
Objective 10.11: *Use the product and quotient rules to find derivatives.*

45. $y = (1 - 3x)(2 + 9x)$

46. $y = 5x^3(x^2 - 3x + 9)$

47. $y = \sqrt{x}(x - 1)^{-1}$

48. $y = 5x\sqrt{9 - x}$

49. $y = \dfrac{1}{(3x^2 + 1)}$ **50.** $y = \dfrac{x + 10}{x - 5}$

51. $y = \dfrac{\sqrt{x}}{5x - 3}$ **52.** $y = \dfrac{x}{\sqrt{x + 3}}$

[10.7]
Objective 10.12: *Find the derivative of a function using the generalized power rule.*

53. $f(x) = (5x + 9)^4$ **54.** $f(x) = (4 - 3x)^8$

55. $y = (5x^2 - 3x)^{1/5}$ **56.** $y = \dfrac{(3x + 7)^5}{x^2 - 5}$

SAMPLE TEST

The following sample test (45 minutes) is intended to review the main ideas of this chapter.

Find the limits in Problems 1–4.

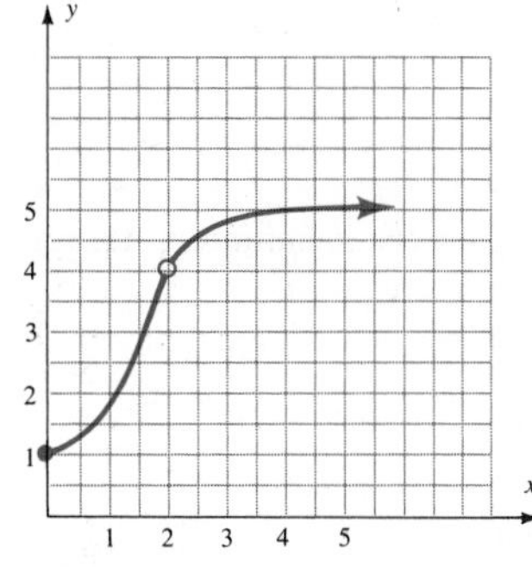

1. $\lim\limits_{x\to 2} f(x)$

2. $\lim\limits_{x\to 2} \dfrac{2 - 5x + 2x^2}{x - 2}$

3. $\lim\limits_{x\to 2} \dfrac{6x + 1}{3 - x}$

4. $\lim\limits_{|x|\to\infty} \dfrac{5x^2 - 3x + 1}{8x^2 - 5x + 4}$

5. Find the point(s) of discontinuity, if any.

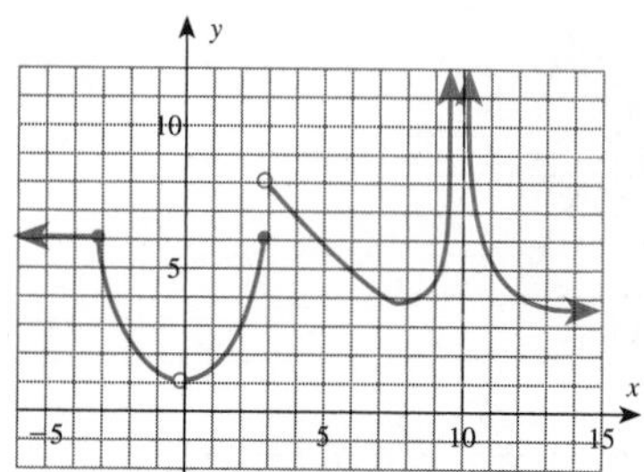

6. Find the point(s) of discontinuity, if any.

$$f(x) = \begin{cases} \dfrac{2x^2 - 7x - 15}{x - 5} & \text{if } -10 \le x \le 10,\ x \ne 5 \\ 13 & \text{if } x = 5 \end{cases}$$

7. Find the average rate of change for the function $y = 5 + x - 2x^2$ for $x = 1$ to $x = 3$.

8. Find the instantaneous rate of change for the function in Problem 7 at $x = 1$.

9. State the definition of derivative.

10. Use the definition of derivative to find the derivative of $y = 1/x$.

Find dy/dx for the functions in Problems 11–18.

11. $y = \dfrac{29}{125}$

12. $y = 3x^2 - 5x + 3$

13. $y = 5x^2 - 2x^{-1}$

14. $y = 2(x - 5)(3x + 2)$

15. $y = \dfrac{15}{4x - 5}$

16. $y = (3x^2 - 5x)^{1/4}$

17. $y = \frac{(x-5)^3}{1-2x}$

18. $y = \frac{5x^3}{\sqrt{x-5}}$

19. Suppose that the profit, P, for a manufacturer is a function of the number of units produced and behaves according to the model

$$P = \frac{1{,}000 - x^2}{100 - x}$$

If the current level of production is 20 units, what is the per unit change in profit if production is increased from 20 to 25 units?

20. Find the marginal profit for the function given in Problem 19.

11

Additional Derivative Topics

CHAPTER OVERVIEW

This chapter enhances and amplifies the idea of the derivative. The discussion should strengthen your understanding of the concept of the derivative by emphasizing that it is a rate of change.

PREVIEW

First, we are introduced to finding a derivative without actually solving for y; this is called *implicit differentiation.* Then meaning is given to dy and dx in the concept called a *differential*. Differentials, derivatives, and implicit differentiation are used in building some business models. The chapter closes by using derivatives to relate rates.

PERSPECTIVE

The concept of a derivative is such an important idea in mathematics that we need time to develop an appreciation of some of the many different places it can be used. This chapter forms a bridge between the definition and manipulation involved in finding a derivative (developed in the last chapter) and the applications of derivatives (introduced in the next chapter).

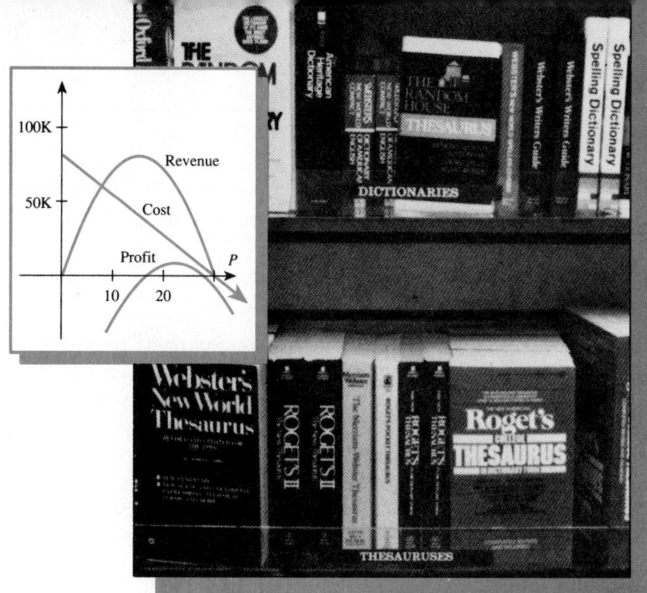

MODELING APPLICATION 9

Publishing: An Economic Model

Karlin Press sells its *World Dictionary* at a list price of \$20 and presently sells 5,000 copies per year. Suppose you are being considered for an editorial position and are asked to do an analysis of the price and sales of this book in order to assess your competency for the position.

After you have finished this chapter, write a paper based on this modeling application. This Modeling Application is continued on page 573.

APPLICATIONS

Management *(Business, Economics, Finance, and Investments)*
Rate of change of price with respect to number of items (11.1, Problems 38–39)
Relationship between sales and advertising cost (11.2, Problem 31)
Average cost (11.2, Problem 32)
Changes in revenue and profit (11.2, Problems 37–38)
Demand equation (11.3, Problem 22)
Revenue, cost, or profit functions (11.3, Problems 23–24, 30, 37–40; 11.5, Problem 25; 11.5, Test Problem 10)
Marginal revenue, cost, or profit (11.3, Problems 25–29, 31, 33–35; 11.5, Problems 25–27)
Average marginal revenue, cost, or profit (11.3, Problems 16–18, 32, 36)
Profit and loss (11.3, Problem 37)
Price elasticity of demand (11.3, Problems 41–44)
Rate of change of profit (11.4, Problems 10–11)
Positioning of a robot arm in an assembly-line conveyor belt (11.4, Problem 13)

Management *(continued)*
Batching process (11.4, Problems 14–15)
Rate of change of wholesale price of apples (11.4, Problem 25)
Maximize monthly revenue from an apartment complex (11.5, Test Problem 9)

Life Sciences *(Biology, Ecology, Health, and Medicine)*
Concentration of alcohol in the bloodstream (11.2, Problem 33)
Average adult pulse rate as a function of a person's height (11.2, Problem 34)
Area of a circular oil slick (11.2, Problem 35)
Area of a dilated pupil (11.2, Problem 36)
Blood velocity (Poiseuille's Law) (11.4, Problems 16–19)
Oil spill (11.4, Problem 22)
The rate of rabies as a function of the skunk population (11.4, Problem 26)
Treatment of a stomach disorder (11.4, Problems 27–28)
The level of carbon monoxide in a city (11.4, Problem 30)

Social Sciences *(Demography, Political science, Population, Psychology, Society, and Sociology)*
Effect of advertising on voting (11.2, Problem 39)
Effect on learning Spanish vocabulary by changing study time (11.2, Problem 40)
Urban sprawl (11.4, Problem 20)

General Interest
Seismological application with concentric circles (11.1, Problem 40)
Distance of one car from another (11.4, Problem 12)
Ripples in a pond (11.4, Problems 23–24)
Height of a weather balloon (11.4, Problem 29)

Modeling Application—
Publishing: An Economic Model

11.1
Implicit Differentiation

Sometimes we are given a function or an equation in terms of two or more variables, say, x and y. In order to use the derivative formulas developed in Chapter 10 it is necessary to **explicitly** solve for y. For example, consider the equation

$$3x^2 - 2x + 5y - 10 = 0$$

Find the slope of this curve at the point $(4, -6)$. To do this, we need to find the derivative. We solve for y:

$$5y = -3x^2 + 2x + 10$$

$$y = -\frac{3}{5}x^2 + \frac{2}{5}x + 2$$

Therefore

$$\frac{dy}{dx} = -\frac{3}{5}(2)x + \frac{2}{5} = -\frac{6}{5}x + \frac{2}{5}$$

The value at the point $(4, -6)$ is found by substitution of the x value ($x = 4$):

$$y' = -\frac{6}{5}(4) + \frac{2}{5} = -\frac{22}{5}$$

Another way to find the derivative is to use the chain rule to find it **implicitly**—that is, without first solving for y. For this example,

$$3x^2 - 2x + 5y - 10 = 0$$

Think of y as a function of x so that

$$D_x x = 1 \qquad \text{and} \qquad D_x y = \frac{dy}{dx}$$

Now differentiate both sides (with respect to x) to obtain

$$D_x(3x^2 - 2x + 5y - 10) = D_x(0)$$

$$\underbrace{D_x(3x^2)}_{6x} + \underbrace{D_x(-2x)}_{(-2)} + \underbrace{D_x(5y)}_{5\frac{dy}{dx}} + \underbrace{D_x(-10)}_{0} = \underbrace{D_x(0)}_{0}$$

Chain rule since y is a function of x

$$6x - 2 + 5\frac{dy}{dx} = 0$$

Solve for $\frac{dy}{dx}$:

$$5\frac{dy}{dx} = -6x + 2$$

$$\frac{dy}{dx} = -\frac{6}{5}x + \frac{2}{5}$$

We can now find the slope at $(4, -6)$ as before. You might ask, why would we want to find the derivative implicitly? Why not always solve for y and then find the derivative? Sometimes it is inconvenient, or even impossible to solve for y, but nevertheless we can find the derivative. The idea is to take the derivative of both sides of the equation, thus treating y as a function of x and using the chain rule. Remember, the derivative of x with respect to x is 1, but the derivative of y with respect to x is $\frac{dy}{dx}$.

EXAMPLE 1 Find $\frac{dy}{dx}$ for $x^2 + y^2 = 25$ both implicitly and explicitly and find the equation of the tangent line at $(3, 4)$.

Solution First, implicitly (without solving for y):

$$D_x(x^2 + y^2) = D_x(25)$$ Take the derivative of both sides with respect to x.

$$D_x(x^2) + D_x(y^2) = D_x(25)$$ Derivative of a sum

$$2x + 2y\frac{dy}{dx} = 0$$ Derivative of x^2 is $2x$.

Derivative of y^2 is $2y\frac{dy}{dx}$ by the chain rule (since we are assuming that y is some, yet unfound, function of x).

$$2y\frac{dy}{dx} = -2x$$ Subtract $2x$ from both sides.

$$\frac{dy}{dx} = -\frac{x}{y}$$ Divide both sides by $2y$, $y \neq 0$.

Next, find the derivative explicitly (that is, solve for y first):

$$y^2 = 25 - x^2$$ Subtract x^2 from both sides.

$$y = \sqrt{25 - x^2}$$ Square root property; select positive value for y since y is positive for the given point, $(3, 4)$.

$$\frac{dy}{dx} = \frac{1}{2}(25 - x^2)^{-1/2}(-2x)$$ Do not forget the chain rule when finding the derivative.

$$= -\frac{x}{\sqrt{25 - x^2}}$$ Notice this is the same as the answer from the first part since $y = \sqrt{25 - x^2}$.

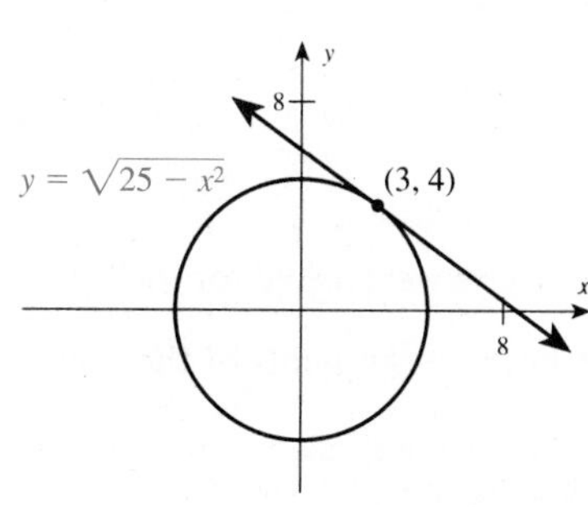

Finally, the equation of the tangent line at $(3, 4)$ is found by finding the slope at $(3, 4)$ using the derivative. The graph is shown in the margin.

$$\frac{dy}{dx} = -\frac{x}{y} = -\frac{3}{4}$$ Evaluate at $(3, 4)$.

Use the point–slope form of the equation of a line, namely

$$y - y_1 = m(x - x_1)$$

or

$$y - 4 = -\frac{3}{4}(x - 3)$$

In standard form this is

$$3x + 4y - 25 = 0$$ ■

Notice from Example 1 that *y is not necessarily a function of x*. When working with implicit differentiation it is often the case for a given value of x that there may be more than one value for y, which violates the definition of a function. We get around this difficulty by saying that if a segment (part) of a graph can be represented by a differentiable function in some vicinity of a point (x, y), then dy/dx will have meaning in that vicinity.*

We now summarize the procedure for implicit differentiation.

Implicit Differentiation

Given an equation involving x and y, where y is assumed to be a differentiable function of x *in some vicinity* of a point (x, y), we can find dy/dx as follows:

1. Take the derivative of both sides with respect to x.
2. Solve for dy/dx:
   a. Collect all terms involving dy/dx on the left side and all terms not involving dy/dx on the right side.
   b. Factor dy/dx from the left side.
   c. Solve for dy/dx by dividing both sides of the equation by the left-hand factor that does not contain dy/dx.

EXAMPLE 2 Find $\frac{dy}{dx}$ where $x^5 - y^5 = 211$.

Solution

$$D_x(x^5 - y^5) = D_x(211)$$

1. Take the derivative of both sides with respect to x.

$$5x^4 - 5y^4\frac{dy}{dx} = 0$$

Do not forget chain rule for $D_x(y^5)$.

$$-5y^4\frac{dy}{dx} = -5x^4$$

2. Solve for $\frac{dy}{dx}$.

$$\frac{dy}{dx} = \frac{-5x^4}{-5y^4}$$

$$= \frac{x^4}{y^4}$$ ■

Note that $\frac{dy}{dx}$ is stated in terms of both x and y, so if you were asked for the slope of the curve at some point you would need to know both components of the point,

* This is, of course, not a definition. The idea of a *vicinity*, or as it is sometimes called, a *neighborhood*, of a point requires a rather precise definition, but the intuitive idea should be clear enough.

not just the x value as when we found $\frac{dy}{dx}$ explicitly. In Example 2 the slope of $x^5 - y^5 = 211$ at $(3, 2)$ is

$$\frac{dy}{dx} = \frac{x^4}{y^4} = \frac{81}{16}$$

WARNING Remember that all the usual derivative procedures apply when doing implicit differentiation, especially when using the product rule. Consider Example 3 carefully.

EXAMPLE 3 If $x^2y^3 = 1$, find $\frac{dy}{dx}$.

Solution Use implicit differentiation; do not forget to use the product rule:

$$D_x(\underbrace{x^2y^3}) = D_x(1)$$

$$\overbrace{x^2D_x(y^3) + y^3D_x(x^2)}^{\text{Product rule}} = 0$$

$$x^2\left(3y^2\frac{dy}{dx}\right) + y^3(2x) = 0$$

$$3x^2y^2\frac{dy}{dx} = -2xy^3$$

$$\frac{dy}{dx} = -\frac{2xy^3}{3x^2y^2} = -\frac{2y}{3x}$$

■

EXAMPLE 4 Find the line tangent to $x^2 + xy + y^2 - 7 = 0$ at $(1, 2)$.

Solution First find the slope of the curve at $(1, 2)$, which means find the derivative at that point. Carry out implicit differentiation:

$$2x + x\frac{dy}{dx} + y + 2y\frac{dy}{dx} + 0 = 0$$

$$x\frac{dy}{dx} + 2y\frac{dy}{dx} = -2x - y$$

$$\frac{dy}{dx}(x + 2y) = -2x - y$$

$$\frac{dy}{dx} = \frac{-2x - y}{x + 2y}$$

At $(1, 2)$:

$$\frac{dy}{dx} = \frac{-2(1) - 2}{1 + 2(2)} = \frac{-4}{5}$$

Finally, the equation of the tangent line is found by using the point–slope form:

$$y - y_1 = m(x - x_1)$$

$$y - 2 = \frac{-4}{5}(x - 1)$$

$$5y - 10 = -4x + 4$$

$$4x + 5y - 14 = 0$$

■

11.1 Problem Set

Suppose that x and y are related by the equations in Problems 1–10. Find dy/dx using both implicit and explicit differentiation.

1. $y + 5x^2 + 12 = 0$ **2.** $3x^3 - y + 15 = 0$

3. $xy = 5$ **4.** $5xy = x^2$

5. $x^2 - 3xy = 50$ **6.** $10x = xy$

7. $x^2 + y^2 = 4, y > 0$ **8.** $x^3 - y^3 = 9$

9. $2x^4 - 5y^3 = 7$ **10.** $x^5y^3 = -1$

Suppose that x and y are related by the equations in Problems 11–20 and use implicit differentiation to find dy/dx.

11. $x^2 - xy + y^2 = 1$

12. $4x^2 - 3xy + 9y^2 = 4$

13. $(2x + 1)^2 + (3y - 5)^2 = 41$

14. $(5x - 2)^2 - (2y + 1)^2 = 30$

15. $\dfrac{x^2 - y^2}{2x + y^2} = 10$

16. $\dfrac{3x^2 + y^2}{x + y} = 100$

17. $3x^2y^3 - 3xy^2 + 5xy = 2$

18. $5x^5y^2 - 5xy^2 - 8x = 100$

19. $x^3 + 2x^2 + xy - 4 = 0$

20. $x^3 - 3x^2 + 2xy + 3 = 0$

Find the standard form of the equation of the line tangent to the curves in Problems 21–31 at the indicated point.

21. $x^2 + y^2 - 4 = 0$ at $(0, 2)$

22. $x^3 - y^3 - 9 = 0$ at $(2, -1)$

23. $2x^4 - 5y^3 - 7 = 0$ at $(-1, -1)$

24. $(2x + 1)^2 + (3y - 5)^2 = 41$ at $(2, 3)$

25. $x^2 + xy + y^2 - 7 = 0$ at $(1, -3)$

26. $x^2 + xy + y^2 - 12 = 0$ at $(2, -4)$

27. $(x - 2)^2 + (y - 1)^2 = 9$ at $x = 2$

28. $(x - 3)^2 + (y - 1)^2 = 16$ at $x = 3$

29. $x^2 - 2xy + y^2 = 0$ at $x = -1$

30. $\dfrac{(x - 1)^2}{9} + \dfrac{(y + 1)^2}{16} = 1$ at $x = 1$

31. $\dfrac{(x + 1)^2}{25} + \dfrac{(y - 1)^2}{4} = 1$ at $x = -1$

APPLICATIONS

Many graphs are famous enough to have been named for a variety of different applications. Problems 32–37 are examples of such curves. Find the slope of the tangent line at the indicated point.

32. Circle: $x^2 + y^2 = 5x + 4y$

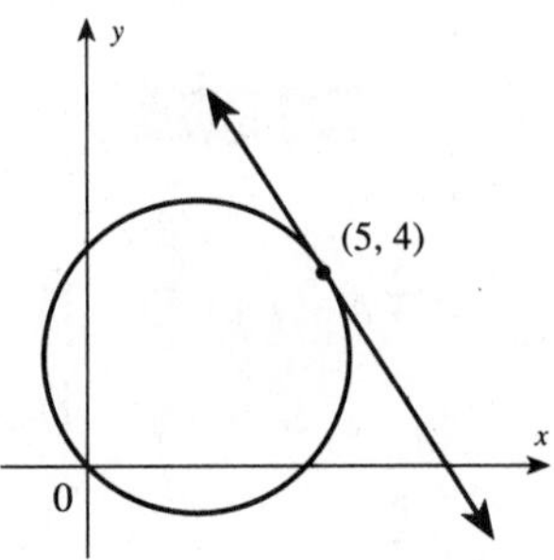

33. Semicubical parabola: $y^2 = 4x^3$

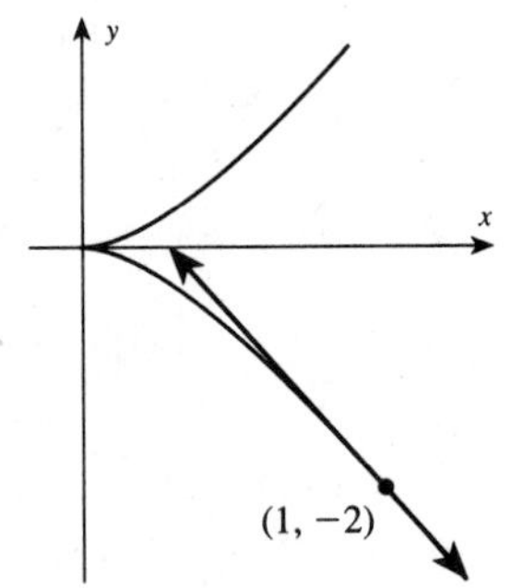

34. Bifolium: $(x^2 + y^2)^2 = 4x^2y$

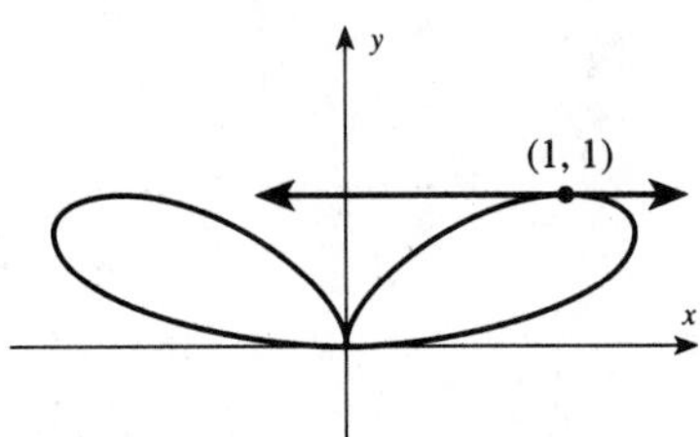

35. Folium of Descartes: $2(x^3 + y^3) = 9xy$

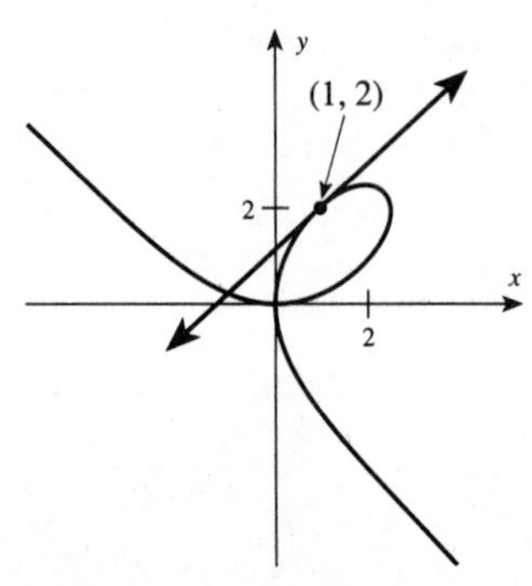

36. Two-leaved rose: $8(x^2 + y^2)^2 = 100(x^2 - y^2)$

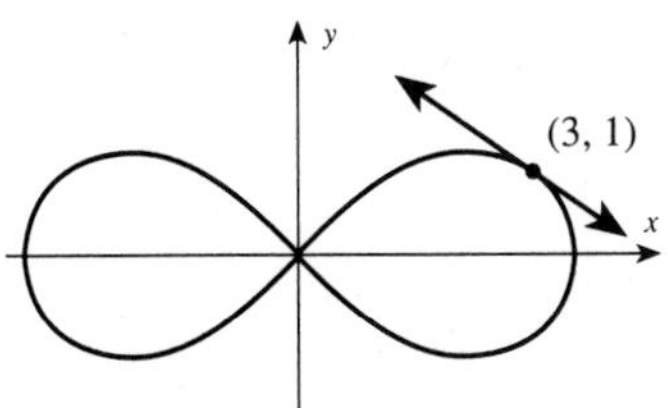

37. Lemniscate of Bernoulli: $12(x^2 + y^2)^2 = 625xy$
(Give slope to nearest tenth.)

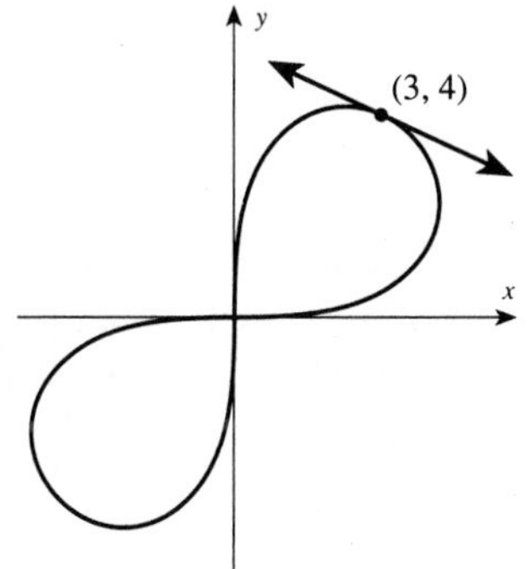

38. Find the rate of change of the price, p, in dollars, with respect to the number of items, x, if

$$x = \sqrt{5{,}000 - p^2}$$

39. Find the rate of change of the price, p, in dollars, with respect to the number of items, x, if

$$x = p^2 - 5p + 500$$

40. Seismologists sometimes use equations of circles around seismological stations. Show that the circles with the equations

$$x^2 + y^2 - 12x - 6y + 25 = 0 \quad \text{and}$$
$$x^2 + y^2 + 2x + y - 10 = 0$$

have tangent lines with the same slope at $x = 2$.

11.2 Differentials

Suppose we consider dx and dy from the derivative dy/dx as two separate quantities. In Chapter 2 we used the symbols Δx and Δy to mean the change in x and the change in y, as follows. Let (x_1, y_1) and (x_2, y_2) be any two points on some curve: then

$$\Delta x = x_2 - x_1 \qquad \text{and} \qquad \Delta y = y_2 - y_1$$

We *define* dx, called the **differential of x**, to be an independent variable equal to the change in x. That is, we define dx to be Δx. Then, if f is differentiable at x, we *define* dy, called the **differential of y**, according to the formula

$$dy = f'(x)\,dx$$

Thus, if $dx \neq 0$, then

$$\frac{dy}{dx} = f'(x)$$

There is a very clear geometrical representation of what we have done here, shown in Figure 11.1 on page 552. Note that $dx = \Delta x$ (the change in x), $\Delta y =$ the change in y that occurs for a change of Δx, and $dy =$ the rise of the tangent line relative to $\Delta x = dx$.

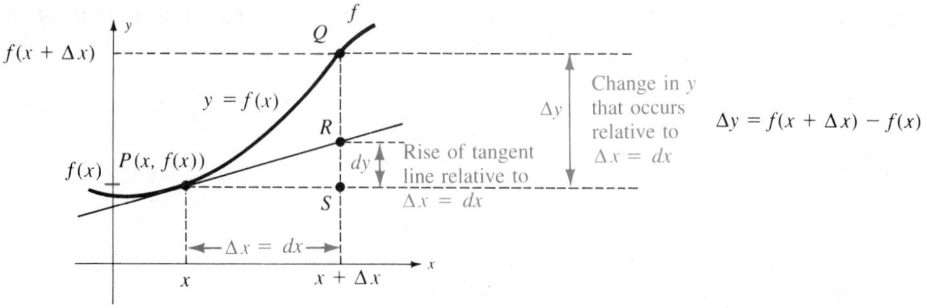

$f(x + \Delta x)$
$y = f(x)$
Q
f
R
$P(x, f(x))$
$f(x)$
S
dy
Rise of tangent line relative to $\Delta x = dx$
Δy
Change in y that occurs relative to $\Delta x = dx$
$\Delta y = f(x + \Delta x) - f(x)$
$\Delta x = dx$
x
$x + \Delta x$

FIGURE 11.1
Geometrical definition of dx and dy

EXAMPLE 1 Use Figure 11.1 to describe Δx, Δy, dx, and dy.

Solution Since $\Delta x = dx$, we see that these are the length of the segment PS; these quantities represent the change in x. The number Δy represents the change in y, which is the length of the segment QS. Finally, the distance dy is defined to be the rise of the tangent line at the point P, so dy is the length of the segment RS. ■

EXAMPLE 2 Find Δy and dy for $f(x) = 6x - 2x^2$ when $x = 2$ and $\Delta x = .1$.

Solution For this example, $y = f(x)$, so

$$x_1 = 2 \qquad \text{and} \qquad x_2 = x_1 + \Delta x = 2 + .1 = 2.1$$
$$y_1 = f(x_1) = f(2) = 6(2) - 2(2)^2 = 4$$
$$y_2 = f(x_2) = f(2.1) = 6(2.1) - 2(2.1)^2 = 3.78$$

Now

$$\begin{aligned} dy &= f'(x)\,dx \\ &= (6 - 4x)(.1) \qquad \text{Since } f'(x) = 6 - 4x \text{ and } \Delta x = .1 \end{aligned}$$

When $x = 2$,

$$dy = [6 - 4(2)](.1) = -.2$$

To find Δy we know that

$$\Delta x = x_2 - x_1 \qquad \text{and} \qquad x_2 = x_1 + \Delta x$$

so that

$$\begin{aligned} \Delta y &= f(x_2) - f(x_1) \\ &= f(x_1 + \Delta x) - f(x_1) \qquad \text{Substitute.} \\ &= f(2 + .1) - f(2) \\ &= f(2.1) - f(2) \\ &= 3.78 - 4 \\ &= -.22 \end{aligned}$$

■

*Approximations Using Differentials**

Note in Example 2 that the numerical values for Δy and dy are almost the same. Since it is generally easier to calculate dy than Δy, it is often convenient to approximate a change in y by using dy. Example 3 illustrates this idea.

EXAMPLE 3 Suppose the demand for a certain product is a function of its price, x, and can be specified according to the formula

$$p(x) = -3x^3 - 2x^2 + 1{,}000$$

How would the demand change as the price changes from \$2 to \$2.10?

Solution The question asks for Δy when $x = 2$ and $\Delta x = .10$. Since

$$\Delta y = p(x + \Delta x) - p(x)$$

you want to find

$$\begin{aligned} \Delta y &= p(2.10) - p(2) \\ &= [-3(2.10)^3 - 2(2.10)^2 + 1{,}000] - [-3(2)^3 - 2(2)^2 + 1{,}000] \\ &= 963.397 - 968 \qquad \text{By calculator} \\ &= -4.603 \end{aligned}$$

■

This means that you would expect the demand to go down about 5 units. Instead of doing this calculation, however, you might approximate Δy by using dy:

$$\begin{aligned} dy &= p'(x)\,dx \\ &= (-9x^2 - 4x)\,dx \end{aligned}$$

Now substitute the values $x = 2$ and $\Delta x = .10$:

$$\begin{aligned} dy &= [-9(2)^2 - 4(2)](.10) \\ &= -4.4 \end{aligned}$$

You would estimate that the demand would decrease by four units. ■

In Example 3 the error introduced by using dy instead of Δy is one unit, but there is a lot less "work" involved. Closer examination shows that the error is only about .203. Example 4 shows how we can estimate the amount of error introduced in an approximation.

* Many books introduce approximations using differentials with problems such as approximating $\sqrt{26}$. To do this, let $f(x) = \sqrt{x}$ and let $x = 25$, $\Delta x = 1$ so

$$\begin{aligned} f(x + \Delta x) &= f(x) + \Delta y & \qquad f(26) = f(25 + 1) &= f(25) + \Delta y \\ &\approx f(x) + dy & &\approx f(25) + dy \\ &= f(x) + f'(x)\,dx & &= f(25) + f'(25)\,dx \end{aligned}$$

Since the functions $f(25)$ and $f'(25)$ are easy to evaluate, and since $dx = \Delta x = 1$, we see that differentials can be used to approximate $\sqrt{26}$. However, with the widespread use of calculators this does not seem like a good motivational problem for using differentials, so we will concentrate on other important applications of the differential.

EXAMPLE 4 What is the error introduced by using dy instead of Δy in Example 3 for an increase in price from 2 to $2 + h$ dollars?

Solution For an increase from 2 to $2 + h$,

$$\begin{aligned}\Delta y &= p(x + \Delta x) - p(x)\\ &= [-3(2 + h)^3 - 2(2 + h)^2 + 1{,}000] - [-3(2)^3 - 2(2)^2 + 1{,}000]\end{aligned}$$

This is a formidable calculation (the details are left for you). After simplification we find that

$$\Delta y = -44h - 20h^2 - 3h^3$$

However, if we approximate this by dy, we see that

$$\begin{aligned}dy &= p'(x)\,dx\\ &= (-9x^2 - 4x)\,dx\end{aligned}$$

At $x = 2$ and $dx = h$,

$$dy = [-9(2)^2 - 4(2)]h = -44h$$

By comparing these two calculations we see that the error introduced by using dy rather than Δy is

$$-20h^2 - 3h^3$$

If h is small (such as .1), we see that the error will be small ($-.203$ in this case). ■

These examples lead us to the following summary:

Differential Approximation

If $f'(x)$ exists, then, for small Δx,

$$\Delta y \approx dy$$

and

$$\begin{aligned}f(x + \Delta x) &= f(x) + \Delta y\\ &\approx f(x) + dy\\ &= f(x) + f'(x)\,dx\end{aligned}$$

EXAMPLE 5 Colortex manufactures pigments for tinting various paints. Its profit function is given by

$$P(x) = 10x - \frac{x^2}{1{,}000} - 3{,}000$$

where x is the number of tubes of pigment produced. What is the expected change in profit if production is changed from 2,000 to 2,010 tubes of pigment?

Solution The change in profit is ΔP and the change in production is Δx. The change in production, $\Delta x = 10$, is given, and since it is relatively small we will approximate

ΔP by dP:

$$\begin{aligned} dP &= P'(x)\,dx \\ &= \left(10 - \frac{2x}{1{,}000}\right)dx \\ &= (10 - .002x)\,dx \end{aligned}$$

Since $\Delta x = dx = 10$ at a production level of 2,000, we have

$$dP = [10 - .002(2{,}000)](10) = 60$$

This means that we would expect the profit to increase by \$60 if production were increased by 10 tubes of pigment. ■

Differential Formulas

Each of the derivative rules derived in the last chapter has a corresponding differential form. For example, the generalized power rule states that

$$\text{If} \qquad y = u^n \qquad \text{then} \qquad \frac{dy}{dx} = nu^{n-1} \cdot \frac{du}{dx}$$

By multiplying both sides by dx we obtain a differential form of this same derivative formula:

$$\text{If} \qquad y = u^n \qquad \text{then} \qquad dy = nu^{n-1} \cdot du$$

All of the derivative rules previously discussed are now summarized in differential form:

Differential Rules

If u and v are differentiable functions and c is a constant, then

CONSTANT RULE:	$dc = 0$
POWER RULE:	$du^n = nu^{n-1} \cdot du$
SUM RULE:	$d(u + v) = du + dv$
DIFFERENCE RULE:	$d(u - v) = du - dv$
PRODUCT RULE:	$d(uv) = u\,dv + v\,du$
QUOTIENT RULE:	$d\left(\frac{u}{v}\right) = \frac{v\,du - u\,dv}{v^2}$

EXAMPLE 6 **a.** If $y = \sqrt{x}$, then $\frac{dy}{dx} = \frac{1}{2}x^{-1/2}$, so, in differential form,

$$dy = \frac{1}{2\sqrt{x}}\,dx$$

b. If $u = x^2 + 1$, then $\frac{du}{dx} = 2x$, so, in differential form,

$$du = 2x\,dx$$

■

11.2

Problem Set

Find dy for the functions in Problems 1–20.

1. $y = 5x^3$

2. $y = 35x^4$

3. $y = 10x^{-1}$

4. $y = 100x^{-2}$

5. $y = 100x^3 - 50x + 10$

6. $y = 500x^5 - 40x^2 + 5{,}000$

7. $y = 3\sqrt{x - 1}$

8. $y = 5\sqrt{x^2 + 4}$

9. $y = 5x$

10. $y = 6x$

11. $y = (5x - 3)^2(x^2 - 3)$

12. $y = (2x^2 + 1)(3x - 1)^3$

13. $y = \dfrac{3x + 1}{x - 2}$

14. $y = \dfrac{5x + 2}{4x + 3}$

15. $y = x^2\left(1 - \dfrac{1}{x} + \dfrac{3}{x^2}\right)$

16. $y = x^3\left(1 - \dfrac{5}{x} + \dfrac{10}{x^2}\right)$

17. $y = \left(5 - \dfrac{1}{x^2}\right)\left(1 + \dfrac{1}{x}\right)$

18. $y = \left(6 - \dfrac{10}{x^3}\right)\left(5x - \dfrac{1}{x}\right)$

19. $y = \dfrac{x^2 - 5x + 1}{x^2 + 3x - 2}$

20. $y = \dfrac{2x^2 + 3x - 5}{3x^2 - 2x + 1}$

Evaluate dy and Δy for the values in Problems 21–24.

21. $y = f(x) = x^2 - 2x + 5, \quad x = 10, \Delta x = .1$

22. $y = f(x) = 2x^2 - 3x, \quad x = 5, \Delta x = .2$

23. $y = f(x) = \sqrt{5x}, \quad x = 5, \Delta x = .15$

24. $y = f(x) = \sqrt{3x - 2}, \quad x = 100, \Delta x = 2$

Estimate Δy by using dy in Problems 25–30.

25. $y = \dfrac{2x - 3}{5x + 2}, \quad x = 100, \Delta x = 3$

26. $y = \dfrac{x^2 + 1}{x - 3}, \quad x = 30, \Delta x = 1$

27. $y = 20\left(3 - \dfrac{1}{x^2}\right), \quad x = 10, \Delta x = .02$

28. $y = \dfrac{1 + 3x}{\sqrt{x}}, \quad x = 20, \Delta x = .1$

29. $y = \dfrac{450(x + 300)}{\sqrt{x - 50}}, \quad x = 1{,}000, \Delta x = 10$

30. $y = \dfrac{1{,}000(20 - 30x)}{\sqrt{x^2 + 20}}, \quad x = 50, \Delta x = 5$

Use differential approximations in Problems 31–40.

APPLICATIONS

31. Suppose that sales, S, can be expressed as a function of advertising cost (c, in thousands of dollars) according to the formula

$$S(c) = 500c - c^2$$

Estimate the increase in sales that will result by increasing the advertising budget from $100,000 to $110,000.

32. The average cost (in dollars) to manufacture x items is

$$A(x) = .05x^3 + .1x^2 + .5x + 10$$

Approximate the change in the average cost as x changes from 10 to 11.

33. The concentration of alcohol in the bloodstream x hours after drinking 1 ounce of alcohol is approximately

$$A(x) = \frac{3x}{100 + x^2}$$

Estimate the change in concentration as x changes from 3 to 3.5.

34. The average adult pulse rate, b, in beats per minute can be expressed as a function of the person's height, h (in inches), according to the formula

$$b = \frac{600}{\sqrt{h}}$$

Approximate the change in pulse rate for a change in height from 72 to 73.5 inches.

35. Find the approximate increase in the area of a circular oil slick as its radius increases from 2 to 2.1 miles.

36. The pupil of a patient's eye is nearly circular and will dilate when the patient is given a certain drug. Estimate the increase in the area of a patient's pupil if the radius increases from 4 to 4.1 millimeters.

37. A company manufactures and sells x items per day. If the cost and revenue equations (in dollars) are

$$C(x) = 400 + 30x \quad \text{and} \quad R(x) = 60x$$

find the approximate changes in revenue and profit if production is increased from 100 to 110 items.

38. A company produces and sells x items per day (in hundreds). The cost and revenue equations (in thousands of dollars) are given on the domain $[0, 35]$:

$$C(x) = 3x^2 - 40x + 200 \quad \text{and} \quad R(x) = 500x - \frac{x^2}{10}$$

Find the approximate changes in revenue and profit if production is increased from 10 to 11 items.

39. The number of people, N, expected to vote in the next election is a function of number of hours, x, of television advertising according to the formula

$$N(x) = 25{,}000 + \frac{x^2}{10} - \frac{x^3}{50{,}000}$$

Find the approximate change in N as x changes from 1,000 to 1,100 hours.

40. A student learns y Spanish vocabulary words in x hours according to the formula

$$y = 50\sqrt{x}$$

What is the approximate increase in the number of words learned as x changes from 3 to 3.5 hours?

11.3 Business Models Using Differentiation

Marginal Analysis

The application of derivatives, that is, rates of change, to cost, revenue, and profit is an important component in the decision-making process for business executives and managers. The word *marginal* is used in business and economics to mean rate of change. Some of these ideas have appeared in examples in previous parts of this book, but for completeness and easy reference they are summarized below:

Marginal Analysis

Suppose that x is the number of units sold in some time interval at a price of p dollars and that

$$C(x) = \text{TOTAL COST} \quad \text{and} \quad R(x) = \text{TOTAL REVENUE}$$
$$= (\text{price per item})(\text{number sold})$$
$$= px$$

then

$$\left.\begin{aligned} P(x) &= R(x) - C(x) = \text{TOTAL PROFIT} \\ p(x) &= \text{PRICE (or DEMAND) function} \end{aligned}\right\}$$

Note that capital P and lowercase p denote different functions.

$$C'(x) = \text{MARGINAL COST}$$
$$R'(x) = \text{MARGINAL REVENUE}$$
$$P'(x) = \text{MARGINAL PROFIT}$$

Marginal can be related to the definition of *derivative* as follows:

Marginal cost, also called the *marginal propensity*, is the rate of change in cost per unit change in production at a given output level. This derivative approximates the *extra cost* of producing one additional unit.

Marginal revenue is the rate of change in revenue per unit change in production at a given output level. This derivative approximates the *extra* revenue for selling one additional unit.

Marginal profit is the rate of change of profit per unit change in production at a given output level. This derivative approximates the *extra profit* for selling one additional unit.

In practice, the smallest level of change in production is one unit. This was denoted in the last section by $\Delta x = dx = 1$. The actual changes in cost, revenue, and profit are denoted by ΔC, and ΔR, and ΔP, respectively, and are approximated by dC, dR, and dP. Recall from the last section that

$$dC = C'(x)\,dx$$
$$dR = R'(x)\,dx$$
$$dP = P'(x)\,dx$$

EXAMPLE 1 Suppose it is known that the total cost (in dollars) for a product is given by the equation

$$C(x) = 18{,}500 + 8.45x$$

where x is the number of items sold. Find the marginal cost.

Solution The marginal cost is $C'(x)$:

$$C'(x) = 8.45$$

This means that it costs an additional \$8.45 to produce one more item at all production levels. ■

The marginal cost is not always a constant. If

$$C(x) = 500 + .045x^2$$

then $C'(x) = .09x$, which is a function of x. This means that the marginal cost depends on the production level of x units. For example, if present production is 1,000 items and if we want to know the cost of producing one additional item, then $x = 1{,}000$, $\Delta x = 1$, and

$$\begin{aligned} dC &= C'(x)\,dx \\ &= .09(1{,}000)(1) \qquad \text{Since } C'(x) = .09x \\ &= 90 \end{aligned}$$

This means that to increase production by one unit would lead to an approximate increase in total cost of \$90.

EXAMPLE 2 Suppose that the market demand for the product in Example 1 is linear and that it has been found that at a selling price of \$15 a company will sell 7,500 items per month, but at a \$25 selling price, sales would drop to 7,000 items per month. Graph the demand function. Also, find the marginal profit.

Solution The first step is to decide on the variables. Apparently, the variables are the price, p, and the number of items per month, x. In mathematics it is customary to put the independent variable (which in this problem is p) on the horizontal axis, and the dependent variable (in this problem it is x) on the vertical axis. This is exactly the opposite of the way that economists do it. In 1890, a book by Alfred Marshall, *Principles of Economics*, the classic that is one of the foundation stones of modern price theory, made the break with mathematics and put the dependent variable, x, on the horizontal axis.* Also note that an economist would have called

* It is unfortunate that mathematicians and economists disagree about how to draw the graphs shown in Figure 11.2. Many mathematics professors will want to put p on the horizontal axis, but the simple fact is that Marshall's scheme is now used by everybody, although mathematicians must wonder about the odd ways of economists.

the number of items n instead of x, but by calling the number of items x instead of n, we are set to draw these graphs the same way that they are drawn in economics and at the same time make them more agreeable to mathematicians by using x for the horizontal axis.

DEMAND: It is linear and passes through the points (7500, 15) and (7000, 25). Use the slope–intercept form: $y - y_1 = m(x - x_1)$, or in terms of the variables of this problem:

$$p - p_1 = m(x - x_1) \quad \text{for ordered pairs } (x, p)$$

First, find m: $m = \dfrac{y_2 - y_1}{x_2 - x_1}$ Slope formula

$= \dfrac{p_2 - p_1}{x_2 - x_1}$ Slope formula for the variables in this problem

$= \dfrac{25 - 15}{7{,}000 - 7{,}500} = \dfrac{10}{-500} = -.02$

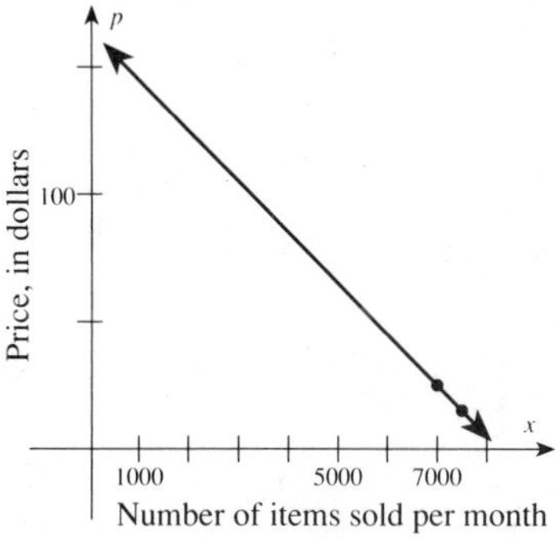

Thus, the equation is

$p - 15 = -.02(x - 7{,}500)$ Substitute (7500, 15) for (x_1, p_1).

$p - 15 = -.02x + 150$

$p = -.02x + 165$

Price, in dollars

Number of items sold per month

FIGURE 11.2

We can draw the demand function as shown in Figure 11.2.

$$\text{TOTAL REVENUE} = xp = x(-.02x + 165) = -.02x^2 + 165x$$

$\text{TOTAL COST} = 18{,}500 + 8.45x$ From Example 1

so

$$\text{TOTAL PROFIT} = \text{TOTAL REVENUE} - \text{TOTAL COST}$$

$$P(x) = (-.02x^2 + 165x) - (18{,}500 + 8.45x)$$

$$= -.02x^2 + 156.55x - 18{,}500$$

The marginal profit is

$$P'(x) = -.04x + 156.55$$

The marginal profit is the rate of change of profit with respect to changes in the number of items. But what does this really mean, and how could it be used? Consider Table 11.1 (page 560) which gives some values for prices from \$100 to \$80.

Notice that as price decreases, the production increases. This makes sense because the higher the price, the less the demand. This production number is determined from the price according to the formula $x = -50p + 8{,}250$.* The profit is calculated from the profit function $P(x) = -.02x^2 + 156.55x - 18{,}500$. Finally take a close look at the profit and marginal profit columns. The maximum profit occurs when the product is priced at \$87. Also notice that as long as the marginal profit was positive the profit was increasing, and when the marginal profit became

* Solve $p = -.02x + 165$ for x.

TABLE 11.1
Comparison of Prices, Production Levels, Profit, and Marginal Profit

p Price	x Production	$P(x)$ Profit	$P'(x)$ Marginal profit
100	3,250	$279,037.50	26.55
99	3,300	$280,315.00	24.55
98	3,350	$281,492.50	22.55
97	3,400	$282,570.00	20.55
96	3,450	$283,547.50	18.55
95	3,500	$284,425.00	16.55
94	3,550	$285,202.50	14.55
93	3,600	$285,880.00	12.55
92	3,650	$286,457.50	10.55
91	3,700	$286,935.00	8.55
90	3,750	$287,312.50	6.55
89	3,800	$287,590.00	4.55
88	3,850	$287,767.50	2.55
87	3,900	$287,845.00	.55
86	3,950	$287,822.50	−1.45
85	4,000	$287,700.00	−3.45
84	4,050	$287,477.50	−5.45
83	4,100	$287,155.00	−7.45
82	4,150	$286,732.50	−9.45
81	4,200	$286,210.00	−11.45
80	4,250	$285,587.50	−13.45

negative, the profit started to decline. It looks like the "turning point" is when the marginal profit is 0. We will show in the next chapter that the maximum value for profit occurs when the marginal profit is zero. ■

EXAMPLE 3 Find the break-even points for the model described in Examples 1–2. Draw graphs for the revenue and the cost. Label the portions of the graph showing both profit and loss, and interpret Table 11.1 in terms of the graph.

Solution The break-even point (from Chapter 2) is when the revenue and cost are equal. Thus

$$-.02x^2 + 165x = 18{,}500 + 8.45x$$

$$.02x^2 - 156.55x + 18{,}500 = 0$$

$$x = \frac{156.55 \pm \sqrt{(-156.55)^2 - 4(.02)(18{,}500)}}{2(.02)}$$

$$\approx \frac{156.55 \pm 151.75}{.04}$$

$$\approx 120 \text{ and } 7{,}707$$

Substitute $x = 120$ into either the revenue or cost equation:

$$18{,}500 + 8.45(120) = 19{,}514; \text{ the point is } (120, 19514)$$

$$18{,}500 + 8.45(7{,}707) = 83{,}628; \text{ the point is } (7707, 83628)$$

The graphs are shown in Figure 11.3.

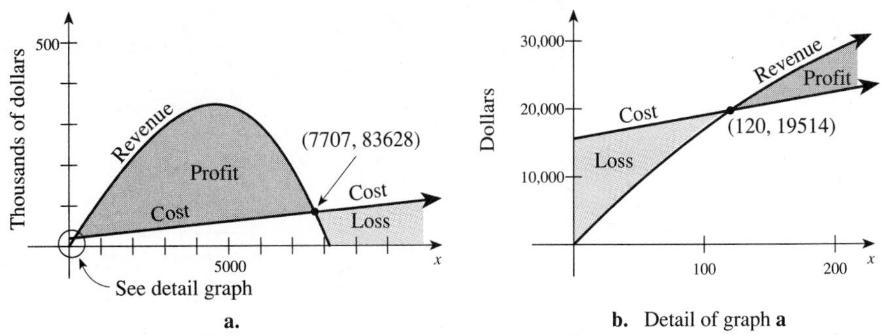

FIGURE 11.3
Cost, revenue, and profit functions

Notice from Figure 11.3 that the profit is positive for $120 < x < 7{,}707$. The maximum profit (from Table 11.1) occurs when $p = 87$ and the production is 3,900. This seems reasonable when we look at Figure 11.3**a** since the largest profit is the part with the biggest bulge. If fewer than 120 or more than 7,707 items are produced, then the profit is negative; this is called a *loss*. ■

Average Marginal Analysis

Sometimes marginal analysis is carried out relative to the average cost, average revenue, or average profit. *Average* here means *cost per unit, revenue per unit,* or *profit per unit.* The average of a number x is denoted by $\bar{x}$.

Average Marginal Analysis

If x is the number of units of a product produced in some time interval, then

$$\bar{C}(x) = \frac{C(x)}{x} = \text{cost per unit} = \text{AVERAGE TOTAL COST}$$

$$\bar{C}'(x) = \text{MARGINAL AVERAGE COST}$$

$$\bar{R}(x) = \frac{R(x)}{x} = \text{revenue per unit} = \text{AVERAGE TOTAL REVENUE}$$

$$\bar{R}'(x) = \text{MARGINAL AVERAGE REVENUE}$$

$$\bar{P}(x) = \frac{P(x)}{x} = \text{profit per unit} = \text{AVERAGE TOTAL PROFIT}$$

$$\bar{P}'(x) = \text{MARGINAL AVERAGE PROFIT}$$

EXAMPLE 4 Find the marginal average cost, marginal average revenue, and marginal average profit for the functions

$$C(x) = 18{,}500 + 8.45x$$
$$R(x) = -.02x^2 + 165x$$
$$P(x) = -.02x^2 + 173.45x - 18{,}500$$

Solution $\bar{C}(x) = \dfrac{C(x)}{x} = \dfrac{18{,}500 + 8.45x}{x} = 18{,}500x^{-1} + 8.45$ This is the average cost or cost per unit.

Marginal average cost: $\bar{C}'(x) = -18{,}500x^{-2}$

$\bar{R}(x) = \dfrac{R(x)}{x} = \dfrac{-.02x^2 + 165x}{x} = -.02x + 165$ This is the average revenue, or revenue per unit.

Marginal average revenue: $\bar{R}'(x) = -.02$

$\bar{P}(x) = \dfrac{P(x)}{x} = \dfrac{-.02x^2 + 173.45x - 18{,}500}{x} = -.02x + 173.45 - 18{,}500x^{-1}$ This is the average profit, or profit per unit.

Marginal average profit: $\bar{P}'(x) = -.02 + 18{,}500\,x^{-2}$

Note that $\bar{P}'(x) = \bar{R}'(x) - \bar{C}'(x)$. ■

11.3 Problem Set

The cost to produce x items is

$$C(x) = 200 + 6x - x^2 + x^3$$

Find the marginal cost for the given value of x in Problems 1–4.

1. $x = 1$ **2.** $x = 0$

3. $x = 4$ **4.** $x = 2$

The revenue function for a product is

$$R(x) = 9x - .001x^2$$

Find the marginal revenue for the given value of x in Problems 5–8.

5. $x = 0$ **6.** $x = 10$

7. $x = 50$ **8.** $x = 100$

The profit function for a product is

$$P(x) = x^3 - 8x^2 + 2x + 50$$

Find the marginal profit for the given value of x in Problems 9–12.

9. $x = 2$ **10.** $x = 1$

11. $x = 4$ **12.** $x = 6$

Find the marginal average cost for the functions given in Problems 13–15.

13. $C(x) = 200 + 6x - x^2 + x^3$

14. $C(x) = .25x^2 + 45x + 225$

15. $C(x) = 5{,}000 + .4x^2$

Find the marginal average revenue for the functions given in Problems 16–18.

16. $R(x) = 9x - .001x^2$ **17.** $R(x) = 50x - .5x^2$

18. $R(x) = 10x - .01x^2$

Find the marginal average profit for the functions given in Problems 19–21.

19. $P(x) = x^3 - 8x^2 + 2x + 50$

20. $P(x) = 5x - .05x^2$

21. $P(x) = x^3 - 50x^2 + 5x + 200$

APPLICATIONS

In Problems 22–27 suppose that the sales of a company are presently 10,000 items per year and that at a list price of \$15 the company will sell 15,000 items but at a list price of \$25 sales will drop to 5,000 items. Use (x, p) where x is the number of items and p is the price.

22. What is the demand equation (assuming it is linear)?

23. What is the revenue function?

24. What is the cost function in terms of price if

$$C(x) = 15{,}000 + 8.5x$$

where x is the total number of items sold?

25. Find and interpret the marginal revenue function.

26. Find and interpret the marginal cost function.

27. Find and interpret the marginal profit function.

Suppose the demand equation for a product is

$$p = 200 - .04x$$

Use (x, p) where x is the total number of items produced and p is the price. Also suppose the cost equation is

$$C(x) = 50{,}000 + 50x$$

Use this information in Problems 28–37.

28. What is the marginal cost in terms of x?

29. What is the marginal average cost in terms of x?

30. What is the revenue equation in terms of x?

31. What is the marginal revenue in terms of x?

32. What is the marginal average revenue in terms of x?

33. Find $R'(1{,}000)$ and $R'(2{,}500)$ and interpret.

34. What is the marginal profit in terms of x?

35. Find $P'(1{,}000)$ and $P'(2{,}500)$ and interpret.

36. What is the marginal average profit in terms of x?

37. Graph the cost and revenue functions on the same coordinate system and show the regions of profit and loss.

38. The total cost to produce x items is

$$C(x) = 1{,}000 + 5x - x^2 + x^3$$

Find the marginal cost.

39. The total revenue for x items is

$$R(x) = 5x - .001x^2$$

Find the marginal revenue.

40. The profit in dollars from the sale of x items is

$$P(x) = x^3 - 3x^2 + 5x + 10$$

Find the marginal profit.

The laws of supply and demand can be used to predict the direction of changes in price and quantity in response to various shifts in supply and demand. However, it is often not enough to know whether quantity rises or falls in response to a change in price; it is also important to know by how much. The relative responsiveness of consumers to a change in the price of an item is called the ***price elasticity of demand****. If $p(x)$ is a differentiable demand function, then the price elasticity of demand is denoted by the Greek letter η (eta) and is defined by the following equation:*

$$\eta = \frac{\textbf{Percentage change in quantity demanded}}{\textbf{Rate of change in price}} = \frac{\frac{p}{x}}{\frac{dp}{dx}}$$

For a given price, if $|\eta| < 1$, the demand is ***inelastic****, and if $|\eta| > 1$, the demand is* ***elastic****. Determine whether the demand functions in Problems 41–44 are elastic, inelastic, or neither at the indicated x-value.*

41. $p(x) = 500 - 4x$; where $x = 50$

42. $p(x) = 10 - .005x$; where $x = 5{,}000$

43. $p(x) = 100 - .5x^2$; where $x = 20$

44. $p(x) = 500(x + 1)^{-1}$; where $x = 24$

*11.4
Related Rates

The concept of derivative has been defined and interpreted. We now use the idea that the derivative represents a rate of change of one variable with respect to another variable. That is, we focus on the fact that

$$\frac{dy}{dx} = \text{the rate of change of } y \text{ with respect to } x$$

$$\frac{dP}{dx} = \text{the rate of change of } P \text{ with respect to } x$$

$$\frac{dz}{dt} = \text{the rate of change of } z \text{ with respect to } t$$

and so on.

* Optional section.

If a formula is given as an equation, we know that we often solve that equation for an unknown value. However, sometimes the information we have at hand is the *rate* at which the variables are changing. We now explore how to use a given formula to find unknown quantities when we *know* the rates of change. For example, if the profit P is given by the equation

$$P = 75x - \frac{x^3}{50{,}000} - 100{,}000$$

for sales of a total of x items, and if you know the rate at which the profit is changing *with respect to time*, you can transform this profit formula into one involving rates by using implicit differentiation and taking the derivative of both sides with respect to time:

$$\begin{aligned}\frac{dP}{dt} &= \frac{d}{dt}\left(75x - \frac{x^3}{50{,}000} - 100{,}000\right)\\ &= 75\frac{dx}{dt} - \frac{3}{50{,}000}x^2\frac{dx}{dt}\end{aligned}$$

Notice that this formula now relates

$$\frac{dP}{dt} = \text{the rate of change of } P \text{ with respect to time } t$$

$$\frac{dx}{dt} = \text{the rate of change of } x \text{ with respect to time } t$$

Since the formula relates rates, the problem is called a **related rate** problem. When working a related rate problem you must distinguish between the general situation and the specific situation. The **general situation** comprises properties that are true at *every* instant of time, while the **specific situation** refers to those properties that are true only at the *particular* instant of time that the problem investigates. When working related rate problems you should first formulate a model that incorporates all the facts that are true at every instant of time (the general situation). Then substitute into the equation from the general situation the facts that represent the particular instant under investigation (the specific situation). Finally, solve the equation from the specific situation for the desired unknown. This process is illustrated by the following examples.

EXAMPLE 1 The total profit, P, is related to the number of items produced, x, by the formula

$$P = 75x - \frac{x^3}{50{,}000} - 100{,}000$$

Suppose that production is increasing by 100 units per week. How fast is the profit increasing when production is at 1,000 items?

Solution *The general situation:* The formula for the general situation is given. We take the derivative of both sides of the given equation with respect to t in order to transform the equation into one on related rates:

$$\frac{dP}{dt} = 75\frac{dx}{dt} - \frac{3}{50{,}000}x^2\frac{dx}{dt}$$

This equation is true for every instant of time in this problem.

The specific situation: At the instant under consideration in this problem (namely, when production is at 1,000 items), we are given that production is increasing at 100 units per week. That is, we are given $dx/dt = 100$. Thus, when production is 1,000 items, we have

$$\begin{aligned}\frac{dP}{dt} &= 75(100) - \frac{3}{50{,}000}(1{,}000^2)(100)\\ &= 1{,}500\end{aligned}$$

This means that the profit is increasing at $1,500 per week when production is 1,000 items and when production is increasing at 100 items per week. ■

EXAMPLE 2 Two conveyor belts are moving away from an assembly point at right angles to one another, as shown in Figure 11.4. If two items simultaneously leave the assembly table, how fast is the distance between them changing after 2 minutes?

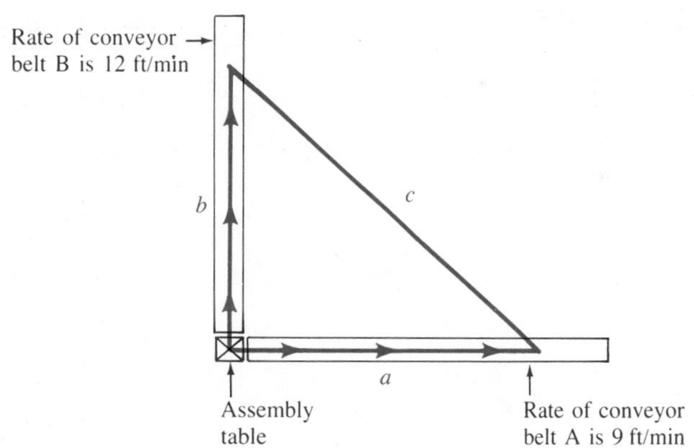

FIGURE 11.4
Assembly point configuration

Solution *The general situation:* Since a, b, and c in Figure 11.4 represent the sides of a right triangle, they are related by the formula

$$c^2 = a^2 + b^2$$

Convert this formula into one involving rates by taking the derivative of both sides with respect to time t:

$$2c\frac{dc}{dt} = 2a\frac{da}{dt} + 2b\frac{db}{dt}$$

The specific situation: At the instant under consideration (namely, at $t = 2$ minutes), you know that $da/dt = 9$ and $db/dt = 12$. Find dc/dt when $t = 2$. When $t = 2$, $a = 18$, $b = 24$, and

$$c^2 = 18^2 + 24^2$$
$$c = 30$$

Substitute these values into the formula relating the rates:

$$2(30)\frac{dc}{dt} = 2(18)(9) + 2(24)(12)$$
$$\frac{dc}{dt} = 15$$

The items are moving apart at 15 feet per minute. ■

EXAMPLE 3 A cylindrical vat with dimensions given in Figure 11.5 is being filled at the rate of 1,000 cubic feet per minute. How fast is the surface rising when the depth is 12 feet?

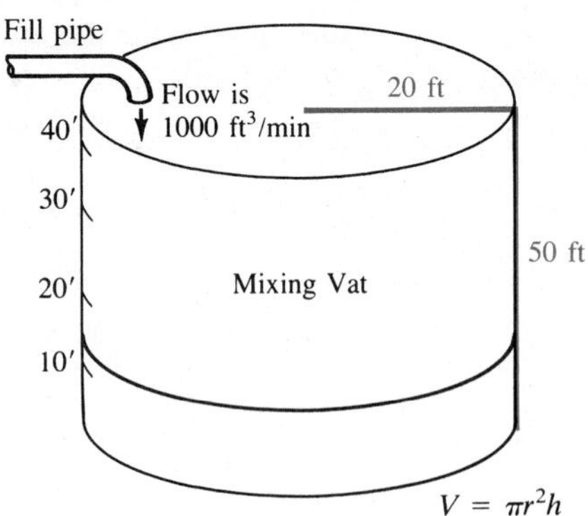

FIGURE 11.5
Mixing vat in an industrial process

Solution *The general situation:* The necessary relationship from Figure 11.5 requires a geometric formula from your previous courses. This is the formula for the volume of a cylinder: $V = \pi r^2 h$. In this problem, $r = 20$ feet (constant), so that the formula we will use is $V = 400\pi h$. Convert this to a formula involving rates:

$$\frac{dV}{dt} = \frac{d}{dt}(400\pi h)$$
$$= 400\pi\frac{dh}{dt}$$

The specific situation: The instant under investigation is when the depth is 12 feet. At this instant, $dV/dt = 1{,}000$ and you want to find dh/dt. Substitute these values into the formula from the general situation:

$$1{,}000 = 400\pi \frac{dh}{dt}$$

$$\frac{dh}{dt} = \frac{1{,}000}{400\pi} \approx .796$$

That is, the height is rising at .796 feet per minute (this is about $9\frac{1}{2}$ inches per minute). ■

The procedure for solving related rate problems is summarized in the box.

Procedure for Solving Related Rate Problems

THE GENERAL SITUATION

1. Draw a figure if appropriate. Use letters to describe the data in the problem. Since rates involve variables and not constants, be careful not to label a quantity with a number unless it *never* changes in the problem.
2. Find a formula relating the variables. (The appropriate formula for most of the problems in this section is given.)
3. Differentiate the equation: usually implicitly and usually with respect to time.

THE SPECIFIC SITUATION

4. List the known quantities; list as unknown the quantity you want to find. If there are other variables or quantities in the problem use the given formula to eliminate those "extra" variables. Substitute in all the values in the formula. The only remaining variable should be the unknown. Solve for the unknown.

EXAMPLE 4 Illustrate steps 3 and 4 of the above procedure to find and interpret dy/dt where $x^3 + 5y^2 = 84$ and $dx/dt = 10$ where $x = 4$. Assume that both x and y are positive.

Solution *The general situation* (*step* 3): You are asked to find the rate at which y is changing with respect to time at *that particular instant* when $x = 4$ if you know that x is changing at a rate of 10 units per unit of time. This means that $dx/dt = 10$ is part of the general situation (it does not change throughout the problem). On the other hand, $x = 4$ and dy/dt are part of the specific situation (this is the instant with which we are concerned). Differentiate the formula implicitly:

$$3x^2\frac{dx}{dt} + 10y\frac{dy}{dt} = 0$$

Known: $dx/dt = 10$ To find: dy/dt

The specific situation (*step* 4): The formula still involves both an x and a y, but we are interested in finding dy/dt at the instant when $x = 4$ (x is not a constant in this problem), so we use the *given* formula to find y *at that instant*:

$$x^3 + 5y^2 = 84$$
$$4^3 + 5y^2 = 84$$
$$5y^2 = 20$$
$$y^2 = 4$$
$$y = 2, -2 \qquad \text{Reject } y = -2 \text{ since } y \text{ must be positive.}$$

Now substitute all values into the formula; the only unknown value should be dy/dt:

$$3x^2\frac{dx}{dt} + 10y\frac{dy}{dt} = 0$$
$$3(4)^2(10) + 10(2)\frac{dy}{dt} = 0$$
$$\frac{dy}{dt} = -24$$

The result says that y is decreasing at the rate of 24 units per unit of time. ■

EXAMPLE 5 The blood in a blood vessel flows faster toward the center of the blood vessel and slower toward the outside of the blood vessel. This flow is described by a formula called *Poiseuille's law*:

$$V = \frac{p}{4Lv}(R^2 - r^2)$$

where V is the velocity of the blood, R is the radius of the blood vessel, r is the distance of the blood from the center of the blood vessel, and p, L, and v are constants related to the blood pressure, length of the blood vessel, and the viscosity of the blood vessel. If a person goes from a warm house into a cold winter night the person's blood vessels will contract at a rate of

$$\frac{dR}{dt} = -.0025 \text{ mm/min}$$

at a place where $R = .01$. Also assume that $r = .005$, $p = 100$, $L = 1$ mm, and $v = .05$ are constants. Find the rate of change of velocity with respect to time at the location where $R = .01$.

Solution *The general situation:* The necessary relationship is given with Poiseuille's law, which is (with the necessary constants)

$$V = \frac{100}{4(1)(.05)}(R^2 - .005^2)$$

Convert this to a formula involving rates:

$$\frac{dV}{dt} = 500\left(2R \cdot \frac{dR}{dt}\right) = 1{,}000R\frac{dR}{dt}$$

The specific situation: The instant under investigation is when $R = .01$ and $\frac{dR}{dt} = -.0025$ is

$$\frac{dV}{dt} = 1{,}000(.01)(-.0025) = -.025$$

The velocity is decreasing at the rate of .025 mm/min. ■

11.4 Problem Set

In Problems 1–9 find the indicated rate, given the other information. Assume that both x and y are positive.

1. Find dy/dt where $x^2 + y^2 = 25$ and $dx/dt = 4$ when $x = 3$.
2. Find dx/dt where $x^2 + y^2 = 25$ and $dy/dt = 2$ when $x = 4$.
3. Find dy/dt where $5x^2 - y = 100$ and $dx/dt = 10$ when $x = 10$.
4. Find dx/dt where $4x^2 - y = 100$ and $dy/dt = -6$ when $x = 1$.
5. Find dx/dt where $y = 2\sqrt{x} - 9$ and $dy/dt = 5$ when $x = 9$.
6. Find dy/dt where $y = 5\sqrt{x + 9}$ and $dx/dt = 2$ when $x = 7$.
7. Find dy/dt where $xy = 10$ and $dx/dt = -2$ when $x = 5$.
8. Find dy/dt where $5xy^2 = 10$ and $dx/dt = -2$ when $x = 1$.
9. Find dx/dt where $x^2 + xy - y^2 = 20$ and $dy/dt = 5$ when $x = 4$ and $y > 0$.

APPLICATIONS

10. The profit, P, in dollars, is related to the number of items produced, x, by the formula

$$P = 125x - \frac{x^2}{200} - 500 \qquad \text{where } 0 \leq x \leq 1{,}000$$

Suppose that production is increasing at five units per week. How fast is the profit changing when production is at 200 items per week?

11. Suppose that production in Problem 10 is decreasing by one unit per week. How fast is the profit changing when production is at 200 items?
12. Two cars start driving from the same point. One goes north at a rate of 30 mph and the other west at 40 mph. How fast is the distance between them changing after 5 hours?
13. A product is moving along a conveyor belt at the rate of 3 feet per second. A robot arm is suspended 8 feet above the belt. At what rate is the distance between the product and the robot arm changing when the product is 9 feet from the base of the arm?

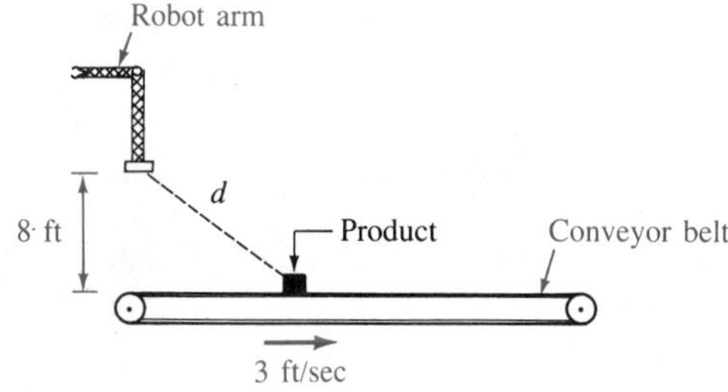

14. How fast is the surface of the vat described in Example 3 rising when the depth is 20 feet?
15. How fast is the surface of the vat described in Example 3 rising if it is being filled at a rate of 500 cubic feet per minute and the other details remain the same?
16. Find the rate of change of blood velocity with respect to time in Example 5 at the instant where $R = .02$.
17. Find the rate of change of blood velocity with respect to time in Example 5 if

$$\frac{dR}{dt} = -.0015 \text{ mm/min}$$

and all the other details of the example remain the same.

18. A person with heart problems needs to take a nitroglycerin tablet which dilates the blood vessels. Suppose the constants are the same as in Example 5 and the nitroglycerin dilates the blood vessel at the rate of

$$\frac{dR}{dt} = .003 \text{ mm/min}$$

What is the rate of change of blood velocity at the instant where $R = .01$?

19. A person with heart problems needs to take a nitroglycerin tablet which dilates the blood vessels. Suppose the constants are the same as in Example 5 and the nitroglycerin

dilates the blood vessel at the rate of

$$\frac{dR}{dt} = .003 \text{ mm/min}$$

What is the rate of change of blood velocity at the instant where $R = .02$?

20. Urban sprawl for a certain city is increasing in a circular manner in such a way that the radius r is increasing at the rate of 3 miles per year. At the moment where $r = 25$ miles, how fast is the area increasing?

21. A supplier services a circular area in such a way that its radius r is increasing at the rate of 1.5 miles per year. How fast is the area increasing at the instant when $r =$ 4 miles?

22. Assume that oil is spilled from a ruptured tanker and forms a circular oil slick whose radius R is increasing at a constant rate of one-half a foot per minute ($dR/dt = .5$). How fast is the area of the spill increasing when the radius of the spill is 100 feet?

23. A pebble is thrown into a still pond and the result is a circular ripple. If the radius of this circle is expanding at 1 foot per second, how fast is the area changing when the radius is 10 feet?

24. How fast is the circumference of the circular ripple in Problem 23 changing?

25. The wholesale price of apples is d dollars per ton, and the daily supply x is related to the price by the formula

$$d = \frac{-5x}{x + 1{,}000} + 70$$

Suppose that there are 2,000 tons available today and that the supply is decreasing at 200 tons per day. At what rate is the price changing?

26. The number of cases of rabies, r, in an area is related to the number of skunks, s, according to the formula

$$r = .0005s^{2/3}$$

A program to eliminate the skunk population is instituted and it is estimated that 200 skunks per day are destroyed and that the present skunk population is 5,000. At what rate is the number of cases of rabies changing?

27. A certain medical procedure requires that a spherical balloon be inserted into the stomach and then inflated. If the radius of the balloon is increasing at the rate of .3 centimeter per minute, how fast is the volume changing when the radius is 4 centimeters? (Use $V = \frac{4}{3}\pi r^3$.)

28. How fast is the surface area of the sphere in Problem 27 increasing? (Use $S = 4\pi r^2$.)

29. A weather balloon is rising vertically at the rate of 10 feet per second. An observer is standing on the ground 300 feet from the point where the balloon was released. At what rate is the distance between the observer and the balloon changing when the balloon is 400 feet high?

30. The level of carbon monoxide (in parts per million, ppm) in a city can be predicted by considering it as a function of the number of registered automobiles in that city according to the formula

$$P = 1 + .2x + .001x^2$$

If the number of automobiles is increasing at 4,000 per year, how is the level of carbon monoxide changing when the city has exactly 25,000 cars?

*11.5 Review

The material of this chapter is reviewed in the following list of objectives. After each objective there are some practice questions. For a sample test select the first question of each set and check your answers. The second question for each objective has no answer given. If you are having trouble with a particular type of problem, look back at the indicated section in the text. When you are finished reviewing these objectives, a sample examination is given at the end of this section.

[11.1]

Objective 11.1: *Find the derivative of y with respect to x implicitly.*

1. $x^5 + 2y^2 + y + 10 = 0$
2. $3x^8y^4 = 1$
3. $(x - 3)^2 - (y + 1)^2 = 1$
4. $x^3 - 3y^2 + 40 = 0$

Objective 11.2: *Find the slope of a tangent line at a given point.*

5. A semicubical parabola

$$y^2 = 8x^3 \text{ at } (2, 8)$$

* Optional section.

6. A semicubical parabola

$$y = x^{2/3} \text{ at } (8, 4)$$

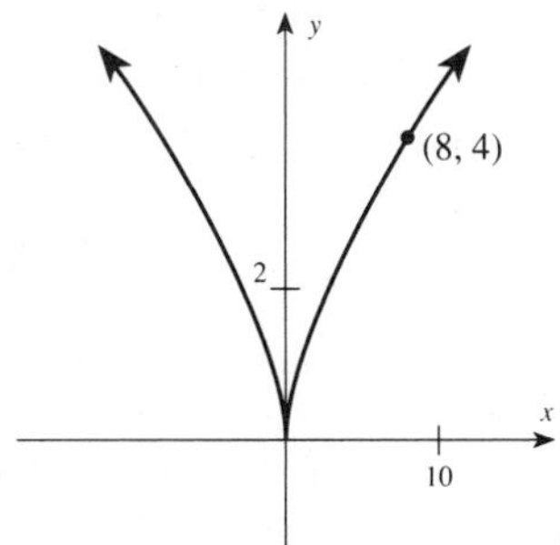

7. $y = 5x^2(6x - 5)^3$ at $x = 1$
8. $x^3 - 3y^2 + 40 = 0$ at $(2, 4)$

Objective 11.3: *Find the standard form equation of a line tangent to a given curve at a given point.*

9. A semicubical parabola

$$y^2 = 8x^3 \text{ at } (2, 8)$$

10. A semicubical parabola

$$y = x^{2/3} \text{ at } (8, 4)$$

11. $u = 5x^2(6x - 5)^3$ at $x = 1$
12. $x^3 - 3y^2 + 40 = 0$ at $(2, 4)$

[11.2]
Objective 11.4: *Find dy for the given functions.*

13. $y = 6\sqrt{3x^2 + 5}$
14. $y = \dfrac{2x + 1}{x - 3}$
15. $y = \dfrac{2x^2 + x - 1}{3x^2 + x - 4}$
16. $y = \left(1 - \dfrac{1}{x^2}\right)\left(2 - \dfrac{1}{x}\right)$

Objective 11.5: *Find dy and Δy for the indicated values.*

17. $y = f(x) = 5x^2, x = 10, \Delta x = .1$
18. $y = f(x) = 2x^2 + 5x - 3, x = 5, \Delta x = .1$
19. $y = f(x) = \sqrt{6x}, x = 6, \Delta x = .2$
20. $y = f(x) = \dfrac{x + 1}{x - 5}, x = 1, \Delta x = .01$

Objective 11.6: *Estimate Δy by using dy.*

21. $y = 5\left(10 - \dfrac{1}{x^3}\right), x = 10, \Delta x = .03$
22. $y = \dfrac{1 - 2x}{\sqrt{x}}, x = 50, \Delta x = .1$
23. $y = \dfrac{x^2 - 1}{x + 2}, x = 20, \Delta x = 1$
24. $y = \dfrac{50(10 - 8x)}{\sqrt{x^2 + 10}}, x = 40, \Delta x = 2$

[11.3]
Objective 11.7: *Solve applied problems based on the preceding objectives. For specific examples of the types of applications look at the list of applications in this chapter on page 545.*

25. **a.** *Profit function.* Suppose that the profit, P, for a manufacturer is a function of the number of units produced and behaves according to the model

$$P(x) = \frac{200 - x^3}{10 - x^2}$$

If the current level of production is 10 units, what is the per unit increase in profit if production is increased from 10 to 15 units?

b. *Marginal profit.* Find the marginal profit for the function given in part **a**.

26. *Cost function.* Suppose that the cost, C, in dollars for producing x items is given by the formula

$$C(x) = 20x^3(5x - 100)^2$$

Find the average rate of change of cost as x increases from 50 to 75 units.

27. *Marginal cost.* Find the marginal cost for the function given in Problem 26.
28. *Rate of change.* An object moving in a straight line travels d centimeters in t minutes according to the formula $d(t) = .005t^3 + 20t$. What is the object's velocity?

[11.4]
Objective 11.8: *Find the indicated rate, given the other information. Assume that both x and y are positive.*

29. Find $\dfrac{dy}{dt}$ where $4x^2 + 9y^2 = 36$ and $\dfrac{dx}{dt} = 3$ when $x = 2$ and $y \geq 0$.
30. Find $\dfrac{dx}{dt}$ where $y = 4x^2$ and $\dfrac{dy}{dt} = -5$ when $x = 2$.
31. Find $\dfrac{dy}{dt}$ where $x^2 + 2xy + y^2 = 49$ and $\dfrac{dx}{dt} = 5$ when $x = 4$.
32. Find $\dfrac{dx}{dt}$ where $xy = 4$ and $\dfrac{dy}{dt} = -4$ when $x = 1$.

SAMPLE TEST

The following sample test (45 minutes) is intended to review the main ideas of this chapter.

Find dy/dx in Problems 1–3 by using implicit differentiation.

1. $y - 10x^2 + 6x - 15 = 0$
2. $6xy = 20x$

3. $x^4 + 5xy - 3x^2y + 9xy^2 - 155 = 0$

4. Find the standard form of the equation of the line tangent to the circle $(x - 2)^2 + y^2 = 25$ at the point $(5, 4)$.

Find the indicated rate, given the information in Problems 5–8.

5. Find dy/dt where $9x^2 + 16y^2 = 145$ and $dx/dt = 4$ when $x = 3$.

6. Find dy/dt where $xy = 9$ and $dx/dt = -8$ when $x = 1$.

7. Find dx/dt where $y = 25x^2$ and $dy/dt = -4$ when $x = 3$.

8. Find dx/dt where $x^2 + 2xy + y^2 = 64$ and $dy/dt = 3$ when $x = 4$.

9. A property management company manages 100 apartments renting for \$500 with all the apartments rented. For each \$50 per month increase in rent there will be 2 vacancies with no possibility of filling them. If x represents the number of \$50 price increases, find the marginal revenue.

10. Suppose that for a company manufacturing lawn chairs, the cost and revenue functions are

$$C = 15{,}000 + 45x$$

$$R = 100x - \frac{x^2}{2{,}000}$$

where the production output is x chairs per week. If production is increasing at a rate of 50 chairs per week when production output is 2,000 chairs per week, find the rate of increase (decrease) in **a.** cost **b.** revenue **c.** profit.

MODELING APPLICATION 9

Publishing: An Economic Model

Karlin Press sells its *World Dictionary* at a list price of $20 and presently sells 5,000 copies per year. Suppose you are being considered for an editorial position and are asked to do an analysis of the price and sales of this book in order to assess your competency for the position. The annual costs associated with this book are summarized in the following table. The dictionary presently has a net cost of 80% of the list (selling) price.

Costs Associated with Publishing
World Dictionary

Cost	Amount, $
Advertising	1,750.00
Author's royalty	0.00
Binding	1.85 per book
Composition	3,600.00
Computer services	750.00
Investment return	3,000.00
Operating overhead	9,000.00*
Printing	4.90 (per book)
Set-up charges	400.00
Storage	.50 (per book)
Taxes	1.20 (per book)

*Salaries and offices are prorated for this title.

Analyze the price and sales of the dictionary. Your analysis should reach a conclusion about pricing the book for maximum revenue and/or maximum profit. Some consideration should also be given to market demand (which is assumed to be linear). Suppose you do some additional market research to find that at a list price of $15, the company will sell 7,500 copies, but at $25 sales would drop to 2,500. You might also want to give some thought to inventory and reprint schedule. To this end you find that the annual inventory (for maximum profit) should be 6,885, but 7,000 copies are allowed because they need some for office and sample copies. For general guidelines about writing this essay, see the commentary accompanying Modeling Application 1 on page 151.

12

Applications and Differentiation

CHAPTER OVERVIEW

The calculus of Chapters 10 and 11 is put to work in this chapter by discussing some important applications of the derivative. The applications demonstrate some of the power and versatility of differential calculus.

PREVIEW

Graphing, aided by the idea of a derivative, leads to curve sketching and the determination of relative maximums and minimums. We then turn to optimization (finding the absolute maximum or minimum), an extremely important concept in the business world.

PERSPECTIVE

We continue to explore the usefulness of calculus in a variety of different real-world settings. The only reason for introducing the derivative in this book is to apply it as a real-world model and to use it to derive additional mathematics to use in real-world models.

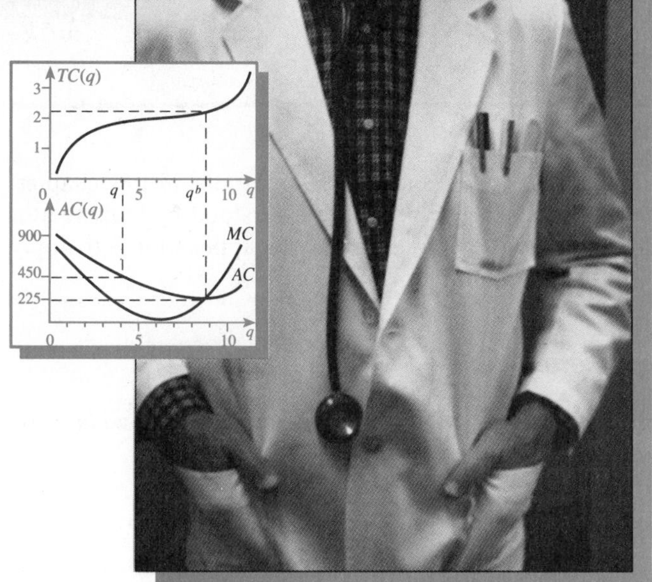

MODELING APPLICATION 10

Health Care Pricing

"Use of surgery and hospital costs vary greatly by region ... many doctors are baffled by this finding. And some fear that politicians and reimbursement officials will seize on data that nobody understands to rationalize budget cuts for Medicare and other health-insurance programs."—The Wall Street Journal, March 5, 1986, page 33

The figure in the margin shows a total cost function for New York hospitals, based on an article, "Financial Management and DRGS," by Steven Ullmann, *Health Services Research*, 1984. Set up another coordinate axis, and carefully construct the graphs of the average cost function and of the marginal cost function. As an example, show how to find the average cost and marginal cost for 175,000 patient-days per year.*

TC(q) Total cost, in dollars

15,000,000

10,000,000

5,000,000

TC

Patient-days per year

0 50,000 100,000 150,000 200,000 q

Write a paper that develops a mathematical model associated with the pricing of health care. For general guidelines about writing this essay, see the commentary for Modeling Application 1 on page 151.

* This modeling application is from Yves Nievergelt, "Graphic Differentiation Clarifies Health Care Pricing," *UMAP Journal*, Volume 9, No. 1. Copyright 1988 by COMAP, Inc.

APPLICATIONS

Management (*Business, Economics, Finance, and Investments*)

Sales as a function of advertising (12.1, Problem 43)
Point of diminishing returns (12.2, Problems 41–48)
Determining when sales are increasing (12.2, Problems 49–50)
Marginal revenue (12.2, Problem 51; 12.3, Problem 36; 12.5, Problem 29)
Cost–benefit model (12.3, Problem 35; 12.5, Problem 30)
Average cost (12.3, Problems 37–38)
Maximum profit (12.4, Problems 21–27, 34–37)
Minimal average cost (12.4, Problems 28, 32–33)

Management (*continued*)

Maximize volume of a box (12.4, Problem 31)
Minimize cost (12.4, Problem 39)
Worker efficiency (12.4, Problem 40)
Largest area in an enclosure (12.4, Problems 41–42)
Tour pricing for group fares (12.4, Problems 43–44)
Property management (12.5, Problem 31)

Life Sciences (*Biology, Ecology, Health, and Medicine*)

Concentration of a drug in the bloodstream (12.3, Problem 39)
Fish swimming to spawn (12.3, Problem 40)
Air pollution (12.4, Problems 29, 38)
Maximize yield per acre (12.4, Problem 45)

Social Sciences (*Demography, Political science, Population, Psychology, Society, and Sociology*)

Learning curve (12.1, Problem 44)
Timing of a political campaign (12.1, Problem 45)
Voting patterns (12.4, Problem 30)

General Interest

Maximum volume of a box (12.4, Problem 31)
Largest area in an enclosure (12.4, Problems 41–42)
Minimize cardboard for a poster (12.5, Problem 32)

Modeling Application—

Health Care Pricing

12.1

First Derivatives and Graphs

We continue our study of the derivatives and their applications in this chapter. Many of the phenomena we will study can be better understood by looking at a graph of the relationship between two variables, so our first application is to apply the idea of the derivative to the graph of a function.

Increasing and Decreasing Functions

If the graph of a function rises from left to right, the function is said to be *increasing*, and if it drops, it is said to be *decreasing*, as is shown in Figure 12.1.

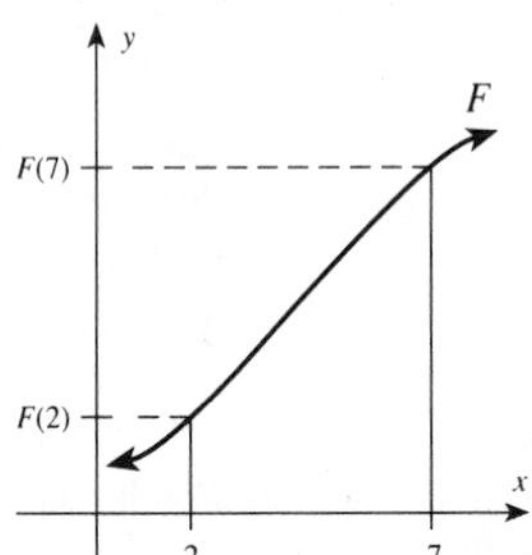

Function F is an increasing function on $(2, 7)$; $F'(x) > 0$ for all x in $(2, 7)$.

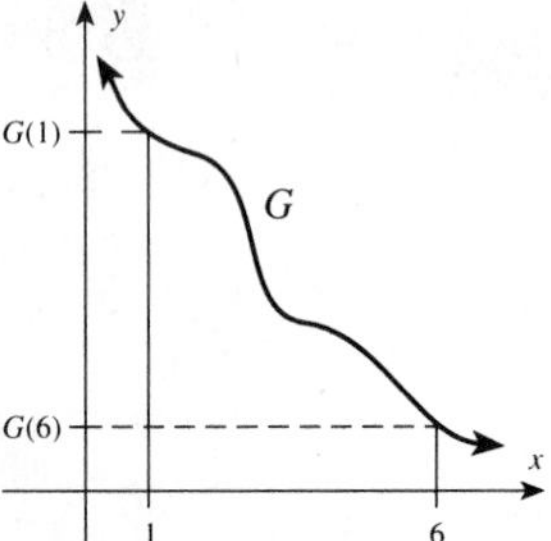

Function G is a decreasing function on $(1, 6)$; $G'(x) < 0$ for all x in $(1, 6)$.

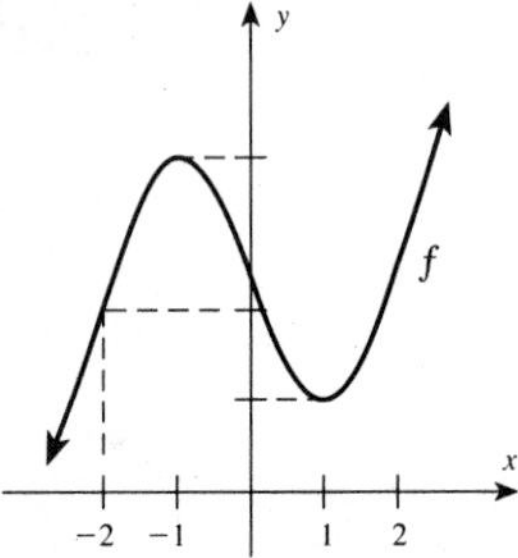

Function f is increasing on $(-2, -1)$, decreasing on $(-1, 1)$, and increasing on $(1, \infty)$.

FIGURE 12.1

Notice that functions are classified as increasing on an open interval (a, b) if for every x_1 and x_2 on the interval such that $x_2 > x_1$, then $f(x_2) > f(x_1)$. If a function is increasing, then we say its graph is rising. Furthermore, if $f'(x) > 0$ for some x, then we say that f is increasing at the point x. Similarly, we say that f is decreasing on some interval I if for every x_1 and x_2 on the interval such that $x_2 > x_1$, then $f(x_2) < f(x_1)$, and is decreasing at a point x if $f'(x) < 0$. If a function is decreasing, then we say its graph is falling. These features of functions are generalized graphically in Table 12.1.

Increasing and Decreasing Functions

If $f'(x) > 0$ for all x on an interval I, then f is **increasing** over I.
If $f'(x) < 0$ for all x on an interval I, then f is **decreasing** over I.

Graphing Parabolas Using Calculus

We will now apply these ideas to parabolas whose equations have the form $y = ax^2 + bx + c$. Section 2.5 promised that this type of parabola would be discussed

TABLE 12.1
Increasing and Decreasing Functions on the Interval (a, b)

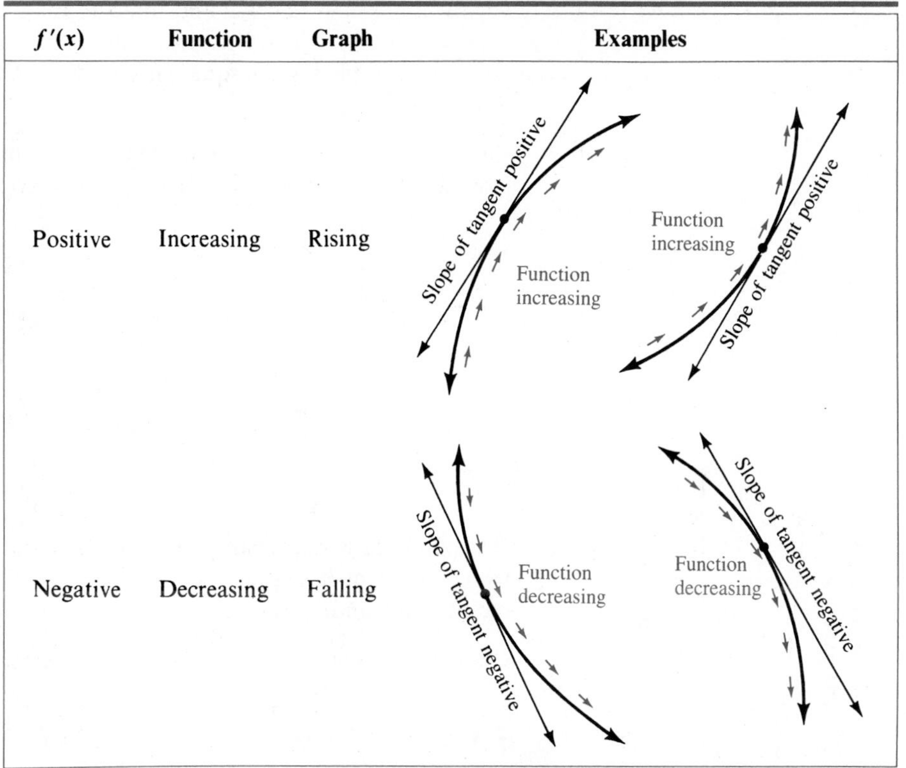

$f'(x)$	Function	Graph	Examples
Positive	Increasing	Rising	
Negative	Decreasing	Falling	

after we learned some calculus techniques. Now recall that parabolas of this form are functions and open upward if $a > 0$ and downward if $a < 0$. Furthermore the vertex of the parabola gives the extreme values for the function—it is a high point if $a < 0$ and a low point if $a > 0$. The slope of the parabola at the vertex must be zero. These possibilities are summarized in Figure 12.2.

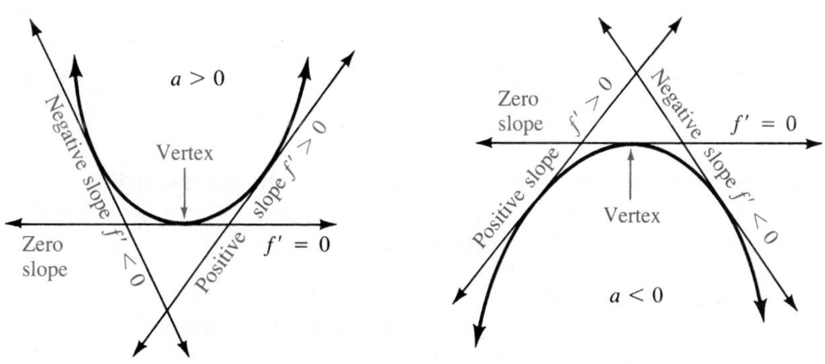

FIGURE 12.2
Graphs of parabolas showing slope

EXAMPLE 1 Graph $y = 3x^2 + 12x + 9$ and indicate the intervals for which the function is increasing and for which it is decreasing. Also compare the methods of finding the vertex by completing the square (as we did in Chapter 2) with that of using calculus (using the derivative).

Solution The function is increasing or decreasing depending on where the derivative is positive and where it is negative. If $f(x) = y = 3x^2 + 12x + 9$, then

$$\begin{aligned} f'(x) &= 6x + 12 \\ &= 6(x + 2) \end{aligned}$$

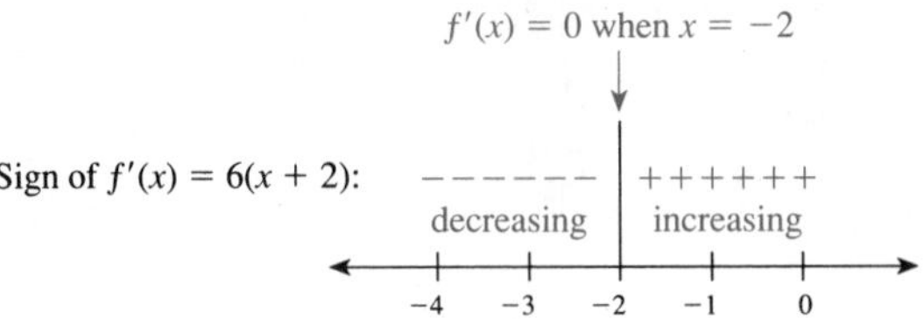

If a parabola is decreasing on $(-\infty, -2)$ and increasing on $(-2, \infty)$ with a horizontal tangent line at $x = -2$, it must open upward. Now we will compare the methods for finding the vertex:

Complete the square (precalculus)	*Use calculus*
$y = 3x^2 + 12x + 9$	$y = 3x^2 + 12x + 9$
$y - 9 = 3(x^2 + 4x)$	$y' = 6x + 12$
$y - 9 + 12 = 3(x^2 + 4x + 2^2)$	$y' = 0$ if $x = -2$
$y + 3 = 3(x + 2)^2$	If $x = -2$,
Vertex is $(-2, -3)$	then $y = 3(-2)2 + 12(-2) + 9 = -3$
	Vertex is $(-2, -3)$

Slope of tangent negative, function decreasing

Slope of tangent positive, function increasing

Horizontal tangent

$f'(x) = 0$ when $x = -2$

The graph is shown in the margin. ■

EXAMPLE 2 Graph $y = 10x - x^2$.

Solution First, find the vertex: $y' = 10 - 2x$

Then, find the horizontal tangent: If $y' = 0$, then $10 - 2x = 0$

$$10 = 2x$$
$$5 = x$$

If $x = 5$, then

$$y = 10(5) - (5)^2 = 25$$

The vertex is the point $(5, 25)$.

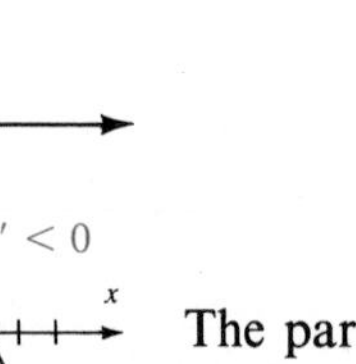

The parabola opens downward ($a = -1$ is less than zero), has the y-intercept at $(0, 0)$, and is symmetric about the vertical line through the vertex. The graph is shown in the margin. ■

Increasing on $(-\infty, 5)$
Decreasing on $(5, \infty)$

Maximums and Minimums

Since we will be graphing functions other than parabolas, some new terminology will be helpful. If f is a continuous function on an interval $[a, b]$, then there will be some point c_1 in that interval so that

$$f(c_1) \geq f(x) \text{ for all } x \text{ on } [a, b]$$

The value $f(c_1)$ is called an **absolute maximum** on $[a, b]$, and we say that the function f has an absolute maximum at $x = c_1$. Similarly, there is another point c_2 on $[a, b]$, not necessarily different from c_1, such that

$$f(c_2) \leq f(x) \text{ for all } x \text{ on } [a, b]$$

The value $f(c_2)$ is called an **absolute minimum** on $[a, b]$, and we say that the function f has an absolute minimum at $x = c_2$.

In addition to these absolute extremes, there may also be points that are higher or lower than the surrounding points. The values of the function at these points are called *local* (or *relative*) *maximums and minimums*. Several of these points are shown in Figure 12.3 and are defined in the following box.

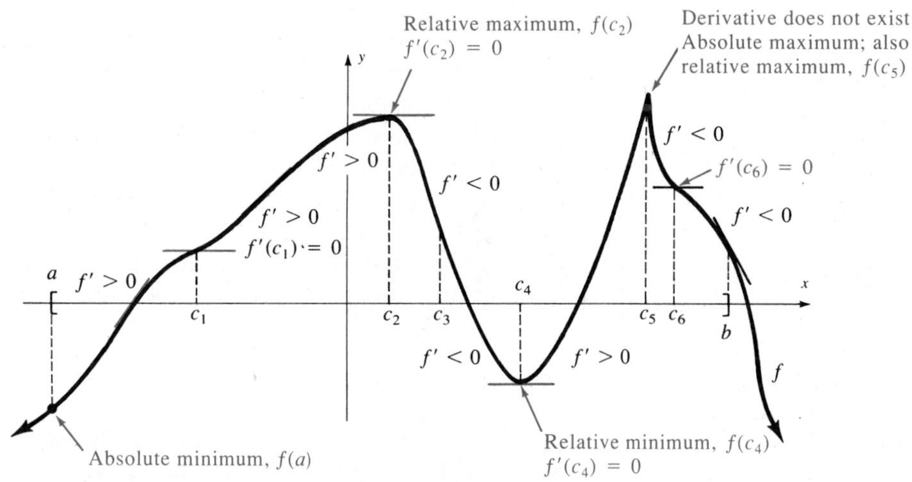

FIGURE 12.3
Absolute and Relative Extremes

Relative Maximum

A function f has a **local** (or **relative**) **maximum** at $x = c$ if there exists some interval (a, b) around c so that $f(x) < f(c)$ for all x (except c) in the interval (a, b). The value M for which $f(c) = M$ is the **relative maximum**.

Relative Minimum

A function f has a **local** (or **relative**) **minimum** at $x = c$ if there exists some interval (a, b) around c so that $f(x) > f(c)$ for all x (except c) in the interval (a, b). The value m for which $f(c) = m$ is the **relative minimum**.

We focus on absolute maximums and minimums in Section 12.3. In this section we use calculus to locate the relative extremes. Notice in Figure 12.3 that each and every relative maximum and relative minimum occurs at values of x for which $f'(x) = 0$ (horizontal tangent lines) or else where $f'(x)$ does not exist (places with sharp points on the graph). In Chapter 1 the term **critical value** was used when solving quadratic inequalities to mean a value that caused a factor or an expression to be zero. In the context of derivatives, a critical value is not only a value that causes a factor of the derivative to be zero (to have a horizontal tangent line), but also a value for which the derivative does not exist.

Critical Value

A **critical value** for a function is an interior point c of its domain at which the function has a horizontal tangent, or at which the derivative does not exist. That is, c is a critical value if

$$f'(c) = 0 \quad \text{or} \quad f'(c) \text{ does not exist.}$$

Of what significance are the critical values? Suppose f is a continuous function defined over an interval $[a, b]$, with critical values $c_1, c_2, \ldots, c_n$. Then everything that is interesting about the curve f will occur either at the endpoints a, b, or at one of these critical values. Everywhere else, the curve is either increasing or decreasing. We call this the **function behavior principle**. Be careful about the way you read this principle. It does *not* say that if c is a critical value, then $f(c)$ is a relative maximum or a relative minimum. It does say that $f(c)$ is a candidate to be a relative maximum or a relative minimum. We must, therefore, develop a strategy that will lead us to identify any relative maximums and minimums from the list of candidates (i.e., critical values). We use two derivative tests for this purpose. We first discuss a derivative test that works in all cases. In the next section, we discuss an easier, second-derivative test, which will work most of the time. For those cases in which this easier second-derivative test does not work, we will need to revert back to the first-derivative test discussed in this section.

First-Derivative Test

The first test, appropriately called the **first-derivative test**, makes use of the fact that the derivative gives the slope of a tangent line of a curve at a particular point. Suppose that c is a critical value. Let c^- be a point to the left of c (but to the right of any other critical value) and let c^+ be a point to the right of c (but to the left of any other critical value). There are then four possibilities as shown in Figure 12.4.

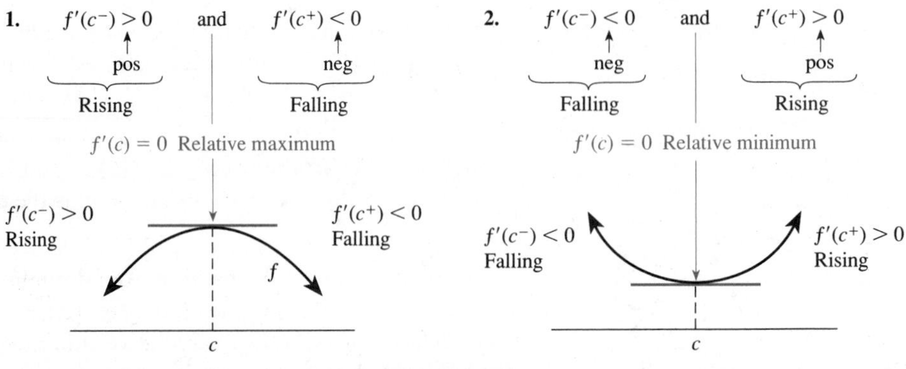

FIGURE 12.4
First-derivative test

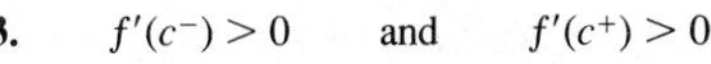

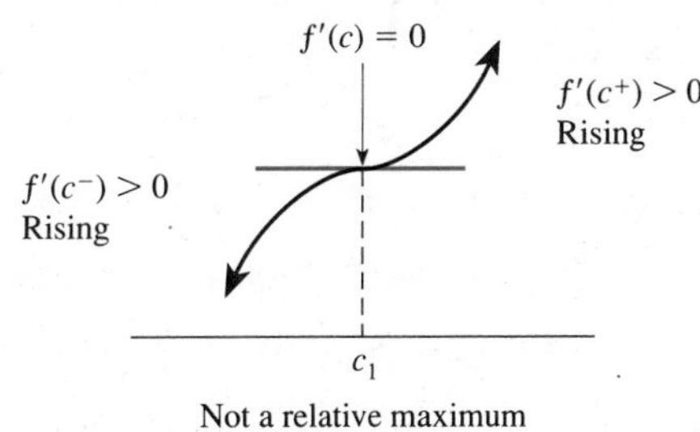

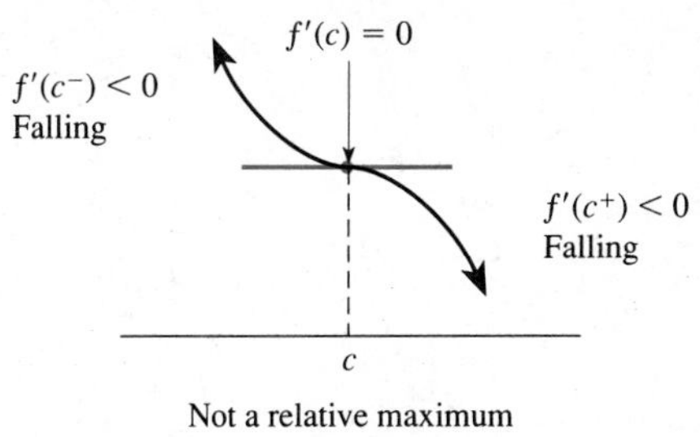

FIGURE 12.4 *(continued)*
First-derivative test

First-Derivative Test

1. Find the critical values:
   a. Find all values c such that $f'(c) = 0$.
   b. Find all values c for which the derivative does not exist but $f(c)$ is defined.
2. **a.** $f(c)$ is a **relative maximum** if

$$f'(c^-) > 0 \quad \text{(rising)} \qquad \text{and} \qquad f'(c^+) < 0 \quad \text{(falling)}$$

   b. $f(c)$ is a **relative minimum** if

$$f'(c^-) < 0 \quad \text{(falling)} \qquad \text{and} \qquad f'(c^+) > 0 \quad \text{(rising)}$$

EXAMPLE 3 Find the critical values for $f(x) = 5x^3 + 4x^2 - 12x - 25$.

Solution Critical values are values in the domain for which $f'(x) = 0$ or $f'(x)$ is not defined.

$$f'(x) = 15x^2 + 8x - 12$$

We see that $f'(x)$ is defined for all x, so we look for x values for which

$$\begin{aligned} f'(x) &= 0 \\ 15x^2 + 8x - 12 &= 0 \\ (3x - 2)(5x + 6) &= 0 \end{aligned}$$

The critical values are $x = \frac{2}{3}$ and $x = -\frac{6}{5}$. ■

If we want to graph the curve whose equation is given in Example 3 we can use the first-derivative test. We want to know where $f'(x)$ is positive and where it is negative.

Look at the factored form of $f'(x)$:

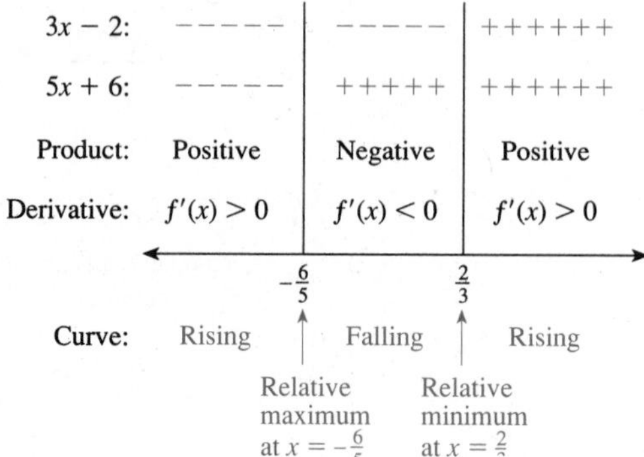

By plotting some points you can graph this curve, as shown in Figure 12.5.

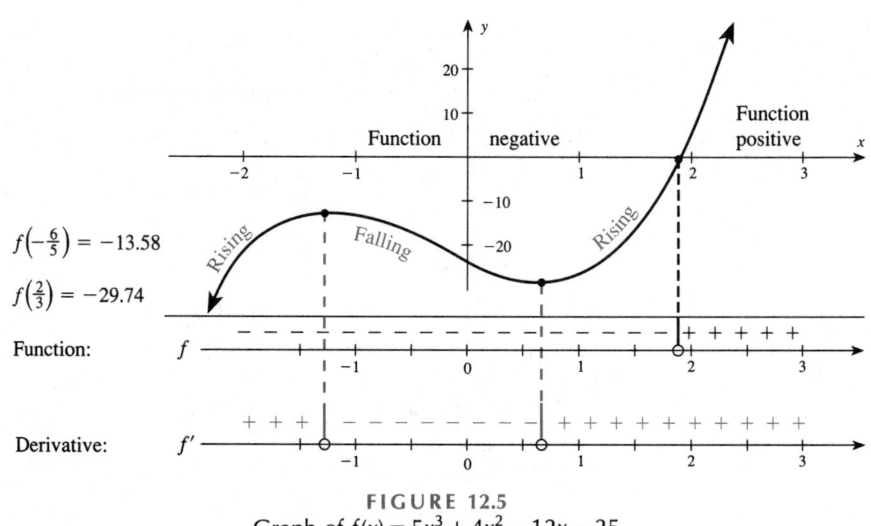

FIGURE 12.5
Graph of $f(x) = 5x^3 + 4x^2 - 12x - 25$

$f(x)$ values found by calculator

x	$f(x)$	$f'(x)$	
$-\frac{6}{5}$	-13.58	0	Relative maximum
$\frac{2}{3}$	-29.74	0	Relative minimum
2	7	>0	Rising } Same for
-2	-25	>0	Rising } each interval
0	-25	<0	Falling } by function behavior principle

EXAMPLE 4 Find the critical values for $f(x) = |x + 1|$.

Solution If $x > -1$:

$$f(x) = x + 1 \quad \text{and} \quad f'(x) = 1$$

If $x < -1$:

$$f(x) = -x - 1 \quad \text{and} \quad f'(x) = -1$$

At $x = -1$,

$$\lim_{h \to -1^+} \frac{|-1 + h + 1| - |-1 + 1|}{h} = \lim_{h \to -1^+} 1 = 1$$

$$\lim_{h \to -1^-} \frac{|-1 + h + 1| - |-1 + 1|}{h} = \lim_{h \to -1^-} -1 = -1$$

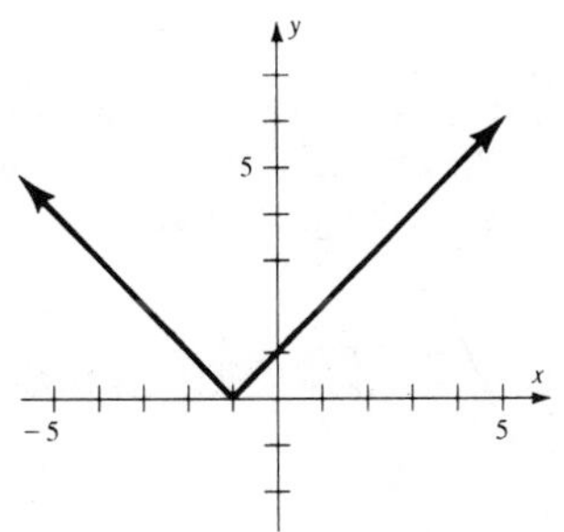

FIGURE 12.6
Graph of $f(x) = |x + 1|$

Thus the derivative does not exist at $x = -1$ and $x = -1$ is in the domain, so $x = -1$ is a critical value.

Apply the first-derivative test. Let

$$f'(c^-) = -1 \quad \text{since} \quad c^- < -1 \quad \text{(falling)}$$
$$f'(c^+) = 1 \quad \text{since} \quad c^+ > -1 \quad \text{(rising)}$$

There is a relative minimum at $f(-1) = 0$. The graph is shown in Figure 12.6.

12.1 Problem Set

Graph the curves in Problems 1–16 by using the derivative to find the vertex.

1. $f(x) = 8x^2$
2. $f(x) = -12x^2$
3. $f(x) = -20x^2$
4. $2x^2 + 5y = 0$
5. $5y + 15x^2 = 0$
6. $5x^2 + 4y = 20$
7. $f(x) = 5x^2 - 20x + 2$
8. $f(x) = 9 + 24x - 12x^2$
9. $x^2 + 4y - 3x + 1 = 0$
10. $x^2 - 4x + 10y + 13 = 0$
11. $9x^2 + 6x + 18y - 23 = 0$
12. $9x^2 + 6y + 18x - 23 = 0$
13. $y = 5x^2 - 3x + 1$
14. $y = 2x^2 + 5x - 8$
15. $y = 7x^2 + 2x - 3$
16. $y = 2x^2 - 5x + 3$

Identify the intervals for which the functions in Problems 17–22 are increasing, decreasing, or constant. Also identify the places where there is a horizontal tangent.

17.

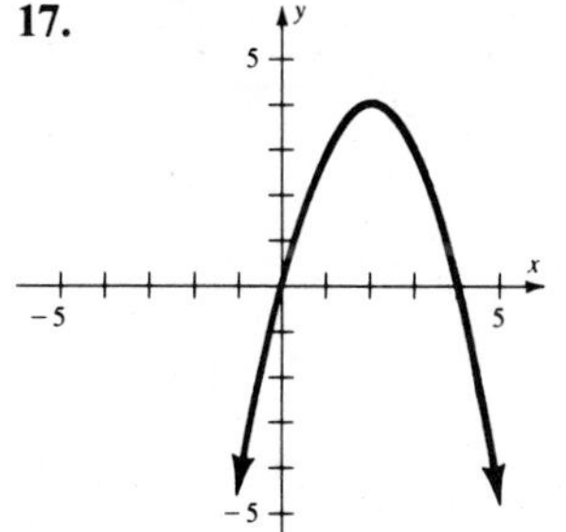

18.

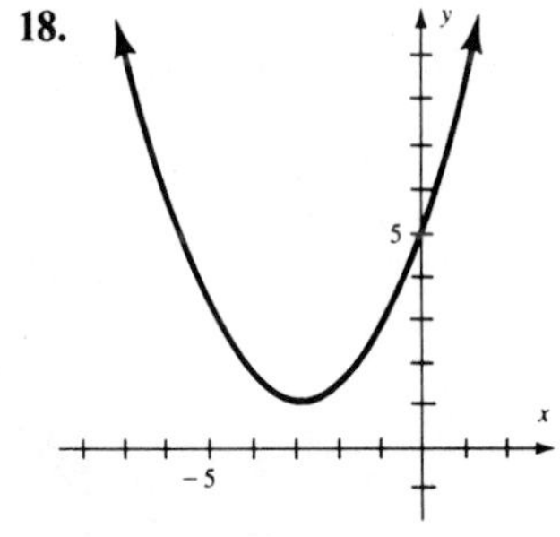

19. 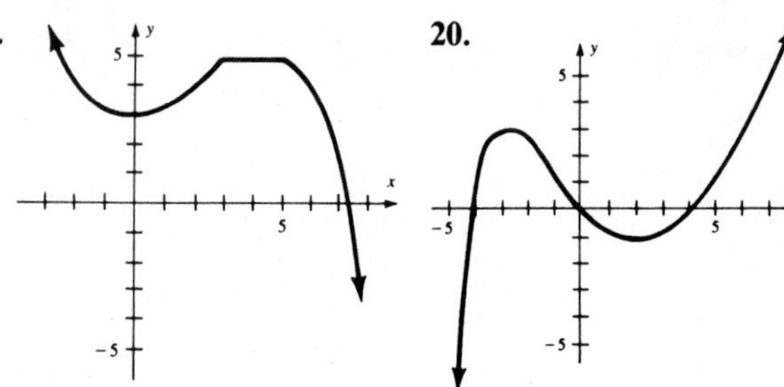

20.

21. 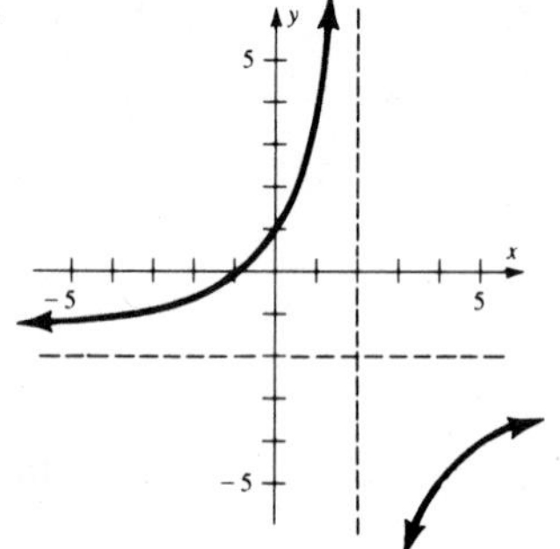

22. 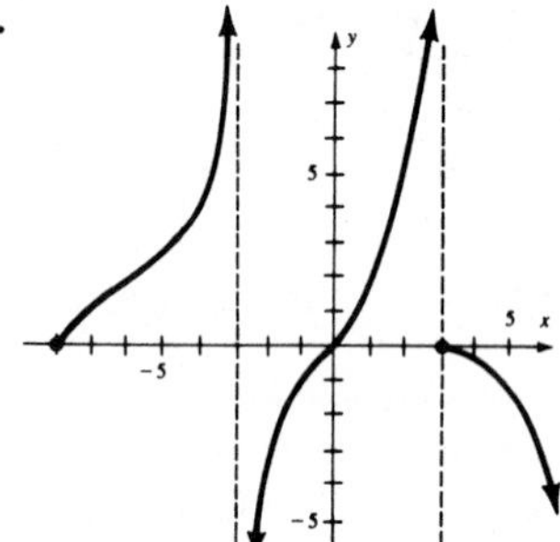

23. The graph of a function f is shown. Draw the tangent lines and measure the slope of f at the points labeled A, B, C, D, and E. Then set up another coordinate system and plot the corresponding points to draw the graph of f'.

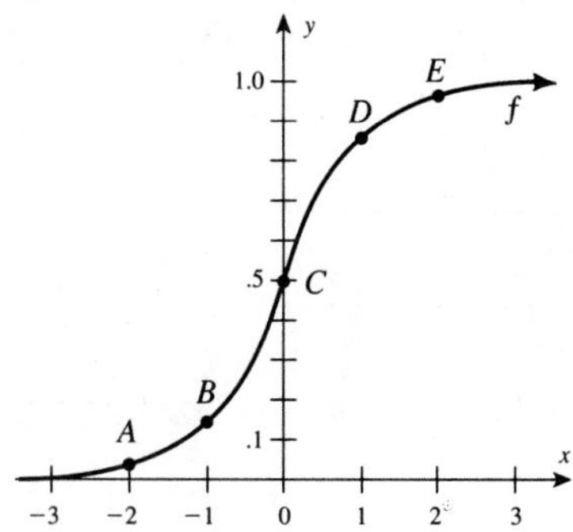

24. The graph of a function f is shown. Draw the tangent lines and measure the slope of f at the points labeled A, B, C, D, and E. Then set up another coordinate system and plot the corresponding points to draw the graph of f'.

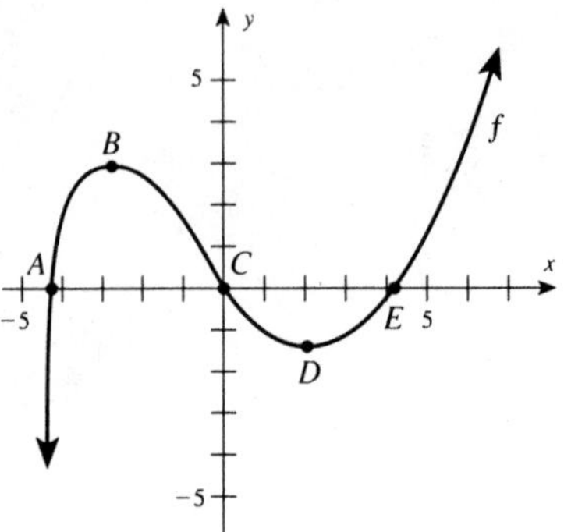

25. The graph of the derivative g' is shown, but it does not show the function g. Sketch the general shape of a function g that may have as its derivative the given graph g'.

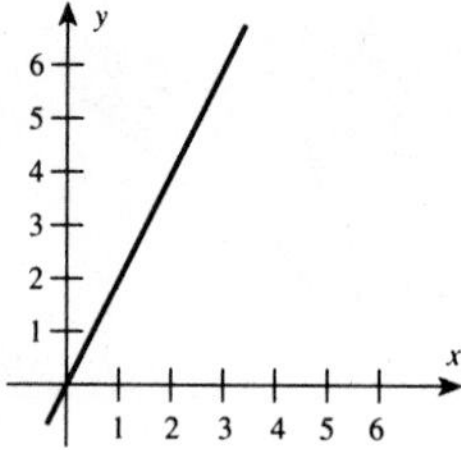

26. The graph of the derivative g' is shown, but it does not show the function g. Sketch the general shape of a function g that may have as its derivative the given graph g'.

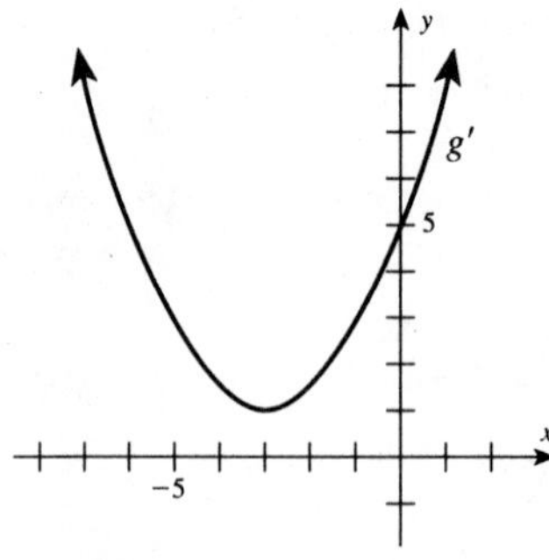

*For the functions in Problems 27–42, (**a**) find the critical values and (**b**) determine the intervals for which the functions are increasing and decreasing.*

27. $f(x) = 5 + 10x - x^2$

28. $f(x) = 10 + 6x - x^2$

29. $f(x) = x^2 - 12x + 6$

30. $f(x) = x^2 + 5x + 6$

31. $g(x) = 2x^3 - 3x^2 - 36x + 4$
32. $g(x) = x^3 + 3x^2 - 24x - 4$
33. $y = x^3 + 5x^2 + 8x - 5$
34. $y = \frac{8}{3}x^3 - 5x^2 - 3x + 10$
35. $y = 3x^4 + 2x^3 - 9x^2 + 12$
36. $y = 3x^4 + 8x^3 - 18x^2 + 5$
37. $f(x) = 3x^5 - 25x^3 - 540x + 90$
38. $f(x) = 3x^5 - 85x^3 + 240x - 200$
39. $y = \dfrac{x+1}{x-2}$
40. $y = \dfrac{x^2}{x-1}$
41. $f(x) = 1 + \dfrac{1}{x} + \dfrac{1}{x^2}$
42. $f(x) = x - 4\sqrt{x}$

APPLICATIONS

43. If a company has sales of

$$S(x) = 10{,}000 + 5{,}000x - 25x^2 - x^3$$

where x is the amount spent on advertising in thousands of dollars, when are the sales increasing?

44. The time t, in minutes, that it takes to learn a list of x items is given by the formula

$$t(x) = 5x\sqrt{x - 10} \qquad \text{for } x \geq 10$$

For what values of x is t increasing?

45. Suppose you are the campaign manager for a candidate, and also suppose that a function showing the percentage of people who are aware of your candidate is

$$f(x) = \frac{10x}{x^2 + 50} + .1$$

where x is the number of months after you begin campaigning. If the election is held on November 2, when should you begin campaigning if you want the maximum number of people to be aware of your candidate at election time?

12.2 Second Derivatives and Graphs

Higher-Order Derivatives

We have used the derivative to find out where f is increasing and where it is decreasing. Since the derivative of f is also a function we can repeat the process and use the derivative of f' to determine where f' is increasing and where it is decreasing. The derivative of a derivative is called its **second derivative** and is denoted by f''. Taking still another derivative gives a function f''' called the **third derivative**. Successive derivatives are denoted by

$$f', f'', f''', f^{(4)}, f^{(5)}, \ldots, f^{(n)}$$

Note the use of parentheses for fourth derivatives and higher. This is to avoid confusion with powers f^4, f^5, and so on.

The graph of a function and its successive derivatives is shown in Figure 12.7.

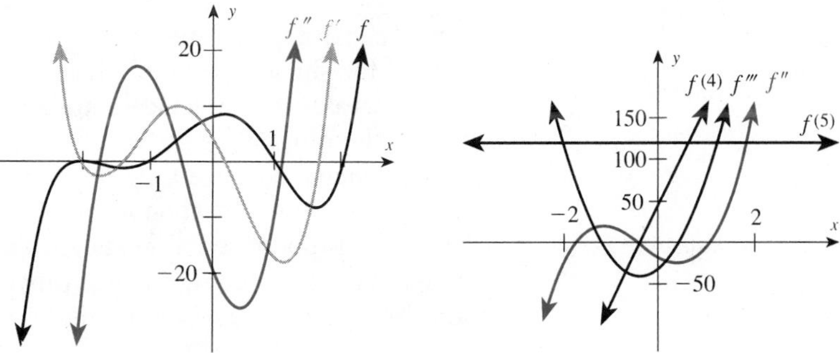

FIGURE 12.7
Comparison of graphs of a function and its derivatives

COMPUTER COMMENT

Use the *Function plotter* on the software accompanying this book to draw the graph of a function. You can also graph relations using the *Implicit relation plotter*.

EXAMPLE 1 Find all successive derivatives of $f(x) = x^5 + 2x^4 - 5x^3 - 10x^2 + 4x + 8$.

Solution

$$f'(x) = 5x^4 + 8x^3 - 15x^2 - 20x + 4$$
$$f''(x) = 20x^3 + 24x^2 - 30x - 20$$
$$f'''(x) = 60x^2 + 48x - 30$$
$$f^{(4)}(x) = 120x + 48$$
$$f^{(5)}(x) = 120$$
$$f^{(6)}(x) = f^{(7)}(x) = \cdots = 0$$

■

Another commonly used notation for higher derivatives uses the dy/dx notation as follows:

$$\frac{dy}{dx}, \frac{d^2y}{dx^2}, \frac{d^3y}{dx^3}, \frac{d^4y}{dx^4}, \ldots, \frac{d^ny}{dx^n}$$

EXAMPLE 2 If $y = 8x^{1/2}$, find the first three derivatives.

Solution

$$\frac{dy}{dx} = 8\left(\frac{1}{2}\right)x^{1/2-1} = 4x^{-1/2}$$
$$\frac{d^2y}{dx^2} = \frac{d}{dx}\left[\frac{dy}{dx}\right] = \frac{d}{dx}[4x^{-1/2}] = 4\left(-\frac{1}{2}\right)x^{-1/2-1} = -2x^{-3/2}$$
$$\frac{d^3y}{dx^3} = \frac{d}{dx}\left[\frac{d^2y}{dx}\right] = \frac{d}{dx}[-2x^{-3/2}] = -2\left(-\frac{3}{2}\right)x^{-3/2-1} = 3x^{-5/2}$$

■

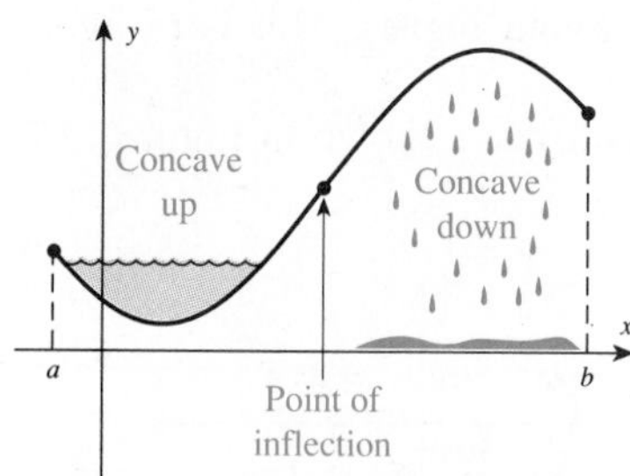

FIGURE 12.8
Concavity and inflection points

Concavity

Now we can use the second derivative to develop a second-derivative test to inform us about the shape of a graph. If a curve lies above its tangent line at each point on an interval (a, b), it is **concave upward** over (a, b); if a curve lies below its tangent line at each point on the interval (a, b), it is **concave downward**. A point on a graph that separates a concave downward portion of a curve from a concave upward portion is called an **inflection point**. Inflection points can occur where $f''(x) = 0$ or where $f''(x)$ is undefined. Figure 12.8 illustrates these ideas.

We can now use the second derivative to obtain additional information about a graph. We consider two possibilities: $f''(x) > 0$ and $f''(x) < 0$.

Case I: If $f''(x) > 0$ on (a, b), then f' is increasing on (a, b).
What are the possibilities for f when f' is increasing?

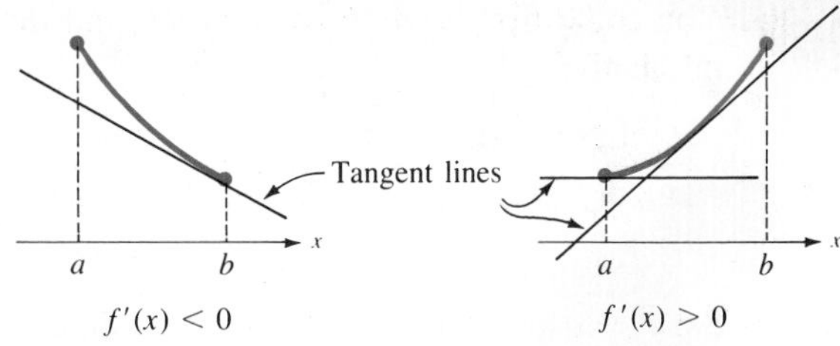

$f'(x) < 0$ $f'(x) > 0$

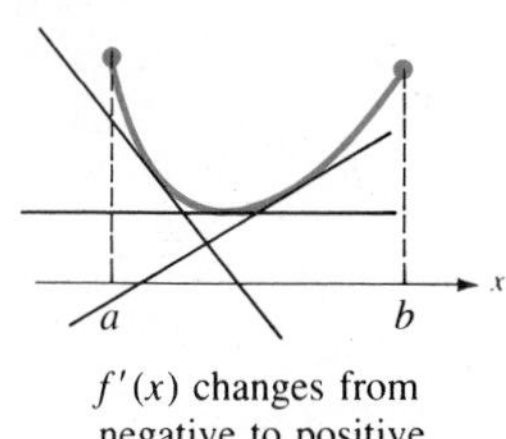

$f'(x)$ changes from negative to positive

$f'(x)$ is increasing
$f''(x) > 0$ on (a, b)
Concave upward

f' is increasing and is negative; the curve is falling
f' is increasing and is positive; the curve is rising
f' is increasing and changes from negative to positive

Curves having the shape illustrated by $f''(x) > 0$ are concave upward.

Case II: If $f''(x) < 0$ on (a, b), then f' is decreasing on (a, b). What are the possibilities for f when f' is decreasing?

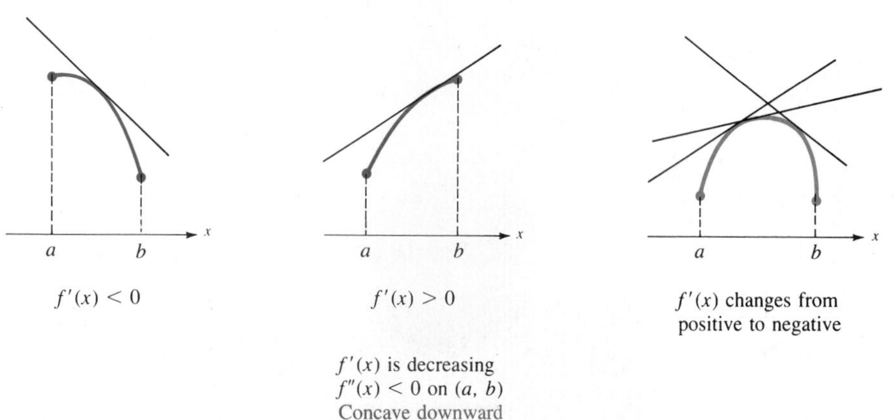

$f'(x) < 0$ $f'(x) > 0$ $f'(x)$ changes from positive to negative

$f'(x)$ is decreasing
$f''(x) < 0$ on (a, b)
Concave downward

f' is decreasing and negative; the curve is falling
f' is decreasing and positive; the curve is rising
f' is decreasing and changes from positive to negative

Curves having the shape illustrated by $f''(x) < 0$ are concave downward.

EXAMPLE 3 Given: $4x^3 - 5x^2 - 8x - 2y + 20 = 0$

a. Find the critical values.
b. Find the intervals for which the function is increasing or decreasing.
c. Find the interval for which the function is concave upward or concave downward.

Solution **a.** You could first solve for y and then find the derivative, but instead you find y' implicitly:

$$12x^2 - 10x - 8 - 2y' = 0$$

$$2y' = 12x^2 - 10x - 8$$

$$y' = 6x^2 - 5x - 4$$

Set $y' = 0$ and solve for the critical values.

$$6x^2 - 5x - 4 = 0$$

$$(3x - 4)(2x + 1) = 0$$

The critical values are $x = \frac{4}{3}$ and $x = -\frac{1}{2}$.

b. The intervals on which the function is increasing or decreasing can be found by finding where the derivative is positive or negative. The procedure is identical to that used when solving inequalities (see Section 1.6).

$$6x^2 + 5x - 4 = (3x - 4)(2x + 1)$$

Plot the critical values and solve on a number line.

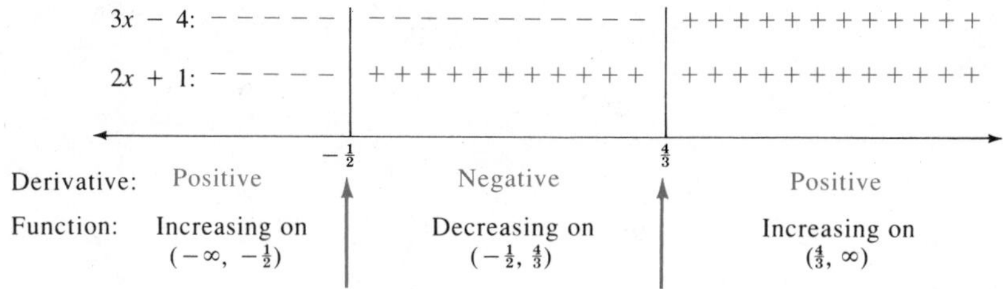

c. For the concavity, look at the second derivative:

$$y'' = 12x - 5$$

Set $y'' = 0$ and solve.

$$12x - 5 = 0$$

$$x = \frac{5}{12}$$

The first derivative is not zero at $x = \frac{5}{12}$, so

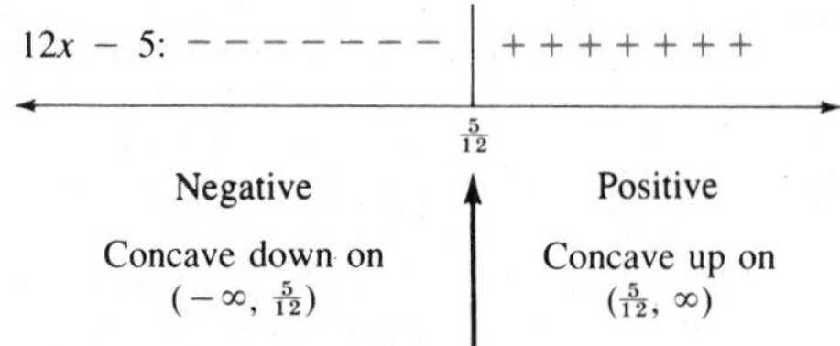

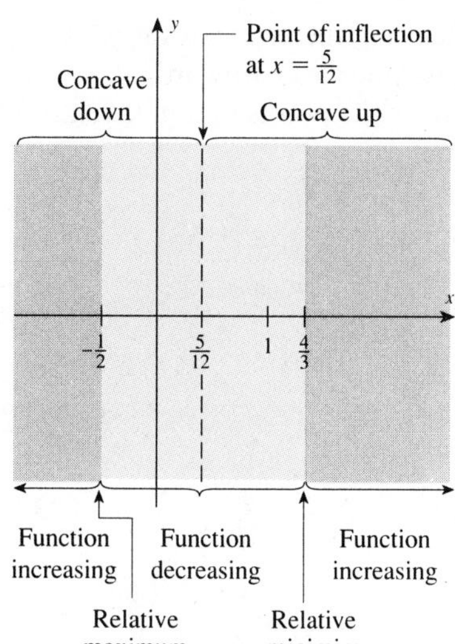

In preparation for graphing such curves we summarize all of this information on a coordinate grid as follows:

Color screen shows portions where the function is decreasing.
Gray screen shows portions where the function is increasing.
These screens are separated by points on the graph that are relative maximums or relative minimums.

Concavity is also marked, and places where concavity changes are marked by a point of inflection.

COMPUTER COMMENT

All of the work of this example can be summarized using the *Function plotter*.

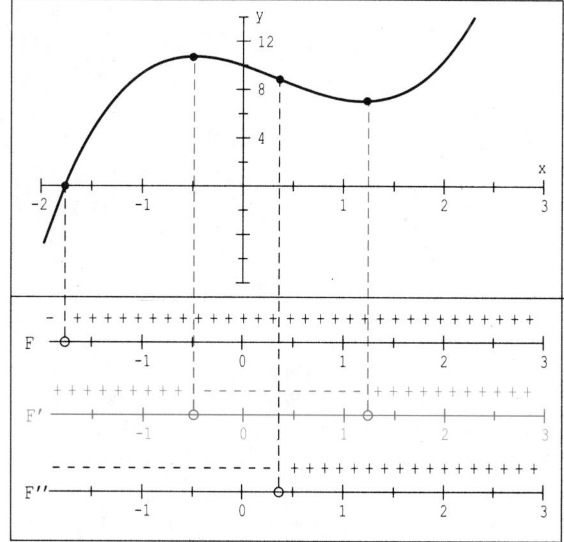

Notice that F shows where the function is positive and where it is negative. F′ shows where the function is increasing and where it is decreasing. F″ shows where the function is concave down and where it is concave up.
Where the curve crosses the axis the value of F is zero;
where the curve reaches a relative maximum or minimum, the value of F′ is zero;
where the curve changes concavity, the value of F″ is zero.

EXAMPLE 4 Identify the intervals for which the function is increasing, the intervals for which it is decreasing, where it is concave upward, where it is concave downward, as well as places where there is a horizontal tangent line. Identify the points that give relative maximum and relative minimum values for the function.

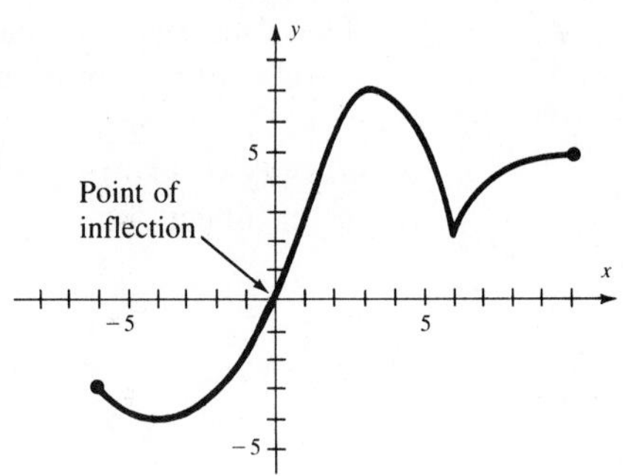

Solution

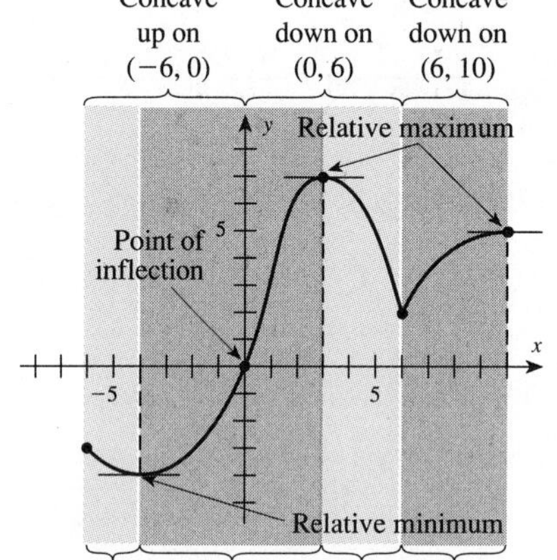

The horizontal tangents are at $x = -4$, $x = 3$, and $x = 10$.

(There is no horizontal tangent at $x = 6$, since there is not even a derivative defined at $x = 6$.)

Second-Derivative Test

For most of your work you will use what is called the **second-derivative test**. Figure 12.9 illustrates the two possibilities that give a relative maximum or a relative minimum.

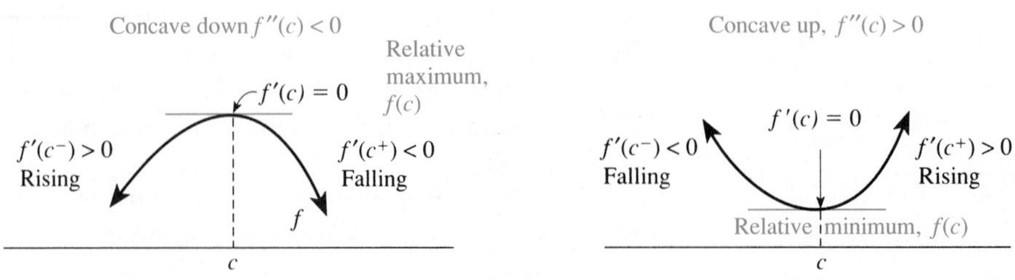

FIGURE 12.9

If a curve has a horizontal tangent line at $x = c$ and is concave upward ($f'' > 0$) in an interval containing the point of tangency, then $f(c)$ is a relative minimum. On the other hand, if the curve is concave downward ($f'' < 0$), then $f(c)$ is a relative maximum. This observation leads us to the second-derivative test:

Second-Derivative Test

If f is a continuous function on an interval (a, b),

1. Find the critical values of f. Suppose that c is a critical value of f.
2. **a.** If $f''(c) > 0$, then $f(c)$ is a relative minimum.
   Note: Greater than zero, concave *up*; thus a relative minimum.
   b. If $f''(c) < 0$, then $f(c)$ is a relative maximum.
   Note: Less than zero, concave *down*; thus a relative maximum.
   c. If $f''(c) = 0$, then the second-derivative test fails and gives no information, so the first-derivative test must be used.

EXAMPLE 5 Graph the parabola defined by $f(x) = x^2 - 8x + 7$ by using the second-derivative test.

Solution $f(x) = x^2 - 8x + 7$
$f'(x) = 2x - 8$; critical value $x = 4$
$f''(x) = 2$; thus, there is a relative minimum at $x = 4$ since $f''(4) = 2 > 0$. This minimum value is

$$f(4) = 16 - 8(4) + 7 = -9$$

■

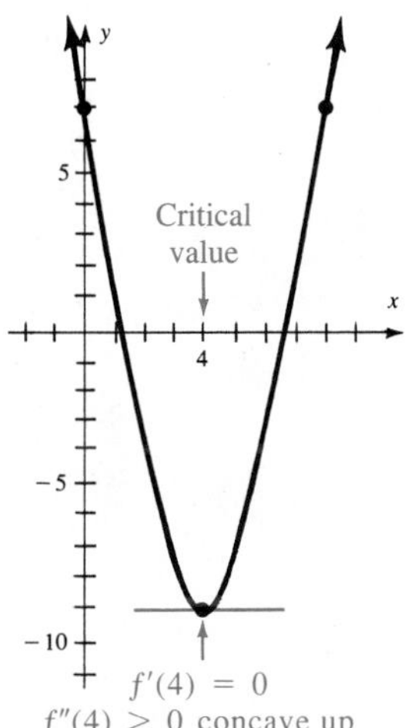

EXAMPLE 6 Graph $f(x) = 3x^4 + 8x^3 - 6x^2 - 24x + 6$.

Solution You need to keep track of the function for points on the curve, the derivative for slope and critical values, and the second derivative for concavity and points of inflection. We will keep track of this with the following table.

Function	First derivative	Second derivative
$f(x) = 3x^4 + 8x^3 - 6x^2 - 24x + 6$	$f'(x) = 12x^3 + 24x^2 - 12x - 24$	$f''(x) = 36x^2 + 48x - 12$
	Critical values: This is a polynomial, so the derivative is defined everywhere. $12x^3 + 24x^2 - 12x - 24 = 0$ $x^3 + 2x^2 - x - 2 = 0$ $x^2(x+2) - (x+2) = 0$ $(x^2 - 1)(x+2) = 0$ $(x-1)(x+1)(x+2) = 0$ $x = 1, -1, -2$	
Evaluate function to plot points:		**Second-derivative test:**
$f(1) = 3(1)^4 + 8(1)^3 - 6(1)^2 - 24(1) + 6$ $= -13$	Check $x = 1$	$f''(1) = 36(1)^2 + 48(1) - 12$ $= 72 > 0$
Point $(1, -13)$		Relative minimum $f(1) = 13$ at $x = 1$
$f(-1) = 3 - 8 - 6 + 24 + 6$ $= 19$	Check $x = -1$	$f''(-1) = 36 - 48 - 12$ $= -24 < 0$
Point $(-1, 19)$		Relative maximum $f(-1) = 19$ at $x = -1$
$f(-2) = 14$		$f''(-2) > 0$
Point $(-2, 14)$	Check $x = -2$	Relative minimum $f(-2) = 14$ at $x = -2$
		Check concavity: $36x^2 + 48x - 12 = 0$ $3x^2 + 4x - 1 = 0$ $x = \dfrac{-2 \pm \sqrt{7}}{3}$ $\approx .22, -1.55$
$f(.22) \approx .52$	$f'(.22) \approx -25$	Check $x = \dfrac{-2 + \sqrt{7}}{3}$ $\approx .22$
Point $(.22, .52)$	decreasing	Point of inflection at $x \approx .22$
$f(-1.55) \approx 16.31$	$f'(-1.55) \approx 7.6$	Check $x = \dfrac{-2 - \sqrt{7}}{3}$ ≈ -1.55
Point $(-1.55, 16.31)$	increasing	Point of inflection at $x \approx -1.55$

If there were any other relative maximums or minimums, they would have shown up in the list of critical values (which they did not), so it is easy to complete the graph. The graph is shown in Figure 12.10. ■

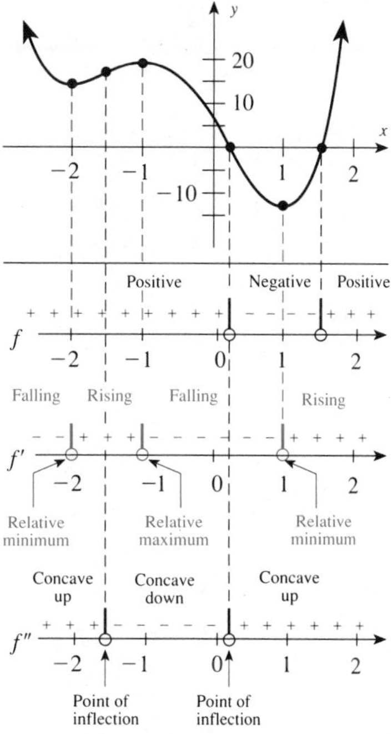

FIGURE 12.10
Graph of $f(x) = 3x^4 + 8x^3 - 6x^2 - 24x + 6$

EXAMPLE 7 Graph $f(x) = x^4$.

Solution

Function	First derivative	Second derivative
$f(x) = x^4$	$f'(x) = 4x^3$	$f''(x) = 12x^2$
	Critical values: $4x^3 = 0$ $x = 0$	
$f(0) = 0$ Point $(0, 0)$	Check $x = 0$	**Second-derivative test:** $f''(x) = 0$ Test fails.
$f(-1) = 1$ Point $(-1, 1)$ $f(1) = 1$ Point $(1, 1)$ Point $(0, 0)$ Plot some points: $f(2) = 16$ $f(-2) = 16$	**First-derivative test:** Choose $c^- = -1$ $f'(-1) < 0$ decreasing Choose $c^+ = 1$ $f'(1) > 0$ increasing relative minimum at $x = 0$	

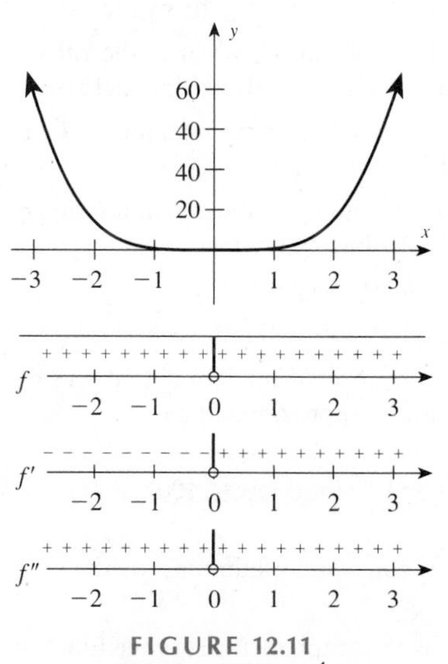

FIGURE 12.11
Graph of $f(x) = x^4$

12.2 Problem Set

Find the indicated derivatives for the functions in Problems 1–12.

1. $f(x) = 2x^5 - 3x^4 + x^3 - 5x^2 + 19x - 120;\ f'''(x)$
2. $f(x) = 25 + x - 3x^2 + 4x^3 - x^4 + 3x^5;\ f^{(5)}(x)$
3. $y = \sqrt{5x};\ \frac{d^4y}{dx^4}$
4. $y = 18\sqrt{3x};\ \frac{d^3y}{dx^3}$
5. $g(x) = 4x^3 - 2x^{-1};\ g^{(4)}(x)$
6. $g(x) = 5x^{-2} - 3x;\ g'''(x)$
7. $y = 3x^{5/3};\ \frac{d^3y}{dx^3}$
8. $y = -4x^{-3/2};\ \frac{d^4y}{dx^4}$
9. $y = \frac{3}{x-1};\ \frac{d^3y}{dx^3}$
10. $y = \frac{-5}{2x+1};\ \frac{d^3y}{dx^3}$
11. $y = \frac{x^2-1}{x+4};\ \frac{d^2y}{dx^2}$
12. $y = \frac{3x^2+5}{x-9};\ \frac{d^2y}{dx^2}$

Find all relative maximums and minimums in Problems 13–30 by using either the first- or the second-derivative test as appropriate. You do not need to graph these functions.

13. $f(x) = (2x + 1)^4$
14. $f(x) = (3x - 5)^4$
15. $f(x) = \frac{x^2+9}{x}$
16. $f(x) = \frac{x^2+1}{x}$
17. $f(x) = \sqrt{x^2+1}$
18. $f(x) = \sqrt{x^2+4}$
19. $f(x) = 8x^2 - x^4$
20. $f(x) = 2x^2 - x^4$
21. $f(x) = x^3 + 5x^2 - 8x + 10$
22. $f(x) = 6x^3 - 21x^2 - 36x + 15$
23. $y = 4x^3 - 27x^2 - 30x - 6$
24. $y = 2x^3 + 7x^2 - 40x + 5$
25. $g(x) = x + \frac{2}{x}$
26. $g(x) = x + \frac{4}{x}$
27. $g(x) = \frac{x^2}{x-1}$
28. $g(x) = \frac{x^2}{x-9}$
29. $f(x) = (x + 1)^{2/3}$
30. $f(x) = (x - 3)^{2/3}$

For the functions in Problems 31–40, **(a)** *find the critical values;* **(b)** *determine the intervals for which the functions are increasing and for which they are decreasing; and* **(c)** *determine the intervals for which the functions are concave up or concave down.*

31. $f(x) = 3 + 12x - x^2$
32. $f(x) = 8 + 6x - x^2$
33. $g(x) = x^3 - 7x^2 - 5x + 8$
34. $g(x) = x^3 - 5x^2 - 8x + 10$
35. $y = x^3 + 11x^2 - 45x + 125$
36. $y = x^3 - 2x^2 - 15x - 75$
37. $12x^3 - 5x^2 - 4x - 2y + 14 = 0$
38. $2x^3 - 12x^2 - 30x - 6y + 5 = 0$
39. $f(x) = (2 - x)^3 - 8$
40. $f(x) = (1 + x)^3 - 1$

APPLICATIONS

If $S(x)$ is the number of units of a product sold after spending x dollars on advertising, then the ***point of diminishing returns*** *is that point of inflection for S at which the rate of change of sales changes from positive to negative.*

41. A company estimates that it will sell $S(x)$ units of a product after spending x thousands of dollars on advertising, according to the formula

$$S(x) = 4{,}000 - x^3 + 45x^2 + 60x$$

for $10 \le x \le 35$. What is the point of diminishing returns?

42. A company estimates that it will sell $S(x)$ units of a product after spending x thousands of dollars on advertising, according to the formula

$$S(x) = 20{,}000 - 4x^3 + 180x^2 - 2{,}400x$$

for $10 \le x \le 30$. What is the point of diminishing returns?

43. For the application in Problem 41, when is the rate of change of sales per unit change in advertising increasing?
44. For the application in Problem 42, when is the rate of change of sales per unit change in advertising increasing?
45. What are the sales corresponding to the maximum rate of change for the sales in Problem 41?
46. What are the sales corresponding to the maximum rate of change for the sales in Problem 42?
47. Graph both S and S' for Problem 41.
48. Graph both S and S' for Problem 42.
49. The marketing research department found that the demand for a product can be approximated by

$$p = 1{,}156 - \frac{1}{12}x^2 \qquad \text{for } 0 \le x \le 100$$

Find the relative maximums and minimums for the revenue function.

50. Over which intervals is the graph of the revenue function in Problem 49 concave upward? Concave downward?

51. A manufacturer has determined that the demand of selling items is given by the formula

$$p(x) = 5 - \left(\frac{x}{100}\right)^2$$

a. Find an expression for the revenue.
b. Find the marginal revenue.
c. Is the marginal revenue increasing or decreasing when $x = 100$ items?

12.3 Curve Sketching—Relative Maximums and Minimums

One of the most important tools in mathematics is the ability to quickly sketch a wide variety of functions. In this section we combine the graphing techniques of calculus with those we used in Chapter 2.*

Table 12.2 presents procedures for graphing functions. It is, of course, not necessary to use all the procedures to graph every function. And, if any of the procedures are too difficult, it is possible to omit that particular procedure. The more you know about a curve, the fewer points you need to plot.

TABLE 12.2
Graphing Strategy

Step	Procedure
Simplify	First, simplify, if possible, the function you wish to graph. That is, combine similar terms, reduce fractions, and simplify radical expressions.
Second-derivative test	Use the second-derivative test to find the relative maximum or minimum values: 1. Find the critical values, c. 2. $f''(c) > 0$, relative minimum $f(c)$ at $x = c$; concave up $f''(c) < 0$, relative maximum $f(c)$ at $x = c$; concave down $f''(c) = 0$, test fails
First-derivative test	Use the first-derivative test if the second-derivative test fails: 1. Relative minimum $f(c)$ if $f'(c^-) < 0$ and $f'(c^+) > 0$ 2. Relative maximum $f(c)$ if $f'(c^-) > 0$ and $f'(c^+) < 0$ The curve is increasing if $f'(x) > 0$ The curve is decreasing if $f'(x) < 0$
Asymptotes	If $f(x) = P(x)/D(x)$, where $P(x)$ and $D(x)$ are polynomial functions with no common factors, then: 1. *Vertical asymptote:* $x = r$ if $D(r) = 0$ 2. *Horizontal asymptote:* If $\lim_{x \to \infty} f(x) = L$ then $y = L$ is a horizontal asymptote.
Intercepts	1. *x-intercept:* Set $y = 0$ and solve for x. 2. *y-intercept:* Set $x = 0$ and solve for y.
Plot points	Plot any additional points necessary to draw the graph.

* If you omitted Chapter 2, it is now a good idea to look at Section 2.6.

EXAMPLE 1 Graph $f(x) = \dfrac{x^2}{x-2}$.

Solution

Function	First derivative	Second derivative
$f(x) = \dfrac{x^2}{x-2}$	$f'(x) = \dfrac{(x-2)(2x) - x^2(1)}{(x-2)^2}$ $= \dfrac{2x^2 - 4x - x^2}{(x-2)^2}$ $= \dfrac{x^2 - 4x}{(x-2)^2} = \dfrac{x(x-4)}{(x-2)^2}$	$f''(x) = \dfrac{(x-2)^2(2x-4) - x(x-4)(2)(x-2)}{(x-2)^4}$ $= \dfrac{8}{(x-2)^3}$ There were several simplification steps left out; can you fill in the details?
	Critical values: $\dfrac{x(x-4)}{(x-2)^2} = 0$ $x(x-4) = 0$ $x = 0, 4$ Critical values	
$f(0) = \dfrac{0^2}{0-2} = 0$ Point $(0, 0)$ $f(4) = \dfrac{4^2}{4-2} = 8$ Point $(4, 8)$	Check $x = 0$ Check $x = 4$	**Second-derivative test** $f''(0) = \dfrac{8}{(0-2)^3} = -1 < 0$ Relative maximum at $x = 0$ $f''(4) = \dfrac{8}{(4-2)^3} = 1 > 0$ Relative minimum at $x = 4$
Intercepts: $x = 0$ found above $y = 0$: $0 = \dfrac{x^2}{x-2}$ $0 = 0$ $(0, 0)$ is only intercept.	**Increasing and Decreasing** x: −− | +++ | +++ | +++ $x - 4$: −− | −−− | −−− | +++ $(x-2)^2$: ++ | +++ | +++ | +++ 0, 2, 4 Derivative: Pos. Neg. Neg. Pos. Function: Rising Falling Falling Rising	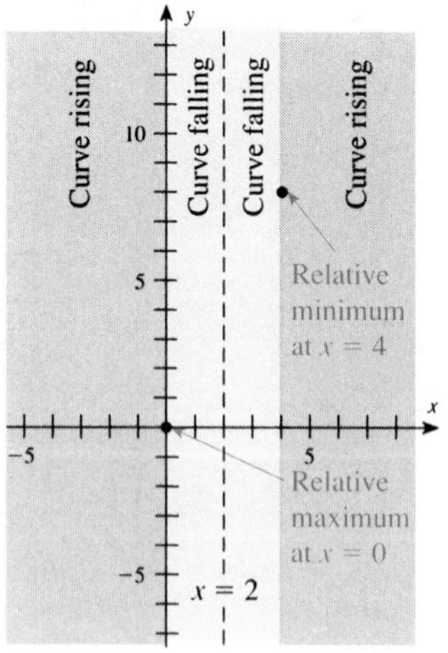 Curve rising Curve falling Curve falling Curve rising Relative minimum at $x = 4$ Relative maximum at $x = 0$ $x = 2$ y, x, 10, 5, −5, 5, −5

Since this is a rational function, we check asymptotes. There is a vertical asymptote $x = 2$. Check for a horizontal asymptote:

$$\lim_{|x|\to\infty} \frac{x^2}{x-2} \text{ does not exist; no horizontal asymptotes.}$$

$$\left[\textit{Note:} \quad \text{Check both } \lim_{x\to\infty} \frac{x^2}{x-2} \text{ and } \lim_{x\to-\infty} \frac{x^2}{x-2}.\right]$$

Draw the asymptote, and plot another value, say $x = 8$, to draw the graph shown in Figure 12.12.

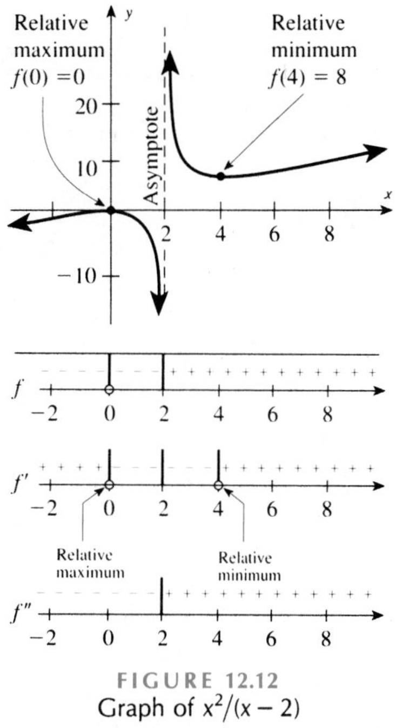

FIGURE 12.12
Graph of $x^2/(x-2)$

EXAMPLE 2 Graph $f(x) = x + \frac{9}{x}$.

Solution The domain excludes $x = 0$.

Function	First derivative	Second derivative
$f(x) = x + 9x^{-1}$	$f'(x) = 1 - 9x^{-2}$	$f''(x) = 18x^{-3}$
	Critical values: $1 - 9x^{-2} = 0$ $x^2 = 9$ $x = \pm 3$ Also, derivative does not exist at $x = 0$, but this value is not in the domain for f.	

(continued)

Function	First derivative	Second derivative
$f(3) = 3 + 9(3)^{-1}$ $= 6$ Point (3, 6) $f(-3) = -3 + 9(-3)^{-1}$ $= -6$ Point $(-3, -6)$	Check $x = 3$ Check $x = -3$	**Second-derivative test:** $f''(3) = 18(3)^{-3} = \frac{2}{3} > 0$ Relative minimum at $x = 3$ $f''(3) = 18(-3)^{-3} = -\frac{2}{3} < 0$ Relative maximum at $x = -3$
Intercepts: $x = 0$: not in the domain; no y-intercept $y = 0$: $0 = x + 9x^{-1}$ $x^2 + 9 = 0$ No solution; no x-intercept **Asymptotes:** Horizontal: $x = 0$ Vertical: $\lim_{\lvert x\rvert \to \infty} f(x)$ does not exist. No vertical asymptotes	**Increasing and decreasing:** *Note:* $f'(x) = 1 - 9x^{-2}$ $= \dfrac{x^2 - 9}{x^2}$ $= \dfrac{(x-3)(x+3)}{x^2}$ Factors: $x - 3$: − − | − − | − − | + + + $x + 3$: − − | + + | + + | + + + x^2: + + | + + | + + | + + + (number line: −3, 0, 3) Derivative: Pos. Neg. Neg. Pos. Function: Rising Falling Falling Rising f'': Neg. Neg. Pos. Pos. Concavity: Down Down Up Up	**Concavity:** $f''(x) = \dfrac{18}{x^3}$ **Points of inflection:** $f''(x) \neq 0$; none (graph labels: y; x; −15; 15; Curve rising; $x = 0$ asymptote; Relative minimum at $x = 3$; Relative maximum at $x = -3$; Curve falling; Curve rising; Concave down; Concave up)

One additional observation that might help with this problem has to do with the limit we considered when looking for a horizontal asymptote:

$$\lim_{|x| \to \infty} x + \frac{9}{x}$$

Although it is true that this limit does not exist, you might also notice that, for large values of $|x|$, the term $9/x$ is negligible, so that the values of f are almost identical to the values on the line $y = x$. This line is called a **slant asymptote**. If you draw the line $y = x$ you can use this line to help you sketch f as shown in Figure 12.13.

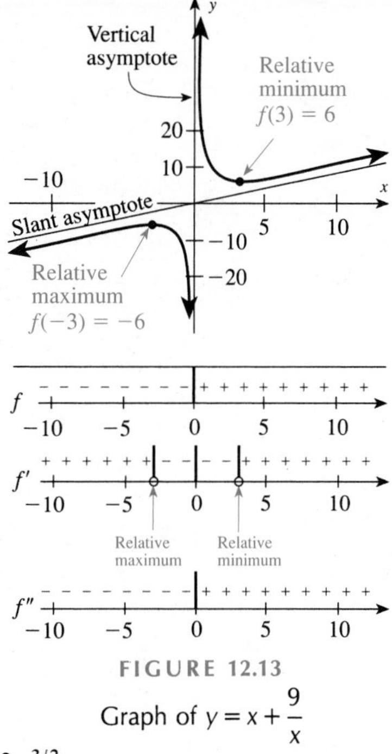

FIGURE 12.13
Graph of $y = x + \frac{9}{x}$

EXAMPLE 3 Graph $f(x) = 12x^{1/2} - 2x^{3/2}$.

Solution The domain is all nonnegative real numbers.

Function	First derivative	Second derivative
$f(x) = 12x^{1/2} - 2x^{3/2}$	$f'(x) = 6x^{-1/2} - 3x^{1/2}$ $= 3x^{-1/2}(2 - x)$	$f''(x) = -3x^{-3/2} - \frac{3}{2}x^{-1/2}$ $= -\frac{3}{2}x^{-3/2}(2 + x)$
	Critical values: First, $f'(x) = 0$ if $x = 2$ Not defined if $x = 0$	
$f(2) = 12(2)^{1/2} - 2(2)^{3/2}$ $= 12\sqrt{2} - 4\sqrt{2}$ $= 8\sqrt{2}$ Point $(2, 8\sqrt{2})$ $f(0) = 12(0)^{1/2} - 2(0)^{3/2}$ $= 0$ Point $(0, 0)$	Check $x = 2$ Check $x = 0$	**Second-derivative test:** $f''(2) = -\frac{3}{2}(2)^{-3/2}(2 + 2)$ < 0 Relative maximum at $x = 2$ $f''(x) = -3(0)^{-3/2} - \frac{3}{2}(0)^{-1/2}$ $= 0$ Test fails
	First-derivative test: Values to the left of 0 not defined	

(continued)

Function	First derivative	Second derivative
There are no asymptotes.	$f(c) > 0$ to the right of critical value $x = 0$; this is a relative minimum	
Intercepts: y-intercept ($x = 0$): $12x^{1/2} - 2x^{3/2}$ $= 12(0)^{1/2} - 2(0)^{3/2}$ $= 0$ Point $(0, 0)$ x-intercept ($y = 0$): $12x^{1/2} - 2x^{3/2} = 0$ $2x^{1/2}(6 - x) = 0$ $x = 0, 6$ These are the points $(6, 0)$ and $(0, 0)$.	**Increasing and Decreasing:** $2 - x$: Not in domain; $+ +$; $- - -$ x: Not in domain; $+ +$; $+ + +$ (0, 2) Derivative: Positive, Negative Function: Rising, Falling f'': Negative, Negative Concavity: Down, Down $f'(6) = 3(6)^{-1/2}(2 - 6)$ < 0 Falling at $x = 6$	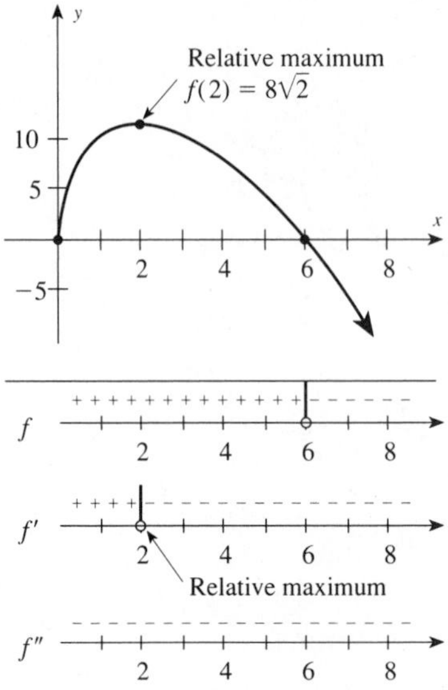 FIGURE 12.14 Graph of $12x^{1/2} - 2x^{3/2}$

The graph is shown in Figure 12.14 ■

12.3 Problem Set

In Problems 1–6, draw a curve on the interval (a, b) satisfying the stated conditions.

1. $f'(x) > 0$ and $f''(x) > 0$
2. $f'(x) > 0$ and $f''(x) < 0$
3. $f'(x) < 0$ and $f''(x) > 0$
4. $f'(x) < 0$ and $f''(x) < 0$
5. Passes through $(-2, 3)$, $(0, 5)$, and $(2, 7)$; $f'(-2) = 0$ and $f'(2) = 0$; $f''(x) > 0$ if $x < 0$, $f''(x) < 0$ if $x > 0$, and $f''(0) = 0$
6. Passes through $(2, 8)$, $(5, 6)$, and $(8, 4)$; $f'(2) = 0$ and $f'(8) = 0$; $f''(x) < 0$ if $x < 5$, $f''(x) > 0$ if $x > 5$, and $f''(5) = 0$

Graph the functions in Problems 7–34.

7. $y = 6x - x^2$
8. $y = 20x - 5x^2$
9. $y = 2x^2 - 3x + 5$
10. $3x^2 + 4x - 2y + 8 = 0$
11. $y = x^3 + 3x^2 - 9x + 5$
12. $y = x^3 - 3x^2 - 24x + 10$
13. $y = x^3 - 48x + 50$
14. $y = x^3 - 75x + 1$
15. $f(x) = x^3 + 1$
16. $f(x) = x^5$
17. $g(x) = x^4 - 1$
18. $g(x) = (x - 1)^4$
19. $f(x) = x^4 - 2x^2$
20. $f(x) = x^4 - 8x^2$

21. $g(x) = x + \frac{4}{x}$

22. $g(x) = x + \frac{1}{x}$

23. $f(x) = 3x^4 + 8x^3 - 6x^2 - 24x - 5$

24. $f(x) = 3x^4 + 20x^3 - 24x^2 - 240x + 20$

25. $y = \frac{x^2}{x - 1}$

26. $y = \frac{x^2}{x + 1}$

27. $y = \frac{x + 1}{x - 1}$

28. $y = \frac{2x + 1}{x - 1}$

29. $y = \frac{x - 5}{x + 5}$

30. $y = \frac{x - 2}{x + 2}$

31. $y = 3x^{2/3}$

32. $y = -3x^{1/3}$

33. $f(x) = 6x^{1/2} - 4x^{3/2}$

34. $f(x) = 6x^{1/2} - 12x^{3/2} + 5$

APPLICATIONS

35. The cost of removing $p\%$ of the pollutants from the atmosphere is given by the formula

$$C(p) = \frac{20{,}000p}{100 - p}$$

Graph $C(p)$ on $[0, 100)$.

36. A manufacturer has determined that the demand of selling x items is given by the formula

$$p(x) = 5 - \left(\frac{x}{100}\right)^2$$

Sketch the marginal revenue function.

37. The cost of manufacturing a certain item is

$$C(x) = 5{,}000 + \frac{1}{2}x^2$$

where x is the number of units produced. Graph the average cost function and the marginal cost function on the same coordinate axes.

38. The cost of manufacturing a product is

$$C(x) = 10{,}000 + .1x^2$$

where x is the number of units produced. Graph the average cost function and the marginal cost function on the same coordinate axes.

39. The concentration C of a drug in the bloodstream t minutes after injection is given by the formula

$$C(t) = \frac{10t}{8 + t^3}$$

Graph $C(t)$.

40. The number of fish swimming upstream to spawn is approximated by the formula

$$S(x) = 13{,}200x - 2x^3 - 45x^2 - 10{,}000$$

where x is the temperature of the water in degrees Fahrenheit $(30 \le x \le 70)$. Graph $S(x)$.

12.4 Absolute Maximum and Minimum

We can now use the derivative in one of its most powerful applications for management, life sciences, and social sciences—that of finding the absolute maximum and absolute minimum of a given function. All of the necessary techniques have now been developed, so the examples in this section illustrate the wide variety of ways these ideas can be applied. Absolute maximum and minimum were defined in the last section, and we only need now to be aware that the **absolute maximum** of a continuous function will occur at the critical value giving the largest value of the function, or else it will occur at an endpoint of a closed interval. On the other hand, the **absolute minimum** will occur at the critical value giving the smallest value of the function or at an endpoint. These situations were illustrated in Figure 12.3, which is repeated at the top of page 602 for easy reference.

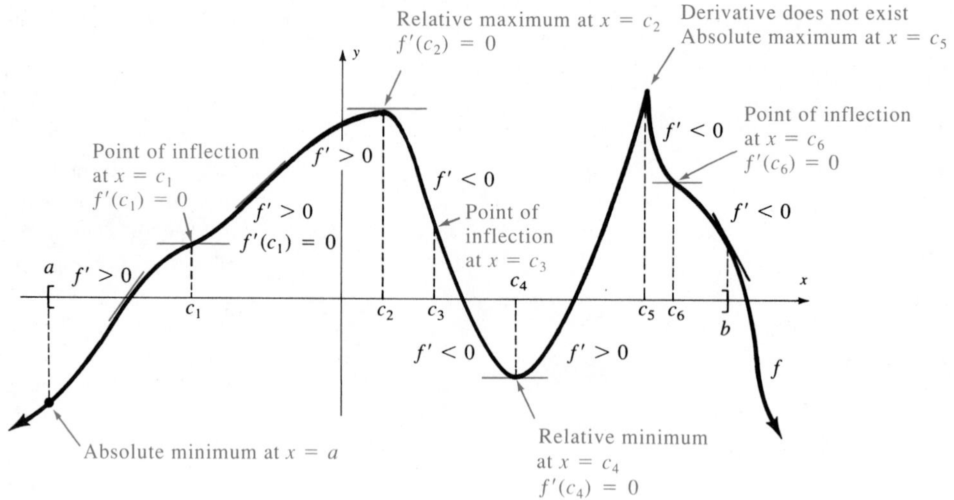

CALCULATOR COMMENT

Graphing calculators and the [trace] key allow you to look at a function on some interval and then to approximate the maximum and minimum values to any reasonable desired degree of accuracy.

Procedure for Finding the Absolute Maximum and Absolute Minimum

Given a continuous function f over an interval $[a, b]$, there exists an absolute maximum and minimum for f on the interval. To find these absolute extremes, carry out the following steps:

1. Find the endpoints and the critical values: $a, c_1, c_2, \ldots, c_n, b$. Remember, the critical values are values that cause the derivative to be either zero or undefined.
2. Evaluate $f(a), f(c_1), f(c_2), \ldots, f(b)$.
3. The **absolute maximum** of f is the largest of the values found in step 2.
4. The **absolute minimum** of f is the smallest of the values found in step 2.

EXAMPLE 1 Find the absolute maximum and minimum for

$$f(x) = 3x^5 - 50x^3 + 135x + 20$$

on $[-2, 4]$.

Solution **1.** $f'(x) = 15x^4 - 150x^2 + 135$

2. Critical values:

$$\begin{aligned} 15x^4 - 150x^2 + 135 &= 0 \\ x^4 - 10x^2 + 9 &= 0 \\ (x^2 - 9)(x^2 - 1) &= 0 \\ (x - 3)(x + 3)(x - 1)(x + 1) &= 0 \\ x &= 3, -3, 1, -1 \end{aligned}$$

3. List: -2, $-1, 1, 3$, 4

Endpoints: -2 and 4

Critical values: $-1, 1, 3$

$x = -3$ is not in the interval $[-2, 4]$, so do not test $x = -3$.

4. Test each of these values:

$$\begin{aligned} f(-2) &= 3(-2)^5 - 50(-2)^3 + 135(-2) + 20 = 54 \\ f(-1) &= 3(-1)^5 - 50(-1)^3 + 135(-1) + 20 = -68 \\ f(1) &= 3(1)^5 - 50(1)^3 + 135(1) + 20 = 108 \\ f(3) &= 3(3)^5 - 50(3)^3 + 135(3) + 20 = -196 \\ f(4) &= 3(4)^5 - 50(4)^3 + 135(4) + 20 = 432 \end{aligned}$$

■

CALCULATOR COMMENT

When graphing a function on a calculator, the most difficult part is determining the correct domain and range. For example, on a TI81 if you input the function in Example 1 using the Y= key and then graph using the standard domain and range you obtain:

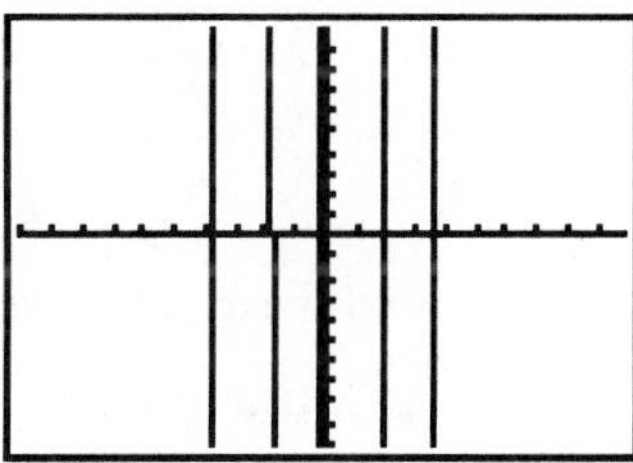

Now, use the RANGE key to set Xmin = −2; Xmax = 10; then use the TRACE key to find X = 4.0631579 Y = 536.84 and X = 2.963158 Y = −194.1321 so that you use the RANGE key to set Ymin = −200;

Ymax = 550. The [GRAPH] now shows:

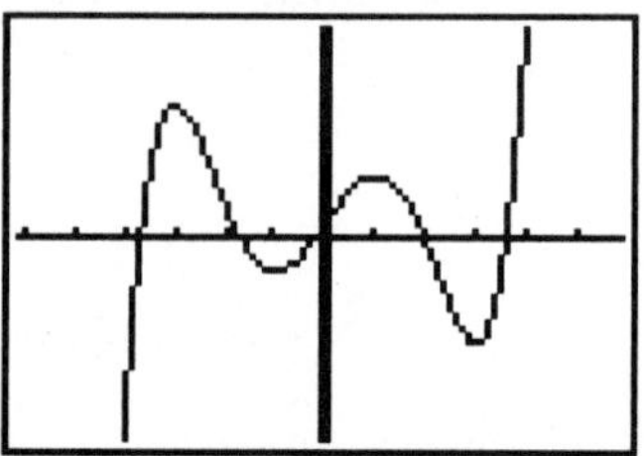

Now the [TRACE] gives an approximation for the extreme values: Min: X = 2.9894737, Y = −195.9604 and the Max: X = 4 Y = 432

■

As noted, the real value of finding absolute maximum and minimum values is in the applications to which these ideas can be applied. The remainder of this section concerns such applications.

EXAMPLE 2 Westel Corporation manufactures telephones and has developed a new cellular phone. Production analysis shows that its price must not be less than \$50; if x units are sold, then the demand is given by the formula

$$p(x) = 150 - x$$

The total cost of producing x units is given by the formula

$$C(x) = 2{,}500 + 30x$$

Find the maximum profit, and determine the price that should be charged to maximize the profit.

Solution The total profit, $P(x)$, is found by

$$\begin{aligned} P(x) &= \text{TOTAL REVENUE} - \text{TOTAL COST} \\ &= (\text{number of items})\overbrace{(\text{price per item})}^{\text{demand function}} - \text{cost} \\ &= x \cdot p(x) - C(x) \\ &= x(150 - x) - (2{,}500 + 30x) \\ &= 150x - x^2 - 2{,}500 - 30x \\ &= -x^2 + 120x - 2{,}500 \end{aligned}$$

Find the absolute maximum value of P:

1. Critical points: $P'(x) = -2x + 120$ and $-2x + 120 = 0$ when $x = 60$. There are no values for which the derivative is not defined, so there is one critical value: $x = 60$.
2. Endpoints: Given $p(x) \geq 50$, we see that

$$\begin{aligned} 150 - x &\geq 50 \\ 100 &\geq x \\ x &\leq 100 \end{aligned}$$

so one endpoint is 100. The other endpoint is merely implied, namely, 0 phones, so that the domain is $[0, 100]$.

3. Find the absolute maximum of P by checking $x = 0$, 60, and 100:

$$P(0) = -2{,}500$$
$$P(60) = -(60)^2 + 120(60) - 2{,}500 = 1{,}100$$
$$P(100) = -(100)^2 + 120(100) - 2{,}500 = -500$$

4. The absolute maximum of P is 1,100 and occurs when $x = 60$.

The answer to the question asked might not be the absolute maximum or minimum value, and you must be careful to answer the question asked. For example, this problem asks for the price. The price needs to be found for $x = 60$. Use the demand function:

$$p(60) = 150 - 60 = 90$$

The phones should be priced at \$90 per unit. ■

The analysis in Example 2 uses the techniques of this chapter, but we were solving this type of problem in Chapter 2. Two additional techniques will simplify the work shown in Example 2. The first relates second-degree problems to the material of this chapter. It is called the **second-derivative test for absolute maximum and minimum**:

Second-Derivative Test for Absolute Maximum and Minimum

If f is a continuous function with only *one* critical value c and if $f''(c)$ exists and is not zero [that is, if $f''(c) = 0$, then this test fails], and if

$f''(c) > 0$ then there is an *absolute minimum* at $x = c$

$f''(c) < 0$ then there is an *absolute maximum* at $x = c$

This test tells us in Example 2 that since there was only one critical value at $x = 60$ and since $P''(x) = -2$, $x = 60$ *must* give the absolute maximum value of P, so we did not need to go through the evaluation of P at 0, 60, and 100.

The second technique involves assuming that the business in question has known cost and revenue functions $C(x)$ and $R(x)$, as shown in Figure 12.15 on page 606. Profit is positive when $R(x)$ exceeds $C(x)$. The points a and b are the break-even points.

The maximum profit occurs at a critical point c_1 of $P(x)$. Assume that $P'(x)$ exists for all x in some interval, usually $[0, \infty)$ and that the critical point c_1 occurs at some x in this interval, so that

$$P'(x) = 0$$

Then

$$P(x) = R(x) - C(x)$$
$$P'(x) = R'(x) - C'(x)$$
$$0 = R'(x) - C'(x)$$
$$R'(x) = C'(x)$$

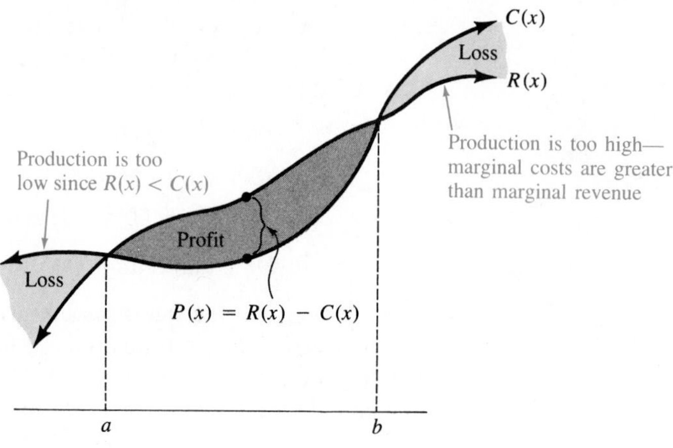

FIGURE 12.15
Maximum value of a profit function

Maximum Profit

> The maximum profit is achieved when the marginal revenue and marginal cost are equal.

Turning, one final time, to Example 2, we see that

$$R(x) = x(150 - x) = 150x - x^2 \qquad \text{and} \qquad C(x) = 2{,}500 + 30x$$

so

$$R'(x) = 150 - 2x \qquad\qquad C'(x) = 30$$

Thus, the maximum profit is found when

$$\begin{aligned} 150 - 2x &= 30 \\ 120 &= 2x \\ 60 &= x \end{aligned}$$

It is well known in economics that the minimal average cost of a product occurs when the average cost is equal to the marginal cost. Although this principle will not be proved here (see Problem 47 in Problem Set 12.4), it is illustrated in Example 3.

EXAMPLE 3 A product has a total cost function given by

$$C(x) = .125x^2 + 20{,}000$$

where x is the number of units produced. Show that the minimal average cost occurs when the average cost is equal to the marginal cost.

Solution First, find the average cost: $\bar{C}(x) = \dfrac{C(x)}{x} = .125x + 20{,}000x^{-1} \qquad x > 0$

Next, find the minimal average cost: $\bar{C}'(x) = .125 - 20{,}000x^{-2}$
Critical values:

$$\begin{aligned} \bar{C}'(x) &= 0 \\ .125 - 20{,}000x^{-2} &= 0 \\ .125 &= 20{,}000x^{-2} \\ .125x^2 &= 20{,}000 \\ x^2 &= 160{,}000 \\ x &= 400, -400 \end{aligned}$$

↑ Reject since $x > 0$.

There is only one critical value, $c = 400$ (see Figure 12.16). Use the second-derivative test.

$$\bar{C}''(x) = 40{,}000x^{-3}$$

and $\bar{C}''(400) > 0$, so by the second-derivative test for absolute maximum and minimum, the absolute minimum occurs at $x = 400$: $\bar{C}(400) = .125x + 20{,}000(400)^{-1} = 100$.

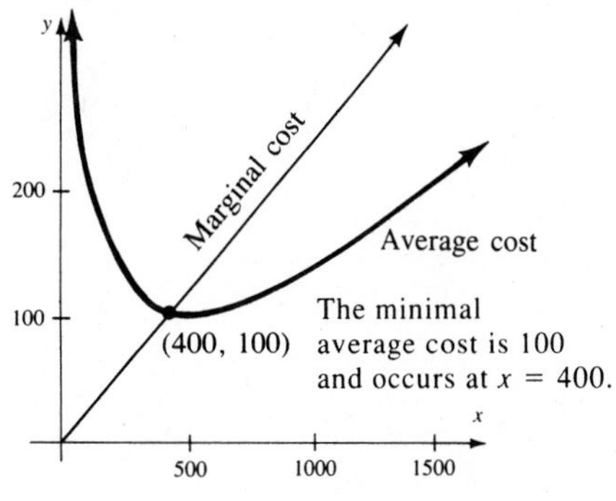

FIGURE 12.16
The minimal average cost of a product occurs when the average cost is equal to the marginal cost.

Now, find the marginal cost: $C'(x) = .25x$. Find the value for which the average cost is equal to the marginal cost:

$$\frac{C(x)}{x} = C'(x)$$

Then

$$\begin{aligned} .125x + 20{,}000x^{-1} &= .25x \\ 20{,}000x^{-1} - .125x &= 0 \\ 20{,}000 - .125x^2 &= 0 &&\text{Multiply both sides by } x\ (x > 0). \\ 160{,}000 - x^2 &= 0 &&\text{Divide both sides by } .125. \\ (400 - x)(400 + x) &= 0 \\ x &= 400, -400 \end{aligned}$$

↑ Reject since $x > 0$ ■

EXAMPLE 4 As more and more industrial areas are constructed, there is a growing need for standards ensuring the control of the pollutants released into the air. Suppose that the air pollution at a particular location is based on the distance from the source of the pollution according to the following principle:

For distances greater than or equal to 1 mile, the concentration of particulate matter (in parts per million, ppm) decreases as the reciprocal of the distance from the source.

This means that if you live 3 miles from a plant emitting 60 ppm, the pollution at your home is $\frac{60}{3} = 20$ ppm. On the other hand, if you live 10 miles from the plant, the pollution at your home is $\frac{60}{10} = 6$ ppm.

Suppose that two plants 10 miles apart are releasing 60 and 240 ppm, respectively, and you want to know the location between them at which the pollution is a minimum.

Solution

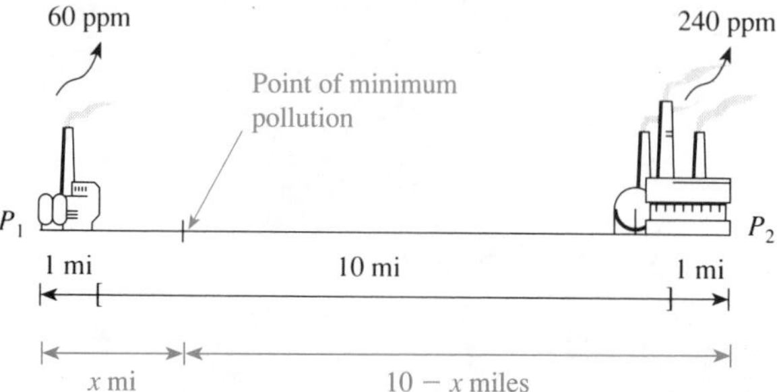

Let x be the distance from plant P_1. The domain is $[1, 9]$ since the given formula implies that you cannot be closer than 1 mile to either plant.

$$\text{Pollution from } P_1 = \frac{60}{x}$$

$$\text{Pollution from } P_2 = \frac{240}{10 - x}$$

The total pollution at any point is the sum of the pollution from the plants:

$$P(x) = \frac{60}{x} + \frac{240}{10 - x}$$

CALCULATOR COMMENT

Graph Y=
60/X+240/(10−X)
RANGE
Xmin=0
Xmax=10
Ymin=0
Ymax=300

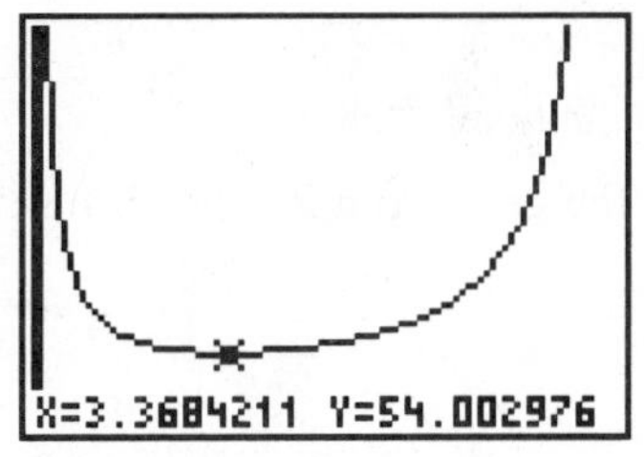

TRACE shows:
X=3.3684211 Y=54.002976

Find the absolute minimum value of P on $[1, 9]$:

1. Critical points:
   a. Values for which $P'(x) = 0$:

$$P(x) = 60x^{-1} + 240(10 - x)^{-1}$$
$$P'(x) = -60x^{-2} - 240(10 - x)^{-2}(-1)$$
$$= -60x^{-2} + 240(10 - x)^{-2}$$

Set $P'(x) = 0$ and solve:

$$\frac{-60}{x^2} + \frac{240}{(10 - x)^2} = 0 \qquad \text{Multiply by } x^2(10 - x)^2.$$
$$-60(10 - x)^2 + 240x^2 = 0 \qquad \text{Divide by 60.}$$
$$-(100 - 20x + x^2) + 4x^2 = 0$$
$$3x^2 + 20x - 100 = 0$$
$$(3x - 10)(x + 10) = 0$$
$$x = \frac{10}{3}, -10$$

Reject this value since it is not in the domain.

   b. Values for which $P'(x)$ do not exist: $x = 0$ and $x = 10$. Reject these since they are not in the domain $[1, 9]$.
2. Endpoints: $x = 1$, $x = 9$
3. Find the absolute minimum of P by checking $x = 1, \frac{10}{3}$, and 9:

$$P(1) = \frac{60}{1} + \frac{240}{9} \approx 87 \text{ ppm}$$
$$P\left(\frac{10}{3}\right) = \frac{60}{\frac{10}{3}} + \frac{240}{\frac{20}{3}} = 54 \text{ ppm}$$
$$P(9) = \frac{60}{9} + \frac{240}{1} \approx 247 \text{ ppm}$$

The minimum pollution is found at $3\frac{1}{3}$ miles from plant P_1. ■

EXAMPLE 5 The voting patterns of a geographical region show that the percent, P, of voters in a national election varies according to age, x, and fits the model predicted by the formula.

$$P(x) = .002x^3 - .195x^2 + 6x \qquad 18 \le x \le 50$$

Graph P and discuss its possible meaning, and then find the absolute maximum and minimum percentage of the population voting in a national election.

Solution $P'(x) = .006x^2 - .390x + 6$

Set $P'(x) = 0$ and solve to find the critical values:

$$.006x^2 - .390x + 6 = 0$$
$$x^2 - 65x + 1{,}000 = 0 \qquad \text{Divide by .006.}$$
$$(x - 25)(x - 40) = 0$$

$P''(x) = .012x - .390$, so

$$P''(25) = .012(25) - .390 < 0 \qquad \text{Relative maximum at } x = 25$$
$$P''(40) = .012(40) - .390 > 0 \qquad \text{Relative minimum at } x = 40$$

Find some values of P (which are needed both for the graph and for the absolute maximum and absolute minimum):

$$P(25) = 59.375$$
$$P(40) = 56$$
$$P(18) = 56.484$$
$$P(50) = 62.5$$

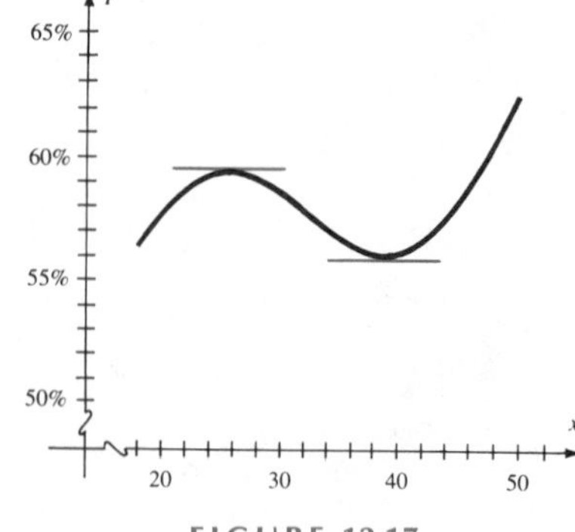

FIGURE 12.17
Graph of voting patterns

From ages 18 to 25 the percentage of the population voting is increasing (derivative positive), and from 25 to 40 it is decreasing, at which time it begins to increase again. From the graph in Figure 12.17 and the calculations we see that the absolute maximum percentage of the population voting is at age 50, the endpoint of the interval [18, 50]. ■

EXAMPLE 6 A shipping box is to be constructed from a piece of corrugated cardboard 17 inches long and 11 inches wide. A square is to be cut out of each corner so that the sides can be folded up to form the box. What is the maximum volume (to the nearest in.^{3}) for the finished box?

Solution Let x be the size of the cutout square. The volume, V, can be calculated as a function of x:

$$V(x) = lwh = (17 - 2x)(11 - 2x)x$$
$$= (187 - 56x + 4x^2)x$$
$$= 4x^3 - 56x^2 + 187x$$

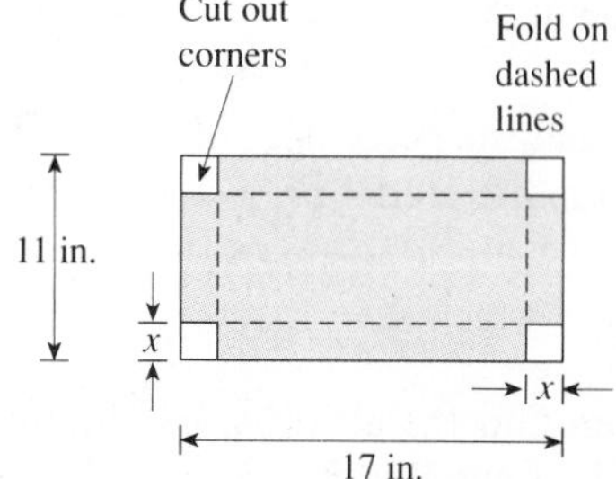

The domain is [0, 5.5] since x cannot be negative, nor can $11 - 2x$. Take the derivative and find the critical values:

$$V'(x) = 12x^2 - 112x + 187$$

$$V'(x) = 0 \quad \text{when} \quad x = \frac{112 \pm \sqrt{(-112)^2 - 4(12)(187)}}{2(12)}$$
$$\approx 7.1555308,\ 2.1778026$$

This first value is not in the domain, so $x \approx 2.18$ inches. The length of the box is $17 - 2x \approx 12.644395$ and the width is $11 - 2x \approx 6.6443948$. The volume is

$$V \approx 2.1778026 \times 12.644395 \times 6.6443948 \approx 182.96668$$

In practical measurements, you should cut $2\frac{1}{8}$ in. from each corner to form a box with dimensions

$$12\tfrac{3}{4} \text{ in. by } 6\tfrac{3}{4} \text{ in. by } 2\tfrac{1}{8} \text{ in.}$$

with a volume of 183 in.^{3}. ■

12.4 Problem Set

Find the absolute maximum and minimum for the functions in Problems 1–20 and also give the x-values for those extrema.

1. $f(x) = x^3 - 3x^2$ on $[-1, 4]$
2. $f(x) = x^3 - 12x$ on $[0, 4]$
3. $f(x) = 3x^5 - 50x^3 + 135x + 750$ on $[-5, 0]$
4. $f(x) = 3x^5 - 50x^3 + 135x + 750$ on $[-3, 3]$
5. $g(x) = x^3 - 2x^2 - 15x + 42$ on $[0, 5]$
6. $g(x) = x^3 + 5x^2 + 3x + 20$ on $[-5, 0]$
7. $h(x) = 3x^5 - 25x^3 + 60x - 200$ on $[0, 3]$
8. $h(x) = 3x^5 - 25x^3 + 60x - 200$ on $[-3, 3]$
9. $f(x) = 1 + x^{1/3}$ on $[-1, 8]$
10. $g(x) = 1 - x^{2/3}$ on $[-1, 8]$
11. $f(x) = \sqrt{x} + x$ on $[0, 4]$
12. $g(x) = x - \sqrt{x}$ on $[0, 9]$
13. $f(x) = (x + 1)^2(x - 3)$ on $[0, 5]$
14. $g(x) = (x - 1)(x + 4)^3$ on $[-5, 1]$
15. $f(x) = 2(x - 3)^{-1}$ on $[0, 5]$
16. $f(x) = 2x(x - 1)^{-1}$ on $[0, 3]$
17. $f(x) = |x|$ on $[-3, 3]$
18. $f(x) = |x + 2|$ on $[-3, 3]$
19. $f(x) = \dfrac{x - 1}{(x + 1)^2}$ on $[0, 3]$
20. $f(x) = \dfrac{x}{(x - 1)^2}$ on $[-2, 0]$

APPLICATIONS

In Problems 21–26 revenue and cost functions are given. Find the number of items, x, which produces a maximum profit. Also find the maximum profit.

21. $R(x) = 35x - .03x^2$; $C(x) = 6{,}000 + 5x$
22. $R(x) = 2{,}000x - .05x^2$; $C(x) = 120{,}000 + 150x$
23. $R(x) = 5{,}000x - .1x^2$; $C(x) = 80{,}000 + 200x$
24. $R(x) = 2x + \dfrac{x}{x - 16}$; $C(x) = \dfrac{2}{3} + x$ on $[0, 15]$
25. $R(x) = 5x + \dfrac{x}{x - 64}$; $C(x) = 10 + x$ on $[0, 62]$
26. $R(x) = 6x + \dfrac{x}{x - 245}$; $C(x) = 50 + x$ on $[0, 240]$
27. Rework Example 2 except that the price must not be less than \$100.
28. Rework Example 3 for $C(x) = .25x^2 + 12{,}100$.
29. Rework Example 4, except assume that the plants are 20 miles apart.
30. Rework Example 5, except assume that the voting formula is

$$P(x) = .002x^3 - .207x^2 + .48x$$

31. Rework Example 6 for a piece of cardboard 10 inches long and 5 inches wide.
32. Show the result of Problem 28 graphically.
33. A manufacturer has the following costs in producing x items ($0 \le x \le 200$): unit cost, \$25; fixed costs, \$500; repairs, $x^2/20$ dollars.
    a. What is the average cost $\bar{C}(x)$ per item if x items are produced?
    b. Find the critical values for $\bar{C}(x)$. Where is this average increasing and where is it decreasing?
    c. What is the minimal average cost?
34. A consulting firm determines that the demand (price equation) for a certain product is $p = 1{,}000 - 10x$ dollars, where x is the number of items produced. The cost function is $C(x) = 5{,}000 + 500x$. Find the maximum profit and determine the price that should be charged to make the maximum profit. [*Note:* Price should be nonnegative.]

35. The price equation for a new product is $50 - x$ dollars, where x is the number of items produced. The cost function is $C(x) = 100 + 20x$. Find the maximum profit and determine the price that should be charged to make the maximum profit. [*Note:* Price must be nonnegative.]

36. Suppose a tax of $2 per item is imposed on the product described in Problem 35. Find the maximum profit and determine the price that should be charged to make the maximum profit.

37. Suppose a tax of $4 per item is imposed on the product described in Problem 35. Find the maximum profit and determine the price that should be charged to make the maximum profit.

38. Suppose that the air pollution at a particular location is based on the distance from the source of the pollution according to the principle that for distances greater than or equal to one mile, the concentration of particulate matter (in parts per million, ppm) decreases as the reciprocal of the distance from the source. Suppose that two plants 50 miles apart are releasing 180 ppm and 300 ppm, respectively. What is the location between the plants where pollution is a minimum?

39. A retailer has determined that the cost C for ordering and storing x units of a product is

$$C(x) = 5x + \frac{50{,}000}{x} \quad \text{on } [1, 200]$$

Find the order size that will minimize the cost if the delivery truck can bring a maximum of 125 per order.

40. An assembly line worker can memorize $p\%$ of a given list of tasks in x continuous hours according to the formula

$$p(x) = 95x - 25x^2 \text{ on } [0, 3]$$

How long should it take a worker to memorize the maximum percentage? What is the maximum percentage?

41. A farmer has 1,000 feet of fence and wishes to enclose a rectangular plot of land. The land borders a river and no fence is required on that side. (Refer to the figure at the top of the next column.) What should the length of the side parallel to the river be in order to include the largest possible area?

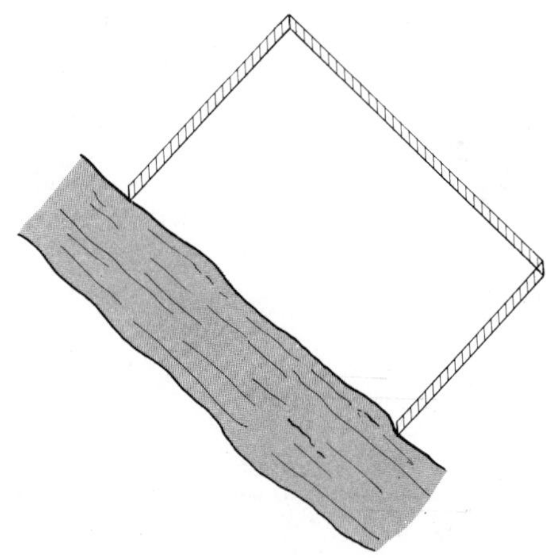

42. A fence must be built around a rectangular area of 1,600 ft^2. The fence along three sides is to be made of a material that costs $12 per foot. The material for the fourth side costs $4 per foot. Find the dimensions (rounded to the nearest foot) of the rectangle that would be the least expensive to build, provided that no side is less than 10 ft long.

43. A tour agency is booking a tour and has 100 people signed up. The price of a ticket is $2,000 per person. The agency has booked a plane seating 150 people at a cost of $125,000. Additional costs to the agency are incidental fees of $500 per person. For each $5 that the price is lowered, a new person will sign up. How much should the price be lowered to maximize the profit for the tour agency?

44. Suppose that the tour agency described in Problem 43 is able to obtain a booking for a plane that will seat 225 people at the same price for the charter as the first airline. Answer the question under these conditions.

45. A viticulturist estimates that if 50 grapevines are planted per acre, each grapevine will produce 150 pounds of grapes. For each additional grapevine planted per acre (up to 20), the average yield per vine drops by 2 pounds. How many grapevines should be planted to maximize the yield per acre?

*12.5 Review

The material of this chapter is reviewed in the following list of objectives. After each objective there are some practice questions. For a sample test select the first question of each set and check your answers. The second question for each objective has no answer given. If you are having trouble with a particular type of problem, look back at the indicated section in the text. When you are finished reviewing these objectives, a sample examination is given at the end of this section.

* Optional section.

[12.1]

Objective 12.1: *Graph a parabola by using calculus to find its vertex and how it opens.*

1. $y = 2x^2 - 8x + 5$
2. $x^2 - 6x + 3y - 4 = 0$
3. $x^2 + 4x + 2y + 3 = 0$
4. $2x^2 - 16x - 3y + 23 = 0$

Objective 12.2: *Graph a curve by finding the critical values, finding when it is increasing or decreasing, and by deciding upon the concavity.*

5. $f(x) = x^3 - 27x$
6. $f(x) = 2x^3 - 3x^2 - 36x$
7. $g(x) = x^3 + 3x^2 - 9x + 5$
8. $g(x) = 4x^3 + 5x^2 + 2x + 3$

[12.2]

Objective 12.3: *Find successive derivatives of a given function.*

9. Find all derivatives of $y = 3x^4 - x^3 + 5x^2 + 79$.
10. Find all derivatives of $y = 1 - 3x^3 + x^5$.
11. Find the first four derivatives of $f(x) = \dfrac{1}{\sqrt{x}}$.
12. Find the first four derivatives of $g(x) = 3x^5 - 3x^{-1}$.

[12.3]

Objective 12.4: *Find all relative maximums and minimums for a given function.*

13. $f(x) = x^3 - x^2 - 5x$
14. $g(x) = x - \dfrac{2}{x}$
15. $t(x) = 4x^2 - x^4$
16. $s(x) = x^3 - 2x^2 - 4x - 8$

Objective 12.5: *Graph a given function.*

17. $f(x) = 4x^3 - 30x^2 + 48x$
18. $g(x) = 3x^5 - 50x^3 + 135x + 12$
19. $y = x^3 - 1$
20. $y = \dfrac{x-2}{x+1}$

[12.4]

Objective 12.6: *Find the absolute maximum and minimum for a function defined on a closed interval.*

21. $y = 3x^5 - 85x^3 + 240x$ on $[-2, 5]$
22. $y = 2x + \dfrac{32}{x}$ on $[1, 9]$
23. $y = (x-1)(x+3)^3$ on $[-5, 5]$
24. $y = 1 - x^{3/2}$ on $[0, 4]$

Objective 12.7: *If you are given revenue and cost functions, find the number of items to produce a maximum profit.*

25. $R(x) = 50x - .05x^2$; $C(x) = 5{,}000 + 2x$
26. $R(x) = 10x - 3x^2 + .01x^3$; $C(x) = 15{,}000 + x$; $x \geq 100$
27. $R(x) = 4x + .345x^2 - .005x^3$; $C(x) = 4{,}000 + x$
28. $R(x) = 10x + \dfrac{x}{(x-49)}$; $C(x) = 3{,}000 + x$ on $[0, 48]$

Objective 12.8: *Solve applied problems based on the preceding objectives.*

29. *Marginal revenue.* A manufacturer has determined that the demand is given by the formula
$$p(x) = 25 - \left(\frac{x}{500}\right)^2$$
Is the marginal revenue increasing or decreasing when $x = 100$ items?
30. *Cost–benefit model.* The cost–benefit model relating the cost, C, of removing $p\%$ of the pollutants from the atmosphere is
$$C(p) = \frac{50{,}000p}{110 - p}$$
where the domain of p is $[0, 100]$. Find the minimum cost if the EPA requires 85% of the pollutants to be removed.
31. *Property management.* A property management company manages 100 apartments renting for \$500 with all the apartments rented. For each \$50 per month increase in rent there will be two vacancies with no possibility of filling them. What rent per apartment will maximize the monthly revenue?
32. A rectangular cardboard poster is to have 208 sq in. for printed matter. It is to have a 3-inch margin at the top and a 2-inch margin at the sides and bottom.

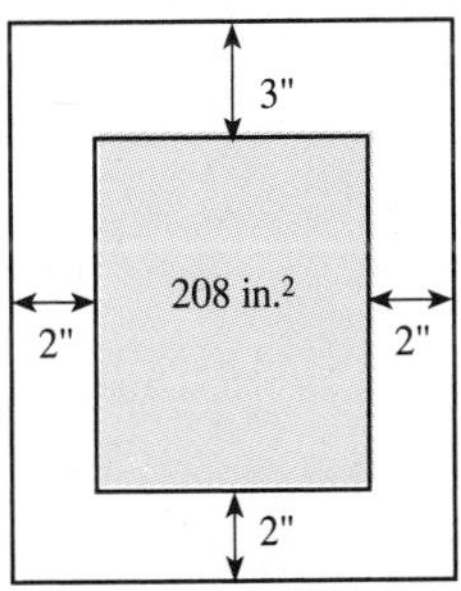

Find the length and width (to the nearest inch) of the poster so that the amount of cardboard used is minimized.

SAMPLE TEST

The following sample test (45 minutes) is intended to review the main ideas of this chapter.

1. **a.** Find all nonzero derivatives of $y = 1 - 3x^3 + x^5$.
   b. Find the first four derivatives of $g(x) = 3x^5 - 3x^{-1}$.

Graph the functions in Problems 2–5.

2. $x^2 - 6x + 3y - 4 = 0$
3. $g(x) = x^3 + 3x^2 - 9x + 5$
4. $y = x + \dfrac{4}{x}$
5. $y = \dfrac{x-2}{x+1}$

Find all the relative maximums and minimums for the functions in Problems 6 and 7.

6. $f(x) = 8x^2 - 2x^4$
7. $g(x) = x + 8/x$
8. Find the absolute maximum and minimum for

$$y = (x-1)(x+3)^3 \text{ on } [-5, 5]$$

9. A manufacturer has determined that the price of selling x items is given by the formula

$$p(x) = 50 - (x/1{,}000)^2$$

Is the marginal revenue increasing or decreasing when $x = .100$ items?

10. A manufacturer needs to package a product in a closed rectangular box. If the sides of the box cost \$4.00 per square foot and the base and top (which are squares) cost \$8.00 per square foot, what are the dimensions of the box with greatest volume that can be constructed for \$96?

Evaluate the limits in Problems 1–5.

1. Find $\lim_{x \to 3} f(x)$ of the accompanying graph.

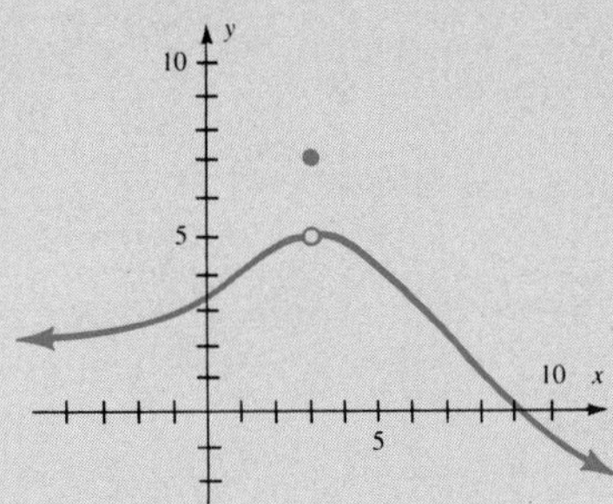

2. Find $\lim_{x \to -4} \dfrac{x+8}{x-4}$.

3. Find $\lim_{x \to 5} \dfrac{2x^2 - 7x - 15}{x - 5}$.

4. Find $\lim_{x \to 2} \dfrac{3x^2 - 5x - 2}{x - 2}$.

5. Find $\lim_{x \to \infty} \dfrac{(3x+1)(5x-2)}{x^2}$.

Find the points of discontinuity over the given domain for the functions in Problems 6–8.

6. $f(x) = \dfrac{x^2 - 15x + 56}{x - 8}$

7. $f(x) = \begin{cases} \dfrac{x^2 - 15x + 56}{x - 8} & \text{if } 0 \le x \le 10, x \ne 8 \\ 7 & \text{if } x = 8 \end{cases}$

8. $f(x) = \begin{cases} \dfrac{x^2 - 15x + 56}{x - 8} & \text{if } 0 \le x \le 10, x \ne 8 \\ 1 & \text{if } x = 8 \end{cases}$

9. Find the instantaneous rate of change for $y = 4 - \dfrac{1}{x}$ at $x = 1$.

10. State the definition of derivative.

Find the derivative of the functions in Problems 11–16.

11. $y = 25 - 12x^2 - 5x^3$

12. $y = 5x\sqrt{9 - x}$

13. $y = (2 - 5x)^5$

14. $y = \dfrac{(2x+5)^8}{x^2 - 5}$

15. $3x^5y^4 = 10$

16. $2x^2 + xy + 3y^2 = 100$

17. Find the equation of the line tangent to $f(x) = 2x - 6x^3$ at the point $(1, -4)$.

18. Graph $2x^2 + 24x - 3y - 3 = 0$ using calculus.

19. Find all derivatives of $y = 2x^4 - x^3 + 3x^2 + 25$.

20. Find all relative maximums and minimums of $f(x) = 4x^2 - x^4$.

21. Find the absolute maximum and minimum of $y = \sqrt{x} - x$ on $[0, 9]$.

22. Find the maximum profit when
$R(x) = 20x + .11x^2 - .01x^3$ and $C(x) = 10{,}000 + x$.

APPLICATIONS

23. A manufacturer has determined that the price of an item is determined by the number of items sold according to the following price formula for x items sold:

$$p(x) = 25 - \frac{x}{500}$$

Is the marginal revenue increasing or decreasing when $x = 100$ items?

24. Find the marginal cost for the function
$C(x) = 20x^3(5x - 100)^2$.

25. An artist's print is 20 in.^{2} and is to be framed so that it has 2 in. of matting on each side, and 4 in. on the top and bottom. What are the dimensions of the frame enclosing the smallest area possible? (That is, the frame should enclose the least area; around your answer to the nearest inch.)

13

Exponential and Logarithmic Functions

CHAPTER OVERVIEW

There are many applications involving growth or decay that cannot be modeled with only the algebraic functions considered thus far. Two additional functions—exponential and logarithmic—are now added to our repertoire. Each is defined and graphed, and we learn how to find their derivatives.

PREVIEW

We are first introduced to exponential functions and then to logarithmic functions. Next, we solve equations involving these functions, and, finally, we learn how to differentiate them.

PERSPECTIVE

Persons needing calculus for management (business, economics, finance, or investments), life sciences (biology, ecology, health, or medicine), or social sciences (demography, political science, population, psychology, society, or sociology) need to understand how things grow (as in money or populations), and how things decay (as in forgetting or inflation). In order to understand these concepts, two of the most useful functions are introduced in this chapter: exponential and logarithmic functions. The material of this chapter will be particularly important in Section 15.2 when solving differential equations.

World record running rates for mile run (meters/minute)

MODELING APPLICATION 11*

World Running Records

World records for footraces at all distances have improved consistently ever since records have been kept. For example, if we consider the mile run, the magic 4-minute mile was broken in 1954 by Roger Bannister of the United Kingdom. Since then, the record has decreased steadily to the present time when the world record is under 3 minutes 47 seconds!

After you have finished this chapter, write a paper that develops a mathematical model to answer the question, Will a 3-minute mile ever be run? Is there an "ultimate" time for a mile run, and, if so, what should we expect that time to be? For general guidelines about writing this essay, see the commentary for Modeling Application 1 on page 151.

* This modeling application is from Joseph Brown, "Predicting Future Improvements in Footracing," *MATYC Journal*, Fall 1980, pp. 173–179.

APPLICATIONS

Management *(Business, Economics, Finance, and Investments)*
Finding the future value in an interest-bearing account (13.1, Problems 40–45)
Number of units sold as a function of advertising expenditures (13.2, Problems 40, 41; 13.5, Test Problem 18)
Determining the time for an interest-bearing account (13.3, Problems 47–52; 13.5, Problem 46)
Monthly payment on an installment loan (13.3, Problems 55–56)
Maximum value of a demand equation (13.4, Problem 41)
Marginal cost (13.4, Problem 42)
Rate at which a money supply is increasing (13.4, Problem 45)
Sales growth pattern (13.4, Problem 48)
Price–demand equation (13.4, Problem 49; 13.5, Test Problem 19)
Price–supply equation (13.4, Problem 50)

Management *(continued)*
Amount spent on advertising (13.5, Problem 48)
Equilibrium for supply/demand (13.5, Problem 51)

Life Sciences *(Biology, Ecology, Health, and Medicine)*
Determining the pH of a substance (13.2, Problems 46, 48)
Petroleum reserves and time until they are exhausted (13.3, Problems 59–60)
Spread of a disease in a town (13.4, Problems 43–44)
Blood pressure of the aorta artery (13.4, Problem 46)
Amount of drug present in the body after being administered (13.4, Problem 47)
Half-life (13.5, Problem 47)

Social Sciences *(Demography, Political science, Population, Psychology, Society, and Sociology)*
Learning curve for a typing test (13.2, Problems 42–45)
World growth rate (13.3, Problems 44–46)
Forgetting curve in psychology (13.3, Problems 53–54)
Dating an artifact in archaeology (13.3, Problems 57–58; 13.5, Problem 47)
Percentage who will hear a presidential announcement (13.5, Problem 52; Test Problem 20)

General Interest
Insurance policy (13.1, Problem 50)
Magnitude of an earthquake on the Richter scale (13.2, Problems 47, 49)
Inflation (13.5, Problem 45)

Modeling Application—
World Running Records

13.1

Exponential Functions

Linear, quadratic, polynomial, and rational functions are examples of what are all called **algebraic functions**. An algebraic function is a function that can be expressed in terms of algebraic operations alone. If a function is not algebraic, it is called a **transcendental function**. In this chapter two examples of transcendental functions—*exponential* and *logarithmic* functions—are discussed.

Definition of Exponential Function

Exponential Function

The function f is an **exponential function** if

$$f(x) = b^x$$

where b is a positive constant other than 1 and x is any real number. The number x is called the **exponent** and b is called the **base**.

Laws of Exponents

Recall that if n is a natural number, then

$$b^n = \underbrace{b \cdot b \cdot b \cdot \cdots \cdot b}_{n \text{ factors}}$$

Furthermore, if $b \neq 0$, then $b^0 = 1$, $b^{-n} = 1/b^n$, and $b^{1/n} = \sqrt[n]{b}$ (for $b \geq 0$ when n is even). Also, $b^{m/n} = (b^{1/n})^m$. These definitions are used in conjunction with five **laws of exponents**:

Laws of Exponents

For a and b positive real numbers and rational numbers p and q, there are five rules that govern the use of exponents. The form 0^0 and division by zero, as well as even roots of negative numbers, are excluded whenever they occur.

First law: $b^p \cdot b^q = b^{p+q}$

Second law: $\dfrac{b^p}{b^q} = b^{p-q}$

Third law: $(b^p)^q = b^{pq}$

Fourth law: $(ab)^p = a^p b^p$

Fifth law: $\left(\dfrac{a}{b}\right)^p = \dfrac{a^p}{b^p}$

EXAMPLE 1 Simplify the following expressions.

a. $16^{1/2} = (4^2)^{1/2}$
$= 4^1$
$= 4$

b. $-16^{1/2} = -(4^2)^{1/2}$
$= -4$

c. $(-16)^{1/2}$ is not defined since b must be greater than or equal to zero for square roots ($x = \frac{1}{2}$ is a square root)

d. $343^{2/3} = (7^3)^{2/3}$
$= 7^2$
$= 49$

e. $25^{-3/2} = (5^2)^{-3/2}$
$= 5^{-3}$
$= 1/5^3$
$= 1/125$ ■

EXAMPLE 2 Use the ordinary rules of algebra to simplify the following expressions.

a. $x(x^{2/3} + x^{1/2}) = x^1x^{2/3} + x^1x^{1/2}$
$= x^{1+2/3} + x^{1+1/2}$
$= x^{5/3} + x^{3/2}$

b. $(x^{1/2} + y^{1/2})(x^{1/2} - y^{1/2}) = x^{1/2}x^{1/2} - x^{1/2}y^{1/2} + x^{1/2}y^{1/2} - y^{1/2}y^{1/2}$
$= x - y$ ■

Graphing Exponential Functions

The next step is to enlarge the domain of x to include all real numbers. This is done with the help of the following property:

Squeeze Theorem for Exponents

> Suppose b is a real number greater than 1. Then for any real number x there is a unique real number b^x. Moreover, if p and q are any two positive, rational numbers such that $p < x < q$, then
>
> $$b^p < b^x < b^q$$

The squeeze theorem gives meaning to expressions such as $2^{\sqrt{3}}$. Since

$$1.732 < \sqrt{3} < 1.733$$

the squeeze theorem says that

$$2^{1.732} < 2^{\sqrt{3}} < 2^{2.733}$$

This means that even though only rational exponents were previously defined, we can extend the definition to any real number exponent by using the squeeze theorem to give us any desired degree of accuracy. The case where $0 < b < 1$ is considered in Problem Set 13.1.

We can now consider an exponential function since this extended definition allows exponents to be any real number.

EXAMPLE 3 Graph the function $f(x) = 2^x$.

Solution Begin by plotting points, as shown in the table and Figure 13.1.

x	$y = f(x) = 2^x$
-3	$2^{-3} = \frac{1}{8}$
-2	$2^{-2} = \frac{1}{4}$
-1	$2^{-1} = \frac{1}{2}$
0	$2^0 = 1$
1	$2^1 = 2$
2	$2^2 = 4$
3	$2^3 = 8$

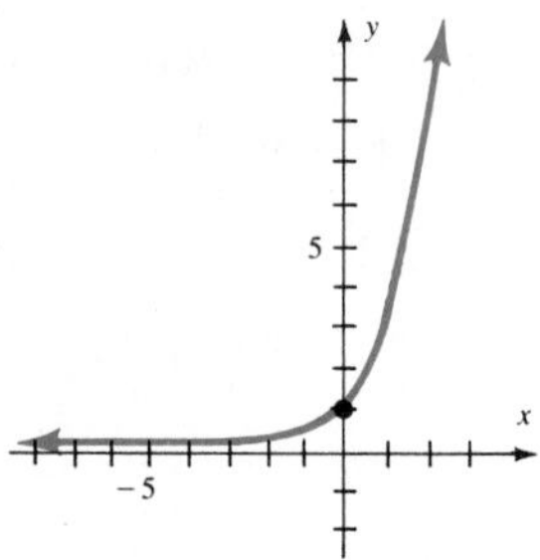

FIGURE 13.1
Graph of $y = 2^x$

Connect these points with a smooth curve to obtain the graph shown in Figure 13.1. Note that the squeeze theorem allows us to draw this as a smooth curve. ■

EXAMPLE 4 Sketch $f(x) = (\frac{1}{2})^x$.

Solution Notice that $y = (\frac{1}{2})^x = (2^{-1})^x$ or 2^{-x}.

x	$y = f(x) = (\frac{1}{2})^x$
-3	$(2^{-1})^{-3} = 8$
-2	$(2^{-1})^{-2} = 4$
-1	$(2^{-1})^{-1} = 2$
0	$(2^{-1})^0 = 1$
1	$(\frac{1}{2})^1 = \frac{1}{2}$
2	$(\frac{1}{2})^2 = \frac{1}{4}$
3	$(\frac{1}{2})^3 = \frac{1}{8}$

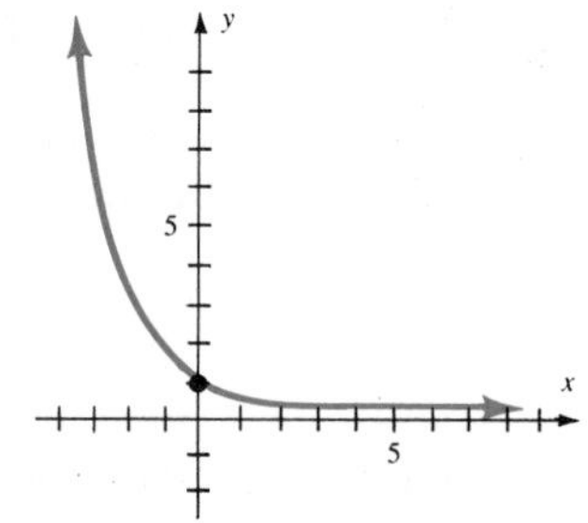

FIGURE 13.2
Graph of $f(x) = (\frac{1}{2})^x$

Connect the points shown in the table with a smooth graph, as shown in Figure 13.2. ■

The exponential function can be used as a model for certain types of growth or decay. If $b > 1$ (as in Example 3), then the exponential function is called a **growth function**, and if $0 < b < 1$ (as in Example 4), then it is called a **decay function**.

*Interest**

One of the most common examples of growth is the way money grows in a bank account. This is a very important concept not only in business but in personal money management. If a sum of money, called the **principal** or **present value**, is

* This is a review of material in Chapter 5. It is included here for those who skipped Chapter 5.

denoted by P and invested at an annual interest rate of r for t years, then the *future amount* of money present is denoted by A and is found by

$$A = P + I$$

where I denotes the amount of interest. **Interest** is an amount of money paid for the use of another's money. **Simple interest** is found by multiplication:

$$I = Prt$$

For example, \$1,000 invested for 3 years at 15% simple interest generates interest of

$$I = \$1{,}000(.15)(3) = \$450$$

so the future amount in 3 years is

$$A = \$1{,}000 + \$450 = \$1{,}450$$

Most businesses, however, pay interest on the interest as well as on the principal, and when this is done, it is called **compound interest**. For example, \$1,000 invested at 15% annual interest compounded annually for 3 years can be found as follows:

First year:

$$\begin{aligned} A &= P + I \\ &= P \cdot 1 + Pr && I = Prt \text{ and } t = 1 \\ &= P(1 + r) && \text{For this example, } A = \$1{,}000(1 + .15) \\ & && = \$1{,}150 \end{aligned}$$

Second year:

$$\begin{aligned} A &= P(1 + r) + I && \text{The total amount from the first year becomes the principal for the second year.} \\ &= P(1 + r) \cdot 1 + P(1 + r) \cdot r \\ &= P(1 + r)(1 + r) \\ &= \mathbf{P(1 + r)^2} && \text{For this example, } A = \$1{,}000(1 + .15)^2 \\ & && = \$1{,}322.50 \end{aligned}$$

Third year:

$$\begin{aligned} A &= \mathbf{P(1 + r)^2} \cdot 1 + \mathbf{P(1 + r)^2} \cdot r \\ &= \mathbf{P(1 + r)^2}(1 + r) \\ &= P(1 + r)^3 && \text{For this example, } A = \$1{,}000(1 + .15)^3 \\ & && = \$1{,}520.88 \end{aligned}$$

The pattern illustrated in Example 3 can be generalized:

Compound Interest or Future Value Formula

If a principal of P dollars is invested at an interest rate of i per period for a total of N periods, then the future amount A is given by the formula

$$A = P(1 + i)^N$$

Compound interest is usually stated in terms of an annual interest rate r and a given number of years t. The frequency of compounding (that is, the number of compoundings per year) is denoted by n where

$n = 1$ if compounded annually

$n = 2$ if compounded semiannually

$n = 4$ if compounded quarterly

$n = 12$ if compounded monthly

$n = 360$ if compounded daily (ordinary interest)

$n = 365$ if compounded daily (exact interest)

Therefore, $i = r/n$ and $N = nt$, as illustrated in Examples 5 and 6.

EXAMPLE 5 If \$12,000 is invested for 5 years at 18% compounded semiannually, what is the amount present at the end of 5 years?

Solution $P = \$12{,}000$; $r = .18$; $i = \frac{.18}{2} = .09$; $t = 5$; $N = 2(5) = 10$; thus

$$\begin{aligned} A &= P(1+i)^N \\ &= \$12{,}000(1+.09)^{10} \\ &\approx \$28{,}408.36 \end{aligned}$$

Use a calculator and round answers involving money to the nearest cent; however, do not round until you are ready to state your final answer. ■

EXAMPLE 6 Rework Example 5 except now the interest is compounded monthly.

Solution $P = \$12{,}000$; $i = \frac{.18}{12} = .015$; $N = 12(5) = 60$. Thus

$$\begin{aligned} A &= \$12{,}000(1+.015)^{60} \\ &\approx \$29{,}318.64 \end{aligned}$$

■

Examples 5 and 6 show how the amount increases as the number of times it is compounded increases. A reasonable extension is to ask what happens if the interest is compounded even more frequently than monthly. Can we compound daily, hourly, every minute, or every split second? The answer is yes; in fact, money can be compounded **continuously**, which means that at every instant the newly accumulated interest is used as part of the principal for the next instant. In order to understand these concepts consider the following contrived example. Suppose \$1 is invested at 100% interest for 1 year compounded at different intervals. The compound interest formula for this example is

$$A = \left(1 + \frac{1}{n}\right)^n$$

The calculations of this formula for different values of n are shown in Table 13.1.

TABLE 13.1
Effect of Compound Interest on a \$1 Investment

Number of periods	Formula	Amount
Annually, $n = 1$	$(1 + \frac{1}{1})^1$	\$2.00
Semiannually, $n = 2$	$(1 + \frac{1}{2})^2$	\$2.25
Quarterly, $n = 4$	$(1 + \frac{1}{4})^4$	\$2.44
Monthly, $n = 12$	$(1 + \frac{1}{12})^{12}$	\$2.61
Daily, $n = 360$	$(1 + \frac{1}{360})^{360}$	\$2.715
Hourly, $n = 8{,}640$	$(1 + \frac{1}{8{,}640})^{8{,}640}$	\$2.7181

If these calculations are continued for even larger n, you will obtain the following:

$n = 10{,}000$	then, the formula yields	2.718145926
$n = 100{,}000$		2.718268237
$n = 1{,}000{,}000$		2.718280469
$n = 10{,}000{,}000$		2.718281828
$n = 100{,}000{,}000$		2.718281828

The calculator cannot distinguish values of $(1 + \frac{1}{n})^n$ for larger n. These values are approaching a particular number. This number, it turns out, is an irrational number, so it does not have a convenient decimal representation. (That is, its decimal representation does not terminate and does not repeat.) Mathematicians have agreed to denote this number by the symbol e, which is defined as a limit:

The Number e

$$\lim_{n\to\infty}\left(1 + \frac{1}{n}\right)^n = e$$

EXAMPLE 7 Find e, e^2, and e^{-3}.

Solution On a calculator, locate a key labeled e^x.

$$e \approx 2.718281828$$
$$e^2 \approx 7.389056099$$
$$e^{-3} \approx .0497870684$$

■

For interest *compounded continuously*, the following formula is used:

$$A = Pe^{rt}$$

You can find e, as well as powers of e, by using Table 10 in Appendix E or by using a calculator.

EXAMPLE 8 Find the future value of the principal in Example 5 if the interest is compounded continuously.

Solution Since $P = \$12{,}000$, $r = .18$, and $t = 5$,

$$\begin{aligned} A &= \$12{,}000e^{(.18)(5)} \\ &\approx \$29{,}515.24 \end{aligned}$$

■

You should memorize at least the first six digits of e:

$$e \approx 2.71828$$

■

13.1 Problem Set

Simplify the expressions in Problems 1–18. Eliminate negative exponents from your answers.

1. $25^{1/2}$
2. $-25^{1/2}$
3. $(-25)^{1/2}$
4. $(-27)^{1/3}$
5. $-27^{1/3}$
6. $27^{1/3}$
7. $7^{1/3} \cdot 7^{2/3}$
8. $8^{4/3} \cdot 8^{-1/3}$
9. $1{,}000^{-1/3}$
10. $.001^{-2/3}$
11. $100^{-3/2}$
12. $.01^{-3/2}$
13. $\dfrac{2^3 \cdot 2^{-4}}{2^5 \cdot 2^{-2}}$
14. $\dfrac{3^{-2} \cdot 3^3}{3^5 \cdot 3^{-4}}$
15. $\dfrac{9^{1/2} \cdot 27^{2/3}}{3^2 \cdot 81}$
16. $\dfrac{4^{1/2} \cdot 8^{2/3}}{16 \cdot 4^3}$
17. $2^{-1} + 3^{-1}$
18. $\dfrac{4^{-1} + 9^{-1}}{36^{-1}}$

Evaluate the expressions in Problems 19–24.

19. e^3
20. e^{-2}
21. $e^{.05}$
22. $e^{-.2}$
23. $e^{.045}$
24. $e^{-4.85}$

Simplify the expressions in Problems 25–30.

25. $x^{1/2}(x^{1/2} + x^{1/2})$
26. $x(x^{1/2} + x^{-1/2})$
27. $x^{2/3}(x^{-2/3} + x^{1/3})$
28. $x^{1/4}(x^{3/4} + x^{-1/4})$
29. $(x^{1/2} + y^{1/2})^2$
30. $(x^{1/2} - y^{1/2})^2$

Sketch the graph of each function in Problems 31–39.

31. $y = 3^x$
32. $y = 4^x$
33. $y = (\frac{1}{3})^x$
34. $y = e^x$
35. $y = e^{-x}$
36. $y = -e^{-x}$
37. $y = e^{x-2}$
38. $y = 10^{x-1}$
39. $y - 2 = 2^{x+1}$

APPLICATIONS

40. If \$1,000 is invested at 12% compounded semiannually, how much money will there be in 10 years?
41. If \$1,000 is invested at 16% compounded continuously, how much money will there be in 25 years?
42. If \$1,000 is invested at 14% compounded continuously, how much money will there be in 10 years?
43. If \$8,500 is invested at 18% compounded monthly, how much money will there be in 4 years?
44. If \$3,600 is invested at 15% compounded daily, how much money will there be in 7 years? (Use a 365-day year; this is *exact interest*.)
45. If \$9,400 is invested at 14% compounded daily, how much money will there be in 6 months? (Use a 360-day year; this is *ordinary interest*.)
46. Find formulas for present value P for interest compounded n times per year and for interest compounded continuously.
47. If you want to have \$10,000 in the bank five years from now and make a deposit compounded semiannually at 11.5%, how much do you need to deposit?
48. Repeat Problem 47 except deposit in a bank paying interest daily. (Assume a 365-day year.)
49. Repeat Problem 47 except deposit in a bank paying continuous interest.
50. Suppose an insurance agent offers you an insurance policy that will pay your beneficiary \$25,000 if you die in the next 25 years, *and* will pay you \$25,000 in 25 years if you live. What is the present value of the \$25,000 you will receive if you live? Assume that the annual inflation rate will be 12% compounded annually for the next 25 years.
51. The hypothesis for the squeeze theorem for exponents requires $b > 1$. What can you say about the case when $0 < b < 1$?

13.2 Logarithmic Functions

Definition of Logarithmic Functions

Suppose the exponent for an exponential function is the unknown. For example, the compound interest formula

$$A = P(1 + i)^N$$

was used in the last section to find A. Now suppose that A, P, and i are known, but the length of time it will take for P to grow to A is not known. If the exponent in an equation is the unknown value, the equation is called an **exponential equation.** Consider

$$A = b^x$$

where $b > 1$. How can we solve this exponential equation for x? Notice that

x is the exponent of base b that yields the value A

This can be rewritten as

x = exponent of base b to get A

It appears that the equation is now solved for x, but we have simply changed the notation. The expression "exponent of b to get A" is called, for historical reasons, "the log of A to the base b." That is,

x = log A to base b

This phrase is shortened to the notation

$$x = \log_b A$$

The term *log* is an abbreviation for **logarithm**, which means **exponent**.

Logarithm

The logarithm

$$x = \log_b A$$

is the **exponent** on a base b that yields the value A. In symbols

$$x = \log_b A \quad \Leftrightarrow \quad b^x = A$$

Logarithmic Function

The function f defined for $x > 0$ by

$$f(x) = \log_b x$$

where $b > 0$, $b \neq 1$, is called the **logarithmic function with base b.**

Remember the definition of logarithm is nothing more than a notational change.

WARNING Remember: logarithm means exponent.

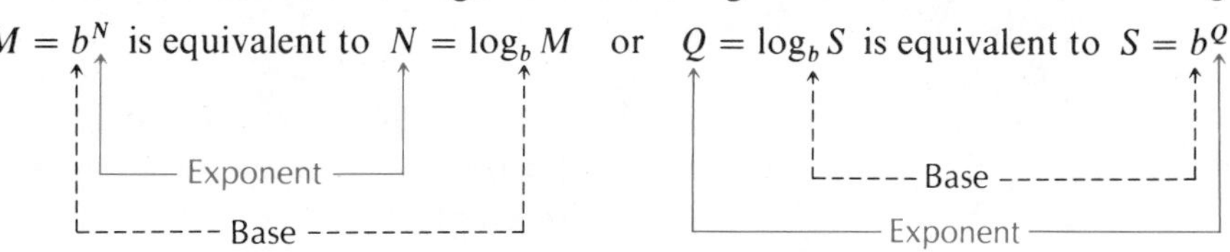

EXAMPLE 1 Change the following expressions from exponential form to logarithmic form.

a. $5^2 = 25$ The base is 5 and the exponent is 2: $\log_5 25 = 2$.

b. $3^2 = 9$ This is the same as $\log_3 9 = 2$.

c. $\frac{1}{8} = 2^{-3}$ In logarithmic form: $\log_2 \frac{1}{8} = -3$.

d. $\sqrt{16} = 4$ Logarithmic form: $\log_{16} 4 = \frac{1}{2}$. (Remember $\sqrt{16} = 16^{1/2}$.) ■

EXAMPLE 2 Change the following expressions from logarithmic form to exponential form.

a. $\log_{10} 100 = 2$ The base is 10 and the exponent is 2: $10^2 = 100$.

b. $\log_{10} 1/1{,}000 = -3$ This is the same as $10^{-3} = 1/1{,}000$.

c. $\log_3 1 = 0$ Exponential form: $3^0 = 1$. ■

Evaluate Logarithms

To **evaluate a logarithm** means to find a numerical value for the given logarithm. The first ones you are asked to evaluate use a property of exponents that says if two bases on two logarithms are the same and the numbers are equal, then the exponents must be equal.

Exponential Property of Equality

For positive real b ($b \neq 1$)

$$b^x = b^y \quad \text{if and only if} \quad x = y$$

EXAMPLE 3 Evaluate the given logarithms.

a. $\log_2 64$ **b.** $\log_3 \frac{1}{9}$ **c.** $\log_9 27$ **d.** $\log_{10} 1$ **e.** $\log_{10} 100$ **f.** $\log_{10} 10$

Solution **a.** Since it is usually necessary to supply a variable to convert to exponential form, we use N in these examples. That is,

$$N = \log_2 64$$

We write this in exponential form: $2^N = 64$

$$2^N = 2^6$$

$$N = 6 \qquad \text{Use the exponential property of equality.}$$

Thus $\log_2 64 = 6$.

b. Let $N = \log_3 \frac{1}{9}$, so $3^N = \frac{1}{9}$

$$3^N = 3^{-2}$$

$$N = -2$$

Thus $\log_3 \frac{1}{9} = -2$.

c. $N = \log_9 27$, so $9^N = 27$

$$3^{2N} = 3^3$$

$$2N = 3$$

$$N = \tfrac{3}{2}$$

Thus $\log_9 27 = \frac{3}{2}$.

d. $\log_{10} 1 = 0$. Can you do this mentally?

e. $\log_{10} 100 = 2$.

f. $\log_{10} 10 = 1$. ■

Suppose you cannot use the exponential property of equality. For example, suppose you want to find $\log_{10} 5.03$. Since 5.03 is between 1 and 10 and

$$10^0 = 1$$

$$10^x = 5.03 \qquad \text{You want to find this } x.$$

$$10^1 = 10$$

The number x should be between 0 and 1, by the squeeze theorem for exponents. There are tables that show approximations for these exponents. However, calculators have, to a large extent, eliminated the need for extensive log tables. Remember, however, that calculator answers are approximate. The key for $\log_{10}$ is simply labeled *log*. Base 10 is fairly common, and if the logarithm is to the base 10 it is called a **common logarithm** and is written without the subscript 10.

EXAMPLE 4 Evaluate the following logarithms correct to four decimal places.

a. $\log 7.68$ **b.** $\log 852$ **c.** $\log .00728$

Solution **a.** DISPLAY: 0.88536122

To four decimal places, $x = .8854$

b. DISPLAY: 2.930439595

To four decimal places, 2.9304.

c. DISPLAY: −2.137868621

To four decimal places, $\log .00728 = -2.1379$. ■

Keep in mind that a logarithm is an exponent. That is, for Example 4a, the answer .8854 is the *exponent* on the base of 10 that gives the given number, 7.68. That is

$$10^{.8854} \approx 7.68$$

In addition to common logarithms, another logarithm is frequently encountered. This is a logarithm to the base e called a **natural logarithm**. Natural logarithms are denoted by

$$\log_e x = \ln x$$

and this is sometimes pronounced as "lawn x."

EXAMPLE 5 Find $\ln 3.49$ to two decimal places.

Solution BY CALCULATOR: If your calculator has a log key, chances are it also has a ln key:

DISPLAY: 1.249901736 ■

EXAMPLE 6 Find $\ln .403$.

Solution DISPLAY: −0.908818717

Thus $\ln .403 \approx -.909$. Always keep in mind that a logarithm is nothing more than an exponent, so this calculator display means that

$$e^{-.909} \approx .403$$ ■

Graphing Logarithmic Functions

To graph a logarithmic function you use the definition of logarithm and then construct a table of values.

EXAMPLE 7 Graph $y = \log_2 x$.

Solution From the definition of logarithm, $2^y = x$. Use this equation to construct a table of values:

y	$x = 2^y$
-3	$2^{-3} = \frac{1}{8}$
-2	$2^{-2} = \frac{1}{4}$
-1	$2^{-1} = \frac{1}{2}$
0	$2^0 = 1$
1	$2^1 = 2$
2	$2^2 = 4$
3	$2^3 = 8$

The graph is shown in Figure 13.3.

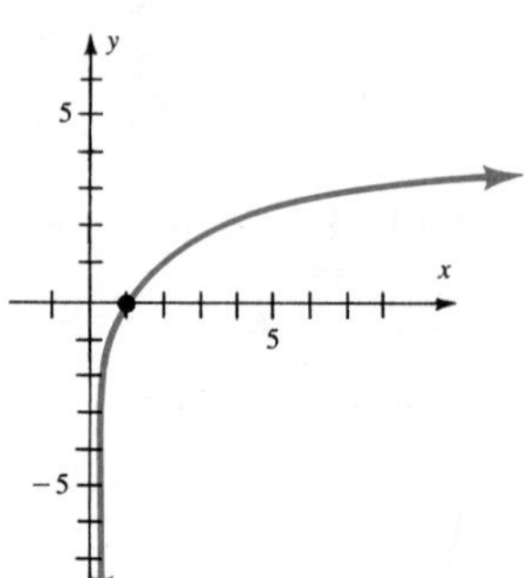

FIGURE 13.3
Graph of $y = \log_2 x$

■

FIGURE 13.4
Graphs of $y = 2^x$, $y = \log_2 x$, and $y = x$

Exponential and Logarithmic Functions Are Inverses

Compare your answer in Example 7 with the graph of $y = 2^x$ shown in Figure 13.4.

Functions whose graphs are symmetric with respect to the line $y = x$ as shown in Figure 13.4 are said to be **inverse functions**. This relationship is needed in order to find e on several brands of calculators. If a calculator has

[ln x] and [INV]

keys, but no e^x key, you can still use the calculator to find e^x. Since

$y = \ln x$ and $y = e^x$

are inverse functions, to find e (or e^1) press

[1] [INV] [ln x] DISPLAY: 2.718281828

These two keys give the inverse of the ln x function; that is, they give e^x when the x value is input just prior to pressing these keys.

If, for example, you need $e^{5.2}$ on such a calculator, press

[5.2] [INV] [ln x] DISPLAY: 181.2722419

EXAMPLE 8 What is the sequence of keys to press to evaluate $\ln e$?*

Solution [1] [INV] [ln x] [ln x] The display is, of course, 1.

This gives e. This gives ln e.

■

* Answers may vary depending on brand and model.

13.2

Problem Set

Write the equations in Problems 1–6 in logarithmic form.

1. $64 = 2^6$
2. $125 = 5^3$
3. $9 = (\frac{1}{3})^{-2}$
4. $\frac{1}{2} = 4^{-1/2}$
5. $a = b^c$
6. $m = n^p$

Write the equations in Problems 7–12 in exponential form.

7. $\log_4 2 = \frac{1}{2}$
8. $\log_2 \frac{1}{8} = -3$
9. $\log .01 = -2$
10. $\log 10{,}000 = 4$
11. $\ln e^2 = 2$
12. $\ln x = .03$

Use the definition of logarithm or a calculator to evaluate the expressions in Problems 13–30.

13. $\log_b b^2$
14. $\log_t t^3$
15. $\log_\pi \sqrt{\pi}$
16. $\ln e^4$
17. $\log 1{,}000$
18. $\log_2 8$
19. $\log_{16} 1$
20. $\log 4.27$
21. $\log 1.08$
22. $\log 8.43$
23. $\log 9{,}760$
24. $\log .042$
25. $\log .321$
26. $\log .0532$
27. $\ln 2.27$
28. $\ln 16.77$
29. $\ln 2$
30. $\ln 13$

Graph the functions in Problems 31–39.

31. $y = \log_3 x$
32. $y = \log_{1/2} x$
33. $y = \ln x$
34. $y = \log x$
35. $y = \ln \sqrt{x}$
36. $y = \log_\pi x$
37. $y = \ln(x + 2)$
38. $y + 2 = \ln x$
39. $y - 1 = \log(x - 2)$

APPLICATIONS

40. An advertising agency conducts a survey and finds that the number of units sold, N, is related to the amount a spent on advertising (in dollars) according to the following formula:

$$N = 1{,}500 + 300 \ln a \qquad a \geq 1$$

a. How many units are sold after spending \$1,000?
b. How many units are sold after spending \$50,000?

41. Graph the sales curve (see Problem 40)

$$N = 1{,}500 + 300 \ln a$$

42. Psychologists have found that in many learning situations a person's rate of learning is rapid at first and then slows down. For the learning curve for learning the touch system of typing, where t is measured in months and P is the number of words typed per minute,

$$P(t) = 80(1 - e^{-.2t})$$

a. What is the expected number of words typed per minute after one month?
b. What is the expected number of words typed per minute after six months?

43. A learning curve describes the rate at which a person learns certain tasks. If a person sets a goal of typing N words per minute (wpm), the length of time, t (in days), to achieve this goal is given by

$$t = -62.5 \ln(1 - N/80)$$

a. How long would it take to learn to type 30 wpm?
b. If we accept this formula, is it possible to learn to type 80 wpm?

44. Graph the learning curve (see Problem 42)

$$P(t) = 80(1 - e^{-.2t})$$

45. Graph the learning curve (see Problem 43)

$$t = -62.5 \ln(1 - N/80)$$

46. The pH of a substance measures its acidity or alkalinity. It is found by the formula

$$\text{pH} = -\log[\text{H}^+]$$

where $[\text{H}^+]$ is the concentration of hydrogen ions in an aqueous solution given in moles per liter.
a. What is the pH (to the nearest tenth) of a lemon for which $[\text{H}^+] = 2.86 \times 10^{-4}$?
b. What is the pH (to the nearest tenth) of rain water for which $[\text{H}^+] = 6.31 \times 10^{-7}$?

47. The Richter scale for measuring earthquakes, developed by Gutenberg and Richter, relates the energy E (in ergs) to the magnitude of the earthquake, M, by the formula

$$M = \frac{\log E - 11.8}{1.5}$$

a. A small earthquake is one that releases 15^{15} ergs of energy. What is the magnitude of such an earthquake on the Richter scale?
b. A large earthquake is one that releases 10^{25} ergs of energy. What is the magnitude of such an earthquake on the Richter scale?

48. Graph the pH formula in Problem 46

$$\text{pH} = -\log[\text{H}^+]$$

for $1 \leq [\text{H}^+] \leq 10^{14}$.
(*Hint:* Let the scale on the x-axis be powers of 10.)

49. Graph the magnitude of an earthquake

$$M = \frac{\log E - 11.8}{1.5}$$

for $10^{12} \le E \le 10^{17}$.
(*Hint:* Let the scale on the x-axis be powers of 10.)

50. Prove that $\log_b 1 = 0$ for all $b > 0, b \neq 1$.

51. Prove that $\log_b b^x = x$ for all x, where $b > 0, b \neq 1$.

52. Prove that $\ln e^x = x$ for all x.

53. Prove that $b^{\log_b x} = x$ for all $x > 0$ where $b > 0, b \neq 1$.

13.3 Logarithmic and Exponential Equations

Logarithmic Equations

We now discuss solving two types of equations involving transcendental functions. The first type is called a **logarithmic equation**. All logarithmic equations fall into one of four categories, as illustrated by Examples 1–4.

EXAMPLE 1 Solve $\log_2 \sqrt{2} = x$ for x.

Solution *The exponent is the unknown*; apply the definition of logarithm:

$$2^x = \sqrt{2}$$
$$2^x = 2^{1/2}$$
$$x = \frac{1}{2} \qquad \text{Exponential property of equality}$$

■

EXAMPLE 2 Solving $\log_x 25 = 2$.

Solution *The base is the unknown*; apply the definition of logarithm:

$$x^2 = 25$$
$$x = \pm 5$$

Be sure the values you obtain are permissible values for the definition of a logarithm. In this case, $x = -5$ is not a permissible value since a logarithm with a negative base is not defined. Therefore the solution is $x = 5$. ■

EXAMPLE 3 Solve $\ln x = 5$.

Solution *The power itself is the unknown*; apply the definition of logarithm:

$$e^5 = x$$

$x \approx 148.41$ DISPLAY: 148.4131591 ■

The first three categories of logarithmic equation problems all use the definition of logarithm. The last category requires a property that follows from the exponential property of equality.

Log of Both Sides Theorem

If A, B, and b are positive real numbers with $b \neq 1$, then

$$\log_b A = \log_b B \quad \text{is equivalent to} \quad A = B$$

EXAMPLE 4 Solve $\log_5 x = \log_5 72$.

Solution Use the log of both sides theorem: $x = 72$. ■

When solving a logarithmic equation, the goal is to write the logarithmic equation with a *single* logarithmic function on either one or both sides of the equation. If the logarithmic equation has the form

$$\log_b A = \log_b B$$

you can use the log of both sides theorem to find the unknown, as Example 4 shows. On the other hand, if the equation has the form

$$\log_b A = N$$

(a log on one side only), then you apply the definition of logarithm to solve the equation, as Examples 1–3 show. In either case, you must first algebraically simplify in order to put the equations into one of these forms. To do this you need some additional theorems about logarithms.

Since logarithms are exponents, we can rewrite the laws of exponents in logarithmic form. For example,

$$b^M \cdot b^N = b^{M+N}$$

This means that if $b^M = A$ and $b^N = B$, then in logarithmic form

$$M = \log_b A \qquad \text{and} \qquad N = \log_b B$$

But if we apply the definition of logarithm to $b^M \cdot b^N = b^{M+N}$, we obtain

$$M + N = \log_b b^M b^N$$

or, by substitution,

$$\log_b A + \log_b B = \log_b AB$$

This is the logarithmic form for the first law of exponents. The second and third laws of exponents can also be written in logarithmic form, as summarized here. You are asked to derive the second and third laws of logarithms in the problem set.

First Law of Logarithms	$\log_b AB = \log_b A + \log_b B$	The log of the product of two numbers is the sum of the logs of those numbers.
Second Law of Logarithms	$\log_b \dfrac{A}{B} = \log_b A - \log_b B$	The log of the quotient of two numbers is the log of the numerator minus the log of the denominator.
Third Law of Logarithms	$\log_b A^p = p \log_b A$	The log of the pth power of a number is p times the log of that number.

Examples 5 and 6 use these laws of logarithms to solve a logarithmic equation that reduces to the type with a logarithm on both sides of the equation. This means that the last step will be to use the log of both sides theorem.

EXAMPLE 5 Solve $\log_8 3 + \frac{1}{2}\log_8 25 = \log_8 x$ for x.

Solution

$$\log_8 3 + \frac{1}{2}\log_8 25 = \log_8 x$$

$$\log_8 3 + \log_8 25^{1/2} = \log_8 x \qquad \text{Third law of logarithms}$$

$$\log_8 3 + \log_8 5 = \log_8 x \qquad 25^{1/2} = \sqrt{25} = 5$$

$$\log_8(3 \cdot 5) = \log_8 x \qquad \text{First law of logarithms}$$

$$15 = x \qquad \text{Log of both sides theorem}$$

The solution is 15. ■

EXAMPLE 6 Solve $\ln x - \frac{1}{2}\ln 2 = \frac{1}{2}\ln(x + 4)$ for x.

Solution

$$\ln x - \frac{1}{2}\ln 2 = \frac{1}{2}\ln(x + 4)$$

$$2\ln x - \ln 2 = \ln(x + 4)$$

$$2\ln x = \ln 2 + \ln(x + 4)$$

$$\ln x^2 = \ln 2(x + 4)$$

$$x^2 = 2x + 8$$

$$x^2 - 2x - 8 = 0$$

$$(x - 4)(x + 2) = 0$$

$$x = 4, -2$$

Since $\ln(-2)$ is not defined, $x = -2$ is an extraneous root. Therefore, the solution is 4. ■

Exponential Equations

The second type of equations are **exponential equations** and they fall into one of three categories.

	Common log	Natural log	Arbitrary base
Base:	10	e	b
Example:	$10^x = 5$	$e^x = 3.456$	$7^x = 3$

Examples 7–9 illustrate each of these types.

EXAMPLE 7 Solve $10^{5x+3} = 195$ for x.

Solution This is a *common log* problem. Use the definition of logarithm to write it in logarithmic form: $\log 195 = 5x + 3$.

Now solve this equation for x: $\log 195 - 3 = 5x$

$$\frac{\log 195 - 3}{5} = x$$

For an approximate answer use a calculator. DISPLAY: -0.1419930777

■

EXAMPLE 8 Solve $e^{-.000425t} = \frac{1}{2}$ for t.

Solution This is a natural logarithm problem. Use the definition to write

$$\ln .5 = -.000425t$$

$$t = \frac{\ln .5}{-.000425}$$

Use a calculator to find $t \approx 1{,}630.934543$; this is about 1,600. ■

Since you do not have tables or calculator keys for bases other than 10 or e, you must proceed differently for an *arbitrary base b*. You should use the log of both sides theorem and the procedure for solving equations. Generally you can choose either base 10 or base e for solving these equations.

EXAMPLE 9 Solve $7^x = 3$ for x.

Solution We will work this two ways: base 10 and base e so it is clear that both produce exactly the same answer.

WARNING $\frac{\log 3}{\log 7} \neq \log \frac{3}{7}$

Base e	*Base* 10	
$\ln 7^x = \ln 3$	$\log 7^x = \log 3$	Log of both sides
$x \ln 7 = \ln 3$	$x \log 7 = \log 3$	Third law of exponents
$x = \frac{\ln 3}{\ln 7}$	$x = \frac{\log 3}{\log 7}$	Divide both sides by the coefficient of x.
$\approx \frac{1.099}{1.946}$	$\approx \frac{.4771}{.8451}$	Evaluate by calculator.
$\approx .5646$	$\approx .5646$	Divide.

■

There is another method for solving the exponential equation of Example 9. This one directly applies the definition of logarithm:

$$7^x = 3 \qquad \text{is the same as} \qquad \log_7 3 = x$$

If there were a logarithm base 7 key on a calculator or a log base 7 table, you would have the answer. Since there are not, you need one final logarithm theorem that changes logarithms from one base to another.

Change of Base Theorem

$$\log_a x = \frac{\log_b x}{\log_b a} \qquad \text{In particular, } \log_a x = \frac{\log x}{\log a} = \frac{\ln x}{\ln a}.$$

Notice that to change from base a to another (possibly more familiar) base b, you simply change the base on the given logarithm from a to b and then divide by the logarithm to the base b of the old base a. The proof of this theorem is identical to the steps outlined in Example 9. That is, if $a^N = x$, then $N = \log_a x$ and

$$\log_b a^N = \log_b x \qquad \text{Take the } \log_b \text{ of both sides.}$$

$$N \log_b a = \log_b x \qquad \text{Third law of exponents}$$

$$N = \frac{\log_b x}{\log_b a} \qquad \text{Divide both sides by } \log_b a.$$

$$\text{Thus, } \log_a x = \frac{\log_b x}{\log_b a} \qquad \text{Substitute } N = \log_a x.$$

EXAMPLE 10 Change $\log_7 3$ to logarithms with base 10 and evaluate.

Solution

$$\log_7 3 = \frac{\log 3}{\log 7}$$

$$\approx \frac{.4771}{.8451} \approx .5646$$

■

EXAMPLE 11 Solve $6^{3x+2} = 200$ for x.

Solution

$$\log_6 200 = 3x + 2 \qquad \text{Use the definition of logarithm.}$$

$$\log_6 200 - 2 = 3x \qquad \text{Solve for } x.$$

$$x = \frac{\log_6 200 - 2}{3}$$

$$= \frac{\dfrac{\log 200}{\log 6} - 2}{3}$$

By calculator, $x \approx .3190157417 \approx .32$. ■

Applications

An important application of exponential equations involves the calculation of human population growth. Human populations grow according to the exponential equation

$$P = P_0 e^{rt}$$

where P_0 is the size of the initial population, r is the growth rate, t is the length of time, and P is the size of the population after time t.

EXAMPLE 12 On April 3, 1987, newspaper headlines proclaimed that the world population reached 5 billion. If the annual growth rate is 2%, when will the world population reach 6 billion?

Solution The initial population P_0 is 5 (billion), P is 6 (billion), $r = .02$, and t is the unknown. Substitute the known values into the population growth formula:

$$6 = 5e^{.02t}$$

$$\frac{6}{5} = e^{.02t}$$

$.02t = \ln 1.2$ Use the definition of logarithm; note that $\frac{6}{5} = 1.2$.*

$$t = \frac{\ln 1.2}{.02} \approx 9.11607784$$

We would expect the world's population to reach 6 billion about 9 years after it reached 5 billion; this would be in 1996. ■

EXAMPLE 13 An almanac lists the 1970 population of San Antonio, Texas, as 654,153, and the 1980 population as 783,296. What was the growth rate of San Antonio for this period?

Solution Since $P_0 = 654{,}153$, $P = 783{,}296$, and $t = 10$, we have

$$P = P_0 e^{rt}$$

$$\frac{P}{P_0} = e^{rt}$$

$rt = \ln\left(\frac{P}{P_0}\right)$ Use the definition of logarithm.*

$$r = \frac{1}{t} \ln \frac{P}{P_0}$$

For this example,

$$r = \frac{1}{10} \ln \frac{783{,}296}{654{,}153} \approx .0180169389$$

The growth rate is about 1.8%. ■

EXAMPLE 14 How long (to the nearest month) will it take for your money to double if you deposit it at a credit union paying 12.5% interest compounded quarterly?

Solution The compound interest formula is

$$A = P(1 + i)^N$$

where $i = \frac{r}{n} = \frac{.125}{4} = .03125$, $A = 2P$, and the unknown is $N = nt = 4t$, where t is the time (in years) to be found to the nearest month.

$$2P = P(1.03125)^{4t}$$

$2 = 1.03125^{4t}$ Divide both sides by P ($P \neq 0$).

$4t = \log_{1.03125} 2$ Use the definition of logarithm.*

$t = \frac{1}{4} \log_{1.03125} 2$ Divide both sides by 4.

≈ 5.6313765 By calculator.

* Many people like to solve equations like these by "taking the log of both sides." This is an acceptable way of proceeding, but unnecessary. It dates back to the time when calculators were not available. The method we use here assumes that you have a calculator, and reinforces the definition of a logarithm as an exponent.

This is 5 years and some months. To find the number of months, subtract 5 and multiply by 12 (to convert .6313765 of a year to months). The answer (to the nearest month) is 5 years 8 months. ■

13.3 Problem Set

Solve Problems 1–43.

1. $\log_5 25 = x$
2. $\log_2 128 = x$
3. $\log(\frac{1}{10}) = x$
4. $\log_x 84 = 2$
5. $\log_x 28 = 2$
6. $\log x = 2$
7. $\ln x = 3$
8. $\ln x = \ln 14$
9. $\ln 9.3 = \ln x$
10. $\log_3 x^2 = \log_3 125$
11. $\ln x^2 = \ln 12$
12. $\log_2 8\sqrt{2} = x$
13. $\log_3 27\sqrt{3} = x$
14. $\log_x 1 = 0$
15. $\log_x 10 = 0$
16. $\log x = 5$
17. $2^x = 128$
18. $8^x = 32$
19. $125^x = 25$
20. $(\frac{2}{3})^x = \frac{9}{4}$
21. $3^{4x-3} = \frac{1}{9}$
22. $27^{2x+1} = 3$
23. $(\frac{3}{2})^x = 10$
24. $1 = 2\log x - \log 1{,}000$
25. $\ln x - \ln e^3 = 1$
26. $\log_a 102 + \log_a 4 - \log_a 3.1 = 1$
27. $\log_b 6 - \log_b 2.8 + \log_b 3.9 = 1$
28. $\log_8 5 + \frac{1}{2}\log_8 9 = \log_8 x$
29. $\log_7 x - \frac{1}{2}\log_7 4 = \frac{1}{2}\log_7(2x - 3)$
30. $\ln x - \frac{1}{2}\ln 3 = \frac{1}{2}\ln(x + 6)$
31. $\frac{1}{2}\ln x = 3\ln 5 - \ln x$
32. $\log_x(5x - 4) = 2$
33. $\log_x(x + 6) = 2$
34. $\frac{1}{2}\log_2 x = 3\log_2 3 - \log_2 x$
35. $\ln 10 - \frac{1}{2}\ln 25 = \ln x$
36. $10^{5-3x} = .041$
37. $10^{2x-1} = 515$
38. $5^{-x} = 8$
39. $4^x = .82$
40. $e^{2x} = 10$
41. $e^{5x} = \frac{1}{4}$
42. $e^{1-2x} = 3$
43. $e^{1-5x} = 15$

APPLICATIONS

44. The world population reached 4 billion on March 18, 1976 and 5 billion on April 3, 1987. What was the growth rate for this period of time (to the nearest hundredth of a percent)?
45. Use the data given in Example 12 to estimate the year the world population will reach 7 billion.
46. Use the data given in Example 12 to estimate the year the world population will reach 8 billion.
47. If \$1,000 is invested at 12% compounded semiannually, how long will it take (to the nearest half-year) for the money to double?
48. Repeat Problem 47, except compound daily and give your answer to the nearest day (365-day year).
49. Repeat Problem 47, except compound continuously and give your answer to the nearest day (365-day year).
50. How long will it take for \$1,000 to triple if it is invested at 12% compounded quarterly? (Give your answer to the nearest quarter.)
51. Repeat Problem 50, except compound continuously and give your answer to the nearest day.
52. If \$1,000 is invested at 12% interest compounded quarterly, how long will it take (to the nearest quarter) for the money to reach \$2,500?
53. Psychologists are concerned with forgetting. In an experiment, students were asked to remember a set of nonsense syllables, such as "nem." They then had to recall the syllables after t seconds. The model used to describe forgetting in this experiment is

$$R = 80 - 27\ln t \qquad t > 1$$

where R is the percentage of students who remember the syllables after t seconds.

   a. What percentage of the students remembered the syllables after 3 seconds?
   b. In how many seconds would only 10% of the students remember the syllables?

54. Solve the forgetting curve formula in Problem 53 for t.
55. If P dollars are borrowed for n months at a monthly interest rate of i, then the monthly payment m is found by the formula

$$m = \frac{Pi}{1 - (1 + i)^{-n}}$$

Use this formula to find the monthly car payment after a down payment of \$2,487 on a new car costing \$12,487. The car is financed for 4 years at 12%. (*Hint:* $P = \$10{,}000$ and $i = .01$.)

56. A home loan is made for \$110,000 at 12% interest for 30 years. What is the monthly payment and what is the

total amount of interest paid over the life of the loan? (*Hint:* Use the formula given in Problem 55.)

57. A formula used for carbon-14 dating in archaeology is

$$A = A_0\left(\frac{1}{2}\right)^{t/5,700} \quad \text{or} \quad P = \left(\frac{1}{2}\right)^{t/5,700}$$

where P is the percentage of carbon-14 present after t years. Solve for t in terms of P.

58. Some bone artifacts found at the Lindemeir site in northeastern Colorado were tested for their carbon-14 content. If 25% of the original carbon-14 was still present, what is the probable age of the artifacts? Use the formula in Problem 57.

59. In 1975 ($t = 0$), the world use of petroleum, P_0, was 19,473 million barrels of oil. If the world reserves at that time were estimated to be 584,600 million barrels and the growth rate for the use of oil is k, then the total amount A used during a time interval $t > 0$ is given by

$$A = \frac{P_0}{k}(e^{kt} - 1)$$

How long will it be before the world reserves are depleted if $k = 8\%$? What does your answer mean?

60. Solve the formula in Problem 59 for t.

61. Prove that $\log_b \frac{A}{B} = \log_b A - \log_b B$.

62. Prove that $\log_b A^p = p \log_b A$.

13.4 Derivatives of Logarithmic and Exponential Functions

Derivative of Logarithmic Function

Rates of change and graphs of logarithmic and exponential functions are fairly common applications in management and the life and social sciences. In Chapters 10 and 11 the notion of a derivative as a limit was defined; that is, if $y = f(x)$, then

$$y' = \lim_{h \to 0} \frac{f(x + h) - f(x)}{h}$$

provided this limit exists. We found a variety of derivatives by using this formula and derived some derivative formulas that made the process very efficient, so that in some problems we no longer had to resort to the definition of derivative. Now we will find the derivatives of the logarithmic and exponential functions. Our first attempt might be to try to apply some of our derivative formulas, but since the logarithmic and exponential functions are transcendental and not algebraic, none of the derivative formulas applies. Whenever we need to find the derivative of a new class of functions we must go back to the definition of derivative.

We begin by finding the derivative of the natural logarithm; that is, we let $y = \ln x$. We need to find

$$\lim_{h \to 0} \frac{\ln(x + h) - \ln x}{h} = \lim_{h \to 0} \frac{\ln\left(\dfrac{x + h}{x}\right)}{h}$$

This is a property of logarithms; do you see which one?

$$= \lim_{h \to 0} \frac{1}{h} \ln\left(\frac{x + h}{x}\right)$$

$$= \lim_{h \to 0} \ln\left(\frac{x + h}{x}\right)^{1/h}$$

Another property of logarithms

$$= \lim_{h \to 0} \ln\left(1 + \frac{h}{x}\right)^{1/h}$$

Let $m = \dfrac{x}{h}$ so that $\dfrac{h}{x} = \dfrac{1}{m}$ and $\dfrac{1}{h} = \dfrac{m}{x}$. Also, as $h \to 0$, $m \to \infty$. Now substitute these changes into the derivative formula:

$$\begin{aligned}
\lim_{h\to 0} \ln\left(1 + \frac{h}{x}\right)^{1/h} &= \lim_{m\to\infty} \ln\left(1 + \frac{1}{m}\right)^{m/x} \\
&= \lim_{m\to\infty} \ln\left[\left(1 + \frac{1}{m}\right)^{m}\right]^{1/x} \\
&= \lim_{m\to\infty} \frac{1}{x} \ln\left(1 + \frac{1}{m}\right)^{m} \\
&= \frac{1}{x} \lim_{m\to\infty} \ln\left(1 + \frac{1}{m}\right)^{m} \\
&= \frac{1}{x} \ln e \qquad \text{Remember the definition of } e. \\
&= \frac{1}{x} \qquad \ln e = 1
\end{aligned}$$

This turns out to be a very pleasing result because of its simplicity! We can also obtain a more general result by applying the chain rule:

Derivative of Logarithm

If $y = \ln|x|$, then $y' = \dfrac{1}{x}$.

If $y = \ln|u|$, where u is a function of x, then $y' = \dfrac{u'}{u}$.

The absolute value bars have been added because the domain of $\ln x$ includes only positive values of x. We are not differentiating $|x|$ here but are simply restricting the domain of x.

EXAMPLE 1 If $y = \ln|3x|$, find y'.

Solution $u(x) = 3x$, so $u'(x) = 3$. Thus

$$y' = \frac{3}{3x} = \frac{1}{x}$$

■

EXAMPLE 2 If $y = \ln|5x^3 + 3x^2 - 4|$, find y'.

Solution $u(x) = 5x^3 + 3x^2 - 4$, so $u'(x) = 15x^2 + 6x$. (This step is usually done mentally.)

$$y' = \frac{15x^2 + 6x}{5x^3 + 3x^2 - 4}$$

■

EXAMPLE 3 If $y = (3x^2 + 2x)\ln|5x - 7|$, find y'.

Solution Use the product rule:

$$y' = (3x^2 + 2x)\left(\frac{5}{5x - 7}\right) + (6x + 2)\ln|5x - 7|$$

$$= \frac{5x(3x + 2)}{5x - 7} + 2(3x + 1)\ln|5x - 7|$$ ■

Properties of logarithms can sometimes simplify the process of finding derivatives. This is especially true when quotients are involved, as shown in Example 4.

EXAMPLE 4 If $y = \ln\left(\frac{x}{x + 1}\right)$ find y' as follows:

a. By taking the derivative of the logarithm

b. By using properties of logarithms before taking the derivative

Solution **a.** $\frac{d}{dx}\left[\ln\left(\frac{x}{x + 1}\right)\right] = \frac{1}{\frac{x}{x+1}}\left[\frac{x}{x + 1}\right]' = \frac{x + 1}{x}\left[\frac{(x + 1)(1) - x(1)}{(x + 1)^2}\right] = \frac{1}{x(x + 1)}$

b. $\frac{d}{dx}\left[\ln\left(\frac{x}{x + 1}\right)\right] = \frac{d}{dx}[\ln x - \ln(x + 1)] = \frac{1}{x} - \frac{1}{x + 1}$

$$= \frac{x + 1 - x}{x(x + 1)} = \frac{1}{x(x + 1)}$$ ■

WARNING ***Take note of the proper notation for logarithms.***

The notation of logarithms can be confusing. Remember, $\ln x$ means $\log_e x$, $\log x$ means $\log_{10} x$. Also, be sure you distinguish between $\ln x^2$, which means $\ln(xx)$, and $(\ln x)^2$, which means $(\ln x)(\ln x)$.

EXAMPLE 5 If $y = \log x$ find $\frac{dy}{dx}$.

Solution The derivative formula we derived applies only to natural logarithms. In order to find the derivative for a common logarithm, we use the change of base rule:

$$y = \log x$$

$$= \frac{\ln x}{\ln 10}$$

$$= \frac{1}{\ln 10}\ln x \qquad \textit{Note: } \frac{1}{\ln 10} \text{ is a constant.}$$

$$\frac{dy}{dx} = \frac{d}{dx}\left(\frac{1}{\ln 10}\ln x\right) \qquad \text{Take the derivative of both sides.}$$

$$= \frac{1}{\ln 10}\frac{1}{x} \qquad \frac{d}{dy}\ln x = \frac{1}{x}$$ ■

EXAMPLE 6 The demand equation of a commodity is

$$p = \frac{200 \ln(x + 1)}{x}$$

and the cost of producing x units is $C(x) = 2x$. Show that the profit function has a relative maximum and not a minimum point.

Solution The profit function $P(x)$ is found by $P(x) = R(x) - C(x)$. We are given the cost function, but we need to find $R(x)$. The revenue function is found by multiplying the number of units times the demand, so $R(x) = xp = 200 \ln(x + 1)$.

$$P(x) = 200 \ln(x + 1) - 2x$$

$$P'(x) = 200\frac{1}{x + 1} - 2$$

$$= 200(x + 1)^{-1} - 2$$

Set $P'(x) = 0$ and solve for x:

$$200(x + 1)^{-1} - 2 = 0$$

$$200 = 2(x + 1)$$

$$99 = x$$

To determine if this is a relative maximum or minimum, check the second derivative:

$$P''(x) = -200(x + 1)^{-2}$$

This function is negative for all values of x, so the profit function has a relative maximum at $x = 99$ and no relative minimum. ■

CALCULATOR COMMENT

When graphing P in Example 6 using a function using a standard domain and range, nothing shows on the screen. Using the TRACE, it appears that the largest value for x is
X=98.947368 Y=723.03401
This leads us to input the following values in the

RANGE:
Xmin=0
Xmax=200
Ymin=0
Ymax=1000
Graph shows:

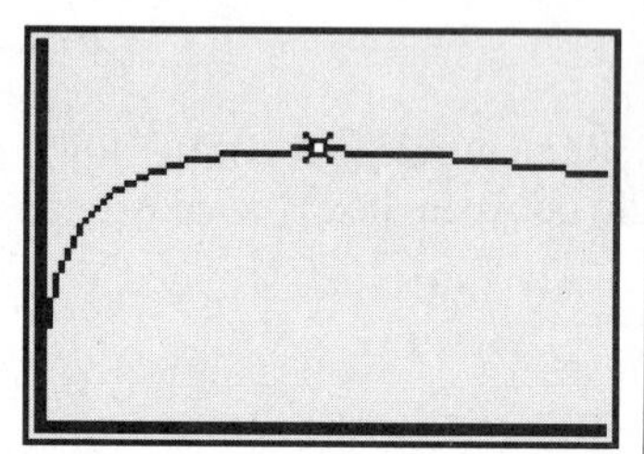

Derivative of Exponential Function

Now we turn our attention to finding the derivative of the exponential function $y = e^x$. To do this, we use the definition of logarithm to write $x = \ln y$. We can now use the derivative rule for $\ln y$ (found above) by using implicit differentiation on the equation $\ln y = x$:

$$\frac{1}{y} \cdot \frac{dy}{dx} = 1$$

We solve for dy/dx:

$$\frac{dy}{dx} = y$$

Since $y = e^x$, we see that the derivative of e^x is e^x; what could be easier than this! This result shows one of the reasons why base e rather than base 10 is used almost exclusively in more advanced work. Again, this result is generalized using the chain rule:

WARNING
$\frac{d}{dx}(e^x) \neq xe^{x-1}$
This is a common mistake.

Derivative of Exponential

If $y = e^x$,	then	$y' = e^x$.
If $y = e^u$,	where u is a function of x, then	$y' = u' \cdot e^u$.

EXAMPLE 7 If $y = e^{10x}$, then $y' = 10e^{10x}$. ■

EXAMPLE 8 If $y = e^{4x^2+5}$, then $y' = 8xe^{4x^2+5}$. ■

EXAMPLE 9 If $y = x^3e^{4x}$, then $y' = x^3(4e^{4x}) + 3x^2e^{4x}$ Use the product rule first.

$$= 4x^3e^{4x} + 3x^2e^{4x} \quad \text{or} \quad x^2e^{4x}(4x + 3)$$ ■

EXAMPLE 10 If $y = \dfrac{e^x}{\ln|x|}$, then $y' = \dfrac{\ln|x| \cdot e^x - e^x \cdot (1/x)}{\ln^2|x|}$

Use the quotient rule first. Note that $[\ln (x)]^2$ is written more simply as $\ln^2 x$.

$$= \frac{\frac{e^x}{x}(x\ln|x| - 1)}{\ln^2|x|}$$

$$= \frac{e^x(x\ln|x| - 1)}{x\ln^2|x|}$$ ■

EXAMPLE 11 Suppose you study the spread of a disease introduced into a small town of 2,000 persons. Assume that everyone in the town has an equal chance of contracting the disease. A model for predicting the number of people, N, contracting the disease is given by

$$N(t) = \frac{2{,}000}{1 + 1{,}999e^{-.5t}}$$

where t is the number of days since the disease was introduced into the community. What is the rate at which members of the community are contracting the disease?

Solution The requested function is dN/dt. Write $N(t) = 2{,}000(1 + 1{,}999e^{-.5t})^{-1}$. Then,

$$\frac{dN}{dt} = 2{,}000(-1)(1 + 1{,}999e^{-.5t})^{-2}(-.5)1{,}999e^{-.5t}$$

$$= \frac{2{,}000(-1)(-.5)1{,}999e^{-.5t}}{(1 + 1{,}999e^{-.5t})^2}$$

$$= \frac{1{,}999{,}000e^{-.5t}}{(1 + 1{,}999e^{-.5t})^2}$$

This means, for example, that on day 0, $N(0) = 2{,}000/(1 + 1{,}999) = 1$ person has the disease, but dN/dt at $t = 0$ is

$$\frac{1{,}999{,}000}{2{,}000^2} = .49975$$

so the rate means that nearly one person is contracting the disease every 2 days. On the other hand, after 10 days, $N(10) \approx 138$, so about 138 people have the disease and the rate at which new people are now contracting the disease is dN/dt at $t = 10$:

$$\frac{1{,}999{,}000e^{-5}}{(1 + 1{,}999e^{-5})^2} \approx 64.33598929$$

which means that at the time $t = 10$, new people are contracting the disease at the rate of about 64 people per day. ■

Graphing Techniques for Logarithmic and Exponential Functions

EXAMPLE 12 Graph $f(x) = 10x - 5x \ln x$.

Solution Domain is $x > 0$.

Function	First derivative	Second derivative
$f(x) = 10x - 5x \ln x$	$f'(x) = 5 - 5 \ln x$ Detail of work: $f'(x) = 10 - (5x \cdot \frac{1}{x} + 5 \ln x)$ $= 10 - 5 - 5 \ln x$ $= 5 - 5 \ln x$ Critical values: $5 - 5 \ln x = 0$ $\ln x = 1$ $x = e$	$f''(x) = -5x^{-1}$
$f(e) = 10e - 5e \ln e$ $= 10e - 5e$ $= 5e$ ≈ 13.59	Check $x = e$ If $x < e$, $f'(x) > 0$ increasing If $x > e$, $f'(x) < 0$ decreasing	$f''(e) = -5e^{-1} < 0$ Relative maximum at $x = e$ No inflection point Concave downward for $x > 0$

The graph is shown in Figure 13.5.

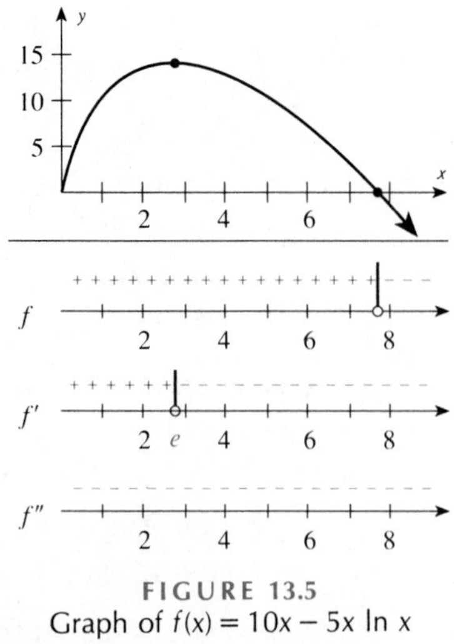

FIGURE 13.5
Graph of $f(x) = 10x - 5x \ln x$ ■

EXAMPLE 13 Graph $f(x) = xe^x$.

Solution Domain is $(-\infty, \infty)$.

Function	First derivative	Second derivative
$f(x) = xe^x$	$f'(x) = e^x(x + 1)$ Detail of work: $f'(x) = x\dfrac{d}{dx}e^x + e^x\dfrac{d}{dx}(x)$ $= xe^x + e^x$ $= e^x(x + 1)$ Critical values: $e^x(x + 1) = 0$ Since $e^x \neq 0$, the only critical value is $x = -1$.	$f''(x) = e^x(x + 2)$
$f(-1) = -e^{-1}$	If $x < -1$, $f'(x) < 0$ decreasing If $x > -1$, $f'(x) > 0$ increasing	$f''(-1) = e^{-1} > 0$ Relative minimum at $x = -1$
$f(0) = 0$		Points of inflection: $f''(x) = 0$ if $x = -2$ If $x < -2$ Concave downward If $x > -2$ Concave upward

The graph is shown in Figure 13.6.

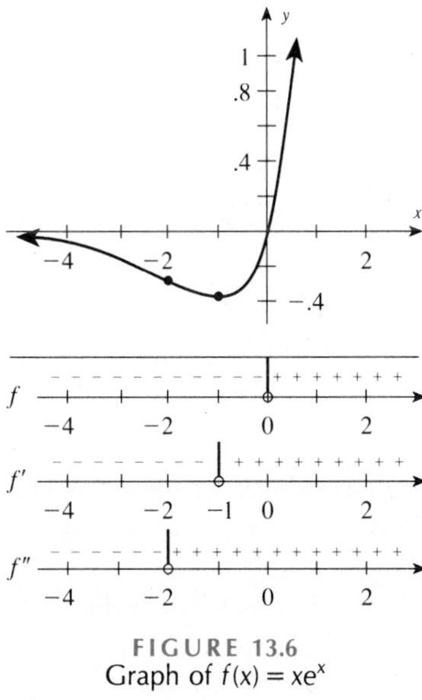

FIGURE 13.6
Graph of $f(x) = xe^x$

13.4

Problem Set

Find the derivatives of the functions in Problems 1–34.

1. $y = e^{2x}$
2. $y = e^{8x}$
3. $y = e^{-x}$
4. $y = e^{-3x}$
5. $f(x) = xe^x$
6. $f(x) = x^2e^{3x}$
7. $f(x) = (2x^3 + 1)e^{5x^2}$
8. $f(x) = (5x^2 - x)e^{-3x^2}$
9. $y = -2e^{5x}(3x^2 + 5)$
10. $y = (3x + 2e^{5x})^4$
11. $y = \ln|5 - x|$
12. $y = \ln|3 - x^2|$
13. $y = (2x^3 + e^{5x})^4$
14. $y = (e^{6x} - 5x^4)^3$
15. $y = \ln x^4$
16. $y = (\ln x)^4$
17. $y = \ln\sqrt{x}$
18. $y = \sqrt{\ln|x|}$
19. $y = \dfrac{\ln x^3}{e^x}$
20. $y = \dfrac{e^{3x}}{\ln x^2}$
21. $f(x) = \ln\left|\dfrac{x}{x+2}\right|$
22. $f(x) = \ln\left|\dfrac{x-3}{x^2}\right|$
23. $f(t) = e^{t^2}(1 - e^{-t})$
24. $f(t) = e^{(t+1)/t}$
25. $y = (e^x - e^{-x})^2$
26. $y = (e^{-x} + e^x)^3$
27. $y = \dfrac{\ln|x|}{5x + 2}$
28. $y = \dfrac{3x^2 - 7}{\ln|x|}$
29. $f(x) = \dfrac{e^x}{\ln|3x|}$
30. $f(x) = \dfrac{e^{5x}}{\ln|5x|}$
31. $y = \dfrac{2{,}000}{1 + 6e^{.3x}}$
32. $y = \dfrac{500}{1 - 30e^{.2x}}$
33. $y = \dfrac{100\ln|x|}{1 + 40e^{-.2x}}$
34. $y = \dfrac{900\ln|x|}{1 - 70e^{-.3x}}$

Graph the functions in Problems 35–40.

35. $f(x) = 2x - 5x\ln x$
36. $f(x) = x - \ln x$
37. $f(x) = \dfrac{\ln x}{x}, \quad x \geq 1$
38. $f(x) = \dfrac{\ln x^2}{x^2}, \quad x \geq 1$
39. $y = x^2e^{3x}$
40. $y = e^x - e^{-x}$

APPLICATIONS

41. The demand equation of a commodity is

$$d = \frac{500\ln(x + 10)}{x^2}$$

where x is the number of units produced ($1 \leq x \leq 20$). The cost of producing x units is $C(x) = 2x^2$. Find the marginal revenue.

42. Find the marginal cost for the information in Problem 41.

43. The spread of a disease in a town of 5,000 people is described by the model

$$N(t) = \frac{5{,}000}{1 + 4{,}999e^{-.1t}}$$

where N is the number of people contracting the disease after t days.
   a. How many people have the disease on day 0?
   b. What is the rate at which people are getting the disease on day 0?

44. a. How many people in Problem 43 have the disease on the tenth day?
   b. What is the rate at which people are getting the disease on the tenth day?

45. The amount of money, P, invested at 12% interest and compounded continuously for t years, has a future value A according to the formula $A = Pe^{.12t}$. At what rate is the amount of money A increasing after 1 year? After 5 years?

46. The blood pressure in the aorta changes between beats according to the formula

$$P(t) = e^{-kt}$$

for an appropriate constant k, where t is the time in milliseconds since the last beat. What is the rate at which the pressure is changing with respect to time after 3 milliseconds if $k = .025$?

47. The amount of a drug present in the body is a function of the amount administered and the length of time t (in hours) since it was administered. For 5 milliliters of a certain drug, the amount of drug present behaves according to the formula

$$A(t) = 5e^{-.03t}$$

What is the rate at which the amount of drug present is changing expressed as a function of time?

48. Sales often follow a growth pattern of rapid initial growth with some leveling off after a period of time. This pattern is described by the formula

$$S(t) = 25{,}000 - 10{,}000e^{-.2t}$$

   a. How many items will be sold initially?
   b. What is the rate of change of sales initially?
   c. What is the rate of change of sales after t years?

49. The price–demand equation for x units of a commodity is given by

$$d(x) = 500e^{-.1x}$$

Find the marginal revenue.

50. The price–supply equation for x units of the commodity described in Problem 49 is given by

$$s(x) = 50e^{.05x}$$

Find the marginal supply.

51. If $y = \log x$, show that

$$y' = \frac{\log e}{x}$$

52. If $y = \log_b x$, show that

$$y' = \frac{\log_b e}{x}$$

53. If $y = \log_b u$, where u is a function of x, show that

$$\frac{dy}{dx} = \frac{\log_b e}{u} \cdot \frac{du}{dx}$$

*13.5 Review

The material of this chapter is reviewed in the following list of objectives. After each objective there are some practice questions. For a sample test select the first question of each set and check your answers. The second question for each objective has no answer given. If you are having trouble with a particular type of problem, look back at the indicated section in the text. When you are finished reviewing these objectives, a sample examination is given at the end of this section.

[13.1]

Objective 13.1: *Simplify expressions with positive, negative, and fractional exponents.*

1. $125^{2/3}$
2. $2^{1/2} \cdot 3^{1/3}$
3. $\dfrac{27^{2/3}}{27^{1/2}}$
4. $(x^{1/2} - y^{1/2})(x^{1/2} + y^{1/2})$

Objective 13.2: *Sketch the graph of exponential functions.*

5. $y = (\frac{1}{2})^x$ 6. $y = -2^x$
7. $y = 2^{-x}$ 8. $y = e^{-x/2}$

Objective 13.3: *Evaluate expressions with the natural base.*

9. e^1 10. e^4
11. $e^{1.05}$ 12. $e^{-.005}$

[13.2]

Objective 13.4: *Write an exponential equation in logarithmic form.*

13. $10^{.5} = \sqrt{10}$ 14. $e^0 = 1$
15. $9^3 = 729$ 16. $(\sqrt{2})^3 = 2\sqrt{2}$

Objective 13.5: *Write a logarithmic equation in exponential form.*

17. $\log 1 = 0$ 18. $\ln \frac{1}{e} = -1$
19. $\log_2 64 = 6$ 20. $\log_\pi \pi = 1$

Objective 13.6: *Evaluate common and natural logarithms.*

21. $\ln 3$ 22. $\log 3$
23. $\log .0021$ 24. $\ln .013$

Objective 13.7: *Graph logarithmic functions.*

25. $y = \log x$ 26. $y = \ln x$
27. $y = \log_5 x$ 28. $y = \log_{1/5} x$

[13.3]

Objective 13.8: *Solve logarithmic equations.*

29. $\log_5 25 = x$ 30. $\log_x(x + 6) = 2$
31. $3 \log 3 - \frac{1}{2} \log 3 = \log \sqrt{x}$
32. $2 \ln \frac{e}{\sqrt{7}} = 2 - \ln x$

Objective 13.9: *Solve exponential equations.*

33. $10^{x+2} = 125$ 34. $5^{2x-3} = .5$
35. $e^{4-3x} = 15$ 36. $10^{-x^2} = .45$

[13.4]

Objective 13.10: *Find the derivatives of exponential functions.*

37. $y = 4.9e^{-.05x^2}$ 38. $y = 3{,}500x - 500e^{3x}$
39. $y = x^3e^{x^4}$ 40. $y = \dfrac{1{,}000}{x - 999e^{-.2x}}$

Objective 13.11: *Find the derivatives of logarithmic functions.*

41. $y = \ln x^2$ 42. $y = \ln|x^4(x^2 - 5)|$
43. $y = \dfrac{\ln x^2}{x + 3}$ 44. $y = 250 \ln|3x^2 - 5|$

* Optional section.

Objective 13.12: *Solve applied problems based on the preceding objectives.*

45. If a person's present salary is \$30,000 per year, use the formula $A = Pe^{rt}$ to determine the salary necessary to equal this salary in 15 years if you assume the 1991 annual inflation rate of 8.2% compounded continuously.

46. If \$5,500 is invested at 13.5% compounded daily, how long will it take for this to grow to \$10,000? (Use a 365-day year.)

47. *Archaeology.* The half-life formula for carbon-14 dating is

$$A = 10\left(\frac{1}{2}\right)^{t/5{,}700}$$

where A is the amount present (in milligrams) after t years. Solve this equation for t.

48. *Advertising.* An advertising agency conducted a survey and found that the number of units sold, N, is related to the amount spent on advertising (in dollars) by the formula

$$A = 1{,}500 + 300 \ln a \qquad (a \geq 1)$$

What is the rate at which N is changing at the instant that $a = \$10{,}000$?

49. The atmospheric pressure P in pounds per square inch (psi) is given by

$$P = 14.7e^{-.21a}$$

where a is the altitude above sea level (in miles). As a hot-air balloon is rising the pressure is constantly changing. How fast is the pressure changing when the balloon is one mile above sea level?

50. When a satellite has an initial radioisotope power supply of 50 watts, its power output in watts is given by the equation

$$P = 50e^{-t/250}$$

where t is the time in days. Solve for t.

51. *Equilibrium point.* Suppose the price–demand and price–supply equations for x thousands of units of a commodity are given by

DEMAND: $p(x) = 300e^{-.3x}$

SUPPLY: $s(x) = 30e^{.1x}$

Find the value of x for the equilibrium point.

52. Suppose the president makes a major policy announcement. A model to predict the percentage, N, of the population that will have heard the announcement is a function of the time, t (in days), after the announcement. Suppose a suitable model is given by the formula

$$N = 1 - e^{-2.5t}$$

How long will it take for 90% of the population ($N = .9$) to hear of the announcement?

SAMPLE TEST

The following sample test (45 minutes) is intended to review the main ideas of this chapter.

Solve Problems 1–8.

1. $\log_6 36 = x$
2. $\log_x(2x + 15) = 2$
3. $2\log 2 - \frac{1}{2}\log 2 = \log\sqrt{x}$
4. $2\ln\frac{e}{\sqrt{5}} = 1 - \ln x$
5. $10^{x-1} = 250$
6. $3^{1-2x} = .5$
7. $e^{2x+3} = 10$
8. $10^{-x^2} = .75$

Find the derivatives in Problems 9–16.

9. $y = 5.5e^{-.5x^2}$
10. $y = (x^2 - 1)e^{3x}$
11. $y = \ln x^5$
12. $y = (\ln x)^5$
13. $y = \ln|x^2(4 - x^3)|$
14. $y = e^x \ln x$
15. $y = \dfrac{\ln x^2}{4 - x}$
16. $y = \dfrac{1{,}000\ln|x^3|}{2e^x}$
17. The inflation formula (continuous)

$$A = Pe^{rt}$$

gives the future value after t years of an inflation rate of r. Solve this equation for r.

18. An advertising agency conducts a survey that finds that the number of units sold, N, is related to the amount spent on advertising (in dollars) by the formula

$$N = 2{,}500 + 200 \ln a \qquad (a \geq 1)$$

What is the rate at which N is changing at the instant that $a = \$20{,}000$?

19. The price–demand and price–supply equations for x thousands of units of a commodity are given by

Demand: $p(x) = 200e^{-.2x}$

Supply: $s(x) = 20e^{.1x}$

Find the number of units that should be manufactured in order to achieve equilibrium.

20. The president makes a major policy announcement. The percentage of the population that will have heard about the announcement is a function of the time t (in days) after the announcement according to the formula

$$N = 1 - e^{-1.5t}$$

where N is the percentage written as a decimal. How long will it take for 90% of the population to hear about the announcement?

14

The Integral

CHAPTER OVERVIEW

The last main idea of calculus is introduced in this chapter—the idea of an integral and a procedure called integration.

PREVIEW

The integral is first introduced as an antiderivative, and then an indefinite integral is defined. The idea of a definite integral is applied to finding areas, and finally we give a formal definition of the definite integral and state the Fundamental Theorem of Calculus.

PERSPECTIVE

The skills you learn in this chapter will be put to use in the next two chapters when applications of integration are considered. As you will see in this chapter, the ability to "reverse" the process of differentiation is important since you quite often know the derivative but need to know the original function. The indefinite integral is used for that purpose.

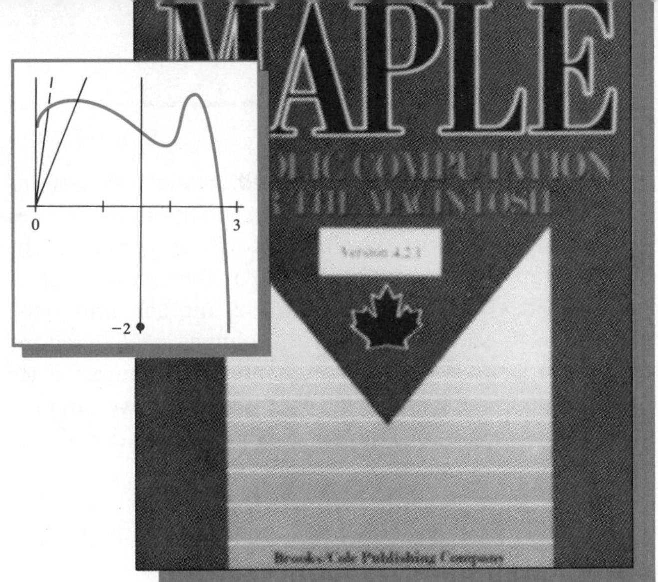

MODELING APPLICATION 12

Computers in Mathematics

In the 1980s, mathematics instruction was forever altered with the advent of powerful, inexpensive calculators and computer programs that would do symbolic manipulation and function graphing. This new software allows both instructors and students to focus on concepts and processes rather than on tedious and complicated calculations involved in realistic problems.

DERIVE, the successor to MUMATH, is a powerful computer algebra system for PC-compatible computers. It is available from The Soft Warehouse, 3615 Harding Avenue, Suite 505, Honolulu, HI 96816.

MATHEMATICA is advertised as more than a program—it is a *system* for doing mathematics. It is available from Wolfram Research, Inc., P.O. Box 6059, Champaign, IL 61821. It runs on both Macintosh and IBM-compatible computers.

MAPLE also does algebraic manipulation in an easy-to-use format for Macintosh computers. It is available from Brooks/Cole Publishing, 555 Forest Lodge Road, Pacific Grove, CA 93950.

After you have finished this chapter, write a paper on a computer program that does symbolic manipulation. Your paper should describe the program, as well as evaluate it.

APPLICATIONS

Management (*Business, Economics, Finance, and Investments*)
Finding the cost function given the marginal cost (14.1, Problems 37–40)
Finding the profit function for Turbo Plus, Inc. (14.1, Problem 43)
Finding the cost of producing 100,000 processors (14.1, Problem 44)
Determining the output after t hours of work (14.2, Problem 42)
Finding the profit when the marginal profit is known (14.2, Problems 43–44)
Finding the sales (14.2, Problem 45)
Determining the accumulated sales (14.3, Problems 45–46)
Finding the Gini index to measure money flow (14.3, Problems 47–48)
Accumulated cost and revenue (14.3, Problem 49; 14.4, Problems 42–43)
Total cost of production (14.4, Problem 36)
Automobile sales from a graph in the *Wall Street Journal* (14.5, Problems 37–38)
Total paper and paperboard production in the U.S. (14.5, Problem 39)

Management (*continued*)
Light output vs. time for a flashbulb (14.6, Problem 38)
Treasury bond rate (14.7, Problem 60)
Determining if the eucalyptus tree grows fast enough to make commercial growing a profitable enterprise (14.7, Test Problem 17)

Life Sciences (*Biology, Ecology, Health, and Medicine*)
Infection by a new strain of influenza (14.1, Problems 45–46)
Flu epidemic (14.2, Problem 41)
Predicting oil consumption (petroleum) given present production and rate of consumption (14.2, Problems 48–49; 14.3, Problems 35–39; 14.7, Problem 19)
Toxic dumping (14.3, Problems 40–44; 14.7, Problem 59)
Rate of healing (14.3, Problem 50)
Amount of pollutants dumped into the Russian River (14.4, Problem 37)
Chemical pollution (14.4, Problem 39)

Life Sciences (*continued*)
Growing rate of a culture (14.4, Problem 40)
Rabbit population (14.5, Problem 41)

Social Sciences (*Demography, Political science, Population, Psychology, Society, and Sociology*)
Predicting population given growth rate (14.1, Problems 41–42; 14.2, Problem 47)
Predicting the world population (14.2, Problem 46; 14.5, Problem 42)
Growth rate of Chicago, Illinois (14.7, Test Problem 18)

General Interest
Velocity of an object thrown off Hoover Dam (14.1, Problems 47–48)
Temperature variations (14.5, Problems 43–44)
Estimating areas of swimming pools and parcels (14.5, Problems 45–46; 14.6, Problem 37)

Modeling Application—
Computers in Mathematics

14.1

The Antiderivative

Calculus is divided into two broad categories. The first, **differential calculus**, involves the definition of derivative and related applications of instantaneous rates of change, marginal costs, profits, and curve sketching, as well as finding maximums and minimums. In this chapter, we begin our study of the second part of calculus, **integral calculus**, which deals with the definition of another limit—the **definite integral**. Integral calculus is used to find areas, volumes, and functions when we have information about the function's rate of change. For example, if we know the present world population and the rate at which it is growing, we can produce a formula (subject to certain assumptions) that predicts the population size at any future time or estimates its size at some time in the past.

EXAMPLE 1 Multiplex Corporation knows that the marginal cost (in thousands of dollars) to produce x items (in thousands) is given by the formula

$$f(x) = 2x + 4$$

Find the cost of producing 50,000 items ($x = 50$) if the fixed costs are \$25,000.

Solution In Section 11.3 marginal cost was defined to be the rate of change in cost per unit change in production at an output level of x units. This means that if F is the cost function, then the marginal cost, f, is a function that can be found by taking the derivative of F at x:

$$F'(x) = f(x)$$

Find a function F so that $F'(x) = f(x)$:

$$F'(x) = 2x + 4$$

By trial and error (we will develop better methods very shortly), we see that

$$F(x) = x^2 + 4x$$

is such a function, but this function is not unique:

If $F_1(x) = x^2 + 4x + 6$, then $F_1'(x) = 2x + 4$;
if $F_2(x) = x^2 + 4x - 100$, then $F_2'(x) = 2x + 4$;
if $F_3(x) = x^2 + 4x + 14.8$, then $F_3'(x) = 2x + 4; \ldots$

There are many functions that have $2x + 4$ as a derivative, but they all differ by a constant since the derivative of a constant is 0. Thus

$$F(x) = x^2 + 4x + C$$

for any constant C. The cost of producing 50,000 items is given by $F(50)$, but first we need to find C, the **constant of integration**. The fixed costs are the costs that are present even if $x = 0$. This means that $F(0) = 25$, so

$$F(0) = 0^2 + 4(0) + C = 25$$
$$C = 25$$

Thus

$$F(x) = x^2 + 4x + 25$$

is the cost equation for our original problem, so

$$F(50) = 50^2 + 4(50) + 25 = 2{,}725$$

The cost of producing 50,000 items is \$2,725,000. ■

In Example 1 we went through a (trial-and-error) process that might be called the process of **antidifferentiation**. A more common name for the process of antidifferentiation is **integration**. If $F(x)$ is an antiderivative of $f(x)$, then $F(x) + C$ is called the **indefinite integral** of $f(x)$. The adjective *indefinite* is used because the constant C is arbitrary or indefinite.

Indefinite Integral

If F is an antiderivative of f, then we write

$$\int f(x)\,dx = F(x) + C$$

and say, "The indefinite integral of $f(x)$ with respect to x is $F(x) + C$." The symbol $\int$ is called the **integral symbol**, $f(x)$ is called the **integrand**, and C is called the **constant of integration**.

NOTE:

Variable of integration

$$\int f(x)\,dx = F(x) + C$$

Integrand

Constant of integration

Antiderivative of $f(x)$

We will say more about the symbol dx later, but for now remember that the dx in the indefinite integral identifies the independent variable of integration, as illustrated by Example 2.

EXAMPLE 2 $\int 3x\,dx$ is read "the indefinite integral of $3x$ with respect to x."

$\int 9y^2\,dy$ is read "the indefinite integral of $9y^2$ with respect to y."

$\int (5z^2 + 3z + 2)\,dz$ is read "the indefinite integral of $5z^2 + 3z + 2$ with respect to z."

$\int x^2y^3\,dx$ is read "the indefinite integral of x^2y^3 with respect to x." ■

EXAMPLE 3 Show that $F(x) = x^6$ is an antiderivative of $f(x) = 6x^5$.

Solution Use the power rule for derivatives:

$$F(x) = x^6$$
$$F'(x) = 6x^5$$

■

EXAMPLE 4 Find an antiderivative of $f(x) = x^5$.

Solution By trial and error (the power rule for derivatives in reverse),

$$F(x) = \frac{x^6}{6} + C$$

Check: $F'(x) = \dfrac{6x^5}{6} + 0 = x^5$ ■

Integral notation makes it easy to state some integration formulas. For example, the power rule for derivatives says if $f(x) = x^n$, then $f'(x) = nx^{n-1}$, so if this process is reversed we find a corresponding integration formula to be

$$\text{If} \quad f(x) = x^n, \quad \text{then} \quad F(x) = \int x^n\,dx = \frac{x^{n+1}}{n+1} + C \qquad n \neq -1$$

($n \neq -1$ keeps the denominator from being 0).

To prove this formula, find the derivative of

$$F(x) = \frac{x^{n+1}}{n+1} + C \qquad n \neq -1$$

$$F'(x) = \frac{(n+1)x^{n+1-1}}{n+1} + 0$$

$$= x^n$$

EXAMPLE 5

a. $\int x^8\,dx = \frac{x^9}{9} + C$

b. $\int x^3\,dx = \frac{x^4}{4} + C$

c. $\int \sqrt{x}\,dx = \int x^{1/2}\,dx$

$$= \frac{x^{3/2}}{\frac{3}{2}} + C = \frac{2}{3}x^{3/2} + C$$

d. $\int dx = \int x^0\,dx = \frac{x^1}{1} + C = x + C$

e. $\int \frac{dx}{x^3} = \int x^{-3}\,dx$

$$= \frac{x^{-2}}{-2} + C = -\frac{1}{2}x^{-2} + C$$

■

As you see, if the derivatives of two functions are equal, then the functions differ by at most a constant. This fact is summarized by adding C in our answers for Example 5. This result is summarized in the following box.

Functions with Equal Derivatives

If F and G are differentiable functions on an interval (a, b) such that

$$F'(x) = G'(x)$$

then

$$F(x) = G(x) + C$$

for some constant C.

If we quickly review the derivative formulas from Chapters 10 and 13, we should be able to restate each as an integration formula.

Review of Derivative Formulas

POWER RULE:	If $y = x^n$, then $y' = nx^{n-1}$.
CONSTANT RULE:	If $y = k$, then $y' = 0$.
CONSTANT TIMES A FUNCTION RULE:	If $y = kf$, then $y' = kf'$.
SUM RULE:	If $y = f + g$, then $y' = f' + g'$.
DIFFERENCE RULE:	If $y = f - g$, then $y' = f' - g'$.
LOGARITHMIC RULE:	If $y = \ln\lvert x\rvert$, then $y' = \dfrac{1}{x}$.
EXPONENTIAL RULE:	If $y = e^x$, then $y' = e^x$.

(We will consider the product, quotient, and chain rules later.)

These derivative formulas can be divided into two types: two *direct integration forms* and three *procedural types.* For now, all our examples can be reduced to one of the direct integration types by first using the procedural types.

Integration Formulas

DIRECT INTEGRATION FORMULAS

$$\int x^n\,dx = \begin{cases} \dfrac{x^{n+1}}{n+1} + C & n \neq -1 \\ \ln\lvert x\rvert + C & n = -1 \end{cases}$$

Power rule (This includes the constant rule as the case where $n = 0$.) ($n \neq -1$)

Logarithmic rule ($n = -1$)

$$\int e^x\,dx = e^x + C$$

Exponential rule

PROCEDURAL INTEGRATION FORMULAS

$$\int kf(x)\,dx = k\int f(x)\,dx$$

The **integral of a constant times a function** is that constant times the integral.

$$\int [f(x) + g(x)]\,dx = \int f(x)\,dx + \int g(x)\,dx$$

The **integral of a sum** is the sum of the integrals.

$$\int [f(x) - g(x)]\,dx = \int f(x)\,dx - \int g(x)\,dx$$

The **integral of a difference** is the difference of the integrals.

EXAMPLE 6

$$\int (3x^2 - 4x + 5)\,dx = \int 3x^2\,dx - \int 4x\,dx + \int 5\,dx \qquad \text{Sum and difference formulas}$$

$$= 3\int x^2\,dx - 4\int x\,dx + 5\int dx \qquad \text{Constant formula}$$

$$= 3\left(\frac{x^3}{3} + C_1\right) - 4\left(\frac{x^2}{2} + C_2\right) + 5(x + C_3) \qquad \text{Power formula}$$

$$= x^3 - 2x^2 + 5x + (3C_1 - 4C_2 + 5C_3)$$

$$= x^3 - 2x^2 + 5x + C$$

■

In Example 6 since $3C_1 - 4C_2 + C_3$ represents a sum of arbitrary constants, they can be combined into a single constant, C. Also, the steps in Example 6 are shown in great detail so you can relate them to the proper formulas, but in practice your work would look like this:

$$\int (3x^2 - 4x + 5)\,dx = \frac{3x^3}{3} - \frac{4x^2}{2} + 5x + C$$

$$= x^3 - 2x^2 + 5x + C$$

When we carry out the process of writing an integral in functional notation (or as a numerical value), we say that we are **evaluating** the integral.

EXAMPLE 7 Evaluate the given integrals.

a. $$\int (5x^4 + 2x^3 + \sqrt[3]{x})\,dx = \frac{5x^5}{5} + \frac{2x^4}{4} + \frac{x^{4/3}}{\frac{4}{3}} + C \qquad \text{Remember, } \sqrt[3]{x} = x^{1/3}$$

$$= x^5 + \frac{1}{2}x^4 + \frac{3}{4}x^{4/3} + C$$

b. $$\int \frac{1}{x}\,dx = \int x^{-1}\,dx = \ln|x| + C$$

c. $$\int (x - e^x)\,dx = \frac{x^2}{2} - e^x + C$$

d. $$\int \frac{x^3 + x + 1}{x}\,dx = \int \left(\frac{x^3}{x} + \frac{x}{x} + \frac{1}{x}\right)dx \qquad \text{Simplify first, then integrate.}$$

$$= \int (x^2 + 1 + x^{-1})\,dx$$

$$= \frac{x^3}{3} + x + \ln|x| + C$$

■

As shown in Example 1, it is possible to find C if some value of the original function is known. The most commonly known value is the one when the variable is 0. That is, if

$$F'(x) = f(x)$$

and $f(0)$ is known, then this known value is called the **initial value**.

EXAMPLE 8 The population of Santa Rosa, California, is growing at the rate $3{,}000 + 500t^{1/4}$, where t is measured in years. The present population is 92,500. Predict the population in 5 years.

Solution The population function for time t is found by knowing

$$P'(t) = 3{,}000 + 500t^{1/4}$$

Thus, by integrating with respect to t,

$$P(t) = \int (3{,}000 + 500t^{1/4})\,dt = 3{,}000t + \frac{500t^{1/4+1}}{\frac{5}{4}} + C$$

$$= 3{,}000t + 400t^{5/4} + C$$

The initial condition ($t = 0$) is 92,500, so

$$P(0) = 3{,}000(0) + 400(0)^{5/4} + C = 92{,}500$$

$$C = 92{,}500$$

In 5 years,

$$P(5) = 3{,}000(5) + 400(5)^{5/4} + 92{,}500$$

$$\approx 110{,}490.6976$$

The population in 5 years will be about 110,000. ■

14.1 Problem Set

Evaluate the indefinite integrals in Problems 1–32. You can check each answer by differentiating your answers.

1. $\int x^7\,dx$
2. $\int x^{10}\,dx$
3. $\int 4x^3\,dx$
4. $\int 12x^5\,dx$
5. $\int 3\,dx$
6. $\int 8\,dx$
7. $\int (5x + 7)\,dx$
8. $\int (5 - 2x)\,dx$
9. $\int dx$
10. $\int (9x^2 - 4x + 3)\,dx$
11. $\int (18x^2 - 6x + 5)\,dx$
12. $\int \frac{dx}{5}$
13. $\int \frac{dx}{9}$
14. $\int \frac{3\,dx}{x}$
15. $\int \frac{5\,dx}{x^2}$
16. $\int -3e^x\,dx$
17. $\int (1 - e^x)\,dx$
18. $\int (x - \sqrt{x})\,dx$
19. $\int (\sqrt[3]{x} + \sqrt{2})\,dx$
20. $\int (\sqrt[3]{5} + \sqrt{2})\,dx$
21. $\int \frac{3x - 1}{\sqrt{x}}\,dx$
22. $\int \frac{5x^2 + 2x + 3}{\sqrt{x}}\,dx$
23. $\int \frac{x}{\sqrt[3]{x}}\,dx$
24. $\int \frac{x^2}{\sqrt[3]{x^2}}\,dx$
25. $\int \frac{x^2 + x + 1}{x}\,dx$
26. $\int \frac{3x^2 + 2x + 1}{x}\,dx$
27. $\int y^3\sqrt{y}\,dy$
28. $\int (3y^{-2/3} - 4y^{1/2})\,dy$
29. $\int (1 + z^2)^2\,dz$
30. $\int (3z + 1)^2\,dz$
31. $\int (3u^2 - u^{-1} + e^u)\,du$
32. $\int (5u^4 - 3u^{-1} + 4e^u)\,du$

In Problems 33–36 find the antiderivative $F(x)$ for the given function that satisfies the initial conditions.

33. $f(x) = 3x^2 + 4x + 1$; $F(0) = 10$
34. $f(x) = 3x^2 + 4x + 1$; $F(1) = -8$
35. $f(x) = \dfrac{5x + 1}{x}$; $F(1) = 5{,}000$
36. $f(x) = \dfrac{3x^2 + 2x + 1}{x}$; $F(1) = -50$

APPLICATIONS

In Problems 37–40 find the cost function for the given marginal cost functions.

37. $C'(x) = 4x + 8$; fixed cost = \$5,000
38. $C'(x) = 128x + 300$; fixed cost = \$32,000
39. $C'(x) = .009x^2$; fixed cost = \$18,500
40. $C'(x) = x + 100e^x$; fixed cost = \$1,900
41. The population of New Haven, Connecticut, is growing at the rate of $450 + 600\sqrt{t}$, where t is measured in years, and the present population is 420,000. Predict the population in 5 years.
42. The population of Charlotte, North Carolina, is growing at the rate of $1{,}000 + 500t^{3/4}$, where t is measured in years, and the present population is 330,000. Predict the population in 10 years.
43. The marginal profit of Turbo Plus is

$$P'(x) = 200x - 10{,}000$$

where x is the sales in thousands of items. Find the profit function if there is a \$50,000 loss (−\$50,000 profit) when no items are produced.

44. The marginal cost for producing x new 32-bit microprocessors is

$$C'(x) = 100x^{-2/3} + .00001\,x$$

If the fixed costs are \$120,000, what is the cost of producing 100,000 processors?

45. A new strain of influenza is known to be spreading at the rate of $240t - 3t^2$ cases per day, where t is the number of days measured from the first recorded outbreak. How many people will be affected on the tenth day, if there were 50 cases on the first day?
46. Repeat Problem 45 for the fifth day, if there were five cases on the first day.
47. It is known that the acceleration of a freely falling body is a constant, 32 feet per second per second. Use the formula

$$a(t) = -32$$

to derive a formula for the velocity (at time t) for an object thrown from the top of Hoover Dam (726 feet) with an initial velocity of −72 feet per second. [*Hint:* If v is the velocity function, then $v'(t) = a(t)$; it is negative because the object was thrown downward.]

48. Derive a formula for the height of the object thrown from Hoover Dam in Problem 47. [*Hint:* If h is the height function, then $h'(t) = v(t)$.]
49. The slope of the tangent line to a curve is given by

$$f'(x) = 3x^2 + 5$$

If the point (1, 2) is on the curve, find the equation of the curve.

50. Find the equation of a curve whose slope is $\sqrt{x}$ if the point (9, 19) is on the curve.

14.2 Integration by Substitution

One of the most important integration techniques comes from the *generalized power rule*. For example, consider

$$\int x(x^2 + 1)^2\,dx$$

We could proceed by writing

$$\int x(x^4 + 2x^2 + 1)\,dx = \int (x^5 + 2x^3 + x)\,dx = \int \frac{x^6}{6} + \frac{2x^4}{4} + \frac{x^2}{2} + C$$

$$= \frac{1}{6}x^6 + \frac{1}{2}x^4 + \frac{1}{2}x^2 + C$$

However, it is often not practical (or possible) to multiply [as with $x(x^2 + 1)^{12}$, for example]. We will proceed, instead, by *substitution*. The complicating ingredient is the extra factor as a result of the chain rule (Section 10.7). In order to handle this factor, we need to use differentials (discussed in Section 11.2).

Let us take another look at the preceding example. This time, instead of multiplying, we will make a substitution. Look at the expression raised to a power, namely, $x^2 + 1$. Let

$$u = x^2 + 1$$

then

$$\frac{du}{dx} = 2x$$

Now, solve for dx:

$$du = 2x\,dx$$

$$dx = \frac{du}{2x} \qquad (x \neq 0)$$

If substitution eliminates the variable x (leaving only the variable u), the integrand may be in a simpler, more easily integrable, form.

$$\int x(x^2+1)^2\,dx = \int xu^2 \frac{du}{2x}$$

$$= \frac{1}{2}\int u^2\,du \qquad \textit{Note:} \text{ The } x\text{'s cancel and } \tfrac{1}{2} \text{ is a constant.}$$

$$= \frac{1}{2}\frac{u^3}{3} + C \qquad \text{Use the power rule on } u.$$

$$= \frac{(x^2+1)^3}{6} + C \qquad \text{Write } u \text{ in terms of } x \text{ for the final answer.}$$

By using substitution we are able to integrate this function without a lot of multiplication. This has tremendous advantages, as shown in Example 1.*

EXAMPLE 1 Evaluate $\int x(x^2+1)^{12}\,dx$.

Solution Let $u = x^2 + 1$; then $\frac{du}{dx} = 2x$, so $dx = \frac{du}{2x}$ $(x \neq 0)$. Substitute to obtain

$$\int x\underbrace{(x^2+1)^{12}}_{u}\underbrace{dx}_{\frac{du}{2x}} = \int xu^{12}\frac{du}{2x} = \frac{1}{2}\int u^{12}\,du$$

↑ Bring constant out in front of integration.

$$= \frac{1}{2}\frac{u^{13}}{13} + C = \frac{1}{26}\underbrace{(x^2+1)^{13}}_{\uparrow} + C$$

Replace u with original variable. ■

* If you did expand, you would see that the result agrees with that shown here.

$$\frac{1}{6}(x^6 + 3x^4 + 3x^2 + 1) + C = \frac{1}{6}x^6 + \frac{1}{2}x^4 + \frac{1}{2}x^2 + \frac{1}{6} + C$$

EXAMPLE 2 Evaluate $\int x^2\sqrt{x^3 - 2}\, dx$.

Solution Let $u = x^3 - 2$; $\dfrac{du}{dx} = 3x^2$, so $dx = \dfrac{du}{3x^2}$. Substitute

$$\begin{aligned}\int x^2\sqrt{x^3 - 2}\, dx &= \int x^2\sqrt{u}\,\frac{du}{3x^2} \\ &= \frac{1}{3}\int u^{1/2}\, du \\ &= \frac{1}{3}\,\frac{u^{3/2}}{\frac{3}{2}} + C \\ &= \frac{2}{9}u^{3/2} + C \\ &= \frac{2}{9}(x^3 - 2)^{3/2} + C\end{aligned}$$

■

The procedure is generalized below:

Integration by Substitution

1. Make a choice for u, say, $u = f(x)$.
   (u is usually inside parentheses, or an exponent, or under a radical.)
2. Find $\dfrac{du}{dx} = f'(x)$.
3. Make the substitution*

   $$u = f(x) \quad \text{and} \quad du = f'(x)\, dx$$

WARNING **Note Step 4.** ⟶

4. *Everything* must be in terms of u with no remaining x's. If this is not possible, try another substitution.
5. Evaluate the integral.
6. Replace u by $f(x)$ so the final answer is in terms of x.

When using the substitution method, make sure that every x is eliminated. If this cannot be done, try a different choice, but remember that not every function *can* be integrated by this technique. This means that *after* substitution and simplification, you *must* have one of the following forms:

$$\int u^n\, du = \begin{cases} \dfrac{u^{n+1}}{n+1} + C & n \neq -1 \\ \ln|u| + C & n = -1 \end{cases}$$

or

$$\int e^u\, du = e^u + C$$

* In making the substitution $du = f'(x)\, dx$ it is often easiest to assume that values that cause division by 0 are excluded from the domain, and then solve $du = f'(x)\, dx$ for dx before making the substitution. If you do this, pay particular attention to the WARNING on step 4.

EXAMPLE 3 Evaluate $\int \frac{x^2\,dx}{(x^3-2)^5}$.

Solution Let u be the value in parentheses; that is, let $u = x^3 - 2$. Then

$$du = 3x^2\,dx \qquad \text{so} \qquad dx = \frac{du}{3x^2}$$

$$\int \frac{x^2\,dx}{(x^3-2)^5} = \int \frac{x^2}{u^5}\cdot\frac{du}{3x^2} = \int \frac{du}{3u^5}$$

All x's must be eliminated:

$(x^3-2)^5 = u^5$ and $dx = \frac{du}{3x^2}$

$$= \frac{1}{3}\int u^{-5}\,du = \frac{1}{3}\,\frac{u^{-4}}{-4} + C$$

$$= \frac{-(x^3-2)^{-4}}{12} + C$$

■

EXAMPLE 4 Evaluate $\int e^{3x+5}\,dx$.

Solution Let u be the exponent; that is, $u = 3x + 5$, so

$$du = 3\,dx \qquad \text{and} \qquad dx = \frac{du}{3}$$

Substitute:

$$\int e^{3x+5}\,dx = \int e^u \frac{du}{3}$$

$$= \frac{1}{3}\int e^u\,du$$

$$= \frac{1}{3}e^u + C$$

$$= \frac{1}{3}e^{3x+5} + C$$

■

EXAMPLE 5 Evaluate $\int x^3\sqrt{3x^2-1}\,dx$.

Solution Let u be the expression under the radical; that is, $u = 3x^2 - 1$, so

$$du = 6x\,dx \qquad \text{and} \qquad dx = \frac{du}{6x}$$

Substitute:

$$\int x^3\sqrt{u}\frac{du}{6x} = \frac{1}{6}\int x^2\underbrace{\sqrt{u}\,du}$$

You can handle this.

WARNING

x^2 is a variable, so do not try to treat it like a constant. That is, do not bring out the x^2 in front of the integral sign.

Only constants may be moved across an integral.

All x's must be eliminated; note:

$$u = 3x^2 - 1$$

$$u + 1 = 3x^2$$

$$\frac{u+1}{3} = x^2$$

$$= \frac{1}{6}\int \frac{u+1}{3}\sqrt{u}\,du$$

$$= \frac{1}{18}\int (u^{3/2} + u^{1/2})\,du$$

$$= \frac{1}{18}\left(\frac{u^{5/2}}{\frac{5}{2}} + \frac{u^{3/2}}{\frac{3}{2}}\right) + C$$

$$= \frac{1}{45}u^{5/2} + \frac{1}{27}u^{3/2} + C$$

$$= \frac{1}{45}(3x^2 - 1)^{5/2} + \frac{1}{27}(3x^2 - 1)^{3/2} + C$$

■

Different parts of a problem may require different substitutions.

EXAMPLE 6 Evaluate $\int\left(\frac{x^2}{1-x^3} + xe^{x^2}\right)dx$.

Solution

$$\int\left(\frac{x^2}{1-x^3} + xe^{x^2}\right)dx = \int \frac{x^2}{1-x^3}\,dx + \int xe^{x^2}\,dx$$

$u = 1 - x^3$
$du = -3x^2\,dx$

$v = x^2$
$dv = 2x\,dx$

$$= \int \frac{x^2}{u}\frac{du}{-3x^2} + \int xe^v\frac{dv}{2x}$$

$$= -\frac{1}{3}\int u^{-1}\,du + \frac{1}{2}\int e^v\,dv$$

$$= -\frac{1}{3}\ln|u| + \frac{1}{2}e^v + C$$

$$= -\frac{1}{3}\ln|1 - x^3| + \frac{1}{2}e^{x^2} + C$$

■

EXAMPLE 7 The TexRite Company has found that the marginal profit for a product is

$$P'(x) = \frac{1{,}600x}{\sqrt[3]{(8x^2 - 33{,}344)^2}}$$

where x is the number of units sold on the domain $[75, 5000]$. If the break-even point is 100 units [that is, $P(100) = 0$], what is the approximate total profit for 1,000 items?

Solution

$$P(x) = \int \frac{1{,}600x\,dx}{\sqrt[3]{(8x^2 - 33{,}344)^2}} = \int \frac{1{,}600x}{\sqrt[3]{u^2}}\frac{du}{16x}$$

Let $u = 8x^2 - 33{,}344$

$du = 16x\,dx$

$$= \int 100u^{-2/3}\,du$$

$$= 100\frac{u^{1/3}}{\frac{1}{3}} + C$$

$$= 300(8x^2 - 33{,}344)^{1/3} + C$$

Now, $P(100) = 0$, so

$$300(8 \cdot 100^2 - 33{,}344)^{1/3} + C = 0$$

$$C = -10{,}800$$

$$P(x) = 300(8x^2 - 33{,}344)^{1/3} - 10{,}800$$

$$P(1{,}000) \approx 49{,}116.52389$$

The profit is about \$49,000. ■

14.2 Problem Set

Evaluate the indefinite integrals in Problems 1–32.

1. $\int (5x + 3)^3\,dx$

2. $\int (1 - 5x)^4\,dx$

3. $\int \frac{5}{(5 - x)^4}\,dx$

4. $\int \frac{6}{(x + 8)^2}\,dx$

5. $\int \sqrt{3x + 5}\,dx$

6. $\int \sqrt[5]{9 - 4x}\,dx$

7. $\int 6x(3x^2 + 1)\,dx$

8. $\int 2x(4 - 5x^2)\,dx$

9. $\int \frac{2x + 5}{\sqrt{x^2 + 5x}}\,dx$

10. $\int \frac{2x + 3}{\sqrt{x^2 + 3x}}\,dx$

11. $\int \frac{dx}{6 + 5x}$

12. $\int \frac{dx}{1 - 4x}$

13. $\int \frac{x\,dx}{1 - 3x^2}$

14. $\int \frac{2x\,dx}{x^2 + 5}$

15. $\int e^{5x}\,dx$

16. $\int e^{1 - 3x}\,dx$

17. $\int 5x^2e^{4x^3}\,dx$

18. $\int 3xe^{x^2 + 5}\,dx$

19. $\int \frac{2x - 1}{(4x^2 - 4x)^2}\,dx$

20. $\int \frac{2x - 1}{(x - x^2)^3}\,dx$

21. $\int \frac{4x^3 - 4x}{x^4 - 2x^2 + 3}\,dx$

22. $\int \frac{x^3 - x}{(x^4 - 2x^2 + 3)^2}\,dx$

23. $\int \frac{\ln|x|}{x}\,dx$

Hint: Let $u = \ln|x|$.

24. $\int \frac{\ln|x + 1|}{x + 1}\,dx$

Hint: Let $u = \ln|x + 1|$.

25. $\int \ln e^x\,dx$

26. $\int \ln e^{x^2}\,dx$

27. $\int\left(\frac{3x}{4x^2+1} + x^2e^{x^3}\right)dx$

28. $\int\left(\frac{4x+6}{\sqrt{x^2+3x}} + xe^{3x^2}\right)dx$

29. $\int x\sqrt[3]{(x^2+1)^2}\,dx$

30. $\int (x^2-4x+4)^{2/5}\,dx$

31. $\int \frac{x\,dx}{\sqrt{x+1}}$

32. $\int x^2\sqrt{5-x}\,dx$

In Problems 33–36 consider the following integrals.

a. $\int (x^2+1)^n\,dx$

b. $\int (x^2+1)^n x\,dx$

c. $\int (x^2+1)^n x^2\,dx$

d. $\int (x^2+1)^n x^3\,dx$

33. Integrate each part for $n = 1$.

34. Integrate each part for $n = 2$.

35. Integrate each part for $n = 3$.

36. Which parts can you easily integrate for an arbitrary n? What causes these parts to be comparatively easy to integrate and why?

In Problems 37–40 consider the following integrals.

a. $\int (x^3+1)^n\,dx$

b. $\int (x^3+1)^n x\,dx$

c. $\int (x^3+1)^n x^2\,dx$

d. $\int (x^3+1)^n x^3\,dx$

37. Integrate each part for $n = 1$.

38. Integrate each part for $n = 2$.

39. Integrate each part for $n = 3$.

40. Which parts can you easily integrate for an arbitrary n? What causes these parts to be comparatively easy to integrate and why?

APPLICATIONS

41. A flu epidemic is spreading at the rate of

$$P'(t) = 40t(5-t^2)^2$$

people per day, where t is the number of days since the first outbreak. If P is a function representing the number of sick people, and if $P(0) = 10$, find P.

42. Let $P(t)$ be the total output after t hours of work. Find P if the rate of production at time t $(1 < t < 40)$ is

$$P'(t) = \frac{50}{50-t}$$

and production (to the nearest unit) is 71 when $t = 1$.

43. If the marginal profit (in dollars) for a product is

$$P'(x) = \frac{100x}{\sqrt[3]{(x^2-36)^2}}$$

where x is the number of units sold $(x \geq 7)$, find the profit for 100 items if the break-even point is 10 units.

44. The marginal profit for a product is

$$P'(x) = \frac{20x}{\sqrt{x^2-16}}$$

where x is the number of units sold $(x \geq 5)$. If the break-even point is 5 units, what is the profit for 25 items?

45. Sales are changing at a rate given by the formula

$$S'(t) = 2{,}000e^{-.2t}$$

where t is the time in years and $S(t)$ is the sales. What are the sales in 3 years if initial sales are 15,000 units?

46. In 1976 the world population reached 4 billion and was growing at a rate approximated by the formula

$$P'(t) = .072e^{.018t}$$

where t is measured in years since 1976 and $P(t)$ gives the population. What is the predicted population in the year 2000?

47. In 1985 the growth rate of San Antonio, Texas, was given by the formula

$$P'(t) = 14{,}000e^{.0175t}$$

where t is measured in years since 1980 and $P(t)$ gives the population. If the population in 1984 was 842,779, predict the population in the year 2000.

48. Between 1980 and 1985 worldwide oil consumption dropped from 2,400 million barrels to 1,950 million barrels. This rate of consumption is given by the formula

$$R(t) = 78e^{-.04t}$$

where t is measured in years since 1985. The total consumption since 1985 is given by $T(t)$, where $T'(t) = R(t)$. Find the consumption from 1985 to the year 2000.

49. If we consider worldwide oil consumption from 1925 to 1985, the rate of consumption is given by the formula

$$R(t) = 32.5e^{.048t}$$

where t is measured in years since 1985. If the worldwide consumption from 1925 to 1985 was 1,950 million barrels, predict the consumption from 1985 to the year 2000 if the total consumption is given by $T(t)$, where $T'(t) = R(t)$.

14.3 The Definite Integral

We will introduce the definite integral with an example, and our approach will be informal and intuitive. These ideas will be formalized in Section 14.5 when we introduce a notion called Riemann sums and the Fundamental Theorem of Calculus.

The Definite Integral

In 1986 world oil prices plummeted. The price drop was related to overproduction and declining consumption. Suppose that between 1980 and 1985 the rate of consumption (in billions of barrels) is given by the formula

$$R(t) = 78e^{-.04t}$$

where t is measured in years after 1985.* If we use this consumption rate, we can predict the amount of oil that will be consumed from 1987 to 1995. The total consumption since 1985 is given by $T(t)$, where $T'(t) = R(t)$. Then

$$\begin{aligned} T(t) &= \int T'(t)\,dt = \int R(t)\,dt \\ &= \int (78e^{-.04t})\,dt \\ &= \frac{78e^{-.04t}}{-.04} + C \\ &= -1{,}950e^{-.04t} + C \end{aligned}$$

We can find C because if $t = 0$, then oil consumption is also zero. Thus

$$\begin{aligned} T(0) = -1{,}950e^{-.04t} + C &= 0 \\ -1{,}950 + C &= 0 \\ C &= 1{,}950 \end{aligned}$$

Thus

$$T(2) = -1{,}950e^{-.04(2)} + 1{,}950 \approx 150$$ Total consumption from 1985 to 1987

$$T(10) = -1{,}950e^{-.04(10)} + 1{,}950 \approx 643$$ Total consumption from 1985 to 1995

We can find the total consumption from 1987 to 1995 by subtraction:

$$T(10) - T(2) = 643 - 150 = 493$$

We would predict approximately 493 billion barrels to be consumed, assuming the rate of consumption does not change. However, if prices continue to drop, we would expect the rate of consumption to again turn upward (see Problems 35–39 in Problem Set 14.3).

* Equations like this are derived in Section 15.2.

Let us take a closer look at what we have done. Since $T(t)$ is an antiderivative of $R(t)$ over the interval from 1987 to 1995, we see that $T(10) - T(2)$ is the *net change* of the function T over this interval. In general, if $F(x)$ is any function and a and b are real numbers with $a < b$, then the *net change of* $F(x)$ *over the interval* $[a, b]$ is the number

$$F(b) - F(a)$$

The quantity $F(b) - F(a)$ is often abbreviated by the symbol

$$F(x)\Big|_a^b$$

This discussion suggests the following formula:

Definite Integral

Let f be a function defined over the interval $[a, b]$. Then the **definite integral** of f over this interval is denoted by

$$\int_a^b f(x)\,dx$$

and is the net change of an antiderivative of f over that interval. Thus, if $F(x)$ is an antiderivative of $f(x)$, then

$$\int_a^b f(x)\,dx = F(x)\Big|_a^b = F(b) - F(a)$$

WARNING The definite integral $\int_a^b f(x)\,dx$ is a real number. The indefinite integral $\int f(x)\,dx$ is a set of functions—namely, all of those functions whose derivative is $f(x)$.

The function f is called the **integrand** and the constants a and b are called the **limits of integration**. The number x is called a **dummy variable** because the definite integral is a fixed number and not a function of x.

Evaluating Definite Integrals

Both the integral sign and the symbol dx are dropped when the antiderivative is found. Also, when using the definite integral it is not necessary to take into account a constant of integration. [Remember, in the introductory example, the constant of integration (1,950) had no bearing on the final answer since it was first added and then subtracted.] The process of finding the value of a definite integral is called **evaluating the integral**.

EXAMPLE 1 Evaluate the given integrals.

a. $\int_2^3 6x^2\,dx$ **b.** $\int_{-2}^2 x^3\,dx$ **c.** $\int_1^{10} \frac{dx}{x}$

Solution **a.** $$\int_2^3 6x^2\,dx = \frac{6x^3}{3}\Big|_2^3 = \frac{6(3)^3}{3} - \frac{6(2)^3}{3} = 2(27) - 2(8) = 38$$

Evaluate at 3 first.

Antiderivative of $6x^2$

Subtract the value of the antiderivative at 2.

b. $\int_{-2}^{2} x^3\,dx = \left.\frac{x^4}{4}\right|_{-2}^{2} = \frac{2^4}{4} - \frac{(-2)^4}{4} = 0$

c. $\int_{1}^{10} \frac{dx}{x} = \ln|x|\Big|_{1}^{10} = \ln|10| - \ln|1| = \ln 10 - 0 \approx 2.3$ ■

Remember, $\ln 1 = 0$.

Properties of the Definite Integral

There are many properties of the definite integral, several of which are analogous to those we stated for the indefinite integral:

Properties of the Definite Integral

1. $\int_a^a f(x)\,dx = 0$ where a is in the domain of f
2. $\int_a^b dx = b - a$
3. $\int_a^b f(x)\,dx = -\int_b^a f(x)\,dx$
4. $\int_a^b f(x)\,dx = \int_a^c f(x)\,dx + \int_c^b f(x)\,dx$ for any points a, b, and c in the closed interval $[a, b]$
5. $\int_a^b kf(x)\,dx = k\int_a^b f(x)\,dx$
6. $\int_a^b [f(x) \pm g(x)]\,dx = \int_a^b f(x)\,dx \pm \int_a^b g(x)\,dx$

EXAMPLE 2 Evaluate the given integrals.

a. $\int_{10}^{10} (3x^2 - \sqrt{x})\,dx$

b. $\int_{4}^{3} (3x^2 - \sqrt{x})\,dx$

c. $\int_{-2}^{\pi/\sqrt{2}} e^t\,dt + \int_{\pi/\sqrt{2}}^{2} e^t\,dt$ (correct to the nearest tenth)

Solution **a.** $\int_{10}^{10} (3x^2 - \sqrt{x})\,dx = 0$

b.
$$\int_{4}^{3} (3x^2 - \sqrt{x})\,dx = -\int_{3}^{4} (3x^2 - \sqrt{x})\,dx = -\left[\frac{3x^3}{3} - \frac{x^{3/2}}{\frac{3}{2}}\right]\Bigg|_{3}^{4}$$
$$= -(4^3 - \tfrac{2}{3}\cdot 4^{3/2}) + (3^3 - \tfrac{2}{3}\cdot 3^{3/2})$$
$$= -(64 - \tfrac{16}{3}) + (27 - 2\sqrt{3})$$
$$= \tfrac{16}{3} - 37 - 2\sqrt{3}$$
$$= \frac{-95 - 6\sqrt{3}}{3}$$

c. $\int_{-2}^{\pi/\sqrt{2}} e^t\,dt + \int_{\pi/\sqrt{2}}^{2} e^t\,dt = \int_{-2}^{2} e^t\,dt = e^t\Big|_{-2}^{2} = e^2 - e^{-2} \approx 7.2$ ■

Substitution with the Definite Integral

Always be careful about the evaluation of the variable when you use substitution in order to find the antiderivative.

EXAMPLE 3 Evaluate $\int_1^{\sqrt{5}} \frac{6x\,dx}{\sqrt{1+3x^2}}$.

Solution

$$\begin{aligned}\int_1^{\sqrt{5}} \frac{6x\,dx}{\sqrt{1+3x^2}} &= \int_{x=1}^{x=\sqrt{5}} \frac{du}{\sqrt{u}} \\ &= \int_{x=1}^{x=\sqrt{5}} u^{-1/2}\,du \\ &= 2u^{1/2}\Big|_{x=1}^{x=\sqrt{5}} \\ &= 2\sqrt{1+3x^2}\Big|_1^{\sqrt{5}} \\ &= 2(4) - 2(2) \\ &= 4\end{aligned}$$

Note that we need to say $x=\sqrt{5}$ and $x=1$ are the limits of integration so that they are not confused with the variable u.

Also notice: $u = 1 + 3x^2$
$du = 6x\,dx$

■

Instead of substituting back to the original variable, we can change the limits of integration. For Example 3, since $u = 1 + 3x^2$, we can substitute $x = 1$ (lower limit):

$$u = 1 + 3(1)^2 = 4$$

and $x = \sqrt{5}$ (upper limit):

$$u = 1 + 3(\sqrt{5})^2 = 16$$

Then we can write

If $x = \sqrt{5}$, then $u = 16$.

Note the change in the limits of integration.

$$\begin{aligned}\int_1^{\sqrt{5}} \frac{6x\,dx}{\sqrt{1+3x^2}} &= \int_4^{16} u^{-1/2}\,du \\ &= 2u^{1/2}\Big|_4^{16} \\ &= 2(\sqrt{16} - \sqrt{4}) \\ &= 4\end{aligned}$$

If $x = 1$, then $u = 4$.

Substitution with the Definite Integral

If f is continuous on the set of values taken on by g, and g' is continuous on $[a, b]$, and if f has an antiderivative on that interval, then

$$\int_a^b f[g(x)]g'(x)\,dx = \int_{g(a)}^{g(b)} f(u)\,du \qquad u = g(x), \quad du = g'(x)\,dx$$

provided these integrals exist.

EXAMPLE 4 Evaluate $\int_1^2 e^{-3t}\,dt$ correct to three decimal places.

Solution

$$\int_1^2 e^{-3t}\,dt = -\frac{1}{3}\int_{-3}^{-6} e^u\,du = -\frac{1}{3}e^u\Big|_{-3}^{-6} = -\frac{1}{3}(e^{-6} - e^{-3}) \approx .016$$

Let $u = -3t$; if $t = 2$, then $u = -6$
$du = -3dt$; if $t = 1$, then $u = -3$

■

Applications

If you are given some function and want to find the rate at which the function is changing at some point, then we know that the derivative will give us this rate. On the other hand, it is not uncommon to know a function which *itself* represents the rate of change. For example, you might know the divorce rate, or rate of flow of blood, or velocity (which is the rate of change of distance with respect to time), or rate of discharge of pollutants into a water supply, or marginal cost (rate of change of cost with respect to number of items). We could cite many other examples in which you might know the rate and wish to find the original function. It should be clear that to find the original function, whose rate of change is known, requires an integral, but it might not be clear what this integral represents. Some examples are given in Table 14.1.

TABLE 14.1
Examples of Given Rates and Their Associated Integrals

Given rate	Associated definite integral
Divorce rate over some interval of time	Total accumulated number of divorces for the given time interval
Rate of flow of blood over some time interval	Total accumulated flow of blood for the given time interval
Velocity	Distance (or position) function
Rate of discharge over some time interval	Total amount of discharge over the given time interval
Marginal cost	Total cost function
Marginal revenue	Total revenue function
Marginal profit	Total profit function
Rate of depreciation with respect to time, t	The value of the item after t years
Rate of consumption after some time, t	The total consumption after t years

EXAMPLE 5 A factory has been dumping pollutants into a lake since 1985. The rate of change in the concentration of the pollutant (in parts per million) at time t is given by the formula

$$P(t) = 150t^{3/2}$$

where t is the number of years since 1985. Ecologists have established that when the total pollution level reaches 24,000 ppm (parts per million), all bass in the lake will be killed. How much longer could the factory operate before all bass in the lake are killed?

Solution We know the rate and wish to find the total accumulation. This is a process that leads to a definite integral. The total accumulation of pollution dumped into the lake in x years since 1985 is given by

$$\int_0^x 150t^{3/2}\,dt = \frac{150t^{5/2}}{\frac{5}{2}}\Bigg|_0^x = 60t^{5/2}\Big|_0^x = 60x^{5/2}$$

Now, we need to solve the equation $60x^{5/2} = 24{,}000$:

$$\begin{aligned}60x^{5/2} &= 24{,}000\\ x^{5/2} &= 400\\ (x^{5/2})^{2/5} &= (400)^{2/5}\\ x &\approx 10.985605\end{aligned}$$

The bass in the lake will all be dead approximately 11 years after 1985—that is, in 1996. ■

14.3 Problem Set

Evaluate the definite integrals in Problems 1–30.

1. $\int_2^3 x^2\,dx$

2. $\int_{-2}^3 x^3\,dx$

3. $\int_1^4 \sqrt{x}\,dx$

4. $\int_1^{100} dx$

5. $\int_{-4}^{-1} \sqrt{1-2x}\,dx$

6. $\int_{-3}^1 \sqrt{1-x}\,dx$

7. $\int_0^2 e^{2x}\,dx$

8. $\int_0^3 e^{-x}\,dx$

9. $\int_1^{10} x^{-1}\,dx$

10. $\int_1^{e^3} \frac{dx}{x}$

11. $\int_1^{32} x^{-2/5}\,dx$

12. $\int_4^9 x^{3/2}\,dx$

13. $\int_3^8 (x-2)^{-1}\,dx$

14. $\int_5^{10} (x-3)^{-1}\,dx$

15. $\int_{-1}^2 x(1+x^4)\,dx$

16. $\int_{-1}^1 x^3(1+x^4)\,dx$

17. $\int_2^2 x(1+x)^4\,dx$

18. $\int_{-1}^1 x^2(1+x)^4\,dx$

19. $\int_0^1 \frac{dx}{(2x+1)^2}$

20. $\int_{-1}^0 \frac{dx}{(1-3x)^3}$

21. $\int_1^2 \left(x^{-2} - \frac{2}{x} + x^3\right)dx$

22. $\int_1^2 \left(x^2 + \frac{5}{x} - \sqrt{x}\right)dx$

23. $\int_6^6 \left(x^2 + \frac{\sqrt{5x}}{3} - 5\right)dx$

24. $\int_0^2 \left(x^2 + \frac{\sqrt{5x}}{3} - 5\right)dx$

25. $\int_{-1}^1 \frac{x\,dx}{\sqrt{1+3x^2}}$

26. $\int_{-1}^1 x\sqrt{1+3x^2}\,dx$

27. $\int_{-1}^0 5x^2(x^3+1)^{10}\,dx$

28. $\int_{-1}^0 3x^4(1-x^5)^3\,dx$

29. $\int_1^e 3x^{-1}\ln^3 x\,dx$

30. $\int_1^e x^{-1}\ln^2 x\,dx$

31. If f and g are continuous on $[-3, 2]$ and

$$\int_{-3}^{0} f(x)\,dx = 4, \quad \int_{0}^{2} f(x)\,dx = 10,$$

and $\int_{-3}^{2} g(x) = -5$, find

a. $\int_{-3}^{2} f(x)\,dx$ **b.** $\int_{2}^{-3} g(x)\,dx$

c. $\int_{-3}^{2} [2f(x) - 3g(x)]\,dx$

32. If f and g are continuous on $[5, 10]$ and

$$\int_{5}^{8} f(x)\,dx = -3, \quad \int_{8}^{10} f(x)\,dx = 4,$$

and $\int_{5}^{10} g(x)\,dx = 2$, find

a. $\int_{5}^{10} f(x)\,dx$ **b.** $\int_{7}^{7} [f(x) + g(x)]\,dx$

c. $\int_{10}^{5} g(x)\,dx$

33. If f is continuous on $[1, 5]$ and

$$\int_{1}^{3} f(x)\,dx = 9 \quad \text{and} \quad \int_{1}^{5} f(t)\,dt = 15, \quad \text{find} \int_{3}^{5} f(y)\,dy.$$

34. If g is continuous on $[-3, 2]$ and

$$\int_{-3}^{2} g(x)\,dx = 7 \quad \text{and} \quad \int_{0}^{2} g(t)\,dt = 4, \quad \text{find} \int_{-3}^{0} g(y)\,dy.$$

APPLICATIONS

35. Suppose that the rate of oil consumption (in billions of barrels) since 1985 is given by the formula

$$R(t) = 78e^{-.04t}$$

where t is measured in years since 1985. Predict the total number of barrels of oil consumed between 1987 and 2000.

36. Suppose that because of the 1986 crash in oil prices the demand for oil (in billions of barrels) changed so that it conforms to the formula

$$R(t) = 32.5e^{.048t}$$

where t is measured in years since 1986. Predict the total number of barrels consumed between 1988 and 1996 using 1986 as $t = 0$.

37. Repeat Problem 36 for the years 1987 to 2000.

38. If the known oil reserves in 1985 were 670.3 billion barrels, use the information in Problem 35 to estimate the length of time before all known reserves are depleted.

39. If the known oil reserves in 1985 were 670.3 billion barrels, use the information in Problem 36 to estimate the length of time before all known reserves are depleted.

40. What is the total amount of pollutants dumped into the lake in Example 5 between the years 1985 and 1990? Recall that the formula for the concentration of the pollutants (in ppm) at time t is given by the formula

$$P(t) = 150t^{3/2}$$

41. A factory has been dumping pollutants into a river since 1990. The rate of concentration of the pollutants (in ppm) at time t is given by the formula

$$P(t) = 350t^{5/2}$$

where t is the number of years since 1990. What is the total amount of pollutants dumped into the river between 1990 and 1995?

42. What is the total amount of pollutants dumped into the lake in Example 5 in 1992? Recall that the formula for the concentration of the pollutants (in ppm) at time t is given by the formula

$$P(t) = 150t^{3/2}$$

43. What is the total amount of pollutants dumped into the river in Problem 41 in 1992? Recall that the formula for the concentration of the pollutants (in ppm) at time t is given by the formula

$$P(t) = 350t^{5/2}$$

44. It is known that all life within 20 miles of a dumping site will die if the total concentration of the pollutants reaches 28,000 ppm. If dumping started in 1990, when will the pollution level be high enough to kill all life within the 20-mile radius of the dumping site? Assume that the formula for the concentration of pollutants (in ppm) at time t is given by $P(t) = 350t^{5/2}$.

45. The sales of Thornton Publishing are continuously growing according to the function

$$S(t) = 50e^{t}$$

where $S(t)$ is the rate of sales in dollars in the tth year. What are the accumulated sales after 10 years?

46. When will the accumulated sales in Problem 45 pass $1 million?

47. In economics, the Gini index, G, can be used to measure how money is distributed among the population according to the formula

$$G = 1 - \int_{0}^{1} 2f(x)\,dx$$

The closer G is to zero, the more the money is spread evenly throughout the population. Find G for the function $f(x) = x^3$.

48. Find the Gini index, G, for the function $f(x) = x^{3/2}$. The Gini index is a measure for the amount of money spread throughout the population (see Problem 47) and is given by

$$G = 1 - \int_0^1 2f(x)\,dx$$

The closer this number is to zero, the more the money is spread evenly throughout the population.

49. The total accumulated cost and revenue for a piece of business machinery is

$$C'(t) = \frac{t}{20} \quad \text{and} \quad R'(t) = 10te^{-t^2}$$

where t is the time in years. Find the cost and revenue functions if the initial cost is \$25,000 and the initial revenue is \$0.

50. The rate of healing for a wound is given by the formula

$$A'(t) = -.85e^{-.1t}$$

where A is the number of square centimeters of unhealed skin after t days. How much will the area change in the first week if the initial wound has an area of 10 cm^2?

14.4 Area Between Curves

Area Function

One of the most common applications of definite integrals involves the area under a curve.

Let f be a continuous nonnegative function on $[a, b]$. Let $A(x)$ be the **area function** that denotes the area of the region of $[a, b]$ bounded by f on top and the x-axis on the bottom and the vertical lines x and a, as shown in Figure 14.1.

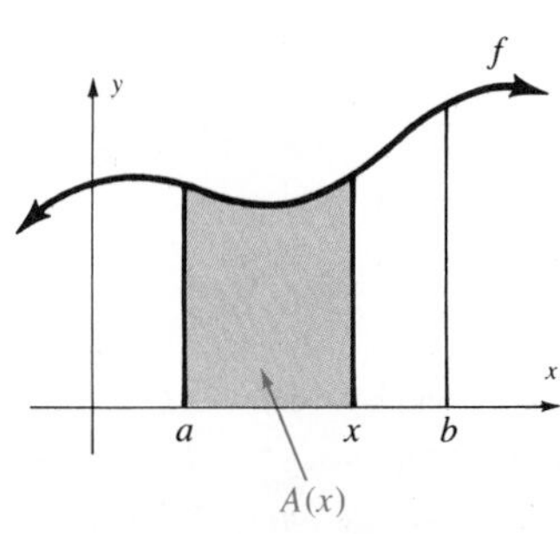

FIGURE 14.1
The area function

EXAMPLE 1 Let $a = 0$ and find $A(x)$ for each of the given functions.

a. $f(x) = 4$ **b.** $f(x) = 4x$ **c.** $f(x) = 4x + 3$

Solution **a.**

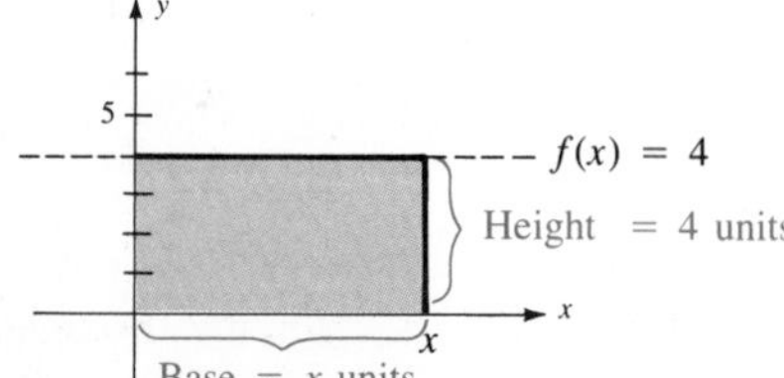

The figure formed is a rectangle, so

$$\begin{aligned} A(x) &= bh \\ &= x(4) \\ &= 4x \end{aligned}$$

b.

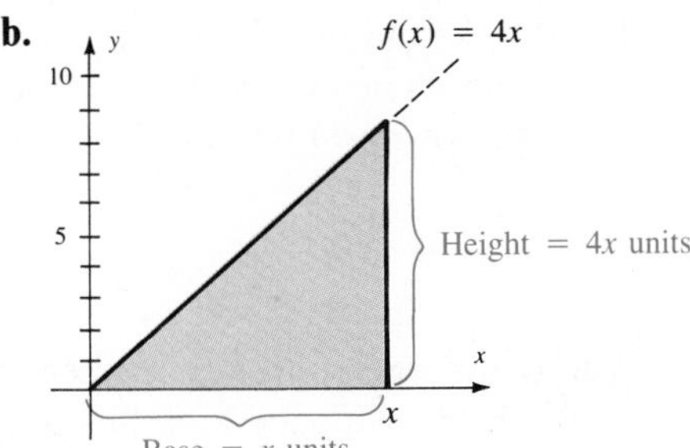

The figure formed is a triangle, so

$$\begin{aligned} A(x) &= \frac{1}{2}bh \\ &= \frac{1}{2}(x)(4x) \\ &= 2x^2 \end{aligned}$$

c.

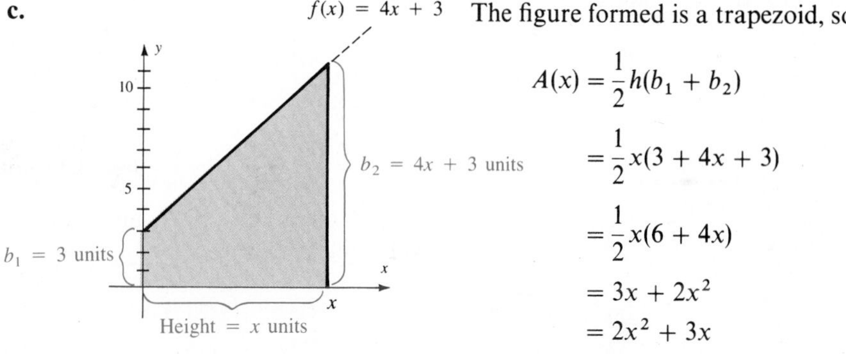

The figure formed is a trapezoid, so

$$\begin{aligned} A(x) &= \frac{1}{2}h(b_1 + b_2) \\ &= \frac{1}{2}x(3 + 4x + 3) \\ &= \frac{1}{2}x(6 + 4x) \\ &= 3x + 2x^2 \\ &= 2x^2 + 3x \end{aligned}$$

■

The area function for each part of Example 1 is the antiderivative of f where $C = 0$. This is true whenever $a = 0$. That is,

Function	Area function	Antiderivative
$f(x) = 4$	$A(x) = 4x$	$F(x) = 4x + C$
$f(x) = 4x$	$A(x) = 2x^2$	$F(x) = 2x^2 + C$
$f(x) = 4x + 3$	$A(x) = 2x^2 + 3x$	$F(x) = 2x^2 + 3x + C$

Area Under a Curve

The fact that the area of the figures bounded by f in Example 1 turned out to be the antiderivatives of f is not a coincidence. In elementary mathematics, areas are defined in terms of the area of a square (i.e., square units). However, when we want to find the area of a region with a curved boundary, the situation is much more complicated. Even the formula for the area of a circle that you remember from elementary school is a result of the calculus. In order to find the area of regions bounded by curves (or functions in general), we make the following definition.

Area Under a Curve

If f is a continuous nonnegative function on $[a, b]$, then the **area** of the region bounded by the graph of f and the x-axis and the vertical lines $x = a$ and $x = b$ is given (exactly) by

$$A = \int_a^b f(x)\,dx$$

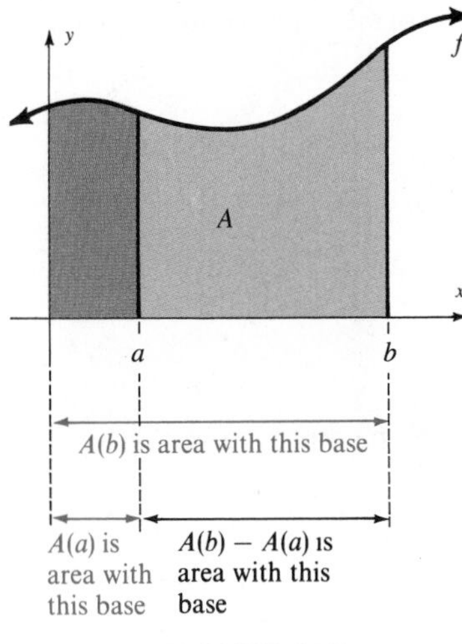

FIGURE 14.2
$A = A(b) - A(a)$

The area A in this definition is shown with the color screen in Figure 14.2. We see that the idea of area as defined by the integral in the box can be justified in terms of area functions (shown in Figure 14.1)—namely, $A(b) - A(a)$:

$$A = \int_a^b f(x)\,dx = A(b) - A(a)$$

Next, we wish to find a relationship between the function f and the area functions $A(b) - A(a)$. We begin with the area function $A(x)$ and shall show that $A'(x) = f(x)$, which means that $A(x)$ is an antiderivative of $f(x)$. From the definition of derivative,

$$A'(x) = \lim_{h \to 0} \frac{A(x+h) - A(x)}{h}$$

Consider $\dfrac{A(x+h) - A(x)}{h}$ in Figure 14.3. Notice that $A(x+h) - A(x)$ is the region shaded in color. This area is approximated by the area of the rectangle with base h and height $f(x)$. Thus

$$A(x+h) - A(x) \approx f(x) \cdot h$$
$$\frac{A(x+h) - A(x)}{h} \approx f(x)$$

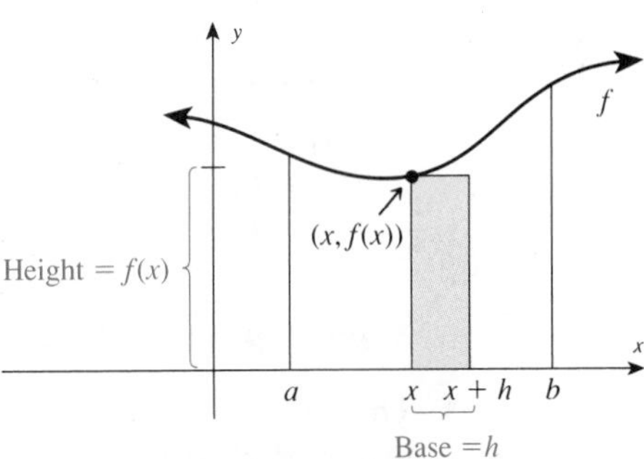

FIGURE 14.3
Area of shaded rectangle is $f(x)h$, which approximates $A(x+h) - A(x)$, as shown in color.

Now if we let $h \to 0$, then the left side has $A'(x)$ as a limit, which is equal to the right side. Thus,

$$A'(x) = f(x)$$

It is therefore reasonable to denote $A(x) = F(x)$, where F is an antiderivative of f. We can now restate the area definition:

$$A = \int_a^b f(x)\,dx = F(x)\Big|_a^b = F(b) - F(a)$$

We will now see by examples that this idea of area under a curve compares exactly with your previous knowledge of area from plane geometry.

EXAMPLE 2 Find the areas of the rectangle, triangle, and trapezoid from Example 1 for $a = 0$ and $b = 10$ using both the results of Example 1 and the area under a curve definition.

a. $f(x) = 4$ **b.** $f(x) = 4x$ **c.** $f(x) = 4x + 3$

Solution **a.** From Example 1a, $A(x) = 4x$, so $A(10) = 40$ and $A(0) = 0$. Using calculus,

$$\int_0^{10} 4dx = 4x\Big|_0^{10} = 4(10 - 0) = 40$$

b. From Example 1b, $A(x) = 2x^2$, so $A(10) = 2(10)^2 = 200$ and $A(0) = 0$. Using calculus,

$$\int_0^{10} 4x\,dx = 2x^2\Big|_0^{10} = 2(10)^2 - 2(0)^2 = 200$$

c. From Example 1c, $A(x) = 2x^2 + 3x$, so $A(10) = 2(10)^2 + 3(10) = 230$ and $A(0) = 0$. Using calculus,

$$\int_0^{10} (4x + 3)\,dx = (2x^2 + 3x)\Big|_0^{10} = [2(10)^2 + 30] - (2 \cdot 0^2 + 0) = 230$$ ■

EXAMPLE 3 Find the area under the curve $y = x^2$ on $[2, 4]$.

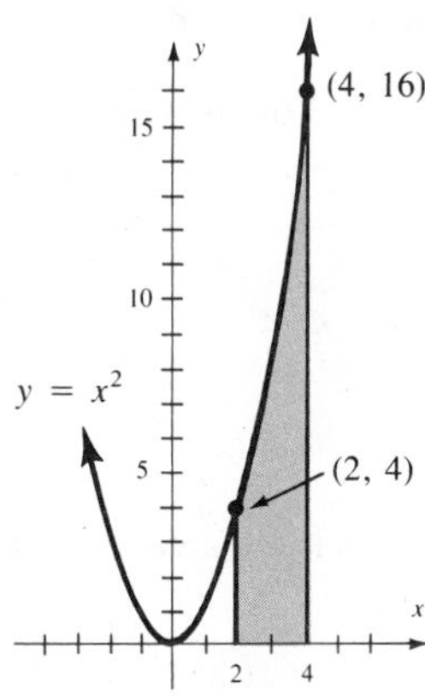

Solution $$\int_2^4 x^2\,dx = \frac{x^3}{3}\Big|_2^4 = \frac{4^3}{3} - \frac{2^3}{3} = \frac{64}{3} - \frac{8}{3} = \frac{56}{3} = 18\frac{2}{3}$$ ■

Area Between a Curve and the x-Axis

There is an important restriction on the conditions of the area under a curve definition. This condition is that the function f be nonnegative. Another way of looking at this is to say that the definite integral does not always give the area! You must, therefore, be careful about the way you state your conclusions. Consider the next example carefully.

EXAMPLE 4 Evaluate $\int_{-2}^{2} x^3\,dx$.

Solution It would **not** be correct to say that this integral is the area bounded by the curve $f(x) = x^3$, the x-axis, and the lines $x = -2$ and $x = 2$. If you graph $f(x) = x^3$ (shown in the margin) we see that there is an area, but if we evaluate

$$I = \int_{-2}^{2} x^3\,dx = \frac{x^4}{4}\bigg|_{-2}^{2} = \frac{2^4}{4} - \frac{(-2)^4}{4} = 0$$

the definite integral is not the area. ■

Compare the areas shown by the two graphs in Figure 14.4.

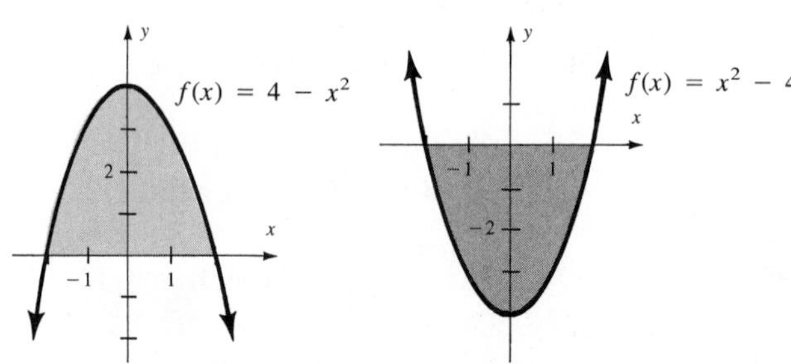

FIGURE 14.4
Areas bounded by two curves

For the curve bounded by $f(x) = 4 - x^2$, the integral **is** the area.

$$\int_{-2}^{2} (4 - x^2)\,dx = 4x - \frac{x^3}{3}\bigg|_{-2}^{2}$$

$$= 8 - \frac{8}{3} - \left(-8 + \frac{8}{3}\right)$$

This is the area. $= 16 - \frac{16}{3} = \frac{32}{3}$

For the curve bounded by $f(x) = x^2 - 4$, the integral is **not** the area.

$$\int_{-2}^{2} (x^2 - 4)\,dx = \frac{x^3}{3} - 4x\bigg|_{-2}^{2}$$

$$= \frac{8}{3} - 8 - \left(-\frac{8}{3} + 8\right)$$

This is not the area. $= \frac{16}{3} - 16 = -\frac{32}{3}$

It appears that if f is negative over $[a, b]$, then the area is the opposite of the given integral. Consider the function shown in Figure 14.5.

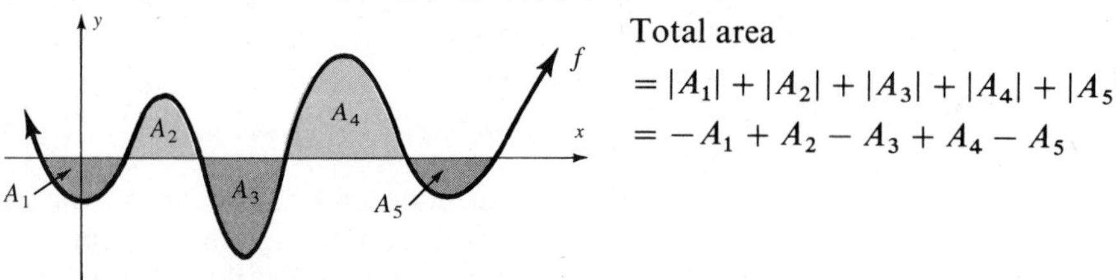

Total area

$$= |A_1| + |A_2| + |A_3| + |A_4| + |A_5|$$

$$= -A_1 + A_2 - A_3 + A_4 - A_5$$

FIGURE 14.5
Area defined by a function over $[a, b]$. A_2 and A_4 are positive numbers representing areas and A_1, A_3, and A_5 are opposites of the enclosed areas.

EXAMPLE 5 Find the area bounded by the curve $f(x) = x^3$, the x-axis, and the lines $x = -2$ and $x = 2$. See Example 4.

Solution We need to know which part of this curve is above the x-axis and which is below. We solve $f(x) = 0$, which tells us where the curve crosses the x-axis; in this example the curve crosses at $x = 0$. Let A be the desired area.

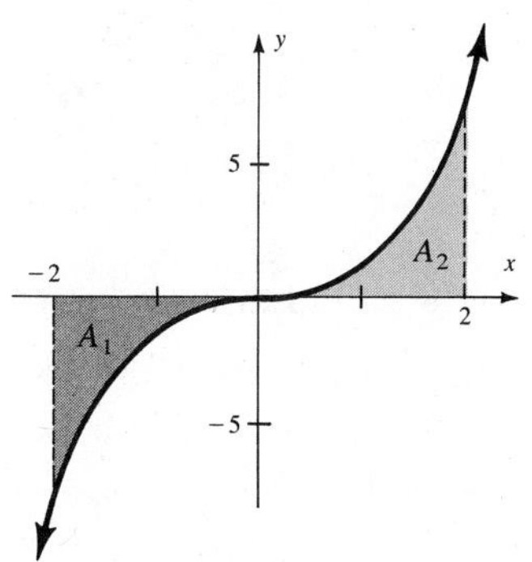

$$A = |A_1| + |A_2| \quad \text{where} \quad A_1 = \int_{-2}^{0} x^3\,dx = \frac{x^4}{4}\Big|_{-2}^{0} = 0 - \frac{16}{4} - 0 = -4$$

$$\text{and} \quad A_2 = \int_{0}^{2} x^3\,dx = \frac{x^4}{4}\Big|_{0}^{2} = 4$$

Thus, $A = -A_1 + A_2 = -(-4) + 4 = 8$. ■

Area Between Curves

The difficulties encountered when dealing with functions like the one in Example 5 which have both positive and negative values (shown in Figure 14.5) can be simplified if we find the area *between* two curves.

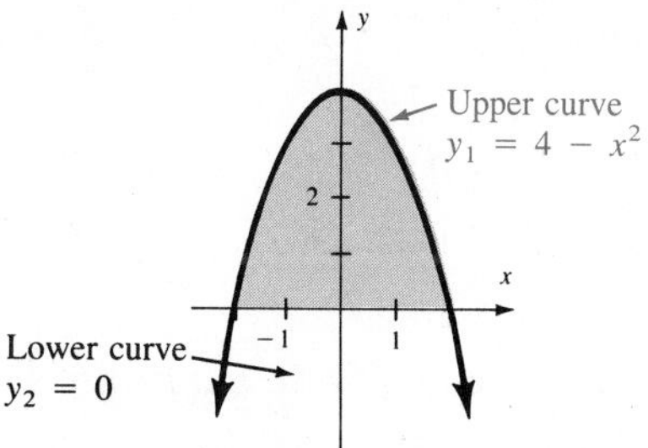

Difference between two curves:

$$y_1 - y_2 = (4 - x^2) - 0 = 4 - x^2$$

$$\text{Area} = \int_a^b (y_1 - y_2)\,dx$$

Upper curve: y_1; Lower curve: y_2

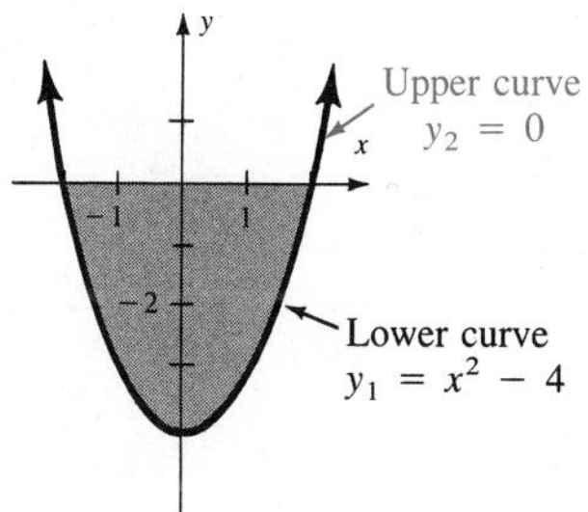

Difference between two curves:

$$y_2 - y_1 = 0 - (x^2 - 4) = 4 - x^2$$

$$\text{Area} = \int_a^b (y_2 - y_1)\,dx$$

Upper curve: y_2; Lower curve: y_1

Area Between Two Curves

Suppose that f and g are continuous functions on $[a, b]$ and that $f(x) \geq g(x)$ for $a \leq x \leq b$.* The area of the region bounded by f above, g below, $x = a$ on the left, and $x = b$ on the right is

$$A = \int_a^b [\underset{\substack{\uparrow \\ \text{Upper} \\ \text{curve}}}{f(x)} - \underset{\substack{\uparrow \\ \text{Lower} \\ \text{curve}}}{g(x)}]\, dx$$

EXAMPLE 6 Find the area between the curves $y = x^2$ and $y = 6 - x$ from $x = -3$ to $x = 2$.

Solution First sketch the curves. The top curve is $y = 6 - x$ and the bottom curve is $y = x^2$. Thus the area between the curves is given by the definite integral

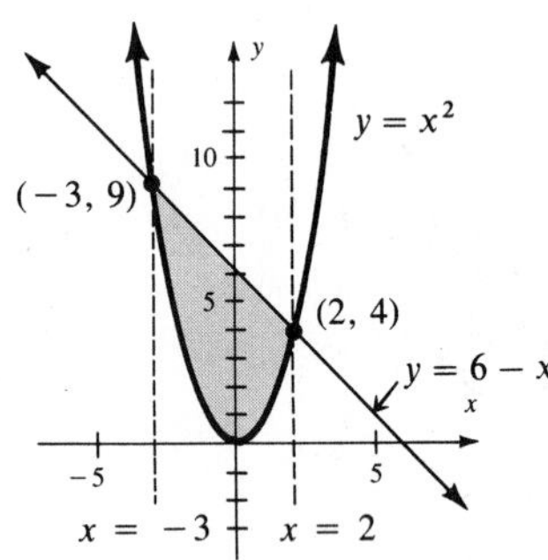

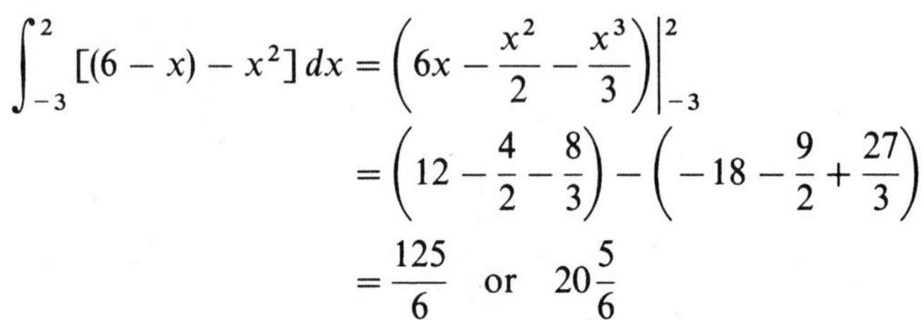

$$\int_{-3}^{2} [(6 - x) - x^2]\, dx = \left(6x - \frac{x^2}{2} - \frac{x^3}{3}\right)\Bigg|_{-3}^{2}$$

$$= \left(12 - \frac{4}{2} - \frac{8}{3}\right) - \left(-18 - \frac{9}{2} + \frac{27}{3}\right)$$

$$= \frac{125}{6} \quad \text{or} \quad 20\frac{5}{6}$$

■

Area Between Curves That Intersect

Sometimes the limits of integration for problems like Example 6 are implied instead of given. In such cases you need to solve the system of equations formed by the boundaries in the problem:

$$\begin{cases} y = x^2 \\ y = 6 - x \end{cases}$$

By substitution,

$$\begin{aligned} 6 - x &= x^2 \\ x^2 + x - 6 &= 0 \\ (x + 3)(x - 2) &= 0 \\ x &= -3, 2 \end{aligned}$$

The points of intersection for the curves are now found:

If $x = -3$, then $y = 6 - (-3) = 9$. Point: $(-3, 9)$
If $x = 2$, then $y = 6 - 2 = 4$. Point: $(2, 4)$

You would use these values for the limits of integration as shown in Example 6.

A condition of the formula for the area between two curves is that you subtract the lower curve from the upper curve. This means you must know whether the curves intersect over the interval of integration. If they do intersect, then you must divide the problem into separate integrals as illustrated by Example 7.

* This means that the curve f is never below g anywhere on the interval from a to b.

EXAMPLE 7 **a.** Evaluate $\int_0^2 (x^3 - x)\,dx$.

b. Find the area of the region bounded by $y = x^3$ and $y = x$ on the interval [0, 2] and compare with the answer you found in part **a**.

Solution **a.** Integral $= \int_0^2 (x^3 - x)\,dx = \left(\frac{x^4}{4} - \frac{x^2}{2}\right)\Big|_0^2 = \left(\frac{16}{4} - \frac{4}{2}\right) - \left(\frac{0}{4} - \frac{0}{2}\right) = 2$

b. When finding the area you must make sure that you identify the upper and lower curves on the interval. To see whether the curves cross we can solve the system

$$\begin{cases} y = x^3 \\ y = x \end{cases}$$

Use substitution to find

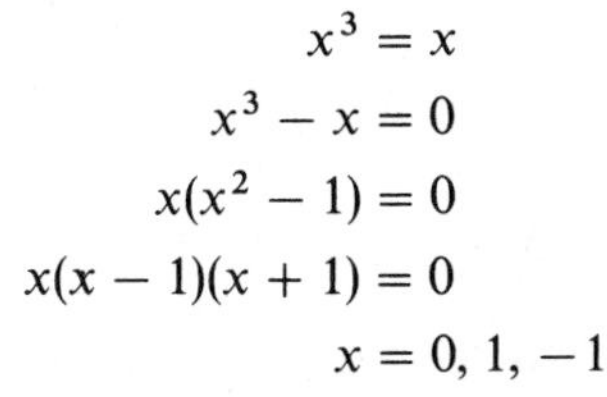

$$\begin{aligned} x^3 &= x \\ x^3 - x &= 0 \\ x(x^2 - 1) &= 0 \\ x(x - 1)(x + 1) &= 0 \\ x &= 0, 1, -1 \end{aligned}$$

The graph is shown in the margin. In order to find the area, we must break the integral into two parts in order to make sure that the equation for the lower curve is subtracted from the equation of the upper curve.

$$\begin{aligned} \text{Area} &= \int_0^1 (x - x^3)\,dx + \int_1^2 (x^3 - x)\,dx \\ &= \left(\frac{x^2}{2} - \frac{x^4}{4}\right)\Big|_0^1 + \left(\frac{x^4}{4} - \frac{x^2}{2}\right)\Big|_1^2 \\ &= \left(\frac{1}{2} - \frac{1}{4}\right) - 0 + \left[\left(\frac{16}{4} - \frac{4}{2}\right) - \left(\frac{1}{4} - \frac{1}{2}\right)\right] \\ &= \frac{5}{2} \end{aligned}$$

Notice that the area is not the same as the integral over the interval [0, 2]. ■

EXAMPLE 8 Find the area between the curve $y = x^2 - 9$, the x-axis, and the lines $x = 1$ and $x = 9$.

Solution Check to see whether the curves $y = x^2 - 9$ and $y = 0$ cross by solving

$$\begin{aligned} x^2 - 9 &= 0 \\ (x - 3)(x + 3) &= 0 \\ x &= 3, -3 \end{aligned}$$

↑ Not an interval

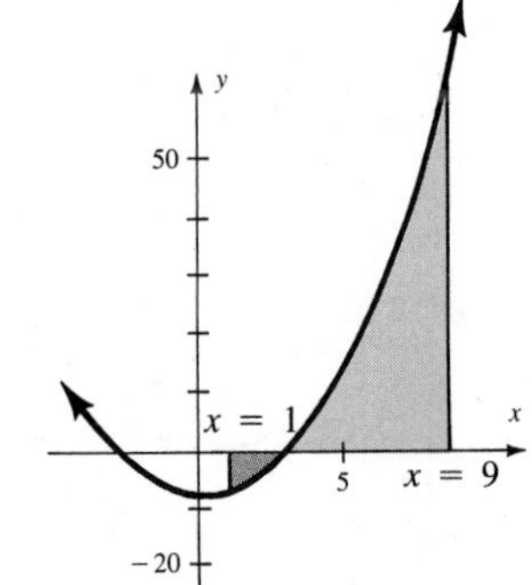

The curves cross at $x = 3$.

The area is found by dividing the interval into two different integrals:

$$\int_1^3 [0-(x^2-9)]\,dx + \int_3^9 [(x^2-9)-0]\,dx = \left(-\frac{x^3}{3}+9x\right)\Bigg|_1^3 + \left(\frac{x^3}{3}-9x\right)\Bigg|_3^9$$

Top curve (labels 0 in the first integral); Top curve (labels x^2-9 in the second integral)

$$= (-9+27) - \left(-\frac{1}{3}+9\right) + (243-81) - (9-27)$$

$$= \frac{568}{3} \quad\text{or}\quad 189\frac{1}{3}$$

■

EXAMPLE 9 The 1986 crash of oil prices brought about a change in the worldwide consumption rate of petroleum products (*Wall Street Journal*, February 11, 1986). We can measure the effect of that change by finding the area between two curves. For example, suppose that

pre-1985 rate of consumption: $R_1(t) = 78e^{-.04t}$ $(0 \le t \le 5)$

post-1985 rate of consumption: $R_2(t) = 50.2e^{.048t}$ $(t \ge 5)$

in billions of barrels for a base year of 1980 ($t = 0$). How much extra oil would be consumed between 1985 and 1990?

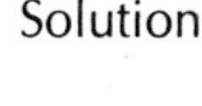

Solution

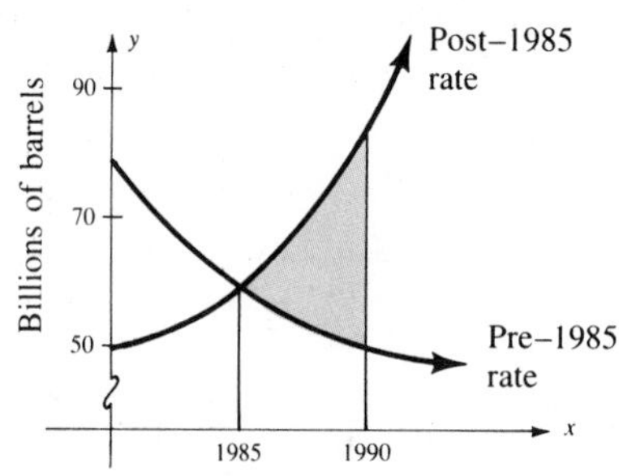

$$\int_5^{10} (50.2e^{.048t} - 78e^{-.04t})\,dt$$

$$= \frac{50.2}{.048}e^{.048t} - \frac{78}{-.04}e^{-.04t}\Bigg|_5^{10}$$

$$\approx 2{,}997.27 - 2{,}926.04$$

$$\approx 71.2$$

The additional oil consumed would be about 71.2 billion barrels. ■

14.4 Problem Set

Write integrals to express the shaded region in Problems 1–6. Both positive regions (color) and negative regions (gray) should be included.

1.

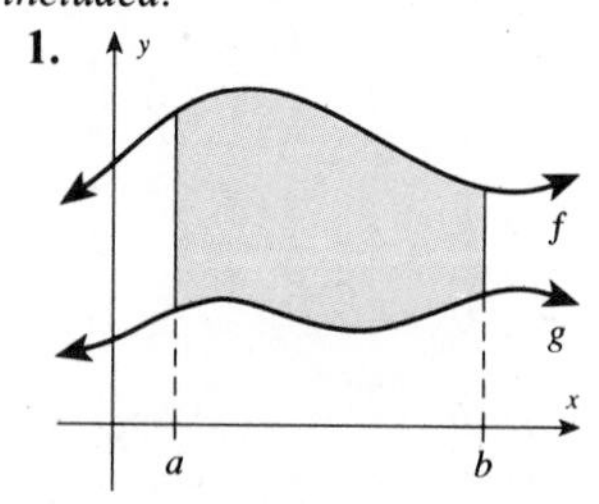

2.

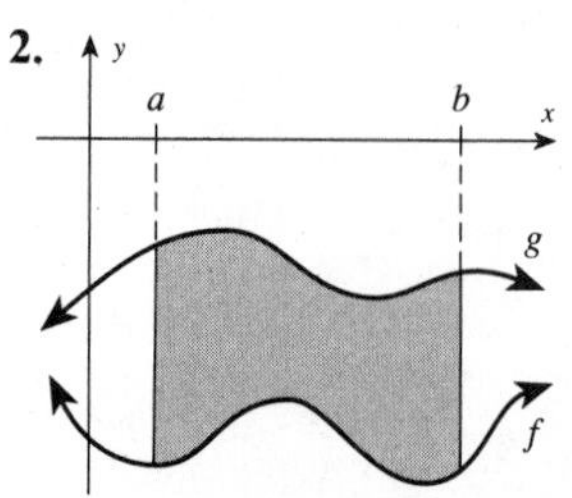

3.

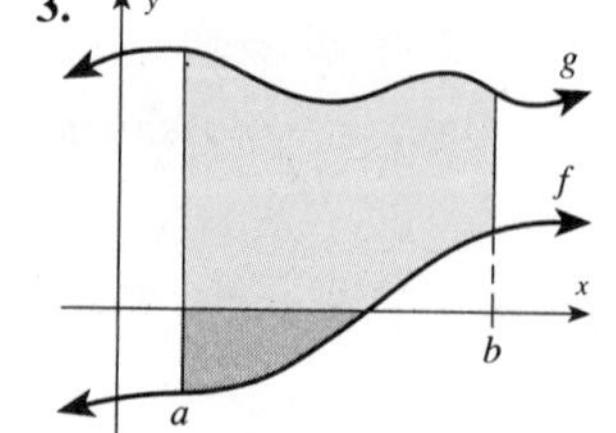

4.

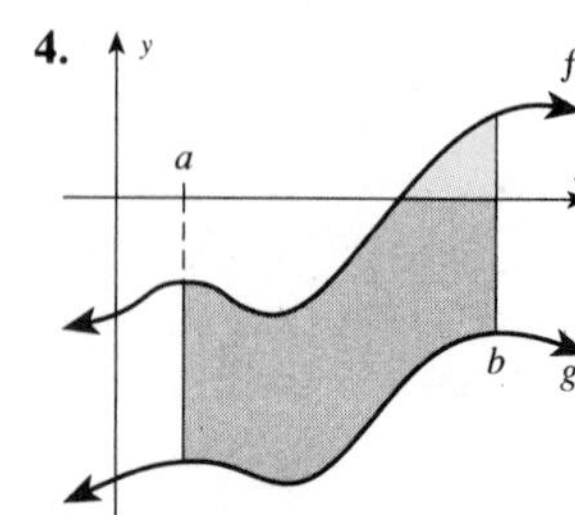

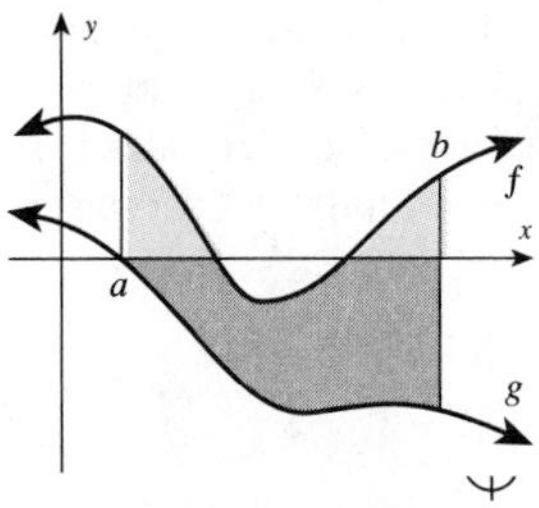

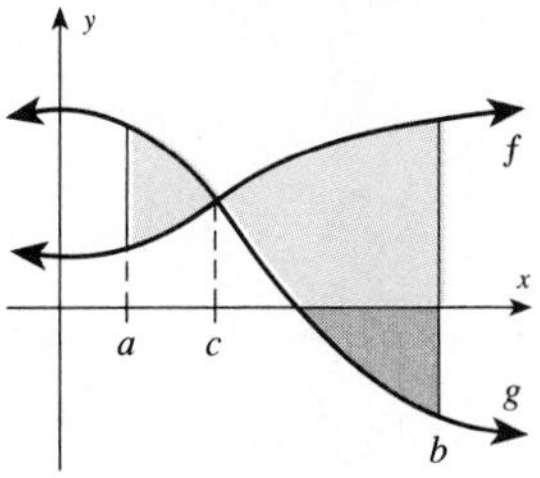

Find the area bounded by the conditions given in Problems 7–35.

7. $y = 6x^2$, the x-axis, and the lines $x = 1$, $x = 5$
8. $y = 4x^3$, the x-axis, and the lines $x = 1$, $x = 4$
9. $y = 3x^2 + 4$, $y = 0$, $x = -1$, and $x = 2$
10. $y = 12x^2 - 2x + 5$, $y = 0$, $x = 0$, $x = 2$
11. $y = 4x^3 - 3x^2 + 2x$, the x-axis on $[0, 2]$
12. $y = e^x$, the x-axis on $[-1, 2]$
13. $y = e^{-x}$, the x-axis on $[-2, 1]$
14. $y = x^2 - 4$ and the x-axis
15. $y = x^2 - 9$ and the x-axis
16. $y = x^2$ and $y = 6 - x$ on $[-3, 2]$
17. $y = x^2$ and $y = 6 - x$ on $[3, 5]$
18. $y = x^3$ and $y = x^2$ on $[0, 1]$
19. $y = x^3$ and $y = x^2$ on $[0, 2]$
20. $y = x^2$ and $y = 6x$ on $[0, 5]$
21. $y = x^2$ and $y = 2x + 4$ on $[0, 1]$
22. $y = x^2 + 4$ and $y = 2x + 4$ on $[-1, 1]$
23. $y = x^2 + 4$ and $y = 2x + 4$ on $[1, 5]$
24. $y = x^3$ and $y = x$
25. $y = x$ and $y = x^4$
26. $y = x^2 - x$ and $y = 6$
27. $y = x^2 - 5x$ and $y = 6$
28. $y = x^{-1}$ on $[.5, 1]$
29. $y = x^{-1}$ on $[.01, 1]$
30. $y = x^2$ and $y = \sqrt{x}$
31. $y = x$ and $y = \sqrt{x}$
32. $y = \sqrt{x}$ and $y = 4$
33. $y = \sqrt{x}$ and $y = 9$
34. $y = x^{-1}$ and $y = e^x$ on $[1, 2]$
35. $y = e^{.5x}$ and $y = x^{-1}$ on $[1, 2]$

APPLICATIONS

36. The Sorite Corporation has found that production costs are determined by the function

$$C(x) = \frac{100}{\sqrt{x + 1}} + 50$$

where $C(x)$ is the cost to produce the xth item. What is the approximate total cost of producing 2,000 items?

37. The EPA (Environmental Protection Agency) estimates that pollutants are being dumped into the Russian River according to the formula

$$P(t) = \frac{1{,}000}{\sqrt[3]{(t + 5)^2}}$$

where t is the year and $P(t)$ is pollutants in tons. What is the total amount (to the nearest ton) of pollutants that will be dumped into the river in the next 25 years?

38. The marginal revenue and marginal cost (in dollars) per day are given in terms of the tth month by the formulas

$$R'(t) = 500e^{.01t} \qquad \text{and} \qquad C'(t) = 50 - .1t$$

where $R(0) = C(0) = 0$, and t is the month. Find the total profit for the first year.

39. A chemical spill is known to kill fish according to the formula

$$N(t) = \frac{100{,}000}{\sqrt{t + 250}}$$

where N is the number of fish killed on the tth day since the spill. What is the approximate total number of fish killed in 30 days?

40. A culture is growing at a rate of

$$R'(x) = 200e^{.05t} \qquad \text{for } 0 \le t \le 10$$

per hour. Find the area between the graph of this equation and the t-axis. What do you think this area represents?

41. The rate of change of the demand for a product with respect to time (in weeks) is given by the equation

$$D'(x) = 20 + .012t^2 \qquad \text{for } 0 \le t \le 25$$

Find the area between the graph of this equation and the t-axis. What do you think this area represents?

42. Suppose the accumulated cost of a piece of equipment is $C(t)$ and the accumulated revenue is $R(t)$, where both of these are measured in thousands of dollars and t is the number of years after the piece of equipment was installed. If it is known that

$$C'(t) = 1 \quad \text{and} \quad R'(t) = 2e^{-.1t}$$

find the area (to the nearest unit) between the graphs of C' and R' (do not forget $t \geq 0$). What do you think this area represents?

43. Suppose the accumulated cost of a piece of equipment is $C(t)$ and the accumulated revenue is $R(t)$, where both of these are measured in thousands of dollars and t is the number of years after the piece of equipment was installed. If it is known that

$$C'(t) = 18 \quad \text{and} \quad R'(t) = 21e^{-.01t}$$

find the area (to the nearest unit) between the graphs of C' and R' (do not forget $t \geq 0$). What do you think this area represents?

14.5 The Fundamental Theorem of Calculus

In this section we will formulate a very important theorem in calculus—so important, in fact, that it carries the name *The Fundamental Theorem of Calculus.* In order to integrate a function f we need to find an antiderivative F. What if we cannot find such an antiderivative? For example,

$$\int_1^5 \frac{x}{x+1}\,dx$$

certainly should have a value, but with the techniques of integration considered thus far, we cannot find an antiderivative of

$$f(x) = \frac{x}{x+1}$$

We will evaluate this integral in Example 2, but we begin with a simpler example, one which we can evaluate.

EXAMPLE 1 Find the area bounded by $f(x) = x^2$, the x-axis, and the lines $x = 1$ and $x = 5$.

Solution The region is shown in Figure 14.6.

$$\int_1^5 x^2\,dx = \left.\frac{x^3}{3}\right|_1^5 = \frac{125}{3} - \frac{1}{3} = \frac{124}{3}$$

■

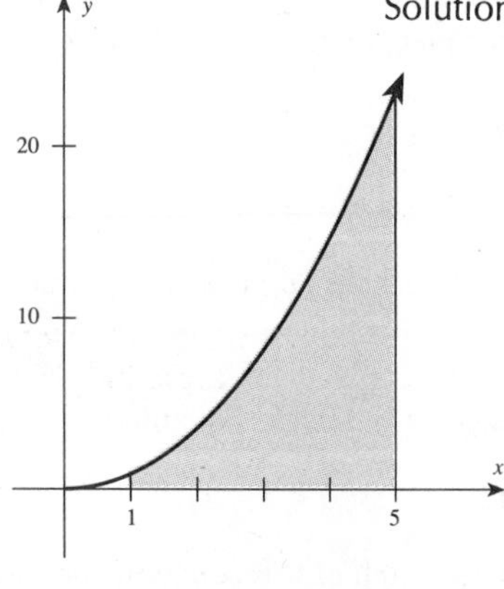

FIGURE 14.6

Suppose that we were not able to find the antiderivative of $f(x) = x^2$ in Example 1. We could approximate the area by dividing the area bounded by f, the x-axis, and the lines $x = 1$, $x = 5$ into one, two, and four rectangles, as shown in Figure 14.7.

COMPUTER APPLICATION

These computer-generated drawings show the area under the curve $y = x^2$ as approximated by rectangles using left endpoints. The Riemann sum, defined later in this section, approximates the actual area.

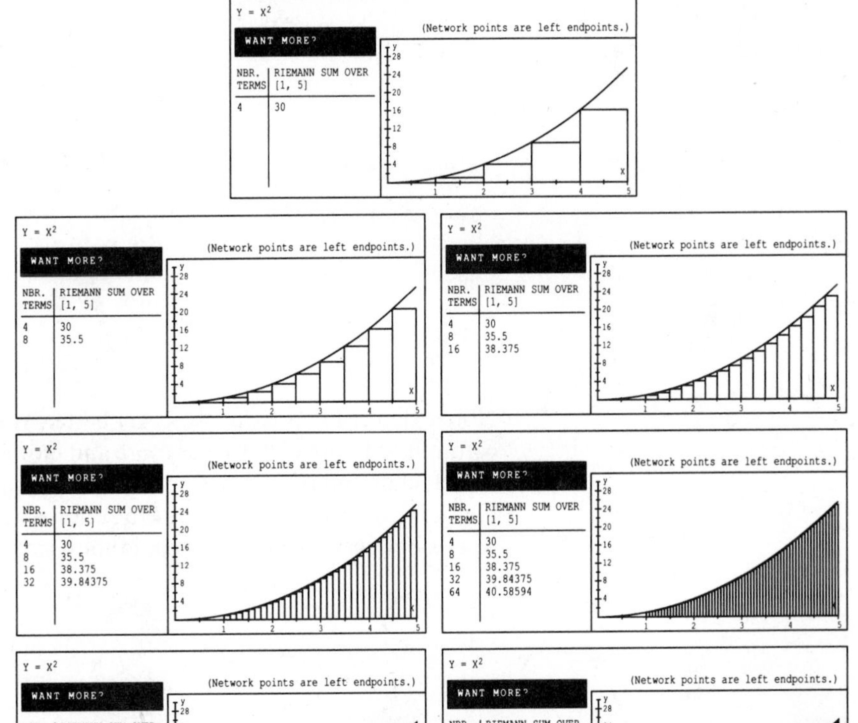

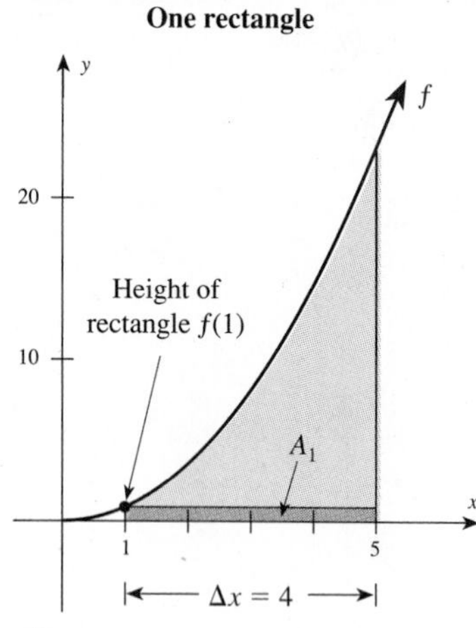

$$\int_1^5 x^2\,dx \approx A_1$$
$= \Delta x \cdot f(1)$
$= 4 \cdot 1^1$
$= 4$

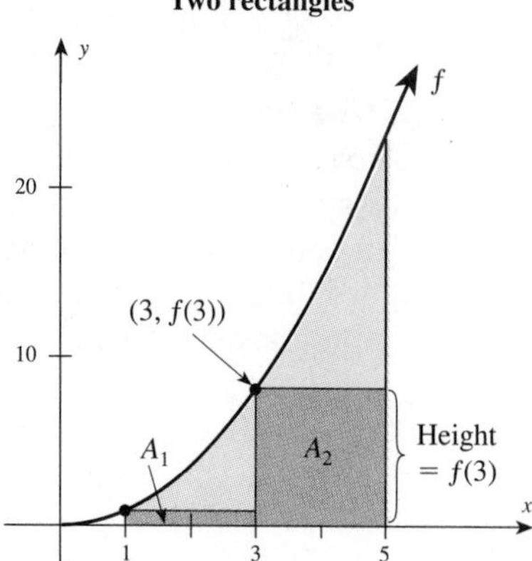

$$\int_1^5 x^2\,dx \approx A_1 + A_2$$
$= \Delta x \cdot f(1) + \Delta x \cdot f(3)$
$= 2 \cdot 1^2 + 2 \cdot 3^2$
$= 20$

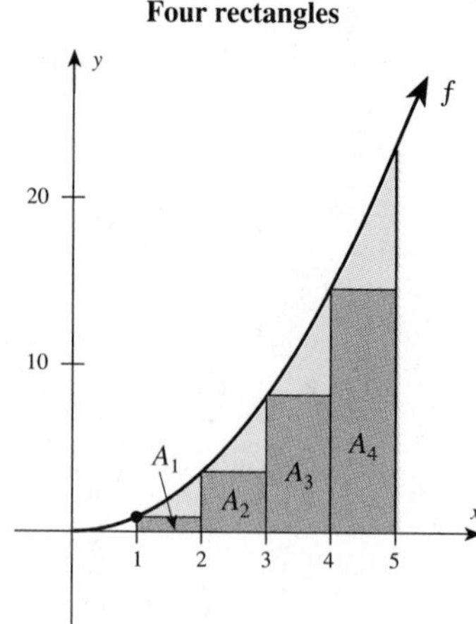

$$\int_1^5 x^2\,dx \approx A_1 + A_2 + A_3 + A_4$$
$= \Delta x \cdot f(1) + \Delta x \cdot f(2) + \Delta x \cdot f(3) + \Delta x \cdot f(4)$
$= 1 \cdot 1^2 + 1 \cdot 2^2 + 1 \cdot 3^2 + 1 \cdot 4^2$
$= 30$

FIGURE 14.7

Notice that the approximations get better. We would expect the approximations to continue to improve as we use more and more rectangles. Also notice that we picked the left endpoint of each subinterval. We could have selected the right endpoint, the midpoint, or any point c_k in the kth subinterval. The following calculations show the results obtained by picking the midpoints for the subintervals shown in Figure 14.7.

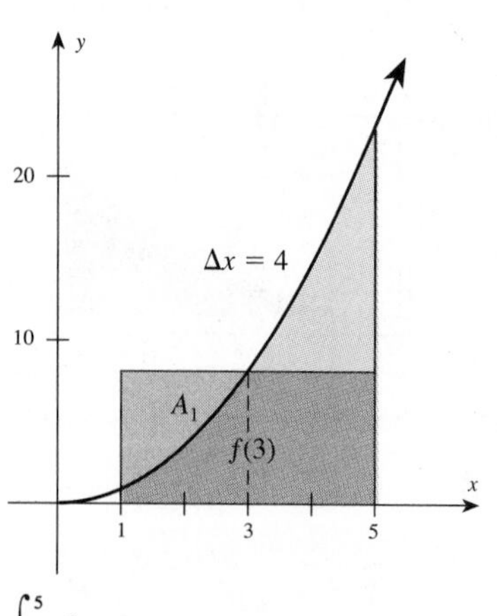

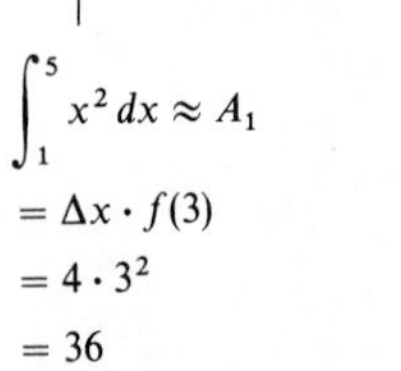

$$\int_1^5 x^2\,dx \approx A_1$$
$= \Delta x \cdot f(3)$
$= 4 \cdot 3^2$
$= 36$

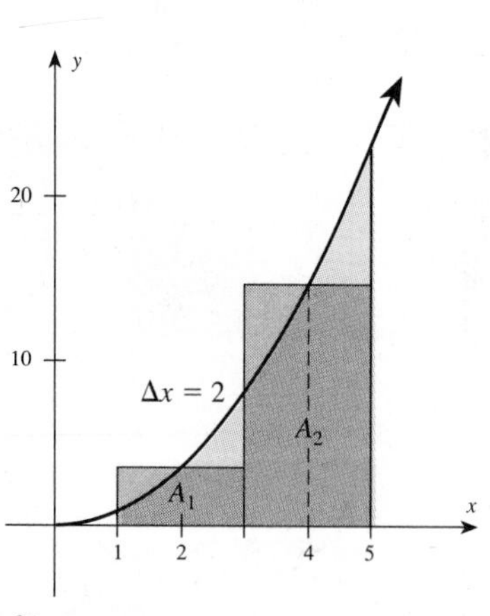

$$\int_1^5 x^2\,dx \approx A_1 + A_2$$
$= \Delta x \cdot f(2) + \Delta x \cdot f(4)$
$= 2 \cdot 2^2 + 2 \cdot 4^2$
$= 40$

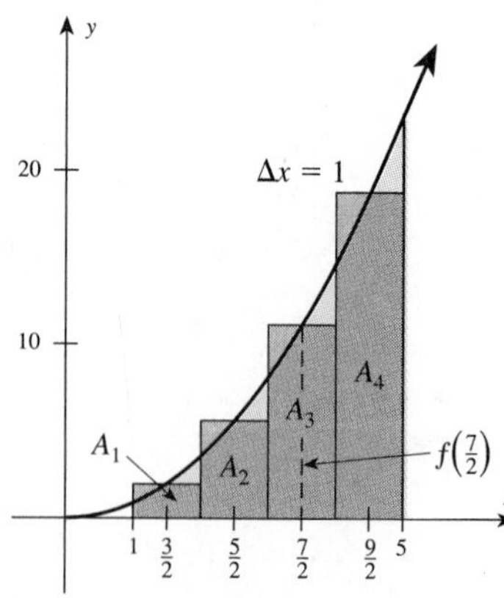

$$\int_1^5 x^2\,dx \approx A_1 + A_2 + A_3 + A_4$$
$= \Delta x \cdot f(\tfrac{3}{2}) + \Delta x \cdot f(\tfrac{5}{2}) + \Delta x \cdot f(\tfrac{7}{2}) + \Delta x \cdot f(\tfrac{9}{2})$
$= 1 \cdot (\tfrac{3}{2})^2 + 1 \cdot (\tfrac{5}{2})^2 + 1 \cdot (\tfrac{7}{2})^2 + 1 \cdot (\tfrac{9}{2})^2$
$= 41$

As you can see, selecting the midpoint of each subinterval usually approximates the area with fewer approximations. However, if you take enough equal subintervals the rectangles are so small that it does not matter which point c_k you choose in the kth subinterval. This leads us to what we call the rectangular approximation to the integral, which is true for any continuous function f.

Rectangular Approximation

Divide the interval from $x = a$ to $x = b$ into n equal subintervals, each with length

$$\Delta x = \frac{b - a}{n}$$

Let c_k be *any* point in the kth subinterval. Then,

$$\begin{aligned}\int_a^b f(x)\,dx &\approx \Delta x \cdot f(c_1) + \Delta x \cdot f(c_2) + \cdots + \Delta x \cdot f(c_n) \\ &= \Delta x[f(c_1) + f(c_2) + \cdots + f(c_n)] \\ &= \Delta x \sum_{k=1}^{n} f(c_k)\end{aligned}$$

EXAMPLE 2 Evaluate $\int_1^5 \frac{x}{x+1}\,dx$. Approximate this integral (to two decimal places) by using the rectangular approximation with $n = 4$.

Solution We want the value of this integral. This is not an area problem; a rectangular approximation exists for any continuous function. If f is not positive on $[a, b]$, then neither the integral nor the sum represents an area. In this problem, we want the value of the integral.

In order to use the rectangular approximation for this integral, we must make sure that f is continuous over $[1, 5]$. The only suspicious point is $x = -1$ which is not in the interval, so f is continuous on $[1, 5]$.

The four subintervals are

$$\begin{array}{ccccc}[a, a + \Delta x], & [a + \Delta x, a + 2\Delta x], & [a + 2\Delta x, a + 3\Delta x], & \text{and} & [a + 3\Delta x, a + 4\Delta x] \\ [1, 2], & [2, 3], & [3, 4], & \text{and} & [4, 5]\end{array}$$

As a check, you should note that $a + 4\Delta x = b$, the right endpoint. In general,

$$\begin{aligned}a + n\Delta x &= a + n\left(\frac{b - a}{n}\right) \\ &= a + b - a \\ &= b\end{aligned}$$

See Appendix D for a discussion of spreadsheets.

SPREADSHEET PROGRAM

	A	B	C	D
1	What is a?			
2	What is b?			
3	What is n?		Delta x is	+(B2-B1)/B3
4				
5	c values	f values	Sum of f values	
6	+B1+D3	[enter function here]		
7	replicate	replicate		
8				
9			@sum(B6..B9)	
10				
11				
12				
13				

```
What is a?    1
What is b?    5
What is n?    4            Delta x is 1

c values     f values      Sum of f values
    1.5              0.6
    2.5      0.71428571429
    3.5      0.77777777778
    4.5      0.81818181818 2.91024531
```

Choose the midpoints of each of these intervals:

Intervals:	$[1,2]$	$[2,3]$	$[3,4]$	$[4,5]$
Midpoints:	$c_1 = \frac{3}{2}$	$c_2 = \frac{5}{2}$	$c_3 = \frac{7}{2}$	$c_4 = \frac{9}{2}$
Heights:	$f\left(\frac{3}{2}\right) = \frac{3}{5}$	$f\left(\frac{5}{2}\right) = \frac{5}{7}$	$f\left(\frac{7}{2}\right) = \frac{7}{9}$	$f\left(\frac{9}{2}\right) = \frac{9}{11}$
Bases:	$\Delta x = 1$	$\Delta x = 1$	$\Delta x = 1$	$\Delta x = 1$

Then, the rectangular approximation for $n = 4$ is

$$\begin{aligned}\int_1^5 \frac{x}{x+1}\,dx &\approx \Delta x[f(c_1) + f(c_2) + \cdots + f(c_n)] \\ &= (1)\left[\frac{3}{5} + \frac{5}{7} + \frac{7}{9} + \frac{9}{11}\right] \\ &\approx 2.91\end{aligned}$$

■

Areas with Tabular Functions

It is possible to find the area of an irregularly shaped region by using a rectangular approximation and a function defined by a table rather than by a formula. A function defined by tabular values is called a **tabular function**.

EXAMPLE 3 Find the surface area of a swimming pool whose dimensions are shown in the following figure.

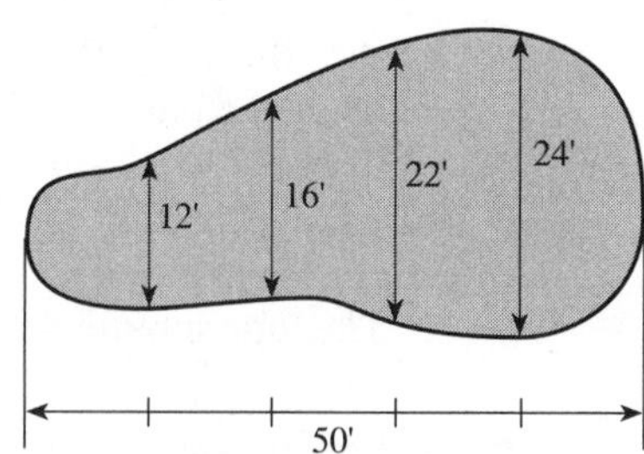

Solution We can represent the information in the sketch in tabular form by letting x be the horizontal distance (in feet) and f be the width of the pool at x (we are using the right-hand endpoints of each subinterval).

x	10	20	30	40	50
$f(x)$	12	16	22	24	0

$$\int_0^{50} f(x)\,dx \approx \Delta x[f(10) + f(20) + f(30) + f(40) + f(50)]$$

$$= 10(12 + 16 + 22 + 24 + 0) = 740$$

We would estimate the surface area of the swimming pool to be 740 ft^2. ■

The Definite Integral as a Limit of a Sum

In Example 2 we found the area under the curve $y = f(x)$ over the closed interval $[a, b]$, by partitioning the interval into n subintervals of equal width $\Delta x = \dfrac{b-a}{n}$. Then we evaluated f at the midpoint of the kth subinterval for $k = 1, 2, 3, 4$. We then formed the approximating sum S_4 of the areas of the rectangles. Instead of taking 4 rectangles, consider n rectangles and the sum S_n. Since we expect the estimates S_n to improve as Δx decreases, we *define* the area A under the curve, above the x-axis, bounded by the lines $x = a$ and $x = b$ to be the limit of S_n as $\Delta x \to 0$.

This approach to the area problem contains the essentials of integration, but there is no compelling reason for the partition points to be evenly spaced or to insist on evaluating f at midpoints. These conventions are for convenience of computation, and to accommodate applications other than area, it is necessary to consider a more general type of approximating sum and to specify what is meant by the limit of such a sum. The approximating sums that occur in integration problems are called **Riemann sums**, and the following definition contains a step-by-step description of how such sums are formed.

Riemann Sum

Suppose a continuous function f is given, along with a closed interval $[a, b]$ on which f is defined. Then:

Step 1: Partition the interval $[a, b]$ into n subintervals by choosing points $\{x_0, x_1, \ldots, x_n\}$ arranged so that

$$a = x_0 < x_1 < x_2 < \cdots < x_{n-1} < x_n = b$$

Call this partition P. For $k = 1, 2, \ldots, n$, the kth subinterval width is $\Delta x_k = x_k - x_{k-1}$. The largest of these widths is called the **norm** of the partition P and is denoted by $\|P\|$; that is,

$$\|P\| = \max_{k=1,2,\ldots,n} \{\Delta x_k\}$$

Step 2: Choose a number arbitrarily from each subinterval. For $k = 1, 2, \ldots, n$, the number x_k^* chosen from the kth subinterval is called the kth *subinterval representative* of the partition P.

Step 3: Form the sum:

$$R_n = f(x_1^*)\Delta x_1 + f(x_2^*)\Delta x_2 + \cdots + f(x_n^*)\Delta x_n$$

$$= \sum_{k=1}^{n} f(x_k^*)\Delta x_k$$

This is the **Riemann sum** associated with f, the given partition P, and the chosen subinterval representatives $x_1^*, x_2^*, \ldots, x_n^*$.

EXAMPLE 4 Suppose the interval $[-2, 1]$ is partitioned into 6 subintervals with subdivision points

$$a = x_0 = -2$$
$$x_1 = -1.6$$
$$x_2 = -.93$$
$$x_3 = -.21$$
$$x_4 = .35$$
$$x_5 = .82$$
$$x_6 = 1 = b$$

Find the norm of this partition P and the Riemann sum associated with the function $f(x) = 2x$, the given partition, and the subinterval representatives $x_1^* = -1.81$, $x_2^* = -1.12$, $x_3^* = -.55$, $x_4^* = -.17$, $x_5^* = .43$, $x_6^* = .94$.

Solution Before we can find the norm of the partition or the required Riemann sum, we must compute the subinterval width Δx_k and evaluate f at each subinterval representative x_k^*. These values are shown in Figure 14.8 and the computations are shown at the top of the next page.

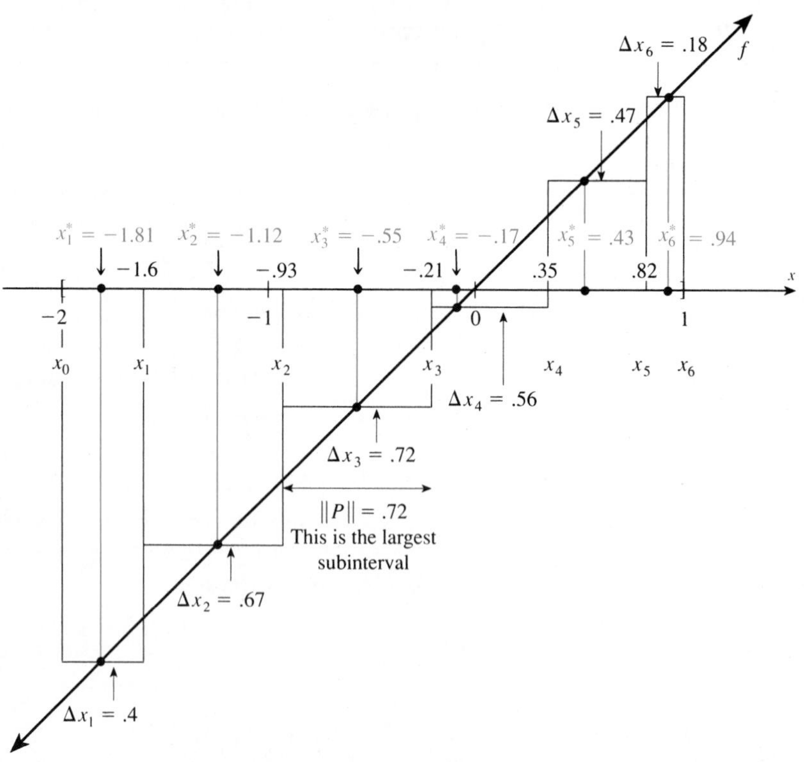

FIGURE 14.8
Graphical representation of a Riemann sum

k	$\overbrace{x_k - x_{k-1}}^{\text{Given}} = \Delta x_k$	$\overset{\text{Given}}{\downarrow}$ x_k^*	$f(x_k^*) = \overbrace{2x_k^*}^{\text{Given}}$
1	$-1.6 - (-2) = .40$	-1.81	$f(-1.81) = -3.62$
2	$-.93 - (-1.6) = .67$	-1.12	$f(-1.12) = -2.24$
3	$-.21 - (-.93) = .72$	$-.55$	$f(-.55) = -1.10$
4	$.35 - (-.21) = .56$	$-.17$	$f(-.17) = -.34$
5	$.82 - .35 = .47$	$.43$	$f(.43) = .86$
6	$1.00 - .82 = .18$	$.94$	$f(.94) = 1.88$

From the table, we see that the largest subinterval width is $\Delta x_3 = .72$, so the partition has norm $\|P\| = .72$. Finally, by using the definition, we find the Riemann sum:

$$R_6 = (-3.62)(.4) + (-2.24)(.67) + (-1.10)(.72) + (-.34)(.56) + (.86)(.47) + (1.88)(.18)$$
$$= -3.1886$$

■

WARNING Notice from Example 4 that the Riemann sum does not necessarily represent an area. The sum found is negative (and areas must be nonnegative). Also notice from Figure 14.8 that the nonnegativity requirement for the function in the definition of area is not met.

Definition of Definite Integral as a Riemann Sum

By comparing the formula for Riemann sum with that of area at the beginning of this section, we recognize that the sum S_n used to approximate area is actually a special kind of Riemann sum which has

$$\Delta x_k = \Delta x = \frac{b-a}{n} \qquad \text{and} \qquad x_k^* = a + k\,\Delta x$$

for $k = 1, 2, \ldots, n$. Since each subinterval in the partition P associated with S_n has width Δx, the norm of the partition is

$$\|P\| = \Delta x = \frac{b-a}{n}$$

This kind of partition is called a **regular partition**. When we express the area under the curve $y = f(x)$ as $A = \lim_{\Delta x \to 0} S_n$, we are actually saying that A can be estimated to any desired accuracy by finding a Riemann sum of the form S_n with norm

$$\|P\| = \frac{b-a}{n}$$

sufficiently small. We use this interpretation as a model for the following definition.

Definite Integral

If f is defined on the closed interval $[a, b]$ we say f is **integrable** on $[a, b]$ if

$$I = \lim_{\|P\| \to 0} f(x_1)\Delta x_1 + f(x_2)\Delta_2 + \cdots + f(x_n)\Delta x_n$$

$$= \lim_{\|P\| \to \infty} \sum_{k=1}^{n} f(x_k^*)\Delta x_k$$

exists. This limit is called the **definite integral** of f from a to b. The definite integral is denoted by

$$I = \int_a^b f(x)\,dx$$

In other words, if the limit of the Riemann sum of f exists, then we say that f is *integrable.* In particular, it means that the number I can be approximated to any prescribed degree of accuracy by any Riemann sum with norm sufficiently small.

The last step in putting together this Riemann sum definition and the ideas developed earlier in this chapter, is a result that forms the basis for much of the calculus. It reformulates the informal definition of definite integral given in Section 14.3 as a Riemann sum, and is known as the **Fundamental Theorem of Calculus**.

Fundamental Theorem of Calculus

If a function f is continuous on an interval $[a, b]$, then

$$\int_a^b f(x)\,dx = F(b) - F(a)$$

where F is any antiderivative of f.

The fundamental theorem of calculus is the most important theorem of calculus. It ties together the ideas of limits, derivatives, areas, and antiderivatives. What good is all of this? Just as the derivative was motivated by looking at a rate of change, but was then found to be useful in many settings, so we see that the limit

$$\lim_{\|P\| \to 0} [\Delta x_1 \cdot f(c_1) + \Delta x_2 \cdot f(c_2) + \cdots + \Delta x_n \cdot f(c_n)] = \lim_{\|P\| \to 0} \sum_{k=1}^{n} f(c_k)\Delta x_k$$

occurs in many practical problems. Just as the derivative is not easy to find from the definition, so is the integral not easy to find from the definition. However, once we recognize a particular application as a derivative, or as an integral, we can apply the appropriate derivative or integral rules. The following application shows how we can do this for a Riemann sum.

Average Value of a Function

Suppose f is a continuous function over some interval $[a, b]$, and we wish to find the average value of f over this interval. For example, suppose a typist's speed in words

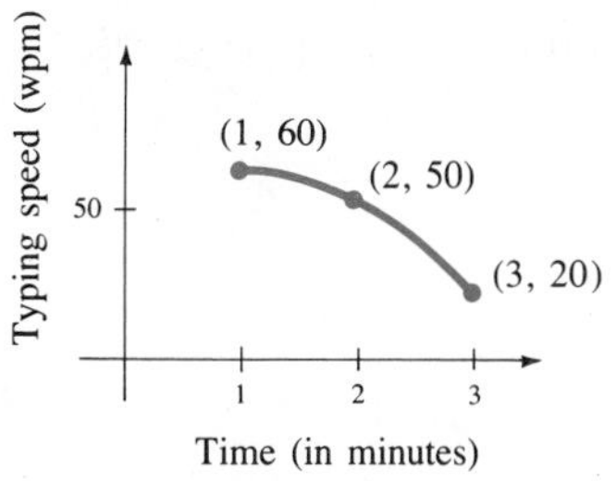

FIGURE 14.9
Typing speed as a function of the length of time of the typing test

per minute is a function of the length of a typing test which is administered, as shown in Figure 14.9. What is the typist's average speed for the three timed tests?

$$\text{Average} = \frac{60 + 50 + 20}{3} = \frac{130}{3}\text{ wpm}$$

However, this average does not accurately represent the situation because there are infinitely many values for t over the interval $[1, 3]$. That is, suppose that typing speed $w(t)$ is represented by the continuous function

$$w(t) = -10t^2 + 20t + 50 \qquad \text{for } t \text{ in } [1, 3]$$

Suppose we divide the interval into n subintervals, and let c_k be any point in the kth subinterval. Then the average is given by

$$\text{Average} = \frac{1}{n}[w(c_1) + w(c_2) + \cdots + w(c_n)]$$

Now the limit as $n \to \infty$ is the average speed over the time interval $[1, 3]$. This looks like the formula for a Riemann sum except it does not have the Δx_k. For this example, we are looking for

$$\Delta t = \frac{b - a}{n}$$

Multiply the expression on the right by 1 written as $\frac{b - a}{b - a}$:

$$\begin{aligned}\text{Average} &= \frac{1}{n} \cdot \frac{b - a}{b - a}[w(c_1) + w(c_2) + \cdots + w(c_n)] \\ &= \frac{1}{b - a} \cdot \frac{b - a}{n}[w(c_1) + w(c_2) + \cdots + w(c_n)] \\ &= \frac{1}{b - a}\Delta t[w(c_1) + w(c_2) + \cdots + w(c_n)]\end{aligned}$$

Thus, over the interval $[a, b]$,

$$\begin{aligned}\text{Average} &= \lim_{\|P\| \to 0} \frac{1}{b - a}\Delta t[w(c_1) + w(c_2) + \cdots + w(c_n)] \\ &= \frac{1}{b - a}\lim_{n \to \infty} \Delta t[w(c_1) + w(c_2) + \cdots + w(c_n)]\end{aligned}$$

We see that this is a Riemann sum, so by the definition of a definite integral we have

$$\text{Average value over the interval } [a, b] = \frac{1}{b - a}\int_a^b w(t)\,dt$$

The following definition of average value is motivated by the previous discussion.

Average Value

> The **average value** of a continuous function f over $[a, b]$ is
>
> $$\frac{1}{b - a}\int_a^b f(x)\,dx$$

EXAMPLE 5 If the typist's speed in the above discussion is given by the continuous function

$$w(t) = -10t^2 + 20t + 50 \qquad \text{for } t \text{ in } [1, 3]$$

find the average typing speed in words per minute.

Solution The average is

$$\begin{aligned}\frac{1}{3-1}\int_1^3 (-10t^2 + 20t + 50)\,dt &= \frac{1}{2}\left[\frac{-10t^3}{3} + \frac{20t^2}{2} + 50t\right]_1^3 \\ &= \frac{1}{2}\left[(-90 + 90 + 150) - \left(-\frac{10}{3} + 10 + 50\right)\right] \\ &\approx 47 \text{ wpm}\end{aligned}$$

■

EXAMPLE 6 Suppose the cost function for producing x items is

$$C(x) = 5{,}000 + 8x$$

a. What is the total cost of producing 100 items?

b. What is the average cost per unit if 100 items are produced?

c. What is the average value of the cost function over the interval [0, 100]?

Solution **a.** The total cost of producing 100 items is $C(100) = 5{,}000 + 8(100) = 5{,}800$.

b. The average cost per unit is $\dfrac{c(100)}{100} = \dfrac{5{,}800}{100} = 58$.

c. The average value of the cost function over [0, 100] is

$$\frac{1}{100-0}\int_0^{100} (5{,}000 + 8x)\,dx = \frac{1}{100}(5{,}000x + 4x^2)\Big|_0^{100} = 5{,}400$$

■

Notice that the average value of the cost function (part **c** of Example 6) is considerably different from the average cost of an item (part **b**). This is because the cost value gives the average value of producing 100 items.

EXAMPLE 7 Given the demand function

$$p(x) = 50e^{.03x}$$

find the average value of the price (in dollars) over the interval [0, 100].

Solution Average value of the price

$$\begin{aligned}&= \frac{1}{100-0}\int_0^{100} 50e^{.03x}\,dx \\ &= \frac{1}{2}\int_0^{100} e^{.03x}\,dx \\ &= \frac{e^{.03x}}{.06}\Big|_0^{100} \\ &= \frac{50}{3}(e^{.03(100)} - e^{.03(0)}) \\ &= \frac{50}{3}(e^3 - 1) \approx 318.09\end{aligned}$$

The average value is approximately \$318. ■

14.5

Problem Set

In Problems 1–6 find the area bounded by the x-axis, the curve $y = f(x)$, and the given vertical lines by

a. *using an antiderivative.*
b. *sketching the curve, and then by doing a rectangular approximation (correct to two decimal places) by choosing $n = 4$, and c_k to be the left endpoint of each subinterval.*

1. $y = x^2 + 1, x = -2, x = 2$
2. $y = x^2 + 2, x = -2, x = 2$
3. $y = 3x^2, x = 1, x = 4$
4. $y = 6x^2, x = 1, x = 4$
5. $y = 4 - x^2, x = 0, x = 2$
6. $y = 9 - x^2, x = -1, x = 3$

Evaluate the integrals in Problems 7–12 by using a rectangular approximation. Give your answer to two decimal places by using a rectangular approximation with $n = 4$ and with Δ_k the midpoint of each subinterval.

7. $\int_1^5 \frac{1}{(2x+1)^2}\,dx$
8. $\int_1^5 \frac{1}{2x+1}\,dx$
9. $\int_0^2 e^{.5x}\,dx$
10. $\int_1^5 \frac{\ln x}{x}\,dx$
11. $\int_0^{1/2} \frac{x}{\sqrt{1+3x^2}}\,dx$
12. $\int_0^{1/2} \frac{1}{\sqrt{1+3x^2}}\,dx$

Evaluate the integrals for the tabular functions in Problems 13–16.

13. $\int_0^6 f(x)\,dx$

x	1	3	5
$f(x)$	2.4	5.5	6.1

14. $\int_1^9 f(x)\,dx$

x	2	4	6	8
$f(x)$	4.9	5.1	4.3	6.2

15. $\int_5^{45} f(x)\,dx$

x	10	20	30	40
$f(x)$	130	180	110	150

16. $\int_0^{.6} f(x)\,dx$

x	.1	.2	.3	.4	.5
$f(x)$	.32	12.1	8.2	2.0	.55

Find the average value of each function in Problems 17–24 over the indicated interval.

17. $f(x) = 2x^2 + 1, \quad [0,4]$
18. $f(x) = 3x^2 + 2, \quad [2,5]$
19. $f(x) = 300 - 4x, \quad [0,10]$
20. $f(x) = 4x - 5x^3, \quad [-1,2]$
21. $f(x) = \sqrt[3]{x}, \quad [1,8]$
22. $f(x) = \sqrt{2x+3}, \quad [3,11]$
23. $f(x) = e^{-.5x}, \quad [0,10]$
24. $f(x) = 128e^{.04x}, \quad [0,10]$

In Problems 25–36 find $F'(x)$.

25. $F(x) = \int_0^x (t+5)\,dt$
26. $F(x) = \int_0^x t(t^2+1)^3\,dt$
27. $F(x) = \int_{-3}^x (t+5)^2\,dt$
28. $F(x) = \int_1^x (t^2+4)\,dt$
29. $F(x) = \int_4^x \sqrt{t}\,dt$
30. $F(x) = \int_1^x e^{3t}\,dt$
31. $F(x) = \int_0^x (t^3 - 3t^2 + 5t - 7)\,dt$
32. $F(x) = \int_{-3}^x (t^2 - 2t + 5)\,dt$
33. $F(x) = \int_8^x \sqrt[3]{t}\,dt$
34. $F(x) = \int_0^x \sqrt[4]{t+3}\,dt$
35. $F(x) = \int_0^x t^2(t+1)^3\,dt$
36. $F(x) = \int_2^x t(3t^4-1)^2\,dt$

APPLICATIONS

37. The graph below, from the March 5, 1986, issue of the *Wall Street Journal*, shows the rate of U.S. automobile sales. Estimate the total sales for July–December.

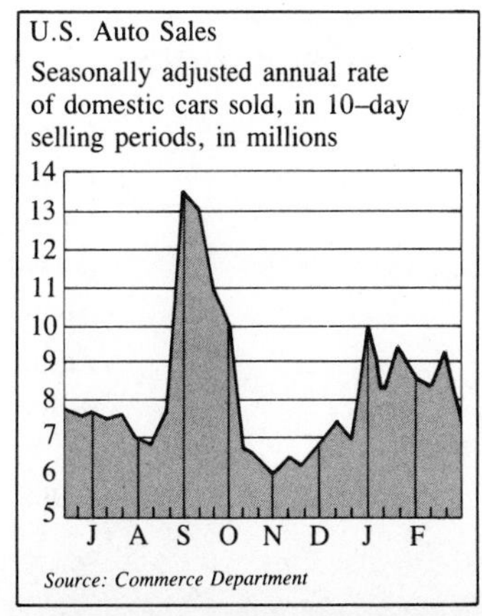

38. Use the graph in Problem 37 to estimate the total automobile sales for the last quarter of the year (Oct. 1 to Dec. 31).

39. The graph below shows the rate of paper and paperboard production in the United States as reported by the American Paper Institute. Estimate the total production for 1985.

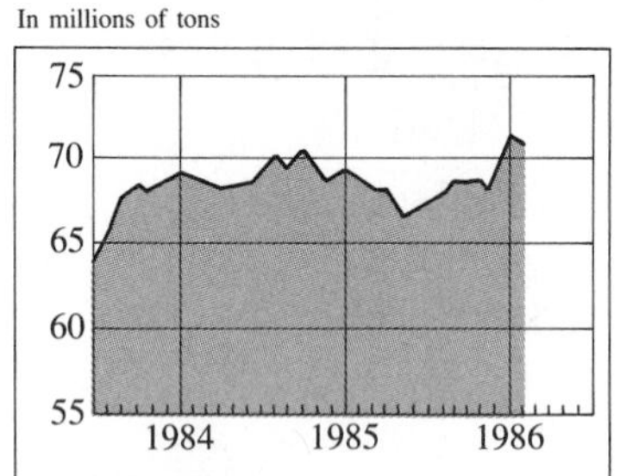

40. If the supply function is

$$p = S(x) = 10(e^{.03x} - 1)$$

what is the average price (in dollars) over the interval $[10, 50]$?

41. The number of rabbits in a limited geographical area (such as an island) is approximated by

$$P(t) = 500 + t - .25t^2$$

for the number t of years on the interval $[0, 5]$. What is the average number of rabbits in the area over the 5-year time period?

42. The world population (in billions) is given by

$$P(t) = 5e^{.03t} \qquad (t = 0 \text{ in } 1987)$$

What is the average population of the earth over the next 30 years?

43. The temperature (in degrees Fahrenheit) varies according to the formula

$$F(t) = 6.44t - .23t^2 + 30$$

Find the average daily temperature if t is the time of day (in hours) measured from midnight.

44. Repeat Problem 43 for the daylight hours—that is, for t on the interval $[6, 20]$.

45. Estimate the surface area of the following swimming pool.

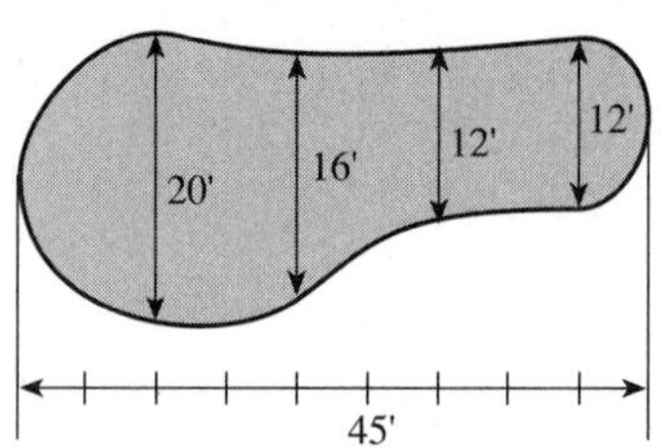

46. A surveyor made the following measurements on a parcel of land.

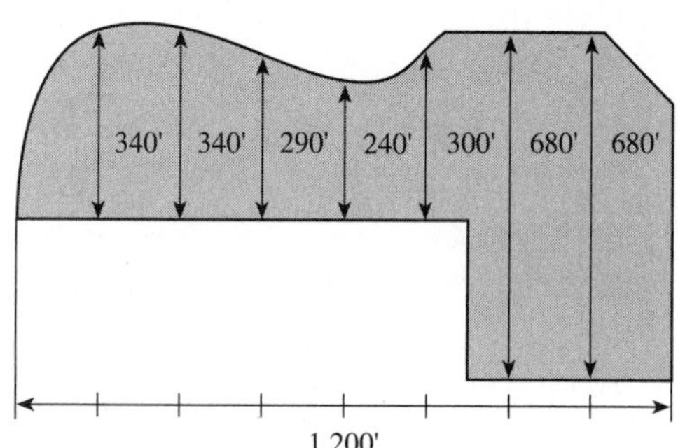

Estimate the area. (One acre is 43,560 ft^2.)

*14.6 Numerical Integration

Calculators or computers are being used more and more to evaluate definite integrals (see the Modeling Application at the beginning of Chapter 14). **Numerical integration** is the process of finding a numerical approximation for a definite integral *without actually carrying out the antidifferentiation process.* This means that you can easily write a computer or calculator program to evaluate definite integrals if you understand numerical integration. Numerical integration is also useful when it is difficult or impossible to find an antiderivative to carry out the integration process.

* Optional section.

Suppose a function f is continuous on $[a, b]$. Then there are three common approximations to the definite integral:

Approximations for the Definite Integral

$\int_a^b f(x)\,dx$ can be approximated by

RECTANGLES: $A_n = \Delta x[f(x_0) + f(x_1) + \cdots + f(x_{n-1})]$

TRAPEZOIDS: $T_n = \dfrac{\Delta x}{2}[f(x_0) + 2f(x_1) + 2f(x_2) + \cdots + 2f(x_{n-1}) + f(x_n)]$

PARABOLAS: $P_n = \dfrac{\Delta x}{3}[f(x_0) + 4f(x_1) + 2f(x_2) + \cdots + 4f(x_{n-1}) + f(x_n)]$
(true for n an even integer)

where $\Delta x = \dfrac{b-a}{n}$ for some positive integer n. (Δx is read "delta x.")

Even though there are many other approximation schemes, these will suffice for our needs. While we will not prove these approximations, it is instructive to see where these formulas come from.

Rectangular Approximation

We considered this method in the preceding section. Recall that we need to let f be a function that is continuous on a closed interval $[a, b]$. Suppose further that $a < b$, and that f is nonnegative on the interval, as shown in Figure 14.10.

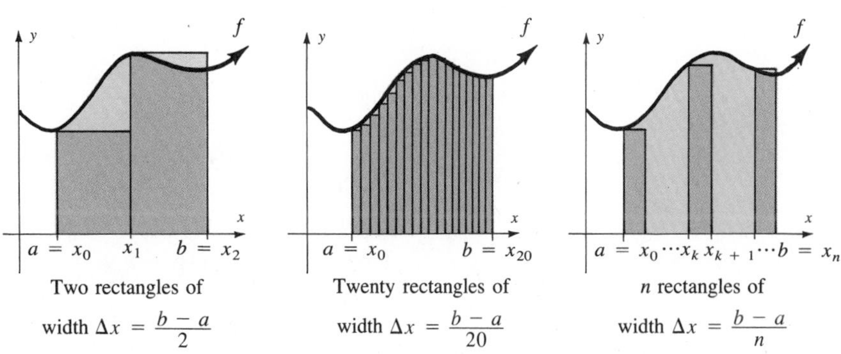

FIGURE 14.10
Rectangular approximation of the definite integral

Divide the interval into n subintervals, each of length $\Delta x = \dfrac{b-a}{n}$. The endpoints of these subintervals are

$$a, \quad a + \Delta x, \quad a + 2\Delta x, \quad a + 3\Delta x, \quad \ldots, \quad a + n\Delta x$$

which are, for convenience, labeled $x_0, x_1, x_2, \ldots, x_n$, respectively. The area of the kth rectangle is

$$f(x_{k-1})\Delta x$$

If you add the areas of all the n rectangles, you clearly obtain A_n.

EXAMPLE 1 Let $f(x) = x^\pi + 1$. Find an approximation for

$$\int_1^3 (x^\pi + 1)\,dx$$

by using a rectangular approximation for $n = 1, 2, 4$, and 8. (We pick 1, 2, 4, 8, ... only as a matter of convenience; *any* subdivision will suffice.)

Solution *First approximation, $n = 1$:*

$$\begin{aligned} A_1 = \int_1^3 (x^\pi + 1)\,dx &\approx f(x_0)\Delta x \\ &= f(1)(2) \\ &= (1^\pi + 1)(2) \\ &= 4 \end{aligned}$$

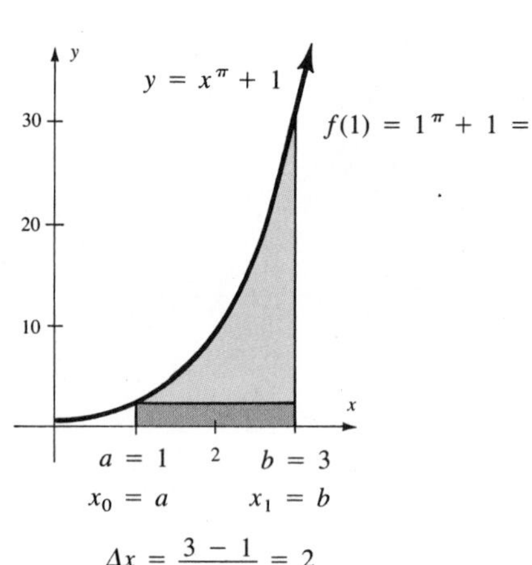

Second approximation, $n = 2$:

$$\begin{aligned} A_2 &= \int_1^3 (x^\pi + 1)\,dx \\ &\approx f(x_0)\Delta x + f(x_1)\Delta x \\ &= f(1)(1) + f(2)(1) \\ &= (1^\pi + 1) + (2^\pi + 1) \\ &\approx 2 + 9.82 \\ &= 11.82 \end{aligned}$$

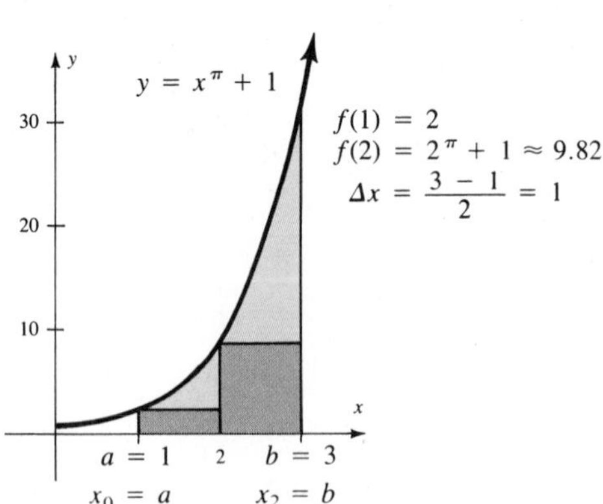

Third approximation, $n = 4$:

$$A_4 = \int_1^3 (x^\pi + 1)\,dx$$
$$\approx f(1)(.5) + f(\tfrac{3}{2})(.5) + f(2)(.5) + f(\tfrac{5}{2})(.5)$$
$$\approx 2(.5) + 4.57(.5) + 9.82(.5) + 18.79(.5)$$
$$\approx 17.59$$

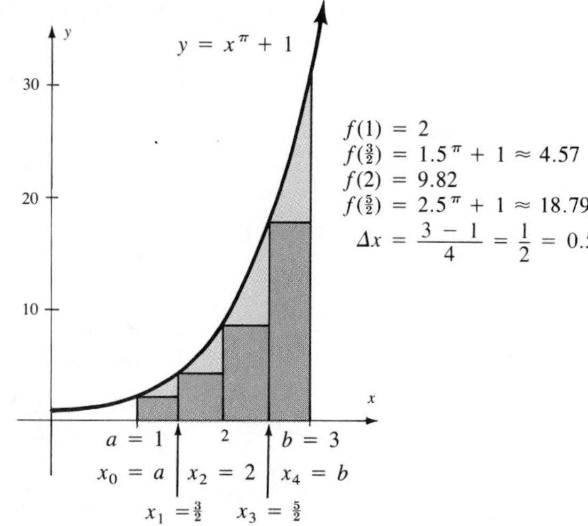

Notice that as $n \to \infty$, the approximate area (gray) is more closely approximating the actual area (color).

Fourth approximation, $n = 8$:

$$A_8 = \int_1^3 (x^\pi + 1)\,dx \approx f(1)(.25) + f(1.25)(.25) + f(1.5)(.25) + \cdots + f(2.75)(.25)$$
$$\approx .25[f(1) + f(1.25) + f(1.5) + f(1.75) + f(2) + f(2.25) + f(2.5) + f(2.75)]$$
$$\approx .25(83.78)$$
$$\approx 20.95$$

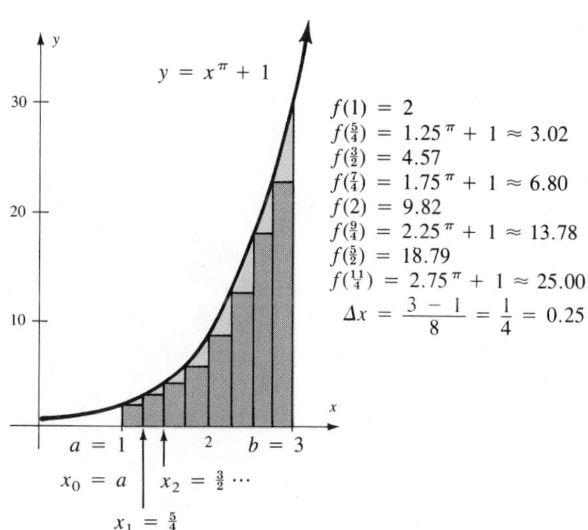

■

COMPUTER APPLICATION

One of the most useful computer programs in integral calculus is called *Numerical integration.* This program will find rectangular approximations for left endpoints, right endpoints, midpoints, as well as trapezoidal and Simpson's approximations. In addition, look at Appendix D for a spreadsheet application of these approximation methods.

Trapezoidal Approximation

In order to carry out a trapezoidal approximation, you simply approximate the area represented by the definite integral using trapezoids instead of rectangles, as shown in Figure 14.11. To derive the trapezoidal approximation, use the formula for the area of a trapezoid:

$$A = \left(\frac{b_1 + b_2}{2}\right)h$$

for bases b_1 and b_2 and height h. From this formula you can find the area of the kth trapezoid:

$$\frac{f(x_{k-1}) + f(x_k)}{2}\Delta x$$

$\Delta x = h$ in the trapezoidal formula $b_1 = f(x_{k-1})$ and $b_2 = f(x_k)$

Therefore, the sum of the areas of all the trapezoids is

$$T_n = \left[\frac{f(x_0) + f(x_1)}{2} + \frac{f(x_1) + f(x_2)}{2} + \cdots + \frac{f(x_{n-1}) + f(x_n)}{2}\right]\Delta x$$

$$= \frac{\Delta x}{2}[f(x_0) + 2f(x_1) + 2f(x_2) + \cdots + 2f(x_{n-1}) + f(x_n)]$$

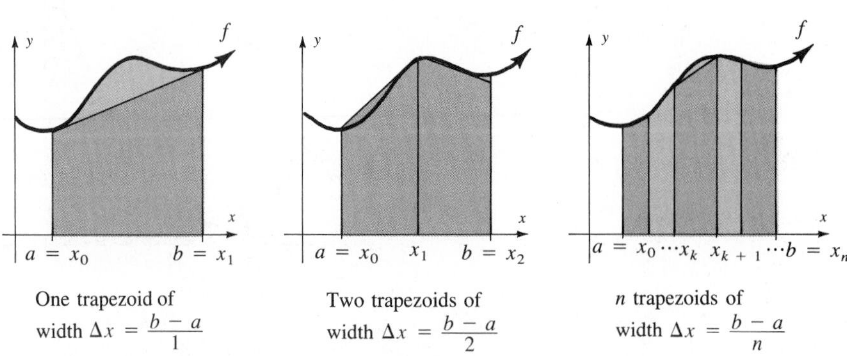

FIGURE 14.11
Trapezoidal approximation of the definite integral

EXAMPLE 2 Let $f(x) = x^\pi + 1$. Find an approximation for

$$\int_1^3 (x^\pi + 1)\,dx$$

by using a trapezoidal approximation for $n = 1$ and $n = 4$.

Solution *First approximation, $n = 1$:*

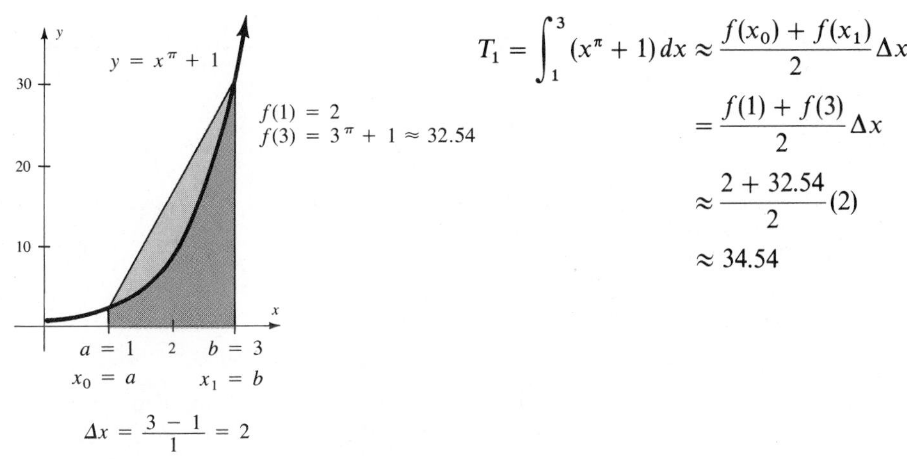

$$T_1 = \int_1^3 (x^\pi + 1)\,dx \approx \frac{f(x_0) + f(x_1)}{2}\Delta x$$
$$= \frac{f(1) + f(3)}{2}\Delta x$$
$$\approx \frac{2 + 32.54}{2}(2)$$
$$\approx 34.54$$

Second approximation, $n = 4$:

$$T_4 = \int_1^3 (x^\pi + 1)\,dx \approx \frac{.5}{2}[f(1) + 2f(1.5) + 2f(2) + 2f(2.5) + f(3)]$$
$$\approx .25(2 + 9.14 + 19.65 + 37.58 + 32.54)$$
$$\approx .25(100.92)$$
$$\approx 25.23$$

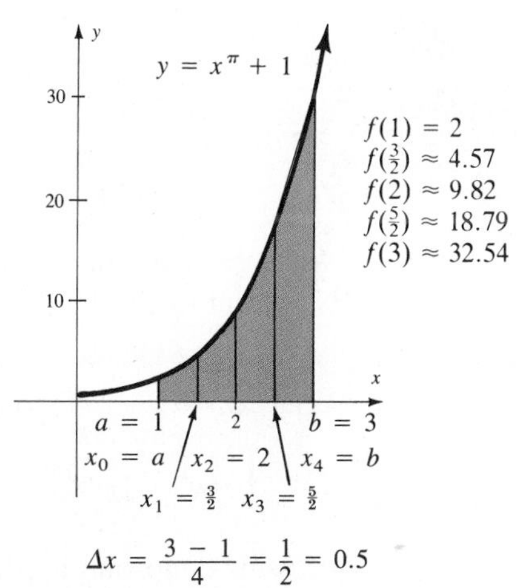

■

COMPUTER APPLICATION

In the *Numerical integration* program accompanying this book, you have the following options:

Rectangular approximation: It uses midpoints.
Trapezoidal approximation: Apply Trapezoidal rule.
Parabola approximation: Apply Simpson's rule.

For Example 2 we obtain:

n	Trapezoid	Midpoint	Simpson
1	34.54428070	19.64995565	24.61473066
2	27.09711817	23.36399996	24.60837270
4	25.23055907	24.29665927	24.60795920
8	24.76360917	24.53009496	24.60793303
16	24.64685206	24.58847105	24.60793138
32	24.61766155	24.60306615	24.60793128

Parabolic Approximation

The parabolic approximation, also known as **Simpson's rule**, uses parabolas to approximate the area under the graph of a function, as shown in Figure 14.12.

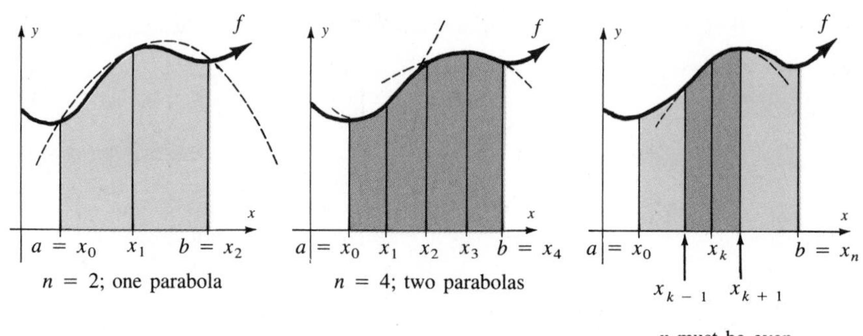

FIGURE 14.12
Parabolic approximation of the definite integral

We will not attempt to derive Simpson's rule, but instead illustrate it in Example 3. Note that to use Simpson's rule, n must be even.

EXAMPLE 3 Let $f(x) = x^{\pi} + 1$. Find an approximation for

$$\int_1^3 (x^{\pi} + 1)\,dx$$

by using Simpson's approximation for $n = 6$.

Solution First, gather the necessary values,

$$\Delta x = \frac{3-1}{6} = \frac{1}{3}$$

and $f(1) = 2$, $f(\frac{4}{3}) \approx 3.47$, $f(\frac{5}{3}) \approx 5.98$, $f(2) \approx 9.82$, $f(\frac{7}{3}) \approx 15.32$, $f(\frac{8}{3}) \approx 22.79$, $f(3) \approx 32.54$. Thus the integral is approximated by

$$P_6 \approx \frac{\frac{1}{3}}{3}[2 + 4(3.47) + 2(5.98) + 4(9.82) + 2(15.32) + 4(22.79) + 32.54]$$

$$\approx \frac{1}{9}(221.46) \approx 24.61$$

■

If you compare Examples 1–3 with the exact value of the definite integral, you will note that each successive approximation is better than the previous one and that the larger the value for n the more accurate the results. For the definite integral in Examples 1–3 we see that

$$\int_1^3 (x^\pi + 1)\,dx = \left.\frac{x^{\pi+1}}{\pi+1} + x\right|_1^3 \approx 24.60793128$$

EXAMPLE 4 Use Simpson's rule with $n = 4$ to approximate the value of

$$\int_0^1 \frac{dx}{x+1}$$

to four decimal places. Check your answer by finding the exact value of the definite integral.

Solution $\Delta x = \dfrac{b-a}{n} = \dfrac{1-0}{4} = .25$

Therefore, since $f(x) = \dfrac{1}{x+1}$,

$$P_4 = \int_0^1 \frac{dx}{x+1} = \frac{.25}{3}[f(x_0) + 4f(x_1) + 2f(x_2) + 4f(x_3) + f(x_4)]$$

$$= \frac{1}{12}[f(0) + 4f(.25) + 2f(.5) + 4f(.75) + f(1)]$$

$$\approx \frac{1}{12}[1 + 4(.8) + 2(.6) + 4(.5714285714) + .5]$$

$$\approx \frac{1}{12}(8.319047619)$$

$$\approx .6932539683 \quad \text{or} \quad .6933 \quad \text{(to four decimal places)}$$

The exact value of the definite integral is found as follows:

$$\int_0^1 \frac{dx}{x+1} = \left|\ln(x+1)\right|_0^1 = \ln 2 - \ln 1 = \ln 2$$

(This is .6931 to four decimal places.)

■

14.6 Problem Set

Consider $\int_1^6 \sqrt{x+3}\,dx$. *Then*

$$\int_1^6 \sqrt{x+3}\,dx = \frac{2}{3}(x+3)^{3/2}\Big|_1^6 = \frac{2}{3}(27) - \frac{2}{3}(8) = \frac{38}{3}$$

Approximate this integral by finding the requested values in Problems 1–12.

1. A_1
2. A_2
3. A_3
4. A_4
5. T_1
6. T_2
7. T_3
8. T_4
9. P_2
10. P_4
11. P_6
12. T_8

Consider $\int_2^4 \frac{x\,dx}{(1+2x)^2}$. *Then*

$$\int_2^4 \frac{x\,dx}{(1+2x)^2} = \frac{1}{4}\left[\ln|1+2x| + \frac{1}{1+2x}\right]_2^4$$
$$= .25\left(\ln 9 + \frac{1}{9}\right) - .25\left(\ln 5 + \frac{1}{5}\right)$$
$$\approx .5770839221 - .4523594781$$
$$\approx .124724444$$

Approximate this integral by finding the requested values in Problems 13–24.

13. A_1
14. A_2
15. A_3
16. A_4
17. T_1
18. T_2
19. T_3
20. T_4
21. P_2
22. P_4
23. P_6
24. P_8

Approximate the integrals in Problems 25–28 by using a rectangular approximation (pick left endpoints). Round your answer to two decimal places using $n = 4$.

25. $\int_3^7 \frac{\sqrt{1+x}}{x^3}\,dx$
26. $\int_1^4 \ln x\,dx$
27. $\int_1^3 x^x\,dx$
28. $\int_0^2 e^{-x^2}\,dx$

Approximate the integrals in Problems 29–32 by using a trapezoidal approximation. Round your answer to two decimal places using $n = 4$.

29. $\int_3^7 \frac{\sqrt{1+x}}{x^3}\,dx$
30. $\int_1^4 \ln x\,dx$
31. $\int_1^3 x^x\,dx$
32. $\int_0^2 e^{-x^2}\,dx$

Approximate the integrals in Problems 33–36 by using Simpson's rule. Round your answer to two decimal places using $n = 4$.

33. $\int_3^7 \frac{\sqrt{1+x}}{x^3}\,dx$
34. $\int_1^4 \ln x\,dx$
35. $\int_1^3 x^x\,dx$
36. $\int_0^2 e^{-x^2}\,dx$

APPLICATIONS

37. A landscape architect needs to estimate the number of cubic yards of fill necessary to level the area shown below. Use Simpson's rule with $\Delta x = 5$ feet and y values equal to the distances measured in the figure to find the surface area. Assume that the depth averages 3 feet. (*Note:* 1 cubic yard equals 27 cubic feet.)

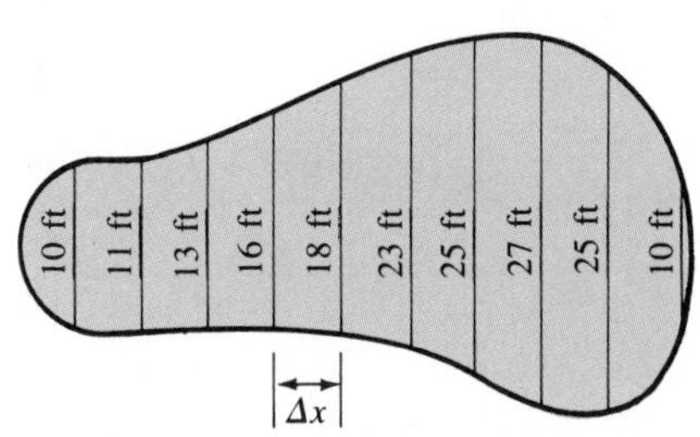

Each contour line is 5 feet from the others

38. The rate at which flashbulbs give off light varies during the flash. For some bulbs, the light output, measured in lumens, reaches a peak and fades quickly (as shown in part **a** of the figure below) and for others the light output stays level for a relatively longer period of time (shown in part **b**).

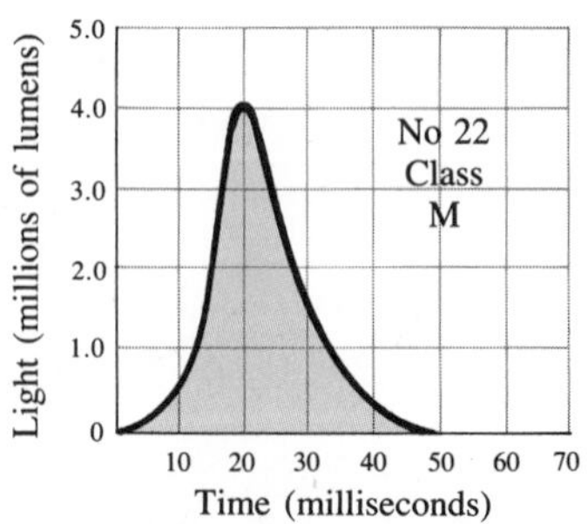

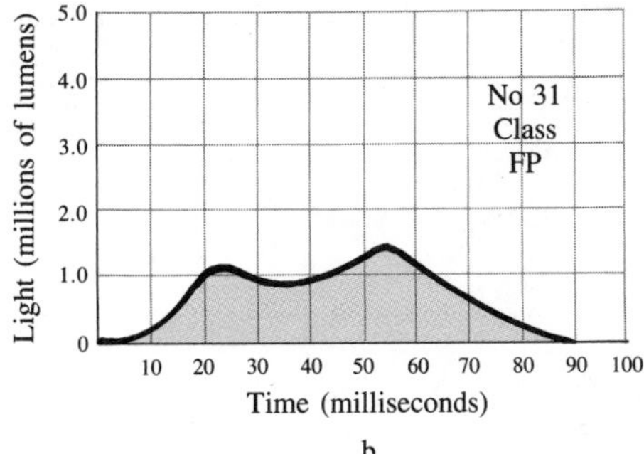

b

The amount A of light emitted by the flashbulb in the interval from 20 to 70 milliseconds after the button is pressed is given by the formula

$$A = \int_{20}^{70} L(t)\,dt \text{ lumens-milliseconds}$$

where $L(t)$ is the lumen output of the bulb as a function of time. Use the trapezoidal rule and the numerical data from the table below to determine which bulb gets more light to the film.*

* From W. U. Walton et al., *Integration* (Newton, MA: Project CALC, Education Development Center, 1975), p. 83. Data in the table from *Photographic Lamp and Equipment Guide*, P4–15P, General Electric Company, Cleveland, Ohio. The material is quoted from Thomas and Finney, *Calculus and Analytic Geometry*, 6th ed. (Reading, MA: Addison-Wesley Publishing Company, 1984), pp. 311–312.

Light Output Versus Time

Time after ignition (in milliseconds)	Light output (#22 bulb)	Light output (#31 bulb)
0	0	0
5	.2	.1
10	.5	.3
15	2.6	.7
20	4.2	1.0
25	3.0	1.2
30	1.7	1.0
35	.7	.9
40	.35	1.0
45	.2	1.1
50		1.3
55		1.4
60		1.3
65		1.0
70		.8
75		.6
80		.3
85		.2
90		0

*14.7 Review

The material of this chapter is reviewed in the following list of objectives. After each objective there are some practice questions. For a sample test select the first question of each set and check your answers. The second question for each objective has no answer given. If you are having trouble with a particular type of problem, look back at the indicated section in the text. When you are finished reviewing these objectives, a sample examination is given at the end of this section.

[14.1]
Objective 14.1: *Evaluate an indefinite integral using one of the three direct integration formulas.*

1. $\int x^6\,dx$ **2.** $\int \frac{dx}{x}$

3. $\int e^u\,du$ **4.** $\int u^{-5}\,du$

Objective 14.2: *Evaluate an indefinite integral using one of the integration formulas.*

5. $\int (6x^2 - 2x - 5)\,dx$ **6.** $\int \frac{du}{12}$

7. $\int \frac{(x-2)\,dx}{3x^2 - 5x - 2}$ **8.** $\int \left(e^{5x+1} + \frac{1}{e}\right)dx$

Objective 14.3: *Find an antiderivative that satisfies given initial conditions.*

9. $f(x) = e^x + 9;\quad F(0) = 10$

10. $f(x) = 3x^2 - 9;\quad F(10) = 0$

* Optional section.

11. $f(x) = \frac{1}{x}$; $F(1) = 0$

12. $f(x) = \frac{x^2 - 9}{x + 3}$; $F(10) = 100$

Objective 14.4: *Find the cost function, given the marginal cost function.*

13. $C'(x) = 9x^2 + 1$; fixed cost, \$25,000

14. $C'(x) = 250 - 8x$; fixed cost, \$8,000

15. $C'(x) = 10 - \frac{1}{\sqrt{x}}$; fixed cost, \$1,500

16. $C'(x) = .005x^3$; fixed cost, \$35,000

[14.2]

Objective 14.5: *Find indefinite integrals by substitution.*

17. $\int\left(e^{-3x} - \frac{1}{e}\right)dx$

18. $\int(2 + 3x^2)^5 x\,dx$

19. $\int \frac{2x - 5}{\sqrt{(x^2 - 5x)^3}}\,dx$

20. $\int x^2\sqrt{10 - x}\,dx$

[14.3]

Objective 14.6: *Evaluate definite integrals.*

21. $\int_1^2 \frac{x^3 + 2}{x^2}\,dx$

22. $\int_0^2 x\sqrt{2x^2 + 1}\,dx$

23. $\int_1^8 x^{1/3}(1 + x^{4/3})^3\,dx$

24. $\int_{\sqrt{17}}^5 x\sqrt{x^2 - 16}\,dx$

[14.4]

Objective 14.7: *Find the area bounded by the x-axis, the curve $y = f(x)$, and two given vertical lines.*

25. $y = \sqrt{5x + 1}$; $x = 3, x = \frac{8}{5}$

26. $y = 5x^4 - 20$; $x = -1; x = 2$

27. $y = \frac{1}{(1 - 2x)^3}$; $x = 1, x = 5$

28. $y = \frac{\ln(3x + 4)}{3x + 4}$; $x = 1; x = 3$

Objective 14.8: *Write an integral to express a shaded region.*

29.

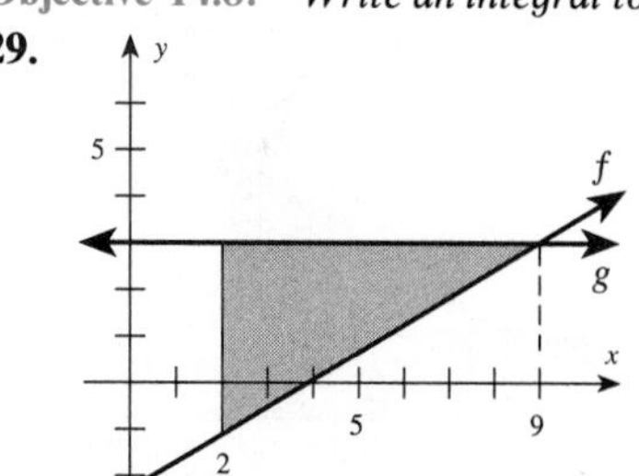

30.

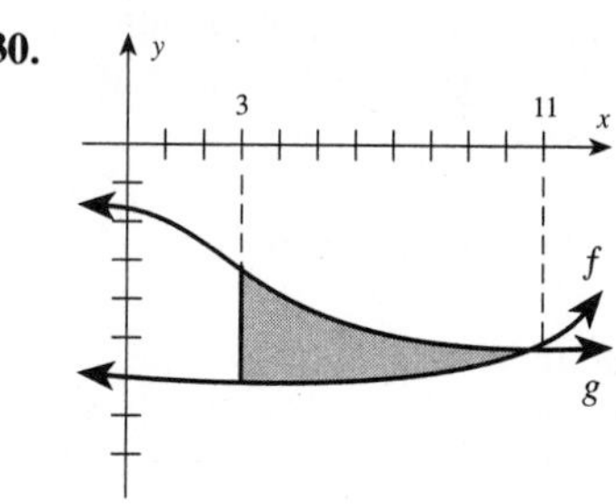

31.

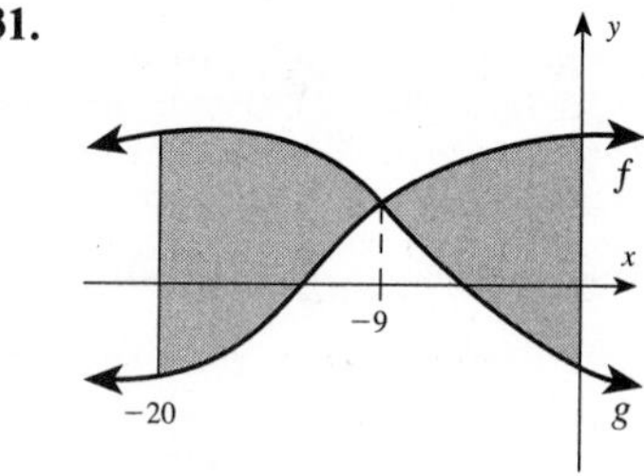

32.

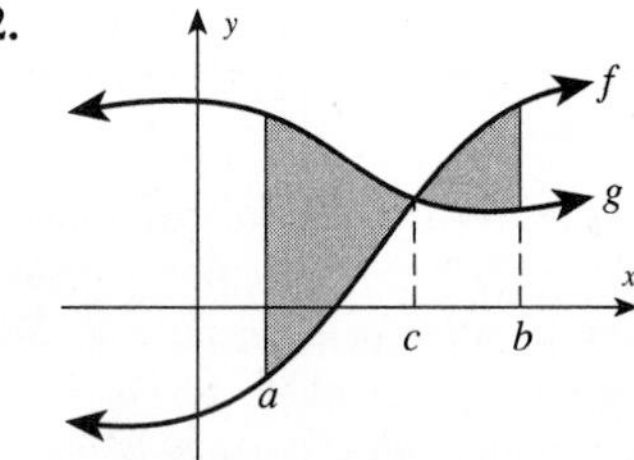

Objective 14.9: *Find the area between two given curves.*

33. $y = \frac{1}{2}x^2$ and $y = 4 - x$ from $x = -2$ to $x = 3$

34. $y = x^2 - 16$ and the x-axis

35. $y = x^2 - 3x$ and $x - 4y = 0$

36. $y = x^3$ and $y = x$

[14.5]

Objective 14.10: *Approximate integrals using a rectangular approximation. Use left endpoints to find*

$$\int_4^{25} \ln\sqrt{x}\,dx$$

correct to two decimal places where n is specified in Problems 37–40.

37. $n = 1$ **38.** $n = 2$

39. $n = 4$ **40.** $n = 8$

Objective 14.11: *Evaluate integrals for tabular functions.*

41. $\int_0^{10} f(x)\,dx$

x	1	3	5	7	9
$f(x)$	63	49	55	82	80

42. $\int_0^{2} f(x)\,dx$

x	.5	1	1.5	2
$f(x)$	8.4	12.1	16.4	11.4

43. The graph shown here from the *Wall Street Journal* presents the rate of domestic cars sold from June 1 to February 28. Estimate the sales for the 4th quarter (September 1–December 31). (*Hint:* Imagine a rectangle 30 days wide with an area equal to the part under the curve for that month.)

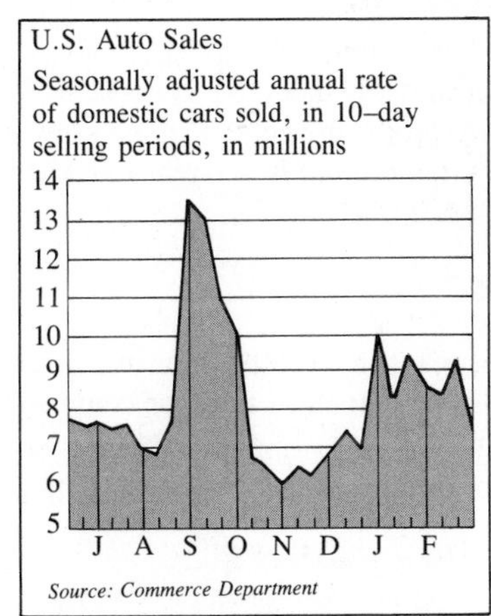

44. Using the graph in Problem 43, estimate the accumulated sales from July 1 to November 1.

Objective 14.12: *Find the average value for a given function.*

45. $f(x) = 6x^2 - 2x + 5$ on $[-2, 1]$

46. $f(x) = \dfrac{3x - 2}{x^2}$ on $[1, 4]$

47. $f(x) = xe^{x^2}$ on $[1, 3]$

48. $f(x) = x\sqrt{9 - 2x^2}$ on $[0, 2]$

[14.6]

Objective 14.13: *Approximate integrals using a trapezoidal approximation. Find*

$$\int_0^2 \frac{dx}{1 + x^2}$$

correct to two decimal places where n is specified in Problems 49–52.

49. $n = 1$ **50.** $n = 2$

51. $n = 4$ **52.** $n = 8$

Objective 14.14: *Approximate integrals using Simpson's rule. Find*

$$\int_0^2 \sqrt{4 - x^2}\,dx$$

correct to two decimal places where n is specified in Problems 53–56.

53. $n = 1$ **54.** $n = 2$

55. $n = 8$ **56.** $n = 4$

Objective 14.15: *Solve applied problems based on the preceding objectives.*

57. *Marginal profit.* The marginal profit of Campress Press-Ons is

$$P'(x) = 250x - 5{,}000$$

where x is the number of items. Find the profit function if there is a \$1,500 loss when no items are produced.

58. *Marginal cost.* The marginal cost of Campress Press-Ons is

$$C'(x) = \frac{\sqrt{x}}{1{,}000} + .0001x$$

If the fixed costs are \$1,500, what is the cost of producing 5,000 items (to the nearest thousand dollars)?

59. Industrial waste is pumped into a 1,000-gallon holding tank at the rate of 1 gal/min and the well-stirred mixture is removed at the same rate. If the concentration of waste in the tank is changing at a rate of

$$c'(t) = .001e^{-.001t}$$

after t minutes, find the concentration of waste in the tank after one hour.

60. The graph below (from the *Wall Street Journal*, February 26, 1986) shows the 10-year Treasury Bond rate for the years 1980–1985. Estimate the area under this curve using a rectangular approximation where $n = 6$.

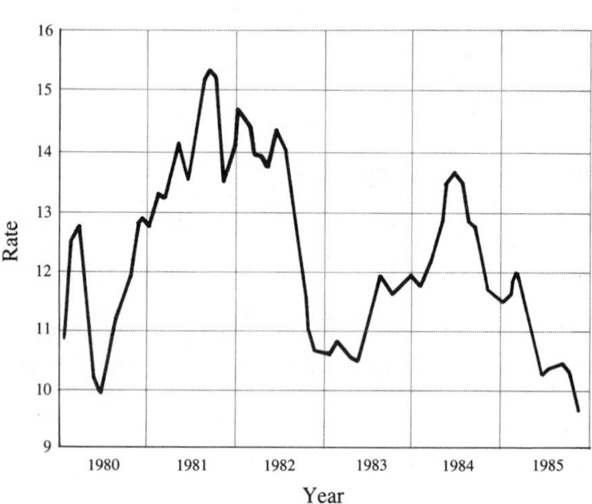

SAMPLE TEST

The following sample test (45 minutes) is intended to review the main ideas of this chapter.

Evaluate the integrals in Problems 1–12. Be sure to include the constant of integration when working with indefinite integrals.

1. $\int x^5\,dx$

2. $\int \frac{dx}{x}$

3. $\int e^u\,du$

4. $\int u^{-4}\,du$

5. $\int (18x^2 - 6x - 3)\,dx$

6. $\int \frac{(x+1)}{2x^2+4x+5}\,dx$

7. $\int \left(e^{3x+1} + \frac{1}{e}\right)dx$

8. $\int \frac{6x^2 - 3x + 2}{x}\,dx$

9. $\int_{-1}^{3} \frac{x^3+2}{x^2}\,dx$

10. $\int_{2}^{2\sqrt{3}} x\sqrt{2x^2+1}\,dx$

11. $\int_{1}^{8} [x^{1/3}(1 + x^{4/3})^3]\,dx$

12. $\int_{3\sqrt{2}}^{5} x\sqrt{x^2-9}\,dx$

13. Find the area bounded by the x-axis, the curve $y = \sqrt{3x+4}$, and the lines $x = 4$ and $x = \frac{5}{3}$.

14. Find the area bounded by the curves $y = x^2$, $y = 6 - x$, and the lines $x = 0$ and $x = 4$.

15. Find the area bounded by the curves $y = x^3$ and $y = x$.

16. The marginal profit of Campress Press-Ons is

$$P'(x) = 250x - 5{,}000$$

where x is the number of items. Find the profit function if there is a \$2,500 loss when no items are produced.

17. One of the fastest growing trees is the eucalyptus. In order to determine whether it is economically feasible to harvest it as a crop for firewood, it must grow at least 6 feet in 5 years ($t = 6$, since it is 1-year-old when planted). If the growth rate is

$$1.2 + 5t^{-4} \qquad t \geq 1$$

feet per year, where t is the time in years, is it economically feasible to plant these trees for commercial harvesting if they are 12 inches tall when planted (that is, when $t = 1$)?

18. In 1986 the growth rate of Chicago, Illinois, was

$$P'(t) = -.035e^{-.011t}$$

where t is measured in years after 1980 and $P(t)$ gives the population in millions. If the population in 1984 was 3,000,000, predict the population at the turn of the century using the 1986 growth rate.

19. The rate of consumption (in billions of barrels per year) for oil conforms to the formula

$$R(t) = 32.4e^{.048t}$$

for t years after 1985. If the total oil still left in the earth is estimated to be 670 billion barrels, estimate the length of time before all available oil is consumed if the rate does not change.

20. Find a decimal approximation for $\int_0^3 \sqrt{9 - x^2}\, dx$ correct to two decimal places. Use the left endpoints for a rectangular approximation where $n = 4$.

15

Applications and Integration

CHAPTER OVERVIEW

The calculus you learned in Chapter 14 is put to work in this chapter as we look at some important applications in order to show some of the power and versatility of integral calculus. In addition, our study of integration is expanded by the introduction of two new integration techniques: by parts, and by table. It might be noted that more and more emphasis today is given to integration by using tables. Although techniques of integration are important, new computer technology has made numerical integration and integration by table more important than ever before.

PREVIEW

We can now use the techniques of integration in a variety of interesting and important applications in mathematics and business. These applications include total value, money flow, consumers' and producers' surplus, lease payments, depreciation, optimum time to overhaul equipment, net investment flow, and capital formation. We learn the counterpart of the product rule for differentiation in a process called *integration by parts.* Probabilistic models are also considered (in an optional section), the most important of these being the normal curve, or normal density function.

PERSPECTIVE

This chapter completes your introduction to elementary calculus. The ideas of limit, the derivative, and the integral have been introduced and some of their applications discussed. If you have mastered these ideas, you have the skill to build and solve mathematical models that will help you analyze, understand, and forecast many real-life situations. The next chapter considers an additional useful topic: functions of several variables.

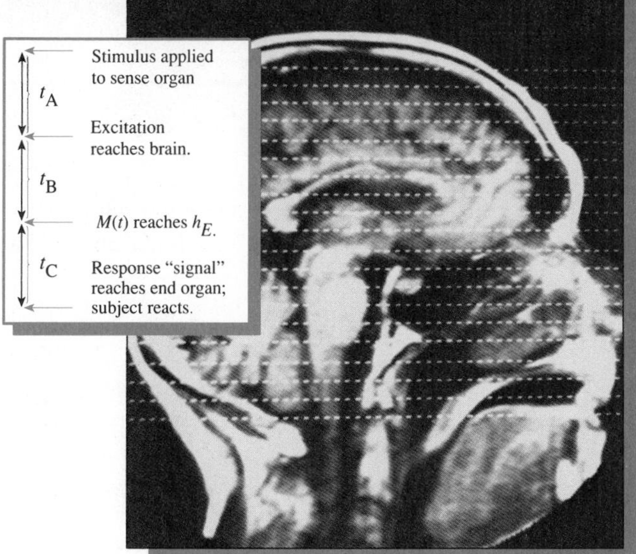

MODELING APPLICATION 13

Modeling the Nervous System

The nervous system has the ability to respond to a variety of stimuli by generating electrical impulses and transmitting them through nerve cells. Nerve cells vary in length from a few micrometers to more than a meter. Nerve cells that transmit impulses from sense organs to the spinal cord and brain are called *sensory* or *afferent neurons*. Those that transmit messages to the muscles and glands are called *motor* or *efferent neurons*. The central nervous system (CNS) consists of the brain and spinal cord; the peripheral nervous system (PNS) consists of the sensory and motor neurons; and the autonomic nervous system (ANS) consists of neurons that regulate processess not under conscious control.

Do some reading in the following sources and write a paper discussing a model that describes reaction time in the central nervous system. For general guidelines about writing this essay, see the commentary for Modeling Application 1 on page 151.*

*Horelick, Brindell, Sinan Koont, and Sheldon F. Gottleif. *Modeling the Nervous System: Reaction Time and the Central Nervous System.* © 1983 COMAP, Inc., 271 Lincoln Street, Suite 4, Lexington, MA 02173. COMAP is a nonprofit corporation engaged in research and development in mathematics education.

Erlanger, J., and H. S. Gasser. *Electrical Signs of Nervous Activity* (Philadelphia: University of Pennsylvania Press, 1937).

APPLICATIONS

Management (*Business, Economics, Finance, and Investments*)
Total value/total income (15.1, Problems 1–12)
Money flow (15.1, Problems 13–18)
Finding total sales (15.1, Problems 19–24)
Monthly lease payments (15.1, Problems 27–28)
Net excess profit (15.1, Problems 29–30)
Time period for a piece of equipment to pay for itself (15.1, Problems 33–34)
Consumer's/producer's surplus (15.1, Problems 35–38; 15.7, Problems 17–20)
Depreciation (15.1, Problems 39–40; 15.7, Problem 70; Test Problem 20)
Optimum time to overhaul equipment (15.1, Problem 41)
Net investment flow (15.1, Problem 42)
Capital value (15.1, Problem 43; 15.5, Problems 23–25; 15.7, Problem 71; Test Problem 18)
Domar's capital expansion model (15.2, Problem 30)

Management (*continued*)
Effect of advertising (15.2, Problems 39–40)
Total profit (15.3, Problem 38)
Accumulated sales (15.4, Problem 67)
Total cost (15.5, Problem 66)
Endowment fund (15.5, Problem 31)

Life Sciences (*Biology, Ecology, Health, and Medicine*)
Reaction time of a drug (15.1, Problems 25–26)
Growth rate of bacteria (15.2, Problem 31; 15.3, Problem 39)
Carbon-14 dating (15.2, Problems 34–35)
Blood pressure (15.2, Problem 36)
Drug dose (15.2, Problems 37–38)
Pollutants dumped into a river (15.4, Problem 65; 15.5, Problem 32)
Oil production (15.5, Problem 33)
Seed germination (15.6, Problem 37)
Rainfall in Ferndale, CA (15.6, Problem 40)
Spread of a disease (15.7, Problem 69)
Oil consumption (15.7, Problem 71)

Social Sciences (*Demography, Political science, Population, Psychology, Society, and Sociology*)
Estimating the crime rate (15.2, Problem 29)
Grading on a curve (15.7, Problem 73)
Rate of learning (15.6, Problem 36; 15.7, Test Problem 19)

General Interest
Future value (15.1, Problems 31–32)
Maintenance costs (15.3, Problem 41)
Radioactive materials in the atmosphere (15.5, Problems 26–28)
Breaking strength (15.6, Problems 38–39)
Life of a lightbulb (15.6, Problems 41–43; 15.7, Problem 72)
Telephone waiting time (15.6, Problem 44)
Battery life (15.6, Problem 45)

Modeling Application—
Modeling the Nervous System

15.1

Business Models Using Integration

Rates

Suppose Wayne Savick introduced a new batching process computer with an annual rate of sales t years after it was first introduced given by the function $S(t)$. He found that

$$S(t) = 5 + 15t^2$$

for $0 \leq t \leq 3$. What is the rate of sales at $t = 2$ years? Consider

$$S(2) = 5 + 15(2)^2 = 65$$

How should this be interpreted? Can we say 65 computers sold during the second year? Not really; notice that

$$S(1) = 5 + 15(1)^2 = 20$$

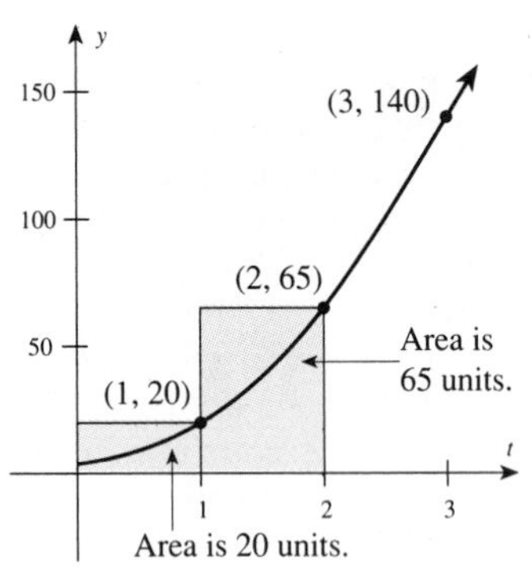

FIGURE 15.1
Graph of the sales of a new batching process computer

Since the rate of sales is 20 units when $t = 1$ and 65 units when $t = 2$, is seems reasonable that the number of units sold at different times during the second year is between 20 and 65. Remember, you cannot assume that the annual rate of sales is constant on a yearly basis. The graph of S is shown in Figure 15.1.

A better approximation for finding the total number of computers sold during the entire second year is found when we calculate

$$S\left(\frac{3}{2}\right) = 5 + 15(1.5)^2 = 38.75$$

Thus, during the first half of the second year, the number of units produced should be at least

$$\frac{1}{2}(20) = 10$$

and during the second half of the second year it should be

$$\frac{1}{2}(38.75) = 19.375$$

This approximation is shown in Figure 15.2.

The number of square units in the area of the rectangles in Figure 15.2 represents the total sales if sales during the first half of the year remain at a constant 20 units and during the second half at a constant 38.75 units. However, sales are not constant. So we now see (by using Riemann sums) that the desired answer is

$$\begin{aligned}\int_1^2 (5 + 15t^2)\,dt &= 5t + \frac{15t^3}{3}\Bigg|_1^2 \\ &= [10 + 5(2)^3] - [5 + 5(1)^3] \\ &= 40\end{aligned}$$

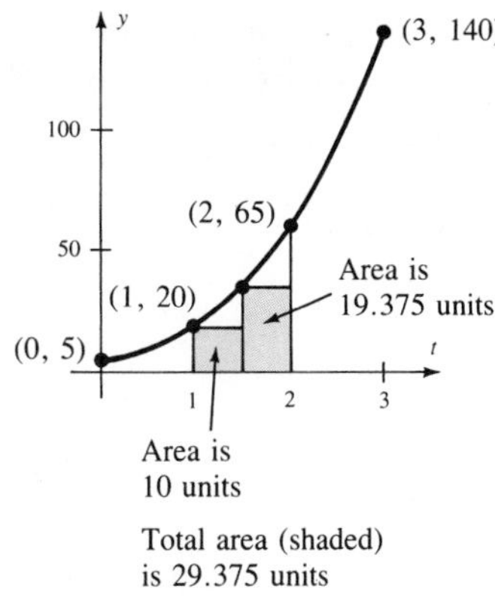

FIGURE 15.2

EXAMPLE 1 During the first year for which a new product is in the market, $S(t)$ is the rate of sales after t months and is given by the formula

$$S(t) = 90\sqrt{t} + 150$$

Find the total sales for the first 9 months.

Solution

$$\int_0^9 (90\sqrt{t} + 150)\,dt = \frac{90t^{3/2}}{\frac{3}{2}} + 150t\Big|_0^9$$

$$= 60t^{3/2} + 150t\Big|_0^9$$

$$= 60(9)^{3/2} + 150(9) = 2{,}970$$

This means that about 2,970 units are sold in the first 9 months. ■

The definite integral we used in Example 1 is called the **total value** over a period of time.

Total Value of a Function over Time

The **total value** of an integrable function f over $[a, b]$ is given by

$$\int_a^b f(x)\,dx$$

A common variation of total value is to determine the length of time it will take for a piece of new equipment to pay for itself.

EXAMPLE 2 A solar heater is being considered as a replacement for an electric heater. The cost of the solar heater is bid at \$38,500, and the rate of savings in operating costs by using a solar instead of an electric heater is given by

$$S(x) = 8{,}000x + 3{,}000$$

where $S(x)$ is in dollars per year and x is the number of years the solar heater will be in use. How long will it take the solar heater to pay for itself?

Solution Let t be the number of years the solar heater is used until the operating cost savings is equal to the initial replacement cost. Then

$$\int_0^t (8{,}000x + 3{,}000)\,dx = 38{,}500$$

$$\frac{8{,}000x^2}{2} + 3{,}000x\Big|_0^t = 38{,}500$$

$$\frac{8{,}000t^2}{2} + 3{,}000t = 38{,}500$$

$$4{,}000t^2 + 3{,}000t - 38{,}500 = 0$$

$$8t^2 + 6t - 77 = 0$$

$$(4t - 11)(2t + 7) = 0$$

$$t = \frac{11}{4}, -\frac{7}{2}$$

Reject negative in this application.

Thus it would take about $\frac{11}{4}$ years or 2 years and 9 months for the solar heater to pay for itself. ■

Continuous Income Stream

The growth rate of money with respect to time is stated as a compound interest formula or interest with continuous compounding. If this growth rate is denoted by $f'(t)$, then the integral of that rate of growth is the function $f(t)$ that gives the accumulated money. The value of this function at $t = k$ is the total amount at this time, and the value at $t = 0$ is the amount at the beginning of the time period. Thus, the total growth is $f(k) - f(0)$ or

$$\int_0^k f'(x)\,dx$$

This value denotes what is called a **continuous income stream**. For example, suppose money flows continuously into a computer video game and grows at a rate given by

$$f(t) = 10e^{.08t}$$

where t is in hours and $0 \le t \le 12$. Find the total amount of money that accumulates in the machine during the 12-hour period:

$$\int_0^{12} 10e^{.08t}\,dt = 10\frac{e^{.08t}}{.08}\Big|_0^{12} \approx 201.46$$

The function f is called the **rate of flow** and even though in reality the income would be collected at specific intervals (each month, each week, or each day), we assume that the income is actually received continuously. That is, we say that the income is received in a **continuous stream**, and the total value formula for this application is referred to as **total income**.

EXAMPLE 3 If the rate of flow for an investment is

$$f(x) = 25{,}000e^{.09t}$$

find the total income produced in one year and in ten years.

Solution The total income for each part is found with these respective integrals:

$$\int_0^1 25{,}000e^{.09t}\,dt = 25{,}000\frac{e^{.09t}}{.09}\Bigg|_0^1 = \$26{,}159.52$$

$$\int_0^{10} 25{,}000e^{.09t}\,dt = 25{,}000\frac{e^{.09t}}{.09}\Bigg|_0^{10} = \$405{,}445.31$$

The total income produced in one year is \$26,159.52 and in 10 years is \$405,445.31. ■

You might recognize the formula used in Example 3 as the formula for the continuous compounding of \$25,000 invested at 9% per year. If we calculate $A(t) = 25{,}000e^{.09t}$ for $t = 1$ we obtain \$27,354.36. How does this relate to the number we found in Example 3? This represents the future value of a \$25,000 deposit. In other words, of this \$27,354.36, \$25,000 is the initial investment and \$2,354.36 is the interest. On the other hand, when we integrate as we did in Example 3, we are finding the *total income for the year*—namely, \$26,159.52. These differences are even more remarkable if we let $t = 10$:

$A(t) = 25{,}000e^{.09t}$ for $t = 10$ gives \$61,490.08.

This means that the future value for a \$25,000 deposit at 9% in 10 years compounded continuously is \$61,490.08, while the *total income* for the same period is given by the rate of flow formula, which yields \$405,445.31.

If the rate of flow formula $f(x)$ in the total value of a function formula is the continuous interest formula, namely $f(t) = Pe^{rt}$, then we have still another name for the total value formula—**total money flow**.

Total Money Flow

The **total money flow** for P dollars over a period of T years at a rate r is given by the formula

$$F = \int_0^T Pe^{rt}\,dt$$

The total money flow is sometimes called **the amount of an annuity**. An *annuity* (see Section 5.3) is a sequence of equal periodic payments into an account. If the interest is earned continuously at an annual rate of r, and a payment of m dollars is made n times a year for t years, then the above formula for N payments ($N = nt$) is

$$A = \int_0^N me^{it}\,dt$$

where $i = \dfrac{r}{n}$.

EXAMPLE 4 If a person places \$50 a month into an account paying 9% compounded continuously, how much will be in the account after 5 years?

Solution $m = 50$, $n = 12$, $N = 5(12) = 60$, $r = .09$, and $i = \frac{.09}{12} = .0075$

$$A = \int_0^{60} 50e^{.0075t}\,dt = 50\frac{e^{.0075t}}{.0075}\Bigg|_0^{60} = \frac{50}{.0075}(e^{.45} - 1) = 3{,}788.747903$$

There would be \$3,788.75 in the account. ■

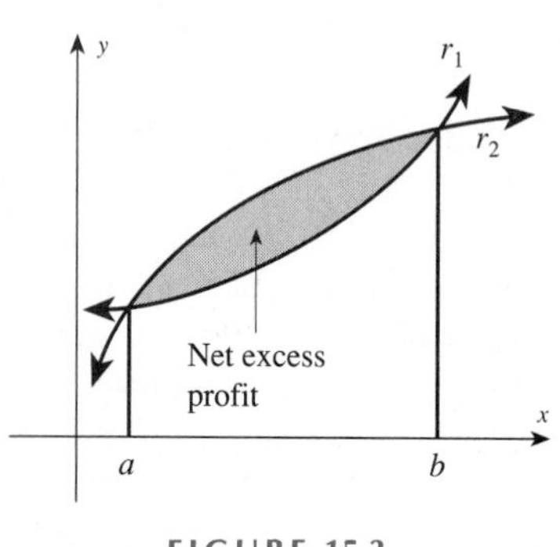

FIGURE 15.3

There are three business applications of integration that involve the area between two curves which we will now consider. These are *net excess profit*, *consumers' surplus*, and *producers' surplus*.

Net Excess Profit

Suppose you are interested in comparing the profit from two investments t years from now and you know that the investments are generating profits at the rates of $r_1(t)$ and $r_2(t)$ dollars per year. Suppose also that you know that for the next N years $r_2(t) > r_1(t)$. The *net excess profit* is defined to be the area between the curves representing these rates as shown in Figure 15.3.

Net Excess Profit

If profit is generated by the rates $r_1(t)$ and $r_2(t)$ at time t over a closed interval of time $[0, N]$ where $r_2(t) > r_1(t)$, then the **net excess profit** is defined by the formula

$$\int_0^N [r_2(t) - r_1(t)]\,dt$$

EXAMPLE 5 Suppose that t years from now, two investment plans will be generating profits at the rates of $10 + t^2$ and $50 + 6t$ in thousands of dollars per year, respectively.

a. When will these investments be generating the same rate of profit?

b. Which is the better investment for the interval of time $[0, N]$ if N is the length of time until the investments generate the same rate of profit? See Figure 15.4.

c. What is the net excess profit?

Solution **a.**

$$10 + t^2 = 50 + 6t$$
$$t^2 - 6t - 40 = 0$$
$$(t - 10)(t + 4) = 0$$
$$t = 10, -4$$

Since t is a length of time, reject $t = -4$, so $t = 10$; it will be 10 years before these investments are generating the same rate of profit; that is $N = 10$.

b. On the interval $[0, 10]$ we have $50 + 6t \geq 10 + t^2$, so $r_2(t) = 50 + 6t$ and $r_1(t) = 10 + t^2$.

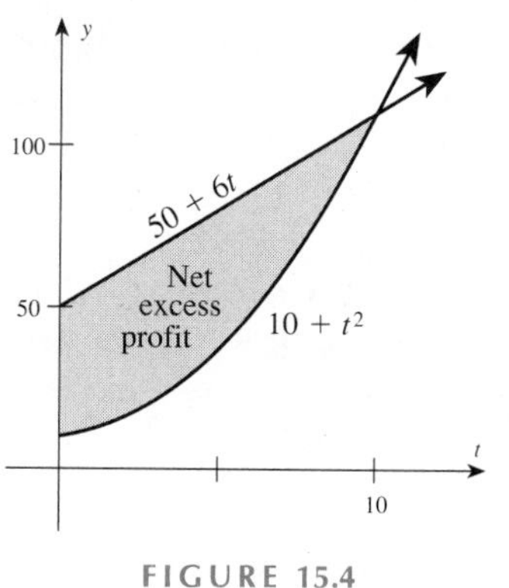

FIGURE 15.4

c. NET EXCESS PROFIT $= \int_0^N [r_2(t) - r_1(t)]\,dt$

$$= \int_0^{10} [(50 + 6t) - (10 + t^2)]\,dt = \int_0^{10} (-t^2 + 6t + 4)\,dt$$

$$= \left(\frac{-t^3}{3} + 3t^2 + 4t\right)\Bigg|_0^{10}$$

$$= \frac{20}{3}$$

The excess profit is \$6,667. (This is $\frac{20}{3}$ thousands of dollars.) ■

Consumers' and Producers' Surplus

An important business application of the area between curves involves the concepts of **consumers' surplus** and **producers' surplus**. The intersection of the demand and supply functions provides an *equilibrium price*, or point, as shown in Figure 15.5.

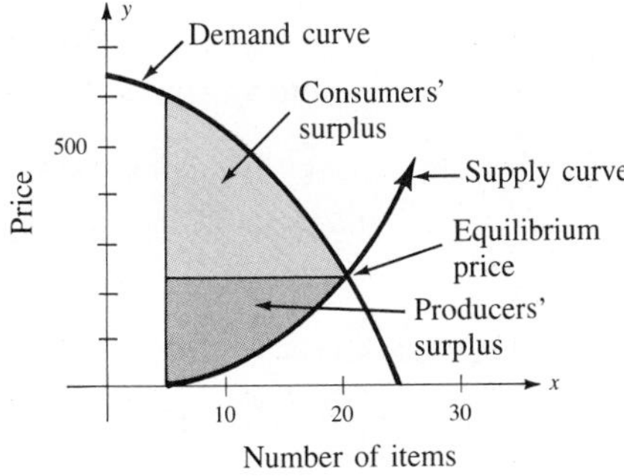

FIGURE 15.5
Consumers' and producers' surplus

The demand function, $D(x)$, gives the price per unit that consumers are willing to pay to get the xth item of the same commodity. For example, a company might be willing to spend \$1,000 for a personal computer. However, once they have the first computer, they might be willing to pay only \$800 for the second computer, and \$500 for a third computer. The price that consumers are willing to pay to get one additional unit usually decreases as the number of units already bought increases. This demand function, which in economics is called the **marginal willingness to spend**, can be thought of as the rate of change with respect to x of the *total* amount consumers are willing to spend for x units. Thus, if $A(x)$ is the total amount (in dollars) that consumers are willing to spend to get x units of the commodity, and if $A(x)$ is differentiable, then

$$D(x) = \frac{dA}{dx}$$

Thus, the total amount consumers are willing to spend to get n units of the commodity is the definite integral

$$A(n) = A(n) - A(0) = \int_0^n \frac{dA}{dx}\,dx = \int_0^n D(x)\,dx$$

In a competitive economy, the total amount that consumers *actually spend* on a commodity is generally less than the total amount they would be *willing* to spend. The difference between these two amounts is known as the **consumers' surplus** because it can be thought of as a savings realized by the consumer.

CONSUMERS' SURPLUS

= TOTAL AMOUNT CONSUMERS WOULD BE WILLING TO SPEND

− ACTUAL CONSUMER EXPENDITURE

In order to understand this example, go back to the company that was willing to purchase a computer. Reconsider the fact that the company was willing to spend \$1,000 on the first computer, \$800 on the second computer, and \$500 for the third. Also suppose that the marked price is \$800. Then the company would buy two computers for a total of

2(\$800) = \$1,600

This is less than the company would have been willing to spend to get two computers—namely, \$1,000 + \$800 = \$1,800. The savings

\$1,800 − \$1,600 = \$200

is the company's consumer surplus. This information is summarized in Figure 15.6.

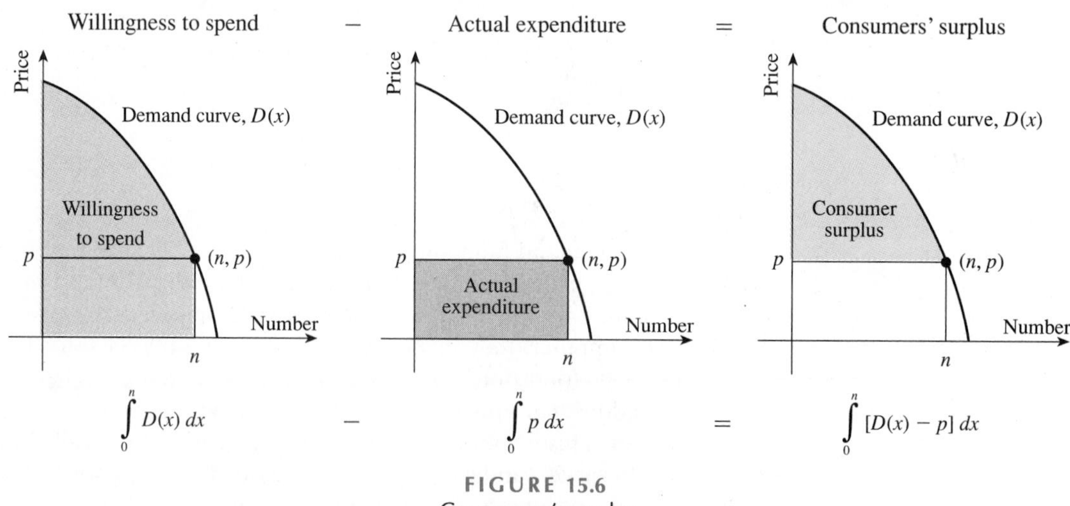

FIGURE 15.6
Consumers' surplus

On the other hand, suppose $S(x)$ is the price–supply function that represents the price at which a producer would be willing to sell the xth item of the same commodity. For example, a company might be willing to sell a single cut glass for

\$100, but would be willing to sell a dozen of the same cut glasses for \$75 each. The **producers' surplus** is the difference between the price (the actual consumer expenditure), p, and the total amount the producers would have been willing to accept for the sale of their product. These ideas are summarized in the following box.

Consumers' Surplus

If n items are supplied at a price of p dollars per item for a demand function $D(x)$, then

$$\text{CONSUMERS' SURPLUS} = \int_0^n [D(x) - p]\,dx$$

Producers' Surplus

If n items are produced at a price of p dollars per item for a supply function $S(x)$, then

$$\text{PRODUCERS' SURPLUS} = \int_0^n [p - S(x)]\,dx$$

The most common situation is for (n, p) to be the equilibrium point for the demand and supply functions.

EXAMPLE 6 The price, in dollars, for a product is

$$D(x) = 650 - x - x^2$$

where x is the number of items in the domain $[0, 25]$. The supply curve, in dollars, is given by

$$S(x) = x^2 - 9x + 10$$

where x is in the domain $[10, 25]$. Find the consumers' and producers' surplus.

Solution First, find the equilibrium price.

$$D(x) = 650 - x - x^2$$
$$S(x) = x^2 - 9x + 10$$

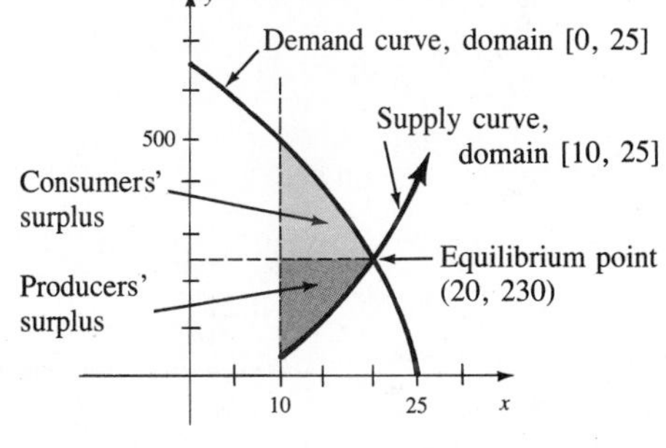

Find x so that $D(x) = S(x)$:

$$650 - x - x^2 = x^2 - 9x + 10$$
$$2x^2 - 8x - 640 = 0$$
$$x^2 - 4x - 320 = 0$$
$$(x - 20)(x + 16) = 0$$
$$x = 20, -16$$

Reject the negative value

The domain for the surplus is the intersection of the domain for the supply and demand; in this example it is given as $[10, 25]$. For $x = 20$, the price is 230, so the equilibrium point is $(20, 230)$.

$$\begin{aligned}
\text{Consumers' surplus} &= \int_{10}^{20} [(650 - x - x^2) - 230]\,dx \\
&= \int_{10}^{20} (420 - x - x^2)\,dx \\
&= \left(420x - \frac{x^2}{2} - \frac{x^3}{3}\right)\Big|_{10}^{20} \\
&= 8{,}400 - 200 - \frac{8{,}000}{3} - 4{,}200 + 50 + \frac{1{,}000}{3} \\
&\approx 1{,}716.67 \\
\text{Producers' surplus} &= \int_{10}^{20} [230 - (x^2 - 9x + 10)]\,dx \\
&= \int_{10}^{20} (220 + 9x - x^2)\,dx \\
&= \left(220x + \frac{9x^2}{2} - \frac{x^3}{3}\right)\Big|_{10}^{20} \\
&\approx 1{,}216.67
\end{aligned}$$

The consumers' surplus is \$1,716.67, and the producers' surplus is \$1,216.67. ■

15.1
Problem Set

APPLICATIONS

Find the total value of the functions in Problems 1–6.

1. $f(x) = \sqrt{x} + 100$ on $[4, 9]$
2. $S(x) = 50\sqrt[3]{x} - 30x$ on $[1, 8]$
3. $m(x) = \dfrac{x-1}{x+1}$ on $[0, 2]$
4. $t(x) = \dfrac{2x+1}{x-2}$ on $[3, 5]$
5. $f(t) = 65e^{-5t}$ on $[0, 3]$
6. $f(t) = 1{,}000e^{.005t}$ on $[0, 2]$

Find the total income produced in the first 5 years for the rate of flow functions given in Problems 7–12.

7. $f(x) = 500$
8. $f(x) = 4{,}000$
9. $f(x) = 200e^{.08t}$
10. $f(x) = 500e^{.11t}$
11. $f(x) = 150e^{-.05t}$
12. $f(x) = 850e^{-.09t}$

Determine the total money flow for Problems 13–18.

13. \$8,000 at 8% for 6 years
14. \$83,000 at 10% for 2 years
15. \$2,500 at 5% for 1 year
16. \$125,000 at 11% for 15 years
17. \$6,500 at 7% for 5 years
18. \$4 million at 10% for 8 years
19. A new product has a rate of sales given by

$$S(t) = 60 + 24t^2 \qquad \text{for } 0 \le t \le 3$$

where t is the number of years since the product was introduced. Find the number of items sold in the second year.

20. Repeat Problem 19 for the third year.
21. Repeat Problem 19 for the total 3-year period.
22. During the first year a new commodity has been on the market, y units per month were sold after x months elapsed, according to the formula

$$y = 300\sqrt[3]{x} + 50 \qquad 0 \le x \le 12$$

Find the total sales during the first 6 months.

23. Repeat Problem 22 for the second 6 months.
24. Repeat Problem 22 for the entire first year.

25. The rate of reaction to a certain drug is measured in minutes after administration according to the formula

$$R(t) = x^{-.5}$$

Evaluate the total reaction in the first 10 minutes.

26. Repeat Problem 25 for the first hour.

27. An equipment leasing company determines that the rate of maintenance cost in dollars per year on a piece of new equipment is

$$M(x) = 150(1 + \sqrt[3]{x^2})$$

where x is the number of years the equipment has been leased. What should be the total amount for maintenance costs for a 4-year lease? What amount should be added to the monthly payment to pay for all the maintenance costs for a 4-year lease?

28. Repeat Problem 27 for a 3-year lease.

29. Find the net excess profit for the functions $r_1(x) = x^2 + 30$ and $r_2(x) = 15 + 8x$ on $[3, 5]$.

30. Find the net excess profit for the functions $r_1(x) = x^2 + 20$ and $r_2(x) = 24 + 3x$ on $[0, 4]$.

31. If a person deposits \$2,250 each year into a tax-sheltered retirement account paying 9% compounded continuously, how much will be in the account in 20 years?

32. If a person deposits \$25 per month into an account paying 7% compounded continuously, how much will be in the account in 8 years?

33. Suppose that a new piece of equipment costs \$15,000. The rate of operating cost savings is $S(x)$ in dollars per year and is given by the formula

$$S(x) = 2{,}400x + 10{,}500$$

where x is the number of years the equipment has been used. How long will it take the piece of equipment to pay for itself?

34. If a piece of machinery has a rate of operating cost savings in dollars per year given by

$$S(x) = 4{,}800x + 22{,}500$$

how long it will take for the piece of machinery to pay for itself if x is the number of years the equipment has been used and the original cost was \$5,100?

35. The price, in dollars, for a product is

$$D(x) = 2{,}000 - 15x - x^2$$

where x is the number of items in the domain $[10, 30]$. The supply function is given by

$$S(x) = x^2 - 3x + 560$$

Find the producers' surplus.

36. Find the consumers' surplus for the information in Problem 35.

37. Find the producers' surplus if the demand and supply functions are

$$D(x) = \frac{236 - 107x}{(x-3)^2} \quad \text{and} \quad S(x) = x + 20$$

for x on $[0, 3)$, where x is the number of items (in thousands).

38. Find the consumers' surplus for the functions given in Problem 37.

39. If the total depreciation at the end of t years is represented by $f(t)$, then the depreciation rate is $f'(t)$ over an interval $[0, t]$. Since $f'(t)$ is usually known, the total depreciation can be found by using the formula

$$f(t) = \int_0^t f'(x)\,dx$$

Suppose a \$38,000 piece of equipment is depreciated over a 10-year period using *straight-line depreciation.* That is, each year the depreciation is

$$\frac{38{,}000}{10} = 3{,}800$$

So $f'(x) = 3{,}800$. Find the total depreciation for the first 3 years.

40. Many items do not depreciate at a constant rate (see Problem 39). Automobiles or computers, for example, depreciate much more quickly in the early years and more slowly toward the end of the time interval. Suppose an automobile depreciates according to the formula

$$f'(x) = 3{,}000\sqrt{5 - x}$$

Find the total depreciation for the first 3 years.

41. A piece of machinery requires an overhaul after time t. If $E(t)$ represents the expense connected with the equipment, we see that

$$E(t) = C + \text{total depreciation}$$

where C is the cost of overhaul. Now, from Problem 39,

$$E(t) = C + \int_0^t f'(x)\,dx$$

(*Also note:* $E'(t) = f'(t)$ since the derivative of C is zero.) The average expense, $A(t)$, is found by dividing by the number of years t:

$$A(t) = \frac{E(t)}{t}$$

If no other factors are involved, the best time to overhaul the equipment is at the value of t for which E has a relative minimum:

$$A'(t) = \frac{tE'(t) - E(t)}{t^2} \quad \text{provided } E'(t) \text{ exists}$$

Set $A'(t) = 0$ and solve for $E'(t)$ to find the critical value:

$$\frac{tE'(t) - E(t)}{t^2} = 0$$

$$tE'(t) - E(t) = 0 \quad \text{Multiply both sides by } t^2$$

$$\underbrace{E'(t)}_{} = \frac{E(t)}{t} \leftarrow \text{Average expense}$$

$E'(t) = f'(x)$ is the rate of depreciation

This critical value is a relative minimum (this is left for you to verify), so you see that the best time to overhaul occurs when the rate of depreciation equals the average expense. If a piece of equipment depreciates according to the formula

$$f'(x) = 1{,}000\sqrt{x} \quad \text{and} \quad C = 500$$

when should the equipment be overhauled to minimize the average expense?

42. If P dollars is invested at a rate of r percent per year compounded continuously for t years, then

$$A = Pe^{rt}$$

is the future amount. The rate of change of A with respect to time t is called the *net investment flow* and is found by

$$\frac{dA}{dt} = Pe^{rt}(r) = Pre^{rt}$$

Find the net investment flow for $P = \$50$ and $r = .09$.

43. The process by which a corporation increases its accumulated wealth is called *capital formation.* If the net investment flow (see Problem 42) is given by a function $f(x)$, then the increase in capital over the interval $[a, b]$ is

$$\int_a^b f(t)\,dt = F(b) - F(a)$$

Suppose Intel Corporation has a net investment flow approximated by the function $f(t) = \sqrt{t}$, where t is in years and f is in millions of dollars per year. What is the amount of capital formation over the next 5 years?

15.2 Differential Equations

Many of the applied integration problems we have worked on have involved knowing the derivative of a function and needing to find the original function. The process of finding this original function is called *antidifferentiation,* and we have used integration in order to find the desired result. More generally, if an equation involves an unknown function (often denoted by y) and one or more of its derivatives, it is called a **differential equation**. If the known derivative is dy/dx, it is called a **first-order differential equation.**

Differential Equations with No y Terms

The simplest type of differential equation is one in which there are no y terms. These differential equations can be solved by integration. We use this type of equation to introduce the terminology we will use when solving differential equations. Then we will turn to differential equations with both x and y terms in which the variables can be separated.

The first type of differential equation is solved by antidifferentiation, which is essentially what we have been doing since Chapter 14.

Differential Equation Theorem 1

A differential equation of the type

$$\frac{dy}{dx} = f(x)$$

has the solution

$$y = \int f(x)\,dx + C$$

where C is an arbitrary constant.

This result is easily derived by integrating, as shown in Example 1.

EXAMPLE 1 Solve the differential equation $y' = 6x$.

Solution

$$\frac{dy}{dx} = 6x \qquad \text{First set up the problem by writing } y' \text{ as } \frac{dy}{dx}.$$

$$dy = 6x\,dx \qquad \text{Separate variables (multiply both sides by } dx\text{).}$$

$$\int dy = \int 6x\,dx \qquad \text{Integrate both sides.}$$

$$y + C_1 = \frac{6x^2}{2} + C_2 \qquad \text{Evaluate the integrals.}$$

$$y = 3x^2 + (C_2 - C_1) \qquad \text{Combine constants.}$$

$$y = 3x^2 + C$$

■

The last form in Example 1 is called the **general solution**. It combines all the constants into one arbitrary constant. In practice, though, your work for Example 1 should look like this:

$$\frac{dy}{dx} = 6x$$

$$y = \int 6x\,dx + C = 3x^2 + C$$

If we take different values for C, we will obtain **particular solutions**. For example,

$$3x^2 \qquad 3x^2 + 5 \qquad \text{and} \qquad 3x^2 - 10$$

are all particular solutions. Thus, if you know the value of a function at a particular point, you can solve for C, as shown by Example 2.

EXAMPLE 2 Solve $y' = 4x - e^{-x} - \sqrt{x}$ if $y = 50$ when $x = 0$.

Solution
$$y = \int (4x - e^{-x} - \sqrt{x})\,dx$$
$$= \frac{4x^2}{2} - \frac{e^{-x}}{-1} - \frac{x^{3/2}}{\frac{3}{2}} + C = 2x^2 + e^{-x} - \frac{2}{3}x\sqrt{x} + C$$

Now, if $x = 0$, then $y = 50$, so by substitution you obtain

$$50 = 2(0)^2 + e^{-0} - \frac{2}{3}(0)\sqrt{0} + C$$
$$50 = 1 + C$$
$$49 = C$$

Thus the particular solution is

$$y = 2x^2 + e^{-x} - \frac{2}{3}x\sqrt{x} + 49$$

■

If the known value for the independent variable is 0, then the values $x = 0$, $y = 50$ are called an **initial condition** or a **boundary condition**.

You can verify that a function is a solution of a differential equation by differentiation. You might wish to check the results of Example 2 in this fashion.

EXAMPLE 3 Derive the formula for P dollars invested at an annual rate of r compounded continuously for t years.

Solution Let A be the amount in the account at any time t. Then

$$\frac{dA}{dt}$$

is the rate of growth of A with respect to time t; this means

$$\frac{dA}{dt} = rA \qquad \text{with initial conditions } A(0) = P; \quad A > 0, \quad P > 0$$

We want to find a function $A = A(t)$ that satisfies these conditions. Solve the equation for r and integrate both sides with respect to t:

$$\frac{dA}{dt} = rA$$
$$\frac{1}{A}\frac{dA}{dt} = r$$
$$\int \frac{1}{A}\frac{dA}{dt}\,dt = \int r\,dt$$
$$\int A^{-1}\,dA = \int r\,dt$$

$$\begin{aligned} \ln|A| &= rt + C \\ A &= e^{rt+C} \\ &= e^C e^{rt} \end{aligned}$$

Since $A(0) = P$, we evaluate $A(t) = e^C e^{rt}$ at $t = 0$ to find

$$\begin{aligned} A(0) &= e^C e^{r(0)} \\ P &= e^C \end{aligned}$$

Thus, the desired formula is

$$A = Pe^{rt}$$

■

Separation of Variables

Sometimes the differential equation involves two variables, say x and y, and it is possible to separate those variables into the product of two separate functions, say s and t, as follows:

$$\begin{aligned} \frac{dy}{dx} &= s(x) \cdot t(y) \\ \frac{1}{t(y)} dy &= s(x)\,dx \qquad t(y) \neq 0 \\ [t(y)]^{-1}\,dy &= s(x)\,dx \\ [t(y)]^{-1}\,dy - s(x)\,dx &= 0 \end{aligned}$$

We say this equation is of the *form*

$$g(y)\,dy + f(x)\,dx = 0$$

If we integrate both sides of this equation, we have the following result.

Differential Equation Theorem 2

If a differential equation can be written in the form

$$g(y)\,dy + f(x)\,dx = 0$$

then the general solution is given by

$$\int g(y)\,dy + \int f(x)\,dx = C$$

where C is an arbitrary constant.

EXAMPLE 4 Find the general solution of

$$\frac{dy}{dx} = x^2 y$$

and check by differentiating.

Solution Algebraically **separate the variables** by dividing each side by y and multiplying by dx:

$$y^{-1}\,dy = x^2\,dx$$
$$y^{-1}\,dy - x^2\,dx = 0$$
$$\int y^{-1}\,dy - \int x^2\,dx = C$$
$$\ln|y| - \frac{x^3}{3} = C$$

Check: Take the derivative of both sides with respect to x:

$$\frac{1}{y}y' - \frac{3x^2}{3} = 0$$

Solve for $y' = dy/dx$:

$$\frac{y'}{y} = x^2$$
$$y' = x^2 y$$

so it checks. ■

Note in Example 4 that we did not solve for y. Sometimes, however, we will want to solve for y. For Example 4,

$$\ln|y| = \frac{1}{3}x^2 + C$$
$$|y| = e^{(1/3)x^2 + C}$$
$$= e^{(1/3)x^2}e^C$$
$$y = \pm e^C e^{(1/3)x^2}$$
$$= Me^{(1/3)x^2}$$

where the constant $\pm e^C$ is rewritten as the constant M. You can also check by finding the derivative of this result to see if it gives the original differential equation.

EXAMPLE 5 Find the general solution of $dy/dx = ky$, where k is a constant.

Solution Separate the variables:

$$y^{-1}\,dy - k\,dx = 0$$
$$\int y^{-1}\,dy - \int k\,dx = C$$
$$\ln|y| - kx = C$$
$$\ln|y| = kx + C$$
$$|y| = e^{kx+C} = e^{kx}e^C$$
$$y = Me^{kx}$$

where the constant $\pm e^C$ is written as M. ■

Examples 4 and 5 are so common that the result of these examples is stated as a theorem:

Differential Equation Theorem 3

> If the rate of change of a variable y with respect to x is linear, then we say that $dy/dx = ky$ for some constant k. That is, the rate of change of y is directly proportional to the quantity itself. The solution of this differential equation is
>
> $$y = Me^{kx}$$

EXAMPLE 6 If the divorce rate has been a constant 5% (1975–1990) and if the number of divorces in 1987 was 1,157,000, estimate the number of divorces in 1992.

Solution Since the rate is a constant, we can write

$$\frac{dy}{dt} = .05y$$

where y is the number of divorces and t is the time. From Theorem 3,

$$y = Me^{.05t}$$

The initial condition is $y = 1{,}157{,}000$ for $t = 0$ (1987):

$$1{,}157{,}000 = Me^{.05(0)}$$
$$1{,}157{,}000 = M$$

Thus

$$y = 1{,}157{,}000e^{.05t}$$

Now, for 1992, $t = 5$, so

$$y = 1{,}157{,}000e^{.05(5)} \approx 1{,}486{,}000 \qquad \text{(DISPLAY: 1485617.4)}$$ ■

Growth and Decay

The last examples of this section lead to some of the most important applications in growth and decay. Suppose we begin with the assumption that the rate of change is linear.

Nobel laureate Willard Libby, from Sebastopol, CA, found, in 1946, that all living plants and animals have a constant level of carbon-14 in their tissues. However, once the plant or animal dies, the radioactive carbon-14 present decays at a rate proportional to the amount present. Thus, if $P(t)$ is the amount of carbon-14 present in an artifact, then

$$\frac{dP}{dt} = rP \qquad \text{where } P(0) = P_0\text{, the amount of carbon-14 present at death}$$

Libby found that the decay rate for carbon-14 is the constant $r = -.0001216$ (This gives a half-life for carbon-14 of 5,750 years.)

TABLE 15.1 Summary of Growth Models

P = population size; r = growth rate; t = time; L = limiting value of population; M = constant initial population

	Growth model	Equation and solution	Sample applications
J-curve 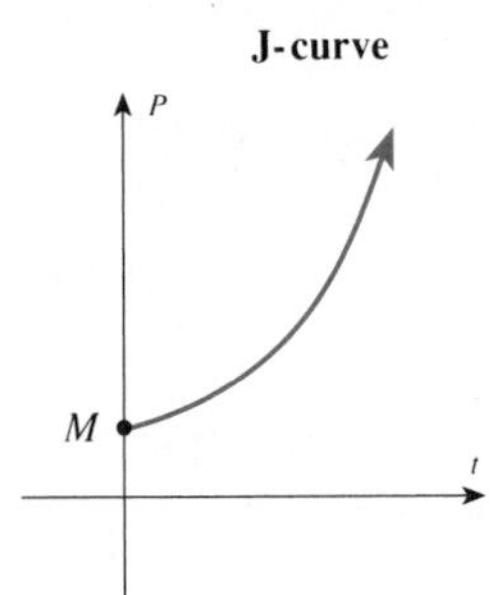	**Uninhibited growth** ($r > 0$) Rate is proportional to the amount present.	$\frac{dP}{dt} = rP$ Solution: $P = Me^{rt}$	Exponential growth; short-term population growth; interest compounded continuously; inflation; price–supply curves
L-curve 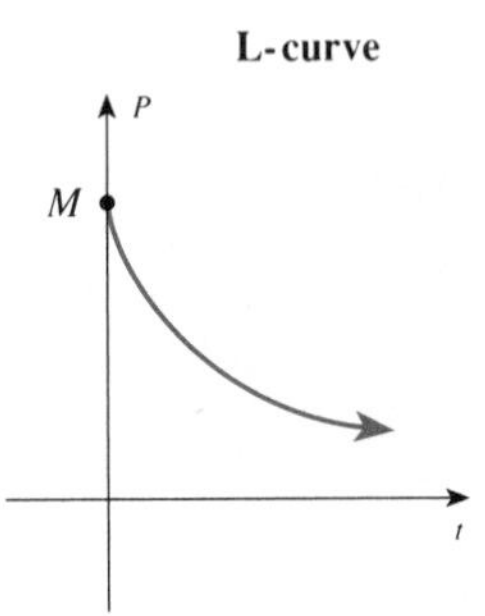	**Uninhibited decay** ($k > 0$) Rate is proportional to the amount present.	$\frac{dP}{dt} = -kP$ Solution: $P = Me^{-kt}$	Radioactive decay; depletion of natural resources; price–demand curves
S-curve 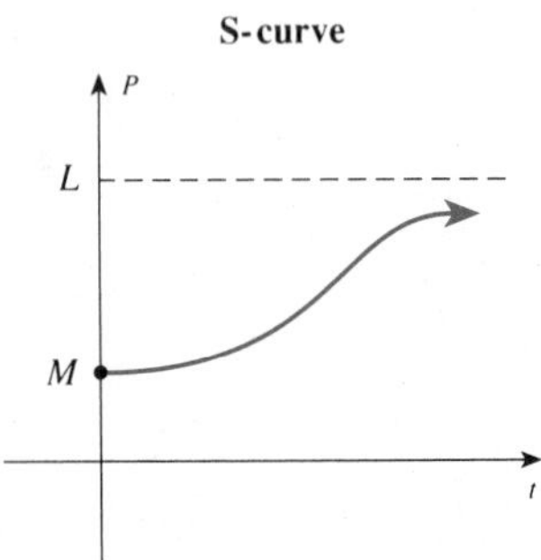	**Inhibited growth or Logistic growth** ($r > 0$) Rate is proportional to the amount present and to the difference between the amount present and a fixed amount.	$\frac{dP}{dt} = rP(L - P)$ Solution: $P = \frac{ML}{M + (L - M)e^{-Lrt}}$	Long-term population growth (with a limiting value); spread of a disease in a population; sales fads (for example, singing flowers or dancing Coke cans); growth of a business
C-curve 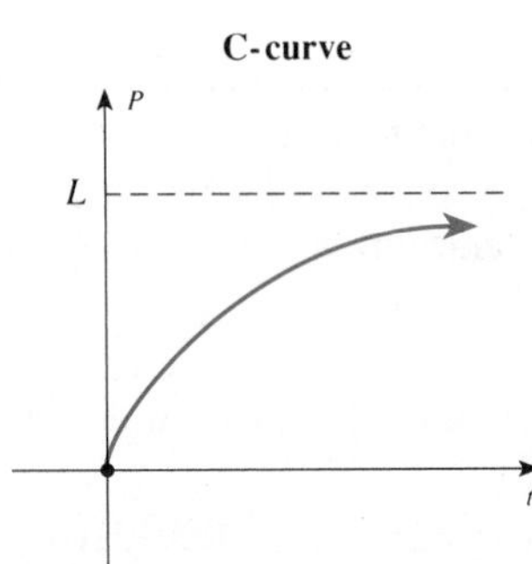	**Limited growth** ($r > 0$) Rate is proportional to the difference between the amount present and a fixed limit.	$\frac{dP}{dt} = r(L - P)$ Solution: $P = L(1 - e^{-rt})$	Learning curve; diffusion of information by mass media; intravenous infusion of a medication; Newton's law of cooling; depreciation; sales of new products; growth of a business

EXAMPLE 7 Some bone artifacts were found at the Lindenmeier site in northeastern Colorado and tested for their carbon-14 content. If 12.5% of the original carbon-14 was still present, what is the probable age of the artifacts?

Solution $\frac{dP}{dt} = -.0001216P$

From Theorem 2,

$$P = Me^{-.0001216t}$$

From the initial condition, when $t = 0$, $P = P_0$:

$$P_0 = Me^0$$
$$P_0 = M$$

Thus,

$$P = P_0e^{-.0001216t}$$

Since $P/P_0 = 12.5\%$, we can divide both sides by P_0 and substitute this constant:

$$.125 = e^{-.0001216t}$$

Solve for t:

$$-.0001216t = \ln .125$$
$$t \approx 17{,}100$$

The artifact is about 17,100 years old. ■

Four common growth and decay models are summarized in Table 15.1. They only begin to represent a topic that could fill an entire course.

As you work through the problems of this section, be aware that we have discussed only the simplest techniques for solving differential equations. You are still not able to solve most differential equations, so the problem set comprises a very select set of differential equations for you to solve. The general solution of differential equations is substantial enough to constitute a separate mathematics course.

15.2 Problem Set

Solve the differential equations in Problems 1–20.

1. $\frac{dy}{dx} = x^2$

2. $\frac{dy}{dx} = 5$

3. $\frac{dy}{dx} = 8x - 10$

4. $\frac{dy}{dx} = 5 - x$

5. $y' = 4x^3 - 3x^2 - 5$

6. $y' = 10 - 9x^2 + 2x$

7. $5\frac{dy}{dx} = e^x$

8. $8\frac{dy}{dx} = \sqrt{x}$

9. $12\frac{dy}{dx} = \sqrt{5x+1}$

10. $\frac{3}{2}\frac{dy}{dx} = e^{4x-3}$

11. $yy' = x$

12. $yy' = 5x^2$

13. $y^2\frac{dy}{dx} = x^3 - 3$

14. $y\frac{dy}{dx} = \sqrt{x} + e^{3x} + 4$

15. $\frac{dy}{dx} = 2xy$

16. $\frac{dy}{dx} = 4x^2y - 3xy + y$

17. $\frac{dP}{dt} = .02P$

18. $\frac{dP}{dt} = 1{,}250P$

19. $\frac{dN}{dt} = .001N$

20. $\frac{dN}{dt} = .15N$

Solve the differential equations in Problems 21–28 for a particular solution.

21. $\frac{dy}{dx} = x^2y^{-2}$ for $y = 5$ when $x = 0$

22. $x^2\frac{dy}{dx} = y$ for $y = 2$ when $x = 1$

23. $\frac{dP}{dt} = .02P$ for $P = e^3$ when $t = 0$

24. $\frac{dN}{dt} = .12N$ for $N = 2.3$ million when $t = 0$ (estimate to two places)

25. $x\frac{dy}{dx} - y\sqrt{x} = 0$ for $y = 1$ when $x = 1$

26. $5xy - 3y = \frac{dy}{dx}$ for $y = e^2$ when $x = 2$

27. $\frac{dy}{dx} = \frac{xy}{1+x^2}$ for $y = 2$ when $x = -1$

28. $\frac{dy}{dx} = \frac{1+x^2}{xy}$ for $y = 4$ when $x = 2$

APPLICATIONS

29. You read a newspaper story that says that the crime rate in a certain area is increasing at a constant rate of 3% per year. If 3,500 major crimes were reported in 1986, estimate the number of major crimes in 1996.

30. Domar's capital expansion model is based on the present value of an investment, P, the investment productivity (a constant h), the marginal productivity to consume (a constant k), and time, t:

$$\frac{dP}{dt} = hkP$$

Solve this equation for P where the initial investment is \$150,000.

31. Suppose that the growth rate of bacteria in milk is a constant 10% per hour and that the maximum number (in millions) permitted is 1,000. How long (to the nearest hour) will the milk be acceptable if we assume that initially there are 12 million bacteria?

32. If the marginal price

$$\frac{dp}{dx}$$

for x units of demand per day is proportional to the price, p, and if the demand at \$5 is 400 while the demand at \$25 is 0, find the price–demand equation.

33. If the marginal price

$$\frac{dp}{dx}$$

for x units of supply per day is proportional to the price, p, and if at a price of \$10 there is no supply, but at a price of \$20 the daily supply is 100, find the price–supply equation.

34. An artifact was found at the Lindenmeier site with 11.8% of the original carbon-14 present. What is the probable age of the artifact?

35. A skull was discovered at the Debert site in Nova Scotia. Tests showed that 28% of the original carbon-14 was still present. What is the probable age of the artifact?

36. Suppose that the blood pressure in the aorta changes between beats with respect to time t according to the formula

$$\frac{dP}{dt} = -kt$$

for some constant k. Solve $P = P(t)$ so that $P(0) = P_0$.

37. Suppose that the amount of a drug injected into a patient decreases at a rate proportional to the amount present according to the formula

$$\frac{dQ}{dt} = -.05t$$

where t is the time and $Q(0) = 5$ milliliters. Solve the equation $Q = Q(t)$.

38. How many milliliters of the drug described in Problem 37 are in the body after 4 hours?

39. Suppose the rate of television advertising exposure to a new product is proportional to the number of those who have not seen the product out of L possible viewers. If no

one is aware of the product at the start of the advertising campaign and 10% of L are aware of the product after two weeks, solve

$$\frac{dP}{dt} = r(L - P)$$

for $P = P(t)$, the number of people who are aware of the product after t days of advertising.

40. How many people are aware of the product described in Problem 39 after 30 days? Assume that there are 2 million possible viewers.

15.3 Integration by Parts

In Section 14.1 we were able to restate integration formulas corresponding to the power, sum, and difference differentiation rules. In Section 14.2 we used substitution to reverse the chain rule. Now we develop an integration formula from the product rule for differentiation.

Suppose that u and v are differentiable functions of x. Then

$$\frac{d}{dx}(uv) = u\frac{dv}{dx} + v\frac{du}{dx}$$

or, using differentials,

$$d(uv) = u\,dv + v\,du$$

Integrate both sides to obtain

$$\int d(uv) = \int u\,dv + \int v\,du$$

$$uv = \int u\,dv + \int v\,du$$

Solve this equation for $\int u\,dv$ and obtain a formula called **integration by parts:**

Integration by Parts

$$\int u\,dv = uv - \int v\,du$$

EXAMPLE 1 Evaluate $\int xe^x\,dx$ using integration by parts.

Solution To use integration by parts, you must choose u and dv so that the new integral is easier to integrate than the original.

Integrate by parts:

Formula:

$$\int u \; dv = uv - \int v \; du$$

$$\int xe^x\,dx = xe^x - \int e^x\,dx$$

$$= xe^x - e^x + C$$

Let $u = x$ and $dv = e^x\,dx$;

then $du = dx$ $\quad v = \int e^x\,dx = e^x$

* See note below

Integration by parts is often confusing the first time you try to do it because there is no absolute choice for u and dv. Experience will help you in deciding. In the previous example, you might have chosen

$u = e^x$ and $dv = x\,dx$

$du = e^x\,dx$ $\quad v = \int x\,dx = \dfrac{x^2}{2}$

Then

$$\int e^x x\,dx = e^x\frac{x^2}{2} - \int \frac{x^2}{2} e^x\,dx$$

$$\int u \; dv = u \; v - \int v \; du$$

$$= \frac{1}{2}x^2 e^x - \frac{1}{2}\int x^2 e^x\,dx$$

* *Note:* When we wrote $v = \int e^x\,dx = e^x$ above, we should have written $v = \int e^x\,dx = e^x + K$. Then our work would have looked like

$$\int xe^x\,dx = x(e^x + K) - \int (e^x + K)\,dx$$

$$= xe^x + xK - \int e^x\,dx - \int K\,dx$$

$$= xe^x + xK - \int e^x\,dx - Kx + C_1$$

$$= xe^x - \int e^x\,dx + C_1$$

$$= xe^x - e^x + C_2 + C_1$$

$$= xe^x - e^x + C \qquad \text{where} \qquad C = C_1 + C_2$$

The intermediate constant K in integration by parts will always drop out, so we usually omit the constant when calculating v from dv and include only the constant C at the end.

Note, however, that this choice of u and dv leads to a more complicated form than the original. Therefore, when you are integrating by parts, if you make a choice for u and dv that leads to a more complicated form than when you started, consider going back and making another choice for u and dv.

Sometimes you may have to integrate by parts more than once, as illustrated by Example 2.

EXAMPLE 2 Evaluate $\int 5x^2 e^{3x}\,dx$ using integration by parts.

Solution

Let $u = x^2$ and $dv = e^{3x}\,dx$

$du = 2x\,dx$ $\quad v = \int e^{3x}\,dx = \frac{1}{3}e^{3x}$

Integrate by parts:

$$\int 5\underset{u}{x^2}\,\underset{dv}{e^{3x}\,dx} = 5\left[\underset{u}{x^2}\underset{v}{\left(\frac{1}{3}e^{3x}\right)} - \int \underset{v}{\frac{1}{3}e^{3x}}\,\underset{du}{2x\,dx}\right]$$

$$= \frac{5}{3}x^2e^{3x} - \frac{10}{3}\int xe^{3x}\,dx$$

$$= \frac{5}{3}x^2e^{3x} - \frac{10}{3}\int xe^{3x}\,dx$$ Now use integration by parts again.

$u = x$ and $dv = e^{3x}\,dx$

$du = dx$ $\quad v = \int e^{3x}\,dx = \frac{1}{3}e^{3x}$

$$= \frac{5}{3}x^2e^{3x} - \frac{10}{3}\left[x\left(\frac{1}{3}e^{3x}\right) - \int \frac{1}{3}e^{3x}\,dx\right]$$

$$= \frac{5}{3}x^2e^{3x} - \frac{10}{9}xe^{3x} + \frac{10}{9}\int e^{3x}\,dx$$

$$= \frac{5}{3}x^2e^{3x} - \frac{10}{9}xe^{3x} + \frac{10}{9}\frac{e^{3x}}{3} + C$$

$$= \frac{5}{3}x^2e^{3x} - \frac{10}{9}xe^{3x} + \frac{10}{27}e^{3x} + C$$ ■

EXAMPLE 3 Evaluate $\int \ln x\, dx$ using integration by parts.

Solution

Let $u = \ln x$	and	$dv = dx$
$du = \frac{1}{x}\,dx$		$v = \int dx = x$

$$\int \ln x\, dx = x \ln x - \int x\left(\frac{1}{x}\right) dx$$
$$= x \ln x - x + C$$

■

If you use integration by parts with the definite integral, be sure to evaluate the first part at both of the limits of integration, as illustrated by Example 4.

WARNING: Do not forget this evaluation.

EXAMPLE 4 $\displaystyle\int_0^1 xe^{2x}\, dx = \frac{1}{2}xe^{2x}\Big|_0^1 - \frac{1}{2}\int_0^1 e^{2x}\, dx$

$u = x$	and	$dv = e^{2x}\, dx$
$du = dx$		$v = (\frac{1}{2})e^{2x}$

It is usually easier to simplify the algebra and then do one evaluation here at the end of the problem.

$$= \left(\frac{1}{2}xe^{2x} - \frac{1}{4}e^{2x}\right)\Big|_0^1$$
$$= \left(\frac{1}{2}e^2 - \frac{1}{4}e^2\right) - \left(0 - \frac{1}{4}\right)$$
$$= \frac{1}{4}e^2 + \frac{1}{4}$$
$$\approx 2.097264$$

■

Remember that integration by parts, as with other integration methods, is not a method that "always works." There are many functions whose integrals cannot be found by the methods described in this book. In these cases you can use approximate integration for definite integrals as discussed in Section 14.6.

COMPUTER APPLICATION

Do not forget that approximate (or computer) integration can work for definite integrals that might be difficult to integrate to a closed form. For Example 4, we found the exact value to be $.25e^2 + .25$, which can be approximated as 2.097264. We have shown the *Riemann sum* program for this same example.

```
Y = Xe^(2X)

How many rectangles? 512

a = 0      b = 1.00000

norm = 1.953E-4

Riemann sum = 2.09726
```

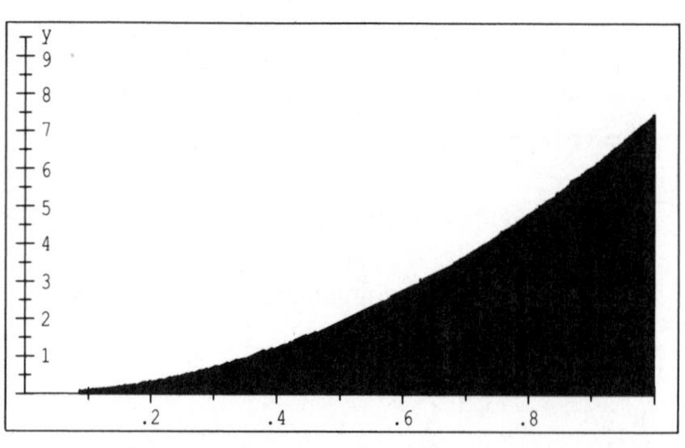

15.3 Problem Set

Evaluate the integrals in Problems 1–34.

1. $\int 12xe^{3x}\,dx$
2. $\int 60xe^{5x}\,dx$
3. $\int_0^1 xe^{-x}\,dx$
4. $\int_0^{1/2} 9xe^{-2x}\,dx$
5. $\int x\sqrt{1-x}\,dx$
6. $\int 3x\sqrt{1-2x}\,dx$
7. $\int x(x+2)^3\,dx$
8. $\int x(3x-1)^4\,dx$
9. $\int \ln(x+1)\,dx$
10. $\int \ln(2x-1)\,dx$
11. $\int \frac{x\,dx}{\sqrt{1-x}}$
12. $\int \frac{x\,dx}{\sqrt{2-3x}}$
13. $\int 6x^2e^{2x}\,dx$
14. $\int 8x^2e^{-2x}\,dx$
15. $\int x^2\sqrt{1-2x}\,dx$
16. $\int 5x^2\sqrt{1-x}\,dx$
17. $\int x^3\sqrt[3]{1-x^2}\,dx$
18. $\int x\ln x\,dx$
19. $\int x^2\ln x\,dx$
20. $\int x^2\ln 3x\,dx$
21. $\int 6x^2e^{4x}\,dx$
22. $\int 24x^2e^{3x}\,dx$
23. $\int \frac{x^3\,dx}{\sqrt{1-x^2}}$
24. $\int \frac{x^3\,dx}{\sqrt{x^2+1}}$
25. $\int x^3e^{x^2}\,dx$
26. $\int x^3\sqrt{x^2+1}\,dx$
27. $\int_1^2 \ln 3x\,dx$
28. $\int_1^3 \ln 2x\,dx$

29. $\int_0^1 (x - 4)e^x\,dx$

30. $\int_0^2 (2x - 3)e^x\,dx$

31. $\int_e^{e^2} \frac{dx}{x \ln x}$

32. $\int_e^{e^2} \sqrt{x} \ln x\,dx$

33. $\int_1^e (\ln x)^2\,dx$

34. $\int_{1/3}^e 3(\ln 3x)^2\,dx$

35. Derive the formula

$$\int x^m e^x\,dx = x^m e^x - m\int x^{m-1} e^x\,dx$$

36. Derive the formula

$$\int (\ln x)^m\,dx = x(\ln x)^m - m\int (\ln x)^{m-1}\,dx$$

37. **a.** Evalute the following integral by parts:

$$\int \frac{x^3}{x^2 + 1}\,dx$$

b. Evaluate the integral in part **a** by first dividing the integrand.

PPLICATIONS

38. The marginal cost and revenue equations (in millions of dollars) for a product are given by

$$R'(x) = \frac{x}{\sqrt{x - 1}} \quad \text{and} \quad C'(x) = .2x$$

where x is the time in years. The area between the graphs of the marginal functions for a time period where $R'(x) > C'(x)$ is the total accumulated profit bounded by two values of x. What is the total profit for years 2 through 5?

39. A contaminated water supply is treated, and the rate of harmful bacteria t days after the treatment is given by

$$N'(t) = te^{-.05t} \qquad 0 \le t \le 10$$

where $N(t)$ is the number of millions of bacteria per liter of water. The initial count is 1 million.

a. Find $N(t)$.

b. Find the total bacteria count after 7 days.

40. If the marginal profit is given by

$$P'(t) = \frac{2t}{(1 + 3t)^{2/3}}$$

where t is the time in years, find P if the profit at time 0 is $0.

41. Maintenance costs for a certain building can be approximated by

$$M'(x) = 100x^2 + \frac{5{,}000x}{x^2 + 1}$$

where x is the age of the building in years and $M(x)$ is the total accumulated cost of maintenance for x years. Find the total maintenance costs for the period from 5 to 10 years after the building was built.

15.4 Using Tables of Integrals

Professional mathematicians and scientists often integrate functions by using tables of integrals. One of the most common sources of integral tables is the Chemical Rubber Company's *Standard Mathematical Tables*, which is much more extensive than what we need for this course. We have, however, a shortened version on the endpapers (the inside covers) of this book.

To use a Brief Integral Table, first classify the integral by type. The types listed in our table include the following:

Forms containing $a + bx$
Forms containing $\sqrt{x^2 \pm a^2}$ or $\sqrt{a^2 - x^2}$
Logarithmic forms
Exponential forms

More extensive tables of integrals are available, but these four fundamental types will serve our purposes, and if you learn how to use this integral table you will be able to use more extensive tables.

After deciding which form applies, match the individual type with the problem at hand by making appropriate choices for the constants. More than one form may apply, but the results derived by using different formulas will be the same (except for the constant).

Take a few moments to look at the **Brief Integral Table**. Notice that this integral table has two basic types of **integration formulas**. The first gives a formula that is the antiderivative, while the second simply rewrites the integral in another form. Examples 1 and 2 illustrate each of these types.

EXAMPLE 1 Evaluate $\int x^2(3-x)^5\,dx$.

Solution This is an integral of the form $a + bx$; from the Brief Integral Table, we see that Formula 11 applies, where $a = 3$, $b = -1$, and $n = 5$:

$$\int x^2(3-x)^5\,dx = \frac{1}{(-1)^3}\left[\frac{(3-x)^{5+3}}{5+3} - 2(3)\frac{(3-x)^{5+2}}{5+2} + 3^2\frac{(3-x)^{5+1}}{5+1}\right] + C$$

$$= -\left[\frac{(3-x)^8}{8} - 6\frac{(3-x)^7}{7} + 9\frac{(3-x)^6}{6}\right] + C$$

$$= -\frac{1}{8}(3-x)^8 + \frac{6}{7}(3-x)^7 - \frac{3}{2}(3-x)^6 + C$$

$$= -\frac{1}{112}(3-x)^6[14(3-x)^2 - 96(3-x) + 168] + C$$

$$= -\frac{1}{112}(3-x)^6[14(9-6x+x^2) - 288 + 96x + 168] + C$$

$$= -\frac{1}{112}(3-x)^6(14x^2 + 12x + 6) + C$$

$$= -\frac{1}{56}(3-x)^6(7x^2 + 6x + 3) + C$$

■

EXAMPLE 2 Evaluate $\int \ln^4 x\,dx$.

Solution This is an integral in logarithmic form; from the Brief Integral Table we see that Formula 34 applies. Notice that this form involves another integral.

$$\int \ln^4 x\,dx = x\ln^4 x - 4\int \ln^{4-1} x\,dx$$

$$= x\ln^4 x - 4\left(x\ln^3 x - 3\int \ln^{3-1} x\,dx\right) \qquad \text{Formula 34, again}$$

$$= x\ln^4 x - 4x\ln^3 x + 12\int \ln^2 x\,dx$$

Now use Formula 33:

$$= x\ln^4 x - 4x\ln^3 x + 12(x\ln^2 x - 2x\ln x + 2x) + C$$

$$= x\ln^4 x - 4x\ln^3 x + 12x\ln^2 x - 24x\ln x + 24x + C$$

■

It is often necessary to make substitutions before using one of the integration formulas, as shown in Example 3.

EXAMPLE 3 Evaluate $\displaystyle\int \frac{x\,dx}{\sqrt{8-5x^2}}$.

Solution This is an integral of the form $\sqrt{a^2-x^2}$, but it does not exactly match any of the Formulas 15–29. Note, however, that, except for the coefficient of 5, it is like Formula 24. Let $u=\sqrt{5}\,x$ (so $u^2=5x^2$); then $du=\sqrt{5}\,dx$:

$$x=\frac{u}{\sqrt{5}} \qquad dx=\frac{du}{\sqrt{5}}$$

$$\int \frac{x\,dx}{\sqrt{8-5x^2}}=\int \frac{\frac{u}{\sqrt{5}}\cdot\frac{du}{\sqrt{5}}}{\sqrt{8-u^2}}=\frac{1}{5}\int \frac{u\,du}{\sqrt{8-u^2}}$$

Now apply Formula 24, where $a^2=8$:

$$\begin{aligned}\frac{1}{5}\int \frac{u\,du}{\sqrt{8-u^2}}&=\frac{1}{5}\left(-\sqrt{8-u^2}\right)+C\\&=-\frac{1}{5}\sqrt{8-5x^2}+C\end{aligned}$$

■

As you can see from Example 3, using an integral table is not a trivial task. In fact, other methods of integration are often preferable, if possible. For Example 3, you can let $u=8-5x^2$ and integrate by substitution:

$$u=8-5x^2 \qquad \text{and} \qquad du=-10x\,dx \qquad \text{so} \qquad dx=\frac{du}{-10x}$$

Substitute:

$$\begin{aligned}\int \frac{x\,dx}{\sqrt{8-5x^2}}&=\int \frac{x\cdot\frac{du}{-10x}}{\sqrt{u}}\\&=\frac{-1}{10}\int u^{-1/2}\,du\\&=\frac{-1}{10}2u^{1/2}+C\\&=\frac{-1}{5}\sqrt{8-5x^2}+C\end{aligned}$$

This answer is the same, of course, as the one we obtained in Example 3. The point of this calculation is to emphasize that you should try simple methods of integration before turning to the table of integrals.

Also, since sometimes more than one formula can be used, it is to our advantage to pick the one that best simplifies the algebra, as shown by Example 4.

EXAMPLE 4 Evaluate $\int \sqrt{1-3x}\,dx$.

Solution *Method* I: Substitution (without integral tables). Let $u = 1 - 3x$, then $du = -3\,dx$, and

$$\begin{aligned}\int \sqrt{1-3x}\,dx &= \int u^{1/2}\left(\frac{du}{-3}\right)\\ &= -\frac{1}{3}\frac{u^{3/2}}{\frac{3}{2}} + C\\ &= -\frac{2}{9}u^{3/2} + C\\ &= -\frac{2}{9}(1-3x)^{3/2} + C\end{aligned}$$

Method II: Use Formula 9, with $a = 1$, $b = -3$, and $n = \frac{1}{2}$.

$$\begin{aligned}\int \sqrt{1-3x}\,dx &= \frac{(1-3x)^{3/2}}{(\frac{1}{2}+1)(-3)} + C\\ &= -\frac{2}{9}(1-3x)^{3/2} + C\end{aligned}$$

Method III: Use Formula 13, with $a = 1$, $b = -3$, and $m = 0$:

$$\begin{aligned}\int \sqrt{1-3x}\,dx &= \frac{2}{-3(2\cdot 0+3)}[x^0\sqrt{(1-3x)^3} - 0] + C\\ &= -\frac{2}{9}\sqrt{(1-3x)^3} + C\\ &= -\frac{2}{9}(1-3x)^{3/2} + C\end{aligned}$$

Note: Even though this integral is 0, you must not forget the constant of integration, which was incorporated into the formula.

■

EXAMPLE 5 Evaluate $\int 5x^2\sqrt{3x^2+1}\,dx$.

Solution This is similar to Formula 25, but you must take care of the 5 (using procedural Formula 1) and the 3 (by making a substitution). Let $u = \sqrt{3}\,x$, then $du = \sqrt{3}\,dx$, and

$$\int 5x^2\sqrt{3x^2+1}\,dx = \int 5\left(\frac{u^2}{3}\right)\sqrt{u^2+1}\,\frac{du}{\sqrt{3}}$$

$dx = du/\sqrt{3}$

$3x^2 = u^2$

$x^2 = u^2/3$

$$= \frac{5}{3\sqrt{3}}\int u^2\sqrt{u^2+1}\,du$$

$$= \frac{5}{3\sqrt{3}}\left[\frac{u}{4}\sqrt{(u^2+1)^3} - \frac{1^2u}{8}\sqrt{u^2+1} - \frac{1^4}{8}\ln|u+\sqrt{u^2+1}|\right] + C$$

Note the use of the $\mp$ symbol in conjunction with the $\pm$ symbol in Formula 25. If you use the top symbol in one place, then you must use the top symbol throughout.

$$= \frac{5u}{12\sqrt{3}}(u^2+1)^{3/2} - \frac{5u}{24\sqrt{3}}(u^2+1)^{1/2} - \frac{5}{24\sqrt{3}}\ln|u+\sqrt{u^2+1}| + C$$

$$= \frac{5\sqrt{3}\,x}{12\sqrt{3}}(3x^2+1)^{3/2} - \frac{5\sqrt{3}\,x}{24\sqrt{3}}(3x^2+1)^{1/2} - \frac{5}{24\sqrt{3}}\ln|\sqrt{3}\,x+\sqrt{3x^2+1}| + C$$

$$= \frac{5x}{12}(3x^2+1)^{3/2} - \frac{5x}{24}(3x^2+1)^{1/2} - \frac{5}{24\sqrt{3}}\ln|\sqrt{3}\,x+\sqrt{3x^2+1}| + C \quad ■$$

15.4 Problem Set

Integrate the expressions in Problems 1–12 using the Brief Integral Table. Indicate the formula used.

1. $\int (1+bx)^{-1}\,dx$ **2.** $\int (a+bx)^5\,dx$

3. $\int \frac{x\,dx}{\sqrt{x^2+a^2}}$ **4.** $\int \frac{x\,dx}{\sqrt{a^2-x^2}}$

5. $\int \frac{dx}{x^2\sqrt{x^2-a^2}}$ **6.** $\int \frac{dx}{x^2\sqrt{a^2-x^2}}$

7. $\int x\ln x\,dx$ **8.** $\int x^2\ln x\,dx$

9. $\int x^{-1}\ln x\,dx$ **10.** $\int x^5\ln x\,dx$

11. $\int xe^{ax}\,dx$ **12.** $\int \frac{dx}{a+be^{2x}}$

Evaluate the integrals in Problems 13–64. If you use the Brief Integral Table, state the formula used.

13. $\int \frac{x^2dx}{\sqrt{x^2+1}}$ **14.** $\int \frac{x^2\,dx}{\sqrt{x^3+1}}$

15. $\int \frac{dx}{x^2\sqrt{x^2+16}}$ **16.** $\int \frac{dx}{x^2\sqrt{16x^2+1}}$

17. $\int \frac{x\,dx}{\sqrt{4x^2+1}}$ **18.** $\int \frac{x^3\,dx}{\sqrt{4x^4+1}}$

19. $\int \frac{dx}{x\sqrt{1-9x^2}}$ **20.** $\int \frac{\sqrt{4x^2+1}}{x}\,dx$

21. $\int \frac{x\,dx}{\sqrt{x^2+4}}$ **22.** $\int \frac{dx}{x\sqrt{x^2+4}}$

23. $\int (1 + x)^3\,dx$

24. $\int (1 + 5x)^3\,dx$

25. $\int x(1 + x)^3\,dx$

26. $\int 4x(1 - 6x)^3\,dx$

27. $\int x\sqrt{1 + x}\,dx$

28. $\int x\sqrt{1 + 3x}\,dx$

29. $\int xe^{4x}\,dx$

30. $\int xe^{-5x}\,dx$

31. $\int x\ln 2x\,dx$

32. $\int x\ln 3x\,dx$

33. $\int x^2\ln 5x\,dx$

34. $\int x^3\ln x\,dx$

35. $\int \frac{dx}{3 + 5e^x}$

36. $\int \frac{dx}{1 + e^{5x}}$

37. $\int \frac{dx}{1 + e^{2x}}$

38. $\int \frac{e^{2x}\,dx}{1 + e^{2x}}$

39. $\int \frac{dx}{\sqrt{1 + x}}$

40. $\int \frac{x^3 + 2x + 1}{x}\,dx$

41. $\int x^2(1 + x)^3\,dx$

42. $\int x^3(1 + x)^3\,dx$

43. $\int 5x^2(1 - x)^3\,dx$

44. $\int 3x^2(3 - 2x)^3\,dx$

45. $\int 2x\sqrt{1 - 2x}\,dx$

46. $\int 3x\sqrt{4 - 3x}\,dx$

47. $\int x^2(2 + 3x)^3\,dx$

48. $\int x^2\sqrt{4x^2 + 1}\,dx$

49. $\int x^2\sqrt{4x^3 + 1}\,dx$

50. $\int x^2e^x\,dx$

51. $\int x^2e^{3x}\,dx$

52. $\int \sqrt{2 + 9x^2}\,dx$

53. $\int \frac{\sqrt{2 + 9x}}{x^2}\,dx$

54. $\int \frac{\sqrt{2 + 9x}}{x}\,dx$

55. $\int \frac{\sqrt{2 + 9x^2}}{x}\,dx$

56. $\int 3\sqrt{5 + 4x^2}\,dx$

57. $\int \frac{1}{\sqrt{2 + 9x^2}}\,dx$

58. $\int x\sqrt{9x^2 - 5}\,dx$

59. $\int 5x\sqrt{9 - 16x^2}\,dx$

60. $\int \ln^5 x\,dx$

61. $\int \frac{\sqrt{x^2 - 1}}{x^2}\,dx$

62. $\int x^3(x^4 + 1)^8\,dx$

63. $\int x^2\sqrt{3 + 10x}\,dx$

64. $\int (\ln 3x + e^{5x} + \sqrt{x^2 + 1})\,dx$

APPLICATIONS

65. A local citizens' group estimates the rate at which pollutants are being dumped into the Eel River according to the formula

$$P(t) = \frac{5{,}000}{t\sqrt{100 + t^2}} \qquad t \geq 1$$

where t is the month and $P(t)$ is the number of tons of pollutants per month. What is the total amount of pollutants that will be dumped into the river in the next 11 months (that is, from $t = 1$ to $t = 12$)?

66. Davis Industries has production costs for a product determined by the function

$$C(x) = \frac{100x}{\sqrt{x^2 + 100}} + 50$$

where $C(x)$ is the cost to produce the xth item. What is the approximate total cost of producing 1,000 items?

67. The sales of Simon's Simple Tool Kit are growing according to the function

$$S(t) = te^{.1t} \qquad t \geq 12$$

where $S(t)$ is the sales in thousands of dollars in the tth month. What are the accumulated sales for the second year (that is, for $t = 12$ to $t = 24$)?

15.5 Improper Integrals

The graph of the function $f(x) = e^{-x}$ is shown in Figure 15.7**a** on page 738. Note that the region between the graph of $f(x) = e^{-x}$ and the x-axis over the interval $[0, \infty]$ is unbounded. It is possible to define the area of this unbounded region. To do this, we consider a vertical line $x = b$, as shown in Figure 15.7**b**. We find the

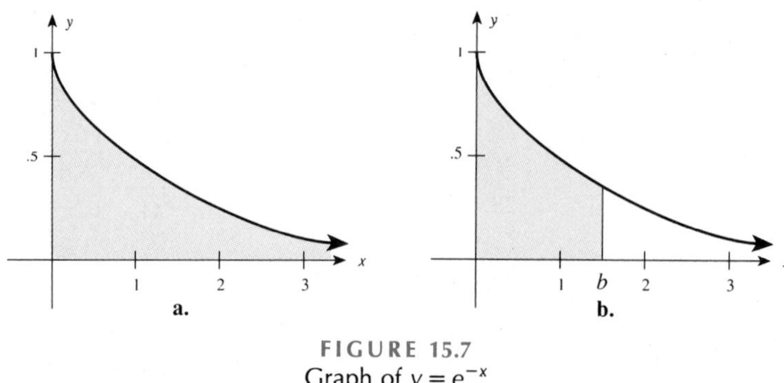

FIGURE 15.7
Graph of $y = e^{-x}$

area of the shaded region:

$$A(b) = \int_0^b e^{-x}\,dx = -e^{-x}\Big|_0^b = (-e)^{-b} - (-e^0) = 1 - e^{-b}$$

Now we let the vertical line $x = b$ slide to the right; that is, suppose $b \to \infty$. The area changes according to

$$\lim_{b\to\infty} (1 - e^{-b}) = 1 \quad \text{since} \lim_{b\to\infty} e^{-b} = 0$$

Now

$$\lim_{b\to\infty} A(b) = \lim_{b\to\infty} (1 - e^{-b}) = 1$$

This limit is defined to be the area of the shaded region in Figure 15.7**a**. The

$$\lim_{b\to\infty} \int_0^b e^{-x}\,dx \quad \text{is denoted by} \quad \int_0^\infty e^{-x}\,dx$$

and is called an **improper integral**.

Improper Integral

If f is continuous over the indicated interval and the limit exists, then the following integrals are called **improper integrals:**

$$\int_a^\infty f(x)\,dx = \lim_{b\to\infty} \int_a^b f(x)\,dx \qquad \text{if the limit exists}$$

$$\int_{-\infty}^b f(x)\,dx = \lim_{a\to-\infty} \int_a^b f(x)\,dx \qquad \text{if the limit exists}$$

$$\int_{-\infty}^\infty f(x)\,dx = \int_{-\infty}^c f(x)\,dx + \int_c^\infty f(x)\,dx$$

where c is any point on $(-\infty, \infty)$ where both improper integrals on the right exist

If the indicated limit exists, then the improper integral is said to **converge**; if the limit does not exist, then the improper integral is said to **diverge**.

EXAMPLE 1 Evaluate the following integrals.

a. $\int_1^{\infty} 2x^{-3}\,dx$ **b.** $\int_{-\infty}^{-1} \frac{dx}{x}$ **c.** $\int_{-\infty}^{\infty} xe^{-x^2}\,dx$

Solution **a.** $\int_1^{\infty} 2x^{-3}\,dx = \lim_{b\to\infty} \int_1^b 2x^{-3}\,dx = \lim_{b\to\infty}(-x^{-2})\Big|_1^b = \lim_{b\to\infty}(-b^{-2}+1) = 1$

b. $\int_{-\infty}^{-1} \frac{dx}{x} = \lim_{a\to-\infty} \int_a^{-1} x^{-1}\,dx = \lim_{a\to-\infty} (\ln|x|)\Big|_a^{-1} = \lim_{a\to-\infty} (0 - \ln|a|)$

This limit does not exist, so we say that this integral diverges.

c.
$$\begin{aligned}
\int_{-\infty}^{\infty} xe^{-x^2}\,dx &= \int_{-\infty}^{0} xe^{-x^2}\,dx + \int_0^{\infty} xe^{-x^2}\,dx \\
&= \lim_{a\to-\infty} \int_a^0 xe^{-x^2}\,dx + \lim_{b\to\infty} \int_0^b xe^{-x^2}\,dx \\
&= \lim_{a\to-\infty} \int_{x=a}^{x=0} \frac{-1}{2} e^u\,du + \lim_{b\to\infty} \int_{x=0}^{x=b} \frac{-1}{2} e^u\,du \\
&= \lim_{a\to-\infty} (-.5e^{-x^2})\Big|_a^0 + \lim_{b\to\infty}(-.5e^{-x^2})\Big|_0^b \\
&= \lim_{a\to-\infty} (-.5 + .5e^{-a^2}) + \lim_{b\to\infty}(-.5e^{-b^2} + .5) \\
&= -.5 + 0 + 0 + .5 = 0
\end{aligned}$$

Let $u = -x^2$
$du = -2x\,dx$

■

EXAMPLE 2 The **capital value**, V, of a property over T years is given by

$$V = \int_0^T Re^{-rt}\,dt$$

where R is the annual rent, or income, and r is the current interest rate. Suppose the current interest rate is 8% and you have an indeterminant lease (no date of termination) paying \$60,000 per year. What is the capital value?

Solution
$$\begin{aligned}
V &= \int_0^{\infty} 60{,}000e^{-.08t}\,dt = \lim_{b\to\infty} \int_0^b 60{,}000e^{-.08t}\,dt = 60{,}000 \lim_{b\to\infty} \int_0^b e^{-.08t}\,dt \\
&= 60{,}000 \lim_{b\to\infty} \frac{e^{-.08t}}{-.08}\Bigg|_0^b = -750{,}000 \lim_{b\to\infty}(e^{-.08b} - 1) \\
&= -750{,}000(0-1) = 750{,}000
\end{aligned}$$

The capital value for the lease is \$750,000. ■

15.5 Problem Set

Evaluate the integrals in Problems 1–22 that converge.

1. $\int_0^{\infty} e^{-x}\,dx$
2. $\int_{-\infty}^{0} e^{-x}\,dx$
3. $\int_0^{\infty} e^{-x/2}\,dx$
4. $\int_1^{\infty} \frac{dx}{x^4}$
5. $\int_1^{\infty} \frac{dx}{x^3}$
6. $\int_1^{\infty} \frac{dx}{\sqrt{x}}$
7. $\int_1^{\infty} \frac{dx}{\sqrt[3]{x}}$
8. $\int_1^{\infty} \frac{dx}{x^{.9}}$

9. $\int_1^\infty \frac{dx}{x^{1.1}}$

10. $\int_1^\infty \frac{x\,dx}{\sqrt{x^2+2}}$

11. $\int_{-\infty}^0 \frac{2x\,dx}{x^2+1}$

12. $\int_1^\infty \frac{x\,dx}{(1+x^2)^2}$

13. $\int_{-\infty}^\infty x^2\,dx$

14. $\int_{-\infty}^\infty (x^2+1)^{-1/2}\,dx$

15. $\int_{-\infty}^\infty \frac{3x\,dx}{(3x^2+2)^3}$

16. $\int_e^\infty \frac{dx}{x\ln x}$

17. $\int_e^\infty \frac{dx}{x(\ln x)^2}$

18. $\int_1^\infty \ln x\,dx$

19. $\int_1^\infty \frac{e^{-\sqrt{x}}}{\sqrt{x}}\,dx$

20. $\int_1^\infty \frac{x^2}{\sqrt[3]{x^3+2}}\,dx$

21. $\int_1^\infty \frac{dx}{x^2\sqrt{x^2+4}}$

22. $\int_{-\infty}^1 \frac{dx}{x^2\sqrt{4-x^2}}$

APPLICATIONS

23. Suppose the current interest rate is 6% and you have an indeterminant lease paying \$20,200 per year. What is the capital value?

24. Suppose the current interest rate is 11% and you have an indeterminant lease paying \$3,600 per year. What is the capital value?

25. Show that the capital value of a rental property with an indeterminant lease paying an annual rent of \$$R$ with a current interest rate of r percent is R/r.

26. The amount of radioactive material being released into the atmosphere annually—that is, the amount present at time T—is given by

$$A = \int_0^T Pe^{-rt}\,dt$$

If a recent United Nations publication estimates that $r = .002$ and $P = 200$ millirems, estimate the total future buildup of radioactive material in the atmosphere.

27. Suppose that $r = .05$ in Problem 26. Estimate the total future buildup of radioactive material in the atmosphere.

28. Show that the buildup of radioactive material in the atmosphere as reported in Problem 26 approaches a limiting value of P/r.

29. Find the area of the unbounded region between the x-axis and the curve $y = \frac{2}{(x-4)^3}$ for $x \geq 6$.

30. Find the area of the unbounded region between the x-axis and the curve $y = \frac{2}{(x-4)^3}$ for $x \leq 2$.

31. Suppose that an endowment produces a perpetual income with a rate of income (in dollars per year)

$$f(t) = 25{,}000e^{.08t}$$

The capital value formula for a variable rate of income is the same as the one in Example 2 with the constant R replaced by the rate of flow function. Find the capital value at 12% compounded continuously.

32. The rate at which a chemical is being released into a river at time t is given by $455e^{-.02t}$ tons per year. Find the total amount of chemical that will be released into the river in the indefinite future.

33. Suppose that an oil well produces $P(t)$ thousand barrels of crude oil per month according to the formula

$$P(t) = 100e^{-.02t} - 100e^{-.1t}$$

where t is the number of months that the well has been in production. What is the total amount of oil produced by the oil well?

*15.6 Probability Density Functions

The models we have considered so far have been deterministic models; now we turn to a **probabilistic model**, used for situations that are random in character and that attempts to predict the outcomes of these events with a certain stated or known degree of accuracy. For example, if we toss a coin, it is impossible to predict in advance whether the outcome will be a head or a tail. Our intuition tells us that the outcome is equally likely to be a head or a tail, and somehow we sense that if we repeat the experiment of tossing a coin a large number of times, heads will occur

* The material from Chapters 6–8 needed for this section is reviewed here, so that it is not necessary to have studied those chapters. If these concepts are new to you, this section will take extra time. This is an optional section.

"about half the time." To check this out, I recently flipped a coin 1,000 times and obtained 460 heads and 540 tails. The percentage of heads is $\frac{460}{1,000} = .46 = 46\%$, which is called the **relative frequency**.

Relative Frequency of a Repeated Experiment

If an experiment is repeated n times and an event occurs m times, then

$$\frac{m}{n}$$

is called the **relative frequency** of the event.

Our task is to create a model that will assign a number p, called the *probability of an event*, which will represent the relative frequency. This means that for a *sufficiently large number of repetitions* of an experiment

$$p \approx \frac{m}{n}$$

Probabilities can be obtained in one of three ways:

1. *Theoretical probabilities* (also called *a priori* models) are obtained by logical reasoning. For example, the probability of rolling a die and obtaining a 3 is $\frac{1}{6}$ because there are six possible outcomes, each with an equal chance of occurring, so a 3 should appear $\frac{1}{6}$ of the time.
2. *Empirical probabilities* (also called *a posteriori* models) are obtained from experimental data. For example, an assembly line producing brake assemblies for General Motors produces 1,500 brakes per day. The probability of a defective brake can be obtained by experimentation. Suppose the 1,500 brakes are selected at random and are tested; 3 are found to be defective. Then the relative frequency, or probability, is

$$\frac{3}{1,500} = .002 \text{ or } .2\%$$

3. *Subjective probabilities* are obtained from experience and indicate a measure of "certainty." For example, a TV reporter studies the satellite maps and issues a prediction about tomorrow's weather based on experience under similar circumstances: "80% chance of rain tomorrow."

A probability measure must conform to these different ways of using the word *probability*. We begin by defining some terms. An **experiment** is the observation of any physical occurrence. A **sample space** of an experiment is the set of all possible outcomes. An **event** is a subset of the sample space. If an event is the empty set, it is called the **impossible event**; and if it has only one element, it is called a **simple event**.

EXAMPLE 1 List the sample space for each experiment, and then list one event for each of the sample spaces and state whether it is a simple event.

a. Experiment 1: recording the results of the UCLA football team in a given season

b. Experiment 2: simultaneously tossing a coin and rolling a die

c. Experiment 3: installing a computer chip and recording the time to failure

Solution **a.** Experiment 1: $S = \{w, l, t\}$
where w, l, and t denote win, lose, and tie, respectively. An example of an event E is that "UCLA wins," which is $E = \{w\}$. This is a simple event. Another example is event F, "UCLA does not lose," which is $F = \{w, t\}$; this is not a simple event.

b. Experiment 2: $S = \{1H, 1T, 2H, 2T, 3H, 3T, 4H, 4T, 5H, 5T, 6H, 6T\}$
An example of an event is "obtaining an even number or a head," which is $E = \{1H, 2H, 2T, 3H, 4H, 4T, 5H, 6H, 6T\}$. E is not a simple event.

c. Experiment 3: $S = \{t \mid t \geq 0\}$
An example of an event is "the chip lasts more than 1,000 hours," which is $E = \{t \mid t > 1{,}000\}$. E is not a simple event. ■

A **random variable** X is a function that has the following properties: the domain of X is contained in a set S, certain of whose subsets correspond to events for which there is an associated *probability function* (several of which will be defined in this section); the range of the function is a real number. A **discrete random variable** is one that has only a finite number of values, each of which is associated with a nonnegative probability, the sum of these probabilities being 1. If X can assume infinitely many values arranged in a sequence, it is called an **infinite discrete random variable**; and if X can assume any real value on an interval, then it is called a **continuous random variable.**

Suppose the heights of 60 students in a mathematics class are recorded (to the nearest inch), as shown in Table 15.2. The results in the table are often referred to as a **frequency distribution** because such results show the distribution of the numbers for each occurrence.

TABLE 15.2 Frequency Distribution of Student Heights

X = height (in inches)	61 or smaller	62	63	64	65	66	67	68	69	70	71	72	73 or taller
Number of occurrences (frequency)	2	2	2	3	5	9	14	10	5	3	2	1	2
Relative frequency	$\frac{2}{60}$	$\frac{2}{60}$	$\frac{2}{60}$	$\frac{3}{60}$	$\frac{5}{60}$	$\frac{9}{60}$	$\frac{14}{60}$	$\frac{10}{60}$	$\frac{5}{60}$	$\frac{3}{60}$	$\frac{2}{60}$	$\frac{1}{60}$	$\frac{2}{60}$

Probability Distribution

A **probability distribution** is the collection of all values that a random variable assumes along with the probabilities that correspond to these values. Furthermore,

1. $P(X = x_1) + P(X = x_2) + \cdots + P(X = x_n) = 1$
2. $0 \leq P(X = x_i) \leq 1$ for every $1 \leq i \leq n$

It is very useful to display the information in a probability distribution graphically in a bar graph called a **histogram**.

EXAMPLE 2 Represent the information in Table 15.2 in a histogram.

Solution Use the horizontal axis to delineate the values of the random variable (the heights) and the vertical axis to represent the relative frequencies. Draw each bar so that the width is 1 unit. The histogram is shown in Figure 15.8.

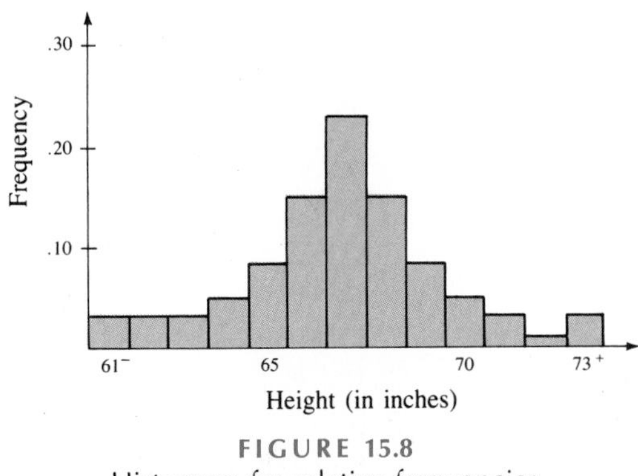

FIGURE 15.8
Histogram for relative frequencies ■

There is a very important relationship between the area of the rectangles in a histogram and the probabilities. Since the width of each bar is 1 unit, the **area of each bar is the probability of occurrence for that random variable**. Therefore the **sum of the areas of the rectangles is 1**. This property is very important in the study of probability distributions.

It is possible to record the heights of the students more accurately. In fact, theoretically, the height of a student could be any positive real number in some domain (say, 0 to 100 inches). We can therefore consider the graph of the relative frequencies as a continuous curve, as shown in Figure 15.9.

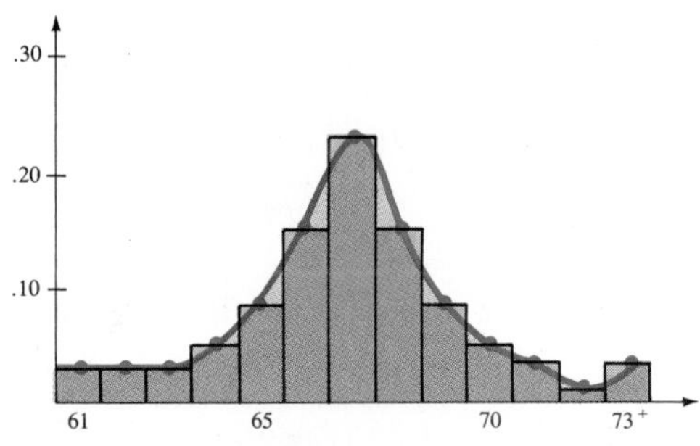

FIGURE 15.9
Relative frequencies as a continuous curve (this curve is drawn by connecting the points at the top of the bars and smoothing the resulting polygon into a curve)

If we now define the **probability** of an event as the relative frequency of occurrence of the event, and if the random variable is assumed to be continuous, then we can use the area under the curve between two values to find the probability that the random variable is between those values. For example,

$$P(65 \le X \le 71)$$

is the probability that a student's height is between 65 and 71 inches and is shown by the shaded region in Figure 15.10.

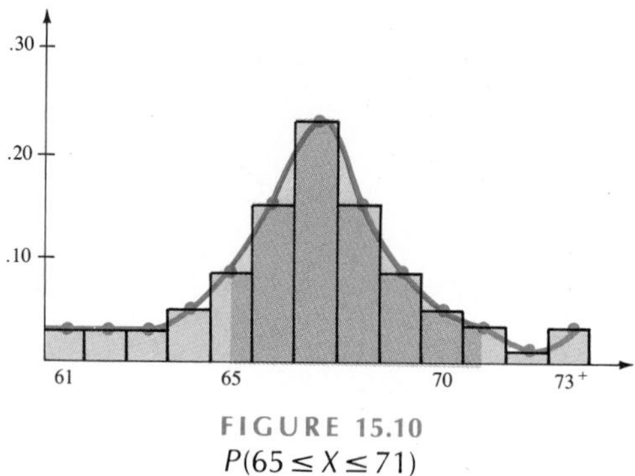

FIGURE 15.10
$P(65 \le X \le 71)$

Since the definite integral can be used to find the area under the graph of $f(x)$ from $x = a$ to $x = b$, if we can find a function f to describe a relative frequency curve, then the probability that a continuous random variable X associated with f will be between a and b is

$$P(a \le X \le b) = \int_a^b f(x)\,dx$$

A function f that can be used to describe a relative frequency curve is called a **probability density function**. Such a function must satisfy certain conditions, as summarized below:

Probability Density Function

If X is a continuous random variable associated with a function f, then f is a **probability density function** if

1. $f(x) \ge 0$ for all x in the domain $(-\infty, \infty)$
2. $\int_{-\infty}^{\infty} f(x)\,dx = 1$
3. If $[a, b]$ is a subinterval of $(-\infty, \infty)$, then

$$P[a \le X \le b] = \int_a^b f(x)\,dx$$

EXAMPLE 3 Show that $f(x) = \dfrac{x}{6} - \dfrac{1}{12}$ is a probability density function on $[1, 4]$.

Solution

1. $\displaystyle\int_1^4 \frac{x}{6} - \frac{1}{12}\,dx = \frac{x^2}{12} - \frac{x}{12}\bigg|_1^4 = \frac{16}{12} - \frac{4}{12} - \frac{1}{12} + \frac{1}{12} = \frac{12}{12} = 1$
2. If x is between 1 and 4, then f is between $\frac{1}{12}$ and $\frac{7}{12}$, so $f(x) \ge 0$ for all X on $[a, b]$.

Thus, f is a probability density function. ■

EXAMPLE 4 Suppose that the life of a computer chip is described by the probability density function

$$f(x) = \frac{72}{35x^3}$$

where x is the number of months it will function properly. The domain for x is the interval $[1, 6]$. What is the probability that the chip will last longer than 4 months?

Solution The probability that the chip will last longer than 4 months is

$$P(4 \le X \le 6) = \int_4^6 \frac{72}{35x^3}\,dx = \frac{72}{35}\frac{x^{-2}}{-2}\Bigg|_4^6 = -\frac{36}{35x^2}\Bigg|_4^6$$

$$= -\frac{1}{35} + \frac{9}{140} = .036$$

■

Sometimes it is necessary to construct a probability density function. For example, suppose $f(x) = x^2$ on $[1, 4]$. Then

$$\int_1^4 x^2 = \frac{x^3}{3}\Bigg|_1^4 = \frac{64}{3} - \frac{1}{3} = \frac{63}{3} = 21$$

If you want to construct a probability density function using f, simply multiply by the reciprocal of the value of the integral, in this case, $\frac{1}{21}$, so that the value of the integral will be 1:

$$f(x) = \frac{1}{21}x^2$$

is a probability density function.

Uniform Probability Distribution

If a density function is a constant (the graph is a horizontal line), then the appropriate model is the **uniform probability distribution**. Suppose the given interval is $[1, 5]$. Then the length of the interval is $5 - 1 = 4$, so if the probability density function is a constant it must be $\frac{1}{4}$ (so that the area is 1), as shown in Figure 15.11 on page 746. Notice that if the interval is $[4, 10]$, then $f(x) = \frac{1}{6}$; if the interval is $[2, 4]$, then $f(x) = \frac{1}{2}$. This observation leads us to the following definition of the uniform probability distribution:

Uniform Distribution

A continuous random variable X is said to be **uniformly distributed** over an interval $[a, b]$ if it has a probability density function

$$f(x) = \frac{1}{b - a} \qquad \text{for } a \le x \le b$$

The graph is shown in Figure 15.11.

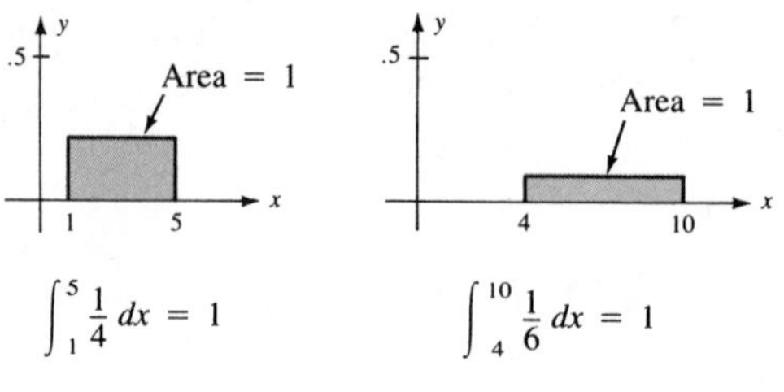

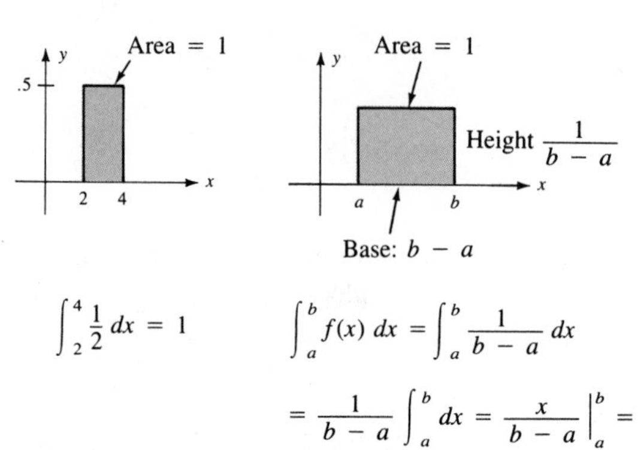

FIGURE 15.11
Some uniform distributions

EXAMPLE 5 Cholesterol levels are artificially introduced into blood samples so that the resulting mixtures are uniformly distributed over the interval [200, 300]. What is the probability that the cholesterol level is between 240 and 265?

Solution The probability density function for x is given by

$$f(x) = \frac{1}{300 - 200} = \frac{1}{100} = .01 \qquad \text{for } 200 \le x \le 300$$

Then the desired probability is

$$\int_{240}^{265} .01\,dx = .01x\Big|_{240}^{265}$$
$$= .01(265 - 240)$$
$$= .25$$

■

Exponential Probability Distribution

A probability density function will often follow an exponential curve. We find this curve in response to the question, "How long do you need to wait if you are observing a sequence of events occurring in time in order to observe the first occurrence of the event?"

Exponential Distribution

A continuous random variable X is said to be **exponentially distributed** over an interval $[0, \infty)$ if it has a probability density function

$$f(x) = ke^{-kx} \qquad \text{for } k > 0$$

The graph is shown in Figure 15.12.

EXAMPLE 6 The useful life of a machine part is given by

$$f(x) = .015e^{-.015t} \qquad 0 \le t < \infty$$

where t is the number of months to failure. What is the probability that the part will fail in the first year?

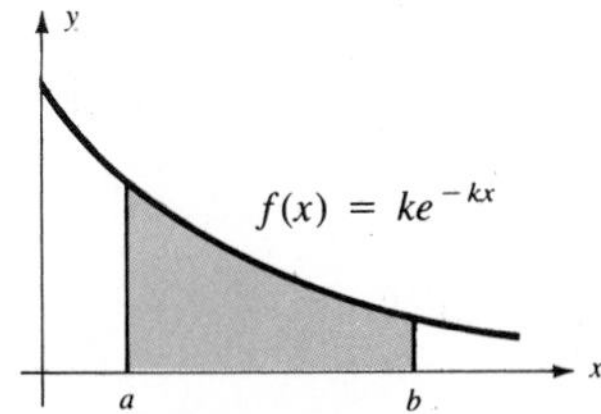

FIGURE 15.12
Exponential distribution

Solution

$$\int_0^{12} .015e^{.015}\,dt = \left.\frac{.015e^{-.015t}}{-.015}\right|_0^{12}$$
$$= -e^{-.015(12)} + e^{-.015(0)}$$
$$\approx .16$$

■

Normal Probability Distribution

A very common probability distribution is the one associated with the so-called bell-shaped curve. Suppose we survey the results of 100,000 IQ scores and obtain the frequency distribution shown in Figure 15.13.

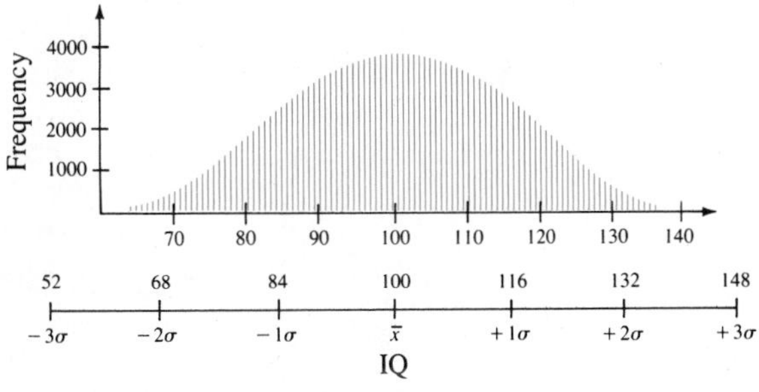

FIGURE 15.13
Frequencies of IQ scores

If we connect the endpoints of the bars in Figure 15.13 by drawing a smooth curve, we obtain a curve very close to a curve called the *normal distribution curve*, or simply the **normal curve**, as shown in Figure 15.14.

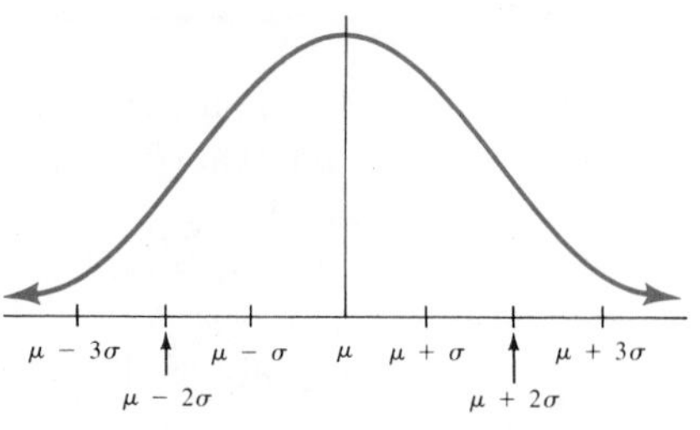

$$f(x) = \frac{e^{-(x-\mu)^2/(2\sigma^2)}}{\sigma\sqrt{2\pi}}$$

FIGURE 15.14
Normal distribution curve

Normal Distribution

A continuous random variable X is said to be **normally distributed** over an interval $(-\infty, \infty)$ if it has a probability density function

$$f(x) = \frac{e^{-(x-\mu)^2/(2\sigma^2)}}{\sigma\sqrt{2\pi}}$$

where μ and σ are real numbers with $\sigma \geq 0$. The number μ is called the **mean** and the number σ is called the **standard deviation**. The graph is shown in Figure 15.14.

Since the normal distribution is a probability distribution, we know the area under this curve is 1. Therefore we can relate the area to probabilities as follows for a random variable X (see Figure 15.15).

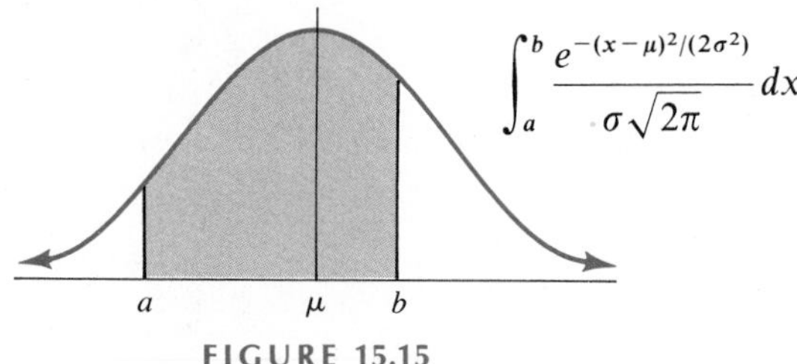

FIGURE 15.15

$P(a \leq X \leq b)$ is the area under the associated normal curve between a and b.
$P(X > \mu) = P(X < \mu) = \frac{1}{2}$; that is, the curve is symmetric about the mean.
$P(X = x) = 0$ for any real number x. (Since there are infinitely many possibilities, the probability of a particular value is 0.)
$P(X < x) = P(X \leq x)$ for any real number x.
$P(X > x) = 1 - P(X \leq x)$ for any real number x.
If $\mu = 0$ and $\sigma = 1$, then the curve is called the **standard normal curve**.

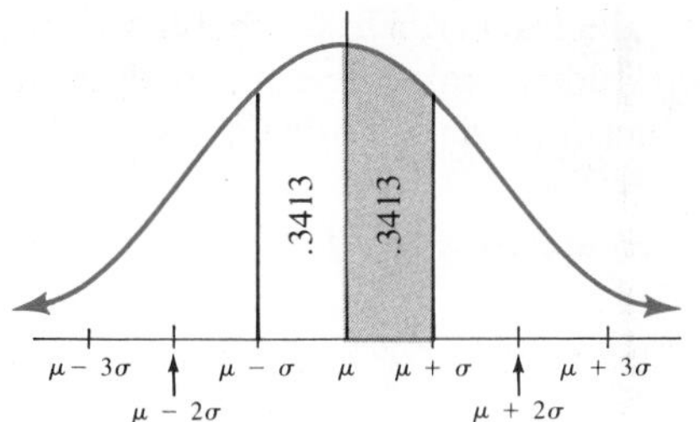

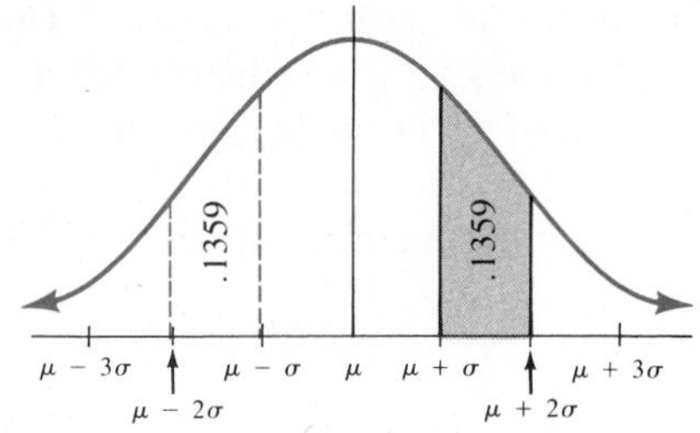

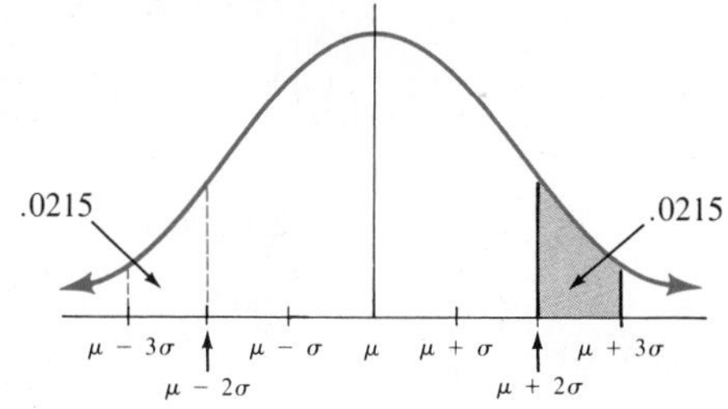

Since the normal distribution is so important for so many applications, and since the calculation of the integral for the probability density function is so complicated, tables of approximate values of the definite integral of the standard normal curve have been calculated. Table 4 in Appendix E is such a table. It contains values of

$$P(X \leq b) = \int_{-\infty}^{b} \frac{1}{\sqrt{2\pi}} e^{-x^2/2}\, dx$$

EXAMPLE 7 Find $P(X \leq .57)$, $P(X > -.13)$, and $P(-.05 < X < .93)$.

Solution

$P(X \leq .57) = .7157$ From Table 4

$P(X > -.13) = 1 - P(X \leq -.13)$ From Table 4

$= 1 - .4483$

$= .5517$

$P(-.05 < X < .93) = .8238 - .4801$ From Table 4

$= .3437$ ■

What if you are not working with a standard normal curve? That is, suppose the curve is a normal curve but does not have a mean of 0. You can then use the information in Figure 15.16.

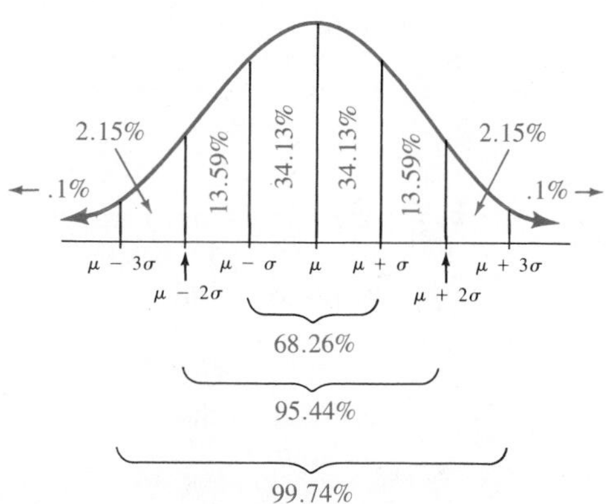

FIGURE 15.16
Areas under a normal curve. Notice that approximately 34% (.3413) lies between μ and $\mu + \sigma$; 14% (.1359) lies between $\mu + \sigma$ and $\mu + 2\sigma$; and 2% (.0215) lies between $\mu + 2\sigma$ and $\mu + 3\sigma$.

EXAMPLE 8 A teacher claims to grade "on a curve." That is, the teacher believes that the scores on a given test are normally distributed. If 200 students take the exam, with mean 73 and standard deviation 9, how would the teacher grade the students?

Solution First, draw a normal curve with a mean 73 and standard deviation 9, as shown in Figure 15.17.

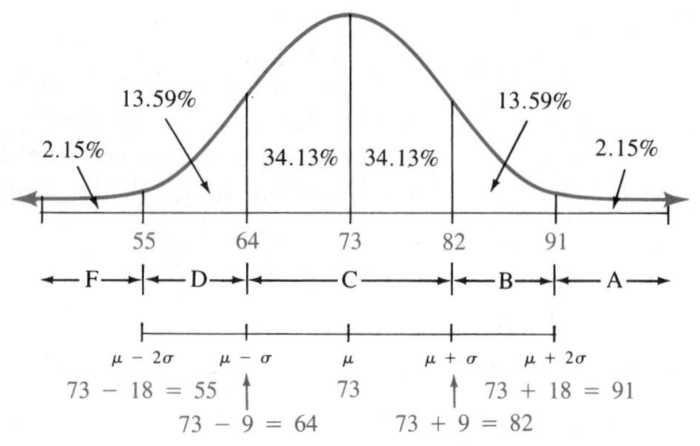

FIGURE 15.17

The interval $[73, 82]$ $(73 + 1\sigma = 82)$ contains about 34% of the class, and the interval $[82, 91]$ $(73 + 2\sigma = 91)$ contains about 14%. Finally, about 2% of the class will score above 91 or below 55:

Score	Letter grade	Number	Percent
92–100	A	4	2%
83–91	B	28	14%
64–82	C	136	68%
55–63	D	28	14%
0–54	F	4	2%

■

EXAMPLE 9 The Ridgemont Light Bulb Company tests a new line of light bulbs and finds their lifetimes to be normally distributed, with a mean life of 98 hours and a standard deviation of 13.

a. What percentage of bulbs will last less than 72 hours?

b. What is the probability that a bulb selected at random will last more than 111 hours?

Solution Draw a normal curve with mean 98 and standard deviation 13. (See the figure at the top of page 751.)

a. $P(X < 72) = .02$

b. $P(X > 111) = .1359 + .0215 \approx .1574$

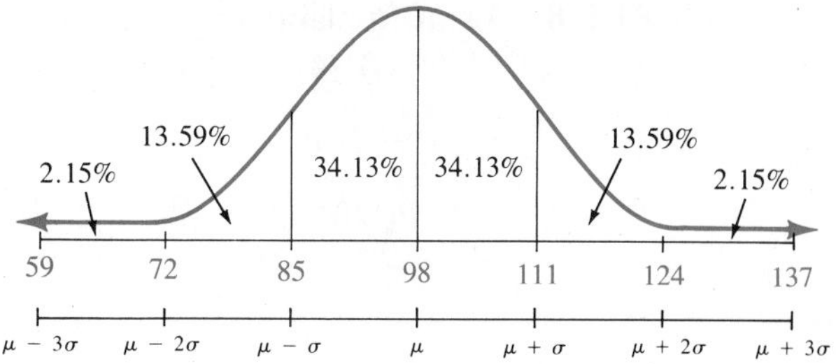

z-Scores

If a division of standard deviations finer than 1, 2, or 3 is needed, a ***z*-score** is used. The z-score translates any normal curve into a standard normal curve by using the simple calculation given below:

***z*-Score**

The area to the left of a value x under a normal curve with mean μ and standard deviation σ is the same as the area under a standard normal curve to the left of the following z-value:

$$z = \frac{x - \mu}{\sigma}$$

z = 1
z = 2
Area between z = 1 and z = 2 is shaded
μ − 3σ
μ − 2σ
μ − σ
μ
μ + σ
μ + 2σ
μ + 3σ
0 z = 1
z = 2
z = 3

EXAMPLE 10 Find the probability that one of the light bulbs described in Example 9 will last between 110 and 120 hours.

Solution From Example 9, $\mu = 98$ and $\sigma = 13$. Let $x = 110$; then

$$z = \frac{110 - 98}{13} \approx .92$$

Now, look up $z = .92$ in Table 4 and find

$$P(X < 110) = P(z < .92) = .8212$$

For $x = 120$,

$$z = \frac{120 - 98}{13} \approx 1.69$$

Again, from Table 4,

$$P(X < 120) = P(z < 1.69) = .9545$$

Thus

$$P(110 \le X \le 120) = .9545 - .8212 = .1333$$

■

15.6 Problem Set

APPLICATIONS

Describe the sample space for the experiments in Problems 1–6. Define a random variable and characterize it as a discrete or continuous random variable.

1. Ten people are asked if they graduated from college, and the number of people responding yes is recorded.
2. An Everready Long-Life® battery is tested, and the life of the battery is recorded.
3. The number of words in which an error is made on a typing test consisting of 500 words is recorded.
4. A person is randomly selected for a public opinion poll and asked two questions: Are you male (m) or female (f)? Do you consider yourself a Democrat (d), Republican (r), Independent (i), or are you not registered (n) to vote?
5. A psychologist is studying sibling relationships in families with three children. Each family is asked about the sex of the children in the family and the results are recorded. If a family has a girl, then a boy, and finally another boy, this would be recorded as {gbb}.
6. A sample of five radios is selected from an assembly line and tested. The number of defective radios is recorded.
7. The heights of 30 students are (numbers are rounded to the nearest inch): 66, 68, 64, 70, 67, 67, 68, 64, 65, 66, 64, 70, 72, 71, 69, 64, 63, 70, 71, 63, 68, 67, 65, 69, 65, 67, 66, 69, 69, 67. Give the frequency distribution, define the random variable, and draw a histogram.
8. The wages of employees of a small accounting firm are (in thousands of dollars): 10, 15, 15, 15, 20, 20, 25, 40, 60, 8, 8, 8, 8, 8, 6, 4. Give the frequency distribution, define the random variable, and draw a histogram.
9. The times that a bank teller spent on each transaction were recorded as follows (rounded to the nearest minute): 8, 7, 3, 6, 5, 7, 3, 4, 5, 5, 4, 2, 1, 5, 2, 1, 3, 3, 4, 6, 4, 4, 2, 5, 4, 6, 4, 5, 4, 5. Give the frequency distribution, define the random variable, and draw a histogram.

Determine whether the functions in Problems 10–17 are probability density functions.

10. $f(x) = 6x$ on $[1, 5]$
11. $f(x) = \dfrac{x^2}{72}$ on $[1, 5]$
12. $f(x) = 6x$ on $[0, \sqrt{3}/3]$
13. $f(x) = \frac{1}{4}$ on $[1, 5]$
14. $f(x) = \frac{1}{4}$ on $[3, 6]$
15. $f(x) = 4$ on $[1, 4]$
16. $f(x) = \frac{3}{125}x^2$ on $[0, 5]$
17. $f(x) = \frac{3}{64}x^2$ on $[1, 4]$

Find k such that each function in Problems 18–27 is a probability density function over the given interval.

18. $f(x) = kx$ on $[2, 6]$
19. $f(x) = kx$ on $[-2, 3]$
20. $f(x) = kx^2$ on $[-1, 3]$
21. $f(x) = kx^2$ on $[0, 5]$
22. $f(x) = 3k$ on $[1, 4]$
23. $f(x) = \dfrac{k}{4}$ on $[1, 7]$
24. $f(x) = kx^{2/3}$ on $[0, 8]$
25. $f(x) = kx^{1/2}$ on $[1, 4]$
26. $f(x) = k/x$ on $[1, 6]$
27. $f(x) = k/x^2$ on $[1, 4]$

Find the area under the standard normal curve satisfying the conditions in Problems 28–33.

28. $X < -2$
29. $X < 1.23$
30. $X > 1$
31. $X > 1.69$
32. $.5 < X < 1.61$
33. $-2.8 \le X \le -.46$
34. The price of an item in dollars is a continuous random variable with a probability density function of $f(x) = 2$ for $1.25 \le x \le 1.75$. What is the probability that the price is more than \$1.50?
35. A number is selected at random from the interval $[0, 100]$. The probability density function is $f(x) = .01$. Find the probability that the number selected is between 35 and 75.
36. The number of minutes to learn a certain task is a random variable with a probability density function of
$$f(t) = .1e^{-.1t}$$
Find the probability that the task is learned in less than 10 minutes.
37. The seeds of many plants are dispersed by the wind, and the distance x (in feet) that the seed travels is given by a probability density function
$$d(x) = .5e^{-.5x}$$
Find the probability that the seeds will be dispersed from 5 to 10 feet away.
38. The breaking strength of a rope (in pounds) is normally distributed, with a mean of 100 pounds and a standard deviation of 16. What is the probability that the rope will break will a force of 132 pounds or less?

39. The diameter of an electric cable is normally distributed, with a mean of .9 inch and a standard deviation of .01. What is the probability that the diameter will exceed .91 inch?

40. The annual rainfall in Ferndale, California, is known to be normally distributed, with a mean of 35.5 inches and a standard deviation of 2.5. What is the probability that the rainfall will exceed 32 inches?

41. If the life of a light bulb is normally distributed with a mean of 250 hours and a standard deviation of 25 hours, what is the probability that the bulb with burn for less than 210 hours?

42. What is the probability that the bulb described in Problem 41 will burn for more than 270 hours?

43. What is the probability that the bulb described in Problem 41 will burn between 240 and 280 hours?

44. At Caltex Corporation the probability density function for the length of time (in minutes) that a randomly selected telephone customer must wait to be helped is given by

$$f(t) = .25e^{-t/4}$$

What is the probability that a customer is kept waiting more than 3 minutes?

45. For a particular type of battery, the probability density function for the life (in hours) of a battery selected at random is given by

$$f(t) = .01e^{-.01t}$$

What is the probability that the battery will last at least 50 hours?

46. The probability density function for the time (in months) after it is purchased that a new telephone will need servicing is given by

$$f(x) = .05e^{-x/20}$$

If the phone is guaranteed for a year, what is the probability that a customer selected at random will have a phone that is no longer covered by the guarantee but that needs servicing?

*15.7

Review

The material of this chapter is reviewed in the following list of objectives. After each objective there are some practice questions. For a sample test select the first question of each set and check your answers. The second question for each objective has no answer given. If you are having trouble with a particular type of problem, look back at the indicated section in the text. When you are finished reviewing these objectives, a sample examination is given at the end of this section.

[15.1]

Objective 15.1: *Find the total value of a given function.*

1. $f(x) = \sqrt[3]{2x + 5}$ on $[1, 40]$
2. $f(x) = .25e^{-.12x}$ on $[0, 4]$
3. $f(x) = \dfrac{x}{2x - 3}$ on $[2, 4]$
4. $f(x) = \ln x$ on $[1, 10]$

Objective 15.2: *Find net excess profit for the given functions over the given intervals.*

5. $r_1(x) = 10 + x^2$; $r_2(x) = 34 + 5x$ on $[0, 8]$
6. $r_1(x) = 15 + x^2$; $r_2(x) = 25 + 3x$ on $[0, 5]$
7. $r_1(x) = 40 + x^2$; $r_2(x) = 32 + 6x$ on $[2, 4]$
8. $r_1(x) = 30 + x^2$; $r_2(x) = 12 + 9x$ on $[3, 6]$

Objective 15.3: *Find the total income for a given rate of flow of a continuous income stream over a given interval* $[a, b]$.

9. $f(t) = 1{,}000$, $[0, 5]$ 10. $f(t) = 100e^{.09t}$ $[0, 3]$
11. $f(t) = 250e^{-.11t}$, $[0, 4]$ 12. $f(t) = 100 + t^2$, $[0, 2]$

Objective 15.4: *Find the total money flow for P dollars over a period of T years at a given rate.*

13. \$3,900 at 8% for 30 years
14. 1,000 e^{rt} for a rate of 8% and time 4 years
15. Find the amount of an annuity with an annual deposit of \$2,500 for 25 years into an account paying 7.5% compounded continuously.
16. If a person places \$125 a month into an account paying 8% compounded continuously, how much will be in the account in 15 years?

Objective 15.5: *Find the consumers' or producers' surplus.*

17. The price, in dollars, for a product is given by the demand function

$$D(x) = 250 + 10x - x^2$$

where x is the number of items (in thousands) in the domain $[10, 20]$. The supply function is given by

$$S(x) = x^2 - 20x + 250$$

Find the consumers' surplus.

* Optional section.

18. The price, in dollars, for a product is given by the demand function

$$D(x) = 4(25 - x^2)$$

where x is the number of items (in thousands) in the domain [0, 10]. The supply function is given by

$$S(x) = x^2 + 5x + 40$$

Find the consumers' surplus.

19. Using the information in Problem 17, find the producers' surplus.

20. Using the information in Problem 18, find the producers' surplus.

[15.2]

Objective 15.6: *Solve first-order differential equations.*

21. $\dfrac{dy}{dx} = 8x^3 - 2x^2 + 1$
22. $\dfrac{dy}{dx} = \sqrt{x} + 5$
23. $3y' = 2x - 5$
24. $6y' - 5 = 11x$
25. $4yy' = x + 5$
26. $yy' = e^{2x+1} - x$
27. $\dfrac{dy}{dx} = 5xy$
28. $\dfrac{dy}{dx} = 3x^3y^2 - 2xy^2 - y^2$

Objective 15.7: *Solve first-order differential equations for a particular solution.*

29. $\dfrac{dy}{dx} = x^3y^{-2}$ for $y = -5$ when $x = 0$
30. $y'x^2 = y$ for $y = 1$ when $x = 2$
31. $2xy = y'$ for $y = e^2$ when $x = 1$
32. $\dfrac{dy}{dx} = \dfrac{xy}{5 - y^2}$ for $y = 1$ when $x = 0$

[15.3]

Objective 15.8: *Evaluate indefinite integrals by parts.*

33. $\displaystyle\int \frac{x}{e^{2x}}\,dx$
34. $\displaystyle\int \frac{x^5}{(x^3 + 1)^2}\,dx$
35. $\displaystyle\int \ln\sqrt{2x}\,dx$
36. $\displaystyle\int x^2\sqrt{10 - x}\,dx$

[15.4]

Objective 15.9: *Evaluate indefinite integrals by using the Brief Integral Table.*

37. $\displaystyle\int x^2\sqrt{x^2 - 16}\,dx$
38. $\displaystyle\int (3 - 4x)^6\,dx$
39. $\displaystyle\int \ln^3 3x\,dx$
40. $\displaystyle\int x^2\sqrt{10 - x}\,dx$

Objective 15.10: *Evaluate indefinite integrals using any appropriate method.*

41. $\displaystyle\int x^2e^{-2x}\,dx$
42. $\displaystyle\int \frac{3dx}{1 - e^{3x}}$
43. $\displaystyle\int \frac{8x + 5}{4x^2 + 5x - 3}\,dx$
44. $\displaystyle\int \frac{5\,dx}{2 - e^{-x}}$

[15.5]

Objective 15.11: *Evaluate improper integrals that converge.*

45. $\displaystyle\int_0^\infty xe^{-x^2}\,dx$
46. $\displaystyle\int_{-\infty}^1 \frac{x\,dx}{x^2 + 1}$
47. $\displaystyle\int_0^\infty .5e^{-.5x}\,dx$
48. $\displaystyle\int_e^\infty \ln x\,dx$

*[15.6]

Objective 15.12: *Describe the sample space for a given experiment. Define a random variable for the experiment and be able to tell if it is a discrete or a continuous random variable.*

49. A pair of dice is rolled and the sum of the top faces is recorded.

50. A coin is tossed three times and the sequence of heads and tails is recorded.

51. The number of minutes it takes a rat to make its way through a maze is recorded; compare with Problem 52.

52. The number of rats that can make their way through a maze is recorded; compare with Problem 51.

Objective 15.13: *Given a set of data, prepare a frequency distribution and draw a histogram.*

53. A pair of dice is rolled and the sum of the spots on the tops of the dice is recorded as follows: 3, 2, 6, 5, 3, 8, 8, 7, 10, 9, 7, 5, 12, 9, 6, 8, 11, 11, 8, 7, 7, 7, 10, 7, 9, 7, 9, 6, 6, 9, 4, 4, 6, 3, 4, 10, 6, 9, 6, 11.

54. Blane, Inc. a consulting firm was employed to perform an efficiency study at National City Bank. As part of the study, they found the number of times per day that people were waiting in line to be the following: 2 were waiting 20 times; 3 were waiting 15 times; 4, 7 times; 5, 5 times; 6, 2 times; 7, 1 time; and there were never more than 7 people in line at any time during the day.

55. Blane, Inc., the consulting firm described in Problem 54, also noted the transaction times (rounded to the nearest minute) for customers at National City Bank, as follows: 1 minute, 10 times; 2 minutes, 12 times; 3 minutes, 18 times; 4 minutes, 25 times; 5 minutes, 16 times; 6 minutes, 10 times; 7 minutes, 6 times; 8 minutes, 1 time; 9 minutes, none; and 10 minutes, 2 times. No transaction took more than 10 minutes.

56. Three coins are tossed onto a table and the following frequencies are noted: 0 heads, 17 times; 1 head, 59 times; 2 heads, 56 times; and 3 heads, 18 times.

* Optional section.

Objective 15.14: *Determine whether a given function is a probability density function.*

57. $f(x) = \frac{3}{63}x^2$ on $[1, 4]$ **58.** $f(x) = \frac{x^2}{3}$ on $[-1, 1]$

59. $f(x) = \frac{2}{3}$ on $[2, 5]$ **60.** $f(x) = .1$ on $[0, 10]$

Objective 15.15: *Find a constant k in order to define a probability density function over a given interval.*

61. $f(x) = kx^{1/4}$ on $[0, 16]$ **62.** $f(x) = k$ on $[-4, 0]$

63. $f(x) = 5kx$ on $[1, 10]$ **64.** $f(x) = k\sqrt{2x}$ on $[2, 8]$

Objective 15.16: *Find the area under the standard normal curve.*

65. $x < 0$ **66.** $-1.03 < x < 1.59$

67. $x > -.11$ **68.** $-.5 \le x \le 1.5$

Objective 15.17: *Solve applied problems based on the preceding objectives.*

69. *Spread of a disease.* If a disease is spreading at a rate of $40t - 6t^2$ cases per day where t is the number of days measured from the first outbreak, give the number of people affected on the seventh day. Assume that there were 4 cases recorded on the first day.

70. *Depreciation.* If the total depreciation at the end of t years is given by $f(t)$ and the depreciation rate is $f'(t) = 200\sqrt{10 - x}$, find the total depreciation for the first five years.

71. *Capital value.* Suppose that the current interest rate is 12% and the British Embassy in Washington, D. C., has an inderminant lease paying \$576,000 per year. What is the capital value of this lease?

***72.** If the life of a light bulb is normally distributed with a mean of 250 hours and a standard deviation of 25 hours, find the following probabilities:

a. $P(X > 250)$ **b.** $P(X < 220)$
c. $P(200 < X < 300)$ **d.** $P(220 \le X \le 320)$

***73.** *Grading on a curve.* Suppose that for a certain exam a teacher grades on a curve. It is known that the mean is 50 and the standard deviation is 5. There are 45 students in the class.

a. How many students should receive a C?
b. How many students should receive an A?
c. What score would be necessary to obtain an A?
d. If an exam paper is selected at random, what is the probability that it will be a failing paper?

SAMPLE TEST

The following sample test (75 minutes) is intended to review the main ideas of this chapter.

* Optional section.

Evaluate the integrals in Problems 1–9. Use substitution, integration by parts, or the Brief Integral Table.

1. $\int (2x - 1)^3\,dx$ **2.** $\int x(2x - 1)^3\,dx$

3. $\int x^2(2x - 1)^3\,dx$ **4.** $\int x(2x^2 - 1)^3\,dx$

5. $\int \frac{8x + 5}{4x^2 + 5x - 3}\,dx$ **6.** $\int \frac{4x^2 + 5x - 3}{x}\,dx$

7. $\int \frac{3}{1 - e^{3x}}\,dx$ **8.** $\int \ln^2 x\,dx$

9. $\int \ln^3 x\,dx$

State the formula you would use from the Brief Integral Table to approximate the integrals in Problems 10–12. You do not need to actually carry out the integration.

10. $\int x^2(20 - 5x)^4\,dx$ **11.** $\int \frac{\sqrt{100 - x}}{x^5}\,dx$

12. $\int_{.5}^{2} t^4 e^{-t}\,dt$

Evaluate the improper integrals in Problems 13–14 that converge.

13. $\int_1^{\infty} x^2 e^{-x^3}\,dx$ **14.** $\int_{-\infty}^{1} \frac{x^2\,dx}{x^3 + 1}$

Solve the differential equations in Problems 15–16.

15. $4y' = 6 + 10x$

16. $\frac{dy}{dx} = \frac{xy}{5 - y^2}$ for $y = 1$ when $x = 0$

17. Determine the total money flow for \$35,000 at 12% for 6 years.

18. Suppose that the current interest rate is 10% and the French Embassy in Washington, D.C., has a 50-year lease paying \$682,000 per year. What is the capital value of this lease?

19. The rate of learning a certain task is estimated by the formula

$$\frac{dN}{dt} = 5e^{-.1t}$$

Find an expression for N if it is known that at time $t = 0$, $N = 0$.

20. The total depreciation at the end of t years is given by $f(t)$ and the depreciation rate is

$$f'(t) = 200\sqrt{10 - x}$$

Find the total depreciation (to the nearest \$10) for the first 5 years.

1. Graph $y = e^x$.

2. Graph $y = \ln\sqrt{x}$.

Solve the equations in Problems 3–6.

3. $\log\left(\frac{x-5}{6}\right) = 2$

4. $3\ln\frac{e}{\sqrt[3]{5}} = 3 - \ln x$

5. $10^{-x} = .5$

6. $e^{1-x} = 105$

7. Fill in the blanks to complete the properties of the definite integral.
   a. $\int_a^a f(x)\,dx =$ ______________
   b. $\int_a^b dx =$ ______________
   c. $\int_a^b f(x)\,dx =$ ______________ where $b < a$
   d. $\int_a^c f(x)\,dx + \int_c^b f(x)\,dx =$ ______________ where $a < c < b$
   e. $\int_a^b kf(x)\,dx =$ ______________
   f. $\int_a^b [f(x) \pm g(x)]\,dx =$ ______________
   g. $\int_{x=a}^{x=b} u\,dv =$ ______________

Evaluate the integrals in Problems 8–15.

8. **a.** $\int \frac{du}{u}$ **b.** $\int du$

9. **a.** $\int (5x^4 + 3x^2 + 5)\,dx$ **b.** $\int e^{2x}\,dx$

10. $\int x^2(5 - 2x^3)^4\,dx$

11. $\int \ln^4 5x\,dx$

12. $\int \frac{100\,dx}{5 - e^{-x}}$

13. $\int \frac{8x - 3}{4x^2 - 3x + 2}\,dx$

14. $\int_1^{\infty} x^{-3/2}\,dx$

15. $\int_{-\infty}^{\infty} \frac{x\,dx}{(x^2 + 1)^2}$

16. Write a formula for the area bounded above by the curve $y = 1/\sqrt{x^2 + 4}$, below by the x-axis, on the left by the y-axis, and on the right by the line $x = t$.

17. Find the area bounded by $y = x^2$ and $y = 32 - x^2$.

18. Find $\int_1^2 \ln x^2\,dx$ correct to two decimal places using one (or all) of the following approximations for $n = 4$.
   a. Rectangular approximation
   b. Trapezoidal approximation
   c. Simpson's rule

APPLICATIONS

19. A piece of replacement equipment costs \$48,000. The rate of operating cost savings is $S(x)$ in dollars and is given by the formula

$$S(x) = 5{,}000x + 2{,}000$$

where x is the number of years the piece of equipment will be used. How long will it take the piece of equipment to pay for itself?

20. In the course of any year, the number y of cases of a disease is reduced by 10% according to the growth formula

$$y = P_0 e^{-.1t}$$

for t years with P_0 cases reported today. If there are 100,000 cases today, how long will it take to reduce the number of cases to less than 10,000?

21. Use the formula $A = P(1 + i)^N$ to find out how long after depositing \$1,000 at 8% compounded daily (365-day year) you must wait in order to have \$5,000.

22. In 1986 the growth rate of Houston, Texas, was given by the formula

$$P'(x) = 1.6e^{.025t}$$

where t is measured in years since 1980 and $P(t)$ is the population in millions. If the 1986 population was 1,860,000, predict the population at the turn of the century. Comment on this result and growth rate.

23. A businessperson receives a shipment of Christmas trees on November 20. The sales pattern is such that the inventory moves slowly at the beginning but as Christmas approaches, the demand increases so that x days after

November 20 the inventory is y trees, where

$$y = 2{,}450 - 2x^2 \qquad \text{for } 0 \le x \le 35$$

What is the average inventory for the first 30 days?

24. Let X be a normally distributed random variable with mean 55 and standard deviation 10. Find the following probabilities.

a. $P(X \ge 55)$ **b.** $P(X > 60)$
c. $P(40 \le X \le 50)$

25. The wait time (in seconds) for a response at a particular terminal in a time-share network is a continuous random variable (X) with a probability density function

$$f(x) = .01e^{-x/100}$$

What is the probability that a user must wait more than 1 minute for a response?

16

Functions of Several Variables

CHAPTER OVERVIEW

In order to build realistic mathematical models for real-life situations it is necessary to consider functions of more than one variable. This chapter defines and discusses such functions.

PREVIEW

The important concept of this chapter is that of a partial derivative. We use it in maximum–minimum applications and the method of least squares. Another method of maximization, Lagrange multipliers, is introduced in Section 16.4. Multiple integration applications, including finding volumes, are developed in Section 16.5.

PERSPECTIVE

Even though this chapter may be considered optional because of time constraints, it is very important in serious model building. The concept of a function of two, three, or more variables is an easy one; unfortunately, the graphical representation of these functions is not. If you do not have time to consider the ideas of this chapter in class, it is useful to study this chapter on your own after you have completed this course.

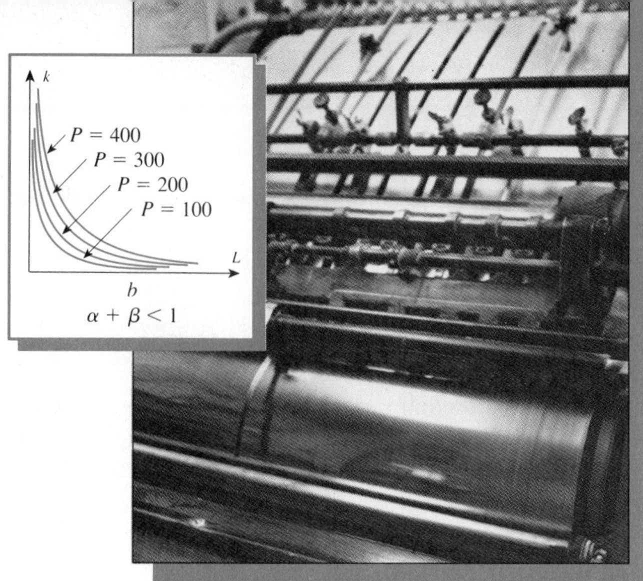

MODELING APPLICATION 14

The Cobb–Douglas Production Function

Karlin Corporation manufactures only one product, which is sold at the price P_0. The firm employs a labor force L, which must be paid an average wage p_1. The firm also requires capital K in terms of tools, buildings, and so forth. The cost of using one unit of capital is p_2. Karlin wishes to maximize its profit. Determine what data need to be collected and construct a model that will accomplish Karlin's goal of maximizing profit.

Developing a mathematical model is no easy task. As you have realized by now, there is usually a great deal of work involved. Every good model should include listing the assumptions, translating the assumptions into mathematical notation, building the model, and then checking the model against tabulated data. Read "The Cobb–Douglas Production Function" by Robert Geitz (The UMAP Journal, Unit 509. © 1981 Education Development Center, Inc.) This paper is a perfect illustration of the modeling process. After reading it, develop a model to accomplish Karlin Corporation's goal.

APPLICATIONS

Management *(Business, Economics, Finance, and Investments)*

Cost function
(16.1, Problems 43, 45–46; 16.2, Problem 50; 16.6, Problem 45; Test Problem 15)
Revenue function
(16.1, Problem 44; 16.2, Problem 51)
Profit (16.2, Problem 49)
Rate of change of an investment
(16.2, Problems 55–56)
Rate of change of an annuity
(16.2, Problem 57)
Maximize profit (16.3, Problems 19, 21)
Minimize labor cost for a function of two variables (16.3, Problem 20)
Minimize cost of shipping container
(16.3, Problem 22)
Least material to construct a shipping container (16.3, Problem 23)
Maximum yield on farm production
(16.4, Problem 15)

Management *(continued)*

Minimize cost relative to supply
(16.4, Problem 16)
Maximum area for a fenced enclosure, given fixed costs (16.4, Problem 17)
Minimum surface area for a standard size Coke can (16.4, Problem 18)
Marginal cost and revenue of a function of two variables
(16.4, Problem 20; 16.6, Problems 46–48)
Average value of a function
(16.5, Problems 55–56)
Cobb–Douglas production function
(16.5, Problem 57)

Life Sciences *(Biology, Ecology, Health, and Medicine)*

Amount of blood flow as a function of blood vessel size (16.1, Problem 47)
Poiseuille's law for blood flow
(16.2, Problem 52)

Life Sciences *(continued)*

Surface area of a human body
(16.2, Problems 53–54)
Supplying a 1,000 calorie diet while minimizing the cost (16.4, Problem 19)

Social Sciences *(Demography, Political science, Population, Psychology, Society, and Sociology)*

Intelligence quotient (IQ) (16.1, Problem 42)

General Interest

Cost function for finishing a room
(16.1, Problems 45–46)
Largest volume that can be mailed in the U.S.
(16.3, Problem 24)

Modeling Application—
The Cobb–Douglas Production Function

16.1

Three-Dimensional Coordinate System

Many real-life models involve more than one variable. Suppose, for example, we consider one of the most fundamental applications, that of the total cost of producing an item. If Ballad Corporation produces a single record with fixed costs of \$2,000 and a unit cost of \$.35, then

$$C(x) = 2{,}000 + .35x$$

for x records produced. However, if a second record is produced with additional fixed costs of \$500 and a unit cost of \$.30, then the total cost of producing x records of the first type and y records of the second type requires what we call a **function of two independent variables** x and y:

$$C(x, y) = 2{,}500 + .35x + .30y$$

Function of Two or More Variables

Suppose D is a collection of ordered n-tuples of real numbers $(x_1, x_2, \ldots, x_n)$. Then a function f with **domain** D is a rule that assigns a number

$$z = f(x_1, x_2, \ldots, x_n)$$

to each n-tuple in D. The function's **range** is the set of z values the function assumes. The symbol z is called the **dependent variable** of f, and f is said to be a **function of the n independent variables** $x_1, x_2, \ldots, x_n$.

You have already considered many examples of functions of several variables, as Example 1 shows.

EXAMPLE 1

Area of a rectangle: $K(l, w) = lw$

Volume of a box: $V(l, w, h) = lwh$

Simple interest: $I(P, r, t) = P(1 + rt)$

Compound interest: $A(P, r, t, n) = P(1 + \frac{r}{n})^{nt}$

Find each of the requested values and interpret your results in terms of what you know about each of these formulas.

a. $K(25, 15)$ **b.** $V(5, 20, 30)$

c. $I(100000, .08, 15)$ **d.** $A(450000, .09, 30, 12)$

Solution

a. $K(25, 15) = 25(15) = 375$; the area of a 25 by 15 rectangle is 375 square units.

b. $V(5, 20, 30) = 5(20)(30) = 3{,}000$; the volume of a 5 by 20 by 30 box is 3,000 cubic units.

c. $I(100000, .08, 15) = 100{,}000[1 + .08(15)] = 100{,}000[2.2] = 222{,}000$; future value of a \$100,000 investment at 8% simple interest for 15 years is \$220,000.

d. $A(450000, .09, 30, 12) = 450{,}000(1 + \frac{.09}{12})^{30(12)} = 450{,}000(1.0075)^{360} \approx 6{,}628{,}759.26$; the future value of a \$450,000 investment at 9% compounded monthly is approximately \$6,628,759.26. ■

Even though a function of several variables has been defined for the general case and Example 1 shows functions of several variables, this chapter focuses primarily on functions of two variables. That is, $z = f(x, y)$ is the notation used for z, a function of two independent variables x and y. In order to graph such a function we need to consider **ordered triplets** (x, y, z) and a **three-dimensional coordinate system**, just as we have already considered ordered pairs (x, y) and a two-dimensional coordinate system. We draw a coordinate system with three mutually perpendicular axes, as shown in Figure 16.1.

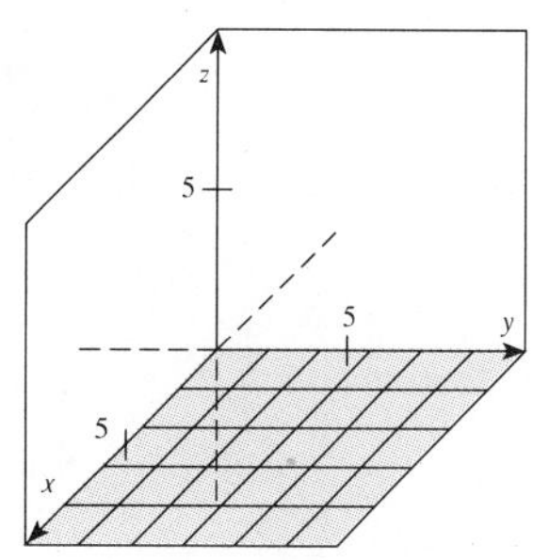

FIGURE 16.1
Three-dimensional coordinate system

Think of the x-axis and the y-axis as the floor and the z-axis as a line perpendicular to the floor. All of the graphs we have done up to now in this book would now be drawn on the "floor." Example 2 shows how to plot points in three dimensions.

EXAMPLE 2 Graph the following ordered triplets.

Solution **a.** $(10, 20, 10)$ **b.** $(-12, 6, 12)$

c. $(-12, -18, 6)$ **d.** $(20, -10, 18)$

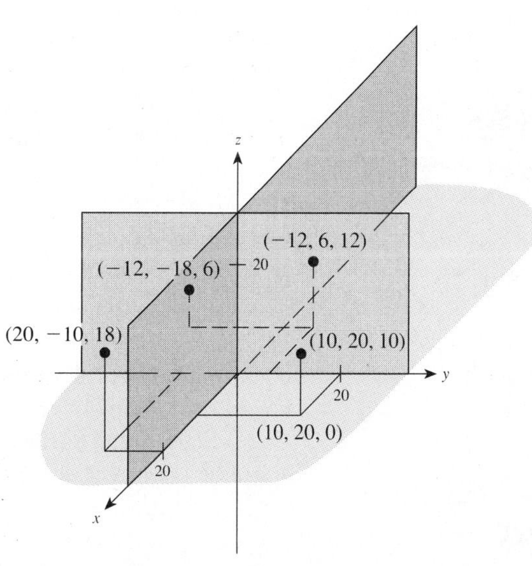

■

If you orient yourself in a room (your classroom, for example), as shown in Figure 16.2, you will notice certain important planes:

FIGURE 16.2
A typical classroom; assume the dimensions are 25 by 30 feet with an 8 foot ceiling.

Floor: ***xy*-plane**
Equation is $z = 0$.
Ceiling: a plane parallel to the xy-plane
Equation is $z = 8$.
Front wall: ***yz*-plane**
Equation is $x = 0$.
Back wall: plane parallel to the yz-plane
Equation is $x = 30$.
Left-side wall: ***xz*-plane**
Equation is $y = 0$.
Right-side wall: plane parallel to the xz-plane
Equation is $y = 25$.

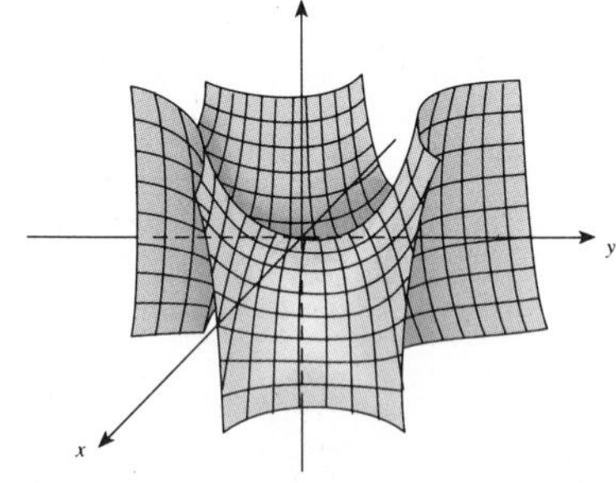

a. $z = x^3 - 3xy^2$

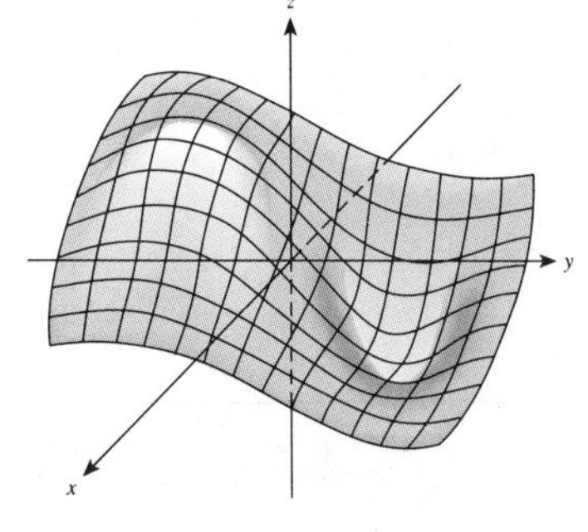

b. $z = \dfrac{-1}{x^2 + y^2 + 1}$

FIGURE 16.3
Graphs of surfaces in three dimensions

The xy-, xz-, and yz-planes are called the **coordinate planes**. Name the coordinates of several objects in the figure.

Just as points in the plane are associated with ordered pairs satisfying an equation in two variables, points in space are associated with ordered triplets satisfying an equation. The graph of any function of the form $z = f(x, y)$ is called a **surface**. It is beyond the scope of this course to have you spend a great deal of time graphing three-dimensional surfaces, but you should be aware that computer programs have simplified the task of graphing surfaces, as shown in Figure 16.3.

The remainder of this section is a brief introduction to some of the more common three-dimensional surfaces.

Planes

The graph of $ax + by + cz = d$ is a **plane** if $a, b, c,$ and d are real numbers ($a, b,$ and c not all zero).

EXAMPLE 3 Graph the planes defined by the given equations.

a. $x + 3y + 2z = 6$

b. $y + z = 5$

c. $x = 4$

Solution It is customary to show only the portion of the graph that lies in the **first octant** (that is, where x, y, and z are all positive). To graph a plane, find some ordered triplets satisfying the equation. The best ones to use are often those on one of the coordinate axes.

a. Let $x = 0$ and $y = 0$; then $z = 3$; point is $(0, 0, 3)$.

Let $x = 0$ and $z = 0$; then $y = 2$; point is $(0, 2, 0)$.

Let $y = 0$ and $z = 0$; then $x = 6$; point is $(6, 0, 0)$.

Plot these points as shown in Figure 16.4a and use them to draw the plane.

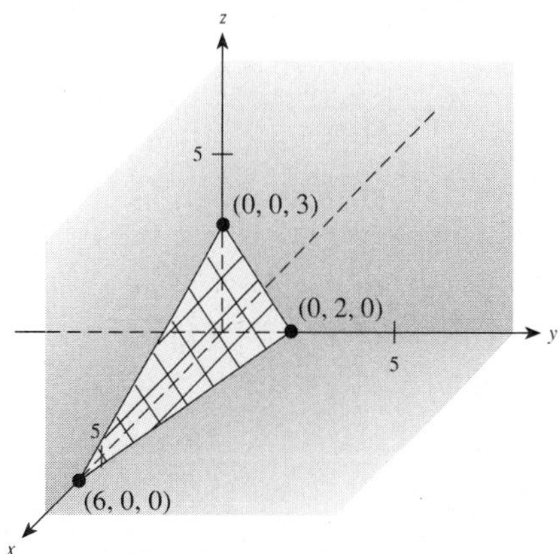

FIGURE 16.4a
Graphing planes

b. If exactly two of the coefficients a, b, c are not zero (i.e., one of the variables is missing from the equation of a plane), then that plane is parallel to the axis corresponding to the missing variable; in this case it is parallel to the x-axis. Draw the line $y + z = 5$ on the yz-plane, and then complete the plane as shown in Figure 16.4b on page 764.

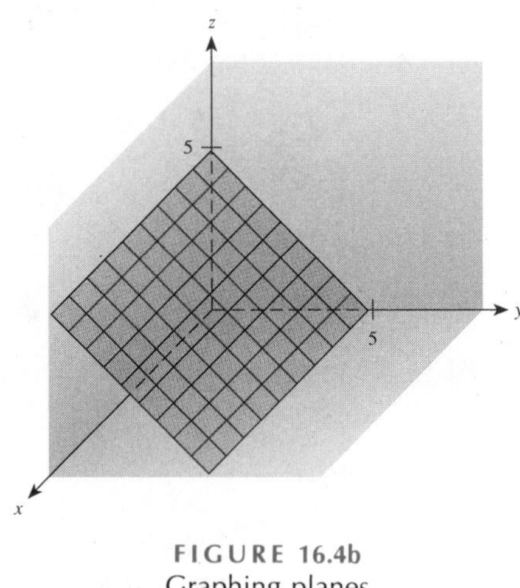

FIGURE 16.4b
Graphing planes

c. If exactly one of a, b, c is nonzero (i.e., two variables are missing), then the plane is parallel to the plane of the two variables missing in the equation, as shown in Figure 16.4c.

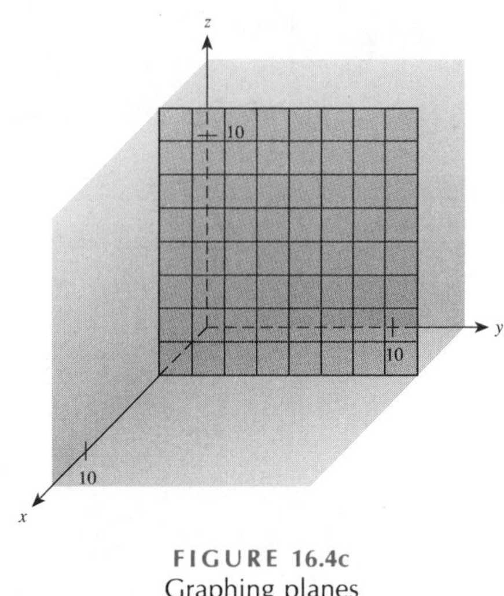

FIGURE 16.4c
Graphing planes

Quadric Surfaces

The graph of the equation

$$Ax^2 + By^2 + Cz^2 + Dxy + Exz + Fyz + Gy + Hy + Iz + J = 0$$

is called a **quadric surface**. The **trace** of a curve is found by setting one of the variables equal to a constant and then graphing the resulting curve. If $x = k$ (k a constant), then the resulting curve is drawn in the plane $x = k$, which is parallel to the yz-plane; similarly, if $y = k$, then the curve is drawn in the plane $y = k$, which is parallel to the xz-plane; and if $z = k$, then the curve is drawn in the plane $z = k$, which is parallel to the xy-plane. Table 16.1 shows the quadric surfaces.

TABLE 16.1 Quadric Surfaces

Surface	Description	Surface	Description		
Elliptic cone 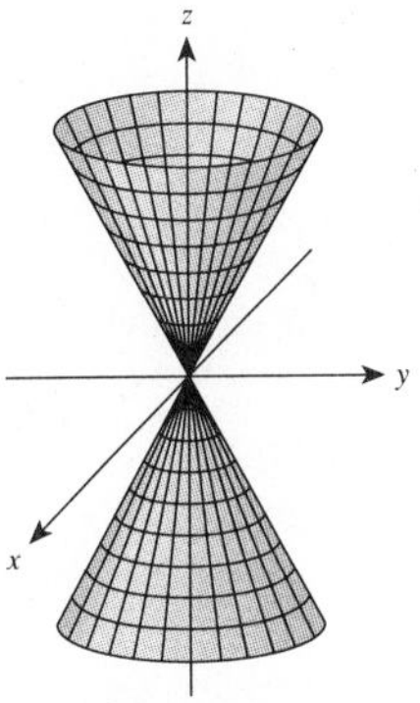	The trace in the xy-plane is a point; in planes parallel to the xy-plane it is an ellipse. Traces in the xz- and yz-planes are intersecting lines; in planes parallel to these they are hyperbolas $z^2 = \frac{x^2}{a^2} + \frac{y^2}{b^2}$	Elliptic paraboloid 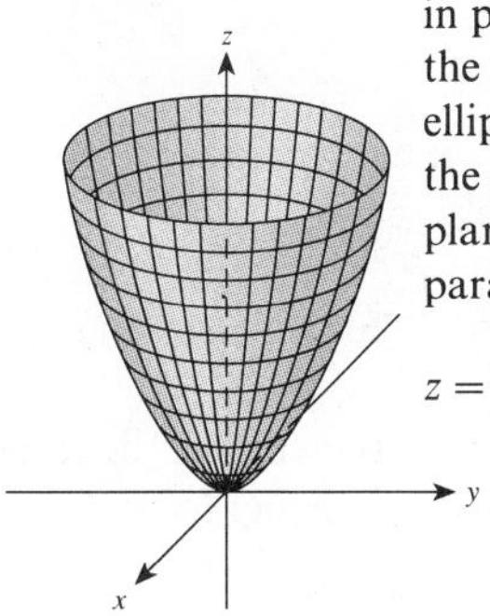	The trace in the xy-plane is a point; in planes parallel to the xy-plane it is an ellipse. Traces in the xz- and yz-planes are parabolas $z = \frac{x^2}{a^2} + \frac{y^2}{b^2}$	Ellipsoid or sphere	The traces in the coordinate planes are ellipses. $\frac{x^2}{a^2} + \frac{y^2}{b^2} + \frac{z^2}{c^2} = 1$ 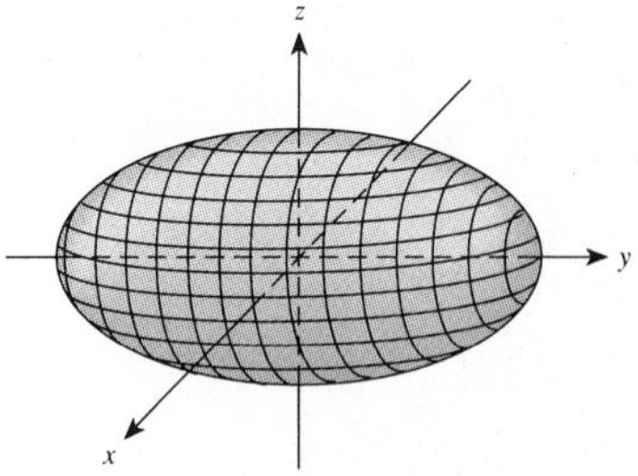
Hyperboloid of one sheet 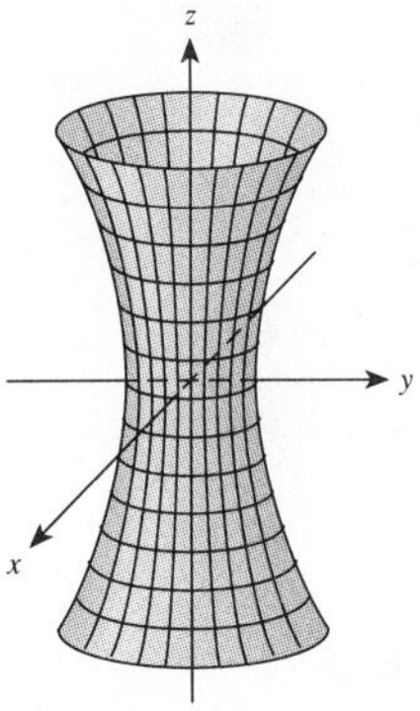	The trace in the xy-plane is an ellipse; in the xz- and yz-planes the traces are hyperbolas $\frac{x^2}{a^2} + \frac{y^2}{b^2} - \frac{z^2}{c^2} = 1$	Hyperboloid of two sheets 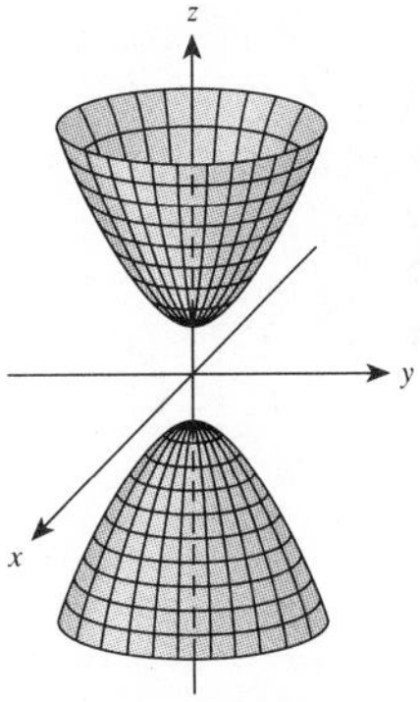	There is no trace in the xy-plane. In planes parallel to the xy-plane, which intersect the surface, the traces are ellipses. Traces in the xz- and yz-planes are the hyperbolas $\frac{x^2}{a^2} + \frac{y^2}{b^2} - \frac{z^2}{c^2} = -1$		If $a^2 = b^2 = c^2 = r^2$, then the graph is a sphere $x^2 + y^2 + z^2 = r^2$ 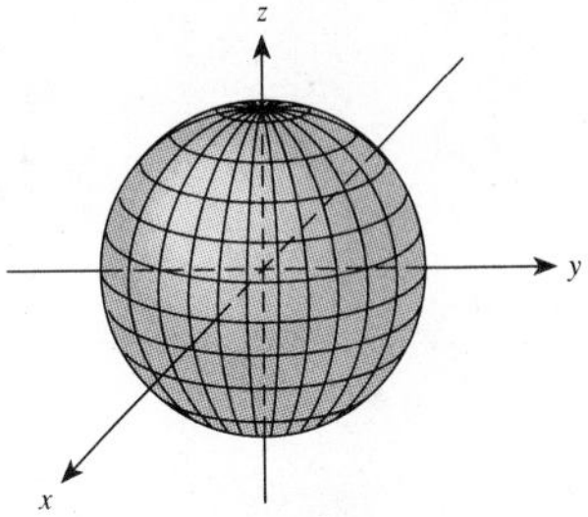
Hyperbolic paraboloid	The trace in the xy-plane is two intersecting lines; in planes parallel to the xy-plane the traces are hyperbolas. Traces in the xz- and yz-planes are parabolas $z = \frac{y^2}{b^2} - \frac{x^2}{a^2}$	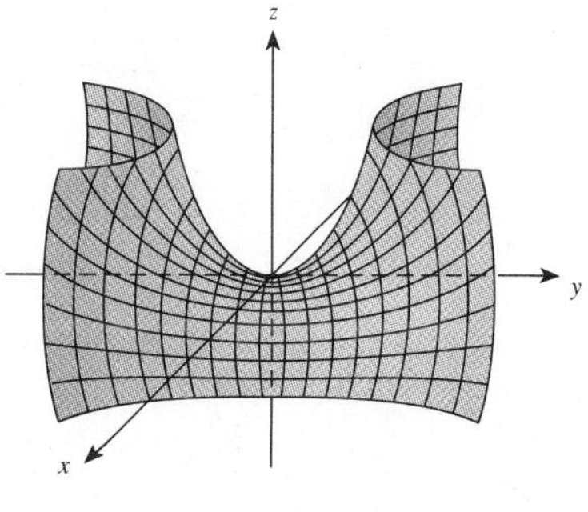			

Circular Cylinders

The graphs of

$$y^2 + z^2 = r^2 \qquad x^2 + z^2 = r^2 \qquad \text{and} \qquad x^2 + y^2 = r^2$$

are **right circular cylinders** of radius r, parallel to the x-axis, y-axis, and z-axis, respectively.

EXAMPLE 4 Graph the following equations.

a. $x^2 + y^2 = 9$ **b.** $y^2 + z^2 = 16$ **c.** $x^2 + z^2 = 25$

Solution **a.** This is a cylinder parallel to the z-axis (the z variable is missing), as shown in Figure 16.5a.

b. This is a cylinder parallel to the x-axis, as shown in Figure 16.5b.

c. This is a cylinder parallel to the y-axis, as shown in Figure 16.5c.

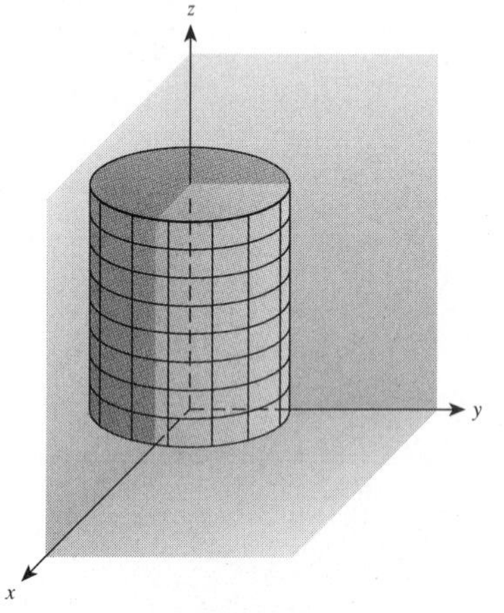

a. Graph of $x^2 + y^2 = 9$

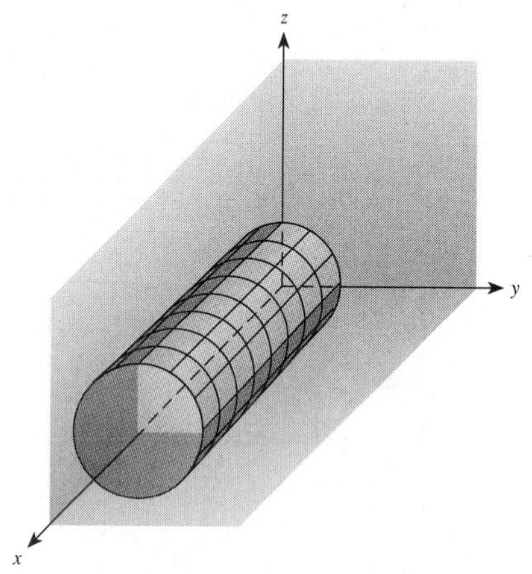

b. Graph of $y^2 + z^2 = 16$

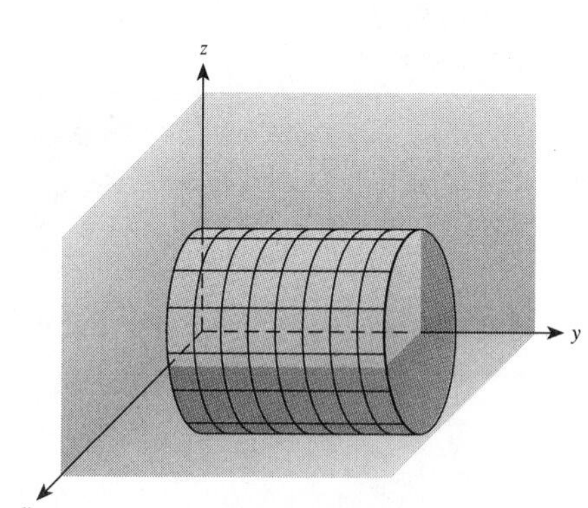

c. Graph of $x^2 + z^2 = 25$

FIGURE 16.5
Graphs of right circular cylinders

■

16.1 Problem Set

Evaluate the functions K, V, I, and A from Example 1 in Problems 1–10, and interpret your results.

1. $K(15, 35)$
2. $K(45, 90)$
3. $V(3, 5, 8)$
4. $V(15, 25, 8)$
5. $I(500, .05, 3)$
6. $I(1250, .08, 12)$
7. $A(2500, .12, 6, 4)$
8. $A(5500, .18, 5, 6)$
9. $A(110000, .09, 30, 12)$
10. $A(250000, .15, 15, 12)$

Evaluate $f(x, y) = x^2 - 2xy + y^2$ *for the values in Problems 11–14.*

11. $f(2, 3)$
12. $f(-1, 4)$
13. $f(-2, 5)$
14. $f(0, 6)$

Evaluate $g(x, y) = \dfrac{2x - 4y}{x^2 + y^2}$ *for the values in Problems 15–18.*

15. $g(2, 1)$
16. $g(-3, 2)$
17. $g(5, -3)$
18. $g(-3, -4)$

Evaluate $h(x, y) = \dfrac{e^{xy}}{\sqrt{x^2 + y^2}}$ *for the values in Problems* 19–22. *Round answers to the nearest hundredth.*

19. $h(0, 5)$ **20.** $h(-2, 3)$

21. $h(-2, -3)$ **22.** $h(1, 1)$

Graph the ordered triplets in Problems 23–25.

23. a. $(1, 2, 3)$ **b.** $(-3, 2, 4)$ **c.** $(1, -4, 3)$ **d.** $(-5, -9, -8)$

24. a. $(2, 4, 3)$ **b.** $(-3, 2, 4)$ **c.** $(10, -20, -5)$ **d.** $(-1, -2, -3)$

25. a. $(10, 5, 20)$ **b.** $(5, -15, -5)$ **c.** $(3, 2, -4)$ **d.** $(-5, -1, 3)$

Graph the surfaces in Problems 26–41.

26. $2x + y + 3z = 6$ **27.** $x + 2y + 5z = 10$

28. $x + y + z = 1$ **29.** $3x - 2y - z = 12$

30. $z^2 = \dfrac{x^2}{4} + \dfrac{y^2}{9}$ **31.** $z = \dfrac{x^2}{4} + \dfrac{y^2}{9}$

32. $\dfrac{x^2}{1} + \dfrac{y^2}{4} + \dfrac{z^2}{9} = 1$ **33.** $\dfrac{x^2}{9} + \dfrac{y^2}{4} + \dfrac{z^2}{25} = 1$

34. $x^2 + y^2 + z^2 = 9$ **35.** $z = x^2 + y^2$

36. $\dfrac{x^2}{9} - \dfrac{y^2}{1} + \dfrac{z^2}{4} = 1$ **37.** $\dfrac{x^2}{9} + \dfrac{y^2}{1} - \dfrac{z^2}{4} = -1$

38. $y^2 + z^2 = 25$ **39.** $x^2 + y^2 = 36$

40. $x^2 + z^2 = 4$ **41.** $y^2 + z^2 = 20$

APPLICATIONS

42. The intelligence quotient (IQ) is defined as $Q(x, y) = \dfrac{100x}{y}$ where Q is the IQ, x is a person's mental age as measured on a standardized test, and y is a person's chronological age measured in years. Find (to the nearest unit) and interpret
a. $Q(15, 13)$ **b.** $Q(6, 9)$
c. $Q(15, 15)$ **d.** $Q(10.5, 9.8)$

43. A company manufactures two types of golf carts. The first has a fixed cost of \$2,500, a variable cost of \$800, and x are produced. The second has a fixed cost of \$1,200, a variable cost of \$550, and y are produced. Write a cost function $C(x, y)$ and find
a. $C(10, 15)$ **b.** $C(5, 25)$
c. $C(15, 10)$ **d.** $C(0, 30)$

44. If the revenue function for the golf carts in Problem 43 is $R(x, y) = 1{,}500x + 900y$ find
a. $R(10, 15)$ **b.** $R(5, 25)$
c. $R(15, 10)$ **d.** $R(0, 30)$

45. If the dimensions of the room in Figure 16.2 are x feet wide, y feet long, and z feet high, and if ceiling material is \$2 per square foot, wall material is \$.75 per square foot, and floor material is \$1.25 per square foot, write a cost function for the ceiling, floor, and wall (assuming no doors or windows).

46. If $C(x, y, z)$ is the cost function for Problem 45, find
a. $C(25, 30, 8)$ **b.** $C(12, 14, 8)$
c. $C(15, 20, 10)$

47. The amount of blood flowing in a blood vessel measured in milliliters is given by $F(l, r) = .002l/r^4$ where l is the length of the blood vessel and r is the radius. Find
a. $F(3.1, .002)$ **b.** $F(15.3, .001)$
c. $F(6, .005)$

16.2 Partial Derivatives

One of the most important and useful concepts in mathematics is that of a derivative. In this section we consider the derivative of a function of several variables. We begin with a geometric interpretation and then apply this interpretation to some particular examples.

Consider a surface $z = f(x, y)$, as shown in Figure 16.6a on page 768. Hold one of the variables constant, say, $y = b$. This gives a curve $z = f(x, b)$, which is the intersection of the plane $y = b$ and the surface. The slope of this curve is called the **partial derivative** of f with respect to x and is denoted by f_x or $\partial z/\partial x$ (see Figure 16.6b). Similarly, if we let $x = a$ (a constant), then the curve that is the intersection of this plane and the surface has slope f_y or $\partial z/\partial y$, which is called the **partial derivative** of f with respect to y (see Figure 16.6c).

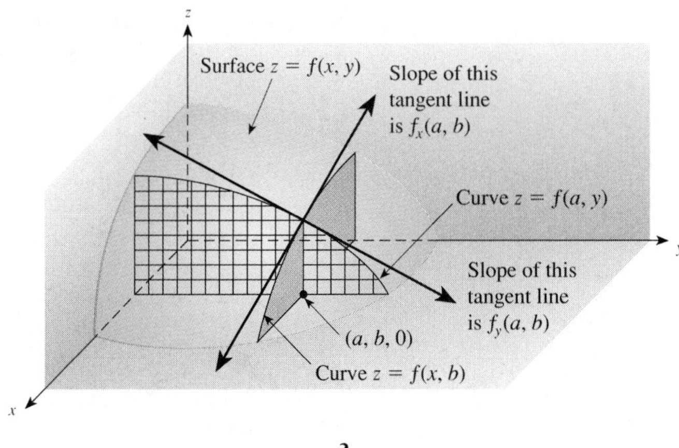

a.

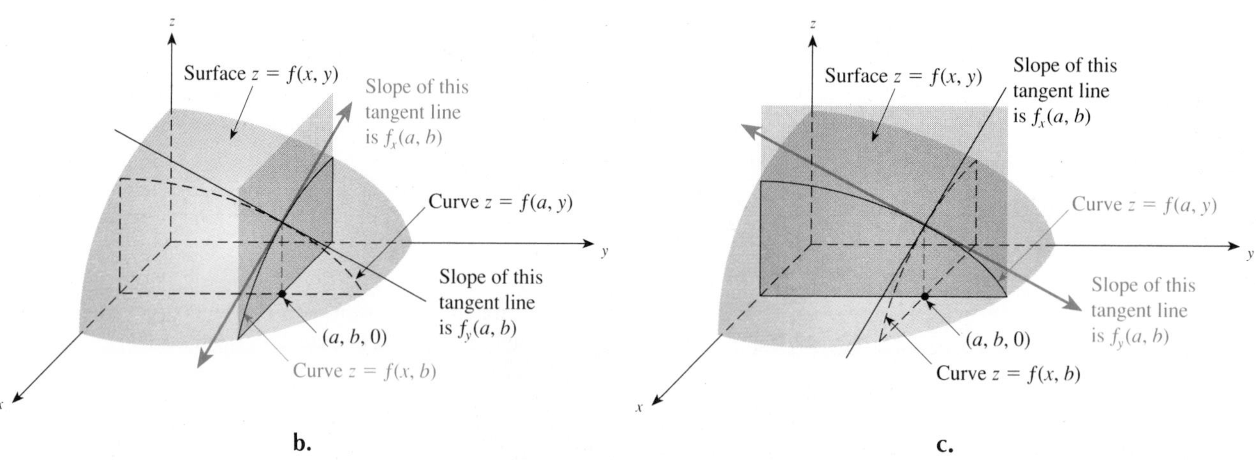

FIGURE 16.6
Geometric interpretation of a partial derivative

Partial Derivative

If $z = f(x, y)$, then the **partial derivative of f with respect to x**, denoted by $\partial z/\partial x$, f_x, or $f_x(x, y)$, is defined by

$$\frac{\partial z}{\partial x} = \lim_{h \to 0} \frac{f(x + h, y) - f(x, y)}{h} \qquad \text{Hold } y \text{ constant.}$$

Also, the **partial derivative of f with respect to y**, denoted by $\partial z/\partial y$, f_y, or $f_y(x, y)$, is defined by

$$\frac{\partial z}{\partial y} = \lim_{h \to 0} \frac{f(x, y + h) - f(x, y)}{h} \qquad \text{Hold } x \text{ constant.}$$

provided these limits exist.

This definition says that these are the usual derivatives found by holding y or x, respectively, constant. Similar definitions could be formulated for functions with three or more independent variables.

EXAMPLE 1 Let $f(x, y) = 3xy^2 - 2x^3y + 5x$ and $g(x, y) = (3x - 5y)^4$. Find the requested partial derivatives.

a. f_x **b.** f_y **c.** g_x **d.** g_y

Solution **a.** Treat y as a constant and find the derivative with respect to x:

$$f_x = \frac{\partial}{\partial x} f(x, y) = 3y^2 - 6x^2y + 5$$

b. Treat x as a constant and find the derivative with respect to y:

$$f_y = \frac{\partial}{\partial y} f(x, y) = 6xy - 2x^3$$

c. Treat y as a constant:

$$g_x = \frac{\partial}{\partial x} g(x, y) = 4(3x - 5y)^3(3) = 12(3x - 5y)^3$$

d. Treat x as a constant:

$$g_y = \frac{\partial}{\partial y} g(x, y) = 4(3x - 5y)^3(-5) = -20(3x - 5y)^3$$

■

EXAMPLE 2 Suppose that x is the inventory (in thousands of dollars) and y is the number of employees of a dress store, where $80 \le x \le 250$ and $3 \le y \le 8$. Also, suppose the weekly profit function (in dollars) is

$$P(x, y) = 5{,}000 + 10x - 15xy + 10x^2 - 3(y - 8)^2$$

At the present time the inventory is \$105,000 ($x = 105$) and there are four employees ($y = 4$). Approximate the rate of change of P per unit change in x if y remains fixed at 4.

Solution This is the partial derivative of P with respect to x at $(105, 4)$:

$$P_x = (10 - 15y + 20x)\Big|_{(105,4)} = 10 - 15(4) + 20(105) = 2{,}050$$

This says that the instantaneous rate of change of the profit is changing \$2,050 per \$1,000 change in x when y remains fixed at four employees. ■

It is possible to find the partial derivative of a partial derivative. The idea is very straightforward. Study the following definition of **higher-order partial derivatives**.

Higher-Order Partial Derivatives

If $z = f(x, y)$, then

$$\frac{\partial^2 z}{\partial x^2} = \frac{\partial}{\partial x}\left(\frac{\partial z}{\partial x}\right) = f_{xx}(x, y) = f_{xx}$$

$$\frac{\partial^2 z}{\partial y^2} = \frac{\partial}{\partial y}\left(\frac{\partial z}{\partial y}\right) = f_{yy}(x, y) = f_{yy}$$

$$\frac{\partial^2 z}{\partial x\, \partial y} = \frac{\partial}{\partial x}\left(\frac{\partial z}{\partial y}\right) = f_{yx}(x, y) = f_{yx}$$

$$\frac{\partial^2 z}{\partial y\, \partial x} = \frac{\partial}{\partial y}\left(\frac{\partial z}{\partial x}\right) = f_{xy}(x, y) = f_{xy}$$

For the mixed partial derivative, $\frac{\partial^2 z}{\partial x\, \partial y} = f_{yx}$, start with z and first differentiate with respect to y (keep x constant) and then with respect to x. For $\frac{\partial^2 z}{\partial y\, \partial z} = f_{xy}$, start with z and first differentiate with respect to x (keep y constant) and then with respect to y. That is, read f_{yx} from left to right; y first, then x.

EXAMPLE 3 For $z = f(x, y) = 5x^2 - 2xy + 3y^3$, find the requested higher-order partial derivatives. Pay particular attention to the notation—part of what this example is illustrating is the variety in notation that can be used for higher-order partial derivatives.

a. $\frac{\partial^2 z}{\partial x\, \partial y}$ **b.** $\frac{\partial^2 z}{\partial y\, \partial x}$ **c.** $\frac{\partial^2 z}{\partial x^2}$ **d.** $f_{xy}(3, 2)$

e. $\frac{\partial^2}{\partial x\, \partial y}(2x^4 - 3x^2y^2 + 5y^3 + 25)\Big|_{(-2,5)}$

Solution **a.** First differentiate with respect to y:

$$\frac{\partial z}{\partial y} = -2x + 9y^2$$

Then differentiate with respect to x:

$$\begin{aligned}\frac{\partial^2 z}{\partial x\, \partial y} &= \frac{\partial}{\partial x}\left(\frac{\partial z}{\partial y}\right)\\ &= \frac{\partial}{\partial x}(-2x + 9y^2)\\ &= -2\end{aligned}$$

b. First differentiate with respect to x, then with respect to y:

$$\frac{\partial z}{\partial x} = 10x - 2y \quad \text{and} \quad \frac{\partial^2 z}{\partial y\, \partial x} = \frac{\partial}{\partial y}(10x - 2y) = -2$$

c. Differentiate with respect to x twice:

$$\frac{\partial z}{\partial x} = 10x - 2y \qquad \text{and} \qquad \frac{\partial^2 z}{\partial x^2} = \frac{\partial}{\partial x}(10x - 2y) = 10$$

d. Differentiate first with respect to x and then with respect to y; finally, evaluate at $(3, 2)$:

$$f_x(x, y) = 10x - 2y \qquad \text{and} \qquad f_{xy}(x, y) = -2$$

At $(3, 2)$ the value is -2. Notice that since the value is a constant, it is -2 at all points.

e. First differentiate with respect to y, then with respect to x; finally, evaluate at $(-2, 5)$:

$$\frac{\partial^2}{\partial x\,\partial y}(2x^4 - 3x^2y^2 + 5y^3 + 25) = \frac{\partial}{\partial x}(-6x^2y + 15y^2) = -12xy$$

At the point $(-2, 5)$:

$$\frac{\partial^2}{\partial x\,\partial y}(2x^4 - 3x^2y^2 + 5y^3 + 25)\bigg|_{(-2,5)} = -12xy\bigg|_{(-2,5)} = -12(-2)(5) = 120 \quad \blacksquare$$

Notice from parts **a** and **b** that

$$\frac{\partial^2 z}{\partial x\,\partial y} = \frac{\partial^2 z}{\partial y\,\partial x}$$

but in general *this is not true*. However, for all of the functions in this book it will be true.

16.2 Problem Set

Let

$$z = f(x, y) = 5x^2 - 3x^3y^4 + 2y^3 - 15 \quad \textit{and}$$
$$w = g(x, y) = (4x - 3y)^5$$

Find the derivatives in Problems 1–20.

1. f_x
2. g_x
3. g_y
4. f_y
5. $f_x(1, 2)$
6. $g_x(2, -1)$
7. $g_y(3, -1)$
8. $f_y(-2, 3)$
9. $\dfrac{\partial^2 z}{\partial x\,\partial y}$
10. $\dfrac{\partial^2 z}{\partial y\,\partial x}$
11. $\dfrac{\partial^2 w}{\partial y\,\partial x}$
12. $\dfrac{\partial^2 w}{\partial x\,\partial y}$
13. $f_{xx}(0, 2)$
14. $f_{xy}(1, 2)$
15. $f_{yx}(-1, 0)$
16. $f_{yy}(2, -1)$
17. $g_{xx}(0, 2)$
18. $g_{xy}(1, 2)$
19. $g_{yx}(-1, 0)$
20. $g_{yy}(2, -1)$

Let

$$z = f(x, y) = e^{3x+2y} \quad \textit{and}$$
$$w = g(x, y) = \sqrt{x^2 - 3y^2}$$

Find the derivatives in Problems 21–40.

21. f_x
22. g_x
23. g_y
24. f_y
25. $f_x(1, 2)$
26. $g_x(-1, 2)$
27. $g_y(3, -2)$
28. $f_y(-2, 3)$

29. $\dfrac{\partial^2 z}{\partial x\,\partial y}$

30. $\dfrac{\partial^2 z}{\partial y\,\partial x}$

31. $\dfrac{\partial^2 w}{\partial y\,\partial x}$

32. $\dfrac{\partial^2 w}{\partial x\,\partial y}$

33. $f_{xx}(0, 2)$

34. $f_{xy}(1, 2)$

35. $f_{yx}(-1, 0)$

36. $f_{yy}(2, -1)$

37. $g_{xx}(2, 0)$

38. $g_{xy}(2, 1)$

39. $g_{yx}(-1, 0)$

40. $g_{yy}(2, -1)$

Find f_x, f_y, and f_λ for the functions in Problems 41–44.

41. $f(x, y, \lambda) = x + 2xy + \lambda(xy - 10)$

42. $f(x, y, \lambda) = 2x + 2y + \lambda xy$

43. $f(x, y, \lambda) = x^2 + y^2 - \lambda(3x + 2y - 6)$

44. $f(x, y, \lambda) = x^2 - y^2 - \lambda(5x - 3y + 10)$

Find $\partial f/\partial b$ and $\partial f/\partial m$ in Problems 45–48.

45. $f(b, m) = (10m + 5b)^2 + (2m + b)$

46. $f(b, m) = (m + b - 4)^2 + (2m + 2b - 8)^2$

47. $f(b, m) = (m + b + 1)^2 + (2m + 2b + 2)^2 + (3m + 3b + 3)^2$

48. $f(b, m) = (2m - b - 3)^3 + (2m - b - 3)^2 + (2m - b - 3)$

APPLICATIONS

49. Hartwell Corporation sells microwave ovens and finds that its profit is a function of the price, p, of the oven as well as the amount spent on advertising, a, according to the function

$$P(p, a) = 2ap + 50p - 10p^2 - .1a^2p - 100$$

a. $\dfrac{\partial P}{\partial a}$ is the rate of change of profit as a function of the change in advertising spending. Find $\dfrac{\partial P}{\partial a}$.

b. $\dfrac{\partial P}{\partial p}$ is the rate of change of profit as a function of a change in price. Find $\dfrac{\partial P}{\partial p}$.

50. Ritetex is producing x units of one item and y units of another. The cost function is

$$C(x, y) = 3{,}700 + 2{,}500x + 550y$$

Find and interpret $C_x(x, y)$ and $C_y(x, y)$.

51. The revenue function for the items produced by Ritetex in Problem 50 is

$$R(x, y) = 1{,}500x + 900y$$

If P is the profit function, find $P_x(5, 10)$ and $P_y(5, 10)$ and interpret what these numbers mean.

52. The amount of blood flowing in a blood vessel (in milliliters) is given by

$$F(l, r) = .002l/r^4$$

where l is the length of the blood vessel and r is the radius. (This formula is called *Poiseuille's law.*) Find $\partial F/\partial r$ and $\partial F/\partial l$ and interpret.

53. The number of square inches of a person's body area is a function of the person's height and weight:

$$A(w, h) = 40.5w^{.425}h^{.725}$$

where A is the surface area in square inches, w is the weight in pounds, and h is the height in feet. Find your own surface area.

54. Find $A_w(180, 6)$ and $A_h(180, 6)$ for the function A of Problem 53 and interpret the results.

55. If \$100 is invested at an annual rate of r for t years, the future value, A, is given by the formula

$$A = 100(1 + r)^t$$

What is the instantaneous rate of change of A per unit change in r if t remains fixed at 5?

56. If \$25 is deposited monthly to an account paying an annual rate of r compounded monthly for 5 years, the future value, A, is given by the formula

$$A = 25\left[\frac{(1 + i)^{60} - 1}{i}\right]$$

where $i = \dfrac{r}{12}$. What is the instantaneous rate of change of A per unit change of i?

57. If P is the present value of an ordinary annuity of equal payments of \$25 per month for 5 years at an annual interest rate of r, then

$$P = 25\left[\frac{1 - (1 + i)^{-60}}{i}\right]$$

where $i = \dfrac{r}{12}$. What is the instantaneous rate of change of P per unit change in r if t remains fixed at 5?

16.3

Maximum–Minimum Applications

We have seen how important optimization is for functions of a single variable. In this section we consider relative maximums and minimums of functions of two variables—that is, of the type $z = f(x, y)$. In this discussion we assume that all second-order partial derivatives exist. Geometrically, the existence of all second-order partial derivatives guarantees that the surface has no tears, ruptures, sharp points, edges, or corners, as shown in Figure 16.7. We will be looking for bulges or dents, which are called **relative maximums** and **minimums**.

Relative Maximum and Relative Minimum

Let a function be defined by $f(x, y)$ for each point in some region of the xy-plane containing the x- and y-axes. Let (a, b) be some point in this region. If there exists a circular region with center at (a, b) such that for all (x, y) in that region,

$f(x, y) \leq f(a, b)$ then $f(a, b)$ is a **relative maximum**

$f(x, y) \geq f(a, b)$ then $f(a, b)$ is a **relative minimum**

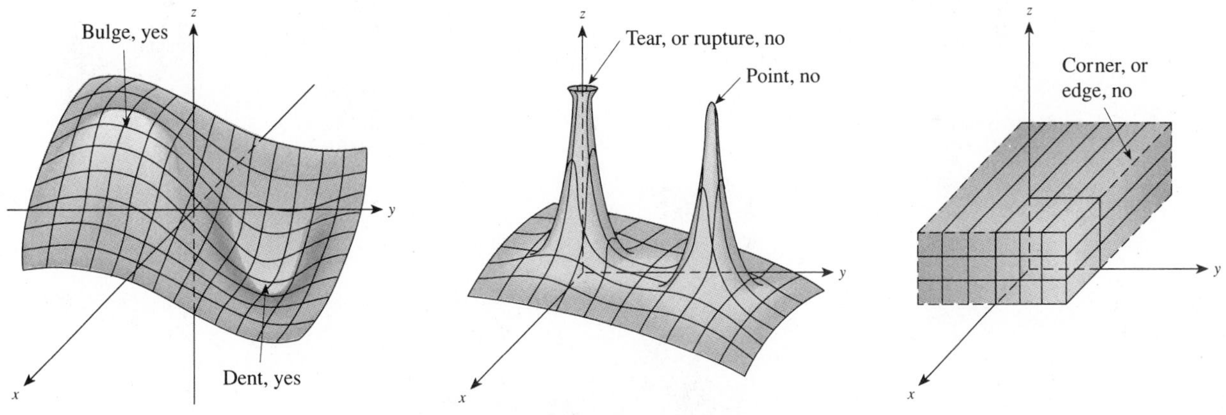

FIGURE 16.7
Surfaces that have or do not have partial derivatives existing at all points

The top of the bulge in Figure 16.7 is a high point which gives a relative maximum for the function and the bottom of the dent is a low point which gives a relative minimum. In this section we are not concerned with boundary points or absolute maximum–minimum theory but, nevertheless, we will be able to consider a great many maximum and minimum problems.

Remember when you did maximums and minimums for a function of a single variable? You first found critical values $x = c$ (values of x where $f'(x) = 0$ or $f'(x)$

does not exist); $f(c)$ was not necessarily a maximum or a minimum but just a possibility. You needed to test further to determine if $f(c)$ was a maximum, minimum, or neither by using the first- or second-derivative tests. We do the same for functions of two variables. We find **critical points** (a, b). These are points for which the partial derivatives f_x and f_y are equal to zero.

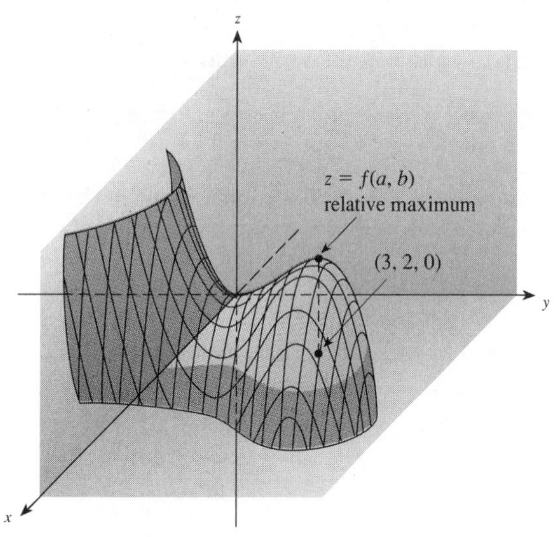

a. Relative maximum at $z = f(a, b)$ at the point $(3, 2, f(3, 2))$

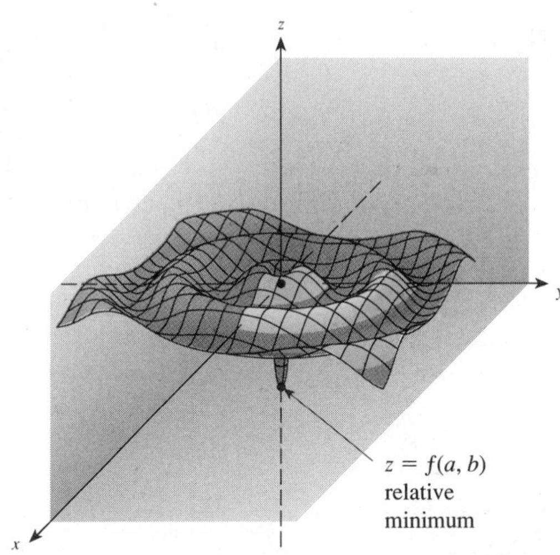

b. Relative minimum $z = f(a, b)$ at $(0, 0, f(0, 0))$

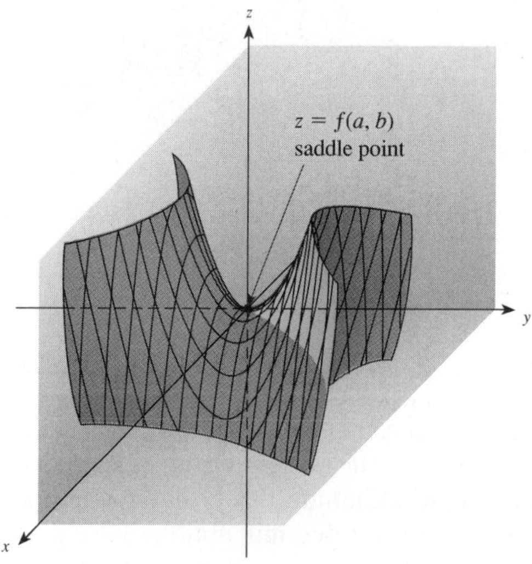

c. Saddle point $(0, 0, 0)$

FIGURE 16.8
Critical points

Critical Points

> Let $z = f(x, y)$ have a relative maximum or a relative minimum at the point (a, b). If both f_x and f_y exist at (a, b), then
>
> $$f_x(a, b) = 0 \qquad \text{and} \qquad f_y(a, b) = 0$$
>
> On the other hand, if (a, b) is a point for which
>
> $$f_x(a, b) = 0 \qquad \text{and} \qquad f_y(a, b) = 0$$
>
> then (a, b) is called a **critical point** for the function f. The value $f(a, b)$ is *not necessarily* a relative maximum or a minimum, but is a candidate.

If you find a point (a, b) that is a critical point, then there are three possibilities, as shown in Figure 16.8. Notice that a saddle point is a point for which the two first partial derivatives are zero but is not itself a relative maximum or a relative minimum. For example, $z = x^2 - y^2$ has both $f_x(0, 0) = 0$ and $f_y(0, 0) = 0$ and $(0, 0, f(0, 0))$ is a saddle point. A **saddle point** gives neither a maximum nor a minimum value for the function.

EXAMPLE 1 Find critical points for the function

$$f(x, y) = 3x^2 + 4y^2 - 2xy + 22x + 30$$

Solution First find the partial derivatives:

$$f_x(x, y) = 6x - 2y + 22 \qquad \text{and} \qquad f_y(x, y) = 8y - 2x$$

Next, set these partials equal to zero:

$$6x - 2y + 22 = 0 \qquad \text{and} \qquad 8y - 2x = 0$$

Finally, solve the system

$$\begin{cases} 6x - 2y = -22 & \longleftarrow \text{This is } 6x - 2y + 22 = 0. \\ -2x + 8y = 0 & \longleftarrow \text{This is } 8y - 2x = 0. \end{cases} \quad 3$$

$$+ \begin{cases} 6x - 2y = -22 \\ -6x + 24y = 0 \end{cases}$$

$$22y = -22$$

$$y = -1$$

To find x, substitute into either of the equations:

$$8y - 2x = 0 \qquad \text{so} \qquad 8(-1) - 2x = 0$$

$$x = -4$$

A critical point is thus $(-4, -1)$. This says that if f has any maximums or minimums they must occur at $(x, y) = (-4, -1)$. ■

EXAMPLE 2 Find the critical points for the surface defined by

$$f(x, y) = 6xy - 4x^3 - 4y^3 - 10$$

Solution $f_x = 6y - 12x^2$ and $f_y = 6x - 12y^2$

Set these equal to zero and solve the system:

$$\begin{cases} 6y - 12x^2 = 0 \\ 6x - 12y^2 = 0 \end{cases}$$

Solve by substitution; from the first equation,

$$6y = 12x^2$$
$$y = 2x^2$$

Substitute into the second equation:

$$6x - 12(2x^2)^2 = 0$$
$$6x - 48x^4 = 0$$
$$x - 8x^4 = 0$$
$$x(1 - 8x^3) = 0$$
$$x(1 - 2x)(1 + 2x + 4x^2) = 0$$

Solve by setting each factor equal to zero:

$x = 0$ $\quad 1 - 2x = 0$ $\quad 1 + 2x + 4x^2 = 0$

$x = \frac{1}{2}$ $\quad$ No real solution (the discriminant is negative)

Finally, find y:

$y = 2x^2$ so if $x = 0$, $y = 0$

if $x = \frac{1}{2}$, $y = 2\left(\frac{1}{2}\right)^2 = \frac{1}{2}$

The critical points are $(0, 0)$ and $(\frac{1}{2}, \frac{1}{2})$. This means that if f has any maximums or minimums they must occur at $(x, y) = (0, 0)$ or $(x, y) = (\frac{1}{2}, \frac{1}{2})$. ■

Since there is no guarantee that $f(a, b)$ is a maximum or a minimum at the critical point (a, b), we need a further test, called a **second-derivative test**, to find the relative maximums or minimums.

Second-Derivative Test for Functions of Two Variables

If *all* of the following conditions hold:

1. $z = f(x, y)$
2. All second-order partial derivatives exist in some circular region containing (a, b) as the center.
3. (a, b) is a critical point [that is, $f_x(a, b) = 0$ and $f_y(a, b) = 0$].
4. $A = f_{xx}(a, b)$, $B = f_{xy}(a, b)$, $C = f_{yy}(a, b)$, $D = B^2 - AC$

Also if:

$D < 0$ and $A < 0$, then $f(a, b)$ is a **relative maximum**.
$D < 0$ and $A > 0$, then $f(a, b)$ is a **relative minimum**.
$D > 0$, then f has a **saddle point** at (a, b) and $f(a, b)$ is neither a relative maximum nor a relative minimum.
$D = 0$, then the *test fails*.

EXAMPLE 3 Apply the second-derivative test for the critical values found in Examples 1 and 2.

Solution **a.** For Example 1, $f(x, y) = 3x^2 + 4y^2 - 2xy + 22x + 30$ has critical point $(-4, -1)$. $A = f_{xx} = 6$, $B = f_{xy} = -2$, and $C = f_{yy} = 8$, so $D = B^2 - AC = 4 - 6(8) < 0$. Thus, $D < 0$ and $A > 0$, so $f(-4, -1)$ is a relative minimum. Even though you would not be expected to graph this surface, it is shown in Figure 16.9a.

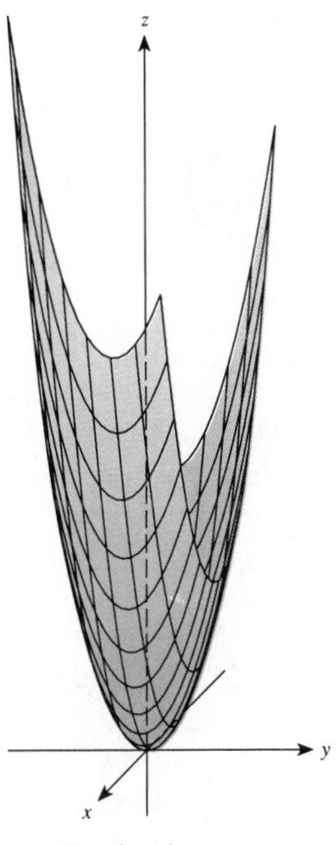

a. Graph of
$f(x, y) = 3x^2 + 4y^2 - 2xy + 22x + 30$

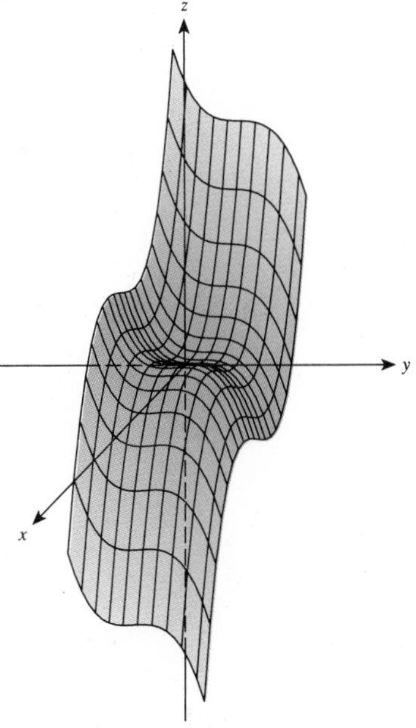

b. Graph of
$f(x, y) = 6xy - 4x^3 - 4y^3 - 10$

FIGURE 16.9
Graphs of surfaces from Examples 1 and 2

b. For Example 2, $f(x, y) = 6xy - 4x^3 - 4y^3 - 10$ has critical points $(0, 0)$ and $(\frac{1}{2}, \frac{1}{2})$. For point $(0, 0)$:

$$
\begin{aligned}
f_{xx}(x, y) &= -24x \quad &&\text{so} \quad A = f_{xx}(0, 0) = 0\\
f_{xy}(x, y) &= 6 \quad &&\text{so} \quad B = f_{xy}(0, 0) = 6\\
f_{yy}(x, y) &= -24y \quad &&\text{so} \quad C = f_{yy}(0, 0) = 0\\
D &= B^2 - AC = 36 - 0 > 0
\end{aligned}
$$

Thus $D > 0$, so there is a saddle point at $(0, 0)$ and $f(0, 0)$ is neither a relative maximum nor a relative minimum. This means that when $(x, y) = (0, 0)$,

$$
\begin{aligned}
z = f(0, 0) &= 6(0)(0) - 4(0)^3 - (0)^3 - 10\\
&= -10
\end{aligned}
$$

Thus the saddle point is $(0, 0, -10)$.

For point $(\frac{1}{2}, \frac{1}{2})$:

$$A = f_{xx}\left(\frac{1}{2}, \frac{1}{2}\right) = -12$$
$$B = f_{xy}\left(\frac{1}{2}, \frac{1}{2}\right) = 6$$
$$C = f_{yy}\left(\frac{1}{2}, \frac{1}{2}\right) = -12$$
$$D = B^2 - AC = 36 - 144 < 0$$
$$f\left(\frac{1}{2}, \frac{1}{2}\right) = 6\left(\frac{1}{2}\right)\left(\frac{1}{2}\right) - 4\left(\frac{1}{2}\right)^3 - 4\left(\frac{1}{2}\right)^3 - 10$$
$$= -\frac{19}{2}$$

Thus $D < 0$ and $A < 0$, so there is a relative maximum of -9.5 at $(\frac{1}{2}, \frac{1}{2})$. Even though you are not expected to graph this surface, it is shown in Figure 16.9b so that you can see the high point and the saddle point. ■

EXAMPLE 4 A manufacturer makes two models of widgets, standard and deluxe. Market research shows the following profit equations:

$$\text{Standard} = [44{,}000 + 500(2y - 5x)]x$$
$$\text{Deluxe} = [500(x - y)]y$$

where x is the price of the standard model and y the price of the deluxe model. How should the widgets be priced to maximize the profit?

Solution Let P = total profit, then

$$P(x, y) = \text{Profit from standard widget} + \text{profit from deluxe widget}$$
$$= [44{,}000 + 500(2y - 5x)]x + [500(x - y)]y$$
$$= 44{,}000x + 1{,}000xy - 2{,}500x^2 + 500xy - 500y^2$$
$$= 44{,}000x + 1{,}500xy - 2{,}500x^2 - 500y^2$$

$$P_x(x, y) = 44{,}000 + 1{,}500y - 5{,}000x \quad \text{and} \quad P_y(x, y) = 1{,}500x - 1{,}000y$$

Set these equal to zero and solve:

$$y = \frac{3}{2}x \qquad \text{From } P_y(x, y) = 0$$

Substitute this into the equation for $P_x(x, y) = 0$:

$$44{,}000 + 1{,}500\left(\frac{3}{2}x\right) - 5{,}000x = 0$$
$$-2{,}750x = -44{,}000$$
$$x = 16$$

If $x = 16$, then $y = (\frac{3}{2})16 = 24$; a critical point is thus $(16, 24)$. Finally, apply the second-derivative test:

$$A = P_{xx} = -5{,}000$$
$$B = P_{xy} = 1{,}500$$
$$C = P_{yy} = -1{,}000$$
$$D = B^2 - AC = (1{,}500)^2 - (-5{,}000)(-1{,}000) < 0$$

Thus, $D < 0$ and $A < 0$, so $(16, 24)$ is a relative maximum. Also, since this is the *only* critical value, we conclude this must also be the absolute maximum. This is analogous to the second-derivative test in two dimensions (see Section 12.4). This means that the profit is maximized if the standard model is priced at \$16 and the deluxe model is priced at \$24. ■

16.3 Problem Set

Find the relative maximums, relative minimums, and saddle points in Problems 1–18. If the second-derivative test fails, simply say so and do not continue with further analysis.

1. $f(x, y) = 3x^2 + 5xy + y^2$
2. $f(x, y) = x^2 + xy - 3x + 2y + 5$
3. $f(x, y) = x^2 + xy + y^2 - 3y$
4. $f(x, y) = x^2 + xy + 2x - 3y + 1$
5. $f(x, y) = x^3 - 3xy - y^3$
6. $f(x, y) = x^3 + y^3 - 3x - 3y$
7. $f(x, y) = 3xy - 5x^2 - y^2 + 3x - 5y - 4$
8. $f(x, y) = 4xy - 6x^2 - y^2 + 2x - 6y - 4$
9. $f(x, y) = xy + 2x - 3y - 4$
10. $f(x, y) = xy - x^2 - 2y^2 + x - y - 5$
11. $f(x, y) = x^2 - y + e^x$
12. $f(x, y) = x^2 + y - e^y$
13. $f(x, y) = xe^y$
14. $f(x, y) = ye^x$
15. $f(x, y) = 4xy - x^4 - y^4$
16. $f(x, y) = xy - x^4 - y^2$
17. $f(x, y) = x^3 + 3xy + y^3$
18. $f(x, y) = x^3 - 3xy + y^3 + 15$

APPLICATIONS

19. Miltex Corporation finds that its profit (in thousands of dollars) is given by the function

$$P(a, n) = -3a^2 - 5n^2 + 34a - 2n + 2an + 40$$

where a is the amount spent on advertising (in thousands of dollars), and n is the number of items (in thousands). Find the maximum value of P and the values of a and n that yield this maximum.

20. The labor cost (per item) for Miltex Corporation is approximated by the function

$$L(x, y) = .5x^2 + 5y^2 - 8x - 26y + 3xy + 55$$

where x is the number of hours of machine time, and y is the number of hours of finishing time. Fine the minimum labor cost and the values of x and y that yield this minimum.

21. A manufacturer makes two models of class rings, standard and deluxe. Market research has shown the following profit equation for x standard rings and y deluxe rings:

$$P(x, y) = 3{,}950x + 150xy - 150x^2 - 50y^2 - 110y$$

How many of each ring should be manufactured in order to maximize the profit?

22. A closed rectangular box with a volume of 8 cubic feet is made from two kinds of material. The top and bottom are made of material costing \$.25 per square foot and the sides require special reinforcing that raises the cost to \$.50 per square foot. What are the cost and dimensions of the box (to the nearest inch) so that the cost of materials is minimized?

23. What are the dimensions of a rectangular box, open at the top, having a volume of 64 cubic feet, and using the least amount of material for its construction?

24. The U.S. Postal Service states that a package cannot have a combined length and girth (distance around) exceeding 120 inches. What are the dimensions of the largest (in volume) box that can be mailed?

25. The second-derivative test does not mention the case where $D < 0$ and $A = 0$. Show that this case is not possible.

16.4 Lagrange Multipliers

In the last section we found the relative maximums and relative minimums of functions of two variables. Sometimes the function to be maximized or minimized has one or more secondary conditions. These conditions are called **constraints**, and the function to be maximized or minimized is called the **objective function.**

The method we will present in this section was first presented by Joseph Lagrange (1736–1813) in a paper he wrote when he was 19, and is called the **method of Lagrange multipliers.**

Method of Lagrange Multipliers

The relative maximums and minimums of the function $z = f(x, y)$ subject to the constraint $g(x, y) = 0$ will be attained at those points (x_0, y_0) which can be found as follows (provided all the partial derivatives exist):

1. Formulate the problem in the form of the objective and constraint functions.
2. Write the equation $F(x, y, \lambda) = f(x, y) + \lambda g(x, y)$. The variable λ is called the *Lagrange multiplier.*
3. Find the partial derivatives F_x, F_y, and F_λ and solve the system

$$\begin{cases} F_x(x, y, \lambda) = 0 \\ F_y(x, y, \lambda) = 0 \\ F_\lambda(x, y, \lambda) = 0 \end{cases}$$

   Solutions (x_0, y_0, λ_0) of this system are called **critical points** for F.
4. Evaluate $z = f(x, y)$ at each critical point. The relative maximums and minimums will be among this list of values.

In addition, if the endpoints of the constraint curve (if any) are also included in the list found in step 3, then the largest value yields the maximum of z subject to the constraint $g(x, y) = 0$ and the smallest value yields the minimum of z subject to the constraint $g(x, y) = 0$.

Keep in mind that this is an alternate method to the second-derivative test introduced in the last section. We begin by considering an example.

EXAMPLE 1 A university extension agricultural service concludes that on a particular farm the yield of wheat per acre (measured in bushels) is a function of the number of acre-feet of water, x, applied and the pounds of fertilizer, y, applied during the growing season according to the formula

$$f(x, y) = 140 - x^2 - 2y^2$$

where f is the yield function. Suppose that water costs \$20 per acre-foot, fertilizer costs \$12 per pound, and the farmer will invest \$236 per acre for water and fertilizer. How much water and fertilizer should the farmer buy to maximize the yield?

Solution In this problem the objective function is f and the constraint is

$$20x + 12y = 236$$

(Note that a more realistic constraint would be that the farmer would spend no more than \$236, in which case this constraint is $20x + 12y \leq 236$; this type of problem was considered in Chapter 4.)

Step 1: Formulate the problem in the form of objective and constraint functions.

$$\begin{aligned} &\text{Maximize:} && f(x, y) = 140 - x^2 - 2y^2 \\ &\text{Subject to:} && g(x, y) = 20x + 12y - 236 = 0 \end{aligned}$$

Step 2: Write the function $F(x, y, \lambda)$ by introducing the Lagrange multiplier λ.

$$\begin{aligned} F(x, y, \lambda) &= f(x, y) + \lambda g(x, y) \\ &= 140 - x^2 - 2y^2 + \lambda(20x + 12y - 236) \\ &= 140 - x^2 - 2y^2 + 20\lambda x + 12\lambda y - 236\lambda \end{aligned}$$

Step 3: Find the partial derivatives F_x, F_y, and F_λ and solve the system $F_x = 0$, $F_y = 0$, $F_\lambda = 0$. Solutions to this system are called *critical points for F.*

$$\begin{aligned} F_x &= -2x + 20\lambda \\ F_y &= -4y + 12\lambda \\ F_\lambda &= 20x + 12y - 236 \end{aligned}$$

These partial derivatives lead to the system of equations

$$\begin{cases} -2x + 20\lambda = 0 \\ -4y + 12\lambda = 0 \\ 20x + 12y - 236 = 0 \end{cases} \quad \text{or} \quad \begin{cases} x - 10\lambda = 0 \\ y - 3\lambda = 0 \\ 5x + 3y = 59 \end{cases}$$

Begin by solving the first pair of equations simultaneously:

$$\begin{matrix} 3 \\ -10 \end{matrix} \begin{cases} x - 10\lambda = 0 \\ y - 3\lambda = 0 \end{cases} \quad + \quad \begin{cases} 3x - 30\lambda = 0 \\ -10y + 30\lambda = 0 \end{cases}$$
$$3x - 10y = 0$$

Use this result along with the third equation of the original system to complete the solution:

$$\begin{matrix} -5 \\ 3 \end{matrix} \begin{cases} 3x - 10y = 0 \\ 5x + 3y = 59 \end{cases} \quad + \quad \begin{cases} -15x + 50y = 0 \\ 15x + 9y = 177 \end{cases}$$
$$59y = 177$$
$$y = 3$$

If $y = 3$, then $3x - 10(3) = 0$ implies that $x = 10$. Also $y - 3\lambda = 0$. This means that $3 - 3\lambda = 0$ or $\lambda = 1$. This system has only one critical point: $(x, y, \lambda) = (10, 3, 1)$.

Step 4: Evaluate $z = f(x, y)$ at each critical point. The relative maximum or relative minimum values of $f(x, y)$ will be among these values in the problem. In this example there is only one critical point, and since there is no minimum, it follows that this point must provide an absolute maximum.

Since $f(x, y) = 140 - x^2 - 2y^2$, we see

$$f(10, 3) = 140 - 10^2 - 2(3)^2 = 22$$

This says that the maximum yield per acre is 22 bushels. ■

EXAMPLE 2 Minimize $f(x, y) = x^2 + y^2$ subject to $x + y - 1 = 0$.

Solution *Step 1:* This was done for us in the statement of the problem:

Minimize: $f(x, y) = x^2 + y^2$

Subject to: $g(x, y) = x + y - 1 = 0$

Step 2:

$$F(x, y, \lambda) = f(x, y) + \lambda g(x, y) = x^2 + y^2 + \lambda x + \lambda y - \lambda$$

Step 3:

$$F_x = 2x + \lambda$$
$$F_y = 2y + \lambda$$
$$F_\lambda = x + y - 1$$

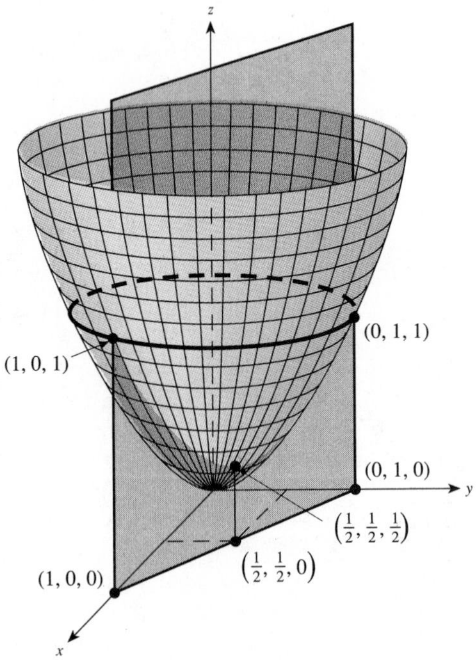

FIGURE 16.10
Graph of $f(x, y) = x^2 + y^2$ showing that the minimum value on this surface satisfying $x + y = 1$ is .5.

Solve the system

$$\begin{cases} 2x + \lambda = 0 \\ 2y + \lambda = 0 \\ x + y - 1 = 0 \end{cases}$$

to find $x = .5$, $y = .5$, $\lambda = -1$, or the critical point: $(.5, .5, -1)$.

Step 4: $f(.5, .5) = .5^2 + .5^2 = .5$

By checking other points near $(.5, .5)$, you can see that this point is a minimum. This situation is illustrated quite nicely in Figure 16.10. ■

EXAMPLE 3 Find two numbers whose sum is 100 and whose product is a maximum.

Solution *Step* 1: Let x and y be the two numbers. Then:

Maximize: $f(x, y) = xy$ — That is, maximize the product.

Subject to: $g(x, y) = x + y - 100 = 0$ — This is the constraint; the sum must be 100.

Step 2:

$$\begin{aligned} F(x, y, \lambda) &= f(x, y) + \lambda g(x, y) \\ &= xy + \lambda(x + y - 100) \\ &= xy + \lambda x + \lambda y - 100\lambda \end{aligned}$$

Step 3:

$$\begin{aligned} F_x &= y + \lambda \\ F_y &= x + \lambda \\ F_\lambda &= x + y - 100 \end{aligned}$$

By solving the system

$$\begin{cases} y + \lambda = 0 \\ x + \lambda = 0 \\ x + y = 100 \end{cases}$$

simultaneously, we find that $x = 50$, $y = 50$, and $\lambda = -50$.

Step 4: Since $f(x, y) = xy$, we see that $f(50, 50) = 50(50) = 2{,}500$. Nearby values of x and y yield smaller products, so we conclude that the maximum product is 2,500, which occurs for the numbers 50 and 50. ■

The method of Lagrange multipliers can easily be extended to functions of more than two variables, as illustrated by Example 4.

EXAMPLE 4 Suppose the temperature T at any point (x, y, z) in a region of space is given by the formula $T = 5{,}000 - (xy + xz + yz)$. Find the lowest temperature on the plane $x + y + z = 100$.

Solution *Step* 1: Minimize: $f(x, y, z) = 5{,}000 - xy - xz - yz$

Subject to: $g(x, y, z) = x + y + z - 100 = 0$

Step 2:
$$
\begin{aligned}
F(x, y, z, \lambda) &= f(x, y, z) + \lambda g(x, y, z) \\
&= 5{,}000 - xy - xz - yz + \lambda(x + y + z - 100) \\
&= 5{,}000 - xy - xz - yz + \lambda x + \lambda y + \lambda z - 100\lambda
\end{aligned}
$$

Step 3:
$$
\begin{aligned}
F_x &= -y - z + \lambda \\
F_y &= -x - z + \lambda \\
F_z &= -x - y + \lambda \\
F_\lambda &= x + y + z - 100
\end{aligned}
$$

Solve the system

$$
\begin{cases}
y + z - \lambda = 0 \\
x + z - \lambda = 0 \\
x + y - \lambda = 0 \\
x + y + z = 100
\end{cases}
$$

From the first two equations: $x - y = 0$
From the first and third equations: $x - z = 0$
From the fourth equation: $z = 100 - x - y$; substitute this result into the equation $x - z = 0$ to find

$$
\begin{aligned}
x - (100 - x - y) &= 0 \\
2x + y &= 100
\end{aligned}
$$

Finally,

$$
\begin{cases}
2x + y = 100 \\
x - y = 0
\end{cases}
$$
$$
\begin{aligned}
3x &= 100 \\
x &= \frac{100}{3}
\end{aligned}
$$

By substitution, $y = \frac{100}{3}$, $z = \frac{100}{3}$, and $\lambda = \frac{200}{3}$.

Step 4: $f(x, y, z) = 5{,}000 - xy - xz - yz$

So

$$
\begin{aligned}
f\left(\frac{100}{3}, \frac{100}{3}, \frac{100}{3}\right) &= 5{,}000 - \left(\frac{100}{3}\right)\left(\frac{100}{3}\right) - \left(\frac{100}{3}\right)\left(\frac{100}{3}\right) - \left(\frac{100}{3}\right)\left(\frac{100}{3}\right) \\
&= 5{,}000 - 3\left(\frac{100}{3}\right)^2 = 5{,}000 - \frac{10{,}000}{3} \\
&= \frac{5{,}000}{3} \approx 1{,}667
\end{aligned}
$$

By checking other nearby points on the plane, you see that this is a minimum. Thus the minimum temperature is about 1,667 degrees. ■

16.4
Problem Set

Find the relative maximums for $f(x, y)$ in Problems 1–6.

1. $f(x, y) = xy$ subject to $x + y = 20$
2. $f(x, y) = 2xy - 5$ subject to $x + y = 12$
3. $f(x, y) = -2x^2 - 3y^2$ subject to $x + 2y = 24$
4. $f(x, y) = 16 - x^2 - y^2$ subject to $x + 2y = 6$
5. $f(x, y) = x^2y$ subject to $x + 2y = 14$
6. $f(x, y) = 4xy^2$ subject to $x - 4y = 16$

Find the relative minimums for $f(x, y)$ in Problems 7–12.

7. $f(x, y) = x^2 + y^2$ subject to $x + y = 24$
8. $f(x, y) = x^2 + y^2$ subject to $x + y = 8$
9. $f(x, y) = x^2 + y^2 - xy - 4$ subject to $x + y = 6$
10. $f(x, y) = x^2 + y^2$ subject to $x + y = 9$
11. $f(x, y) = x^2 + y^2$ subject to $x + y = 16$
12. $f(x, y) = x^2 + y^2$ subject to $2x + y = 20$
13. Find two numbers whose sum is 10 and whose product is a maximum.
14. Find two numbers whose sum is 120 and whose product is a maximum.

APPLICATIONS

15. How would the farmer of Example 1 maximize the yield if the amount spent is \$100 instead of \$236?
16. A wholesaler supplies two types of radio-controlled airplanes, models A and B. Suppose x units of model A and y units of model B can be supplied at a cost of

$$C(x, y) = 6x^2 + 18y^2$$

If the supplier is limited to shipping no more than 100 models, how much of each item should be supplied in order to minimize the cost? (Assume that the number shipped is exactly 100 models.)

17. A rancher needs to build a rectangular fenced enclosure. Because of a difference in terrain, one width costs \$3 per foot and the other costs \$9 per foot. The length costs \$6 per foot. What is the maximum area that can be enclosed if the rancher has \$4,000 to spend?
18. A can of Classic Coke® holds about 25 cubic inches. Find the minimum surface area of a can with a volume of 25 cubic inches. This gives the appropriate minimum amount of material required. Then measure a can of Coke to compare your answer with the size of the actual can.
19. A patient is put on a 1,000 calorie diet. Let us suppose (for purposes of this problem) that these calories will be supplied by two food types, meat and vegetables. Let x be the number of ounces of meat costing \$.15 per ounce and y the number of ounces of vegetables costing \$.03 per ounce. If the two foods produce

$$C(x, y) = 20xy$$

calories, determine what mixture of meat and vegetables should be supplied to minimize the cost.

20. A company buys x items of new equipment and uses y hours of labor at a cost of $C(x, y)$ dollars and a revenue of $R(x, y)$ dollars. The partial derivatives C_x and R_y are the marginal cost and marginal revenue with respect to x, and the partial derivatives C_y and R_y are the marginal cost and marginal revenue with respect to y. Show that if a Lagrange multiplier is used to find the maximum of $R(x, y)$ subject to the constraint $C(x, y) = \text{constant}$, then the Lagrange multiplier is the ratio of the marginal revenue and the marginal cost with respect to each variable at the maximum.

Problems 21–24 involve functions of three variables.

21. Find the maximum value of

$$f(x, y, z) = x - y + z$$

on the sphere

$$x^2 + y^2 + z^2 = 100$$

22. Find the minimum value of

$$f(x, y, z) = x - y + z$$

on the sphere

$$x^2 + y^2 + z^2 = 100$$

23. The temperature T at point (x, y, z) in a region of space is given by the formula

$$T = 100 - xy - xz - yz$$

Find the lowest temperature on the plane

$$x + y + z = 10$$

24. Find the largest product of numbers x, y, and z such that their sum is 24.

*16.5 Multiple Integrals

In this chapter we have defined functions with two or more independent variables, defined the derivatives of such functions, and then looked at some applications of the derivatives. It seems reasonable to now ask how we integrate functions of two or more variables.

Integration was originally motivated by looking at antiderivatives as the reverse of the process of differentiation. With functions of several variables the question becomes one of reversing the process of partial differentiation. We write $\int f(x, y)\,dx$ to indicate that we are to find the antiderivative with respect to x while holding y fixed, and $\int f(x, y)\,dy$ to indicate that we are to antidifferentiate $f(x, y)$ with respect to y, holding x fixed.

EXAMPLE 1 **a.** Evaluate $\int (12x^2y^3 + 2x - 6y^2)\,dx$.

b. Evaluate $\int (12x^2y^3 + 2x - 6y^2)\,dy$.

Check your results by using partial derivatives.

Solution **a.**

$$\int (12x^2y^3 + 2x - 6y^2)\,dx = \frac{12y^3x^3}{3} + \frac{2x^2}{2} - 6y^2x + C(y)$$
$$= 4x^3y^3 + x^2 - 6xy^2 + C(y)$$

Integrate with respect to x; note that the constant of integration is a *function of y* since the derivative of a function of y with respect to x is 0 (remember to think of y as a constant in this context).

Check by taking the partial derivative with respect to x:

$$\frac{\partial}{\partial x}[4x^3y^3 + x^2 - 6xy^2 + C(y)]$$
$$= 4(3)x^2y^3 + 2x - 6y^2 + 0$$
$$= 12x^2y^3 + 2x - 6y^2$$

b.

$$\int (12x^2y^3 + 2x - 6y^2)\,dy = \frac{12x^2y^4}{4} + 2xy - \frac{6y^3}{3} + C(x)$$
$$= 3x^2y^4 + 2xy - 2y^3 + C(x)$$

This time integrate with respect to y (and treat x as a constant).

Check: $\frac{\partial}{\partial y}[3x^2y^4 + 2xy - 2y^3 + C(x)] = 12x^2y^3 + 2x - 6y^2$ ■

* Optional section.

This indefinite integration of a function of two variables extends quite easily to definite integration, as shown in Example 2.

EXAMPLE 2 Evaluate

$$\int_{-1}^{4} (x^2 - xy + y^2)\,dx = \frac{x^3}{3} - \frac{x^2 y}{2} + xy^2 \Big|_{x=-1}^{x=4}$$

$$= \left(\frac{64}{3} - 8y + 4y^2\right) - \left[\frac{-1}{3} - \frac{1}{2}y + (-1)y^2\right]$$

$$= \frac{65}{3} - \frac{15}{2}y + 5y^2$$

■

Notice in Example 2 that integrating and evaluating a definite integral, which is a function $f(x, y)$ with respect to x, produces a function of y alone (including constants). This integral, in turn, can be integrated with respect to y:

$$\int_{1}^{3} \left(\frac{65}{3} - \frac{15}{2}y + 5y^2\right) dy = \frac{65}{3}y - \frac{15}{4}y^2 + \frac{5y^3}{3}\Big|_{y=1}^{y=3}$$

$$= \left(65 - \frac{135}{4} + 45\right) - \left(\frac{65}{3} - \frac{15}{4} + \frac{5}{3}\right)$$

$$= \frac{170}{3}$$

What we have done here is a process called **double integration** that can be summarized using the notation

$$\iint_R f(x, y)\,dA$$

where, in this example, $f(x, y) = x^2 - xy + y^2$, dA indicates that this is an integral over a two-dimensional region, and R is a region on which the function is defined—in this example,

$$-1 \le x \le 4$$
$$1 \le y \le 3$$

Geometrically, this process can be viewed as the volume of a solid under the surface $f(x, y)$ (where $f(x, y) \ge 0$) and directly over the rectangle R, as shown in Figure 16.11.

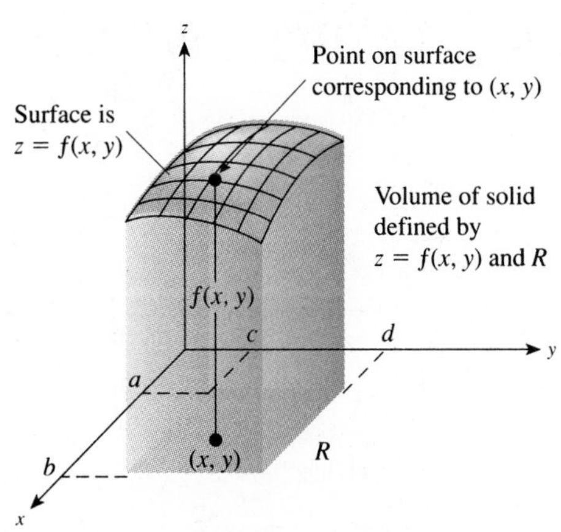

FIGURE 16.11
Geometric interpretation of a double integral

$$V = \iint_R (x^2 - xy + y^2)\,dA$$

Double Integral

The **double integral** of a function $f(x, y)$ over a rectangle R defined by the boundaries

$$a \le x \le b \quad \text{and} \quad c \le y \le d$$

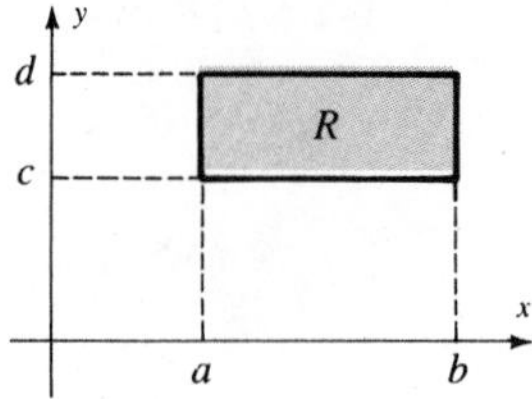

is

$$\iint_R f(x, y)\, dA = \int_a^b \int_c^d f(x, y)\, dy\, dx = \int_c^d \int_a^b f(x, y)\, dx\, dy$$

The function $f(x, y)$ is called the **integrand** and R is called the **region of integration**. The expression dA indicates that it is an integral over a two-dimensional region (i.e., an area). The integrals at the right are called **iterated integrals** and the order in which dx and dy are written indicates the order of integration as follows:

$$\int_c^d \int_a^b f(x, y)\, dx\, dy = \int_c^d \left[\int_a^b f(x, y)\, dx \right] dy$$

Integrate first with respect to x, then with respect to y.

$$\int_a^b \int_c^d f(x, y)\, dy\, dx = \int_a^b \left[\int_c^d f(x, y)\, dy \right] dx$$

Integrate first with respect to y, then with respect to x.

EXAMPLE 3 Evaluate

a. $\displaystyle\int_0^3 \int_0^1 (12x^2y^3 - 4xy)\, dy\, dx$ **b.** $\displaystyle\int_0^1 \int_0^3 (12x^2y^3 - 4xy)\, dx\, dy$

Solution **a.**

$$\begin{aligned}\int_0^3 \int_0^1 (12x^2y^3 - 4xy)\, dy\, dx &= \int_0^3 \left(\frac{12x^2y^4}{4} - \frac{4xy^2}{2} \right)\Bigg|_0^1 dx \\ &= \int_0^3 (3x^2 - 2x)\, dx \\ &= \frac{3x^3}{3} - \frac{2x^2}{2}\Bigg|_0^3 \\ &= 27 - 9 = 18\end{aligned}$$

Integrate first with respect to y.

b.

$$\begin{aligned}\int_0^1 \int_0^3 (12x^2y^3 - 4xy)\, dx\, dy &= \int_0^1 \left(\frac{12x^3y^3}{3} - \frac{4x^2y}{2} \right)\Bigg|_0^3 dy \\ &= \int_0^1 (108y^3 - 18y)\, dy \\ &= \frac{108y^4}{4} - \frac{18y^2}{2}\Bigg|_0^1 \\ &= 27 - 9 = 18\end{aligned}$$

Integrate first with respect to x.

■

Notice that the integrals in parts **a** and **b** of Example 3 are the same except for the order of integration. The fact that the answers are the same is no coincidence. We will be reversing the order of integration later in this section.

Double Integrals over General Regions

The limits of integration in Examples 1–3 are constant, which corresponds to integration over a rectangular region. We now consider variable limits of integration. There are four types of regions with variable limits:

I. Variable limits of integration for y: $\displaystyle\iint_R f(x, y)\,dA = \int_a^b \int_{g_1(x)}^{g_2(x)} f(x, y)\,dy\,dx$

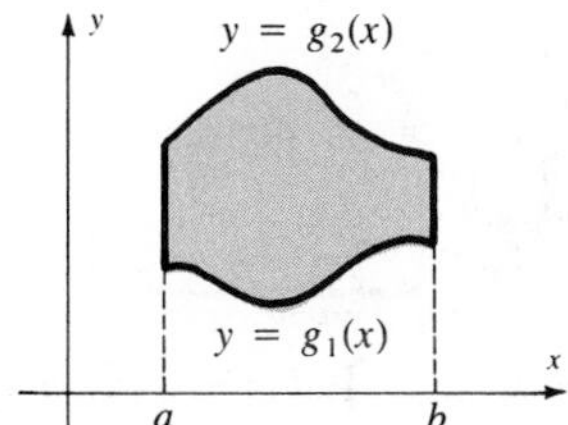

II. Variable limits of integration for x: $\displaystyle\iint_R f(x, y)\,dA = \int_c^d \int_{h_1(y)}^{h_2(y)} f(x, y)\,dx\,dy$

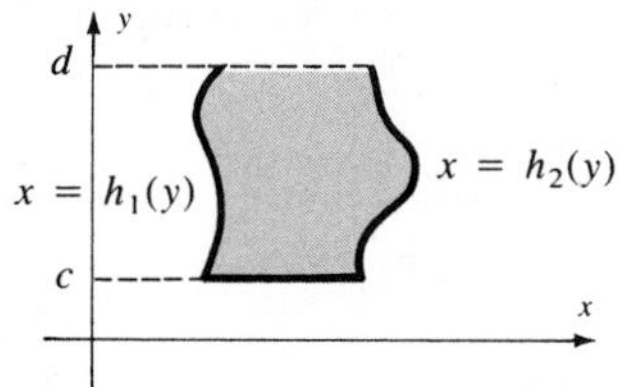

III. Variable limits of integration for either x or y:

$$\iint_R f(x, y)\,dA = \int_a^b \int_{g_2(x)}^{g_1(x)} f(x, y)\,dy\,dx = \int_c^d \int_{h_2(y)}^{h_1(y)} f(x, y)\,dx\,dy$$

IV. A region that does not have boundaries specified as functions. A region like this, which does not have a boundary that can easily be expressed as a function of x or a function of y, can be broken up into smaller regions that fit one of the previous three types. However, regions such as this are beyond the scope of this book.

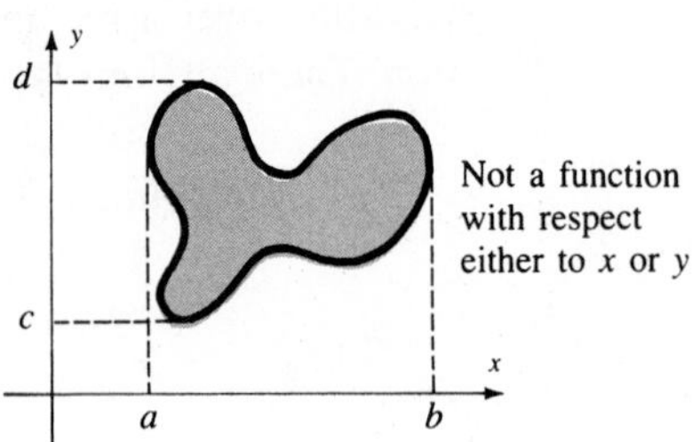

Double Integrals over Variable Regions

Let $z = f(x, y)$ be a function of two variables. If R is a region defined by $g_1(x) \le y \le g_2(x)$, $a \le x \le b$ (type I), then

$$\iint_R f(x, y)\, dA = \int_a^b \int_{g_1(x)}^{g_2(x)} f(x, y)\, dy\, dx$$

If R is a region defined by $h_1(y) \le x \le h_2(y)$, $c \le y \le d$ (type II), then

$$\iint_R f(x, y)\, dA = \int_c^d \int_{h_1(y)}^{h_2(y)} f(x, y)\, dx\, dy$$

WARNING When setting up these double integrals, remember that the variable limits of integration are on the inner integral and the constant limits of integration are on the outer integral.

EXAMPLE 4 Evaluate the double integral of $f(x, y) = 6xy$ over the region shown in the figure.

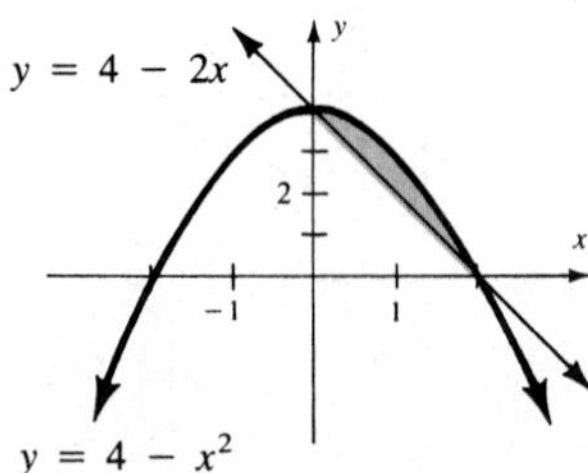

Solution This is a type I region with variable limits for y, so we can set up the following integral:

$$\begin{aligned}
\iint_R f(x,y)\,dA &= \int_0^2 \int_{4-2x}^{4-x^2} 6xy\,dy\,dx \\
&= \int_0^2 3xy^2 \Big|_{4-2x}^{4-x^2} dx \\
&= \int_0^2 3x(4-x^2)^2 - 3x(4-2x)^2\,dx \\
&= \int_0^2 3x(16-8x^2+x^4) - 3x(16-16x+4x^2)\,dx \\
&= \int_0^2 (3x^5 - 36x^3 + 48x^2)\,dx \\
&= \left(\frac{3x^6}{6} - \frac{36x^4}{4} + \frac{48x^3}{3}\right)\Bigg|_0^2 \\
&= 32 - 144 + 128 - 0 = 16
\end{aligned}$$

■

EXAMPLE 5 Integrate $f(x, y) = x - y$ over the region shown in the figure.

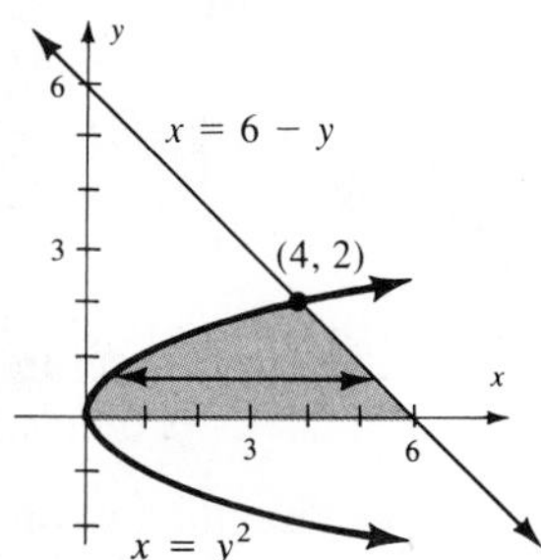

Solution This is a type II region with variable limits for x:

$$\begin{aligned}
\int_0^2 \int_{y^2}^{6-y} (x-y)\,dx\,dy &= \int_0^2 \left(\frac{x^2}{2} - xy\right)\Bigg|_{y^2}^{6-y} dy \\
&= \int_0^2 \left[\frac{1}{2}(6-y)^2 - (6-y)y - \frac{1}{2}(y^2)^2 + y^3\right] dy \\
&= \int_0^2 \left(-\frac{1}{2}y^4 + y^3 + \frac{3}{2}y^2 - 12y + 18\right) dy \\
&= \left(-\frac{y^5}{10} + \frac{y^4}{4} + \frac{3y^3}{6} - \frac{12y^2}{2} + 18y\right)\Bigg|_0^2 \\
&= -\frac{32}{10} + \frac{16}{4} + \frac{24}{6} - \frac{48}{2} + 36 - 0 \\
&= 16\frac{4}{5}
\end{aligned}$$

■

One of the hardest parts in evaluating a double integral is deciding which variable to integrate with respect to first, and which second. After that, the next step is deciding the limits of integration. These steps are illustrated for the regions we used in Examples 4 and 5.

Problem: Evaluate $\iint_R f(x, y)\, dA$.

Example 4

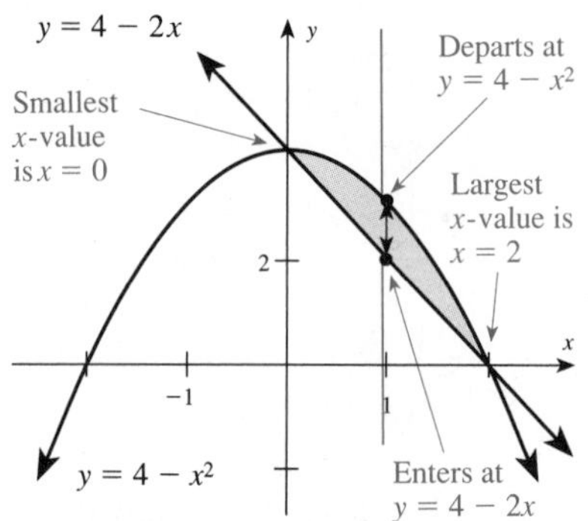

Type I: vertical line; integrate with respect to y first:

$$\int_0^2 \int_{4-2x}^{4-x^2} f(x, y)\, dy\, dx$$

Example 5

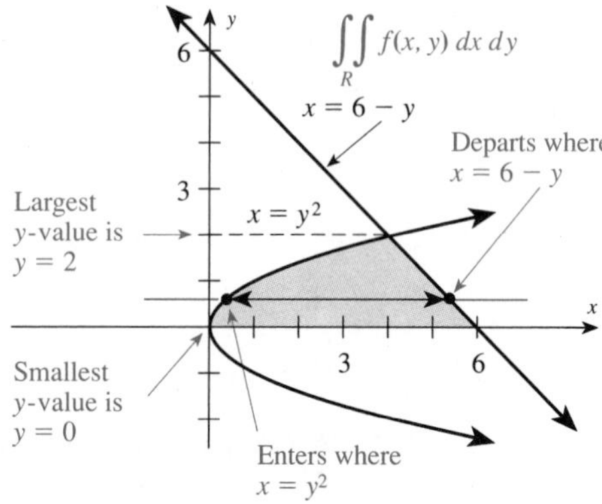

Type II: horizontal line; integrate with respect to x first:

$$\int_0^2 \int_{y^2}^{6-y} f(x, y)\, dx\, dy$$

Type I. Integrate first with respect to y, then with respect to x:

1. Imagine a vertical line L cutting through R in the direction of increasing y.
2. Integrate from the y value where L enters R to the y value where L departs R (these should be functions of x; that is, solve for y).
3. Choose the x limits that include all the vertical lines that pass through R.

Type II. Integrate first with respect to x, then with respect to y:

1. Imagine a horizontal line L cutting through R in the direction of increasing x.
2. Integrate from the x value where L enters R to the x value where L departs R (these should be functions of y; that is, solve for x).
3. Choose the y limits that include all the horizontal lines that pass through R.

EXAMPLE 6 Evaluate $f(x, y) = x + y$ two different ways over the region shown in the figure.

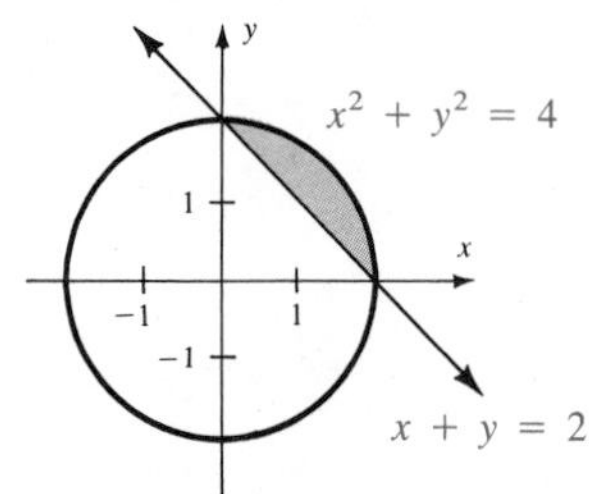

Solution This is a type III region with variable limits.

I. Integrate *first with respect to* y, then x; consider a **vertical line**, as shown in the figure at the right:

$$\int_0^2 \int_{2-x}^{\sqrt{4-x^2}} (x+y)\,dy\,dx$$

$$= \int_0^2 \left(xy + \frac{1}{2}y^2\right)\Big|_{2-x}^{\sqrt{4-x^2}} dx$$

$$= \int_0^2 \left[x\sqrt{4-x^2} + \frac{1}{2}(4-x^2) - x(2-x) - \frac{1}{2}(2-x)^2\right] dx$$

$$= \int_0^2 x\sqrt{4-x^2}\,dx = -\frac{1}{2}\int_4^0 u^{1/2}\,du$$

Let $u = 4 - x^2$	If $x = 0$, then $u = 4$
$du = -2x\,dx$	If $x = 2$, then $u = 0$

$$= -\frac{1}{2}\left(\frac{2}{3}\right)u^{3/2}\Big|_4^0 = \frac{8}{3}$$

II. Integrate *first with respect to* x, then y; consider a **horizontal line**, as shown in the figure at the right:

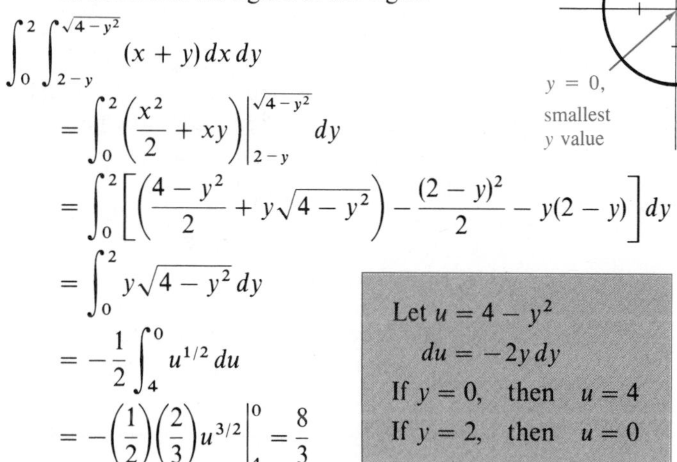

$$\int_0^2 \int_{2-y}^{\sqrt{4-y^2}} (x+y)\,dx\,dy$$

$$= \int_0^2 \left(\frac{x^2}{2} + xy\right)\Big|_{2-y}^{\sqrt{4-y^2}} dy$$

$$= \int_0^2 \left[\left(\frac{4-y^2}{2} + y\sqrt{4-y^2}\right) - \frac{(2-y)^2}{2} - y(2-y)\right] dy$$

$$= \int_0^2 y\sqrt{4-y^2}\,dy$$

$$= -\frac{1}{2}\int_4^0 u^{1/2}\,du$$

$$= -\left(\frac{1}{2}\right)\left(\frac{2}{3}\right)u^{3/2}\Big|_4^0 = \frac{8}{3}$$

Let $u = 4 - y^2$	
$du = -2y\,dy$	
If $y = 0$, then	$u = 4$
If $y = 2$, then	$u = 0$

■

Reversing the Order of Integration

As Examples 4–6 show, sometimes we integrate first with respect to y (using a vertical line), sometimes with respect to x (using a horizontal line), and other times it is possible to do it either way. If the integration can be done either way, we can change the order of integration. Over a rectangular region, it is simply a matter of interchanging the numerical limits of integration, but when interchanging the order of integration with variable limits, we need to look at the region R more carefully, as Example 7 shows.

EXAMPLE 7 Reverse the order of integration for

$$\int_2^{11}\int_{\sqrt{y-2}}^{3} f(x, y)\,dx\,dy$$

Solution You can see that this integral is integrated first with respect to x (so a horizontal line should be used). This means that to find R, you write

$$x = 3 \qquad \text{and} \qquad x = \sqrt{y-2}$$

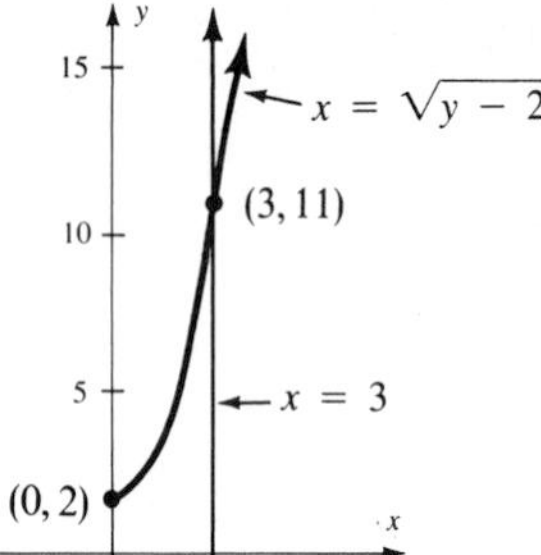

and graph these curves as shown in the margin.

Next, install the limits of integration for y, namely,

$$y = 2 \qquad \text{and} \qquad y = 11$$

and graph these as shown at the right. Shade R and reverse the order of integration. Then solve each equation for y and draw a vertical line to find the limits of integration.

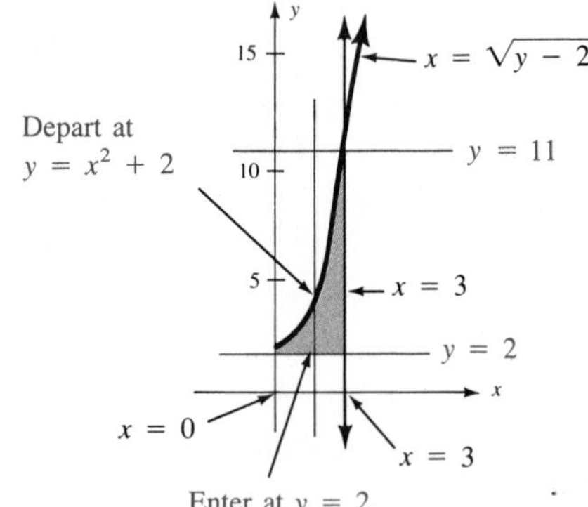

$$x = \sqrt{y-2}$$
$$x^2 = y - 2$$
$$y = x^2 + 2$$

It enters at $y = 2$ and departs at $y = x^2 + 2$; the least value for x is 0 and the largest value for x is 3, so the limits can now be reversed:

$$\int_0^{3}\int_2^{x^2+2} f(x, y)\,dy\,dx$$

■

Volume

As stated earlier, a geometrical interpretation of double integration is that of a **volume** under a surface. That idea is now formalized:

Volume

Let $z = f(x, y)$ be a nonnegative surface over a region R, as illustrated by Figure 16.12. The volume of the solid under the surface of f and over the region R is

$$\iint_R f(x, y)\,dA$$

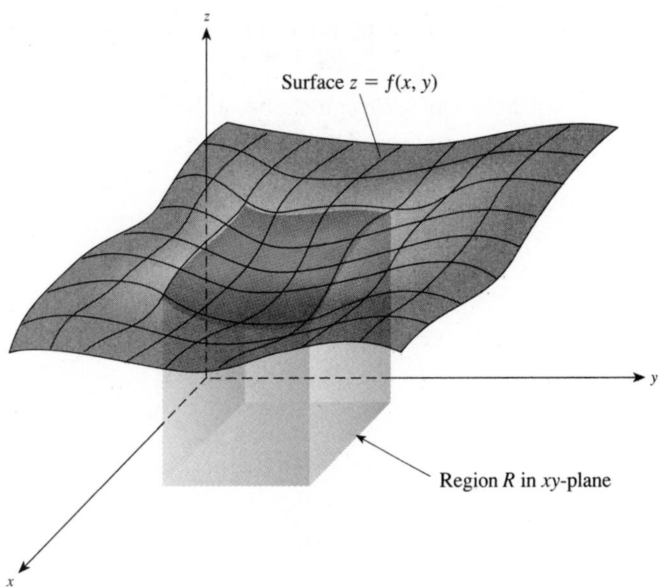

FIGURE 16.12

EXAMPLE 8 Find the volume of the solid bounded on the bottom by the region bounded by the curve with equation $y = x^2$ and the line $2x - y = 0$ and on the top by the plane $z = 10$.

Solution First draw the region R in two dimensions. (The three-dimensional figure is for reinforcement only; you do not need to draw it for your work.)

We will integrate with respect to y first (vertical line).

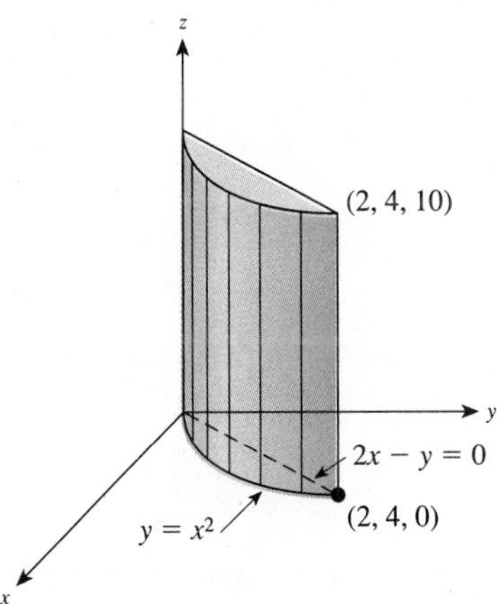

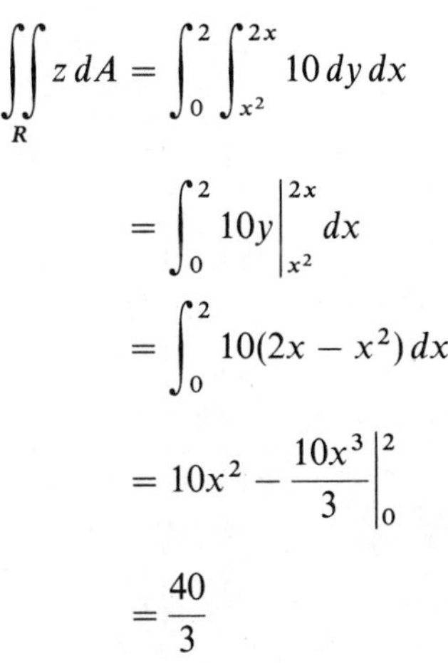

$$\iint_R z\,dA = \int_0^2 \int_{x^2}^{2x} 10\,dy\,dx$$

$$= \int_0^2 10y\Big|_{x^2}^{2x} dx$$

$$= \int_0^2 10(2x - x^2)\,dx$$

$$= 10x^2 - \frac{10x^3}{3}\Big|_0^2$$

$$= \frac{40}{3}$$

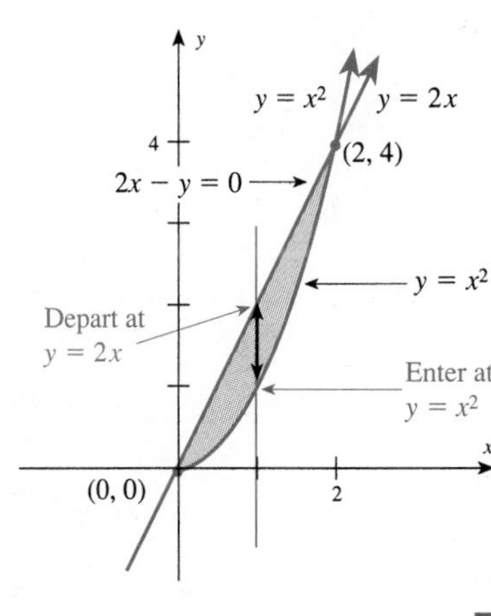

■

EXAMPLE 9 Find the volume of the function $z = x + y$ over the region R, where $0 \le x \le 2$ and $1 \le y \le 3$.

Solution We can choose either order of iteration. As a check, we will evaluate both ways.

$$\iint_R (x + y)\,dA$$

Method I:

$$\int_1^3 \int_0^2 (x + y)\,dx\,dy = \int_1^3 \left(\frac{x^2}{2} + xy\right)\Bigg|_0^2 dy = \int_1^3 (2 + 2y)\,dy = 2y + y^2\Big|_1^3 = 6 + 9 - 2 - 1 = 12$$

Method II:

$$\int_0^2 \int_1^3 (x + y)\,dy\,dx = \int_0^2 \left(xy + \frac{y^2}{2}\right)\Bigg|_1^3 dx = \int_0^2 \left(3x + \frac{9}{2} - x - \frac{1}{2}\right) dx = \int_0^2 (2x + 4)\,dx = x^2 + 4x\Big|_0^2 = 4 + 8 - 0 - 0 = 12$$

■

16.5 Problem Set

Evaluate the integrals in Problems 1–12.

1. $\int_0^2 (x^2y + y^2)\,dy$
2. $\int_0^2 (x^2y + y^2)\,dx$
3. $\int_1^4 (xy^2 - x)\,dx$
4. $\int_1^4 (xy^2 - x)\,dy$
5. $\int_0^4 x\sqrt{x^2 + 2y}\,dy$
6. $\int_0^4 x\sqrt{x^2 + 2y}\,dx$
7. $\int_1^2 3e^{x+2y}\,dx$
8. $\int_1^2 3e^{x+2y}\,dy$
9. $\int_1^5 xe^{x^2+3y}\,dx$
10. $\int_1^5 xe^{x^2+3y}\,dy$
11. $\int_1^{e^x} \frac{x}{y}\,dy$
12. $\int_1^{e^x} \frac{x}{y}\,dx$

Evaluate the iterated integrals in Problems 13–24, and draw the region R defined by the double integral.

13. $\int_1^4 \int_0^3 dy\,dx$
14. $\int_{-2}^3 \int_0^4 dy\,dx$
15. $\int_{-3}^2 \int_2^4 dx\,dy$
16. $\int_2^5 \int_{-1}^3 dx\,dy$
17. $\int_1^3 \int_0^y (x + 3y)\,dx\,dy$
18. $\int_0^2 \int_{y^2}^{\frac{1}{2}y+1} xy\,dx\,dy$
19. $\int_0^2 \int_0^{y^2} xy\,dx\,dy$
20. $\int_0^1 \int_{y^2}^{(y+3)^{1/2}} 2xy\,dx\,dy$
21. $\int_0^5 \int_0^x (2x + y)\,dy\,dx$
22. $\int_1^3 \int_1^{x^2} 3xy\,dy\,dx$
23. $\int_0^1 \int_{x^3}^{x^2} x^2y\,dy\,dx$
24. $\int_{-.5}^2 \int_{x^2}^{\frac{3}{2}x+1} 4x\,dy\,dx$

Evaluate the integrals in Problems 25–30 by first reversing the order of integration. (Note that these are the odd-numbered Problems 13–23.)

25. $\int_1^4 \int_0^3 dy\,dx$
26. $\int_{-3}^2 \int_2^4 dx\,dy$
27. $\int_1^3 \int_0^y (x + 3y)\,dx\,dy$
28. $\int_0^2 \int_0^{y^2} xy\,dx\,dy$
29. $\int_0^5 \int_0^x (2x + y)\,dy\,dx$
30. $\int_0^1 \int_{x^3}^{x^2} x^2y\,dy\,dx$

Evaluate each double integral over the region R in Problems 31–38.

31. $\iint_R (x^2 + y)\,dx\,dy$ $\quad 0 \le x \le 2$, $0 \le y \le 3$

32. $\iint_R (x + 2y)\,dy\,dx \quad \begin{matrix} -1 \le x \le 2 \\ 1 \le y \le 3 \end{matrix}$

33. $\iint_R \frac{dy\,dx}{x} \quad \begin{matrix} 1 \le x \le 3 \\ 0 \le y \le 1 - x \end{matrix}$

34. $\iint_R (3 - 3x^2)\,dy\,dx \quad \begin{matrix} 0 \le x \le 2 \\ 0 \le y \le -\frac{9}{2}x \end{matrix}$

35. $\iint_R y^3 e^{xy}\,dx\,dy \quad \begin{matrix} 0 \le x \le y^2 \\ 1 \le y \le 2 \end{matrix}$

36. $\iint_R e^{x+y}\,dx\,dy \quad \begin{matrix} 0 \le x \le 2y \\ 0 \le y \le 1 \end{matrix}$

37. $\iint_R xy\sqrt{x^2 + y^2}\,dx\,dy \quad \begin{matrix} 1 \le x \le 3 \\ 2 \le y \le 4 \end{matrix}$

38. $\iint_R xy\sqrt{x^2 + y^2}\,dy\,dx \quad \begin{matrix} 0 \le x \le 4 \\ 1 \le y \le 3 \end{matrix}$

Set up the integrals in Problems 39–44 *two ways.*

39. $\iint_R (x + y)\,dA$, R bounded by $x = y^2$ and $y = x^2$

40. $\iint_R xy\,dA$, R bounded by $y = x^2$ and $y = 2x$

41. $\iint_R 2xy\,dA$, R bounded by $x + y = 5$, $x = 0$, and $y = 0$

42. $\iint_R (2x + y)\,dA$, R bounded by $y = x$, $y = 2 - x$, and the x-axis

43. $\iint_R dA$, R bounded by $y = x^2$, $y = 8 - x^2$, and the y-axis

44. $\iint_R 3xy^2\,dA$, R bounded by $y = 2x$, $y = 3 - x$, and $y = 0$

Find the volume of the solid bounded by the surface $z = f(x, y)$ about the region R as specified in Problems 45–54.

45. $z = 5$; $R: 1 \le x \le 5, -3 \le y \le 2$

46. $z = x + y$; $R: -1 \le x \le 2, 0 \le y \le 4$

47. $z = xy\sqrt{x^2 + y^2}$; $R: 0 \le x \le 1, 0 \le y \le 1$

48. $z = x^2 y$; $R: 0 \le x \le 2, 0 \le y \le 4$

49. $z = 5xy$; R bounded by $x = y^2$ and $y = x^2$

50. $z = 2xy$; R bounded by $y = x^2$ and $y = x$

51. $z = 6$; R bounded by $y = x^2$ and $y = 3x$

52. $z = 5$; R bounded by $x = y^2$ and $x = 2y$

53. $z = xy^2$; R bounded by $y = x^2$ and $y = x$

54. $z = \sqrt{\frac{x}{y}}$; R bounded by $y = x^2$ and $y = x$

55. The *average value* of the function $z = f(x, y)$ over the rectangle $R: a \le x \le b, c \le y \le d$ is

$$\frac{1}{(b - a)(d - c)} \iint_R f(x, y)\,dA$$

Find the average value of the surface $z = f(x, y) = x - 2y$ over the rectangle $0 \le x \le 2, 0 \le y \le 3$.

56. Find the average value of $f(x, y) = 4 - x - y$ over the rectangle $0 \le x \le 2$, $0 \le y \le 1$. (See Problem 55 for the definition of average value.)

APPLICATION

57. If an industry invests L thousand labor-hours ($0 \le L \le 5$) and K million dollars ($0 \le K \le 2$) in the production of P thousand units of a commodity, then P is given by

$$P(L, K) = L^{.75} K^{.25}$$

This is called a *Cobb–Douglas production function.** Find the average number of units produced for the indicated ranges of x and y. (See Problem 55 for a definition of average value.)

* You are asked to do some research on this model in the modeling application at the beginning of this chapter.

*16.6 Review

The material of this chapter is reviewed in the following list of objectives. After each objective there are some practice questions. For a sample test select the first question of each set and check your answers. The second question for each objective has no answer given. If you are having trouble with a particular type of problem, look back at the indicated section in the text. When you are finished reviewing these objectives, a sample examination is given at the end of this section.

* Optional section.

[16.1]

Objective 16.1: *Evaluate functions of several variables.*

Evaluate the functions in Problems 1–4 for the point $(5, -12)$.

1. $f(x, y) = 3x^2 - 2xy + y^2$
2. $g(x, y) = e^{x/y}$
3. $K(l, w) = lw$
4. $P(L, K) = L^{.5}K^{.3}$

Objective 16.2: *Plot points in three dimensions.*

5. **a.** $(5, 0, 0)$ **b.** $(3, -5, 5)$
6. **a.** $(0, 1, 0)$ **b.** $(-1, 2, 1)$
7. **a.** $(0, 0, 10)$ **b.** $(5, 10, -10)$
8. **a.** $(2, 5, 7)$ **b.** $(-2, -5, -4)$

Objective 16.3: *Graph surfaces in space.*

9. $x^2 + y^2 = 9$
10. $z = x^2 + y^2$
11. $x + y + 2z = 10$
12. $x^2 + y^2 - z^2 = 0$

[16.2]

Objective 16.4: *Find partial derivatives.*

13. Find f_x for $f(x, y) = (2x - 5y)^{12}$.
14. Find $g_y(1, 3)$ for $g(x, y) = e^{5xy}$.
15. If $z = T(x, y)$ find $\dfrac{\partial z}{\partial y}$ for $T(x, y) = \dfrac{1{,}000}{\sqrt{x^2 + y^2}}$.
16. If $f(x, y, \lambda) = 2xy + \lambda(2x + 3y - 100)$ find f_λ.

Objective 16.5: *Find higher-order partial derivatives.*

Let $z = f(x, y) = 3x^5 - 5x^4y^3 + 2x^2 - 150$ for Problems 17–20.

17. $f_{xx}(x, y)$
18. $f_{yx}(1, -1)$
19. $\dfrac{\partial^2 z}{\partial y\, \partial x}$
20. $\dfrac{\partial^2 z}{\partial y^2}$

[16.3]

Objective 16.6: *Find the relative maximums, relative minimums, and saddle points of a function of two variables.*

21. $g(x, y) = \dfrac{y}{x}$
22. $f(x, y) = xy - 4x + 3y + 120$
23. $f(x, y) = 2x^2 - 3xy + y^2$
24. $g(x, y) = e^{xy}$

[16.4]

Objective 16.7: *Find relative maximums or minimums using the method of Lagrange multipliers.*

25. Find the relative maximum of $f(x, y) = 12 - x^2 - y^2$ subject to $x + y = 10$.
26. Find the relative minimum of $g(x, y) = x^2 + 4y^2$ subject to $x + 2y = 12$.
27. Find two numbers whose sum is 250 and whose product is a maximum.
28. Find the smallest value for a product of two numbers if their difference must be 10.

[16.5]

Objective 16.8: *Evaluate integrals that are functions of two variables.*

29. $\displaystyle\int_1^3 x^2y^3\, dx$
30. $\displaystyle\int_{-1}^4 5e^{x+2y}\, dy$
31. $\displaystyle\int_0^1 x^2y^2\sqrt{x^3 + 8}\, dx$
32. $\displaystyle\int_1^{e^\pi} \frac{x^2}{y}\, dy$

Objective 16.9: *Evaluate iterated integrals.*

33. $\displaystyle\int_0^9 \int_{\sqrt{x}}^{\sqrt{x}+2} y\, dy\, dx$
34. $\displaystyle\int_0^{\ln 3} \int_0^x e^y dy\, dx$
35. $\displaystyle\int_1^2 \int_0^{1/y} y^3\, dx\, dy$
36. $\displaystyle\int_0^3 \int_0^{y\sqrt{9-y^2}} dx\, dy$

Objective 16.10: *Evaluate integrals by reversing the order of integration.*

37. $\displaystyle\int_1^8 \int_{2/y}^{\sqrt{2y}} dx\, dy$ (correct to the nearest tenth)
38. $\displaystyle\int_0^4 \int_{2y}^8 dx\, dy$
39. $\displaystyle\int_0^2 \int_{x^2}^4 dy\, dx$
40. $\displaystyle\int_1^{e^2} \int_{\ln y}^2 dx\, dy$

Objective 16.11: *Find a volume between a given surface and a given region by evaluating a double integral.*

41. $\displaystyle\iint_R (x + 2y)\, dA$, R bounded by $y = x^2$ and $y = 5x$
42. $\displaystyle\iint_R y^{-1}\, dy\, dx$ $\quad 1 \le x \le 2$, $1 \le y \le e^3$
43. $\displaystyle\iint_R x^2y^3\, dx\, dy$ $\quad 0 \le x \le 2$, $0 \le y \le 1$
44. $\displaystyle\iint_R xy^2\, dA$, R bounded by $x = y^2$ and $y = x^2$

Objective 16.12: *Solve applied problems based on the preceding objectives.*

45. A company produces two types of skateboards, standard and competition models. The weekly demand and cost equations are given where p is the price (in dollars) of the standard skateboard and q is the price (in dollars) of the competition model:

$$p = 50 - .05x + .001y$$
$$q = 130 + .01x - .04y$$

for x the weekly demand (in hundreds) for standard skateboards and y the weekly demand (in hundreds) for competition skateboards. If the cost function is

$$C(x, y) = 90 + 20x + 90y$$

find $C_x(5, 8)$ and $C_y(5, 8)$ and interpret each of these.

46. Karlin Enterprises employs between 100 and 500 employees and has a capital investment of between 3 and 5 million dollars. A research company has determined that Karlin's productivity (units per employee per day) is approximated by the formula

$$z = P(x, y) = 4xy - 3x^2 - y^3$$

where x is the size of the labor force (in hundreds) and y is the capital investment (in millions of dollars). Find the marginal productivity of labor when $x = 2$ and $y = 4$ and interpret your answer.

47. Find the marginal productivity of capital in Problem 46 when $x = 5$ and $y = 2$ and interpret your answer.

48. Find the maximum productivity for Karlin Enterprises (see Problem 46) in terms of labor force and capital investment.

SAMPLE TEST

The following sample test (45 minutes) is intended to review the main ideas of this chapter.

1. Evaluate $P(L, K) = L^{.4}K^{.2}$ at the point (10, 50).

2. Plot the following points in three dimensions.
a. (0, 3, 0) **b.** (4, −2, 5)
c. (2, 6, 3)

3. Graph the surface $2x + y + z = 8$.

4. Graph the surface $y^2 + z^2 - x^2 = 0$.

Find the requested derivatives or integrals in Problems 5–10.

5. f_x for $f(x, y) = (x - 3y)^{10}$

6. f_λ for $f(x, y, \lambda) = 4xy + \lambda(2x + 3y - 50)$

7. f_{xx} for $f(x, y) = 4x^4 - 3x^3y^2 + 2x^2 - 250$

8. $\int_0^1 x^2y^2\sqrt{x^3 + 8}\,dx$

9. $\int_0^9 \int_{\sqrt{x}}^{\sqrt{x}+2} dy\,dx$

10. $\int_0^3 \int_0^{y\sqrt{9-y^2}} dx\,dy$

11. Find the relative maximums, relative minimums, and saddle points of $f(x, y) = 2x^2 - 3xy + y^2$.

12. Use the method of Lagrange multipliers to find two numbers whose sum is 250 and whose product is a maximum.

13. Evaluate

$$\int_1^{e^2} \int_{\ln y}^{2} dx\,dy$$

by reversing the order of integration.

14. Find the volume of the region R bounded by $y = x^2$ and $y = 5x$.

$$\iint_R (x + 2y)\,dA$$

15. A company produces two types of skateboards, standard and competition. The weekly demand and cost equations are:

$$p = 50 - .05x + .001y$$
$$q = 130 + .01x - .04y$$

where p is the price of the standard skateboard and q is the price of the competition model for x, the weekly demand (in hundreds) for standard skateboards, and y, the weekly demand (in hundreds) for competition skateboards. If the cost function is

$$C(x, y) = 90 + 20x + 90y$$

find the revenue function, R, and evaluate $C(5, 8)$ and $R(5, 8)$.

CUMULATIVE REVIEW
for chapters 10–16

1. In your own words, discuss the meaning of limit. Use examples and graphs.
2. In your own words, discuss the definition of derivative. Include in your discussion reasons why the derivative is important, as well as some of the principal applications of derivative.
3. In your own words, discuss the definition of integral. Include in your discussion reasons why the integral is important, as well as some of the principal applications of the integral.
4. State the fundamental theorem of integral calculus. Discuss why you think this theorem is fundamental.

Find the derivatives of the functions in Problems 5–7.

5. $y = 5x^{-1} + 3\sqrt{x}$
6. $y = \dfrac{-3}{1 - 2x}$
7. $y = \ln|1 - 11x|$

Evaluate the integrals in Problems 8–10.

8. $\int (3x^2 - 2x)\,dx$
9. $\int_0^2 4x\,dx$
10. $\int_1^3 5x^{-1}\,dx$
11. Evaluate $\int_0^1 \int_{e^x}^{e} \frac{1}{\ln y}\,dy\,dx$ by reversing the order of integration.

APPLICATIONS

12. Changes in oxygen pressure (P_{O_2}) have been recorded on the graph below for time (in seconds) on the interval $[0, 20]$.

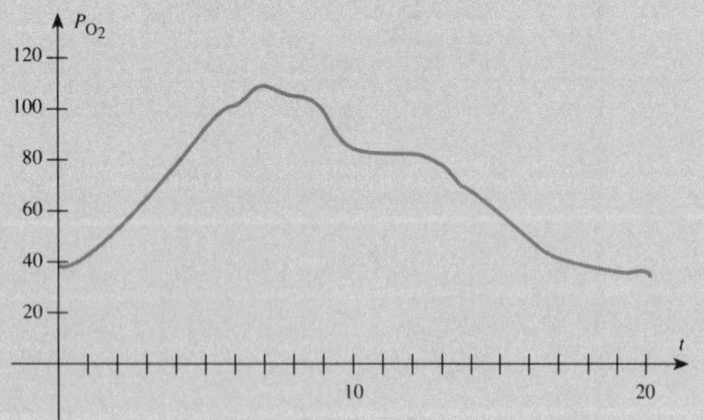

Is P_{O_2} increasing, decreasing, or constant for
a. $t < 3$?
b. $10 \le t \le 13$?
c. $t \ge 15$?

13. The cost of producing x units of a product is $C(x)$, where

$$C(x) = 5x + 8{,}000$$

The product sells for \$130 per unit.
a. What is the break-even point?
b. What revenue will the company receive if it sells just that number of units?

14. A company finds its sales are related to the amount spent on training programs by

$$T(x) = \frac{50 + 25x}{x + 5}$$

where T is the sales in hundreds of thousands of dollars when x thousand dollars are spent on training. Find the rate of change of sales when $x = 5$.

15. Suppose the profit in thousands of dollars from an item is $P(x) = x^3 - 13x^2 + 40x$, where x is the price in dollars. Find the maximum possible profit on $[0, 7]$.
16. The rate of sales of a brand of razor blades, in thousands, is given by $S(x) = 3x^2 + 4x$, where x is the time in months that the new product has been on the market. Find the total sales after 10 months.
17. Find the area of the region enclosed by the curves

$$y = 2x - 4 \quad \text{and} \quad y^2 = 4x \quad \text{for } x \ge 1$$

(*Note:* $y^2 = 4x$ can be broken into two functions, $y = 2\sqrt{x}$ and $y = -2\sqrt{x}$.)

18. A manufacturer finds it costs her $x^2 + 5x + 7$ dollars to produce x tons of dulconite. At production levels above 3 tons, she must hire additional workers, and her costs increase by $3x - 9$ dollars on the total production. If the price she receives is \$13 per ton regardless of how much she manufactures, and if her plant capacity is 10 tons, what level of output maximizes profits?
19. Graph $f(x) = \frac{1}{4}x^4 - \frac{3}{2}x^2$ showing the relative maximums, relative minimums, and points of inflection.
20. The functional life of a timing device selected at random is determined by the probability density function

$$f(t) = .005e^{-.005t}$$

where t is the number of months it has been in operation. What is the probability that the device will operate for more than 5 years ($t = 60$ months)?

Appendixes

Logic is a method of reasoning that accepts only inescapable conclusions. This is possible because of the strict way every concept is *defined*, or accepted without definition. So, we begin by defining a crucial term—a **statement**:

Statement

A **statement** is a meaningful sentence that is either true or false, but not both true and false.

Statements can be either **simple** or **compound**. Simple statements are those without **connectives**—words such as *not*, *and*, *or*, *neither . . . nor*, *if . . . then*, *either . . . or*, *unless*, *because*, and so on, while compound statements have at least one connective. A compound statement is formed by using connectives to combine simple statements. Because of the definition of a statement, the **truth value** of *any* statement labels it as either true (T) or false (F). The true value of a *compound* statement depends on the truth values of its component parts.

Letters such as p, q, r, s, . . . are used to denote simple statements, and then certain connectives are defined. There are three **fundamental connectives: conjunction** (*and*), **disjunction** (*or*), and **negation** (*not*). These connectives are defined by a **truth table**, which lists all possibilities. The truth values for each connective are chosen to correspond to everyday usage.

Fundamental Connectives

Conjunction:	*p and q*	symbolized by	$p \wedge q$
Disjunction:	*p or q*	symbolized by	$p \vee q$
Negation:	*not p*	symbolized by	$\sim p$

Truth Table Definition

p	q	$p \wedge q$	$p \vee q$	$\sim p$
T	T	T	T	F
T	F	F	T	F
F	T	F	T	T
F	F	F	F	T

The definition of the fundamental connectives, along with a truth table, are used to determine the truth or falsity of a compound statement.

EXAMPLE 1 Construct a truth table for the compound statement:

Alfie did ***not*** *come last night* ***and*** *he did* ***not*** *pick up his money.*

Solution Let

p: Alfie came last night

q: Alfie picked up his money

Then the statement can be written $\sim p \wedge \sim q$. To begin, list all the possible combinations of truth values for the simple statements p and q:

p	q	
T	T	
T	F	
F	T	
F	F	

Insert the truth values for $\sim p$ and $\sim q$:

p	q	$\sim p$	$\sim q$	
T	T	F	F	
T	F	F	T	
F	T	T	F	
F	F	T	T	

Finally, insert the truth values for $\sim p \wedge \sim q$:

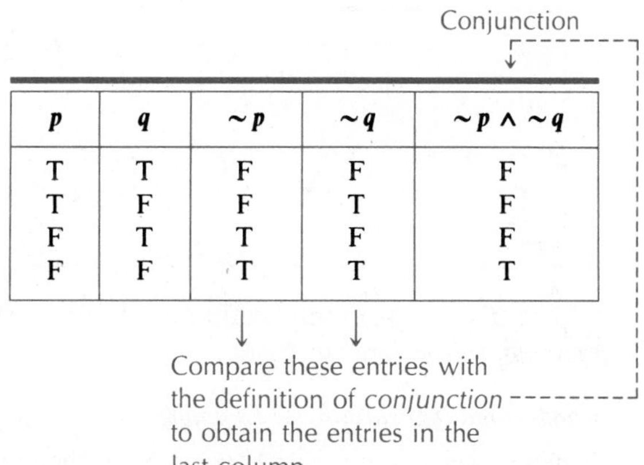

p	q	$\sim p$	$\sim q$	$\sim p \wedge \sim q$
T	T	F	F	F
T	F	F	T	F
F	T	T	F	F
F	F	T	T	T

The only time the compound statement is true is when *both* p and q are false. ■

EXAMPLE 2 Construct a truth table for $\sim(\sim p)$.

Solution

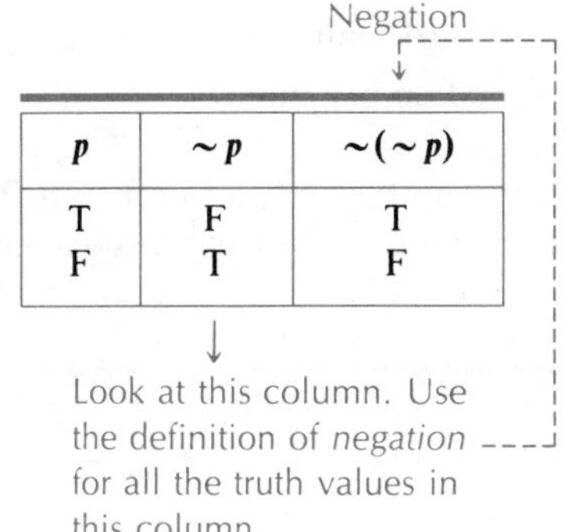

p	$\sim p$	$\sim(\sim p)$
T	F	T
F	T	F

Note that $\sim(\sim p)$ and p have the same truth values. If two statements have the same truth values, one can replace the other in any logical expression. This means that the double negative of a statement is the same as the original statement.

Law of Double Negation

$\sim(\sim p)$ may be replaced by p in any logical expression.

Truth tables can be used to prove certain useful results and to introduce some additional operators. The first one we will consider is called the **conditional**. The statement "if p, then q" is called a *conditional statement*. It is symbolized by $p \to q$; p is called the **antecedent**, and q is called the **consequent**.

Definition of the Conditional

The *conditional* is defined by the following truth table:

p	q	$p \to q$
T	T	T
T	F	F
F	T	T
F	F	T

The *if* part of an implication need not be stated first. All the following statements have the same meaning:

Conditional translation	*Example*
If p, then q	If you are 18, then you can vote.
q, if p	You can vote, if you are 18.
p, only if q	You are 18 only if you can vote.
$p \subseteq q$	p is a subset of q.

Some statements, although not originally written as a conditional, can be put into if–then form. For example, "All ducks are birds" can be rewritten as "If it is a duck, then it is a bird." Thus we add one more form to the list:

All p are q. All 18-year-olds can vote.

EXAMPLE 3 Translate the following sentence (which appeared on the 1986 federal income tax form) into symbolic form:

If you received capital gains distributions for the year and you do not need Schedule D to report any other gains or losses, do not file that schedule.

Solution *Step 1:* Isolate the simple statements and assign them variables. Let

c: You received capital gains distributions for the year.

g: You need Schedule D to report other gains for the year.

l: You need Schedule D to report other losses for the year.

f: You need to file Schedule D.

Your choice of variables is, of course, arbitrary, but you must be careful not to let variables represent compound statements.

Step 2: Rewrite the sentence, making substitutions for the variables.

If c and ($\sim g$ or $\sim l$), then $\sim f$.

Step 3: Complete the translation to symbols:

$$[c \wedge (\sim g \vee \sim l)] \rightarrow \sim f$$

■

Related to the conditional $p \rightarrow q$ are other statements, which are defined below.

Converse, Inverse, and Contrapositive

Given the conditional $p \rightarrow q$, we define:

Converse: $q \rightarrow p$

Inverse: $\sim p \rightarrow \sim q$

Contrapositive: $\sim q \rightarrow \sim p$

Not all these statements are equivalent, as shown by Table A.1.

TABLE A.1

p	q	$\sim p$	$\sim q$	Statement $p \rightarrow q$	Converse $q \rightarrow p$	Inverse $\sim p \rightarrow \sim q$	Contrapositive $\sim q \rightarrow \sim p$
T	T	F	F	T	T	T	T
T	F	F	T	F	T	T	F
F	T	T	F	T	F	F	T
F	F	T	T	T	T	T	T

Note that, if the conditional is true, the converse and inverse are not necessarily true. However, the contrapositive is always true if the original conditional is true.

Law of Contraposition

> A conditional may always be replaced by its contrapositive without having its truth value affected.

EXAMPLE 4 Consider the following:

p: You obey the law.
q: You will go to jail.

Given $p \to \sim q$, write the converse, inverse, and contrapositive.

Solution *Statement:* $p \to \sim q$ *Converse:* $\sim q \to p$
Inverse: $\sim p \to q$ *Contrapositive:* $q \to \sim p$

The double negative $\sim(\sim q)$ is replaced by q in the inverse and contrapositive. Make this simplification whenever possible.

Statement:	$p \to \sim q$	If you obey the law, then you will not go to jail.
Contrapositive:	$q \to \sim p$	If you go to jail, then you did not obey the law.
Converse:	$\sim q \to p$	If you do not go to jail, then you obey the law.
Inverse:	$\sim p \to q$	If you do not obey the law, then you will go to jail.

■

The power of logic is in drawing conclusions not explicitly stated or known. There are three main types of reasoning that allow us to draw such conclusions. These are *direct reasoning*, *indirect reasoning*, and *transitivity*.

Direct reasoning is the drawing of a conclusion from two premises. The following example, called a **syllogism**, is an illustration:

$p \to q$	If you receive an A on the final, then you will pass the course.
p	You receive an A on the final.
q	You pass the course.

This argument consists of two *premises*, or *hypotheses*, and a *conclusion*; the argument is valid if

$$[(p \to q) \land p] \to q$$

is always true, as shown in Table A.2.

TABLE A.2
Truth Table for Direct Reasoning

p	q	$p \to q$	$(p \to q) \land p$	$[(p \to q) \land p] \to q$
T	T	T	T	T
T	F	F	F	T
F	T	T	F	T
F	F	T	F	T

The pattern of argument is illustrated below:

Direct Reasoning

Major premise:	$p \rightarrow q$	
Minor premise:	p	
Conclusion:	$\therefore q$	Three dots (∴) are used to mean *therefore*.

This reasoning is also sometimes called *modus ponens*, *law of detachment*, or *assuming the antecedent*.

EXAMPLE 5

If you play chess, then you are intelligent.
You play chess.
∴ You are intelligent. ■

EXAMPLE 6

If you are a logical person, then you will understand this example.
You are a logical person.
∴ You understand this example. ■

We say that these arguments are *valid* since we recognize them as direct reasoning. The following syllogism illustrates what we call **indirect reasoning**:

$p \rightarrow q$	If you receive an A on the final, then you will pass the course.
$\sim q$	You did not pass the course.
$\therefore \sim p$	∴ You did not receive an A on the final.

We can prove this result is valid by direct reasoning as follows:

$$[(\sim q \rightarrow \sim p) \wedge \sim q] \rightarrow \sim p$$

Also, since $\sim q \rightarrow \sim p$ is logically equivalent to the contrapositive $p \rightarrow q$, we have

$$[(p \rightarrow q) \wedge \sim q] \rightarrow \sim p$$

We can also prove that indirect reasoning is valid by using a truth table, and you are asked to do this in the problem set. The pattern of argument is illustrated below:

Indirect Reasoning

Major premise:	$p \rightarrow q$
Minor premise:	$\sim q$
Conclusion:	$\therefore \sim p$

This method is also known as reasoning by *denying the consequent* or as *modus tollens*.

EXAMPLE 7

If the cat takes the rat, then the rat will take the cheese.
The rat does not take the cheese.

∴The cat does not take the rat. ■

EXAMPLE 8

If you received an A on the test, then I am Napoleon.
I am not Napoleon.

∴You did not receive an A on the test. ■

Sometimes we must consider some extended arguments. **Transitivity** allows us to reason through several premises to some conclusion. The argument form is as follows:

Transitivity

Premise: $p \to q$
Premise: $q \to r$
Conclusion: $\therefore p \to r$

Transitivity is proved by a truth table, as Table A.3 shows.

TABLE A.3 Truth Table for Transitivity

p	q	r	$p \to q$	$q \to r$	$p \to r$	$(p \to q) \wedge (q \to r)$	$[(p \to q) \wedge (q \to r)] \to (p \to r)$
T	T	T	T	T	T	T	T
T	T	F	T	F	F	F	T
T	F	T	F	T	T	F	T
T	F	F	F	T	F	F	T
F	T	T	T	T	T	T	T
F	T	F	T	F	T	F	T
F	F	T	T	T	T	T	T
F	F	F	T	T	T	T	T

EXAMPLE 9

If you attend class, then you will pass the course. $p \to q$
If you pass the course, then you will graduate. $q \to r$

∴If you attend class, then you will graduate. $\therefore p \to r$

Transitivity can be extended so that a chain of several if–then sentences is connected together. For example, we could continue:

If you graduate, then you will get a good job.
If you get a good job, then you will meet the right people.
If you meet the right people, then you will become well known.

∴If you attend class, then you will become well known. ■

Several of these argument forms may be combined into one argument. Remember:

Procedure for Drawing Logical Conclusions

1. Translate into symbols.
2. Simplify the symbolic argument. Replace a statement by its contrapositive, use direct or indirect reasoning, or use transitivity.
3. Translate the conclusion back into words.

EXAMPLE 10 Form a valid conclusion using all these statements:

1. If I receive a check for \$500, then we will go on vacation.
2. If the car breaks down, then we will not go on vacation.
3. The car breaks down.

Solution First, change into symbolic form:

1. $c \to v$ where $c =$ I receive a \$500 check.
2. $b \to \sim v$ $v =$ We will go on vacation.
3. b $b =$ Car breaks down.

Next, simplify the symbolic argument. In this example the premises must be rearranged:

2. $b \to \sim v$ Second premise
1. $c \to v$ First premise
3. b

Now $c \to v$ is the same as $\sim v \to \sim c$ (replace the statement by its contrapositive). That is, if 1 is replaced by 1′, $\sim v \to \sim c$, the following argument is obtained:

2.	$b \to \sim v$	Second premise
1′.	$\sim v \to \sim c$	Contrapositive of first premise
	$\therefore b \to \sim c$	Transitive
3.	b	Third premise
	$\therefore \sim c$	Direct reasoning

Finally, translate the conclusion back into words: "I did not receive a check for \$500." ■

A Problem Set

According to the definition, which of the examples in Problems 1–4 are statements?

1. a. Hickory, Dickory, Dock, the mouse ran up the clock.
   b. $3 + 5 = 9$
   c. Is John ugly?
   d. John has a wart on the end of his nose.
2. a. March 13, 1984, is Monday.
   b. Division by zero is impossible.
   c. Logic is not as difficult as I had anticipated.
   d. $4 - 6 = 2$
3. a. $6 + 9 \neq 7 + 8$
   b. Thomas Jefferson was the twenty-third president.
   c. Sit down and be quiet!
   d. If wages continue to rise, then prices will also rise.

4. a. John and Mary were married on August 3, 1979.
   b. $6 + 12 \neq 10 + 8$
   c. Do not read this sentence.
   d. Do you have a cold?

Translate the statements in Problems 5–15 into symbols. For each simple statement, be sure to indicate the meanings of the symbols you use. Answers are not unique.

5. W. C. Fields is eating, drinking, and having a good time.
6. Sam will not seek and will not accept the nomination.
7. Jack will not go tonight and Rosamond will not go tomorrow.
8. Fat Albert lives to eat and does not eat to live.
9. The decision will depend on judgment or intuition, and not on who paid the most.
10. Everything happens to everybody sooner or later if there is time enough. (G. B. Shaw)
11. We are not weak if we make a proper use of those means which the God of Nature has placed in our power. (Patrick Henry)
12. A useless life is an early death. (Goethe)
13. All work is noble. (Thomas Carlyle)
14. Everything's got a moral if only you can find it. (Lewis Carroll)
15. You don't have to itemize deductions on Schedule A or complete the worksheet if you have earned income of $5,300 or more. (1990 tax form)

Construct a truth table for the statements in Problems 16–39.

16. $\sim p \vee q$
17. $\sim p \wedge \sim q$
18. $\sim(p \wedge q)$
19. $\sim r \vee \sim s$
20. $\sim(\sim r)$
21. $(r \wedge s) \vee \sim s$
22. $p \wedge \sim q$
23. $\sim p \vee \sim q$
24. $(\sim p \wedge q) \vee \sim q$
25. $(p \wedge \sim q) \wedge p$
26. $(\sim p \vee q) \wedge (q \wedge p)$
27. $(p \vee q) \vee (p \wedge \sim q)$
28. $p \vee (p \rightarrow q)$
29. $p \rightarrow (\sim p \rightarrow q)$
30. $(p \wedge q) \rightarrow p$
31. $\sim p \rightarrow \sim(p \wedge q)$
32. $(p \wedge q) \wedge (p \rightarrow \sim q)$
33. $(p \rightarrow p) \rightarrow (q \rightarrow \sim q)$
34. $(p \rightarrow \sim q) \rightarrow (q \rightarrow \sim p)$
35. $(p \rightarrow q) \rightarrow (\sim q \rightarrow \sim p)$
36. $[p \wedge (p \vee q)] \rightarrow p$
37. $(p \wedge q) \wedge \sim r$
38. $[(p \vee q) \wedge \sim r] \wedge r$
39. $[p \wedge (q \vee \sim p)] \vee r$
40. Prove indirect reasoning by using a truth table.

Name the type of reasoning illustrated by the arguments in Problems 41–48.

41. If I inherit $1,000, I will buy you a cookie.
    I inherit $1,000.
    Therefore, I will buy you a cookie.
42. All snarks are fribbles.
    All fribbles are ugly.
    Therefore, all snarks are ugly.
43. If you understand a problem, it is easy.
    The problem is not easy.
    Therefore, you do not understand the problem.
44. If I do not get a raise in pay, I will quit.
    I do not get a raise.
    Therefore, I quit.
45. If Fermat's last theorem is ever proved, then my life is complete.
    My life is not complete.
    Therefore, Fermat's last theorem is not proved.
46. All mathematicians are eccentrics.
    All eccentrics are rich.
    Therefore, all mathematicians are rich.
47. No students are enthusiastic.
    You are enthusiastic.
    Therefore, you are not a student.
48. If you like beer, you will like Bud.
    You do not like Bud.
    Therefore, you do not like beer.

In Problems 49–65 form a valid conclusion using all the statements for each argument.

49. If you can learn mathematics, then you are intelligent. If you are intelligent, then you understand human nature.
50. If I am idle, then I become lazy.
    I am idle.
51. All trebbles are frebbles.
    All frebbles are expensive.
52. If we interfere with the publication of false information, we are guilty of suppressing the freedom of others.
    We are not guilty of suppressing the freedom of others.
53. If a nail is lost, then a shoe is lost.
    If a shoe is lost, then a horse is lost.
    If a horse is lost, then a rider is lost.
    If a rider is lost, then a battle is lost.
    If a battle is lost, then a kingdom is lost.
54. If you climb the highest mountain, you will feel great.
    If you feel great, then you are happy.
55. If $b = 0$, then $a = 0$.
    $a \neq 0$.
56. If $a \cdot b = 0$, then $a = 0$ or $b = 0$.
    $a \cdot b = 0$.
57. If I eat that piece of pie, I will get fat.
    I will not get fat.

58. If we win first prize, we will go to Europe.
If we are ingenious, we will win first prize.
We are ingenious.

59. If I can earn enough money this summer, I will attend college in the fall.
If I do not participate in student demonstrations, then I will not attend college.
I earned enough money this summer.

60. If I am tired, then I cannot finish my homework.
If I understand the material, then I can finish my homework.

61. If you go to college, then you will get a good job.
If you get a good job, then you will make a lot of money.
If you do not obey the law, then you will not make a lot of money.
You go to college.

62. All puppies are nice.
This animal is a puppy.
No nice creatures are dangerous.

63. Babies are illogical.
Nobody is despised who can manage a crocodile.
Illogical persons are despised.

64. Everyone who is sane can do logic.
No lunatics are fit to serve on a jury.
None of your sons can do logic.

65. No ducks waltz.
No officers ever decline to waltz.
All my poultry are ducks.

APPENDIX B
Mathematical Induction

Mathematical induction is an important method of proof in mathematics, allowing us to prove results involving the set of positive integers. Mathematical induction is not the scientific method or the inductive logic used in the experimental sciences; it is a form of deductive logic in which conclusions are inescapable.

The first step in establishing a result by mathematical induction is often the observation of a pattern. Let us begin with a simple example. Suppose we wish to know the sum of the first n odd integers. We could begin by looking for a pattern:

$$\begin{aligned}
&1 &&= 1\\
&1 + 3 &&= 4\\
&1 + 3 + 5 &&= 9\\
&1 + 3 + 5 + 7 &&= 16\\
&1 + 3 + 5 + 7 + 9 &&= 25
\end{aligned}$$

Do you see a pattern? It appears that the sum of the first n odd numbers is n^2 since the sum of the first three odd numbers is 3^2, of the first four odd numbers is 4^2, and so on. We now wish to *prove* deductively that

$$1 + 3 + 5 + \cdots + \underbrace{(2n - 1)}_{\substack{\uparrow\\ n\text{th odd number}}} = n^2$$

is true for all positive integers n. How can we proceed? We prove certain propositions about the positive integers. The proposition is denoted by $P(n)$. For example, in the above problem we let

$$P(n) = 1 + 3 + 5 + \cdots + (2n - 1) = n^2$$

This means that

$$\begin{aligned}
&P(1): && 1 = 1^2\\
&P(2): && 1 + 3 = 2^2\\
&P(3): && 1 + 3 + 5 = 3^2\\
&P(4): && 1 + 3 + 5 + 7 = 4^2\\
& && \vdots\\
&P(100): && 1 + 3 + 5 + \cdots + 199 = 100^2\\
& && \vdots\\
&P(x - 1): && 1 + 3 + 5 + \cdots + (2x - 3) = (x - 1)^2\\
&P(x): && 1 + 3 + 5 + \cdots + (2x - 1) = x^2\\
&P(x + 1): && 1 + 3 + 5 + \cdots + (2x + 1) = (x + 1)^2
\end{aligned}$$

Now we need to show that $P(n)$ is true for all n (n a positive integer).

Principle of Mathematical Induction (PMI)

> If a given proposition $P(n)$ is true for $P(1)$ and if the truth of $P(k)$ implies the truth of the proposition for $P(k+1)$, then $P(n)$ is true for all positive integers.

Thus, for proof by mathematical induction, we need to

1. Prove $P(1)$ is true.
2. Assume $P(k)$ is true.
3. Prove $P(k+1)$ is true.
4. Conclude that $P(n)$ is true for all positive integers n.

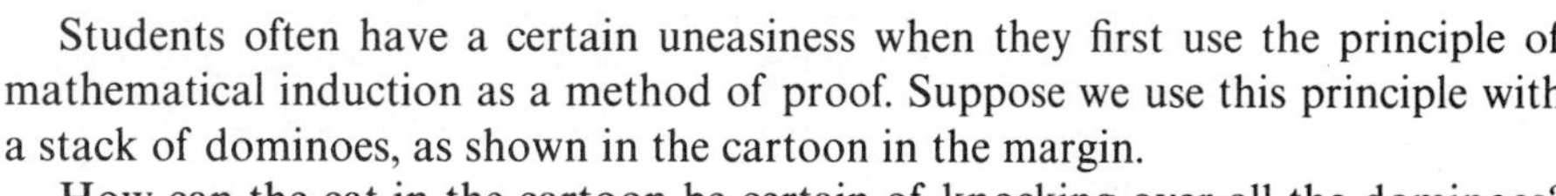

Students often have a certain uneasiness when they first use the principle of mathematical induction as a method of proof. Suppose we use this principle with a stack of dominoes, as shown in the cartoon in the margin.

How can the cat in the cartoon be certain of knocking over all the dominoes? The cat would have to be able to knock over the first one. Also, the dominoes would be arranged so that *if* the kth domino falls, then the next one, the $(k+1)$st, will also fall. That is, each domino is set up so that if it falls, it causes the next one to fall. This is a kind of "chain reaction." The first domino falls; this knocks over the next one (the second domino); the second one knocks over the next one (the third domino); the third one knocks over the next one; this continues until all the dominoes are knocked over.

Let us now return to the example to prove (for all positive integers n)

$$1 + 3 + 5 + \cdots + (2n-1) = n^2$$

Step 1: Prove $P(1)$ is true: $1 = 1^2$ is true.

Step 2: Assume $P(k)$ is true: $1 + 3 + 5 + \cdots + (2k-1) = k^2$

Step 3: Prove $P(k+1)$ is true.

TO PROVE: $1 + 3 + 5 + \cdots + [2(k+1) - 1] = (k+1)^2$

This is found by substituting $(k+1)$ for n in the original statement we want to prove. Next, we simplify. This is so we will know when we are finished with step 3 of the proof.

$$\begin{aligned} 1 + 3 + 5 + \cdots + [2(k+1) - 1] &= (k+1)^2 \\ 1 + 3 + 5 + \cdots + (2k + 2 - 1) &= (k+1)^2 \\ 1 + 3 + 5 + \cdots + (2k+1) &= (k+1)^2 \end{aligned}$$

The procedure for step 3 is to begin with the hypothesis (from step 2) and *prove*

$$1 + 3 + 5 + \cdots + (2k+1) = (k+1)^2$$

Statements	**Reasons**
1. $1 + 3 + 5 + \cdots + (2k-1) = k^2$	1. By hypothesis (step 2)
2. $1 + 3 + 5 + \cdots + (2k-1) + (2k+1) = k^2 + (2k+1)$	2. Add $(2k+1)$ to both sides.
3. $1 + 3 + 5 + \cdots + (2k-1) + (2k+1) = k^2 + 2k + 1$	3. Associative
4. $1 + 3 + 5 + \cdots + (2k-1) + (2k+1) = (k+1)^2$	4. Factoring (distributive)

Step 4: The proposition is true for all positive integers by the principle of mathematical induction (PMI).

EXAMPLE 1 Prove or disprove: $2 + 4 + 6 + \cdots + 2n = n(n + 1)$.

Proof *Step* 1: Prove $P(1)$ true:

$2 \stackrel{?}{=} 1(1 + 1)$

$2 = 2$

True.

Step 2: Assume $P(k)$. HYPOTHESIS: $2 + 4 + 6 + \cdots + 2k = k(k + 1)$

Step 3: Prove $P(k + 1)$. TO PROVE: $2 + 4 + 6 + \cdots + 2(k + 1) = (k + 1)(k + 2)$

Statements		Reasons
1. $2 + 4 + 6 + \cdots + 2k$	$= k(k + 1)$	1. Hypothesis (step 2)
2. $2 + 4 + 6 + \cdots + 2k + 2(k + 1)$	$= k(k + 1) + 2(k + 1)$	2. Add $2(k + 1)$ to both sides.
3.	$= (k + 1)(k + 2)$	3. Factor

Step 4: The proposition is true for all positive integers by PMI. ■

EXAMPLE 2 Prove or disprove: $n^3 + 2n$ is divisible by 3.

Proof *Step* 1: Prove $P(1)$: $1^3 + 2 \cdot 1 = 3$, which is divisible by 3.

Step 2: Assume $P(k)$. HYPOTHESIS: $k^3 + 2k$ is divisible by 3.

Step 3: Prove $P(k + 1)$. TO PROVE: $(k + 1)^3 + 2(k + 1)$ is divisible by 3.

Statements	Reasons
1. $(k + 1)^3 + 2(k + 1) = k^3 + 3k^2 + 3k + 1 + 2k + 2$	1. Distributive, associative, and commutative axioms
2. $\quad = (3k^2 + 3k + 3) + (k^3 + 2k)$	2. Commutative and associative axioms
3. $\quad = 3(k^2 + k + 1) + (k^3 + 2k)$	3. Distributive axiom
4. $3(k^2 + k + 1)$ is divisible by 3	4. Definition of divisibility by 3
5. $k^3 + 2k$ is divisible by 3	5. Hypothesis
6. $(k + 1)^3 + 2(k + 1)$ is divisible by 3	6. Both terms are divisible by 3 and therefore the sum is divisible by 3.

Step 4: The proposition is true for all positive integers n by PMI. ■

Example 3 shows that *even though* we make an assumption in step 2, it is not going to help if the proposition is not true.

EXAMPLE 3 Prove or disprove: $n + 1$ is prime.

Proof *Step* 1: Prove $P(1)$: $1 + 1 = 2$ is a prime.

Step 2: Assume $P(k)$. HYPOTHESIS: $k + 1$ is a prime.

Step 3: Prove $P(k + 1)$. TO PROVE: $(k + 1) + 1$ is a prime. This is not possible since $(k + 1) + 1 = k + 2$, which is not prime whenever k is an even positive integer.

Step 4: Any conclusions? We cannot conclude that the statement is false, only that induction does not work. But this statement is, in fact, false, and a counterexample is $n = 3$ since $n + 1 = 4$ is not prime. ■

EXAMPLE 4 Prove or disprove: $1 \cdot 2 \cdot 3 \cdot 4 \cdot \cdots \cdot n < 0$.

Proof Students often slip into the habit of skipping either the first or second step in a proof by mathematical induction. This is dangerous, and it is important to check every step. Suppose you do not verify the first step.

Step 2: Assume $P(k)$. HYPOTHESIS: $1 \cdot 2 \cdot 3 \cdot \cdots \cdot k < 0$

Step 3: Prove $P(k + 1)$. TO PROVE: $1 \cdot 2 \cdot 3 \cdot \cdots \cdot k \cdot (k + 1) < 0$
$1 \cdot 2 \cdot 3 \cdot \cdots \cdot k < 0$ by hypothesis.
$k + 1$ is positive since k is a positive integer. Then, since we know that the product of a negative and a positive is negative,

$$\underbrace{1 \cdot 2 \cdot 3 \cdot \cdots \cdot k}_{\text{Negative}} \cdot \underbrace{(k + 1)}_{\text{Positive}} < 0$$

Step 3 is proved.

Step 4: The proposition is not true for all positive integers since the first step, $1 < 0$, does not hold. ■

B Problem Set

1. Prove: $1 + 2 + 3 + \cdots + n = \dfrac{n(n + 1)}{2}$
for all positive integers n.

2. Prove: $1^2 + 2^2 + 3^2 + \cdots + n^2 = \dfrac{n(n + 1)(2n + 1)}{6}$
for all positive integers n.

3. Prove:
$$1^2 + 3^2 + 5^2 + \cdots + (2n - 1)^2 = \frac{n(2n - 1)(2n + 1)}{3}$$
for all positive integers n.

4. Prove: $1^3 + 2^3 + 3^3 + \cdots + n^3 = \dfrac{n^2(n + 1)^2}{4}$
for all positive integers n.

5. Prove:
$$2^2 + 4^2 + 6^2 + \cdots + (2n)^2 = \frac{2n(n + 1)(2n + 1)}{3}$$
for all positive integers n.

6. Prove:
$$1 \cdot 2 + 2 \cdot 3 + 3 \cdot 4 + \cdots + n(n + 1) = \frac{n(n + 1)(n + 2)}{3}$$
for all positive integers n.

7. Prove:
$$1 \cdot 3 + 2 \cdot 4 + 3 \cdot 5 + \cdots + n(n + 2) = \frac{n(n + 1)(2n + 7)}{6}$$
for all positive integers n.

8. Prove: $1 + r + r^2 + \cdots + r^n = \dfrac{r^{n+1} - 1}{r - 1}$
for all positive integers n.

9. Prove: $n^5 - n$ is divisible by 5 for all positive integers n.

10. Prove: $n(n + 1)(n + 2)$ is divisible by 6 for all positive integers n.

11. Prove: $(1 + n)^2 \geq 1 + n^2$ for all positive integers n.

12.
$$1^3 = 1^2$$
$$1^3 + 2^3 = 3^2$$
$$1^3 + 2^3 + 3^3 = 6^2$$
$$1^3 + 2^3 + 3^3 + 4^3 = 10^2$$

Make a conjecture based on the above pattern and then prove or disprove your conjecture.

13.
$$1 = 1$$
$$1 + 4 = 5$$
$$1 + 4 + 7 = 12$$
$$1 + 4 + 7 + 10 = 22$$

Make a conjecture based on the above pattern and then prove or disprove your conjecture.

14. Prove: $\binom{k}{r} + \binom{k}{r-1} = \binom{k+1}{r}$.

Use the formula $\binom{n}{m} = \dfrac{n!}{m!(n-m)!}$ and not induction. You will need this result in Problem 15.

15. The binomial theorem can be proved for any positive integer n by using mathematical induction. Fill in the missing steps and reasons.

TO PROVE:

$$(a+b)^n = \sum_{j=0}^{n} \binom{n}{j} a^{n-j} b^j$$
$$= \binom{n}{0} a^n + \binom{n}{1} a^{n-1} b + \binom{n}{2} a^{n-2} b^2 + \cdots$$
$$+ \binom{n}{r} a^{n-r} b^r + \cdots + \binom{n}{n-1} a b^{n-1} + \binom{n}{n} b^n$$

Step 1: Prove the binomial theorem is true for $n = 1$.
a. Fill in these details.

Step 2: Assume the theorem is true for $n = k$.
b. Fill in the statement of the hypothesis.

Step 3: Prove the theorem is true for $n = k + 1$.

TO PROVE: $(a+b)^{k+1} = \sum_{j=0}^{k+1} \binom{k+1}{j} a^{k+1-j} b^j$

BY HYPOTHESIS:
c. Fill in the statement of the hypothesis.
d. Fill in the details; the final simplified form is

$$(a+b)^{k+1} = a^{k+1} + \cdots$$
$$+ \left[\binom{k}{r} + \binom{k}{r-1}\right] a^{k-r+1} b^r + \cdots + b^{k+1}$$

e. Use Problem 14 to complete the proof.

This appendix provides a brief introduction to calculators and their use. In the last few years, pocket calculators have been one of the fastest selling items in the United States. This is probably because most people (including mathematicians!) do not like to do arithmetic and a good calculator can be purchased for less than $20.

This book was written with the assumption that you have or will have a calculator. This appendix is included to help you choose a calculator and understand the calculator comments in this book.

Calculators are classified by the types of problems they are equipped to handle, as well as by the type of logic for which they are programmed. The problem of selecting a calculator is compounded by the multiplicity of brands from which to choose.

The different types of calculators are distinguished primarily by their price.

1. *Four-function calculators.* These calculators have a keyboard consisting of the numerals and the four arithmetic operations, or functions: addition [+], subtraction [−], multiplication [×], and division [÷]. These calculators are obsolete.
2. *Four-function calculators with memory* ($10–$20). Usually these are no more expensive than four-function calculators, and offer a memory register: [M], [STO], or [M⁺]. The more expensive models may have more than one memory register. Memory registers allow you to store partial calculations for later recall. Some models will even remember the total when they are turned off.
3. *Scientific calculators* ($20–$50). These calculators have additional mathematical functions, such as square root [√], trigonometric [sin], [cos], and [tan], and logarithmic [log] and [exp]. Depending on the particular brand, a scientific model may have other keys as well.
4. *Special-purpose calculators* ($40–$400). Special-use calculators for business, statistics, surveying, medicine, or even gambling and chess are available.
5. *Programmable calculators* ($50–$600). With these calculators you can enter a *sequence* of steps for the calculator to repeat on your command. Some of these calculators allow the insertion of different cards that "remember" the sequence of steps for complex calculations.
6. *Algebraic (Symbolic) Manipulation and Graphing Calculators* ($75–$250). These calculators represent the next generation of calculators. They are capable of graphing curves and doing algebraic manipulation. They will, no doubt, revolutionize the way mathematics is taught in the future.

Calculator Usage in This Book

The calculators used in preparing this book were a *Sharp* EL531A, for which I paid less than $10, and a *Texas Instruments* TI-81, which cost me less than $100. There are several types of logic used on calculators: *arithmetic*, *algebraic*, *RPN*, and the graphing calculators use what I will call WYSIWYG ("What you see is what you get") logic, in which you input usual mathematical notation. In this book, I have divided calculator usage into two categories. The first assumes a scientific calculator, and **it is assumed that you have a scientific calculator**. No special designation is made in this book for problems that require a scientific calculator. The logic on a scientific calculator will be *arithmetic*, *algebraic*, or *RPN*. To determine the type of logic used by one of these calculators, try this test problem:

If the answer shown is 20, it is an arithmetic-logic calculator. If the answer is 14 (the correct answer), then it is an algebraic-logic calculator. If the calculator has no equal key [=] but has an [ENTER] or [SAVE] key, then it is an RPN-logic calculator. An RPN-logic calculator will give the answer as 14. In algebra you learn to perform multiplication before addition, so that the correct value for

$$2 + 3 \times 4$$

is 14 (multiply first). An algebraic calculator will "know" this fact and will give the correct answer, whereas an arithmetic calculator will simply work from left to right to obtain the incorrect answer, 20. Therefore, if you have an arithmetic-logic calculator, you will need to be careful about the order of operations. Some arithmetic-logic calculators provide parentheses [(] [)] so that operations can be grouped as in

[2] [+] [(] [3] [×] [4] [)] [=]

but then you must remember to insert the parentheses.

With an RPN calculator, the operation symbol is entered after the numbers have been entered. These three types of logic can be illustrated by the problem $2 + 3 \times 4$:

Arithmetic logic	*Algebraic logic*	*RPN logic*	*WYSIWYG logic*
[3]	[2]	[2]	[2]
[×]	[+]	[ENTER]	[+]
[4]	[3]	[3]	[(]
[=]	[×]	[ENTER]	[3]
[+]	[4]	[4]	[×]
[2]	[=]	[×]	[4]
[=]		[+]	[)]
			[=]

Regardless of the type of logic your calculator uses, it is a good idea to check your owner's manual for each type of problem illustrated in the text because there are many different brands of calculators on the market, and many have slight variations in keyboards.

There is also a limit to the accuracy of your calculator. You may have a calculator with a 6-, 8-, or 10-digit display. Test its accuracy with Example 1.

EXAMPLE 1 Find $2 \div 3$.

Solution

Algebraic:	2	÷	3	=
RPN:	2	ENTER	3	÷
Display:	2	2.	3	.6666666667

■

There may be a discrepancy between this answer and the one you obtain on your calculator. Some machines will not round the answer as shown here but will show the display

.6666666666

Others will show a display such as

6.6666 -01

This is a number in scientific notation and should be interpreted as

$$6.6666 \times 10^{-1} \qquad \text{or} \qquad .66666$$

Most calculators will also use scientific notation when the numbers become larger than that allowed by their display register. Example 2 shows how your calculator handles large numbers. Some calculators simply show an overflow and will not accept larger numbers. You need a calculator that will accept and handle large and small numbers.

EXAMPLE 2 Find 50^6.

Solution Check your owner's manual to find out how to use an exponent key. Most calculators will work as shown below.

Algebraic:	50	y^x	6	
RPN:	50	ENTER	6	y^x
WYSIWYG:	50	∧	6	ENTER

When the maximum size of the display has been reached, the calculator should automatically switch to scientific notation. The point at which a calculator will do this varies from one type or brand to another. The answer for this example is 1.5625 10 , which means

$$1.5625 \times 10^{10} = 15{,}625{,}000{,}000$$

■

A Note on Graphing Calculators

Although I have not assumed that you have a graphing calculator for this book, I have included several CALCULATOR COMMENT boxes. For the last few years I have used a *Casio fx*-7000*G* calculator and a *Hewlett-Packard* 28*S*. However, this year I started using a *Texas Instruments* TI-81, and ever since have used it almost continuously. The GRAPH, TRACE, and MATRIX keys proved invaluable in preparing this edition.

The major difference in *using* a graphing calculator rather than a scientific calculator is the way of using the log, LN, $\sqrt{\ }$ keys as well as the way of inputting exponents. Consider the calculations in Example 3.

EXAMPLE 3 Compare the following calculations on a graphing calculator and a scientific calculator by showing typical keystrokes.

a. $\sqrt{45.3}$ **b.** $\sqrt{4^2 + 5^2}$ **c.** $\log 85$ **d.** $5^{1/3}$ **e.** $\frac{1}{7}$

Solution

	SCIENTIFIC CALCULATOR	GRAPHING CALCULATOR
a.	[45.3] [$\sqrt{x}$]	[$\sqrt{\ }$] [45.3]
b.	[4] [x^2] [+] [5] [x^2] [=] [$\sqrt{x}$]	[$\sqrt{\ }$] [(] [4] [x^2] [+] [5] [x^2] [)] [ENTER]
c.	[85] [log]	[LOG] [85]
d.	[5] [y^x] [(] [1] [÷] [3] [)] [=]	[5] [^] [3] [x^{-1}] [ENTER]
e.	[7] [1/x]	[7] [x^{-1}] [ENTER] ■

C

Problem Set

Use a calculator to evaluate each of the expressions in Problems 1–31.

1. $(14)(351)$

2. $(218)(263)$

3. $(4{,}158)(.00456)$

4. $(3.00)^3(182)$

5. $(2.00)^4(1{,}245)(277)$

6. $(6.00)^5(1{,}456)(288)$

7. $\dfrac{(1{,}979)(1{,}356)}{452}$

8. $\dfrac{(515)(20{,}600)}{200}$

9. $\dfrac{(618)(460)(125)}{650}$

10. $[.14 + (197)(25.08)]19$

11. $(990)(1{,}117)(342) - 89$

12. $\dfrac{1.00}{.005 + .020}$

13. $\dfrac{1.500 \times 10^4 + (7.000)(67.00)}{20{,}000}$

14. $(6.28)^{1/2}(4.85)$

15. $(8.23)^{1/2}(6.14)$

16. $\dfrac{1.00}{\sqrt{8.48} - \sqrt{21.3}}$

17. $\dfrac{1.00}{\sqrt{4.83} + \sqrt{2.51}}$

18. $[(4.083)^2(4.283)^3]^{-2/3}$

19. $[(6.128)^4(3.412)^2]^{-1/2}$

20. $\dfrac{16^2 + 25^2 - 9^2}{(2.0)(16)(25)}$

21. $\dfrac{216^2 + 418^2 - 315^2}{(2.00)(216)(418)}$

22. $\dfrac{4.82^2 + 6.14^2 - 9.13^2}{(2.00)(4.82)(6.14)}$

23. $\sqrt{\dfrac{(51)(36)}{212}}$

24. $\sqrt{\dfrac{25 + 49}{1 + 4(51)}}$

25. $\sqrt{\dfrac{45 + 156}{2 + 51(19)}}$

26. $\dfrac{18.361^2 + 15.215^2 - 13.815^2}{(2.0000)(18.361)(15.215)}$

27. $\dfrac{17.813^2 + 13.451^2 - 19.435^2}{2(17.813)(13.451)}$

28. $\dfrac{(2.51)^2 + (6.48)^2 - (2.51)(6.48)(.3462)}{(2.51)(6.48)}$

29. $\dfrac{241^2 + 568^2 - (241)(568)(.5213)}{(241)(568)}$

30. $1{,}500\left(1 + \dfrac{.105}{12}\right)^{8(12)}$

Software Accompanying Text

Microcomputers are changing the curriculum today as much as calculators changed the curriculum in the last decade. The personal computer is accessible to a great many students, and more and more schools and colleges have computer labs in which computer programs can be used.

The history of the marriage of software and textbooks is still in a state of transition. At first, software was written to accompany specific textbooks, but this proved less than desirable for at least two reasons. First, it is costly to develop *good* software, so many books claimed "tailor-made" software. However, it was not very general, and for the most part left both instructor and student wanting for better and more useful software. The second reason was that computer technology was changing so quickly that the software written for a textbook would either be released long after the textbook or else it became dated by the time the book came out. For these reasons I am recommending a stand-alone software program that I found *extremely* useful in understanding the material in this book. It is called The MATH LAB by Chris Avery and Charles B. Barker. I have included several COMPUTER APPLICATION boxes in this book to direct you to particular aspects of this program. MATH LAB comes in IBM format. Here is a quick summary of how you can use MATH LAB to help you with this course.

Finite Mathematics

A. Straight-line plotter
B. Matrix product
C. Row reduction
D. Matrix inverse
E. Matrix equations and inverses
F. Simplex method
G. Counting problems
H. Probability problems
I. Probability properties
J. Statistics
K. Binomial coefficients
L. Correlation and regression
M. Markov chains
N. Two-person zero-sum game
O. Financial formulas
P. Finance problems
Q. Loan amortization

Calculus

A. Function plotter
B. Parametric equations
C. Implicit relations
D. Limits
E. Riemann sums
F. Numerical integration
G. Double integration
H. Sequences and series
I. Surface plotter
J. Differential equations

SPREADSHEET PROGRAM		
	A	B
1	x	y=1/(3x^2-5x-2)
2	-3	@1/(3*A2^2-5*A2-2)
3	+A2+0.25	replicate
4	replicate	

x	y=1/(3x^2-5x-2)
-3	0.025
-2.75	0.029038113
-2.5	0.034188034
-2.25	0.040920716
-2	0.05
-1.75	0.062745098
-1.5	0.081632653
-1.25	0.111888112
-1	0.166666667
-0.75	0.290909091
-0.5	0.8
-0.25	-1.77777778
0	-0.5
0.25	-0.32653061
0.5	-0.26666667
0.75	-0.24615385
1	-0.25
1.25	-0.28070175
1.5	-0.36363636
1.75	-0.64
2	ERR
2.25	0.516129032
2.5	0.235294118
2.75	0.144144144
3	0.1

Spreadsheet Programs

Many of us now have access to a home computer as well as software for word processing, data bases, and spreadsheets. Spreadsheets provide a powerful, easy tool for processing data. The most commonly used spreadsheets today are versions of Excel and Lotus 1–2–3.* Although it is not appropriate to develop the techniques of spreadsheets in this book, suffice it to say that *it is very easy* to learn enough about spreadsheets to make them a useful tool in mathematics.

The basic format for a spreadsheet is the arrangement of data into **cells** which are referenced by a row number (1, 2, 3, . . .) and a column number (A, B, C, . . .). The first spreadsheet shown in the text is reproduced in the margin. Look at the part labeled "SPREADSHEET PROGRAM." Cell A1 contains the letter x and cell A2 contains the numeral -3. What is contained in cell A3? (*Answer:* the formula "+A2+0.25") A cell can contain a word or a letter (such as x), a number (such as -3), or a formula (such as +A2+0.25). This means that 0.25 is to be added to the contents of cell A2. The plus sign precedes a cell reference to distinguish the contents of a cell from a word. Almost any formula can be put into a cell. Notice the formula in cell B2:

Spreadsheet notation	*Algebraic notation*
1/(3∗A2^2−5∗A2−2)	$\dfrac{1}{3(x^2-5x-2)}$

In the program (cell A4) you see the word **replicate**. As we will use it in this book, it means "copy the cell right above, and repeat for an entire column of entries." Thus, you see the first column of output; these entries were entered by using a COPY command on cell A3 and then replicated for the entire column. This means cell A4 contains "+A3+0.25", cell A5 contains "+A4+0.25", and cell A6 contains "+A5+0.25", ... The changing of the cell that is referenced is automatically carried out by the program when using the replicate command. Do you see how all of the entries in the first column are generated with a single formula and the replicate command? In turn, the second column is evaluating a function (in preparation to graphing it in Chapter 1).

Finally, notice in the program that the entry in cell C3 has a formula preceded by @. The way you communicate the type of entry you are putting into a cell is as follows:

First entry is a letter or " ' " to indicate a word or text.
First entry is "+" or a numeral to indicate a number or to reference a cell.
First entry is "@" to indicate a formula.

From time to time in this book you will find inserts showing a spreadsheet program and its associated output.

* In preparing this book we use a Lotus 1–2–3-compatible spreadsheet called *Quattro-Pro*.

TABLE 1 Pascal's Triangle—Combinatorics

n	$\binom{n}{0}$	$\binom{n}{1}$	$\binom{n}{2}$	$\binom{n}{3}$	$\binom{n}{4}$	$\binom{n}{5}$	$\binom{n}{6}$	$\binom{n}{7}$	$\binom{n}{8}$	$\binom{n}{9}$	$\binom{n}{10}$
0	1										
1	1	1									
2	1	2	1								
3	1	3	3	1							
4	1	4	6	4	1						
5	1	5	10	10	5	1					
6	1	6	15	20	15	6	1				
7	1	7	21	35	35	21	7	1			
8	1	8	28	56	70	56	28	8	1		
9	1	9	36	84	126	126	84	36	9	1	
10	1	10	45	120	210	252	210	120	45	10	1
11	1	11	55	165	330	462	462	330	165	55	11
12	1	12	66	220	495	792	924	792	495	220	66
13	1	13	78	286	715	1287	1716	1716	1287	715	286
14	1	14	91	364	1001	2002	3003	3432	3003	2002	1001
15	1	15	105	455	1365	3003	5005	6435	6435	5005	3003
16	1	16	120	560	1820	4368	8008	11440	12870	11440	8008
17	1	17	136	680	2380	6188	12376	19448	24310	24310	19448
18	1	18	153	816	3060	8568	18564	31824	43758	48620	43758
19	1	19	171	969	3876	11628	27132	50388	75582	92378	92378
20	1	20	190	1140	4845	15504	38760	77520	125970	167960	184756

NOTE: $\binom{n}{m} = \dfrac{n(n-1)(n-2)\cdot\cdots\cdot(n-m+1)}{m(m-1)(m-2)\cdot\cdots\cdot 3\cdot 2\cdot 1}$; $\binom{n}{0} = 1$; $\binom{n}{1} = n$

For coefficients missing from the above table, use the relation

$$\binom{n}{m} = \binom{n}{n-m}$$

For example,

$$\binom{20}{11} = \binom{20}{9} = 167{,}960$$

TABLE 2
Squares and Square Roots

n	n^2	$\sqrt{n}$	$\sqrt{10n}$	n	n^2	$\sqrt{n}$	$\sqrt{10n}$
1	1	1.000	3.162	51	2601	7.141	22.583
2	4	1.414	4.472	52	2704	7.211	22.804
3	9	1.732	5.477	53	2809	7.280	23.022
4	16	2.000	6.325	54	2916	7.348	23.238
5	25	2.236	7.071	55	3025	7.416	23.452
6	36	2.449	7.746	56	3136	7.483	23.664
7	49	2.646	8.367	57	3249	7.550	23.875
8	64	2.828	8.944	58	3364	7.616	24.083
9	81	3.000	9.487	59	3481	7.681	24.290
10	100	3.162	10.000	60	3600	7.746	24.495
11	121	3.317	10.488	61	3721	7.810	24.698
12	144	3.464	10.954	62	3844	7.874	24.900
13	169	3.606	11.402	63	3969	7.937	25.100
14	196	3.742	11.832	64	4096	8.000	25.298
15	225	3.873	12.247	65	4225	8.062	25.495
16	256	4.000	12.649	66	4356	8.124	25.690
17	289	4.123	13.038	67	4489	8.185	25.884
18	324	4.243	13.416	68	4624	8.246	26.077
19	361	4.359	13.784	69	4761	8.307	26.268
20	400	4.472	14.142	70	4900	8.367	26.458
21	441	4.583	14.491	71	5041	8.426	26.646
22	484	4.690	14.832	72	5184	8.485	26.833
23	529	4.796	15.166	73	5329	8.544	27.019
24	576	4.899	15.492	74	5476	8.602	27.203
25	625	5.000	15.811	75	5625	8.660	27.386
26	676	5.099	16.125	76	5776	8.718	27.568
27	729	5.196	16.432	77	5929	8.775	27.749
28	784	5.292	16.733	78	6084	8.832	27.928
29	841	5.385	17.029	79	6241	8.888	28.107
30	900	5.477	17.321	80	6400	8.944	28.284
31	961	5.568	17.607	81	6561	9.000	28.460
32	1024	5.657	17.889	82	6724	9.055	28.636
33	1089	5.745	18.166	83	6889	9.110	28.810
34	1156	5.831	18.439	84	7056	9.165	28.983
35	1225	5.916	18.708	85	7225	9.220	29.155
36	1296	6.000	18.974	86	7396	9.274	29.326
37	1369	6.083	19.235	87	7569	9.327	29.496
38	1444	6.164	19.494	88	7744	9.381	29.665
39	1521	6.245	19.748	89	7921	9.434	29.833
40	1600	6.325	20.000	90	8100	9.487	30.000
41	1681	6.403	20.248	91	8281	9.539	30.166
42	1764	6.481	20.494	92	8464	9.592	30.332
43	1849	6.557	20.736	93	8649	9.644	30.496
44	1936	6.633	20.976	94	8836	9.695	30.659
45	2025	6.708	21.213	95	9025	9.747	30.822
46	2116	6.782	21.448	96	9216	9.798	30.984
47	2209	6.856	21.679	97	9409	9.849	31.145
48	2304	6.928	21.909	98	9604	9.899	31.305
49	2401	7.000	22.136	99	9801	9.950	31.464
50	2500	7.071	22.361	100	10000	10.000	31.623

TABLE 3
Binomial Probabilities

This table computes the probability of exactly k successes in n independent binomial trials with the probability of success on a single trial equal to p. It thus contains the individual terms for specified choices of k, n, and p.* For entries 0+, the probability is less than .0005, but greater than 0.

$$\binom{n}{k} p^k (1 - p)^{n-k}$$

		p											
n	k	.01	.05	.10	.15	.20	.25	.30	.35	.40	.45	.50	k
2	0	.980	.903	.810	.723	.640	.563	.490	.423	.360	.303	.250	0
	1	.020	.095	.180	.255	.320	.375	.420	.455	.480	.495	.500	1
	2	0+	.003	.010	.023	.040	.063	.090	.122	.160	.202	.250	2
3	0	.970	.857	.729	.614	.512	.422	.343	.275	.216	.166	.125	0
	1	.029	.135	.243	.325	.384	.422	.441	.444	.432	.408	.375	1
	2	0+	.007	.027	.057	.096	.141	.189	.239	.288	.334	.375	2
	3	0+	0+	.001	.003	.008	.016	.027	.043	.064	.091	.125	3
4	0	.961	.815	.656	.522	.410	.316	.240	.179	.130	.092	.063	0
	1	.039	.171	.292	.368	.410	.422	.412	.384	.346	.299	.250	1
	2	.001	.014	.049	.098	.154	.211	.265	.311	.346	.368	.375	2
	3	0+	0+	.004	.011	.026	.047	.076	.111	.154	.200	.250	3
	4	0+	0+	0+	.001	.002	.004	.008	.015	.026	.041	.062	4
5	0	.951	.774	.590	.444	.328	.237	.168	.116	.078	.050	.031	0
	1	.048	.204	.328	.392	.410	.396	.360	.312	.259	.206	.156	1
	2	.001	.021	.073	.138	.205	.264	.309	.336	.346	.337	.313	2
	3	0+	.001	.008	.024	.051	.088	.132	.181	.230	.276	.312	3
	4	0+	0+	0+	.002	.006	.015	.028	.049	.077	.113	.156	4
	5	0+	0+	0+	0+	0+	.001	.002	.005	.010	.018	.031	5
6	0	.941	.735	.531	.377	.262	.178	.118	.075	.047	.028	.016	0
	1	.057	.232	.354	.399	.393	.356	.303	.244	.187	.136	.094	1
	2	.001	.031	.098	.176	.246	.297	.324	.328	.311	.278	.234	2
	3	0+	.002	.015	.041	.082	.132	.185	.235	.276	.303	.313	3
	4	0+	0+	.001	.005	.015	.033	.060	.095	.138	.186	.234	4
	5	0+	0+	0+	0+	.002	.004	.010	.020	.037	.061	.094	5
	6	0+	0+	0+	0+	0+	0+	.001	.002	.004	.008	.016	6
7	0	.932	.698	.478	.321	.210	.133	.082	.049	.028	.015	.008	0
	1	.066	.257	.372	.396	.367	.311	.247	.185	.131	.087	.055	1
	2	.002	.041	.124	.210	.275	.311	.318	.298	.261	.214	.164	2
	3	0+	.004	.023	.062	.115	.173	.227	.268	.290	.292	.273	3
	4	0+	0+	.003	.011	.029	.058	.097	.144	.194	.239	.273	4
	5	0+	0+	0+	.001	.004	.012	.025	.047	.077	.117	.164	5
	6	0+	0+	0+	0+	0+	.001	.004	.008	.017	.032	.055	6
	7	0+	0+	0+	0+	0+	0+	0+	.001	.002	.004	.008	7
8	0	.923	.663	.430	.272	.168	.100	.058	.032	.017	.008	.004	0
	1	.075	.279	.383	.385	.336	.267	.198	.137	.090	.055	.031	1
	2	.003	.051	.149	.238	.294	.311	.296	.259	.209	.157	.109	2
	3	0+	.005	.033	.084	.147	.208	.254	.279	.279	.257	.219	3
	4	0+	0+	.005	.018	.046	.087	.136	.188	.232	.263	.273	4
	5	0+	0+	0+	.003	.009	.023	.047	.081	.124	.172	.219	5
	6	0+	0+	0+	0+	.001	.004	.010	.022	.041	.070	.109	6
	7	0+	0+	0+	0+	0+	0+	.001	.003	.008	.016	.031	7
	8	0+	0+	0+	0+	0+	0+	0+	0+	.001	.002	.004	8

* For $p > .50$, the value of $\binom{n}{k} p^k (1 - p)^{n-k}$ is found by using the table entry for $\binom{n}{n-k} (1 - p)^{n-k} p^k$.

TABLE 3 *(continued)*

		p											
n	***k***	**.01**	**.05**	**.10**	**.15**	**.20**	**.25**	**.30**	**.35**	**.40**	**.45**	**.50**	***k***
9	0	.914	.630	.387	.232	.134	.075	.040	.021	.010	.005	.002	0
	1	.083	.299	.387	.368	.302	.225	.156	.100	.060	.034	.018	1
	2	.003	.063	.172	.260	.302	.300	.267	.216	.161	.111	.070	2
	3	0+	.008	.045	.107	.176	.234	.267	.272	.251	.212	.164	3
	4	0+	.001	.007	.028	.066	.117	.172	.219	.251	.260	.246	4
	5	0+	0+	.001	.005	.017	.039	.074	.118	.167	.213	.246	5
	6	0+	0+	0+	.001	.003	.009	.021	.042	.074	.116	.164	6
	7	0+	0+	0+	0+	0+	.001	.004	.010	.021	.041	.070	7
	8	0+	0+	0+	0+	0+	0+	0+	.001	.004	.008	.018	8
	9	0+	0+	0+	0+	0+	0+	0+	0+	0+	.001	.002	9
10	0	.904	.599	.349	.197	.107	.056	.028	.013	.006	.003	.001	0
	1	.091	.315	.387	.347	.268	.188	.121	.072	.040	.021	.010	1
	2	.004	.075	.194	.276	.302	.282	.233	.176	.121	.076	.044	2
	3	0+	.010	.057	.130	.201	.250	.267	.252	.215	.166	.117	3
	4	0+	.001	.011	.040	.088	.146	.200	.238	.251	.238	.205	4
	5	0+	0+	.001	.008	.026	.058	.103	.154	.201	.234	.246	5
	6	0+	0+	0+	.001	.006	.016	.037	.069	.111	.160	.205	6
	7	0+	0+	0+	0+	.001	.003	.009	.021	.042	.075	.117	7
	8	0+	0+	0+	0+	0+	0+	.001	.004	.011	.023	.044	8
	9	0+	0+	0+	0+	0+	0+	0+	.001	.002	.004	.010	9
	10	0+	0+	0+	0+	0+	0+	0+	0+	0+	0+	.001	10
11	0	.895	.569	.314	.167	.086	.042	.020	.009	.004	.001	0+	0
	1	.099	.329	.384	.325	.236	.155	.093	.052	.027	.013	.005	1
	2	.005	.087	.213	.287	.295	.258	.200	.140	.089	.051	.027	2
	3	0+	.014	.071	.152	.221	.258	.257	.225	.177	.126	.081	3
	4	0+	.001	.016	.054	.111	.172	.220	.243	.236	.206	.161	4
	5	0+	0+	.002	.013	.039	.080	.132	.183	.221	.236	.226	5
	6	0+	0+	0+	.002	.010	.027	.057	.099	.147	.193	.226	6
	7	0+	0+	0+	0+	.002	.006	.017	.038	.070	.113	.161	7
	8	0+	0+	0+	0+	0+	.001	.004	.010	.023	.046	.081	8
	9	0+	0+	0+	0+	0+	0+	.001	.002	.005	.013	.027	9
	10	0+	0+	0+	0+	0+	0+	0+	0+	.001	.002	.005	10
	11	0+	0+	0+	0+	0+	0+	0+	0+	0+	0+	0+	11
12	0	.886	.540	.282	.142	.069	.032	.014	.006	.002	.001	0+	0
	1	.107	.341	.377	.301	.206	.127	.071	.037	.017	.008	.003	1
	2	.006	.099	.230	.292	.283	.232	.168	.109	.064	.034	.016	2
	3	0+	.017	.085	.172	.236	.258	.240	.195	.142	.092	.054	3
	4	0+	.002	.021	.068	.133	.194	.231	.237	.213	.170	.121	4
	5	0+	0+	.004	.019	.053	.103	.158	.204	.227	.222	.193	5
	6	0+	0+	0+	.004	.016	.040	.079	.128	.177	.212	.226	6
	7	0+	0+	0+	.001	.003	.011	.029	.059	.101	.149	.193	7
	8	0+	0+	0+	0+	.001	.002	.008	.020	.042	.076	.121	8
	9	0+	0+	0+	0+	0+	0+	.001	.005	.012	.028	.054	9
	10	0+	0+	0+	0+	0+	0+	0+	.001	.002	.007	.016	10
	11	0+	0+	0+	0+	0+	0+	0+	0+	0+	.001	.003	11
	12	0+	0+	0+	0+	0+	0+	0+	0+	0+	0+	0+	12
13	0	.878	.513	.254	.121	.055	.024	.010	.004	.001	0+	0+	0
	1	.115	.351	.367	.277	.179	.103	.054	.026	.011	.004	.002	1
	2	.007	.111	.245	.294	.268	.206	.139	.084	.045	.022	.010	2
	3	0+	.021	.100	.190	.246	.252	.218	.165	.111	.066	.035	3
	4	0+	.003	.028	.084	.154	.210	.234	.222	.184	.135	.087	4
	5	0+	0+	.006	.027	.069	.126	.180	.215	.221	.199	.157	5
	6	0+	0+	.001	.006	.023	.056	.103	.155	.197	.217	.209	6
	7	0+	0+	0+	.001	.006	.019	.044	.083	.131	.177	.209	7
	8	0+	0+	0+	0+	.001	.005	.014	.034	.066	.109	.157	8
	9	0+	0+	0+	0+	0+	.001	.003	.010	.024	.050	.087	9
	10	0+	0+	0+	0+	0+	0+	.001	.002	.006	.016	.035	10
	11	0+	0+	0+	0+	0+	0+	0+	0+	.001	.004	.010	11
	12	0+	0+	0+	0+	0+	0+	0+	0+	0+	0+	.002	12
	13	0+	0+	0+	0+	0+	0+	0+	0+	0+	0+	0+	13

TABLE 3 *(continued)*

n	k	.01	.05	.10	.15	.20	.25	.30	.35	.40	.45	.50	k
14	0	.869	.488	.229	.103	.044	.018	.007	.002	.001	0+	0+	0
	1	.123	.359	.356	.254	.154	.083	.041	.018	.007	.003	.001	1
	2	.008	.123	.257	.291	.250	.180	.113	.063	.032	.014	.006	2
	3	0+	.026	.114	.206	.250	.240	.194	.137	.085	.046	.022	3
	4	0+	.004	.035	.100	.172	.220	.229	.202	.155	.104	.061	4
	5	0+	0+	.008	.035	.086	.147	.196	.218	.207	.170	.122	5
	6	0+	0+	.001	.009	.032	.073	.126	.176	.207	.209	.183	6
	7	0+	0+	0+	.002	.009	.028	.062	.108	.157	.195	.209	7
	8	0+	0+	0+	0+	.002	.008	.023	.051	.092	.140	.183	8
	9	0+	0+	0+	0+	0+	.002	.007	.018	.041	.076	.122	9
	10	0+	0+	0+	0+	0+	0+	.001	.005	.014	.031	.061	10
	11	0+	0+	0+	0+	0+	0+	0+	.001	.003	.009	.022	11
	12	0+	0+	0+	0+	0+	0+	0+	0+	.001	.002	.006	12
	13	0+	0+	0+	0+	0+	0+	0+	0+	0+	0+	.001	13
	14	0+	0+	0+	0+	0+	0+	0+	0+	0+	0+	0+	14
15	0	.860	.463	.206	.087	.035	.013	.005	.002	0+	0+	0+	0
	1	.130	.366	.343	.231	.132	.067	.031	.013	.005	.002	0+	1
	2	.009	.135	.267	.286	.231	.156	.092	.048	.022	.009	.003	2
	3	0+	.031	.129	.218	.250	.225	.170	.111	.063	.032	.014	3
	4	0+	.005	.043	.116	.188	.225	.219	.179	.127	.078	.042	4
	5	0+	.001	.010	.045	.103	.165	.206	.212	.186	.140	.092	5
	6	0+	0+	.002	.013	.043	.092	.147	.191	.207	.191	.153	6
	7	0+	0+	0+	.003	.014	.039	.081	.132	.177	.201	.196	7
	8	0+	0+	0+	.001	.003	.013	.035	.071	.118	.165	.196	8
	9	0+	0+	0+	0+	.001	.003	.012	.030	.061	.105	.153	9
	10	0+	0+	0+	0+	0+	.001	.003	.010	.024	.051	.092	10
	11	0+	0+	0+	0+	0+	0+	.001	.002	.007	.019	.042	11
	12	0+	0+	0+	0+	0+	0+	0+	0+	.002	.005	.014	12
	13	0+	0+	0+	0+	0+	0+	0+	0+	0+	.001	.003	13
	14	0+	0+	0+	0+	0+	0+	0+	0+	0+	0+	0+	14
	15	0+	0+	0+	0+	0+	0+	0+	0+	0+	0+	0+	15
16	0	.851	.440	.185	.074	.028	.010	.003	.001	0+	0+	0+	0
	1	.138	.371	.329	.210	.113	.053	.023	.009	.003	.001	0+	1
	2	.010	.146	.275	.277	.211	.134	.073	.035	.015	.006	.002	2
	3	0+	.036	.142	.229	.246	.208	.146	.089	.047	.022	.009	3
	4	0+	.006	.051	.131	.200	.225	.204	.155	.101	.057	.028	4
	5	0+	.001	.014	.056	.120	.180	.210	.201	.162	.112	.067	5
	6	0+	0+	.003	.018	.055	.110	.165	.198	.198	.168	.122	6
	7	0+	0+	0+	.005	.020	.052	.101	.152	.189	.197	.175	7
	8	0+	0+	0+	.001	.006	.020	.049	.092	.142	.181	.196	8
	9	0+	0+	0+	0+	.001	.006	.019	.044	.084	.132	.175	9
	10	0+	0+	0+	0+	0+	.001	.006	.017	.039	.075	.122	10
	11	0+	0+	0+	0+	0+	0+	.001	.005	.014	.034	.067	11
	12	0+	0+	0+	0+	0+	0+	0+	.001	.004	.011	.028	12
	13	0+	0+	0+	0+	0+	0+	0+	0+	.001	.003	.009	13
	14	0+	0+	0+	0+	0+	0+	0+	0+	0+	.001	.002	14
	15	0+	0+	0+	0+	0+	0+	0+	0+	0+	0+	0+	15
	16	0+	0+	0+	0+	0+	0+	0+	0+	0+	0+	0+	16

(Column headings .01 through .50 are values of p.)

TABLE 3 *(continued)*

		p											
n	***k***	**.01**	**.05**	**.10**	**.15**	**.20**	**.25**	**.30**	**.35**	**.40**	**.45**	**.50**	***k***
17	0	.843	.418	.167	.063	.023	.008	.002	.001	0+	0+	0+	0
	1	.145	.374	.315	.189	.096	.043	.017	.006	.002	.001	0+	1
	2	.012	.158	.280	.267	.191	.114	.058	.026	.010	.004	.001	2
	3	.001	.041	.156	.236	.239	.189	.125	.070	.034	.014	.005	3
	4	0+	.008	.060	.146	.209	.221	.187	.132	.080	.041	.018	4
	5	0+	.001	.017	.067	.136	.191	.208	.185	.138	.087	.047	5
	6	0+	0+	.004	.024	.068	.128	.178	.199	.184	.143	.094	6
	7	0+	0+	.001	.007	.027	.067	.120	.168	.193	.184	.148	7
	8	0+	0+	0+	.001	.008	.028	.064	.113	.161	.188	.185	8
	9	0+	0+	0+	0+	.002	.009	.028	.061	.107	.154	.185	9
	10	0+	0+	0+	0+	0+	.002	.009	.026	.057	.101	.148	10
	11	0+	0+	0+	0+	0+	.001	.003	.009	.024	.052	.094	11
	12	0+	0+	0+	0+	0+	0+	.001	.002	.008	.021	.047	12
	13	0+	0+	0+	0+	0+	0+	0+	.001	.002	.007	.018	13
	14	0+	0+	0+	0+	0+	0+	0+	0+	0+	.002	.005	14
	15	0+	0+	0+	0+	0+	0+	0+	0+	0+	0+	.001	15
	16	0+	0+	0+	0+	0+	0+	0+	0+	0+	0+	0+	16
	17	0+	0+	0+	0+	0+	0+	0+	0+	0+	0+	0+	17
18	0	.835	.397	.150	.054	.018	.006	.002	0+	0+	0+	0+	0
	1	.152	.376	.300	.170	.081	.034	.013	.004	.001	0+	0+	1
	2	.013	.168	.284	.256	.172	.096	.046	.019	.007	.002	.001	2
	3	.001	.047	.168	.241	.230	.170	.105	.055	.025	.009	.003	3
	4	0+	.009	.070	.159	.215	.213	.168	.110	.061	.029	.012	4
	5	0+	.001	.022	.079	.151	.199	.202	.166	.115	.067	.033	5
	6	0+	0+	.005	.030	.082	.144	.187	.194	.166	.118	.071	6
	7	0+	0+	.001	.009	.035	.082	.138	.179	.189	.166	.121	7
	8	0+	0+	0+	.002	.012	.038	.081	.133	.173	.186	.167	8
	9	0+	0+	0+	0+	.003	.014	.039	.079	.128	.169	.185	9
	10	0+	0+	0+	0+	.001	.004	.015	.038	.077	.125	.167	10
	11	0+	0+	0+	0+	0+	.001	.005	.015	.037	.074	.121	11
	12	0+	0+	0+	0+	0+	0+	.001	.005	.015	.035	.071	12
	13	0+	0+	0+	0+	0+	0+	0+	.001	.004	.013	.033	13
	14	0+	0+	0+	0+	0+	0+	0+	0+	.001	.004	.012	14
	15	0+	0+	0+	0+	0+	0+	0+	0+	0+	.001	.003	15
	16	0+	0+	0+	0+	0+	0+	0+	0+	0+	0+	.001	16
	17	0+	0+	0+	0+	0+	0+	0+	0+	0+	0+	0+	17
	18	0+	0+	0+	0+	0+	0+	0+	0+	0+	0+	0+	18

TABLE 3 *(continued)*

		p											
n	***k***	**.01**	**.05**	**.10**	**.15**	**.20**	**.25**	**.30**	**.35**	**.40**	**.45**	**.50**	***k***
19	0	.826	.377	.135	.046	.014	.004	.001	0+	0+	0+	0+	0
	1	.159	.377	.285	.153	.068	.027	.009	.003	.001	0+	0+	1
	2	.014	.179	.285	.243	.154	.080	.036	.014	.005	.001	0+	2
	3	.001	.053	.180	.243	.218	.152	.087	.042	.017	.006	.002	3
	4	0+	.011	.080	.171	.218	.202	.149	.091	.047	.020	.007	4
	5	0+	.002	.027	.091	.164	.202	.192	.147	.093	.050	.022	5
	6	0+	0+	.007	.037	.095	.157	.192	.184	.145	.095	.052	6
	7	0+	0+	.001	.012	.044	.097	.153	.184	.180	.144	.096	7
	8	0+	0+	0+	.003	.017	.049	.098	.149	.180	.177	.144	8
	9	0+	0+	0+	.001	.005	.020	.051	.098	.146	.177	.176	9
	10	0+	0+	0+	0+	.001	.007	.022	.053	.098	.145	.176	10
	11	0+	0+	0+	0+	0+	.002	.008	.023	.053	.097	.144	11
	12	0+	0+	0+	0+	0+	0+	.002	.008	.024	.053	.096	12
	13	0+	0+	0+	0+	0+	0+	.001	.002	.008	.023	.052	13
	14	0+	0+	0+	0+	0+	0+	0+	.001	.002	.008	.022	14
	15	0+	0+	0+	0+	0+	0+	0+	0+	.001	.002	.007	15
	16	0+	0+	0+	0+	0+	0+	0+	0+	0+	0+	.002	16
	17	0+	0+	0+	0+	0+	0+	0+	0+	0+	0+	0+	17
	18	0+	0+	0+	0+	0+	0+	0+	0+	0+	0+	0+	18
	19	0+	0+	0+	0+	0+	0+	0+	0+	0+	0+	0+	19
20	0	.818	.358	.122	.039	.012	.003	.001	0+	0+	0+	0+	0
	1	.165	.377	.270	.137	.058	.021	.007	.002	0+	0+	0+	1
	2	.016	.189	.285	.229	.137	.067	.028	.010	.003	.001	0+	2
	3	.001	.060	.190	.243	.205	.134	.072	.032	.012	.004	.001	3
	4	0+	.013	.090	.182	.218	.190	.130	.074	.035	.014	.005	4
	5	0+	.002	.032	.103	.175	.202	.179	.127	.075	.036	.015	5
	6	0+	0+	.009	.045	.109	.169	.192	.171	.124	.075	.037	6
	7	0+	0+	.002	.016	.055	.112	.164	.184	.166	.122	.074	7
	8	0+	0+	0+	.005	.022	.061	.114	.161	.180	.162	.120	8
	9	0+	0+	0+	.001	.007	.027	.065	.116	.160	.177	.160	9
	10	0+	0+	0+	0+	.002	.010	.031	.069	.117	.159	.176	10
	11	0+	0+	0+	0+	0+	.003	.012	.034	.071	.119	.160	11
	12	0+	0+	0+	0+	0+	.001	.004	.014	.035	.073	.120	12
	13	0+	0+	0+	0+	0+	0+	.001	.004	.015	.037	.074	13
	14	0+	0+	0+	0+	0+	0+	0+	.001	.005	.015	.037	14
	15	0+	0+	0+	0+	0+	0+	0+	0+	.001	.005	.015	15
	16	0+	0+	0+	0+	0+	0+	0+	0+	0+	.001	.005	16
	17	0+	0+	0+	0+	0+	0+	0+	0+	0+	0+	.001	17
	18	0+	0+	0+	0+	0+	0+	0+	0+	0+	0+	0+	18
	19	0+	0+	0+	0+	0+	0+	0+	0+	0+	0+	0+	19
	20	0+	0+	0+	0+	0+	0+	0+	0+	0+	0+	0+	20

TABLE 4 Standard Normal Cumulative Distribution

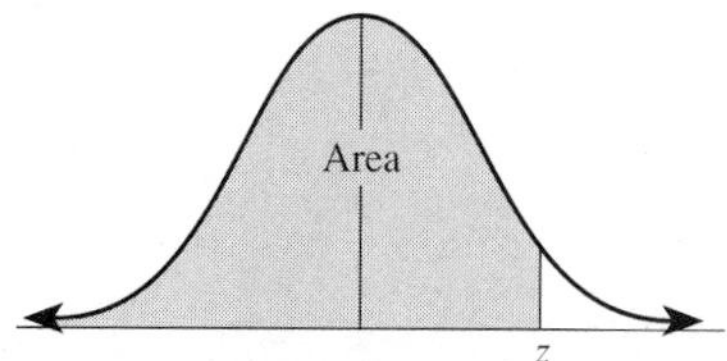

z	.00	.01	.02	.03	.04	.05	.06	.07	.08	.09
−3.4	.0003	.0003	.0003	.0003	.0003	.0003	.0003	.0003	.0003	.0002
−3.3	.0005	.0005	.0005	.0004	.0004	.0004	.0004	.0004	.0004	.0003
−3.2	.0007	.0007	.0006	.0006	.0006	.0006	.0006	.0005	.0005	.0005
−3.1	.0010	.0009	.0009	.0009	.0008	.0008	.0008	.0008	.0007	.0007
−3.0	.0013	.0013	.0013	.0012	.0012	.0011	.0011	.0011	.0010	.0010
−2.9	.0019	.0018	.0017	.0017	.0016	.0016	.0015	.0015	.0014	.0014
−2.8	.0026	.0025	.0024	.0023	.0023	.0022	.0021	.0021	.0020	.0019
−2.7	.0035	.0034	.0033	.0032	.0031	.0030	.0029	.0028	.0027	.0026
−2.6	.0047	.0045	.0044	.0043	.0041	.0040	.0039	.0038	.0037	.0036
−2.5	.0062	.0060	.0059	.0057	.0055	.0054	.0052	.0051	.0049	.0048
−2.4	.0082	.0080	.0078	.0075	.0073	.0071	.0069	.0068	.0066	.0064
−2.3	.0107	.0104	.0102	.0099	.0096	.0094	.0091	.0089	.0087	.0084
−2.2	.0139	.0136	.0132	.0129	.0125	.0122	.0119	.0116	.0113	.0110
−2.1	.0179	.0174	.0170	.0166	.0162	.0158	.0154	.0150	.0146	.0143
−2.0	.0228	.0222	.0217	.0212	.0207	.0202	.0197	.0192	.0188	.0183
−1.9	.0287	.0281	.0274	.0268	.0262	.0256	.0250	.0244	.0239	.0233
−1.8	.0359	.0352	.0344	.0336	.0329	.0322	.0314	.0307	.0301	.0294
−1.7	.0446	.0436	.0427	.0418	.0409	.0401	.0392	.0384	.0375	.0367
−1.6	.0548	.0537	.0526	.0516	.0505	.0495	.0485	.0475	.0465	.0455
−1.5	.0668	.0655	.0643	.0630	.0618	.0606	.0594	.0582	.0571	.0559
−1.4	.0808	.0793	.0778	.0764	.0749	.0735	.0722	.0708	.0694	.0681
−1.3	.0968	.0951	.0934	.0918	.0901	.0885	.0869	.0853	.0838	.0823
−1.2	.1151	.1131	.1112	.1093	.1075	.1056	.1038	.1020	.1003	.0985
−1.1	.1357	.1335	.1314	.1292	.1271	.1251	.1230	.1210	.1190	.1170
−1.0	.1587	.1562	.1539	.1515	.1492	.1469	.1446	.1423	.1401	.1379
−.9	.1841	.1814	.1788	.1762	.1736	.1711	.1685	.1660	.1635	.1611
−.8	.2119	.2090	.2061	.2033	.2005	.1977	.1949	.1922	.1894	.1867
−.7	.2420	.2389	.2358	.2327	.2296	.2266	.2236	.2206	.2177	.2148
−.6	.2743	.2709	.2676	.2643	.2611	.2578	.2546	.2514	.2483	.2451
−.5	.3085	.3050	.3015	.2981	.2946	.2912	.2877	.2843	.2810	.2776
−.4	.3446	.3409	.3372	.3336	.3300	.3264	.3228	.3192	.3156	.3121
−.3	.3821	.3783	.3745	.3707	.3669	.3632	.3594	.3557	.3520	.3483
−.2	.4207	.4168	.4129	.4090	.4052	.4013	.3974	.3936	.3897	.3859
−.1	.4602	.4562	.4522	.4483	.4443	.4404	.4364	.4325	.4286	.4247
−.0	.5000	.4960	.4920	.4880	.4840	.4801	.4761	.4721	.4681	.4641

(continued)

TABLE 4 *(continued)*

z	.00	.01	.02	.03	.04	.05	.06	.07	.08	.09
0.0	.5000	.5040	.5080	.5120	.5160	.5199	.5239	.5279	.5319	.5359
.1	.5398	.5438	.5478	.5517	.5557	.5596	.5636	.5675	.5714	.5753
.2	.5793	.5832	.5871	.5910	.5948	.5987	.6026	.6064	.6103	.6141
.3	.6179	.6217	.6255	.6293	.6331	.6368	.6406	.6443	.6480	.6517
.4	.6554	.6591	.6628	.6664	.6700	.6736	.6772	.6808	.6844	.6879
.5	.6915	.6950	.6985	.7019	.7054	.7088	.7123	.7157	.7190	.7224
.6	.7257	.7291	.7324	.7357	.7389	.7422	.7454	.7486	.7517	.7549
.7	.7580	.7611	.7642	.7673	.7704	.7734	.7764	.7794	.7823	.7852
.8	.7881	.7910	.7939	.7967	.7995	.8023	.8051	.8078	.8106	.8133
.9	.8159	.8186	.8212	.8238	.8264	.8289	.8315	.8340	.8365	.8389
1.0	.8413	.8438	.8461	.8485	.8508	.8531	.8554	.8577	.8599	.8621
1.1	.8643	.8665	.8686	.8708	.8729	.8749	.8770	.8790	.8810	.8830
1.2	.8849	.8869	.8888	.8907	.8925	.8944	.8962	.8980	.8997	.9015
1.3	.9032	.9049	.9066	.9082	.9099	.9115	.9131	.9147	.9162	.9177
1.4	.9192	.9207	.9222	.9236	.9251	.9265	.9278	.9292	.9306	.9319
1.5	.9332	.9345	.9357	.9370	.9382	.9394	.9406	.9418	.9429	.9441
1.6	.9452	.9463	.9474	.9484	.9495	.9505	.9515	.9525	.9535	.9545
1.7	.9554	.9564	.9573	.9582	.9591	.9599	.9608	.9616	.9625	.9633
1.8	.9641	.9649	.9656	.9664	.9671	.9678	.9686	.9693	.9699	.9706
1.9	.9713	.9719	.9726	.9732	.9738	.9744	.9750	.9756	.9761	.9767
2.0	.9772	.9778	.9783	.9788	.9793	.9798	.9803	.9808	.9812	.9817
2.1	.9821	.9826	.9830	.9834	.9838	.9842	.9846	.9850	.9854	.9857
2.2	.9861	.9864	.9868	.9871	.9875	.9878	.9881	.9884	.9887	.9890
2.3	.9893	.9896	.9898	.9901	.9904	.9906	.9909	.9911	.9913	.9916
2.4	.9918	.9920	.9922	.9925	.9927	.9929	.9931	.9932	.9934	.9936
2.5	.9938	.9940	.9941	.9943	.9945	.9946	.9948	.9949	.9951	.9952
2.6	.9953	.9955	.9956	.9957	.9959	.9960	.9961	.9962	.9963	.9964
2.7	.9965	.9966	.9967	.9968	.9969	.9970	.9971	.9972	.9973	.9974
2.8	.9974	.9975	.9976	.9977	.9977	.9978	.9979	.9979	.9980	.9981
2.9	.9981	.9982	.9982	.9983	.9984	.9984	.9985	.9985	.9986	.9986
3.0	.9987	.9987	.9987	.9988	.9988	.9989	.9989	.9989	.9990	.9990
3.1	.9990	.9991	.9991	.9991	.9992	.9992	.9992	.9992	.9993	.9993
3.2	.9993	.9993	.9994	.9994	.9994	.9994	.9994	.9995	.9995	.9995
3.3	.9995	.9995	.9995	.9996	.9996	.9996	.9996	.9996	.9996	.9997
3.4	.9997	.9997	.9997	.9997	.9997	.9997	.9997	.9997	.9997	.9998

TABLE 5 Compound Interest

Compounded amount of $1 *for N periods at i% per period.*

N \ i	1%	1½%	2%	2½%	3%	3½%	4%	N
1	1.010000	1.015000	1.020000	1.025000	1.030000	1.035000	1.040000	1
2	1.020100	1.030225	1.040400	1.050625	1.060900	1.071225	1.081600	2
3	1.030301	1.045678	1.061208	1.076891	1.092727	1.108718	1.124864	3
4	1.040604	1.061364	1.082432	1.103813	1.125509	1.147523	1.169859	4
5	1.051010	1.077284	1.104081	1.131408	1.159274	1.187686	1.216653	5
6	1.061520	1.093443	1.126162	1.159693	1.194052	1.229255	1.265319	6
7	1.072135	1.109845	1.148686	1.188686	1.229874	1.272279	1.315932	7
8	1.082857	1.126493	1.171659	1.218403	1.266770	1.316809	1.368569	8
9	1.093685	1.143390	1.195093	1.248863	1.304773	1.362897	1.423312	9
10	1.104622	1.160541	1.218994	1.280085	1.343916	1.410599	1.480244	10
11	1.115668	1.177949	1.243374	1.312087	1.384234	1.459970	1.539454	11
12	1.126825	1.195618	1.268242	1.344889	1.425761	1.511069	1.601032	12
13	1.138093	1.213552	1.293607	1.378511	1.468534	1.563956	1.665074	13
14	1.149474	1.231756	1.319479	1.412974	1.512590	1.618695	1.731676	14
15	1.169069	1.250232	1.345868	1.448298	1.557967	1.675349	1.800944	15
16	1.172579	1.268986	1.372786	1.484506	1.604706	1.733986	1.872981	16
17	1.184304	1.288020	1.400241	1.521618	1.652848	1.794676	1.947900	17
18	1.196147	1.307341	1.428246	1.559659	1.702433	1.857489	2.025817	18
19	1.208109	1.326951	1.456811	1.598650	1.753506	1.922501	2.106849	19
20	1.220190	1.346855	1.485947	1.638616	1.806111	1.989789	2.191123	20
25	1.282432	1.450945	1.640606	1.853944	2.093778	2.363245	2.665836	25
30	1.347849	1.563080	1.811362	2.097568	2.427262	2.806794	3.243398	30
35	1.416603	1.683881	1.999890	2.373205	2.813862	3.333590	3.946089	35
40	1.488864	1.814018	2.208040	2.685064	3.262038	3.959260	4.801021	40
45	1.564811	1.954213	2.437854	3.037903	3.781596	4.702359	5.841176	45
50	1.644632	2.105242	2.691588	3.437109	4.383906	5.584927	7.106683	50
55	1.728525	2.267944	2.971731	3.888773	5.082149	6.633141	8.646367	55
60	1.816697	2.443220	3.281031	4.399790	5.891603	7.878091	10.519627	60
65	1.909366	2.632042	3.622523	4.977958	6.829983	9.356701	12.798735	65
70	2.006763	2.835456	3.999558	5.632103	7.917822	11.112825	15.571618	70
75	2.109128	3.054592	4.415835	6.372207	9.178926	13.198550	18.945255	75
80	2.216715	3.290663	4.875439	7.209568	10.640891	15.675738	23.049799	80
85	2.329790	3.544978	5.382879	8.156964	12.335709	18.617859	28.043605	85
90	2.448633	3.818949	5.943133	9.228856	14.300467	22.112176	34.119333	90
95	2.573538	4.114092	6.561699	10.441604	16.578161	26.262329	41.511386	95
100	2.704814	4.432046	7.244646	11.813716	19.218632	31.191408	50.504948	100

N \ i	4½%	5%	5½%	6%	6½%	7%	7½%	N
1	1.045000	1.050000	1.055000	1.060000	1.065000	1.070000	1.075000	1
2	1.092025	1.102500	1.113025	1.123600	1.134225	1.144900	1.155625	2
3	1.141166	1.157625	1.174241	1.191016	1.207950	1.225043	1.242297	3
4	1.192519	1.215506	1.238825	1.262477	1.286466	1.310796	1.335469	4
5	1.246182	1.276282	1.306960	1.338226	1.370087	1.402552	1.435629	5
6	1.302260	1.340096	1.378843	1.418519	1.459142	1.500730	1.543302	6
7	1.360862	1.407100	1.454679	1.503630	1.553987	1.605781	1.659049	7
8	1.422101	1.477455	1.534687	1.593848	1.654996	1.718186	1.783478	8
9	1.486095	1.551328	1.619094	1.689479	1.762570	1.838459	1.917239	9
10	1.552969	1.628895	1.708144	1.790848	1.877137	1.967151	2.061032	10
11	1.622853	1.710339	1.802092	1.898299	1.999151	2.104852	2.215609	11
12	1.695881	1.795856	1.901207	2.012196	2.129096	2.252192	2.381780	12
13	1.772196	1.885649	2.005774	2.132928	2.267487	2.409845	2.560413	13
14	1.851945	1.979932	2.116091	2.260904	2.414874	2.578534	2.752444	14
15	1.935282	2.078928	2.232476	2.396558	2.571841	2.759032	2.958877	15
16	2.022370	2.182875	2.355263	2.540352	2.739011	2.952164	3.180793	16
17	2.113377	2.292018	2.484802	2.692773	2.917046	3.158815	3.419353	17
18	2.208479	2.406619	2.621466	2.854339	3.106654	3.379932	3.675804	18
19	2.307860	2.526950	2.765647	3.025600	3.308587	3.616528	3.951489	19
20	2.411714	2.653298	2.917757	3.207135	3.523645	3.869684	4.247851	20
25	3.005434	3.386355	3.813392	4.291871	4.827699	5.427433	6.098340	25
30	3.745318	4.321942	4.983951	5.743491	6.614366	7.612255	8.754955	30
35	4.667348	5.516015	6.513825	7.686087	9.062255	10.676581	12.568870	35
40	5.816365	7.039989	8.513309	10.285718	12.416075	14.974458	18.044239	40
45	7.248248	8.985008	11.126554	13.764611	17.011098	21.002452	25.904839	45
50	9.032636	11.467400	14.541961	18.420154	23.306679	29.457025	37.189746	50
55	11.256308	14.635631	19.005762	24.650322	31.932170	41.315001	53.390690	55
60	14.027408	18.679186	24.839770	32.987691	43.749840	57.946427	76.649240	60
65	17.480702	23.839901	32.464587	44.144972	59.941072	81.272861	110.039897	65
70	21.784136	30.426426	42.429916	59.075930	82.124463	113.989392	157.976504	70
75	27.146996	38.832686	55.454204	79.056921	112.517632	159.876019	226.795701	75
80	33.830096	49.561441	72.476426	105.795993	154.158907	224.234388	325.594560	80
85	42.158455	63.254353	94.723791	141.578904	211.211062	314.500328	467.433099	85
90	52.537105	80.730365	123.800206	189.464511	289.377460	441.102980	671.060665	90
95	65.470792	103.034676	161.801918	253.546255	396.472198	618.669748	963.394370	95
100	81.588518	131.501258	211.468600	339.302084	543.201271	867.716326	1383.077210	100

TABLE 5 *(continued)* Future and Present Value

Use this table to find the future value (multiply principal by table number) and to find present value (divide future values by table number).

N \ i	8%	9%	10%	11%	12%	13%	14%	N
1	1.080000	1.090000	1.100000	1.110000	1.120000	1.130000	1.140000	1
2	1.166400	1.188100	1.210000	1.232100	1.254400	1.276900	1.299600	2
3	1.259712	1.295029	1.331000	1.367631	1.404928	1.442897	1.481544	3
4	1.360489	1.411582	1.464100	1.518070	1.573519	1.630474	1.688960	4
5	1.469328	1.538624	1.610510	1.685058	1.762342	1.842435	1.925415	5
6	1.586874	1.677100	1.771561	1.870415	1.973823	2.081952	2.194973	6
7	1.713824	1.828039	1.948717	2.076160	2.210681	2.352605	2.502269	7
8	1.850930	1.992563	2.143589	2.304538	2.475963	2.658444	2.852586	8
9	1.999005	2.171893	2.357948	2.558037	2.773079	3.004042	3.251949	9
10	2.158925	2.367364	2.593742	2.839421	3.105848	3.394567	3.707221	10
11	2.331639	2.580426	2.853117	3.151757	3.478550	3.835861	4.226232	11
12	2.518170	2.812665	3.138428	3.498451	3.895976	4.334523	4.817905	12
13	2.719624	3.065805	3.452271	3.883280	4.363493	4.898011	5.492411	13
14	2.937194	3.341727	3.797498	4.310441	4.887112	5.534753	6.261349	14
15	3.172169	3.642482	4.177248	4.784589	5.473566	6.254270	7.137938	15
16	3.425943	3.970306	4.594973	5.310894	6.130394	7.067326	8.137249	16
17	3.700018	4.327633	5.054470	5.895093	6.866041	7.986078	9.276464	17
18	3.996019	4.717120	5.559917	6.543553	7.689966	9.024268	10.575169	18
19	4.315701	5.141661	6.115909	7.263344	8.612762	10.197423	12.055693	19
20	4.660957	5.604411	6.727500	8.062312	9.646293	11.523088	13.743490	20
25	6.848475	8.623081	10.834706	13.585464	17.000064	21.230542	26.461916	25
30	10.062657	13.267678	17.449402	22.892297	29.959922	39.115898	50.950159	30
35	14.785344	20.413968	28.102437	38.574851	52.799620	72.068506	98.100178	35
40	21.724521	31.409420	45.259256	65.000867	93.050970	132.781552	188.883514	40
45	31.920449	48.327286	72.890484	109.530242	163.987604	244.641402	363.679072	45
50	46.901613	74.357520	117.390853	184.564827	289.002190	450.735925	700.232988	50
55	68.913856	114.408262	189.059142	311.002466	509.320606	830.451725	1348.238807	55
60	101.257064	176.031292	304.481640	524.057242	897.596933	1530.053473	2595.918660	60
65	148.779847	270.845963	490.370725	883.066930	1581.872491	2819.024345	4998.219642	65
70	218.606406	416.730086	789.746957	1488.019132	2787.799828	5193.869624	9623.644985	70
75	321.204530	641.190893	1271.895371	2507.398773	4913.055841	9569.368113	18529.506390	75
80	471.954834	986.551668	2048.400215	4225.112750	8658.483100	17630.940454	35676.981807	80
85	693.456489	1517.932029	3298.969030	7119.560696	15259.205681	32483.864937	68692.981028	85
90	1018.915089	2335.526582	5313.022612	11996.873812	26891.934223	59849.415520	132262.467379	90
95	1497.120549	3593.497147	8556.676047	20215.430053	47392.776624	110268.668614	254660.083396	95
100	2199.761256	5529.040792	13780.612340	34064.175270	83522.265727	203162.874228	490326.238126	100

N \ i	15%	16%	17%	18%	19%	20%	N
1	1.150000	1.160000	1.170000	1.180000	1.190000	1.200000	1
2	1.322500	1.345600	1.368900	1.392400	1.416100	1.440000	2
3	1.520875	1.560896	1.601613	1.643032	1.685159	1.728000	3
4	1.749006	1.810639	1.873887	1.938778	2.005339	2.073600	4
5	2.011357	2.100342	2.192448	2.287758	2.386354	2.488320	5
6	2.313061	2.436396	2.565164	2.699554	2.839761	2.985984	6
7	2.660020	2.826220	3.001242	3.185474	3.379315	3.583181	7
8	3.059023	3.278415	3.511453	3.758859	4.021385	4.299817	8
9	3.517876	3.802961	4.108400	4.435454	4.785449	5.159780	9
10	4.045558	4.411435	4.806828	5.233836	5.694684	6.191736	10
11	4.652391	5.117265	5.623989	6.175926	6.776674	7.430084	11
12	5.350250	5.936027	6.580067	7.287593	8.064242	8.916100	12
13	6.152788	6.885791	7.698679	8.599359	9.596448	10.699321	13
14	7.075706	7.987518	9.007454	10.147244	11.419773	12.839185	14
15	8.137062	9.265521	10.538721	11.973748	13.589530	15.407022	15
16	9.357621	10.748004	12.330304	14.129023	16.171540	18.488426	16
17	10.761264	12.467685	14.426456	16.672247	19.244133	22.186111	17
18	12.375454	14.462514	16.878953	19.673251	22.900518	26.623333	18
19	14.231772	16.776517	19.748375	23.214436	27.251616	31.948000	19
20	16.366537	19.460759	23.105599	27.393035	32.429423	38.337600	20
25	32.918953	40.874244	50.657826	62.668627	77.388073	95.396217	25
30	66.211772	85.849877	111.064650	143.370638	184.675312	237.376314	30
35	133.175523	180.314073	243.503474	327.997290	440.700607	590.668229	35
40	267.863546	378.721158	533.868713	750.378345	1051.667507	1469.771568	40
45	538.769269	795.443826	1170.479411	1716.683879	2509.650603	3657.261988	45
50	1083.657442	1670.703804	2566.215284	3927.356860	5988.913902	9100.438150	50
55	2179.622184	3509.048796	5626.293659	8984.841120	14291.666609	22644.802257	55
60	4383.998746	7370.201365	12335.356482	20555.139966	34104.970919	56347.514353	60
65	8817.787387	15479.940952	27044.628088	47025.180900	81386.522174	140210.646915	65
70	17735.720039	32513.164839	59293.941729	107582.222368	194217.025056	348888.956932	70
75	35672.867976	68288.754533	129998.886072	246122.063716	463470.508558	868147.369314	75
80	71750.879401	143429.715890	285015.802412	563067.660386	1106004.544354	2160228.462010	80
85	144316.646994	301251.407222	624882.336142	1288162.407650	2639317.992285	5375339.686589	85
90	290272.325206	632730.879999	1370022.050417	2947003.540121	6298346.150529	13375565.248934	90
95	583841.327636	1328551.025313	3003702.153303	6742030.208228	15030081.387632	33282686.520228	95
100	1174313.450700	2791651.199375	6585460.885837	15424131.905453	35867089.727971	82817974.522015	100

TABLE 6 Ordinary Annuity

Amount of ordinary annuity of \$1 at compound interest. Remember, an ordinary annuity requires payment at the end of the period.

N \ i	1%	1½%	2%	2½%	3%	3½%	4%	N
1	1.000000	1.000000	1.000000	1.000000	1.000000	1.000000	1.000000	1
2	2.010000	2.015000	2.020000	2.025000	2.030000	2.035000	2.040000	2
3	3.030100	3.045225	3.060400	3.075625	3.090900	3.106225	3.121600	3
4	4.060401	4.090903	4.121608	4.152516	4.183627	4.214943	4.246464	4
5	5.101005	5.152267	5.204040	5.256329	5.309136	5.362466	5.416323	5
6	6.152015	6.229551	6.308121	6.387737	6.468410	6.550152	6.632975	6
7	7.213535	7.322994	7.434283	7.547430	7.662462	7.779408	7.898294	7
8	8.285671	8.432839	8.582969	8.736116	8.892336	9.051687	9.214226	8
9	9.368527	9.559332	9.754628	9.954519	10.159106	10.368496	10.582795	9
10	10.462213	10.702722	10.949721	11.203382	11.463879	11.731393	12.006107	10
11	11.566835	11.863262	12.168715	12.483466	12.807796	13.141992	13.486351	11
12	12.682503	13.041211	13.412090	13.795553	14.192030	14.601962	15.025805	12
13	13.809328	14.236830	14.680332	15.140442	15.617790	16.113030	16.626838	13
14	14.947421	15.450382	15.973938	16.518953	17.086324	17.676986	18.291911	14
15	16.096896	16.682138	17.293417	17.931927	18.598914	19.295681	20.023588	15
16	17.257864	17.932370	18.639285	19.380225	20.156881	20.971030	21.824531	16
17	18.430443	19.201355	20.012071	20.864730	21.761588	22.705016	23.697512	17
18	19.614748	20.489376	21.412312	22.386349	23.414435	24.499691	25.645413	18
19	20.810895	21.796716	22.840559	23.946007	25.116868	26.357180	27.671229	19
20	22.019004	23.123667	24.297370	25.544658	26.870374	28.279682	29.778079	20
25	28.243200	30.063024	32.030300	34.157764	36.459264	38.949857	41.645908	25
30	34.784892	37.538681	40.568079	43.902703	47.575416	51.622677	56.084938	30
35	41.660276	45.592088	49.994478	54.928207	60.462082	66.674013	73.652225	35
40	48.886373	54.267894	60.401983	67.402554	75.401260	84.550278	95.025516	40
45	56.481075	63.614201	71.892710	81.516131	92.719861	105.781673	121.029392	45
50	64.463182	73.682828	84.579401	97.484349	112.796867	130.997910	152.667084	50
55	72.852457	84.529599	98.586534	115.550921	136.071620	160.946890	191.159173	55
60	81.669670	96.214652	114.051539	135.991590	163.053437	196.516883	237.990685	60
65	90.936649	108.802772	131.126155	159.118330	194.332758	238.762876	294.968380	65
70	100.676337	122.363753	149.977911	185.284114	230.594064	288.937865	364.290459	70
75	110.912847	136.972781	170.791773	214.888297	272.630856	348.530011	448.631367	75
80	121.671522	152.710852	193.771958	248.382713	321.363019	419.306787	551.244977	80
85	132.978997	169.665226	219.143939	286.278570	377.856952	503.367394	676.000123	85
90	144.863267	187.929900	247.156656	329.154253	443.348904	603.205027	827.983334	90
95	157.353755	207.606142	278.084960	377.664154	519.272026	721.780816	1012.784648	95
100	170.481383	228.803043	312.232306	432.548654	607.287733	862.611657	1237.623705	100

N \ i	4½%	5%	5½%	6%	6½%	7%	7½%	N
1	1.000000	1.000000	1.000000	1.000000	1.000000	1.000000	1.000000	1
2	2.045000	2.050000	2.055000	2.060000	2.065000	2.070000	2.075000	2
3	3.137025	3.152500	3.168025	3.183600	3.199225	3.214900	3.230625	3
4	4.278191	4.310125	4.342266	4.374616	4.407175	4.439943	4.472922	4
5	5.470710	5.525631	5.581091	5.637093	5.693641	5.750739	5.808391	5
6	6.716892	6.801913	6.888051	6.975319	7.063728	7.153291	7.244020	6
7	8.019152	8.142008	8.266894	8.393838	8.522870	8.654021	8.787322	7
8	9.380014	9.549109	9.721573	9.897468	10.076856	10.259803	10.466371	8
9	10.802114	11.026564	11.256260	11.491316	11.731852	11.977989	12.229849	9
10	12.288209	12.577893	12.875354	13.180795	13.494423	13.816448	14.147087	10
11	13.841179	14.206787	14.583498	14.971643	15.371560	15.783599	16.208119	11
12	15.464032	15.917127	16.385591	16.869941	17.370711	17.888451	18.423728	12
13	17.159913	17.712983	18.286798	18.882138	19.499808	20.140643	20.805508	13
14	18.932109	19.598632	20.292572	21.015066	21.767295	22.550488	23.365921	14
15	20.784054	21.578564	22.408663	23.275970	24.182169	25.129022	26.118365	15
16	22.719337	23.657492	24.641140	25.672528	26.754010	27.888054	29.077242	16
17	24.741707	25.840366	26.996403	28.212880	29.493021	30.840217	32.258035	17
18	26.855084	28.132385	29.481205	30.905653	32.410067	33.999033	35.677388	18
19	29.063562	30.539004	32.102671	33.759992	33.516722	37.378965	39.353192	19
20	31.371423	33.065954	34.868318	36.785591	38.825309	40.995492	43.304681	20
25	44.565210	47.727099	51.152588	54.864512	58.887679	63.249038	67.977862	25
30	61.007070	66.438848	72.435478	79.058186	86.374864	94.460786	103.399403	30
35	81.496618	90.320307	100.251364	111.434780	124.034690	138.236878	154.251606	35
40	107.030323	120.799774	136.605614	154.761966	175.631916	199.635112	227.256520	40
45	138.849965	159.700156	184.119165	212.743514	246.324587	285.749311	332.064515	45
50	178.503028	209.347996	246.217476	290.335905	343.179672	406.528929	482.529947	50
55	227.917959	272.712618	327.377486	394.172027	475.879533	575.928593	698.542534	55
60	289.497954	353.583718	433.450372	533.128181	657.689842	813.520383	1008.656538	60
65	366.237831	456.798011	572.083392	719.082861	906.785722	1146.755161	1453.865297	65
70	461.869680	588.528511	753.271204	967.932170	1248.068666	1614.134174	2093.020048	70
75	581.044362	756.653718	990.076429	1300.948680	1715.655875	2269.657419	3010.609352	75
80	729.557699	971.228821	1299.571387	1746.599891	2356.290874	3189.062680	4327.927467	80
85	914.632336	1245.087069	1704.068919	2342.981741	3234.016343	4478.576120	6219.107984	85
90	1145.269007	1594.607301	2232.731017	3141.075187	4436.576302	6287.185427	8934.142195	90
95	1432.684259	2040.693529	2923.671235	4209.104250	6084.187663	8823.853541	12831.924930	95
100	1790.855956	2610.025157	3826.702467	5638.368059	8341.558016	12381.661794	18427.696132	100

TABLE 6 *(continued)*

N \ i	8%	9%	10%	11%	12%	13%	14%	N
1	1.000000	1.000000	1.000000	1.000000	1.000000	1.000000	1.000000	1
2	2.080000	2.090000	2.100000	2.110000	2.120000	2.130000	2.140000	2
3	3.246400	3.278100	3.310000	3.342100	3.374400	3.406900	3.439600	3
4	4.506112	4.573129	4.641000	4.709731	4.779328	4.849797	4.921144	4
5	5.866601	5.984711	6.105100	6.227801	6.352847	6.480271	6.610104	5
6	7.335929	7.523335	7.715610	7.912860	8.115189	8.322706	8.535519	6
7	8.922803	9.200435	9.487171	9.783274	10.089012	10.404658	10.730491	7
8	10.636628	11.028474	11.435888	11.859434	12.299693	12.757263	13.232760	8
9	12.487558	13.021036	13.579477	14.163972	14.775656	15.415707	16.085347	9
10	14.486562	15.192930	15.937425	16.722009	17.548735	18.419749	19.337295	10
11	16.645487	17.560293	18.531167	19.561430	20.654583	21.814317	23.044516	11
12	18.977126	20.140720	21.384284	22.713187	24.133133	25.650178	27.270749	12
13	21.495297	22.953385	24.522712	26.211638	28.029109	29.984701	32.088654	13
14	24.214920	26.019189	27.974983	30.094918	32.392602	34.882712	37.581065	14
15	27.152114	29.360916	31.772482	34.405359	37.279715	40.417464	43.842414	15
16	30.324283	33.003399	35.949730	39.189948	42.753280	46.671735	50.980352	16
17	33.750226	36.973705	40.544703	44.500843	48.883674	53.739060	59.117601	17
18	37.450244	41.301338	45.599173	50.395936	55.749715	61.725138	68.394066	18
19	41.446263	46.018458	51.159090	56.939488	63.439681	70.749406	78.969235	19
20	45.761964	51.160120	57.274999	64.202832	72.052442	80.946829	91.024928	20
25	73.105940	84.700896	98.347059	114.413307	133.333870	155.619556	181.870827	25
30	113.283211	136.307539	164.494023	199.020878	241.332684	293.199215	356.786847	30
35	172.316804	215.710755	271.024368	341.589555	431.663496	546.680819	693.572702	35
40	259.056519	337.882445	442.592556	581.826066	767.091420	1013.704243	1342.025099	40
45	386.505617	525.858734	718.904837	986.638559	1358.230032	1874.164630	2590.564800	45
50	573.770156	815.083556	1163.908529	1668.771152	2400.018249	3459.507117	4994.521346	50
55	848.923201	1260.091796	1880.591425	2818.204240	4236.005047	6380.397885	9623.134336	55
60	1253.213296	1944.792133	3034.816395	4755.065839	7471.641112	11761.949792	18535.133283	60
65	1847.248083	2998.288474	4893.707253	8018.790272	13173.937422	21677.110345	35694.426015	65
70	2720.080074	4619.223180	7687.469566	13518.355744	23223.331897	39945.150956	68733.178463	70
75	4002.556624	7113.232148	12708.953714	22785.443391	40933.798673	73602.831635	132346.474212	75
80	5886.935428	10950.574090	20474.002146	38401.025004	72145.692501	135614.926571	254828.441480	80
85	8655.706112	16854.800326	32979.690296	64714.188149	127151.714005	249868.191823	490657.007341	85
90	12723.938616	25939.184247	53120.226118	109053.398293	224091.118528	460372.427073	944724.766995	90
95	18701.506857	39916.634964	85556.760466	183767.545936	394931.471864	848212.835490	1818993.452831	95
100	27484.515704	61422.675465	137796.123398	309665.229724	696010.547721	1562783.647911	3502323.129475	100

N \ i	15%	16%	17%	18%	19%	20%	N
1	1.000000	1.000000	1.000000	1.000000	1.000000	1.000000	1
2	2.150000	2.160000	2.170000	2.180000	2.190000	2.200000	2
3	3.472500	3.505600	3.538900	3.572400	3.606100	3.640000	3
4	4.993375	5.066496	5.140513	5.215432	5.291259	5.368000	4
5	6.742381	6.877135	7.014400	7.154210	7.296598	7.441600	5
6	8.753738	8.977477	9.206848	9.441968	9.682952	9.929920	6
7	11.066799	11.413873	11.772012	12.141522	12.522713	12.915904	7
8	13.726819	14.240093	14.773255	15.326996	15.902028	16.499085	8
9	16.785842	17.518508	18.284708	19.085855	19.923413	20.798902	9
10	20.303718	21.321469	22.393108	23.521309	24.708862	25.958682	10
11	24.349276	25.732904	27.199937	28.755144	30.403546	32.150419	11
12	29.001667	30.850169	32.823926	34.931070	37.180220	39.580502	12
13	34.351917	36.786196	39.403993	42.218663	45.244461	48.496603	13
14	40.504705	43.671987	47.102672	50.818022	54.840909	59.195923	14
15	47.580411	51.659505	56.110126	60.965266	66.260682	72.035108	15
16	55.717472	60.925026	66.648848	72.939014	79.850211	87.442129	16
17	65.075093	71.673030	78.979152	87.068036	96.021751	105.930555	17
18	75.836357	84.140715	93.405608	103.740283	115.265884	128.116666	18
19	88.211811	98.603230	110.284561	123.413534	138.166402	154.740000	19
20	102.443583	115.379747	130.032936	146.627970	165.418018	186.688000	20
25	212.793017	249.214024	292.104856	342.603486	402.042491	471.981083	25
30	434.745146	530.311731	647.439118	790.947991	966.712169	1181.881569	30
35	881.170156	1120.712955	1426.491022	1816.651612	2314.213721	2948.341146	35
40	1779.090308	2360.757241	3134.521839	4163.213027	5529.828982	7343.857840	40
45	3585.128460	4965.273911	6879.290650	9531.577105	13203.424228	18281.309940	45
50	7217.716277	10435.648773	15089.501673	21813.093666	31515.336327	45497.190750	50
55	14524.147893	21925.304976	33089.962703	49910.228445	75214.034786	113219.011287	55
60	29219.991638	46057.508533	72555.038129	114189.666478	179494.583786	281732.571766	60
65	58778.582580	96743.380952	159080.165226	261245.449442	428344.853547	701048.234576	65
70	118231.466926	203201.030246	348782.010169	597673.457599	1022189.605560	1744439.784661	70
75	237812.453171	426798.465828	764693.447483	1367339.242866	2439313.202939	4340731.846568	75
80	478332.529343	896429.474315	1676557.661247	3128148.113254	5821071.286073	10801137.310052	80
85	962104.313290	1882815.045139	3675772.565539	7156452.264725	13891142.064658	26876693.432947	85
90	1935142.168042	3954561.749997	8058947.355395	16372236.334003	33149185.002785	66877821.244672	90
95	3892268.850907	8305937.658205	17668830.313546	37455717.823488	79105686.250695	166413427.601142	95
100	7828749.671335	17445313.746092	38737999.328452	85689616.141407	188774151.199846	414089867.610073	100

TABLE 7 Annuity Due

Amount of annuity due for $1 at compound interest. Remember, annuity due requires payment at the beginning of the period.

N \ i	1%	1½%	2%	2½%	3%	3½%	4%	N
1	1.010000	1.015000	1.020000	1.025000	1.030000	1.035000	1.040000	1
2	2.030100	2.045225	2.060400	2.075625	2.090900	2.106225	2.121600	2
3	3.060401	3.090903	3.121608	3.152516	3.183627	3.214943	3.246464	3
4	4.101005	4.152267	4.204040	4.256329	4.309136	4.362466	4.416323	4
5	5.152015	5.229551	5.308121	5.387737	5.468410	5.550152	5.632975	5
6	6.213535	6.322994	6.434283	6.547430	6.662462	6.779408	6.898294	6
7	7.285671	7.432839	7.582969	7.736116	7.892336	8.051687	8.214226	7
8	8.368527	8.559332	8.754628	8.954519	9.159106	9.368496	9.582795	8
9	9.462213	9.702722	9.949721	10.203382	10.463879	10.731393	11.006107	9
10	10.566835	10.863262	11.168715	11.483466	11.807796	12.141992	12.486351	10
11	11.682503	12.041211	12.412090	12.795553	13.192030	13.601962	14.025805	11
12	12.809328	13.236830	13.680332	14.140442	14.617790	15.113030	15.626838	12
13	13.947421	14.450382	14.973938	15.518953	16.086324	16.676986	17.291911	13
14	15.096896	15.682138	16.293417	16.931927	17.598914	18.295681	19.023588	14
15	16.257864	16.932370	17.639285	18.380225	19.156881	19.971030	20.824531	15
16	17.430443	18.201355	19.012071	19.864730	20.761588	21.705016	22.697512	16
17	18.614748	19.489376	20.412312	21.386349	22.414435	23.499691	24.645413	17
18	19.810895	20.796716	21.840559	22.946007	24.116868	25.357180	26.671229	18
19	21.019004	22.123667	23.297370	24.544658	25.870374	27.279682	28.778079	19
20	22.239194	23.470522	24.783317	26.183274	27.676486	29.269471	30.969202	20
25	28.525631	30.513969	32.670906	35.011708	37.553042	40.313102	43.311745	25
30	35.132740	38.101762	41.379441	45.000271	49.002678	53.429471	58.328335	30
35	42.076878	46.275969	50.994367	56.301413	62.275944	69.007603	76.598314	35
40	49.375237	55.081912	61.610023	69.087617	77.663298	87.509537	98.826536	40
45	57.045885	64.568414	73.330564	83.554034	95.501457	109.484031	125.870568	45
50	65.107814	74.788070	86.270989	99.921458	116.180773	135.582837	158.773767	50
55	73.580982	85.797543	100.558264	118.439694	140.153768	166.580031	198.805540	55
60	82.486367	97.657871	116.332570	139.391380	167.945040	203.394974	247.510313	60
65	91.846015	110.434814	133.748679	163.096289	200.162741	247.119577	306.767116	65
70	101.683100	124.199209	152.977469	189.916217	237.511886	299.050690	378.862077	70
75	112.021975	139.027372	174.207608	220.260504	280.809781	360.728561	466.576621	75
80	122.888237	155.001515	197.647397	254.592280	331.003909	433.982524	573.294776	80
85	134.308787	172.210204	223.526818	293.435534	389.192660	520.985253	703.133728	85
90	146.311900	190.748849	252.099789	337.383110	456.649371	624.317203	861.102667	90
95	158.927293	210.720235	283.646659	387.105758	534.850186	747.043145	1053.296034	95
100	172.186197	232.235089	318.476952	443.362370	625.506365	892.803065	1287.128653	100

N \ i	4½%	5%	5½%	6%	6½%	7%	7½%	N
1	1.045000	1.050000	1.055000	1.060000	1.065000	1.070000	1.075000	1
2	2.137025	2.152500	2.168025	2.183600	2.199225	2.214900	2.230625	2
3	3.278191	3.310125	3.342266	3.374616	3.407175	3.439943	3.472922	3
4	4.470710	4.525631	4.581091	4.637093	4.693641	4.750739	4.808391	4
5	5.716892	5.801913	5.888051	5.975319	6.063728	6.153291	6.244020	5
6	7.019152	7.142008	7.266894	7.393838	7.522870	7.654021	7.787322	6
7	8.380014	8.549109	8.721573	8.897468	9.076856	9.259803	9.446371	7
8	9.802114	10.026564	10.256260	10.491316	10.731852	10.977989	11.229849	8
9	11.288209	11.577893	11.875354	12.180795	12.494423	12.816448	13.147087	9
10	12.841179	13.206787	13.583498	13.971643	14.371560	14.783599	15.208119	10
11	14.464032	14.917127	15.385591	15.869941	16.370711	16.888451	17.423728	11
12	16.159913	16.712983	17.286798	17.882138	18.499808	19.140643	19.805508	12
13	17.932109	18.598632	19.292572	20.015066	20.767295	21.550488	22.365921	13
14	19.784054	20.578564	21.408663	22.275970	23.182169	24.129022	25.118365	14
15	21.719337	22.657492	23.641140	24.672528	25.754010	26.888054	28.077242	15
16	23.741707	24.840366	25.996403	27.212880	28.493021	29.840217	31.258035	16
17	25.855084	27.132385	28.481205	29.905653	31.410067	32.999033	34.677388	17
18	28.063562	29.539004	31.102671	32.759992	34.516722	36.378965	38.353192	18
19	30.371423	32.065954	33.868318	35.785591	37.825309	39.995492	42.304681	19
20	32.783137	34.719252	36.786076	38.992727	41.348954	43.865177	46.552532	20
25	46.570645	50.113454	53.965981	58.156383	62.715378	67.676470	73.076201	25
30	63.752388	69.760790	76.419429	83.801677	91.989230	101.073041	111.154358	30
35	85.163966	94.836323	105.765189	118.120867	132.096945	147.913460	165.820476	35
40	111.846688	126.839763	144.118923	164.047684	187.047990	213.609570	244.300759	40
45	145.098214	167.685164	194.245719	225.508125	262.335685	305.751763	356.969354	45
50	186.535665	219.815396	259.759438	307.756059	365.486351	434.985955	518.719693	50
55	238.174268	286.348249	345.383247	417.822348	506.811702	616.243594	750.933224	55
60	302.525362	371.262904	457.290142	565.115872	700.439682	870.466810	1084.305779	60
65	382.718533	479.637912	603.547978	762.227832	965.726794	1227.028022	1562.905195	65
70	482.653815	617.954936	794.701120	1026.008100	1329.193129	1727.123566	2249.996552	70
75	607.191358	794.486404	1044.530633	1379.005601	1827.173507	2428.533438	3236.405054	75
80	762.387795	1019.790262	1371.047813	1851.395885	2509.449781	3412.297067	4652.522027	80
85	955.790791	1307.341422	1797.792710	2483.560646	3444.227405	4792.076448	6685.541082	85
90	1196.806112	1674.337666	2355.531223	3329.539698	4724.953761	6727.288407	9604.202860	90
95	1497.155051	2142.728205	3084.473153	4461.650505	6479.659861	9441.523288	13794.319300	95
100	1871.444474	2740.526415	4037.171102	5976.670142	8883.759287	13248.378119	19809.773342	100

TABLE 7 *(continued)*

N \ i	8%	9%	10%	11%	12%	13%	14%	N
1	1.080000	1.090000	1.100000	1.110000	1.120000	1.130000	1.140000	1
2	2.246400	2.278100	2.310000	2.342100	2.374400	2.406900	2.439600	2
3	3.506112	3.573129	3.641000	3.709731	3.779328	3.849797	3.921144	3
4	4.866601	4.984511	5.105100	5.227801	5.352847	5.480271	5.610104	4
5	6.335929	6.523335	6.715610	6.912860	7.115189	7.322706	7.535519	5
6	7.922803	8.200435	8.487171	8.783274	9.089012	9.404658	9.730491	6
7	9.636628	10.028474	10.435888	10.859434	11.299693	11.757263	12.232760	7
8	11.487558	12.021036	12.579477	13.163972	13.775656	14.415707	15.085347	8
9	13.486562	14.192930	14.937425	15.722009	16.548735	17.419749	18.337295	9
10	15.645487	16.560293	17.531167	18.561430	19.654583	20.814317	22.044516	10
11	17.977126	19.140720	20.384284	21.713187	23.133133	24.650178	26.270749	11
12	20.495297	21.953385	23.522712	25.211638	27.029109	28.984701	31.088654	12
13	23.214920	25.019189	26.974983	29.094918	31.392602	33.882712	36.581065	13
14	26.152114	28.360916	30.772482	33.405359	36.279715	39.417464	42.842414	14
15	29.324283	32.003399	34.949730	38.189948	41.753280	45.671735	49.980352	15
16	32.750226	35.973705	39.544703	43.500843	47.883674	52.739060	58.117601	16
17	36.450244	40.301338	44.599173	49.395936	54.749715	60.725138	67.394066	17
18	40.446263	45.018458	50.159090	55.939488	62.439681	69.749406	77.969235	18
19	44.761964	50.160120	56.274999	63.202832	71.052442	79.946829	90.024928	19
20	49.422921	55.764530	63.002499	71.265144	80.698736	91.469917	103.768418	20
25	78.954415	92.323977	108.181765	126.998771	149.333934	175.850098	207.332743	25
30	122.345868	148.575217	180.943425	220.913174	270.292606	331.315113	406.737006	30
35	186.102148	235.124723	298.126805	379.164406	483.463116	617.749325	790.672881	35
40	279.781040	368.291865	486.851811	645.826934	859.142391	1145.485795	1529.908613	40
45	417.426067	573.186021	790.795321	1095.168801	1521.217636	2117.806032	2953.243872	45
50	619.671769	888.441076	1280.299382	1852.335979	2688.020438	3909.243042	5693.754335	50
55	916.837058	1373.500057	2068.650567	3128.206707	4744.325653	7209.849610	10970.373143	55
60	1353.470360	2119.823425	3338.298035	5278.123082	8368.238046	13291.003265	21130.051943	60
65	1995.027929	3268.134436	5383.077978	8900.857202	14754.809912	24495.134690	40691.645657	65
70	2937.686480	5034.953266	8676.216525	15005.374875	26010.131725	45138.020581	78355.823448	70
75	4322.761154	7753.423041	13979.849085	25291.842164	45845.854514	83171.199747	150874.980601	75
80	6357.890263	11936.125758	22521.402360	42625.137755	80803.175601	153244.867025	290504.423288	80
85	9348.162601	18371.732355	36277.659326	71832.748846	142409.919685	282351.056760	559348.988369	85
90	13741.853705	28273.710829	58432.248730	121049.272105	250982.052751	520220.842593	1076986.234374	90
95	20197.627405	43509.132110	94112.436513	203981.975989	442323.248488	958480.504103	2073652.536227	95
100	29683.276961	66950.716257	151575.735738	343728.404994	779531.813448	1765945.522139	3992648.367601	100

N \ i	15%	16%	17%	18%	19%	20%	N
1	1.150000	1.160000	1.170000	1.180000	1.190000	1.200000	1
2	2.472500	2.505600	2.538900	2.572400	2.606100	2.640000	2
3	3.993375	4.066496	4.140513	4.215432	4.291259	4.368000	3
4	5.742381	5.877135	6.014400	6.154210	6.296598	6.441600	4
5	7.753738	7.977477	8.206848	8.441968	8.682952	8.929920	5
6	10.066799	10.413873	10.772012	11.141522	11.522713	11.915904	6
7	12.726819	13.240093	13.773255	14.326996	14.902028	15.499085	7
8	15.785842	16.518508	17.284708	18.085855	18.923413	19.798902	8
9	19.303718	20.321469	21.393108	22.521309	23.708862	24.958682	9
10	23.349276	24.732904	26.199937	27.755144	29.403546	31.150419	10
11	28.001667	29.850169	31.823926	33.931070	36.180220	38.580502	11
12	33.351917	35.786196	38.403993	41.218663	44.244461	47.496603	12
13	39.504705	42.671987	46.102672	49.818022	53.840909	58.195923	13
14	46.580411	50.659505	55.110126	59.965266	65.260682	71.035108	14
15	54.717472	59.925026	65.648848	71.939014	78.850211	86.442129	15
16	64.075093	70.673030	77.979152	86.068036	95.021751	104.930555	16
17	74.836357	83.140715	92.405608	102.740283	114.265884	127.116666	17
18	87.211811	97.603230	109.284561	122.413534	137.166402	153.740000	18
19	101.443583	114.379747	129.032936	145.627970	164.418018	185.688000	19
20	117.810120	133.840506	152.138535	173.021005	196.847442	224.025600	20
25	244.711970	289.088267	341.762681	404.272113	478.430565	566.377300	25
30	499.956918	615.161608	757.503768	933.318630	1150.387481	1418.257883	30
35	1013.345680	1300.027028	1668.994496	2143.648902	2753.914328	3538.009375	35
40	2045.953854	2738.478399	3667.390552	4912.591372	6580.496488	8812.629408	40
45	4122.897729	5759.717737	8048.770061	11247.260984	15712.074831	21937.571928	45
50	8300.373719	12105.352576	17654.716957	25739.450526	37503.250230	54596.628900	50
55	16702.770077	25433.353773	38715.256362	58894.069565	89504.701395	135862.813544	55
60	33602.990383	53426.709898	84889.394611	134743.806444	213598.554705	338079.086119	60
65	67595.369968	112222.321904	186123.793315	308269.630342	509730.375721	841257.881492	65
70	135966.186964	235713.195085	408074.951898	705254.679967	1216405.630617	2093327.741593	70
75	273484.321146	495086.220361	894691.333555	1613460.306582	2902782.711497	5208878.215881	75
80	550082.408744	1039858.190205	1961572.463659	3691214.773639	6927074.830427	12961364.772062	80
85	1106419.960284	2184065.452361	4300653.901680	8444613.672375	16530459.056942	32252032.119537	85
90	2225413.493248	4587291.629996	9428968.405813	19319238.874124	39447530.153314	80253385.493606	90
95	4476109.178543	9634887.683518	20672531.466849	44197747.031716	94135766.638327	199696113.121370	95
100	9003062.122036	20236563.945467	45323459.214289	101113747.046860	224641239.927817	496907841.132087	100

TABLE 8 Present Value of an Annuity

Amount needed to deposit today to equal the future value of the annuity; it is the amount you can borrow with a given monthly payment on an installment loan.

N \ i	1%	1½%	2%	2½%	3%	3½%	4%	N
1	.990099	.985222	.980392	.975610	.970874	.966184	.961538	1
2	1.970395	1.955883	1.941561	1.927424	1.913470	1.899694	1.886095	2
3	2.940985	2.912200	2.883883	2.856024	2.828611	2.801637	2.775091	3
4	3.901966	3.854385	3.807729	3.761974	3.717098	3.673079	3.629895	4
5	4.853431	4.782645	4.713460	4.645828	4.579707	4.515052	4.451822	5
6	5.795476	5.697187	5.601431	5.508125	5.417191	5.328553	5.242137	6
7	6.728195	6.598214	6.471991	6.349391	6.230283	6.114544	6.002055	7
8	7.651678	7.485925	7.325481	7.170137	7.019692	6.873956	6.732745	8
9	8.566018	8.360517	8.162237	7.970866	7.786109	7.607687	7.435332	9
10	9.471305	9.222185	8.982585	8.752064	8.530203	8.316605	8.110896	10
11	10.367628	10.071118	9.786848	9.514209	9.252624	9.001551	8.760477	11
12	11.255077	10.907505	10.575341	10.257765	9.954004	9.663334	9.385074	12
13	12.133740	11.731532	11.348374	10.983185	10.634955	10.302738	9.985648	13
14	13.003703	12.543382	12.106249	11.690912	11.296973	10.920520	10.563123	14
15	13.865053	13.343233	12.849264	12.381378	11.937935	11.517411	11.118387	15
16	14.717874	14.131264	13.577709	13.055003	12.561102	12.094117	11.652296	16
17	15.562251	14.907649	14.291872	13.712198	13.166118	12.651321	12.165669	17
18	16.398269	15.672561	14.992031	14.353364	13.753513	13.189682	12.659297	18
19	17.226008	16.426168	15.678462	14.978891	14.323799	13.709837	13.133939	19
20	18.045553	17.168639	16.351433	15.589162	14.877475	14.212403	13.590326	20
25	22.023156	20.719611	19.523456	18.424376	17.413148	16.481515	15.622080	25
30	25.807708	24.015838	22.396456	20.930293	19.600441	18.392045	17.292033	30
35	29.408580	27.075595	24.998619	23.145157	21.487220	20.000661	18.664613	35
40	32.834686	29.915845	27.355479	25.102775	23.114772	21.355072	19.792774	40
45	36.094508	32.552337	29.490160	26.833024	24.518713	22.495450	20.720040	45
50	39.196118	34.999688	31.423606	28.362312	25.729764	23.455618	21.482185	50
55	42.147192	37.271467	33.174788	29.713979	26.774428	24.264053	22.108612	55
60	44.955038	39.380269	34.760887	30.908656	27.675564	24.944734	22.623490	60
65	47.626608	41.337786	36.197466	31.964577	28.452892	25.517849	23.046682	65
70	50.168514	43.154872	37.498619	32.897857	29.123421	26.000397	23.394515	70
75	52.587051	44.841600	38.677114	33.722740	29.701826	26.406689	23.680408	75
80	54.888206	46.407323	39.744514	34.451817	30.200763	26.748776	23.915392	80
85	57.077676	47.860722	40.711290	35.096215	30.631151	27.036804	24.108531	85
90	59.160881	49.209855	41.586929	36.665768	31.002407	27.279316	24.267278	90
95	61.142980	50.462201	42.380023	36.169171	31.322656	27.483504	24.397756	95
100	63.028879	51.624704	43.098352	36.614105	31.598905	27.655425	24.504999	100

N \ i	4½%	5%	5½%	6%	6½%	7%	7½%	N
1	.956938	.952381	.947867	.943396	.938967	.934579	.930233	1
2	1.872668	1.859410	1.846320	1.833393	1.820626	1.808018	1.795565	2
3	2.748964	2.723248	2.697933	2.673012	2.648476	2.624316	2.600526	3
4	3.587526	3.545951	3.505150	3.465106	3.425799	3.387211	3.349326	4
5	4.389977	4.329477	4.270284	4.212364	4.155679	4.100197	4.045885	5
6	5.157872	5.075692	4.995530	4.917324	4.841014	4.766540	4.693846	6
7	5.892701	5.786373	5.682967	5.582381	5.484520	5.389289	5.296601	7
8	6.595886	6.463213	6.334566	6.209794	6.088751	5.971299	5.857304	8
9	7.268790	7.107822	6.952195	6.801692	6.656104	6.515232	6.378887	9
10	7.912718	7.721735	7.537626	7.360087	7.188830	7.023582	6.864081	10
11	8.528917	8.306414	8.092536	7.886875	7.689042	7.498674	7.315424	11
12	9.118581	8.863252	8.618518	8.383844	8.158725	7.942686	7.735278	12
13	9.682852	9.393573	9.117079	8.852683	8.599742	8.357651	8.125640	13
14	10.222825	9.898641	9.589648	9.294984	9.013842	8.745468	8.489154	14
15	10.739546	10.379658	10.037581	9.712249	9.402669	9.107914	8.827120	15
16	11.234015	10.837770	10.462162	10.105895	9.767764	9.446649	9.141507	16
17	11.707191	11.274066	10.864609	10.477260	10.110577	9.763223	9.433960	17
18	12.159992	11.689587	11.246074	10.827603	10.432466	10.059087	9.706009	18
19	12.593294	12.085321	11.607654	11.158116	10.734710	10.335595	9.959078	19
20	13.007936	12.462210	11.950382	11.469921	11.018507	10.594014	10.194491	20
25	14.828209	14.093945	13.413933	12.783356	12.197877	11.653583	11.146946	25
30	16.288889	15.372451	14.533745	13.764831	13.058676	12.409041	11.810386	30
35	17.461012	16.374194	15.390552	14.498246	13.686957	12.947672	12.272511	35
40	18.401584	17.159086	16.046125	15.046297	14.145527	13.331709	12.594409	40
45	19.156347	17.774070	16.547726	15.455832	14.480228	13.605522	12.818629	45
50	19.762008	18.255925	16.931518	15.761861	14.724521	13.800746	12.974812	50
55	20.248021	18.633472	17.225170	15.990543	14.902825	13.939939	13.083602	55
60	20.638022	18.929290	17.449854	16.161428	15.032966	14.039181	13.159381	60
65	20.950979	19.161070	17.621767	16.289123	15.127953	14.109940	13.212165	65
70	21.202112	19.342677	17.753304	16.384544	15.197282	14.160389	13.248933	70
75	21.403634	19.484970	17.853947	16.455848	15.247885	14.196359	13.274543	75
80	21.565345	19.596460	17.930953	16.509131	15.284818	14.222005	13.292383	80
85	21.695110	19.683816	17.989873	16.548947	15.311775	14.240291	13.304809	85
90	21.799241	19.752262	18.034954	16.578699	15.331451	14.253328	13.313464	90
95	21.882800	19.805891	18.069447	16.600932	15.345812	14.262623	13.319493	95
100	21.949853	19.847910	18.095839	16.617546	15.356293	14.269251	13.323693	100

TABLE 8 (continued)

N \ i	8%	9%	10%	11%	12%	13%	14%	N
1	.925926	.917431	.909091	.900901	.892857	.884956	.677193	1
2	1.783265	1.759111	1.735537	1.712523	1.690051	1.668102	1.646661	2
3	2.577097	2.531295	2.486852	2.443715	2.401831	2.361153	2.321632	3
4	3.312127	3.239720	3.169865	3.102446	3.037349	2.974471	2.913712	4
5	3.992710	3.889651	3.790787	3.695897	3.604776	3.517231	3.433081	5
6	4.622880	4.485919	4.355261	4.230538	4.111407	3.997550	3.888668	6
7	5.206370	5.032953	4.868419	4.712196	4.563757	4.422610	4.288305	7
8	5.746639	5.534819	5.334926	5.146123	4.967640	4.798770	4.638864	8
9	6.246888	5.995247	5.759024	5.537048	5.328250	5.131655	4.946372	9
10	6.710081	6.417658	6.144567	5.889232	5.650223	5.426243	5.216116	10
11	7.138964	6.805191	6.495061	6.206515	5.937699	5.686941	5.452733	11
12	7.536078	7.160725	6.813692	6.492356	6.194374	5.917647	5.660292	12
13	7.903776	7.486904	7.103356	6.749870	6.423548	6.121812	5.842362	13
14	8.244237	7.786150	7.366687	6.981865	6.628168	6.302488	6.002072	14
15	8.559479	8.060688	7.606080	7.190870	6.810864	6.462379	6.142168	15
16	8.851369	8.312558	7.823709	7.379162	6.973986	6.603875	6.265060	16
17	9.121638	8.543631	8.021553	7.548794	7.119630	6.729093	6.372859	17
18	9.371887	8.755625	8.201412	7.701617	7.249670	6.839905	6.467420	18
19	9.603599	8.950115	8.364920	7.839294	7.365777	6.937969	6.550369	19
20	9.818147	9.128546	8.513564	7.963328	7.469444	7.024752	6.623131	20
25	10.674776	9.822580	9.077040	8.421745	7.843139	7.329985	6.872927	25
30	11.257783	10.273654	9.426914	8.693793	8.055184	7.495653	7.002664	30
35	11.654568	10.566821	9.644159	8.855240	8.175504	7.585572	7.070045	35
40	11.924613	10.757360	9.779051	8.951051	8.243777	7.634376	7.105041	40
45	12.108402	10.881197	9.862808	9.007910	8.282516	7.660864	7.123217	45
50	12.233485	10.961683	9.914814	9.041653	8.304498	7.675242	7.132656	50
55	12.318614	11.013993	9.947106	9.061678	8.316972	7.683045	7.137559	55
60	12.376552	11.047991	9.967157	9.073562	8.324049	7.687280	7.140106	60
65	12.415983	11.070087	9.979607	9.080614	8.328065	7.689579	7.141428	65
70	12.442820	11.084449	9.987338	9.084800	8.330344	7.690827	7.142115	70
75	12.461084	11.093782	9.992138	9.087283	8.331637	7.691504	7.142472	75
80	12.473514	11.099849	9.995118	9.088757	8.332371	7.691871	7.142657	80
85	12.481974	11.103791	9.996969	9.089632	8.332787	7.692071	7.142753	85
90	12.487732	11.106354	9.998118	9.090151	8.333023	7.692179	7.142803	90
95	12.491651	11.108019	9.998831	9.090459	8.333157	7.692238	7.142829	95
100	12.494318	11.109102	9.999274	9.090642	8.333234	7.692270	7.142843	100

N \ i	15%	16%	17%	18%	19%	20%	N
1	.869565	.862069	.854701	.847458	.840336	.833333	1
2	1.625709	1.605232	1.585214	1.565642	1.546501	1.527778	2
3	2.283225	2.245890	2.209585	2.174273	2.139917	2.106481	3
4	2.854978	2.798181	2.743235	2.690062	2.638586	2.588735	4
5	3.352155	3.274294	3.199346	3.127171	3.057635	2.990612	5
6	3.784483	3.684736	3.589185	3.497603	3.409777	3.325510	6
7	4.160420	4.038565	3.922380	3.811528	3.705695	3.604592	7
8	4.487322	4.343591	4.207163	4.077566	3.954366	3.837160	8
9	4.771584	4.606544	4.450566	4.303022	4.163332	4.030967	9
10	5.018769	4.833227	4.658604	4.494086	4.338935	4.192472	10
11	5.233712	5.028644	4.836413	4.656005	4.486500	4.327060	11
12	5.420619	5.197107	4.988387	4.793225	4.610504	4.439217	12
13	5.583147	5.342334	5.118280	4.909513	4.714709	4.532681	13
14	5.724476	5.467529	5.229299	5.008062	4.802277	4.610567	14
15	5.847370	5.575456	5.324187	5.091578	4.875863	4.675473	15
16	5.954235	5.668497	5.405288	5.162354	4.937700	4.729561	16
17	6.047161	5.748704	5.474605	5.222334	4.989664	4.774634	17
18	6.127966	5.817848	5.533851	5.273164	5.033331	4.812195	18
19	6.198231	5.877455	5.584488	5.316241	5.070026	4.843496	19
20	6.259331	5.928841	5.627767	5.352746	5.100862	4.869580	20
25	6.464149	6.097092	5.766234	5.466906	5.195148	4.947587	25
30	6.565980	6.177198	5.829390	5.516806	5.234658	4.978936	30
35	6.616607	6.215338	5.858196	5.538618	5.251215	4.991535	35
40	6.641778	6.233497	5.871335	5.548152	5.258153	4.996598	40
45	6.654293	6.242143	5.877327	5.552319	5.261061	4.998633	45
50	6.660515	6.246259	5.880061	5.554141	5.262279	4.999451	50
55	6.663608	6.248219	5.881307	5.554937	5.262790	4.999779	55
60	6.665146	6.249152	5.881876	5.555285	5.263004	4.999911	60
65	6.665911	6.249596	5.882135	5.555437	5.263093	4.999964	65
70	6.666291	6.249808	5.882254	5.555504	5.263131	4.999986	70
75	6.666480	6.249908	5.882308	5.555533	5.263147	4.999994	75
80	6.666574	6.249956	5.882332	5.555546	5.263153	4.999998	80
85	6.666620	6.249979	5.882344	5.555551	5.263156	4.999999	85
90	6.666644	6.249990	5.882349	5.555554	5.263157	5.000000	90
95	6.666655	6.249995	5.882351	5.555555	5.263158	5.000000	95
100	6.666661	6.249998	5.882352	5.555555	5.263158	5.000000	100

TABLE 9 Mortality Table Based on 100,000 Persons Living at Age 0. l_x = *Number of Living;* d_x = *Number of Deaths;* p_x = *Probability of Living;* q_x = *Probability of Dying, for Age x from 0 to 99*

x	l_x	d_x	p_x	q_x	x	l_x	d_x	p_x	q_x
0	100,000	708	.9929	.0071	50	87,624	729	.9917	.0083
1	99,292	175	.9982	.0018	51	86,895	792	.9909	.0091
2	99,117	151	.9985	.0015	52	86,103	858	.9900	.0100
3	98,966	144	.9986	.0015	53	85,245	928	.9891	.0109
4	98,822	138	.9986	.0014	54	84,317	1,003	.9881	.0119
5	98,684	133	.9987	.0014	55	83,314	1,083	.9870	.0130
6	98,551	128	.9987	.0013	56	82,231	1,168	.9858	.0142
7	98,423	124	.9987	.0013	57	81,063	1,260	.9845	.0156
8	98,299	121	.9988	.0012	58	79,803	1,357	.9830	.0170
9	98,178	119	.9988	.0012	59	78,446	1,458	.9814	.0186
10	98,059	119	.9988	.0012	60	76,988	1,566	.9797	.0204
11	97,940	120	.9988	.0012	61	75,422	1,677	.9778	.0222
12	97,820	123	.9988	.0013	62	73,745	1,793	.9757	.0243
13	97,697	129	.9987	.0013	63	71,952	1,912	.9734	.0266
14	97,568	136	.9986	.0014	64	70,040	2,034	.9710	.0291
15	97,432	142	.9986	.0015	65	68,006	2,159	.9683	.0318
16	97,290	150	.9985	.0016	66	65,847	2,287	.9653	.0347
17	97,140	157	.9984	.0016	67	63,560	2,418	.9620	.0381
18	96,983	164	.9983	.0017	68	61,142	2,548	.9583	.0417
19	96,819	168	.9983	.0017	69	58,594	2,672	.9544	.0456
20	96,651	173	.9982	.0018	70	55,922	2,784	.9502	.0498
21	96,478	177	.9982	.0018	71	53,138	2,877	.9459	.0542
22	96,301	179	.9982	.0019	72	50,261	2,948	.9414	.0587
23	96,122	182	.9981	.0019	73	47,313	2,993	.9368	.0633
24	95,940	183	.9981	.0019	74	44,320	3,019	.9319	.0681
25	95,757	185	.9981	.0019	75	41,301	3,030	.9266	.0734
26	95,572	187	.9981	.0020	76	38,271	3,030	.9208	.0792
27	95,385	190	.9980	.0020	77	35,241	3,020	.9143	.0857
28	95,195	193	.9980	.0020	78	32,221	2,998	.9070	.0931
29	95,002	198	.9979	.0021	79	29,223	2,957	.8988	.1012
30	94,804	202	.9979	.0021	80	26,266	2,888	.8901	.1100
31	94,602	207	.9978	.0022	81	23,378	2,790	.8807	.1194
32	94,395	212	.9978	.0023	82	20,588	2,659	.8709	.1292
33	94,183	218	.9977	.0023	83	17,929	2,499	.8606	.1394
34	93,965	226	.9976	.0024	84	15,430	2,314	.8500	.1500
35	93,739	235	.9975	.0025	85	13,116	2,113	.8389	.1611
36	93,504	247	.9974	.0027	86	11,003	1,901	.8272	.1728
37	93,257	261	.9972	.0028	87	9,102	1,685	.8149	.1851
38	92,996	280	.9970	.0030	88	7,417	1,470	.8018	.1982
39	92,716	301	.9968	.0033	89	5,947	1,263	.7876	.2124
40	92,415	326	.9965	.0035	90	4,684	1,068	.7720	.2280
41	92,089	354	.9962	.0039	91	3,616	888	.7544	.2456
42	91,735	383	.9958	.0042	92	2,728	725	.7342	.2658
43	91,352	414	.9955	.0045	93	2,003	579	.7109	.2891
44	90,938	447	.9951	.0049	94	1,424	450	.6840	.3160
45	90,491	484	.9947	.0054	95	974	341	.6499	.3501
46	90,007	525	.9942	.0058	96	633	253	.6003	.3997
47	89,482	569	.9937	.0064	97	380	185	.5132	.4869
48	88,913	618	.9931	.0070	98	195	129	.3385	.6615
49	88,295	671	.9924	.0076	99	66	66	.0000	1.0000

Based on the 1958 CSO Mortality Table prepared in cooperation with the National Association of Insurance Commissioners. Courtesy of the Society of Actuaries, Chicago, Illinois.

TABLE 10 Powers of e

x	e^x	e^{-x}	x	e^x	e^{-x}	x	e^x	e^{-x}
.00	1.000	1.000	.50	1.649	.607	1.00	2.718	.368
.01	1.010	.990	.51	1.665	.600	1.01	2.746	.364
.02	1.020	.980	.52	1.682	.595	1.02	2.773	.361
.03	1.031	.970	.53	1.699	.589	1.03	2.801	.357
.04	1.041	.961	.54	1.716	.583	1.04	2.829	.353
.05	1.051	.951	.55	1.733	.577	1.05	2.858	.350
.06	1.062	.942	.56	1.751	.571	1.06	2.886	.346
.07	1.073	.932	.57	1.768	.566	1.07	2.915	.343
.08	1.083	.923	.58	1.786	.560	1.08	2.945	.340
.09	1.094	.914	.59	1.804	.554	1.09	2.974	.336
.10	1.105	.905	.60	1.822	.549	1.10	3.004	.333
.11	1.116	.896	.61	1.840	.543	1.11	3.034	.330
.12	1.127	.887	.62	1.859	.538	1.12	3.065	.326
.13	1.139	.878	.63	1.878	.533	1.13	3.096	.323
.14	1.150	.869	.64	1.896	.527	1.14	3.127	.320
.15	1.162	.861	.65	1.916	.522	1.15	3.158	.317
.16	1.174	.852	.66	1.935	.517	1.16	3.190	.313
.17	1.185	.844	.67	1.954	.512	1.17	3.222	.310
.18	1.197	.835	.68	1.974	.507	1.18	3.254	.307
.19	1.209	.827	.69	1.994	.502	1.19	3.287	.304
.20	1.221	.819	.70	2.014	.497	1.20	3.320	.301
.21	1.234	.811	.71	2.034	.492	1.21	3.353	.298
.22	1.246	.803	.72	2.054	.487	1.22	3.387	.295
.23	1.259	.795	.73	2.075	.482	1.23	3.421	.292
.24	1.271	.787	.74	2.096	.477	1.24	3.456	.289
.25	1.284	.779	.75	2.117	.472	1.25	3.490	.287
.26	1.297	.771	.76	2.138	.468	1.26	3.525	.284
.27	1.310	.763	.77	2.160	.463	1.27	3.561	.281
.28	1.323	.756	.78	2.182	.458	1.28	3.597	.278
.29	1.336	.748	.79	2.203	.454	1.29	3.633	.275
.30	1.350	.741	.80	2.226	.449	1.30	3.669	.273
.31	1.363	.733	.81	2.248	.445	1.31	3.706	.270
.32	1.377	.726	.82	2.270	.440	1.32	3.743	.267
.33	1.391	.719	.83	2.293	.436	1.33	3.781	.264
.34	1.405	.712	.84	2.316	.432	1.34	3.819	.262
.35	1.419	.705	.85	2.340	.427	1.35	3.857	.259
.36	1.433	.698	.86	2.363	.423	1.36	3.896	.257
.37	1.448	.691	.87	2.387	.419	1.37	3.935	.254
.38	1.462	.684	.88	2.441	.415	1.38	3.975	.252
.39	1.477	.677	.89	2.435	.411	1.39	4.015	.249
.40	1.492	.670	.90	2.460	.407	1.40	4.055	.247
.41	1.507	.664	.91	2.484	.403	1.41	4.096	.244
.42	1.522	.657	.92	2.509	.399	1.42	4.137	.242
.43	1.537	.651	.93	2.535	.395	1.43	4.179	.239
.44	1.553	.644	.94	2.560	.391	1.44	4.221	.237
.45	1.568	.638	.95	2.586	.387	1.45	4.263	.235
.46	1.584	.631	.96	2.612	.383	1.46	4.306	.232
.47	1.600	.625	.97	2.638	.379	1.47	4.349	.230
.48	1.616	.619	.98	2.664	.375	1.48	4.393	.228
.49	1.632	.613	.99	2.691	.372	1.49	4.437	.225

TABLE 10 *(continued)*

x	e^x	e^{-x}	x	e^x	e^{-x}	x	e^x	e^{-x}
1.50	4.482	.223	2.00	7.389	.135	2.50	12.182	.082
1.51	4.527	.221	2.01	7.463	.134	2.51	12.305	.081
1.52	4.572	.219	2.02	7.538	.133	2.52	12.429	.080
1.53	4.618	.217	2.03	7.614	.131	2.53	12.554	.080
1.54	4.665	.214	2.04	7.691	.130	.254	12.680	.079
1.55	4.712	.212	2.05	7.768	.129	2.55	12.807	.078
1.56	4.759	.210	2.06	7.846	.127	2.56	12.936	.077
1.57	4.807	.208	2.07	7.925	.126	2.57	13.066	.077
1.58	4.855	.206	2.08	8.004	.125	2.58	13.197	.076
1.59	4.904	.204	2.09	8.085	.124	2.59	13.330	.075
1.60	4.953	.202	2.10	8.166	.122	2.60	13.464	.074
1.61	5.003	.200	2.11	8.248	.121	2.61	13.599	.074
1.62	5.053	.198	2.12	8.331	.120	2.62	13.736	.073
1.63	5.104	.196	2.13	8.415	.119	2.63	13.874	.072
1.64	5.155	.194	2.14	8.499	.118	2.64	14.013	.071
1.65	5.207	.192	2.15	8.585	.116	2.65	14.154	.071
1.66	5.259	.190	2.16	8.671	.115	2.66	14.296	.070
1.67	5.312	.188	2.17	8.758	.114	2.67	14.440	.069
1.68	5.366	.186	2.18	8.846	.113	2.68	14.585	.069
1.69	5.420	.185	2.19	8.935	.112	2.69	14.732	.068
1.70	5.474	.183	2.20	9.025	.111	2.70	14.880	.067
1.71	5.529	.181	2.21	9.116	.110	2.71	15.029	.067
1.72	5.585	.179	2.22	9.207	.109	2.72	15.180	.066
1.73	5.641	.177	2.23	9.300	.108	2.73	15.333	.065
1.74	5.697	.176	2.24	9.393	.106	2.74	15.487	.065
1.75	5.755	.174	2.25	9.488	.105	2.75	15.643	.064
1.76	5.812	.172	2.26	9.583	.104	2.76	15.800	.063
1.77	5.871	.170	2.27	9.679	.103	2.77	15.959	.063
1.78	5.930	.169	2.28	9.777	.102	2.78	16.119	.062
1.79	5.989	.167	2.29	9.875	.101	2.79	16.281	.061
1.80	6.050	.165	2.30	9.974	.100	2.80	16.445	.061
1.81	6.110	.164	2.31	10.074	.099	2.81	16.610	.060
1.82	6.172	.162	2.32	10.176	.098	2.82	16.777	.060
1.83	6.234	.160	2.33	10.278	.097	2.83	16.945	.059
1.84	6.297	.159	2.34	10.381	.096	2.84	17.116	.058
1.85	6.360	.157	2.35	10.486	.095	2.85	17.288	.058
1.86	6.424	.156	2.36	10.591	.094	2.86	17.462	.057
1.87	6.488	.154	2.37	10.697	.093	2.87	17.637	.057
1.88	6.553	.153	2.38	10.805	.093	2.88	17.814	.056
1.89	6.619	.151	2.39	10.913	.092	2.89	17.993	.056
1.90	6.686	.150	2.40	11.023	.091	2.90	18.174	.055
1.91	6.753	.148	2.41	11.134	.090	2.91	18.357	.054
1.92	6.821	.147	2.42	11.246	.089	2.92	18.541	.054
1.93	6.890	.145	2.43	11.359	.088	2.93	18.728	.053
1.94	6.959	.144	2.44	11.473	.087	2.94	18.916	.053
1.95	7.029	.142	2.45	11.588	.086	2.95	19.016	.052
1.96	7.099	.141	2.46	11.705	.085	2.96	19.298	.052
1.97	7.171	.139	2.47	11.822	.085	2.97	19.492	.051
1.98	7.243	.138	2.48	11.941	.084	2.98	19.688	.051
1.99	7.316	.137	2.49	12.061	.083	2.99	19.886	.050
						3.00	20.086	.050

N	0	1	2	3	4	5	6	7	8	9
1.0	.0000	.0043	.0086	.0128	.0170	.0212	.0253	.0294	.0334	.0374
1.1	.0414	.0453	.0492	.0531	.0569	.0607	.0645	.0682	.0719	.0755
1.2	.0792	.0828	.0864	.0899	.0934	.0969	.1004	.1038	.1072	.1106
1.3	.1139	.1173	.1206	.1239	.1271	.1303	.1335	.1367	.1399	.1430
1.4	.1461	.1492	.1523	.1553	.1584	.1614	.1644	.1673	.1703	.1732
1.5	.1761	.1790	.1818	.1847	.1875	.1903	.1931	.1959	.1987	.2014
1.6	.2041	.2068	.2095	.2122	.2148	.2175	.2201	.2227	.2253	.2279
1.7	.2304	.2330	.2355	.2380	.2405	.2430	.2455	.2480	.2504	.2529
1.8	.2553	.2577	.2601	.2625	.2648	.2672	.2695	.2718	.2742	.2765
1.9	.2788	.2810	.2833	.2856	.2878	.2900	.2923	.2945	.2967	.2989
2.0	.3010	.3032	.3054	.3075	.3096	.3118	.3139	.3160	.3181	.3201
2.1	.3222	.3243	.3263	.3284	.3304	.3324	.3345	.3365	.3385	.3404
2.2	.3424	.3444	.3464	.3483	.3502	.3522	.3541	.3560	.3579	.3598
2.3	.3617	.3636	.3655	.3674	.3692	.3711	.3729	.3747	.3766	.3784
2.4	.3802	.3820	.3838	.3856	.3874	.3892	.3909	.3927	.3945	.3962
2.5	.3979	.3997	.4014	.4031	.4048	.4065	.4082	.4099	.4116	.4133
2.6	.4150	.4166	.4183	.4200	.4216	.4232	.4249	.4265	.4281	.4298
2.7	.4314	.4330	.4346	.4362	.4378	.4393	.4409	.4425	.4440	.4456
2.8	.4472	.4487	.4502	.4518	.4533	.4548	.4564	.4579	.4594	.4609
2.9	.4624	.4639	.4654	.4669	.4683	.4698	.4713	.4728	.4742	.4757
3.0	.4771	.4786	.4800	.4814	.4829	.4843	.4857	.4871	.4886	.4900
3.1	.4914	.4928	.4942	.4955	.4969	.4983	.4997	.5011	.5024	.5038
3.2	.5051	.5065	.5079	.5092	.5105	.5119	.5132	.5145	.5159	.5172
3.3	.5185	.5198	.5211	.5224	.5237	.5250	.5263	.5276	.5289	.5302
3.4	.5315	.5328	.5340	.5353	.5366	.5378	.5391	.5403	.5416	.5428
3.5	.5441	.5453	.5465	.5478	.5490	.5502	.5514	.5527	.5539	.5551
3.6	.5563	.5575	.5587	.5599	.5611	.5623	.5635	.5647	.5658	.5670
3.7	.5682	.5694	.5705	.5717	.5729	.5740	.5752	.5763	.5775	.5786
3.8	.5798	.5809	.5821	.5832	.5843	.5855	.5866	.5877	.5888	.5899
3.9	.5911	.5922	.5933	.5944	.5955	.5966	.5977	.5988	.5999	.6010
4.0	.6021	.6031	.6042	.6053	.6064	.6075	.6085	.6096	.6107	.6117
4.1	.6128	.6138	.6149	.6160	.6170	.6180	.6191	.6201	.6212	.6222
4.2	.6232	.6243	.6253	.6263	.6274	.6284	.6294	.6304	.6314	.6325
4.3	.6335	.6345	.6355	.6365	.6375	.6385	.6395	.6405	.6415	.6425
4.4	.6435	.6444	.6454	.6464	.6474	.6484	.6493	.6503	.6513	.6522
4.5	.6532	.6542	.6551	.6561	.6571	.6580	.6590	.6599	.6609	.6618
4.6	.6628	.6637	.6646	.6656	.6665	.6675	.6684	.6693	.6702	.6712
4.7	.6721	.6730	.6739	.6749	.6758	.6767	.6776	.6785	.6794	.6803
4.8	.6812	.6821	.6830	.6839	.6848	.6857	.6866	.6875	.6884	.6893
4.9	.6902	.6911	.6920	.6928	.6937	.6946	.6955	.6964	.6972	.6981
5.0	.6990	.6998	.7007	.7016	.7024	.7033	.7042	.7050	.7059	.7067
5.1	.7076	.7084	.7093	.7101	.7110	.7118	.7126	.7135	.7143	.7152
5.2	.7160	.7168	.7177	.7185	.7193	.7202	.7210	.7218	.7226	.7235
5.3	.7243	.7251	.7259	.7267	.7275	.7284	.7292	.7300	.7308	.7316
5.4	.7324	.7332	.7340	.7348	.7356	.7364	.7273	.7380	.7388	.7396
N	0	1	2	3	4	5	6	7	8	9

N	0	1	2	3	4	5	6	7	8	9
5.5	.7404	.7412	.7419	.7427	.7435	.7443	.7451	.7459	.7466	.7474
5.6	.7482	.7490	.7497	.7505	.7513	.7520	.7528	.7536	.7543	.7551
5.7	.7559	.7566	.7574	.7582	.7589	.7597	.7604	.7612	.7619	.7627
5.8	.7634	.7642	.7649	.7657	.7664	.7672	.7679	.7686	.7694	.7701
5.9	.7709	.7716	.7723	.7731	.7738	.7745	.7752	.7760	.7767	.7774
6.0	.7782	.7789	.7796	.7803	.7810	.7818	.7825	.7832	.7839	.7846
6.1	.7853	.7860	.7868	.7875	.7882	.7889	.7896	.7903	.7910	.7917
6.2	.7924	.7931	.7938	.7945	.7952	.7959	.7966	.7973	.7980	.7987
6.3	.7993	.8000	.8007	.8014	.8021	.8028	.8035	.8041	.8048	.8055
6.4	.8062	.8069	.8075	.8082	.8089	.8096	.8102	.8109	.8116	.8122
6.5	.8129	.8136	.8142	.8149	.8156	.8162	.8169	.8176	.8182	.8189
6.6	.8195	.8202	.8209	.8215	.8222	.8228	.8235	.8241	.8248	.8254
6.7	.8261	.8267	.8274	.8280	.8287	.8293	.8299	.8306	.8312	.8319
6.8	.8325	.8331	.8338	.8344	.8351	.8357	.8363	.8370	.8376	.8382
6.9	.8388	.8395	.8401	.8407	.8414	.8420	.8426	.8432	.8439	.8445
7.0	.8451	.8457	.8463	.8470	.8476	.8482	.8488	.8494	.8500	.8506
7.1	.8513	.8519	.8525	.8531	.8537	.8543	.8549	.8555	.8561	.8567
7.2	.8573	.8579	.8585	.8591	.8597	.8603	.8609	.8615	.8621	.8627
7.3	.8633	.8639	.8645	.8651	.8657	.8663	.8669	.8675	.8681	.8686
7.4	.8692	.8698	.8704	.8710	.8716	.8722	.8727	.8733	.8739	.8745
7.5	.8751	.8756	.8762	.8768	.8774	.8779	.8785	.8791	.8797	.8802
7.6	.8808	.8814	.8820	.8825	.8831	.8837	.8842	.8848	.8854	.8859
7.7	.8865	.8871	.8876	.8882	.8887	.8893	.8899	.8904	.8910	.8915
7.8	.8921	.8927	.8932	.8938	.8943	.8949	.8954	.8960	.8965	.8971
7.9	.8976	.8982	.8987	.8993	.8998	.9004	.9009	.9015	.9020	.9025
8.0	.9031	.9036	.9042	.9047	.9053	.9058	.9063	.9069	.9074	.9079
8.1	.9085	.9090	.9096	.9101	.9106	.9112	.9117	.9122	.9128	.9133
8.2	.9138	.9143	.9149	.9154	.9159	.9165	.9170	.9175	.9180	.9186
8.3	.9191	.9196	.9201	.9206	.9212	.9217	.9222	.9227	.9232	.9238
8.4	.9243	.9248	.9253	.9258	.9263	.9269	.9274	.9279	.9284	.9289
8.5	.9294	.9299	.9304	.9309	.9315	.9320	.9325	.9330	.9335	.9340
8.6	.9345	.9350	.9355	.9360	.9365	.9370	.9375	.9380	.9385	.9390
8.7	.9395	.9400	.9405	.9410	.9415	.9420	.9425	.9430	.9435	.9440
8.8	.9445	.9450	.9455	.9460	.9465	.9469	.9474	.9479	.9484	.9489
8.9	.9494	.9499	.9504	.9509	.9513	.9518	.9523	.9528	.9533	.9538
9.0	.9542	.9547	.9552	.9557	.9562	.9566	.9571	.9576	.9581	.9586
9.1	.9590	.9595	.9600	.9605	.9609	.9614	.9619	.9624	.9628	.9633
9.2	.9638	.9643	.9647	.9652	.9657	.9661	.9666	.9671	.9675	.9680
9.3	.9685	.9689	.9694	.9699	.9703	.9708	.9713	.9717	.9722	.9727
9.4	.9731	.9736	.9741	.9745	.9750	.9754	.9759	.9763	.9768	.9773
9.5	.9777	.9782	.9786	.9791	.9795	.9800	.9805	.9809	.9814	.9818
9.6	.9823	.9827	.9832	.9836	.9841	.9845	.9850	.9854	.9859	.9863
9.7	.9868	.9872	.9877	.9881	.9886	.9890	.9894	.9899	.9903	.9908
9.8	.9912	.9917	.9921	.9926	.9930	.9934	.9939	.9943	.9948	.9952
9.9	.9956	.9961	.9965	.9969	.9974	.9978	.9983	.9987	.9991	.9996
N	0	1	2	3	4	5	6	7	8	9

APPENDIX F
Answers to Odd-Numbered Problems

CHAPTER 1

1.1 Algebra Pretest (page 2)

Part I

1. E **3.** A **5.** C **7.** D **9.** D **11.** C **13.** B **15.** D **17.** F **19.** F **21.** F **23.** F **25.** T **27.** T **29.** T

Part II

1. B **3.** C **5.** C **7.** A **9.** B **11.** D **13.** C **15.** A **17.** D

1.2 Real Numbers (page 7)

1. a. I, Q, R **b.** Q', R **3. a.** N, W, I, Q, R **b.** N, W, I, Q, R **5. a.** Q, R **b.** Q, R **7. a.** Q, R **b.** Q', R **9. a.** Undefined. **b.** W, I, Q, R **11.** Since all prime numbers are natural numbers P is drawn inside N. **13.** All even numbers are whole numbers but all are not natural numbers (0 is even but not natural); so E is drawn inside W but not entirely inside N. **15.** False **17.** False **19.** True **21.** True **23.** True **25.** False **27.** True **29.** True **31.** True **33.** False **35.** False **37.** False **39.** True **41.** 9 **43.** 19 **45.** $-\pi$ **47.** $\pi - 2$ **49.** $10 - \pi$ **51.** $\sqrt{20} - 4$ **53.** $8 - \sqrt{50}$ **55.** $7 - 2\pi$ **57.** $|x + 3| = -(x + 3)$ **59.** $|y - 5| = y - 5$ **61.** Therefore, $|4 + 3t| = -(4 + 3t)$ **63.** 10 **65.** 11 **67.** $\pi - 2$ **69.** $2 - \sqrt{3}$

1.3 Algebraic Expressions (page 13)

1. 13 **3.** 11 **5.** 36 **7.** -25 **9.** -49 **11.** 33 **13.** 6 **15.** 9 **17.** 4 **19.** 1 **21.** 0 **23.** 42 **25.** -71 **27.** -1 **29. a.** $x^2 + 4x - 12$ **b.** $x^2 + x - 20$ **31. a.** $x^2 - 8x + 15$ **b.** $x^2 - x - 12$ **33. a.** $2y^2 - y - 1$ **b.** $2y^2 - 5y + 3$ **35. a.** $6y^2 + 5y - 6$ **b.** $6y^2 + 13y + 6$ **37. a.** $x^2 - y^2$ **b.** $a^2 - b^2$ **39. a.** $x^2 + 4x + 4$ **b.** $x^2 - 4x + 4$ **41. a.** $a^2 + 2ab + b^2$ **b.** $a^2 - 2ab + b^2$ **43.** $3x - 4z$ **45.** $4x - 2y + z$ **47.** $-2y$ **49.** $2x + 2y$ **51.** $-x + 4y - 9$ **53.** $4x^2 - 8x - 1$ **55.** $-3x^2 + 3x - 9$ **57.** $2x^2 - x - 3$ **59.** $3x^2 - 4x - 11$ **61.** $x^3 + 3x^2 + 3x + 1$ **63.** $x^3 - x^2 - 4x - 2$ **65.** $3x^2 - 10x + 1$ **67.** $15x^3 - 22x^2 + 5x + 2$

1.4 Factoring (page 19)

1. $4x(5y - 3)$ **3.** $2(3x - 1)$ **5.** $x(y + z^2 + 3)$ **7.** $a^2 + b^2$ does not factor. **9.** $(m - n)^2$ **11.** $x^{-3}(1 + x^2 + x^5)$ **13.** $(4x - 1)(x + 3)$ **15.** $(2x - 3)(x + 5)$ **17.** $(3x + 1)(x - 2)$ **19.** $2x^{-3}(x - 8)(x + 3)$ **21.** $x^{-2}(2x - 3)(x - 1)$ **23.** $x^{1/3}(x^2 + x + 1)$ **25.** $(x - y - 1)(x - y + 1)$ **27.** $x^{-1/2}(x + 3)^2$ **29.** $(a + b - x - y)(a + b + x + y)$ **31.** $(2x - 3)(x + 2)$ **33.** $(6x - 1)(x + 8)$ **35.** $(6x + 1)(x + 8)$ **37.** $(4x - 3)(x + 4)$ **39.** $(9x - 2)(x - 6)$ **41.** $-(x - 1)(x + 1)(x - 3)(x + 3)$ **43.** $\frac{1}{36}(2x - 1)(2x + 1)(3x - 1)(3x + 1)$ **45.** $(x^2 - 3x - 8)(x - 4)(x + 1)$ **47.** $(2x + 2y + a + b)(x + y - 3a - 3b)$ **49.** $18x^2(x - 5)^2(2x - 5)$ **51.** $5x(3x + 1)^2(15x + 2)$ **53.** $2x(4x + 3)^2(10x + 3)$ **55.** $(x + 1)^2(x - 2)^3(7x - 2)$ **57.** $6(2x - 1)^2(3x + 2)(5x + 1)$ **59.** $2(x + 5)^2(x^2 - 2)^2(5x^2 + 15x - 4)$

1.5 Linear Equations and Inequalities (page 24)

1. $x = 2$ **3.** $x = -5$ **5.** $x = -3$ **7.** $x = -3$ **9.** $x = -2$ **11.** $x = 2$ **13.** $x = 3$ **15.** $x = 2$ **17.** $x \leq 3$ or $(-\infty, 3]$ **19.** $x < -5$ or $(-\infty, -5)$ **21.** $x \geq -4$ or $[-4, \infty)$ **23.** $x > -3$ or $(-3, \infty)$ **25.** $5 < x < 9$ or $(5, 9)$ **27.** $-1 < x < 4$ or $(-1, 4)$ **29.** $-7 < x < -4$ or $(-7, -4)$ **31.** $y' = \dfrac{x^4}{y^4}$ **33.** $y' = \dfrac{-2y}{x}$ **35.** $y' = \dfrac{2x - y}{x - 2y}$ **37.** $y' = \dfrac{-3x^2 - 4x - y}{x}$ **39.** $y' = \dfrac{-4(x + 1)}{25(y - 1)}$ **41.** $y' = \dfrac{3y^2 - 6xy^3 - 5y}{9x^2y^2 - 6xy + 5x}$ **43.** $x = 3$ **45.** $x < -1$ or $(-\infty, -1)$ **47.** $x \leq 10$ or $(-\infty, 10]$ **49.** $x < -2$ **51.** $x \leq 1$ or $(-\infty, 1]$ **53.** $x = 1$ **55.** No solution **57.** $x = 0$ **59.** Any real number is a solution.

1.6 Quadratic Equations and Inequalities (page 30)

1. $x = -5$ or $x = 3$ **3.** $x = -9$ or $x = 2$ **5.** $x = -\frac{1}{2}$ or $x = \frac{4}{5}$ **7.** $x = 1$ or $x = -6$ **9.** $x = 5$
11. $x = -\frac{2}{3}$ or $x = \frac{1}{4}$ **13.** No real solutions **15.** $x = \pm\frac{\sqrt{5}}{2}$ **17.** $x = 0$ or $x = \frac{7}{3}$ **19.** $x = -\frac{1}{3}$ or $x = 2$
21. $[2, 6]$ **23.** $(-3, 0)$ **25.** $(-\infty, -2] \cup [8, \infty)$ **27.** $(-\infty, \frac{1}{3}) \cup (4, \infty)$ **29.** $[-\infty, -3] \cup [3, \infty)$
31. There is no real solution. **33.** $(-\infty, -2) \cup (3, \infty)$ **35.** $[2, 3]$ **37.** $[-5, 1]$ **39.** $(-\infty, -\frac{1}{6}) \cup (10, \infty)$

1.7 Rational Expressions (page 33)

1. $\frac{2 + 3x + 3y}{x + y}$ **3.** $\frac{x - y}{2}$ **5.** $\frac{3x + 2}{x + 2}$ **7.** $\frac{11}{2(x + y)}$ **9.** 1 **11.** $3x^2 - 17x - 6$ **13.** $\frac{x^2 + 2x + 3}{x^2}$
15. $\frac{3}{x - 1}$ **17.** $\frac{3 + 5x - x^2}{x^2}$ **19.** $\frac{x^2 + 2xy + y^2}{xy} = \frac{(x + y)^2}{xy}$ **21.** $\frac{x^2 - 2xy + y^2}{xy} = \frac{(x - y)^2}{xy}$
23. $\frac{x^2 + 3x^4y^3 + 3y}{3x^3y^2}$ **25.** $\frac{-2}{x + 7}$ **27.** $\frac{1 - x^2}{x^2 + 1} = \frac{(1 - x)(1 + x)}{x^2 + 1}$ **29.** $\frac{4(x - 1)}{(x + 2)(x - 2)}$

1.8 Review (page 34)

Objective Questions

1.

Number	Natural	Whole	Integer	Rational	Irrational	Real
			Set			
-8	No	No	Yes	Yes	No	Yes
$\frac{5}{6}$	No	No	No	Yes	No	Yes
$.\overline{23}$	No	No	No	Yes	No	Yes
$\sqrt{169}$	Yes	Yes	Yes	Yes	Yes	Yes
$\frac{3}{0}$	No	No	No	No	No	No

3. -5 **5.** $4 - x$ **7.** $2\pi + 4$ **9. a.** -64 **b.** 64 **c.** -64 **d.** -1 **11.** 3 **13.** -1 **15. a.** -1
b. 1 **17. a.** $x^2 + 5x - 14$ **b.** $x^2 - 9$ **19. a.** $x^2 + xy - 2y^2$ **b.** $x^2 - 9y^2$ **21.** $3x - 4z$ **23.** $-2x^2 - 7$
25. $(x - 2)(x - 3)$ **27.** $(5 - x)(5 + x)$ **29.** $(3x + 2)(3x - 2)(x + 2)(x - 2)$ **31.** $x = 20$ **33.** $x = 6$
35. $(7, \infty)$ **37.** $(-3, \infty)$ **39.** $[-13, -8]$ **41.** $(-10, -6)$ **43.** $x = \frac{1}{3}, x = -\frac{2}{5}$ **45.** $x = -\frac{5}{3}, -\frac{3}{2}$
47. $3 \pm \sqrt{2}$ **49.** $(5, 7)$ **51.** $(-\infty, -\frac{2}{3}) \cup (\frac{5}{2}, \infty)$ **53.** $\frac{8x - 5y}{60}$ **55.** $\frac{6x + 1}{(3x - 1)^2}$

Sample Test

1.

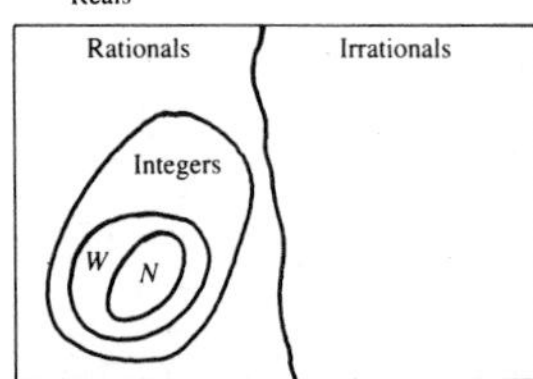

3. $5 - \sqrt{10}$ **5.** -306 **7.** $x^2 + 2xy + y^2$ **9.** $(6x + 1)(x - 5)$
11. $x = 6$ **13.** $(-\infty, -\frac{4}{5})$ **15.** $x = 0$ or $x = \frac{1}{5}$ **17.** $(\frac{2}{3}, 1)$
19. $\frac{14x - 5y}{60}$

CHAPTER 2

2.1 Functions and Graphs Pretest (page 38)

1. A **3.** C **5.** C **7.** B **9.** C **11.** A **13.** A **15.** B **17.** B **19.** B

2.2 Functions (page 45)

1. Answers vary. **3.** Answers vary. **5. a.** y is a function of x, since for each year, x, there will be one and only one closing price, y. **b.** y is not a function of x, since for each closing price, x, we may not be able to uniquely determine the year, y.
7. a. $.29 **b.** $.53 **9. a.** 11 **b.** -11 **11. a.** 7 **b.** 47 **c.** -78 **d.** 497 **13. a.** 6 **b.** -26
c. 34 **d.** -394 **15. a.** 1 **b.** -1 **c.** -1 **d.** 1 **17. a.** All real numbers **b.** All real numbers
19. a. All real numbers except 5 **b.** $[\frac{1}{2}, \infty)$. **21. a.** $5t - 2$ **b.** $5w - 2$ **23. a.** $5t + 5h - 2$ **b.** $5s + 5t - 2$
25. $5t + 5h + 38$ **27.** $2h^2 + 8h + 1$ **29.** $10x^2 - 2$ **31.** $50x^2 - 60x + 11$ **33.** $250x^2 - 100x - 27$
35. $8x^4 - 32x^3 - 16x^2 + 96x + 65$ **37.** 2 **39.** $4x + 2h$ **41.** $4x + 2h$ **43.** $2x + h - 2$ **45. a.** $.92

b. \$.45 **47.** \$1.14 **49. a.** \$.51 **b.** $e(1984) - e(1944)$ **51. a.** \$.03 **b.** The average change per year in the price of gasoline from 1944 to 1984. **53. a.** \$.02 **b.** \$.01 **c.** \$.02 **d.** \$.03 **e.** $\dfrac{s(1944 + h) - s(1944)}{h}$

55. The population will be about 356,500,000. **57.** 208,900,000 **59. a.** $-14{,}800$ **b.** The average change in number of marriages from 1982 to year 1982 + h.

2.3 Graphs (page 50)

1. A picture of the ordered pairs for the function. **3.** The x-intercepts are the points where the graph crosses (or touches) the x-axis. To find the x-intercept, set $y = 0$. The y-intercept is the point where the graph crosses the y-axis. To find the y-intercept, set $x = 0$. **5.** $(2, f(2))$ **7.** $(x_0, f(x_0))$ **9.** $(x_0, g(x_0))$ **11.** $(3, h(3))$ **13.** $(x_0 + t, h(x_0 + t))$ **15.** Domain is $x = -3$; range is $(-\infty, \infty)$; x-intercept $(-3, 0)$; not a function. **17.** Domain is $-2 \le x \le 5$; the range is $-5 \le y \le 3$; x-intercepts are $(-\frac{3}{2}, 0)$ and $(\frac{15}{4}, 0)$; y-intercept is $(0, 3)$; a function. **19.** The domain and the range are all real numbers; x-intercepts are $(-3, 0)$, $(-1, 0)$ and $(2, 0)$; y-intercept is $(0, \frac{5}{2})$; the relation is a function.

21.

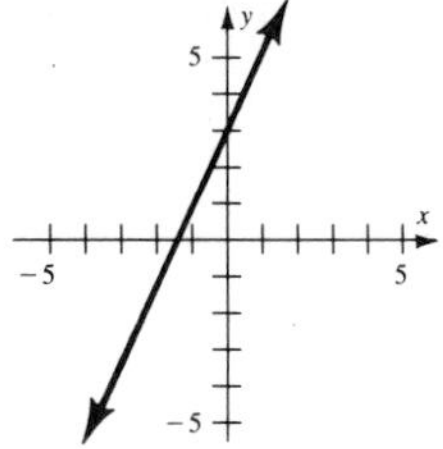

23.

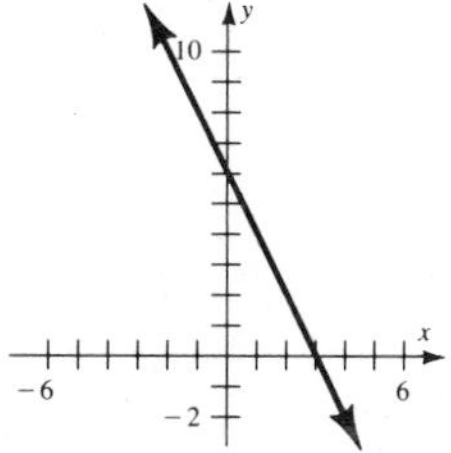

25.

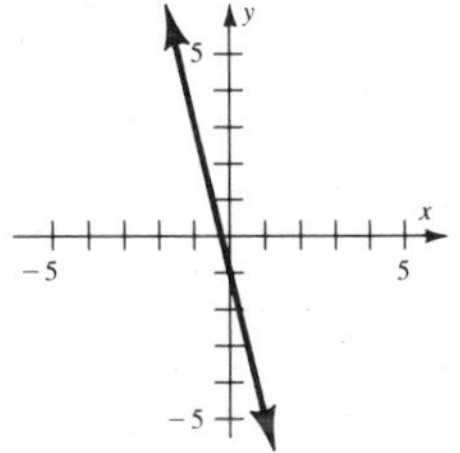

27.

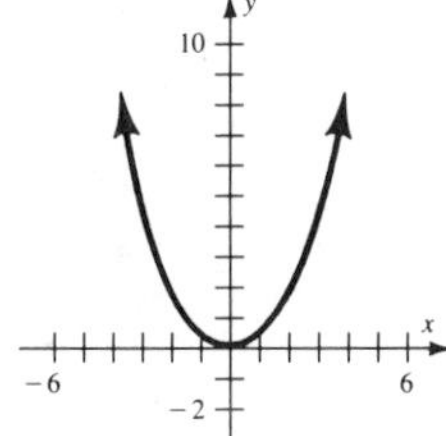

29.

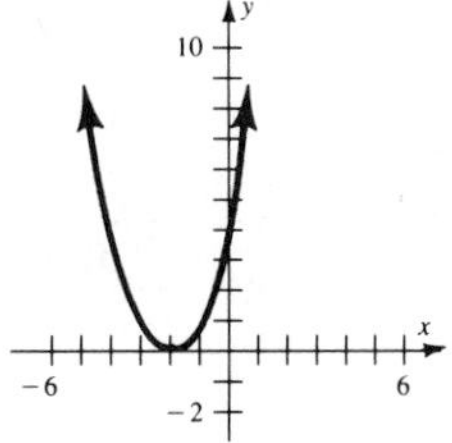

31.

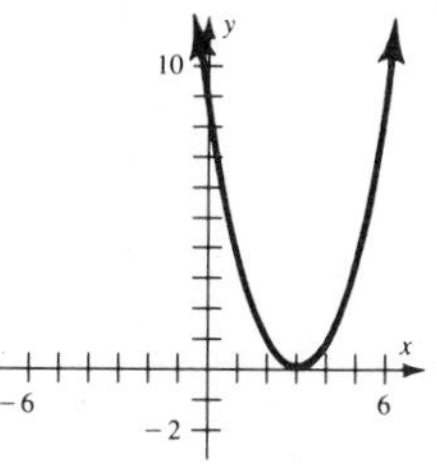

33.

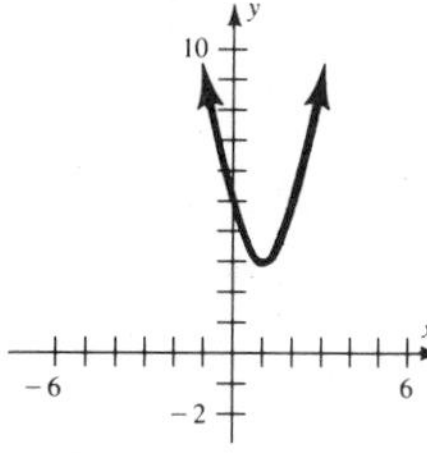

35.

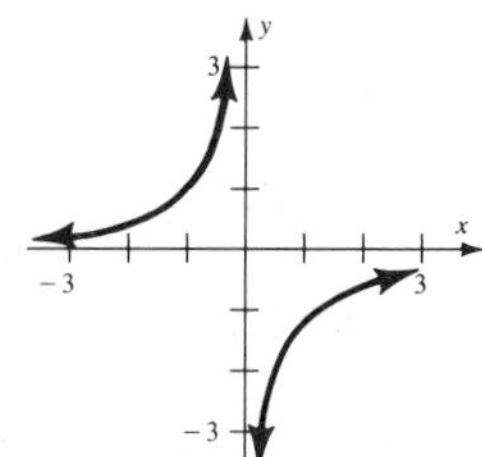

37.

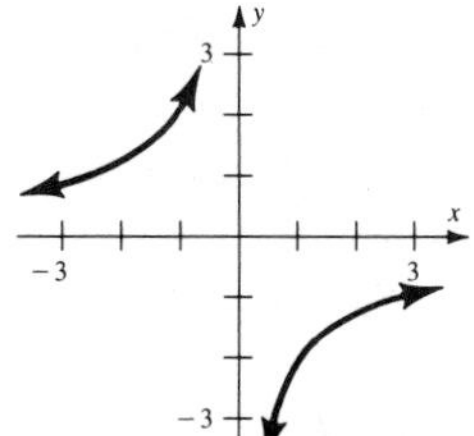

39.

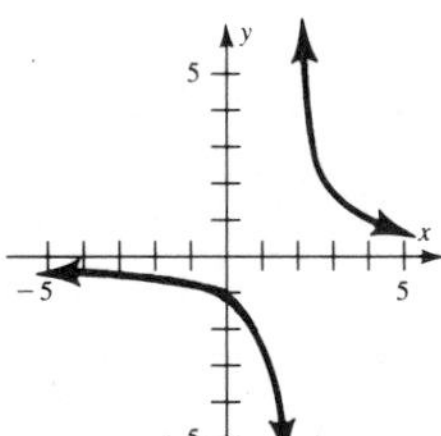

41.

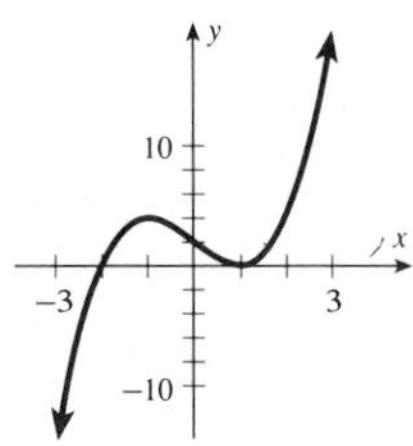

43.

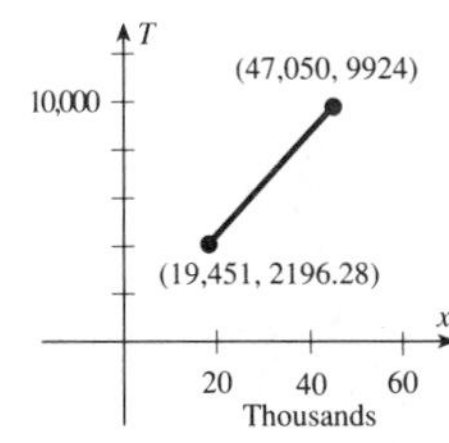

45.

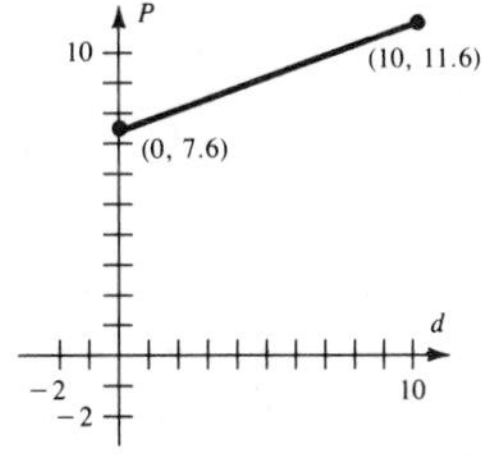

47.

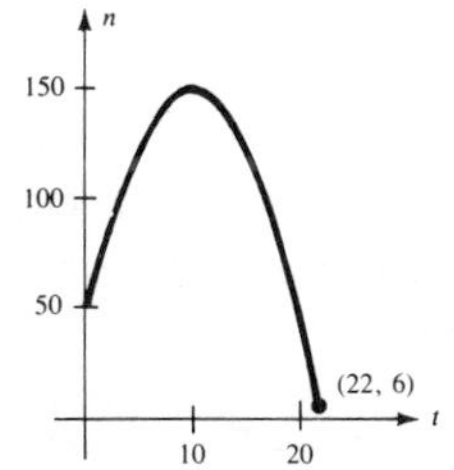

2.4 Linear Functions (page 62)

1. The x-intercept is $(-2, 0)$; the y-intercept is $(0, 4)$. **3.** The x-intercept is $(-1, 0)$; the y-intercept is $(0, -\frac{4}{3})$. **5.** The x-intercept is $(-5, 0)$; the y-intercept is $(0, 2)$. **7.** The y-intercept is $(0, -2)$. **9.** $\frac{1}{3}$ **11.** $\frac{5}{7}$ **13.** 1

15. The slope is $m = 2$; the y-intercept is $b = 4$ or $(0, 4)$. **17.** The slope is $m = 9$; the y-intercept is $b = 1$ or $(0, 1)$.
19. The slope is $m = \frac{2}{3}$; the y-intercept is $b = \frac{5}{3}$ or $(0, \frac{5}{3})$. **21.** The slope is $m = 0$; the y-intercept is $b = 5$ or $(0, 5)$.
23. This is a vertical line; hence it has no slope and no y-intercept.
25. Slope $= 2$; y-intercept is -5

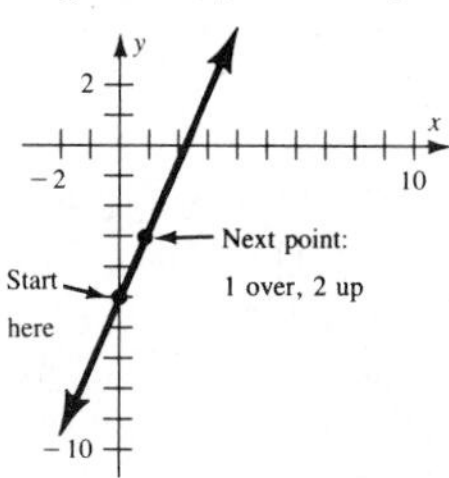

27. Slope $= -\frac{1}{4}$; y-intercept is 2

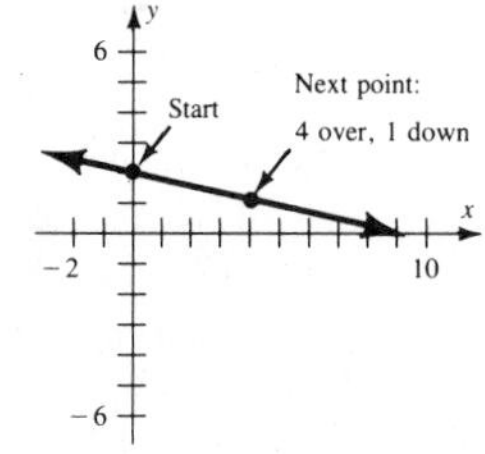

29. Slope $= \frac{3}{5}$; y-intercept is $\frac{2}{5}$

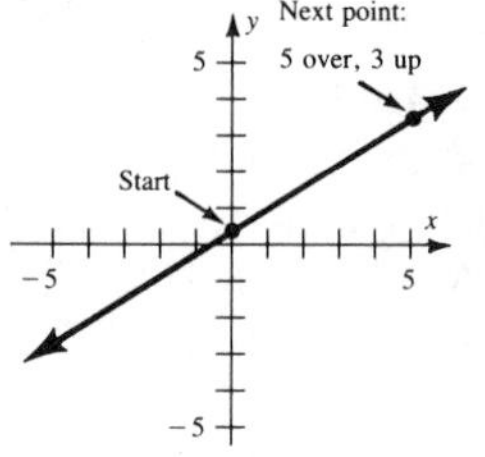

31. Slope is $-\frac{1}{3}$; y-intercept is 3

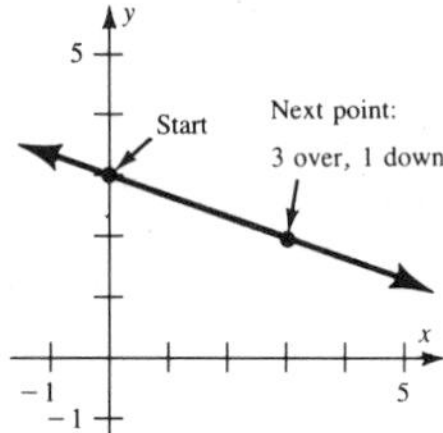

33. Slope $= \frac{2}{3}$; y-intercept is 0

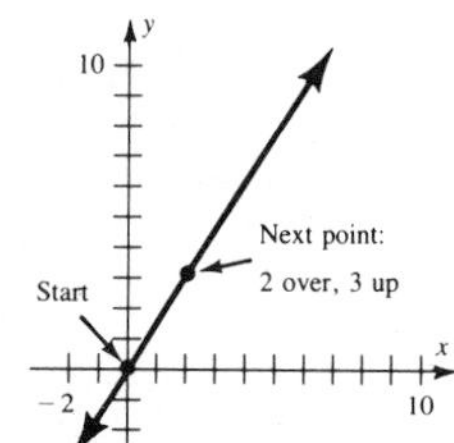

35. (No slope and no y-intercept) $x = -\frac{5}{2}$

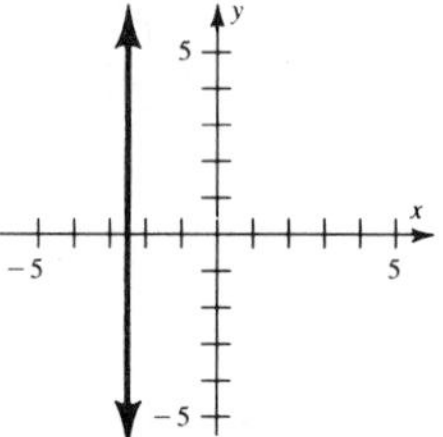

37.

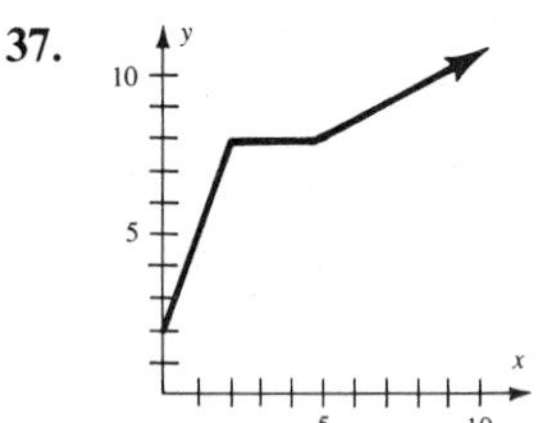

39.

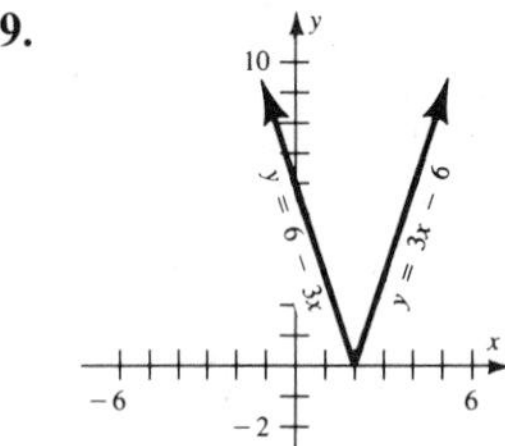

41.

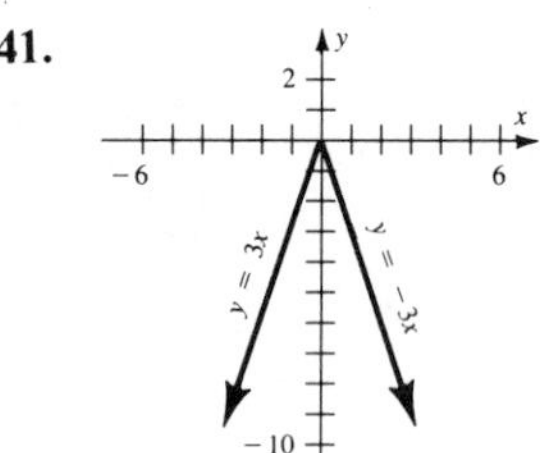

43.

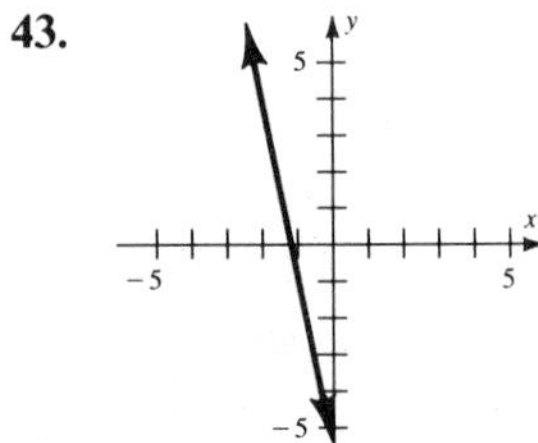

45.

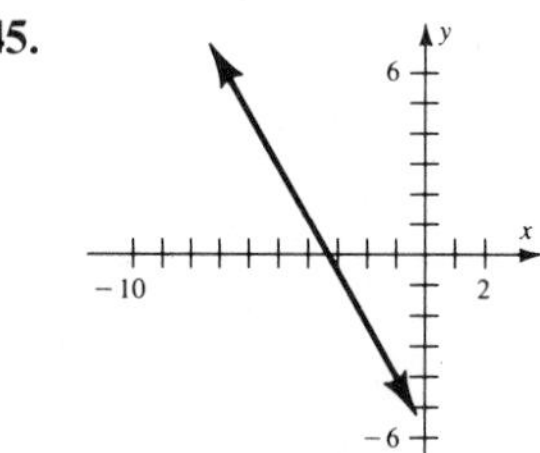

47.

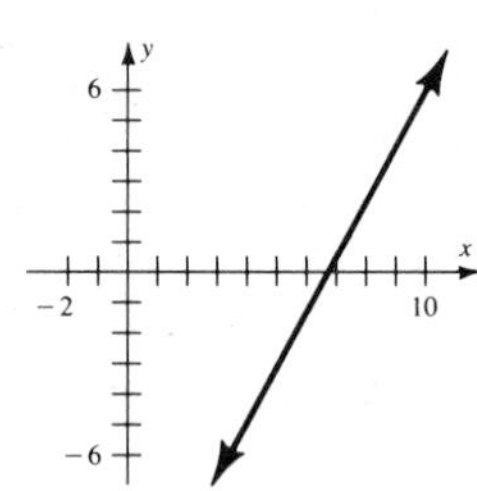

49. $2x + y + 3 = 0$ **51.** $y - 4 = 0$ **53.** $x + y - 2 = 0$ **55.** $3x - 5y - 27 = 0$ **57.** $2x - y - 4 = 0$
59. $y - 6 = 0$ **61.** $2x - y = 0$; four boxes could be supplied. **63.** $3x - 10y + 82 = 0$; 20.2 (million).
65. $y - 60 = 0$ **67.** The expected cost is \$535. **69.** The expected percentage is 63.0%.
71. Let x = amount on Form 1040, line 37; $T = .15x$

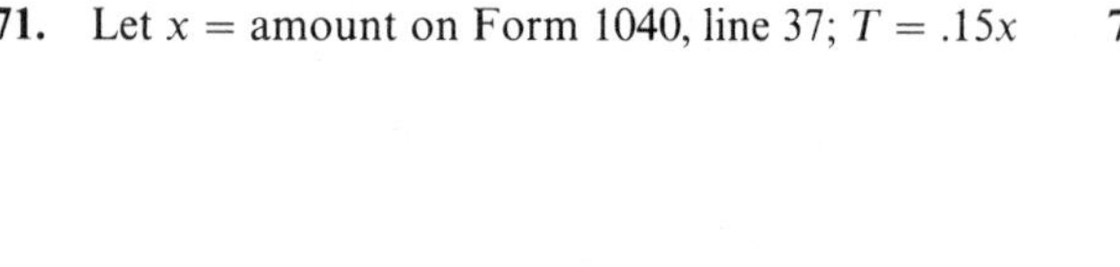

73.

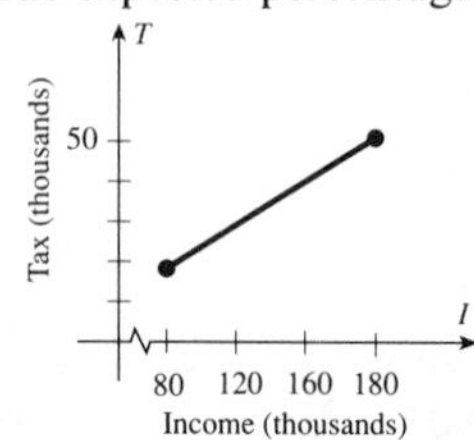

75. Let x = amount on Form 1040, line 37; $T = \begin{cases} .15x & \text{for } 0 < x \le 32{,}450 \\ .28x + 4{,}867.50 & \text{for } 32{,}450 < x \le 78{,}401 \\ .33x + 17{,}733.50 & \text{for } 78{,}401 < x \le 185{,}730 \end{cases}$

2.5 *Quadratic and Polynomial Functions* (page 72)

1.

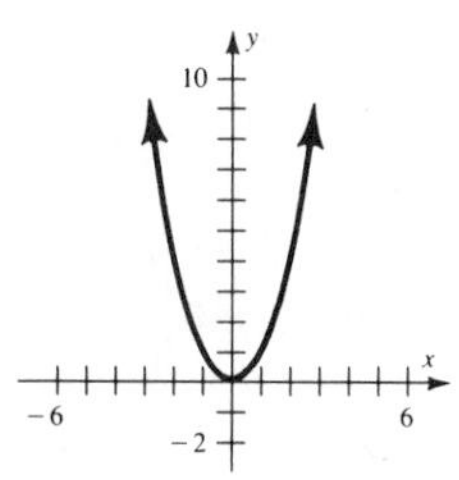

3.

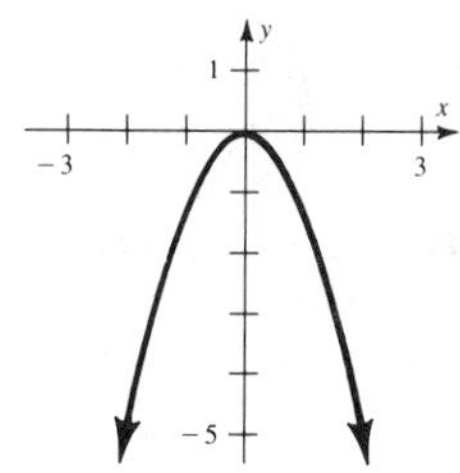

5.

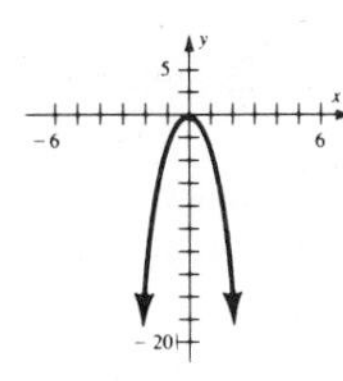

7.

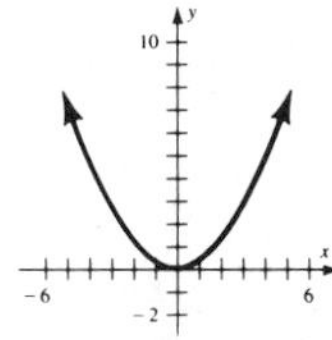

9.

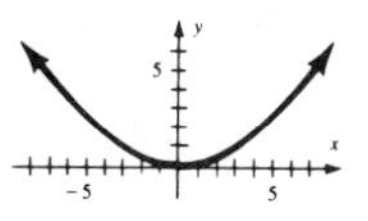

11.

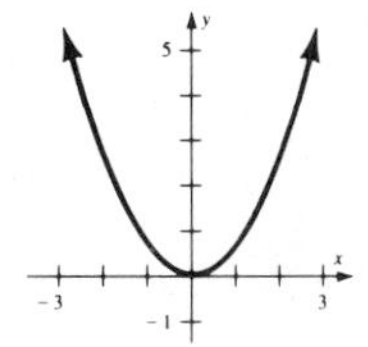

13. Vertex is $(1, 0)$.

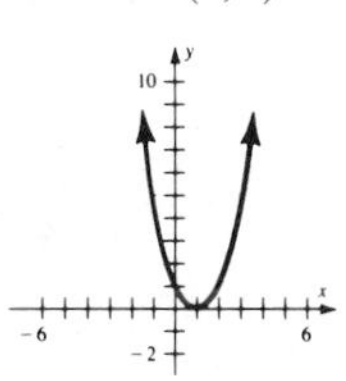

15. Vertex is $(-3, 0)$.

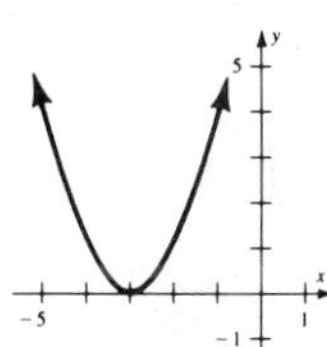

17. Vertex is $(1, 0)$.

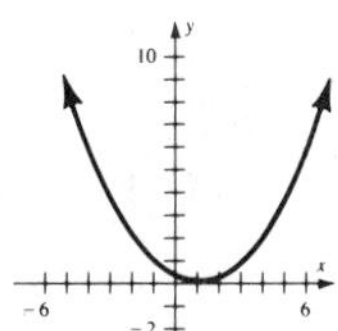

19. Vertex is $(1, 2)$.

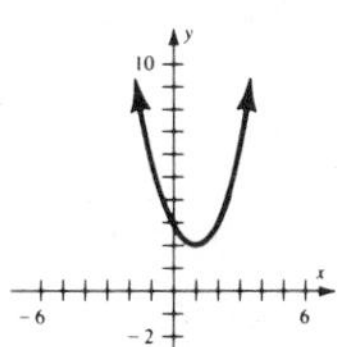

21. Vertex is $(1, 2)$.

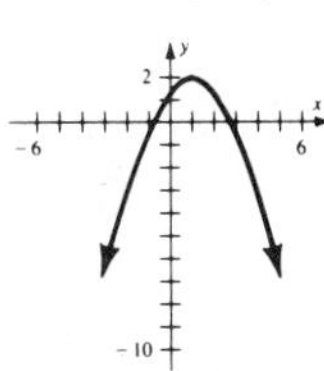

23. Vertex is $(-\frac{1}{3}, -\frac{2}{3})$.

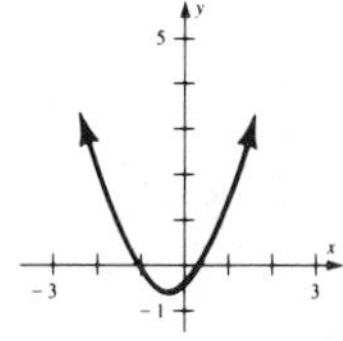

25. Vertex is $(\frac{3}{5}, -\frac{2}{5})$.

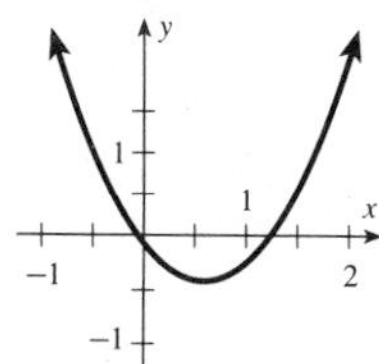

27. $y = (x - 2)^2$

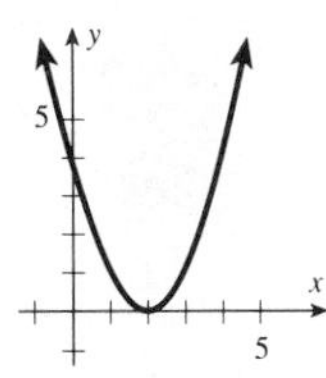

29. $y = -2(x + 1)^2$

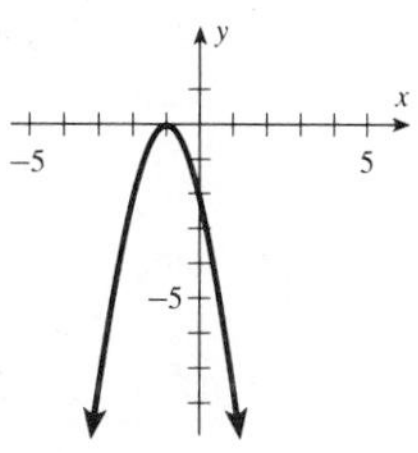

31. $y - 1 = 3(x + 2)^2$

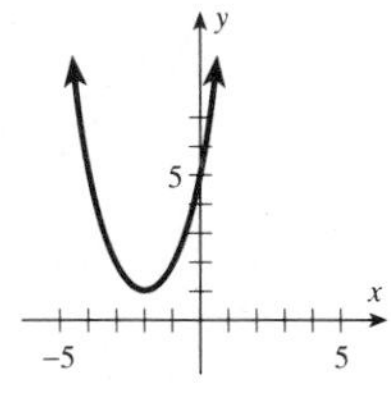

33. $y + 2 = 2(x + 1)^2$

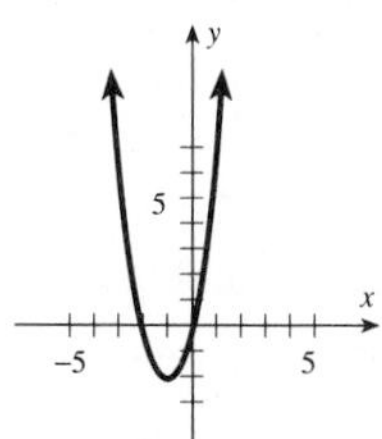

35. $y - 3 = -2(x + 1)^2$

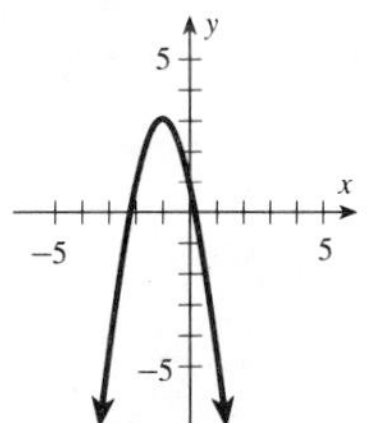

37. $y + 4 = -3(x - 2)^2$

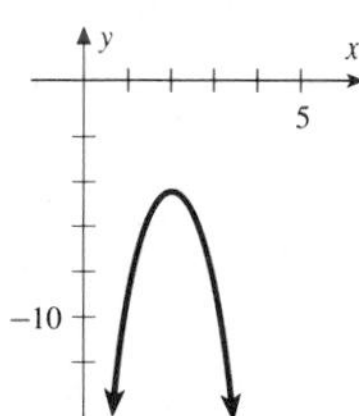

39. The maximum value of y is 3, occurring when $x = -1$. **41.** The maximum value of y is 1,250, occurring when $x = 450$.
43. The minimum value of y is $-1{,}400$, occurring when $x = 560$. **45.** The maximum value of y is -14, occurring

when $x = -3$. **47.** The minimum value for y is 13, occurring at $x = -3$. **49.** The maximum is $y = -4$, which occurs when $x = 1$. **51. a.** $-10x^2 + 1{,}040x$ **b.** The break-even points are $(19, 16150)$ and $(35, 24150)$.
53. a. The maximum profit is \$1,156,250, which occurs when 375 boats are produced; that is, 375 should be produced.
b. The loss is \$250,000. **c.** The maximum profit is \$1,156,150. **55.** The maximum profit is \$25, which occurs when $x = 15$. **57.** They intersect at $(30, 600)$ and $(40, 400)$. **59.** The maximum height was 3,456 units.

2.6 Rational Functions (page 80)

1. A rational function is a function of the form $\frac{P(x)}{Q(x)}$ where $P(x)$ and $Q(x)$ are polynomials and $Q(x) \neq 0$.
3. A horizontal asymptote will be present if: (a) the degrees of the numerator and denominator are equal, in which case the ratio of leading coefficients gives the value of the asymptote or (b) the degree of the numerator is less than that of the denominator, in which case the x-axis will be the asymptote. **5.** Vertical asymptote is $x = \frac{3}{2}$; horizontal asymptote is $y = 0$ (the x-axis). **7.** Deleted point at $x = -3$. Vertical asymptote $x = 1$. The horizontal asymptote is $y = 0$ (the x-axis).
9. The vertical asymptotes are $x = -\frac{3}{2}$ and $x = \frac{2}{3}$; the horizontal asymptote is $y = 0$ (the x-axis). **11.** There is a vertical asymptote at $x = \dfrac{-2 \pm \sqrt{2}}{2}$; the horizontal asymptote is the ratio of leading coefficients: $y = \frac{5}{2}$.
13. No vertical asymptote; the horizontal asymptote is $y = 0$. **15.** The vertical asymptotes are $x = -2$ and $x = 1$; the horizontal asymptote is $y = 0$ (the x-axis).

17.
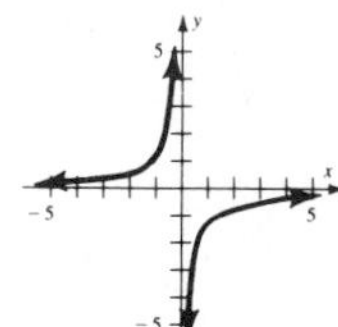

19.
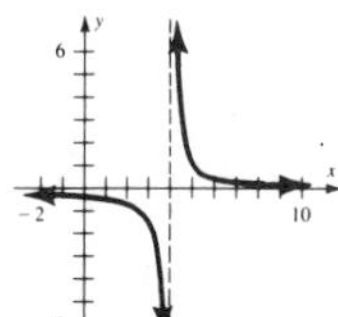

21.
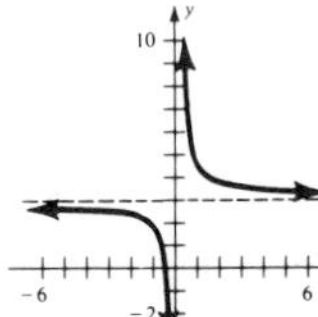

23.
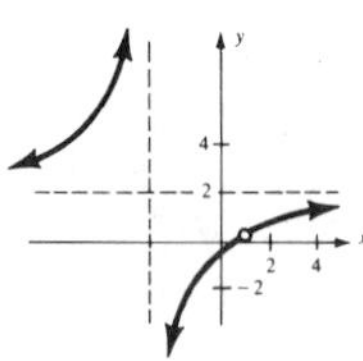

25.
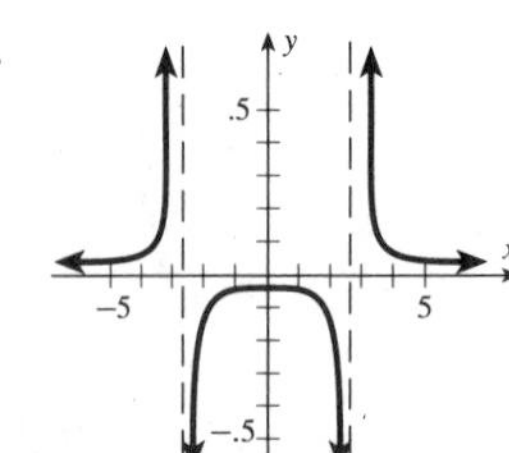

27.
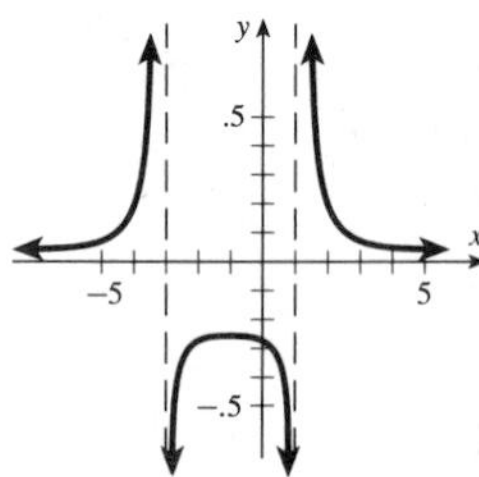

29. \$106,667

31.
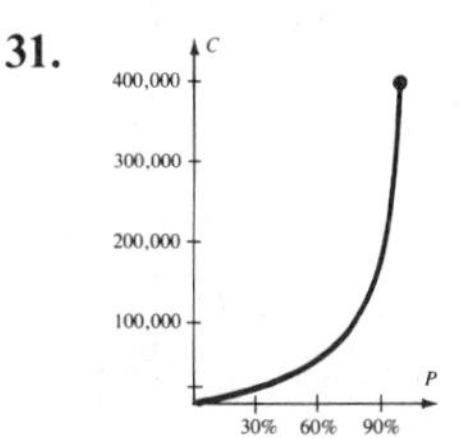

33. \$72,000

35.

37.
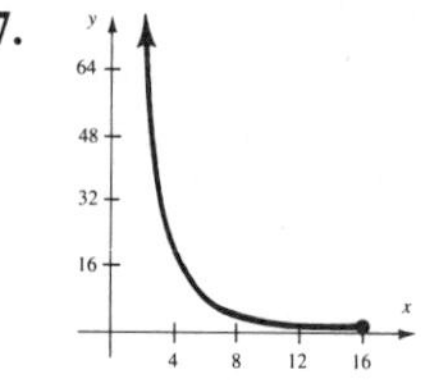

39.
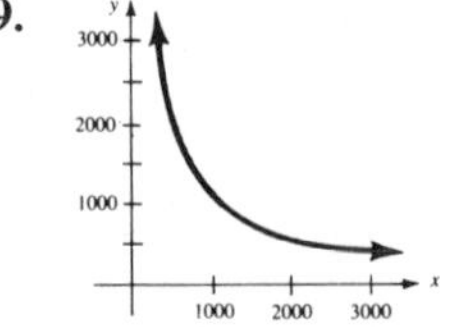

2.7 Review (page 81)

Objective Questions

1. 47 **3.** 15 **5.** $3s^2 + 18s + 27$ **7.** $6xh + 3h^2$ **9.** The coordinates of A are $(2, 3)$; the domain is all real numbers; the range is $y \leq 4$. The x-intercepts are -1 and 3, and the y-intercept is 3. **11.** Coordinates of C are $(x_0 + h, f(x_0 + h))$. The domain is $s \leq x \leq t$; the range is $r \leq y \leq q$. The x-intercept is at d; the y-intercept is at q.

13.

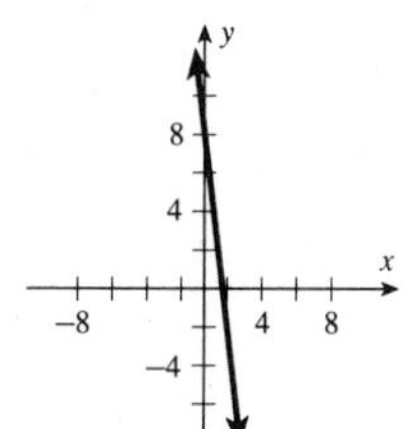

15.

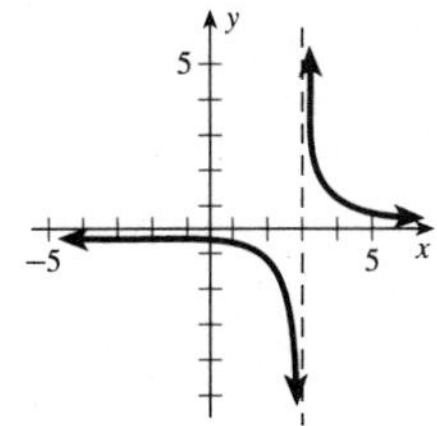

17. x-intercept $(\frac{4}{5}, 0)$; y-intercept $(0, -4)$ **19.** The x-intercept is $(-4, 0)$; the y-intercept is $(0, 3)$. **21.** $\frac{3}{7}$ **23.** -5

25.

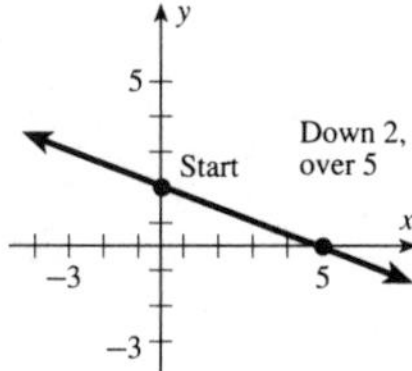

27.

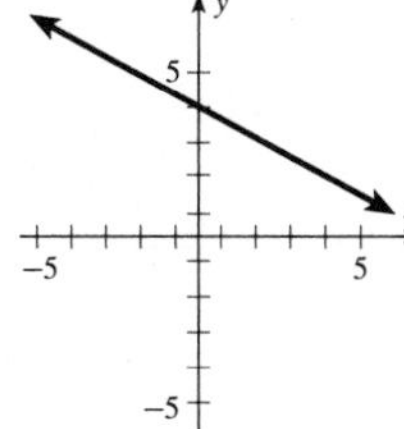

29.

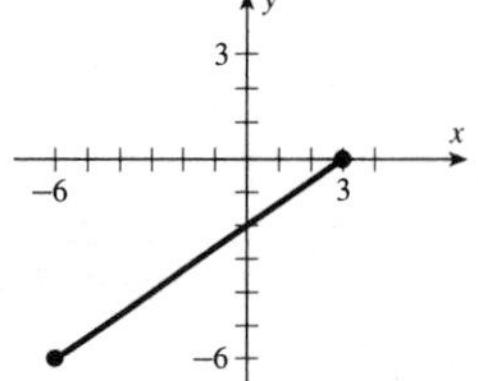

31.

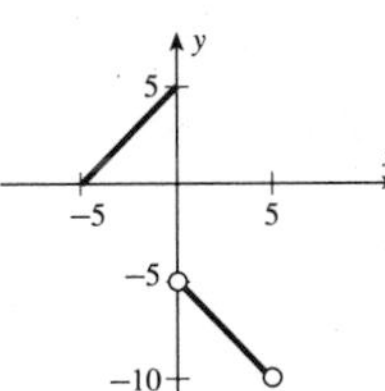

33. $5x - y = 0$ **35.** $x - 5y - 20 = 0$

37.

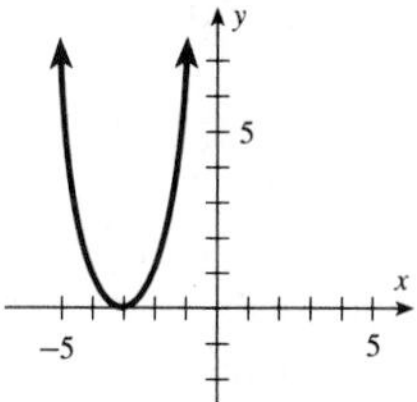

39.

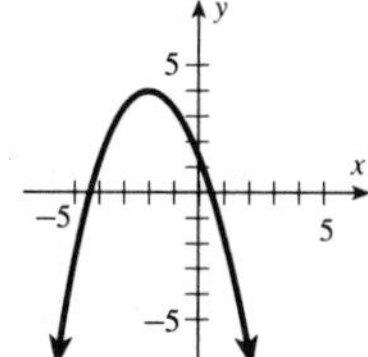

41. Maximum: $y = -250$; occurs when $x = 1{,}300$. **43.** Minimum: $y = 250$; occurs when $x = 3$.
45. There is a vertical asymptote at $x = 5$; there is a horizontal asymptote at $y = 0$.
47. The vertical asymptotes are at $x = 3$ and $x = -2$; horizontal asymptote at $y = 0$.

49.

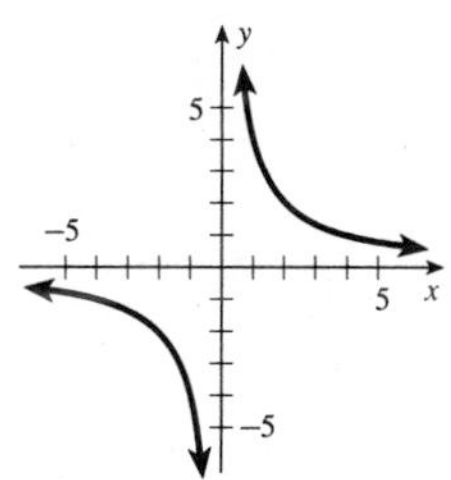

51.

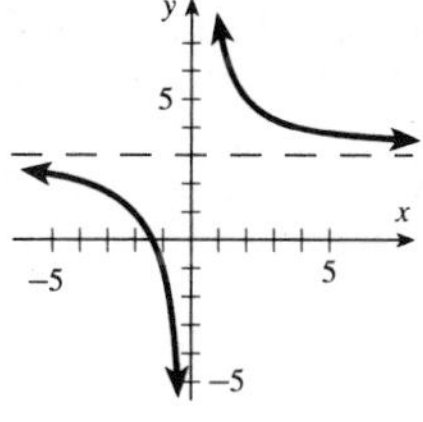

53. $\dfrac{Z(1985) - Z(1980)}{5}$ **55.** Call $S(x)$ the supply function: $S(x) = 30x - 4{,}000$

57. Break-even points occur at $(10, 35000)$ and $(50, 75000)$. **59.** \$380,000

Sample Test

1. 14 **3.** $-4t-3$ **5.** -4 **7.**

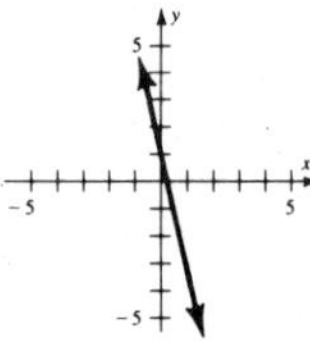

9. Slope $= -8$

11.

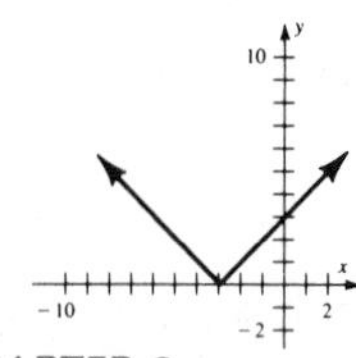

13. $5x + y - 170 = 0$ **15.** $x + 3 = 0$
17. The maximum value of y is 550, which occurs when $x = 40$.
19. \$4,500

CHAPTER 3

3.1 *Systems of Equations* (page 95)

1.

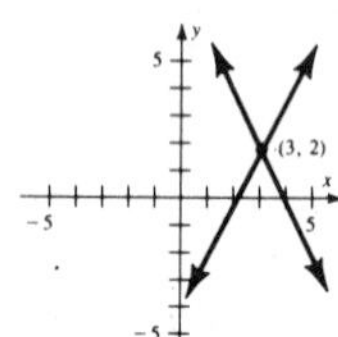

3.

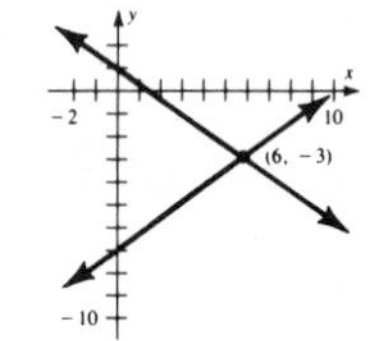

5. Dependent system

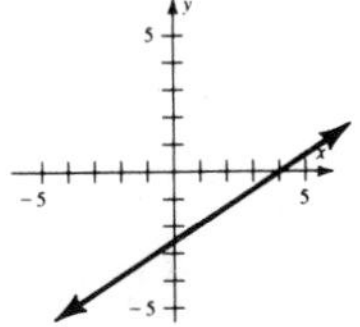

7. The solution is $(a, b) = (2, 3)$. **9.** The solution is $(m, n) = (-3, 6)$. **11.** The system is dependent.
13. The solution is $(c, d) = (1, 1)$. **15.** $(q_1, q_2) = (1, 0)$ **17.** $(\frac{2}{7}, 3)$ **19.** $(\frac{3}{5}, \frac{1}{2})$ **21.** $(-6, 2)$ **23.** $(-\frac{8}{5}, -\frac{21}{5})$
25. The solution is $(x, y) = (31, 19)$. **27.** $x = \dfrac{a}{a^2 - b^2}$; $y = \dfrac{-b}{a^2 - b^2}$ **29.** $(1800, 200)$
31. $A(0,0)$, $B(10,0)$, $C(6,12)$, $D(0,15)$ **33.** $B(12,0)$, $C(4,4)$, $D(0,20)$ **35.** $A(0,0)$, $B(22,0)$, $C(22,16)$, $D(14,30)$, $E(0,40)$
37. $B(90,0)$, $C(40,10)$, $D(10,20)$, $E(0,70)$ **39.** $A(0,0)$, $B(42,0)$, $C(30,36)$, $D(18,45)$, $E(0,48)$
41. $A(0,0)$; $B(\frac{7}{2},0)$; $C(\frac{5}{2},2)$; $D(0,3)$ **43.** $A(0,0)$; $B(\frac{8}{3},0)$; $C(\frac{5}{3},\frac{5}{3})$; $D(0,\frac{8}{3})$ **45.** $A(0,0)$; $B(2,0)$; $C(\frac{7}{4},\frac{1}{2})$; $D(1,\frac{5}{4})$; $E(0,\frac{3}{2})$
47. When the mileage is 150, both rates are the same. **49.** When the mileage is 800, both rates are the same.
51. The equilibrium point is $(2000, 18)$. **53.** The price should be \$15. **55. a.** Supply: $x = 125$; Demand: $x = 75$
b. $p = \$200$ **c.** $p = \$400$ **d.** $(x, p) = (\frac{250}{3}, \frac{700}{3})$; The equilibrium price is $\frac{700}{3} \approx \$233.33$.
e. See (**d**). The number of units is $\frac{250}{3}$, or about 83. **57.** For a price of \$4, both supply and demand are 400 shirts.
59. Cost and revenue are both \$2,000 when 1,000 sunglasses are produced. **61.** If 10,000 disks are produced, both cost and revenue will be 30,000. **63.** If 3,733 cards are produced, cost and revenue will be about \$5,600.
65. If 2,240 cards are produced, both cost and revenue will be \$4,480. **67.** $(x, y) = (500, 200)$
69. The mixture should contain 60 units of grade A fertilizer and 20 units of grade B fertilizer.

3.2 *Introduction to Matrices* (page 111)

1. a. 3×3, square **b.** 3×1, column **c.** 3×2 **d.** 4×1, column **e.** 2×3 **f.** 1×4, row
3. Matrices are equal if and only if all corresponding entries are equal. **a.** $y = 4$ and $x = 2$ **b.** $z = 6$, $x = 4$, and $y = -9$.

5.

	Part 1	Part 2	Part 3	Part 4
Factory A	25	42	193	0
Factory B	16	39	150	0
Factory C	50	50	50	50
Factory D	0	0	0	320

7.

	A	B	C	D
A	2	1	1	0
B	1	2	2	0
C	1	2	0	1
D	0	0	1	0

9. a. $3 \times 3, c_{12}$ **b.** $3 \times 3, c_{32}$ **c.** $3 \times 3, c_{21}$
d. $2 \times 2, c_{11}$ **e.** $2 \times 2, c_{21}$ **11. a.** $[-13 \quad 8 \quad -1]$; order 1×3 **b.** $[-13 \quad 0 \quad -9]$; order 1×3

13. $\begin{bmatrix} 2 & 4 & 2 \\ 6 & -2 & 4 \\ 2 & 2 & 5 \end{bmatrix}$ **15.** $\begin{bmatrix} 0 & -4 & 2 \\ 0 & 0 & 0 \\ 6 & 0 & -5 \end{bmatrix}$ **17.** Not conformable **19.** $\begin{bmatrix} 3 & 4 & 4 \\ 9 & -3 & 6 \\ 6 & 3 & 5 \end{bmatrix}$ **21.** Not conformable

23. $\begin{bmatrix} 16 \\ 1 \\ 12 \end{bmatrix}$ **25.** $\begin{bmatrix} 10 & 5 & 8 \\ 9 & 3 & 11 \\ 6 & 11 & 7 \end{bmatrix}$ **27.** $\begin{bmatrix} 10 & 5 & 8 \\ 9 & 3 & 11 \\ 6 & 11 & 7 \end{bmatrix}$ **29.** $\begin{bmatrix} 13 & -4 & 10 \\ 8 & 3 & 4 \\ 21 & 4 & -2 \end{bmatrix}$ **31.** $\begin{bmatrix} 35 & 5 & 18 \\ 46 & 2 & 16 \\ 39 & -7 & 26 \end{bmatrix}$

33. $\begin{bmatrix} -1 & 37 & 32 \\ 4 & 14 & 45 \\ 27 & 9 & 28 \end{bmatrix}$ **35.** $(AB)C = \begin{bmatrix} 34 & 117 & 44 \\ 45 & 143 & 97 \\ 109 & 100 & 151 \end{bmatrix}$ **37.** $AB + AC = \begin{bmatrix} 13 & 25 & 20 \\ 25 & 31 & 23 \\ 42 & 24 & 33 \end{bmatrix}$

39. $(B + C)A = \begin{bmatrix} 48 & 1 & 28 \\ 54 & 5 & 20 \\ 60 & -3 & 24 \end{bmatrix}$ **41.** Not conformable **43.** $\begin{bmatrix} 1 & 2 \\ 4 & 0 \\ -1 & 3 \\ 2 & 1 \end{bmatrix}$

45. Not conformable **47.** Not conformable (for multiplication) **49.** Not conformable **51.** $B^3 = \begin{bmatrix} 42 & 70 \\ -35 & 7 \end{bmatrix}$

53. **a.** $T = \begin{array}{c} \\ A \\ B \\ C \\ D \\ E \end{array} \begin{array}{c} \begin{array}{ccccc} A & B & C & D & E \end{array} \\ \begin{bmatrix} 0 & 1 & 0 & 1 & 1 \\ 0 & 0 & 1 & 0 & 0 \\ 1 & 0 & 0 & 0 & 0 \\ 1 & 0 & 0 & 0 & 0 \\ 1 & 0 & 0 & 0 & 0 \end{bmatrix} \end{array}$ **b.** $T^2 = \begin{array}{c} \\ A \\ B \\ C \\ D \\ E \end{array} \begin{array}{c} \begin{array}{ccccc} A & B & C & D & E \end{array} \\ \begin{bmatrix} 2 & 0 & 1 & 0 & 0 \\ 1 & 0 & 0 & 0 & 0 \\ 0 & 1 & 0 & 1 & 1 \\ 0 & 1 & 0 & 1 & 1 \\ 0 & 1 & 0 & 1 & 1 \end{bmatrix} \end{array}$ **c.** $T^3 = \begin{array}{c} \\ A \\ B \\ C \\ D \\ E \end{array} \begin{array}{c} \begin{array}{ccccc} A & B & C & D & E \end{array} \\ \begin{bmatrix} 1 & 2 & 0 & 2 & 2 \\ 0 & 1 & 0 & 1 & 1 \\ 2 & 0 & 1 & 0 & 0 \\ 2 & 0 & 1 & 0 & 0 \\ 2 & 0 & 1 & 0 & 0 \end{bmatrix} \end{array}$

55. **a.** $\begin{bmatrix} 28 & 56 & 24 \\ 20 & 32 & 4 \end{bmatrix}$ **b.** $\begin{bmatrix} 12 & 6 & 12 \\ 8 & 0 & 12 \end{bmatrix}$ **c.** Total revenue is \$7,560. **d.** Revenue: $\begin{bmatrix} 12 & 32 & 16 \\ 12 & 20 & 0 \end{bmatrix} \begin{bmatrix} 90 \\ 110 \\ 120 \end{bmatrix}$;

Profit: $\begin{bmatrix} 12 & 32 & 16 \\ 12 & 20 & 0 \end{bmatrix} \begin{bmatrix} 27 \\ 33 \\ 36 \end{bmatrix}$ **57.** Using two intermediaries, U.S. can communicate with the USSR in 4 ways, with Cuba in only 1 way, and with Mexico in 3 ways. **59.** False **61.** True **63.** True

3.3 Gauss–Jordan Elimination (page 127)

1. **a.** $\left[\begin{array}{cc|c} 4 & 5 & -16 \\ 3 & 2 & 5 \end{array}\right]$ **b.** $\left[\begin{array}{ccc|c} 1 & 1 & 1 & 4 \\ 3 & 2 & 1 & 7 \\ 1 & -3 & 2 & 0 \end{array}\right]$ **c.** $\left[\begin{array}{cccc|c} 1 & 3 & 1 & 1 & 3 \\ 1 & 0 & -2 & 2 & 0 \\ 0 & 0 & 1 & 5 & -14 \\ 0 & 1 & -3 & -1 & 2 \end{array}\right]$ **3.** **a.** $\begin{cases} 2x_1 + x_2 + 4x_3 = 3 \\ 6x_1 + 2x_2 - x_3 = -4 \\ -3x_1 - x_2 = 1 \end{cases}$

b. $\begin{cases} x_1 = 5 \\ x_2 = -3 \\ x_3 = 4 \end{cases}$ **c.** $\begin{cases} x_1 = 3 \\ x_2 = 2 \\ x_3 = -8 \\ 0 = 1 \end{cases}$ **5.** **a.** $\left[\begin{array}{ccc|c} 1 & 5 & 3 & -2 \\ -1 & 3 & -2 & 3 \\ -2 & 2 & 4 & 8 \end{array}\right]$ **b.** $\left[\begin{array}{ccc|c} -2 & 2 & 4 & 8 \\ -2 & 6 & -4 & 6 \\ 1 & 5 & 3 & -2 \end{array}\right]$ **c.** $\left[\begin{array}{ccc|c} -1 & 1 & 2 & 4 \\ -1 & 3 & -2 & 3 \\ 1 & 5 & 3 & -2 \end{array}\right]$

d. $\left[\begin{array}{ccc|c} 0 & 12 & 10 & 4 \\ -1 & 3 & -2 & 3 \\ 1 & 5 & 3 & -2 \end{array}\right]$ **e.** $\left[\begin{array}{ccc|c} -2 & 2 & 4 & 8 \\ -2 & -2 & -5 & 5 \\ 1 & 5 & 3 & -2 \end{array}\right]$ **7.** $x = 1, y = 2$ **9.** $x = -6, y = -4$ **11.** $\begin{cases} x = \frac{10}{3} + 2t \\ y = 3t \end{cases}$

13. Inconsistent system (no simultaneous solution) **15.** Inconsistent system (no simultaneous solution)
17. Inconsistent system (no simultaneous solution) **19.** $x = 1, y = 2, z = 3$ **21.** $x = 3, y = -1, z = 2$
23. $x = 3, y = 2, z = 5$ **25.** $x = 2, y = -1, z = 0$ **27.** $x = 6, y = 2, z = -9$ **29.** $x = 1, y = 1, z = 2$

31. $\begin{cases} x = \frac{8}{5} - 7t \\ y = -\frac{19}{5} + t \\ z = 5t \end{cases}$ **33.** Inconsistent system (no simultaneous solution) **35.** $x = -7 + 11t, y = -2 + 7t, z = 3t$

37. $w = \frac{18}{5} + 3s - t, x = \frac{7}{5} + 7s - 14t, y = 5s, z = 5t$ **39.** $w = \frac{11}{2} - 6s, x = 2 - 4s + t, y = 3s, z = 3t$ **41.** $x_1 = -\frac{3}{2} - 2t, x_2 = 1, x_3 = -\frac{1}{4} + 3t, x_4 = 12t$ **43.** $x_1 = 4, x_2 = -1, x_3 = 1, x_4 = -3$ **45.** $x_1 = 1, x_2 = -1, x_3 = 1, x_4 = 2$ **47.** $x_1 = \frac{3}{2} + 51t, x_2 = \frac{1}{3} - 94t, x_3 = -\frac{1}{3} - 62t, x_4 = -\frac{3}{2} + 33t, x_5 = 6t$ **49.** Three containers of spray I and four containers of spray II are needed. **51.** Four units of candy I, 5 units of candy II, and 6 units of candy III **53.** No. The best bargain is the large scoop at $\frac{1.40}{6} \approx \$.23$ per ounce.

3.4 Inverse Matrices (page 136)

1. Yes **3.** Yes **5.** Yes **7.** No **9.** Yes **11.** No **13.** The inverse is $\begin{bmatrix} 2 & 7 \\ 1 & 4 \end{bmatrix}$

15. The inverse is $\begin{bmatrix} 2 & -5 \\ -1 & 3 \end{bmatrix}$ **17.** The inverse is $\begin{bmatrix} 0 & \frac{1}{2} \\ \frac{1}{3} & -\frac{1}{6} \end{bmatrix} = \frac{1}{6}\begin{bmatrix} 0 & 3 \\ 2 & -1 \end{bmatrix}$ **19.** The inverse is $\frac{1}{22}\begin{bmatrix} 2 & -3 \\ 1 & 4 \end{bmatrix}$

21. The inverse is $\begin{bmatrix} 4 & 3 \\ 2 & 2 \end{bmatrix}$ **23.** The inverse is $\begin{bmatrix} 3 & -17 & -20 \\ 3 & -18 & -20 \\ -1 & 6 & 7 \end{bmatrix}$ **25.** The inverse is $\begin{bmatrix} 3 & 3 & -1 \\ -2 & -2 & 1 \\ -4 & -5 & 2 \end{bmatrix}$

27. The inverse is $\frac{1}{12}\begin{bmatrix} -2 & 2 & 2 \\ 8 & -8 & 4 \\ 7 & -1 & -1 \end{bmatrix}$ **29.** The inverse is $\frac{1}{10}\begin{bmatrix} -12 & 0 & 8 & 1 \\ 6 & 0 & -4 & 2 \\ 2 & 0 & 2 & -1 \\ 0 & 10 & 0 & 0 \end{bmatrix}$ **31.** $x = 3, y = 2$

33. $x = 5, y = -4$ **35.** $x = -1, y = 4$ **37.** $x = -4, y = 6$ **39.** $x = 3, y = 1$ **41.** $x = -2, y = 2$ **43.** $x = 23, y = -30$ **45.** $x = -5, y = 4$ **47.** $x = 5, y = 6, z = 1$ **49.** $x = 1, y = -2, z = 3$ **51.** $x = 5, y = 4, z = -1$ **53.** $x = -3, y = 2, z = 7$ **55.** $w = 3, x = -2, y = 1, z = -4$ **57.** 6 bags of grain I, 3 bags of grain II, and 1 bag of grain III. **59.** 22 bags of grain I, 8 bags of grain II, 11 bags of grain III. **61.** Ten ounces of each food should be supplied.

3.5 Leontief Models (page 145)

1. A one-parameter solution is: $x = t, y = 3t, z = 2t$. **3.** A one-parameter solution is: $x = -6t, y = 2t, z = 3t$. **5.** A one-parameter solution is: $x_1 = 154t, x_2 = 89t, x_3 = 30t$. **7.** Answers vary. One possible solution is: $x_1 = \$50{,}000$; $x_2 = \$47{,}000$; $x_3 = \$30{,}000$. **9.** Let x_1, x_2, x_3, and x_4 represent the payments from the farmer, builder, tailor, and merchant, respectively. One possibility is to choose $x_4 = \$900$; then $x_1 = \$900, x_2 = \$1{,}125, x_3 = \$1{,}125$. **11.** The required output in 3 years is about \$166 for farming, \$158 for construction, and \$147 for clothing. **13.** The required output in 5 years would be about \$144 for farming, \$205 for construction, and \$99 for clothing. **15.** $\begin{bmatrix} 1.41 & .14 & .34 \\ .33 & 1.41 & .39 \\ .17 & .07 & 1.16 \end{bmatrix}$

17. The output needed would be about \$347, \$252, and \$214 for manufacturing, agriculture, and transportation, respectively.

19. $\begin{bmatrix} 1.414 & 0 & 0 \\ .0267 & 1.261 & .027 \\ .0797 & .016 & 1.276 \end{bmatrix}$

3.6 Review (page 146)

Objective Questions

1. (2, 2)

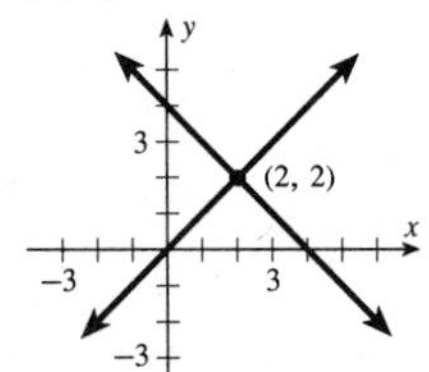

3. (5, 4)

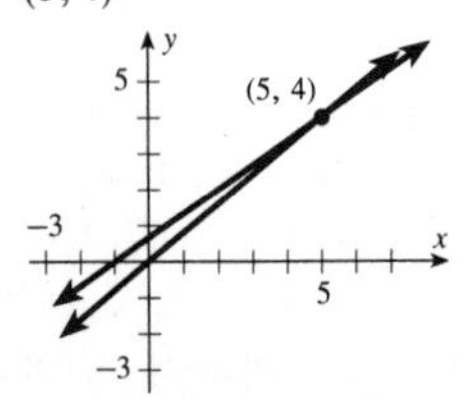

5. (3, 7)

7. $(1, 2)$ **9.** $(5, 2)$ **11.** $(\frac{1}{2}, \frac{5}{6})$ **13.** 2×5 **15.** 5×1 **17.** Cannot be equal since $0 \neq 1$.
19. Cannot be equal since first matrix is 1×1 and the second is 1×3. **21.** Answers vary.

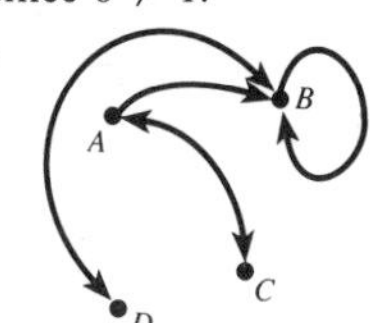

23. $\begin{array}{c} \\ A \\ B \\ C \\ D \end{array}\begin{array}{c} \begin{array}{cccc} A & B & C & D \end{array} \\ \begin{bmatrix} 0 & 1 & 1 & 0 \\ 0 & 0 & 0 & 1 \\ 1 & 1 & 0 & 0 \\ 0 & 1 & 0 & 1 \end{bmatrix} \end{array}$ **25.** $\begin{bmatrix} 1 & 3 & 1 \\ -7 & 1 & 51 \end{bmatrix}$ **27.** $\begin{bmatrix} 11 & 4 & 2 \\ 5 & 1 & 1 \\ 2 & 14 & 1 \end{bmatrix}$ **29.** $[-1 \quad 12 \quad -4]$ **31.** $\begin{bmatrix} -7 \\ -12 \\ -3 \\ 11 \end{bmatrix}$

33. $\begin{bmatrix} 12 & 3 & 9 \\ 18 & -6 & 15 \\ 0 & 27 & -18 \end{bmatrix}$ **35.** $\begin{bmatrix} .5 \\ .9 \\ 1 \end{bmatrix}$ **37.** $\begin{bmatrix} -15 & 8 & -8 & 1 \\ 31 & -15 & 12 & -48 \\ -10 & 34 & -2 & -11 \end{bmatrix}$ **39.** $\begin{bmatrix} 93 & -9 & 72 & -30 \\ -48 & 30 & -45 & 33 \\ 27 & -33 & 18 & -27 \end{bmatrix}$

41. $\begin{bmatrix} 29 & 5 & -21 \\ 11 & 2 & 39 \end{bmatrix}$ **43.** Not conformable **45.**

	Ct. Hse. Square	Student's Inn	College
Ct. Hse. Square	1	2	1
Student's Inn	1	0	1
College	1	1	0

47.

	Ct. Hse. Square	Student's Inn	College
Ct. Hse. Square	28	27	21
Student's Inn	16	17	11
College	16	16	12

49. $\left[\begin{array}{ccc|c} 3 & -2 & 0 & -3 \\ 1 & 0 & 4 & 7 \end{array}\right]$ **51.** $\left[\begin{array}{cc|c} 3 & 2 & -1 \\ 5 & 3 & 0 \\ 2 & -4 & 26 \end{array}\right]$

53. $\begin{cases} x_1 + 4x_2 + 9x_3 = -1 \\ 2x_1 + 3x_2 + x_3 = 0 \\ 4x_1 - x_2 + 2x_3 = 3 \\ 3x_1 - 4x_2 = 1 \end{cases}$ **55.** $\begin{cases} x_1 + 2x_4 = -3 \\ x_2 + 5x_3 = -1 \end{cases}$ **57.** $x_1 = 1, x_2 = 2$ **59.** $x = 3, y = -5$

61. $x = 6 - t, y = -1 + 2t, z = 5t$ **63.** $x = \frac{1}{2} - \frac{13}{16}z, y = 1 + \frac{7}{8}z$. Let $z = 16t$; then $x = \frac{1}{2} - 13t, y = 1 + 14t, z = 16t$
65. The mixture should contain 20 bags of grain A, 8 bags of grain B, and 12 bags of grain C.
67. The number of married adults the following year: $.9(38{,}000) + .3(14{,}000) = 38{,}400$. The number of single adults the following year: $.1(38{,}000) + .7(14{,}000) = 13{,}600$. **69.** Yes **71.** Yes **73.** The inverse is $\begin{bmatrix} 1 & -1 & 1 \\ 0 & 2 & -1 \\ 2 & 3 & 0 \end{bmatrix}$.

75. The inverse is $\frac{1}{14}\begin{bmatrix} 1 & 3 \\ -4 & 2 \end{bmatrix}$. **77.** $x = -2, y = 10, z = 19$ **79.** $x = 5, y = 4$ **81.** $(58t, 43t, 18t)$ **83.** $(t, 3t, 4t)$
85. $x_1 = \frac{23}{19}x_3, x_2 = \frac{23}{19}x_3; x_1 = \$36{,}316; x_2 = \$50{,}526; x_3 = \$30{,}000$ **87.** $x_1 = \frac{4}{3}x_3, x_2 = \frac{5}{4}x_3; x_1 = \$40{,}000; x_2 = \$37{,}500;$ $x_3 = \$30{,}000$ **89.** $T = \begin{bmatrix} .1 & .2 & .4 \\ .1 & .167 & .33 \\ .12 & .12 & .28 \end{bmatrix}; I - T = \begin{bmatrix} .9 & -.2 & -.4 \\ -.1 & .833 & -.33 \\ -.12 & -.12 & .72 \end{bmatrix}$ **91.** 349 for M_1, 367 for M_2, 453 for M_3

Sample Test

1. a. Adding; $(\frac{-25}{7}, \frac{64}{21})$ **b.** The graphs of these two linear equations intersect at $(3, 2)$. **c.** $x = 5, y = 8$

3. a. $\begin{bmatrix} 8 & 3 & 7 \\ -2 & 16 & 17 \end{bmatrix}$ **b.** Not conformable **5. a.** $\begin{bmatrix} 11 & 8 & -3 \\ 3 & 8 & -6 \\ 11 & -6 & 14 \end{bmatrix}$ **b.** $\begin{bmatrix} 5 & 2 & 0 \\ 3 & 0 & 0 \\ 2 & 3 & 5 \end{bmatrix}$ **7. a.** $x = 3, y = -2$

b. $x = 2, y = -5$ **c.** $x = \frac{7}{5} + 2t, y = -\frac{24}{5} + 11t, z = 5t$ **9. a.** $A^{-1} = \begin{bmatrix} 1 & -1 & 1 \\ 0 & 2 & -1 \\ 2 & 3 & 0 \end{bmatrix}$ **b.** $\begin{bmatrix} 12 \\ -5 \\ 13 \end{bmatrix}$

CHAPTER 4

4.1 Systems of Linear Inequalities (page 157)

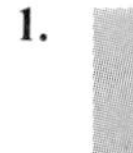

1. 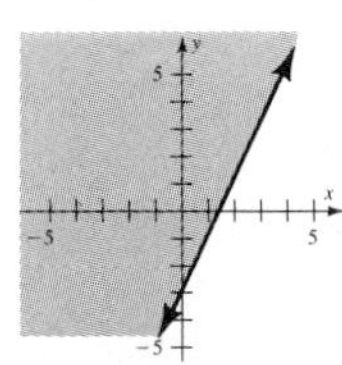

3. 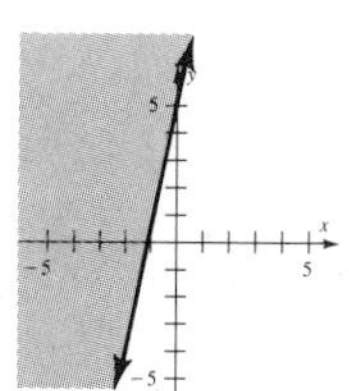

5. 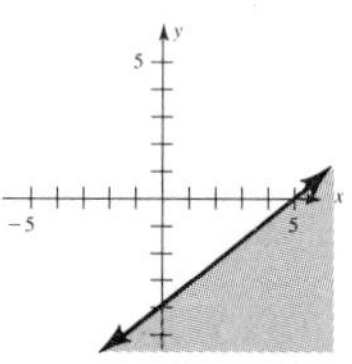

7. 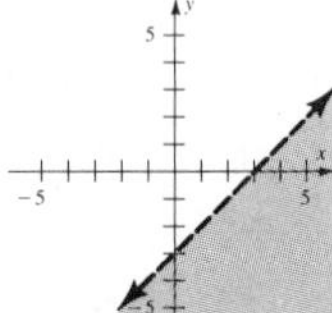

9. 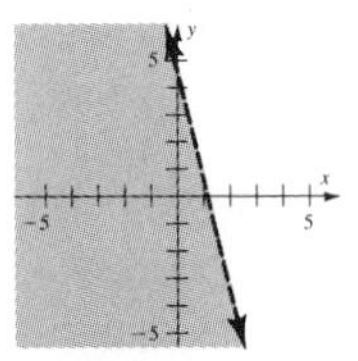

11. 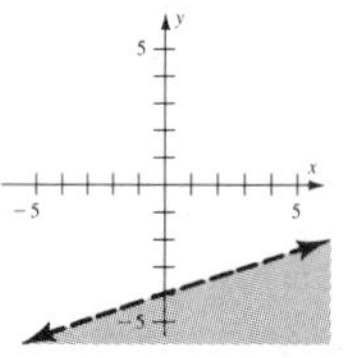

13. 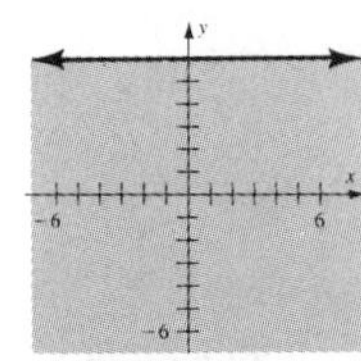

15. 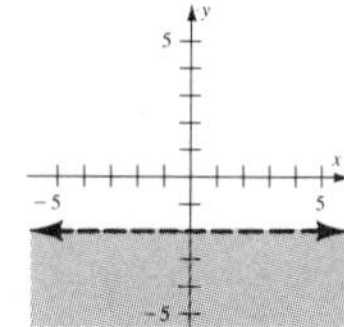

17. 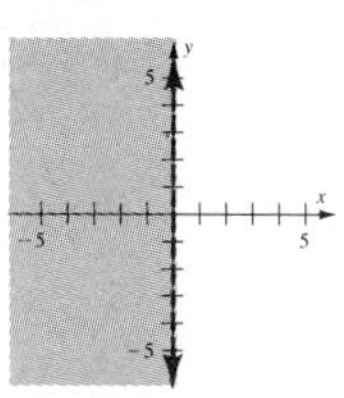

19. 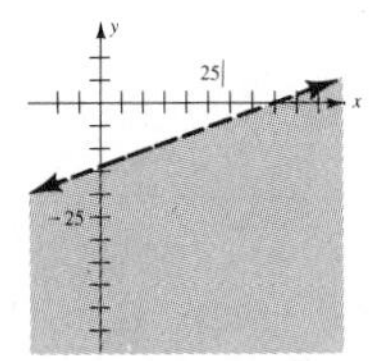

21. 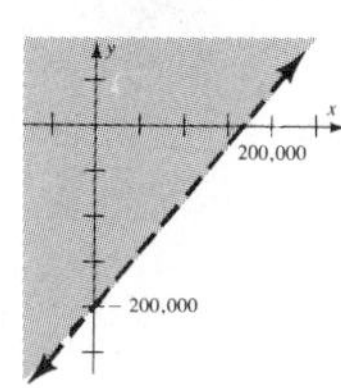

23. 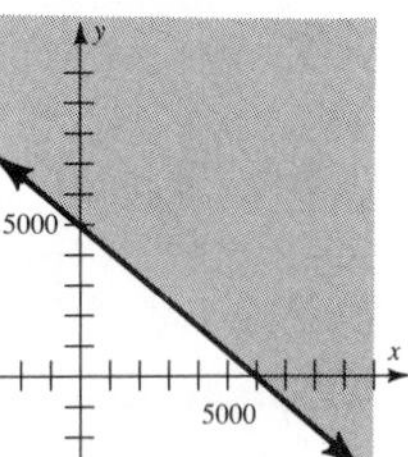

25. 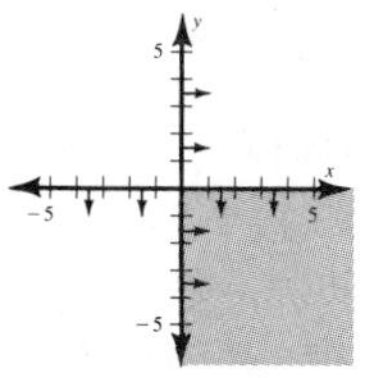

27. 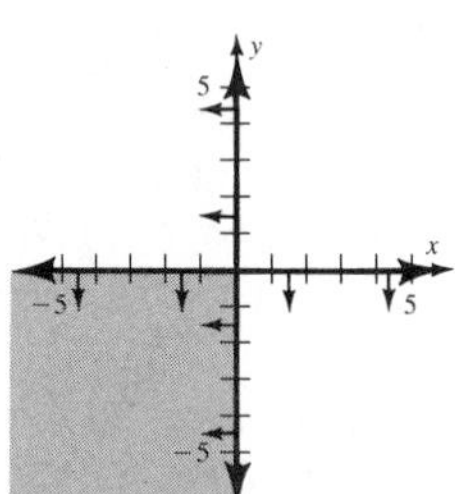

29. 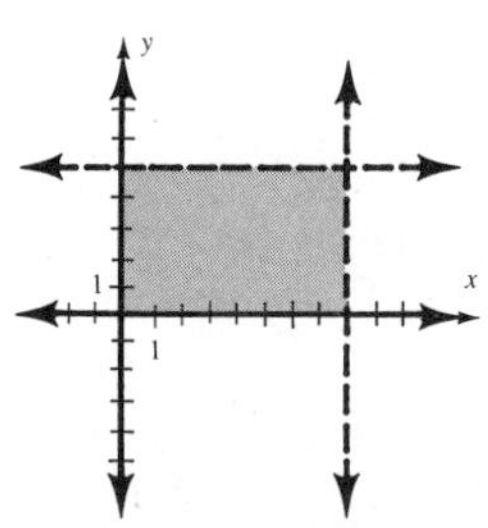

31. 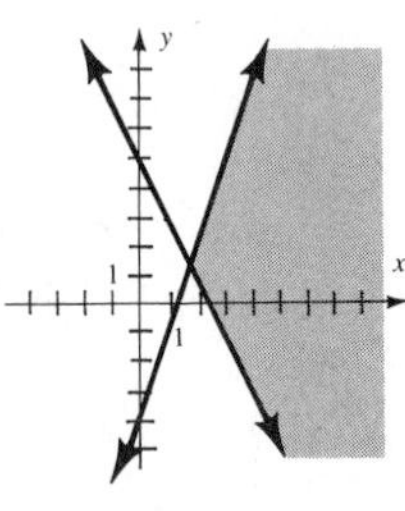

33. 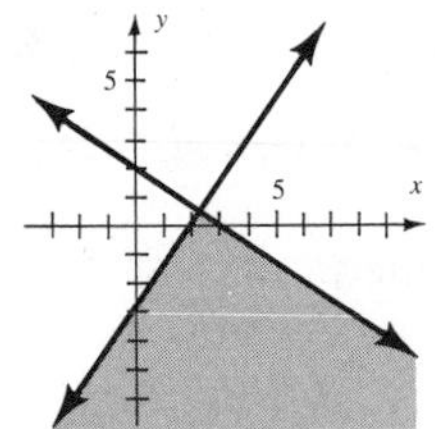

35. 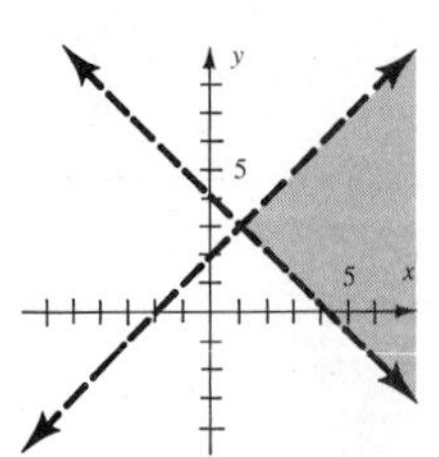

37. 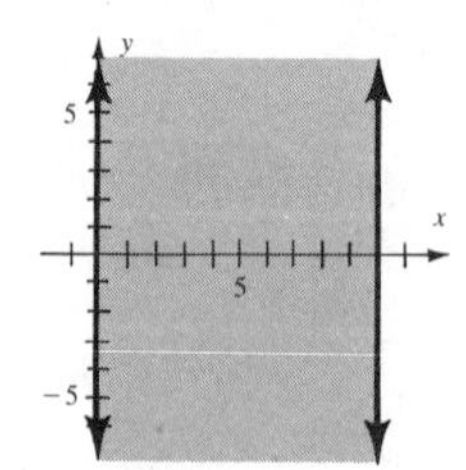

39. 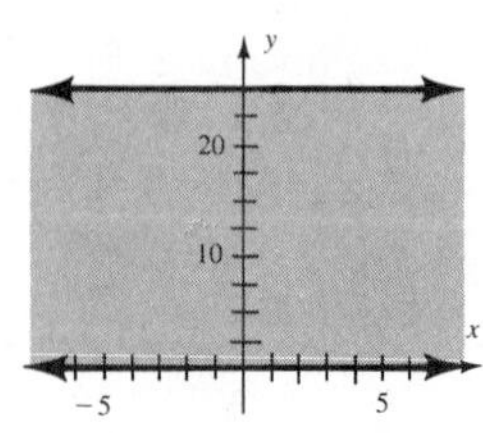

41.
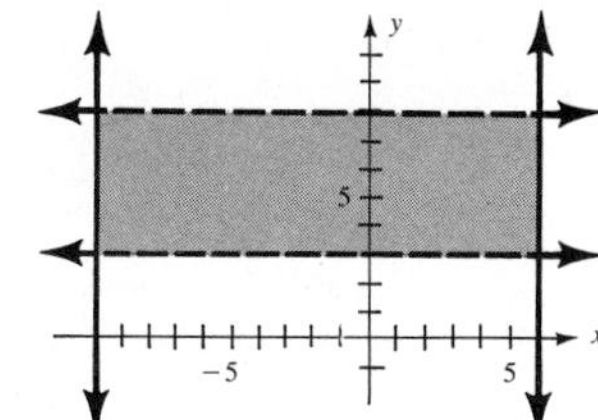

43.
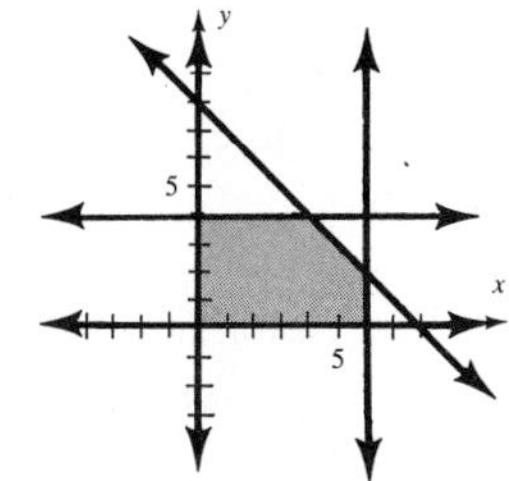

45.
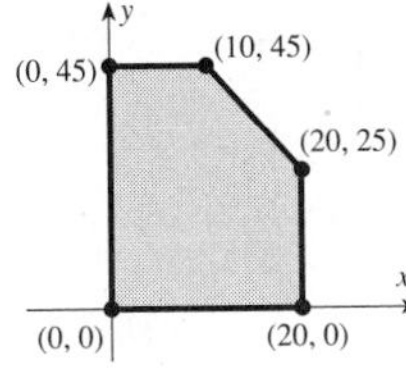

47.
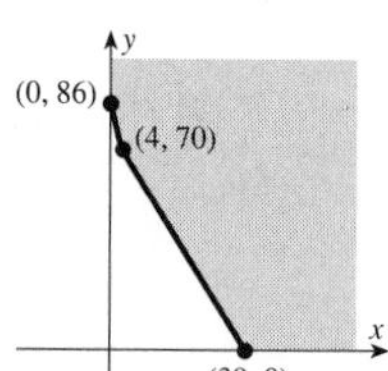

49.
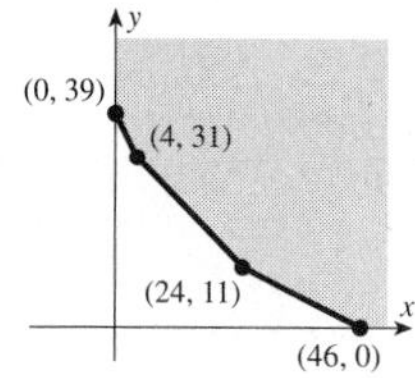

51.
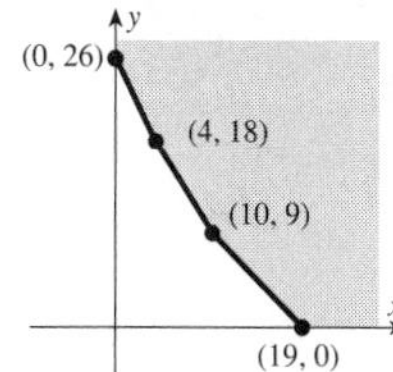

53.
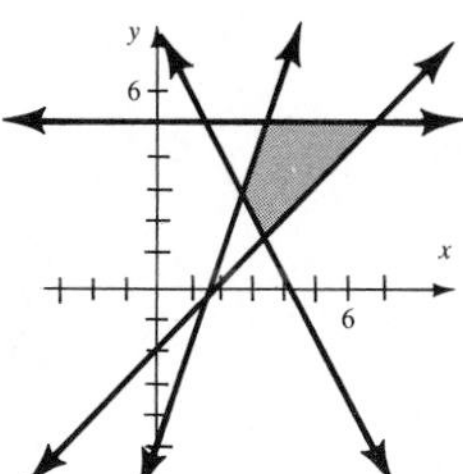

55.
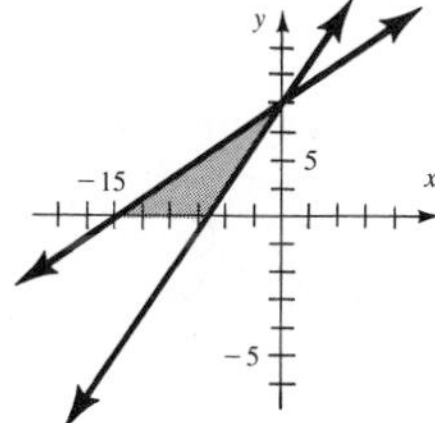

57.
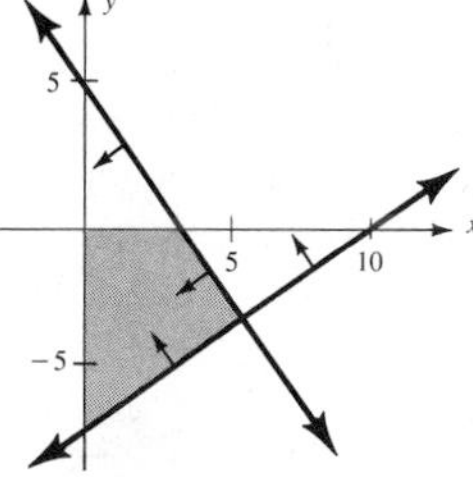

59.
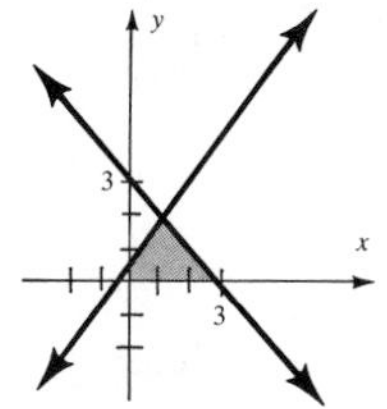

4.2 Formulating Linear Programming Models (page 166)

1. Let x = number of acres of corn
y = number of acres of wheat
Maximize $P = (1.20)(100x) + (2.50)(40y) = 120x + 100y$

Subject to $\begin{cases} x \geq 0, \quad y \geq 0 \\ 120x + 60y \leq 24{,}000 \\ 100x + 40y \geq 20{,}000 \\ x + y \leq 500 \end{cases}$

3. Let x = the number of standard items produced
y = the number of economy items produced
Maximize $P = 45x + 30y$

Subject to $\begin{cases} x \geq 0, \quad y \geq 0 \\ 3x + 3y \leq 1{,}500 \\ 2x + 0y \geq 800 \end{cases}$

5. Let x = ounces of Corn Flakes and
y = ounces of Honeycombs
Minimize $C = .07x + .19y$

Subject to $\begin{cases} x \geq 0, \quad y \geq 0 \\ 23x + 14y \geq 322 \\ 7x + 17y \geq 119 \end{cases}$

7. Let x = amount invested in Pertec stock
y = amount invested in Campbell Municipal Bonds
Maximize $R = .20x + 10y$

Subject to $\begin{cases} x \geq 0, \quad y \geq 0 \\ x + y \leq 100{,}000 \\ x \leq 70{,}000 \\ y \geq 20{,}000 \\ y \leq 3x \end{cases}$

9. Let x_1 = no. shipped from New Orleans to Chicago
x_2 = no. shipped from New Orleans to L.A.
y_1 = no. shipped from Atlanta to Chicago
y_2 = no. shipped from Atlanta to L.A.
Minimize $C = 120x_1 + 180x_2 + 100y_1 + 200y_2$

Subject to $\begin{cases} x_1 \geq 0, \quad x_2 \geq 0, \quad y_1 \geq 0, \quad y_2 \geq 0 \\ x_1 + x_2 \leq 1{,}200 \\ y_1 + y_2 \leq 1{,}500 \\ x_1 + y_1 \geq 1{,}500 \\ x_2 + y_2 \geq 900 \end{cases}$

11. Let x = no. of dozen of cream-filled candies
y = no. of dozen of solid chews
Maximize $P = 6x + 4.8y$

Subject to $\begin{cases} x \geq 0, \quad y \geq 0 \\ x \geq 25 \\ y \geq 50 \\ x + y \leq 125 \end{cases}$

13. Let x = number of Glassbelt tires produced
y = number of Rainbelt tires produced
Maximize $R = 52x + 36y$

Subject to $\begin{cases} x \geq 0, \quad y \geq 0 \\ 3x + y \leq 200 \\ 2x + 2y \leq 400 \end{cases}$

15. Let x = number of commercial guests
y = number of other guests
Maximize $P = 4.50x + 3.50y$

Subject to $\begin{cases} x \geq 0, \quad y \geq 0 \\ .40x + .20y \leq 50 \\ x + y \leq 200 \end{cases}$

17. Let x = number of Alpha units produced
y = number of Beta units produced
Maximize $P = 5x + 8y$

Subject to $\begin{cases} x \geq 0, \quad y \geq 0 \\ x \geq 700 \\ x + 3y \leq 1{,}200 \\ x + 2y \leq 1{,}000 \end{cases}$

19. Let x = number of A1 animals
y = number of A2 animals
z = number of A3 animals
Maximize $T = x + y + z$

Subject to $\begin{cases} x \geq 0, \quad y \geq 0, \quad z \geq 0 \\ 12x + 15y + 20z \leq 8{,}000 \\ 6x + 20y + 10z \leq 10{,}000 \\ 16x + 5y + 10z \leq 6{,}000 \end{cases}$

21. Let x = number of tabies manufactured
y = number of shelves manufactured
P = profit
Maximize $P = 4x + 2y$

Subject to $\begin{cases} x \geq 30 \\ y \geq 25 \\ 2x + y \leq 200 \\ 2x + 2y \leq 240 \\ 2x + 3y \leq 300 \end{cases}$

23. Let x = number of hours she jogs
y = number of hours she plays handball
z = number of hours she dances
E = exercise time (total)
Minimize $E = x + y + z$

Subject to $\begin{cases} x \geq 3 \\ y \geq 2 \\ z \geq 5 \\ x \geq y \\ 900x + 600y + 800z \geq 9{,}000 \end{cases}$

25. Let x, y, and z represent units produced on day, swing, and graveyard shift.
Minimize labor costs $C = 100x + 150y + 180z$

Subject to $\begin{cases} x \geq 20, \quad y + z \geq 50, \quad x \geq 0 \\ x \leq 50, \quad 60x \leq 1{,}800, \quad y \geq 0 \\ y \leq 50, \quad 60y \leq 1{,}500, \quad z \geq 0 \\ z \leq 50, \quad 60z \leq 3{,}000 \end{cases}$

27. Let x = number of cases shipped from L.A. to Chicago
y = number of cases shipped from L.A. to Dallas
z = number of cases shipped from Seattle to Chicago
w = number of cases shipped from Seattle to Dallas

Minimize $C = 9x + 7y + 7z + 8w$

$$\text{Subject to } \begin{cases} x + y \le 90 & x \ge 0 \\ z + w \le 130 & y \ge 0 \\ x + z \ge 80 & z \ge 0 \\ y + w \ge 110 & w \ge 0 \end{cases}$$

29. Let x = no. of units of copper produced
y = no. of units of lead produced
z = no. of units of zinc produced

Maximize $P = 90x + 100y + 70z$

$$\text{Subject to } \begin{cases} x \ge 0, \quad y \ge 0, \quad z \ge 0 \\ 2x + 3y + z \le 10 \\ 2x + 2y + z \le 12 \\ x + y + 3z \le 8 \end{cases}$$

4.3 Graphical Solution of Linear Programming Problems (page 179)

1. **a.** D **b.** A **3.** **a.** None **b.** C

5. $(0, 4\frac{1}{2}), (6, 0), (5, 2), (0, 0)$

7. $(0, 0), (0, 4), (4, 0), (2, 3)$

9. $(6, 0), (6, 2), (4, 4), (0, 4), (0, 0)$

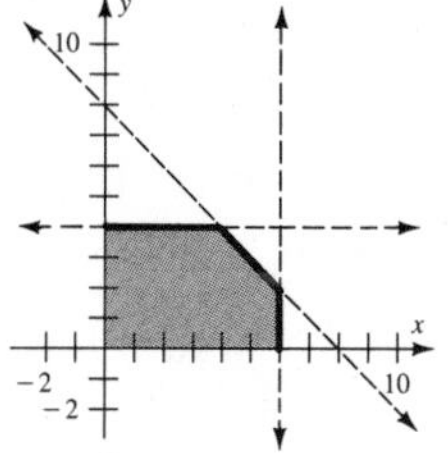

11. $(\frac{24}{13}, \frac{16}{13}), (0, 4), (0, \frac{8}{5})$

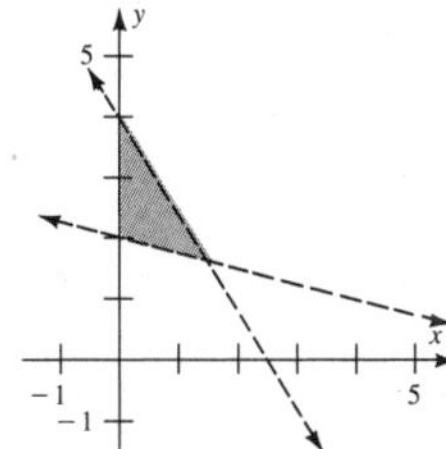

13. $(0, 0), (10, 0), (6, 6), (10, 2)$

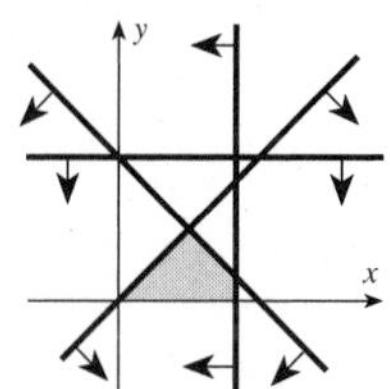

15. $(0, 40), (50, 0), (8, 24), (\frac{200}{7}, \frac{60}{7})$

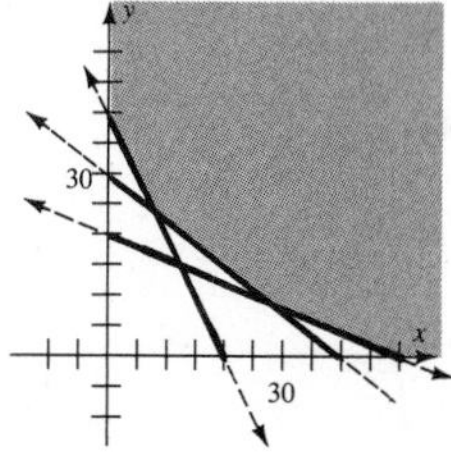

17. $(3, 2), (5, 5), (7, 5), (\frac{10}{3}, \frac{4}{3})$

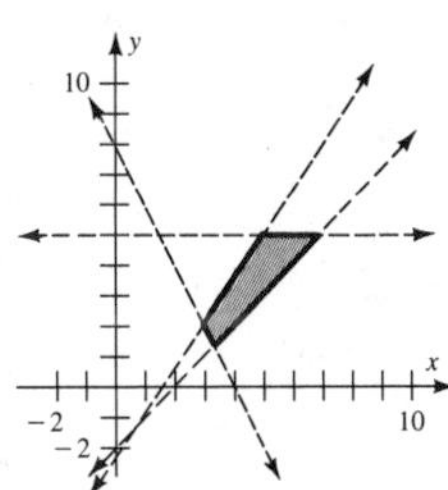

19. Maximum of 190 occurs at $(5, 2)$. **21.** Maximum of 600 occurs at $(6, 0)$. **23.** Maximum of 500 occurs at $(2, 3)$. **25.** Maximum of 72 occurs at $(0, 6)$. **27.** Maximum of 12 occurs at $(6, 0)$. **29.** Maximum of 120 occurs at $(0, 6)$. **31.** Minimum of $\frac{7,360}{7}$ occurs at $(\frac{200}{7}, \frac{60}{7})$. **33.** Maximum of 45 occurs at $(6, 3)$. **35.** Minimum of $20\frac{2}{3}$ occurs at $(\frac{10}{3}, \frac{4}{3})$. **37.** Minimum of 560 occurs at $(4, 0)$. **39.** $P = \$8,000$ occurs at $(\frac{400}{11}, 0)$. **41.** The farmer should plant 100 acres of corn and 200 acres of wheat (leaving 200 acres unplanted). **43.** They should manufacture 400 standard models and 100 economy models. **45.** The minimum cost is \$1.19. **47.** You should invest \$70,000 in Pertec stock and \$30,000 in Campbell bonds.

4.4 Slack Variables and the Pivot (*page 190*)

1. Maximize $z = 30x_1 + 20x_2$

Subject to $\begin{cases} 2x_1 + x_2 \le 12 \\ 5x_1 + 8x_2 \le 40 \\ x_1 \ge 0 \\ x_2 \ge 0 \end{cases}$

3. Maximize $z = 100x_1 + 100x_2$

Subject to $\begin{cases} 3x_1 + 2x_2 \le 12 \\ x_1 + 2x_2 \le 8 \\ x_1 \ge 0 \\ x_2 \ge 0 \end{cases}$

5. Conditions 1 and 3 are violated.

7. Conditions 2 and 3 are violated.

9. Condition 3 is violated.

11. Condition 2 is violated.

13.
$$\begin{array}{cccccc} x_1 & x_2 & y_1 & y_2 & z & \end{array}$$
$$\left[\begin{array}{ccccc|c} 3 & 1 & 1 & 0 & 0 & 300 \\ 2 & 2 & 0 & 1 & 0 & 400 \\ \hline -2 & -3 & 0 & 0 & 1 & 0 \end{array}\right]$$

15.
$$\begin{array}{cccccc} x_1 & x_2 & y_1 & y_2 & z & \end{array}$$
$$\left[\begin{array}{ccccc|c} 1 & 1 & 1 & 0 & 0 & 200 \\ 4 & 2 & 0 & 1 & 0 & 500 \\ \hline -45 & -35 & 0 & 0 & 1 & 0 \end{array}\right]$$

17.
$$\begin{array}{ccccccc} x_1 & x_2 & y_1 & y_2 & y_3 & z & \end{array}$$
$$\left[\begin{array}{cccccc|c} 12 & 150 & 1 & 0 & 0 & 0 & 1{,}200 \\ 6 & 200 & 0 & 1 & 0 & 0 & 1{,}200 \\ 16 & 50 & 0 & 0 & 1 & 0 & 800 \\ \hline -1 & -1 & 0 & 0 & 0 & 1 & 0 \end{array}\right]$$

19.
$$\begin{array}{ccccccc} x_1 & x_2 & y_1 & y_2 & y_3 & z & \end{array}$$
$$\left[\begin{array}{cccccc|c} 8 & 5 & 3 & 1 & 0 & 0 & 1{,}000 \\ 5 & 1 & 3 & 0 & 1 & 0 & 800 \\ \hline -12 & -7 & -5 & 0 & 0 & 1 & 0 \end{array}\right]$$

21.
$$\begin{array}{cccccccccc} x_1 & x_2 & x_3 & x_4 & y_1 & y_2 & y_3 & y_4 & z & \end{array}$$
$$\left[\begin{array}{ccccccccc|c} 1 & 1 & 0 & 0 & 1 & 0 & 0 & 0 & 0 & 90 \\ 0 & 0 & 1 & 1 & 0 & 1 & 0 & 0 & 0 & 130 \\ 1 & 0 & 1 & 0 & 0 & 0 & 1 & 0 & 0 & 80 \\ 0 & 1 & 0 & 1 & 0 & 0 & 0 & 1 & 0 & 110 \\ \hline -9 & -7 & -7 & -8 & 0 & 0 & 0 & 0 & 1 & 0 \end{array}\right]$$

23.
$$\begin{array}{cccccc} x_1 & x_2 & y_1 & y_2 & z & \end{array}$$
$$\left[\begin{array}{ccccc|c} 3 & 0 & 1 & 0 & 0 & 90 \\ 1 & \frac{1}{3} & 0 & \frac{1}{6} & 0 & 3 \\ \hline -12 & -6 & 0 & 0 & 1 & 0 \end{array}\right]$$
Divide row 2 by 6.

$$\begin{array}{cccccc} x_1 & x_2 & y_1 & y_2 & z & \end{array}$$
$$\left[\begin{array}{ccccc|c} 0 & -1 & 1 & -\frac{1}{2} & 0 & 81 \\ 1 & \frac{1}{3} & 0 & \frac{1}{6} & 0 & 3 \\ \hline 0 & -2 & 0 & 2 & 1 & 36 \end{array}\right]$$
Add -3(row 2) to row 1; add 12(row 1) to row 3.

25.
$$\begin{array}{cccccc} x_1 & x_2 & y_1 & y_2 & z & \end{array}$$
$$\left[\begin{array}{ccccc|c} 1 & 4 & \frac{1}{3} & 0 & 0 & 430 \\ 0 & 1 & 1 & \frac{1}{2} & 0 & 81 \\ \hline 0 & -6 & 4 & 0 & 1 & 360 \end{array}\right]$$
Divide row 2 by 2.

$$\begin{array}{cccccc} x_1 & x_2 & y_1 & y_2 & z & \end{array}$$
$$\left[\begin{array}{ccccc|c} 1 & 0 & -\frac{11}{3} & 2 & 0 & 106 \\ 0 & 1 & 1 & \frac{1}{2} & 0 & 81 \\ \hline 0 & 0 & 10 & 3 & 1 & 846 \end{array}\right]$$
Add -4(row 2) to row 1 and add 6(row 2) to row 3.

27.

$$\begin{array}{ccccc|c} x_1 & x_2 & y_1 & y_2 & z & \\ \end{array}$$
$$\left[\begin{array}{ccccc|c} 2 & 1 & \frac{1}{4} & 0 & 0 & 10 \\ 12 & 6 & 0 & 1 & 0 & 600 \\ \hdashline -8 & -10 & 0 & 0 & 1 & 0 \end{array}\right]$$

Divide row 1 by 4.

$$\begin{array}{ccccc|c} x_1 & x_2 & y_1 & y_2 & z & \\ \end{array}$$
$$\left[\begin{array}{ccccc|c} 2 & 1 & \frac{1}{4} & 0 & 0 & 10 \\ 0 & 1 & -\frac{3}{2} & 1 & 0 & 540 \\ \hdashline 12 & 0 & \frac{5}{2} & 0 & 1 & 100 \end{array}\right]$$

Add -6(row 1) to row 2; add 10(row 1) to row 3.

29.

$$\begin{array}{cccccc|c} x_1 & x_2 & y_1 & y_2 & y_3 & z & \\ \end{array}$$
$$\left[\begin{array}{cccccc|c} 2 & 3 & 1 & 0 & 0 & 0 & 200 \\ 1 & 3 & 0 & \frac{1}{2} & 0 & 0 & 50 \\ 3 & 2 & 0 & 0 & 1 & 0 & 300 \\ \hdashline -10 & -5 & 0 & 0 & 0 & 1 & 0 \end{array}\right]$$

Divide row 2 by 2.

$$\begin{array}{cccccc|c} x_1 & x_2 & y_1 & y_2 & y_3 & z & \\ \end{array}$$
$$\left[\begin{array}{cccccc|c} 0 & -3 & 1 & -1 & 0 & 0 & 100 \\ 1 & 3 & 0 & \frac{1}{2} & 0 & 0 & 50 \\ 0 & -7 & 0 & -\frac{3}{2} & 1 & 0 & 150 \\ \hdashline 0 & 25 & 0 & 5 & 0 & 1 & 500 \end{array}\right]$$

Add -2(row 2) to row 1.
Add -3(row 2) to row 3.
Add 10(row 2) to row 4.

31.

$$\begin{array}{cccccc|c} x_1 & x_2 & y_1 & y_2 & y_3 & z & \\ \end{array}$$
$$\left[\begin{array}{cccccc|c} 0 & -5 & 1 & 0 & 0 & 0 & 300 \\ \frac{1}{2} & 1 & 0 & \frac{1}{2} & 0 & 0 & 10 \\ 0 & 8 & 0 & -2 & 1 & 0 & 180 \\ \hdashline 0 & -4 & 0 & 6 & 0 & 1 & 300 \end{array}\right]$$

Divide row 2 by 2.

$$\begin{array}{cccccc|c} x_1 & x_2 & y_1 & y_2 & y_3 & z & \\ \end{array}$$
$$\left[\begin{array}{cccccc|c} \frac{5}{2} & 0 & 1 & \frac{5}{2} & 0 & 0 & 350 \\ \frac{1}{2} & 1 & 0 & \frac{1}{2} & 0 & 0 & 10 \\ -4 & 0 & 0 & -6 & 1 & 0 & 100 \\ \hdashline 2 & 0 & 0 & 8 & 0 & 1 & 340 \end{array}\right]$$

Add 5(row 2) to row 1.
Add -8(row 2) to row 3.
Add 4(row 2) to row 4.

33.

$$\begin{array}{ccccccc|c} x_1 & x_2 & x_3 & y_1 & y_2 & y_3 & z & \\ \end{array}$$
$$\left[\begin{array}{ccccccc|c} 1 & 2 & 4 & 1 & 0 & 0 & 0 & 80 \\ 1 & 4 & 3 & 0 & 1 & 0 & 0 & 60 \\ 4 & 2 & 1 & 0 & 0 & \frac{1}{2} & 0 & 5 \\ \hdashline -2 & -3 & -4 & 0 & 0 & 0 & 1 & 0 \end{array}\right]$$

Divide row 3 by 2.

$$\begin{array}{ccccccc|c} x_1 & x_2 & x_3 & y_1 & y_2 & y_3 & z & \\ \end{array}$$
$$\left[\begin{array}{ccccccc|c} -15 & -6 & 0 & 1 & 0 & -2 & 0 & 60 \\ -11 & -2 & 0 & 0 & 1 & -\frac{3}{2} & 0 & 45 \\ 4 & 2 & 1 & 0 & 0 & \frac{1}{2} & 0 & 5 \\ \hdashline 14 & 5 & 0 & 0 & 0 & 2 & 1 & 20 \end{array}\right]$$

Add -4(row 3) to row 1.
Add -3(row 3) to row 2.
Add 4(row 3) to row 4.

35. **a.**

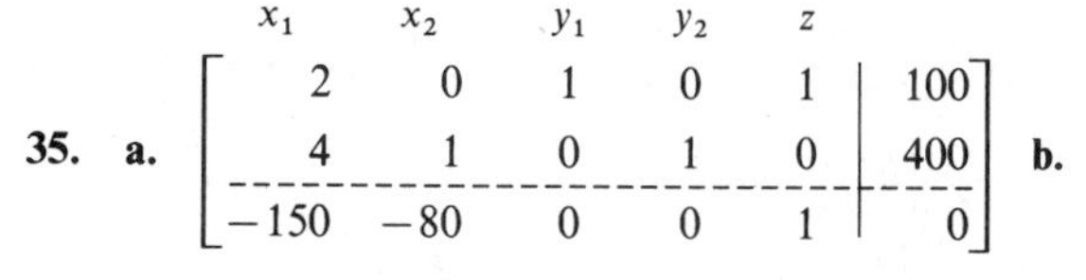

x_1	x_2	y_1	y_2	z	
2	0	1	0	1	100
4	1	0	1	0	400
−150	−80	0	0	1	0

b.

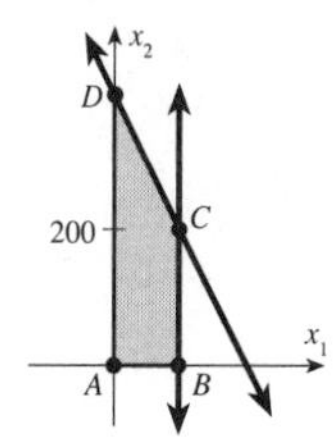

c. Point $A(0,0)$

x_1	x_2	y_1	y_2	z
0	0	**100**	**400**	**0**

System values

$$\begin{aligned} 2(0) + 0(0) + 1(\mathbf{100}) + 0(400) + 0(0) &= 100 \\ 4(0) + 1(0) + 0(100) + 1(\mathbf{400}) + 0(0) &= 400 \\ -150(0) - 80(0) + 0(100) + 0(400) + 1(\mathbf{0}) &= 0 \end{aligned}$$

↑ ↑ ↑

For the first equation to be true, this value must be 100.
For the second equation to be true, this value must be 400.
For the third equation to be true, this value must be 0.

Point $B(50,0)$

x_1	x_2	y_1	y_2	z
50	0	**0**	**200**	**7,500**

System values

$$\begin{aligned} 2(50) + 0(0) + 1(\mathbf{0}) + 0(200) + 0(7{,}500) &= 100 \\ 4(50) + 1(0) + 0(0) + 1(\mathbf{200}) + 0(7{,}500) &= 400 \\ -150(50) - 80(0) + 0(0) + 0(200) + 1(\mathbf{7{,}500}) &= 0 \end{aligned}$$

↑ ↑ ↑

For the first equation to be true, this value must be 0.
For the second equation to be true, this value must be 200.
For the third equation to be true, this value must be 7,500.

Point $C(50,200)$

x_1	x_2	y_1	y_2	z
50	200	**0**	**0**	**23,500**

System values

$$\begin{aligned} 2(50) + 0(200) + 1(\mathbf{0}) + 0(0) + 0(23{,}500) &= 100 \\ 4(50) + 1(200) + 0(0) + 1(\mathbf{0}) + 0(23{,}500) &= 400 \\ -150(50) - 80(200) + 0(0) + 0(0) + 1(\mathbf{23{,}500}) &= 0 \end{aligned}$$

↑ ↑ ↑

For the first equation to be true, this value must be 0.
For the second equation to be true, this value must be 0.
For the third equation to be true, this value must be 23,500.

Point $D(0,400)$

x_1	x_2	y_1	y_2	z
0	400	**100**	**0**	**32,000**

System values

$$\begin{aligned} 2(0) + 0(400) + 1(\mathbf{100}) + 0(0) + 0(32{,}000) &= 100 \\ 4(0) + 1(400) + 0(100) + 1(\mathbf{0}) + 0(32{,}000) &= 400 \\ -150(0) - 80(400) + 0(100) + 0(0) + 1(\mathbf{32{,}000}) &= 0 \end{aligned}$$

↑ ↑ ↑

For the first equation to be true, this value must be 100.
For the second equation to be true, this value must be 0.
For the third equation to be true, this value must be 32,000.

The maximum value is 32,000 at $x_1 = 0$ and $x_2 = 400$.

37. **a.**

$$\begin{array}{ccccc|c} x_1 & x_2 & y_1 & y_2 & z & \\ 0 & 5 & 1 & 0 & 1 & 300 \\ 6 & 1 & 0 & 1 & 0 & 240 \\ \hline -15 & -25 & 0 & 0 & 1 & 0 \end{array}$$

b.

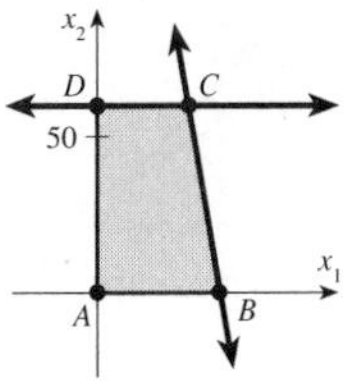

Point $A(0,0)$ System values

x_1	x_2	y_1	y_2	z
0	0	**300**	**240**	**0**

$$\begin{aligned} 0(0) + 5(0) + 1(\mathbf{300}) + 0(240) + 0(0) &= 300 \\ 6(0) + 1(0) + 0(300) + 1(\mathbf{240}) + 0(0) &= 240 \\ -15(0) - 25(0) + 0(300) + 0(240) + 1(\mathbf{0}) &= 0 \end{aligned}$$

↑ ↑ ↑

For the first equation to be true, this value must be 300.
For the second equation to be true, this value must be 240.
For the third equation to be true, this value must be 0.

Point $B(40,0)$ System values

x_1	x_2	y_1	y_2	z
40	0	**300**	**0**	**600**

$$\begin{aligned} 0(40) + 5(0) + 1(\mathbf{300}) + 0(0) + 0(600) &= 300 \\ 6(40) + 1(0) + 0(300) + 1(\mathbf{0}) + 0(600) &= 240 \\ -15(40) - 25(0) + 0(300) + 0(0) + 1(\mathbf{600}) &= 0 \end{aligned}$$

↑ ↑

For the first equation to be true, this value must be 300.
For the second equation to be true, this value must be 0.
For the third equation to be true, this value must be 600.

Point $C(30,60)$ System values

x_1	x_2	y_1	y_2	z
30	60	**0**	**0**	**1,950**

$$\begin{aligned} 0(30) + 5(60) + 1(\mathbf{0}) + 0(0) + 0(1{,}950) &= 300 \\ 6(30) + 1(60) + 0(0) + 1(\mathbf{0}) + 0(1{,}950) &= 240 \\ -15(30) - 25(60) + 0(0) + 0(0) + 1(\mathbf{1{,}950}) &= 0 \end{aligned}$$

↑ ↑ ↑

For the first equation to be true, this value must be 0.
For the second equation to be true, this value must be 0.
For the third equation to be true, this value must be 1,950.

Point $D(0,60)$ System values

x_1	x_2	y_1	y_2	z
0	60	**0**	**180**	**1,500**

$$\begin{aligned} 0(0) + 5(60) + 1(\mathbf{0}) + 0(180) + 0(1{,}500) &= 300 \\ 6(0) + 1(60) + 0(0) + 1(\mathbf{180}) + 0(1{,}500) &= 240 \\ -15(0) - 25(60) + 0(0) + 0(180) + 1(\mathbf{1{,}500}) &= 0 \end{aligned}$$

↑ ↑ ↑

For the first equation to be true, this value must be 0.
For the second equation to be true, this value must be 180.
For the third equation to be true, this value must be 1,500.

The maximum value is 1,950 at $x_1 = 30$, $x_2 = 60$.

39. **a.**

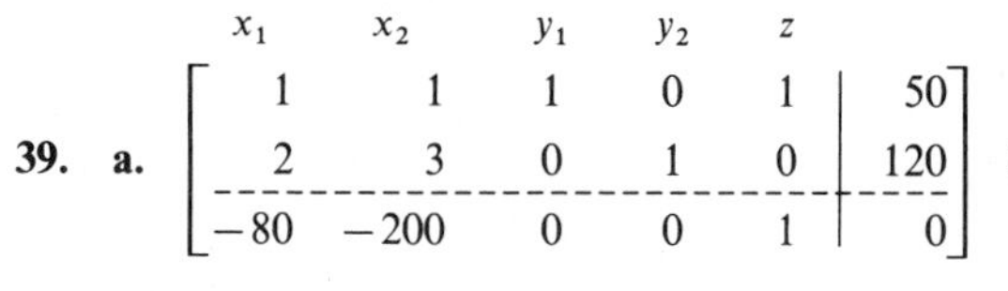

$$\begin{array}{ccccc|c} x_1 & x_2 & y_1 & y_2 & z & \\ 1 & 1 & 1 & 0 & 1 & 50 \\ 2 & 3 & 0 & 1 & 0 & 120 \\ \hline -80 & -200 & 0 & 0 & 1 & 0 \end{array}$$

b.

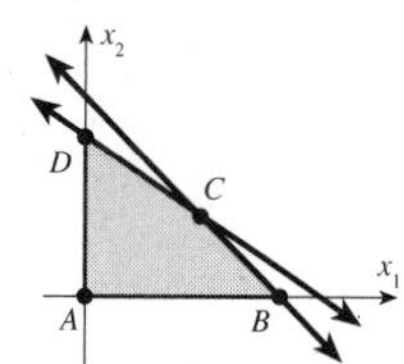

Point $A(0,0)$ System values

x_1	x_2	y_1	y_2	z
0	0	**50**	**120**	**0**

$$\begin{array}{rl} & x_1 \qquad x_2 \qquad y_1 \qquad y_2 \qquad z \\ 1(0) + & 1(0) + 1(\mathbf{50}) + 0(120) + 0(0) = 50 \\ 2(0) + & 3(0) + 0(50) + 1(\mathbf{120}) + 0(0) = 120 \\ -80(0) - & 200(0) + 0(50) + 0(120) + 1(\mathbf{0}) = 0 \end{array}$$

↑ ↑ ↑

For the first equation to be true, this value must be 50.

For the second equation to be true, this value must be 120.

For the third equation to be true, this value must be 0.

Point $B(50,0)$ System values

x_1	x_2	y_1	y_2	z
50	0	**0**	**20**	**4,000**

$$\begin{array}{rl} & x_1 \qquad x_2 \qquad y_1 \qquad y_2 \qquad z \\ 1(50) + & 1(0) + 1(\mathbf{0}) + 0(20) + 0(4{,}000) = 50 \\ 2(50) + & 3(0) + 0(0) + 1(\mathbf{20}) + 0(4{,}000) = 120 \\ -80(50) - & 200(0) + 0(0) + 0(20) + 1(\mathbf{4{,}000}) = 0 \end{array}$$

↑ ↑ ↑

For the first equation to be true, this value must be 0.

For the second equation to be true, this value must be 20.

For the third equation to be true, this value must be 4,000.

Point $C(30,20)$ System values

x_1	x_2	y_1	y_2	z
30	20	**0**	**0**	**6,400**

$$\begin{array}{rl} & x_1 \qquad x_2 \qquad y_1 \qquad y_2 \qquad z \\ 1(30) + & 1(20) + 1(\mathbf{0}) + 0(0) + 0(6{,}400) = 50 \\ 2(30) + & 3(20) + 0(0) + 1(\mathbf{0}) + 0(6{,}400) = 120 \\ -80(30) - & 200(20) + 0(0) + 0(0) + 1(\mathbf{6{,}400}) = 0 \end{array}$$

↑ ↑ ↑

For the first equation to be true, this value must be 0.

For the second equation to be true, this value must be 0.

For the third equation to be true, this value must be 6,400.

Point $D(0,40)$ System values

x_1	x_2	y_1	y_2	z
0	40	**10**	**0**	**8,000**

$$\begin{array}{rl} & x_1 \qquad x_2 \qquad y_1 \qquad y_2 \qquad z \\ 1(0) + & 1(40) + 1(\mathbf{10}) + 0(0) + 0(8{,}000) = 50 \\ 2(0) + & 3(40) + 0(10) + 1(\mathbf{0}) + 0(8{,}000) = 120 \\ -80(0) - & 200(40) + 0(10) + 0(0) + 1(\mathbf{8{,}000}) = 0 \end{array}$$

↑ ↑ ↑

For the first equation to be true, this value must be 10.

For the second equation to be true, this value must be 0.

For the third equation to be true, this value must be 8,000.

The maximum value is 800 at $x_1 = 0$ and $x_2 = 40$.

4.5 Maximization by the Simplex Method (*page 203*)

1. **a.** $x_1 = 0, x_2 = 0, y_1 = 30, y_2 = 50, z = 0$ **b.** $x_1 = 0, x_2 = 0, y_1 = 120, y_2 = 180, z = 0$ **3.** **a.** $x_1 = 0, x_2 = 0, x_3 = 0, y_1 = 60, y_2 = 30, y_3 = 40, z = 0$ **b.** $x_1 = 0, x_2 = 0, x_3 = 0, y_1 = 80, y_2 = 50, y_3 = 60, y_4 = 90, z = 0$
5. **a.** $x_1 = 10, y_1 = 12, y_3 = 20, x_2 = y_2 = 0, z = 32$. Final tableau. **b.** $x_1 = 120, x_3 = 80, y_1 = 20, x_2 = y_2 = y_3 = 0, z = 360$. Because the bottom row has negative entries, this is not the final tableau.

7.

$$\begin{array}{ccccc|c} x_1 & x_2 & y_1 & y_2 & z & \\ 6 & 3 & 1 & 0 & 0 & 20 \\ 2 & \textcircled{4} & 0 & 1 & 0 & 4 \\ \hline -4 & -20 & 0 & 0 & 1 & 0 \end{array}$$

The pivot is circled.

$$\begin{array}{ccccc|c} x_1 & x_2 & y_1 & y_2 & z & \\ 6 & 3 & 1 & 0 & 0 & 20 \\ \frac{1}{2} & 1 & 0 & \frac{1}{4} & 0 & 1 \\ \hline -4 & -20 & 0 & 0 & 1 & 0 \end{array}$$

Divide row 2 by the pivot.

$$\begin{array}{ccccc|c} x_1 & x_2 & y_1 & y_2 & z & \\ \frac{9}{2} & 0 & 1 & -\frac{3}{4} & 0 & 17 \\ \frac{1}{2} & 1 & 0 & \frac{1}{4} & 0 & 1 \\ \hline 6 & 0 & 0 & 5 & 1 & 20 \end{array}$$

Carry out pivoting process.

9.

$$\begin{array}{cccccc|c} x_1 & x_2 & y_1 & y_2 & y_3 & z & \\ 2 & 3 & 1 & 0 & 0 & 0 & 50 \\ 45 & \textcircled{15} & 0 & 1 & 0 & 0 & 30 \\ 4 & 3 & 0 & 0 & 1 & 0 & 70 \\ \hline -5 & -20 & 0 & 0 & 0 & 1 & 0 \end{array}$$

The pivot is circled.

$$\begin{array}{cccccc|c} x_1 & x_2 & y_1 & y_2 & y_3 & z & \\ 2 & 3 & 1 & 0 & 0 & 0 & 50 \\ 3 & 1 & 0 & \frac{1}{15} & 0 & 0 & 2 \\ 4 & 3 & 0 & 0 & 1 & 0 & 70 \\ \hline -5 & -20 & 0 & 0 & 0 & 1 & 0 \end{array}$$

Divide row 2 by the pivot.

$$\begin{array}{cccccc|c} x_1 & x_2 & y_1 & y_2 & y_3 & z & \\ -7 & 0 & 1 & -\frac{1}{5} & 0 & 0 & 44 \\ 3 & 1 & 0 & \frac{1}{15} & 0 & 0 & 2 \\ -5 & 0 & 0 & -\frac{1}{5} & 1 & 0 & 64 \\ \hline 55 & 0 & 0 & \frac{4}{3} & 0 & 1 & 40 \end{array}$$

Carry out pivoting process.

11.

$$\begin{array}{ccccccc|c} x_1 & x_2 & x_3 & y_1 & y_2 & y_3 & z & \\ 5 & -1 & -3 & 1 & 0 & 0 & 0 & 20 \\ 2 & 1 & 9 & 0 & 1 & 0 & 0 & 10 \\ 1 & 3 & \textcircled{9} & 0 & 0 & 1 & 0 & 2 \\ \hline -4 & -5 & -9 & 0 & 0 & 0 & 1 & 0 \end{array}$$

The pivot is circled.

$$\begin{array}{ccccccc|c} x_1 & x_2 & x_3 & y_1 & y_2 & y_3 & z & \\ 5 & -1 & -3 & 1 & 0 & 0 & 0 & 20 \\ 2 & 1 & 9 & 0 & 1 & 0 & 0 & 10 \\ \frac{1}{9} & \frac{1}{3} & 1 & 0 & 0 & \frac{1}{9} & 0 & \frac{2}{9} \\ \hline -4 & -5 & -9 & 0 & 0 & 0 & 1 & 0 \end{array}$$

Divide row 3 by the pivot.

$$\begin{array}{ccccccc|c} x_1 & x_2 & x_3 & y_1 & y_2 & y_3 & z & \\ \frac{16}{3} & 0 & 0 & 1 & 0 & \frac{1}{3} & 0 & \frac{62}{3} \\ 1 & -2 & 0 & 0 & 1 & -1 & 0 & 8 \\ \textcircled{\frac{1}{9}} & \frac{1}{3} & 1 & 0 & 0 & \frac{1}{9} & 0 & \frac{2}{9} \\ \hdashline -3 & -2 & 0 & 0 & 0 & 9 & 1 & 2 \end{array}$$

Carry out pivoting process. (The next pivot is circled.)

$$\begin{array}{ccccccc|c} x_1 & x_2 & x_3 & y_1 & y_2 & y_3 & z & \\ \frac{16}{3} & 0 & 0 & 1 & 0 & \frac{1}{3} & 0 & \frac{62}{3} \\ 1 & -2 & 0 & 0 & 1 & -1 & 0 & 8 \\ 1 & 3 & 9 & 0 & 0 & 1 & 0 & 2 \\ \hdashline -3 & -2 & 0 & 0 & 0 & 1 & 1 & 2 \end{array}$$

Divide row 3 by the pivot.

$$\begin{array}{ccccccc|c} x_1 & x_2 & x_3 & y_1 & y_2 & y_3 & z & \\ 0 & -16 & -48 & 1 & 0 & -5 & 0 & 10 \\ 0 & -5 & -9 & 0 & 1 & -2 & 0 & 6 \\ 1 & 3 & 9 & 0 & 0 & 1 & 0 & 2 \\ \hdashline 0 & 7 & 27 & 0 & 0 & 4 & 1 & 8 \end{array}$$

Carry out pivoting process.

13. Maximum of $z = 600$ occurs when $x_1 = 0, x_2 = 200$. **15.** Maximum of 480 occurs when $x_1 = 0$ and $x_2 = 30$. **17.** Maximum of 7,500 occurs when $x_1 = 50$ and $x_2 = 150$. **19.** Maximum of 4,700 occurs when $x_1 = 700$ and $x_2 = 150$. **21.** Maximum of $z = 100$ occurs when $x_1 = 50$ and $x_2 = 0$. **23.** Maximum of 400 occurs when $x_1 = 100$ and $x_2 = 0$. **25.** Maximum of 6,000 occurs when $x_1 = 500$ and $x_2 = 0$. **27.** Maximum of $z = 1{,}600$ occurs when $x_1 = 80$ and $x_4 = 110$. **29.** 700 Alpha products and 150 Beta products should be manufactured. Then the (maximum) profit will be $4,700. **31.** 700 Alpha products and 150 Beta products should be manufactured. Then the (maximum) profit will be $4,700. (The condition added to Problem 29 is superfluous.) **33.** Rosenberg's would place 3 advertisements in the *Press Democrat*, 5 advertisements in the *News Herald*, and 5 advertisements in *U.S.A. Today*. **35.** The maximum profit is $1,600 when 50 model 2 shoes, 250 model 3 shoes, and no model 1 shoes are manufactured.

4.6 Nonstandard Linear Programming Problems (page 213)

1. The solution shows a maximum value of 400 when $x_1 = 0$ and $x_2 = 8$.

3.

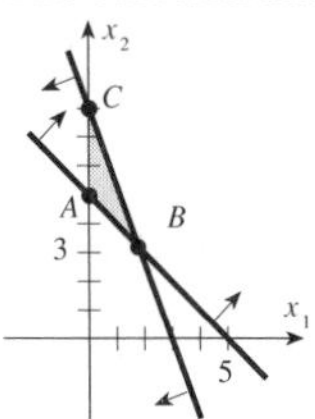

Maximum value is 400 at (0, 8).

Check corner points:	Objective function $20x_1 + 50x_2$
A: (0, 5)	$20(0) + 50(5) = 250$
B: $(\frac{9}{5}, \frac{16}{5})$	$20(0) + 50(8) = 400$
C: (0, 8)	$20(\frac{9}{5}) + 50(\frac{16}{5}) = 196$

Reconcile with simplex method (Problem 1):

1st tableau (0, 0) is not feasible (see negative in right-hand column).
2nd tableau (5, 0) is not feasible (see negative in right-hand column).
3rd tableau $(\frac{9}{5}, \frac{16}{5})$ is point B.
Result tableau (0, 8) is point C.

5. The minimum value is $z' = -z = 300$ when $x_1 = 2$ and $x_2 = \frac{8}{3}$.

7. 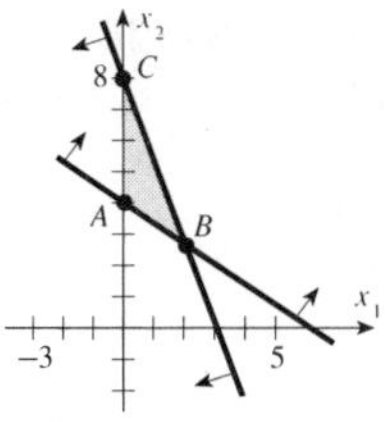

Minimum value is 300 at $(2, \frac{8}{3})$.

Check corner points:	Objective function $30x_1 + 90x_2$
A: $(0,4)$	$30(0) + 90(4) = 360$
B: $(2,\frac{8}{3})$	$20(2) + 90(\frac{8}{3}) = 300$
C: $(0,8)$	$30(0) + 90(8) = 720$

Minimum value is 300 at $(2, \frac{8}{3})$.
Reconcile with simplex method (Problem 5):
1st tableau point $(0, 0)$ is not feasible.
2nd tableau $(6, 0)$ is not feasible.
3rd tableau $(2, \frac{8}{3})$ is the point that gives the minimum value.

9. Maximum of 400 occurs when $x_1 = 0$ and $x_2 = 10$. **11.** Maximum of 600 occurs when $x_1 = 20$ and $x_2 = 0$. **13.** Maximum of $\frac{405}{2}$ occurs when $x_1 = \frac{15}{4}$ and $x_2 = -\frac{9}{2}$. **15.** Maximum of 400 occurs when $x_1 = 8$ and $x_2 = 0$. **17.** Maximum of 12 occurs when $x_1 = 0$ and $x_2 = 4$. **19.** Maximum of $z' = 40$, so minimum of -40 occurs when $x_1 = 2$ and $x_2 = 0$. **21.** Maximum of $z' = -125$, so minimum of 125 occurs when $x_1 = 3$ and $x_2 = 2$. **23.** Maximum of $z' = 30$, so minimum of $z = z' + 50 = 80$ occurs when $x_1 = 15$, $x_2 = 0$, and $x_3 = 10 - x_2 = 10$. **25.** Minimum cost of \$57.00 occurs when no units of A, 57 units of B, and no units of C are mixed. **27.** Any ordered pair on the line segment with endpoints (80, 40) and (87.5, 25) will yield the maximum, \$400, when substituted into the objective function. Since we desire integral solutions, we could start at (80, 40) and use the idea of the slope of the line segment, which is $-\frac{2}{1}$, to generate other integral solutions: (81, 38), (82, 36), (83, 34), (84, 32), (85, 30), (86, 28), and (87, 26). **29.** Answers vary, but the key idea is that the simplex method always gives the maximum, *but only one* set of values where it occurs. The occurrence of a maximum will not be unique if and only if the objective function has the same slope as one of the sides of the (polygonal) feasibility region. In this case, we might have noticed that the objective function and the first constraint both have slopes of -2. Perhaps in such a situation, one should resort to the graphing method when possible. **31.** Sebastopol Winery should ship 600 from Windsor to Santa Rosa, 300 from Windsor to San Rafael, and 700 from Graton to San Rafael. **33.** The final tableau shows 6 trainees, 15 regular employees and 1 supervisor. The cost is \$1,026.

4.7 Duality (page 222)

1. a. $A^T = \begin{bmatrix} 6 & 4 \\ 9 & 8 \end{bmatrix}$ **b.** $B^T = \begin{bmatrix} 5 & 3 \\ 6 & 8 \end{bmatrix}$ **c.** $C^T = \begin{bmatrix} 4 & 6 \\ 9 & 1 \\ 1 & 4 \end{bmatrix}$ **3. a.** $G^T = \begin{bmatrix} 1 \\ 3 \\ 5 \end{bmatrix}$ **b.** $H^T = [4 \quad 9 \quad 6]$ **c.** $J^T = [1 \quad 0 \quad 3 \quad 2]$ **5.** Maximize $z' = 10y_1 + 30y_2$

Subject to $\begin{cases} 2y_1 + 3y_2 \le 3 \\ 8y_1 + 5y_2 \le 4 \\ y_1 \ge 0, \quad y_2 \ge 0 \end{cases}$

7. Maximize $z' = 10y_1 + 2y_2 + y_3 + 15y_4$

$\begin{cases} y_1 + 2y_2 + 3y_4 \le 3 & y_1 \ge 0 \\ y_1 + y_3 \le 2 & y_2 \ge 0 \\ y_1 + 3y_2 + 2y_3 + 5y_4 \le 5 & y_3 \ge 0 \\ & y_4 \ge 0 \end{cases}$

9. Minimum value is 24 when $x_1 = 0$ and $x_2 = 6$. **11.** Minimum value is 342 when $x_1 = 15$ and $x_2 = 3$. **13.** Minimum value is 441 when $x_1 = 3$ and $x_2 = 15$. **15.** Minimum value is 528 when $x_1 = 12$ and $x_2 = 4$. **17.** Minimum value is 320 when $x_1 = 8$ and $x_2 = 4$. **19.** Minimum value is 210 when $x_1 = 15$ and $x_2 = 5$. **21.** Minimum value is 206 when $x_1 = 9$, $x_2 = 5$, and $x_3 = 1$. **23.** Minimum value is 792 when $x_1 = 12$, $x_2 = 6$, and $x_3 = 6$. **25.** Minimum value is 850 when $x_1 = 5$, $x_2 = 10$, and $x_3 = 10$. **27.** Minimum value is 25 when $x_1 = 5$, $x_2 = 5$, and $x_3 = 0$. **29.** Minimum of 660 occurs when $x_1 = 0$, $x_2 = 0$, and $x_3 = 33$. **31.** Minimum of 72 occurs when $x_1 = 0$ and $x_2 = 6$. **33.** Minimum of 120 occurs when $x_1 = 0$ and $x_2 = 6$. **35.** Minimum of 23 occurs when $x_1 = 4$ and $x_2 = 1$. **37.** Minimum of 560 occurs when $x_1 = 4$ and $x_2 = 0$. **39.** The Gainesville plant should operate 30 days and the Sacramento plant should operate 20 days. The minimum cost is \$900,000. **41.** She should spend 5 hours jogging, 2 hours playing handball, and 5 hours dancing per week. **43.** 30 units, 25 units, and 50 units should be produced on day, swing, and graveyard shifts, respectively. **45.** Ship 30 sets from Burlingame to Hillsborough and 35 sets from San Jose to Palo Alto. **47.** Ship 15 from San Antonio to Ft. Worth, 10 from San Antonio to Houston, and 25 from Dallas to Houston.

4.8 Review (page 226)

Objective Questions

1.

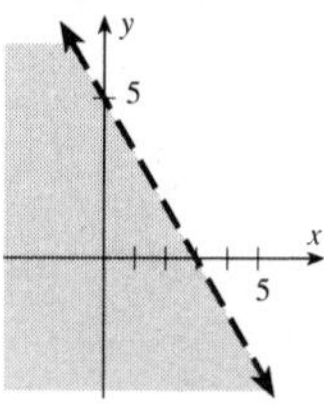

3.

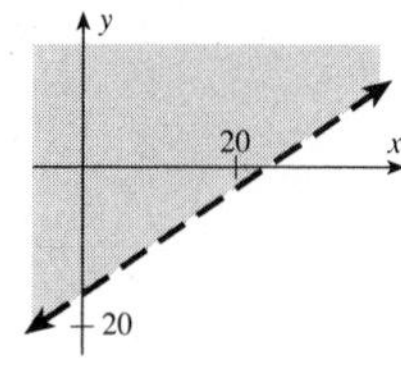

5.

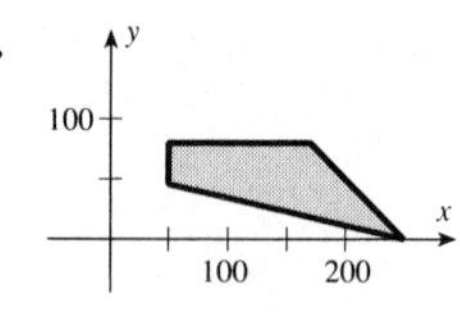

7.

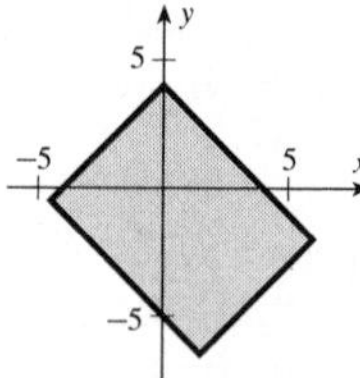

9. Let a = ounces of food A
b = ounces of food B
Minimize $C = .14a + .06b$

$$\text{Subject to } \begin{cases} a \geq 0, \quad b \geq 0 \\ 6a + 2b \geq 100 \\ 3a + 2b \geq 60 \\ a + 2b \geq 40 \end{cases}$$

11. Let x = number of four-person branches
y = number of six-person branches
Maximize income $I = 50{,}000x + 65{,}000y$

$$\text{Subject to } \begin{cases} x \geq 0, \quad y \geq 0 \\ 175{,}000x + 200{,}000y \leq 1{,}275{,}000 \\ 4x + 6y \leq 32 \\ x + y \leq 10 \end{cases}$$

13.

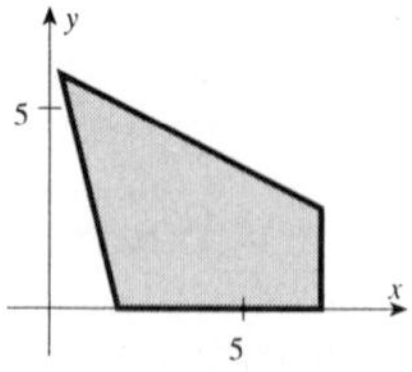

15.

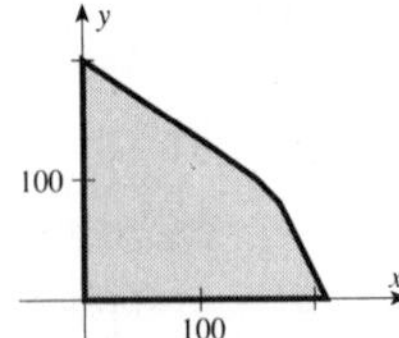

17. $(\frac{2}{7}, \frac{41}{7})$, $(\frac{7}{4}, 0)$, $(7, 0)$, $(7, \frac{5}{2})$ **19.** $(0,0)$, $(210,0)$, $(170,80)$, $(150,100)$, $(0,200)$ **21.** Maximum of 52.5 occurs at $(x, y) = (7, \frac{5}{2})$. **23.** The maximum is 1,170 which occurs at $(x, y) = (170, 80)$. **25.** The corner points of the feasibility region which is not bounded are $(0, 50)$, $(12, 14)$, and $(40, 0)$. The minimum is \$2.52 and occurs when 12 ounces of food A and 14 ounces of food B are bought. **27.** The maximum is \$380,000 and occurs with 5 four-person branches and 2 six-person branches. **29.** The maximum profits will occur when 50 commercial and 150 other guests are served. **31.** Maximize $z = 60x_1 + 20x_2$

$$\text{Subject to } \begin{cases} x_1 \geq 0 \\ x_2 \geq 0 \\ 3x_1 + 2x_2 \leq 52 \\ 6x_1 - 4x_2 \leq 100 \end{cases}$$

33. Violates condition 3 **35.**

$$\begin{array}{c} \begin{array}{ccccccc|c} x_1 & x_2 & x_3 & y_1 & y_2 & y_3 & z & \end{array} \\ \left[\begin{array}{ccccccc|c} 2 & 0 & 1 & 1 & 0 & 0 & 0 & 50 \\ 0 & 4 & 1 & 0 & 1 & 0 & 0 & 90 \\ 3 & 4 & 0 & 0 & 0 & 1 & 0 & 100 \\ \hdashline -6 & -25 & -3 & 0 & 0 & 0 & 1 & 0 \end{array}\right] \end{array}$$

37.

$$\begin{array}{c} \begin{array}{ccccccc|c} x_1 & x_2 & x_3 & y_1 & y_2 & y_3 & z & \end{array} \\ \left[\begin{array}{ccccccc|c} 3 & 2 & 0 & 1 & 0 & 0 & 0 & 120 \\ 0 & 3 & 1 & 0 & 1 & 0 & 0 & 90 \\ 0 & 0 & 1 & 0 & 0 & 1 & 0 & 100 \\ \hdashline -15 & -5 & -1 & 0 & 0 & 0 & 1 & 0 \end{array}\right] \end{array}$$

39.

$$\begin{array}{c} \begin{array}{cccccc|c} x_1 & x_2 & y_1 & y_2 & y_3 & z & \end{array} \\ \left[\begin{array}{cccccc|c} 0 & -3 & 1 & -\frac{4}{3} & 0 & 0 & 20 \\ 1 & 2 & 0 & \frac{1}{3} & 0 & 0 & 4 \\ 0 & 1 & 0 & -\frac{1}{3} & 1 & 0 & 16 \\ \hdashline 0 & 15 & 0 & \frac{10}{3} & 0 & 1 & 40 \end{array}\right] \end{array}$$

41.
$$\begin{array}{ccccccc|c} x_1 & x_2 & x_3 & y_1 & y_2 & y_3 & z & \\ \hline -2 & -5 & 0 & 1 & -\frac{4}{3} & 0 & 0 & \frac{20}{3} \\ 1 & 2 & 1 & 0 & \frac{1}{3} & 0 & 0 & \frac{40}{3} \\ 4 & -1 & 0 & 0 & -\frac{1}{3} & 1 & 0 & \frac{50}{3} \\ \hline 3 & 10 & 0 & 0 & \frac{8}{3} & 0 & 1 & \frac{320}{3} \end{array}$$

43. $x_1 = x_2 = x_3 = 0, y_1 = 30, y_2 = 20, y_3 = 10, z = 0$

45. $x_1 = y_2 = y_3 = 0, x_2 = 38, x_3 = 19, y_1 = 60, z = 120$

47.
$$\begin{array}{ccccc|c} x_1 & x_2 & y_1 & y_2 & z & \\ \hline 0 & -2 & 1 & -2 & 0 & 4 \\ 1 & 2 & 0 & \frac{1}{2} & 0 & 5 \\ \hline 0 & 5 & 0 & \frac{3}{2} & 1 & 15 \end{array}$$

49.
$$\begin{array}{ccccccc|c} x_1 & x_2 & x_3 & y_1 & y_2 & y_3 & z & \\ \hline 1 & 1 & 3 & 1 & 0 & 0 & 0 & 2 \\ 0 & -2 & -5 & -3 & 1 & 0 & 0 & 24 \\ 0 & -2 & -3 & -2 & 0 & 1 & 0 & 16 \\ \hline 0 & 2 & 13 & 5 & 0 & 0 & 1 & 95 \end{array}$$
Final simplex tableau

51. Maximum of 610 occurs when $x_1 = 10$, $x_2 = 22$, and $x_3 = 0$. **53.** Maximum of $z = 690$ occurs when $x_1 = 40$, $x_2 = 0$, and $x_3 = 90$. **55.** The maximum population of 72 occurs with 40 of A1 species and 32 of A2 species.

57. They should produce 80 tables and 40 shelves to reach the maximum profit of \$5,600.

59. Standard form

$$\begin{array}{ccccc|c} x_1 & x_2 & y_1 & y_2 & z & \\ \hline 1 & 3 & 1 & 0 & 0 & 30 \\ 2 & -1 & 0 & 1 & 0 & 24 \\ \hline -5 & -2 & 0 & 0 & 1 & 0 \end{array}$$

61. Not in standard form

$$\begin{array}{ccccccc|c} x_1 & x_2 & x_3 & y_1 & y_2 & y_3 & z & \\ \hline 1 & 0 & 1 & 1 & 0 & 0 & 0 & 10 \\ 0 & 1 & 1 & 0 & 1 & 0 & 0 & 30 \\ -1 & -2 & 0 & 0 & 0 & 1 & 0 & 10 \\ \hline -1 & -2 & -5 & 0 & 0 & 0 & 1 & 0 \end{array}$$

63. Maximum of 65 occurs when $x_1 = 7$ and $x_2 = 10$. **65.** The maximum value is 300 when $x_1 = 3$ and $x_2 = 4$.

67. $\begin{bmatrix} 6 & -4 & 3 \\ 8 & 3 & 4 \\ 1 & -2 & 5 \\ 9 & 0 & 7 \end{bmatrix}$ **69.** $\begin{bmatrix} 8 \\ 1 \\ 4 \end{bmatrix}$

71. Maximize $z' = 100y_1 + 600y_2$

Subject to $\begin{cases} y_1 + 2y_2 \le 30 \\ 5y_1 + 3y_2 \le 25 \\ y_1 \ge 0, \quad y_2 \ge 0 \end{cases}$

73. Maximize $z' = 70y_1 + 60y_2 + 50y_3$

Subject to $\begin{cases} 3y_1 + 2y_2 + 5y_3 \le 100 \\ 5y_1 + 3y_2 + 2y_3 \le 200 \\ y_1 \ge 0, \quad y_2 \ge 0, \quad y_3 \ge 0 \end{cases}$

75. Minimum is 5,000 which occurs when $x_1 = 0$ and $x_2 = 200$. **77.** The minimum is 3,000 which occurs when $x_1 = 30$ and $x_2 = 0$. **79.** Minimum cost \$2 occurs when 2 pounds of peanuts, and 0 pounds of cashews are bought.

81. Minimum of 1 hour occurs when $\frac{1}{2}$ hour is spent collecting cabbage and $\frac{1}{2}$ hour is spent collecting peanuts.

Sample Test

1.
$$\begin{array}{ccccccc|c} x_1 & x_2 & x_3 & y_1 & y_2 & y_3 & z & \\ \hline 2 & 0 & 1 & 1 & 0 & 0 & 0 & 50 \\ 0 & \textcircled{4} & 1 & 0 & 1 & 0 & 0 & 100 \\ 3 & 4 & 0 & 0 & 0 & 1 & 0 & 109 \\ \hline -2 & -20 & -3 & 0 & 0 & 0 & 1 & 0 \end{array}$$

For extra practice solve: Maximum of 506 occurs when $x_1 = 3$, $x_2 = 25$, and $x_3 = 0$.

3. The dual is maximize $z' = 18y_1 + 36y_2 + 30y_3$

Subject to $\begin{cases} 9y_1 + \ \ 3y_2 + 2y_3 \le 50 \\ \ \ y_1 + 12y_2 + 3y_3 \le 80 \end{cases}$

$$\begin{array}{cccccc} y_1 & y_2 & y_3 & x_1 & x_2 & z' \end{array}$$

$$\left[\begin{array}{cccccc|c} 9 & 3 & 2 & 1 & 0 & 0 & 50 \\ 1 & 12 & 3 & 0 & 1 & 0 & 80 \\ \hline -18 & -36 & -30 & 0 & 0 & 1 & 0 \end{array}\right] \begin{array}{l} \text{Initial tableau} \\ \text{Pivot is row 2, column 2.} \\ \\ \end{array}$$

For extra practice, solve:

$$\begin{array}{cccccc} y_1 & y_2 & y_3 & x_1 & x_2 & z' \end{array}$$

$$\left[\begin{array}{cccccc|c} \frac{9}{2} & \frac{3}{2} & 1 & \frac{1}{2} & 0 & 0 & 25 \\ -\frac{25}{2} & \frac{15}{2} & 0 & -\frac{3}{2} & 1 & 0 & 5 \\ \hline 117 & 9 & 0 & 15 & 0 & 1 & 750 \end{array}\right] \text{Final tableau}$$

The minimum value is 750 when $x_1 = 15$ and $x_2 = 0$.

5. C **7.** C **9.** C

CHAPTER 5

5.1 Difference Equations (page 245)

1. 6, 16, 26, 36 **3.** 1, 4, 16, 64 **5.** 10, −50, 250, −1,250 **7.** 4, 11, 25, 53 **9.** 1, 5, 13, 29 **11.** 3, 13, 63, 313

For Problems 13–39 refer to the difference equation $x_{n+1} = ax_n + b$, *which has solution* $x_n = \dfrac{b}{1-a} + \left(x_0 - \dfrac{b}{1-a}\right)a^n$, $a \ne 1$.

13. $b = 8$; $x_4 = 32$ **15.** $x_0 = 1, a = 3, b = 0, n = 4$: $x_4 = 81$ **17.** 10 **19.** $x_0 = 0, a = 5, b = 2, n = 4$: $x_4 = 312$ **21.** $x_0 = 8, a = 2, b = 1, n = 4$: $x_4 = 143$ **23.** $x_0 = 0, a = -\frac{1}{2}, b = 1, n = 4$: $x_4 = \frac{5}{8}$ **25.** $x_n = 5 + 25n$ **27.** $x_n = 4 - 2n$ **29.** $x_n = 4n$ **31.** $a = 3, b = 0, x_0 = 1$: $x_n = 3^n$ **33.** $a = 2, b = 0, x_0 = 1$: $x_n = 2^n$ **35.** $a = 2, b = 3, x_0 = 0$: $x_n = 3 - 3(2^n)$ **37.** $a = 5, b = 9, x_0 = 2$: $x_n = -\frac{9}{4} + \frac{17}{4}(5^n)$ **39.** $a = 4, b = 2, x_0 = 1$: $x_n = -\frac{2}{3} + \frac{5}{3}(4^n)$ **41.** 20 **43.** 49 **45.** 11 **47.** 80 **49.** 75 **51.** In four years the salary will be $56,965. **53.** $x_n = \dfrac{2{,}500}{1 - 1.05} + \left(38{,}000 - \dfrac{2{,}500}{1 - 1.05}\right)1.05^n = 88{,}000(1.05)^n - 50{,}000$ **55.** 188,470 **57.** 2^{24} (8 hours = 24 20-minute periods) **59.** $T_0 = 1$; $T_1 = T_0 + 2$; $T_2 = T_1 + 3$; …, $T_n = T_{n-1} + (n + 1)$

5.2 Interest (page 255)

1. Simple interest = $400. Compounded annually, interest = $469.33. **3.** Simple interest = $720. Compounded annually, interest = $809.86. **5.** Simple interest = $12,000. Compounded annually, interest = $43,231.47. **7.** Simple interest amount = $1,400. Compounded annually, the amount = $1,469.33. **9.** Simple interest = $2,720. Compounded annually, amount = $2,809.86. **11.** Simple interest = $17,000. Compounded annually, amount = $48,231.47. **13.** $i = 9\%$; $N = 5$; $A = \$1{,}538.62$; $I = \$538.62$ **15.** $i = 8\%$; $N = 3$; $A = \$629.86$; $I = \$129.86$ **17.** $i = 2\%$; $N = 12$; $A = \$634.12$; $I = \$134.12$ **19.** $i = 4\frac{1}{2}\%$; $N = 40$; $A = \$29{,}081.82$; $I = \$24{,}081.82$ **21.** $i = 5\%$; $N = 40$; $A = \$35{,}199.94$; $I = \$30{,}199.94$ **23.** $i = 2\%$; $N = 60$; $A = \$13{,}124.12$; $I = \$9{,}124.12$ **25.** $i = 4\%$; $N = 5$; $A = \$1{,}520.82$; $I = \$270.82$ **27.** $24,067.32 **29.** $19,898.24 **31.** $29,960.19 **33.** 6.17% **35.** 12.75% **37.** $16,889.10 **39.** $153,278.42 **41.** $5 **43.** $12 **45.** $2,691 **47.** $256 **49.** $168,187 **51.** $1,177,312 **53.** $3,656 **55.** $131,444 **57.** $2,219 **59.** The second offer is better.

5.3 Annuities (page 260)

1. $56,642 **3.** $59,498 **5.** $22,620 **7.** $61,173 **9.** $61,878 **11.** $23,299 **13.** $4,883 **15.** $220,498 **17.** $2,930 **19.** $72,433 **21.** $14,897 **23.** $310,478 **25.** $988,458 **27.** Ordinary annuity: $24,297. **29.** Annuity due: $1,374,583. **31.** Answers vary.

5.4 Amortization (page 266)

1. $5,628.89 **3.** $5,655.87 **5.** $6,934.43 **7.** $4,313.07 **9.** $1,181.41 **11.** $5,250.32 **13.** $44.42 **15.** $272.19 **17.** $111.52 **19.** $133.78 **21.** $887.30 **23.** $4,802.66 **25.** $131,022 **27.** $1,780.28

29. Present value of the annuity is $25,906, which has greater value than the $25,000. **31.** You can finance about $9,300. If you add to this amount your down payment and the value of your trade-in, you will know the amount you can spend. **33.** $96,800 down; about a 27% down payment. **35.** Price range of $187,500 to $219,000. **37.** Price range from $170,500 to $198,700.

5.5 Sinking Funds (page 269)

1. $1,193.20 **3.** $1,896.70 **5.** $4,485.45 **7.** $6,631.19 **9.** $2,076.83 **11.** $351.35 **13.** $5,426.39 **15.** $255,310 **17. a.** Present value **b.** $56,255 **19. a.** Future value **b.** $A = \$140{,}522$ **21. a.** Present value **b.** $131,444 **23. a.** Future value **b.** $292 **25. a.** Present value **b.** $165,135 **27. a.** Installment payments **b.** $1,317 **29. a.** Ordinary annuity **b.** $175,611

5.6 Review (page 272)

Objective Questions

1. 10, 30, 90, 270 **3.** 4,700 **5.** $x_n = 50 + 250n$ **7.** $x_n = 5 + 10n$ **9.** $156 **11.** $360,000 **13.** $177.16 **15.** $404,928 **17.** Simple, $556; compounded annually, $577.16 **19.** Simple, $1,360,000; compounded, $1,404,928 **21.** $589.55 **23.** $A = \$1{,}433{,}244.61$ **25.** $651.29 **27.** $i = \frac{.16}{4} = .04$; $N = 4(3) = 12$; $A = 900(1.601032) = \$1{,}440.93$ **29.** 10.47% **31.** 10.52% **33.** $428.67 **35.** $3,506.90 **37.** $6,476.72 **39.** $25,461.86 **41.** $6,552.28 **43.** $26,034.75 **45.** $1,610.14 **47.** $40,920.31 **49.** $1,007.14 **51.** $246.42 **53.** Annuity due; $4,882.89 **55.** Ordinary annuity; $80,834.93 **57.** $146.93 **59.** $353.61 **61.** Future value ($1,404.93) **63.** Present value ($5,674.27) **65.** Sinking fund ($1,574.10) **67.** Future value ($13,540.81) **69.** Ordinary annuity ($86,297.39) **71.** Sinking fund ($81.83) **73.** Sinking fund ($1,237.26)

Sample Test

1. Future value; $1,347.85 **3.** Present value; $5,536.76 **5.** Sinking fund; $1,670.92 **7.** Future value; $13,540.81 **9.** Annuity due; $57,149.45 **11.** Present value; $15,098.48 **13.** Sinking fund; $139.62 **15.** Present value of an annuity; $3,433,081 **17.** Simple interest + Sinking fund; $13,628,876 **19.** C

CHAPTER 6

6.1 Sets and Set Operations (page 284)

1. Answers vary. **3.** Answers vary. **5.** { }; cardinality 0 **7.** {2}, {4}, {2, 4}, { }; cardinality 2 **9.** {2}, {4}, {6}, {8}, {2, 4}, {2, 6}, {2, 8}, {4, 6}, {4, 8}, {6, 8}, {2, 4, 6}, {2, 4, 8}, {2, 6, 8}, {4, 6, 8}, {2, 4, 6, 8}, { }; cardinality 4 **11.** 1,024 **13.** 100; 2^{100} or $2^{100} - 1$ if you exclude the empty set **15.** $\{(a, 1), (a, 3), (a, 5), (b, 1), (b, 3), (b, 5)\}$ **17.** $\{(a, a), (a, b), (b, a), (b, b)\}$ **19.** Answers vary. **21.** {6, 8} **23.** {1, 2, 3, 4, 5, 6, 7} **25.** ∅ **27.** {3, 4, 6, 8, 10} **29.** {1, 2, 3, 4, 5} **31.** {1, 2, 3, 4, 5, 7} **33.** {1, 2} **35.** {5} **37.** {3, 4, 7} **39. a.** F **b.** T **c.** T **d.** T **e.** F

41.

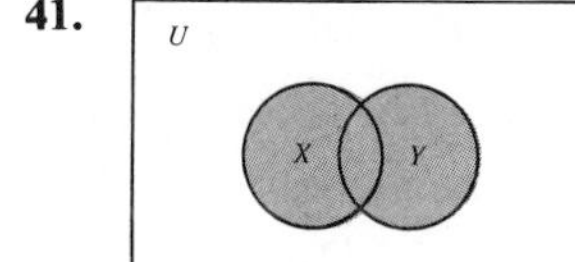

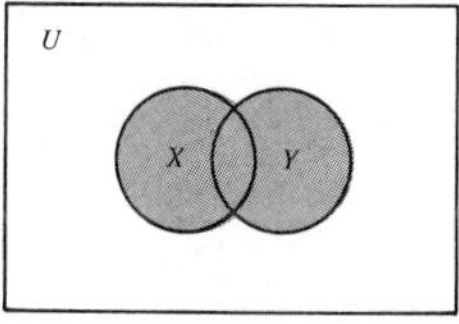

43.

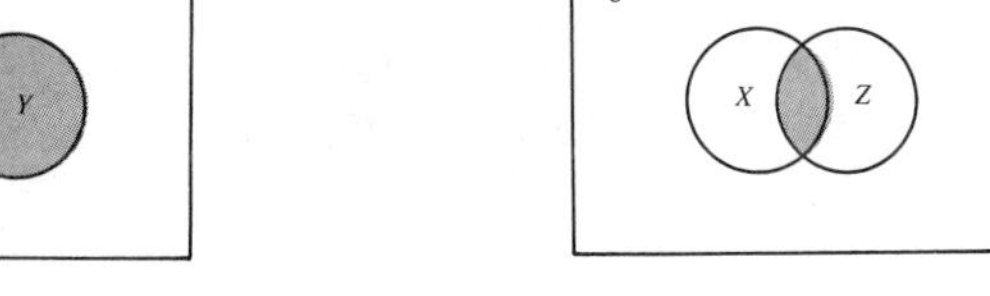

45.

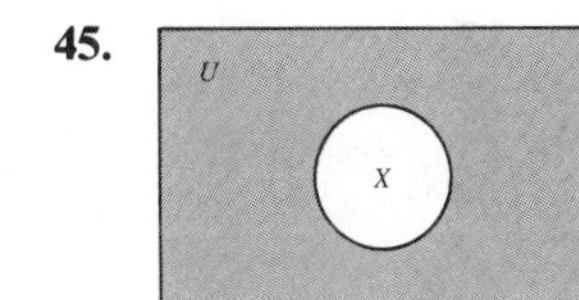

47.

49.

51.

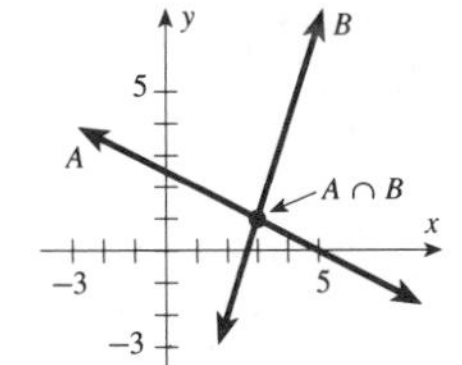

53.

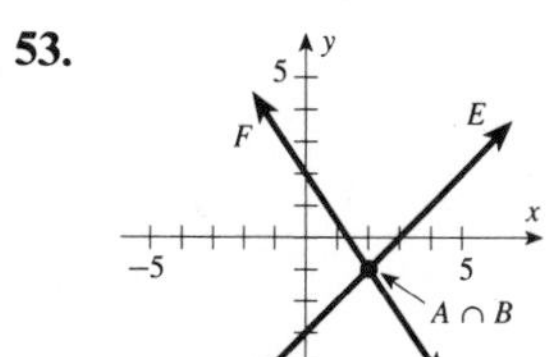

55. **a.** $A \cup B = \{$Bob Wisner, Joan Marsh, Craig Barth, Phyllis Niklas, Shannon Smith, Christy Anton$\}$; union
b. $A \cap B = \{$Phyllis Niklas, Craig Barth$\}$; intersection **57.** Answers vary. There should be 16 disjoint regions.

6.2 *Combined Operations with Sets* *(page 291)*

1.

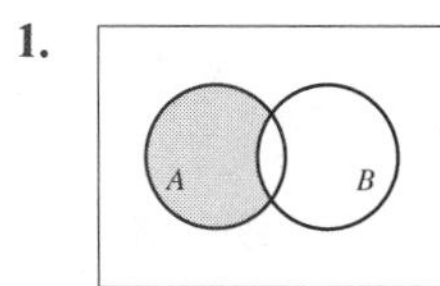

3. 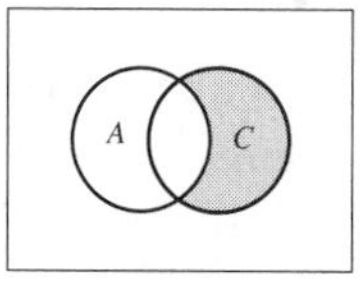 **5.** 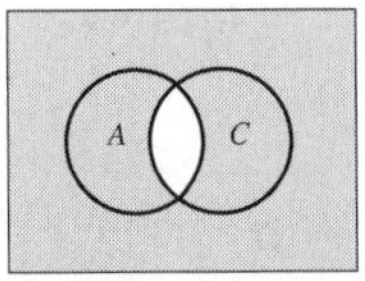 **7.**

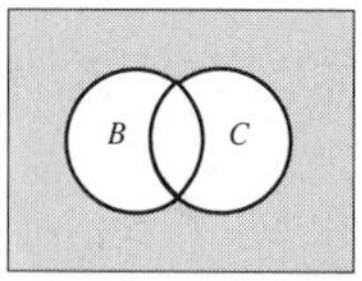

9.

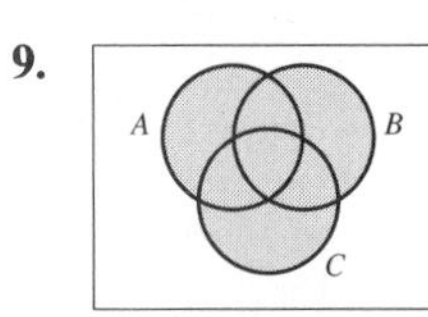

11.

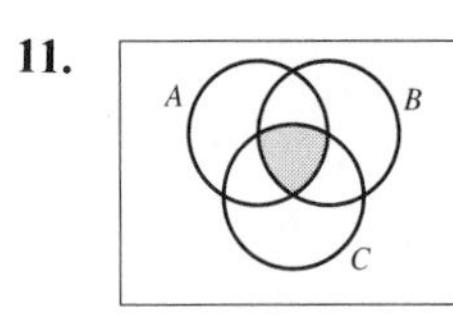

13.

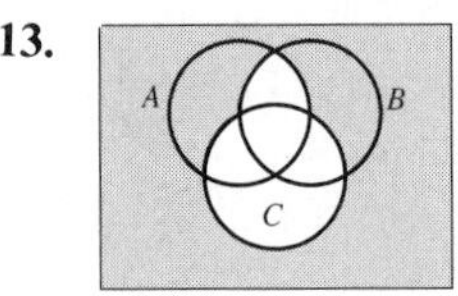

15.

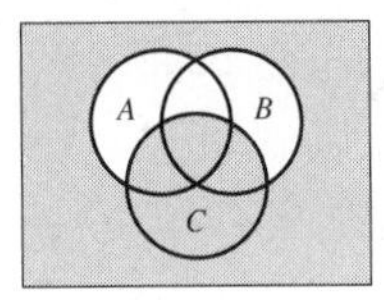

17. $\varnothing$ **19.** A or $\{2, 4, 6, 8\}$ **21.** $\{4, 6\}$ **23.** $\bar{A}$ or $\{1, 3, 5, 7, 9, 10\}$ **25.** F **27.** T **29.** F
31. 72 persons can travel in 2 buses. **33.** There are 34 people playing. **35.** 144 **37.** 70

39.

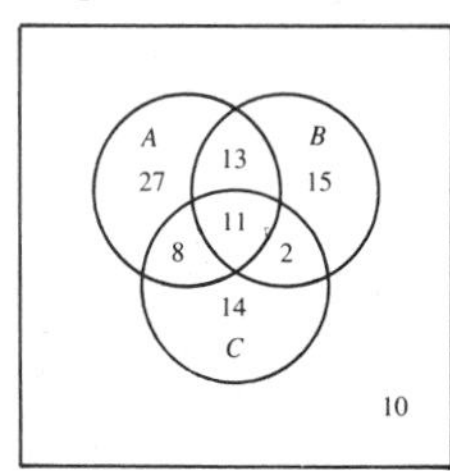

41. **a.** 12 **b.** 7 **c.**

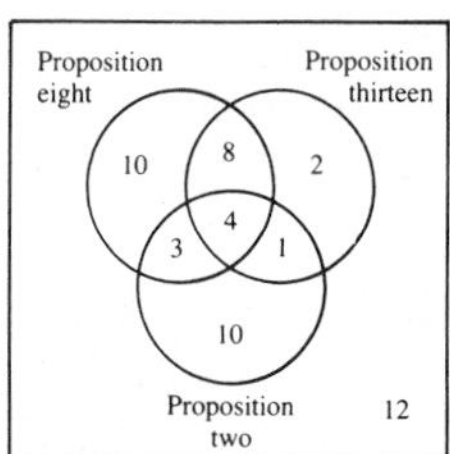

6.3 *Fundamental Counting Principle* *(page 298)*

1. 696 **3.** 40,200 **5.** 24 **7.** 144 **9.** 3,628,800 **11.** 72 **13.** 11,880 **15.** 126 **17.** 330
19. 15 **21.** 210 **23.** 24 **25.** 8 **27.** 18 **29.** 60 **31.** 900,000,000 **33.** 2,600,000 **35.** 786,240
37. 1,024 **39.** **a.** 10,000 **b.** 456,976 **c.** 1,679,616 **41.** 120 **43.** 27,000 **45.** 15

6.4 *Permutations* *(page 304)*

1. **a.** 9 **b.** 72 **c.** 504 **d.** 3,024 **e.** 1 **3.** **a.** 95,040 **b.** 60 **c.** 1,680 **d.** 1 **e.** $\dfrac{g!}{(g-h)!}$ **5.** **a.** 210
b. 120 **c.** $\dfrac{50!}{2}$ **d.** 25 **e.** $\dfrac{m!}{(m-3)!} = m(m-1)(m-2)$ **7.** 5,040 **9.** 360 **11.** 34,650 **13.** 4,989,600
15. 831,600 **17.** 6 possibilities **19.** 120 **21.** 336 **23.** 5,040 **25.** 362,880 **27.** 56 **29.** 3,603,600
31. It will take almost 17 years. **33.** 360 ways **35.** 210 ways **37.** 3,628,800 ways **39.** 15,120 ways

6.5 *Combinations* *(page 312)*

1. **a.** 9 **b.** 36 **c.** 84 **d.** 126 **e.** 1 **3.** **a.** 35 **b.** 1 **c.** 1,225 **d.** 25 **e.** $\dfrac{g!}{(g-h)!h!}$ **5.** **a.** 2,520
b. 1 **c.** 45 **d.** $\dfrac{n!}{(n-4)!} = n(n-1)(n-2)(n-3)$ **e.** $\dfrac{n!}{4!(n-4)!}$ **7.** **a.** 1 **b.** 23,426 **c.** 1 **d.** 5,040
e. $\dfrac{n!}{(n-5)!} = n(n-1)(n-2)(n-3)(n-4)$ **9.** **a.** 20 **b.** 420 **11.** **a.** 2,520 **b.** 1,680 **13.** 75,287,520 **15.** 6
17. 1,287 **19.** Permutation; ${}_8P_5 = 6{,}720$ **21.** Permutation; ${}_5P_5 = 120$ **23.** Combination; $\dbinom{100}{6} = 1{,}192{,}052{,}400$
25. Permutation; ${}_6P_6 = 720$ **27.** Neither; $3 \cdot 3 \cdot 2 \cdot 2 \cdot 1 \cdot 1 = 36$ ways **29.** Combination; $\dbinom{5}{3} = 10$

31. Permutation; ${}_6P_5 = 720$ **33.** $\binom{5}{2}\binom{25}{5}\binom{18}{5} = 4{,}552{,}178{,}400$ **35.** $\dfrac{50!}{15!10!25!}$ **37.** $\binom{50}{7, 5, 38} = 961{,}149{,}000{,}000$

39. $\binom{10}{4, 5, 1} = 1{,}260$ **41. a.** $\binom{10}{3} = 120$ **b.** $\binom{11}{2, 8, 1} = 495$ **c.** $\binom{10}{1}\binom{11}{2} = 550$ **d.** $\binom{10}{1}\binom{6}{1}\binom{5}{1} = 300$

e. $\binom{6}{2}\binom{5}{1} = 75$ **43.** 26,400 **45.** Answers vary.

6.6 Binomial Theorem (*page 318*)

1. 8 **3.** 28 **5.** 56 **7.** 12 **9.** 1 **11.** 153 **13.** $x^5 + 5x^4y + 10x^3y^2 + 10x^2y^3 + 5xy^4 + y^5$

15. $x^4 + 4x^3y + 6x^2y^2 + 4xy^3 + y^4$ **17.** $x^5 - 5x^4y + 10x^3y^2 - 10x^2y^3 + 5xy^4 - y^5$

19. $x^5 + 10x^4 + 40x^3 + 80x^2 + 80x + 32$ **21.** $16x^4 + 96x^3y + 216x^2y^3 + 216xy^4 + 81y^4$

23. $1 - 10x + 45x^2 - 120x^3 + 210x^4 - 252x^5 + 210x^6 - 120x^7 + 45x^8 - 10x^9 + x^{10}$

25. $\binom{15}{0}x^{15} - \binom{15}{1}x^{14}y + \cdots + (-1)^r\binom{15}{r}x^{15-r}y^r + \cdots + \binom{15}{14}xy^{14} - \binom{15}{15}y^{15}$

27. $\binom{20}{0}x^{20} + \binom{20}{1}x^{19}y + \cdots + \binom{20}{r}x^{20-r}y^r + \cdots + \binom{20}{19}xy^{19} + \binom{20}{20}y^{20}$

29. Term: $\binom{11}{7}a^4(-b)^7$; coefficient: $\binom{11}{7}(-1)^7 = -330$ **31.** Term: $\binom{14}{4}a^{10}(3b)^4$; coefficient: $\binom{14}{4}(3)^4 = 81{,}081$

33. Term: $\binom{8}{1}(x^2)^7(-2y)^1$; coefficient: $\binom{8}{1}(-2) = -16$ **35.** 2^{100} **37.** $\binom{10}{6} = 210$ **39.** Answers vary.

41. Answers vary.

6.7 Review (*page 319*)

Objective Questions

1. $\{3\}$ **3.** $\{4, 5, 6\}$ **5.** $\varnothing, \{1, 2, 3\}, \{1\}, \{2\}, \{3\}, \{1, 2\}, \{1, 3\}, \{2, 3\}$ **7.** $\varnothing, \{h, m\}, \{h\}, \{m\}$ **9.** $\{5, 6, 7, 8, 9, 10\}$

11. $\{4, 6, 7, 8, 9, 10\}$ **13.** $\{10\}$ **15.** $\{6\}$ **17.** $\{1, 2, 3, 4, 6, 7, 8, 9\}$ **19.** $\{1, 2, 3, 5, 7, 8, 9, 10\}$

21. U R S

23. 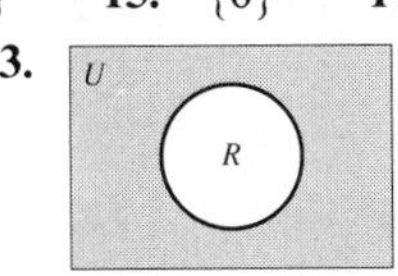

25. 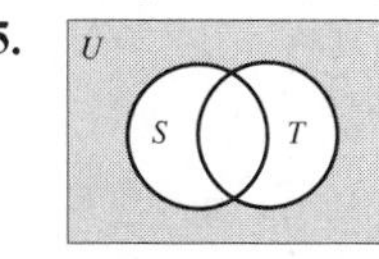

27. 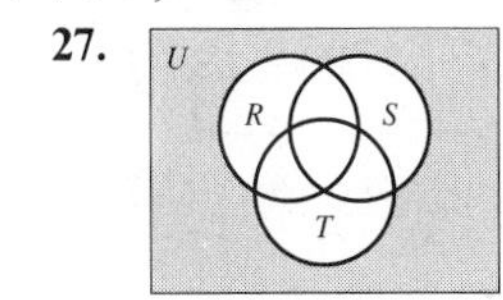

29. $\{6, 8, 10, 12, 16\}$ **31.** $\bar{A} \cap \bar{C} = \{2, 4, 10, 14\} \cap \{2, 4, 6\} = \{2, 4\}$ **33.** 241 **35.** 51 **37.** 210 **39.** 18,090

41. 360 **43.** 144 **45.** 6,720 **47.** 1,560 **49.** 20 **51.** 60 **53.** 56 **55.** 780 **57.** 1,024

59. 75,287,520 **61.** $x^7 + 7x^6y + 21x^5y^2 + 35x^4y^3 + 35x^3y^4 + 21x^2y^5 + 7xy^6 + y^7$

63. $32x^5 + 80x^4y + 80x^3y^2 + 40x^2y^3 + 10xy^4 + y^5$ **65.** $-2{,}002$ **67.** -560 **69.** 64 **71.** 6

Sample Test

1. a. $\{2, 3, 4, 8, 9, 10, 12\}$ **b.** $\{2, 3, 4, 5, 6, 8, 9, 10, 11, 12\}$ **c.** $\{5, 6\}$ **d.** $\{2, 3, 4, 7, 8, 9, 10, 11, 12\}$ **3. a.** 5,040

b. 5,016 **c.** 6 **d.** 9,900 **5. a.** 28 **b.** 168 **7.** 42 **9.** 505,785,627,648 **11.** 1,024

13. $T^6 + 6T^5H + 15T^4H^2 + 20T^3H^3 + 15T^2H^4 + 6TH^5 + H^6$ **15.**

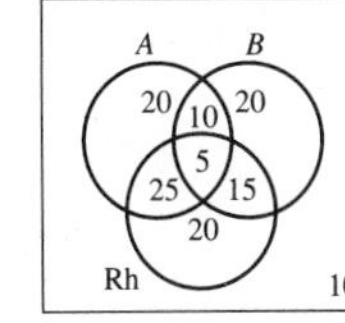

a. Sum of all regions shown = 125. **b.** Region 7 has 5 people. **c.** Region 4 has 25 people. **d.** Region 8 has 10 people.

e. Regions 1, 2, and 3 add up to 60 people.

CHAPTER 7

7.1 *Probability Models* (page 329)

1. $S = \{2, 3, 4, 5, 6, 7, 8, 9, 10, 11, 12\}$ **3.** $\{0, 1, 2, 3, 4, 5, 6, 7, 8, 9, 10\}$ **5.** $\{4, 5, 6, 7\}$
7. {nonnegative integers less than or equal to 500} **9.** $\{bbb, bbg, bgb, bgg, gbb, gbg, ggb, ggg\}$
11. Not mutually exclusive **13.** Mutually exclusive **15.** Not mutually exclusive **17.** Not mutually exclusive
19. Not mutually exclusive **21.** $L \cup E$ = {roll a number less than 3 or roll an even number} = $\{1, 2, 4, 6\}$
23. $E \cap L$ = {roll a number less than 3 and roll an even} = $\{2\}$ **25.** $\bar{L}$ = {roll a 3 or larger} = $\{3, 4, 5, 6\}$
27. $\bar{E} \cap \bar{L}$ = {roll an odd number which is also at least a 3} = $\{3, 5\}$ **29.** Yes; answers vary. **31.** $\frac{3}{11}$ **33.** $\frac{5}{11}$
35. 0 **37.** $\frac{8}{11}$ **39.** Yes; answers vary. **41.** $\frac{1}{9}$ **43.** $\frac{5}{9}$ **45.** 0 **47.** $\frac{8}{9}$

7.2 *Probability of Equally Likely Events* (page 336)

1. a. $\frac{3}{8}$ **b.** $\frac{1}{4}$ **c.** $\frac{3}{8}$ **3. a.** $\frac{1}{4}$ **b.** $\frac{1}{4}$ **c.** $\frac{1}{2}$ **5. a.** $\frac{7}{20}$ **b.** $\frac{7}{20}$ **c.** $\frac{3}{10}$ **7.** About 35 times **9. a.** $\frac{1}{13}$
b. $\frac{1}{4}$ **c.** $\frac{1}{52}$ **d.** $\frac{4}{13}$ **11. a.** $\frac{7}{24}$ **b.** $\frac{7}{24}$ **c.** $\frac{7}{12}$ **d.** $\frac{7}{12}$ **e.** 0 **13.** Answers vary; $P(B) = \frac{1}{2}$ **15.** .05
17. .19 **19. a.** .66 **b.** .56 **21.** No **23.** $\frac{1}{9}$ **25.** $\frac{1}{6}$ **27.** $\frac{1}{9}$ **29.** $\frac{1}{18}$ **31.** $\frac{7}{36}$ **33.** $\frac{1}{2}$ **35.** $\frac{2}{9}$
37. $\frac{1}{36}$ **39.** There are 16 possible rolls. **a.** $\frac{1}{16}$ **b.** $\frac{1}{8}$ **c.** $\frac{3}{16}$

41.

	8	7	6	5	4	3	2	1
8	8, 8	7, 8	6, 8	5, 8	4, 8	3, 8	2, 8	1, 8
7	8, 7	7, 7	6, 7	5, 7	4, 7	3, 7	2, 7	1, 7
6	8, 6	7, 6	6, 6	5, 6	4, 6	3, 6	2, 6	1, 6
5	8, 5	7, 5	6, 5	5, 5	4, 5	3, 5	2, 5	1, 5
4	8, 4	7, 4	6, 4	5, 4	4, 4	3, 4	2, 4	1, 4
3	8, 3	7, 3	6, 3	5, 3	4, 3	3, 3	2, 3	1, 3
2	8, 2	7, 2	6, 2	5, 2	4, 2	3, 2	2, 2	1, 2
1	8, 1	7, 1	6, 1	5, 1	4, 1	3, 1	2, 1	1, 1

43. a. $\frac{9}{196}$ **b.** $\frac{1}{196}$ **c.** $\frac{1}{14}$ **d.** $\frac{1}{49}$

7.3 *Calculated Probabilities* (page 345)

1. $\frac{5}{8}$ **3.** $\frac{1}{4}$ **5.** $\frac{3}{4}$ **7.** $\frac{6}{13}$ **9.** $\frac{3}{13}$ **11.** $\frac{10}{13}$ **13.** The *odds* are 1 to 3. **15.** $\frac{9}{10}$
17. The *odds* are 1 to 12. **19.** $\frac{1}{4}$ **21.** .00815 **23.** $\frac{2}{17}$ **25.** $\frac{1}{4}$ **27.** $\frac{3}{8}$ **29.** $\frac{5}{16}$ **31.** $\frac{5}{16}$ **33.** 2 to 7
35. .38 **37.** .05 **39.** .07 **41.** .08 **43.** .31 **45.** .23 **47.** .01 **49.** .33 **51.** .14 **53.** .01
55. .21 **57.** .06 **59.** .00001 385 **61. a.** .027136 **b.** .116948 **c.** .411438 **d.** .475695 **e.** .507297
f. .538344 **g.** .706316 **h.** .891232 **i.** .970374 **63.** First game, P(one 6) = .51775; second game, P(one 12) = .4914

7.4 *Conditional Probability* (page 356)

1. $\frac{3}{8}$ **3.** $\frac{1}{8}$ **5.** $\frac{3}{7}$ **7.** $\frac{7}{8}$ **9.** $\frac{1}{4}$ **11.** $\frac{1}{4}$ **13.** $\frac{2}{5}$ **15.** $\frac{1}{6}$ **17.** $\frac{2}{11}$ **19.** The second branch of path 3
21. $P(A_1 \cap D)$ or $P(D \mid A_1)P(A_1)$ **23.** .3 **25.** .36 **27.** .19 **29.** .53 **31.** .32 **33.** .59
35. $P(E \mid F) = .50$, $P(F \mid E) = .20$ **37.** $P(E \mid F) = \frac{1}{3}$, $P(F \mid E) = \frac{1}{2}$ **39.** $P(E \mid F) = .50$, $P(F \mid E) = .80$ **41.** 60%
43. a. $\frac{1}{13}$ **b.** $\frac{1}{17}$ **c.** $\frac{4}{663}$ **d.** $\frac{8}{663}$ **45.** $\frac{7}{30}$ **47.** .004 **49.** .0205

7.5 *Independent Events, Intersections, and Unions* (page 364)

1. Independent **3.** Independent **5.** Not independent **7.** Independent **9.** Independent
11. Not independent **13.** Not independent **15.** Not independent **17.** $\frac{1}{2}$ **19.** $\frac{5}{6}$ **21.** $\frac{1}{12}$ **23.** $\frac{2}{3}$
25. $\frac{4}{9}$ **27.** $\frac{11}{12}$ **29.** $\frac{1}{3}$ **31.** $\frac{5}{9}$ **33.** $\frac{35}{36}$ **35.** $\frac{1}{4}$ **37.** 0 **39.** $\frac{1}{2}$ **41. a.** $\frac{1}{2}$ **b.** $\frac{1}{3}$ **c.** $\frac{1}{6}$
43. a. $\frac{5}{6}$ **b.** $\frac{11}{12}$ **c.** $\frac{17}{18}$ **45.** $\frac{1}{2}$ **47. a.** $\frac{1}{2}$ **b.** $\frac{1}{6}$ **c.** $\frac{1}{2}$ **49.** $\frac{1}{4}$ **51.** $\frac{1}{2}$ **53.** $\frac{7}{12}$ **55.** $\frac{1}{6}$ **57.** 1
59. $\frac{1}{2}$ **61.** $\frac{9}{64}$ **63.** $\frac{15}{64}$ **65.** $\frac{5}{14}$ **67.** $\frac{15}{28}$ **69.** $\frac{41}{56}$ **71.** .6137

7.6 *Bayes' Theorem* (page 373)

1. .06 **3.** .12 **5.** .3 **7.** .2 **9.** .1 **11.** .18 **13.** .57 **15.** .175 **17.** $\frac{1}{2}$ **19.** $\frac{2}{5}$ **21.** $\frac{3}{4}$

23.

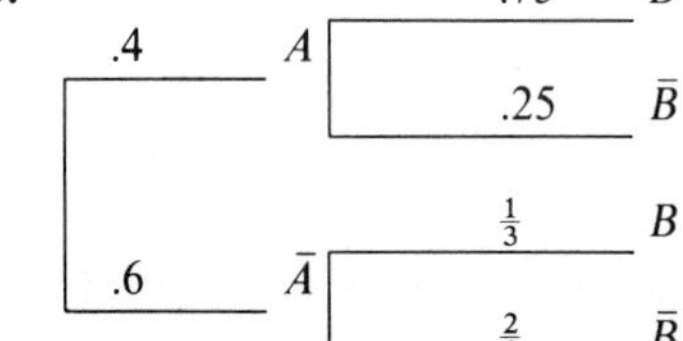

25. $\frac{2}{13}$ **27.** $\frac{7}{20}$ **29.** $\frac{7}{10}$ **31.** .05 **33.** .04 **35.** .05 **37.** $\frac{1}{4}$ **39.** $\frac{4}{17}$ **41.** $\frac{1}{4}$ **43.** .04 **45.** .04 **47.** .05 **49.** .25 **51.** .44 **53.** .59 **55.** $\frac{2}{15}$ **57.** $\frac{30}{37}$ **59.** $\frac{7}{37}$ **61.** .30 **63.** .42 **65.** .81 **67.** $\frac{2}{3}$ **69.** **a.** .22 **b.** .09 **c.** .70 **71.** $\frac{5}{12}$

7.7 Review (page 375)

Objective Questions

1. $S = \{$all integers between 0 and 1.44 MB, inclusive$\}$ **3.** {blue–blue, blue–black, blue–brown, black–brown, black–black, brown–brown} **5.** Mutually exclusive **7.** Mutually exclusive **9.** Answers vary. **11.** Not a probability model **13.** $\frac{3}{7}$ **15.** $\frac{5}{7}$ **17.** $\frac{5}{7}$ **19.** 1 **21.** $\frac{1}{13}$ **23.** $\frac{4}{13}$ **25.** Answers vary. **27.** Answers vary. **29.** .013 **31.** .099 **33.** .89 **35.** .28 **37.** $\frac{1}{3}$ **39.** $\frac{26}{31}$ **41.** .225 **43.** 1 **45.** .41 **47.** .44 **49.** .30 **51.** .25 **53.** .75 **55.** $\frac{60}{121} \approx .50$ **57.** .18 **59.** .27 **61.** .73 **63.** .2 **65.** .32 **67.** .286 **69.** .64

Sample Test

1. **a.** For a sample space S and event E associate a real number $P(E)$ such that: (i) $0 \le P(E) \le 1$; (ii) $P(S) = 1$; (iii) If E and F are mutually exclusive events, then $P(E \cup F) = P(E) + P(F)$ **b.** Their intersection is empty. **c.** One event does not affect the probability of the other. **3.** No. The probabilities of A, B, and C are $\frac{1}{4}$, $\frac{1}{4}$, and $\frac{1}{2}$, respectively. **5.** **a.** .6 **b.** .625 **c.** .85 **d.** .475 **7.** **a.** .333 **b.** .208 **c.** .24 **d.** .625 **9.** **a.** $\frac{1}{3}$ **b.** $\frac{4}{15}$ **c.** $\frac{11}{15}$

CHAPTER 8

8.1 Random Variables (page 384)

1.

Inches	*Tally*	*Frequency*
63	|	2
64	||||	4
65	|||	3
66	|||	3
67	𝍸	5
68	|||	3
69	||||	4
70	|||	3
71	||	2
72	|	1

(to the nearest inch) where $X = 63, 64, \ldots, 72$.

3.

No. of Spots	*Tally*	*Frequency*
2	|	1
3	|||	3
4	𝍸	5
5	||	2
6	𝍸 ||||	9
7	𝍸 |||	8
8	𝍸 |	6
9	𝍸 |	6
10	||||	4
11	𝍸	5
12	|	1

Let X = sum of dice = 2, 3, 4, ..., 11, 12.

5.

No. of Years	*Frequency*	*Relative Frequency*
13	1	$\frac{1}{25} = .04$
14	2	$\frac{2}{25} = .08$
15	1	$\frac{1}{25} = .04$
16	3	$\frac{3}{25} = .12$
17	3	$\frac{3}{25} = .12$
18	0	0
19	3	$\frac{3}{25} = .12$
20	3	$\frac{3}{25} = .12$
21	2	$\frac{2}{25} = .08$
22	4	$\frac{4}{25} = .16$
23	1	$\frac{1}{25} = .04$
24	1	$\frac{1}{25} = .04$
25	1	$\frac{1}{25} = .04$
	25	1

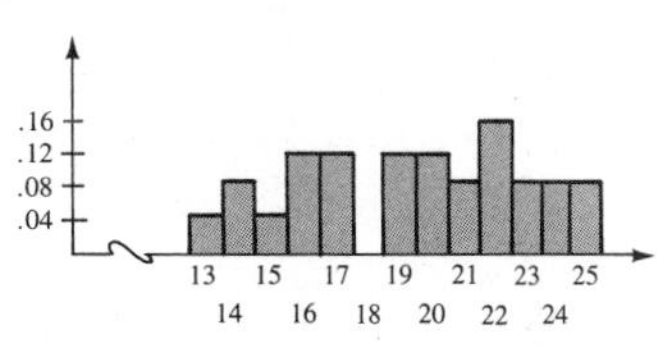

7.

No. Waiting	*Frequency*	*Relative Frequency*
2	20	$\frac{20}{50} = .40$
3	15	$\frac{15}{50} = .30$
4	7	$\frac{7}{50} = .14$
5	5	$\frac{5}{50} = .10$
6	2	$\frac{2}{50} = .04$
7	1	$\frac{1}{50} = .02$
	50	1

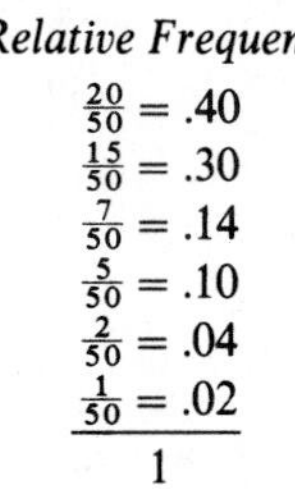

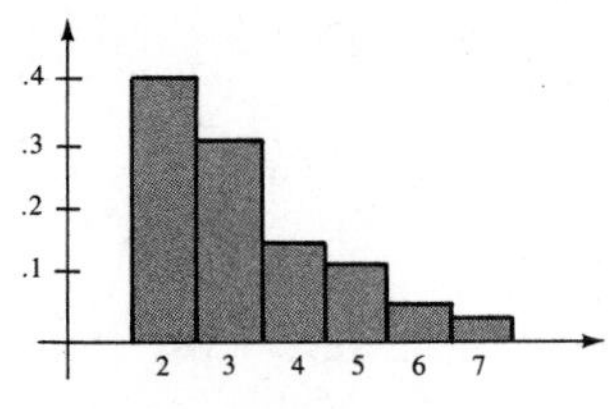

9.

No. of Heads	*Frequency*	*Relative Frequency*
3	18	$\frac{18}{150} = .12$
2	56	$\frac{56}{150} = .373$
1	59	$\frac{59}{150} = .393$
0	17	$\frac{17}{150} = .113$
	150	1

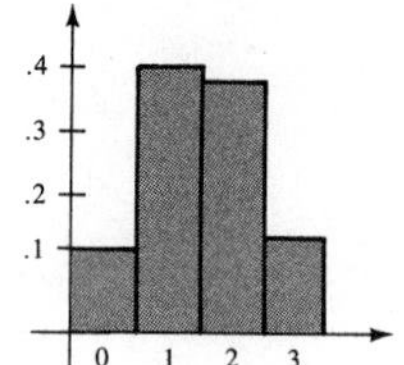

11.

No. of Heads	*Relative Frequency*
0	$\frac{1}{16} = .0625$
1	$\frac{4}{16} = .25$
2	$\frac{6}{16} = .375$
3	$\frac{4}{16} = .25$
4	$\frac{1}{16} = .0625$
	1

13.

No. of Aces	*Relative Frequency*
0	$\frac{1,128}{1,326} = .851$
1	$\frac{192}{1,326} = .145$
2	$\frac{6}{1,326} = .005$
	1

15.

No. of Defectives	*Relative Frequency*
0	$\frac{64}{125} = .512$
1	$\frac{48}{125} = .384$
2	$\frac{12}{125} = .096$
3	$\frac{1}{125} = .008$
	1

17.

No. of Hearts	*Relative Frequency*
4	$\frac{{}_{13}C_4}{{}_{52}C_4} \approx .002641$
3	$\frac{{}_{13}C_3 \cdot {}_{39}C_1}{{}_{52}C_4} \approx .0412004$
2	$\frac{{}_{13}C_2 \cdot {}_{39}C_2}{{}_{52}C_4} \approx .2134934$
1	$\frac{{}_{13}C_1 \cdot {}_{39}C_3}{{}_{52}C_4} \approx .4388475$
0	$\frac{{}_{39}C_4}{{}_{52}C_4} \approx .3038175$
	$.9999988 \approx 1$

19.

No. of Almond Joys	*Relative Frequency*
0	$\frac{595}{1,225} \approx .486$
1	$\frac{525}{1,225} \approx .429$
2	$\frac{105}{1,225} \approx .086$
	1

21.

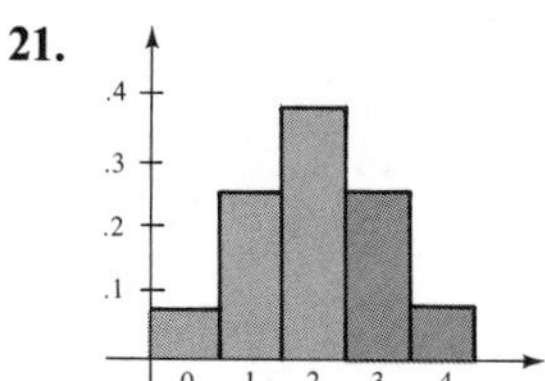

23.

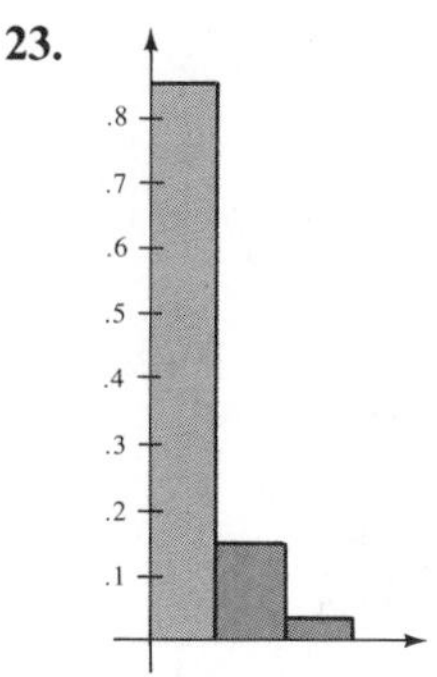

25.

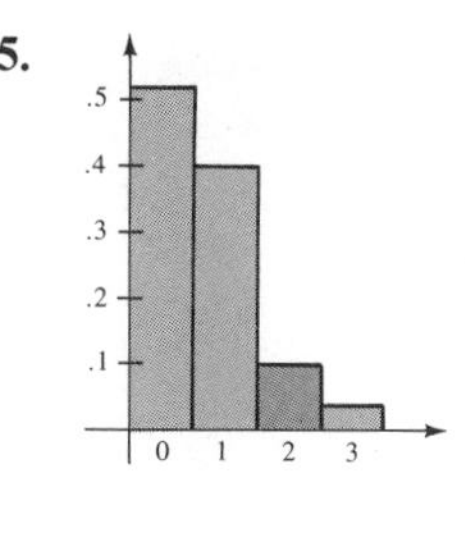

27.

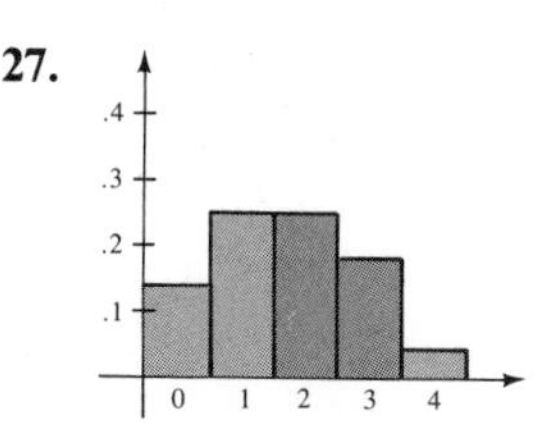

29.

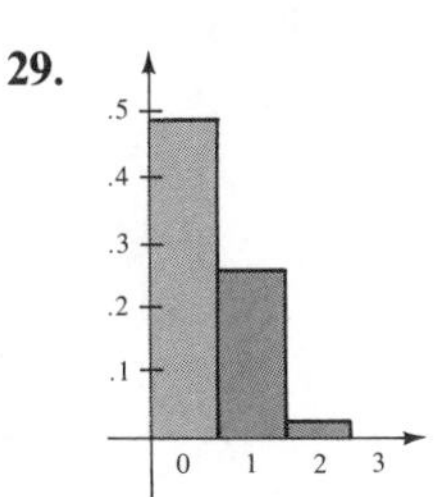

31.

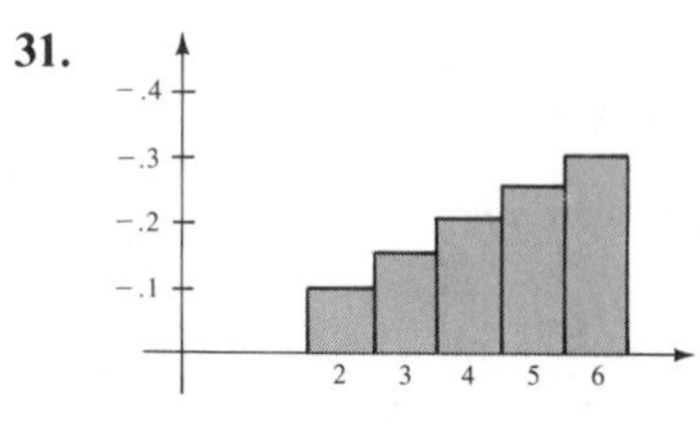

8.2 Analysis of Data (page 393)

1. Mean: 3; Median: 3; Mode: none; Range: 4; Variance: 2.5; Standard deviation: 1.58 **3.** Mean: 105; Median: 105; Mode: none; Range: 4; Variance: 2.5; Standard deviation: 1.58 **5.** Mean: 6; Median: 7; Mode: 7; Range: 8; Variance: $\frac{20}{3}$; Standard deviation: 2.58 **7.** Mean: 10; Median: 8; Mode: none; Range: 18; Variance: 52; Standard deviation: 7.21 **9.** Mean: 91; Median: 95; Mode: 95; Range: 17; Variance: 50.5; Standard deviation: 7.11 **11.** Mean: 3; Median: 3; Mode: 3; Range: 4; Variance: 1.67; Standard deviation: 1.29 **13.** Answers vary. **15.** Mean: \$15,800; Median: \$16,000; Mode: \$10,000 **17.** Mean: 68.1; Median: 70; Mode: 70 **19.** Mean: 56; Median: 65; Mode: 70; Range: 90 **21.** Variance: $s^2 = 141\frac{2}{3}$; Standard deviation: $s = \sqrt{141\frac{2}{3}} \approx 11.9$ **23.** Variance: $s^2 = \dfrac{371{,}600{,}000}{4} \approx 41{,}288{,}888.9$; Standard deviation: $s = \sqrt{41{,}288{,}888.9} \approx 6{,}425.6$ **25.** Mean: $\frac{3}{4}$; Variance: $\frac{11}{16}$; Standard deviation: .83 **27.** Mean: 3.4; Variance: .84; Standard deviation: .92 **29.** Mean: 2.85; Variance: 2.3275; Standard deviation: 1.526 **31.** Answers vary. **33.** Answers vary. **35.** Answers vary. **37.** Answers vary. **39.** 9.1596 **41.** 9.5736 **43.** 10.4198

8.3 The Binomial Distribution (page 400)

1. .512 **3.** .096 **5.** .488 **7.** .6561 **9.** .0001 **11.** .9477 **13.** .132 **15.** .128 **17.** .016 **19.** 0 + ($\approx$.0002) **21.** Mean: 1.5; Variance: 1.05; Standard deviation: 1.02 **23.** Mean: 7.8; Variance: 2.73; Standard deviation: 1.65 **25.** Mean: 3.0; Variance: 1.50; Standard deviation: 1.22 **27.** Mean: .7; Variance: .63; Standard deviation: .79 **29.** Mean: 6; Variance: 2.4; Standard deviation: 1.55 **31.** .4812 **33.** .5443 **35.** .4838 **37.** The smallest number of missiles is 16. **39.** Mean: 1.2; Standard deviation: .98 **41.** Mean: 30; Standard deviation: 3.46 **43.** **a.** .109375 **b.** .1143 **45.** .3125 **47.** $n2^{1-n}$

8.4 The Normal Distribution (page 406)

1. .0228 **3.** .8413 **5.** .2709 **7.** .9772 **9.** .8413 **11.** .0455 **13.** .2548 **15.** .7926 **17.** 34 persons **19.** 8 persons **21.** 25 **23.** 1 student **25.** 2.3% **27.** 97.8% **29.** 40.5 inches **31.** .50 **33.** .1151 **35.** .8823 **37.** .0228 **39.** .0668

8.5 Normal Approximation to the Binomial (page 411)

1. Yes **3.** No **5.** Yes **7.** Yes **9.** No **11.** .8485 **13.** .0010 **15.** .1515 **17.** .6793 **19.** .0909 **21.** .0005 **23.** .6293 **25.** .9278 **27.** .6985 **29.** .1856 **31.** .0028 **33.** .0306 **35.** .9997 **37.** .52 **39.** .8665 **41.** .9998 **43.** .0032 **45.** .0223 **47.** .00351 **49.** .06797 **51.** .0132

8.6 Correlation and Regression (page 418)

1. Yes **3.** No **5.** Yes **7.** No **9.** No **11.** No

13. .995

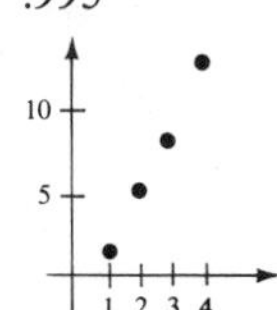

There is a linear correlation at 5%.

15. −.9847

There is a linear correlation at 1%.

17. .970

There is a linear correlation at 1%.

19. $y' = 3.9x - 3$ **21.** $y' = .56x + 41.58$ **23.** $y' = -2.71x + 30.06$ **25.** .842, significant at 1%
27. $y' = 2.45x + 6.09$ **29.** $r = .358$, no significant correlation **31.** $r = -.936$, significant at 1%
33. $y' = .468x + 19.235$ **35.** $y' = -6.186x + 87.179$

8.7 Review (page 419)

Objective Questions

1.

Scores	*Frequency*
40	1
45	0
50	0
55	2
60	3
65	5
70	4
75	5
80	2
85	2
90	1
	25

3.

Wages	*Frequency*
14,000	1
16,000	1
18,000	5
20,000	2
25,000	1
30,000	1
40,000	2
45,000	3
	16

5.

Scores	*Frequency*	*Relative Frequency*
40	1	$\frac{1}{25} = .04$
45	0	$\frac{0}{25} = 0$
50	0	$\frac{0}{25} = 0$
55	2	$\frac{2}{25} = .08$
60	3	$\frac{3}{25} = .12$
65	5	$\frac{5}{25} = .20$
70	4	$\frac{4}{25} = .16$
75	5	$\frac{5}{25} = .20$
80	2	$\frac{2}{25} = .08$
85	2	$\frac{2}{25} = .08$
90	1	$\frac{1}{25} = .04$
		1

7.

Wages	*Frequency*	*Relative Frequency*
14,000	1	$\frac{1}{16} = .0625$
16,000	1	$\frac{1}{16} = .0625$
18,000	5	$\frac{5}{16} = .3125$
20,000	2	$\frac{2}{16} = .125$
25,000	1	$\frac{1}{16} = .0625$
30,000	1	$\frac{1}{16} = .0625$
40,000	2	$\frac{2}{16} = .125$
45,000	3	$\frac{3}{16} = .1875$
		1

9.

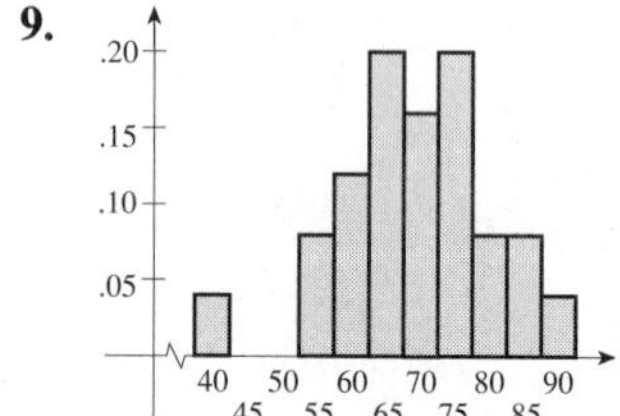

11.

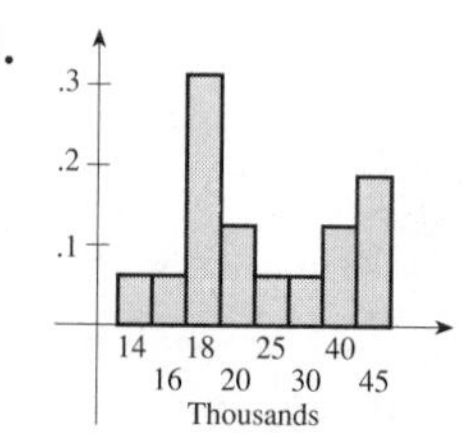

13. Mean: 16; Median: 15; Mode: 24 **15.** Mean: 28,400; Median: 21,000; Mode: 21,000 **17.** Mean: 69.2; Median: 70; Mode: 75 and 65 (bimodal) **19.** Mean: \$26,875; Median: \$20,000; Mode: \$18,000 **21.** Range: 16; Variance: $42\frac{1}{3}$; Standard deviation: 6.51 **23.** Range: 5; Variance: $\frac{35}{12}$; Standard deviation: 1.7078 **25.** .0098 **27.** .211 **29.** Mean: 9.8; Variance: 9.70396; Standard deviation: 3.1290 **31.** Mean = 12.5; Variance = 9.375; Standard deviation ≈ 3.06 **33.** .0668 **35.** .3830 **37.** .3085 **39.** .4972 **41.** 1 A, 7 B's, 34 C's, 7 D's, and 1 F **43.** .0262 **45.** .0287

47.

49.

51. $r = -.914$, significant at both 1% and 5% **53.** $y' = -.12x + 30.4$ **55.** $y' = .218x + 81.821$

Sample Test

1.

Roll	2	3	4	5	6	7	8	9	10	11	12
Frequency	1	3	4	3	8	9	6	6	4	5	1

3.

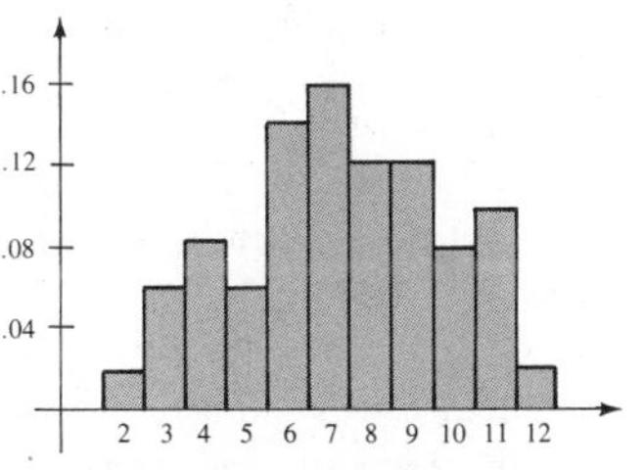

5. Range: 10; Variance: 5.9424; Standard deviation ≈ 2.4377 **7.** Mean: 8; Variance: 1.6; Standard deviation: 1.26 **9.** .003 **11.** $y' = 9.23x + 114.7$

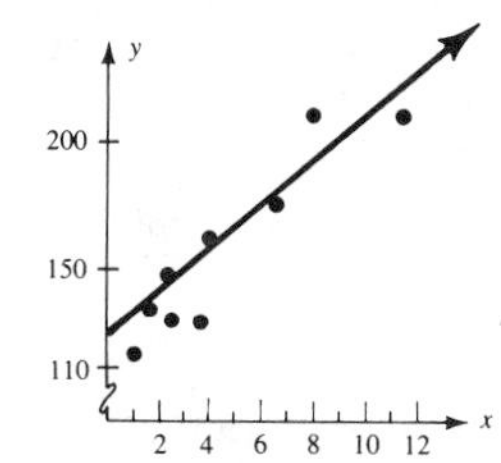

13. D **15.** A

CHAPTER 9

9.1 Introduction to Markov Chains (page 430)

1. a. No **b.** Yes **c.** No **3. a.** No **b.** Yes **c.** No **5. a.** Yes **b.** No

7. $T^2 = \begin{bmatrix} \frac{3}{8} & \frac{5}{8} \\ \frac{5}{16} & \frac{11}{16} \end{bmatrix}$; $T^3 = \begin{bmatrix} \frac{11}{32} & \frac{21}{32} \\ \frac{21}{64} & \frac{43}{64} \end{bmatrix}$ **9.** $T^2 = \begin{bmatrix} .452 & .548 \\ .3151 & .6849 \end{bmatrix}$; $T^3 = \begin{bmatrix} .39724 & .60276 \\ .346587 & .653413 \end{bmatrix}$

11. $T^2 = \begin{bmatrix} .2075 & .605 & .1875 \\ .265 & .3675 & .3675 \\ .425 & .385 & .19 \end{bmatrix}$; $T^3 = \begin{bmatrix} .3524 & .4136 & .234 \\ .2603 & .4889 & .2509 \\ .2753 & .393 & .3318 \end{bmatrix}$ **13.** $\begin{array}{c c} & \begin{array}{cc} 0 & 1 \end{array} \\ \begin{array}{c} 0 \\ 1 \end{array} & \begin{bmatrix} .3 & .7 \\ .6 & .4 \end{bmatrix} \end{array}$

15.

		Next day: Increase	Next day: Decrease
Present	Increase	.3	.7
	Decrease	.8	.2

17. a. No. The bottom row is not a probability vector. **b.** The probability that a person who rented a car from SFO will return it to SFO is .8. **c.** SJC

19. $T^2 = \begin{bmatrix} .69 & .17 & .15 \\ .33 & .53 & .15 \\ .46 & .29 & .41 \end{bmatrix}$ **21.** $T^2 = \begin{bmatrix} .22 & .52 & .26 \\ .13 & .70 & .17 \\ .13 & .34 & .53 \end{bmatrix}$ **23.** $\frac{3}{4}$ **25.** $\frac{39}{64}$ **27.** $\frac{9}{16}$ **29.** $\frac{7}{16}$ **31.** .9891 **33.** .98901171 **35.** .9891 **37.** .0109

9.2 Regular Markov Chains (page 437)

For Problems 1–11 *let the given matrix be T.*

1. $T^n \approx \begin{bmatrix} \frac{1}{4} & \frac{3}{4} \\ \frac{1}{4} & \frac{3}{4} \end{bmatrix}$ for large n. **3.** Not regular **5.** $T^n \approx \begin{bmatrix} \frac{4}{13} & \frac{9}{13} \\ \frac{4}{13} & \frac{9}{13} \end{bmatrix}$ for large n. **7.** $T^n \approx \begin{bmatrix} \frac{5}{8} & \frac{3}{8} \\ \frac{5}{8} & \frac{3}{8} \end{bmatrix}$ for large n.

9. $T^n \approx \begin{bmatrix} \frac{2}{9} & \frac{4}{9} & \frac{1}{3} \\ \frac{2}{9} & \frac{4}{9} & \frac{1}{3} \\ \frac{2}{9} & \frac{4}{9} & \frac{1}{3} \end{bmatrix}$ for large n. **11.** $T^n \approx \begin{bmatrix} \frac{1}{5} & \frac{1}{2} & \frac{3}{10} \\ \frac{1}{5} & \frac{1}{2} & \frac{3}{10} \\ \frac{1}{5} & \frac{1}{2} & \frac{3}{10} \end{bmatrix}$ for large n. **13. a.** No **b.** Yes; $[\frac{1}{2} \;\; \frac{1}{2}]$ **c.** No, because the matrix is not regular. **15. a.** $0 \le a \le 1$ **b.** $[\frac{1}{2} \;\; \frac{1}{2}]$ **17.** In the long run, she will buy canned 40% of the time, packaged 45% of the time, and dry 15% of the time. **19.** $\begin{array}{c} \\ A \\ B \\ C \end{array} \begin{array}{c} \begin{array}{ccc} A & B & C \end{array} \\ \begin{bmatrix} .7 & .1 & .2 \\ .1 & .8 & .1 \\ .1 & .4 & .5 \end{bmatrix} \end{array}$

21. In the long run, 25% ($\frac{1}{4}$) of the market will be for Alpha, 54% ($\frac{13}{24}$) for Better Bean, and 21% ($\frac{5}{24}$) for Carolyn's.
23. The long-range voting pattern is 57% ($\frac{4}{7}$) Democrat, 12% ($\frac{6}{49}$) Republican, and 31% ($\frac{15}{49}$) Independent.
25. The distribution after 2 generations will be 22.5% Jersey, 27.5% Guernsey, and 50% white-faced.
27. The first company can expect 40% of the market, and the second can expect 60%. **29.** The probability is $\frac{1}{3}$.

9.3 Absorbing Markov Chains (page 443)

1. Not absorbing **3.** Not absorbing **5.** Not absorbing **7.** All three states are absorbing. **9.** States 1 and 2 are absorbing. **11.** State 3 is absorbing. **13.** If the system begins in state 1, the probability that it will end up in absorbing state 2 is .375; the probability it will end up in absorbing state 3 is .625. **15.** If the system begins in state 3, the probability that it will end up in absorbing state 1 is .30, the probability it will end up in absorbing state 2 is .70.
17. Since there is only one absorbing state, the probability of ending in that state is 100% regardless of the beginning state.
19. If the system begins in state 2, the probability that it will end up in absorbing state 1 is $\frac{1}{2}$, and the probability it will end up in absorbing state 4 is also $\frac{1}{2}$. If the system begins in state 3, the situation is identical. **21.** If the system is in state 2, there is a probability of $\frac{43}{65}$ that it will end up in absorbing state 1 and a probability of $\frac{22}{65}$ that it will end up in absorbing state 3. If the system begins in state 4, there are probabilities of $\frac{28}{65}$ and $\frac{37}{65}$ that it will end up in absorbing states 1 and 3, respectively.
23. If the system begins in state 2, 3, or 4, there is an equal chance of ending up in either absorbing state 1 or absorbing state 5.

25. a. $\begin{array}{c} \\ 0 \\ 1 \\ 2 \end{array} \begin{array}{c} \begin{array}{ccc} 0 & 1 & 2 \end{array} \\ \begin{bmatrix} 1 & 0 & 0 \\ \frac{1}{2} & 0 & \frac{1}{2} \\ 0 & 0 & 1 \end{bmatrix} \end{array}$ **b.** $[\frac{1}{2} \;\; \frac{1}{2}]$ **c.** 50% **27.** The probability of a returned item ending up being passed is $\frac{7}{8}$.
29. The probability of ending up in room 1 after starting in room 5 is $\frac{3}{8}$.

9.4 Expectation (page 448)

1. 2 **3.** 6 **5.** 2.05 **7.** 8¢ **9.** 2¢ **11.** $12 **13.** $318 **15.** $85 **17.** −5¢ **19.** −5¢ **21.** −5¢ **23.** −8¢ **25.** −5¢ **27.** $E = -\$8{,}125$; they should not sink the test well. **29.** $0 **31.** It is not a good deal for the player. **33.** Two wells should be tried. **35.** The optimal number of cars is 12.

9.5 Game Theory (page 458)

1. The optimal strategies are $[1 \;\; 0]$ and $\begin{bmatrix} 0 \\ 1 \end{bmatrix}$; the value is 0. **3.** The optimal strategies are $[\frac{1}{2} \;\; \frac{1}{2}]$ and $\begin{bmatrix} \frac{1}{2} \\ \frac{1}{2} \end{bmatrix}$; the value is 0.

5. The optimal strategies are $[1 \;\; 0]$ and $\begin{bmatrix} 0 \\ 1 \end{bmatrix}$; the value is −2. **7.** Strictly determined, $[1 \;\; 0]$ and $\begin{bmatrix} 0 \\ 1 \end{bmatrix}$; the value is −1.

9. The optimal strategies are $[\frac{1}{3} \;\; \frac{2}{3}]$ and $\begin{bmatrix} \frac{1}{6} \\ \frac{5}{6} \end{bmatrix}$; the value is $-\frac{1}{6}$. **11.** Strictly determined, $[1 \;\; 0 \;\; 0]$ and $\begin{bmatrix} 1 \\ 0 \\ 0 \end{bmatrix}$; the value is 2.

13. The optimal strategies are $[\frac{1}{6} \;\; 0 \;\; \frac{5}{6}]$ and $\begin{bmatrix} 0 \\ \frac{2}{3} \\ \frac{1}{3} \end{bmatrix}$; the value is $-\frac{1}{3}$. **15.** Strictly determined, $[0 \;\; 1 \;\; 0]$ and $\begin{bmatrix} 0 \\ 1 \\ 0 \end{bmatrix}$; the value is -1. **17.** The optimal strategies are $[\frac{1}{4} \quad 0 \quad \frac{3}{4}]$ and $\begin{bmatrix} \frac{1}{4} \\ \frac{3}{4} \\ 0 \end{bmatrix}$; the value is $-\frac{1}{4}$. **19.** Not strictly determined.

21. Not strictly determined. **23.** The optimal strategies are $[\frac{1}{5} \quad \frac{4}{5}]$ and $\begin{bmatrix} \frac{1}{10} \\ \frac{9}{10} \end{bmatrix}$; the value is $\frac{11}{5}$. **25.** The optimal strategies are $[\frac{3}{5} \quad \frac{2}{5}]$ and $\begin{bmatrix} \frac{2}{5} \\ \frac{3}{5} \end{bmatrix}$; the value is -40.

9.6 *m* × *n* Matrix Games (page 463)

1. The optimal strategies are $[\frac{5}{6} \quad \frac{1}{6} \quad 0]$ and $\begin{bmatrix} \frac{1}{6} \\ \frac{5}{6} \end{bmatrix}$; the value is $\frac{7}{6}$. **3.** The optimal strategies are $[\frac{1}{3} \quad \frac{2}{3} \quad 0 \quad 0]$ and $\begin{bmatrix} \frac{2}{3} \\ \frac{1}{3} \end{bmatrix}$; the value is $\frac{1}{3}$. **5.** The optimal strategies are $[\frac{1}{2} \quad \frac{1}{2}]$ and $\begin{bmatrix} \frac{5}{8} \\ 0 \\ \frac{3}{8} \end{bmatrix}$; the value is $-\frac{1}{2}$. **7.** The optimal strategies are $[\frac{1}{3} \quad 0 \quad \frac{2}{3}]$ and $\begin{bmatrix} \frac{2}{3} \\ 0 \\ \frac{1}{3} \end{bmatrix}$; the value is $\frac{5}{3}$. **9.** The optimal strategies are $[0 \quad 1 \quad 0]$ and $\begin{bmatrix} 0 \\ 0 \\ 1 \end{bmatrix}$; the value is -1. **11.** The optimal strategies are $[0 \quad \frac{1}{6} \quad \frac{5}{6} \quad 0]$ and $\begin{bmatrix} 0 \\ \frac{1}{6} \\ 0 \\ \frac{5}{6} \end{bmatrix}$; the value is $\frac{13}{6}$. **13.** The optimal strategies are $[\frac{1}{2} \quad \frac{1}{2} \quad 0]$ and $\begin{bmatrix} \frac{1}{2} \\ \frac{1}{2} \\ 0 \end{bmatrix}$; the value is $\frac{1}{2}$.

9.7 Review (page 463)

Objective Questions

1. $T = \begin{matrix} & \begin{matrix} 0 & 1 & 2 \end{matrix} \\ \begin{matrix} 0 \\ 1 \\ 2 \end{matrix} & \begin{bmatrix} .12 & .63 & .25 \\ .47 & .15 & .38 \\ .17 & .42 & .41 \end{bmatrix} \end{matrix}$ **3.** Present $\begin{matrix} & \text{Next} \\ & \begin{matrix} S & L & G \end{matrix} \\ \begin{matrix} S \\ L \\ G \end{matrix} & \begin{bmatrix} 0 & .60 & .40 \\ .20 & .50 & .30 \\ 0 & .50 & .50 \end{bmatrix} \end{matrix}$ **5.** $T^3 = \begin{matrix} & \begin{matrix} L & G \end{matrix} \\ \begin{matrix} L \\ G \end{matrix} & \begin{bmatrix} .688 & .312 \\ .624 & .376 \end{bmatrix} \end{matrix}$ Probability is .312.

7. Probability is .72. **9.** The expected number of Buzza-Cardoza customers is 2,500,000. **11.** 33% with A, 38% with B, 29% with C **13.** Regular **15.** Not regular

For Problems 17 and 19, let V be the fixed probability vector.

17. $V = [\frac{1}{3} \quad \frac{2}{3}]$ **19.** $V = [\frac{5}{49} \quad \frac{25}{49} \quad \frac{19}{49}]$ **21.** The long-term expectation is $\frac{1}{3}$ of the market for Hallmark and $\frac{2}{3}$ for Buzza-Cardoza (regardless of the initial shares). **23.** The percentage of business is 33% for A, 39% for B, and 28% for C. **25.** Yes, third state is absorbing. **27.** Not a transition matrix. **29.** Since state 2 is the only absorbing state, there is a 100% probability that the system will end up in state 2 regardless of the beginning state. **31.** If a system begins in state 2, there is a 67% ($\frac{2}{3}$) chance it will end up in absorbing state 1 and a 33% ($\frac{1}{3}$) chance it will end up in absorbing state 4. If it begins in state 3, there is a 33% ($\frac{1}{3}$) chance it will end up in absorbing state 1 and a 67% ($\frac{2}{3}$) chance it will end up in state 4.

33. a. States 0 and 4 are absorbing states. **b.** Favors the second player (60% chance of losing) **35. a.** States 1 and 6 are absorbing states. **b.** $Q = \begin{matrix} & \begin{matrix} 2 & 3 & 4 & 5 \end{matrix} \\ \begin{matrix} 2 \\ 3 \\ 4 \\ 5 \end{matrix} & \begin{bmatrix} \frac{1}{4} & \frac{1}{4} & \frac{1}{4} & \frac{1}{8} \\ \frac{1}{4} & \frac{1}{2} & 0 & 0 \\ \frac{1}{4} & 0 & \frac{1}{2} & 0 \\ 1 & 0 & 0 & 0 \end{bmatrix} \end{matrix}$ **c.** $I_4 - Q = \begin{bmatrix} \frac{3}{4} & -\frac{1}{4} & -\frac{1}{4} & -\frac{1}{8} \\ -\frac{1}{4} & \frac{1}{2} & 0 & 0 \\ -\frac{1}{4} & 0 & \frac{1}{2} & 0 \\ -1 & 0 & 0 & 1 \end{bmatrix}$ **37.** 12¢ **39.** 44¢

41. Contract III has the highest expectation. **43.** House II has the highest expectation. **45.** Not strictly determined.

47. Strictly determined, the value is 1. **49.** Strictly determined, the value is -2. Strategies are $[1 \quad 0]$ and $\begin{bmatrix} 0 \\ 1 \end{bmatrix}$.

51. The optimal strategies are $[\frac{1}{3}\ \ \frac{2}{3}]$ and $\begin{bmatrix}\frac{1}{6}\\ \frac{5}{6}\end{bmatrix}$; the value is $-\frac{2}{3}$. **53.** The optimal strategies are $[\frac{1}{6}\ \ 0\ \ \frac{5}{6}]$ and $\begin{bmatrix}0\\ \frac{2}{3}\\ \frac{1}{3}\end{bmatrix}$; the value is $-\frac{4}{3}$. **55.** The strategies are $[\frac{6}{11}\ \ \frac{5}{11}\ \ 0]$ and $\begin{bmatrix}\frac{5}{11}\\ \frac{6}{11}\\ 0\end{bmatrix}$; the value is $-\frac{20}{11}$. **57.** The optimal strategies are $[\frac{6}{11}\ \ \frac{5}{11}]$ and $\begin{bmatrix}\frac{15}{22}\\ \frac{7}{22}\end{bmatrix}$; the value is $-\frac{100}{11}$. **59.** The strategies are $[\frac{1}{5}\ \ \frac{4}{5}]$ and $\begin{bmatrix}0\\ \frac{2}{5}\\ 0\\ 0\\ \frac{3}{5}\end{bmatrix}$; the value is $-\frac{1}{5}$. This is a loss of \$200.

61. The optimal strategies are $[\frac{7}{12}\ \ \frac{5}{12}\ \ 0]$ and $\begin{bmatrix}0\\ \frac{7}{12}\\ \frac{5}{12}\\ 0\end{bmatrix}$; the value is $-\frac{11}{12}$. **63.** The optimal strategies are $[\frac{1}{2}\ \ \frac{1}{2}\ \ 0]$ and $\begin{bmatrix}\frac{1}{2}\\ \frac{1}{2}\\ 0\end{bmatrix}$; the value is $\frac{1}{2}$.

Sample Test

1. Present (rows) / Next (columns):

$$\begin{array}{c|ccc} & S & L & G\\ \hline S & .20 & .75 & .05\\ L & .10 & .60 & .30\\ G & 0 & .40 & .60\end{array}$$

3. This game has solution $v_1 = \frac{5}{8}$, $v_2 = \frac{3}{8}$, the fixed probability vector is $[\frac{5}{8}\ \ \frac{3}{8}]$.

5. **a.** 0 and 5 are absorbing states. **b.** The game does not favor either player. **c.** $\frac{1}{5}$ or .2 **7.** The strategies of the original game are $[\frac{3}{4}\ \ 0\ \ \frac{1}{4}]$ and $\begin{bmatrix}\frac{1}{4}\\ \frac{3}{4}\\ 0\end{bmatrix}$, the value of the game is $\frac{5}{4}$. **9.** The optimal number is 23 cars.

Cumulative Review Chapters 3–9 (page 470)

1. **a.** $\begin{bmatrix}14 & -1 & 12\\ 20 & -8 & 5\\ -13 & 9 & 6\end{bmatrix}$ **b.** $\begin{bmatrix}18 & -1 & 11\\ 23 & -10 & 6\\ -13 & 12 & 8\end{bmatrix}$ **c.** $\begin{bmatrix}12 & -1 & 1\\ 3 & -3 & 4\\ 2 & 3 & -4\end{bmatrix}$ **d.** $\frac{1}{5}\begin{bmatrix}-1 & 3 & 2\\ -6 & 8 & 7\\ -9 & 12 & 8\end{bmatrix}$ **3.** (6, 13)

5. $(-3, 5, -2)$ **7.**

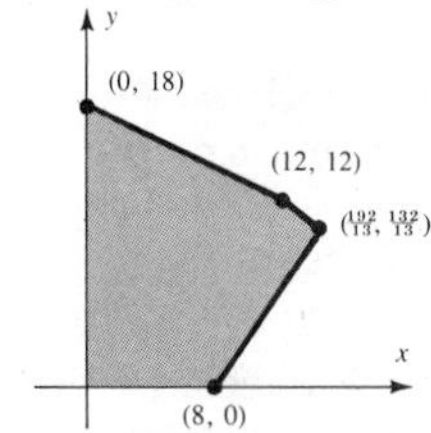

9. **a.** {2} **b.** {1, 3, 5, 7, 9} **c.** {2, 4, 6, 8, 10} **d.** { } **e.** {1, 4, 6, 8, 9, 10} **11.** **a.** 32 **b.** 720 **13.** **a.** .0625 **b.** .125 **c.** .0625 **d.** .25 **e.** .0004 **15.** .1

17. **a.**

Data	40	45	50	55	60	65	70	75	80	85	90	95
Frequency	1	0	0	1	0	2	4	1	2	1	1	2

b.
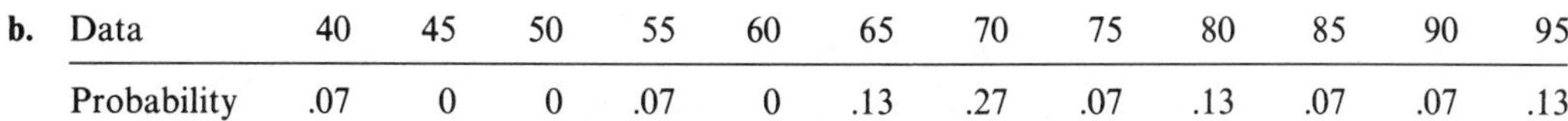

Data	40	45	50	55	60	65	70	75	80	85	90	95
Probability	.07	0	0	.07	0	.13	.27	.07	.13	.07	.07	.13

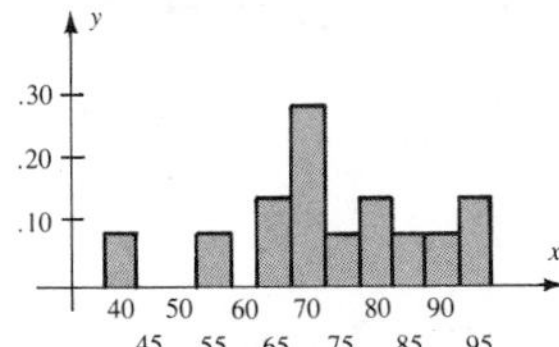

c. Mean = 73.7, median = 70, mode = 70 **d.** Range = 55, variance = 204.9, standard deviation = 14.3
19. **a.** $r = -.6013$; no significant correlation at 1% or 5% **b.** $y' = -.16x + 41.6$ **21.** Maximum number of animals is 400 (they are all species *C*). **23.** The "mixture" should consist of 2 pounds of peanuts and no pounds of cashews.
25. *C* **27.** *A* **29.** *C* **31.** *C* **33.** *B* **35.** Answers vary. **37.** Answers vary. **39.** After two months, companies *A* and *B* would each have 37.3% of the market and company *C* would have 25.3%. **41.** $[\frac{6}{11} \quad \frac{5}{11}]$ **43.** *A*

45. Strictly determined game with value -1; strategies are $[0 \quad 0 \quad 1]$ and $\begin{bmatrix} 1 \\ 0 \\ 0 \end{bmatrix}$. **47.** \$6,574.44 **49.** \$17,624.76

CHAPTER 10

10.1 Limits (page 492)

1. 0 **3.** 8 **5.** 7 **7.** 6 **9.** 2 **11.** 2 **13.** 15 **15.** Limit does not exist. **17.** 0 **19.** 5
21. $-\frac{1}{6}$ **23.** 1 **25.** 0 **27.** $\frac{1}{2}$ **29.** 4 **31.** 8 **33.** $\frac{1}{6}$ **35.** 0 **37.** Limit does not exist.
39. 0 **41.** $\frac{1}{4}$ **43.** 2 **45.** $\frac{3}{2}$ **47.** 5 **49.** Limit does not exist. **51.** 1
53. The function is decreasing without bound. **55.** $\frac{4}{3}$ **57.** 3 **59.** 5 **61.** 10 **63.** $y = 3$ **65.** $y = \frac{2}{3}$
67. $y = 2$ **69.** **a.** \$7.00 **b.** \$12.00 **c.** \$7.00 **d.** Limit does not exist. **71.** **a.** Limit does not exist. **b.** 80%
c. 70% **d.** 50% **73.** $\frac{1}{2}$ **75.** $-\frac{1}{9}$

10.2 Continuity (page 501)

1. Suspicious point at $x = 4$; $x = 4$ is a point of discontinuity. **3.** Suspicious point at $x = 2$; $x = 2$ is a point of discontinuity.
5. Suspicious points at $x = 1$ and $x = 4$; $x = 4$ is a point of discontinuity. **7.** Suspicious points at $x = 1$, $x = 6$; $x = 1$ and $x = 6$ are points of discontinuity. **9.** Suspicious points at $x = 2$, $x = -2$; $x = -2$ and $x = 2$ are points of discontinuity.
11. There are no suspicious points; hence the function is continuous for $x > 0$. **13.** Yes **15.** Yes
17. Since there are no abrupt jumps or undefined values, it is continuous. The domain is $0 \le x < 24$, where x is the hour of the day.
19. Not continuous (unless the price remains constant). The domain is $0 \le x < 24$, where x is the hour of the day.
21. Not continuous; the domain is $t > 0$, where t is the number of minutes. **23.** Continuous on $[-5, 5]$
25. Continuous on $[5, 10]$ **27.** Discontinuities at $x = 1$ and $x = -1$ **29.** $x = -2$ is the only discontinuity on $[-5, 5]$.
31. Continuous on $[0, 2]$. **33.** Continuous on $[0, 5]$. **35.** $x = 3$ is a discontinuity. **37.** Discontinuous at $x = -2$.
39. Continuous on $[0, 5]$ **41.** $x = -2$ is the only suspicious point; $x = -2$ is a discontinuity.
43. Continuous on $[-5, 5]$ **45.** Continuous on $[-5, 5]$ **47.** **a.** Not continuous **b.** \$.52, \$.75, \$.75
c.
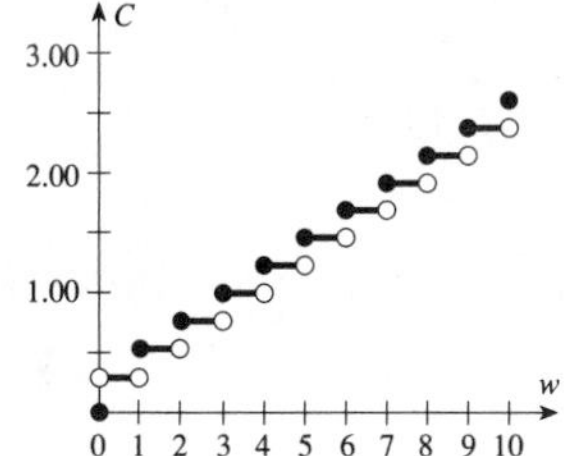

49. Answers vary.

10.3 Rates of Change (page 512)

1. $\frac{20}{3}$ ft/sec **3.** 40 ft/sec **5.** $\frac{10}{7}$ units/worker **7.** $\frac{5}{6}$ units/worker **9.** $-10\ \frac{\text{points}}{\text{year}}$ **11.** $-9\ \frac{\text{points}}{\text{year}}$

13. .1486 trillion/yr **15.** .2564 trillion/yr **17.** $59.3\ \frac{\text{miles}}{\text{hour}}$ **19.** $36\ \frac{\text{miles}}{\text{hour}}$ **21.** Slope $\to 0$

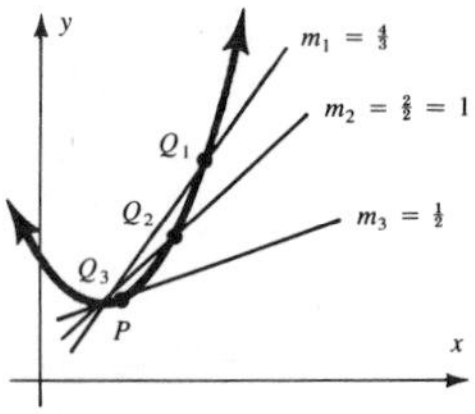

23. Slope $= 2$ for *each* value of Q **25.** Slope $\to 0$ **27.** -3 **29.** 0 **31.** 12 **33.** -9 **35.** $\frac{1}{6}$
37. -3 **39.** 0 **41.** 6 **43.** -3 **45.** 2 **47.** 0 **49.** 9 **51.** 8,900 **53.** 5,930
55. $60x + 30h - 100$ **57.** $60x - 100$ **59.** The instantaneous rate of change of cost at a given x.

10.4 Definition of Derivative (page 519)

1. Derivative does not exist at $x = 1$. **3.** The derivative does not exist at $x = 3$. **5.** The derivative does not exist at $x = 4$.

7. $4x$ **9.** $-6x$ **11.** -5 **13.** $6x + 4$ **15.** $-6x - 50$ **17.** $6x^2$ **19.** $\frac{3}{x^2}$ **21.** $\frac{\sqrt{x}}{x}$

23. $30x + y + 45 = 0$ **25.** $5x + y - 4 = 0$ **27.** $8x - y + 6 = 0$ **29.** The marginal cost is \$0.
31. The actual cost of producing an extra item is \$1. **33.** $2x - 100$ **35.** The marginal cost is \$20.
37. The actual cost is \$21. **39.** -188 **41.** $\frac{31}{5}$ **43.** 20 miles per min

10.5 Differentiation Techniques, Part I (page 525)

1. $7x^6$ **3.** $12x^{11}$ **5.** $-5x^{-6}$ **7.** 0 **9.** $32x^{-9}$ **11.** $\frac{-\sqrt{x}}{4x}$ **13.** $-40x^{-9}$ **15.** $15x^{1/4}$

17. $6x + 1$ **19.** $4x - 5$ **21.** $15x^2 - 10x + 4$ **23.** $\frac{2x^5 - x^2 - 3}{x^4}$ **25.** $8x^7 + 16x^3$ **27.** $\frac{-2x - 10}{x^3}$

29. $-35x^6 + x^{-1/2} + 3x^{-2}$ **31.** $1 - 2x^{-3}$ **33.** $\frac{9}{2}x^{7/2} + \frac{15}{2}x^{3/2}$ **35.** $\frac{1}{3}x^{-2/3}$ **37.** π **39.** $4\pi r^2$
41. $\frac{C}{2\pi}$ **43.** $f'(x) = 5x^4$; $f'(1) = 5$ **45.** $f'(x) = -\frac{1}{2}x^{-3/2}$; $f'(1) = -\frac{1}{2}$ **47.** $f'(x) = 6x - 4$; $f'(0) = -4$
49. $f'(x) = 2x + 2$; $f'(0) = 2$; $2x - y + 1 = 0$ **51.** $f'(x) = 4x^{-3}$; $f'(1) = 4$; $4x - y - 4 = 0$
53. $f'(x) = \frac{1}{3}x^{-2/3}$; $f'(1) = \frac{1}{3}$; $x - 3y + 2 = 0$ **55.** Marginal cost is $C'(x) = 40x + 500$ **57.** $\frac{dm}{dt} = \frac{-400}{t^2}$
59. The rate of change of earnings is 25.6 (thousand dollars per year). **61.** The instantaneous rate of change of number of tasks (N) per hour (unit change in x) is $N' = \frac{25}{2\sqrt{x}}$. At the end of the fifth hour, the subjects are learning at the rate of 6 tasks per hour.
63. The rate of change of P with respect to x is about 20 foxes per rabbit. **65. a.** The average rate of change in the CPI from 1990 to 1994 was 4.8. **b.** 4.9 **67.** Answers vary. **69.** Answers vary.

10.6 Differentiation Techniques, Part II (page 530)

1. $20x^3 - 60x$ **3.** $2x - 1$ **5.** $24x^3 - 10x$ **7.** $60x^5 - 125x^4 + 20x^3$ **9.** $\frac{-3}{(x-3)^2}$ **11.** $\frac{-8}{(x-3)^2}$

13. $\frac{-16x}{(x^2-5)^2}$ **15.** $\frac{-1}{(x+2)^2}$ **17.** $10x^4 + 16x^3 - 3x^2 - 16x$ **19.** $20x^4 - 44x^3 - 9x^2 + 104x + 13$

21. $42x^6 - 20x^4 + 100x^3 - 81x^2 - 20x + 43$ **23.** $-\frac{25}{3}x^{-4/3} + 25x^{2/3}$ **25.** $30x^{3/2} - 15x^{-1/2}$

27. $\frac{-10x^2 + 6}{(5x^2 - 11x + 3)^2}$ **29.** $-\frac{35}{8}$ **31.** $\frac{83}{4}$ **33.** $-\frac{3}{2}$ **35.** $x + y = 0$ **37.** $83x - 4y - 180 = 0$

39. $x + 9y + 2 = 0$ **41.** $\dfrac{-200{,}000x - 1{,}500{,}000}{(x^2 + 15x + 25)^2}$ **43.** When the price is \$15, the demand is decreasing by about 20 items per unit change in price. The slope of the graph at $x = 15$ is -20. **45.** $N'(x) = 575x^{-2}$ **47.** The slope of the graph at $x = 10$ is 5.75. The number of items is increasing at the point $x = 10$ since N' is positive. **49.** $.01(.01 + .005t)^{-2}$
51. $\dfrac{-20t^2 + 10}{(2t^2 + 1)^2}$

10.7 The Chain Rule (page 539)

1. $9(3x + 2)^2$ **3.** $20(5x - 1)^3$ **5.** $3(2x^2 + x)^2(4x + 1)$ or $3x^2(2x + 1)^2(4x + 1)$ **7.** $2(2x^2 - 3x + 2)(4x + 3)$
9. $4(x^3 + 5x)^3(3x^2 + 5)$ or $4x^3(x^2 + 5)^3(3x^2 + 5)$ **11.** $-2(2x^2 - 5x)^{-3}(4x - 5)$ or $-2x^{-3}(2x - 5)^{-3}(4x - 5)$
13. $-(x^4 + 3x^3)^{-2}(4x^3 + 9x^2)$ or $-x^{-4}(x + 3)^{-2}(4x + 9)$ **15.** $15(4x^3 + 3x^2)^2(12x^2 + 6x)$ or $90x^5(4x + 3)^2(2x + 1)$
17. $\frac{1}{4}(x^2 - 3x)^{-3/4}(2x - 3)$ or $\dfrac{2x - 3}{4(x^2 - 3x)^{3/4}}$ **19.** $x(x^2 + 16)^{-1/2}$ or $\dfrac{x\sqrt{x^2 + 16}}{x^2 + 16}$ **21.** $\frac{15}{2}x^2(x^3 + 8)^{-1/2}$
23. $\frac{1}{2}(3x + 1)^{-1/2}(9x + 2)$ **25.** $\frac{2}{3}(2x + 5)^{2/3}$ **27.** $-5(5x + 3)^{-2}$ **29.** $-4x(x^2 + 3)^{-3}$ **31.** $60x - y - 35 = 0$
33. $3x - 8y + 11 = 0$ **35.** $7x - 2y - 25 = 0$ **37.** $f'(x) = 0$ if $x = 0, 3, \frac{3}{2}$ **39.** $f'(x) = 0$ if $x = \frac{3}{2}$
41. $f'(x) = 0$ if $x = \frac{5}{2}$ **43. a.** $(2x + 1)^2[3(3x + 2)^2 3] + [2(2x + 1)(2)](3x + 2)^3$ **b.** $(2x + 1)(3x + 2)^2(30x + 17)$
45. a. $(5x + 1)^2[-1(4x + 3)^{-2}(4)] + [2(5x + 1)(5)](4x + 3)^{-1}$ **b.** $2(5x + 1)(4x + 3)^{-2}(10x + 13)$
47. a. $\dfrac{(5x + 3)[2(2x - 5)(2)] - 5(2x - 5)^2}{(5x + 3)^2}$ **b.** $\dfrac{(2x - 5)(10x + 37)}{(5x + 3)^2}$ **49. a.** $\dfrac{(2x - 5)^2[4(x + 5)^3] - (x + 5)^4[4(2x - 5)]}{(2x - 5)^4}$
b. $\dfrac{4(x + 5)^3(x - 10)}{(2x - 5)^3}$ **51. a.** $3(x^2 + 1)^{-1/2} - 3x^2(x^2 + 1)^{-3/2}$ **b.** $3(x^2 + 1)^{-3/2}$ **53.** The rate of change at any t is: $-5{,}400 \cdot 10^9$ bacteria per minute. **55.** $\dfrac{dA}{dr} = 100{,}000(1 + r)^9$ **57.** $\dfrac{dP}{dr} = -100{,}000(1 + r)^{-11}$ **59. a.** The present enrollment is 4,000. **b.** The enrollment in 10 years will be about 5,691. **c.** $400(1 + .2t)^{-3/2}$ **d.** The annual enrollment is increasing by about 400 students per year. **e.** The annual enrollment is increasing by about 77 students per year.
61. Answers vary; $\dfrac{f}{g} = fg^{-1}$. Using the product rule,

$$\begin{aligned}(fg^{-1})' &= f'(g^{-1}) + f(g^{-1})' \\ &= f'g^{-1} + f(-g^{-2}g') \\ &= \frac{f'}{g} - \frac{fg'}{g^2} \\ &= \frac{fg'}{g^2} - \frac{fg'}{g^2} \\ &= \frac{gf' - fg'}{g^2}\end{aligned}$$

10.8 Chapter Review (page 540)

Objective Questions

1. 10 **3.** The limit does not exist. **5.** The limit does not exist. **7.** The limit is -5. **9.** 7 **11.** 0
13. Suspicious at $x = -3$ and $x = 4$; discontinuous at $x = 4$ **15.** Suspicious at $x = 7$; discontinuous at $x = 7$
17. Discontinuous **19.** Continuous **21.** $f(x)$ is continuous on $[-5, 5]$ **23.** Suspicious at $x = 8$; discontinuous at $x = 8$ **25.** 16 **27.** $\dfrac{\sqrt{2}}{2}$ **29.** 10 **31.** $\frac{1}{2}$ **33.** For a function $f(x)$, the derivative is $f'(x) = \lim\limits_{h \to 0} \dfrac{f(x + h) - f(x)}{h}$ provided this limit exists. **35.** $\dfrac{-1}{(x - 5)^2}$ **37.** $14x^{13}$ **39.** $-\frac{7}{9}x^{-16/9}$ **41.** 0
43. $6x^2 - 10x$ **45.** $3 - 54x$ **47.** $-\frac{1}{2}x^{-1/2}(x - 1)^{-2}(x + 1)$ **49.** $\dfrac{-6x}{(3x^2 + 1)^2}$ **51.** $\dfrac{-5x - 3}{2x^{1/2}(5x - 3)^2}$
53. $20(5x + 9)^3$ **55.** $y' = \frac{1}{5}(5x^2 - 3x)^{-4/5}(10x - 3)$

Sample Test

1. 4 **3.** $\lim_{x \to 2} \frac{6x+1}{3-x} = 13$ **5.** Look for "jumps, holes, or poles" (vertical asymptotes). There is a hole at $x = 0$, a jump at $x = 3$, and a pole at $x = 10$. **7.** Let $f(x) = y = 5 + x - 2x^2$. The average rate of change is $\frac{f(3)-f(1)}{3-1} = \frac{f(3)-f(1)}{3-1} = \frac{(-10)-(4)}{3-1} = -7$. **9.** The derivative of $f(x)$ is $f'(x) = \lim_{h \to 0} \frac{f(x+h)-f(x)}{h}$ provided this limit exists. **11.** 0 **13.** $y' = 10x + 2x^{-2}$ **15.** $y' = \frac{-60}{(4x-5)^2}$ **17.** $y' = \frac{-(x-5)^2(4x+7)}{(1-2x)^2}$

19. The profit is decreasing by \$.50 per unit.

CHAPTER 11

11.1 Implicit Differentiation (page 550)

1. Explicitly: $\frac{dy}{dx} = -10x$; implicitly: $\frac{dy}{dx} = -10x$ **3.** Explicitly: $\frac{dy}{dx} = -5x^{-2}$; implicitly: $\frac{dy}{dx} = \frac{-y}{x}$

5. Explicitly: $\frac{dy}{dx} = \frac{1}{3} + \frac{50}{3}x^{-2}$; implicitly: $\frac{dy}{dx} = \frac{2}{3} - \frac{y}{x}$ **7.** Explicitly: $\frac{dy}{dx} = -x(4-x^2)^{-1/2}$; implicitly: $\frac{dy}{dx} = -\frac{x}{y}, y > 0$ **9.** Explicitly: $\frac{dy}{dx} = \frac{8}{15}x^3\left(\frac{2}{5}x^4 - \frac{7}{5}\right)^{-2/3}$; implicitly: $\frac{dy}{dx} = \frac{8x^3}{15y^2}$ **11.** $\frac{dy}{dx} = \frac{2x-y}{x-2y}$

13. $\frac{dy}{dx} = \frac{-2(2x+1)}{3(3y-5)}$ **15.** $\frac{dy}{dx} = \frac{x-10}{11y}$ **17.** $\frac{dy}{dx} = \frac{-6xy^3+3y^2-5y}{9x^2y^2-6xy+5x}$ **19.** $\frac{dy}{dx} = \frac{-3x^2-4x-y}{x}$

21. $y - 2 = 0$ **23.** $8x + 15y + 23 = 0$ **25.** $x + 5y + 14 = 0$ **27.** $y - 4 = 0$ and $y + 2 = 0$ **29.** $x - y = 0$

31. $y - 3 = 0$ and $y + 1 = 0$ **33.** At $(1, -2)$, $\frac{dy}{dx} = -3$ **35.** At $(1, 2)$, $\frac{dy}{dx} = 2$ **37.** At $(3, 4)$, $\frac{dy}{dx} \approx -.2$

39. $\frac{dp}{dx} = \frac{1}{2p-5}$

11.2 Differentials (page 556)

1. $dy = 15x^2\,dx$ **3.** $dy = -10x^{-2}\,dx$ **5.** $dy = (300x^2 - 50)\,dx$ **7.** $dy = \frac{3}{2}(x-1)^{-1/2}\,dx$ **9.** $dy = 5\,dx$

11. $dy = 2(5x-3)(10x^2-3x-15)\,dx$ **13.** $dy = -7(x-2)^{-2}\,dx$ **15.** $dy = (2x-1)\,dx$

17. $dy = (-5x^{-2} + 2x^{-3} + 3x^{-4})\,dx$ **19.** $dy = \frac{8x^2-6x+7}{(x^2+3x-2)^2}dx$ **21.** $dy = 1.8, \Delta y = 1.81$

23. $dy = .075, \Delta y \approx .0744457825$ **25.** $\Delta y \approx dy = \frac{57}{502^2} \approx .0002261869$ **27.** $\Delta y \approx \frac{40}{10^3}(.02) = .0008$

29. $\Delta y \approx dy \approx 46.105036$ **31.** $\Delta S \approx dS = 3{,}000$, so the sales will increase by about 3,000 units.

33. The alcohol will increase by about .01 percent. **35.** The area of a circle is 1.256637 square miles.

37. The change in revenue is $\Delta R \approx dR = \$600$. The change in profit is $\Delta P \approx dP = \$300$. **39.** 14,000 votes

11.3 Business Models Using Differentiation (page 562)

1. 7 **3.** 46 **5.** 9 **7.** 8.9 **9.** −18 **11.** −14 **13.** $\bar{C}(x) = 200x^{-1} + 6 - x + x^2$; $\bar{C}'(x) = -200x^{-2} - 1 + 2x$ **15.** $\bar{C}(x) = 5{,}000x^{-1} + .4x$; $\bar{C}'(x) = -5{,}000x^{-2} + .4$ **17.** $\bar{R}(x) = 50 - .5x$; $\bar{R}'(x) = -.5$

19. $\bar{P}(x) = x^2 - 8x + 2 + \frac{50}{x}$; $\bar{P}'(x) = 2x - 8 - \frac{50}{x^2}$ **21.** $\bar{P}(x) = x^2 - 50x + 5 + \frac{200}{x}$; $\bar{P}'(x) = 2x - 50 - \frac{200}{x^2}$

23. $R(x) = -.001x^2 + 30x$ **25.** $R'(x) = -.002x + 30$ This is approximately the change of revenue relative to a unit change in x. **27.** $P'(x) = -.002x + 21.5$; this is the change in profit relative to a unit change in x. **29.** $\bar{C}'(x) = 50{,}000x^{-2}$ **31.** $R'(x) = 200 - .08x$ **33.** At a production level of 1,000 items, the revenue is increasing at a rate of \$84 per item. At a production level of 2,500 items the revenue does not change per unit change in the number of items. **35.** At a production level of 1,000 items, the profit is increasing at \$70 per item. At 2,500 items the profit is decreasing at \$50 per item.

37.

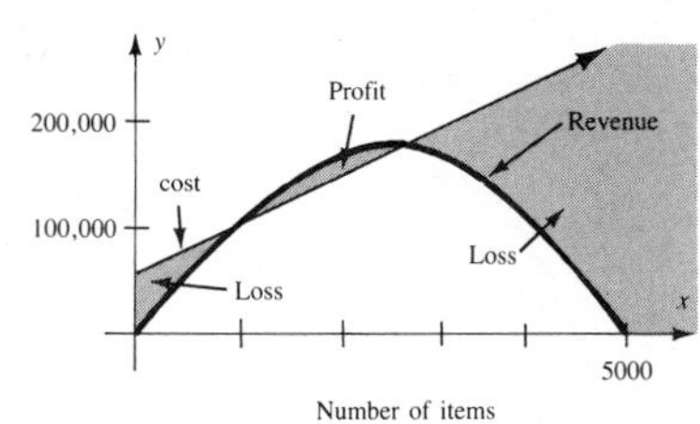

39. $R'(x) = 5 - .002x$ **41.** Elastic demand **43.** Inelastic demand

11.4 Related Rates (page 569)

1. $\frac{dy}{dt} = -3$ **3.** $\frac{dy}{dt} = 1{,}000$ **5.** $\frac{dx}{dt} = 15$ **7.** $\frac{dy}{dt} = \frac{4}{5}$ **9.** $\frac{dx}{dt} = 0$ **11.** Profit is decreasing at a rate of \$123 per week. **13.** The distance is changing at about 2.24 feet per second. **15.** $\frac{dh}{dt} \approx .398$ **17.** $\frac{dV}{dt} = -.015$ **19.** $\frac{dV}{dt} = -.06$ **21.** The area is increasing at a rate of about 37.7 square miles per year. **23.** The area is growing at a rate of about 62.8 square feet per second. **25.** The price is increasing at \$.11 per day. **27.** $\frac{dV}{dt} \approx 60.32$ cm^3 per minute **29.** $\frac{ds}{dt} = 8$ feet per second

11.5 Review (page 570)

Objective Questions

1. $\frac{dy}{dx} = \frac{-5x^4}{4y+1}$ **3.** $\frac{dy}{dx} = \frac{x-3}{y+1}$ **5.** $f'(1) = 6$ **7.** $f'(1) = 100$ **9.** $6x - y - 4 = 0$ **11.** $100x + y + 95 = 0$ **13.** $dy = 18x(3x^2+5)^{-1/2}\,dx$ **15.** $dy = \frac{-x^2 - 10x - 3}{(3x^2 + x - 4)^2}\,dx$ **17.** $dy = 10; \Delta y = 10.05$ **19.** $dy = .1; \Delta y \approx .09918$ **21.** $\Delta y \approx .000045$ **23.** $y = \frac{x^2-1}{x+2}; \Delta y \approx dy \approx .9938$ **25. a.** \$1.18 **b.** $\frac{x^4 - 30x^2 + 400x}{(10 - x^2)^2}$ **27.** $2{,}500x^2(x-20)(x-12)$ **29.** $-\frac{4}{5}\sqrt{5}$ **31.** $\frac{dy}{dt} = -5$

Sample Test

1. $\frac{dy}{dx} = 20x - 6$ **3.** $\frac{dy}{dx} = \frac{4x^3 - 6xy + 9y^2 + 5y}{3x^2 - 18xy - 5x}$ **5.** $\frac{dy}{dt} = \pm\frac{27}{8}$ **7.** $\frac{dx}{dt} = \frac{-2}{75}$

9. The marginal revenue is $4{,}000 - 200x$.

CHAPTER 12

12.1 First Derivatives and Graphs (page 583)

1.

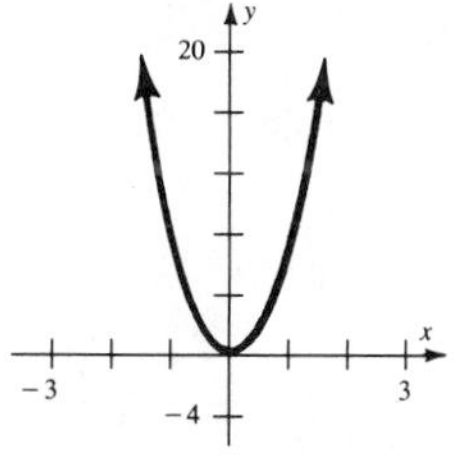

3.

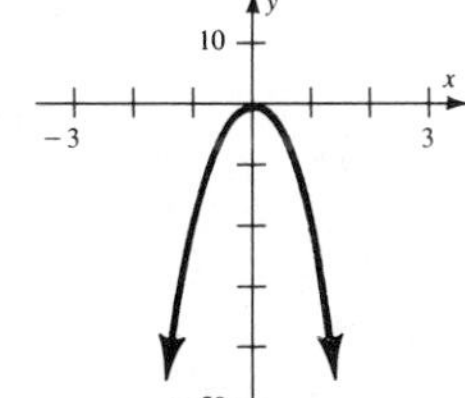

5.

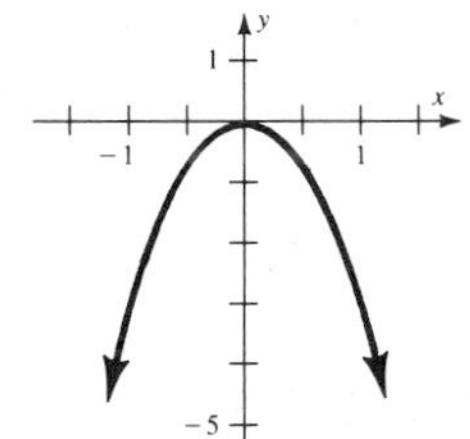

7. 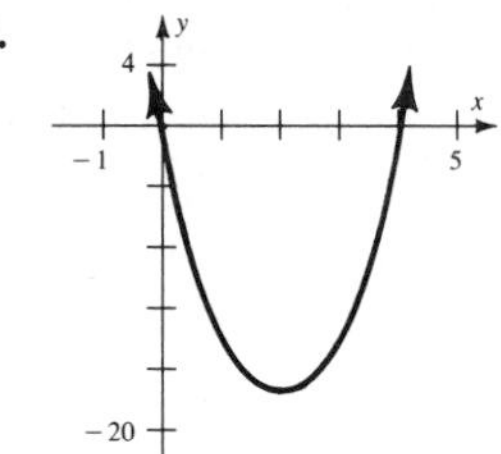

9. 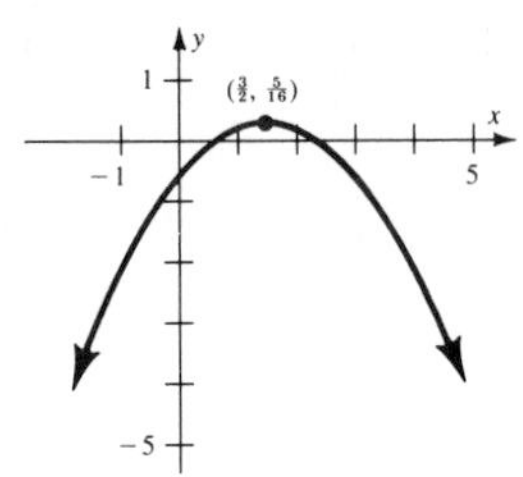

11. 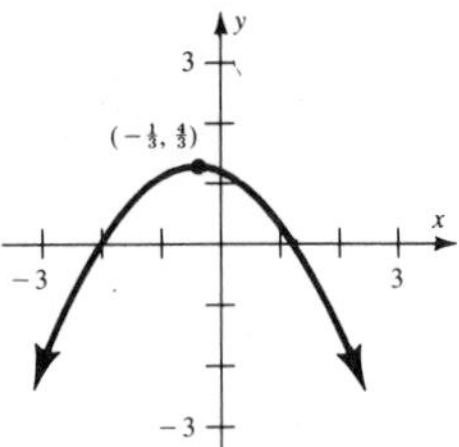

13. 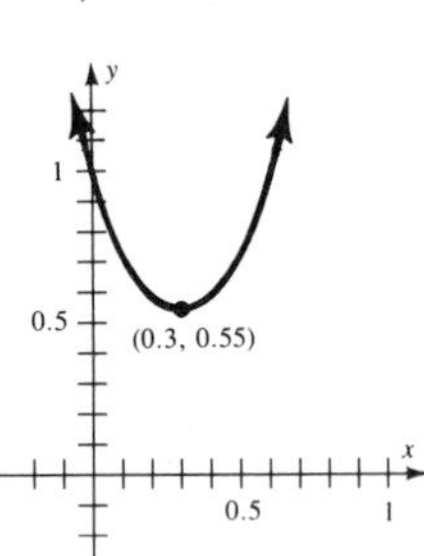

15. 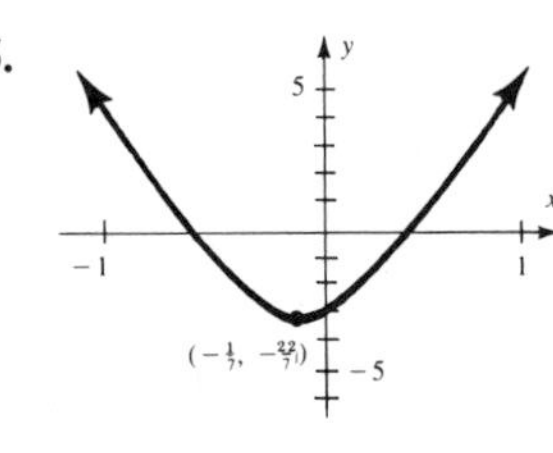

17. Increasing on $(-\infty, 2)$ and decreasing on $(2, \infty)$; horizontal tangent at $(2, 4)$. **19.** Decreasing on $(-\infty, 0)$ and on $(5, \infty)$; increasing on $(0, 3)$; constant on $(3, 5)$; horizontal tangent at $(0, 3)$ and for all $3 < x < 5$ **21.** Increasing on $(-\infty, 2)$ and $(2, \infty)$; no horizontal tangents **23.**

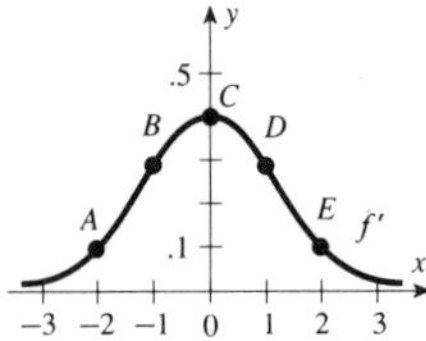

25.

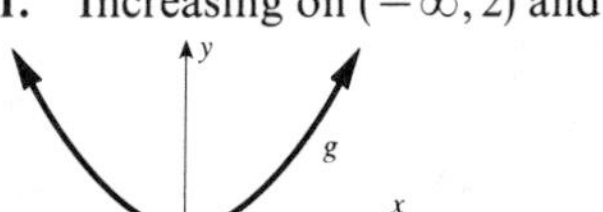

27. $f'(x) = 10 - 2x$ **a.** Critical value is 5. **b.** $f(x)$ is increasing on $(-\infty, 5)$ and decreasing on $(5, \infty)$.
29. $f'(x) = 2x - 12$ **a.** Critical value is 6. **b.** $f(x)$ is increasing on $(6, \infty)$ and decreasing on $(-\infty, 6)$.
31. $g' = 6x^2 - 6x - 36$ **a.** Critical values: $3, -2$ **b.** $g(x)$ is increasing on $(-\infty, -2)$ and $(3, \infty)$ and decreasing on $(-2, 3)$.
33. $y' = 3x^2 + 10x + 8$ **a.** Critical values: $-\frac{4}{3}, -2$ **b.** y is increasing on $(-\infty, -2)$ and $(-\frac{4}{3}, \infty)$ and decreasing on $(-2, -\frac{4}{3})$. **35.** $y' = 12x^3 + 6x^2 - 18x$ **a.** Critical values: $0, -\frac{3}{2}, 1$ **b.** y is increasing on $(-\frac{3}{2}, 0)$ and $(1, \infty)$ and decreasing on $(-\infty, -\frac{3}{2})$ and $(0, 1)$. **37.** $f'(x) = 15x^4 - 75x^2 - 540$ **a.** Critical values: $3, -3$ **b.** $f(x)$ is increasing on $(-\infty, -3)$ and $(3, \infty)$ and decreasing on $(-3, 3)$. **39.** $y' = \dfrac{-3}{(x-2)^2}$ **a.** No critical values **b.** y is increasing on $(-\infty, 2)$ and $(2, \infty)$. **41.** $f'(x) = -x^{-2} - 2x^{-3}$ **a.** Critical values: $0, -2$ **b.** $f(x)$ is increasing on $(-2, 0)$ and decreasing elsewhere. **43.** $S(x)$ is increasing when the amount spent on advertising is under \$33,333. **45.** Begin campaign April 1.

12.2 Second Derivatives and Graphs (page 594)

1. $f'(x) = 10x^4 - 12x^3 + 3x^2 - 10x + 19$; $f''(x) = 40x^3 - 36x^2 + 6x - 10$; $f'''(x) = 120x^2 - 72x + 6$ **3.** $y = \sqrt{5}\,x^{1/2}$; $y' = \dfrac{\sqrt{5}}{2}x^{-1/2}$; $y'' = -\dfrac{\sqrt{5}}{4}x^{-3/2}$; $y''' = \dfrac{3\sqrt{5}}{8}x^{-5/2}$; $\dfrac{d^4y}{dx^4} = \dfrac{-15\sqrt{5}}{16}x^{-7/2}$ **5.** $g'(x) = 12x^2 + 2x^{-2}$; $g''(x) = 24x - 4x^{-3}$; $g'''(x) = 24 + 12x^{-4}$; $g^{(4)}(x) = -48x^{-5}$ **7.** $y' = 5x^{2/3}$; $y'' = \dfrac{10}{3}x^{-1/3}$; $\dfrac{d^3y}{dx^3} = y''' = \dfrac{-10}{9}x^{-4/3}$ **9.** $y = 3(x-1)^{-1}$; $y' = -3(x-1)^{-2}$; $y'' = 6(x-1)^{-3}$; $\dfrac{d^3y}{dx^3} = y''' = -18(x-1)^{-4}$ **11.** $y' = \dfrac{x^2 + 8x + 1}{(x+4)^2}$; $\dfrac{d^2y}{dx^2} = y'' = \dfrac{30}{(x+4)^3}$
13. Relative minimum 0 at $x = -\frac{1}{2}$ **15.** Relative minimum 6 at $x = 3$; relative maximum -6 at $x = -3$
17. Relative minimum 1 at $x = 0$ **19.** Relative minimum 0 at $x = 0$; relative maximum 16 at $x = -2$; relative maximum 16 at $x = 2$ **21.** Relative minimum $\frac{194}{27}$ at $x = \frac{2}{3}$; relative maximum 58 at $x = -4$ **23.** Relative maximum $\frac{7}{4}$ at $-\frac{1}{2}$; relative minimum -331 at $x = 5$ **25.** Relative minimum $2\sqrt{2}$ at $x = \sqrt{2}$; relative maximum $-2\sqrt{2}$ at $x = -\sqrt{2}$

27. Relative maximum 0 at $x = 0$; relative minimum 4 at $x = 2$ **29.** A relative minimum 0 occurs at $x = -1$. **31. a.** $x = 6$ **b.** increasing for $x < 6$; decreasing for $x > 6$. **c.** concave down for all x **33. a.** $x = -\frac{1}{3}$ or 5 **b.** curve is increasing left of $x = -\frac{1}{3}$, decreasing for $-\frac{1}{3} < x < 5$, and increasing right of $x = 5$ **c.** curve is concave down for $x < \frac{7}{3}$ **35. a.** $x = \frac{5}{3}$ or -9 **b.** curve increasing left of -9, decreasing for $-9 < x < \frac{5}{3}$, and increasing right of $\frac{5}{3}$ **c.** curve is concave up for $x > -\frac{11}{3}$, and concave down for $x > -\frac{11}{3}$ **37. a.** $x = -\frac{2}{9}$ or $\frac{1}{2}$ **b.** curve is increasing left of $-\frac{2}{9}$, decreasing $-\frac{2}{9} < x < \frac{1}{2}$, and increasing right of $\frac{1}{2}$. **c.** curve is concave up for $x > \frac{5}{36}$ and is concave down for $x < \frac{5}{36}$ **39. a.** $x = 2$ **b.** graph decreasing on $(-\infty, 2)$ and $(2, \infty)$ **c.** concave up on $(-\infty, 2)$, concave down on $(2, \infty)$ **41.** The point of diminishing returns is at $x = 15$. **43.** $[0, 15)$ **45.** $x = 15$ is the point of diminishing returns with sales of 11,650 **47.** Relative maximum at $x = \dfrac{30 + \sqrt{980}}{2} \approx 30.65$; $x < 15$ concave up; $x > 15$ concave down

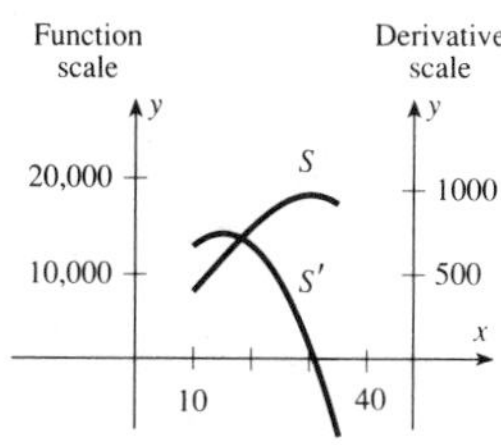

49. Relative maximum at $x = 68$; no relative minimums **51.** Revenue = (price)(number of items) **a.** $R(x) = 5x - \frac{1}{10{,}000}x^3$ **b.** $R'(x) = 5 - \frac{3}{10{,}000}x^2$ **c.** The marginal revenue is decreasing.

12.3 Curve Sketching—Relative Maximums and Minimums (page 600)

1. Answers vary.

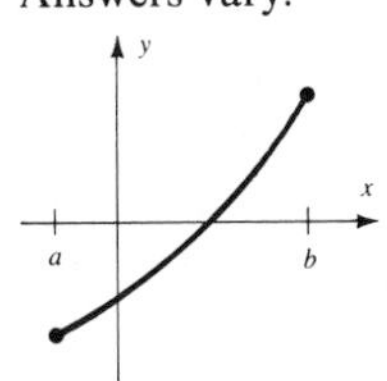

3. Answers vary.

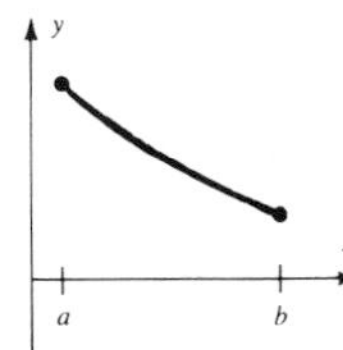

5. Answers vary.

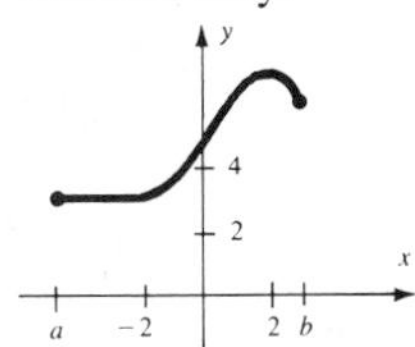

7.

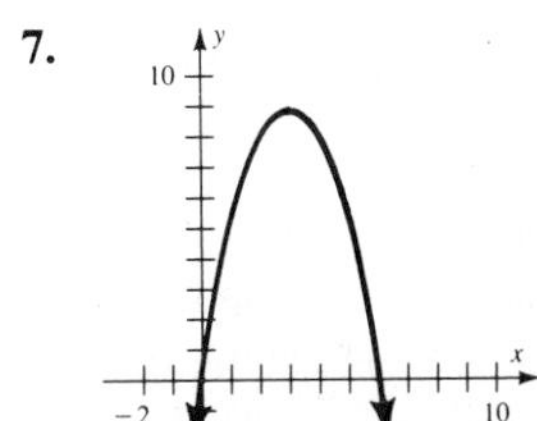

9.

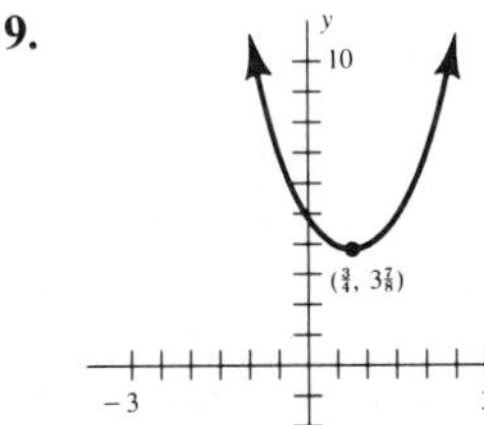

11.

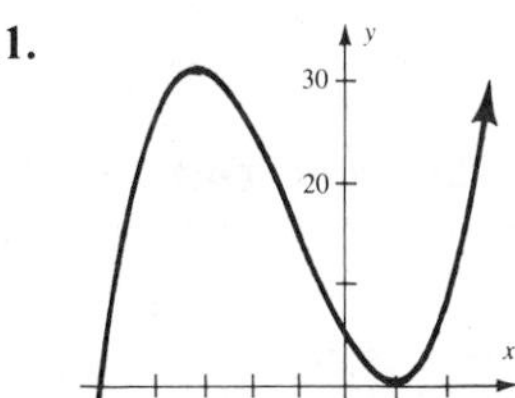

13.

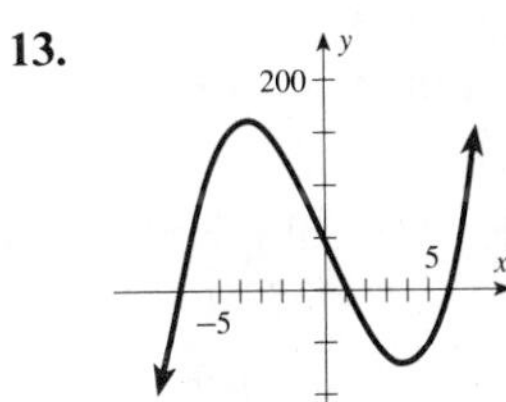

15.

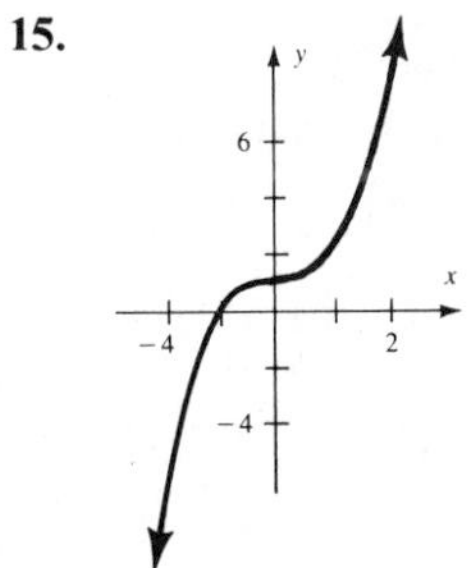

17.

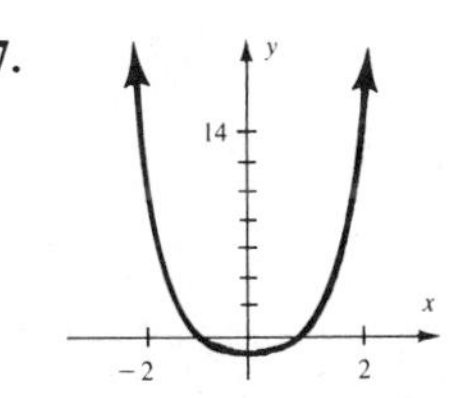

19. 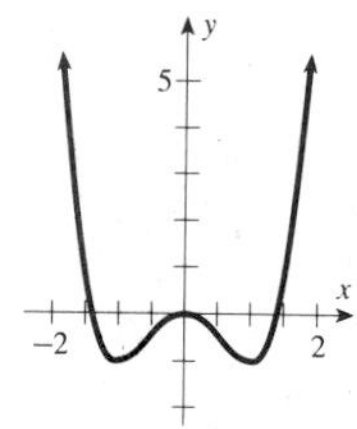

21. 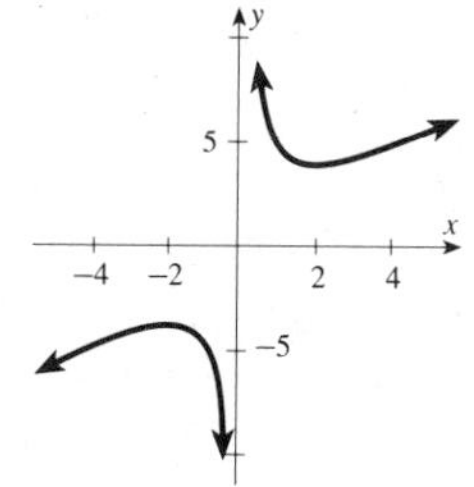

23. 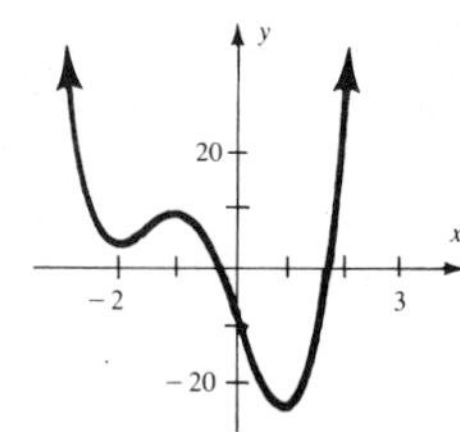

25. 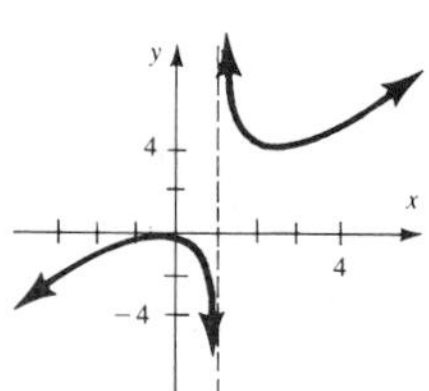

27. 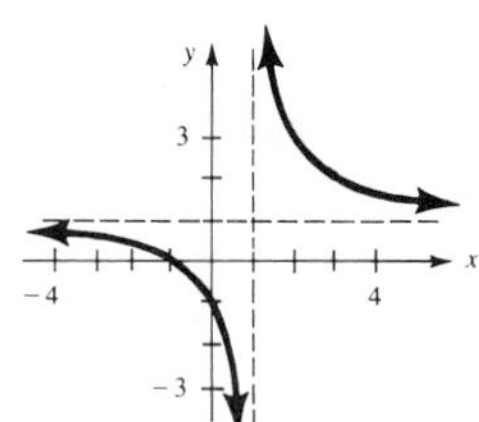

29. 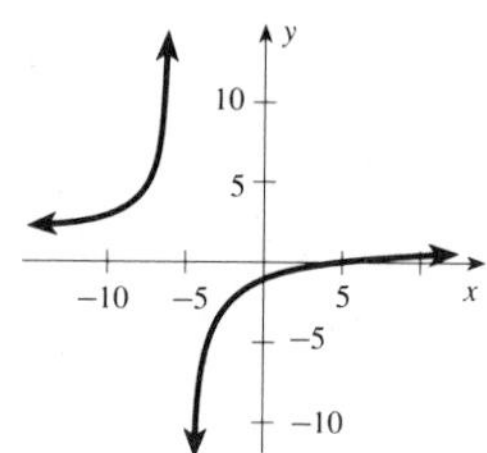

31. 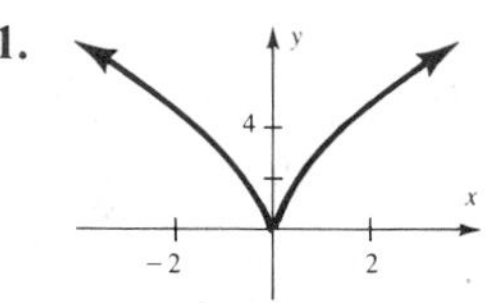

33. 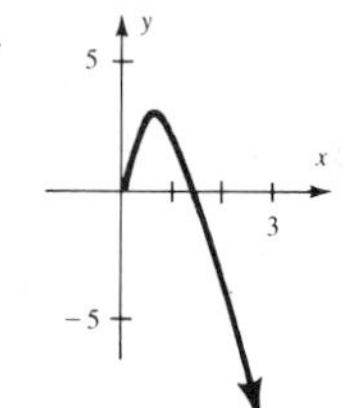

35. 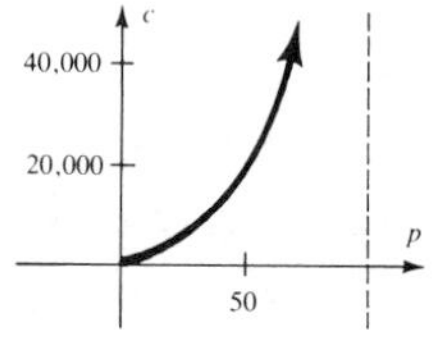

37. 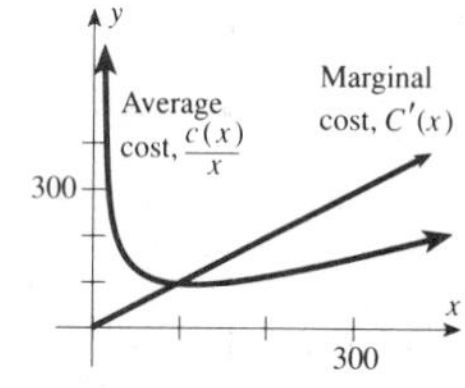

39. 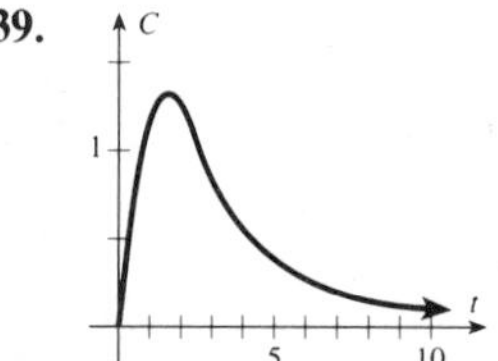

12.4 Absolute Maximum and Minimum *(page 611)*

1. Maximum = 16 at $x = 4$ and minimum = -4 occurs at $x = 2$ and $x = -1$. **3.** Maximum = 966 occurs at $x = -3$; minimum = $-3{,}050$ occurs at $x = -5$. **5.** Maximum = 42 occurs at $x = 0$ and $x = 5$; minimum = 6 occurs at $x = 3$. **7.** Maximum = 34 occurs at $x = 3$; minimum = -200 occurs at $x = 0$. **9.** Maximum = 3 occurs at $x = 8$; minimum = 0 occurs at $x = -1$. **11.** Maximum = 6 occurs at $x = 4$; minimum = 0 occurs at $x = 0$. **13.** Maximum = 72 occurs at $x = 5$; minimum = $-\frac{256}{27}$ occurs at $x = \frac{5}{3}$. **15.** No maximums or minimums **17.** Maximum = 3 occurs at $x = 3$ and $x = -3$; minimum = 0 occurs at $x = 0$. **19.** Maximum = $\frac{1}{8}$ occurs at $x = 3$; minimum = -1 occurs at $x = 0$. **21.** $P(500) = 1{,}500$ **23.** $P(24{,}000) = 57{,}520{,}000$ **25.** $P(60) = 215$ **27.** Absolute maximum at endpoint: $x = 50$; profit is 1,000 at a price of \$100. **29.** The minimum is 27, which occurs $6\frac{2}{3}$ miles from plant P_1. **31.** The box has dimension 1 in. × 3 in. × 8 in. with a volume of 24 in.^{3}. **33. a.** $25 + \dfrac{500}{x} - \dfrac{x}{20}$ **b.** Critical values 0, 100; decreasing on (0, 100) and increasing on (100, 200) **c.** Minimal average cost is \$25 at $x = \$100$ **35.** Maximum profit is \$125 when $x = 15$. The price is \$35. **37.** The maximum profit is \$69 when $x = 13$. The price is \$37. **39.** The minimum cost is \$1,000 when $x = 100$. **41.** The largest area occurs when the side parallel to the river is 500 ft. **43.** Maximum is \$62,500 which occurs when 50 sign up. **45.** Plant 62 vines per acre.

12.5 Review (page 612)

Objective Questions

1.

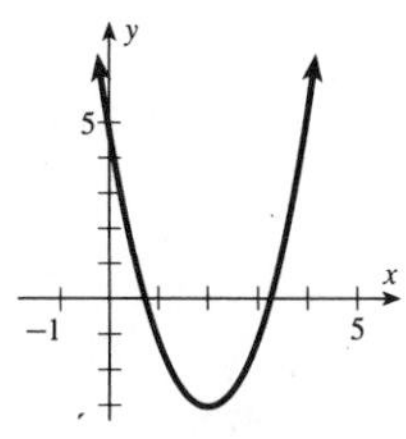

3.

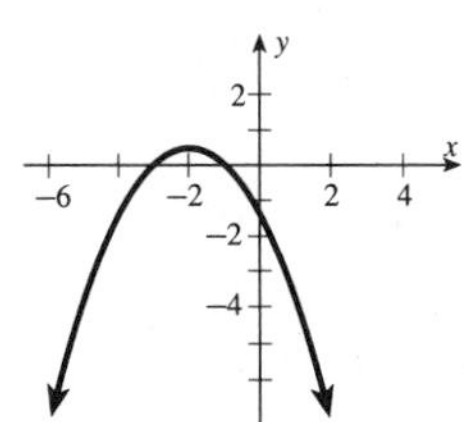

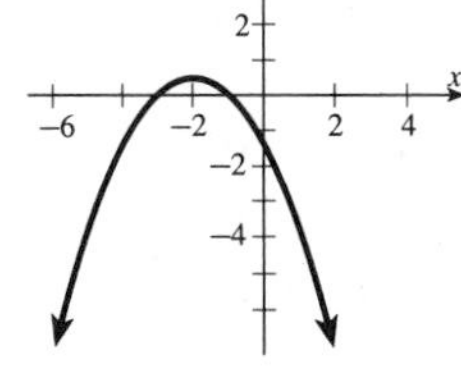

5.

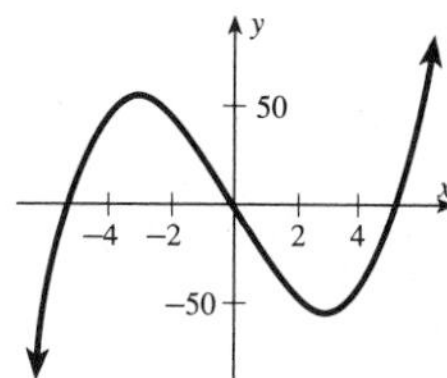

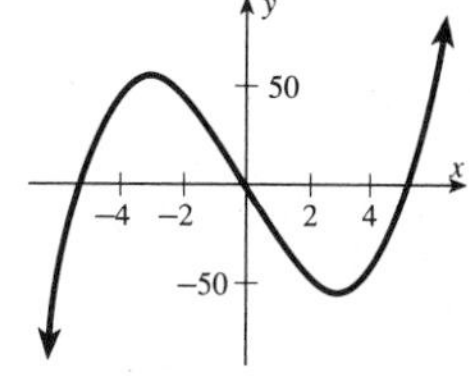

7.

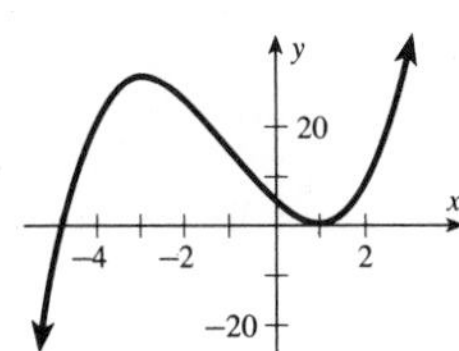

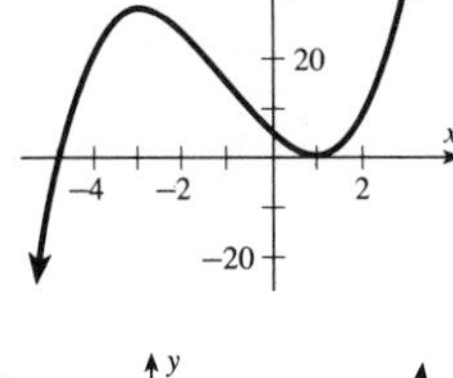

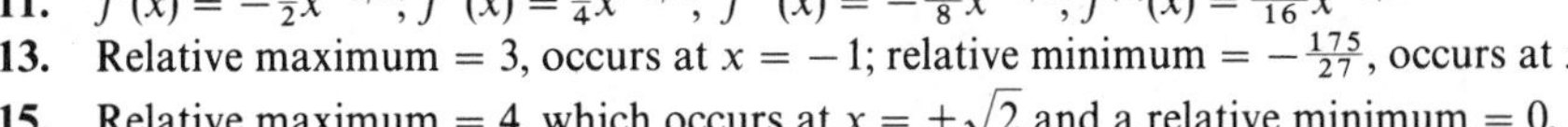

9. $y' = 12x^3 - 3x^2 + 10x$; $y'' = 36x^2 - 6x + 10$; $y''' = 72x - 6$; $y^{(4)} = 72$; $y^{(N)} = 0$ for $N > 4$
11. $f'(x) = -\frac{1}{2}x^{-3/2}$; $f''(x) = \frac{3}{4}x^{-5/2}$; $f'''(x) = -\frac{15}{8}x^{-7/2}$; $f^{(4)}(x) = \frac{105}{16}x^{-9/2}$
13. Relative maximum = 3, occurs at $x = -1$; relative minimum = $-\frac{175}{27}$, occurs at $x = \frac{5}{3}$
15. Relative maximum = 4, which occurs at $x = \pm\sqrt{2}$ and a relative minimum = 0, which occurs at $x = 0$

17.

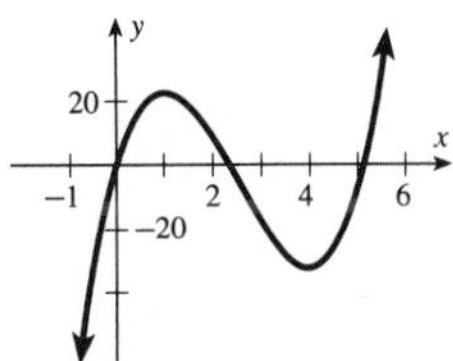

19.

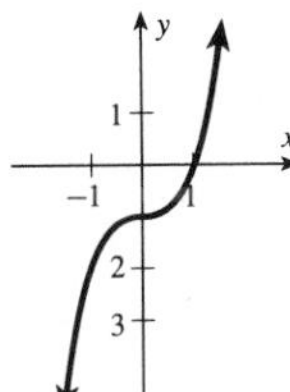

21. The maximum = 158 which occurs at $x = 1$; the minimum = $-1{,}408$ which occurs at $x = 4$. **23.** Absolute minimum -27 at $x = 0$; absolute maximum 2,048 at $x = 5$ **25.** 480 items **27.** Produce 50 items **29.** The marginal revenue is increasing at $x = 100$. **31.** The rent should be $1,500 per month (60 will be rented with 40 vacancies).

Sample Test

1. a. $y' = -9x^2 + 5x^4$; $y'' = -18x + 20x^3$; $y''' = -18 + 60x^2$; $y^{(4)} = 120x$; $y^{(5)} = 120$; $y^{(6)} = y^{(7)} = \cdots = 0$
b. $g' = 15x^4 + 3x^{-2}$; $g'' = 60x^3 - 6x^{-3}$; $g''' = 180x^2 + 18x^{-4}$; $g^{(4)} = 360x - 72x^{-5}$

3.

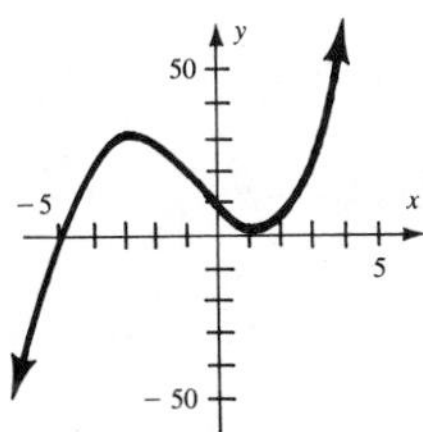

5.

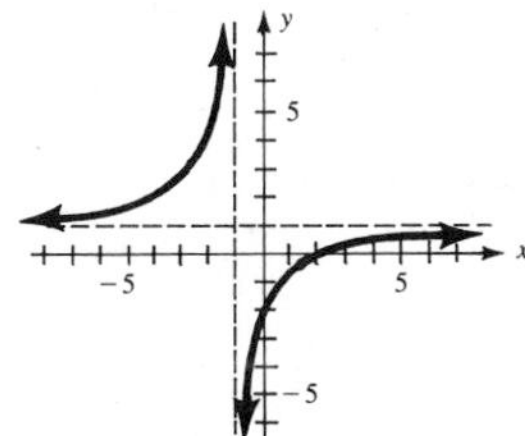

7. There is a relative maximum at $x = -2\sqrt{2}$ and a relative minimum at $x = 2\sqrt{2}$.
9. The marginal revenue is decreasing when $x = 100$.

Cumulative Review Chapters 10–12 (page 615)

1. The limit is 5. **3.** 13 **5.** 15 **7.** Discontinuous at $x = 8$ **9.** At $x = 1$, $\frac{dy}{dx} = 1$ **11.** $y' = -24x - 15x^2$ **13.** $y' = -25(2 - 5x)^4$ **15.** $y' = \frac{-5y}{4x}$ **17.** $16x + y - 12 = 0$ **19.** $y' = 8x^3 - 3x^2 + 6x$; $y'' = 24x^2 - 6x + 6$; $y''' = 48x - 6$; $y^{(4)} = 48$; $y^{(5)} = y^{(6)} = \cdots = 0$ **21.** The maximum is $\frac{1}{4}$, which occurs at $x = \frac{1}{4}$; the minimum is -6, which occurs at $x = 9$. **23.** Marginal revenue is decreasing at $x = 100$. **25.** The frame should be 6 in. by 14 in.

CHAPTER 13

13.1 Exponential Functions (page 624)

1. 5 **3.** Not a real number **5.** -3 **7.** 7 **9.** $\frac{1}{10}$ **11.** $\frac{1}{1,000}$ **13.** 2^{-4} **15.** 3^{-3} **17.** $\frac{5}{6}$
19. 20.085537 **21.** 1.0512711 **23.** 1.0460279 **25.** $2x$ **27.** $1 + x$ **29.** $x + 2x^{1/2}y^{1/2} + y$

31.
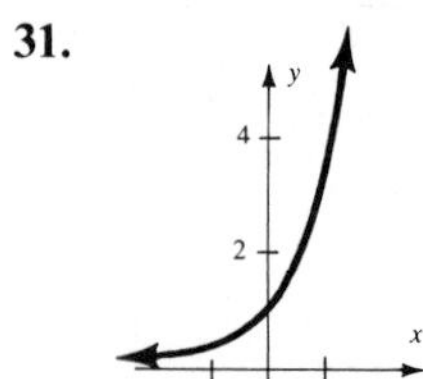

33.
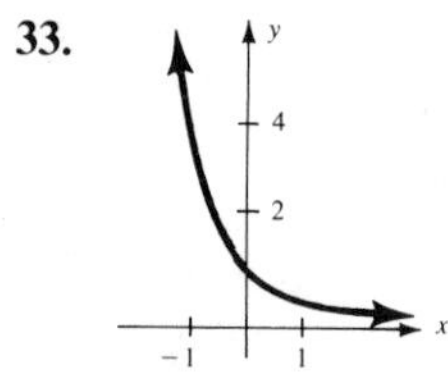

35.
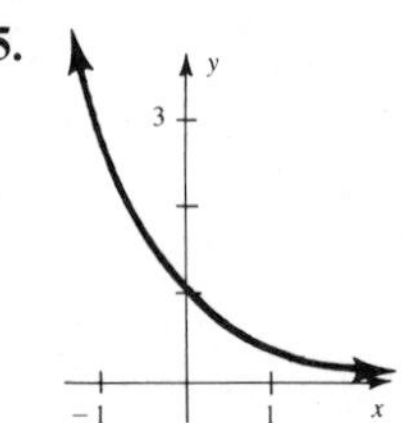

37.
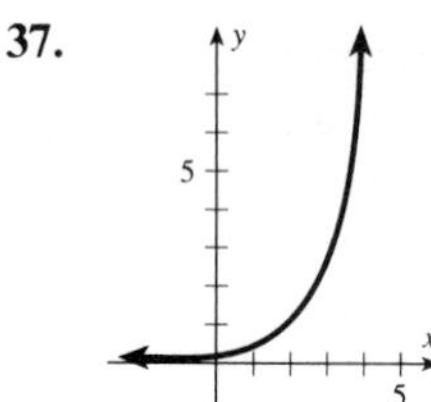

39.
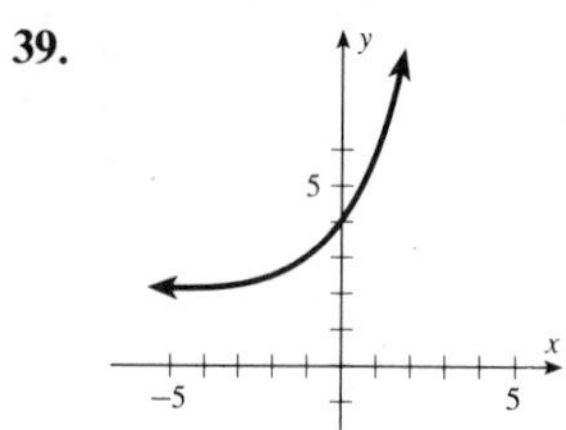

41. \$54,598.15 **43.** \$17,369.57 **45.** \$10,081.44 **47.** \$5,717.37 **49.** \$5,627.05
51. If $p < x < q$ and $0 < b < 1$, then $b^q < b^x < b^p$.

13.2 Logarithmic Functions (page 629)

1. $\log_2 64 = 6$ **3.** $\log_{1/3} 9 = -2$ **5.** $\log_b a = c$ **7.** $4^{1/2} = 2$ **9.** $10^{-2} = .01$ **11.** $e^2 = e^2$ **13.** 2
15. $\frac{1}{2}$ **17.** 3 **19.** 0 **21.** .033423755 **23.** 3.989449818 **25.** $-.493494967$ **27.** .81977983
29. .69314718

31.
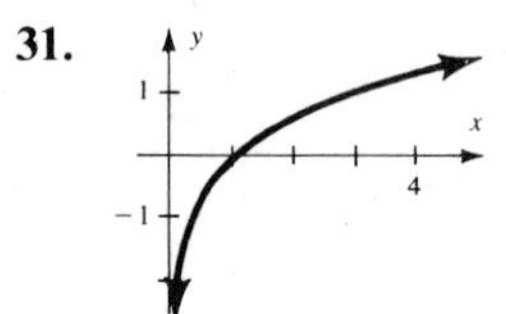

33.
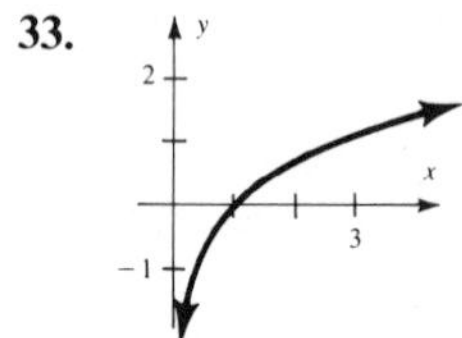

35.
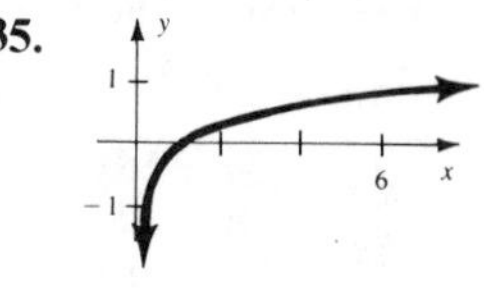

37.
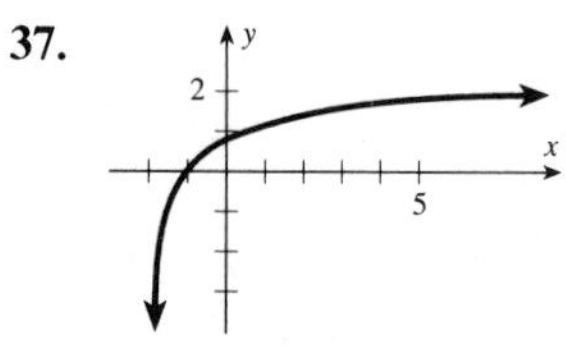

39.
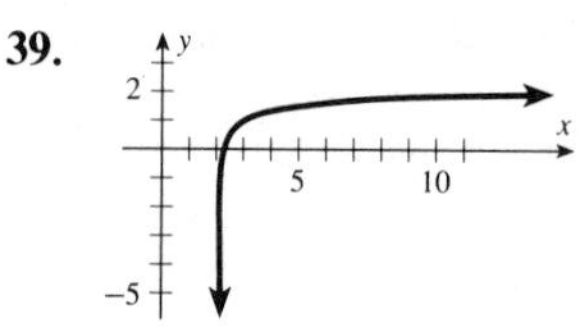

41.
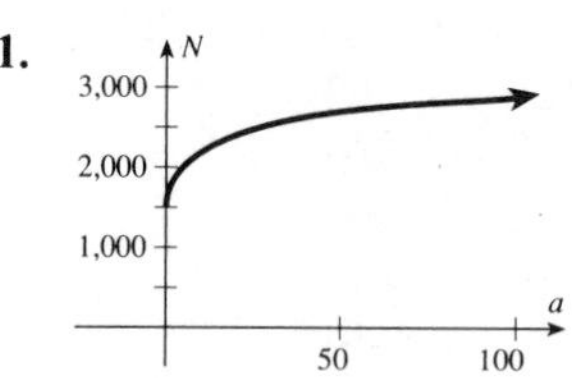

43. a. $t = -62.5 \ln(1 - \frac{30}{80}) \approx 29$; it would take 1 month. **b.** No, since $N = 80$ gives $\ln 0$ which is not defined.

45.
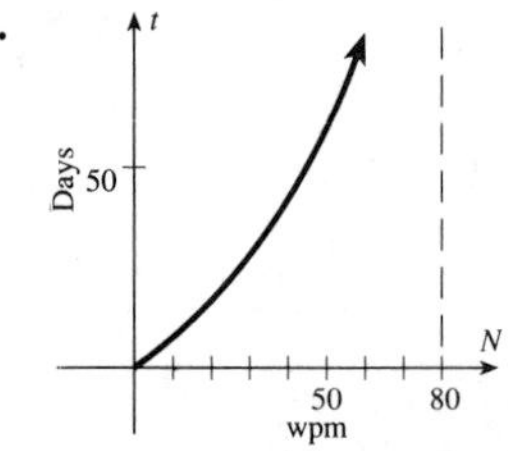

47. a. $M \approx 3.9$ **b.** $M = 8.8$

49. Because of the large numbers, it is difficult to graph (E, M); let $X = \log E$ so that $M = \dfrac{\log 10^x - 11.8}{1.5} = \dfrac{2}{3}X - 7.867$ for $12 \le X \le 17$ as shown at the right

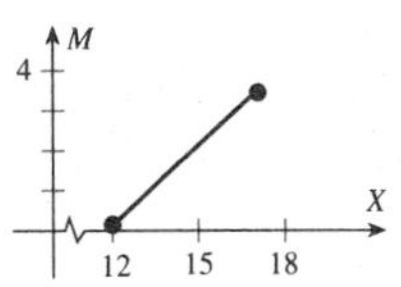

51. Since $b^x = b^x$, then $x = \log_b b^x$ by definition of logarithm.
53. Since $\log_b x = \log_b x$, then $b^{\log_b x} = x$ by definition of logarithm.

13.3 Logarithmic and Exponential Equations (page 636)

1. 2 **3.** -1 **5.** $2\sqrt{7}$ **7.** $e^3 \approx 20.08553692$ **9.** 9.3 **11.** $\pm 2\sqrt{3}$ **13.** 3.5 **15.** No solution
17. 7 **19.** $\frac{2}{3}$ **21.** $\frac{1}{4}$ **23.** 5.6788736 **25.** e^4 **27.** 8.3571428 **29.** 2, 6 **31.** 25 **33.** 3 **35.** 2
37. 1.855903615 **39.** $-.143152$ **41.** $-.2772588$ **43.** $-.34161$ **45.** About 2004
47. To the nearest half-year, it would take 6 years. **49.** 5 years, 283 days **51.** 9.1551 years or 9 years 57 days.
53. a. 50% **b.** About 13 seconds **55.** \$263.34 **57.** $5700 \log_{1/2} P = t$
59. 15.3 years **61.** Answers vary. Let $A = b^x$ and $B = b^y$, which means $\log_b A = x$ and $\log_b B = y$.
$\log_b \dfrac{A}{B} = \log_b \dfrac{b^x}{b^y} = \log_b b^{x-y} = x - y = \log_b A - \log_b B$

13.4 Derivatives of Logarithmic and Exponential Functions (page 644)

1. $2e^{2x}$ **3.** $-e^{-x}$ **5.** $xe^x + e^x$ **7.** $2xe^{5x^2}[10x^3 + 3x + 5]$ **9.** $-2e^{5x}(6x + 15x^2 + 25)$ **11.** $\dfrac{1}{x-5}$

13. $4(2x^3 + e^{5x})^3(6x^2 + 5e^{5x})$ **15.** $\dfrac{4}{x}$ **17.** $\dfrac{1}{2x}$ **19.** $\dfrac{3 - 3x\ln x}{xe^x}$ **21.** $\dfrac{2}{x(x+2)}$ **23.** $e^{t^2-t}(1 + 2te^t - 2t)$

25. $2(e^{2x} - e^{-2x})$ **27.** $\dfrac{5x + 2 - 5x\ln|x|}{x(5x+2)^2}$ **29.** $\dfrac{e^x(x\ln|3x| - 1)}{x\ln^2|3x|}$ **31.** $\dfrac{-3{,}600e^{.3x}}{(1 + 6e^{.3x})}$ **33.** $\dfrac{100(e^{.2x} + 40 + 8x\ln|x|)}{xe^{.2x}(1 + 40e^{-.2x})^2}$

35.

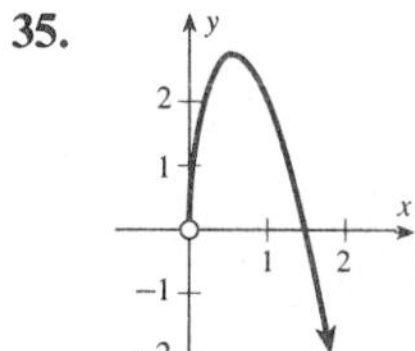

37.

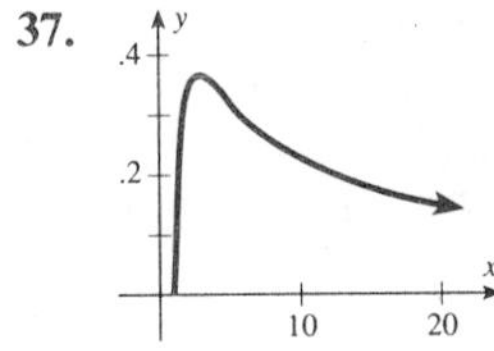

39.

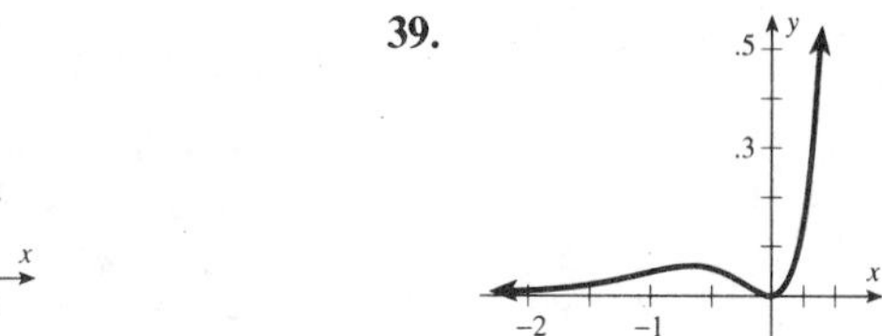

41. $R'(x) = 500\left[\dfrac{x - (x+10)\ln(x+10)}{x^2(x+10)}\right]$ **43. a.** $N(0) = 1$ **b.** .09998 (people per day) **45.** After 1 year: $.1353P$;
after 5 years: $.2187P$ **47.** $A'(t) = -.15e^{-.03t}$ **49.** $R'(x) = -50xe^{-.1x} + 500e^{-.1x} = 50e^{-.1x}(10 - x)$
51. Answers vary. **53.** Answers vary.

13.5 Review (page 645)

Objective Questions

1. 25 **3.** $\sqrt{3}$ **5.**

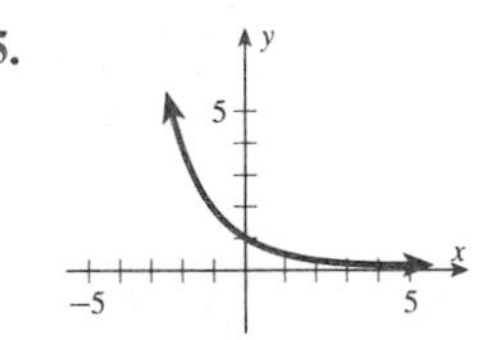

7.

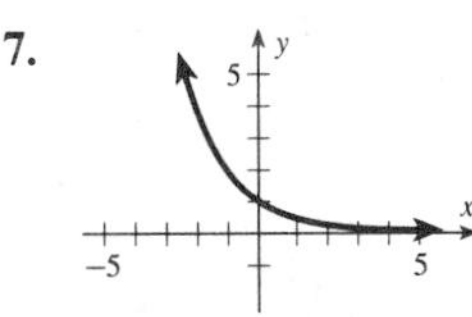

9. $e \approx 2.71828$ **11.** 2.8576511 **13.** $\log\sqrt{10} = .5$ **15.** $\log_9 729 = 3$ **17.** $10^0 = 1$ **19.** $2^6 = 64$

21. 1.0986123 **23.** -2.6777807 **25.** **27.**

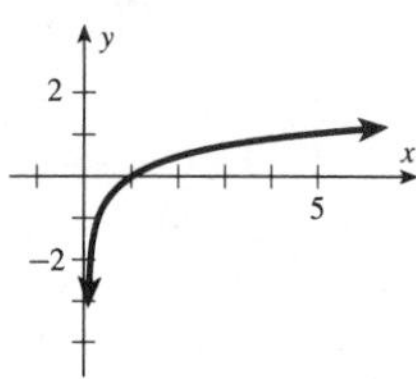

29. $x = 2$ **31.** $x = 3^5$ **33.** $x \approx .09691$ **35.** $x \approx .43065$ **37.** $\dfrac{dy}{dx} = -.49xe^{-.05x^2}$ **39.** $y' = x^2e^{x^4}(4x^4 + 3)$

41. $y' = \dfrac{2}{x}$ **43.** $\dfrac{2x + 6 - x\ln x^2}{x(x + 3)^2}$ **45.** \$102,637.89 **47.** $t = 5{,}700\log_{1/2}\dfrac{A}{10}$ **49.** $P'(1) \approx -2.5023$ psi per mile

51. Equilibrium point at 5,756 items

Sample Test

1. $x = 2$ **3.** $x = 8$ **5.** $x = 3.39794$ **7.** $x \approx -.348707$ **9.** $y' = -5.5xe^{-.5x^2}$ **11.** $y' = \dfrac{5}{x}$

13. $y' = \dfrac{8 - 5x^3}{x(4 - x^3)}$ **15.** $y' = \dfrac{8 - 2x + x\ln x^2}{x(4 - x)^2}$ **17.** $t = \dfrac{1}{t}\ln\dfrac{A}{P}$ **19.** Equilibrium when $x = 7{,}675$ units

CHAPTER 14

14.1 The Antiderivative (page 655)

1. $\frac{x^8}{8} + C$ **3.** $x^4 + C$ **5.** $3x + C$ **7.** $\dfrac{5x^2}{2} + 7x + C$ **9.** $x + C$ **11.** $6x^3 - 3x^2 + 5x + C$ **13.** $\frac{1}{9}x + C$

15. $-5x^{-1} + C$ **17.** $x - e^x + C$ **19.** $\frac{3}{4}x^{4/3} + \sqrt{2}x + C$ **21.** $2x^{3/2} - 2x^{1/2} + C$ **23.** $\frac{3}{5}x^{5/3} + C$

25. $\dfrac{x^2}{2} + x + \ln|x| + C$ **27.** $\frac{2}{9}y^{9/2} + C$ **29.** $z + \dfrac{2z^3}{3} + \dfrac{z^5}{5} + C$ **31.** $u^3 - \ln|u| + e^u + C$

33. $F(x) = x^3 + 2x^2 + x + 10$ **35.** $F(x) = 5x + \ln|x| + 4{,}995$ **37.** $C(x) = 2x^2 - 8x + 5{,}000$

39. $C(x) = .003x^3 + 18{,}500$ **41.** Since $P(t) = 450t + 400t^{3/2} + 420{,}000$, in 5 years the projected population is 426,722.

43. $P(x) = 100x^2 - 10{,}000x - 50{,}000$ **45.** $N(t) = 120t^2 - t^3 - 69$, and on the tenth day $N(10) = 10{,}931$ (cases).

47. $v(t) = -32t - 72$ **49.** $f(x) = x^3 + 5x - 4$

14.2 Integration by Substitution (page 661)

1. $\dfrac{(5x + 3)^4}{20} + C$ **3.** $\dfrac{5}{3(5 - x)^3} + C$ **5.** $\dfrac{2(3x + 5)^{3/2}}{9} + C$ **7.** $\dfrac{(3x^2 + 1)^2}{2} + C$ **9.** $2\sqrt{x^2 + 5x} + C$

11. $\frac{1}{5}\ln|6 + 5x| + C$ **13.** $-\frac{1}{6}\ln|1 - 3x^2| + C$ **15.** $\frac{1}{5}e^{5x} + C$ **17.** $\frac{5}{12}e^{4x^3} + C$ **19.** $-\dfrac{1}{4(4x^2 - 4x)} + C$

21. $\ln(x^4 - 2x^2 + 3) + C$ **23.** $\dfrac{\ln^2|x|}{2} + C$ **25.** $\dfrac{x^2}{2} + C$ **27.** $\frac{3}{8}\ln(4x^2 + 1) + \frac{1}{3}e^{x^3} + C$ **29.** $\frac{3}{10}(x^2 + 1)^{5/3} + C$

31. $\frac{2}{3}(x + 1)^{3/2} - 2\sqrt{x + 1} + C$ **33.** **a.** $\dfrac{x^3}{3} + x + C$ **b.** $\dfrac{x^4}{4} + \dfrac{x^2}{2} + C$ **c.** $\dfrac{x^5}{5} + \dfrac{x^3}{3} + C$ **d.** $\dfrac{x^6}{6} + \dfrac{x^4}{4} + C$

35. **a.** $\dfrac{x^7}{7} + \dfrac{3x^5}{5} + x^3 + x + C$ **b.** $\frac{1}{8}(x^2 + 1)^4 + C$ **c.** $\dfrac{x^9}{9} + \dfrac{3x^7}{7} + \dfrac{3x^5}{5} + \dfrac{x^3}{3} + C$ **d.** $\dfrac{(x^2 + 1)^5}{10} - \dfrac{(x^2 + 1)^4}{8} + C$

37. **a.** $\dfrac{x^4}{4} + x + C$ **b.** $\dfrac{x^5}{5} + \dfrac{x^2}{2} + C$ **c.** $\dfrac{x^6}{6} + \dfrac{x^3}{3} + C$ **d.** $\dfrac{x^7}{7} + \dfrac{x^4}{4} + C$ **39.** **a.** $\dfrac{x^{10}}{10} + \dfrac{3x^7}{7} + \dfrac{3x^4}{4} + x + C$

b. $\dfrac{x^{11}}{11} + \dfrac{3x^8}{8} + \dfrac{3x^5}{5} + \dfrac{x^2}{2} + C$ **c.** $\dfrac{(x^3 + 1)^4}{12} + C$ **d.** $\dfrac{x^{13}}{13} + \dfrac{3x^{10}}{10} + \dfrac{3x^7}{7} + \dfrac{x^4}{4} + C$

41. $P(t) = -\frac{20}{3}(5 - t^2)^3 + \frac{2{,}530}{3} = \frac{20}{3}(t^2 - 5)^3 + \dfrac{2{,}530}{3}$ **43.** \$2,627.77 **45.** \$19,511.88 **47.** 1,120,026 people

49. 714 million barrels

14.3 The Definite Integral (page 668)

1. $\frac{19}{3}$ **3.** $\frac{14}{3}$ **5.** $9-\sqrt{3}$ **7.** $\frac{1}{2}(e^4-1)$ **9.** $\ln 10$ **11.** $\frac{35}{5}$ **13.** $\ln 6$ **15.** 12 **17.** 0 **19.** $\frac{1}{3}$
21. $\frac{17}{4}-\ln 4$ **23.** 0 **25.** 0 **27.** $\frac{5}{33}$ **29.** $\frac{3}{4}$ **31. a.** 14 **b.** 5 **c.** 43 **33.** 6
35. The total consumption is 730 billion barrels. **37.** The total consumption is 615 billion barrels.
39. About 14 years from 1985 to 1999. **41.** The pollution level is 28,000 ppm. **43.** The pollution level is 3,500 ppm.
45. $1,101,273.29 **47.** $\frac{1}{2}$ **49.** $C(t)=\frac{t^2}{40}+25{,}000$ and $R(t)=-5e^{-t^2}+5$

14.4 Area Between Curves (page 678)

1. Since $f(x)>g(x)$ on $[a,b]$, the area $=\int_a^b [f(x)-g(x)]\,dx$ **3.** Since $g(x)>f(x)$ on $[a,b]$, the area $=\int_a^b [g(x)-f(x)]\,dx$
5. Since $f(x)>g(x)$ on $[a,b]$, the area $=\int_a^b [f(x)-g(x)]\,dx$ **7.** 248 **9.** 21 **11.** 12 **13.** e^2-e^{-1} **15.** 36
17. $\frac{86}{3}$ **19.** $\frac{3}{2}$ **21.** $\frac{14}{3}$ **23.** $\frac{56}{3}$ **25.** $\frac{3}{10}$ **27.** $\frac{373}{6}$ **29.** 4.6051702 **31.** $\frac{1}{6}$ **33.** 243
35. 1.4459739 **37.** 4,192 tons **39.** About 184,000 fish **41.** 562.5; this area represents the total demand of 562.5 during the first 25 weeks. **43.** 22.528776; the area represents the accumulated profit.

14.5 The Fundamental Theorem of Calculus (page 691)

1. a. $\frac{28}{3}$ **b.** 10 **3. a.** 63 **b.** $\frac{3{,}006}{64}\approx 46.97$ **5. a.** $\frac{16}{3}$ **b.** $\frac{25}{4}$ **7.** .12 **9.** 3.43 **11.** .1080 **13.** 28
15. 5,700 **17.** $\frac{35}{3}$ **19.** 280 **21.** $\frac{45}{28}$ **23.** $-\frac{1}{5}(e^{-5}-1)\approx .199$ **25.** $x+5$ **27.** $(x+5)^2$ **29.** $x^{1/2}$
31. x^3-3x^2+5x-7 **33.** $\sqrt[3]{x}$ **35.** $x^2(x+1)^3$ **37.** 3.9 million **39.** About 68 million tons
41. 500 rabbits **43.** The average temperature is about 63°F. **45.** 600 ft^2

14.6 Numerical Integration (page 700)

1. $\Delta x=5;\ A_1=10$ **3.** $\Delta x=\frac{5}{3};\ A_3=11.81415$ **5.** $\Delta x=5;\ T_1=12.5$ **7.** $\Delta x=\frac{5}{3};\ T_3\approx 12.64748$
9. $\Delta x=\frac{5}{3};\ P_2\approx 12.66503$ **11.** $\Delta x=\frac{5}{6};\ P_6\approx 12.66664$ **13.** $\Delta x=2;\ A_1=.16$ **15.** $\Delta x=\frac{2}{3};\ A_3\approx .13546$
17. $\Delta x=2;\ T_1\approx .12938$ **19.** $\Delta x=\frac{2}{3};\ T_3\approx .12526$ **21.** $\Delta x=1;\ P_2\approx .12476$ **23.** $\Delta x=\frac{1}{3};\ P_6\approx .12473$
25. $\Delta x=1;\ A_4\approx .14$ **27.** $\Delta x=\frac{1}{2};\ A_4\approx 8.36$ **29.** $\Delta x=1;\ T_4\approx .11$ **31.** $\Delta x=\frac{1}{2};\ T_4\approx 14.86$
33. $\Delta x=1;\ P_4\approx .10$ **35.** $\Delta x=\frac{1}{2};\ P_4\approx 13.81$ **37.** Area ≈ 890 square feet. About 100 yd^3 of fill will be needed.

14.7 Review (page 701)

Objective Questions

1. $\frac{x^7}{7}+C$ **3.** e^u+C **5.** $2x^3-x^2-5x+C$ **7.** $\frac{1}{3}\ln|3x+1|+C$ **9.** $F(x)=e^x+9x+9$
11. $F(x)=\ln|x|$ **13.** $C(x)=3x^3+x+25{,}000$ **15.** $C(x)=10x-2\sqrt{x}+1{,}500$ **17.** $-\frac{e^{-3x}}{3}-\frac{x}{e}+C$
19. $\frac{-2}{\sqrt{x^2-5x}}+C$ **21.** $\frac{5}{2}$ **23.** 15,657 **25.** $\frac{74}{15}$ **27.** $\frac{20}{81}$ **29.** $\int_2^9 [g(x)-f(x)]\,dx$
31. $\int_{-20}^{9} [g(x)-f(x)]\,dx+\int_{-9}^{0} [f(x)-g(x)]\,dx$ **33.** 15 **35.** $\frac{2{,}197}{384}\approx 5.72$ **37.** 14.56
39. $\Delta x=5.25;\ A\approx 24.33$ **41.** 658 **43.** About $2\frac{3}{4}$ million cars were sold in the last quarter. **45.** 12
47. $\frac{1}{4}(e^9-e)\approx 2{,}025.091411$ **49.** 1.2 **51.** $T\approx 1.10$ **53.** n must be *even* for Simpson's rule. **55.** $P=3.08$
57. $P(x)=125x^2-5{,}000x-1{,}500$ **59.** The concentration of waste in the tank is 5.8%.

Sample Test

1. $\frac{x^6}{6}+C$ **3.** e^u+C **5.** $6x^3-3x^2-3x+C$ **7.** $\frac{1}{3}e^u+\frac{1}{e}x+C=\frac{1}{3}e^{3x+1}+\frac{1}{e}x+C$ **9.** $\frac{4}{3}$
11. 15,657.1875 **13.** $\frac{74}{9}$ **15.** $\frac{1}{2}$ **17.** It will be economically feasible. **19.** $t=14.4$ years

CHAPTER 15

15.1 Business Models Using Integration (page 716)

1. $512\frac{2}{3}$ **3.** −.197 **5.** 13 **7.** $2,500 **9.** $1,229.56 **11.** $663.60 **13.** $61,607.44 **15.** $2,563.55
17. $38,913.42 **19.** 116 **21.** 396 **23.** $4,028.35 **25.** 6.32 **27.** $1,507.14; additional monthly cost $31.40

29. $\frac{4}{3}$ **31.** \$126,241.19 **33.** It will take $1\frac{1}{4}$ years. **35.** The producers' surplus is \$3,495.33.
37. The producers' surplus is \$2. **39.** \$11,400 **41.** 1.31 years **43.** \$7,453,560

15.2 Differential Equations (page 725)

1. $y = \frac{x^3}{3} + C$ **3.** $y = 4x^2 - 10x + C$ **5.** $y = x^4 - x^3 - 5x + C$ **7.** $y = \frac{1}{5}e^x + C$ **9.** $y = \frac{1}{90}(5x + 1)^{3/2} + C$
11. $y^2 = x^2 + C$ **13.** $y^3 = \frac{3x^4}{4} - 9x + C$ **15.** $y = Me^{x^2}$ **17.** $50 \ln|P| = t + C$ or $P = Me^{.02t}$
19. $1{,}000 \ln|N| = t + C$ or $N = Me^{.001t}$ **21.** $y^3 = x^3 + 125$ **23.** $P = e^{.02t+3}$ or $\ln P = 3 + .02t$
25. $\ln|y| = 2\sqrt{x} - 2$ **27.** $\ln|y| = \frac{1}{2}\ln(1 + x^2) + \frac{1}{2}\ln 2$ or $y = \sqrt{2x^2 + 2}$ **29.** 4,725 **31.** 44.2 hours
33. $k = .0069314$ **35.** The skull is about 10,500 years old. **37.** $Q(t) = -.025t^2 + 5$ **39.** $P(t) = L(1 - e^{-.075t})$

15.3 Integration by Parts (page 731)

1. $4xe^{3x} - \frac{4}{3}e^{3x} + C$ **3.** $1 - 2e^{-1}$ **5.** $-\frac{2}{3}x(1 - x)^{3/2} + \frac{2}{3}[-\frac{2}{5}(1 - x)^{5/2}] + C$ **7.** $\frac{x(x + 2)^4}{4} - \frac{(x + 2)^5}{20} + C$

9. $(x + 1)\ln(x + 1) - \int dx = (x + 1)\ln(x + 1) - x + C$ **11.** $-2x(1 - x)^{1/2} - \frac{4}{3}(1 - x)^{3/2} + C$

13. $3x^2e^{2x} - 3xe^{2x} + \frac{3}{2}e^{2x} + C$ **15.** $-\frac{x^2(1 - 2x)^{3/2}}{3} - \frac{2x(1 - 2x)^{5/2}}{15} - \frac{2(1 - 2x)^{7/2}}{105} + C$

17. $-\frac{3}{8}(1 - x^2)^{4/3} + \frac{3}{14}(1 - x^2)^{7/3} + C$ or $-\frac{3x^2(1 - x^2)^{4/3}}{8} - \frac{9(1 - x^2)^{7/3}}{56} + C$ **19.** $\frac{x^3 \ln x}{3} - \frac{x^3}{9} + C$
21. $\frac{3}{2}x^2e^{4x} - \frac{3}{4}xe^{4x} + \frac{3}{16}e^{4x} + C$ **23.** $-(1 - x^2)^{1/2} + \frac{1}{3}(1 - x^2)^{3/2} + C$ **25.** $\frac{1}{2}x^2e^{x^2} - \frac{1}{2}e^{x^2} + C$
27. $2\ln 6 - \ln 3 - 1 = \ln 12 - 1$ **29.** $5 - 4e$ **31.** $\ln 2$ **33.** $e - 2$ **35.** Answers vary.
37. **a.** $-\frac{1}{2}\ln|x^2 + 1| + \frac{1}{2}x^2 + C$ **b.** $\frac{x^2}{2} - \frac{1}{2}\ln|x^2 + 1| + C$ **39.** **a.** $N(t) = 401 - 20e^{-.05t}(t + 20)$ **b.** 20.5 million/liter
41. \$32,559.20

15.4 Using Tables of Integrals (page 736)

1. Using Formula 9, $\int (1 + bx)^{-1}\,dx = \frac{1}{b}\ln|1 + bx| + C$ **3.** Using Formula 23, $\int \frac{x\,dx}{\sqrt{x^2 + a^2}} = \sqrt{x^2 + a^2} + C$

5. Using Formula 26, $\int \frac{dx}{x^2\sqrt{x^2 - a^2}} = \frac{\sqrt{x^2 - a^2}}{a^2x} + C$ **7.** Using Formula 30, $\int x \ln x\,dx = \frac{x^2}{2}\ln x - \frac{x^2}{4} + C$

9. Using Formula 32 ($m = -1$), $\int x^{-1}\ln x\,dx = \frac{1}{2}\ln^2 x + C$ **11.** Using Formula 35, $\int xe^{ax}\,dx = \frac{e^{ax}}{a^2}(ax - 1) + C$

13. Using Formula 29 ($a = 1$), $\int \frac{x^2}{\sqrt{x^2 + 1}}\,dx = \frac{x}{2}\sqrt{x^2 + 1} - \frac{1}{2}\ln|x + \sqrt{x^2 + 1}| + C$ **15.** Using Formula 26 ($a = 4$), $\int \frac{dx}{x^2\sqrt{x^2 + 16}} = -\frac{\sqrt{x^2 + 16}}{16x} + C$ **17.** Let $u = 4x^2 + 1$, then $du = 8x\,dx$: $\int \frac{x\,dx}{\sqrt{4x^2 + 1}} = \frac{1}{4}\sqrt{4x^2 + 1} + C$

19. Using Formula 20 $\left(a = \frac{1}{3}\right)$, $\int \frac{dx}{x\sqrt{1 - 9x^2}} = -\ln\left|\frac{1 + \sqrt{1 - 9x^2}}{3x}\right| + C$ **21.** Using Formula 23, $\int \frac{x\,dx}{\sqrt{x^2 + 4}} = \sqrt{x^2 + 4} + C$ **23.** Using Formula 9, with $a = 1$, $b = 1$, and $n = 3$: $\int (1 + x)^3\,dx = \frac{(1 + x)^4}{4} + C$

25. Using Formula 10 with $a = 1$, $b = 1$, and $n = 3$: $\int x(1 + x)^3\,dx = \frac{1}{5}(1 + x)^5 - \frac{1}{4}(1 + x)^4 + C$ **27.** Using Formula 13 with $m = 1$, $a = 1$, $b = 1$: $\int x\sqrt{1 + x}\,dx = \frac{-2(2 - 3x)\sqrt{(1 + x)^3}}{15} + C$ **29.** Using Formula 35 with $a = 4$: $\int xe^{4x}\,dx = \frac{e^{4x}}{16}(4x - 1) + C$ **31.** Let $u = 2x$, then $x = \frac{u}{2}$ and $du = 2\,dx$. Using Formula 30, $\int x \ln 2x\,dx = \frac{x^2}{2}\ln 2x - \frac{x^2}{4} + C$

33. Let $u = 5x$, then $du = 5\,dx$. Then use Formula 31: $\int x^2 \ln 5x\,dx = \frac{x^3}{3}\ln 5x - \frac{x^3}{9} + C$ **35.** Using Formula 39 with

$a = 3, b = 5$, and $m = 1$: $\displaystyle\int \frac{dx}{3 + 5e^x} = \frac{x}{3} - \tfrac{1}{3}\ln|3 + 5e^x| + C$ **37.** Using Formula 39 with $a = 1, b = 1$, and $m = 2$: $\displaystyle\int \frac{dx}{1 + e^{2x}} = x - \tfrac{1}{2}\ln|1 + e^{2x}| + C$ **39.** Let $u = 1 + x$, then $du = dx$. Using Formula 2: $\displaystyle\int \frac{dx}{\sqrt{1 + x}} = 2(1 + x)^{1/2} + C$

41. Using Formula 11 with $a = 1, b = 1$, and $n = 3$: $\displaystyle\int x^2(1 + x)^3\,dx = \frac{(1 + x)^6}{6} - \frac{2(1 + x)^4}{5} + \frac{(1 + x)^4}{4} + C$

43. Using Formula 11 with $a = 1, b = -1$, and $n = 3$: $\displaystyle\int 5x^2(1 + x)^3\,dx = \frac{-5(1 + x)^6}{6} + 2(1 - x)^5 - \frac{5(1 + x)^4}{4} + C$

45. Using Formula 13 with $m = 1, a = 1$, and $b = 2$: $\displaystyle\int 2x\sqrt{1 - 2x}\,dx = \frac{-2(1 + 3x)\sqrt{(1 - 2x)^3}}{15} + C$ **47.** Using Formula 11 with $a = 2, b = 3$, and $n = 3$: $\displaystyle\int x^2(2 + 3x)^3\,dx = \frac{1}{27}\left[\frac{(2 + 3x)^6}{6} - \frac{4(2 + 3x)^5}{5} + (2 + 3x)^4\right] + C$ **49.** Let $U = 4x^3 + 1$, then $du = 12x^2\,dx$. Using Formula 2, $\displaystyle\int x^2\sqrt{4x^3 + 1}\,dx = \tfrac{1}{18}(4x^3 + 1)^{3/2} + C$ **51.** Using Formula 36 with $m = 2$ and $a = 3$: $\displaystyle\int x^2e^{3x}\,dx = \frac{x^2e^{3x}}{3} - \frac{2}{3}\int xe^{3x}\,dx$. This last integral can be done using Formula 35 with $a = 3$: $\displaystyle\frac{e^{3x}}{27}(9x^2 - 6x + 2) + C$

53. Using Formula 14 (twice): $\displaystyle\frac{-\sqrt{(2 + 9x)^3}}{2x} + \frac{9}{4}\left[2\sqrt{2 + 9x} + \sqrt{2}\ln\left|\frac{\sqrt{2 + 9x} - \sqrt{2}}{\sqrt{2 + 9x} + \sqrt{2}}\right|\right] + C$ **55.** Let $u = 3x$, then $du = 3\,dx$: $\displaystyle\int \frac{\sqrt{2 + 9x^2}}{x}\,dx = \int \frac{\sqrt{2 + u^2}}{u}\,du$. Using Formula 21 with $a = \sqrt{2}$: $\displaystyle\sqrt{9x^2 + 2} - \sqrt{2}\ln\left|\frac{\sqrt{2} + \sqrt{9x^2 + 2}}{3x}\right| + C$

57. Let $u = 3x$, then $du = 3\,dx$: $\displaystyle\int \frac{1}{\sqrt{2 + 9x^2}}\,dx = \int \frac{1}{\sqrt{2 + u^2}}\,\frac{du}{3}$. Using Formula 16 with $a = \sqrt{2}$: $\tfrac{1}{3}\ln|3x + \sqrt{9x^2 + 2}| + C$

59. Let $u = 9 - 16x^2$, then $du = -32x\,dx$: $\displaystyle\int 5x\sqrt{9 - 16x^2}\,dx = -\tfrac{5}{48}(9 - 16x^2)^{3/2} + C$ **61.** Using Formula 28 with $a = 1$: $\displaystyle\int \frac{\sqrt{x^2 - 1}}{x^2}\,dx = -\frac{\sqrt{x^2 - 1}}{x} + \ln|x + \sqrt{x^2 - 1}| + C$ **63.** Using Formula 11 with $n = \tfrac{1}{2}$, $a = 3$, and $b = 10$: $\displaystyle\int x^2(3 + 10x)^{1/2}\,dx = \frac{(3 + 10x)^{7/2}}{3{,}500} - \frac{3(3 + 10x)^{5/2}}{1{,}250} + \frac{3(3 + 10x)^{3/2}}{500} + C$. If instead of Formula 11 you use Formula 13, you will obtain the following (equivalent) form: $\displaystyle\frac{6 - 30x + 125x^2}{4{,}375}(3 + 10x)^{3/2} + C$ **65.** The total amount will be $\displaystyle\int_1^{12} \frac{5{,}000}{t\sqrt{100 + t^2}}\,dt$. Using Formula 19 with $a = 10$: $\displaystyle\int_1^{12} \frac{5{,}000}{t\sqrt{100 + t^2}}\,dt = -500\left[\left|\frac{10 + \sqrt{244}}{12}\right| - \ln\left|\frac{10 + \sqrt{101}}{1}\right|\right] \approx 1{,}120$ tons.

67. The accumulated sales are $\displaystyle\int_{12}^{24} te^{.1t}\,dt$. Using Formula 35 (or integration by parts) with $a = .1$: $\displaystyle\frac{e^{2.4}}{.01}(2.4 - 1) - \frac{e^{1.2}}{.01}(1.2 - 1) \approx 1{,}476.842$ thousands or about \$1,480,000.

15.5 Improper Integrals (*page 739*)

1. 1 **3.** 2 **5.** $\tfrac{1}{2}$ **7.** Diverges **9.** 10 **11.** Diverges **13.** Diverges **15.** 0 **17.** 1 **19.** $2e^{-1}$ **21.** $\dfrac{\sqrt{5} - 1}{4}$ **23.** \$336,666.67 **25.** Answers vary. **27.** 4,000 millirems **29.** $\tfrac{1}{4}$ **31.** \$625,000
33. 4,000 (in thousands of barrels) or 4 million barrels

15.6 Probability Density Functions (*page 752*)

1. Let X = number of people who responded "yes," which is a discrete random variable. The sample space is $\{0, 1, 2, \ldots, 10\}$.
3. Let X = number of words with an error, which is a discrete random variable. The sample space is $\{0, 1, 2, \ldots, 499, 500\}$.

5. Let x = number of boys (or the number of girls), which is a discrete random variable. The sample space is $\{bbb, bbg, bgb, bgg, gbb, gbg, ggb, ggg\}$.

7. Let X = height in inches.

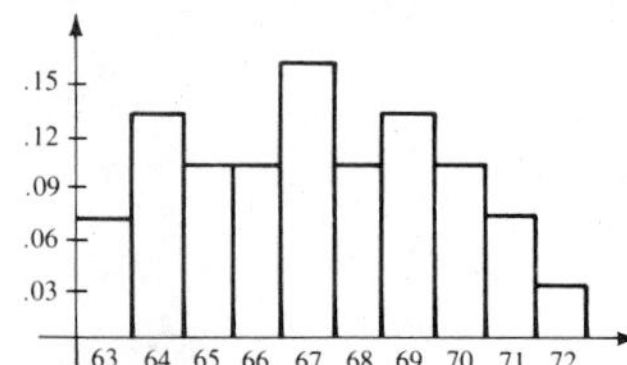

9. Let X = time (in minutes) spent on each transaction.

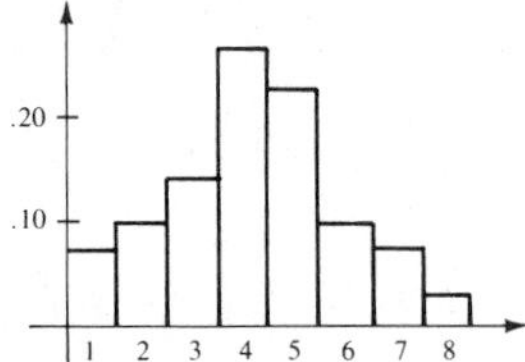

11. $f(x)$ is not a probability density function.
13. This is a probability density function. **15.** $f(x)$ is not a probability density function. **17.** $f(x)$ is not a probability density function. **19.** No value of k will make $f(x)$ a probability density function. **21.** $\frac{3}{125}$ **23.** $\frac{2}{3}$ **25.** $\frac{3}{14}$ **27.** $\frac{4}{3}$ **29.** .8907 **31.** .0455 **33.** .3202 **35.** .40 **37.** .075 **39.** .1587 **41.** .0548 **43.** .5403 **45.** .6065

15.7 Review *(page 753)*

Objective Questions

1. 135.13 **3.** 2.207 **5.** \$181.33 **7.** \$1.33 **9.** 5,000 **11.** \$809.01 **13.** \$488,629.85 **15.** \$184,027.30 **17.** \$208.33 **19.** \$83.33 **21.** $y = 2x^4 - \frac{2}{3}x^3 + x + C$ **23.** $y = \frac{1}{3}x^2 - \frac{5}{3}x + C$ **25.** $y^2 = \frac{1}{4}x^2 + \frac{5}{2}x + C$ **27.** $y = Me^{(5/2)x^2}$ **29.** $y^3 = \frac{3}{4}x^4 - 125$ **31.** $y = e^{x^2+1}$ **33.** $-\frac{e^{2x}}{4}(2x + 1) + C$
35. $x\ln\sqrt{2x} - \frac{1}{2}x + C$ **37.** Use Formula 25 with $a^2 = 16$: $\frac{x}{4}\sqrt{(x - 16)^3} + 2x\sqrt{x^2 - 16} - 32\ln|x + \sqrt{x^2 - 16}| + C$
39. Let $u = 3x$, then $du = 3\,dx$. Use Formula 34, then 33: $x\ln^3 3x - 3x\ln^2 3x + 6x\ln 3x - 6x + C$
41. $-\frac{1}{2}x^2e^{-2x} - \frac{1}{4}e^{-2x}(2x + 1) + C$ **43.** Let $u = 4x^2 + 5x - 3$, then $du = (8x + 5)\,dx$: $\ln|4x^2 + 5x - 3| + C$ **45.** $\frac{1}{2}$
47. 1 **49.** $\{2, 3, 5, \ldots, 12\}$; X = sum of the top faces; discrete. **51.** $[0, \infty)$; X = number of minutes a rat takes to go through a maze; continuous.

53.

Roll	2	3	4	5	6	7	8	9	10	11	12
Frequency	1	3	3	2	7	7	4	6	3	3	1
Rel. Frequency	$\frac{1}{40}$	$\frac{3}{40}$	$\frac{3}{40}$	$\frac{2}{40}$	$\frac{7}{40}$	$\frac{7}{40}$	$\frac{4}{40}$	$\frac{6}{40}$	$\frac{3}{40}$	$\frac{3}{40}$	$\frac{1}{40}$

55.

Number of Minutes	1	2	3	4	5	6	7	8	9	10
Rel. Frequency	$\frac{10}{100}$	$\frac{12}{100}$	$\frac{18}{100}$	$\frac{25}{100}$	$\frac{16}{100}$	$\frac{10}{100}$	$\frac{6}{100}$	$\frac{1}{100}$	$\frac{0}{100}$	$\frac{2}{100}$

57. This is a probability density function.

59. This is not a probability density function. **61.** $k = \frac{5}{128}$ **63.** $k = \frac{2}{495}$ **65.** .5 **67.** .7926 **69.** 276
71. \$4,800,000 **73.** **a.** 31 **b.** 1 **c.** 60 **d.** 2.3%

Sample Test

1. $\frac{1}{8}(2x - 1)^4 + C$ **3.** By parts: $\frac{x^2(2x - 1)^4}{8} - \frac{x(2x - 1)^5}{40} + \frac{(2x - 1)^6}{480} + C$. If instead, you use Formula 11, you will obtain the following (equivalent) form: $\frac{(2x-1)^6}{48} + \frac{(2x-1)^5}{40} + \frac{(2x-1)^4}{32} + C$ **5.** $\ln|4x^2 + 5x - 3| + C$

7. $3x - \ln|1 - e^{3x}| + C$ **9.** $x\ln^3 x - 3x\ln^2 x + 6x\ln x - 6x + C$ **11.** Formula 14 **13.** $\frac{e}{3} \approx .9061$

15. $y = \frac{5}{4}x^2 + \frac{3}{2}x + C$ **17.** \$307,543.02 **19.** $50(1 - e^{-.1t})$

Cumulative Review for Chapters 13–15 (page 756)

1. **3.** $605 = x$ **5.** .301 **7. a.** 0 **b.** $b - a$ **c.** $-\int_b^a f(x)\,dx$ **d.** $\int_a^b f(x)\,dx$

e. $k\int_a^b f(x)\,dx$ **f.** $\int_a^b f(x)\,dx \pm \int_a^b g(x)\,dx$

g. $u(x)v(x)\Big|_{x=a}^{x=b} - \int_{x=a}^{x=b} v\,du = u(b)v(b) - u(a)v(a) - \int_{x=a}^{x=b} v\,du$ where u and v are functions of x.

9. a. $x^5 + x^3 + 5x + C$ **b.** $\frac{1}{2}e^{2x} + C$ **11.** $x\ln^4 5x - 4x\ln^3 5x + 12x\ln^2 5x - 24x\ln 5x + 24x + C$

13. $\ln|4x^2 - 3x + 2| + C$ **15.** 0 **17.** $\frac{512}{3}$ **19.** $t = 4$ years **21.** 20.12 years **23.** 1,850 trees

25. .5488; about $\frac{1}{2}$ second response time

CHAPTER 16

16.1 Three-Dimensional Coordinate System (page 766)

1. The area of a rectangle with dimensions 15 by 35 is $K(15, 35) = (15)(35) = 525$ square units. **3.** The volume of a box 3 by 5 by 8 is $V(3, 5, 8) = (3)(5)(8) = 120$ cubic units. **5.** The simple interest from \$500 investment at 5% for 3 years is $I(500, .05, 3) = 500[1 + (.05)(3)] = 500(1.15) = \$575$. **7.** The future value of a \$2,500 investment at 12% compounded quarterly for 6 years is $A(2500, .12, 6, 4) = 2{,}500(1 + \frac{.12}{4})^{(4)(6)} = 2{,}500(1.03)^{24} \approx 2{,}500(2.0327941) = \$5{,}081.99$

9. The future value of a \$110,000 investment at 9% compounded monthly for 30 years is $A(110{,}000, .09, 30, 12) = 110{,}000(1 + \frac{.09}{12})^{(30)(12)} = 110{,}000(1.0075)^{360} \approx 110{,}000(14.730577) = \$1{,}620{,}363$

11. $f(2, 3) = 2^2 - 2(2)(3) + 3^2 = 1$ **13.** $f(-2, 5) = (-2)^2 - 2(-2)(5) + 5^2 = 49$ **15.** $g(2, 1) = \frac{2(2) - 4(1)}{2^2 + 1^2} = 0$

17. $g(5, -3) = \frac{2(5) - 4(-3)}{5^2 + (-3)^2} = \frac{11}{17}$ **19.** $h(0, 5) = \frac{e^{(0)(5)}}{\sqrt{0^2 + 5^2}} = \frac{1}{5}$ or .2

21. $h(-2, -3) = \frac{e^{(-2)(-3)}}{\sqrt{(-2)^2 + (-3)^2}} = \frac{e^6}{\sqrt{13}} \approx \frac{403.42879}{3.6055513} = 111.89$

23.

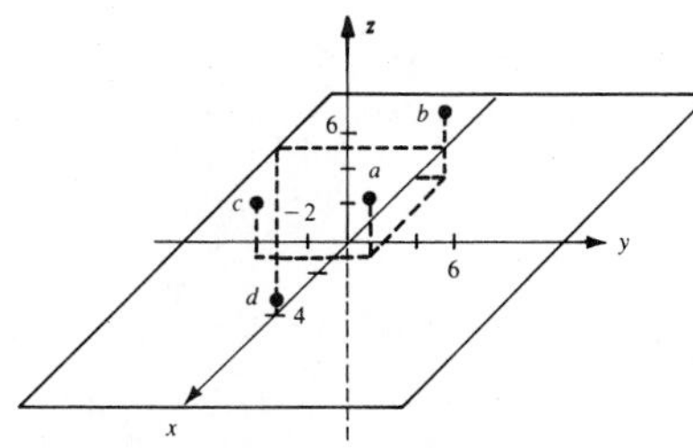

25.

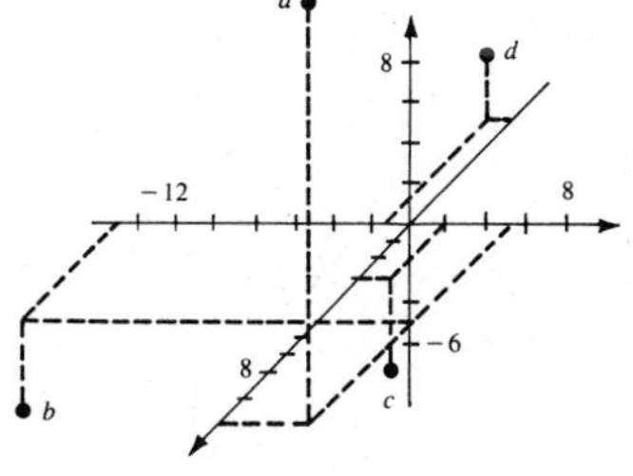

27.

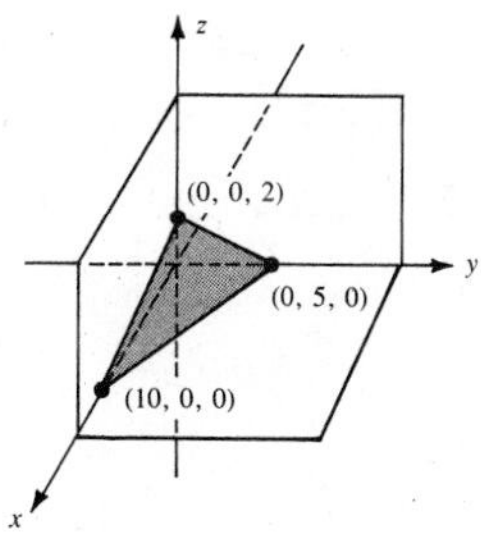

29.

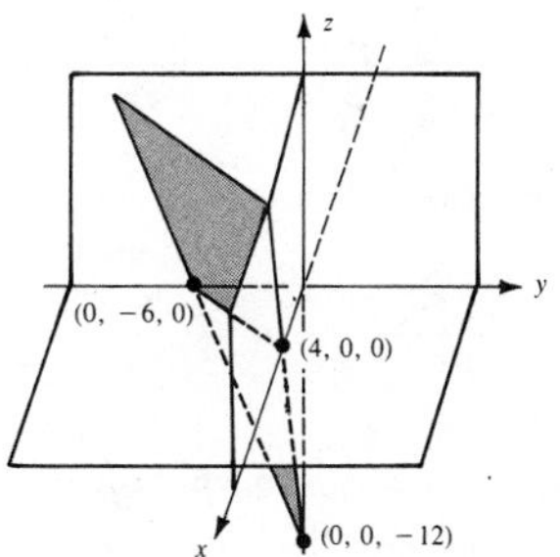

31.

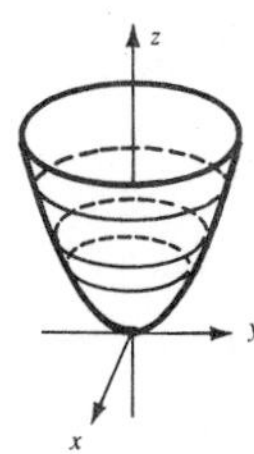

33.

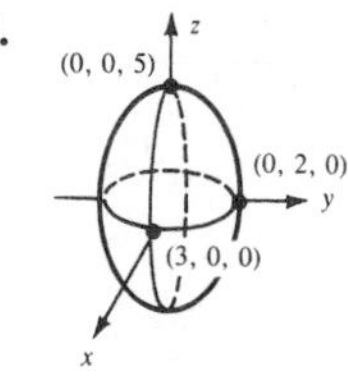

35.

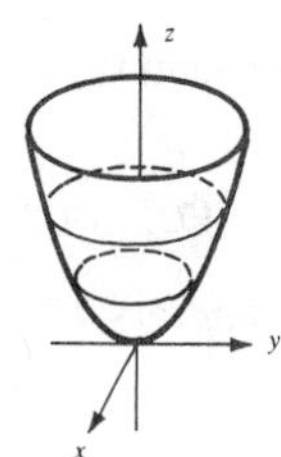

37.

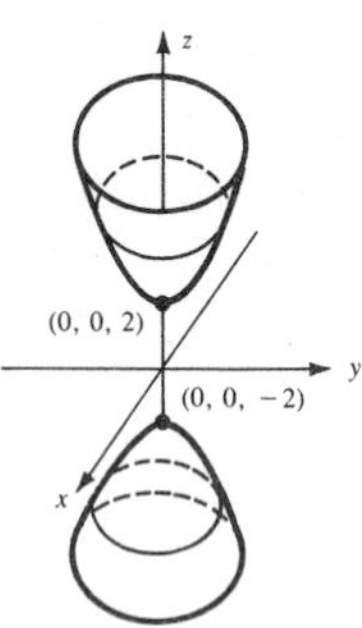

39.

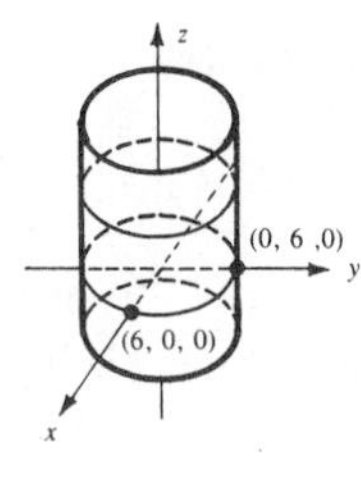

41.

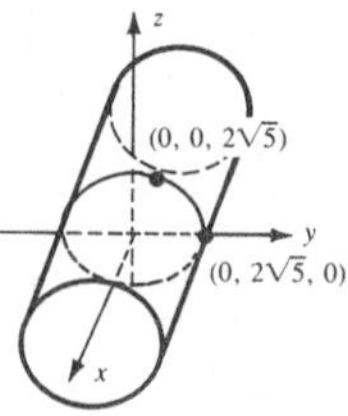

43. **a.** \$19,950 **b.** \$21,450 **c.** \$21,200 **d.** \$20,200 **45.** The total cost is $C(x, y, z) = 2xy + 1.25xy + 1.5xz + 1.5yz$
47. **a.** 3.875×10^8 mℓ **b.** 3.06×10^{10} mℓ **c.** 1.92×10^7 mℓ

16.2 Partial Derivatives *(page 771)*

1. $10x - 9x^2y^4$ **3.** $-15(4x - 3y)^4$ **5.** -134 **7.** $-759{,}375$ **9.** $-36x^2y^3$ **11.** $-240(4x - 3y)^3$

13. 10 **15.** 0 **17.** $-69{,}120$ **19.** 15,360 **21.** $3e^{3x+2y}$ **23.** $\dfrac{-3y}{\sqrt{x^2 - 3y^2}}$ **25.** $3e^7$

27. $g_y(3, -2)$ does not exist. **29.** $6e^{3x+2y}$ **31.** $\dfrac{3xy}{(\sqrt{x^2 - 3y^2})^3}$ **33.** $9e^4$ **35.** $\dfrac{6}{e^3}$ **37.** 0 **39.** 0

41. $f_x = 1 + 2y + \lambda y;\ f_y = 2x + \lambda x;\ f_\lambda = xy - 10$ **43.** $f_x = 2x - 3\lambda;\ f_y = 2y - 2\lambda;\ f_\lambda = -3x - 2y + 6$

45. $\dfrac{\partial f}{\partial b} = 100m + 50b + 1;\ \dfrac{\partial f}{\partial m} = 200m + 100b + 2$

47. It might be easier to rewrite $f(b, m)$:

$$f(b,m) = (m + b + 1)^2 + 4(m + b + 1)^2 + 9(m + b + 1)^2 = 14(m + b + 1)^2$$

$$\frac{\partial f}{\partial b} = 28(m + b + 1) \cdot \frac{\partial}{\partial b}(m + b + 1) = 28(m + b + 1)$$

$$\frac{\partial f}{\partial m} = 28(m + b + 1) \cdot \frac{\partial}{\partial m}(m + b + 1) = 28(m + b + 1)$$

49. **a.** $\dfrac{\partial P}{\partial a} = 2p - .2ap$ **b.** $\dfrac{\partial P}{\partial p} = 2a + 50 - 20p - .1a^2$. At a fixed level of advertising the rate of change of profit per unit change in price is $2a + 50 - 20p - .1a^2$. **51.** The rate of decrease in profit is \$1,000 per unit increase in x, assuming y stays constant, $P_y = 350$, $P_y(5, 10) = 350$. The rate of increase in profit is \$350 per unit increase in y, assuming x stays constant.

53. For most people answers will vary between 900 and 1,700 square feet. **55.** $A_r = \dfrac{\partial A}{\partial r} = 500(1 + r)^4$

57. $P_r = 300\left[\dfrac{rt(1 + \frac{r}{12})^{-12t-1} - 1 + (1 + \frac{r}{12})^{-12t}}{r^2}\right]$. For $t = 5$: $P_r = -\dfrac{300}{r^2}[1 - 5r(1 + \frac{r}{12})^{-61} - (1 + \frac{r}{12})^{-60}]$

16.3 Maximum–Minimum Applications *(page 779)*

1. Saddle point at $(0, 0, f(0, 0)) = (0, 0, 0)$ **3.** Relative minimum at $(-1, 2, f(-1, 2)) = (-1, 2, -3)$
5. Saddle point at $(0, 0, f(0, 0)) = (0, 0, 0)$; relative maximum at $(-1, 1, f(-1, 1)) = (-1, 1, 1)$

7. Relative maximum at $(-\frac{9}{11}, -\frac{41}{11}, \frac{45}{11})$ **9.** Saddle point at $(3, -2, 2)$ **11.** No relative maximums, minimums, or saddle points **13.** No relative maximums, minimums, or saddle points **15.** Saddle point at $(0, 0, 0)$; relative maximums at $(-1, -1, 2)$ and $(1, 1, 2)$ **17.** Saddle point at $(0, 0, 0)$; relative maximum at $(-1, -1, 1)$
19. The maximum is 141 (thousand dollars), which occurs at $(a, n) = (6, 1)$. **21.** Profit will be maximized when about 50 standard and 75 deluxe rings are sold. **23.** The box should therefore be 5.04 by 5.04 by approximately 2.52 feet.
25. Answers vary.

16.4 Lagrange Multipliers (page 785)

1. 100 **3.** $-\frac{3,456}{11} \approx -314.18$ **5.** $\frac{5,488}{27} \approx 203.26$ **7.** $f(12, 12) = 288$ is a relative minimum.
9. $f(3, 3) = 5$ is a relative minimum. **11.** $f(8, 8) = 128$ is a relative minimum. **13.** Both numbers are 5.
15. 119 is a relative maximum; he should apply $\frac{250}{59}$ acre-feet of water and $\frac{75}{59}$ pounds of fertilizer.
17. The rancher should build his fence $166\frac{2}{3}$ by $166\frac{2}{3}$ for a maximum area of 27,778 square feet.
19. \$.95 is a maximum; the optimum mixture is $\sqrt{10} \approx 3.16$ ounces of meat and $5\sqrt{10} \approx 15.81$ ounces of vegetables.
21. $10\sqrt{3}$ is a maximum. **23.** $T(\frac{10}{3}, \frac{10}{3}, \frac{10}{3}) = \frac{200}{3}$

16.5 Multiple Integrals (page 796)

1. $2x^2 + \frac{8}{3}$ **3.** $\frac{15}{2}(y^2 - 1)$ **5.** $\frac{x}{3}(x^2 + 8)^{3/2} - \frac{x^4}{3}$ **7.** $3e^{2y+2} - 3e^{2y+1}$ **9.** $\frac{e^{3y+25}}{2} - \frac{e^{3y+1}}{2}$ **11.** x^2

13. 9 **15.** 10 **17.** $\frac{91}{3}$ **19.** $\int_0^2 \int_0^{y^2} xy\,dx\,dy = \frac{16}{3}$ **21.** $\int_0^5 \int_0^x (2x + y)\,dy\,dx = \frac{625}{6}$

23. $\int_0^1 \int_{x^3}^{x^2} x^2y\,dy\,dx = \frac{1}{63}$ **25.** $\int_0^3 \int_1^4 dy\,dx = 9$ **27.** $\int_0^1 \int_1^3 (x + 3y)\,dx\,dy + \int_1^3 \int_x^3 (x + 3y)\,dy\,dx = \frac{91}{3}$

29. $\int_0^5 \int_y^5 (2x + y)\,dy\,dx = \frac{625}{6}$ **31.** $\int_0^3 \int_0^2 (x^2 + y)\,dx\,dy = 17$ **33.** $\int_0^3 \int_0^{1-x} \frac{1}{x}\,dy\,dx = \ln 3 - 2$

35. $\int_1^2 \int_0^{y^2} y^3e^{xy}\,dx\,dy = \frac{e^8}{3} - \frac{e}{3} - \frac{7}{3}$ **37.** $\int_2^4 \int_1^3 xy\sqrt{x^2 + y^2}\,dx\,dy \approx 92$ **39.** **A.** $\int_0^1 \int_2^{\sqrt{x}} (x + y)\,dy\,dx = \frac{3}{10}$

B. $\int_0^1 \int_{y^2}^{\sqrt{y}} (x + y)\,dx\,dy = \frac{3}{10}$ **41.** **A.** $\int_0^5 \int_0^{5-x} 2xy\,dy\,dx = \frac{625}{12}$ **B.** $\int_0^5 \int_0^{5-x} 2xy\,dx\,dy = \frac{625}{12}$

43. **a.** $\int_0^2 \int_{x^2}^{8-x^2} dy\,dx = \frac{32}{3}$ **b.** $\int_0^4 \int_0^{\sqrt{y}} dx\,dy + \int_4^8 \int_0^{\sqrt{8-y}} dx\,dy = \frac{32}{3}$ **45.** $\int_{-3}^2 \int_1^5 5\,dx\,dy = 100$

47. $\int_0^1 \int_0^1 xy\sqrt{x^2 + y^2}\,dx\,dy = \frac{1}{15}(2^{5/2} - 2) \approx .2437902833$ **49.** $\int_0^1 \int_{x^2}^{\sqrt{x}} 5xy\,dy\,dx = \frac{5}{12}$ **51.** $\int_0^3 \int_{x^2}^{3x} 6\,dy\,dx = 27$

53. $\int_0^1 \int_{x^2}^x xy^2\,dy\,dx = \frac{1}{40}$ **55.** -2 **57.** $\frac{16}{35}\sqrt[4]{250} \approx 1.81776166$

16.6 Review (page 797)

Objective Questions

1. 339 **3.** -60 **5.**

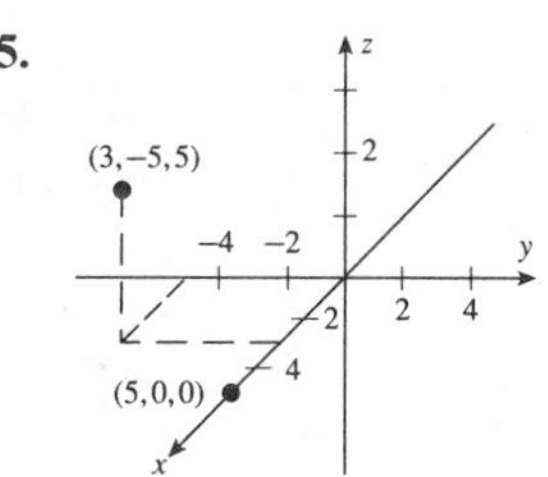

7.

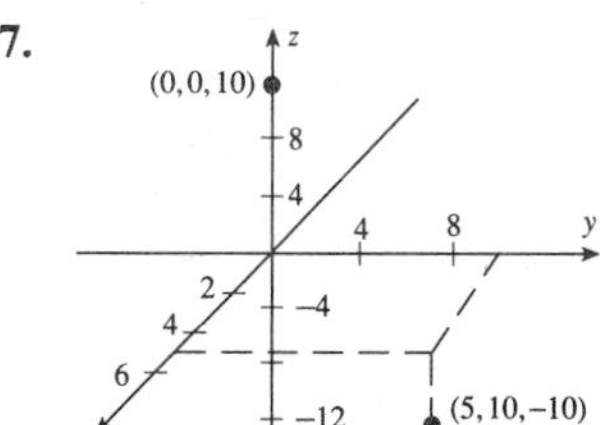

9.

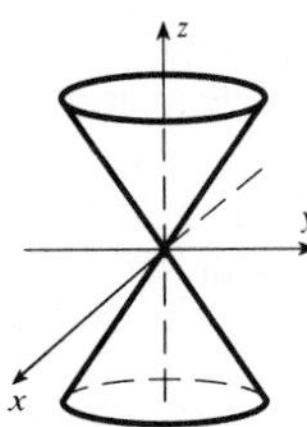

11.

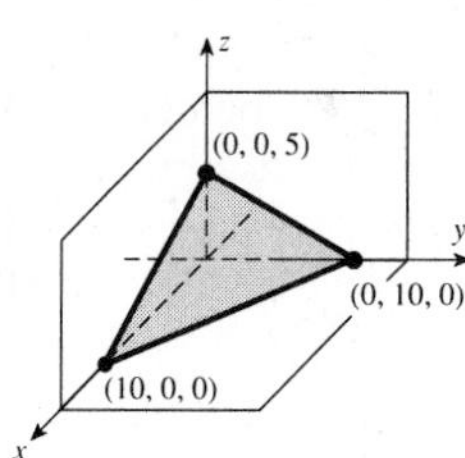

13. $24(2x - 5y)^{11}$

15. $-1{,}000y(x^2 + y^2)^{-3/2}$ **17.** $60x^3 - 60x^2y^3 + 4$ **19.** $-60x^3y^2$ **21.** No relative maximums or minimums
23. A saddle point at $(0, 0, 0)$ **25.** $f(5, 5) = -38$ **27.** The solution is $x = 125$, $y = 125$ ($\lambda = -125$); maximum is 15,625.
29. $\dfrac{26y^3}{3}$ **31.** $(6 - \frac{32}{9}\sqrt{2})y^2$ **33.** 54 **35.** $\displaystyle\int_1^2\int_0^{1/y} y^3\,dx\,dy = \frac{7}{3}$ **37.** 16.2 **39.** $\frac{16}{3}$ **41.** $\dfrac{1{,}875}{4} = 468.75$
43. $\displaystyle\int_0^1\int_0^2 x^2y^3\,dx\,dy = \frac{2}{3}$ **45. a.** 20 **b.** 90 **47.** 8

Sample Test

1. $P(10, 50) = 10^{.4}50^{.2} \approx 5.49$

3.

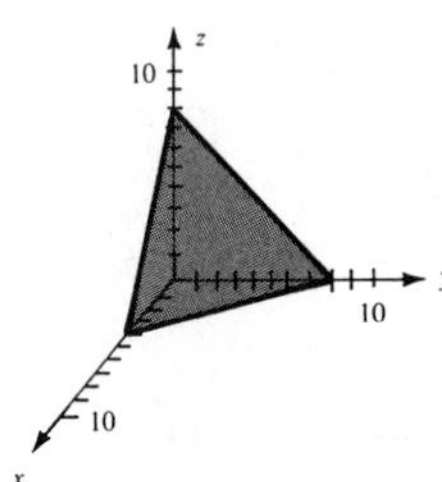

5. $f_x = 10(x - 3y)^9$

7. $f_x = 16x^3 - 9x^2y^2 + 4x$; $f_{xx} = 48x^2 - 18xy^2 + 4$ **9.** 18 **11.** Since $D > 0$, $(0, 0, 0)$ is a saddle point.
13. $\displaystyle\int_0^2\int_1^{e^x} dy\,dx = e^2 - 3 \approx 4.39$ **15.** $R(5, 8) = 910$; $C(5, 8) = 1{,}286.63$

Cumulative Review for Chapters 10–16 (page 800)

1. Answers vary. The limit of a function $f(x)$, as x approaches a, is the number L if the value of $f(x)$ approaches L as x gets closer to a. The values of x get closer and closer (but never reach) a, the values of $f(x)$ get closer and closer (and may or may not) reach L. **3.** Answers vary. Integration is the inverse process of differentiation; it is antidifferentiation. $\int f(x)\,dx = F(x)$ means that $F'(x) = f(x)$. That is, when looking for the integral of $f(x)$ we are seeking a function, $F(x)$, whose derivative is $f(x)$. The definite integral is used to find areas, as well as to sum or total functional values. Finding areas under a curve can be used in calculating consumer and producer surplus, as well as certain probabilities if we are given a probability density function.

5. $-5x^{-2} + \frac{3}{2}x^{-1/2}$ **7.** $\dfrac{-11}{1 - 11x}$ **9.** $\displaystyle\int_0^2 4x\,dx = \frac{4x^2}{2}\bigg|_0^2 = 8$ **11.** $\displaystyle\int_0^1\int_{e^x}^{e} \frac{1}{\ln y}\,dy\,dx = \int_1^e\int_0^{\ln y} \frac{1}{\ln y}\,dx\,dy$

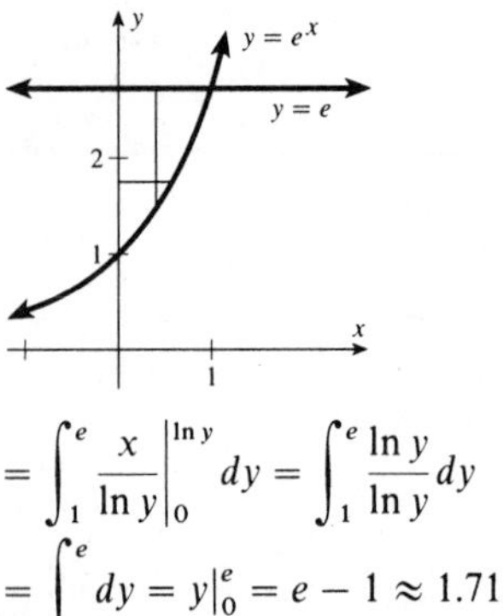

$$= \int_1^e \frac{x}{\ln y}\bigg|_0^{\ln y} dy = \int_1^e \frac{\ln y}{\ln y}\,dy$$
$$= \int_1^e dy = y\big|_0^e = e - 1 \approx 1.718$$

13. **a.** 64 units **b.** \$8,320 **15.** The maximum is 36 (thousand) when the price is \$2.00. **17.** 9

19.

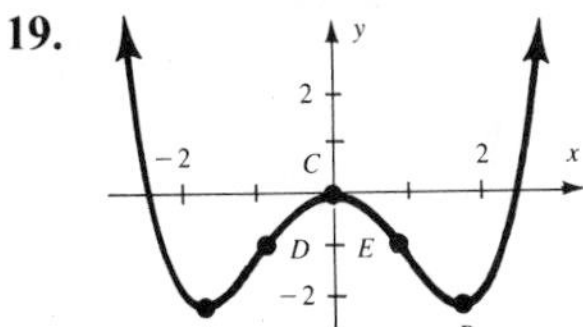

APPENDIXES

Appendix A: Deductive Logic (page 809)

1. *a*, *b*, and *d* are statements. **3.** *a*, *b*, and *d* are statements. **5.** $e \wedge d \wedge g$; where *e*: W. C. Fields is eating, *d*: W. C. Fields is drinking, *g*: W. C. Fields is having a good time. **7.** $\sim J \wedge \sim R$; where *J*: Jack will go tonight; *R*: Rosamond will go tomorrow. **9.** $(J \vee I) \wedge \sim P$; where *J*: The decision will depend on judgment; *I*: The decision will depend on intuition; *P*: The decision will depend on who paid the most. **11.** $P \rightarrow \sim w$; *p*: We make a proper use of those means which the God of nature has placed in our power; *w*: We are weak. **13.** $w \rightarrow n$; where *w*: It is work; *n*: It is noble. **15.** $(t \vee m) \rightarrow (\sim i \vee w)$; *i*: You must itemize deductions on Schedule A; *w*: You must complete the worksheet; *t*: You have earned an income of \$2,300; *m*: You have earned an income of more than \$2,300.

17.

p	q	$\sim p$	$\sim q$	$\sim p \wedge \sim q$
T	T	F	F	F
T	F	F	T	F
F	T	T	F	F
F	F	T	T	T

19.

r	s	$\sim r$	$\sim s$	$\sim r \vee \sim s$
T	T	F	F	F
T	F	F	T	T
F	T	T	F	T
F	F	T	T	T

21.

r	s	$r \wedge s$	$\sim s$	$(r \wedge s) \vee \sim s$
T	T	T	F	T
T	F	F	T	T
F	T	F	F	F
F	F	F	T	T

23.

p	q	$\sim p$	$\sim q$	$\sim p \vee \sim q$
T	T	F	F	F
T	F	F	T	T
F	T	T	F	T
F	F	T	T	T

25.

p	q	$\sim p$	$p \wedge \sim q$	$(p \wedge \sim q) \wedge p$
T	T	F	F	F
T	F	T	T	T
F	T	F	F	F
F	F	T	F	F

27.

p	q	$\sim q$	$p \vee q$	$p \wedge \sim q$	$(p \vee q) \vee (p \wedge \sim q)$
T	T	F	T	F	T
T	F	T	T	T	T
F	T	F	T	F	T
F	F	T	F	F	F

29.

p	q	$\sim p$	$(\sim p \rightarrow q)$	$p \rightarrow (\sim p \rightarrow q)$
T	T	F	T	T
T	F	F	T	T
F	T	T	T	T
F	F	T	F	T

31.

p	q	$\sim p$	$(p \wedge q)$	$\sim(p \wedge q)$	$\sim p \rightarrow \sim(p \wedge q)$
T	T	F	T	F	T
T	F	F	F	T	T
F	T	T	F	T	T
F	F	T	F	T	T

33.

p	q	$\sim q$	$p \rightarrow p$	$q \rightarrow \sim q$	$(p \rightarrow p) \rightarrow (q \rightarrow \sim q)$
T	T	F	T	F	F
T	F	T	T	T	T
F	T	F	T	F	F
F	F	T	T	T	T

35.

p	q	$\sim p$	$\sim q$	$p \rightarrow q$	$\sim q \rightarrow \sim p$	$(p \rightarrow q) \rightarrow (\sim q \rightarrow \sim p)$
T	T	F	F	T	T	T
T	F	F	T	F	F	T
F	T	T	F	T	T	T
F	F	T	T	T	T	T

37.

p	q	r	$\sim r$	$(p \wedge q)$	$(p \wedge q) \wedge \sim r$
T	T	T	F	T	F
T	T	F	T	T	T
T	F	T	F	F	F
T	F	F	T	F	F
F	T	T	F	F	F
F	T	F	T	F	F
F	F	T	F	F	F
F	F	F	T	F	F

39.

p	q	r	$\sim p$	$(q \vee \sim p)$	$p \wedge (q \vee \sim p)$	$[p \wedge (q \vee \sim p)] \vee r$
T	T	T	F	T	T	T
T	T	F	F	T	T	T
T	F	T	F	F	F	T
T	F	F	F	F	F	F
F	T	T	T	T	F	T
F	T	F	T	T	F	F
F	F	T	T	T	F	T
F	F	F	T	T	F	F

41. Direct **43.** Indirect **45.** Indirect **47.** Indirect **49.** If you can learn mathematics, then you understand human nature. **51.** All trebbles are expensive. **53.** If a nail is lost, then the kingdom is lost. **55.** $b \neq 0$ **57.** I will not eat that piece of pie. **59.** I will participate in student demonstrations. **61.** You obey the law. **63.** Babies cannot manage crocodiles. **65.** If they are my poultry, then they are not officers.

Appendix B: Mathematical Induction (page 815)

Answers in this section vary.

Appendix C: Calculators (page 820)

1. 4,914 **3.** 18.96048 **5.** 5,517,840 **7.** 5,937 **9.** 54,669.23077 **11.** 378,193,771 **13.** .77345 **15.** 17.61441762 **17.** .2644086935 **19.** .0078046506 **21.** .6764741715 **23.** 2.942851909 **25.** .454975922 **27.** .251485404 **29.** 2.259842248

Index

TO THE OWNER OF THIS BOOK:

We hope that you have found *College Mathematics and Calculus with Applications to Management, Life and Social Sciences*, 2nd Edition, useful. So that this book can be improved in a future edition, would you take the time to complete this sheet and return it? Thank you.

School and address: ______________________________

____

Department: ______________________________

Instructor's name: ______________________________

1. What I like most about this book is: ______________________________

____

2. What I like least about this book is: ______________________________

____

3. My general reaction to this book is: ______________________________

____

4. The name of the course in which I used this book is: ______________________________

____

5. Were all of the chapters of the book assigned for you to read? Yes No

   If not, which ones weren't? ______________________________

6. What specific suggestions do you have for improving this book?

____

____

____

____

____

____

Optional:

Your name: ______________________________ Date: ____________________

May Brooks/Cole quote you, either in promotion for *College Mathematics and Calculus with Applications to Management, Life and Social Sciences*, 2nd Edition, or in future publishing ventures?

Yes: _______ No: _______

Sincerely,
Karl J. Smith

FOLD HERE

NO POSTAGE NECESSARY IF MAILED IN THE UNITED STATES

BUSINESS REPLY MAIL

FIRST CLASS PERMIT NO. 358 PACIFIC GROVE, CA

POSTAGE WILL BE PAID BY ADDRESSEE

ATT: *Karl J. Smith*

Brooks/Cole Publishing Company
511 Forest Lodge Road
Pacific Grove, California 93950-9968

FOLD HERE

Forms Containing $\sqrt{x^2 \pm a^2}$, $\sqrt{a^2 - x^2}$

15. $\displaystyle\int \sqrt{x^2 \pm a^2}\, dx = \frac{x}{2}\sqrt{x^2 \pm a^2} \pm \frac{a^2}{2}\ln|x + \sqrt{x^2 \pm a^2}| + C$

16. $\displaystyle\int \frac{dx}{\sqrt{x^2 \pm a^2}} = \ln|x + \sqrt{x^2 \pm a^2}| + C$

17. $\displaystyle\int x\sqrt{x^2 \pm a^2}\, dx = \frac{1}{3}\sqrt{(x^2 \pm a^2)^3} + C$

18. $\displaystyle\int x\sqrt{a^2 - x^2}\, dx = -\frac{1}{3}\sqrt{(a^2 - x^2)^3} + C$

19. $\displaystyle\int \frac{dx}{x\sqrt{x^2 + a^2}} = -\frac{1}{a}\ln\left|\frac{a + \sqrt{x^2 + a^2}}{x}\right| + C$

20. $\displaystyle\int \frac{dx}{x\sqrt{a^2 - x^2}} = -\frac{1}{a}\ln\left|\frac{a + \sqrt{a^2 - x^2}}{x}\right| + C$

21. $\displaystyle\int \frac{\sqrt{x^2 + a^2}\, dx}{x} = \sqrt{x^2 + a^2} - a\ln\left|\frac{a + \sqrt{x^2 + a^2}}{x}\right| + C$

22. $\displaystyle\int \frac{\sqrt{a^2 - x^2}\, dx}{x} = \sqrt{a^2 - x^2} - a\ln\left|\frac{a + \sqrt{a^2 - x^2}}{x}\right| + C$

23. $\displaystyle\int \frac{x\, dx}{\sqrt{x^2 \pm a^2}} = \sqrt{x^2 \pm a^2} + C$

24. $\displaystyle\int \frac{x\, dx}{\sqrt{a^2 - x^2}} = -\sqrt{a^2 - x^2} + C$

25. $\displaystyle\int x^2\sqrt{x^2 \pm a^2}\, dx = \frac{x}{4}\sqrt{(x^2 \pm a^2)^3} \mp \frac{a^2x}{8}\sqrt{x^2 \pm a^2} - \frac{a^4}{8}\ln|x + \sqrt{x^2 \pm a^2}| + C$

26. $\displaystyle\int \frac{dx}{x^2\sqrt{x^2 \pm a^2}} = \mp\frac{\sqrt{x^2 \pm a^2}}{a^2x} + C$

27. $\displaystyle\int \frac{dx}{x^2\sqrt{a^2 - x^2}} = -\frac{\sqrt{a^2 - x^2}}{a^2x} + C$

28. $\displaystyle\int \frac{\sqrt{x^2 \pm a^2}}{x^2}\, dx = -\frac{\sqrt{x^2 \pm a^2}}{x} + \ln|x + \sqrt{x^2 \pm a^2}| + C$

29. $\displaystyle\int \frac{x^2\, dx}{\sqrt{x^2 \pm a^2}} = \frac{x}{2}\sqrt{x^2 \pm a^2} \mp \frac{a^2}{2}\ln|x + \sqrt{x^2 \pm a^2}| + C$